Edited by
Christoph Wittmann and James C. Liao

Industrial Biotechnology

Related Titles

Mozzi, F., Raya, R.R., Vignolo, G.M. (eds.)

Biotechnology of Lactic Acid Bacteria – Novel Applications 2e

2nd Edition
2015
Print ISBN: 978-1-118-86840-9

Gupta, V.K., Mach, R., Sreenivasaprasad, S. (eds.)

Fungal Biomolecules - Sources, Applications and Recent Developments

2015
Print ISBN: 978-1-118-95829-2

Sharma, N.K., Stal, L.J., Rai, A.K. (eds.)

Cyanobacteria - An Economic Perspective

2013
Print ISBN: 978-1-119-94127-9

(Further Volumes of the "Advanved Biotechnology" Series:)

Published:

Villadsen, J. (ed.)

Fundamental Bioengineering

2016
Print ISBN: 978-3-527-33674-6

Love, J. Ch. (ed.)

Micro- and Nanosystems for Biotechnology

2016
Print ISBN: 978-3-527-33281-6

Coming soon:

Yoshida, T. (ed.)

Applied Bioengineering

2017
Print ISBN: 978-3-527-34075-0

Wittmann, Ch., and Liao, J.C. (eds.)

Industrial Biotechnology
Products and Processes

2017
Print ISBN: 978-3-527-34181-8

Planned:

Systems Biology

J. Nielsen & S. Hohmann (Chalmers University, Sweden)

2017
Print ISBN: 9783527335589

Emerging Areas in Bioengineering

H. N. Chang (KAIST, South Korea)

2017
Print ISBN: 9783527340880

Edited by Christoph Wittmann and James C. Liao

Industrial Biotechnology

Microorganisms

Volume 1

WILEY-VCH
Verlag GmbH & Co. KGaA

Volume Editors

Christoph Wittmann
Saarland University
Institute of Systems Biotechnology
Campus A1.5
66123 Saarbrücken
Germany

James C. Liao
University of California
Chemistry & Biochemistry
5531 Boelter Hall
420 Westwood Plaza
Los Angeles, CA 90095
United States

Series Editors

Sang Yup Lee
KAIST
373-1 Guseong-Dong
291 Daehak-ro, Yuseong-gu
305-701 Daejon
South Korea

Jens Nielsen
Chalmers University
Department of Chemical and Biological Engineering
Kemivägen 10
412 96 Göteborg
Sweden

Gregory Stephanopoulos
Massachusetts Institute of Technology
Department of Chemical Engineering
77 Massachusetts Avenue
Cambridge, MA 02139
United States

Cover credit: Artwork in the circle
© fotoliaxrender

Library of Congress Card No.: applied for

British Library Cataloguing-in-Publication Data
A catalogue record for this book is available from the British Library.

Bibliographic information published by the Deutsche Nationalbibliothek
The Deutsche Nationalbibliothek lists this publication in the Deutsche Nationalbibliografie; detailed bibliographic data are available on the Internet at <http://dnb.d-nb.de>.

Print ISBN: 978-3-527-34179-5
ePDF ISBN: 978-3-527-80778-9
ePub ISBN: 978-3-527-80780-2
Mobi ISBN: 978-3-527-80781-9
oBook ISBN: 978-3-527-80779-6

Cover Design Adam-Design, Weinheim, Germany
Typesetting SPi Global, Chennai, India
Printing and Binding Markono Print Media Pte Ltd, Singapore
Printed on acid-free paper

To Heike, Isabelle, Felix, and Florian, with thanks from Christoph for their love, support, and inspiration.

To Kelly, Carol, and Clara, with thanks from Jim.

Contents to Volume 1

Contents to Volume 2

List of Contributors

Michael W.W. Adams
University of Georgia
Department of Biochemistry and
Molecular Biology
Life Sciences Bldg.
Athens, GA 30602–7229
USA

Hal Alper
Department of Chemical
Engineering
The University of Texas at Austin
200 E Dean Keeton Street
Stop C0400
Austin, TX 78712
USA

Andriy Luzhetskyy
Helmholtz Institute for
Pharmaceutical Research,
Actinobacteria Metabolic
Engineering Group
Universitätscampus E8
66123 Saarbrücken
Germany

and

University of Saarland
Department of Pharmaceutical
Biotechnology
UdS Campus C2.3
66123 Saarbrücken
Germany

William S. Ansari
University of California
California Center for Algae
Biotechnology, Division of
Biological Sciences
9500 Gilman Drive
San Diego, La Jolla, CA 92093
USA

Judith Becker
Saarland University
Institute of Systems
Biotechnology
Campus A 15
66123 Saarbrücken
Germany

Frank R. Bengelsdorf
Universität Ulm
Institut für Mikrobiologie und
Biotechnologie
Albert-Einstein-Allee 11
89081 Ulm
Germany

Dhananjay Beri
Dartmouth College
Thayer School of Engineering
14 Engineering Drive
Hanover, NH 03755
USA

and

BioEnergy Science Center
Oak Ridge, TN
USA

Oksana Bilyk
Helmholtz Institute for
Pharmaceutical Research,
Actinobacteria Metabolic
Engineering Group
Universitätscampus E8
66123 Saarbrücken
Germany

Juliana Bleckwedel
Centro de Referencia para
Lactobacilos (CERELA)-
CONICET
Chacabuco 145
San Miguel de Tucumán 4000
Argentina

José M. Borrero-de Acuña
Universidad Andrés Bello
Center for Bioinformatics and
Integrative Biology
Biosystems Engineering
Laboratory
Faculty of Biological Sciences
Av. República 239
8340176 Santiago de Chile
Chile

Klaus Buchholz
Technical University
Braunschweig
Institute of Chemical
Engineering
Hans-Sommer-Str. 10
38106 Braunschweig
Germany

Derrick S.W. Chuang
University of California
Department of Chemical and
Biomolecular Engineering
420 Westwood Plaza
5531 Boelter Hall
Los Angeles, CA 90095
USA

John Collins
Science historian
Leipziger Straße 82A
38124 Braunschweig
Germany

Jonathan M. Conway
North Carolina State University
Department of Chemical and
Biomolecular Engineering
EB-1, 911 Partners Way
Raleigh, NC 27695-7905
USA

James A. Counts
North Carolina State University
Department of Chemical and
Biomolecular Engineering
EB-1, 911 Partners Way
Raleigh, NC 27695-7905
USA

Arnold L. Demain
Drew University
Charles A. Dana Research
Institute for Scientists Emeriti
(R.I.S.E.)
36, Madison Ave
Madison, NJ 07940
USA

Fabienne Duchoud
University of California
Department of Chemical and
Biomolecular Engineering
420 Westwood Plaza
5531 Boelter Hall
Los Angeles, CA 90095
USA

Quentin M. Dudley
Northwestern University
Department of Chemical and
Biological Engineering
2145 Sheridan Road
Evanston, IL 60208
USA

and

Northwestern University
Chemistry of Life Processes
Institute
2170 Campus Drive
Evanston, IL 60208
USA

Peter Dürre
Universität Ulm
Institut für Mikrobiologie und
Biotechnologie
Albert-Einstein-Allee 11
89081 Ulm
Germany

Michael Egermeier
BOKU – University of Natural
Resources and Life Sciences
Department of Biotechnology
Muthgasse 18
1190 Vienna
Austria

and

BOKU – University of Natural
Resources and Life Sciences
CD-Laboratory for Biotechnology
of Glycerol
Muthgasse 18
1190 Vienna
Austria

Maria Eugenia Ortiz
Centro de Referencia para
Lactobacilos
(CERELA)-CONICET
Chacabuco 145
San Miguel de Tucumán 4000
Argentina

Stefanie K. Flitsch
Universität Ulm
Institut für Mikrobiologie und
Biotechnologie
Albert-Einstein-Allee 11
89081 Ulm
Germany

Brigitte Gasser
BOKU – University of Natural Resources and Life Sciences
Department of Biotechnology
Muthgasse 18
1190 Vienna
Austria

and

Austrian Centre of Industrial Biotechnology (ACIB GmbH)
Muthgasse 18
1190 Vienna
Austria

Matthias P. Gerstl
Austrian Centre of Industrial Biotechnology (ACIB)
Muthgasse 11
1190 Vienna
Austria

Javier A. Gimpel
Centre for Biotechnology and Bioengineering
Department of Chemical Engineering and Biotechnology
Universidad de Chile
851 Beaucheff, Santiago
Chile

Edward Green
CHAIN Biotechnology Limited
Imperial College Incubator
Imperial College London
Level 1 Bessemer Building
London SW7 2AZ
UK

Adam M. Guss
BioEnergy Science Center
Oak Ridge, TN
USA

and

Oak Ridge National Laboratory
Biosciences Division
1 Bethel Valley Road
Oak Ridge, TN 37831
USA

Michael Hanscho
Austrian Centre of Industrial Biotechnology (ACIB)
Muthgasse 11
1190 Vienna
Austria

Chris Herring
Enchi Corporation
Hanover, NH 03755
USA

Michael E. Himmel
BioEnergy Science Center
Oak Ridge, TN
USA

and

Biosciences Center, National Renewable Energy Laboratory
15013 Denver West Parkway
Golden, CO 80401
USA

Hans-Peter Hohmann
Nutrition Innovation Center
R&D Biotechnology
DSM Nutritional Products Ltd
Wurmisweg 576
4303 Kaiseraugst
Switzerland

Evert K. Holwerda
Dartmouth College
Thayer School of Engineering
14 Engineering Drive
Hanover, NH 03755
USA

and

BioEnergy Science Center
Oak Ridge, TN
USA

Michael C. Jewett
Northwestern University
Department of Chemical and
Biological Engineering
2145 Sheridan Road
Evanston, IL 60208
USA

and

Northwestern University
Chemistry of Life Processes
Institute
2170 Campus Drive
Evanston, IL 60208
USA

and

Northwestern University
Robert H. Lurie Comprehensive
Cancer Center
676 North St. Clair
Chicago, IL 60611
USA

and

Northwestern University
Simpson Querrey Institute for
Bionanotechnology
303 E. Superior
Chicago, IL 60611
USA

Christian Jungreuthmayer
TGM - Technologisches
Gewerbemuseum
Wexstraße 19-23
1200 Vienna
Austria

and

Austrian Centre of Industrial
Biotechnology (ACIB)
Muthgasse 11
1190 Vienna
Austria

Ashty S. Karim
Northwestern University
Department of Chemical and
Biological Engineering
2145 Sheridan Road
Evanston, IL 60208
USA

and

Northwestern University
Chemistry of Life Processes
Institute
2170 Campus Drive
Evanston, IL 60208
USA

Prema S. Karunanithi
University of California
California Center for Algae
Biotechnology, Division of
Biological Sciences
9500 Gilman Drive
San Diego, La Jolla, CA 92093
USA

Matthew W. Keller
University of Georgia
Department of Biochemistry and
Molecular Biology
Life Sciences Bldg.
Athens, GA 30602–7229
USA

Robert M. Kelly
North Carolina State University
Department of Chemical and
Biomolecular Engineering
EB-1, 911 Partners Way
Raleigh, NC 27695-7905
USA

Piyum A. Khatibi
North Carolina State University
Department of Chemical and
Biomolecular Engineering
EB-1, 911 Partners Way
Raleigh, NC 27695-7905
USA

Michael Kohlstedt
Saarland University
Institute of Systems Biology
Biosciences Campus A1.5
66123 Saarbrücken
Germany

Preben Krabben
Green Biologics Limited
45A Western Avenue
Milton Park
Abingdon
Oxfordshire OX14 4RU
UK

Laxmi Krishnappa
University of Groningen,
University Medical Center
Groningen
Department of Medical
Microbiology
Hanzeplein 1
9700 RB Groningen
The Netherlands

Laura L. Lee
North Carolina State University
Department of Chemical and
Biomolecular Engineering
EB-1, 911 Partners Way
Raleigh, NC 27695-7905
USA

James C. Liao
University of California
Department of Chemical
and Biomolecular Engineering
& Departmant of Bioengineering
420 Westwood Plaza
5531 Boelter Hall
Los Angeles, CA, 90095
USA

Sonja Linder
Universität Ulm
Institut für Mikrobiologie und
Biotechnologie
Albert-Einstein-Allee 11
89081 Ulm
Germany

Gina L. Lipscomb
University of Georgia
Department of Biochemistry and Molecular Biology
Life Sciences Bldg.
Athens, GA 30602–7229
USA

Andrew J. Loder
North Carolina State University
Department of Chemical and Biomolecular Engineering
EB-1, 911 Partners Way
Raleigh, NC 27695-7905
USA

Lee R. Lynd
Dartmouth College
Thayer School of Engineering
14 Engineering Drive
Hanover, NH 03755
USA

and

Dartmouth College
Department of Biological Sciences
Hanover, NH
USA

and

Enchi Corporation
Hanover, NH 03755
USA

and

BioEnergy Science Center
Oak Ridge, TN
USA

Hans Marx
BOKU – University of Natural Resources and Life Sciences
Department of Biotechnology
Muthgasse 18
1190 Vienna
Austria

and

BOKU – University of Natural Resources and Life Sciences
CD-Laboratory for Biotechnology of Glycerol
Muthgasse 18
1190 Vienna
Austria

Diethard Mattanovich
BOKU – University of Natural Resources and Life Sciences
Department of Biotechnology
Muthgasse 18
1190 Vienna
Austria

and

Austrian Centre of Industrial Biotechnology (ACIB GmbH)
Muthgasse 18
1190 Vienna
Austria

Stephen P. Mayfield
University of California
California Center for Algae Biotechnology, Division of Biological Sciences
9500 Gilman Drive
San Diego, La Jolla, CA 92093
USA

Nigel Minton
University of Nottingham
BBSRC/EPSRC Synthetic Biology
Research Centre (SBRC)
School of Life Sciences
University Park
Nottingham NG7 2RD
UK

Fernanda Mozzi
Centro de Referencia para
Lactobacilos (CERELA)-
CONICET
Chacabuco 145
San Miguel de Tucumán 4000
Argentina

Rolf Müller
Saarland University, Helmholtz
Centre for Infection Research
and Pharmaceutical
Biotechnology
Department of Microbial Natural
Products, Helmholtz-Institute
for Pharmaceutical Research
Saarland, Saarland University
Campus, Building E8.1,
66123 Saarbrücken
Germany

Stefan Müller
Austrian Academy of Sciences
Johann Radon Institute for
Computational and Applied
Mathematics
Altenberger Straße 69
4040 Linz
Austria

Sean J. Murphy
Dartmouth College
Thayer School of Engineering
14 Engineering Drive
Hanover, NH 03755
USA

and

BioEnergy Science Center
Oak Ridge, TN
USA

Govind Nair
Department of Biotechnology
University of Natural Resources
and Life Sciences
Vienna, Muthgasse 18
1190 Vienna
Austria

and

Austrian Centre of Industrial
Biotechnology (ACIB)
Muthgasse 11
1190 Vienna
Austria

Pablo I. Nikel
Systems and Synthetic Biology
Program
National Spanish Center for
Biotechnology (CNB-CSIC)
Calle Darwin, 3
28049 Madrid
Spain

Daniel G. Olson
Dartmouth College
Thayer School of Engineering
14 Engineering Drive
Hanover, NH 03755
USA

and

BioEnergy Science Center
Oak Ridge, TN
USA

Julie Paye
Dartmouth College
Thayer School of Engineering
14 Engineering Drive
Hanover, NH 03755
USA

and

BioEnergy Science Center
Oak Ridge, TN
USA

and

Novo Nordisk
West Lebanon, NH
USA

Micaela Pescuma
Centro de Referencia para
Lactobacilos (CERELA)-
CONICET
Chacabuco 145
San Miguel de Tucumán 4000
Argentina

Jennifer Pfizenmaier
University of Stuttgart
Institute of Biochemical
Engineering
Allmandring 31
70569 Stuttgart
Germany

Ignacio Poblete-Castro
Universidad Andrés Bello
Center for Bioinformatics and
Integrative Biology
Biosystems Engineering
Laboratory
Faculty of Biological Sciences
Av. República 239
8340176 Santiago de Chile
Chile

Anja Poehlein
Georg-August University
Genomic and Applied
Microbiology and Göttingen
Genomics Laboratory
Grisebachstr. 8
37077 Göttingen
Germany

Zoltán Prágai
Nutrition Innovation Center
R&D Biotechnology
DSM Nutritional Products Ltd
Wurmisweg 576
4303 Kaiseraugst
Switzerland

Georg Regensburger
Johannes Kepler University Linz
Institute for Algebra
Altenberger Straße 69
4040 Linz
Austria

Amanda M. Rhaesa
University of Georgia
Department of Biochemistry
and Molecular Biology
Life Sciences Bldg.
Athens, GA 30602–7229
USA

Gabe M. Rubinstein
University of Georgia
Department of Biochemistry
and Molecular Biology
Life Sciences Bldg.
Athens, GA 30602–7229
USA

Luciana Ruiz-Rodríguez
Centro de Referencia para
Lactobacilos (CERELA)-
CONICET
Chacabuco 145
San Miguel de Tucumán 4000
Argentina

Thomas Rydzak
BioEnergy Science Center
Oak Ridge, TN
USA

and

Oak Ridge National Laboratory
Biosciences Division
Oak Ridge, TN
USA

Michael Sauer
BOKU – University of Natural
Resources and Life Sciences
Department of Biotechnology
Muthgasse 18
1190 Vienna
Austria

and

Austrian Centre of Industrial
Biotechnology (ACIB GmbH)
Muthgasse 18
1190 Vienna
Austria

and

BOKU – University of Natural
Resources and Life Sciences,
CD-Laboratory for
Biotechnology of Glycerol
Muthgasse 18
1190 Vienna
Austria

Bettina Schiel-Bengelsdorf
Universität Ulm
Institut für Mikrobiologie und
Biotechnologie
Albert-Einstein-Allee 11
89081 Ulm
Germany

Israel M. Scott
University of Georgia
Department of Biochemistry
and Molecular Biology
Life Sciences Bldg.
Athens, GA 30602–7229
USA

Xiongjun Shao
Dartmouth College
Thayer School of Engineering
14 Engineering Drive
Hanover, NH 03755
USA

and

BioEnergy Science Center
Oak Ridge, TN
USA

Elizabeth A. Specht
University of California
California Center for Algae
Biotechnology, Division of
Biological Sciences
9500 Gilman Drive
San Diego, La Jolla, CA 92093
USA

Benjamin A. Stegmann
Universität Ulm
Institut für Mikrobiologie und
Biotechnologie
Albert-Einstein-Allee 11
89081 Ulm
Germany

Christopher T. Straub
North Carolina State University
Department of Chemical and
Biomolecular Engineering
EB-1, 911 Partners Way
Raleigh, NC 27695-7905
USA

Jie Sun
Department of Chemical
Engineering
The University of Texas at Austin
200 E Dean Keeton Street
Stop C0400
Austin, TX 78712
USA

Ralf Takors
University of Stuttgart
Institute of Biochemical
Engineering
Allmandring 31
70569 Stuttgart
Germany

Matthew Theisen
University of California
Department of Chemical
and Biomolecular Engineering
& Department of Bioengineering
420 Westwood Plaza
5531 Boelter Hall,
Los Angeles, CA, 90095
USA

Liang Tian
Dartmouth College
Thayer School of Engineering
14 Engineering Drive
Hanover, NH 03755
USA

and

BioEnergy Science Center
Oak Ridge, TN
USA

Erick J. Vandamme
Ghent University
Department of Biochemical and
Microbial Technology
Block B, 2nd floor Coupure links
6539000 Ghent
Belgium

Jan M. van Dijl
University of Groningen
University Medical Center
Groningen
Department of Medical
Microbiology
Hanzeplein 1
9700 RB Groningen
The Netherlands

Nicholas P. Vitko
North Carolina State University
Department of Chemical and
Biomolecular Engineering
EB-1, 911 Partners Way
Raleigh, NC 27695-7905
USA

Silke C. Wenzel
Saarland University
Helmholtz Centre for Infection
Research and Pharmaceutical
Biotechnology
Department of Microbial Natural
Products
Helmholtz-Institute for
Pharmaceutical Research
Saarland, Saarland University
Campus, Building E8.1,
66123 Saarbrücken
Germany

Christoph Wittmann
Saarland University
Institute of Systems Biology,
Biosciences, Campus A1.5
66123 Saarbrücken
Germany

Robert Worthen
Dartmouth College
Thayer School of Engineering
14 Engineering Drive
Hanover, NH 03755
USA

and

BioEnergy Science Center
Oak Ridge, TN
USA

Ying Zhang
University of Nottingham
BBSRC/EPSRC Synthetic Biology
Research Centre (SBRC)
School of Life Sciences
University Park
Nottingham NG7 2RD
UK

Jürgen Zanghellini
Department of Biotechnology
University of Natural Resources
and Life Sciences
Vienna, Muthgasse 18
1190 Vienna
Austria

and

Austrian Centre of Industrial
Biotechnology (ACIB)
Muthgasse 11
1190 Vienna
Austria

Benjamin M. Zeldes
North Carolina State University
Department of Chemical and
Biomolecular Engineering
EB-1, 911 Partners Way
Raleigh, NC 27695-7905
USA

About the Series Editors

Sang Yup Lee is Distinguished Professor at the Department of Chemical and Biomolecular Engineering at the Korea Advanced Institute of Science and Technology. At present, Prof. Lee is the Director of the Center for Systems and Synthetic Biotechnology, Director of the BioProcess Engineering Research Center, and Director of the Bioinformatics Research Center. He has published more than 500 journal papers, 64 books, and book chapters, and has more than 580 patents (either registered or applied) to his credit. He has received numerous awards, including the National Order of Merit, the Merck Metabolic Engineering Award, the ACS Marvin Johnson Award. Charles Thom Award, Amgen Biochemical Engineering Award, Elmer Gaden Award, POSCO TJ Park Prize, and HoAm Prize. He is Fellow of American Association for the Advancement of Science, the American Academy of Microbiology, American Institute of Chemical Engineers, Society for Industrial Microbiology and Biotechnology, American Institute of Medical and Biological Engineering, the World Academy of Science, the Korean Academy of Science and Technology, and the National Academy of Engineering of Korea. He is also Foreign Member of National Academy of Engineering, USA. In addition, he is honorary professor of the University of Queensland (Australia), honorary professor of the Chinese Academy of Sciences, honorary professor of Wuhan University (China), honorary professor of Hubei University of Technology (China), honorary professor of Beijing University of Chemical Technology (China), and advisory professor of the Shanghai Jiaotong University (China). Apart from his academic associations, Prof. Lee is the editor-in-chief of the *Biotechnology Journal* and is also contributing to numerous other journals as associate editor and board member. Prof. Lee is serving as a member of Presidential Advisory Committee on Science and Technology (South Korea).

Jens Nielsen is Professor and Director to Chalmers University of Technology (Sweden) since 2008. He obtained an MSc degree in chemical engineering and a PhD degree (1989) in biochemical engineering from the Technical University of Denmark (DTU) and after that established his independent research group and was appointed full professor there in 1998. He was Fulbright visiting professor at MIT in 1995–1996. At DTU, he founded and directed the Center for Microbial Biotechnology. Prof. Nielsen has published more than 350 research papers and coauthored more than 40 books, and he is inventor of more than 50 patents. He has founded several companies that have raised more than 20 million in venture capital. He has received numerous Danish and international awards and is member of the Academy of Technical Sciences (Denmark), the National Academy of Engineering (USA), the Royal Danish Academy of Science and Letters, the American Institute for Medical and Biological Engineering and the Royal Swedish Academy of Engineering Sciences.

Gregory Stephanopoulos is the W.H. Dow Professor of Chemical Engineering at the Massachusetts Institute of Technology (MIT, USA) and Director of the MIT Metabolic Engineering Laboratory. He is also Instructor of Bioengineering at Harvard Medical School (since 1997). He received his BS degree from the National Technical University of Athens and his PhD from the University of Minnesota (USA). He has coauthored about 400 research papers and 50 patents, along with the first textbook on metabolic engineering. He has been recognized by numerous awards from the American Institute of Chemical Engineers (AIChE) (Wilhelm, Walker and Founders awards), American Chemical Society (ACS), Society of Industrial Microbiology (SIM), BIO (Washington Carver Award), the John Fritz Medal of the American Association of Engineering Societies, and others. In 2003, he was elected member of the National Academy of Engineering (USA) and in 2014 President of AIChE.

Preface

Over the past 100 years, industrial biotechnology has grown into a multibillion dollar market, which now has even begun to include large parts of chemical, material, and fuel production in a rapidly growing bioeconomy. Through decades of research and discovery, industrial biotechnology offers a cornucopia of possibilities. This volume is focused on the microorganisms, which are at the very heart of industrial production and determine its success through their biocatalytic efficiency. Optimized and streamlined through billion years of evolution, microbial cells exhibit networks of hundreds to thousands of biochemical conversions. Embedded into a remarkable architecture of fine-tuned control and regulation, their cellular networks can operate at highest efficiency, versatility, selectivity, vitality, and robustness. This allows top efficiency and quality production processes, provided we are able tame and orchestrate this huge natural power. It is therefore more than worthwhile to collect and compile our current knowledge on the most relevant industrial cell factories, including state-of-the-art strategies from systems biology, systems metabolic engineering, and synthetic biology to design, improve, and upgrade their properties, as has been done in this volume.

Organized in six thematic parts, this volume comprises 19 well-elaborated chapters by leading experts in the field and provides a most comprehensive view on important aspects of industrial microorganisms, including the latest trends in research and development: the level of global analysis, design, and engineering of biological systems.

Part A "From Pioneers to Visionary" covers the period from the early days of industrial biotechnology (Chapter 1) till the time of development of novel concepts of strain design and production, including synthetic biology, genome-scale modeling, and cell-free production (Chapters 2–4) and bridges 100 years of discovery and innovation.

Part B "Multipurpose Bacterial Cell Factories" offers four chapters on some of the most widely used industrial microorganisms. *Escherichia coli* (Chapter 5), *Corynebacterium glutamicum* (Chapter 6), *Bacillus subtilis* (Chapter 7), and *Pseudomonas putida* (Chapter 8) have evolved into synthetic platforms with a broad range of applications. Their product portfolios include fine chemicals, bulk chemicals, drugs, flavors and fragrances, materials, fuels, therapeutic and diagnostic proteins, and enzymes, among others. This progress has taken place largely as a result of several decades of intensive research so that these bacteria belong to the best-characterized biological systems today – a vast knowledge base for further exploration.

Part C "Exploiting Anaerobic Synthetic Power" deals with microorganisms that live without oxygen. As the aeration of the production vessels is expensive and challenging at large scale, this lifestyle offers interesting applications. Chapters 9 and 10 focus on different *Clostridium* strains that are specifically suited to produce solvents such as acetone and butanol, but also discussing the use of thermostable cellulolytic enzymes toward consolidated bioprocessing: an elegant way to couple the decomposition of polymeric raw materials with conversion of the formed sugar into the product of interest in one operation and thereby streamlining production. Chapter 11 introduces lactic acid bacteria, which are well-accepted strains applied in human nutrition and other industrial areas.

Part D "Microbial Treasure Chests for High-Value Molecules" touches the world of natural products – complex molecule structures with biologically unique properties and high value as therapeutics to fight infections, cancer, and other threatening diseases, but also for many other applications, including herbicides, insecticides, or fungicides in agriculture. Chapters 12 and 13 highlight two of the most important bacterial groups that supply these high-value products: *Myxobacteria* and *Streptomyces.*

Part E discusses the use of microorganisms as novel sustainable feedstocks in industrial biotechnology. Chapter 14 sheds light on thermophilic bacteria, whereas Chapters 15 and 16 deal with autotrophic systems that enable production simply from sunlight and carbon dioxide: cyanobacteria and algae.

Part F discusses eukaryotic cell factories of high industrial relevance. Mammalian cells offer great potential to synthesize complex proteins with high therapeutic value, but are far more complex and more difficult to grow, hence requiring specific handling (Chapter 17). Two interesting chapters (Chapter 18 and 19) introduce yeasts, including *Saccharomyces cerevisiae* and *Pichia pastoris.*

Thanks to the leading experts and their excellent contributions, which are greatly appreciated, this volume – together with its sister volume "Industrial Biotechnology Products and Processes" – hopefully sets a milestone of perpetual value. Finally, we would like to thank Claudia Ley and Waltraud Wuest at Wiley for their assistance in the production.

Saarbrücken, July 2016 — *Christoph Wittmann*
Los Angeles, July 2016 — *James C. Liao*

Part I
Industrial Biotechnology: From Pioneers to Visionary

Industrial Biotechnology: Microorganisms, First Edition. Edited by Christoph Wittmann and James C. Liao.

1
History of Industrial Biotechnology

Arnold L. Demain, Erick J. Vandamme, John Collins, and Klaus Buchholz

1.1
The Beginning of Industrial Microbiology

Microbes have been extremely important for life on Earth. They are the progenitors of all life on Earth and are the preeminent system to study evolution. They provide rapid generation times, genetic flexibility, unequaled experimental scale, and manageable study systems. Estimates indicate 5×10^{31} microbial cells exist with a weight of 50 quadrillion metric tons. More photosynthesis is accomplished by microbes than by green plants. More than 60% of the earth's biomass is that of microbes. Over 90% of the cells in human bodies are microorganisms. Sterile animals are less healthy than those colonized by microbes.

Long before their discovery, microorganisms were exploited to serve the needs and desires of humans, that is, to preserve milk, fruits, and vegetables, and to enhance the quality of life with the resultant beverages, cheeses, bread, pickled foods, and vinegar. The use of yeasts dates back to ancient days. The oldest fermentation know-how, the conversion of sugar to alcohol by yeasts, was used to make beer in Sumeria and Babylonia before 7000 BC. By 4000 BC, the Egyptians had discovered that carbon dioxide generated by the action of brewer's yeast could leaven bread. Ancient peoples made cheese with molds and bacteria. Wine was made in China as early as in 7000 BC [1] and in Assyria in 3500 BC. Reference to wine can be found in the Book of Genesis, where it is noted that Noah consumed a bit too much of the beverage. According to the Talmud, "a man without salt and vinegar is a lost man." The Assyrians treated chronic middle ear diseases with vinegar, and Hippocrates treated patients with it in 400 BC. According to the New Testament, vinegar was offered to Jesus on the cross. For thousands of years, moldy cheese, meat, and bread were employed in folk medicine to heal wounds. By 100 BC, ancient Rome had over 250 bakeries which were making leavened bread. As a method of preservation, milk was fermented to lactic acid to make yogurt and also converted into kefyr and kumiss using the *Kluyveromyces* species in Asia. The use of molds to saccharify rice in the Koji process dates back at least to 700 AD. By the fourteenth century AD, the distillation of alcoholic spirits from fermented

Industrial Biotechnology: Microorganisms, First Edition. Edited by Christoph Wittmann and James C. Liao.

grain, a practice thought to have originated in China or the Middle East, was common in many parts of the world. Vinegar manufacture began in Orleans, France, at the end of the fourteenth century, the surface technique being referred to as the *Orleans method*.

Antonie van Leeuwenhoek, in the Netherlands in the seventeenth century, turning his simple lens to the examination of water, decaying matter, and scrapings from his teeth, reported on the presence of tiny "animalcules," that is, moving organisms less than 1/1000th the size of a grain of sand. He was a Dutch merchant with no university training but his spare time interest was the construction of microscopes. This lack of university connection might have caused his discoveries to go unknown, had it not been for the Royal Society in England and its secretary, Henry Oldenburg, who corresponded with European science amateurs. From 1673 to 1723, Leeuwenhoek's great powers as a microscopist were communicated to the Society in a series of letters. Thus the practice of industrial biotechnology has its roots deep in antiquity.

In these early days, most scientists thought that microbes arose spontaneously from nonliving matter. What followed was an argument over spontaneous generation, aptly called the *War of the Infusions* lasting 100 years. Proponents had previously claimed that maggots were spontaneously created from decaying meat; however, this was discredited by Redi. By this time, the theory of spontaneous generation, originally postulated by Aristotle among others, was discredited with respect to higher forms of life, so the proponents concentrated their arguments on bacteria. The theory did seem to explain how a clear broth became cloudy via growth of large numbers of such "spontaneously generated microorganisms" as the broth aged. However, others believed that microorganisms only came from previously existing microbes and that their ubiquitous presence in air was the reason that they would develop in organic infusions after gaining access to these rich liquids. Three independent investigators, Charles Cagniard de la Tour of France, Theodor Schwann, and Friedrich Traugott Kützing of Germany, proposed that the products of fermentation, chiefly ethanol and carbon dioxide, were created by a microscopic form of life. This concept was bitterly opposed by the leading chemists of the period (such as Jöns Jakob Berzelius, Justus von Liebig, and Friedrich Wöhler), who believed fermentation to be strictly a chemical reaction; they maintained that the yeast in the fermentation broth was lifeless, decaying matter. Organic chemistry was flourishing at the time, and these opponents of the living microbial origin were initially quite successful in putting forth their views. Interest in the mechanisms of these fermentations resulted in later investigations by Louis Pasteur, which not only advanced microbiology as a distinct discipline, but also led to the development of vaccines and concepts of hygiene, which revolutionized the practice of medicine.

In 1850, Davaine detected rod-shaped objects in the blood of anthrax-infected sheep and was able to produce the disease in healthy sheep by inoculation of such blood. In the next 25 years, Pasteur of France and John Tyndall of Britain demolished the concept of spontaneous generation and proved that existing microbial life came from preexisting life. In the 1850s, Pasteur detected two distinct forms of amyl alcohol, that is, D and L, able to polarize light in different directions (opticals isomers or enantiomers) but he was not able to separate the two. He found that only one of the two optical isomers (e.g., for tartaric acid) were produced by living microbes carrying out fermentation. Pasteur concluded in 1857 that fermentation was a living process of yeast. In 1861, he proved the presence of microbes in air and discredited the theory of spontaneous generation of microbes. It was at this point that microbiology was born, but it took almost two decades, until 1876, to disprove the chemical hypothesis of Berzelius, Liebig, and Wöhler, that is, that fermentation was the result of contact with decaying matter.

In 1876, the great German microbiologist, Robert Koch, proved that bacteria from anthrax infections were capable of causing the disease. His contributions involving the growth of microbes in pure culture led to the decline of the pleomorphism theory, that is, that one form of bacteria developed into another. It was mainly the work of Koch that led to the acceptance of the idea that specific diseases were caused by specific organisms, each of which had a specific form and function. In 1884, his students, Gaffky and Loeffler, were able to confirm the etiologic role of infectious bacteria in the cases of typhoid fever and diphtheria and, in 1894, Alexandre Yersin, Louis Pasteur's student, for bubonic plague. Yersin also confirmed the presence of the disease organism in the animal vector, rats.

The distillers of Lille in France called upon Pasteur to find out why the contents of their fermentation vats were turning sour. He noted through his microscope that the fermentation broth contained not only yeast cells but also bacteria that could produce lactic acid. He was able to prevent such souring by a mild heat treatment, which later became known as *pasteurization*. One of his greatest contributions was to establish that each type of fermentation was mediated by a specific microorganism. Furthermore, in a study undertaken to determine why French beer was inferior to German beer, he demonstrated the existence of strictly anaerobic life, that is, life in the absence of air. Interest in the mechanisms of these fermentations resulted in the later investigations by Pasteur, which not only advanced microbiology as a distinct discipline, but also led to the development of vaccines and concepts of hygiene, which revolutionized the practice of medicine. With the establishment of the germ theory of disease by Pasteur and Koch, the latter half of the nineteenth century was characterized by the fight against disease and the attention of microbiologists was directed to the medical and sanitation aspects of microbiology. Owing to the work of Pasteur and Koch, it became evident that the body's own defenses played a great part in fighting pathogenic microbes. It was found that when a bacterium invaded the body of a human or an animal, proteins (i.e., antibodies) were formed in the bloodstream. These could specifically neutralize the invading parasite. The science of immunology was thus founded. By injecting either dead forms or attenuated forms of the disease-producing bacterium, Pasteur could render the individual immune to the disease. The production of these vaccines occupied much of the early research in microbiology.

The application of antiseptics materialized t the time of the contributions made by Pasteur. It had been shown in 1846 by Semmelweis that chlorine could control infection, and in 1865, Joseph Lister showed that the same could be done with carbolic acid. Later, Paul Ehrlich used synthetic dyes and established the concept of the "magic bullet." Toward the end of the nineteenth century, Ehrlich began testing many synthetic compounds. He achieved success in 1909, curing relapsing fever, syphilis, and trypanosomiasis with an arsenical product called *Salvarsan* or *Compound 606*, because it was his 606th attempt to produce an arsenical compound which killed the syphilis bacterium *in vivo* without harming the host. This was the first chemotherapeutic drug ever discovered and he coined the term *chemotherapy*. This use of drugs selectively toxic to the parasite but not damaging to the host opened an entirely new field for the curing of human diseases. In 1927, this work was continued by Gerhard Domagk in Germany along with his collaborators Mietzsch and Klarer. They were working at the I.G. Farbenindustrie which was the result of a 1924 merger between Bayer and BASF. Their work resulted in the development of the red-colored molecule Prontosil rubrum. This compound was active in mice against streptococci but strangely was not active *in vitro*. Then in 1935, Trefouel and co-workers in France discovered that the red dye was broken down in the animal to the colorless and inhibitory sulfanilamide. This discovery of the first "pro-drug" also established the important concept that chemicals could kill or inhibit bacteria without toxicity to humans. Although the Nazi government refused to permit Domagk to accept the Nobel Prize in 1939, he later accepted it in 1947. Other synthetic chemotherapeutic drugs gained wide use over the years, including isonicotinic acid hydrazide and *para*-aminosalicylic acid, both for tuberculosis.

For thousands of years, moldy cheese, meat, and bread had been employed in folk medicine to heal wounds. In the 1870s, Tyndall, Pasteur, and William Roberts, a British physician, directly observed the antagonistic effects of one microorganism on another. Pasteur, with his characteristic foresight, suggested that the phenomenon might have some therapeutic potential. During the ensuing 50 years, various microbial preparations were tried as medicines, but they were either too toxic or inactive in live animals. This led to the momentous moment in microbiological history, when, in 1927, Alexander Fleming discovered penicillin (see Section 1.3).

In 1877, Moritz Traube proposed that (i) proteinlike materials catalyzed fermentation and other chemical reactions and (ii) they were not destroyed by such activities. This was the beginning of the concept of what we call enzymology today. He also proposed that fermentation was carried out via multistage reactions in which the transfer of oxygen occurred from one part of a sugar molecule to another, finally forming some oxidized compound such as carbon dioxide and a reduced compound such as alcohol. The field of biochemistry became established in 1897 when Eduard Buchner found that cell-free yeast extracts, lacking whole cells, could convert sucrose into ethanol. Thus, the views of Pasteur were modified and it became understood that fermentation could also be carried out in the absence of living cells.

During World War I, the need for glycerol, used to manufacture ammunition, resulted in the application of yeast to convert sugars into glycerol. This development led to an exhaustive study after the war of the mechanisms involved in these reactions and those converting sugars to ethanol by Neuberg. This was followed by the studies of the Dutch in Delft dealing with oxidation/reduction reactions and the kinetics of enzyme-catalyzed reactions.

Also during World War I, Chaim Weizmann of the United Kingdom applied the butyric acid bacteria, used for centuries for the retting of flax and hemp, for production of acetone and butanol. His use of *Clostridium* during World War I to produce acetone and butanol was the first nonfood fermentation developed for large-scale production; with it came the problems of viral and microbial contamination that had to be solved. Although use of this fermentation faded because it could not compete with chemical means for solvent production, it did provide a base of experience for the development of large-scale cultivation of fungi for production of citric acid. Soon after World War I, an aerobic process was devised in which *Aspergillus niger* was used (see Section 1.2). Not too many years later, the discoveries of penicillin and streptomycin and their commercial development heralded the start of the antibiotic era (see Section 1.3).

1.2 Primary Metabolites and Enzymes

1.2.1 Birth, Rise, and Decline of the Term "Biotechnology" in the Period 1900–1940

The word "biotechnology" was coined around 1919 by the Hungarian agricultural engineer Karoly Ereky, who used the term in the title of his book "Biotechnologie der Fleish-, Fett-, und Milcherzeugung im Landwirtschaflichen Grossbetriebe" (Biotechnology of meat, fat, and milk production in large-scale agricultural industries). Ereky, who later became Hungarian Minister of Food, had established a large intensive pig-rearing farm and processing plant close to Budapest, Hungary, where pigs (called *biotechnological working machines*) converted agro- and waste streams into meat, fat, and leather. In the previous decades, chemistry had merged with technology and had resulted in a novel fast-growing industry, the chemical industry. Erecky envisaged agriculture and biology combined with engineering to lead to a new industrial revolution. His vision, soon to be applied to microorganisms, rather than macroorganisms, became popular among agrobiologists, chemists, and engineers. On the basis of his perception, by fermenting cheap and abundant agricultural produce and waste, both the farmers and the chemical industry became beneficiaries. This vision led scientists and engineers to produce a range of "bio"-chemicals (solvents, alcohols, organic acids, and enzymes) using starch/sugar-fermenting microbes. Although inspirational to many scientists and engineers, his new term *biotechnology* was hardly used at all and was almost forgotten until 1975–1980, whereas the then existing terms such as industrial fermentation and industrial microbiology remained widely used till the late 1980s [2].

1.2.2
Influential Scholars Boosting Industrial Fermentation from 1900 to 1940

In the late nineteenth century, several renowned scientists believed that the emerging industrial application of microbiology would form a new type of industry, differing from the then rapidly growing (petro)chemical industry. This idea was, at least in Europe, based on the huge importance and value of the German beer industry at the turn of the nineteenth century; it was second only to machinery building and surpassed metallurgy and coal mining. Indeed, on the basis of Pasteur's theories and practical findings in France, combined with those of Koch and Cohn in Germany, Lister in the United Kingdom, and Emil Christian Hansen in Denmark, brewing had evolved from an art into a controlled and well-understood malting, mashing, and yeast fermentation process. Also at that time, yeast culture collections were established in Prague, Delft, Berlin, and fermentation and brewing research institutes were founded (Pasteur Institute, Paris; Carlsberg Institute, Copenhagen; Institut fur Gärungsgewerbe (Institute for Fermentation Industries), Berlin). They soon gained impact and fame and still continue to function today, although under other names. In 1898, an English translation appeared of Franz Lafar's famous two-volume handbook in German, "Technical Mycology: The Utilization of Micro-organisms in the Arts and Manufactures." Lafar, the first director of the Vienna Technical Institute, became famous for his improvements of alcohol fermentation and distillery practice.

World War I brought on innovative fermentation applications. In the United Kingdom, Chaim Weizmann, who was trained at the Institut Pasteur in Paris, worked at Manchester University closely together with a brewing equipment manufacturer, R. Seligman, who had introduced the plate heat exchanger. In 1915, Weizmann developed a suitable method to ferment potato starch and grain with anaerobic bacteria to produce the chemical, acetone, on a large scale. Acetone was essential for the manufacture of much-needed ammunition for the British Army. In Germany, in 1915, W. Connstein and K. Lüdecke developed fermentation processes for glycerol, lactic acid, and yeast for animal feed under the pressure of World War I. In the 1920s, Ereky's "Biotechnologie" vision was soon applied to microorganisms (rather than to pigs) by the German microbiologist Paul Lindner, a pupil of Koch, based at the Inst. Gärungsgewerbe, Berlin. This trend was followed up especially in Czechoslovakia, The Netherlands, the United Kingdom, and the United States.

In the 1930s, at the Charles University in Prague, Prof. Konrad Bernhauer became a fervent promoter of the fermentation-based chemistry. His classic textbook of 1936, "Gärungschemisches Praktikum" (Practical Chemistry of Fermentation), condensed the knowledge of fermentation in Europe and the United States. After World War II, he became an important mentor of German scholars at the Inst. Gärungsgewerbe. Ereky and Bernhauer can be considered the prewar advocates of industrial uses of (micro)biology; however, their Nazi links caused their names to become forgotten in later years.

In the Netherlands, at the Technical University Delft, the fermentation metabolism group of J.A. Kluyver became influential in the 1920s and 1930s in providing basic insights of microbial growth, metabolism, and production potential. In 1921, Kluyver became the Chair of General and Applied Microbiology, upon Martinus Beijerinck's retirement. In 1924, he investigated the production of sorbose by *Acetobacter suboxydans* and collaborated with the "Nederlandse Gist-en Spiritus Fabriek" (The Dutch Yeast and Alcohol Manufacturing Company). Over the next few years, he described chemical transformations performed with microbes in a scientific way, including oxidations, fermentations, and incomplete oxidations. By 1926, he had published his famous paper, "Unity and Diversity in the Metabolism of Microorganisms," and explained the term *facultative anaerobes.* He also made a valuable and industrially relevant contribution by developing the technique of submerged culture of molds, later to be widely used in the fermentation industry. His PhD student, C.B. Van Niel discovered and developed the aroma compound of butter, that is, diacetyl, important as a bioflavor for the growing margarine industry. Van Niel left Delft toward the end of 1928 to accept an offer to become Professor at Stanford University's Hopkins Marine Station, in Pacific Grove, CA, USA. His research on photosynthetic bacteria revolutionized the concept of the biochemistry of photosynthesis.

In the United States, especially at the US Department of Agriculture (USDA), employing about 600 chemists in 1915, research boosted the industrial use of biology, especially due to the dislocation of chemical supplies as a result of World War I. The dairy chemist, James Currie, worked on the production of citric acid with the mold *A. niger*. He persuaded, in 1923, the then small, New York-based Chas. Pfizer and Co. to support him. He developed the surface fermentation process in shallow trays to convert sugar into citric acid, which until then, had to be extracted from lemons and other citrus fruits. In 1929, Pfizer switched to submerged fermentation based on the research of Bernhauer. Also in the 1930s, well before World War II, the Research Director of the Dow Chemical company, William J. Hale, promoted heavily the use of chemicals, including ethanol (called agricrude-alcohol), made from cheap farm produce. This principle was named "Chemurgy" by him. He advocated the creation of "Agricenters" for processing of farm products into industrial end products and of raw materials for other process industries. In the meantime, USDA researchers in Washington, DC, developed microbial processes for the production of other organic acids from sugar and starch (i.e., lactic acid, gluconic acid, and others) for use in the food and other industries. However, their laboratory was abandoned to make space for construction of the Pentagon. Four new regional laboratories were set up, including the Northern Regional Research Laboratories (NRRL) in Peoria, Illinois during 1939–1940. It was there that on July 14, 1941, Florey and Heatley arrived from Oxford University with Fleming's penicillin fungus *Penicillium notatum* in their coat pockets! (see Section 1.3).

These developments in biotechnology, during the 1900–1930s, occurred along with those in petrochemical engineering to form a novel and separate field of science and technology. The German chemical companies (e.g., Bayer, BASF, and Hoechst) and several oil companies (BP, Shell, and Standard Oil) were set up to become established firms, although they were not as important then as they became later. In retrospect, the oil crises in 1973 and 1979 forced the chemical and oil industries again to reorientate and this also boosted renewed interest in Ereky's "Biotechnologie" and Hale's "Chemurgy" concepts. The term *industrial biotechnology* surfaced again in the 1980s.

1.2.3 Milestone Achievements in Industrial Fermentation Technology

1.2.3.1 The Acetone–Butanol–Ethanol (ABE) Fermentation Process

By the start of the twentieth century, shortages of natural rubber activated interest in alternative feedstocks and in chemical routes to produce synthetic rubber. This attracted the attention of the young chemist, Chaim Weizmann, who was assistant to Prof. W.H. Perkin at Manchester University, UK. The chemical company Strange and Graham Ltd (London) had also shown interest in a process to prepare butadiene or isoprene, building blocks of rubber by oxidation of *n*-butanol or isoamylalcohol, both obtainable by fermentation of sugars. They recruited Perkin and Weizmann to work on this project. This joint research project had to be refocused during World War I, because Britain's need for acetone as a solvent for the manufacturing of smokeless explosive cordite became critical. Butanol and acetone had already been reported as fermentation products by Pasteur in 1861 and F. Schardinger in 1905. Weizmann was able to select a superior strain of *Clostridium acetobutylicum*, which produced commercially interesting levels of acetone, butanol, and ethanol (ABE) using cereals as feedstock, and he filed a patent in 1915. Owing to the German blockade, Britain soon experienced a shortage of grain and decided to move the solvent production plants to Canada and India. Also in the United States, the US Air Service and the British War Mission purchased the Commercial and Majestic Whiskey distilleries in Terre Haute, Indiana, and modified them for acetone production using the Weizmann process. The Commercial Solvents Corp. of New York managed the new company. Between May and November 1918, 400 000 l of acetone were produced with 800 000 l of *n*-butanol as a coproduct. These surpluses of *n*-butanol became valuable during the prohibition era in the United States (1920–1933), as it could replace amylacetate in lacquers (for automobiles). Butanol also found use in solvents, plasticizers, paints, and resins. From the 1930s onward, the butanol fermentation process was largely superceded by its petrochemical production route. However, today, it is again gaining commercial interest. Weizmann's research work on acetone and its essential role in the British war period was recognized by the Cabinet Minister of Armament David Lloyd George, who later became Prime Minister. Weizmann, always a fervent proponent of a homeland for the Jewish people, was to later (1948) become the first President of the State of Israel. After 1940, continued interest in solvent fermentations [3, 4] led to the further development of the butanol fermentation [5].

1.2.3.2 A Novel Vitamin C Fermentation Process

Early observations on microbial oxidations of sugar alcohols (polyols) culminated in a novel process for vitamin C (ascorbic acid) in the 1930s. In 1867, Pasteur had observed that certain bacteria, which he called *Mycoderma aceti*, oxidized the alcohol in wine into acetic acid (to make vinegar). In 1886, in the United Kingdom, Adrian J. Brown used *Bacterium aceti* (now *Acetobacter aceti* subsp. *xylinum*) to oxidize mannitol to fructose, *n*-propanol to propionic acid, and ethyleneglycol to glycolic acid. In 1898, G. Bertrand reported on the microbial oxidation of other polyols to ketones, for example, sorbitol to sorbose, using Brown's strain; this also laid the basis for the Bertrand–Hudson rule. Revisiting the work of Bertrand in the early 1930s, Tadeus Reichstein from the Chemistry Department of ETH in Zurich, Switzerland, successfully devised a microbial approach for oxidizing D-sorbitol to L-sorbose, an important intermediate in the chemical synthesis of vitamin C. This bioconversion step worked so efficiently that the company F. Hoffmann-La Roche AG in Basel decided to produce vitamin C via this chemoenzymatic route, rather than extracting it from fruits. They used *A. suboxydans* cultures to convert 20% solutions of D-sorbitol into L-sorbose with yields of up to 97%.

Today, various combinations of chemical and microbiological approaches are still used to meet the high demand for vitamin C as a nutriceutical and an antioxidant [6–8]. Other vitamin processes important today include riboflavin [9–12] and vitamin B12 [13, 14].

1.2.3.3 The Lactic Acid Fermentation Process

In 1857, Pasteur described what he called a *lactic yeast*, responsible for the formation of lactic acid, when advising a distillery experiencing difficulties in the fermentation of sugar beet juice to ethanol. Lactic acid remained a specialty product until 1883, when the young MIT-educated chemist Charles E. Avery built the first lactic acid fermentation plant, the Avery Lactate Company, in Littleton, near Boston, Massachusetts; the fermentation substrate used was hydrolyzed corn starch. Avery's aim was to replace cream of tartar (potassium bitartrate) used as an acidulant in the bakery sector. This project was initially successful but after a fire ruined the plant in 1911, several other US companies replaced Avery's company in this effort. Competition also came from lactic acid producers in Germany, for example, Boehringer Co., Knab and Lindenhayn and E. Merck. They switched later to whey, molasses, or sugar as substrate [15]. During World War I, lactic acid production increased considerably in Germany, to meet the military requirements to replace glycerol Also in the United Kingdom and France, new facilities for the lactic acid fermentation were built.

In the United States, the group of L.A. Rogers at the USDA's Bureau of Dairy Industry introduced the use of pure cultures in the American dairy industry and devised a continuous fermentation process for lactic acid based on whey. In 1936, based on the work of the Rogers group, large-scale operation was realized by Sheffield By-Products Company, Norwich, NY, using wooden fermentation vats to grow *Lactobacillus bulgaricus* and stainless steel equipment to counteract the corrosive properties of lactic acid. Also, American Maize Products Co., DuPont, and Clinton Corn Syrup Refining Co. started to produce lactic acid from glucose-rich starch hydrolysates and corn steep liquor in the late 1930s. Then, in Europe, several new lactic acid plants became operational, for example, Byk-Guldenwerke and C. H. Boehringer Sohn (Germany), Société Normande de Produits Chimiques (France), Bowmans Chemicals (UK), Schiedamsche Melkzuur Fabriek (SMF, later named CCA) (The Netherlands), and Kemisk Vaerk Koge (KVK) (Denmark). Other companies were established in Italy, Hungary, Czechoslovakia, Poland, Romania, and Russia. Also, five companies in Japan (Takeda Chemical Industries, Tanabe, Sankyo, Dai Ichi Seiyaku, Dai Nippon Industries) producing lactic acid were operational in 1937. Synthetic lactic acid, manufactured first by Musashino Chem. Lab. Ltd, Tokyo, then came on the market. Petrochemical technology was also increasingly applied to produce ethanol, acetone, and butanol. However, new applications were developed for lactic acid and its derivatives in the food industry, and the medical, health and technical sectors. This led to a revival of the lactic acid fermentation process [16, 17] with several new companies that are still active today.

1.2.3.4
Fermentative Production of Glycerol

In Germany, during World War I, factories focused on manufacturing glycerol by fermentation, equally needed for their weapons and explosives industry. This was based on the "Protol" process, developed in 1915 by W. Connstein and K. Ludecke. They had found that addition of sodium bisulfite to a yeast ethanol fermentation process using beet sugar diverted it into a process yielding glycerol. In 1919, Carl Neuberg and co-worker J. Hirsh revealed the mechanism of bisulfite action, that is, it prevented alcohol formation and rerouted toward glycerol overproduction. For some time, 24 factories in Germany produced 12 000 tons of glycerol per year for use in the production of explosives. Today, glycerol is available in large quantities as a valuable side product of biodiesel production and of fat hydrolysis.

1.2.3.5
L-(−)-Ephedrine by Fermentation

In 1921, Neuberg and Hirsch discovered that yeasts could condense added benzaldehyde with pyruvate-derived acetaldehyde to form the chiral product, L-(+) phenylacetylcarbinol (also named Neuberg's ketone). This alpha-hydroxyketone (acyloin) can easily be chemically converted into L-(−)-ephedrine, an important bronchodilator still made by a bioprocess.

1.2.3.6
Steroid Transformations

An old paper by Lintner and von Liebig in 1911 on the reduction of furfural to furfurol by yeast attracted the interest in 1937 of Mamoli and Vercellone, former students of C. Neuberg. It inspired them to use yeast to reduce 4-androstenedione to testosterone. This was the first example of a successful microbial steroid transformation, to be followed by many more in the early 1950s.

1.2.3.7
The Citric Acid Fermentation Process

Citric acid was commercially produced from the 1820s until about 1919 from Italian lemons; then, microbial citric acid took over. In 1893, Carl Wehmer, while at the Technical College in Hannover, Germany, became interested in mycology and studied fungal metabolic acids, oxalic acid, and citric acid. Two fungal species of a genus which he called *Citromyces (=Penicillium)*, were able to produce considerable levels of citric acid when grown on the surface of 10% sugar solutions. He recognized the importance of his findings and applied for patents in 1894. However, owing to technical and sterility issues, his project never got beyond the pilot scale. Several years later, in 1917, James N. Currie, a dairy scientist at the USDA, who was aware of Wehmer's papers, examined several other molds. He discovered *A. niger* to be a good producer of citric acid when cultured in media with low pH, high sugar levels, and mineral salts. He informed Chas. Pfizer & Co. Inc., then a major producer of lemon-derived citric acid, of his findings and requested commercial interest toward his microbial process. He was then hired by the company and asked to develop his findings into a commercial process. This formed the basis of the first citric acid plant in the United States, in 1923. The Pfizer plant dominated the citric acid market for many years to come and they also built an overseas plant in the United Kingdom in 1936. Similar industrial processes had started in Belgium (in 1919) and led to companies such as the companies Citrique Belge S.A., John and E. Sturge in the United Kingdom, Montanindustrie J. D. Starck A.G. in Czechoslovakia, and Boehringer and J.A. Benckiser in Germany; they all used the "surface culture" process. Details are not well documented owing to the restriction of information by the manufacturers. The fungal mycelium was grown as a surface mat on liquid beet molasses medium in a large number of shallow trays, stacked in a large room kept under semiaseptic conditions; spores were blown in with a sterile air stream. This process was profitable for many years and was only challenged in the 1940s by the development of submerged fermentation processes for citric acid. Today, the latter is still the main process for fermentative production of citric acid [18, 19] as well as other organic acids [20, 21], including acetic acid [22–25].

1.2.3.8
Gluconic Acid Process

Formation of gluconic acid was first observed by Boutroux in 1880 using the bacterium *M. aceti* b (*A. aceti*). In 1922, Molliard described formation of gluconic acid by the mold *A. niger*, along with citric acid and oxalic acid. A few years later (1924), Bernhauer found an *A. niger* strain that almost exclusively formed gluconic acid, when grown as thin mats on glucose solutions at low temperature. Over the coming decades, this fermentation process was intensively studied and optimized by researchers at the USDA and in Japan, where surface as well as submerged fungal fermentations under increased air pressure and at high glucose levels (up to 35%) were developed. Today, such processes are used for large-scale gluconic acid production. 2-Keto-gluconic acid is also produced by fermentation [26].

1.2.3.9
Other Important Fermentation Processes and Products

Other bacterial-based fermentations such as 2,3-butanediol, acetoin, dihydroxyacetone, keto-gluconate, propionic acid, vinegar, and "old traditional" fermented foods (e.g., cheese, yoghurt, pickles, and sauerkraut) were studied during this period both to gain more basic microbiological and biochemical understanding, as well as to develop large-scale controlled fermentations. During the period 1900–1930, important "traditional" yeast-based fermentations, such as the production of food, baker's and feed yeasts, beer brewing, beverage, industrial, and fuel alcohol, were further optimized and reached high volumes worldwide. From the 1930s onward, industrial and fuel alcohols were increasingly produced by chemical synthesis from petroleum feedstock. In the United States in 1936, about 84% of ethanol was still produced by fermentation of different agro-derived substrates (molasses, grain, sulfite liquor, etc.), while only 16% was made from ethyl sulfate via chemical synthesis; in 1946, the figures changed to 64% versus 36%, respectively. This chemical synthesis trend continued for a while to overtake the use of fermentation. However, owing to high petroleum prices and environmental concerns, the tide turned and industrial and fuel ethanol also began to be made microbiologically [27–31]. Other fermentations based on fungal strains, including itaconic acid, kojic acid, fumaric acid, and gallic acid, have been studied by several research groups in Japan, the United States, and in Europe. Their industrial production became very important after World War II. Also of great significance were the fermentations developed in the late 1900s for amino acids [32] especially those for L-glutamic acid [33–37] and L-lysine, as well as those devised for 5′ nucleotides such as guanylic (GMP) and inosinic (IMP) acids [38–40]. Fermentative production of polymers such as dextran, xanthan [41], polyhydroxy butyrate [42, 43], and polylactic acid (PLA) [44] also became important.

1.2.3.10
Applied Biocatalysis and Industrial Enzymes

Although several practical developments in the field of biocatalysis date from the first half of the nineteenth century (e.g., use of diastase extracted from malted barley in the brewing industry) and Emil Christian Hansen's enzyme preparation, rennet, for cheese making (1874), scientific background on enzymes only emerged later in the nineteenth century. This was based on the findings of Emil Fisher starting in 1894 on enzyme specificity and its "lock and key" action and on the 1897 work of Eduard and Hans Buchner on the pure chemical nature of the alcohol fermentation in the absence of living yeast cells. The soluble agent in yeast press juice was called *zymase*. The work eliminated the "*vis vitalis*" (vital force) paradigm altogether. A further key step toward the "chemical paradigm" was the work of J.B. Sumner in 1926 on the crystallization of jack bean (*Canavalia ensiformis*) urease and on the protein nature of enzymes. In the 1930s, several more enzymes were isolated, purified, and crystallized from plants, animal organs, as well as from yeasts, molds, and bacteria. Technical developments on enzymes started at the onset of the twentieth century with the founding of the Rohm and Haas Company in 1907 in Germany, and the description of several practical enzymatic reactions with crude amylase, lipase, protease, trypsin, pepsin, invertase, and others. The kinetic studies by Michaelis and Menten in 1913 were also very important toward the understanding of the physicochemical nature of enzyme action. The Japanese scientist Jokichi Takamine, working in the United States (Peoria, Illinois), was the first to patent a microbial enzyme product (1894). This "Takamine" process involved extraction with aqueous ethanol of extracellular amylases (named "Taka-diastase") from *Aspergillus oryzae*, growing on bran (similar to the ancient Japanese koji process). Early in the twentieth century, plant lipases were produced by mechanical disruption of ricinus seeds and used to produce fatty acids from oils and fats. It was also found that this reaction is reversible and the enzymatic synthesis of fat from glycerol and fatty acids was described as early as in 1911. Proteolytic enzymes were successfully used in 1911 in the United States for the chillproofing of beer. Wheat diastase was found to interact beneficially with dough making and the addition of malt extract became a common practice in bread baking. Production of pectinases started in Europe in the 1930s for use in the fruit juice sector. For leather manufacturing, early tanners kept the animal skins in a warm suspension of dog and bird dung, not knowing that this unpleasant bating practice was based on the action of enzymes (pepsin, trypsin, lipase, etc.) present in animal dung. Once this mechanism was revealed in 1898, a bacterial bate was developed from *Bacillus erodiens* cultures and commercialized as a bacterial culture ("Erodin") adsorbed on wood meal. In 1907, pancreatic extract was introduced as a bating agent by O. Rohm, who founded his own company in Stuttgart, Germany. With the trade name "Oropon," his product became very successful and he moved production to larger facilities in Darmstadt. Here, a growing market, searching for a new and pleasant technical product, was an important factor in his success. It also led to the increasing knowledge on the principles of enzymatic action. Further development of large-scale submerged fermentation processes for enzymes has led to increased industrial production and applications of enzymes. This happened in the late 1950s, with the emergence of detergent enzymes and use of glucoamylase to produce glucose from starch.

1.3 The Antibiotic Era

1.3.1 Penicillin

The very first recorded observation on microbial "antibiosis" dates back to 1877 [45]. Pasteur and Joubert described slower growth of *Clostridium* sp., in the presence of other bacteria. In 1893, Bartolomeo Gosio, an Italian physician, discovered a compound in the culture filtrate of *Penicillium brevicompactum*, which, in pure crystallized form, inhibited growth of *Bacillus anthracis*; it was later rediscovered and named mycophenolic acid [46]. Although it was never used as an antibiotic, owing to its toxicity, a derivative found use as a new immunosuppressant. In the early 1920s, André Gratia, a microbiologist at the University of Liége, Belgium, studied the lysis of bacteria by products derived from other microorganisms. He was one of the first phage researchers after F. d'Herelle, belonging to the period before the viral nature of bacteriophages became clear. In 1925, Gratia described the bacteriolytic effect of certain fungi, including a *Penicillium* strain that exerted this action on anthrax-causing bacteria. Owing to an illness, Gratia did not further pursue this research topic [47].

The accidental discovery of penicillin by Alexander Fleming in 1929 in England began the golden era of antibiotics. He noted that some of his plates containing *Staphylococcus aureus* were contaminated with the mold, *P. notatum*. Strangely, he observed that none of the bacterial colonies could grow in the vicinity of the mold and concluded that the mold was producing some inhibitory agent. He also noted that filtrates of the mold lyzed the staphylococci and were nontoxic in animals. Because of his earlier discovery and studies on lysozyme, he recognized this as an important phenomenon to pursue. He coined the name penicillin for the antibacterial substance in the mold culture broth, and published his findings in 1929. Since the activity was very unstable and Fleming could get no encouragement from his fellow scientists concerning the usefulness of such material, the project was abandoned by Fleming.

Although Fleming's discovery led to penicillin, the first successful chemotherapeutic agent produced by a microbe, thus initiating the golden age of the wonder drugs, the road to development of penicillin as a successful drug was not an easy one. Attempts to isolate penicillin were made in the 1930s by a number of British chemists, but the instability of the substance frustrated their efforts. For a decade, penicillin remained a laboratory curiosity. With the advent of World War II, and the deaths of many British soldiers on the battlefield from bacterial infections after being wounded, a study of penicillin began in 1939 at the Sir William Dunn School of Pathology of the University of Oxford by Howard W. Florey, Ernst B. Chain, Norman G. Heatley, Edward Abraham, and their colleagues. Chain had searched through the papers of Fleming and others on lysozyme, penicillin, and metabolites with inhibitory effects on other organisms.

In the years 1939–1941, under eminent war pressure, Fleming's *P. notatum* strain was grown as a surface culture in large flasks, fermentation media were optimized, penicillin activity measurements were improved, solvent extraction of penicillin from the broth was optimized, toxicity tests in animals (mice) were conducted, as well as protection tests by penicillin of mice infected with lethal doses of virulent *Streptococcus* sp. This amazing effort led to the successful preparation of a stable form of penicillin and the demonstration of its remarkable antibacterial activity and lack of toxicity in mice. Subsequent clinical trials on humans were very successful and time had come for commercial production. Production of penicillin by the strain of *P. notatum* in use was so slow, however, that it took over a year to accumulate enough material for a clinical test on humans; large-scale production became essential. Since British pharmaceutical companies did not show interest, the Oxford University administration was forced to contact the Rockefeller Foundation in New York that had funded their research. Florey and Heatley were sent to New York in 1941 and met with R.G. Harrison, Chair of the National Research Council, who advised them to contact the USDA. There they met Percy A. Wells, acting Chief of the Bureau of Agriculture and Industrial Chemistry. He was in charge of the four regional research laboratories and, fortunately was a fermentation specialist. He sent Florey and Heatley to the USDA's NRRLs in Peoria, Illinois, resulting in a historical outcome.

Florey and Heatley convinced the NRRL and several American pharmaceutical companies (including Merck, Squibb, and Pfizer), to develop the production of penicillin. Heatley remained for a period at the NRRL to work with Moyer and Coghill and then worked for a while at Merck. Thus began a momentous cooperative effort among university and industrial laboratories in the United States and academic institutions in England, which lasted throughout the war.

Although Fleming's original strain produced only traces of penicillin, "brute force" genetic manipulation made tremendous strides in production ability and led to a whole new technology known as *strain improvement*. These early basic genetic studies concentrated heavily on the production of mutants and the study of their properties. The ease with which "permanent" characteristics of microorganisms could be changed by mutation and the simplicity of the mutation technique had tremendous appeal to microbiologists [48]. A cooperative "strain-selection" program was established between workers at the USDA in Peoria, the Carnegie Institution at Cold Spring Harbor in New York, Stanford University and the University of Wisconsin. Strain selection began with *Penicillium chrysogenum* NRRL-1951, the well-known isolate from a moldy cantaloupe obtained in a Peoria market. This strain was capable of producing 60 $\mu g\,ml^{-1}$. Cultivation of spontaneous sector mutants and single-spore isolations led to higher-producing cultures from NRRL 1951. One of these, NRRL 1951-1325, produced 150 $\mu g\,ml^{-1}$. It was next subjected to X-ray treatment by Demerec of the Carnegie Institution, and the mutant X-1612 was obtained, yieldeding 300 $\mu g\,ml^{-1}$. Workers at the University of Wisconsin obtained ultraviolet-induced mutants of Demerec's strain. One of these, Q-176, which produced 550 $\mu g\,ml^{-1}$, became the ancestor of all of the strains subsequently used in industry. The

"Wisconsin family" of superior strains became well known all over the world, some producing over 1800 μg ml^{-1}. The penicillin improvement effort was the start of a long engagement between genetics and industrial microbiology, which ultimately demonstrated that mutation was the major factor involved in the 100- to 1000-fold increases obtained in production of microbial metabolites.

Thousands of lives were saved, on and off the battlefield. This discovery and development of the β-lactam antibiotics was among the most powerful and successful achievements of modern science and technology. Since Fleming's accidental discovery of the penicillin-producing mold, years of steady progress followed, and today, the β-lactam group of compounds is the most successful example of natural product application and chemotherapy.

The discovery and development of the sulfa drugs and penicillin have been reviewed by Dixon [49] and Bentley [46].

Penicillin had been originally produced in surface culture, but titers were very low. Submerged culture soon became the method of choice. By the use of strain improvement and medium modifications, such as the use of corn steep liquor as additive, the yield of penicillin was increased by 100-fold in just a few years. Much of the understanding of the physiology of *P. chrysogenum*, in relation to penicillin production, was achieved by Professor Marvin Johnson and his students at the University of Wisconsin. Further clinical successes were demonstrated in both England and the United States, and finally in 1943, penicillin was used to treat those wounded in battle.

By the 1950s, it was realized the *P. chrysogenum* could use additional acyl compounds as side-chain precursors (other than phenylacetic acid for penicillin G) and produce new penicillins, and one of these, penicillin V (phenoxymethylpenicillin), achieved commercial success. Its commercial application resulted from its stability toward acid which permitted oral administration, an advantage it held over the accepted article of commerce, penicillin G (benzylpenicillin). Penicillin G and penicillin V became the main penicillins of commerce. In commercial production, the usual medium had been a complex one, composed of glucose, corn steep liquor, side-chain precursor (phenylacetic acid for penicillin G or phenoxyacetic acid for penicillin V), and mineral salts. The earliest recognition that glucose had a negative effect on penicillin biosynthesis was made by Johnson and his students. They found glucose to be excellent for growth but poor for penicillin formation, while lactose showed the opposite pattern. They devised a medium containing both sugars in which growth occurred at the expense of glucose and when it was exhausted, the mass of cells began to produce the antibiotic at the expense of lactose. Unlike glucose, lactose was utilized slowly and did not exert carbon catabolite repression on the process. Then, Davey and Johnson found that intermittent or continuous feeding of the less expensive glucose could replace batch feeding of lactose. This represented the birth of the method of fed-batch fermentation which is commonplace in the fermentation industry today.

The biosynthesis of penicillin from its precursors, L-cysteine, L-valine, and phenylacetic acid, was actively studied during the 1950s, 1960s, and 1970s. Of great interest was the relationship between L-lysine and penicillin formation. In

1947, David Bonner made the observation that 25% of the lysine auxotrophs that he had made from *P. chrysogenum* failed to make penicillin and he predicted that (i) there was some relationship between the antibiotic and the amino acid and (ii) there was a common precursor of the two compounds. That he was absolutely correct was established later when Demain and co-workers found that lysine was a potent inhibitor of penicillin biosynthesis [50, 51]. That the inhibition could be reversed by L-α-aminoadipic acid led to the postulations that (i) L-α-aminoadipic acid was involved in penicillin biosynthesis although it did not end up in the final penicillin molecule, (ii) penicillin was derived from α-ketoglutarate and acetyl-CoA via the fungal lysine biosynthetic pathway, and (iii) lysine inhibition of penicillin biosynthesis was due to feedback inhibition by lysine of its own biosynthetic pathway, thus limiting L-α-aminoadipic acid formation. Independently, Arnstein and colleagues detected the tripeptide α-aminoadipyl-cysteinylvaline (ACV) as an intracellular compound in *P. chrysogenum*. Results in several laboratories established L-α-aminoadipic acid as an important precursor of all penicillins. Soon, α-(L-aminoadipyl)-L-cysteinyl-D-valine (LLD-ACV) was established as the crucial intermediate of penicillin biosynthesis. The reaction, sensitive to feedback inhibition by lysine, was later proven to be the initial step of lysine biosynthesis in fungi, that is, the homocitrate synthase reaction.

The future of penicillins became doubtful in the 1950s as resistant strains of *S. aureus* emerged in hospital populations. The staphylococcal population was building up resistance to penicillin via selection of penicillinase-producing strains, and new drugs were clearly needed to combat these resistant forms. Also, the penicillins described up to this point were solvent-soluble, exhibiting a high degree of activity against gram-positive organisms but were much less active against gram-negatives. Fortunately, two developments occurred which led to a rebirth of interest in the penicillins and related antibiotics. One was the 1959 discovery in Japan by Koichi Kato of the accumulation of the "penicillin nucleus" in *P. chrysogenum* broths to which no side-chain precursor had been added. In England, Batchelor and co-workers isolated the compound (6-aminopenicillanic acid, 6-APA) which was the "penicillin nucleus" discovered by Kato. 6-APA was later used to make "semisynthetic" (i.e., chemical modification of a natural antibiotic) penicillins with the beneficial properties of resistance to penicillinase and to acid, plus broad-spectrum antibacterial activity.

The second development was the discovery of a completely different type of penicillin, a hydrophilic type, showing equivalent activity against both classes of microorganisms (i.e., penicillin N). This compound was discovered independently by two groups of workers. Brotzu in 1948 published his work in an unknown Sardinian journal on the isolation of an antibiotic-producing culture of the fungus *Cephalosporium acremonium* (later reclassified as *Acremonium chrysogenum*) from sewage. Unable to purify the antibiotic material, he sent the culture to Florey at Oxford where some 10 years earlier, *P. notatum* had been at a similar stage in history. While the British were studying the components of this antibiotic complex, workers at the Michigan Department of Health announced that a species of *Tilachlidium* produced a new antibiotic which they called *synnematin*.

After the culture was reclassified as *Cephalosporium salmosynnematum*, it was shown that synnematin was a complex of two components, A and B. Almost nothing was ever published on the A component, and we have no knowledge today about its structural relationship to synnematin B. While the above work was going on in Michigan, the British announced that Brotzu's strain produced two antibiotics, that is, "cephalosporin P," active only against gram-positive organisms and "cephalosporin N," which was active against both gram-positive and gram-negative bacteria. "Cephalosporin P" was found to be of steroidal nature and not a β-lactam at all. "Cephalosporin N," on the other hand, was found by Abraham and co-workers to be a true penicillin possessing an α-D-aminoadipyl side chain, and to be identical to synnematin B. It was renamed "penicillin N." In comparison to penicillin G, it was only 1% as active against gram-positive forms, but had equal to or somewhat greater activity against gram-negative bacteria. The hydrophilic nature of penicillin N and its roughly equivalent activity against gram-positive and gram-negative bacteria were due to the carboxyl group in the side-chain.

The tremendous success attained in the battle against disease with penicillin G not only led to the Nobel Prize being awarded to Fleming, Florey, and Chain, but to a new field of antibiotic research, and a new antibiotic industry. Penicillin opened the way for the development of many other antibiotics, and it still remains the most active and one of the least toxic of these compounds. Today, more than 100 antibiotics are used to combat infections to humans, animals, and plants.

1.3.2
The Cephalosporins

A key breakthrough was the finding in Edward Abraham's laboratory at Oxford mentioned above, that is, that a second antibiotic was produced by Brotzu's strain of *A. chrysogenum*. After his earlier contributions as part of the Florey penicillin team, Abraham had established an independent laboratory at Oxford. Abraham and Newton found the new compound to be related to penicillin N in that it consisted of a β-lactam ring attached to a side chain which was identical to that of penicillin N, that is, D-α-aminoadipic acid. It differed, however, from the penicillins in containing a six-membered dihydrothiazine ring in place of the five-membered thiazolidine ring of the penicillins. It was called *cephalosporin C*. Thus, the era of the cephalosporins was launched.

The nucleus of cephalosporin C was named 7-aminocephalosporanic acid (7-ACA). Cephalosporin C strongly absorbed ultraviolet light, was stable to acid and to penicillin β-lactamase, was non-toxic and had *in vivo* activity in mice. Its mode of action was the same as that of the penicillins; that is, inhibition of bacterial cell wall formation. Although neither penicillin N nor cephalosporin C was ever commercialized, they led to important knowledge on the biosynthesis of these compounds and the development of many powerful semi-synthetic cephalosporins of great use in medicine.

The stability of cephalosporin C to penicillin β-lactamase was a very attractive property. The main disadvantage of the molecule was its weak activity; it had only 0.1% of the activity of penicillin G against sensitive staphylococci, although its activity against gram-negative bacteria equaled that of penicillin G. However, by chemical removal of its D-α-aminoadipic acid side chain and replacement with phenylacetic acid, a penicillinase-resistant semi-synthetic compound was obtained which was 100 times more active than cephalosporin C. Many other new cephalosporins with wide antibacterial spectra were developed in the ensuing years, for example, cephalothin, cephaloridine, and cephaloglycin, making the semi-synthetic cephalosporins the most important group of antibiotics at that time. The stability of the cephalosporins to penicillins was evidently a function of its dihydrothiazine ring because (i) the D-α-aminoadipic acid side chain did not render penicillin N immune to attack and (ii) removal of the acetoxy group from cephalosporin C did not decrease its stability to penicillinase. Cephalosporin C competitively inhibited the action of penicillinase from *Bacillus cereus* on penicillin G. Although it did not have a similar effect on the *S. aureus* enzyme, certain of its derivatives did. Another major advantage was that cephalosporins could be given to some patients who were allergic to penicillins.

From a biosynthetic sense, the relationship of penicillin N and cephalosporin C was of great interest. An important development that led to rapid progress in this area was the subcellular work done by Abraham and his Oxford colleagues in the 1970s [52]. They used protoplast lysates of *A. chrysogenum* to convert labeled valine into a penicillin. This led to the discovery by Kohsaka and Demain at MIT in 1976 [53] of the ring expansion reaction, catalyzed by the “expandase” (deacetoxycephalosporin C synthase; DAOCS) enzyme. For many years, it had been thought that penicillin N and cephalosporin C were products of different biosynthetic branches in *A. chrysogenum*. However, the discovery of the ring expansion enzyme showed that cephalosporins were produced from a penicillin. This was confirmed by Yoshida *et al.* [54] who showed that the ring expansion enzyme converted penicillin N to deacetoxycephalosporin C. Extracts of mutants which fermentatively produced penicillin N but not cephalosporins failed to carry out this reaction, whereas early blocked mutants (negative for production of both penicillin N and cephalosporins) did expand the ring.

From the 1970s through the 1980s, knowledge accumulated concerning fermentation and biosynthesis of cephalosporin C. The following were of major importance: the (i) stimulation by DL-methionine via a regulatory mechanism unrelated to its ability to contribute the sulfur atom to the antibiotic, (ii) use of acetate as precursor to the acetoxy group, (iii) L-cysteine and L-valine as precursors of the nucleus, and (iv) L-α-aminoadipic acid as precursor of the D-α-aminoadipyl side chain of cephalosporin C. An important step forward was provided by Banko and co-workers [55] when they demonstrated that the cell-free activity of the enzyme from *A. chrysogenum* formed the important tripeptide precursor of all penicillins and cephalosporins, that is, LLD-ACV. This enzyme, ACV synthetase, was proven to be a single multifunctional enzyme acting on L-α-aminoadipic acid, L-cysteine, and L-valine to produce LLD-ACV. Also of importance was the isolation by Hollander and co-workers [56] of pure isopenicillin N synthetase ("cyclase") which converted the LLD-ACV to isopenicillin N.

The long-held notion that β-lactams were produced only by fungi was shattered by a report from Merck & Co. that a streptomycete produced penicillin N. This provocative finding was presented at the 1962 American Society for Microbiology Annual Meeting and was only published as an abstract. Although much doubt was cast on this report, two reports from Eli Lilly & Co. [57] and Merck & Co. [58] published some 9–10 years later reported that various species of *Streptomyces* and *Nocardia* produced cephalosporins modified at C7 (=cephamycins) and/or at the side chain attached to C3. The discovery of cephamycin C led to much research on and development of prokaryotic cephalosporins as the presence of the methoxy group on the β-lactam ring made the molecule more active against gram-negative and anaerobic pathogens and more resistant to gram-negative β-lactamases. For the first time in the history of the β-lactams, molecules were available which showed a high degree of stability toward these troublesome enzymes. As with fungal cephalosporin C, cephamycin C was never used clinically but was employed for semisynthesis of many medically useful compounds. A more potent semisynthetic cephamycin, cefoxitin, was rapidly commercialized by Merck, to be followed later by cefmetazole, temocillin, cefotetan, and other semisynthetic cepahlosporins.

In the 1970s and the 1980s, the pathways to the penicillins and the cephalosporins including cephamycin C were worked out especially after cell-free systems became available [59]. Late in the 1970s came reports on the production of β-lactam antibiotics which were neither penicillins nor cephalosporins. The most important was clavulanic acid from streptomycetes, which possessed only weak antibiotic activity but was an excellent inhibitor of β-lactamase [60]. It became a blockbuster compound by being coformulated with broad-spectrum semisynthetic penicillins which are susceptible to β-lactamase, for example, with amoxycillin, the combination is known as *Augmentin*.

Another important development in the history of the β-lactam antibiotics was the discovery of the carbapenems at Merck. The first, called *thienamycin*, was discovered by Kahan and his group with a screening protocol based on inhibition of peptidoglycan synthesis [61]. The antibiotic was produced by *Streptomyces cattleya*, which also made cephamycin C. Carbapenems resembled the penicillins in having a β-lactam ring fused to a five-membered ring. They differed in that the five-membered ring was unsaturated and contained a carbon atom instead of the sulfur. Sulfur was, however, present in another location in all the carbapenems produced by streptomycetes. A large number of carbapenems were reported, but thienamycin was the most important. Indeed, it was the most potent, most broad-spectrum, and nontoxic natural antibacterial agent ever found. It inhibited cell wall synthesis, as did the penicillins and cephalosporins, and was relatively resistant to microbial β-lactamases.

The development of commercial process technology for the penicillin nucleus (6-APA) and the cephalosporin nucleus (7-ACA) opened the way for chemical acylation with various side chains yielding many improved semisynthetic penicillins and cephalosporins with broader antibacterial activity and improved pharmacokinetic properties. Broad-spectrum penicillins and cephalosporins became the best-selling antibacterial agents in the pharmaceutical arena.

1.3.3 The Waksman Era

The advent of penicillin, which signaled the beginning of the antibiotic era, was closely followed in the 1940s by the discoveries made by Selman A. Waksman, a soil microbiologist at Rutgers University. He and his students, especially H. Boyd Woodruff, Albert Schatz, and Hubert Lechevalier, succeeded in discovering many new antibiotics from the filamentous bacteria, the actinomycetes, such as actinomycin, neomycin, and the best-known of these new "wonder drugs," streptomycin. These discoveries on the antibiotic-producing abilities of the actinomycetes occurred long before the developments, described above, of β-lactams produced by these filamentous bacteria. Waksman and Woodruff published in 1940 on the discovery of the actinomycins, which were chromo-oligopeptides. One such compound, actinomycin D, has been used for years to combat the Wilms tumor in children and became a very important tool in the development of molecular biology as an inhibitor of RNA polymerase.

After its momentous discovery in 1944 by Waksman, Schatz, and Bugie as a product of *Streptomyces griseus*, streptomycin was used against tuberculosis caused by *Mycobacterium tuberculosis* and also against gram-negative bacteria; bacterial meningitis was also treated with streptomycin. Its major impact on medicine was recognized by the award of the Nobel Prize to Waksman in 1952. As the first commercially successful antibiotic produced by an actinomycete, this aminoglycoside led the way to the recognition of these organisms as the most prolific producers of antibiotics. Streptomycin also provided a valuable tool for studying cell function. After a period of time during which it was thought to act by altering permeability, its interference with protein synthesis was recognized as its primary effect. Its interaction with ribosomes provided much information on their structure and function; it not only inhibited their action but also caused misreading of the genetic code and was required for the function of ribosomes in streptomycin-dependent mutants. With Lechevalier, Waksman reported on the discovery of neomycin in 1948 and candicidin in 1953. Neomycin, an aminoglycoside produced by *Streptomyces fradiae*, served as a topical antibacterial, and the polyene candicidin, made by *S. griseus*, found use as a topical antifungal antibiotic.

Cooperation on the development of industrial processes between Rutgers University, Princeton University, Columbia University, and Merck & Co., Inc. led to the birth of the field of biochemical engineering. With royalties on streptomycin turned over to Rutgers University by the manufacturer, Merck, Waksman was able to build the world-famed Institute of Microbiology.

The discoveries of the aminoglycosides at Rutgers further advanced the antibiotic era and resulted in the later discovery of many more "wonder drugs" such as chloramphenicol in 1947, tetracyclines in 1948, macrolides such as erythromycin in 1952 [62], glycopeptides such as vancomycin in 1956, additional aminoglycosides such as gentamicin in 1963, β-lactams such as cephamycins in 1970, and carbapenems in 1979, ansamycins such as rifamycin in 1957, and polyene macrolides such as nystatin in 1950. Of 15 000 microbial secondary metabolites discovered, about 12 000 were antibiotics. Their unusual chemical structures included β-lactam rings, cyclic peptides containing "unnatural" and nonprotein amino acids, unusual sugars and nucleosides, polyenes, and large macrolide rings. Although most were useless for humans, being either too toxic or inactive in higher organisms, others were lifesavers. The antibiotics were virtually the only drugs utilized for chemotherapy against pathogenic microorganisms and were crucial in the increase in average life expectancy in the United States from 47 years in 1900 to 74 for males and 80 for women in 2000.

For some reason, the actinomycetes were amazingly prolific in the number of antibiotics they could produce. Roughly 70% of all antibiotics were obtained from these filamentous prokaryotes, and 75% of those were in turn made by a single genus, *Streptomyces.* It is quite amazing that strains of *Streptomyces hygroscopicus* produce over 180 different secondary metabolites. About 10% of the antibiotics are made by unicellular bacteria and about 20% by fungi. New bioactive products from microbes were discovered at an amazing pace: 200–300 per year in the late 1970s, increasing to 500 per year by the 1990s. Accompanying the natural product antibiotics in the pharmaceutical arena were synthetic antimicrobials such as the quinolones and fluoroquinolones. Even these synthetics traced their discovery back to a natural product, that is, quinine. The first quinolone, nalidixic acid, was modeled after quinine. Commercialization of antibiotics, however, slowed down in the 1980s and only three, daptomycin, caspofungin acetate, and the synthetic oxazolidinone were commercialized in the ensuing decades.

1.3.4 Mutagenesis and Strain Improvement

The experiences described above in which penicillin-producing cultures were mutagenized and screened for improved producers led the way to the extensive use of genetics to improve production capability. From the 1950s on, genetic manipulation such as mutagenesis/screening was replaced by mutagenesis/selection/screening in which various selective means were used to decrease the number of strains that had to be screened for improved production. Then, it was found that new derivatives, some better than the parent molecule, could be produced by mutants. This was first discovered by Kelner in 1949 but the more active derivatives were not isolated and identified. However, the medically useful metabolites demethyltetracycline and doxorubicin were discovered later by mutation of the cultures producing tetracycline and daunomycin, respectively. In 1969, University of Illinois Professors Kenneth Rinehart and David Gottlieb, and student W.T. Shier devised and reported the technique of "mutational biosynthesis" (=mutasynthesis) [63]. In this process, a mutant blocked in secondary metabolism was fed analogs of the moiety whose biosynthesis was blocked. If successful, the mutant (called an *idiotroph*) produced a new secondary metabolite. Mutational biosynthesis was used for the discovery of many new secondary metabolites. The most well known was the commercial anthelmintic agent doramectin, production of which employed a mutant of the avermectin producer, *Streptomyces avermitilis* [64].

For strain improvement, genetic recombination was virtually ignored in industry before 1975, mainly owing to the low frequency of recombination, as low as 10^{-6}. However, use of polyethyleneglycol-mediated protoplast fusion in actinomycetes by Okanishi *et al.* [65] changed the situation markedly. Okanishi's work on protoplast formation, fusion, and regeneration accelerated the use of genetic recombination. From then on, there was a heightened interest in the application of genetic recombination to the production of important microbial products. Frequencies of recombination increased to even greater than 10^{-1} in some cases. After 1985, many strain improvement programs routinely included (i) transposition mutagenesis, (ii) targeted deletions and duplications by genetic engineering, and (iii) genetic recombination by protoplast fusion and plasmid transformation. Much was known about genetics and regulation in the actinomycetes owing to the elegant research on *Streptomyces coelicolor* by David Hopwood, Keith Chater, Mervyn Bibb, and their colleagues at the John Innes Institute in Norwich, England (which has been called the Temple of Streptomyces Genetics) [66]. Their favorite organism made at least five secondary metabolites (an A-factor-like molecule, and the antibiotics, actinorhodin, undecylprodigiosin, methylenomycin A, and "calcium-dependent antibiotic" or CDA).

These efforts and others by geneticists in academia and industry throughout the world in the 1970s and 1980s revealed that the genes encoding most antibiotic biosynthetic pathways were clustered into operons, thus facilitating transfer of entire pathways from one organism to another. Such clusters were found to also include regulatory and resistance genes. In 1985, "combinatorial biosynthesis" was born. An international effort from the United Kingdom, Japan, and the United States resulted in the cloning of a pathway from one streptomycete producing the isochromanequinone antibiotic actinorhodin into strains producing granaticin, dihydrogranaticin, and mederomycin (which are also isochromanequinones). This resulted in the discovery of two new hybrid antibiotics, mederrhodin A and dihydrogranatirhodin [67]. Combinational biosynthesis became a widely used technique used for discovery of new hybrid drugs [68] by recombinant DNA (rDNA) technology. New antibiotics were discovered [69–71] including those created by changing the order of the genes of an individual pathway in its native host.

Progress in strain development has recently involved extensive use of new genetic techniques such as (i) metabolic engineering, accomplishing quantification and control of metabolic fluxes, and including inverse metabolic engineering and transcript expression analyses such as association analysis and massive parallel signature sequencing; (ii) directed evolution; (iii) molecular breeding including DNA shuffling and whole genome shuffling; and (iv) combinatorial biosynthesis. These efforts facilitated not only the isolation of improved strains but also the elucidation and identification of new genetic targets to be used in product discovery.

1.3.5
Semisynthetic Antibiotics to Combat Resistant Microbes

Although there was a feeling expressed by many in the late 1970s that the era of product discovery for bacterial disease was ending, the battle against resistant microbes continued and featured some surprising developments. These included the following: (i) semisynthetic variations of old antibiotics (ketolides, clarithromycin, azithromycin, glycylglycines); (ii) older underutilized antibiotics (teicoplanin); (iii) new derivatives of previously undeveloped narrow-spectrum antibiotics (streptogrammins); and (iv) a few newly developed antimicrobial agents (caspofungin, daptomycin, and the synthetic epothilones).

Of great success was the development of semisynthetic erythromycins. These included clarithromycin, roxithromycin, azithromycin, and the ketolide, telithromycin. While the first two showed improved acid stability and bioavailability over erythromycin A, they showed no improvement against resistant strains. On the other hand, azithromycin and telithromycin acted against macrolide-resistant bacteria. All the above semisynthetic erythromycins were effective agents for upper respiratory tract infections and could be administered parentally or orally. Telithromycin was bacteriostatic, active orally, and of great importance for community-acquired respiratory infections. Of particular interest was its low abilities to select for resistance mutations and to induce cross-resistance. It also did not induce Macrolide-Lincosamide-Streptogramin B-resistance (MLSB), a problem with other macrolides. For more than 35 years, the glycopeptides vancomycin and teicoplanin were virtually the only natural antibiotics active against multidrug-resistant gram-positive bacteria. Their use was cut down by the increase in multidrug resistance. To the rescue came a number of new semisynthetic antibiotics, one called Synercid. Synercid was composed of a synergistic (by 100-fold) pair of narrow spectrum streptogramins, that is, quinupristin and dalfopristin, both being semisynthesized from natural compounds made by a single strain of *Streptomyces pristinaespiralis*. The pair was constituted by a (Group A) polyunsaturated macrolactone containing an unusual oxazole ring and a dienylamide fragment and a (Group B) cyclic hexadepsipeptide possessing a 3-hydroxypicolinoyl exocyclic fragment. Although the natural streptogramins were poorly water soluble and could not be used intravenously, the Synercid components were both water soluble. They inhibited protein synthesis and were active against vancomycin-resistant *Enterococcus faecium* (VREF), methicillin-resistant *S. aureus* (MRSA), glycopeptide-resistant *S. aureus*, and β-lactam-resistant *Streptococcus pneumoniae*. Synergistic action of the streptogramins was due to the fact that the B component blocked binding of aminoacyl-tRNA complexes to the ribosome, while the A component inhibited peptide bond formation and distorted the ribosome, promoting the binding of the B component. Synercid was approved by FDA in 1999. A semisynthetic tetracycline, that is, a glycylcycline, was successfully developed for use against tetracycline-resistant bacteria. The 9-*t*-butylglycylamido derivative of minocycline called *tigecycline*, is active against resistant gram-positive, gram-negative, and anaerobic bacteria possessing the ribosomal protection resistance mechanism or the active efflux mechanism.

In addition to antibiotics, many other extremely important secondary metabolites were discovered or developed over the years, such as hypocholesterolemic agents including compactin, mevinolin, pravastatin; enzyme inhibitors, such as lipstatin, clavulanic acid, polyoxins, and siderophores; immunosuppressants, for example, sirolimus (a rapamycin), cyclosporine A, and tacrolimus (FK-506); antitumor agents, such as taxol and ergot alkaloids; bioherbicides, including biolaphos and phosphinothricin; plant growth stimulants, such as the gibberellins; antiparasitic agents, including monensin, polyethers, and avermectins; and bioinsecticides.

1.4 The Biotechnology Era Between 1970 and 2015

1.4.1 Biotechnology in the 1960s and 1970s; Governmental and Political Initiatives

By the 1950s, large-scale production of products such as beer, cheese, citric acid, amino acids, enzymes, and pharmaceuticals such as antibiotics had become well established. Growing recognition of economic relevance followed the success of penicillin and the manufacture of other antibiotics based on applied microbiology and biochemical engineering. The most important products (commodities) were ethanol, starch products, L-glutamate, and other amino acids, citric acid, and other organic acids [72]. Enzyme application became a large, important technology, and big business, with a turnover of €2.5 billion in 2010 [73].

Starting in the 1970s and 1980s, biotechnology attracted the attention of government agencies in Germany, the United Kingdom, Japan, the United States, and other countries as a field with innovative potential, capable of contributing to economic growth. A first enthusiastic report by the German chemical technology organization DECHEMA was produced in 1974 for the German Ministry for Education and Science (Bundesministerium für Bildung und Wissenschaft, BMBW). It was the first systematic approach for biotechnology research funding, emphasizing classical biotechnology and aiming to develop a unified research and development strategy. The German government wanted to accelerate biotechnology research and development to identify and encourage innovations in industry [2, 74–76]. This study has been an intriguing example of interaction between politicians, industrialists, and scientists and was termed a *corporatist approach* by Jasanoff [77]. Subsequent studies on biotechnology in other European countries were initiated, for example, in the United Kingdom, Japan, and France [2, 74].

Most topics, however, were classical or conventional. The focus of interest was on the identification of new products and new or improved products, processes, and services. rDNA methods were not mentioned, as they were still in early development, and there was no example that this was a truly enabling technology at the time of the study (1972–1974). Interdisciplinary communication between microbiologists, chemists, and engineers was difficult owing to different working procedures, specialized scientific and technical languages, and different approaches to this new field. No integration of methods and theoretical approaches took place or was developed seriously. The major industries in Germany, as well as in the rest of Europe, notably the chemical, pharmaceutical, and food industries, were conservative in their attitude toward the new field of biotechnology. University education essentially remained confined within traditional disciplines, faculties, and curricula. Exceptions to this were the specialized research units at a few UK, European, and American universities that offered special courses (e.g., University College London and the Technical University of Berlin). The first biotechnology journal of high reputation was established in 1958 by Elmer Gaden, "Journal of Microbiological and Biochemical Engineering." It later became "Biotechnology and Bioengineering." A few other journals were launched in the 1950s and 1960s (such as "Applied Microbiology," renamed "Environmental and Applied Microbiology," and "Applied Microbiology and Biotechnology").

In contrast to Europe, in the United States, during the 1970s and subsequently in the 1980s, a more profound change took place including innovation based on recombinant technologies. This is evidenced notably by the emergence of biotechnology companies during the 1970s and 1980s and by an OTA (Office of Technology Assessment, USA) study [78]. The emphasis in the OTA study was on genetic engineering and rDNA technology resulting in commercial opportunities and support for the rapid commercial exploitation of scientific results; it was closely associated with the business world: "This report focuses on the industrial use of rDNA, cell fusion, and novel bioprocessing techniques. In the past ten years, dramatic new developments in the ability to select and manipulate genetic material have sparked unprecedented interest in the industrial uses of living organisms." Emphasis was placed on the pharmaceutical sector, which had been most active in commercializing biotechnology. Thus, the first recombinant biotechnology products, such as rDNA-produced human insulin, interferon (IFN), and monoclonal antibody (mAb) diagnostic kits, were a direct result of the basic research that led to these new technologies. The highlights of the US competitive position were a well-developed life-science base, the availability of financing for high-risk ventures, and an entrepreneurial spirit that had led the United States to the forefront of biotechnology commercialization. The political framework made it possible for industrialists and scientists to rapidly capitalize on the results of basic research. The transfer of science to the economic sphere resulted in the continued founding of many new biotechnology companies; and the development and marketing of a considerable range of new important products, many blockbusters, mostly pharmaceuticals which were previously not accessible. It was a type of gold rush to support the "New Biotechnology," as recombinant technology was known in the United States. The story of insulin (see the following) is an example of both a great breakthrough and the delayed recognition of the potential of genetics and recombinant technologies by established pharmaceutical companies. As an anecdotal example of the reticence prevailing in European Industry at that time, John Collins, in 1975 and later, Herbert Boyer together with Robert Swanson (founders of Genentech) presented their positive viewpoint of the impact of rDNA-genetics on the synthesis of human insulin at the Novo Company in Denmark, which supplied 30% of the insulin world market. Their views were taken seriously by young scientists but, initially, not by management as they believed that there was no possible solution to the problems involved in the industrial development of recombinant bacteria.

1.4.2
The Development of New Biotechnology Companies

The turning point in genetics ensued from the establishment of a model for the molecular structure of DNA by James Watson and Francis Crick, based on the crystallography data of Rosalind Franklin. But the "*DNA Revolution*" as Hotchkiss termed it, progressed or penetrated slowly into technology, initially having little effect on traditional processes and products [79]. Significant scientific breakthrough events and technological progress provided a new basis for BT (see Section 1.5). New methods and tools that played a key role in the expansion of recombinant technologies include analysis of DNA, RNA, and proteins, and of their structure, synthesis of short DNA molecules, the identification and purification of DNA molecules that code for pharmaceutically active proteins, introduction of such DNA (also from human sources) into bacteria, and expression of the protein in bacteria.

The story begins in 1972 with the concepts for cloning – transfer of DNA from different sources into bacteria in 1972 (see Section 1.5 for details).

The spark of invention has been vividly remembered by Cohen [80] himself, Boyer, Falkow [81], and two others with the remarkable event when, during an evening in an Hawain Deli, during a conference, the concept for cloning was conceived and written on a napkin: The experiments in the laboratories of both Boyer and Cohen during the following month were successful and published in 1973 in the Proceedings of the National Academy of Sciences USA (PNAS). "Boyer's restriction enzyme (*Eco* RI) became the workhorse of the recombinant DNA revolution." It was given by Boyer to other labs working in the field ([82], p. 59). Among the difficulties was screening at the time, but Boyer's team had been able to identify the few bacterial colonies – literally one in a million – containing the toad DNA. "Herb just said he kissed every colony on the plate, until one turned into a prince" (Falkow, cited by Hall [82], p. 63). This was the basis for the first patent on recombinant technologies by Cohen and Boyer, filed in 1974, acknowledged in 1980. (For the controversial debate on patenting, see Section 1.5).

The key to the founding and the success stories of new BT companies, and the expansion of recombinant technologies, were new tools primarily based on recombinant technologies (for details, see Section 1.5).

The patent office in 1980 acknowledged that the claims by Cohen and Boyer, for the use of plasmids and restriction nucleases constituted a novel invention of recombinant-DNA technology. Stanford's decision to patent the rDNA technique and the breadths of the claims upheld by the Patent Office were widely noted, and they set patterns, both in academia and in industry [83].

In the controversy about patenting biological issues, the pro arguments were that biotechnology companies provided the most effective means of translating advances in science and technology into products for medicine and agriculture [82–84] and, of course, there were the precedents of Pasteur's patent on pure cultures of yeast and Chakrabarty's patent on multiplasmid-containing strains able to degrade xenobiotics (made by classical microbiology techniques, not rDNA) [72].

One of the questions asked in Science and Technology Studies (STS) is how basic science leads to technology and how this is converted into innovation. Biology, unlike engineering, physics, and chemistry had no tradition of basic scientists being routinely involved in innovation. Consequently, the founding of Genentech by Boyer and Swanson, the first biotechnology company based on gene cloning, in 1976, has become a paradigm for just such a transition. Swanson, a venture capitalist, and Boyer, a microbiologist and biochemist, had the joint vision to commercialize rDNA. After ups and downs, they admirably achieved their goal, an example to be followed by many others, and a trend that revolutionized modern medicine. Genentech at the outset had the immediate goal of producing pharmaceutical products and exploiting the capacity of bacteria for the production of hormones, such as human insulin and human growth hormone, which could be put to immediate clinical use.

It is to be noted that the first modern biotechnology company, Cetus, was founded in Berkeley, California in 1972, producing mainly chemicals for industry and agriculture and employing about 250 co-workers in 1980 [85]. Cetus was based on a different type of "cloning technology," namely, on the biophysics of scanning and deflecting droplets (originally the Coulter counter), so that single cell-based colonies (called clones) could be plated on growth media. This latter was combined with microscopic scanning of a large grid, a technology also used later on a microscale for high-throughput DNA sequencing, so that the automated screening of millions of single mutant clones could be carried out. Cetus after it failed owing to a delay in FDA approval of their rDNA product interleukin-2 was later incorporated into the Chiron Corporation (Berkeley, 1991). At Cetus, Kary Mullis succeeded with a highly important innovation, the polymerase chain reaction (PCR) method for amplification of DNA (even from tiny amounts). As a potent analytical tool, it had a tremendous influence on various fields of modern biotechnology, even for the identification of criminals as developed by Alec Jeffreys (University of Leicester, UK). Mullis was awarded the Nobel Prize in 1993 and remains the only employee rather than founder of a biotechnology company to win a Nobel Prize. PCR, in turn, led directly to the large-scale development of the biochip industry by the Chiron spin-off Affymetrix (for the development of microarrays and DNA-CHIPs see [72], section 11.4.3) and now allows isolation of any gene from a known genome sequence.

Boyer and Swanson conceived Genentech, the first gene cloning company, in 1976, when they decided to commercialize rDNA. It was founded in the same year, with the immediate goal of producing pharmaceutical products, and to exploit the capacity of bacteria for the production of hormones, such as human insulin and human growth hormone, which could be put to immediate clinical use. The story will be presented in some detail, as it was this success that finally convinced industry that the new biotechnology was really an approach to be taken seriously, as, in particular, the final product was approved for clinical use in a surprisingly short time after the gene had been cloned and expressed (4 years project development time).

Boyer at Genentech and Arthur Riggs and Keiichi Itakura from the Beckman Research Institute, were the first to successfully express a human gene (for somatostatin, a hypothalamus hormone used for several hormonal disorders, previously made by synthesis) in bacteria in 1977 [72]. Somatostatin's small size made it particularly appealing to Riggs and Itakura to synthesize the gene for it. It could be interesting in itself (although necessary for a relatively rare disorder, gigantism). For Swanson, validating the development model was necessary, that is, providing the gene for a protein (e.g., a hormone) and cloning and expressing it in bacteria. The immediate plan was to go for insulin, in order to enter a big market [82]. In addition, cooperation agreements were made with the University of California, San Francisco (UCSF) and the City of Hope National Medical Center, California. The company strategy comprised to make as many patents as possible, to produce scientific publications in high-ranking journals, and to present results at international meetings, a strategy that should, and did, attract many brilliant young scientists.

Involved in the race for insulin, and the "DNA gold rush," were some of the most outstanding scientists in molecular biology, namely, Walter Gilbert of Harvard, Howard Goodman and William Rutter of UCSF, Boyer with David Goeddel, Axel Ullrich, and Peter Seeburg of Genentech [82, 83]. Goeddel, described as a "kamikaze scientist" by colleagues became the first full-time Genentech employee, followed, with some delay, owing to their commitment to their academic careers as postdoctoral fellows at UCSF, by Axel Ullrich, Peter Seeburg (both from Germany) and John Shine (from Australia). They became a formidable, highly qualified, and engaged staff trained in rDNA cloning. Ullrich was the first to clone rat insulin, while Seeburg succeeded in cloning the rat growth hormone at UCSF, but they had been excluded as inventors on patents. In early 1980, the Genentech staff already amounted to 110 employees, 80 of whom were scientists. The years, 1978–1982, were golden ones in the Genentech laboratories, during which bright minded spirits, working furiously, "won many of the cloning races" [83]. The "mad race" won by Goeddel and Ullrich at Genentech over Walter Gilbert at Harvard, for cloning the gene for human insulin, has been recounted engagingly by Hall [82].

Genentech helped to develop 7 of the 13 types of biotechnology-based biopharmaceuticals on the market in 1993; its scientists produced 250 papers annually and 1200 granted patents by 1993. The first important contract was with AB Kabi (Sweden) for the production of somatostatin [85].

The first product commercially manufactured using gene technology was human insulin, which was approved for clinical use in 1982. A connection between Genentech and the Eli Lilly and Company was crucial. It provided Genentech financial assistance, improved the primitive Genentech expression systems to achieve commercially viable levels, provided the manufacturing facilities for a qualitative drug, and contributed the regulatory expertise, which, in 1982, brought the first rDNA drug to the market [83]. For this development, it was important that the Supreme Court of the United States allowed the patenting of new recombinant bacteria [85].

This marked the beginning of the new rDNA era of biotechnology. It now became possible to produce human proteins, hormones, IFNs, interleukins (ILs), antibodies, and so on, for medical treatment of many diseases, a tremendous advance which had not been possible earlier. It created a new "big business," with the introduction of many blockbuster pharmaceuticals. The demonstration that approval and worldwide marketing of insulin could be achieved in only 4 years after the insulin gene had been cloned belied the considerable skepticism that prevailed within the pharmaceutical industry. In particular, the demonstration that a small bioreactor of some 10 l was sufficient to provide enough material for the entire world market of human growth hormone was a dramatic step and went a long way in stimulating enthusiasm for investment in the new area of biotechnology [72] (see Table 1.3; Section 1.4.4.2).

The following are among the impressive list of innovative recombinant medicinals produced by Genentech (later, Genentech/Roche): human growth hormone (1985), IFN α-2a (1980) (in cooperation with Hoffmann-La Roche, Switzerland), Protropin (1985), Activase (tPA) (touted to be the blockbuster drug of the decade) (1987), Rituxan (1997), and Herceptin (1998); and, in addition, Thymosin α_1, proinsulin, and leucocyte and fibroplast IFNs (U. Behrendt, 2009, personal communication) [85, 86]. Scaling up was important, when, in 1979, Genentech approached the rDNA Advisory Committee for permission to use 750-l-reactors, which was granted.

The founding of Genentech had been achieved with $100 000, which was extended by $950 000 in 1978. In the initial 1980 public stock offering, Genentech raised $35 million. For tPA, Genentech had mobilized a huge force to make and market the drug during the early 1980s; however, it was no longer the house of science. Genentech was taken over by Hoffman La-Roche AG, (CH), in 1990, with a majority share of 60% for $2.1 billion, with full control and integration into the company in 2009 ([83], pp. 200–202).

Other selected biotechnological companies are listed below in chronological order.

Genex Corp. Rockville, Maryland, USA was founded 1977, with a main interest in chemicals for industry and IFN [85].

Biogen S.A., was founded in 1978 by a group of leading molecular biologists in Geneva, Switzerland, among them Heinz Schaller and Walter Fiers. It was first run by Dan Adams, and from 1982 by Walter Gilbert. The company first succeeded in the production of an IFN (Avonex) based on the research of Charles Weismann, Zurich. It generated revenues via licenses to other pharmaceutical companies, for example, Schering Plough (Intron A), and via a number of hepatitis B-based vaccines sold by Smith-Kline Beecham and by Merck [82, 83, 85, 87].

Amgen (named for Applied Molecular Genetics), another success story, was started in 1980 in Thousand Oaks, California on the initiative of a venture capitalist, William K. Bowes, who convinced Winston Salser of the Moleculat Biology Institute at UCLA to participate. Salser immediately set out to assemble a distinguished scientific advisory board, including Eugene Goldwasser of the University of Chicago, creator of valuable small amounts of erythropoietin (EPO). The diverse interests of the scientific advisors were reflected in the many directions Amgen took in the early years, apparently confused and without adequate resources, in contrast to Genentech. The leadership desperately needed finally came when George Rathman accepted the presidency. He started with private financing of $19 million, by far the largest initial equity financing in biotechnology history. He then succeeded in early 1983 to raise $43 million via public offering. An early investment by Schering Plough (which thus entered the biotechnology field) was also important. Amgen succeeded in the development of the first two blockbuster drugs for patients with blood disorders or patients undergoing chemotherapy: (i) Epogen (EPO, erythropoetin) for which the market grew to unforeseen dimensions, with excellent clinical responses in 1987 and (ii) Neupogen (G-CSF, granylocyte colony-stimulating factor), entering clinical trials, also in 1987. Amgen also developed Neulasta, as well as an interleukin-1 (IL-1) receptor antagonist. Two more rounds of financing raised another $150 million. In 2002, Amgen acquired Immunex Corporation, another biopharmaceutical company. By mid-2006, seven of its recombinant products had been approved [83, 87].

Chiron Corp., Emeryville, California, founded in 1978 by William J. Rutter, pursued, unlike other start-ups, a wide menu of pharmaceuticals, including vaccines, diagnostics, and therapeutics. Hepatitis Delta (HDV) was cloned and characterized in Michael Houghton's laboratory in 1986 and in 1989 also Hepatitis C virus (HCV) [72, 83]. This latter along with PCR-based diagnostics allowed much-improved (1000-fold increased sensitivity; also for HIV) safety for products still derived from blood serum, for example, for serum albumen used to stabilize other rDNA products used intravenously.

Large pharmaceutical companies entered the field of recombinant drugs later than the new biotechnology companies. They either acquired new biotechnology companies or established in-house know-how, or combined both strategies. Conservative as they had been, many neglected the potential of recombinant technologies, and lost their original strength in several cases, even undergoing a significant decline ([72], section 17.6.3). Remarkable changes in the structure of the pharmaceutical industry occurred during the last century, with the trend toward "big pharma," mainly in the United States and Britain, by mergers and acquisitions (globalization) in the late 1990s and subsequent years and by the emergence of the new biotechnology companies. Data from Ernst & Young indicate that over the past few years, biotechnology firms have become the primary source of approved new medical entities (NMEs). In 2005, major pharmaceutical companies garnered only 11 NME approvals, whereas biotechnology firms had 18, with research budgets of only about one-quarter the size of that of the pharmaceutical industry [88a,b]. In 2007, the world market for pharmaceuticals was over $600 billion. A list of the largest biotechnology companies and their sales is shown in Table 1.1, with sales.

A most important market is diagnostics, with sales of about $35 billion, dominated by Roche with $9.2 billion in 2008, followed by Abbot, Johnson & Johnson, and Bayer [90, 91]. In one important segment, notably diagnostics, enzymes and mABs products are developed by rDNA tools, pioneered from the early 1980s by Boehringer Mannheim (Germany) (taken over by Roche in 1997). The large-scale production of the first recombinant enzyme was achieved in 1981, as well as an enzyme immunoassay for the identification of potato viruses, and, in 1982, test kits for human diagnosis and food analysis with recombinant enzymes.

The strong performance of the core biotechnology firms, that is, small- or medium-sized enterprises, in drug discovery has been of major importance. A boost in the founding of new companies followed during the 1980s and 1990s,

Table 1.1 (a) The largest pharmaceutical companies, with sales in 2014 and (b) the largest biotechnology companies with sales listed [72, 89a,b].

(a) Major pharmaceutical companies (selling biopharmaceuticals)	Sales ($ billion) 2014	(b) Major biotechnology companies	Sales ($ billion) 2014 (except other data given)
Johnson & Johnson (USA)	74.3	Gilead Sciences Inc. (USA)	24.9
Pfizer (USA)	49.6	Amgen Inc. (USA)	20.0
Roche (CH)	48.0	Novo (DK)	12.9
Sanofi-Aventis (F)	41.0	Genentech (USA)[a)]	10.5 (2008)
Merck & Co. (USA)	42.2	Biogen Idec Inc. (USA)	9.7
AbbVie (USA)[b)]	20.0	Celgene Corp. (USA)	7.7
Eli Lilly (USA)	19.6	Genzyme Corp. [c)] (USA)	4.0 (2010)
Total Big Pharma	430.5	Serono [d)] (CH)	2.8 (2006)
		Chiron (USA) [e)]	1.9 (2006)
		Total sales of large biotech companies	62

New data for 2014 from C&EN 2015, March 9, pp. 25, 26.

a) Taken over by Hoffman La-Roche AG, CH, in 1990, with a majority share of 60%, with full control and integrated into the company in 2009.
b) AbbVie formed by split of former Abbot Laboratories in 2013.
c) Taken over by Sanofi in 2011.
d) Taken over by Merck KGA, Germany, in 2006.
e) Taken over by Novartis, CH, in 2005/2006.

first in the United States, and later in Europe. Start-up companies were founded often on a key technology platform. In 2000, there were around 1273 new biotechnology companies with 162 000 employees and $21 billion turnover in United States. About the same time, there were 1570 new biotechnology companies with 61 000 employees and $7.7 billion turnover in Europe (data from and analysis of companies with a maximum of 500 employees in 2000) [92]. Economic data for the biotechnology firms (not including major pharmaceutical companies) have been assembled by Lähteenmäki and Lawrence [88a,b]. Biotechnology companies, whose shares are traded publicly on stock exchanges, that is, 309 firms, generated almost $47 billion in revenue in 2004; 404 firms generated more than $63 billion in 2005 in the United States.

Recently, a remarkable series of new companies, four within 4 years, have been founded in order to exploit a new tool, a technology to explore genomics and genome editing tools of remarkable precision; two founded by the inventors Jennifer Doudna and Emmanuelle Charpentier. These companies created an investment of $345 million within that short time. However, in the area of personalized medicine, some have ethical concerns with respect to germline alterations (as opposed to somatic tissue alterations; see also the end of Section 1.5) [93–95].

1.4.3
New Bioengineering Tools

The key to the founding and the success stories of new biotechnology companies, and the expansion of their technologies, were new tools, primarily based on recombinant technologies. These are discussed in detail in Section 1.5.

Classical biochemical engineering had the following aims [96, 97]: (i) the quantitative investigation of biotransformations and modeling of the processes; (ii) the development of bioreactors and downstream operations; and (iii) the transformation of laboratory results and theoretical approaches (models) into technical dimensions (scale-up). Sustainability (sustainable development) became another relevant aspect, which has been defined as development that meets the needs of the present without compromising the ability of future generations to meet their own needs. Others define it as the optimal growth path that maintains economic development while protecting the environment, optimizing social conditions, and relying on limited, exhaustible natural resources.

Classical fermentative production of products such as ethanol, amino acids, and antibiotics made significant technical progress, including feeding, mixing systems, and process analytics from the 1950s until the 1970s. The essential technologies for production are fermentation, cell culture technology, media design and preparation, downstream processing, filtration (membrane filtration, virus filtration), membrane chromatography, containment technologies, including single-use processes (single-use bags), fermenters, and downstream technologies [72].

The clinical and commercial success that recombinant proteins have had since the mid-1990s, clearly stimulated development of mammalian cell culture technology (for details, see [72], chapter 15 and [98]). The first recombinant therapeutic protein (tissue plasminogen activator, tPA) made in cultivated mammalian cells obtained market approval in 1986 (see Table 1.3), and made Chinese hamster ovary (CHO) cells the dominant host system for manufacturing more than 60% of all new target proteins in the clinical pipeline. Rapid and often dramatic yield improvements were achieved from a titer range of $5-50\,mg\,l^{-1}$ in the early 1980s to $0.5-5\,g\,l^{-1}$ in 2002–2004, an increase by two orders of magnitude. This made it possible to produce complex recombinant proteins for clinical application in kilogram quantities, or even up to a ton per year. The products exhibit all the necessary secondary (post-translational) modifications that only a higher eukaryote can execute, and that are often essential for biological activity and pharmacokinetics: proper folding, disulfide bridge formation, oligomerization, proteolytic processing, phosphorylation, and glycosylation, among others.

Reactor technology remains diversified, with reactor types ranging from roller bottles to stacked plates, hollow fibers, stirred-tank reactors, or disposables such as single-use bioreactors (SUBs), notably for low volume and speciality applications such as the production of viral vaccines [99]. Both CHO and NSO (mouse myeloma derived) cells grow well in suspension with high cell densities. Modern production systems use stirred-tank systems (U. Behrendt, 2009, personal communication) [86]. SUBs are used in the 15–125 l range for seeding larger stainless steel reactors, particularly for mammalian cell culture.

Another important topic is downstream processing. The production of biopharmaceuticals requires the highest standards of sterility and purity, including GMP [72]. Downstream requirements specific for rDNA products include, for example, the renaturation of "inclusion bodies" – proteins aggregated in bacterial cytoplasm.

Bioengineering developed toward biosystems engineering. With the progress of molecular biology, new tools were developed, often referred to as the *omics*, that is, genomics, transcriptomics, proteomics, metabolomics, fluxomics, and more [72]. Bioinformatics has been developed in order to handle, process, and correlate the huge amount of data generated by ever-faster analytical procedures. In order to combine them, a "holistic" approach known as *biosystems engineering* was created. Its aim comprises the quantitative description and improvement of established or developing novel production processes. When searching for a new product or process, it has been proposed that the ideal strategy to follow is a rational procedure of sequencing genomes of microbial species (bacteria, yeasts, or molds) of particular interest for use in metabolic engineering [100]. The next step is annotation and reconstruction of the metabolism. Transcriptome, proteome, and metabolome investigations are essential steps as the basis for rational optimization of a microorganism. Regulatory control of genes is an essential aspect to be considered. Biosystems engineering must take into account the environment of the cell in the bioreactor, namely, concentrations (and their gradients) of substrates, notably oxygen, products, temperature, and pH, that is, the integration of physiology and fluid dynamics in bioprocesses. A further step should address the integration of all steps, from genetic engineering to downstream operations right from the beginning of process development [101]. The integration of all data using bioinformatic tools should enable the construction of networks and models using the genome and further experimental data obtained, both for production processes and target identification for drug design [100]. The analytical methods and tools used include DNA-arrays in transcriptomics (with up to 15 000 oligonucleotides); 2D-gel electrophoresis for qualitative analysis, ELISA for obtaining quantitative data in proteomics; MFA (metabolic flux analysis via mass balances and including isotope materials), FBA (flux balance analysis via modeling) in fluxomics; GC-MS in metabolomics, and classical analysis of substrates, products, pH, pO_2, CO_2, temperature, time, power input, and so on. Genomics encompasses fast automated sequencing, functional sequence analysis, and genome annotation.

The basis of biosystems engineering resides in the molecular structure of biological processes and structures (Figure 1.1). Problems and challenges of biosystems engineering have been addressed recently by the Stephanopoulos group at M.I.T. Their review [102] includes goals such as the improvement of cellular properties, the intelligent design of biochemical pathways and the eventual design of new phenotypes and the engineering of microbial cell factories to produce fuels, chemicals, and pharmaceuticals.

Biosystems engineering stands in contrast to a completely empirical approach for many microorganisms in which rounds of mutagenesis and genome shuffling are combined to optimize metabolic properties of the organism with respect to biotechnological parameters. This can be directed to a single enzyme substrate specificity, or simultaneously to several genes from related organisms, or entire chromosomes. These approaches are based on the "sexual PCR" (DNA shuffling) method pioneered by Willem "Pim" Stemmer (*1954–2013) founder of Maxygen [103].

Information flux		**Mass flux**	
Storage	**Message**	**Product**	**Metabolite A**
DNA	*mRNA*	*Protein*	
(Gen)		*(Enzyme)*	
ATG·CTC ...	AUG·CUC ...	Met·Leu ...	Metabolite A
5000	4000	4000	500–2000
Genome	Transcriptome	Proteome	Metabolome

Figure 1.1 Basic molecular structure of biological production processes with microorganisms.

Synthetic biology is proceeding toward low cost and high volume with expectations for the market to reach nearly $12 billion in 2018, notably using synthetic DNA in organism design [104].

A number of examples of biosystems engineering applied to industrial biotechnology (or "industrial systems biology") have been summarized by Papini *et al.* [105] and Wittmann [106], including examples for commercially highly important products such as ethanol, butanol, polyhydroxyalkanoates, amino acids, polyketides, and antibodies.

The information stored in DNA is transcribed into messenger molecules (mRNA) that in turn encode the synthesis of proteins on the ribosomes. Most of the proteins produced act as catalysts for the reactions in the metabolic network. The entirety of the DNA information is known as the *genome*; for microorganisms, it typically contains from 1000 up to 5000 identified genes. The proteome comprises about 4000 proteins and roughly 2000 metabolites can be identified as comprising the metabolome [100].

Metabolomics is an extension of proteomics and describes the analysis of the catalytic activities that make up the metabolic activities of a cell. Metabolomics is at the core of biosystems engineering and its purpose is to analyze the metabolome (all metabolites, roughly in the range of 500–2000) and their concentrations in the cell under given physiological conditions, as well as the dynamic response to changing environmental conditions (notably in a bioreactor).

Amino acids, citric acid, lactic acid, propanediol, penicillin G, synthetic drug intermediates, and therapeutic proteins are among the industrially relevant products of fermentation and cell culture that have been targets for metabolic engineering. Some of this work has been adopted by industry (see [72], section 16.4.1). The major aim was to optimize the yields of industrial products, which was efficiently realized with *Corynebacterium glutamicum* for lysine and tryptophan, and at the Dupont company for 1,3-propane diol production [107–109].

Bioinformatics aims to effectively deal with data pools from the omics to provide useful interpretations and models. Bioinformatics provides informational techniques specifically enabling access to and interpretation of large amounts of data generated in different fields of biosystems engineering. It is used to develop tools for annotating large genome sequences, analyzing data produced by sequencing machines within a short timescale, interpreting data obtained from proteomics (two-dimensional gels and MS analysis), calculating mass balances in MFA, and determining the kinetics and dynamics from the analysis of metabolic pathways. It comprises understanding and modeling of genomes, proteomes, cell metabolism, and whole-cell biotransformations, and even incorporates the design of cell factories, in addition to modeling cells for new processes and for the formation of new products [110].

The scientific status of biotechnology can actually be recognized as that of a scientific discipline on its own, this being due to the significant progress in the scientific basis, molecular biology, as well as biochemical and biosystems engineering (see Section 1.4.5.1).

1.4.4 Products

The range of products has increased significantly since the 1970s, in the fields of food, feed, industrial commodities, specialties of the pharmaceutical industry, and agriculture. Moreover, services in environmental protection, notably wastewater and exhaust gas treatment, have developed into a large industry (for an extended overview, see Buchholz and Collins [72], chapter 16). Recombinant technology has led to considerable improvements in processing, product purity, reduced costs and prices, and to an extension of the range of products available, notably of pharmaceuticals and in-plant breeding.

1.4.4.1 Food, Feed, Industrial Commodities, and Specialties

For hundreds of years, industrial biotechnology has been a large, traditional technology concerned with producing beer, wine, and bread (see preceding sections), but has nevertheless been a dynamic field which now makes use of the most advanced tools such as rDNA technologies and systems biotechnology. Industrial biotechnology, often termed *white* biotechnology, comprises a broad spectrum of products and processes:

- Fuel, for example, ethanol and biodiesel, and energy, for example, biogas
- Commodities: organic acids, acrylamide, detergents, biopolymers, for example, biodegradable PLA and polyhydroxybutyrate (PHB), and enzymes
- Food and feed ingredients, textile and paper, for example, amino acids, starch derivatives, and sweeteners
- Fine and speciality chemicals, for example, antibiotics, chiral intermediates for pharmaceuticals, agrochemicals, sugars, and derivatives such as sorbitol, specialty enzymes, vitamins, dyes, fragrances, cosmetics, and polysaccharides.

It also aims at the utilization of renewable resources with the challenge of replacing oil-based chemicals and fuels. Environmental processes play a major role in industrial biotechnology, enabling clean and sustainable production and, furthermore, providing a safe environment most notably in crowded areas, big cities, and industrial areas. These technologies represent big business.

Table 1.2 shows a small selection of products; however, many more are actually marketed. For conventional food products, such as beer and wine (about 195 million tons per annum), cheese (about 19 million tons per annum), bakers yeast, vinegar, and others (see [73], chapter 1 table 1.1).

Although the production of biofuels is of considerable economic relevance, it is also characterized by heated political, ecological, and social debates. Recombinant technologies and second-generation fuels are expected to contribute to a reduction in the dependence on fossil resources and to significantly reduce greenhouse gas emissions ([73], section 12.2). A calculation of emissions yielded 94 for gasoline, 77 for currently available bioethanol, and 11 for cellulosic-based ethanol

Table 1.2 Selected products made by fermentation (worldwide, 2003–2005) [72, 96, 111–114].

Product/process	Production ($t\,a^{-1}$)	Price (€ kg^{-1})	Market value (million€)	Company
Ethanol	37 500 000	0.4	15 000	Diverse
Starch products	>10 000 000	—	—	Diverse
Glucose	40 000 000	—	—	Diverse
L-Glutamate	1 500 000	1.20	1800	Ajinomoto, Tanabe Seiyaku (Japan)
L-Lysine	850 000	2.00	1400	Evonic (D)
Citric acid	1 100 000	0.80	880	Diverse
Enzymes	—	—	1830	Novozymes (DK), Genencor (USA)[a)]
HFCS	8 000 000	0.80	6400	ADS, A.E. Staley, Cargill, CPC (USA)
Isomalt	>100 000	—	—	Suedzucker, Cerestar (D)
PLA	140 000	2.52	315	Evonic (D)
Xanthan	40 000	8.40	336	
Penicillins	45 000	—	Total antibiotics: 19 000	DSM(NL), Bayer (D), Kaneka (Japan), a.o.
Cephalosporins	30 000	—	—	
Riboflavin (B2)	30 000	—	—	BASF(D), DSM (NL)
Vitamin C	80 000	8	640	Roche (CH), BASF (D), Takeda (Japan)

Note the inverse relationship between production volume and price.
ADS, Archer Daniels Midland; CPC, CPC International; HFCS, high fructose corn syrup and/or glucose/fructose syrup sweetner; Isomalt, hydrogenated isomaltulose sweetner; and PLA, polylactic acid.
a) Taken over by DuPont (USA).

(measured in kilogram of CO_2 equivalent per MJ of fuel production and burning) [115]. Ethanol is both currently and traditionally one of the most important biotechnology products to emerge since the nineteenth century, although its production has been variable. A total of 46 million m^3 of ethanol was produced worldwide in 2006 and currently 47 and 28 million m^3 are being produced in Brazil and the United States, respectively. To improve the existing technology, the challenges include the optimization of *Saccharomyces cerevisiae* strains using recombinant technologies and obtaining higher osmotic and alcohol tolerance and sufficiently high fermentation rates (yeast being very sensitive to alcohol concentration of $100\,g\,l^{-1}$). In order to identify and use new raw materials, the concept of integrated biorefineries producing commodities from renewable resources has been envisioned. This concept involves processing various feedstocks derived from a range of different biomolecules, including lingocellulosics such as wood straw and whole crops (e.g., maize), into a variety of useful products. The manufacture of such products, which include fuel, energy and materials, building blocks for chemical synthesis, and chemicals, would make use of the entire biomass including by-products [116, 117]. The concept is still in the phase of pilot studies, and the future development of economic conditions will control its realization.

Methanogenesis was detected by Volta in 1776 but it was over 100 years later that this biological process was first exploited technologically. Actually, biogas could, in principle, provide the solution to energy production from any complex biomass, including the huge reservoir of available residues and wastes, estimated at a potential of some 100 million tons per annum (oil equivalent) [118]. Recently, microbial fuel cells (MFCs) have been developed as a most promising concept that makes use of microbial catabolic activity to directly generate electricity from the degradation of organic matter providing access to cheap and environmentally friendly energy sources [119].

Among traditional bulk products glucose is used for fermentation to ethanol, amino acids, and organic acids, with main applications in the food industries, and so on, and as a sweetener, manufactured on a scale of some 40 million tons per annum. Furthermore, it is isomerized enzymatically to give a glucose/fructose syrup (high fructose corn syrup, HFCS) in a process that had been developed and scaled up to very large dimensions during the 1980s ([73], section 8.4). Similarly, the enzymatic isomerization of sucrose to give Isomalt has been established on a large scale. Amino acids, as additives for food, feed, and organic acids, with main applications in the food industries, have been produced in large quantities for many decades, mainly by fermentation but also by enzymatic processes, in the range of over 1 million tons per annum. The success in improving yields by metabolic engineering has been mentioned in the preceding text. Antibiotics, such as penicillin and cephalosporin and their derivatives (e.g., aminopenicillanic and aminocephalosporanic acids), are considered as bulk products in general, as the quantities produced are high and the prices are rather low, in contrast to most pharmaceuticals. New products entering the market during recent years include lactate and 1,3-propanediol as building blocks for polymer synthesis. Thus, the manufacture of PLA, which is based on lactic acid as the monomer, has been established on a large scale. The enzymatic production of 100 000 tons per annum of acrylamide from acrylonitrile was an important economic breakthrough.

For enzymes, a boost occurred from the 1980s on, with recombinant technologies extending the range of enzymes available. The new techniques considerably improved yields, lowered prices, and markedly extended applications, with modified and/or improved selectivity, effectiveness, and stability (both at elevated temperatures and pH). The current market for technical enzymes is about €2.5 billion, due to a significantly extended range of applications. The most important application has been in detergents (34%) and in starch processing to make sweeteners (some 30%), and, later, biofuels. During recent decades, their applications have extended steadily, owing to their excellent stereoselectivity, into the manufacture of fine chemicals (see the following) ([73], p. 15; for details, see chapters 1, 7, and 8).

A most important innovation came from academia, that is, the immobilization of expensive enzymes to give biocatalysts of high stability, high half-life, and low cost, which enabled continuous processing, a breakthrough for large processes. Remarkably, one of the pioneers, Georg Manecke of the Berlin University, found no interest when he offered a patent to the Bayer Company in the mid 1950s. Some 15 years later, another pioneer, Günter Schmidt-Kastner, developed a process for penicillin hydrolysis, to be used in the synthesis of penicillin derivatives at Bayer. This process was successfully scaled up and went into production on a large scale around 1980; it was also developed at Beecham, working together with Malcolm Lilly of Imperial College London, and by other companies [120]. The first process with immobilized enzymes went into production in Japan in 1969 [121]. The largest enzymatic process is glucose isomerization to give a fructose/glucose syrup with production of some 8 million tons per annum.

Cloning of the first industrial enzymes, penicillin amidase and α-galactosidase, both in *Escherichia coli*, was achieved respectively by Wagner, Mayer, and Collins, Braunschweig and by Bückel, Boehringer Mannheim during the late 1970s [122]. Large-scale production of α-galactosidase was carried out at Boehringer Mannheim and for amylases at Novo in the early 1980s. The creation of modified or new activities via generation and design of mutants, directed evolution, and rational design was very successful in the production of chiral chemicals as synthons for pharmaceuticals and agrochemicals. Enzyme discovery, screening, and selection methods have undergone much progress (Bornscheuer, in [73], chapters 3 and 5) (for the history of enzyme technology, see [120, 123]). Hubert Mayer, in the late 1970s, had tried to clone penicillin acylase (PA) without success, an idea that came from Fritz Wagner; the gene of PA at first was not found in the clones. The key for success was the cosmid technique introduced by John Collins, with which the whole *E. coli* genome could be obtained within 200–400 clones ([72], p. 123; [123]). The cosmid patent was filed immediately in 1978, and was the third rDNA patent to be approved worldwide, an entrepreneurial act received negatively by the academic community. Such expression cloning is still used for "empirical gene mining" using DNA from poorly characterized collections (mixtures) of commensal species from various environments even where individual cultivation of isolated species is not possible.

Around 1980, genetic engineering of amylase production was a test case at Novo for food enzymes, going through the approval process, and finally being marketed in 1984.

Fine chemicals made using biotechnology comprise a wide range of products. The more recent innovations include vitamin B12 and the expanding field of chiral organic building blocks for pharmaceuticals (see [72], section 16.4.3; [73], chapter 4). Since the mid-1970s, a new approach and impetus to systematic studies aimed at application of biocatalysis and biotransformation in organic synthesis was undertaken. It included semisynthetic penicillins, esters, and glycerides [124], as well as optically active compounds [125]. It was recognized that biocatalysts could be used for synthesis in organic chemistry, notably for pharmaceuticals and agrochemicals incorporating chirality, that is, the production of pure compounds consisting of one isomer only as the key requirement. An important economic breakthrough was the enzymatic synthesis of chiral aspartame, an intense sweetener, the annual sales of which are in the region of $850 million [126].

Environmental processes play a major role in industrial biotechnology, both enabling clean and sustainable production in industry and providing a sound environment. They comprise wastewater and exhaust air purification as well as soil remediation, and they represent big business ([127, 128], pp. 309–321). Thus, environmental biotechnology is of major importance, considering that all cities worldwide and every town in western countries is obliged to undertake biological wastewater treatment, and that most industries including not only food, pharma, and chemistry but also car factories have a requirement for biological wastewater treatment and in many cases for biological exhaust air treatment systems as well. In 1988, there were about 27 400 aerobic municipal wastewater treatment plants in the European Union. The cost of wastewater treatment worldwide has been estimated to be $30 billion in 1980 and $68 billion in 1990, with an annual growth rate of 9%.

1.4.4.2
Biopharmaceuticals

The setting up of new biotechnology companies (Section 1.4.2) was the key event for the development of biopharmaceuticals based on recombinant technologies. Antibiotic manufacture had been established before this (Section 1.3). Antibiotics are still the most important strategy against bacterial infections. The world production of antibiotics is estimated at over 60 000 tons per annum, valued at more than $30 billion per year [128, 129]. For the synthesis of semisynthetic penicillins and cephalosporins, the fermentation products penicillin G and cephalosporin, respectively, are hydrolyzed by immobilized enzymes to yield the acid form with the intact β-lactam ring as the active principal: 6-APA, 7-ACA, or 7-aminodeacetoxy cephalosporanic acid (7-ADCA).

Protein-based drugs constitute about a quarter of new approvals with a majority being glycoproteins. A total of 77 recombinant biopharmaceutical products gained approval and were available on the market in 2000 and that number had increased to some 165 biopharmaceutical products by 2006, with a market size estimated at $35 billion in 2006, and reached over $60 billion by 2014 [130] ([72], sections 9.7.2 and 17.4.2). The average annual growth rate was 20% ([87], pp. 8–11; [130, 131] (for an overview see: [98]); see also www.centerwatch.com/patient/drugs). Approved recombinant proteins comprise a number of different biopharmaceuticals, that is, blood factors, thrombolytics, insulins, other hormones, hematopoietic growth factors, bone morphogenetic proteins, IFNs, and ILs, mABs and engineered antibodies, vaccines, therapeutic enzymes, and enzyme inhibitors [97]. Most of the drugs listed in Table 1.3 are blockbusters with sales of $1 billion per year or more. Most of the newly approved medicines have been developed to address indications which are the major causes of disease and mortality in the industrialized world, notably infections,

Table 1.3 Examples of rDNA products for medical use compiled and extended after [72, 87, 131, 133].

Active substance	Medical use	Gene first cloned	FDA: first approval[a]	Market size ($\$ \times 10^9\ yr^{-1}$)
First generation				
Human insulin	Replacement therapy diabetes type 1/2	1978	1982	12.15[b]
Growth hormone (huGH)	Replacement therapy dwarfism	1979	1985	1.02[c]
Factor VII/VIIa	Replacement therapy, hemophilia (Alexander's disease)	1987	1999	0.30[d]
Factor VIII	Replacement therapy, hemophilia A	1988	1993[e]	2.40, 3.00[d]
Factor IX	Replacement therapy, hemophilia B	1983	1996[f]	0.30|[d]
Enzymes[g]				1.30[g]
β-Glucocerebrosidase	Gaucher's disease	1985	1994	0.20[h]
α-Galactosidase	First drug for treating Fabry disease	1985	2003	0.20[i]
Iduronate-2-sulfatase	Replacement therapy mucopolysaccharidosis type II	1990	2006	0.20[j]
Human DNAase	Lung clearance, cystic fibrosis (mucoviscidosis)	1990	1993	0.30[k]
Second generation				
Human interferon-α1	Antiviral, limited antitumor (hairy cell leukemia)	1979	1984	0.60[l]
Human interferon-β	Antiviral (HCV), multiple sclerosis (MS)	1979	2002[a]	3.80[m]
Interleukin-2	Metastatic renal carcinoma	1983	1992[n]	0.08[o]
Tissue plasminogen activator (tPA)	Dissolves blood clots (e.g., after heart attack)	1983	1996[a]	0.85[p]
GM-CSF	Shortens neutropenia after chemotherapy	1985	1991[a]	0.08[q]
G-CSF	Use in blood stem-cell allografts	1986	1991[a]	4.50[r]
Erythropoeitin (EPO)	Replacement therapy (EPO lost during hemodialysis)	1985	1996	3.50[s]
Recombinant vaccines				1.43
Hepatitis B vaccine[t] (HBS antigen)	First anticancer vaccine (80% of hepatic cancer → viral origin)	1982	1986	0.20[u]
Papilloma virus vaccine (HPV)	Anti-cervical cancer vaccine	1980	2006	0.60[v]

Monoclonal antibodies (mAbs)				24.60
Anti-TNFα	Autoimmune disease (plaque psoriasis, rheumatoid arthritis)	1991 (Enbrel)	2002[a] (Humira)	11.00[w]
Anti-HER-2/neu	Metastatic breast cancer, gastric cancer ($HER2^+$)	1984	1998	1.90[x]
Anti-CD20	Non-Hodgkin lymphoma, B-cell leukemias	1994	1997	3.50[y]

a) The first FDA approval may be for other indication(s).
b) Insulins (Lantus, Novolog, Humalog, Levemir, Humilin) plus GLP-1 receptor agonists (Victoza, Byetta, Novolin).
c) Human growth hormone (nutropin, genotropin, humatrope, norditropin).
d) Blood factors total about $3500 million, Factor VII (Novoseven,); Factor VIII (Advate, Refacto AF/Xyntha); and Factor IX (Benefix).
e) Human albumin-free first in 2003.
f) 2013 Version with long serum-half-life approved here.
g) Therapeutic enzymes, total US sales (2012).
h) Cerezyme + Vpriv.
i) Fabrazyme + Replagal.
j) Elaprase.
k) Pulmozyme, the first inhaled protein therapeutic. Interferons/cytokines.
l) Pegasys.
m) Avonex, Rebif, Betaseron.
n) Delay in approval caused failure of Cetus (bought by Chiron); US market data updated to 2012.
o) Proleukin, 2010.
p) Alteplase, worldwide.
q) Granulocyte-macrophage-colony-stimulating factor, leukine.
r) Granulocyte-colony-stimulating factor, Neulasta, neupogen.
s) Epogen, Aranesp, Procrit.
t) N.B.: PCR-based testing of blood and blood-derived products for viral contamination testing from 1986 $>10^6$-fold more sensitive, also for previously unknown viruses (HBV, HIV, HCV, HDV, West Nile Virus, etc.).
u) Pediarix, Recombivax HB.
v) Gardasil; monoclonal antibodies (US sales).
w) Humira (first fully human mAb) + Remicade + Enbrel (fusion protein), *world-wide sales $23 billion.*
x) Herceptin.
y) Rituxan, chimeric murine-human mAb.

with hepatitis B and C as the most targeted indications, diabetes, hemophilia, myocardial infarction, and various cancers [132].

Following a number of scientific breakthroughs, about 10 new products were approved in 2014, some with blockbuster status, comprising antibodies, peptides, and enzymes, for example, for treating cancer [134]. However, R&D costs to bring a single drug to market is approaching $3 billion, a 145% increase over estimated costs in 2003 [135].

1.4.4.3
Plant Products, Seeds

The promise of transgenic plants, or green biotechnology, is to create crops with higher yields, which can grow on less fertile land, to feed a growing, hungry world population. Headlines such as "Genes to gasoline" provoke expectations for second-generation biofuels. Crops should be resistant to pests and need fewer chemicals, notably insecticides, fungicides, herbicides, and fertilizers. The use of genomic analysis and quantitative trait linkage has not only accelerated plant breeding programs, but allows definition of new cultivars that need standards for patent protection, although no rDNA methodology was directly applied, that is, the plant concerned is not a genetically modified organism (GMO). The majority of agricultural scientists are convinced that crops of high yield, high quality, low cost, and low environmental impact can be delivered by the exploitation of the techniques for plant biotechnology, in particular by involving molecular breeding strategies ([136], Foreword).

The most successful, in terms of application, has been herbicide resistance, where reduced utilization of herbicides has been claimed. Resistance against diseases (pests) has been one of the priorities in designing genetically modified (GM) plants. This also implies a decrease in the dependence on chemical pesticides. Genetic modification has been shown to provide an additional and less time-consuming tool, and to be a useful means of creating novel resistance, compared to classic breeding.

Heated controversies and media attention have affected the discussion on GM crops. Few novel technologies have provoked so much opposition, and consequently governmental control, as has plant biotechnology. It has been subjected to major political, economic, social, ethical, and environmental scrutiny, much more than other fields of biotechnology. The discussion of risks in general goes back to the formulation of rDNA guidelines in Asilomar in 1975. The critical public perception and resistance of consumers throughout the world has slowed down, and even prevented applications, notably in Europe.

Applications include food, feed, and renewable resources, with over 15 crops modified for 1 or more of some 47 phenotypic traits commercialized since 2000, most with attributes related to input and yield performance. In 2006, 102 million ha were cultivated with GM crops, and in 2007, there were 114 million ha in 22 countries. Seven countries dominate the global market, namely, the United States, Argentina, Brazil, Canada, India, China, and South Africa. Among the products are soy bean, maize, cotton, rape, cucurbit (squash), tomato, papaya, alfalfa, poplar, petunias, and paprika. The growth of the area cultivated since 1996 is obvious. Application was mainly with the combined traits of insecticide resistance and herbicide tolerance. Four crops account for nearly 100% of the commercially grown transgenic crops, namely, soy bean (64%), cotton (43%), maize (24%), and rape (20%).

1.4.5 Further Aspects

The focus of this section has been on the scientific and engineering foundation, background, and development of biotechnology. Nevertheless, some aspects of political and economic conditions, and of sociology of science should be mentioned. Thus, a considerable number of approaches tend to investigate, and find general interpretations and explanations for this unique technological development. They are not discussed here, but mentioned only as suggestions for further reading.

1.4.5.1 Scientific Status

Since the 1990s, the status of biotechnology can be recognized as that of a scientific discipline of its own. Advanced groups in applied biology, biotechnology, and biochemical engineering tended to cooperate and integrate their efforts, with work oriented to the molecular level and with empirical data being interpreted in terms of coherent theories [137, 138]. The basis was molecular genetics, which introduced a new paradigm based on knowledge of the molecular biology involved, and basic engineering concepts, which was oriented at molecular processes, and the integration of these scientific approaches [139].

1.4.5.2 Political, Institutional, and Socioeconomic Conditions

A general overview on the history of biotechnology, including these topics, was written by Bud [2]. He discussed the role of governmental politics, under the heading "Resistance to New Technology – Regulators of Biotechnology" [74]. A critical comparison of governmental politics in Germany, termed a *corporate approach*, and the United States, with the OTA study, was presented by Jasanoff [77]. Yi [140] discussed how academic institutions, government agencies, and the nascent biotechnology industry contested the legal ownership of rDNA technology in the name of the public interest. It meant the reconfiguration of the ownership of public knowledge in late twentieth century American capitalism. (See Section 1.5 for details of the earlier discussion.)

1.4.5.3 Economic Studies

A range of socioeconomic studies have been published in recent years, including the roles of academic, notably university, and company research, networks, and company strategies, both of big pharma and new biotechnology firms However, many studies suffer from a lack of interdisciplinarity in research teams, where it seems that no scientists or only few scientists involved in biotechnology participated.

The group of Zucker and Darby [141–143] found a positive impact of research universities on nearby firms related to a positive impact on the firms' research productivity, increasing the number of biotechnology patents, more products in development, and more products on the market, as of 1989–1990. Arts and Veugelers [144] and Arts *et al.* [145] discussed indicators derived from patent documents to capture the nature and impact of technological inventions, and to compare and validate these indicators within the field of biotechnology. However, the studies suffer from a lack of interdisciplinarity, which may be considered necessary for such work.

1.4.5.4
Science Studies

What drives science and what motives make up its dynamics? Several periods in the history of biotechnology may be interpreted in terms of a concept developed by Rheinberger [146] ([72], part I), who analyzed the early phases of molecular biology, that is, Zameznik's approach to identify the steps in the translation of DNA into protein, which turned out to be a highly complex sequence of reactions. In this concept, epistemic things may be characterized as a field of certain, yet unidentified objects, experimental techniques, and implicit knowledge, that is, a field of problems of phenomena, reactions, and structures which were not understood unequivocally, nor explained. Adequate methods of working on the topic to give unambiguous results were not available. "To put it as a paradox: epistemic things represent that what remains unknown. They may be characterized by a list of activities and properties." Epistemic things may be considered as those questions, problems, and stimuli that provide the dynamics of research in early phases. No ideas of technical application in molecular biology appeared as long as the field remained unclear and undisclosed; these are, epistemic things in Rheinberger's notation. It was only when structures of DNA and protein and enzymatic reactions appeared with consistent character and logical order that rational construction seemed possible, and pioneers began to conceive concepts for application, that is, cloning concepts for making drugs and other proteins.

1.5
How Pioneering Developments Led to Genetic Engineering

1.5.1
Preamble: People and Principles in Developing Enabling Technologies

Bartneck and Rauterberg [147] have argued that emphasis on discoveries rather than on inventions has moved the Nobel Prize away from its original intention of rewarding the greatest contributions to society. It should be clear from the following descriptions of events in molecular biology, molecular genetics, and cell biology that both novel conceptual insights and technical innovations have led to "enabling technologies" which have, to say the least, transformed biotechnology and modern medicine.

Berg and Metz [148] have presented gene technology as having arisen mainly as a consequence of using established tools in new combinations. This does little to explain why. In the following discussion, we differentiate between developments that have occurred by gradual incremental steps, perhaps involving less intellectual input, and those which arose by more radical conceptual changes, even when they mimicked mechanisms used in nature. Contrary to Berg and Metz, we argue that at the inception of gene cloning, it was not an accepted notion that, in the face of enormous ignorance of the basic differences in gene structure and function in different species, cross-species gene cloning, for example, selecting for expression was a likely possibility. Another category of developments involves surprise findings in the early years of gene cloning that caused redirection and creation of whole new fields of investigation. This led to novel enabling technologies beginning in 1972–1973. A special aspect unusual for the biological sciences at that time was that academic discoveries were immediately translated into biotechnology inventions that were taken up rapidly in industrial settings, producing enormous innovation, often with novel medical applications. This involved, in many cases, academics who became entrepreneurs, a fact that was vehemently condemned by many of their peers as *infra dig* and/or, especially with the patenting of sequences and products (necessary for the due diligence process required by potential investors in start-ups), as incompatible with scientific rigor and/or academic freedom.

1.5.2
Academic Freedom and Patenting: Hindrance to Science or Lever for Innovation?

This latter conflict area, although not suffered by Louis Pasteur (a great model to follow?) has occurred often before. Are there parallels to be seen with the conflict between Humphrey Davy and George Stephenson about the invention and patenting (1815) of the miners' lamp that protected miners from igniting methane explosions? Davy refused to patent and received the Rumford medal for this invention in 1816. However, his lamp was dull and deteriorated in damp mines. Stephenson's patented design invented independently, was bright, worked well, and was preferentially manufactured and used. Perhaps a similarity exists with Paul Berg being awarded a Nobel Prize for being the first to discuss cloning principles and doing *in vitro* DNA joining: In contrast, Cohen and Boyer invented and demonstrated a reproducible DNA cloning method, which they patented, opening up a multibillion dollar industry. James Watson (already a Nobel Laureate), the coordinator of the Human Genome Program at the NIH, was incensed by the National Institutes of Health patenting hundreds of Craig Venter's EST DNA sequences. Watson often referred to Venter as "Hitler," later saying in public that "virtually any monkey" could do that sort of work (EST mapping; [149]). Venter moved genomic analysis into an industrial setting on a grand scale and now uses a fund derived from this to donate more financial support for international health programs than the World Health Organization itself.

1.5.3 Conceptual Leaps and Jumps

Enzymes that had been discovered to be part of cellular defense mechanisms to recognize and destroy foreign DNA were reconceived as tools to ensure its uptake into viral and plasmid vectors, thus allowing its stable perpetuation across species barriers: this was the beginning of the ability to manipulate DNA at will. This was true both during 1972–1973, with the use of restriction endonucleases (R.endos) and, more recently, since 2010, with the use of CRISPR/Cas9 (a type of bacterial adaptive immune system recognizing specific fragments of foreign DNA). This latter system has recently been developed into the most specific method for genetic engineering and site-specific mutagenesis in a wide range of prokaryotic and eukaryotic hosts [150] with or without plasmid or viral vectors.

In the absence of knowledge of gene structure and function in higher organisms, rDNA technology was used as a basis for a new industry to produce pharmaceutical proteins at a time when the pharmaceutical industry thought this would be impossible. In retrospect, this would seem to require that we recognize this step as a conceptual leap, although the individual components needed for constructing rDNA molecules are known. A gene is not just a piece of DNA. It is a *particular* piece of DNA that shows stable inheritance and, when in the *correct chromosomal location* in a *specifically developed cell* containing the *required regulatory molecules and receptors*, can be expressed in a living organism, preferably in a controlled fashion. Its protein product may require specific modification before it can be functional and may require extensive purification before it is suitable for use as a pharmaceutical product.

At several steps on the way to acquiring the human genome sequence and finally putting this information to biotechnological use, a conceptual breakthrough involving several developments were needed. The following are some examples:

- Botstein's analysis of how high-resolution genome mapping could be achieved and used for human genetics and gene isolation [151];
- rapid and cheaper analysis made possible initially by better cloning systems for efficient isolation of larger fragments;
- better computer systems and algorithms for sequence analysis that allowed shotgun whole genome analysis;
- the EST concept which allowed more rapid access to the protein-coding regions of genes (not forgetting Sydney Brenner's novel contribution sequencing *Fugu* the Puffer-fish, whose genes essentially lacked introns → 10-fold less sequencing to get to the gene sequences);
- novel concepts for faster parallel sequencing to remove the requirement for previous cloning of large DNA regions;

- the brilliant idea of the PCR, which was completely unexpected [152, 153] and hardly believed before it became generally established with the introduction of a heat-stable DNA polymerase (Taq1), which became generally available through cloning its gene in 1989 [154]. This method allowed, among other things, direct cloning and sequencing of any region bracketed by known sequences (this is known as simplified gene manipulation), *in vitro* affinity selection of gene and gene product and thus isolation of completely novel ligands with specific affinities or functions;
- affinity selection and affinity maturation that involved novel brilliant concepts in the area of phage display and ligand libraries, mimicking natural processes found in antibody maturation, or seen over long periods of time in evolution. For extensive reviews, particularly for the beginnings of gene cloning, phage display, applications in biotechnology, and recombinant antibody development, see [72, 155–157].

1.5.4
Surprise Discoveries Initiate Novel Areas/Methods of Research

A number of discoveries were completely unexpected:

- *Gene splicing*: Many coding regions in eukaryotic genes and particularly their viruses are often not continuous, but interrupted by intervening sequences that have to be removed by a *splicing* process at the mRNA level before translation takes place. Gene splicing was discovered independently by Sharp [158] and Roberts [159] in 1977.
- *Nonuniversality of the genetic code*: One of the assumptions implicit in doing gene cloning is that the genetic code is universal. Luckily for those who had invested in this new technology, this was generally true. Exceptions existed, however, for example, as discovered in bovine mitochondria cytochrome oxidase II where a stop codon translated to tryptophan insertion (see, e.g., a retrospective by Sanger [160]): Later more exceptions were found: for example, four base-pair suppressor tRNAs were found to be commonplace in yeast; and in bacteria, the formation of seleno-enzymes.
- *RNA enzymes*: Studying splicing also led to the discovery of *RNA-enzymes*, interpreted by many as an insight as to what might have been used as the biochemical basis of life before the evolution of complex protein synthesis machinery. This finding contributed to the development of SELEX technology and "synthetic biology."
- *Gene transfer to plants*: The studies of tumors in plants by Jeff Schell and Marc van Montagu led to using the same system that one finds in nature for transfer of genes from bacteria (*Agrobacterium tumefaciens*) to plants, the beginning of genetic engineering in plants.
- Gene copy number variation (GCNV): the discovery that GCNV between individuals is a major factor in human genetics was unsuspected. This complicated

the genome-sequencing program that was based on mixed DNA from many different individuals. It now influences the way human geneticists screen patients; for instance, DNA chip arrays can be used for this purpose.

- *Missing heredity*: (Low clinical relevance of genetic predisposition): The modest results in finding genes strongly affecting widespread syndromes such as metabolic syndrome and susceptibility to cancer was also not expected. Continued huge funding in this area meets significant criticism [161], although many are still looking through rose-tinted glasses at the $2 billion increase in the US Congress budget for Vice President Biden's "moonshot" to defeat cancer.
- *Epigenetics*: Changes in environment cause changes in gene expression via, for example, alteration of methylation at the DNA level and through modification of histone and other chromatin proteins. These so-called *epigenetic modifications* now represent an important new area of research and diagnostics with high relevance in health care.
- *Unknown microbial flora*: Shotgun sequencing of DNA from entire microbial communities revealed a vast unrecognized microbial variation: microbiologists, until the 1980s, had largely confined their studies to microorganisms which could be cultivated as pure colonies in the laboratory. The astounding finding was that perhaps more than 95% of the world's microbes had not yet been isolated in pure culture! Cell growth is dependent on commensal and symbiotic interactions or, for some, cell growth stops at very low cell density ("quorum sensing"). Amazingly, the *most abundant single species in the world, Pelagibacter ubique*, was first discovered by assembly of the complete genome sequence from sequences found in nearly all ocean samples. The huge potential represented by this wealth of microbial diversity and specialization is a boon both for basic research and for the biotechnological industry concerned with secondary metabolites. It represents a rich area of specialization in terms of gene mining for many years to come. The Earth Microbiome Project (EMP) was launched in August 2010, with the ambitious aim of constructing a global catalog of the uncultured microbial diversity of this planet. The primary vision of the EMB, to process the microbial diversity and functional potential from approximately 200 000 environmental samples, marks it as an undertaking so massive that it was at first considered to be pure folly (as late as 2012, Jonathan Eisen was quoted in Nature as saying, "Knight and Gilbert literally talk about sampling the entire planet. It is ludicrous and not feasible – yet they are doing it" [162]).

1.5.5
Methodology Without Which Gene Technology Would Not Have Been Possible

Before dealing with the specific steps that made gene technology possible as a direct precursor to genetic engineering, we should consider the methodological repertoire that had mostly been developed in the beginning and middle of the twentieth century. The standard laboratory methods are often taken for granted, but a quick review of the most important is included here: their effects on discoveries are listed in Table 1.4.

Table 1.4 Laboratory procedures on which the growth of recombinant DNA was also dependent.

Chromatography (see body text 1930–1960s) • Paper/starch → first analysis of nucleotides and chemical synthesis products; inherited changes in proteins; + antisera → antigen recognition (Ouchterlony method) • Size exclusion polymers and zeolites → pure enzymes • HPLC and gas chromatography → combination of the above were the basic tools of biochemistry combined with structural analysis: sequencing of first simple RNA viruses; peptide sequencing *Electrophoresis: agarose/acrylamide gels* → protein separations; isolation of specific DNA restriction fragments; first DNA sequencing; analysis of transcription products (Northern blotting); or proteins (+Western blotting) • Pulse-field electrophoresis → mapping very large (>30 kb DNA fragments e.g., NotI digest of YAC clones; jumping and linking libraries for genome contig building/mapping • Isoelectrophoresis focusing (IF) → refinement of antigen recognition; protein charge shifts detected • Combining 1D and IF → 2D protein gels → detect post-translational modification; proteomics tool *Centrifugation* → preparation of cell pellets/antibody-antigen complexes • Velocity sedimentation → fractionation of organelles; cellular components, RNA, DNA, membranes • Ultracentrifugation (isopycnic) → DNA strand separation, first pure gene; preparation of plasmid/viral DNA	*Electron microscopy (EMI)* → DNA lengths • EM + DNA, or DNA/RNA heteroduplex → discovery of introns, transposons (Tn), insertion elements (IS) • *Scanning tunneling EM (1981)* → image individual atoms; confirmed DNA structure; future single molecule sequencing? Longer fragment sequences: faster sequencing? *Crystallography* → recognition + separation of mirror symmetry enantiomers • X-ray diffraction analysis → absolute molecular structure. First protein structures combined with powerful computing → protein design *Laser-activated fluorescent technologies* • Flow cytometry (cell sorting, 1972)[a] → first high-throughput screening; → with antibodies follow cellular differentiation; separate specific cell types • Combined with tagged gene expression → follow cell differentiation *in vivo* • Single molecule fluorescence (1992) → follow intracellular trafficking; next-generation single molecule sequencing? Novel methods for measuring intermolecular affinities, for example, in drug screening (Evotec)

a) First developed for electrical impedence (Fulwyler, 1965) → founding of Cetus.

1.5.5.1
Centrifugation: Preparation of Molecules of Different Sizes, Shapes, and/or Densities: Velocity Gradient Centrifugation

Friedrich Miescher is generally given credit for being the first (1869) to isolate a cell organelle by centrifugation [163]. The protease stripped extract of the "nucleus fraction" of human pus, which he called *nucleins*, was later renamed by Altmann as nucleic acid, once its chemical nature had been established.

Centrifugation (Table 1.4), initially applied in the dairy industry, and ultracentrifugation, now a routine method in all biology laboratories became a science through the work of Theodor Svedberg in Uppsala, and the unit used for the sedimentation constant (S = Svedberg = 10^{-13} s), reflecting the relative sizes of molecules, is named after him. To study the size of proteins and RNAs, very high g-forces had to be generated, for example, already in the 1930s, 900 000-fold greater than the standard gravitational field. The older authors of this chapter often experienced rotors exploding in their armored centrifuges during molecular (isopycnic) separations in the 1960s and 1970s.

- *Velocity centrifugation*
  Behrens applied gradient centrifugation, where sugar gradients were used to prevent mixing of the tube contents during acceleration and braking. This was used for the fractionation of cells from blood or tissues and analysis and preparation of subcellular fractions, for example, nuclei, chloroplasts, ribosomes, mitochondria [163]. This early work was instrumental from the 1940s onward in the detection of molecular subunits of complex molecules such as the protein subunits of hemoglobin.
  Continuous preparative ultracentrifugation was used in the mid-1970s for isolating mitochondria, chloroplasts, ribosomes, or RNA species from cell lysates on a large scale.
- *Isopycnic density centrifugation*
  During high speed centrifugation (e.g., several days at >300 000*g*), even the salts of heavy metals, such as cesium, form a density gradient in which large molecules form a discrete sharp band at their specific density. This required that engineers develop a rotor that could sustain such huge forces [164]. With this method, it was shown that transfer of resistance properties between bacteria was sometimes associated with the acquisition of "satellite" or "episomal" DNA bands, differing in density from the chromosome [80]. Single strands of bacterial viruses could be separated in isopycnic gradients after melting the DNA and hybridizing it to synthetic poly (rI:rC). This was instrumental in isolating the first pure gene fragment in 1968, the pre-DNA cloning era (see below) [165].

In 1967, Vinograd was instrumental in analyzing plasmid DNA via ultracentrifugation, including the use of DNA-intercalating dyes such ethidium bromide [166, 167]. This allowed reliable detection and purification of plasmid or viral DNA on a large scale, even when it was of the same specific density (AT/GC ratio dependent) as the host chromosome. This was the main method in use from the early 1970s and in the mid-1980s. The fact that ethidium bromide-DNA complexes are highly fluorescent in UV light allowed sensitive detection of DNA in gels, facilitating restriction mapping and later the analysis of PCR reaction products (e.g., DNA fingerprinting developed by Alec Jeffries in 1985). It also introduced mutations in the DNA.

1.5.5.2
X-ray Crystallography: Understanding Molecular Structure at the Atomic Level

The discovery of diffraction of X-rays by crystals by Max Von Laue in 1912 was an indication that their atoms are arranged in ordered lattices. William Henry Bragg refined the X-ray spectrometer for use with X-rays of defined wavelengths. His son, William Lawrence Bragg, developed the mathematics to deduce the positions of atoms within the crystal. This was initially calculated by hand for simple mineral salts. Extending this to proteins required faster computing, improvement of X-ray beam production (e.g., at the DESY synchrotron), and ever more sensitive high-speed data collection of the diffracted beams. The expense of the instrumentation allowed the establishment of the technology initially in relatively few centers. These became centers of excellence in molecular biology, attracting excellent scientists and resulting in rapid developments, for example, as seen from the fact that, between 1904 and 2012, 41 members of either the Cavendish Laboratory, or the Laboratory for Molecular Biology (LMB) at Cambridge University, UK, became Nobel Laureates (Fred Sanger twice; *recommended further reading*: (i) de Chadarevian [168], (ii) Brownlee [169]).

1.5.5.3
Chromatography with Solvent Motion or Electric Charge: Detection of Mutant Gene Products

Electrophoresis as a major analytical tool began with the work of Arne Tiselius. Tiselius, a student of Svedberg, developed moving boundary electrophoresis in 1937. This was based on Faraday's work on electrolysis in the eighteenth century and on studies of moving boundaries in electrolytes by "schlieren" optics, initially applied to protein analysis by Botho Schwerin in 1914 (Section 1.5.5.4). Oliver Smithies in the 1950s was the first to detect altered protein structure correlated with inherited diseases. In inherited thalassemia, those carrying the trait had altered globin conformation with lower affinity for oxygen that caused it to run at a different speed in zone electrophoresis through starch gels.

1.5.5.4
Protein Sequencing

Earlier developments in protein purification and chromatography had a lot to do with the development of protein sequencing by Fred Sanger starting in 1943 in Chibnall's group, Cambridge, UK. Insulin was already available in a pure form from the pharmaceutical industry. Sanger developed a sequencing protocol involving specific chemical modification of the exposed amino groups. After partial hydrolysis of insulin (e.g., with acid or proteases), he fractionated the products using 2D paper chromatography; in one direction by electrophoresis, and in the other by solvent. Ninhydrin stained the breakdown products for visualization on paper. These "fingerprints" could be interpreted in terms of a sequence, initially only near the amino termini of the peptides. The total sequence, including the position of disulfide bonds linking the two chains, was established by 1955. This work finally established that proteins had definite polypeptide chains of specific sequence and were not just amorphous. These data later, in combination with analysis of RNA and DNA structure, allowed the conceptualization that there is a flow of genetic information, each in the form of linear structures, from DNA to RNA, and from RNA to protein. Following more sequencing and the combined use of oligonucleotide synthesis and *in vitro* protein synthesis, the rules for this high fidelity information flow, the genetic code was solved.

1.5.5.5
Nucleic Acid Sequencing – the Prelude: Phage + Bacterial Genetics and Biochemistry, the Gene Concept

Short synthetic oligonucleotides and RNA viruses were the first, pure, single-stranded nucleic acids available for sequencing before gene cloning. Bacterial genetics in combination with biochemistry, to examine the composition of the protein gene products, allowed an initial analysis of genes and the discovery that some jointly regulated inducible gene clusters (operons) were sequestered on plasmids and bacteriophage which could be transferred (by conjugation, transformation, or transduction) between bacteria.

Proteins controlling gene transcription were isolated by binding to bacterial viruses (bacteriophage or "phage"), for example, for the lactose operon (Qβdlac) or for the phage λ (*lambda*) itself. In the late 1960s, the exact molecular and biochemical interaction controlling gene expression in the bacterial *lac* operon was established, 5 years before gene cloning was developed. A circular genetic map existed already for both the phage λ (*lambda*) and the bacterium *E. coli*. The first gene fragment was isolated in1969.

The first isolation of a pure gene fragment was carried out 4 years before gene cloning [165] at Harvard. This involved S1 single-strand nuclease digestion of DNA heteroduplexes (partial hybrids obtained after melting the dsDNA and reannealing at lower temperature) of transducing phages containing the lac operon in inverse orientation. This was hailed by the *New York Times* as the beginning of a revolution in planned genetics, similar to that portrayed in Aldous Huxley's novel "Brave New World," thus inaugurating self-criticism by the community involved in the research.

The Beginnings of Sequencing By the late 1960s, small RNAs, for example, tRNAs and RNA viruses, had been highly purified. Sequencing of the viruses by partial chemical degradation and chromatography of the products was initiated in Fiers and Weissman's groups in 1969. Sanger's group was instrumental in many of the developments in this early period. The isolation of a tRNA pure enough for sequencing required the newly developed DEAE cellulose ion exchange chromatography and further refinement of countercurrent chromatography. Once this had been achieved, Holley's group [170] was successful in sequencing the 77 base yeast alanine-tRNA, the first RNA molecule sequenced [171]. Studies on *in vitro* translation of linear single-stranded RNA or single-stranded DNA (ssDNA) viruses were instrumental in confirming the degeneracy and universality of the genetic code [172]. Sequences specifying the site of initiation of translation were determined by sequencing fragments of the RNA phages protected by ribosome binding from degradation by ribonuclease [173]. The first gene sequence, from an MS2 RNA virus, came from Walter Fiers' lab in Ghent, Belgium [174] in 1972, as did the first total genome sequence in 1976, again the MS2 RNA virus [175]. Sanger and Weissmann were pioneering oligonucleotide sequencing using highly radioactive ^{32}P-labeled RNA, complete or partial digestion with enzymes, and sedimentation and chromatographic separation of the cleavage products. The chromatographic separation methods were the same as those used by those synthesizing oligonucleotides, for example, in Khorana's laboratory, and these synthetic fragments could be used as reference material [176]. Gilbert and Maxam in 1973, established the 23 bp lacO (operator) dsDNA sequence, separating chemical degradation fragments by chromatography [177]. Weissmann's work pioneered the chain-elongation sequencing technology using labeled nucleotides in reiterated viral RNA synthesis reactions lacking one or more of the four nucleotide-triphosphates.

The first RNA molecule to be sequenced was not a virus, but a tRNA. Two enzymes were required in these analyses: (i) bovine pancreatic ribonuclease, which cleaved after pyrimidines (C or U), became a classic system for scientific studies after Armour & Co., the hotdog Company, purified 1 kg of the enzyme and distributed it to scientists (source Wikipedia) and (ii) takadiastase ribonuclease T1, which cleaved 3′ to a guanosine (G). Each of these small fragments was further analyzed by partial digestion with snake venom diesterase from the 3′-ends. Once again, biochemistry and enzymology led to breakthroughs in chemical analysis.

Chemical Degradation and End-Labeling Oligonucleotide sequencing began with a typical chemical analysis for very short products of chemical oligonucleotide synthesis, for instance, chromatographic separations with different solvents before and after partial chemical degradation in a reaction specific for cleavage of chemical bonds in a base-specific manner (e.g., purine/pyrimidine specific). The amounts of chemicals involved were relatively large (mgs). Flat acrylamide gels were used for separating the reaction products. As noted above, the first RNA sequence, gene sequence, and total viral genome sequence were determined *without* the aid of gene cloning. However, it was not until DNA cloning and R.endo cleavage allowed the isolation of specific DNA fragments that sequencing could be used widely for gene analysis [178, 179]. End-labeling with P^{32} phosphate was dependent on the fact that oligonucleotide dephosphorylase and oligonucleotide kinases had been identified, produced, and purified for wide distribution (see Table 1.5). It was characteristic of this period, before companies had assembled a supply of materials for the burgeoning rDNA laboratories, that each group would produce a few key enzymes or vectors themselves and exchange them for other materials, such as host strains and vectors with other laboratories, both at home and abroad. International restrictions on rDNA, as well as commercial interests in it and its products, later reduced this practice severely. These halcyon days transformed to "Golden Helix" and "Genome Wars" to name just two book titles that describe the 1980–1990s entry of molecular genetics into the commercial world.

Oligonucleotide Primer Extension on a Single-Stranded Template (Sanger Sequencing) Initially, (radionuclide) end-labeled primers were extended in four separate base-specific terminating synthesis reactions in the presence of small amounts of strand-terminating dideoxynucleotide triphosphates, in addition to the normal four deoxynucleotide triphosphates (dT, dA, dG, and dC; dNTPs), a method developed by Sanger and Alan Coulson in 1977 [160, 194, 195]. These two scientists were also driving forces in developing the sequencing center at Cambridge, which made a major contribution to the human genome-sequencing

Table 1.5 Genetic engineering and sequencing required pure well-characterized enzymes.

Enzyme	Application	Discovered/references
Alkaline phosphatase (Calf thymus)	Removal of terminal 5′-phosphates before $P^{32}O_4$-labeling of oligonucleotides. Prevention of insert DNA scrambling	Morten [250]
Polynucleotide kinase (PNK)	End-labeling oligos (synthetic or dephosphorylated) (many sequencing methods; tracking hybridization etc.)	Novogrodsky *et al.* [180]
Haelll restriction endonuclease (representative of hundreds)	Preparing specific DNA fragments with blunt ends	Smith and Wilcox [181]
	Specific cleavage of SV40 DNA by Hae R. endos.	Danna and Nathans [182]
EcoRI restriction endonuclease (representative of hundreds)	Preparing specific DNA fragments with cohesive ends	Boyer, [252]
		Mertz and Davis [183]
Terminal transferase (TdT)	Adding oligonucleotide tails to ds oligonucleotides (joining DNA fragments; cloning cDNA). Detecting apoptotic cells	Bollum [251]
DNA ligase from (*E. coli*/T4 ϕ)	Covalent joining DNA fragments with paired cohesive ends	Gellert [255] and Goulian and Kornberg [254]
Exonuclease I	Trimming single-stranded tails from ds DNA	Lehman and Nussbaum [184]
Exonuclease III (λphage)	Preparing 3′ ssDNA overhangs from 5′ PO_4 dsDNA (pre-tailing)	Sriprakash *et al.* [185]
DNA polymerase I (*Klenow fragment*)	Efficient DNA synthesis on a template (Lacking 3′–5′ exonuclease)	Lehman *et al.* [253]
		Klenow and Henningsen [186]
Reverse transcriptase (AMV)	Allows synthesis of a DNA strand complementary to an mRNA	Temin and Mizutani [187] Baltimore [188]
RNAse H	Digests RNA in RNA/DNA hybrids	Stein and Hausen [189]
S1 single-strand nuclease	Digests single-stranded DNA and ssDNA extension on dsDNA	Ando [190]
	Analysis of DNA/DNA or DNA/RNA heteroduplexes	Vogt [191]
Bal31 ss endonuclease	Progressive deletion from ends of dsDNA (making deletions)	Gray *et al.* [192]
T. aquaticus DNA polymerase	PCR (polymerase chain reaction)	Chien *et al.* [193]
Heat stable	DNA amplification – basis for second-generation sequencing	Lawyer *et al.* [154]

program (now called *The Sanger Center*). Gel columns were used later and formed the basis of separation in Leroy Hood's 1986 automated sequencing prototype that was developed as a workhorse for the final stages of the Human Genome Project in 1999.

Sequencing Automation: Large-Scale Sequencing Methods differ in several parameters such as the length of a sequence (length) that is reliably produced without error (raw error rate), the overall speed (raw base sequence per second), and cost. The increase of speed and reduction in cost of sequencing was, for each, over a million-fold in the last 25 years, a rate of development that is more dramatic than that proposed by "Moore's law" for computer technology: Moore's Law predicts doubling in speed and cutting cost by half every 2 years. The rate constant for sequencing is closer to doubling and halving, respectively, every year. As previously discussed, the development of PCR (particularly emulsion PCR as in the 454 sequencers using pyrosequencing), purification of relevant enzymes used in sequencing reactions, novel chemistries, such as reversibly terminating deoxynucleotides with different colored fluorescence for each type of base, and laser detection methods such as microarray scanning, were all important prerequisites for one or other of the sequencing technologies. The mega-sequencing technology, for example, used in the Illumina and Applied Biosystems, about 2008, simultaneously scans millions of sequencing reactions of randomly spaced "clonal amplicons" immobilized on a surface. Although each sequence length is short, the rate of sequencing more than compensates.

Next-Generation Sequencing Pettersson *et al.* [196] provide an excellent review into the previous and ongoing development of brilliant novel principles and their application in the several generations of sequencing technology up to 2009. Sequencing on single molecules can deliver long sequences as the method is devoid of the rising background noise generated by accumulated errors from each step of stochastic sequencing procedures involving many copies of each sequence.

In January 2016, one DNA sequencer supplier provides a machine that ostensibly delivers 1800 Gbp of sequence from 6000 million reads, each read giving 2×150 bp per run. This is far more than is necessary for sequencing an entire human genome to a very high degree of accuracy. The next paradigm shift required is a new method to deal with the data flood. New scientific journals are being created to deal with this new dimension, for example, "GigaScience" was founded 2011 as an open access journal.

1.5.6 DNA: Its Transfer to and Selection in Living Cells

In 1943, Avery, MacLeod, and McCarty had demonstrated that the transfer of DNA (and not RNA, protein, or lipids) was correlated with inheritance of an altered property (phenotype) in the recipient Streptococcal cells (pneumococci), an effect which could be seen both at the level of colony morphology (polysaccharide production when the cells were grown on solid surfaces) and in terms of pathogenicity of the cells when injected into mice. This stimulated microbiologists and biochemists to investigate DNA in more detail culminating in X-ray crystallography. DNA crystal X-ray diffraction data produced by Rosalind Franklin were prematurely used by Watson and Crick. They built a tentative model of the DNA structure, communicated in a one-page letter to Nature in April 1953. The model cartooned a molecule of symmetrical beauty implying that it could be suitable as a carrier for genetic (inherited) information in the form of a long linear code. The presence of an antiparallel strand was interpreted as being a suitable template that would allow replication, mutation, and repair. The model structure was very close to the truth in its essential features and was rapidly accepted. In the 1970s, DNA transformation in which double-stranded supercoiled DNA was "transformed" into bacteria was established and led to the observation that these plasmids were often carrying antibiotic resistance genes [80, 81].

A parallel development of equal importance to Avery and co-workers in 1943 was the attempt to repeat this "transformation" by DNA transfer to animal cells. To demonstrate this, Szybalski made nonreverting mutants in the purine synthetic pathway (hypoxanthine-guanine phosphoribosyltransferase gene; HGPRT a salvage enzyme preventing xanthine and uric acid synthesis) of human cell lines, so that expression of the active *hgprt*-gene allowed selection, positive or negative, on a special medium (HAT medium). These mutants were, in fact, similar to those discovered later for the Lesch–Nyhan syndrome. Transferring DNA from normal cells to the mutant cell lines allowed selection of stable "wild-type" cells, the DNA from which, in turn, could "transform" other mutant cell lines [197]. This was a breakthrough that opened up human and other eukaryotic genetic analysis *in vitro*, and allowed selection of cell fusions. For example, such *hybridomas* were used to first produce mABs or cell lines carrying one, or a limited number of, human chromosomes. One could in fact perhaps in retrospect reevaluate the importance of this finding, saying that genetic engineering really began at this point in 1963!

The production of selectable vectors for use in animals cells often used Szybalski's system or further developments based on selective protocols directed to the purine synthesis pathway, for example, Bacchetti and Graham [198], who transferred the *Herpes simplex* thymidine kinase into a human cell line, albeit with very low efficiency in 1977; and Mulligan and Berg [199] who developed a selection protocol for the expression of an *E. coli gpt* gene in a eukaryotic vector in 1981.

1.5.7 Gene Cloning (1971–1982) the Era of Modern Biotechnology Based on Molecular Biology Begins

1.5.7.1 A Prerequisite for Cloning: Nucleic Acid, Biochemistry, and Enzymology

The possibilities to label, manipulate, and modify DNA molecules relied to a great extent on the availability of the pure, well-characterized enzymes listed in Table 1.5.

1.5.7.2
Applying Known Methods or a Conceptual Jump: the Details

Plasmids and viruses, particularly, have their own complex multicomponent structures (pilli and viral capsids) for transferring their DNA into host cells, with very limited host range. Often, the plasmids and viruses themselves carried DNA modification and destruction systems to inhibit competition and/or kill their hosts. Lateral DNA transfer (conjugation, transduction) between related bacteria was known, often promiscuous and widespread if strong selection pressure such as antibiotic treatment was applied. In the absence of such selection, it was rarely detectable. The spread of antibiotic resistance was, however, a main cause of concern among hospital staffs and a main topic of research for microbiologists including Stanley Falkow and Stanley Cohen in California who were among the first to have the idea of gene cloning (see Section 1.4). The discussion between Boyer and Cohen in the presence of three others, including Falkow, is assumed to be the crystallization point in the realization of gene cloning and the direct precursor to Boyer, who in 1976, with the venture capitalist Swanson, founded Genentech. At least for Cohen and Falkow, it seemed obvious that the isolation of specific DNA fragments (restriction fragments) and their overproduction in pure form (preparation of the recombinant plasmids from the clones) was a simple way to analyze the structure and function of the plasmids, which were causing promiscuous spread of antibiotic resistance in pathogens [80, 81, 148, 200–203]. The jump to cross species barriers with the aim of making pharmaceutical products was probably more in the forefront of Boyer's thoughts, an approach which allowed him to convince investors, in particular the risk capitalist Swanson, that protein therapeutics could be produced in bacteria. This was *contrary to the general conception of the difficulties that would have to be overcome* as outlined in the following list of the state of the art at the time.

Imponderables to Cross-Species Gene Transfer and Expression Among the imponderables that implied how impracticable it would probably be to move and stably express genes across species barriers, were the following:

- It was not all that clear that the genetic code was completely universal (it is not).
- Protein secretion is often mediated by protein–membrane complexes that recognize specific target proteins.
- There are hundreds of types of post-translational modification (proteolytic processing, addition of chemical groups, side chains) of proteins, often determining protein immunogenicity, function, solubility, and so on. This area was largely unexplored.
- Small peptides are scavenged and degraded in bacteria. *Note*: the statement that somatostatin was the first peptide (14aa) gene to be cloned and expressed in *E. coli* is not technically correct. A large chimeric protein had already been expressed, isolated from the bacteria, and chemically cleaved to yield the peptide fragment [204].
- Little was known about protein folding. Chaperone proteins are often necessary for correct folding of proteins and subunit assembly. In bacteria, there is low

oxygen concentration where disulfide bridges do not form. This is in contrast to the oxidizing condition in the cytoplasm of eukaryotes.

- The highly ordered folding of DNA (chromosomes) only existed in eukaryotes. Its role in gene expression was unknown.
- In eukaryotes, regulated secretion, sequestration to and protein release from organelles, as well as mRNA stabilization, were not understood.
- Some DNA structures are unstable, being able to jump to other locations (deletion and insertion) or invert (discovered in plants in 1951 by Barbara McClintock); rediscovered in bacteria, insertion elements, transposons, homing nucleases, CRISPR, and so on, targeted and destabilized, particularly foreign, DNA.
- Sequences with inverted mirror symmetry, that is, palindromes are extremely unstable (unclonable) in prokaryotes. The stability of direct repeats also differs.
- GC methylation common in eukaryotic DNA leads to DNA degradation in most *E. coli* hosts.
- Genetic load was not understood. Recombinant organisms are normally at a severe disadvantage in a natural environment. Under strong selection, recombinants can be forced to survive in the laboratory. However, often the rDNA is still rapidly lost, for example, where high transcription rates are involved.

With all the above in mind, the general impression in 1971 was that gene transfer, controlled gene expression, and stable inheritance, particularly across species barriers, would not be likely or would lead to transient-crippled unstable hybrids with poor expression that would certainly not be suitable as a source for biotechnology products. The mainstream of the classical pharmaceutical industry management was of this opinion, right up to the announcement that Genentech's recombinant insulin had been approved for clinical use in 1984.

By the end of 1974, it was understood that gene cloning was a generally applicable methodology, at least for academic laboratories *to investigate gene structure and function*. In 1977, DNA sequencing started to spread as a general tool. Armies of researchers battled to discover and deal with the practical consequences of the barriers presented above, many turning to species-specific cloning and expression systems, others continuing to deal with overcoming interspecies expression and protein-folding problems. The complexity of post-translational protein modification such as glycosylation has really only been solved by using tissue culture with cells from higher organisms for production for important pharmaceuticals, such as human tPA, EPO, the IFNs, and mABs [72].

How Gene Cloning Started? It has become commonplace to cite the conceptual beginning of gene cloning as being the 1972 paper from Paul Berg's laboratory [205]. As Berg himself admits [148], the paper in fact contains no novel methodology and most of the key enzymes were given to the group as gifts. It should be pointed out that, at that time, whether or not a method would work depended very much on the purity of the enzymes used, none of which was commercially available. The paper *neither demonstrates the feasibility of using the proposed* *protocol for transferring DNA to a foreign organism, nor is this idea original.* The key principle in the Berg-Lobban method that was declared as novel was to join two oligonucleotides via overlapping hybridization, filling out single-strand region, and ligating with DNA ligase. This had been, in fact, demonstrated in Khorana's laboratory as a step in the synthesis of a completely synthetic gene for a tRNA [176]. The oligonucleotide elongation with Sgaramella's TdT (terminal deoxyribonucleotidyl transferase) as proposed in Berg's paper does not, in fact, work very well in practice for cloning purposes and leads to loss of the original restriction sites, thus complicating further analysis, fragment isolation, sequencing, and/or recloning (see discussion on p. 127 and Figure 1.1 in [155]). It did not, therefore, find much application initially (although it was later attractive as a step in complementary DNA (cDNA) cloning), the simpler and efficient method of Cohen and Boyer [206] being the obvious choice. A key discovery in Berg's laboratory was the closed circular ligation of EcoRI cleaved DNA by Metz that implied that the cohesive ends of the DNA were all the same and probably small and cohesive, melting in the range 4–15 °C, a fact which Boyer's laboratory (which had donated the EcoRI) immediately confirmed by DNA sequencing [207]. Rich Roberts particularly followed the route of Hamilton Smith and Boyer with respect to searching for and characterizing further restriction enzymes and using them in analyzing viral and plasmid DNA structure leading, for instance, to his discovery of gene splicing [208].

Danna and Nathans [182] were the first to cleave SV40 with one of the first type II restriction enzymes. Berg's group made *in vitro* SV40 hybrids with phage DNA; however, they transferred it *neither* to *E. coli nor* to eukaryotic cells, leaving open the question as to whether or not opening the restriction site destroyed function(s) required for the vector replication or gene expression. It was left finally to others to use restriction enzymes to investigate SV40 gene structure and function (e.g., the transformation gene, T, t) after transformation of cleaved viral fragments into an animal cell line [209]). A widely used early protocol for transfer of DNA into animal cell cells was coprecipitation of the DNA with calcium phosphate [210].

In 1976, Goff and Berg [211] selected for an *E. coli gpt* and a yeast *tRNA* gene cloned in an SV40 vector that could replicate in monkey cells. These clones could be developed as vectors for further studies of genes in eukaryotic cells using the foreign genes as selective markers. This was the first demonstration of cloning of bacterial DNA with ensuing cross-species transfer into animal cells. Initially, two types of SV40 vectors were available, (i) those that contained only the SV40 replication origin and could carry some 4–5 kb of foreign DNA and required cocultivation with helper virus and (ii) those with small deletions in the "late gene region" that could carry only small foreign fragments.

Transformation of cells by uptake of foreign DNA into bacteria was known since 1944. It was unknown whether eukaryotic DNA could be expressed in microorganisms, but the assumption that the chance of this happening might not be zero caused Berg and his group to postpone attempting cross-species transformation with their *in vitro*-ligated DNA. Perhaps this direction of thought was kept alive by ghosts of irreproducible work published in 1964 by, for example,

Thomas Trautner on Polyoma virus replication and production in transformed *Bacillus subtilis* (see review in [212]). The discussion initiated by Berg led to an excited meeting at Asilomar and the moratorium on gene cloning until guidelines for safe design of such experiments were established [148]. A more detailed analysis of the state of the art might, however, have arrived at a different conclusion that, initially, the most that could be expected from cloning eukaryotic DNA would be improved deletion analysis (specific induced deletion) of viruses and plasmids without expecting cross-species gene expression, the latter being the feeding ground for horror scenarios and public anxiety.

1.5.8 Genome Mapping: Clone Libraries, Restriction Maps, and RFLPs

1.5.8.1 Prelude: Human Genetics before Genome Sequencing

Barbara McClintock was incredibly far ahead of her time: she was one of the main developers of cytogenetics and the originator of the idea of chromosome meiosis and epigenetic effects in plants in the 1940s. She admonished students in life sciences, first of all, to "know your organism." Having a sequence does not imply that one has any knowledge relevant to your organism. Standing alone, it gives little or no insight and initially one has no idea where genes might be. It is only through the pioneering work of hundreds of human geneticists such as Victor McKusick and Leena Peltonen-Palotie, as well as biochemists and cell biologists who worked with them, that sequence work on the human genome could later be interpreted meaningfully in terms of the causality of heritable diseases.

The work of several generations of biochemists working with physicians insured that by 1973, 93 inborn errors of metabolism had been listed. Most of these detailed discoveries of the biochemical pathways and the enzymes involved in metabolic disease were made between 1957 and 1973 [213].

1.5.8.2 Important Ideas on Gene Mapping in the DNA Age

As DNA clones became available, it was considered how these might be used as "probes" or "links" to locate heritable disease genes on the chromosome. Again, it was McClintock, with her studies of meiosis in 1931, who had shown how the relative position of genes could be ordered along the chromosome. In 1980, a scheme was outlined as to how more-detailed linkage maps might be obtained for the whole human genome, using size variation of restriction fragments from the same region but from various individuals (restriction fragment length polymorphisms, RFLPs) [151]. This was also a central theme for the initial work of Jean Weissenbach at Genethon in Evry near Paris, founded in 1990 with a main agenda to develop diagnostics, and possibly treatments, for rare inherited diseases. One of the major breakthroughs that largely changed the negative attitude toward rDNA work was the identification of the dominant genes for the Lesch–Nyhan syndrome (where affected children die a terrible death at an early age). This knowledge was applied as prenatal diagnostic to dramatically reduce the birth rate of affected children among the Ashkenazi Jews of New York. In 1993, the causal gene for Huntington Chorea, another incurable disease involving neuronal degeneration, was isolated and located to a chromosomal segment designated 4p16.3. This was the first autosomal disease locus found using genetic linkage analysis [214] and is associated with restriction fragment linked polymorphisms (RFLPs) where long stretches of triplet repeats in the gene are unstable. Since the symptoms appear late in life, long after child-bearing age, and are untreatable, (family) diagnostics came in for considerable criticism with the "right not to know" being in the forefront of the arguments.

A variety of developments in cloning technology improved the ability to isolate and analyze long regions of the chromosome in particular, simplifying the ordering of adjacent or overlapping clone fragments and assembly of clusters of overlapping sequences ("contigs"). Figures below in brackets give the number of primary clones needed to give a 95% confidence of covering the whole human genome:

- Initially, cloning was inefficient with plasmids rarely carrying fragments of greater than 5 kb with a bias more in the 1–2 kb range (10^7).
  Using the *lambda in vitro* packaging system, cloning became much more efficient and cloning of exclusively larger DNA fragments was ensured (see also [72]).
- with *lambda* cloning vectors, in which a "buffer" fragment had been deleted, inserts were in the size range of 5 kb, later 10–15 kb (6×10^5) [215]
- with *cosmids*; plasmids carrying the *cos*-site of lambda, which are packageable *in vitro* only if they have picked up at least 25–40 kb of insert DNA (2×10^5). Some designate low copy variants, with an F plasmid origin, as fosmids. The original vectors were amplifiable up to high copy number, facilitating further analysis and reducing contamination by *E. coli* chromosomal DNA [216].
- *M13 filamentous phage*. Viral cloning vectors can be isolated as a dsDNA plasmid-form from the cell and as ssDNA from the phage particles. It was developed for cloning by Messing *et al.* [217]. This latter can serve as template for sequencing by the Sanger primer elongation method [160, 218] using a synthetic DNA primer adjacent to the cloning site. It can also be applied to the introduction of specific mutations into the cloned gene by using a partially mismatched primer. It is the vector type used for developing phage-display technology, where clones can be physically selected on the basis of affinity to an immobilized target molecule of choice [219].

A powerful extension of this latter method was "sexual PCR" (DNA shuffling) developed by Stemmer [220] to (i) generate mutants and recombinants in mutant libraries, on a large scale and (ii) to select for novel ligands, which in combination with phage display, could be applied for ligand "affinity maturation" in analogy to the maturation of B-cells producing antibodies with improved affinity to specific target molecules. This recognized the importance of recombination in addition to point mutants in evolution, a factor also used most effectively for altering enzyme specificity. (*Note*: this *in vitro* recombination system can also be used effectively to generate combinatorial library fragments that can also be selected *in vitro*, in combination with *in vitro* gene expression systems (e.g., [221].)

- Following some 9 years after Struhl's *et al.* [222] 1979 demonstration of cloning in yeast, Burke *et al.* [223] showed that very large fragments could be stably cloned in yeast (yeast auxiliary chromosomes; YACs) with a size range of some 40–700 kb (average 80 kb); (10^5). Although reisolating the intact large insert from the yeast cell presented difficulties, YACs could be used for hybridization mapping and preparing sublibraries for sequencing and contig mapping. The libraries are very difficult to produce and are of limited use in tiling (joining or bridging contigs) because to up to 50% of the inserts are of chimeric origin. Their use was dropped in the latter part of the human genome project.
- *PAC vector DNAs* are packaged in a low-copy-number *E. coli* vector using an *in vitro* P1 phage packaging mix [224].

Each generation of cloning vector inventors praised the improved stability of their vector system. Much of this may be hearsay, although it may also be correlated with improvements made to the host strains, which allowed stabilization of the insert DNA and improved protocols, thus reducing palindrome formation of vectors while preparing the chimeric DNA. Purity of enzymes used in DNA manipulation also helped (e.g., purification of cleaner enzymes after cloning and expressing the gene for the enzyme; see Table 1.5).

- *Human accessory chromosome* (HAC) *vectors* , introduced in 1997 by Harrington *et al.* [225] demonstrated clones carrying up to 10 000 kb of DNA. These types of stable vectors are of more importance as gene transfer vectors to human cells to help annotate the human genome without disruption of the extant chromosomes and without use of potentially dangerous fragments of viral vectors.
- *Shuttle vectors* can be replicated and selected for in two hosts. *E. coli* is usually the host for initial isolation and manipulation of DNA as efficient cloning systems had been established for this host. It is then transferred to a new host by transformation or as in the case of *plants*, first into an *Agrobacterium* host, which then uses the T-DNA transfer system to transfer the DNA fragment to the plant host, for example, 1983 [226]. This latter system, the origin of rDNA work with plants, derived directly from studies of the growth of crown gall tumors where a natural system of cell transformation was discovered in which a tumorigenic DNA fragment was transferred from *A. tumefaciens*. Such vectors are of great use in expression cloning in prokaryotes (gene mining), particularly with respect to polyketide antibiotics; and primary and secondary metabolites.

Cosmids, BACs/PACs were used as the major resources in all the large-scale sequencing projects. Cosmids, particularly, are still used (sometimes as shuttle vectors) in "gene mining" (not the *in silico* form of gene mining which refers to searching for homology in databases) in the search for novel enzyme genes where expression of large genes or operons is required: they carry exclusively appropriate sized DNA inserts, are easy to work with, and allow rapid mapping and sequencing of the cloned genes.

The presence of long repeated sequences throughout the genome makes it not only difficult for shotgun sequencing, it also makes the detection of overlapping fragments extremely difficult. Because of this, it was important to gather information from both ends of long sequences of defined length. With such data from, for example, "jumping and linking libraries," the relative position of two contigs separated by long repetitive sequences could be established. *Even today, there is still some 1% of the human genome sequence that is not established, although the regions in question are mapped to a few positions on the human chromosomes.*

- *Jumping and linking libraries*: For those involved for some 15 years in the1980s and 1990s in assembling a total gene/sequence map of large genomes, before total genome shotgun sequencing, a main task was to place a particular fragment to be sequenced with respect to the rest of the known sequences. This yielded the groups of known overlapping sequences and clones called *contigs*. Bridging contigs, that is, finding out the relative position of the contigs separated by as yet nonsequenced regions, was accomplished by sequencing clone libraries, especially made to contain only the ends of very large size-fractionated fragments that had been circularized *in vitro* ("jumping and linking-libraries") [227–229].

Some libraries were made in plasmid vectors exclusively from cDNAs (synthesized on an mRNA template). Many of these contained the open-reading frame from a spliced mRNA. The sequences of these clones were referred to as expressed sequences (ESs) and were used late in the genome sequence program to preferentially locate possible gene region tags (EST; e.g., [230]).

Since the advent of total shotgun genome sequencing and application of superior computer algorithms and hardware for ordering single sequence stretches within very large data sets [231], the earlier methods of gene fragment ordering before the advent of DNA sequencing, have become largely obsolete. Particularly for organisms where "the" genome has already been sequencing, "resquencing" becomes simplified. The shotgun approach was used first to obtain the total genome sequence of a microorganism (*Haemophilus influenzae Rd* in 1995 [232]). The first complete shotgun sequence of a larger eukaryote genome was that of the *Drosophila melanogaster* genome published by Gerald Rubin and Craig Venter's group in 2000 [233].

The human genome, being more than 3000-fold larger than that of *H. influenza* and 25 times larger than that of *Drosophila*, is considerably more difficult to master, relying exclusively on shotgun sequencing. As Venter himself admits, although his human genome sequencing project at Celera was largely based on the shotgun approach, much faster for accumulating raw data, he had used data from contig-linkage mapping and the EST-linkage method which he himself had helped pioneer. In spite of this, his choice of a single genome (his own) for sequencing simplified the linkage in the repetitive regions that in his case would be far less heterogeneous than that provided by the general sequencing (HUGO) community. The single-genome approach also gave for the first time the actual sequence for *each of the diploid chromosome copies separately*. This approach, that took place first in an industrial format with nonpublic funding at Celera under the leadership of Venter, was much maligned by notable leaders of the human genome program in America who had little empathy with Venter's withholding his data for several weeks or months before releasing it. Many in this competitive field found Venter's reveling in his ability to sequence faster and more cheaply than the rest of the community fairly unbearable. A deplorable aspect of this phase of the genome-sequencing project was the extreme infighting for funding and committee lobbying. Much of this attitude appeared to be a continuation of the extreme competitiveness already extant from the early "gene wars," that is, "the characterization of contestants in the cloning races – Goeddel and Gilbert, and many others in supporting roles, ... portray a sorry picture of personal rivalries in the conduct of science. A scientist can be intensely competitive and even unscrupulous in pursuit of a laudable goal" ([83], pp. 196, 198, 199). Also recommended is [234] for a coverage of the early development of the scientific and social issues in the human genome project, covering the heroes and the hypocrites.

The final outcome of the Celera sequencing effort was cost effective. Critics say it scavenged public research data financed by the taxpayer (true or false; who benefitted most from whom?). Some argue that scientists should have been doing more scientific research rather than having *generations of students* doing rote sequencing. Watson's supporters still maintain that Venter could not have completed the sequence without the data from the open HGP database, and may be correct. Conversely, the sequence was also certainly "finished" (still bits missing?) faster (and cheaper) with the help of the independent approach at Celera. We note that as data-flow increases exponentially, emphasis shifts toward improved and more rapid analysis. A little-recognized milestone was the development of the Genome Assembler algorithm by Granger Gideon Sutton at Celera (a nonprofit Organization led by Craig Venter) which succeeded in 1995 in yielding a complete *H. influenzae* genome (an NIH grant proposal was refused), later the *D. melanogaster* genome, and finally the human genome in 1999 all in the face of aggressive skepticism by the Jim Watson-lead "Genome Consortium" (NCBI) that such a random ("shotgun") assembly was impossible (see also [235]).

We note once more that without the accompanying medical, biochemical, and cell biological research on the *biological function of a sequence*, the sequence alone has no intrinsic value. When Francis Collins was asked at a press conference on the completion of the DNA sequence of a particular chromosome about the special significance of this event he said "a chromosome is an entity"; Craig Venter pithily remarked " a piece of shit is also an entity." Why this flagrant breach of political correctness? Perhaps it was necessary to explain to the Press reporting this incident that the worldwide genome project relied on grouping sequence efforts to groups supplied with clones selected as belonging to a particular chromosome so as to later simplify ordering of the sequences generated, for example, to "contigs" of overlapping clones. Membership in such a group required accepting this regime and was a necessary part of the funding program of which Venter was not a part.

Francis Collins should have admitted that he no choice other than this approach. Venter on the other hand did.

1.5.8.3
DNA Hybridization Chip Sequencing and More

Patrick O'Brown and Ronald Davis played a key role in the development of synthesis of oligonucleotide microarrays and their use for quantifying hybridization with nonradioactive oligos (optical sensors) that became a laboratory tool starting in about 1995. Once a genome has been sequenced in its entirety, it can be scanned for polymorphisms by direct hybridization to a DNA array (chips) carrying nucleotide primers corresponding to known polymorphic sites (sites with known sequence divergence) within PCR-amplified regions. The nucleotides hybridizing to the probe oligos on the chip are fluorescent labeled. This methodology has a very wide range of applications, for example, detecting GCNVs; routine screening for particular viruses or antibiotic resistance genes; whole-genome resequencing; comparative DNA methylation studies; distribution of epigenetically modified chromatin on the DNA (combined with chromatin immunoprecipitation, ChIP); and comparative gene transcription studies among others (section 11.4.5 in [72]). A combination of these latter lead to a better understanding of how genes respond to their environment, as well as ashow cells differentiate during embryogenesis, and how this can be reversed when cancer cells are placed with embryonic tissue.

1.5.8.4
Mega Sequencing: Impact on Biotechnology

The early phase of human genome sequencing gave rapid progress in tracking down the genes involved in many of the thousands of heritable diseases, syndromes, or traits that had been identified by human geneticists. Many of the powerful *biological response modifiers*, such as IFN, lymphokines, and growth and cell-differentiation factors, were first clearly characterized as proteins via sequencing and expressing their genes. This provided, at the same time, methods to reproducibly produce these compounds relatively cheaply in large amounts. This changed medical research and medical diagnostics, and increased the range of biopharmaceutical products (see Section 1.4).

The initial impact related to *first-generation products* such as insulin and growth hormone, also blood-clotting factors, that were already well characterized and already in clinical use. Their production via rDNA methods in bacteria or yeast gave an alternative source for a product that no longer had to be derived from human or animal blood or tissue and wase thus free of the danger of contaminating viruses from such sources (e.g., cytomegalovirus, hepatitis viruses, and AIDS).

First Pathogen-Free Source of Pharmaceuticals A fact that is sometimes overlooked in reviews of this period is the identification, cloning, and sequencing of pathogenic viruses. This was not just of general interest. It allowed for the first time the development of DNA-based PCR testing of products derived from blood or tissues for the presence or absence of these viruses ([72], pp. 141–142; eight examples between 1979 and 1990, e.g., [236]) including the AIDS HIV virus. It also lead to the first rDNA-based vaccine for hepatitis B.

The second generation of products could not be characterized in detail until their genes had been cloned and they had been produced in pure form, and in sufficient amounts to be made available for detailed research, and eventually, in several cases, clinical application. One example is that of GCSF (granulocyte colony stimulating factor), cloned at Amgen in 1986 [237], which could be used since the late 1980s for blood stem-cell transplantation to cure, for instance, chemotherapy-resistant Hodgkin-lymphoma. Without this compound to stimulate blood stem-cell release into the donor blood, this treatment would not be an option (the youngest son of one of the authors of this article is still alive, thanks to this development; see Section 1.4 for more examples).

Welte [238] provided detailed insight into the academic, industrial, and medical collaborations that were involved and required at that time to complete this success story, starting with recognition of a protein function, purification of the protein, partial sequence, use of this information to find a cDNA clone sequence synthesized on a messenger RNA (mRNA), cloning the entire gene, and expression and optimization of production and purification, finally leading to extensive clinical testing (see Table 1.6 for methods involved).

The human genome program and the major advances in high-throughput sequencing has influenced biotechnology in a number of ways:

Table 1.6 Methods for enriching mRNAs/detecting specific clones or gene products.

System	Application	Developed by
Southern blotting (DNA) and Northern blotting (RNA)	Nucleic acid transfer to cellulose: detect homology to DNA or RNA probes (colony or phage plaque hybridization similar: → clone identification)	Southern [256] "Southern" Alwine *et al.* [257] "Northern"
Western blotting (1D or 2D)	Transfer of protein to filter: identify with monoclonal antibody	Symington *et al.* [258]
Hybridoma technology	First method of producing monoclonal antibodies (hybridomas not stable)	Köhler and Milstein [259]
rDNA monoclonal antibodies	Produced in cell lines, yeast, insect larvae, or bacteria (more stable production)	For example, Riechmann *et al.* [260]; Marks *et al.* [261]
Fluorophor- and enzyme-labeling	Protein or nucleic acid tagging with various highly fluorescent tags (e.g., GFP) many applications: DNA sequencing, tissue ID, chromosome painting (mapping) replaces radioactivity in the methods above. Also, enzyme-linked signal amplification (ELISA)	For example, review: Phillips [262]; Lichter, P *et al.* [263]; Ju *et al.* [264]

Diagnosis for Predisposition to Disease The *diagnostic possibilities for predisposition* to common diseases are of more limited clinical relevance than many originally expected. This has limited the predicted immediate growth of personal medicine. Sequence-based clinical diagnostics for the some 4000 hereditary mostly recessive traits known before the HUGO sequencing effort had really begun can now be carried out rapidly and with high precision without involving other family members. However, medical family history relating to inherited disorders (anamnesis) will always play a major role in indicating if a genetic causality should be investigated. Preventive surgery in the case of actress Angelina Jolie, who had a very high predisposition to breast cancer associated with particular BRCA gene alleles (over 70% likelihood rather than the general population 5–12%), has highlighted this diagnostic option in which sequence patenting has also played a historic role.

Genomewide Association Sequencing: Genes Affecting Widespread Chronic Syndromes When referring to "the human genome sequence," one does not usually refer to an individual sequence (apart from those of Venter and later of Watson). "The Sequence" was, and still is, the sum of data from thousands of genomes and hundreds of thousands of partial sequences of variations in particular disease-associated regions. The database provides *the basis for future medical and fundamental research related to human health.* However, many investigators warn against putting too strong a bias on looking for genetic factors relating to cancer, for instance, where the impact has been disappointing. By 2010, the US National Cancer Institute's TCGA had cost $375 million; part of an international cancer genome consortium that aims to sequence 25 000 tumor samples for a total of $1 billion. As stated by Robert Weinberg (MIT), "Sequencing endless more cancer genomes isn't going to tell us more than we already know" (http://news.sciencemag.org/2010/04/updated-skeptic-questions-cancer-genome-projects).

1.5.9
Expressing Genes in Other Organisms: Transgenic Animals Carrying rDNA

This review has glossed over many areas that deserve more detailed description with respect to biotechnological development. One of these is the area of recombinant antibodies. This is an exciting area that has produced some of the most innovative products to be derived from rDNA technology: particularly in bacteria; yeasts, both intracellular and surface expression; in Baculovirus vectors in insect cells; in animal cell lines, and in transgenic animals. These studies involved both affinity maturation and realization of the end-product for clinical use. The successful development of this area required finding solutions to nearly all of the problems encountered in the development of genetic engineering as presented above. These aspects have all been discussed in some detail with a listing of all the clinical applications up to and including 2010, including the history, the science, and the business [72].

The production of transgenic animals carrying rDNA in their genome arose from a combination of advances in embryology [239], understanding the differentiation of stem cells [240] in the developing embryo, and development of viral vectors [241]. This has had a broad range of applications, from production of "knockout" mice for fundamental research to animal models for human disease (e.g., a mouse model for HCV infections [242]).

1.5.10
Future Trends

In the search for new therapeutic ligands, the healthy competition between protein modeling and empirical selection procedures continues (e.g., [243]). This is true also in the search for therapeutic targets where "systems biology" should contribute a more global view of the interplay of intracellular pathways using *in silico* techniques [244]. Synthetic biology provides a new tool, independent of existing structures, where novel enzymes or gene regulators are constructed *de novo* by assembly of "Lego"-like functional building blocks [245]. This latter complements the ongoing refinement of genetic engineering with respect to direct gene replacement or mutation in various organisms. In the area of industrial biotechnology, the science continues to develop as a sustainable alternative to chemical synthesis.

The discovery of the key factors (often referred to as the *Yamanaka factors*) which induce differentiated cells to form pluripotent stem cells ((PiPS – protein-induced pluripotent stem cells) e.g., from adult human fibroblasts) opens up the vision of creating collections of immortalized stem-cell cultures of a huge range of tissue specificity as potential source for medical application over and above the requirement for helping postirradiation or chemotherapy recovery of leukemia patients, or even to use this method as a form of personalized medicine [246]. Although this development was achieved through the application of rDNA technology, the final PiPS technology, a radical advance for biotechnology, no longer requires it.

The latest developments in the CRISPR/Cas9 methods of *in vivo* genetic engineering, represent a great increase in the ability to accurately manipulate genetic material by DNA replacement (recombination) directly in living cells or whole organisms, efficiently and with little or no unwanted genetic side effects. This is expected to revolutionize much of modern medicine and biotechnology in the near future (*Note:* the first genetically engineered animal, salmon, engineered to grow faster and intended for food, has just been approved by FDA) [247a, b]. One innovative example with CRISPR/Cas9 technology can be seen in the work of, for example, George Church's group in which this new method was used to destroy 69 retroviral copies in the genome of a pig in a single experiment. This is the first step to generate pigs which are free of retroviruses that might be harmful to man. Further altering tissue specific antigens could lead to swine capable of serving as a safe harbor for human organs. In North America, some 125 000 patients per year are in need of organ donors [248]. In passing, one can refer to George Church (MIT) as the epitomy of the modern molecular geneticist, who has fully embraced the role of the entrepreneur, being founder of at least 9 genome-related companies and is associated with or advising some 81 companies.

From the foregoing, we see that new tools are developed opening novel opportunities for investigation, often replacing older methods. Investigators, who are patient and those who are not, have made major contributions. Isolated individuals and those who build huge consortiums, those believing in entrepreneurship and those who do not, have all played their roles. Recognizing the way to progress is a privilege of the prepared mind independent of political or peer ranking. Let us keep reporting with honesty.

In the quest for solutions, we can be assured to stumble upon surprises as biological systems have a memory and a potential complexity that exceeds the possibility of the physical matter of the Universe to give each of them substance.

References

1. McGovern, P.E. *et al.* (2004) Fermented beverages of pre- and proto-historic China. *Proc. Natl. Acad. Sci. U.S.A.*, **101**, 17593–17598.
2. Bud, R. (1993) *The Uses of Life: A History of Biotechnology*, Cambridge University Press.
3. Lovitt, R.W., Kim, B.H., Shen, G.-J., and Zeikus, J.G. (1988) Solvent production by microorganisms. *Crit. Rev. Biotechnol.*, 7, 107–186.
4. Mermelstein, L.D., Papoutsakis, E.T., Petersen, D.J., and Bennett, G.N. (1993) Metabolic engineering of *Clostridium acetobutylicum* ATCC 824 for increased solvent production by enhancement of acetone formation enzyme activities using a synthetic acetone operon. *Biotechnol. Bioeng.*, **42**, 1053–1060.
5. Qureshi, N., Li, X.-L., Hughes, S., Saha, B.C. *et al.* (2006) Butanol production from corn fiber xylan using *Clostridium acetobutylicum*. *Biotechnol. Progr.*, **22**, 673–680.
6. Bremus, C., Herrmann, U., Bringer-Meyer, S., and Sahm, H. (2006) The use of microorganisms in L-ascorbic acid production. *J. Biotechnol.*, **124**, 196–205.
7. Xu, A., Yao, J., Yu, L., Lv, S. *et al.* (2004) Mutation of *Gluconobacter oxydans* and *Bacillus megaterium* in a two-step process of L-ascorbic acid manufacture by ion beam. *J. Appl. Microbiol.*, **96**, 1317–1323.
8. Saito, Y., Ishii, Y., Hayashi, H., Imao, Y. *et al.* (1997) Cloning of genes coding for L-sorbose and L-sorbosone dehydrogenases from *Gluconobacter oxydans* and microbial production of 2-keto-gulonate, a precursor of L-ascorbic acid, in a recombinant *G. oxydans* strain. *Appl. Environ. Microbiol.*, **63**, 454–460.
9. Perlman, D. (1979) in *Microbial Technology*, 2nd edn, vol. 1 (eds J.H. Peppler and D. Perlman), Academic Press, New York, pp. 521–527.
10. Koizumi, S., Yonetani, Y., Maruyama, A., and Teshiba, S. (2000) Production of riboflavin by metabolically engineered *Corynebacterium ammoniagenes*. *Appl. Microbiol. Biotechnol.*, **51**, 674–679.
11. Perkins, J.B., Sloma, A., Hermann, T., Theriault, K. *et al.* (1999) Genetic engineering of *Bacillus subtilis* for the commercial production of riboflavin. *J. Ind. Microbiol. Biotechnol.*, **22**, 8–18.
12. Huembelin, M., Griesser, V., Keller, T., Schurter, W. *et al.* (1999) GTP cyclohydrolase II and 3.4-dihydroxy-2-butanone 4-phosphate synthase are the

rate-limiting enzymes in riboflavin synthesis of an industrial *Bacillus subtilis* strain used for riboflavin production. *J. Ind. Microbiol. Biotechnol.*, **22**, 1–7.
13. Scott, A.I. and Roessner, C.A. (2002) Biosynthesis of cobalamin (vitamin B12). *Biochem. Soc. Trans.*, **30**, 613–620.
14. Demain, A.L., Daniels, H.J., Schnable, L., and White, F. (1968) Specificity of the stimulatory effect of betaine on the vitamin B12 fermentation. *Nature*, **220**, 1324–1325.
15. Benninga, H. (1990) *A History of Lactic Acid Making*, Kluwer Academic Publishers.
16. Bai, D.-M., Zhao, X.-M., Li, X.-D., and Xu, S.-M. (2004) Strain improvement and metabolic flux analysis in the wild-type and a mutant *Lactobacillus lactis* strain for L(+)-lactic acid production. *Biotechnol. Bioeng.*, **88**, 681–689.
17. Zhou, S., Causey, T.B., Hasona, A., Shanmugam, K.T. *et al.* (2003) Production of optically pure D-lactic acid in mineral salts medium by metabolically engineered *Escherichia coli* W3110. *Appl. Environ. Microbiol.*, **69**, 399–407.
18. Roehr, M. (1998) A century of citric acid fermentation and research. *Food Technol. Biotechnol.*, **36**, 163–171.
19. Hannan, M.A., Rabbi, F., Faizur Rahman, A.T.M., and Choudhury, N. (1973) Analyses of some mutants of *Aspergillus niger* for citric acid production. *J. Ferment. Technol.*, **51**, 606–608.
20. Miall, L.M. (1978) in *Economic Microbiology: Primary Metabolites*, vol. 1 (ed. A.H. Rose), Academic Press, New York, pp. 47–119.
21. Mattey, M. (1992) The production of organic acids. *Crit. Rev. Biotechnol.*, **12**, 87–132.
22. Fukaya, M., Tayama, K., Tamaki, T., Tagami, H. *et al.* (1989) Cloning of the membrane-bound aldehyde dehydrogenase gene of *Acetobacter polyoxogenes* and improvement of acetic acid production by use of the cloned gene. *Appl. Environ. Microbiol.*, **55**, 171–176.
23. Wang, D.I.C., Fleischaker, R.J., and Wang, G.Y. (1978) A novel route to the production of acetic acid by fermentation. *AIChE Symp. Ser.*, **74**, 105–110.
24. Parekh, S.R. and Cheryan, M. (1994) High concentrations of acetate with a mutant strain of *C. thermoaceticum*. *Biotechnol. Lett.*, **16**, 139–142.
25. Cheryan, M., Parekh, S., Shah, M., and Witjitra, K. (1997) Production of acetic acid by *Clostridium thermoaceticum*. *Adv. Appl. Microbiol.*, **43**, 1–33.
26. Sonoyamana, T., Tani, H., Matsuda, K., Kageyama, B. *et al.* (1982) Production of 2-keto-L-gulonic acid from D-glucose by two-stage fermentation. *Appl. Environ. Microbiol.*, **43**, 1064–1069.
27. Lynd, L.R. (1990) Large scale fuel ethanol from lignocellulose: potential, economics, and research priorities. *Appl. Biochem. Biotechnol.*, **24/25**, 695–719.
28. Ingram, L.O., Conway, T., Clark, D.P., Sewell, G.W., and Preston, J.F. (1987) Genetic engineering of ethanol production in *Escherichia coli*. *Appl. Environ. Microbiol.*, **53**, 2420–2425.
29. Doran, J.B. and Ingram, L.O. (1993) Fermentation of crystalline cellulose to ethanol by *Klebsiella oxytoca* containing chromosomally integrated *Zymomonas mobilis* genes. *Biotechnol. Progr.*, **9**, 533–538.
30. Lynd, L.R. (1989) Production of ethanol from lignocellulosic materials using thermophilic bacteria: critical evaluation of potential and review. *Adv. Biochem. Eng.*, **38**, 1–52.
31. Demain, A.L., Newcomb, M., and Wu, J.H.D. (2005) Cellulase, clostridia and ethanol. *Microbiol. Mol. Biol. Rev.*, **69**, 124–154.
32. Eggeling, L. and Sahm, H. (1999) in *Metabolic Engineering* (eds S.Y. Lee and E.T. Papoutsakis), Marcel Dekker, New York, pp. 153–176.
33. Kitano, K., Sugiyama, Y., and Kanzaki, T. (1972) L-Glutamate fermentation with acetic acid by an oleic acid requiring mutant. II. Inhibitory factors against the extracellular accumulation of L-glutamate. *J. Ferment. Technol.*, **50**, 182–191.

34. Nakao, Y., Kikuchi, M., Suzuki, M., and Doi, M. (1972) Microbial production of L-glutamic acid by glycerol auxotrophs and production of L-glutamic acid from *n*-paraffins. *Agric. Biol. Chem.*, **36**, 490–496.
35. Laneelle, G. and Clement, Y. (1986) Glutamate excretion mechanism in *Corynebacterium glutamicum*. 5th International Symposium on the Genetics of Industrial *Microorganisms*, Part B, pp. 247–252.
36. Eggeling, L. and Sahm, H. (2001) The cell wall barrier of *Corynebacterium glutamicum* and amino acid efflux. *J. Biosci. Bioeng.*, **92**, 201–213.
37. Nampoothiri, K.M., Hoischen, C., Bathe, B., Moeckel, B. *et al.* (2002) Expression of genes of lipid synthesis and altered lipid composition modulates L-glutamate efflux of *Corynebacterium glutamicum. Appl. Microbiol. Biotechnol.*, **58**, 89–96.
38. Demain, A.L., Jackson, M., Vitali, R.A., Hendlin, D. *et al.* (1965) Production of xanthosine-5'-monophosphate and inosine-5'-monophosphate by auxotrophic mutants of a coryneform bacterium. *Appl. Microbiol.*, **13**, 757–761.
39. Demain, A.L., Jackson, M., Vitali, R.A., Hendlin, D. *et al.* (1966) Production of guanosine-5′-monophosphate and inosine-5′-monophosphate by fermentation. *Appl. Microbiol.*, **14**, 821–825.
40. Demain, A.L. (1968) Production of purine nucleotides by fermentation. *Prog. Ind. Microbiol.*, **8**, 35–72.
41. Tseng, Y.H., Ting, W.Y., Chou, H.C., Yang, B.Y. *et al.* (1992) Increase of xanthan production by cloning *xps* genes into wild-type *Xanthomonas campestris. Lett. Appl. Microbiol.*, **14**, 43–46.
42. Dawes, E.A. (1988) Polyhydroxybutyrate: an intriguing biopolymer. *Biosci. Rep.*, **8**, 537–547.
43. Pedros-Alio, C., Mas, J., and Guerrero, R. (1985) The influence of poly-β-hydoxybutyrate accumulation on cell volume and buoyant density in *Alcaligenes eutrophus. Arch. Microbiol.*, **143**, 178–184.
44. Datta, R., Tsai, S.P., Bonsignore, P., Moon, S.H. *et al.* (1995) The technological and economic potential of poly(lactic acid) and its derivatives. *FEMS Microbiol. Rev.*, **16**, 221–231.
45. Foster, W. and Raoult, A. (1974) Early description of antibiosis. *J. R. Coll. Gen. Pract.*, **24** (19), 889–894.
46. Bentley, R. (2001) Bartolomeo Gosio, 1863–1944: an appreciation. *Adv. Appl. Microbiol.*, **48**, 229–250.
47. Gratia, J.P. (2000) André Gratia: a forerunner in microbial and viral genetics. *Genetics*, **156**, 471–476.
48. Raper, K.B. (1994) The development of improved penicillin-producing molds. *Ann. N.Y. Acad. Sci.*, **48**, 41–56.
49. Dixon, B. (2006) Sulfa's true significance. *Microbe*, **1**, 500–501.
50. Demain, A.L. and Masurekar, P.S. (1974) Lysine inhibition of in vivo homocitrate synthesis in *Penicillium chrysogenum. J. Gen. Microbiol.*, **82**, 143–151.
51. Friedrich, C.G. and Demain, A.L. (1977) Effects of lysine analogs on *Penicillium chrysogenum. Appl. Environ. Microbiol.*, **34**, 706–709.
52. Fawcett, P.A., Usher, J.J., and Abraham, E.P. (1976) in 2nd International Symposium on the Genetics of Industrial Microorganisms (ed. K.D. MacDonald), Academic Press, New York, pp. 129–138.
53. Kohsaka, M. and Demain, A.L. (1976) Conversion of penicillin N to cephalosporin(s) by cell-free extracts of *Cephalosporium acremonium. Biochem. Biophys. Res. Commun.*, **70**, 465–473.
54. Yoshida, M., Konomi, T., Kohsaka, M., Baldwin, J.E. *et al.* (1978) Cell-free ring expansion of penicillin N to deacetoxycephalosporin C by *Cephalosporium acremonium* CW-19 and its mutants. *Proc. Natl. Acad. Sci. U.S.A.*, **75**, 6253–6257.
55. Banko, G., Demain, A.L., and Wolfe, S. (1987) α-(L-α-aminoadipyl)-L-cysteinyl-D-valine synthetase (ACV synthetase): a multifunctional enzyme with broad substrate specificity for the synthesis of penicillin and cephalosporin precursors. *J. Am. Chem. Soc.*, **109**, 2858–2860.

56. Hollander, I.J., Shen, Y.Q., Heim, J., Demain, A.L. *et al.* (1984) A pure enzyme catalyzing penicillin biosynthesis. *Science*, **224**, 610–612.
57. Nagarajan, R., Boeck, L.D., Gorman, M., Hamill, R.L. *et al.* (1971) β-Lactam antibiotics from *Streptomyces*. *J. Am. Chem. Soc.*, **93**, 2308–2310.
58. Stapley, E.O., Jackson, M., Hernandez, S., Zimmerman, S.B. *et al.* (1972) Cephamycins, a new family of β-lactam antibiotics. I. Production by actinomycetes, including *Streptomyces lactamdurans* sp. n. *Antimicrob. Agents Chemother.*, **2**, 122–131.
59. Bost, P.E. and Demain, A.L. (1977) Studies on the cell-free biosynthesis of β-lactam antibiotics. *Biochem. J.*, **162**, 681–687.
60. Brown, A.G. (1986) Clavulanic acid, a novel β-lactamase inhibitor-a case study in drug discovery and development. *Drug Des. Dev.*, **1**, 1–21.
61. Kahan, F.S., Kahan, F.M., Goegelman, R.T., Currie, S.A. *et al.* (1979) Thienamycin, a new β-lactam antibiotic. I. Discovery, taxonomy, isolation, and physical properties. *J. Antibiot.*, **32**, 1–12.
62. Berdy, J. (1995) *Proceedings of the 9th International Symposium on the Biology Actinomycetes, Part 1*, Allerton Press, New York, pp. 3–23.
63. Shier, W.T., Rinehart, K.L., and Gottlieb, D. (1969) Preparation of four new antibiotics from a mutant of *Streptomyces fradiae*. *Proc. Natl. Acad. Sci. U.S.A.*, **63**, 198–204.
64. McArthur, H.A.I. (1997) The novel avermectin, Doramectin – A successful application of mutasynthesis. Abstract S13 of the 5th International Conference on the Biotechnology of Microbiology (Products: Novel Pharmacology Agrobiology Activities) Williamsburg, VA, 1997, p. 20.
65. Okanishi, M., Suzuki, K., and Umezawa, H. (1974) Formation and reversion of streptomycete protoplasts: cultural condition and morphological study. *J. Gen. Microbiol.*, **80**, 389–400.
66. Hopwood, D.A. (1988) Towards an understanding of gene switching in *Streptomyces*, the basis of sporulation and antibiotic production. *Proc. R. Soc. London*, **235**, 121–138.
67. Hopwood, D.A., Malpartida, F., Kieser, H.M., Ikeda, H. *et al* (1985) Production of hybrid antibiotics by genetic engineering. *Nature*, **314**, 642–644.
68. Floss, H.G. (2006) Combinatorial biosynthesis – potential and problems. *J. Biotechnol.*, **124**, 242–257.
69. Dougherty, T.J. and Barrett, J.F. (2001) ABT-773: a new ketolide antibiotic. *Expert Opin. Invest. Drugs*, **10**, 343–351.
70. Nichterlein, T., Kretschmar, M., and Hof, H. (1996) RP 59500, a streptogramin derivative, is effective in murine listerosis. *J. Chemother.*, **8**, 107–112.
71. Petersen, P.J., Jacobus, N.V., Weiss, W.J., Sum, P.E. *et al.* (1999) In vitro and in vivo antibacterial activities of a novel glycylcycline, the 9-*t*-butylglycylamido derivative of minocycline (GAR-936). *Antimicrob. Agents Chemother.*, **43**, 738–744.
72. Buchholz, K. and Collins, J. (2010) *Concepts in Biotechnology: History, Science and Business*, Wiley-VCH Verlag GmbH, Weinheim.
73. Buchholz, K., Kasche, V., and Bornscheuer, U. (2012) *Biocatalysts and Enzyme Technology*, Wiley-VCH Verlag GmbH, Weinheim.
74. Bud, R. (1994) in *Resistance to New Technology* (ed. M. Bauer), Cambridge University Press, Cambridge, pp. 293–309.
75. Dechema (1974) *Biotechnologie*, Dechema, Frankfurt.
76. Buchholz, K. (1979) in *Geplante Forschung* (eds W. van den Daele, W. Krohn, and P. Weingart), Suhrkamp Verlag, Frankfurt, pp. 64–116.
77. Jasanoff, S. (1985) Technological innovation in a corporatist state: the case of biotechnology in the Federal Republic of Germany. *Res. Policy*, **14**, 23–38.
78. OTA (US Office of Technology Assessment) (1984) Impact of Applied Genetics, OTA, Washington, DC.
79. Hotchkiss, R.D. (1979) in *The Origins of Modern Biochemistry* (eds P.R. Srinivasan, J.S. Fruton, and J.T. Edsall),

New York Academy of Sciences, New York, pp. 321–342.
80. Cohen, S.N. (2013) DNA cloning: a personal view after 40 years. *Proc. Natl. Acad. Sci. U.S.A.*, **110**, 15521–15529.
81. Falkow, S. (2001) I'll have the chopped liver please, or how I learned to love the clone. A recollection of some of the events surrounding one of the pivotal experiments that opened the era of DNA cloning. *ASM News*, **67**, 555–559.
82. Hall, S.S. (1987) *Invisible Frontiers*, Tempus Books of Microsoft Press.
83. Kornberg, A. (1995) *The Golden Helix*, University Science Books, Sausalito, CA.
84. Lear, J. (1978) *Recombinant DNA: The Untold Story*, Crown Press, New York.
85. Ullrich, A. (1980) Genentech-story. *Nachr. Chem. Tech. Lab.*, **28**, 726–730.
86. Behrendt, U. (2009) *Genentech, Data, Technologietransfer-Workshop*, VBU, Dechema, Frankfurt.
87. Walsh, G. (2007) *Pharmaceutical Biotechnology*, John Wiley & Sons, Ltd., Chichester.
88. (a) Lähteenmäki, R. and Lawrence, S. (2005) Public biotechnology 2004 – the numbers. *Nat. Biotechnol.*, **23**, 663–671; (b) Lähteenmäki, R. and Lawrence, S. (2006) Public biotechnology 2005 – the numbers. *Nat. Biotechnol.*, **24**, 625–634.
89. (a) C&EN: Chemical & Engineering News (2009) vol. 7 (July 6), pp. 30–46; (b) July 4, (2011) pp. 34–39.
90. Roche (2008) *Geschäftsbericht 2008*, F. Hoffmann La-Roche AG, Basel.
91. Läsker, K. (2008) Sueddeutsche Zeitung, 19 May 2008, p. 20; 27 May 2008, p. 23.
92. Ernst & Young (2001) Yearly Report cited in the VCI (Verband der chemische Industrie), Yearly Report (Jahresbericht), Frankfurt, p. 24.
93. Chemical & Engineering News (2015) (Sept. 7), pp. 14–20.
94. Chemical & Engineering News (2015) C&E Supplement, pp. 26–29.
95. Chemical & Engineering News (2015) (June 29), pp. 20–21.
96. Heinzle, E., Biwer, A.P., and Cooney, C.L. (2006) Development of Sustainable Bioprocesses, Modelling and Assessment, John Wiley & Sons, Inc., New York.
97. Weuster-Botz, D., Hekmat, D., Puskeiler, R., and Franco-Lara, E. (2007) Enabling technologies: fermentation and downstream processing. *Adv. Biochem. Eng./Biotechnol.*, **105**, 205–247.
98. (a) Walsh, G. (2005) in *Modern Biopharmaceuticals*, vol. 1 (ed. J. Knaeblein), Wiley-VCH Verlag GmbH, Weinheim, pp. 1–34; (b) Wurm, F.M. (2004) Production of recombinant protein therapeutics in cultivated mammalian cells. *Nat. Biotechnol.*, **22**, 1393–1398.
99. Zhou, T.-C., Zhou, W.-W., Hu, W., and Zhong, J.-J. (2010) *Encyclopedia of Industrial Biotechnology*, John Wiley & Sons, Inc., New York, http://tinyurl.com/wileyEIB (accessed 31 May 2016).
100. Deckwer, W.-D., Jahn, D., Hempel, D., and Zeng, A.-D. (2006) Systems biology approaches to bioprocess development. *Eng. Life Sci.*, **6**, 455–469.
101. Hempel, D.C. (2006) Development of biotechnological processes by integrating genetic and engineering metods. *Eng. Life Sci.*, **6**, 443–447.
102. Klein-Marcuschamer, D., Yadav, V.G., Ghaderi, A., and Stephanopoulos, G.N. (2010) De novo metabolic engineering and the promise of synthetic DNA. *Adv. Biochem. Eng./Biotechnol.*, **120**, 101–131.
103. Silverman, J. *et al.* (2005) Multivalent avimer proteins evolved by exon shuffling of a family of human receptor domains. *Nat. Biotechnol.*, **23**, 1556–1561.
104. Chemical & Engineering News (2015) (Nov. 9), p. 9.
105. Papini, M., Salazar, M., and Nielsen, J. (2010) Systems biology of industrial microorganisms. *Adv. Biochem. Eng./Biotechnol.*, **120**, 51–99.
106. Wittmann, C. (2010) Analysis and engineering of metabolic pathway fluxes in *Corynebacterium glutamicum* for amino acid production. *Adv. Biochem. Eng./Biotechnol.*, **120**, 21–49.
107. Marx, A., Eikmanns, B.J., Sahm, H., de Graaf, A.A., and Eggeling, L. (1999)

Response of the central metabolism in *Corynebacterium glutamicum* to the use of an NADH-98 Hiller, dependent glutamate dehydrogenase. *Metab. Eng.*, **1**, 35–48.

108. Pfefferle, W., Möckel, B., Bathe, B., and Marx, A. (2003) Biotechnological manufacture of lysine. *Adv. Biochem. Eng./Biotechnol.*, **79**, 59–112.
109. Raab, R.M., Tyo, K., and Stephanopoulos, G. (2005) Metabolic engineering. *Adv. Biochem. Eng./Biotechnol.*, **100**, 1–17.
110. Melzer, G., Eslahpazir, M., Franco-Lara, E., and Wittmann, C. (2009) Flux design: in silico design of cell factories based on correlation of pathway fluxes to desired properties. *BMC Syst. Biol.*, **3**, e120.
111. Soetaert, W. and Vandamme, E. (eds) (2010) *Industrial Biotechnology*, Wiley-VCH Verlag GmbH, Weinheim.
112. Elander, R.P. (2003) Industrial production of β-lactam antibiotics. *Appl. Microbiol. Biotechnol.*, **61**, 385–392.
113. RFA (Renewable Fuels Association) (2005) http://www.ethanolrfa.org/industry/locations/ (accessed 31 May 2016).
114. Bruggink, A. (2001) *Synthesis of ß-Lactam Antibiotics*, Kluwer Academic Publishers, Dordrecht.
115. Schulz, W.G. (2007) The costs of biofuels. *Chem. Eng. News*, **85** (51), 12–16 (emissions adapted form *Science* (2006) **311**, 506).
116. U.S. Department of Energy (DEO) (2004) *Energy Efficiency and Renewable Energy, Biomass: Top Value Added Chemicals from Biomass* (eds T. Werpy and G. Petersen), U.S. Department of Commerce, Springfield, VA.
117. Kamm, B. and Kamm, M. (2007) Biorefineries – multi product processes. *Adv. Biochem. Eng./Biotechnol.*, **105**, 175–204.
118. EU (2005) http://ec.europa.eu/research/energy/index_eu.htm.
119. Schröder, U., Niessen, J., and Scholz, F. (2003) A generation of microbial fuel cells with current outputs boosted by more than one order of magnitude. *Angew. Chem.*, **115**, 2986–2989.
120. Bornscheuer, U. and Buchholz, K. (2005) Highlights in biocatalysis – historical landmarks and current trends. *Eng. Life Sci.*, **5**, 309–323.
121. Tosa, T., Mori, T., Fuse, N., and Chibata, I. (1969) Studies on continuous enzyme reactions. 6. Enzymatic properties of DEAE-sepharose-aminoacylase complex. *Agric. Biol. Chem.*, **33**, 1047–1056.
122. (a) Mayer, H., Collins, J., and Wagner, F. (1979) Patent DE2930794; (b) McCain, K.W. (1995) The structure of biotechnology R & D. *Scientometrics*, **32** (2), 153–175.
123. Buchholz, K. and Poulson, P.B. (2000) in *Applied Biocatalysis* (eds A.J.J. Straathof and P. Adlercreutz), Harwood Academic Publishers, Amsterdam, pp. 1–15.
124. Andersen, O. and Poulsen, E.B. (1983) in *Enzyme Technology* (ed. R.M. Lafferty), Springer-Verlag, Berlin, pp. 179–188.
125. Klibanov, A. and Cambou, B. (1987) in *Methods in Enzymology*, vol. 136 (ed. K. Mosbach), Academic Press, Orlando, FL, pp. 117–137.
126. Cheetham, P.S.J. (2000) in *Applied Biocatalysis* (eds A.J.J. Straathof and P. Adlercreutz), Harwood Academic Publishers, Amsterdam, pp. 93–152.
127. Jördening, H.-J. and Winter, J. (2005) *Environmental Biotechnology*, Wiley-VCH Verlag GmbH, Weinheim.
128. Buchholz, K., Kasche, V., and Bornscheuer, U. (2005) *Biocatalysts and Enzyme Technology*, Wiley-VCH Verlag GmbH, Weinheim.
129. Hubschwerlen, C. (2007) in *Comprehensive Medicinal Chemistry II*, vol. 7 (eds J.J. Plattner and M.C. Desai), Elsevier, Amsterdam, pp. 497–517.
130. Jarvis, L.M. (2007) *Chem. Eng. News*, 15–20.
131. Aggarwal, S. (2007) What's fueling the biotech engine? *Nat. Biotechnol.*, **25**, 1097–1104.
132. Melmer, G. (2005) in *Production of Recombinant Proteins. Novel Microbial and Eukariotic Expression Systems* (ed. G. Gellissen), Wiley-VCH Verlag GmbH, Weinheim, pp. 361–383.

133. Aggarwal, S. (2014) What's fueling the biotech engine – 2012–2013. *Nat. Biotechnol.*, **32**, 32–39.
134. Jarvis, L.M. (2015) The year in new drugs. *Chem. Eng. News*, **93** (5), 11–16.
135. Chemical & Engineering News (2014) (Nov. 24), p. 6.
136. Slater, A., Scott, N.W., and Fowler, M.R. (2008) *Plant Biotechnology: The Genetic Manipulation of Plants*, Oxford University Press, Oxford.
137. Stephanopoulos, G. (1999) Metabolic fluxes and metabolic engineering. *Metab. Eng.*, **1**, 1–11.
138. Reuss, M. (2001) Editorial. *Bioprocess Biosyst. Eng.*, **24**, 1.
139. Buchholz, K. (2007) Science or not? The status and dynamics of biotechnology. *Biotechnol. J.*, **2**, 1154–1168.
140. Yi, D. (2011) Who owns what? Private ownership and the public interest in recombinant DNA technology in the 1970s. *Isis*, **102**, 446–474.
141. Zucker, L.G. and Darby, M.R. (1997) Present at the biotechnological revolution: transformation of technological identity. *Res. Policy*, **26**, 429–446.
142. Zucker, L.G., Darby, M.R., and Armstrong, J. (1998) Geographically localized knowledge: spillovers or markets? *Econ. Inq.*, **36**, 65–86.
143. Zucker, L.G. and Darby, M.R. (2001) Capturing technological opportunity via Japan's star scientists: evidence from Japanese firms' biotech patents and products. *J. Technol. Transfer*, **26**, 37–58.
144. Arts, S. and Veugelers, R. (2012) *The Technological Origins and Novelty of Breakthrough Inventions*, KU Leuven, http://ssrn.com/abstract=2230366 (accessed 31 May 2016).
145. Arts, S., Appio, F.P., and Van Looy, B. (2013) Inventions shaping technological trajectories: do existing patent indicators provide a comprehensive picture? *Scientometrics*, **97** (2), 397–419.
146. (a) Rheinberger, H.-J. (1997) *Toward a History of Epistemic Things*, Stanford University Press, Stanford, CA; Also:(b) Rheinberger, H.-J. (2001) *Experimentalsysteme und epistemische Dinge*, Wallstein–Verlag, Göttingen.
147. (a) Bartneck, C. and Rauterberg, M. (2007) Physics Nobels should favour inventions. *Nature*, **448**, 644; (b) Bartneck, C. and Rauterberg, M. (2008) The asymmetry between discoveries and inventions in the Nobel Prize in Physics. *Technoetic Arts*, **6**, 73.
148. Berg, P. and Mertz, J.E. (2010) Personal reflections on the origins and emergence of recombinant DNA technology. *Genetics*, **184**, 9–17.
149. Venter, J.C. (2008) *A Life Decoded: My Genome: My Life*, Penguin Press Science, New York.
150. Bassett, A.R. and Ji-Long Liu, J.-L. (2014) CRISPR/Cas9 and genome editing in Drosophila. *J. Genet. Genomics*, **41**, 7–19. doi: 10.1016/j.jgg.2013.12.004
151. Botstein, D., White, R.L., Skolnick, M. *et al.* (1980) Construction of a genetic linkage map in man using restriction fragment length polymorphisms. *Am. J. Hum. Genet.*, **32**, 314–331.
152. Saiki, R.K. *et al.* (1985) Enzymatic amplification of β-globin genomic sequences and restriction site analysis for diagnosis of sickle cell anemia. *Science*, **230**, 1350–1354.
153. Saiki, R.K. *et al.* (1988) Primer-directed enzymatic amplification of DNA with a thermostable DNA polymerase. *Science*, **239**, 487–491.
154. Lawyer, F.C., Stoffel, S., Saiki, R.K., Myambo, K., Drummond, R., and Gelfand, D.H. (1989) Isolation, characterization, and expression in *Escherichia coli* of the DNA polymerase gene from *Thermus aquaticus*. *J. Biol. Chem.*, **264**, 6427–6437.
155. Collins, J. (1977) Gene cloning with small plasmids. *Curr. Top. Microbiol. Immunol.*, **78**, 122–170.
156. Collins, J. (1997) in *Annual Reports in Combinatorial Chemistry and Molecular Diversity* (eds W.H. Moos, M.R. Pavia, B.M. Kay, and B.A. Ellington), Springer-Verlag, Berlin, New York, pp. 210–267, ISBN: 978-94-017-0738-1 (Print), 978-0-306-46904-6 (Online).
157. Winnacker, E.-L. (1987) *From Genes to Clones*, VCH Verlagsgesellschaft, Weinheim.
158. Berget, S.M., Moore, C., and Sharp, P.A. (1977) Spliced segments at

the 5' terminus of adenovirus 2 late mRNA. *Proc. Natl. Acad. Sci. U.S.A.*, **74**, 3171–3175. PMCID: PMC431482.

159. Chow, L.T., Gelinas, R.E., Broker, T.R., and Roberts, R.J. (1977) An amazing sequence arrangement at the 5' ends of adenovirus 2 messenger RNA. *Cell*, **12**, 1–8.
160. (a) Sanger, F. (1980) *Determination of Nucleic Acid Sequences in DNA*, Nobel Prize Lecture; (b) Shizuya, H., Birren, B., Kim, U.-J., Mancino, V., Slepak, T., Tachiira, Y., and Simon, M. (1992) Cloning and stable maintenance of 300-kilobase-pair fragments of human DNA in *Escherichia coli* using an F-based vector. *Proc. Natl. Acad. Sci. U.S.A.*, **89**, 8794–8797.
161. Maher, B. (2008) Personal genomes: the case of missing hereditability. *Nature*, **456**, 18–21.
162. Gilbert, J.A., Jansson, J.K., and Knight, R. (2014) The earth microbiome project: successes and aspirations. *BMC Biol.*, **12** (1), 69.
163. de Duve, C. and Beaufay, H. (1981) A short history of cell fractionation. *J. Cell Biol.*, **91**, 293–299.
164. Pedersen, K.O. (1976) The development of Svedberg's ultracentrifuge, *Biophys. Chem.*, **5**, 3–18.
165. Shapiro, J., MacHattie, L., Eron, L., Ihler, G., Ippen, K., Beckwith, J., Arditti, R., Reznikoff, W., and MacGillivray, R. (1969) The isolation of pure *lac* operon DNA. *Nature*, **224**, 768–774.
166. Hirt, B. (1967) Selective extraction of polyoma DNA from infected mouse cell cultures. *J. Mol. Biol.*, **26**, 365–369.
167. Radloff, R., Bauer, W., and Vinograd, J. (1967) A dye-buoyant-density method for the detection and isolation of closed circular duplex DNA in HeLa cells. *Proc. Natl. Acad. Sci. U.S.A.*, **57**, 1514–1521.
168. de Chadarevian, S. (2002) Designs for Life. Molecular Biology after World War II, Cambridge University Press.
169. Brownlee, D.G. (2014) Fred Sanger Double Nobel Laureate: A Biography, Cambridge University Press.
170. (a) Holley, R.W. *et al.* (1965) Structure of a ribonucleic acid. *Science*, **147**, 1462–1465; (b) Holley, R.W. (1968) *Alanine Transfer RNA*, Nobel Prize Lecture.
171. Sanger, F. (1971) Nucleotide sequences in bacteriophage ribonucleic acid. The eighth hopkins memorial lecture. *Biochem. J.*, **124**, 833–843.
172. Brescher, M.S. (1968) Direct translation of a circular mRNA. *Nature*, **220**, 1088–1091.
173. Hindley, J., Staples, D.H., Billeter, M.A., and Weissmann, C. (1970) Location of the coat cistron on the RNA of phage Qß. *Proc. Natl. Acad. Sci. U.S.A.*, **67**, 1180–1187.
174. Min Jou, W., Haegeman, G., Ysebaert, M., and Fiers, W. (1972) Nucleotide sequence of the gene coding for the bacteriophage MS2 coat protein. *Nature (London)*, **237**, 82–86.
175. Fiers, W., Contreras, R., Duerinck, F., Haegeman, G., Iserentant, D., Merregaert, J., Min Jou, W., Molemans, F., Raeymaekers, A., Van den Berghe, A., Volckaert, G., and Ysebaert, M. (1976) Complete nucleotide-sequence of bacteriophage MS2-RNA – primary and secondary structure of replicase gene. *Nature*, **260**, 500–507.
176. (a) Kleppe, K., Ohtsuka, E., Kleppe, R., Molineux, I., and Khorana, H.G. (1971) Studies on polynucleotides *1, *2XCVI. Repair replication of short synthetic DNA's as catalyzed by DNA polymerases. *J. Mol. Biol.*, **56**, 341–361; (b) Khorana, H.G. (1979) Total synthesis of a gene. *Science*, **203**, 614–662.
177. Gilbert, W. and Maxam, A. (1973) The nucleotide sequence of the lac operator. *Proc. Natl. Acad. Sci. U.S.A.*, **70**, 3581–3584.
178. Maxam, A.M. and Gilbert, W. (1977) A new method for sequencing DNA. *Proc. Natl. Acad. Sci. U.S.A.*, **74**, 560–564.
179. Gilbert, W. (1980) *DNA Sequencing and Gene Structure*, Nobel Prize Lecture.
180. Novogrodsky, A., Tal, M., Traub, A., and Hurwitz, J. (1966) The enzymatic phosphorylation of ribonucleic acid and deoxyribonucleic acid. II. Further properties of the 5'-hydroxyl polynucleotide kinase. *J. Biol. Chem.*, **241**, 2933–2943.
181. Smith, H.O. and Wilcox, K.W. (1970) A restriction enzyme from *Hemophilus*

influenzae. I. Purification and general properties. *J. Mol. Biol.*, **51**, 379–391.
182. Danna, K. and Nathans, D. (1971) Specific cleavage of simian virus 40 DNA by restriction endonuclease of *Hemophilus influenzae*. *Proc. Natl. Acad. Sci. U.S.A.*, **68**, 2913–2917.
183. Mertz, J.E. and Davis, R.W. (1972) Cleavage of DNA by RI restriction endonuclease generates cohesive ends. *Proc. Natl. Acad. Sci. U.S.A.*, **69**, 3370–3374.
184. Lehman, I.R. and Nussbaum, A.L. (1964) The deoxyribonucleases of *Escherichia Coli*. V. on the specificity of exonuclease I (Phosphodiesterase). *J. Biol. Chem.*, **238**, 2628–2636.
185. Sriprakash, K.S., Lundh, N., Huh, M.M.-O., and Radding, C.M. (1975) The specificity of lambda exonuclease. Interactions with single-stranded DNA. *J. Biol. Chem.*, **250**, 5438–5445.
186. Klenow, H. and Henningsen, I. (1970) Selective elimination of the exonuclease activity of the deoxyribonucleic acid polymerase from *Escherichia coli* B by limited proteolysis. *Proc. Natl. Acad. Sci. U.S.A.*, **65**, 168–175.
187. Temin, H.M. and Mizutani, S. (1970) RNA-dependent DNA polymerase in virions of Rous sarcoma virus. *Nature*, **226**, 1211–1213.
188. Baltimore, D. (1970) RNA-dependent DNA polymerase in virions of RNA tumour viruses. *Nature*, **226**, 1209–1211.
189. Stein, H. and Hausen, P. (1969) Enzyme from calf thymus degrading the RNA moiety of DNA-RNA hybrids: effect on DNA dependent RNA polymerase. *Science*, **166**, 393–395.
190. Ando, T. (1966) A nuclease specific for heat-denatured DNA isolated from a product of *Aspergillus oryzae*. *Biochim. Biophys. Acta*, **114**, 158–168.
191. Vogt, V.M. (1973) Purification and further properties of single-strand-specific nuclease from *Aspergillus oryzae*. *Eur. J. Biochem.*, **33**, 192.
192. Gray, H.B. Jr., Ostrander, D.A., Hodnett, J.L., Legerski, R.J., and Robberson, D.L. (1975) Extracellular nucleases of *Pseudomonas BAL 31*. I. Characterization of single strand-specific deoxyriboendonuclease and double-strand deoxyriboexonuclease activities. *Nucleic Acids Res.*, **2**, 1459–1492.
193. Chien, A., Edgar, D.B., and Trela, J.M. (1976) Deoxyribonucleic acid polymerase from the extreme thermophile *Thermus aquaticus*. *J. Bacteriol.*, **127**, 1550–1557.
194. Sanger, F., Nicklen, S., and Coulson, A.R. (1977) DNA sequencing with chain-terminating inhibitors. *Proc. Natl. Acad. Sci. U.S.A.*, **74**, 5463–5467.
195. García-Sancho, M. (2010) A new insight into Sanger's development of sequencing: from proteins to DNA, 1943–1977. *J. Hist. Biol.*, **43**, 265–323.
196. Pettersson, E., Lundeberg, J., and Ahmadian, A. (2008) Generations of sequencing technologies. *Genomics*, **93**, 105–111.
197. Szybalska, E.H. and Szybalski, W. (1962) Genetics of human cell lines. IV: DNA mediated heritable transformation of a biochemical trait. *Proc. Natl. Acad. Sci. U.S.A.*, **48**, 2026–2034.
198. Bacchetti, S. and Graham, F.L. (1977) Transfer of the gene for thymidine kinase to thymidine kinase-deficient human cells by purified herpes simplex viral DNA. *Proc. Natl. Acad. Sci. U.S.A.*, **74**, 1590–1594.
199. Mulligan, R.C. and Berg, P. (1981) Selection for animal cells that express the *Escherichia coli* gene coding for xanthine-guanine phophoribyl transferase. *Proc. Natl. Acad. Sci. U.S.A.*, **78**, 2072–2076.
200. Silver, R.P. and Falkow, S. (1970) Specific labeling and physical characterization of R-factor deoxyribonucleic acid in *Escherichia coli*. *J. Bacteriol.*, **104**, 331–339.
201. (a) Hughes, S.S. (2009) Regional Oral History Office, Bancroft Library, University of California, Berkeley, CA; (b) Transcripts of Interviews with Cohen, S.N., Berg, P., Boyer, H., Falkow, S., and Heynecker, H. (1995) http://bancroft.berkeley.edu/ROHO/ (accessed 31 May 2016).
202. Cohen, S.N., Chang, A.C., and Hsu, L. (1972) Nonchromosomal antibiotic

resistance in bacteria: genetic transformation of *Escherichia coli* by R-factor DNA. *Proc. Natl. Acad. Sci. U.S.A.*, **69**, 2110–2114.
203. Cohen, S.N. (1982) in *From Genetic Engineering to Biotechnology: The Critical Transition* (eds W.J. Whelan and S. Black), John Wiley & Sons, Inc., New York, pp. 213–216.
204. Itakura, K., Hirose, T., Crea, R., Riggs, A.D., Heyneker, H.L., Bolivar, F., and Boyer, H.W. (1977) Expression in *Escherichia coli* of a chemically synthesized gene for the hormone somatostatin. *Science*, **198**, 1056–1063.
205. Jackson, D.A., Symons, R.H., and Berg, P. (1972) Biochemical method for inserting new genetic information into DNA of Simian Virus 40: circular SV40 DNA molecules containing lambda phage genes and the galactose operon of *Escherichia coli*. *Proc. Natl. Acad. Sci. U.S.A.*, **69**, 2904–2909.
206. Cohen, S., Chang, A., Boyer, H., and Helling, R. (1973) Construction of biologically functional bacterial plasmids *in vitro*. *Proc. Natl. Acad. Sci. U.S.A.*, **70**, 3240–3244.
207. Hedgpeth, J.H., Goodman, M., and Boyer, H.W. (1972) DNA nucleotide sequence restricted by the RI endonuclease. *Proc. Natl. Acad. Sci. U.S.A.*, **69**, 3448–3452.
208. Roberts, R.J. (2005) How restriction enzymes became the workhorses of molecular biology. *Proc. Natl. Acad. Sci. U.S.A.*, **102**, 5905–5908.
209. Abrahams, P.J., Mulder, C., Van De Voorde, A., Warnaar, S.O., and van der Eb, A.J. (1975) Transformation of primary rat kidney cells by fragments of simian virus 40 DNA. *J. Virol.*, **16**, 818–823.
210. Graham, F.L. and van der Eb, A.J. (1973) A new technique for the assay of infectivity of human adenovirus 5 DNA. *Virology*, **52**, 456–467.
211. Goff, S. and Berg, P. (1976) Construction of hybrid viruses containing SV40 and lambda phage DNA segments and their propagation in cultured monkey cells. *Cell*, **9**, 695–705.
212. Jones, D. and Sneath, P.H. (1970) Genetic transfer and bacterial taxonomy. *Bacteriol. Rev.*, **34**, 40–81.
213. Harris, H. (1975) *The Principles of Human Biochemical Genetics*, North-Holland Publishing Co., Amsterdam.
214. Macdonald, M. (1993) A novel gene containing a trinucleotide repeat that is expanded and unstable on Huntington's disease chromosomes. The Huntington's Disease Collaborative Research Group. *Cell*, **72**, 971–983, doi: 10.1016/0092-8674(93)90585-E. PMID 8458085.
215. Hohn, B. and Murray, K. (1977) Packaging recombinant DNA molecules into bacteriophage *lambda* particles *in vitro*. *Proc. Natl. Acad. Sci. U.S.A.*, **74**, 3259–3262.
216. Collins, J. and Hohn, B. (1978) Cosmids: a type of plasmid gene-cloning vector that is packageable *in vitro* in bacteriophage λ heads. *Proc. Natl. Acad. Sci. U.S.A.*, **75**, 4242–4246.
217. Messing, J., Gronenberg, B., Müller-Hill, B., and Hofschneider, P.H. (1977) Filamentous coliphage *M13* as a cloning vehicle: insertion of a *HinII* fragment of the *lac* regulatory region in *M13* replicative form *in vitro*. *Proc. Natl. Acad. Sci. U.S.A.*, **74**, 3542–3646.
218. Viera, J. and Messing, J. (1982) The pUC plasmids. An *M13mp7*-derived system for insertion mutagenesis and sequencing with synthetic universal primers. *Gene*, **19**, 259–268.
219. Smith, G.P. (1985) Filamentous fusion phage: novel expression vectors that display cloned antigens on the virion surface. *Science*, **228**, 1315–1317.
220. Stemmer, W.P. (1994) DNA shuffling by random fragmentation and reassembly: in vitro recombination for molecular evolution. *Proc. Natl. Acad. Sci. U.S.A.*, **91**, 10747–10751.
221. Hannes, J. and Plückthun, A. (1997) *In vitro* selection and evolution of functional proteins by using ribosomal display. *Proc. Natl. Acad. Sci. U.S.A.*, **94**, 4937–4942.
222. Struhl, K., Stinchcomb, D.T., Scherer, S., and Davis, R.W. (1979) High-frequency transformation of yeast: autonomous replication of hybrid DNA

molecules. *Proc. Natl. Acad. Sci. U.S.A.*, **76**, 1035–1039.

223. Burke, D.T., Carle, G.F., and Olson, M.V. (1987) Cloning of large segments of exogenous DNA into yeast by means of artificial chromosome vectors. *Science*, **236**, 806–812.

224. Ioannou, P.A., Amemiya, C.T., Garnes, J., Kroisel, P.M., Shiyuza, H., Chen, C., Batzer, M.A., and de Jong, P.J. (1994) A new bacteriophage P1-derived vector for the propagation of large human DNA fragments. *Nat. Genet.*, **6**, 84–89.

225. Harrington, J.J., Van Bokkelen, G., Mays, R.W., Gustashaw, K., and Huntington, W.F. (1997) Formation of de novo centromeres and construction of first-generation human artificial microchromosomes. *Nat. Genet.*, **15**, 345–355.

226. Herrera-Estrella, L., Depicker, A., van Montagu, M., and Schell, J. (1983) Expression of chimaeric genes transferred into plant cells using a Ti-plasmid-derived vector. *Nature*, **303**, 209–213.

227. Collins, F.S. and Weissman, S.M. (1984) Directional cloning of DNA fragments at a large distance from an initial probe: a circularization method. *Proc. Natl. Acad. Sci. U.S.A.*, **21**, 6812–6816.

228. Poustka, A. and Lehrach, H. (1986) Jumping libraries and linking libraries: the next generation of molecular tools in mammalian genetics. *Trends Genet.*, **2**, 174–179.

229. Poustka, A., Pohl, T.M., Barlow, D.P., Frischauf, A.M., and Lehrach, H. (1987) Construction and use of human chromosome jumping libraries from NotI-digested DNA. *Nature*, **325**, 353–355.

230. Bailey, L.C. Jr., Searls, D.B., and Overton, G.C. (1998) Analysis of EST-driven gene annotation in human genomic sequence. *Genome Res.*, **8**, 362–376.

231. Staden, R. (1979) A strategy of DNA sequencing employing computer programs. *Nucleic Acids Res.*, **6**, 2601–2610.

232. Fleischmann, R.D., Adams, A.D., White, O., Clayton, R.A., Kirkness, E.F., Kerlavage, A.R., Bult, C.J., Tomb, J.F., Dougherty, B.A., Merrick, J.M. *et al.* (1995) Whole-genome random sequencing and assembly of *Haemophilus influenzae Rd. Science*, **269**, 496–512.

233. Adams, M.D. *et al.* (2000) The genome sequence of *Drosophila melanogaster*. *Science*, **287**, 2185–2195.

234. Kevles, D.J. and Hood, L. (eds.) (1992) *The Code of codes: Scientific and social issues of the human genome project*. Harvard University Press, Cambridge, Mass.

235. Levy, S. *et al.* (2007) The diploid genome of an individual human. *PLoS Biol.*, **5**(10), e254.

236. Wang, K.-S., Choo, Q.-L., Weiner, A.J. *et al.* (1986) Structure, sequence and expression of the hepatitis δ viral genome. *Nature*, **323**, 508–514.

237. Souza, L.M., Boone, T.C., Gabrilove, J. *et al* (1986) Recombinant human granulocyte colony stimulating factor: effects on normal and leukemic myeloid cells. *Science*, **232**, 61–65.

238. Welte, K. (2010) in 20 Years of GCSF: Clinical and Non-Clinical Studies Molineux (eds M. Foote and T. Arvedson), Springer-Verlag, Heidelberg, pp. 15–24.

239. Gordon, J.W. and Ruddle, F.H. (1981) Integration and stable germ line transformation of genes injected into mouse pronuclei. *Science*, **214**, 1244–1246.

240. Gossler, A., Doetschman, T., Korn, R., Serfling, E., and Kemler, R. (1986) Transgenesis by means of blastocyst-derived embryonic stem cell line. *Proc. Natl. Acad. Sci. U.S.A.*, **83**, 9065–9069.

241. Jaenisch, R. (1976) Germ-line integration and mendelian transmission of the exogenous Moloney leukemia virus. *Proc. Natl. Acad. Sci. U.S.A.*, **73**, 1260–1264.

242. Dorner, M., Horwitz, J.A., Robbins, J.B., Barry, W.T., Feng, Q., Mu, K., Jones, C.T. *et al.* (2011) A genetically humanized mouse model for hepatitis C virus infection. *Nature*, **474**, 208–211.

243. Kügler, J., Schmelz, S., Gentzsch, J., Haid, S., Pollmann, E., van den Heuvel, J., Franke, R., Pietschmann, T., Heinz, D.W., and Collins, J. (2012) High affinity peptide inhibitors of the hepatitis

C virus NS3-4A protease refractory to common resistant mutants. *J. Biol. Chem.*, **287**, 39224–39232.

244. Chuang, H.-Y., Matan Hofree, M., and Ideker, T. (2010) A decade of systems biology. *Annu. Rev. Cell Dev. Biol.*, **26**, 721–744.

245. Wang, Y.-H., Wei, K.Y., and Smolke, C.D. (2013) Synthetic biology: advancing the design of diverse genetic systems. *Annu. Rev. Chem. Biomol. Eng.*, **4**, 69–102.

246. Takahashi, K., Tanabe, K., Ohnuki, M., Narita, M., Ichisaka, T., Tomoda, K., and Yamanaka, S. (2007) Induction of pluripotent stem cells from adult human fibroblasts by defined factors. *Cell*, **131**, 861–872.

247. (a) Doudna, J.A. and Charpentier, E. (2014) The new frontier of genome engineering with CRISPR-Cas9. *Science*, **346**, 1077, doi: 10.1126/science.1258096; (b). Chemical & Engineering News (2015) (Nov. 23), p. 22.

248. Yang, B.L. *et al* (2015) Genome-wide inactivation of porcine endogenous retroviruses (PERVs). *Science*, **350**, 1101–1104.

249. Fulwyler, M.J. (1965) Electronic separation of biological cells by volume. *Science*, **150**, 910–911.

250. Morton, R.K. (1955) Some properties of alkaline phosphatase of cow's milk and calf intestinal mucosa. *Biochem. J.*, **60**, 573–582.

251. Bollum, F.J. (1960) Calf thymus polymerase. *J. Biol. Chem.*, **235**, 2399–2403.

252. Boyer, H.W. (1971) DNA restriction and modification mechanisms in bacteria. *Annu. Rev. Microbiol.*, **25**, 153–176.

253. Lehman, I.R., Bessman, M.J., Simms, E.S. and Kornberg , (1958) A. Enzymatic synthesis of deoxyribonucleic acid. I. Preparation of substrates and partial purification of an enzyme from Escherichia coli. *J. Biol. Chem.*, **233**, 163–170.

254. Goulian, M. and Kornberg, A. (1967) Enzymatic synthesis of DNA. 23. Synthesis of circular replicative form of phage *phi*-X174 DNA. *Proc. Natl. Acad. Sci. U.S.A.*, **58**, 1723–1730.

255. Gellert, M. (1967) Formation of covalent circles of lambda DNA by *E. coli* extracts. *Proc Natl. Acad. Sci. U.S.A.*, **57**, 148–155.

256. Southern, E.M. (1975) Detection of specific sequences among DNA fragments separated by gel electrophoresis. *J. Mol. Biol.*, **98**, 503–517.

257. Alwine, J.C., Kemp, D.J. and Stark, G.R. (1977) Method for detection of specific RNAs in agarose gels by transfer to diazobenzoyloxymethyl-paper and hybridization with DNA probes. *Proc. Natl. Acad. Sci. U.S.A.*, **74**, 5350–5354.

258. Symington, J., Green, M. and Brackmann, K. (1981) Immunoautoradiographic detection of proteins after electrophoretic transfer from gels to diazo-paper: analysis of adenovirus encoded proteins. *Proc. Natl. Acad. Sci. U.S.A.*, **78**, 177–181.

259. Köhler, G and Milstein, C. (1975) Continuous cultures of fused cells secreting antibody of predefined specificity. *Nature (Lond.)*, **256**, 495–497.

260. Riechmann, L., Clark, M., Waldmann, H. and Winter, G. (1988) Reshaping antibodies for therapy. *Nature (Lond.)* **332**, 323–327.

261. Marks, J.D., Hoogenboom, H.R., Bonnert, T.P., McCafferty, J., Griffiths, A.D. and Winter, G. (1991) By-passing immunization: Human antibodies from V-gene libraries displayed on phage. *J. Mol. Biol.*, **222**, 581–597.

262. Phillips, G.J. (2001) Green fluorescent protein – a bright idea for the study of bacterial protein localization. *FEMS Microbiol. Lett.*, **204**, 9–18.

263. Lichter, P., Tang, C.J., Call, K., Hermanson, G., Evans, G.A., Housman, D. and Ward, D.C. (1990) High resolution mapping of human chromosome 11 by *in situ* hybridization with cosmid clones. *Science*, **247**, 64–69.

264. Ju J., Ruan, C., Fuller, C.W., Glazer, A.N. and Mathies (1995) Fluorescent energy transfer dye-labeled primers for DNA sequencing and analysis. *Proc. Natl. Acad. Sci. U.S.A.*, **92**, 4347–4351.

2
Synthetic Biology: An Emerging Approach for Strain Engineering

Jie Sun and Hal Alper

2.1
Introduction

Owing to nature's inherent complexity, biological systems have traditionally been recalcitrant to both quantitative study as well as engineering. However, recent advances in fields such as metabolic engineering, applied molecular biology, and genetics have resulted in a large number of emerging technologies that have the potential to address worldwide issues such as production of chemicals and fuels, improvement of human health, and efficient utilization of plant biomass. Among these technologies, "synthetic biology" has quickly risen into the spotlight as a paradigm for rewiring cellular systems.

Since the term *synthetic biology* was first used in the early twentieth century [1], it has existed as a paradigm at the interface of disciplines such as molecular biology, metabolic engineering, systems biology, mathematics, and physics with the goals of designing and building novel proteins, genetic circuits, metabolic networks, and multicellular consortia from the ground up [2]. Over the past decades, many research groups have developed and expanded this field to realize applications in the chemical, pharmaceutical, agricultural, and food industries. Even more so, these applications (and the field as a whole) have been advanced by several main technological driving forces: (i) advances in novel biological parts construction, such as precise control of gene expression through promoter engineering [3, 4], (ii) cost-effective DNA synthesis and sequencing [5–7], and (iii) computational protein design coupled with high-throughput screening [8–10]. These technologies have led to advances such as rapid integration of multiple biological parts to reprogram cellular networks, including model-based design of biological systems [11], construction and optimization of biochemical pathways [12, 13], design principles for genetic circuits [14], and engineering multicellular systems that utilize cell-cell communication, including development of quorum-sensing networks and construction of microbial consortia [15, 16]. This chapter provides an update on the field to complement numerous reviews of synthetic biology [3, 17–20].

Industrial Biotechnology: Microorganisms, First Edition. Edited by Christoph Wittmann and James C. Liao.

The field of synthetic biology has progressed rapidly in the past decade. The most well-known breakthroughs in synthetic biology (often heralding the new era of synthetic circuit design) occurred in 2000, the year in which two genetic circuits components – a toggle switch [21] and an oscillator [22] – were synthetically designed to mimic those seen in naturally occurring networks. Two years later, directed evolution was applied to generate a library of genetic regulatory circuits, an approach that enabled the capacity for fine-tuned behavior of synthetic networks [23]. Subsequent to the achievement of synthetic cell-cell communication networks in prokaryotes, in 2005 two artificial cell-cell communication systems were integrated in a eukaryote *Saccharomyces cerevisiae*, using signal synthesizers and receptors from *Arabidopsis thaliana* [24]. In 2006, a successful application of synthetic biology in an engineered yeast strain led to the production of the antimalarial drug precursor artemisinic acid at a high titer (100 mg l^{-1}) [25]. This work was further enhanced by a massive metabolic engineering and synthetic biology effort to increase titers another 250-fold to 25 g l^{-1} [26]. Going beyond pathways, a complete microbial genome was assembled by transformation-associated recombination cloning in the yeast *S. cerevisiae* and confirmed to be functional [27]. This final work demonstrates one of the promises of synthetic biology, namely, the ability to generate bottom-up biological systems.

The growing interest and early successes mentioned above show that synthetic biology has a great potential for enhancing a wide range of industrial applications, such as engineering more robust industrial strains for the production of a variety of high value-added compounds. As indicated in Scheme 2.1, the purpose of this chapter is to demonstrate synthetic biology as a paradigm for bottom-up engineering, starting from single gene-expression control (basic elements), synthetic pathway or circuit design (functional and robust modules), to cell-cell communication engineering (microbial communities). Theoretically, once biological parts are well characterized, any desired performance should be observed consistently after assembly of functional parts in a pathway or more complex network. Of course, this premise is predicated on singular parts functioning in a consistent fashion when integrated into larger networks and circuits. Through development of proper stimuli or quorum-sensing networks, microbial communities with desired properties should be able to be generated in a facile manner. However, complete control of cells remains a lofty goal of synthetic biology. Here, with several typical case studies, we will discuss how the fundamental components of synthetic biology combine to regulate genetic networks and control intercellular interactions, and thus serve as a useful platform for engineering cells.

2.2
Basic Elements

Synthetic biology views the cell as a collection of many biological "parts" (e.g., DNA, RNA, proteins, regulatory circuits) that are assembled to generate complex biological functions [28]. These components exhibit defined functions in a pathway, network, or cell. Efforts to rewire synthetic systems have thus focused on

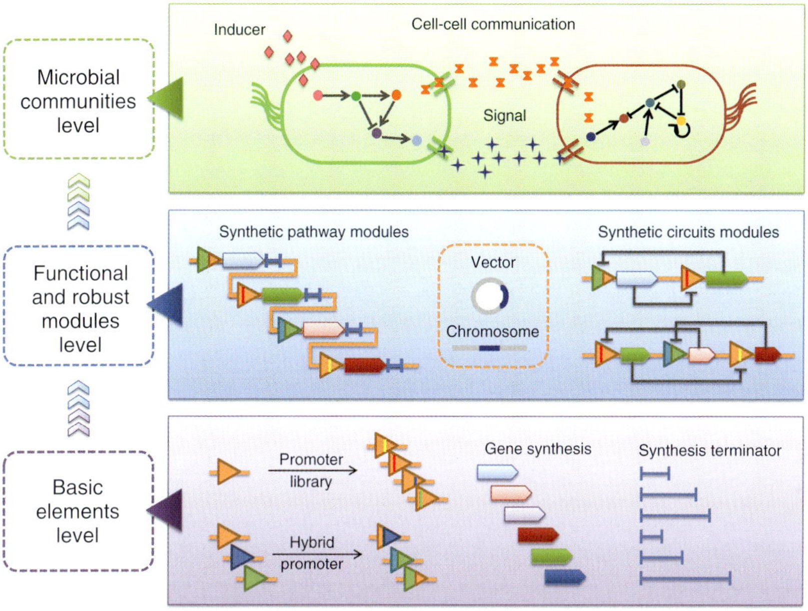

Scheme 2.1 Overview of engineering targets for synthetic biology. Multiple levels of synthetic biology can impact strain engineering. In the bottom panel, basic elements such as synthetic genes and regulatory elements are engineered, where they are combined into synthetic pathways and circuits in the middle panel. All these synthetic modules can enable cell-cell communication as sketched in the top panel.

three layers: transcriptional control, translational control, and protein regulation [17]. This section discusses recent advances in gene synthesis, transcriptional control, and various gene-expression optimization tools to illustrate the importance of engineering these basic elements in isolation and in combination for synthetic networks. Each of these elements is essential for complete control of strain engineering. In a sense, pathway-engineering applications require predictable control of synthetic parts. From the promoters that regulate individual pathway enzymes to the global regulators that can impart bulk physiological traits, well-characterized synthetic parts play a role in each stage of strain engineering.

2.2.1 Gene Synthesis

DNA synthesis is a fundamental technology which has rapidly spurred growth in synthetic biology. The capacity to generate DNA sequences *de novo* enables biology to be "written" as opposed to "copied" from a template. Dozens of gene synthesis companies have emerged in the past decade with the result of translating new advances in DNA synthesis technologies into a continual decrease in synthesis cost [29, 30] that is reminiscent of Moore's law. As a result, it is now highly feasible to synthesize genes rapidly for most synthetic biology applications, such as codon optimization for heterologous gene expression, generation of biosynthetic pathways, and even creation of artificial genomes. The first synthetic DNA sequence was 75 bp long, requiring 20 man-years of labor when it was created in 1970 [31]. Significantly longer DNA sequences, such as those encoding a pathway, are also synthesized by *in vitro* or *in vivo* methods such as Gibson Isothermal Assembly [32], DNA Assembler [12], Overlap Extension PCR (OE-PCR) [33], Ordered Gene Assembly in *Bacillus subtilis* (OGAB) [34], Sequence and Ligation-independent Cloning (SLIC) [35], In-Fusion [36], and Ligase Cycling Reaction (LCR) [37]. These techniques are discussed in more detail in the Section 2.3.1. Moving even further up the scale, it is remarkable to note the first synthesis of a *Mycoplasma genitalium* genome approximately 582 970 bp in length in 2008 by the Craig Venter Institute [27]. In this work, cassettes of 5–7 kb were overlapped and assembled from chemically synthesized oligonucleotides. These fragments were combined to create intermediate clones of approximately 1/8 and 1/4 of the length of the genome. All four 1/4 genomes were further assembled into a complete synthetic genome by transformation-associated recombination cloning in the yeast. A similar strategy was applied to the marine cyanobacteria *Prochlorococcus*, in which the 1.66 Mbp genome of the bacteria was assembled in yeast [38]. Additionally, a synthetic *M. mycoides* genome could be transplanted into *Mycoplasma capricolum* to replace its native genome and form synthetic cells with same phenotypic properties and capability of self-replication as *Mycoplasma mycoides* [39]. Thus, gene synthesis has rapidly transformed our capacity to write DNA – the central code of cells.

Gene synthesis is also very important in the process of codon optimization. Commonly, it is desired to produce valuable compounds using proteins naturally found in rare microorganisms. However, the genetic engineering tools for these organisms are either not well developed or the native expression level from these organisms is too low. Therefore, these valuable proteins are usually expressed in a heterologous manner using well-established hosts such as *Escherichia coli* and yeast for large-scale production. However, evolutionary history has forced different organisms to prefer different codon usage [40]. This bias can manifest itself in heterologous hosts as suboptimal protein expression due to unmatched codon usages. Successful codon optimization strategies usually involve replacing a rare codon(s) in the gene with that of a more frequently used codon in the host [41–44]. As a demonstration of this power, a synthetic library of 154 variants of green fluorescent protein (GFP) varied randomly at synonymous sites was expressed in *E. coli*, resulting in proteins with a 250-fold range of expression level variation across the library [45]. Nonetheless, numerous studies suggest that synonymous codon usage beyond the initiation region can impact expression. Studies with *E. coli* strains overexpressing rare tRNAs were shown to be capable of significantly improved gene expression [44, 46, 47]. Further, Welch *et al.* [48] generated 40 synonymous variants of two different proteins, the DNA polymerase of *Bacillus* phage Φ29 and a synthetic antibody fragment scFv, resulting in more than a 40-fold difference in expression level across each library. Additionally, an *in vitro* programmed microfluidic droplet system was utilized to generate, in an autonomous manner, customized DNA libraries that successfully synthesize libraries of yeast ribosome binding sites and the bacterial Azurine [49]. These recent advances in inexpensive *de novo* DNA synthesis enable creation of a large library of gene variants that can subsequently be screened for expression in a high-throughput manner, thus providing a large pool of basic elements for synthetic biology applications.

2.2.2 Transcriptional Control

Gene transcription is regulated by several factors, such as promoter strength, *cis*- and *trans*-acting factors, cell growth phase, and expression level of RNA polymerase. This section details the typical promoter structure in prokaryotes and eukaryotes, and describes relevant examples of promoter engineering – a synthetic biology approach for tuning expression of genes and generating novel synthetic expression parts. As a complement to well-characterized promoters, optimized gene expression vectors (also mentioned in this section) are critical for fine-tuned transcriptional control.

2.2.2.1 Promoter Engineering

The promoter is a key regulatory part that plays important roles in the performance of a gene, a gene cluster, or a designed gene circuit. A promoter can be thought of as a sequence of DNA, usually located upstream of the genes it controls, that provides an initial binding site for transcription factors and RNA polymerase. In prokaryotes such as *E. coli*, there are two conserved motifs in a "consensus" promoter sequence that are typically located 35 and 10 bp upstream from the transcription start site. The sigma (σ) subunit of prokaryotic RNA polymerase and almost all alternative σ's constitute a set of transcription factors known as σ^{70} *family proteins* in *E. coli* [50]. The σ subunits can interact with core RNA polymerase, recognize promoter DNA, and can direct the process of transcription initiation [51].

In contrast, eukaryotic promoters are organized around a core element and an upstream enhancer element [3]. In this scheme, the upstream enhancer element determines the transcriptional efficiency or promoter strength, while the core element determines the transcription direction and provides a binding site for RNA polymerase. In several reviews [52–54], the core promoter has been defined as the basal structure sufficient to direct transcriptional initiation by RNA polymerase II. The upstream enhancer element in eukaryotic promoters usually contains transcription factor-binding sites (TFBSs) for transcriptional activators or repressors. TFBSs could specifically interact with DNA-binding proteins and RNA polymerase to regulate the expression of genes. In 2013, an open database YEASTRACT (http://www.yeastract.com) contained a large number of TFBSs as well as DNA–protein interactions. The interactions between Transcription Factors (TFs) and an upstream activation sequence (UAS) or an upstream repressive sequence (URS) can increase or reduce transcriptional frequency, respectively. These associations are necessary for transcription via RNA polymerase II.

Most native promoters can be generally categorized into inducible promoters and constitutive promoters that form a large starting pool for synthetic biologists to choose from. Despite the thousands of native promoters found in a given organism, the diversity of promoter strength (as well as the sequences of those commonly used in synthetic biology) is still not sufficient for specific applications, such as strong upregulation or downregulation of certain genes in a pathway for biosynthetic applications and metabolic control analysis. Promoter engineering technologies generally focus on altering promoter strength in both core and enhancer regions to facilitate precise control of gene expression, further expanding the total pool of promoters that can be used for synthetic biology applications [4, 55]. Until recently, the number of promoter elements used in synthetic biology or applied biotechnology applications was very small.

Promoter Strength Characterization Knowledge of promoter strength is critical for synthetic circuit and pathway design. Ultimately, such information is critical to apply the appropriate promoters for expression of multiple genes in a pathway. To this end, many native promoters have been well studied and subsequently diversified into a wide range of promoter strengths. There are several well-characterized reporter genes used to characterize promoter strength, such as GFP [4, 13], β-galactosidase (LacZ) [56, 57], and β-lactamase [58]. However, promoter strengths are not uniform and are influenced by growth conditions, among other factors. For *E. coli*, a set of promoters associated with stationary-phase genes was determined by inserting 300–500 bp DNA fragments of a variety of *E. coli* promoters upstream from the translation initiation codon into a screening vector. Thus, when the culture changes from exponential to stationary phase, expression of growth-related genes is decreased, whereas a number of stationary-phase genes are turned on [59]. In yeast, several reports have compared strengths of constitutive and inducible promoters in great detail. Mumberg *et al.* [56] inserted four constitutive promoters *CYC1*, *ADH*, *GPD*, and *TEF* from yeast into plasmids with low (*CEN/ARS*) or high (2μ) copy number and characterized promoter activity using LacZ. The activity of LacZ was found to vary by approximately three orders of magnitude under different promoter and copy number combinations. The highest expression level was achieved with the 2μ *GPD* construct, while the lowest one was with the *CEN/ARS CYC1* construct. The expression levels for *CYC1* and *ADH* varied between 2.6- and 30-fold, respectively, by moving between *CEN/ARS* and 2μ vectors. Fang *et al.* [60] further showed evidence that using different selective markers in both plasmids and genomic locations could affect protein expression. Partow *et al.* [57] constructed seven yeast-constitutive promoters in an integrative plasmid with a reporter gene *lacZ*. The different reporter systems were stably integrated in a single copy into the genome of *S. cerevisiae* CEN.PK 113-5D at the *URA3* locus, thus avoiding gene copy number variations. When the glucose concentration decreased, the promoters *TEF1* and *PGK1* showed the most constant levels of activity when compared with promoters *ADH1*, *TPI1*, *HXT7*, *TDH3*, and *PYK1*. Promoter *HXT7* was strongest in a continuous glucose-limited culture while promoter *ADH1* was suitable for gene expression at high glucose concentrations. Furthermore, Sun *et al.* [13] cloned 14 constitutive yeast promoters in a 2μ vector with two reporter genes: *gfp* and *lacZ*. The 14 promoters showed similar trends of GFP synthesis on both the translational and transcriptional level. Under four routinely used growth conditions with different concentrations of oxygen and glucose, the relative rankings of promoter strength remained consistent. As a proof-of-concept, 10 of these promoters were further used to drive a 10-gene pathway to successfully convert xylan to a high-value-added compound (zeaxanthin). These examples demonstrate that it is possible to vary expression significantly using promoters and that genetic context is critical for optimal pathway design. Moreover, several factors could play an important role in the characterization of a promoter's strength, such as vector copy number and a variety of environmental stresses, including temperature shock, osmotic stress, oxidative stress, pH, and glucose or nitrogen starvation [61, 62]. Thus, before considering strain engineering applications, a more systematic comparison of promoter strengths should be performed by integrating promoter test cassettes in the genome, where copy number and stability are more tightly controlled. In addition, it is always best to test the function of synthetic parts under the condition of ultimate use (i.e., industrial conditions rather than laboratory conditions).

Promoter Library Synthesis Identification and characterization of native promoters is time consuming, especially for newly identified organisms that may be of biotechnological interest. However, proven workflows enable one to start from a well-studied promoter and vary the expression using library-based screening approaches. The most commonly-evoked technique of directed evolution involves introduction of mutations through methods such as error-prone PCR, DNA shuffling, and saturation mutagenesis. For example, tuning constitutive gene expression in lactic acid bacteria through library-based approaches has progressed over the last decade. First, a library of synthetic promoters with strength spanning over three to four orders of magnitude was constructed in *Lactococcus lactis* [55]. By saturation mutagenesis of the spacer regions between consensus −35 and −10 motifs, a 400-fold change in activity was observed. However, this improvement was highly dependent on the organism, as the strength of promoter variants in *L. lactis* did not correlate well with their respective strengths in *E. coli*. A similar approach was applied to chromosomal genes in *L. lactis* [63]. The phosphofructokinase (pfk) gene was fused to a library of *las* promoters, and an additional gene copy was introduced into a phage attachment site on the chromosome, resulting in a range of *pfk* activities from 1.4- to 11-fold higher than the wild type. The simultaneous modulation of *pfk*, pyruvate kinase (pk), and lactate dehydrogenase (ldh) activities was further investigated by integrating a truncated *pfk* fragment fused with a library of synthetic promoters at the *pfk* locus. The activity of the downstream gene *ldh* was thus altered from 50% to 350% of the wild-type level.

These were some of the first experiments to demonstrate simultaneous, synthetic control of pathway enzymes, thus ushering in the age of synthetic biology for pathway control and engineering. A similar strategy was also developed for *Lactobacillus plantarum* [64] in which a synthetic promoter library was generated, covering 3–4 logs of expression levels. Interestingly, these promoters also behaved similarly in a related, but distinct strain, *Lactobacillus sakei*.

The utility of random mutagenesis was realized in the development of a promoter library for perhaps one of the most commonly used strain engineering backgrounds, *E. coli*. A broad promoter library was developed using error-prone PCR of the constitutive bacteriophage P_L-λ promoter [4]. In this study, nearly 200 random promoter mutants were screened in a quantitative manner. Instead of only measuring the fluorescence of a single reporter gene such as *gfp*, the relative mRNA levels of *gfp*, and the activity of chloramphenicol acetyltransferase (cat) (as measured by resistance to chloramphenicol) were also compared. These three metrics correlated very well at the single-cell level and showed the utility of this approach to characterize a promoter library. Further analysis of a larger set of promoter variants was used to extract critical promoter residues responsible for either enhancing or reducing transcriptional activity [65]. Ultimately, the utility of such a library for tuning expression of pathway enzymes was evidenced by two case studies optimizing expression of *ppc* and *dxs* for growth and pathway flux, respectively. These results demonstrated that optimal expression varied for each gene of interest. Thus, large, diverse libraries are required for strain engineering applications.

This approach was further illustrated in a eukaryotic context through the creation of a library of promoter mutants from the strong yeast promoter, *TEF1*, which also drove a wide dynamic range of yECitrine fluorescent protein production [66]. The 11 *TEF1* mutants covered an activity range between about 8% and 120% of the unmutated *TEF1* promoter. The promoter strengths of these 11 mutants were independent of the carbon source. These promoters were used initially to demonstrate how pathway flux (in this case, glycerol utilization) can be influenced by gene expression. Recently, these *TEF1* promoter variants were used to modulate the expression of graded dominant mutants of a histone acetyltransferase-*GCN5*, thus uncoupling the catalytic function of global yeast GCN5p from its allosteric interactions [67]. The synthetic promoter library concept was also successfully adapted in the context of the minimal promoter *PFY1* from *S. cerevisiae*, which was identified by bioinformatic approaches [68]. Recently, it was shown that computational design of native yeast promoters to reduce nucleosome-binding affinity could improve their activity [69]. The redesigned promoters exhibited higher gene expression in both a plasmid and genomic context. Ultimately, synthetic constitutive promoters across host systems are critical for strain engineering.

Beyond constitutive expression systems, several inducible expression systems are also available, yet require addition of an inducer to initiate promoter activity. Unfortunately, most inducible promoters exhibit an all-or-nothing behavior, and thus are not fit for fine-tuning gene expression. Therefore, random mutagenesis and a multistage flow cytometry screening method were also developed and utilized for the oxygen-regulated *DAN1* promoter in *S. cerevisiae*. Two mutants which obtained activity under less-stringent conditions than the wild-type promoter were isolated, enabling gene expression during the oxygen depletion characteristic of normal fermentation [70]. In contrast, instead of changing the strength of an existing native promoter, error-prone PCR can also be used to construct an artificial promoter that has −35 and −10 box like motifs but may not initially carry promoter activity in *E. coli* [71]. After four rounds of error-prone PCR, the strongest mutant promoter showed 1.94-fold higher activity than the *tac* promoter. Synthetic biology can be applied similarly to other inducible promoter systems to generate a cadre of synthetic tools for pathway applications – from constitutive control to fine-tuned inducible traits.

Hybrid Promoter Approach In contrast to the approaches described above, a new approach was required to consistently obtain higher expression levels. In particular, high transcription is required for heterologous expression of pathway enzymes and protein overproduction. A hybrid promoter approach can best be described as the fusion of two main elements - an enhancer element(s) and a core promoter element - to create a synthetic promoter with novel regulation properties. Most approaches attempt to insert single or tandem enhancer regions to a separate core promoter. Two commonly used bacterial hybrid promoters, *tacI* and *tacII*, were synthesized from sequences of the *trp* and the *lac* UV5 promoters in *E. coli* [72]. The hybrid promoter *tacI* was composed of the DNA sequence upstream of position-20 from the transcriptional start site of the *trp* promoter, and another DNA sequence downstream of position-20 from the *lac* UV5 promoter. The hybrid promoter *tacII* is comprised of two upstream and downstream sequences at position-11 from *trp* and *lac* UV5 promoters as well. These two hybrid promoters increased the transcription 11- and 7-fold, respectively, from the wild-type *lac* UV5 promoter and 2- and 3-fold, respectively, from the *trp* promoter in the absence of the *trp* repressor. These promoters serve as a common starting point in bacterial metabolic engineering efforts.

In yeast, hybrid promoters have also been developed. Several hybrid promoter construction techniques in yeast have been patented [73, 74]. In a nonconventional yeast, *Yarrowia lipolytica*, a hybrid promoter approach developed libraries of tunable promoters through the use of tandem copies of UASs [75]. By doing so, the hybrid promoter library exhibited more than 400-fold variation in mRNA levels, with the strongest promoter exhibiting an eightfold higher gene expression level than wild-type promoters. These promoters ultimately enabled successful pathway and strain engineering efforts in this organism [76]. The importance of UAS regions was also demonstrated in *S. cerevisiae* through the use of multiple, disparate UAS elements in front of the strongest promoter *GPD* (TDH3) [77]. The transcriptional capacity of *GPD* was expanded 2.5-fold as measured by mRNA levels, thus leading to the strongest constitutive promoter in yeast. Additionally, this approach was extended to inducible promoter systems with the development of a galactose-controlled library possessing a 50-fold dynamic range. Moreover, the strongest member of this library had 15% higher transcription levels over the strongest, full *Gal1* promoter [77]. Beyond mutating the native components of promoters, several more rationally designed synthetic hybrid promoters were also developed. The rearrangement of TFBSs with random spacer sequences on a native yeast promoter base enabled a wider variety of choices of available promoters for synthetic biology applications [69]. Furthermore, a workflow for designing robust constitutive and inducible promoters with minimal UAS regions and minimal core elements was established. The size of these synthetic hybrid promoters was reduced by up to 80%, potentially leading to less time-consuming assemblies of single genes and pathways in yeast [78]. Collectively, these hybrid promoter strategies (used in conjunction with the randomization methods described above) can lead to complete control of expression for multiple host organisms. In the future, promoter resource database will allow for even more complicated module construction, such as pathways or gene circuits.

2.2.2.2 Optimization of Gene Expression Vectors

Beyond promoters, expression vectors and genetic context influence synthetic construct behavior. High-copy-number vectors (which usually replicate from tens to hundreds of copies per cell) have long been utilized for recombinant gene expression, particularly because of easy manipulation and high gene expression. However, recent studies have shown that this high overexpression system sometimes causes a metabolic burden in the host cell, and may further be structurally and segregationally unstable [79, 80]. The mRNA stability and copy number of multicopy (pMB1-based) and low-copy (F-based) plasmids were evaluated using an inducible promoter, a *lacZ* reporter gene, and 5′-hairpin structures. Increased inducer significantly decreased cell growth from high-copy plasmid, whereas the inducer concentration had little effect when using a low-copy plasmid [81]. In an experiment with similar results, the isopentenyl diphosphate (IPP) pathway (which includes genes *dxs* and *dxr*) was expressed from several expression vectors under the control of three different promoters and transformed into three different *E. coli* strains [82]. The comparison of different strains revealed that the *dxs* gene under the control of an arabinose-inducible promoter on a medium-copy plasmid resulted in twofold higher lycopene production than under the control of isopropyl β-D-1-thiogalactopyranoside (IPTG)-inducible *trc* and *lac* promoters on medium-copy and high-copy plasmids. When investigating the IPP pathway in pMB1-based plasmids and F-plasmids, the accumulation of metabolites in the stationary phase was enhanced in both cases, although the cell density was 24% lower when using a high-copy plasmid than when using a low-copy plasmid. Overexpression of *dxs* on a high-copy plasmid also significantly decreased cell growth and lycopene production [80]. Thus, the highest copy plasmid is not always the most effective choice for pathway-engineering applications. Sometimes, a low or single-copy plasmid expressing a critical enzyme in a pathway can reduce metabolic burden, resulting in better overall yield of the desired compound.

Beyond copy number, synthetic biology and recombinant DNA technology has led to the development of flexible vectors containing restriction sites in multiple cloning regions (MCSs). As a result, there are usually several base pairs between the upstream promoter and the gene of interest which will appear in the 5′-untranslated region (5′-UTR) of the mRNA. Even without the use of restriction enzyme cloning, a 5′-UTR will still inherently exist owing to spacing in the promoter region between the transcription start site and translation start site. The resulting secondary structure of this 5′-UTR mRNA has been shown to affect protein expression due to translational level control [83–85]. As a result, the composition of the MCS (or any synthetically imposed UTR region) might also influence translation efficiency. As a demonstration of the impact of these regions, the pRS-based yeast shuttle vectors were tested in yeast using yECitrine and a drastic decrease in expression was observed when the gene was placed at different positions along the MCS [86]. This phenomenon was observed for three common promoters-*GPD* (high strength), *TEF* (medium strength), and *CYC1* (low strength). Ultimately, it was determined that the restriction site-dependent inhibition might be related to the thermodynamic free energy of folding of the 5′-UTR. Despite this, the 5′-UTR structure inhibition was shown to be less of a rate-limiting step in the translation of longer genes with rarer codons. Thus, for short, codon-optimized genes (such as those easily generated by DNA synthesis), 5′-UTR regions are critical for determining the output of an expression cassette. To address this, these MCS regions were reengineered with the aid of thermodynamic models to generate sites which exhibited less inhibition of gene expression as the gene of interest was ligated into various positions. A similar model and approach can be used for the development of synthetic expression cassettes to be synthesized.

A standard eukaryotic gene expression cassette is normally comprised of a promoter, a target gene, and a terminator. While numerous studies of the effect of promoter on gene expression have been published, only a few studies have considered the influence of terminators. A recent study correlated the expression level enabled by over 30 terminators in *S. cerevisiae* with the mRNA half-life conferred by these elements [87]. When pairing these expression-enhancing terminators with various promoters, the expression difference was magnified (up to 11-fold with a low promoter). Furthermore, this work demonstrated that pathway flux can be controlled not just by promoter strength but also by terminator choice. Thus, synthetic parts provide a great deal of flexibility for strain and pathway engineering. These vector-mediated optimization studies could be readily extended to other organisms to maximize the utility of cloning technology in synthetic biology. More importantly, these parts can enable newfound control over pathway flux.

Taken together, these studies emphasize the importance of viewing an organism as a collection of basic building blocks, such as genes and transcriptional control elements. Through engineering these basic components, strain engineers can increase the expression of a rate-limiting enzyme or decrease the expression of unnecessary components, thus increasing overall strain productivity.

2.3
Functional and Robust Modules

Assembly of basic elements into functional and robust modules is a pressing challenge in synthetic biology. Modules are not simply combinations of random genes, rather construction of functional modules requires computational- and experimental-based rational design to dictate the composition and arrangement of basic elements. Ultimately, synthetic biology can achieve DNA-level specification of a desired function. In this regard, modules will be developed in a customized fashion with performance as the predescribed set of parameters. This next section discusses recent advances in developing pathway and gene circuit design in synthetic biology with strain engineering as the major application.

2.3.1
Synthetic Pathway Modules

Module design at the pathway level usually includes biological parts such as promoters, genes, proteins, and terminators, as described above. Traditional methods usually modify native pathways or express the partial or entire pathway in a heterologous host. In this regard, the repertoire of parts and enzymes were rather limited and thus the potential products were limited to a small subset of central metabolites and known secondary metabolites. However, recent advances have allowed the design of novel pathways for the production of desired compounds through computational tools [11]. In particular, many databases are designed to help identify the required biological parts and maximize the efficiency of the designed pathway, such as the Kyoto Encyclopedia of Genes and Genomes (KEGG), BRaunschwig ENzyme DAtabase (BRENDA), Universal Protein Resource (Uniprot), and BioCyc Database. For example, an uncharacterized enolase gene *hpbD*, from *Pelagibaca bermudensis*, was identified as an amino acid racemase/epimerase by *in silico* ligand docking of a library of 87 098 metabolites from the KEGG database. Further, with the prediction of several neighboring genes *hpbJ* and *hpbB1* in the genome, it was found that the substrate for HpbD is a group of small betaines and reacts in a 1,1-proton transfer reaction. It was thus found that the entire HpbD genome neighborhood constitutes a catabolic pathway that can degrade betaine tHyp-B into α-ketoglutarate [88].

Although the *de novo* synthesis of a large DNA sequence is possible through the above-mentioned DNA synthesis technologies, current pathway synthesis tools are still focused on developing *in vivo* and *in vitro* construction tools primarily based on restriction enzyme-mediated ligation or homologous recombination of natural or semisynthetic DNA fragments. Nevertheless, several promising technologies have been developed, such as Gibson Isothermal Assembly [32], DNA Assembler [12], and OGAB [34]. These synthetic biology approaches coupled with the rapid growth of these databases will quickly enable metabolite production from more cryptic pathways to become feasible. In this regard, synthetic pathway design of any pathway (no matter how large or how small) is now becoming possible. The potential implications for strain engineering are thus immense.

2.3.1.1
Pathway Assembly Tools

Several common pathway assembly tools have been developed recently to tackle the goals described above. The Gibson assembly technology is based on the two-step thermocycler method which was used to synthesize a complete genome [27]. By using exonuclease III and an antibody-bound *Taq* DNA polymerase, Gibson assembly enabled one-step *in vitro* recombination. Specifically, exonuclease III can remove several nucleotides from the 5′-ends, exposing complementary 3′-ends and allowing these single-stranded DNA overhangs to anneal. Then, the *Taq* DNA polymerase repairs the unpaired sequence to form a double-stranded DNA product. Finally, a ligase joins the 3′-ends and the 5′-ends of homologous DNA fragments into a covalently sealed molecule. This technology can enable construction of DNA molecules as large as 583 kb and construction of clones in *E. coli* as large as 300 kb [89]. For this reason, Gibson assembly has been used in a number of pathway and strain engineering applications recently [90–93].

Another novel one-step cloning method is called the *DNA assembler* [12]. On the basis of traditional homologous recombination mechanism in yeast [94], the DNA assembler allows the assembly of an entire biochemical pathway by combining each gene cassette along with a linearized vector through a single *in vivo* homologous recombination event in *S. cerevisiae*. As a proof of concept, a functional D-xylose utilization pathway (~9 kb, three genes), a functional zeaxanthin biosynthesis pathway (~11 kb, five genes), and a functional D-xylose-to-zeaxanthin pathway (~19 kb, eight genes) were constructed using this method. This technology has even been used to construct pathways up to 45.3 kb in length [95].

Assembly of DNA fragments in a one-step fashion has also been developed for *B. subtilis* plasmids, via a technique known as *Ordered Gene Assembly in B. subtilis* [34]. To demonstrate this technology, five antibiotic resistance genes, together with the backbone plasmid, were generated to have 3 bp protruding sequences at both ends. On the basis of this defined pairing and orientation, the ligation product was successfully generated via a one-step transformation in *B. subtilis*. Later, the OGAB method showed the capability to construct a zeaxanthin biosynthesis pathway in a *B. subtilis*–*E. coli* shuttle vector [96]. Thus, these approaches all lead to new ways to synthesize pathways using non-template-dependent methods.

2.3.1.2
Pathway Metabolic Flux Optimization Approaches

The number of characterized enzymes and enzymes with solved crystal structures is rapidly growing. Still, it is challenging to identify the required enzymes for a synthetic pathway, let alone the exact balance of enzyme activities for optimal pathway performance. In order to maximize product titer and productivities, several groups have attempted to balance metabolic flux through a combinatorial approach, in which the expression level of each gene in the pathway is varied simultaneously to find the optimal pathway configuration [97–100]. However, most of these studies suffer from limited libraries of metabolic pathways or inefficient *in vitro* cloning techniques. A combinatorial assembly of homologous pathway enzymes was developed in *S. cerevisiae*, termed "Customized Optimization of Metabolic PAthways by Combinatorial Transcriptional EngineeRing" (COMPACTER) [97]. This library contained multiple mutant pathways under the control of several mutant promoter and terminator pairs through a single DNA Assembler step. The ability of COMPACTER to generate improved pathways was demonstrated in both model strains and industrial strains. After a single round, a xylose-utilizing industrial strain was generated which exhibited 69% of the xylose consumption rate of the fastest reported xylose-utilizing strain. Also, a cellobiose-utilizing industrial strain was improved through COMPACTER, demonstrating the highest reported cellobiose consumption rate and ethanol productivity. An alternative approach is to rebuild the pathway from the ground up through a technique known as *pathway refactoring*. Under this process, all native regulation is replaced by well-characterized synthetic parts and genes are codon optimized. As an example, the nitrogen fixation pathway from *Klebsiella oxytoca* was refactored into a completely synthetic version of the natural pathway [101]. As a result, the combination of synthetic parts and module development is capable of rapid strain engineering.

At the same time, additional innovative approaches have been developed to better balance metabolic flux at a global level, such as global transcription machinery engineering (gTME) [102, 103], trackable multiplex recombineering (TRMR) [104], multiplex automated genome engineering (MAGE) [105], and RNA interference [106]. The gTME approach reprograms transcription at a global level through mutagenesis of transcriptional regulators. Such a large perturbation to the transcriptome is linked to the mutation of a single gene, in contrast to classical genomewide mutagenesis approaches. This approach was initially demonstrated in prokaryotic cells through engineering σ^{70} factors leading to improved phenotypes such as ethanol tolerance and metabolite overproduction [107]. In eukaryotes, two general transcription factors of the RNA polymerase II: TATA-binding protein (Spt15p) and TATA-binding protein-associated factor (Taf25p) were also used in a gTME approach. Mutagenesis of *spt15* followed by selection led to the desired phenotypes of improved glucose/ethanol tolerance and more efficient ethanol productivity [102]. A second technique to alter phenotypes on a genomewide scale combined molecular barcoding and microarray technologies [104]. By transforming a library of synthetic DNA cassettes into *E. coli* to either increase or decrease downstream gene transcription and translation, expression levels of over 95% of genes could be modified. The incorporation of barcodes into the library of synDNA cassettes enabled tracking each allele within the mixed population via a microarray. Within 1 week, it was possible to map thousands of genes during growth under three media conditions with various inhibitors. Such an innovative approach can rapidly survey for new genetic targets, while at the same time, delivering the desired phenotype. A third technique, using recombination-based genetic engineering, could modify genomic genes and regulatory regions through allelic replacement mediated by single-strand DNA [105]. As a demonstration of this approach, 24 genetic targets in the 1-deoxy-D-xylulose-5-phosphate (DXP) biosynthesis pathway in *E. coli* were modified simultaneously, resulting in the successful overproduction of the downstream isoprenoid product lycopene with more than a fivefold improvement in titer over the parent strain. Finally, optimization of RNA interference in yeast enabled genomewide screens for gene knockdowns improving strain tolerance toward selected environmental pressures [106, 108]. As a result, a pool of tolerant strains with downregulated gene targets could be easily obtained, indicating that this approach is highly suitable for strain engineering. These exciting achievements enabled simultaneous optimization of metabolic pathways and creation of a large library of phenotypes in a less time-consuming manner. In the future, it would be highly desirable to transfer these robust modules into industrial production strains.

2.3.2 Synthetic Circuit Modules

Another major focus in the development of functional and robust modules in synthetic biology is design of gene regulatory circuits at the network level. Despite significant efforts to balance metabolic flux as mentioned above, a metabolic pathway in alternate biological systems can often lead to the accumulation of metabolic intermediates, resulting in toxicity to the cell and causing metabolic burden. In order to optimize cellular bioprocess and precisely control gene expression in response to environmental stimuli, significant efforts have been directed toward engineering responsive gene regulatory circuits. Similar to the construction of electrical circuits, the gene circuit approach treats promoters, repressors, activators, reporter genes, or ribosome binding sites as the nodes in an electrical circuit. While there are inherent flaws in viewing biological components with this paradigm, significant progress has nonetheless been made in creating functional genetic circuits. This section focuses on reviewing the early development of gene circuits and discussing the challenges inherent to next-generation synthetic gene networks and their applications.

2.3.2.1 Examples of Synthetic Circuits

The basic synthetic biology elements discussed above can be used to generate small modules with specified behaviors, such as switches [21, 109–111], oscillators [22, 112], logic gates [92, 113–115], time-delayed circuits [116], and synthetic transcriptional cascades [117–119]. These circuits can be combined with pathways to yield improved biological function of engineered cellular systems.

A genetic toggle switch was constructed in *E. coli* on the basis of the predictions of a mathematical model which predicted a synthetic, bistable gene-regulatory network [21]. This network was composed of two constitutive promoters and two repressors, in which one repressor inhibited the activity of a promoter that drove the expression of the second repressor, which inhibited the activity of the promoter driving the first repressor. This module was found to be the simplest design necessary to achieve robust, bistable behavior.

With the combination of three transcriptional repressor systems, an oscillator was also successfully predicted and experimentally demonstrated in *E. coli* [22]. In this network, the repressor protein Lac inhibited the transcription of repressor gene *tetR* and the repressor protein TetR inhibited the transcription of the repressor gene *cI* from the λ phage. Finally, the repressor protein cI inhibited *lacI* expression, forming a negative feedback loop. The key design feature required for a robust oscillation phenotype was claimed to be a time delay module in the negative feedback loop, specifically as a part of a functional transcription factor in the cascade. The essential requirements for biochemical oscillators have been identified in theoretical studies as (i) a negative feedback loop; (ii) time delay in the feedback signal; (iii) nonlinear kinetic rate laws; and (iv) the appropriate timescale for the designed biochemical reactions to generate oscillatory behavior [120].

Another example of gene circuits which enable complex synthetic programming was reported with two complementary synthetic genetic counters in *E. coli* which function to count inputs [121]. The riboregulated transcriptional cascade (RTC) counters could count up to three input pulse events (the researchers used arabinose pulses in this case). As a result, by coupling the inputs of these RTC counters with cell cycle, it might be possible to program cell death after a specified number of cell divisions. Such a mechanism in engineered strains would be useful for biosensing, bioremediation, or medical purposes.

Alternatively, a gene circuit can be viewed as a set of interchangeable network motifs or components. The inputs to a regulated promoter (usually referring to the protein/inducer pair) could lead to an output as "ON" or "OFF" according to whether the downstream gene is transcribed or not. This features an important concept in gene circuit design, in which the circuit itself is viewed as a collection of logic gates. These gates could consist of "AND," "OR," "NAND," "NOR," or "NOT IF" logic gates based on the functional relationship between the inducers and output readout. For example, higher-order RNA devices have been constructed to function as logic gates and signal filters [113]. It was shown that a molecular input signal was "read" and converted to an observed increase or decrease in gene expression output. Improvement of these devices may allow for the development of more complex gene network designs.

2.3.2.2 Synthetic Circuit Design

The early development of gene circuit design was usually based on a simple mathematical model. To understand and assemble more complex gene networks, more sophisticated computational design tools are being developed. System model optimization in gene circuit design is often carried out through iterative rounds of mathematical design, genetic manipulation, experimental observation, and model refinement [18]. Collectively, these steps comprise the design-build-test cycle described in synthetic biology. As the first step, development of computer-aided tools provide a basis for efficient gene circuit engineering in synthetic biology. Several software packages have been reported to choose precise genetic components, optimize gene expression, and predict the performance of the resulting system. The MIT Registry of Standard Biological Parts (http://partsregistry.org) categorizes circuit components as a database of synthetic or natural network building blocks, such as promoters and ribosome binding sites. With the use of this general way of describing parts, methods such as the Biojade tool (http://web.mit.edu/jagoler/www/biojade/) enable users to drag standard parts onto a canvas and link them using hypothetical wires. Such a graphical design tool supports a library of over 100 parts as well as the analysis of their assembled behavior. Another gene circuit optimization tool, the OptCircuit framework, considers the three key components of genetic elements (promoter, transcripts, and inducers). Through collecting kinetic descriptions of promoter–protein, protein–protein, and protein–inducer pairs, an ensemble of circuits was identified which had the desired behavior [122]. Integrated software, such as Synthetic Biology Software Suite (SynBioSS), allow users to model complete reactions in a gene network that represents almost every step in the process, including transcription, translation, regulation, induction, and degradation [123]. Other gene circuit design tools have been developed which analyze flux balance and sensitivity, write command lines, enable graphical modeling, introduce programming language into gene circuit design, and understand circuit simulation. Each of these tools has been reviewed in detail elsewhere [18, 124].

Although early development of gene circuits was focused on prokaryotes, work has expanded into other host organisms. A positive feedback circuit in *S. cerevisiae* gene circuits was achieved by integrating theory and experiments [125]. By using well-defined promoter elements and transcriptional activators, researchers were able to create a graded response in constitutive systems. The activators bound to enhancers of the promoters in a graded fashion that increased the transcription rate in each cell, resulting in a dose-dependent binary response. The effects of negative feedback were also examined in yeast cells by integrating transcriptional cascades into the chromosome [126]. A TetR-based transcriptional cascade with no feedback showed a bimodal response in yEGFP expression. By including negative autoregulation, a "linearizer" circuit was constructed which reduced the heterogeneity of yEGFP. This autoregulation strategy could reduce the expression noise at intermediate induction by sevenfold. This gene circuit was also highly robust in comparison to different TetR-repressible promoters. The design of synthetic gene circuits has also been established in eukaryotes and thus has important implications for synthetic biology applications in these organisms. Such endeavors can lead to better understanding artificial or natural gene networks and could eventually provide a solution for economic biopharmaceutical manufacturing, more precise gene therapy, and tissue engineering.

2.3.2.3
Next-Generation Synthetic Circuits

Pioneering work has enabled significant progress in designing biological parts and assembling them into functional synthetic circuits by integrating modeling and experiments. However, construction of useful next-generation synthetic gene networks to solve societal issues is still an ongoing challenge. More closely linking synthetic circuit design with endogenous cellular processes has begun, developing a new generation of synthetic circuits [127]. Such complex gene circuits, coupled with more sophisticated integration of fundamental genetic parts, would better mimic natural cellular networks in different organisms. Moreover, it would allow for a better integration and control of both innate and heterologous components. Recent studies have begun to emerge in this direction. Examples include a tunable bandpass filter developed in *E. coli* [128], which will be useful for studying cellular differentiation and development. In addition, recent developments in stem cell biology have started to uncover the genetic networks responsible for uncontrolled population growth and differentiation [129]. Additionally, analog-to-digital and digital-to-analog converters would enable both the activation of genetic pathways with analog inputs as well as the conversion of digital representations of cell metabolism back into analog outputs [121]. Thus, by creating new design and computational tools that take into account biological variability by design, more-reliable synthetic circuits and advanced behaviors can be developed, such as precise control of cell life cycle for maximum productivity, flexible control of gene expression to minimize metabolic burden, and reversible switching of a cell between two metabolic states. The future is thus bright for merging these next-generation synthetic circuits with strain engineering, ranging from process control for chemical production [130], gene therapy [131], programmed microbiome therapy [132], to "intelligent" plants.

2.4
Microbial Communities

Up to this point, we have considered strain engineering in single organisms. However, the idea of consortia-based bioprocessing is gaining interest, especially in biomass processing into fuels [133]. More generally, in either a natural or artificial biosystem, a distributed set of organisms may be necessary to achieve macroscopic performance of complex functions. Synthetic biology has the potential to create control systems capable of stabilizing and harnessing microbial consortia. Previous research has established the capacity for cell-cell communications via synthetic regulatory networks and small molecule signals. In this regard, the development of synthetic gene networks has leveraged well-studied, natural cell-cell communication systems, such as the quorum-sensing phenomenon found in the bacterium *Vibrio fischeri* [134]. For example, a synthetic cell-cell communication circuit was created using an artificial quorum sensor in *E. coli* [135]. The secretion of acetate was used as the signal molecule, and acetyl phosphate (AcP) interacted with two-component regulators which were involved in the phosphate starvation response, nitrogen regulation, and chemotaxis. In this case, GFP expression was driven by the AcP-regulated hybrid promoter based on glnAp2. The signal molecule accumulated as a function of cell density, and once the concentration of the molecule reached a sufficient level, the signal could diffuse across the cell membrane. As a result, a fast transition to high GFP expression was observed in the culture. Through further promoter engineering work, such as mutagenesis of the enhancer region, researchers were able to produce various circuits with different sensitivities. These results emphasize the importance of engineering basic biological parts in synthetic biology. While autoinduction of GFP is clearly a proof of concept, it is not difficult to imagine the utility of these communication and inducer systems in regulating heterologous pathways for secondary metabolite production.

Another sophisticated synthetic network has been developed to reprogram a community of cells in a sophisticated manner [136]. The interaction between multiple signaling pathways and synthetic circuits between cells provided important guidelines to choose most appropriate components for network assembly. In this design, three modules were created: a biosensor module which detects DNA damage, a genetic toggle switch which responds to the biosensor module and generates output signals, and finally a quorum-sensing pathway that converts these signals into a multicellular response. Thus, four *E. coli* strains were engineered that could direct biofilm formation in response to DNA damage and activate protein synthesis in response to cell density. This work represents a significant advance in the development of cells with programmable behaviors and distributed loads in a community.

Another way to adapt existing gene circuits to cell-cell communication is to couple gene expression to cell survival [137]. In this population-control circuit, the LuxI protein from the quorum-sensing system enabled the synthesis of the diffusible acyl-homoserine lactone (AHL) as a signaling molecule during cell growth. At high concentrations, LuxR protein was bound and it activated the LuxR transcriptional regulator, which was engineered to produce a killer gene. Thus, accumulation of the killer protein would cause cell death. In another application, AHL was used as a signal to trigger increased motility and secretion of bactericidal peptides in *E. coli* [138]. In this way, *E. coli* was reprogrammed to autoregulate the abundance of an undesirable organism, in this case, *Pseudomonas aeruginosa*. This concept of communication-regulated cell growth and death could facilitate the cooperation of a natural quorum-sensing module with engineered counterparts and thus confer beneficial functions to engineered cells. Moreover, this approach can help balance populations of cells in bioprocessing consortia.

Synthetic biologists have gained insights into constructing synthetic circuits to study cellular networks. The next level of complexity is to assemble different populations of cellular networks into multicellular systems. Indeed, the study and design of such multicellular systems is still ongoing. Such developments would allow us to precisely control multicellular behaviors, which in turn might be beneficial in solving industrial problems such as reducing the local accumulation of toxic intermediates in a bioreactor and decreasing cell-to-cell variability in gene expression and productivity.

2.5 Conclusions and Future Prospects

The examples described above demonstrate that synthetic biology is an emerging technology that could enable the goal of multicellular optimization in a bottom-up manner. On a basic level, rapid and inexpensive DNA synthesis technology coupled with high-level design principles will expand the capacity to obtain novel DNA products. In addition, fundamental engineering efforts related to improving biological parts, such as promoter and terminator engineering, can generate fine-tuned gene expression and expand the library of pathway construction tools. On the second level, a well-characterized synthetic parts library could be assembled to build functional pathways and synthetic circuits that lead to efficient production of desired compounds or precisely control cellular behaviors. Finally on the long-term level, all of these cutting-edge advancements will make a desired biological system much easier to design and experimentally realize, generating a system which behaves predictably and functions in a manner that is superior to its natural counterpart. Already, with the aid of synthetic biology, living systems have been engineered for many purposes – a wide variety of compounds have been synthesized [139], user-defined physical structures have been generated for tissue regeneration and biomaterial formation, and synthetic multicellular networks can sense, degrade, and produce various compounds. Synthetic biology will represent a quantum leap in industrial biotechnology, which can greatly facilitate the development of new biological products and processes.

Acknowledgments

This work was supported by the DuPont Young Professor Grant and Office of Naval Research Young Investigator Program Award. The authors acknowledge Nathan Crook for critical reading and helpful comments.

References

1. Szybalski, W. (1974) *In vivo* and *in vitro* initiation of transcription. *Adv. Exp. Med. Biol.*, **44** (1), 23–24.
2. McDaniel, R. and Weiss, R. (2005) Advances in synthetic biology: on the path from prototypes to applications. *Curr. Opin. Biotechnol.*, **16** (4), 476–483.
3. Blazeck, J. and Alper, H.S. (2013) Promoter engineering: recent advances in controlling transcription at the most fundamental level. *Biotechnol. J.*, **8** (1), 46–58.
4. Alper, H. *et al.* (2005) Tuning genetic control through promoter engineering. *Proc. Natl. Acad. Sci. U.S.A.*, **102** (36), 12678–12683.
5. Cheng, J.Y. *et al.* (2002) High throughput parallel synthesis of oligonucleotides with 1536 channel

synthesizer. *Nucleic Acids Res.*, **30** (18), e93.

6. Lausted, C. *et al.* (2004) POSaM: a fast, flexible, open-source, inkjet oligonucleotide synthesizer and microarrayer. *Genome Biol.*, **5** (8), R58.
7. Matzas, M. *et al.* (2010) High-fidelity gene synthesis by retrieval of sequence-verified DNA identified using high-throughput pyrosequencing. *Nat. Biotechnol.*, **28** (12), 1291–1294.
8. Saven, J.G. (2011) Computational protein design: engineering molecular diversity, nonnatural enzymes, nonbiological cofactor complexes, and membrane proteins. *Curr. Opin. Chem. Biol.*, **15** (3), 452–457.
9. Pantazes, R.J., Grisewood, M.J., and Maranas, C.D. (2011) Recent advances in computational protein design. *Curr. Opin. Struct. Biol.*, **21** (4), 467–472.
10. Leemhuis, H. *et al.* (2005) New genotype–phenotype linkages for directed evolution of functional proteins. *Curr. Opin. Struct. Biol.*, **15** (4), 472–478.
11. Crook, N. and Alper, H.S. (2012) Model-based design of synthetic, biological systems. *Chem. Eng. Sci.*, **103**, 2–11.
12. Shao, Z. and Zhao, H. (2009) DNA assembler, an in vivo genetic method for rapid construction of biochemical pathways. *Nucleic Acids Res.*, **37** (2), e16.
13. Sun, J. *et al.* (2012) Cloning and characterization of a panel of constitutive promoters for applications in pathway engineering in *Saccharomyces cerevisiae. Biotechnol. Bioeng.*, **109** (8), 2082–2092.
14. Hasty, J., McMillen, D., and Collins, J.J. (2002) Engineered gene circuits. *Nature*, **420** (6912), 224–230.
15. Waters, C.M. and Bassler, B.L. (2005) Quorum sensing: cell-to-cell communication in bacteria. *Annu. Rev. Cell Dev. Biol.*, **21**, 319–346.
16. Eisenhauer, N., Scheu, S., and Jousset, A. (2012) Bacterial diversity stabilizes community productivity. *PLoS One*, **7** (3), e34517.
17. Young, E. and Alper, H. (2010) Synthetic biology: tools to design, build, and optimize cellular processes. *J. Biomed. Biotechnol.*, **2010**, 130781.
18. Purnick, P.E. and Weiss, R. (2009) The second wave of synthetic biology: from modules to systems. *Nat. Rev. Mol. Cell Biol.*, **10** (6), 410–422.
19. Andrianantoandro, E. *et al.* (2006) Synthetic biology: new engineering rules for an emerging discipline. *Mol. Syst. Biol.*, **2**, 2006.0028.
20. Liang, J., Luo, Y., and Zhao, H. (2011) Synthetic biology: putting synthesis into biology. *Wiley Interdiscip. Rev. Syst. Biol. Med.*, **3** (1), 7–20.
21. Gardner, T.S., Cantor, C.R., and Collins, J.J. (2000) Construction of a genetic toggle switch in *Escherichia coli. Nature*, **403** (6767), 339–342.
22. Elowitz, M.B. and Leibler, S. (2000) A synthetic oscillatory network of transcriptional regulators. *Nature*, **403** (6767), 335–338.
23. Yokobayashi, Y., Weiss, R., and Arnold, F.H. (2002) Directed evolution of a genetic circuit. *Proc. Natl. Acad. Sci. U.S.A.*, **99** (26), 16587–16591.
24. Chen, M.T. and Weiss, R. (2005) Artificial cell-cell communication in yeast *Saccharomyces cerevisiae* using signaling elements from *Arabidopsis thaliana. Nat. Biotechnol.*, **23** (12), 1551–1555.
25. Ro, D.K. *et al.* (2006) Production of the antimalarial drug precursor artemisinic acid in engineered yeast. *Nature*, **440** (7086), 940–943.
26. Paddon, C.J. *et al.* (2013) High-level semi-synthetic production of the potent antimalarial artemisinin. *Nature*, **496** (7446), 528–532.
27. Gibson, D.G. *et al.* (2008) Complete chemical synthesis, assembly, and cloning of a *Mycoplasma genitalium* genome. *Science*, **319** (5867), 1215–1220.
28. Endy, D. (2005) Foundations for engineering biology. *Nature*, **438** (7067), 449–453.
29. Caruthers, M.H. (1985) Gene synthesis machines: DNA chemistry and its uses. *Science*, **230** (4723), 281–285.
30. Carr, P.A. and Church, G.M. (2009) Genome engineering. *Nat. Biotechnol.*, **27** (12), 1151–1162.

31. Agarwal, K.L. *et al.* (1970) Total synthesis of the gene for an alanine transfer ribonucleic acid from yeast. *Nature*, **227** (5253), 27–34.
32. Gibson, D.G. *et al.* (2010) Chemical synthesis of the mouse mitochondrial genome. *Nat. Methods*, **7** (11), 901–905.
33. Horton, R.M. *et al.* (1989) Engineering hybrid genes without the use of restriction enzymes: gene splicing by overlap extension. *Gene*, **77** (1), 61–68.
34. Tsuge, K., Matsui, K., and Itaya, M. (2003) One step assembly of multiple DNA fragments with a designed order and orientation in *Bacillus subtilis* plasmid. *Nucleic Acids Res.*, **31** (21), e133.
35. Li, M.Z. and Elledge S.J. (2007) Harnessing homologous recombination *in vitro* to generate recombinant DNA *via* SLIC. *Nat. Methods*, **4** (3), 251–256.
36. Bird, L.E. *et al.* (2014) Application of In-Fusion cloning for the parallel construction of *E. coli* expression vectors. *Methods Mol. Biol.*, **1116**, 209–234.
37. de Kok, S. *et al.* (2014) Rapid and reliable DNA assembly via ligase cycling reaction. *ACS Synth. Biol.*, **3** (2), 97–106.
38. Tagwerker, C. *et al.* (2012) Sequence analysis of a complete 1.66 Mb *Prochlorococcus marinus* MED4 genome cloned in yeast. *Nucleic Acids Res.*, **40** (20), 10375–10383.
39. Gibson, D.G. *et al.* (2010) Creation of a bacterial cell controlled by a chemically synthesized genome. *Science*, **329** (5987), 52–56.
40. Grosjean, H. and Fiers, W. (1982) Preferential codon usage in prokaryotic genes – the optimal codon anticodon interaction energy and the selective codon usage in efficiently expressed genes. *Gene*, **18** (3), 199–209.
41. Kotula, L. and Curtis, P.J. (1991) Evaluation of foreign gene codon optimization in yeast: expression of a mouse Ig kappa chain. *Nat. Biotechnol.*, **9** (12), 1386–1389.
42. Williams, D.P. *et al.* (1988) Design, synthesis and expression of a human interleukin-2 gene incorporating the codon usage bias found in highly expressed *Escherichia coli* genes. *Nucleic Acids Res.*, **16** (22), 10453–10467.
43. Perlak, F.J. *et al.* (1991) Modification of the coding sequence enhances plant expression of insect control protein genes. *Proc. Natl. Acad. Sci. U.S.A.*, **88** (8), 3324–3328.
44. Gustafsson, C., Govindarajan, S., and Minshull, J. (2004) Codon bias and heterologous protein expression. *Trends Biotechnol.*, **22** (7), 346–353.
45. Kudla, G. *et al.* (2009) Coding-sequence determinants of gene expression in *Escherichia coli*. *Science*, **324** (5924), 255–258.
46. Wu, G. *et al.* (2007) SGDB: a database of synthetic genes re-designed for optimizing protein over-expression. *Nucleic Acids Res.*, **35** (Database issue), D76–D79.
47. Del Tito, B.J. Jr., *et al.* (1995) Effects of a minor isoleucyl tRNA on heterologous protein translation in *Escherichia coli*. *J. Bacteriol.*, **177** (24), 7086–7091.
48. Welch, M. *et al.* (2009) Design parameters to control synthetic gene expression in *Escherichia coli*. *PLoS One*, **4** (9), e7002.
49. Yehezkel, T.B. *et al.* (2015) Synthesis and cell-free cloning of DNA libraries using programmable microfluidics. *Nucleic Acids Res.*, **44** (4), e35.
50. Sharp, M.M. *et al.* (1999) The interface of σ with core RNA polymerase is extensive, conserved, and functionally specialized. *Genes Dev.*, **13** (22), 3015–3026.
51. Gruber, T.M. and Gross, C.A. (2003) Assay of *Escherichia coli* RNA polymerase: sigma-core interactions. *Methods Enzymol.*, **370**, 206–212.
52. Juven-Gershon, T. *et al.* (2008) The RNA polymerase II core promoter – the gateway to transcription. *Curr. Opin. Cell Biol.*, **20** (3), 253–259.
53. Butler, J.E. and Kadonaga, J.T. (2002) The RNA polymerase II core promoter: a key component in the regulation of gene expression. *Genes Dev.*, **16** (20), 2583–2592.

54. Smale, S.T. (2001) Core promoters: active contributors to combinatorial gene regulation. *Genes Dev.*, **15** (19), 2503–2508.
55. Jensen, P.R. and Hammer, K. (1998) The sequence of spacers between the consensus sequences modulates the strength of prokaryotic promoters. *Appl. Environ. Microbiol.*, **64** (1), 82–87.
56. Mumberg, D., Muller, R., and Funk, M. (1995) Yeast vectors for the controlled expression of heterologous proteins in different genetic backgrounds. *Gene*, **156** (1), 119–122.
57. Partow, S. *et al.* (2010) Characterization of different promoters for designing a new expression vector in *Saccharomyces cerevisiae. Yeast*, **27** (11), 955–964.
58. Cartwright, C.P. *et al.* (1994) Use of β-lactamase as a secreted reporter of promoter function in yeast. *Yeast*, **10** (4), 497–508.
59. Shimada, T. *et al.* (2004) Classification and strength measurement of stationary-phase promoters by use of a newly developed promoter cloning vector. *J. Bacteriol.*, **186** (21), 7112–7122.
60. Fang, F. *et al.* (2011) A vector set for systematic metabolic engineering in *Saccharomyces cerevisiae. Yeast*, **28** (2), 123–136.
61. Gasch, A.P. *et al.* (2000) Genomic expression programs in the response of yeast cells to environmental changes. *Mol. Biol. Cell*, **11** (12), 4241–4257.
62. Moskvina, E. *et al.* (1998) A search in the genome of *Saccharomyces cerevisiae* for genes regulated via stress response elements. *Yeast*, **14** (11), 1041–1050.
63. Solem, C. and Jensen, P.R. (2002) Modulation of gene expression made easy. *Appl. Environ. Microbiol.*, **68** (5), 2397–2403.
64. Rud, I. *et al.* (2006) A synthetic promoter library for constitutive gene expression in *Lactobacillus plantarum. Microbiology*, **152** (Pt. 4), 1011–1019.
65. Jensen, K. *et al.* (2006) Identifying functionally important mutations from phenotypically diverse sequence data. *Appl. Environ. Microbiol.*, **72** (5), 3696–3701.
66. Nevoigt, E. *et al.* (2006) Engineering of promoter replacement cassettes for fine-tuning of gene expression in *Saccharomyces cerevisiae. Appl. Environ. Microbiol.*, **72** (8), 5266–5273.
67. Lanza, A.M. *et al.* (2012) Linking yeast Gcn5p catalytic function and gene regulation using a quantitative, graded dominant mutant approach. *PLoS One*, **7** (4), e36193.
68. Blount, B.A. *et al.* (2012) Rational diversification of a promoter providing fine-tuned expression and orthogonal regulation for synthetic biology. *PLoS One*, **7** (3), e33279.
69. Curran, K.A. *et al.* (2014) Design of synthetic yeast promoters via tuning of nucleosome architecture. *Nat. Commun.*, **5**, 4002.
70. Nevoigt, E. *et al.* (2007) Engineering promoter regulation. *Biotechnol. Bioeng.*, **96** (3), 550–558.
71. Kagiya, G. *et al.* (2005) Generation of a strong promoter for *Escherichia coli* from eukaryotic genome DNA. *J. Biotechnol.*, **115** (3), 239–248.
72. de Boer, H.A., Comstock, L.J., and Vasser, M. (1983) The *tac* promoter: a functional hybrid derived from the *trp* and *lac* promoters. *Proc. Natl. Acad. Sci. U.S.A.*, **80** (1), 21–25.
73. Rosenberg, S. and Tekamp-Olson, P. (1992) Enhanced yeast transcription employing hybrid GAPDH promoter region constructs. US Patent 5089398 A.
74. Arimura, H. *et al.* (1987) Yeast promoter and process for preparing heterologous protein. US Patent CA1310602 C.
75. Blazeck, J. *et al.* (2011) Tuning gene expression in *Yarrowia lipolytica* by a hybrid promoter approach. *Appl. Environ. Microbiol.*, **77** (22), 7905–7914.
76. Blazeck, J. *et al.* (2013) Heterologous production of pentane in the oleaginous yeast *Yarrowia lipolytica. J. Biotechnol.*, **165** (3–4), 184–194.
77. Blazeck, J. *et al.* (2012) Controlling promoter strength and regulation in *Saccharomyces cerevisiae* using synthetic hybrid promoters. *Biotechnol. Bioeng.*, **109** (11), 2884–2895.

78. Redden, H. and Alper, H.S. (2015) The development and characterization of synthetic minimal yeast promoters. *Nat. Commun.*, **6**, 7810.
79. Chan, K.M. *et al.* (2013) The 2 micron plasmid of *Saccharomyces cerevisiae*: a miniaturized selfish genome with optimized functional competence. *Plasmid*, **70** (1), 2–17.
80. Jones, K.L., Kim, S.W., and Keasling, J.D. (2000) Low-copy plasmids can perform as well as or better than high-copy plasmids for metabolic engineering of bacteria. *Metab. Eng.*, **2** (4), 328–338.
81. Carrier, T., Jones, K.L., and Keasling, J.D. (1998) mRNA stability and plasmid copy number effects on gene expression from an inducible promoter system. *Biotechnol. Bioeng.*, **59** (6), 666–672.
82. Kim, S.W. and Keasling, J.D. (2001) Metabolic engineering of the nonmevalonate isopentenyl diphosphate synthesis pathway in *Escherichia coli* enhances lycopene production. *Biotechnol. Bioeng.*, **72** (4), 408–415.
83. Kozak, M. (2005) Regulation of translation via mRNA structure in prokaryotes and eukaryotes. *Gene*, **361** (21), 13–37.
84. Baim, S.B. and Sherman, F. (1988) mRNA structures influencing translation in the yeast *Saccharomyces cerevisiae*. *Mol. Cell. Biol.*, **8** (4), 1591–1601.
85. Pickering, B.M. and Willis, A.E. (2005) The implications of structured 5′ untranslated regions on translation and disease. *Semin. Cell Dev. Biol*, **16** (1), 39–47.
86. Crook, N.C., Freeman, E.S., and Alper, H.S. (2011) Re-engineering multicloning sites for function and convenience. *Nucleic Acids Res.*, **39** (14), e92.
87. Curran, K.A. *et al.* (2013) Use of expression-enhancing terminators in *Saccharomyces cerevisiae* to increase mRNA half-life and improve gene expression control for metabolic engineering applications. *Metab. Eng.*, **19**, 88–97.
88. Zhao, S. *et al.* (2013) Discovery of new enzymes and metabolic pathways by using structure and genome context. *Nature*, **502** (7473), 698–702.
89. Gibson, D.G. *et al.* (2009) Enzymatic assembly of DNA molecules up to several hundred kilobases. *Nat. Methods*, **6** (5), 343–345.
90. Lan, E.I. and Liao, J.C. (2011) Metabolic engineering of cyanobacteria for 1-butanol production from carbon dioxide. *Metab. Eng.*, **13** (4), 353–363.
91. Esvelt, K.M., Carlson, J.C., and Liu, D.R. (2011) A system for the continuous directed evolution of biomolecules. *Nature*, **472** (7344), 499–503.
92. Moon, T.S. *et al.* (2012) Genetic programs constructed from layered logic gates in single cells. *Nature*, **491** (7423), 249–253.
93. Toettcher, J.E. *et al.* (2011) Light-based feedback for controlling intracellular signaling dynamics. *Nat. Methods*, **8** (10), 837–839.
94. Ma, H. *et al.* (1987) Plasmid construction by homologous recombination in yeast. *Gene*, **58** (2–3), 201–216.
95. Choi, Y.S. *et al.* (2010) Cloning and heterologous expression of the spectinabilin biosynthetic gene cluster from *Streptomyces spectabilis*. *Mol. Biosyst.*, **6** (2), 336–338.
96. Nishizaki, T. *et al.* (2007) Metabolic engineering of carotenoid biosynthesis in *Escherichia coli* by ordered gene assembly in *Bacillus subtilis*. *Appl. Environ. Microbiol.*, **73** (4), 1355–1361.
97. Du, J. *et al.* (2012) Customized optimization of metabolic pathways by combinatorial transcriptional engineering. *Nucleic Acids Res.*, **40** (18), e142.
98. Lu, C. and Jeffries, T. (2007) Shuffling of promoters for multiple genes to optimize xylose fermentation in an engineered *Saccharomyces cerevisiae* strain. *Appl. Environ. Microbiol.*, **73** (19), 6072–6077.
99. Ramon, A. and Smith, H.O. (2010) Single-step linker-based combinatorial assembly of promoter and gene cassettes for pathway engineering. *Biotechnol. Lett.*, **33** (3), 549–555.
100. Bailey, J.E. (1999) Lessons from metabolic engineering for functional

genomics and drug discovery. *Nat. Biotechnol.*, **17** (7), 616–618.
101. Temme, K., Zhao, D., and Voigt, C.A. (2012) Refactoring the nitrogen fixation gene cluster from *Klebsiella oxytoca*. *Proc. Natl. Acad. Sci. U.S.A.*, **109** (18), 7085–7090.
102. Alper, H. *et al.* (2006) Engineering yeast transcription machinery for improved ethanol tolerance and production. *Science*, **314** (5805), 1565–1568.
103. Liu, H. *et al.* (2010) gTME for improved xylose fermentation of *Saccharomyces cerevisiae*. *Appl. Biochem. Biotechnol.*, **160** (2), 574–582.
104. Warner, J.R. *et al.* (2010) Rapid profiling of a microbial genome using mixtures of barcoded oligonucleotides. *Nat. Biotechnol.*, **28** (8), 856–862.
105. Wang, H.H. *et al.* (2009) Programming cells by multiplex genome engineering and accelerated evolution. *Nature*, **460** (7257), 894–898.
106. Crook, N.C., Schmitz, A.C., and Alper, H.S. (2014) Optimization of a yeast RNA interference system for controlling gene expression and enabling rapid metabolic engineering. *ACS Synth. Biol.*, **3** (5), 307–313.
107. Alper, H. and Stephanopoulos, G. (2007) Global transcription machinery engineering: a new approach for improving cellular phenotype. *Metab. Eng.*, **9** (3), 258–267.
108. Drinnenberg, I.A. *et al.* (2009) RNAi in budding yeast. *Science*, **326** (5952), 544–550.
109. Atkinson, M.R. *et al.* (2003) Development of genetic circuitry exhibiting toggle switch or oscillatory behavior in *Escherichia coli*. *Cell*, **113** (5), 597–607.
110. Kramer, B.P. *et al.* (2004) An engineered epigenetic transgene switch in mammalian cells. *Nat. Biotechnol.*, **22** (7), 867–870.
111. Kramer, B.P. and Fussenegger, M. (2005) Hysteresis in a synthetic mammalian gene network. *Proc. Natl. Acad. Sci. U.S.A.*, **102** (27), 9517–9522.
112. Tigges, M., *et al.*, A tunable synthetic mammalian oscillator. *Nature*, 2009. **457**(7227): p. 309–312.
113. Win, M.N. and Smolke, C.D. (2008) Higher-order cellular information processing with synthetic RNA devices. *Science*, **322** (5900), 456–460.
114. Rackham, O. and Chin, J.W. (2005) Cellular logic with orthogonal ribosomes. *J. Am. Chem. Soc.*, **127** (50), 17584–17585.
115. Qi, L.S. *et al.* (2013) Repurposing CRISPR as an RNA-guided platform for sequence-specific control of gene expression. *Cell*, **152** (5), 1173–1183.
116. Weber, W. *et al.* (2007) A synthetic time-delay circuit in mammalian cells and mice. *Proc. Natl. Acad. Sci. U.S.A.*, **104** (8), 2643–2648.
117. Hooshangi, S., Thiberge, S., and Weiss, R. (2005) Ultrasensitivity and noise propagation in a synthetic transcriptional cascade. *Proc. Natl. Acad. Sci. U.S.A.*, **102** (10), 3581–3586.
118. Hussein, R. and Lim, H.N. (2012) Direct comparison of small RNA and transcription factor signaling. *Nucleic Acids Res.*, **40** (15), 7269–7279.
119. Levine, E. *et al.* (2007) Quantitative characteristics of gene regulation by small RNA. *PLoS Biol.*, **5** (9), e229.
120. Novak, B. and Tyson, J.J. (2008) Design principles of biochemical oscillators. *Nat. Rev. Mol. Cell Biol.*, **9** (12), 981–991.
121. Friedland, A.E. *et al.* (2009) Synthetic gene networks that count. *Science*, **324** (5931), 1199–1202.
122. Dasika, M.S. and Maranas, C.D. (2008) OptCircuit: an optimization based method for computational design of genetic circuits. *BMC Syst. Biol.*, **2**, 24.
123. Hill, A.D. *et al.* (2008) SynBioSS: the synthetic biology modeling suite. *Bioinformatics*, **24** (21), 2551–2553.
124. Marchisio, M.A. and Stelling, J. (2009) Computational design tools for synthetic biology. *Curr. Opin. Biotechnol.*, **20** (4), 479–485.
125. Becskei, A., Seraphin, B., and Serrano, L. (2001) Positive feedback in eukaryotic gene networks: cell differentiation by graded to binary response conversion. *EMBO J.*, **20** (10), 2528–2535.
126. Nevozhay, D. *et al.* (2009) Negative autoregulation linearizes the dose–response and suppresses the

heterogeneity of gene expression. *Proc. Natl. Acad. Sci. U.S.A.*, **106** (13), 5123–5128.

127. Nandagopal, N. and Elowitz, M.B. (2011) Synthetic biology: integrated gene circuits. *Science*, **333** (6047), 1244–1248.

128. Sohka, T. *et al.* (2009) An externally tunable bacterial band-pass filter. *Proc. Natl. Acad. Sci. U.S.A.*, **106** (25), 10135–10140.

129. Macarthur, B.D., Ma'ayan, A., and Lemischka, I.R. (2009) Systems biology of stem cell fate and cellular reprogramming. *Nat. Rev. Mol. Cell Biol.*, **10** (10), 672–681.

130. Peralta-Yahya, P.P. *et al.* (2011) Identification and microbial production of a terpene-based advanced biofuel. *Nat. Commun.*, **2**, 483.

131. Loo, L.W. *et al.* (2012) *cis*-Expression QTL analysis of established colorectal cancer risk variants in colon tumors and adjacent normal tissue. *PLoS One*, **7** (2), e30477.

132. Bernard, P. and Couturier, M. (1992) Cell killing by the F plasmid CcdB protein involves poisoning of DNA-topoisomerase II complexes. *J. Mol. Biol.*, **226** (3), 735–745.

133. Shong, J., Jimenez Diaz, M.R., and Collins, C.H. (2012) Towards synthetic microbial consortia for bioprocessing. *Curr. Opin. Biotechnol.*, **23** (5), 798–802.

134. Engebrecht, J., Nealson, K., and Silverman, M. (1983) Bacterial bioluminescence – isolation and genetic-analysis of functions from *Vibrio fischeri*. *Cell*, **32** (3), 773–781.

135. Bulter, T. *et al.* (2004) Design of artificial cell-cell communication using gene and metabolic networks. *Proc. Natl. Acad. Sci. U.S.A.*, **101** (8), 2299–2304.

136. Kobayashi, H. *et al.* (2004) Programmable cells: interfacing natural and engineered gene networks. *Proc. Natl. Acad. Sci. U.S.A.*, **101** (22), 8414–8419.

137. You, L. *et al.* (2004) Programmed population control by cell-cell communication and regulated killing. *Nature*, **428** (6985), 868–871.

138. Hwang, I.Y. *et al.* (2014) Reprogramming microbes to be pathogen-seeking killers. *ACS Synth. Biol.*, **3** (4), 228–237.

139. Curran, K.A. and Alper, H.S. (2012) Expanding the chemical palate of cells by combining systems biology and metabolic engineering. *Metab. Eng.*, **14** (4), 289–297.

3
Toward Genome-Scale Metabolic Pathway Analysis

Jürgen Zanghellini, Matthias P. Gerstl, Michael Hanscho, Govind Nair, Georg Regensburger, Stefan Müller, and Christian Jungreuthmayer

3.1
Introduction

Constraint-based methods (CBMs) are extensively used to study cellular metabolism. These methods mainly rely on the stoichiometry of the biochemical reactions that form a metabolic network – data that can be fairly reliably obtained from annotated genome sequences and metabolic pathway databases. On the basis of this information, genome-scale metabolic networks can be reconstructed [1]. Mathematically, the topology of this network is captured by the (total) stoichiometric matrix that systematically collects the (signed) stoichiometric coefficients of all metabolites in the participating reactions. It is common practice to partition the set of metabolites into external and internal metabolites by defining a systems boundary. This systems boundary may correspond to a physical boundary, like the cell envelope, or to a virtual boundary. Reactions that connect external metabolites with internal ones are known as *exchange reactions* and account for the communication of a network with its environment. Of particular interest is the internal stoichiometric matrix $\boldsymbol{S} \in \mathbb{R}^{m\times r}$ that consists of the (signed) stoichiometric coefficients of the m *internal* metabolites in the r participating reactions (including exchange reactions).

At steady state, owing to mass conservation and thermodynamic feasibility, the metabolic flux vector $\boldsymbol{v} \in \mathbb{R}^r$ fulfills the equations/inequalities

$$\boldsymbol{S}\boldsymbol{v} = \boldsymbol{0}, \tag{3.1a}$$

$$v_i \geq 0 \quad \text{for } i \in I_{\text{irrev}}. \tag{3.1b}$$

The latter inequalities guarantee that all reactions i in the set of irreversible reactions I_{irrev} carry a (steady-state) flux v_i in the thermodynamically feasible forward direction. Typically, as $r > m$, the equations and inequalities (3.1a) and (3.1b) define a set of solutions, rather than a single solution. Characterizing the set of feasible solutions and finding biologically relevant solutions is at the heart of all CBMs.

Elementary flux mode analysis (EFMA) has emerged as a powerful tool in the family of CBMs [2, 3]. The particular power of an EFMA is based on the ability to

Industrial Biotechnology: Microorganisms, First Edition. Edited by Christoph Wittmann and James C. Liao.

unbiasedly decompose a metabolic network into irreducible functional building blocks, called elementary flux modes (EFMs). Three conditions uniquely define the set of EFMs in a metabolic network. An EFM $\boldsymbol{e} \in \mathbb{R}^r$ (i) operates at steady state, $\boldsymbol{Se} = \boldsymbol{0}$, (ii) uses all irreversible reactions in the appropriate direction, $e_i \geq 0$ for $i \in I_{\text{irrev}}$, and (iii) involves a minimal set of active reactions [4]. The first two features are obvious requirements from the equations and inequalities (3.1a) and (3.1b) and apply to all feasible flux distributions, while the characteristic property of an EFM is its support minimality. Recall that the support of a vector is the set of nonzero components. Support minimality of an EFM means that there is no feasible flux distribution having a smaller support, that is, less active reactions. Similar to basis vectors of linear subspaces, EFMs can be seen as generating vectors of a metabolic network as every feasible flux distribution can be represented as a non-negative linear combination [1)]

$$\boldsymbol{v} = \sum_{i=1}^{n} \lambda_i \boldsymbol{e}_i \quad \text{with } \lambda_i \in \mathbb{R} \text{ and } \lambda_i \geq 0, \tag{3.2}$$

where n denotes the number of EFMs in the network. This property allows for numerous applications in basic science and biotechnology. In fact, EFMA has been identified as a useful tool for metabolic engineering [6].

Suppose we are interested in turning a host organism into a growth-coupled cell factory for the production of some (bio)chemical commodity. Is this even possible for a given host and product of interest? The answer is provided by EFMA [7]: growth-coupled (and even partially growth-coupled) production is feasible if and only if there exists at least one growth-coupled EFM. [2)] For an illustration see Figure 3.1, where we plotted the product yield as function of the biomass yield for each EFM in a growth and nongrowth coupled (toy) network. *Escherichia coli* for instance, is in principle capable of producing all central carbon metabolites in a growth-coupled manner even if additional maintenance requirements are taken into account [7]. Thus EFMA provides extremely powerful means of analysis.

Yet, the enumeration of EFMs is computationally hard, which significantly limits the applicability of EFMA. In fact, a complete EFMA is currently limited to medium-scale (metabolic) models. The computational challenge for larger networks is associated with two main problems: (i) the number of EFMs explodes with the size of the metabolic network [9]. This makes a full EFMA of large systems virtually impossible, as even a core metabolic model of *E. coli* may already have several hundred million EFMs [10]. (ii) Even if a full EFMA in large networks was possible, then gazillions of EFMs would need to be stored, processed and further analyzed to gain biological knowledge, which would by far exceed current computational capabilities. For instance, on the basis of a massively parallelized approach, the largest, full EFMA reported to date found

1) Strictly speaking, a biochemically meaningful (thermodynamically feasible) decomposition is a non-negative linear combination of EFMs without cancellations [5].
2) For simplicity, we only discuss the so-called homogeneous case, which is applicable to all systems of the form (3.1). If additionally inhomogeneous constraints, such as minimal maintenance requirements or maximal flux capacities, are considered, the mathematical analysis based on the concept of "elementary flux vectors" yields similar, but more complex decompositions [5, 7, 8].

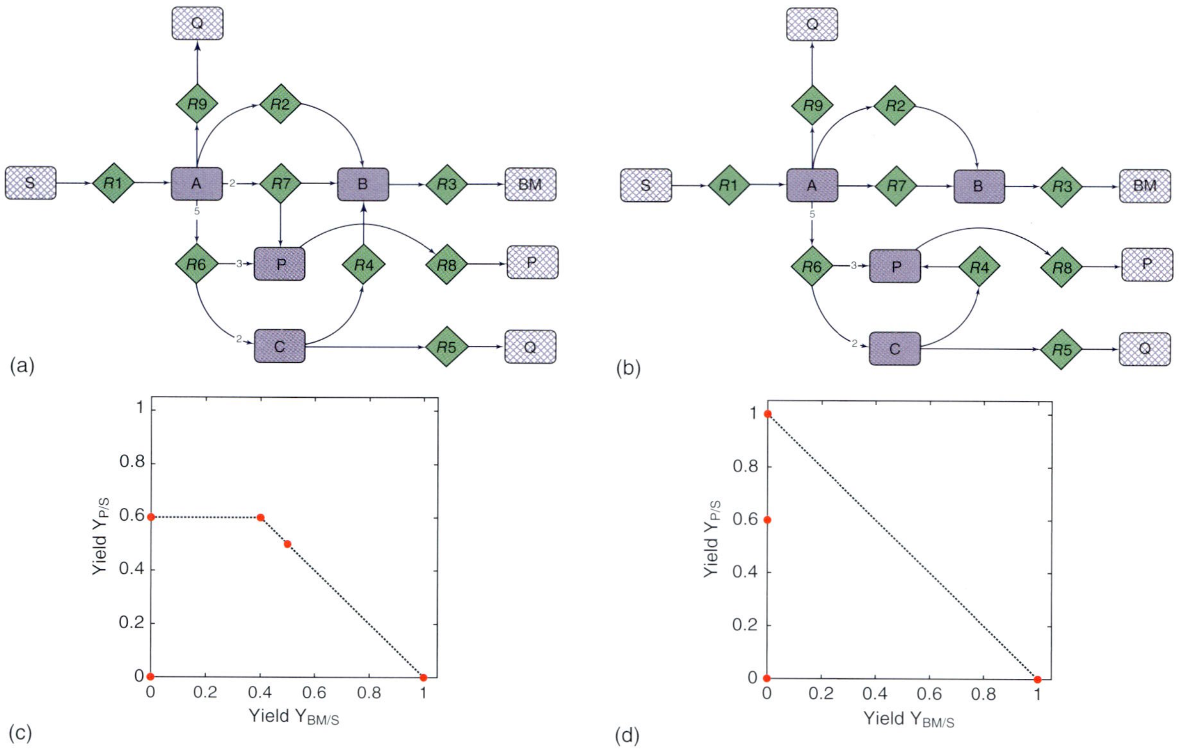

Figure 3.1 Growth- and non-growth-coupled toy metabolic networks (panels a and b, respectively) along with their associated phenotypic yield spaces (panels c and d, respectively). Both networks consist of nine irreversible reactions (diamonds, *R*1–*R*9), four internal metabolites (full rectangles), four external metabolites (checkered rectangles: BM, biomass; *P*, product of interest; *Q* by-product; *S*, substrate; note that the metabolite Q is the product of two reactions–*R*5 and *R*9) and five EFMs. In the phenotypic yield space EFMs are represented by full circles. Note that the point (1/0) in the phenotypic yield space of network B represents two EFMs with identical yields. The feasible yield space is bounded by the two axes and the dashed line. Growth-coupled production of P is achievable only in network A, but not in network B.

2 billion EFMs in a metabolic model of *Phaeodactylum tricornutum* with 318 reactions [11]. These data take up 1.2 TB of storage. Just reading all EFMs one by one takes up to an hour on our standard high-performance workstations. These numbers indicate that it is essential to find ways that identify and select "relevant" EFMs.

In the following, we briefly review currently available methods and then move on to present a novel method for the calculation of "short" EFMs.

One way to overcome the computational problems associated with an EFMA of large-scale networks is to compute only small subsets of EFMs. Subsets may be selected randomly [12, 13] or based on support information [14] or subject to additional constraints [15, 16]. Characterizing a subset of EFMs is also sufficient to address the feasibility problem of growth-coupled production outlined above (see Figure 3.1) as one has to find only one growth-coupled EFM. This is conveniently achieved by two methods developed by David and Bockmayr [17] and Pey and Planes [18], respectively. The latter method has the advantages that additional minimal yield requirements can be considered as well and that the method even remains feasible in genome-scale models. However, even if the analysis is restricted to a particular subset of EFMs (e.g., the subset of growth-coupled EFMs), their computation remains challenging. In fact, it has been shown that a single EFM can be computed in polynomial time, yet computing the complete subset is NP-complete [19].

One very fast EFM enumeration method uses the binary nullspace implementation [20, 21] of the double description (DD) method [22]. This method has proven particularly useful for the complete enumeration of EFMs in metabolic networks [22].

In the following, we will detail the DD method and then outline a new algorithm for the calculation of "short" EFMs that will make the DD method fit for the analysis of genome-scale metabolic models.

3.2 DD Method

We consider a metabolic network at steady state given by the equations and inequalities (3.1a) and (3.1b). After splitting every reversible reaction into two irreversible reactions (with opposite directions), we can assume that all reactions are irreversible. The equations and inequalities (3.1a) and (3.1b) define the so-called flux cone

$$F = \{\boldsymbol{v} \in \mathbb{R}^r \mid \boldsymbol{S}\boldsymbol{v} = \boldsymbol{0} \text{ and } \boldsymbol{v} \geq \boldsymbol{0}\}. \tag{3.3}$$

All thermodynamically feasible steady-state solutions, in particular all EFMs, are elements of this cone. Since we assume that all reactions are irreversible, the flux cone F is pointed and the EFMs coincide with its extreme rays (the "edges" of the flux cone) [23]. The DD method is widely used to enumerate these extreme rays.

The DD method relies on the fact that, due to the Minkowski–Weyl theorem, a polyhedral cone C can be described implicitly by inequalities (by the intersection of half spaces),

$$C = C(\mathbf{A}) = \{\boldsymbol{v} \in \mathbb{R}^r \mid \mathbf{A}\boldsymbol{v} \geq \mathbf{0}\}, \tag{3.4}$$

or explicitly by generators,

$$C = \text{cone}(\mathbf{R}) = \{\boldsymbol{v} \in \mathbb{R}^r \mid \boldsymbol{v} = \mathbf{R}\lambda,\ \lambda \geq \mathbf{0}\}, \tag{3.5}$$

for some matrices $\mathbf{A}$ and $\mathbf{R}$. A pair $(\mathbf{A}, \mathbf{R})$ of corresponding matrices is called a *DD pair*. If the cone C is pointed and the columns of $\mathbf{R}$ are its extreme rays, then $(\mathbf{A}, \mathbf{R})$ is called a *minimal DD pair*.

The DD method iteratively processes the inequalities given by the rows of $\mathbf{A}$. Let $\mathbf{A}_k$ be the submatrix of $\mathbf{A}$ that contains the first k rows of $\mathbf{A}$ and suppose that $(\mathbf{A}_k, \mathbf{R}_k)$ is the corresponding minimal DD pair. The DD method considers the next row of $\mathbf{A}$ and constructs a new minimal DD pair $(\mathbf{A}_{k+1}, \mathbf{R}_{k+1})$.

In the following, we describe one iteration step of the DD method in more detail. (For simplicity, we omit the iteration index.) Given the minimal DD pair $(\mathbf{A}, \mathbf{R})$ and an additional inequality represented by a vector $\boldsymbol{a} \in \mathbb{R}^r$ (i.e., the next unprocessed row of the full matrix), we consider the matrix $\mathbf{A}' = \begin{pmatrix} \mathbf{A} \\ \boldsymbol{a}^t \end{pmatrix}$ defining the cone $C(\mathbf{A}') = C(\mathbf{A}) \cap C(\boldsymbol{a}^t)$. We determine the corresponding matrix $\mathbf{R}'$ as follows: The additional inequality divides the set of column indices J of $\mathbf{R}$ into three sets

$$J^+ = \{j \in J \mid \boldsymbol{a}^t \boldsymbol{r}_j > 0\}, \tag{3.6a}$$

$$J^0 = \{j \in J \mid \boldsymbol{a}^t \boldsymbol{r}_j = 0\}, \tag{3.6b}$$

$$J^- = \{j \in J \mid \boldsymbol{a}^t \boldsymbol{r}_j < 0\}. \tag{3.6c}$$

The first two sets contain the columns of $\mathbf{R}$ that lie in the intersection $C(\mathbf{A}) \cap C(\boldsymbol{a}^t)$, and the third set contains the columns of $\mathbf{R}$ that lie outside. We keep the extreme rays that fulfill the additional inequality and form positive linear combinations of (so-called adjacent) rays such that the new rays lie on the cutting hyperplane $\{\boldsymbol{v} \in \mathbb{R}^r \mid \boldsymbol{a}^t \boldsymbol{v} = 0\}$. Hence, we obtain a matrix $\mathbf{R}' \in \mathbb{R}^{r \times J'}$, where

$$J' = J^+ \cup J^0 \cup J^{\text{adj}}, \tag{3.7a}$$

$$J^{\text{adj}} = \{(j^+, j^-) \in J^+ \times J^- \mid \boldsymbol{r}^{j^+}, \boldsymbol{r}^{j^-} \text{ adjacent}\}, \tag{3.7b}$$

$$\boldsymbol{r}'_j = \boldsymbol{r}_j \text{ for } j \in J^+ \cup J^0, \tag{3.7c}$$

$$\boldsymbol{r}'_{(j^+, j^-)} = (\boldsymbol{a}^t \boldsymbol{r}_{j^+}) \boldsymbol{r}_{j^-} - (\boldsymbol{a}^t \boldsymbol{r}_{j^-}) \boldsymbol{r}_{j^+} \quad \text{for } (j^+, j^-) \in J^{\text{adj}}, \tag{3.7d}$$

that is, for adjacent rays $\boldsymbol{r}^{j^+}, \boldsymbol{r}^{j^-}$. Then, $(\mathbf{A}', \mathbf{R}')$ is a minimal DD pair that satisfies also the additional inequality.

A cone $C(\mathbf{A})$ as in (3.4) is defined by inequalities $\mathbf{A}\boldsymbol{v} \geq \mathbf{0}$. A representation of the flux cone F as in (3.3) by inequalities is given by $F = C(\mathbf{A})$ with

$$\mathbf{A} = \begin{pmatrix} \boldsymbol{S} \\ -\boldsymbol{S} \\ \boldsymbol{I}_r \end{pmatrix}, \tag{3.8}$$

where $\boldsymbol{I}_r \in \mathbb{R}^{r \times r}$ denotes the identity matrix.

Of course, first we need to find an initial minimal DD pair. Following the null space approach [23, 24], we compute a basis of the kernel of the stoichiometric matrix $\boldsymbol{S}$. More specifically, we compute a column-reduced echelon form of the basis and (after a permutation of rows) obtain

$$\mathbf{R}_0 = \begin{pmatrix} \boldsymbol{I}_p \\ \boldsymbol{K} \end{pmatrix}, \tag{3.9}$$

where p is the dimension of the kernel of $\boldsymbol{S}$. By construction, the columns of $\mathbf{R}_0$ satisfy the inequalities coming from the stoichiometric matrix $\boldsymbol{S}$ and the first p non-negativity constraints. The remaining $r - p$ non-negativity constraints are processed consecutively. Explicitly, we consider

$$\mathbf{A}_k = \begin{pmatrix} \boldsymbol{S} & \\ -\boldsymbol{S} & \\ \boldsymbol{I}_{p+k} & \mathbf{0} \end{pmatrix}, \quad k = 0, \ldots, r - p. \tag{3.10}$$

Clearly, the first $p + k$ rows of the corresponding matrices $\mathbf{R}_k$ are non-negative. This fact is the basis for the binary null space approach [23, 24] for which efficient implementations are available [20, 21]. Still, the DD suffers from the combinatorial explosion of intermediate extreme rays resulting from the combination of adjacent rays $\boldsymbol{r}^{j^+}, \boldsymbol{r}^{j^-}$. Thus, a full EFMA is only applicable to medium-scale metabolic networks.

We conclude with the following key observation: at each iteration step $k = 1, \cdots, r - p$, we either keep an extreme ray or compute a positive linear combination of two adjacent rays. In any case, *the number of positive entries within the first $p + k$ components of the resulting rays either remains the same or increases.* In particular, the parents of an extreme ray, that is, the intermediate rays that are combined, never have a larger support than their offspring. (Recall that the support of a vector is the set of nonzero components.) This observation can be exploited computationally, as explained below.

3.3 Calculating Short EFMs in Genome-Scale Metabolic Networks

Suppose we are interested in calculating all EFMs with a maximal cardinality $c_{\max}$ of their support. In this case, we can omit an intermediate extreme ray from the analysis as soon as the number of positive entries in the first $p + k$ components exceeds the maximal cardinality. By omitting intermediate extreme rays, the combinatorial explosion of extreme rays is curbed and the method becomes applicable to larger networks, without compromising the efficiency and speed of the method.

We implemented the maximum cardinality feature as an extension to the open source program *Efmtool* [25]. The software is freely available at the author's webpage [26]. The principal workflow of the algorithm is illustrated in Figure 3.2. This extended version was used throughout the remainder of the study and compared

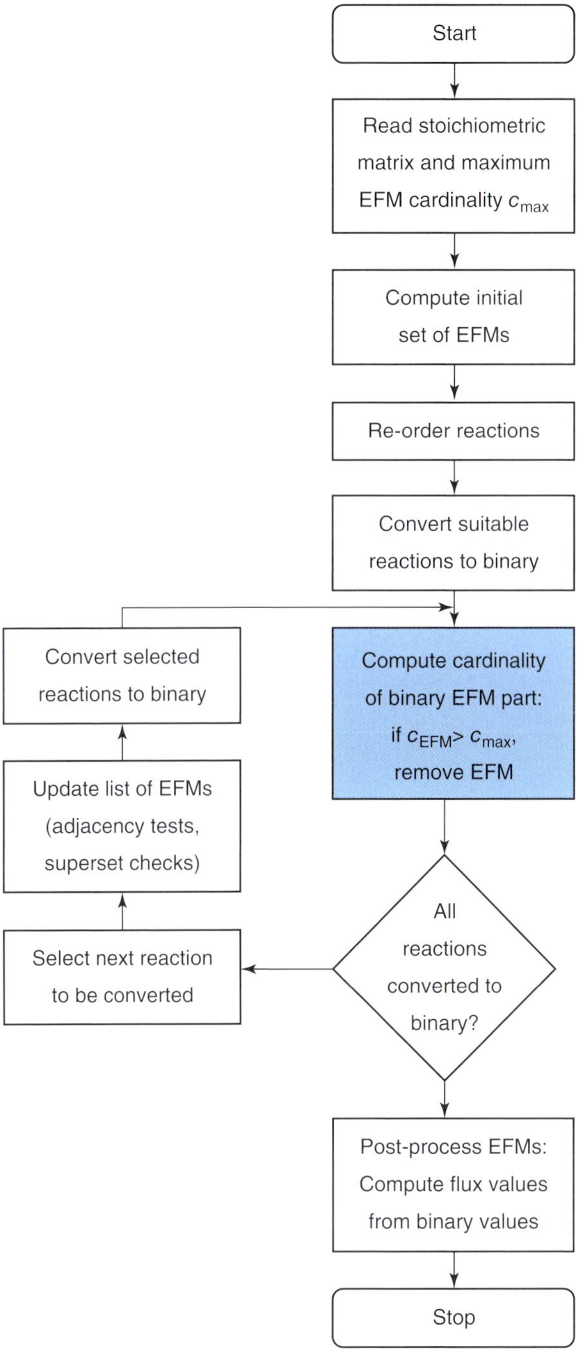

Figure 3.2 Computing the shortest EFMs with the DD method. The box with blue background color highlights the main extension of the maximum cardinality feature over common implementations of the DD method.

to a standard approach of calculating the k-shortest EFMs using a mixed-integer linear program (MILP) [16].

The MILP-based approach was realized by a hybrid code. Performance uncritical parts of the program (e.g., setting up the linear program, input/output handling) were implemented in Perl. C-libraries of the IBM ILOG CPLEX Optimization Studio were used for the computationally expensive task of solving the linear program. CPLEX is a commercial optimization software package for which free academic licenses are available.

We calculated all EFMs up to a cardinality of c_{max} in six metabolic models of various sizes ranging from small-scale to genome-scale. The key properties of these models are listed in Table 3.1. All networks were compressed [23] before the EFMA. We used a computer with two Intel Xeon CPUs (each with six cores, 2.67 GHz) running Ubuntu 14.04. Both programs were allowed to use up to eight parallel threads during the execution.

Figure 3.3 illustrates the runtimes (primary y-axis) for both methods, and the number of enumerated EFMs (secondary y-axis) as function of the EFMs' length for the metabolic networks listed in Table 3.1. Note that both y-axes were scaled logarithmically and that the depicted runtimes did not include pre- or postprocessing steps such as network compression and decompression, which took up to an hour in genome-scale metabolic networks.

Naturally, we found that both, runtimes and the number of EFMs increased with increasing cardinality. Moreover, execution times for both approaches strongly corresponded with the number of EFMs. However, for a fixed number of computed EFMs the execution time varied widely across the models. For instance, in the liver cancer model it took almost 15 h to compute all 307 444 EFMs up to a cardinality of 7, while it took less than 2 h to compute all 990 797 EFMs up to a cardinality of 100 in the *P. tricornutum* model. This indicates not only that the relation between runtime, EFM cardinality, and number of EFM is nontrivial but also that the number of EFMs had a larger effect on the runtime than the cardinalities of the EFMs.

Table 3.1 Main topological properties of the six metabolic models used in this study.

Model	Internal metabolites	External metabolites	Total reactions	Irreversible reactions	EFMs
E. coli core I	69	17	82	35	5 011
E. coli core II	53	15	71	20	429 276
E. coli core III	72	21	95	59	226 269 020
P. tricornutum	327	11	318	103	1 934 729 551
Blattibacteriacae cuenoti Bge	306	43	350	45	Unknown
Liver cancer	1 754	195	2 423	454	Unknown

Note however, that the number of metabolites and the number of reactions has no known functional correlation to the number of EFMs. In fact, previous attempts to find such a correlation failed [27] and the question remains open.

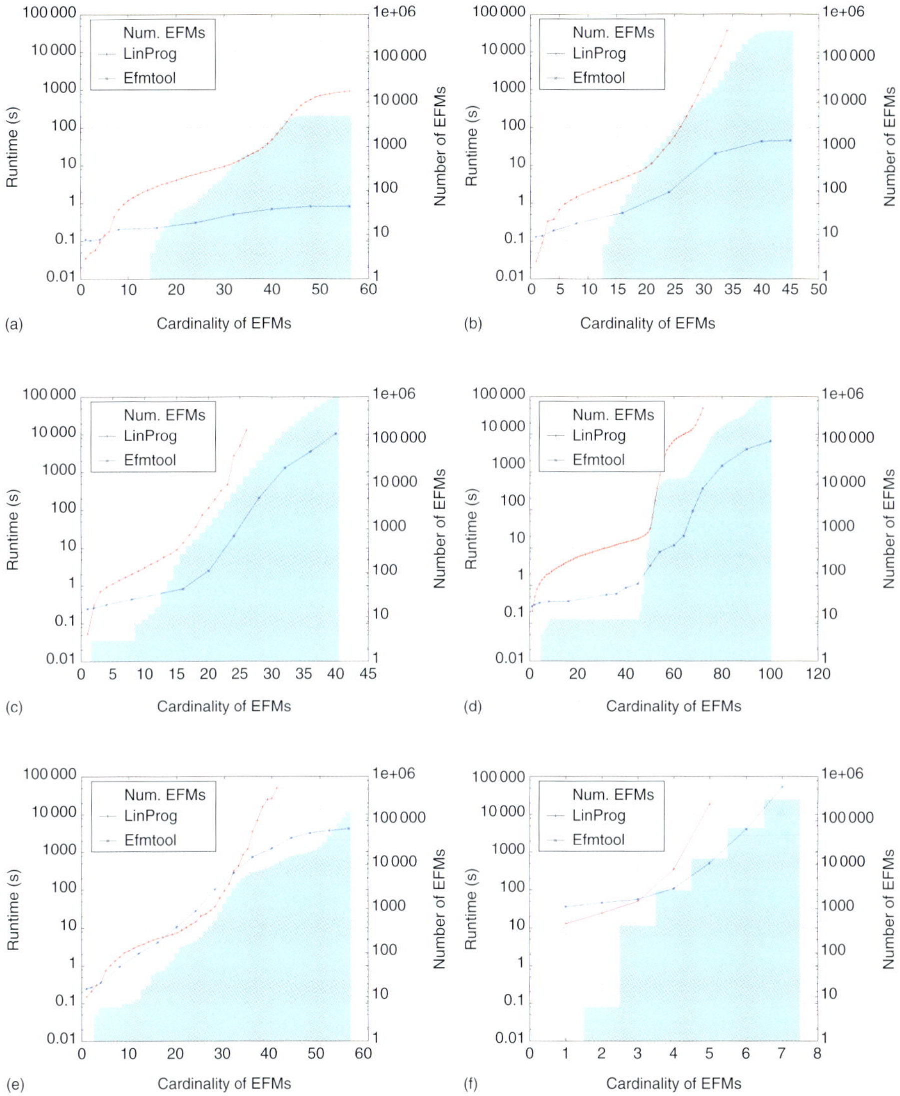

Figure 3.3 Comparison of execution times. Runtime comparison between maximum cardinality EFMA and the MILP based approach in various models (see Table 3.1). (a) *E. coli* core I [28], (b) *E. coli* core II, (c) *E. coli* core III, (d) *P. tricornutum* [11], (e) *Blattibacteriacae cuenoti* Bge [29], and (f) liver cancer [30].

Figure 3.3 shows that for low cardinalities and, therefore, for very low runtimes, maximum cardinality EFMA and the MILP approach exhibit a similar performance. However, Figure 3.3 also clearly demonstrates that for larger cardinalities, maximum cardinality EFMA is much faster than the MILP approach. Importantly, the performance gain grows with increasing EFM cardinalities.

Moreover, in the models (b) to (f) the runtime time requirements for the MILP approach became quickly prohibitively large, while maximum cardinality EFMA successfully continued to output EFMs.

3.4
Conclusions

Here we introduced a maximum cardinality EFMA that efficiently enumerates short EFMs in (large) metabolic networks. Maximum cardinality EFMA is based on the DD method and exploits the fact that new (intermediate) EFMs are always constructed by combining two appropriately selected shorter EFMs. Thus the cardinality of any new (intermediate) EFM is always larger than the cardinality of both of its parent EFMs. An EFM exceeding the user-specified cardinality threshold can therefore safely be removed from further analysis.

Maximum cardinality EFMA was implemented as a minimal invasive extension to *Efmtool* and fully utilizes the computational advantages of the binary null space implementation of the DD method. Both factors, the maximum cardinality EFMA strategy and the binary implementation, result in a major speedup that outperforms other, MILP-based approaches by orders of magnitude.

Maximum cardinality EFMA requires a specific, user-defined maximum cardinality threshold, c_{max}, as input. However, the number of EFMs that will be calculated for the specified threshold is not known *a priori*. In contrast to MILP-based approaches, maximum cardinality EFMA is therefore not able to terminate after it has found a predefined number of EFMs. In practice, this drawback had little effect on the usability of our program. Beginning with low values of c_{max}, we repeatedly executed maximum cardinality EFMA while increasing c_{max} until the desired number of EFMs were calculated. Even if we cumulated all execution times, still maximum cardinality EFMA outperformed the MILP approach in calculating the 10 000 shortest EFMs in all tested models. Conversely, for a given runtime, many more EFMs can be computed by maximum cardinality EFMA than the MILP approach, which in turn results in larger networks that can be analyzed and studied more thoroughly with maximum cardinality EFMA.

In principle, maximum cardinality EFMA is able to fully enumerate all EFMs of a network, as for $c_{max} \rightarrow r$ the ordinary DD method is retrieved. The DD method suffers from a combinatorial explosion in the number of (intermediate) EFMs. The same problem applies to maximum cardinality EFMA. Thus c_{max} has to be sufficiently small in order to remain computationally feasible. Nevertheless, our analysis revealed that in all investigated models maximum cardinality EFMA was able to calculate the 70 000 shortest EFMs within about 1 h on standard workstation computers. Comparable runtimes are not achievable with currently available alternative methods.

Note that the maximum cardinality feature of the DD method as observed above is a very general feature that can be exploited for the integration of -omics data. Suppose an (intermediate) EFM contains two active reactions that together are infeasible. For instance, in *E. coli* the glyoxylate shunt is downregulated under high-glucose conditions [31]. In this case, the glucose uptake and the glyoxylate shunt are not simultaneously active. Any EFM that contains both reactions is therefore infeasible. On the basis of our observation, these EFMs can be discarded without impact on the further analysis. In fact, recent methodological developments exploited this fact and identified thermodynamically feasible [32, 33] or transcriptionally regulated [34, 35] EFMs. However, the underlying principle has never been clearly formulated. Thus by extensively including -omics data based on the principle introduced here, an EFMA of a genome-scale metabolic model seems feasible.

Acknowledgments

CJ, MPG, MH, GMN, and JZ acknowledge support by the Federal Ministry of Science, Research and Economy (BMWFW), the Federal Ministry of Traffic, Innovation and Technology (bmvit), the Styrian Business Promotion Agency SFG, the Standortagentur Tirol, the Government of Lower Austria and ZIT – Technology Agency of the City of Vienna through the COMET-Funding Program managed by the Austrian Research Promotion Agency FFG.

SM and GR were supported by the Austrian Science Fund (FWF), project P28406.

References

1. Thiele, I. and Palsson, B.Ø. (2010) A protocol for generating a high-quality genome-scale metabolic reconstruction. *Nat. Protoc.*, **5** (1), 93–121, doi: 10.1038/nprot.2009.203.
2. Papin, J.A., Price, N.D., Wiback, S.J., Fell, D.A., and Palsson, B.O. (2003) Metabolic pathways in the post-genome era. *Trends Biochem. Sci.*, **28** (5), 250–258.
3. Zanghellini, J., Ruckerbauer, D.E., Hanscho, M., and Jungreuthmayer, C. (2013) Elementary flux modes in a nutshell: properties, calculation and applications. *Biotechnol. J.*, **8** (9), 1009–1016.
4. Schuster, S., Fell, D.A., and Dandekar, T. (2000) A general definition of metabolic pathways useful for systematic organization and analysis of complex metabolic networks. *Nat. Biotechnol.*, **18** (3), 326–332, doi: 10.1038/73786.
5. Müller, S. and Regensburger, G. (2016) Elementary vectors and conformal sums in polyhedral geometry and applications in metabolic pathway analysis, *Front. Genet.*, **7**, 90, doi: 10.3389/fgene.2016.00090.
6. Trinh, C.T. and Thompson, R.A. (2012) Elementary mode analysis: a useful metabolic pathway analysis tool for reprograming microbial metabolic pathways, in *Reprogramming Microbial Metabolic Pathways*, Vol. **64** (eds X. Wang, J. Chen, and P. Quinn), Springer Netherlands, Dordrecht, pp. 21–42.
7. Klamt, S. and Mahadevan, R. (2015) On the feasibility of growth-coupled product synthesis in microbial strains. *Metab. Eng.*, **30**, 166–178, doi: 10.1016/j.ymben.2015.05.006.
8. Urbanczik, R. (2007) Enumerating constrained elementary flux vectors of metabolic networks. *IET Syst. Biol.*, **1** (5), 274–279.
9. Klamt, S. and Stelling, J. (2002) Combinatorial complexity of pathway analysis in metabolic networks. *Mol. Biol. Rep.*, **29** (1-2), 233–236, doi: 10.1023/A:1020390132244.

10. Jungreuthmayer, C., Nair, G., Klamt, S., and Zanghellini, J. (2013) Comparison and improvement of algorithms for computing minimal cut sets. *BMC Bioinformatics*, **14** (1), 318, doi: 10.1186/1471-2105-14-318.
11. Hunt, K.A., Folsom, J.P., Taffs, R.L., and Carlson, R.P. (2014) Complete enumeration of elementary flux modes through scalable demand-based subnetwork definition. *Bioinformatics*, **30** (11), 1569–1578.
12. Machado, D., Soons, Z., Patil, K.R., Ferreira, E.C., and Rocha, I. (2012) Random sampling of elementary flux modes in large-scale metabolic networks. *Bioinformatics*, **28** (18), i515–i521, doi: 10.1093/bioinformatics/bts401.
13. Kaleta, C., De Figueiredo, L.F., Behre, J., and Schuster, S. (2009) EFMEvolver: computing elementary flux modes in genome-scale metabolic networks, in Proceedings *of the German Conference on Bioinformatics, Lecture Notes in Informatics (LNI) P-157*, Gesellschaft für Informatik, Bonn, pp. 179–190.
14. Pey, J., Villar, J.A., Tobalina, L., Rezola, A., García, J.M., Beasley, J.E., and Planes, F.J. (2015) TreeEFM: calculating elementary flux modes using linear optimization in a tree-based algorithm. *Bioinformatics*, **31** (6), 897–904.
15. Kelk, S.M., Olivier, B.G., Stougie, L., and Bruggeman, F.J. (2012) Optimal flux spaces of genome-scale stoichiometric models are determined by a few subnetworks. *Sci. Rep.*, **2**, 580, doi: 10.1038/srep00580.
16. De Figueiredo, L.F., Podhorski, A., Rubio, A., Kaleta, C., Beasley, J.E., Schuster, S., and Planes, F.J. (2009) Computing the shortest elementary flux modes in genome-scale metabolic networks. *Bioinformatics*, **25** (23), 3158–3165, doi: 10.1093/bioinformatics/btp564.
17. David, L. and Bockmayr, A. (2014) Computing elementary flux modes involving a set of target reactions. *IEEE/ACM Trans. Comput. Biol. Bioinf.*, **11** (6), 1099–1107, doi: 10.1109/TCBB.2014.2343964.
18. Pey, J. and Planes, F.J. (2014) Direct calculation of elementary flux modes satisfying several biological constraints in genome-scale metabolic networks. *Bioinformatics*, **30** (15), 2197–2203, doi: 10.1093/bioinformatics/btu193.
19. Acuna, V., Chierichetti, F., Lacroix, V., Marchetti-Spaccamela, A., Sagot, M.F., and Stougie, L. (2009) Modes and cuts in metabolic networks: Complexity and algorithms. *Biosystems*, **95** (1), 51–60.
20. van Klinken, J.B. and van Dijk, K.W. (2016) FluxModeCalculator: an efficient tool for large-scale flux mode computation. *Bioinformatics*, **32** (8), 1265–1266, doi: 10.1093/bioinformatics/btv742.
21. Terzer, M. and Stelling, J. (2008) Large-scale computation of elementary flux modes with bit pattern trees. *Bioinformatics*, **24** (19), 2229–2235, doi: 10.1093/bioinformatics/btn401.
22. Fukuda, K. and Prodon, A. (1996) Double description method revisited, in *Combinatorics and Computer Science, Lecture Notes in Computer Science*, Vol. **1120**, Springer-Verlag, Berlin Heidelberg, pp. 91–111.
23. Gagneur, J. and Klamt, S. (2004) Computation of elementary modes: a unifying framework and the new binary approach. *BMC Bioinformatics*, **5**, 175.
24. Wagner, C. (2004) Nullspace approach to determine the elementary modes of chemical reaction systems. *J. Phys. Chem. B*, **108** (7), 2425–2431, doi: 10.1021/jp034523f.
25. efmtool - elementary flux mode tool, http://www.csb.ethz.ch/tools/software/efmtool.html (accessed 22 June 2016).
26. Regulatory elementary flux mode tool, regEfmtool, http://www.biotec.boku.ac.at/en/arbeitsgruppenresearch-groups/research-group-mattanovich/staff/associated-group-metabolic-modelling/regefmtool/ (accessed 22 June 2016).
27. Yeung, M., Thiele, I., and Palsson, B. (2007) Estimation of the number of extreme pathways for metabolic networks. *BMC Bioinformatics*, **8** (1), 363, doi: 10.1186/1471-2105-8-363.
28. Trinh, C.T., Unrean, P., and Srienc, F. (2008) Minimal *Escherichia coli* cell for the most efficient production of ethanol from hexoses and pentoses. *Appl. Environ. Microbiol.*, **74** (12), 3634–3643.
29. González-Domenech, C.M., Belda, E., Patiño-Navarrete, R., Moya, A., Peretó,

J., and Latorre, A. (2012) Metabolic stasis in an ancient symbiosis: genome-scale metabolic networks from two Blattabacterium cuenoti strains, primary endosymbionts of cockroaches. *BMC Microbiol.*, **12** (Suppl. 1), S5, doi: 10.1186/1471-2180-12-S1-S5.

30. Human Metabolic Atlas, INIT models of cancer cell, http://www.metabolicatlas.com/downloads/initcan (accessed 22 June 2016).
31. Kornberg, H.L. (1966) The role and control of the glyoxylate cycle in *Escherichia coli*. *Biochem. J.*, **99** (1), 1–11.
32. Gerstl, M.P., Ruckerbauer, D.E., Mattanovich, D., Jungreuthmayer, C., and Zanghellini, J. (2015) Metabolomics integrated elementary flux mode analysis in large metabolic networks. *Sci. Rep.*, **5**, 8930.
33. Gerstl, M.P., Jungreuthmayer, C., and Zanghellini, J. (2015) tEFMA: computing thermodynamically feasible elementary flux modes in metabolic networks. *Bioinformatics*, **31** (13), 2232–2234, doi: 10.1093/bioinformatics/btv111.
34. Jungreuthmayer, C., Ruckerbauer, D.E., Hanscho, M., and Zanghellini, J. (2015) Avoiding the enumeration of infeasible elementary flux modes by including transcriptional regulatory rules in the enumeration process saves computational costs. *PLoS ONE*, **10** (6), e0129840.
35. Jungreuthmayer, C., Ruckerbauer, D.E., and Zanghellini, J. (2013) regEfmtool: speeding up elementary flux mode calculation using transcriptional regulatory rules in the form of a three-state logic. *Biosystems*, **113** (1), 37–39.

4
Cell-Free Synthetic Systems for Metabolic Engineering and Biosynthetic Pathway Prototyping

Ashty S. Karim, Quentin M. Dudley, and Michael C. Jewett

4.1
Introduction

Essentially all of today's chemicals and materials are produced from the same petroleum-based precursors: methanol, ethylene, propene, butadiene, benzene, toluene, and xylene. Unfortunately, the use of these petrochemicals has resulted in increased energy demands, shortages of feedstock, higher prices, and added climate change effects. This has motivated a growing need for greener methods of producing fuels, pharmaceuticals, materials, and commodity chemicals [1]. Biology offers an appealing approach to meet this need due to its ability to mimic the specificity and range of the commodity chemical industry's products. In fact, biologically derived substitutes for the aforementioned petroleum-based precursor molecules have already been identified [2, 3].

Natural reaction cascades (i.e., metabolism) in cells offer a plethora of biotransformation networks that can be manipulated to make products from simple monomer starting blocks (CO_2, sugars, etc.). Some of the oldest examples of these processes include the use of yeast for fermentation or leavening of food and alcohol. More recently, our understanding of cellular metabolism and our ability to manipulate these reactions by reading, writing, and editing DNA has vastly increased, giving rise to a growing array of chemical products (e.g., 1,3-propanediol, farnesene, and 1,4-butanediol [4]) that can be harnessed by redirecting cellular metabolism with metabolic engineering [5–8]. The wide range of synthetic biology tools (chassis, vectors, regulatory elements, etc.) is further expanding our ability to engineer biological systems [9, 10].

While the number of metabolic engineering success stories is rapidly growing, there are limitations that currently prohibit a wide distribution of industrial-scale biotechnology. The main problem is that the process of developing and engineering living organisms to make viable production strains is expensive and time consuming [10]. In the past few decades, several notable ventures have established industrial-scale biosynthesis with metabolically engineered cells (Table 4.1). While the design-build-test-learn (DBTL) cycle time is decreasing, the organisms we engineer retain survival as their main objective. Because

Industrial Biotechnology: Microorganisms, First Edition. Edited by Christoph Wittmann and James C. Liao.

Table 4.1 Examples of industrial biotechnology (time and cost).

Target	Institutions	Time to market (yr)	R&D person years
1,3-Propanediol [11]	Genencor, Tate & Lyle, DuPont	15 (1992–2007)	~575
Artemisinin [10]	Amyris, UC Berkeley	13 (2000–2013)	>130
Farnesene [12]	Amyris	4 (2008–2012)	>250

Data from [4, 10, 13] and Z. Serber, Amyris Biotechnologies, DARPA Living Foundries Industry Day (personal communication).

evolution has optimized the metabolism and regulation of cells for growth and adaptation, cells often resist changes that divert resources away from these objectives by altering the catalytic composition of the cell or using embedded regulatory strategies (e.g., feedback inhibition of enzymes). Therefore, it is very difficult to rationally engineer intracellular fluxes to generate high titers from an active synthetic pathway while the machinery of the cell is functioning to maintain cell growth. This leads to several challenges facing the state of the art, namely, cellular fermentations provide us low titers, yields, and productivities along with an essentially limited palate of synthesizable chemicals. Many new computational, genomic, and screening tools are being developed to speed up design of strains for metabolic engineering [5, 14, 15]. Complementary to these efforts are new strategies to eliminate the cell's growth objectives, which can enable high yields and productivities [16, 17].

Cell-free, or *in vitro*, synthetic systems have emerged as a way of decoupling cellular growth objectives from engineering process objectives [18]. The foundational principle of cell-free synthetic systems is that we can conduct precise, complex biomolecular synthesis without using intact cells [19]. Instead, crude cell lysates or purified enzymes are used, which can be accurately monitored and modeled to provide rapid DBTL cycles. Complementing *in vivo* systems, the complexity of cell-free systems range from a few enzymes (2–4 enzymes) to many enzymes (more than 20 enzymes), as highlighted in several reviews [20–23]. In this chapter, we discuss the state of the art in cell-free synthetic systems with regard to cell-free metabolic engineering (CFME). Cell-free protein synthesis (CFPS) is a separate area of development in cell-free biology. CFPS is a useful process for understanding and harnessing biological systems. CFPS is motivating new advances in high-throughput protein production [19, 24–26], clinical manufacture [24], genetic circuit optimization [27], the construction of synthetic ribosomes [28], and incorporation of nonstandard amino acids [29–36]. Perspectives on CFPS are described elsewhere [13, 22, 37–39]. In this review, we begin with a brief introduction of the technological capabilities of the CFME field and its potential benefits for pathway engineering, biomolecular "breadboarding," and production. In the next section, we discuss up-to-date cell-free systems for small-molecule metabolite production and pathway optimization. Finally, we examine challenges and opportunities, providing commentary about new directions in the field.

4.2 Background

Cell-free biology has been used for many decades as a foundational research tool. For example, cell-free fermentations were first discovered in the late nineteenth century/early twentieth century by Eduard Buchner in the attempt to elucidate the mechanism behind alcoholic fermentation of yeast cells [40]. This opened the door to the possibility of using individual enzymes to produce chemicals such as ethanol from reconstituted enzymes [41] and was a seminal study in establishing biochemistry as a field. Specifically, Buchner's work led to discoveries of how intricate metabolisms of different organisms function, as well as opening the aperture to our ability to engineer metabolism for the targeted synthesis of biochemicals. In the 1960s, Nirenberg and Matthaei discovered that protein synthesis, which comprises a complex metabolic system of tens of reactions, could be activated and controlled in crude cell-free extracts [42]. They used this system to discover the genetic code. More recently, Swartz and colleagues demonstrated that cell-free systems could provide highly active energy and cofactor regeneration to fuel high-level ($g\,l^{-1}$) CFPS [19]. Collectively, these advances along with other significant works [21, 23, 43] have motivated the development of capturing cell-free reactions to address the challenges of traditional metabolic engineering. There are two broad classes of cell-free systems typically used for synthesis of small-molecule metabolites: purified enzymes and crude cell lysates (Figure 4.1).

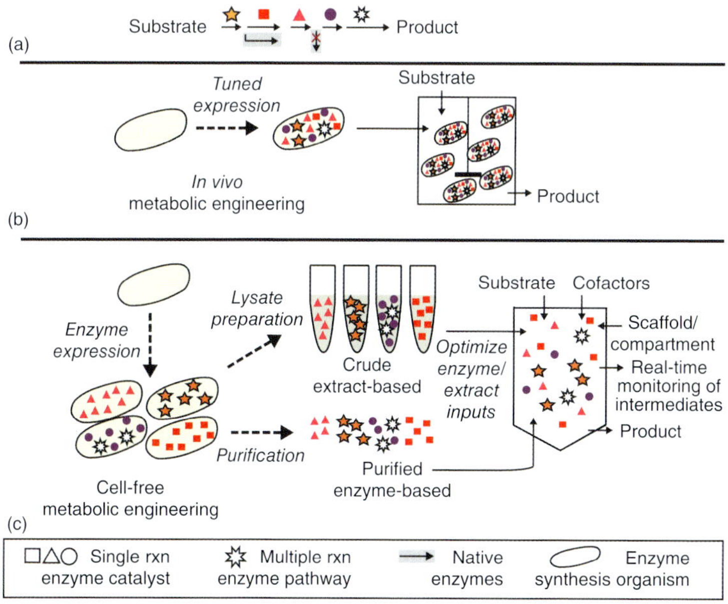

Figure 4.1 Comparison of traditional and cell-free metabolic engineering. (a) Desired biochemical pathway. (b) Methodology for metabolic engineering *in vivo*. (c) Methodology for metabolic engineering *in vitro*.

4.2.1 Purified Enzyme Systems

Purified approaches take advantage of enzymes that have been overexpressed and purified individually, then used as individual biocatalysts or recombined to assemble a pathway of interest. The benefit of these systems is that the reaction network is explicitly defined, which gives exquisite control of reaction conditions and pathway fluxes. Indeed, the clarity of the biosynthetic pathway comes from eliminating unnecessary enzymes and cellular distractions (i.e., growth and other off-pathway metabolites). With respect to CFME, purified enzyme systems are used owing to their distinct advantages over chemical methods, including decreased reaction time, increased product yield, greater product specificity, relatively low cost, and reduced environmental impact [44]; although, few long (with more than four enzymes) pathways have been implemented. From a breadboarding perspective, cell-free systems can troubleshoot pathways difficult to achieve *in vivo* sometimes creating synthetic reaction networks leading to novel pathway design [45]. From a production standpoint, there are already several examples of simple *in vitro* enzyme systems in industrial biocatalysis. For example, immobilized glucose isomerase has been used to harness enzymatic activity for the production of fructose [46] and the industrial production of antibiotics has been pursued with hybrid chemical and biological syntheses [47]. From these commercial examples, and others, a general rule of thumb is that purified enzyme systems are used when the product is of high value or a reaction is not possible using chemistry. Moreover, these systems are used to study the fundamental operation of biosynthetic pathways. Few industrial examples exist with synthetic enzymatic pathways (e.g., with more than 10 purified enzymes), in part because of the high catalyst costs (as a result of purification and stability) [19, 48]. Another challenge with recapitulating pathways from purified enzymes is cofactor cost and regeneration. Despite these challenges, the majority of CFME research to date has utilized purified systems.

4.2.2 Crude Cell Lysate Systems

As described earlier, as early as in the 1900s, scientists have looked to crude lysates as a means to harness and understand biosynthetic pathways [49]. Instead of using purified enzymes, this approach levers the extract of lysed cells as an ensemble of biocatalysts. When compared to the use of purified enzymes, crude cell lysates offer lower system catalyst costs and greater system capabilities (e.g., cofactor regeneration and long-lived biocatalytic activity) [37, 50]. However, use of crude lysates has been limited for metabolic conversions *in vitro*. The presence of hundreds to thousands of active enzymes in crude lysates can make it more challenging to selectively activate and control only desired reactions. For example, the need to bypass off-pathway reactions can limit theoretical yields (an advantage of purified enzyme systems). Also, the cell-like features of these systems are partially as complex as that of whole cells. This comes with the burden of attempting to characterize the contents of the functional activities of the lysate. However, there is a growing set of examples demonstrating that crude cell extracts should not be treated as a "black box," but rather as a set of biochemical reactions that can be activated and controlled [51].

The following sections discuss the most recent successes and advances made in CFME, as well as challenges and opportunities associated with cell-free systems.

4.3 The Benefits of Cell-Free Systems

Table 4.2 summarizes the titers, yields, and productivities of recent CFME reports. These efforts have generated common rules for achieving high metabolic yields and productivities *in vitro*. Optimal metabolic conversions require increased flux through desired pathways, decreased flux through competing reactions, and energy balancing. In short, the same concepts apply to both *in vitro* and *in vivo* metabolic engineering. Thus, development of cell-free systems should be guided by cytoplasmic mimicry to enable highly productive systems [13, 19, 51]. Here, we first describe advances in purified enzyme systems and follow with a discussion of crude lysate-based systems.

Table 4.2 Productivities, yields, and scales of CFME.

Target molecule	Substrate	Final titer (mM)	Productivity (mmol l^{-1} h^{-1})	% of Theoretical yield	References
Purified systems					
Ethanol	Glucose	28.7	1.51	57	[23]
Isobutanol	Glucose	10.3	0.45	53	[23]
Amorpha-4,11-diene	Mevalonate	1.66	0.28	50	[52]
Acetyl phosphate	F6P	~30	20.00	~100	[45]
Hydrogen (H_2)	Starch	N/A	0.44	0	[53]
Hydrogen (H_2)	Cellobiose	N/A	0.48	93	[54]
Hydrogen (H_2)	Cellopentaose	N/A	3.92	68	[54]
Hydrogen (H_2)	Xylose	N/A	1.95	95	[55]
Xylitol (i.e., NAD(P)H)	Xylose/cellobiose	~18	0.38	18	[56]
n-Butanol	Glucose	3.5	0.49	82	[57]
Lactate/lactic acid	Glucose	12.5	1.25	50	[58]
Malate/malic acid	Glucose	5	0.625	86	[59]
Farnesene	Acetyl-CoA	~39	0.260	35	[43]
Isoprene	PEP	~0.5	0.13	~100	[60]
Crude extract systems					
DHAP	Glucose	9.5	19.00	83	[61]
TDHP	Glucose	23	17.25	41	[61]
Ethanol	Glucose	86.8	10.85	20	[62]

4.3.1
Purified Enzyme Systems

Some of the first successes of cell-free systems have been in the use of purified enzymes. In 1985, Welch enabled cell-free ethanol production by reconstituting the yeast glycolytic system *in vitro* [41]. In 1992, Fessner and Walter successfully produced dihydroxyacetone phosphate (DHAP), a key metabolic intermediate in glycolysis and the Calvin cycle [63]. Fessner's system required high-energy phosphoenolpyruvate (PEP) for ATP generation and did not quantify DHAP. That said, similar to Welch's work, it demonstrated the feasibility of a multienzymatic pathway *in vitro*.

With this backdrop, it is clear that combining enzymes to create biosynthetic pathways *in vitro* is not necessarily new, but recently there have been several success stories that utilize more complex systems of enzymes. A pioneering effort in CFME demonstrates the conversion of glucose to biofuels [23]. By developing a system with six and eight purified enzymes for ethanol and isobutanol production, respectively, Guterl and colleagues in the Sieber lab were able to show proof of effective *in vitro* metabolic engineering (Figure 4.2a). The reconstituted glycolytic pathway (nonphosphorylative Entner–Doudoroff pathway) took advantage of promiscuous enzymes to downsize the number of enzymes needed from 10 to 4 with ethanol yields at 57% and isobutanol yields at 53%. On the basis of theoretical activity of individual enzymes and their concentration in the system, the theoretical maximum conversion productivity appears to be 30 $mmol\,l^{-1}\,h^{-1}$; 1.5 $mmol\,l^{-1}\,h^{-1}$ was achieved experimentally. In the success of this pathway, an interesting question to pose is what quantity of enzymes is needed to prepare a feasible cell-free network. Furthermore, this pathway paves the way for cell-free systems to address the capabilities of using multiple cofactors and electron

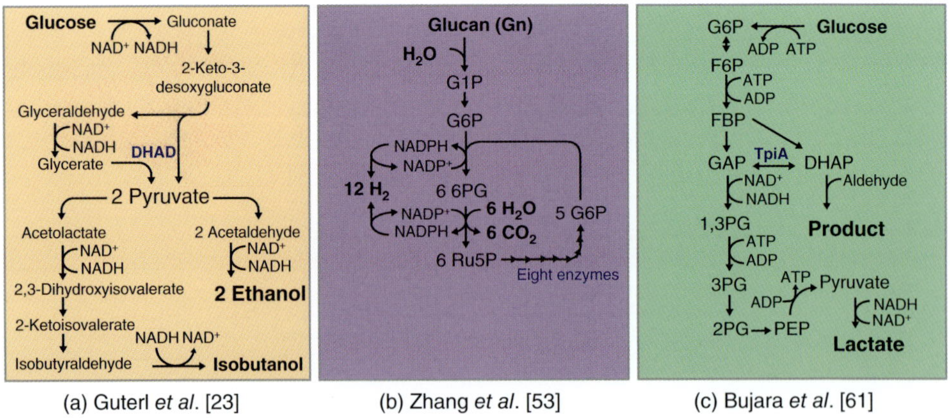

Figure 4.2 Biochemical pathways of key cell-free metabolic engineering achievements. (a) Pathway from [23]. (b) Pathway from [53]. (c) Pathway from [61].

balancing. This example sheds light on the unexplored boundaries of cell-free systems and what they can teach us about *in vivo* biochemical pathways.

In addition to exploring the boundaries of *in vitro* biochemical pathways, cell-free systems are also being used for rapid pathway debugging. Troubleshooting and optimizing metabolic pathways is a pernicious challenge *in vivo.* For example, cases, we have to look no further than one of several success stories (e.g., artemisinin production [10]) that have required more than 100 person years of development (Table 4.1). Cell-free technologies offer some exciting opportunities for screening metabolic pathways because they provide direct access to the reaction conditions and the ability to monitor intermediate reactions. Thus, cell-free systems allow for a desirable *in vivo* pathway to be analyzed and tested for enzymatic activity *in vitro.* It is conceivable that one could take parts or modules of metabolic pathways and "breadboard" them together to test out different pathways [27, 64, 65]. This circuitry approach has value for metabolic engineering and for building complex genetic circuits.

An example of *in vitro* pathway debugging was recently shown for the discovery of a synthetic pathway that mitigates the loss of carbon seen in native glycolysis [45]. Nonoxidative glycolysis (NOG) was conceived as an alternative to the standard Embden–Meyerhof–Parnas (EMP) glycolytic pathway. Using a cell-free purified enzyme system, Bogorad, Lin, and Liao combined 7 His-tag purified enzymes with 7 purchased enzymes to create a 14-enzyme pathway that converts glucose (and xylose) to acetyl-phosphate, a metabolite easily converted to acetyl-CoA, which can enter central metabolism. Through *in vitro* production, they were able to achieve near 100% of theoretical conversion; this resulted in the ability to direct all carbon flux to acetyl-phosphate with no loss to CO_2. Once the *in vitro* system was established, expressing the NOG pathway *in vivo* required less troubleshooting of enzymes used and reaction conditions. The lack of reducing equivalents produced limits the range of products amenable to this pathway, but this exemplifies how a novel pathway can be introduced for microbial production through *in vitro* validation.

Another example demonstrating unique opportunities of the cell-free approach is the synthesis of a precursor to artemisinin, amorpha-4,11-diene from mevalonate [52]. In this study, seven enzymes were His-tag-purified for pathway debugging. Rather than tuning expression levels on individual enzymes *in vivo,* the cell-free approach allowed for direct modulation of pathway enzyme concentrations. Here, a design of experiments method using Taguchi orthogonal arrays explored a broad range of enzyme concentrations. Because this was done in a cell-free, isolated system, the researchers were able to determine that the fifth enzyme in the pathway, farnesyl diphosphate synthase (encoded by *IspA*), had a negative effect on production. Lowering the enzyme concentration, however, lowered the overall yield of amorpha-4,11-diene. This was actually due to precipitation of the intermediate and enzymatic product of the *IspA* enzyme, farnesyl pyrophosphate. By improving the buffer pH and salts in the cell-free environment, they produced high conversion through this multistep pathway. While this study shows how a cell-free system can be relatively easy to manipulate, modulate, and troubleshoot, it also highlights how sometimes new challenges appear *in vitro* (such as protein precipitation) that might not be an issue *in vivo*.

Numerous examples demonstrate the simplicity and usefulness of *in vitro* systems for pathway optimization and achieving high theoretical yields. An instance showing near 100% of theoretical yields is the production of isoprene from PEP [60]. A pathway of 12 enzymes was implemented to achieve high yields. However, by far the largest body of work in CFME is in the production of hydrogen. Hydrogen has been pursued as an alternative energy carrier [20]. Sugar conversion to hydrogen and then to electricity has a higher theoretical process efficiency compared to traditional combustion of fuel and provides an opportunity to address global transportation energy challenges [66]. Recently, the Zhang group produced many reports demonstrating the production of hydrogen from sugars. Starch and water can produce high yields of hydrogen in a cell-free environment [53]. Hydrogen can also be produced cell-free from cellobiose [54] and xylose [55]. Also, *in silico* models of hydrogen production have been created to further expand our cell-free abilities to produce hydrogen and other chemicals [21]. Xylose has been converted to xylitol *in vitro* using oxidation of cellobiose to CO_2 as a source of energy [56]. The cell-free pathway to convert sugars to hydrogen demonstrates complexity difficult to control *in vivo* (Figure 4.2b). Moreover, these examples demonstrate our ability to metabolize C5 sugars that are not favored by bacteria. The Zhang group has also done tangential work to convert cellulose to starch and ethanol using a hybrid purified enzyme/fermentation system; thus, building up the toolset needed to metabolize lignocellulosic biomass [67]. While achieving high theoretical yields, the group is now seeking to improve cofactor turnover, extend peak productivities beyond approximately 8 h, create low-cost cofactor mimics, and reduce catalyst costs to commercialize the technology [22].

Cell-free synthetic systems have also provided a platform to merge the best of synthetic chemistry and synthetic biology to produce hybrid molecules. For example, enzymes can be used to selectively make chiral bonds that would have otherwise required repeated isolation and purification of intermediates by a standard organic synthesis route. This is particularly true in the pharmaceutical industry, where biocatalysis has been shown to enable efficient and economical processes for manufacture of chiral drugs. An illustrative example is the synthesis of stigaliptin, which is traditionally produced by asymmetric hydrogenation of an enamine at high pressure (250 psi) using a rhodium-based chiral catalyst [68]. Among other factors, the stereoselectivity limits the chemistry that can be achieved. Savile *et al.* engineered a transaminase scaffold through directed evolution to develop a biocatalyst with >99.95% preference for the E,E isomer of stigaliptin, a 19% reduction in total waste, and a decrease in manufacturing costs overall. In another example, Bechtold *et al.* were able to produce a precursor to the diabetes treatment, tesaglitazar, with a 100-fold increase over what has been seen *in vivo* [69]. Also, another case shows the production of D-fagomine from glycerol and aldehyde [70].

In summary, purified enzyme systems and synthetic enzymatic pathways are opening up new opportunities for high-yielding bioconversions. It is clear that cell-free synthetic systems provide benefits for validating biosynthetic pathways, building complex metabolic pathways, and expanding the chemical repertoire. While purified enzyme cascades exhibit high theoretical conversions and yields, the regulation of cofactors and stability of enzymes during reaction performance can drastically affect the efficiency of the cell-free system and is currently a limitation, which we discuss later.

4.3.2
Crude Cell Lysate Systems

Crude cell lysates offer an alternative approach to purifying individual enzymes to create a synthetic reaction network. A key difference is that native enzymes in the lysate can enable ATP and cofactor regeneration [13, 71]. While this provides a benefit when compared to purified enzyme systems, the cost adds to the complexity. The emergence of crude extract-based CFME is a new development only in recent years.

One of the first complex pathways activated in cell-free crude lysate systems was the conversion of glucose to DHAP, a key building block from the glycolysis carried out in *Escherichia coli* extracts (Figure 4.2c) [61]. The 10-step pathway was activated *in vitro* utilizing natural cell metabolism, an advantage of a crude lysate system, with the triose-phosphate isomerase enzyme that utilizes DHAP being removed. In other words, native enzymes present in the extract (and endogenous to *E. coli*) converted glucose to DHAP. Cofactors NAD^+ and ATP were added to the direct flux through to DHAP, and similarly AMP nucleosidase was knocked out to inhibit DHAP production from ATP itself (independent of glucose). The authors showed how to modulate cofactors to direct flux in cell-free systems and also how to adapt an *in vivo* system (i.e., endogenous metabolism combined with gene knockouts) to direct an *in vitro* reaction, a step which is of critical importance. However, this system was limited in product yield as the by-product lactate was generated in order to provide a sink for the generated NADH.

While the overall yields were limited because of off-pathway metabolite synthesis, the system subsequently provided the basis for a real-time metabolic analysis of the cell-free reaction through mass spectrometry [72]. A continuously stirred membrane reactor, which contained a membrane to sequester proteins while allowing small-molecule exchange, was used and real-time analysis was carried out directly by ESI-MS at 8 s increments. This elegant and groundbreaking method allowed for the creation of a blueprint for enzyme regulation needed to optimize output of DHAP as well as to determine the limiting reaction in the metabolic pathway. Furthermore, a 2.5-fold increase in output was achieved once it was identified that the ATP-consuming enzymes (HK and PFK) and the enzymes they purified and added to the system, needed close control. While this report provides highly quantitative and time-resolved capabilities to control cell-free reactions for pathway analysis, the precise genomic control needed to upregulate HK and PFK to appropriate levels was not demonstrated.

Beyond understanding pathway operation in a cell-like extract environment, crude lysate systems have also been exploited to create other building blocks. For example, the synthesis of triketone precursors, essential for some antibiotics, have been pursued in cell-free systems [73]. In *E. coli* crude lysates, enzymes from *Bacillus subtilis* were overexpressed to achieve triketide lactone production. While the scientists briefly use this system to characterize enzymes in the pathway, this is a good example of how different properties such as chirality can be achieved through manipulation of enzyme activities. However, this system does not take advantage of any of the benefits of crude lysate and therefore could be performed in a purified system as well. This still remains an important finding in the growing body of CFME literature.

Another key pathway that has been built in crude lysates is the production of acetyl-CoA from CO_2 and H_2 via 3-hydroxypropionate/4-hydroxybutyrate in *Pyrococcus furiosus* extracts [74]. The proposed pathway involves utilizing hyperthermophilic enzymes that can metabolize CO_2 for carbon incorporation and H_2 for redox requirements in a cycle of 13 enzymes. In each turn of the cycle, 2 CO_2 and 1 acetyl-CoA yields 2 acetyl-CoA. Moreover, this method of carbon fixation is more carbon efficient than the Calvin cycle. To ensure flux is driven through the cycle, the authors expended efforts to characterize the first three enzymes describing the reactions from Acetyl-CoA to 3-HP. The growth temperature of *P. furianosis* is 100 °C with high activity at 70 °C; production at this temperature inactivates non-heterologous enzymes, yet the yields are relatively low (up to 60 mg l^{-1} 3-HP) and the cofactor regeneration system (NADPH and H_2) is still expensive.

In summary, lysate-based cell-free systems have so far seen much fewer examples than purified systems, yet the field is growing, mainly because of the potential benefits for cofactor regeneration from endogenous enzymes in the lysate. Looking forward, extracts from different organisms could be combined to create a synergistic cell-free environment for biochemical production.

4.3.3 Variations of Cell-Free Systems

As an alternative to purified enzymes, as well as crude cell lysates, hybrid cell-free systems that straddle the divide between the living and nonliving also exist. For example, permeabilized cells have been explored as partial cell-free systems. This method fixes and perforates the intact membrane to maintain a reaction environment more similar to *in vivo* conditions. While there are many methods for permeabilization of the cell membrane, most methods work by weakening the forces that hold the membrane together allowing for low-weight molecules to diffuse into the cell but large biopolymers remain [75]. These methods have been used to produce secondary metabolites. Flaviolin was produced in permeabilized *E. coli* cells with the benefit of simple reaction conditions and buffers [76]. Permeabilized cells offer benefits of the cellular environment and *in vitro* control. Also, the reusability of permeabilized cells shows that higher yields can be sustained in these systems; reactions converting fumarate to malate were repeated six times using the same cells showing 60 mol% yields [77]. This hybrid approach, while not the best for every biosynthetic pathway, offers an interesting method worthy of further research [75].

Fuel cells are another variation on the cell-free system. The idea here is to use microbial cells to produce electricity by oxidizing metabolites in the feed source; the electricity can be directly used or the reducing power is utilized to power further reactions. In theory, this could virtually eliminate energy challenges in cell-free systems allowing for the production of high-energy molecules at low cost. However, finding the optimal feed is a challenge. Dihydrogen and dioxygen as a feed source has been used for its efficiency, but dihydrogen can be costly; methanol is one alternative that has been tried because it is more readily available [78]. Choosing the right biocatalyst to drive the fuel cell is the next key challenge. Multienzymatic cascades that occur naturally result in high energy densities that are beneficial to these systems; using Kreb's cycle enzymes produces a 4.6-fold power density increase over using an intact mitochondria in a pyruvate/air fuel cell [79]. Furthermore, a higher energy density ($\sim$24 μW cm^{-2}) pyruvate/air fuel cell was made utilizing the Kreb's cycle enzymes cross-linked together [80]. However, enzyme stability is a concern in sustaining fuel cells. One of the largest advances in cell-free fuel cells has been the switch to direct electron transfer and immobilization of enzymes for fuel cell stability [20, 81]. Also, thermostability of enzymes and complete oxidation of glucose are areas that have been addressed to increase stability and maximize energy density [82–84]. In one particular example, Zhang and colleagues showed that a biosynthetic, 13-enzyme pathway in a fuel cell could produce nearly 24 electrons per glucose in maltodextrin [43]. Challenges regarding stability and energy storage are still not well characterized to produce capable fuel cells, but the ease of engineering cell-free systems over whole cells provides potential to create high-activity biofuel cells. Hybrid fuel cells that combine the best components of cell-free systems with enzyme engineering could address these challenges [20].

4.4 Challenges and Opportunities in Cell-Free Systems

With many lab-scale successes and demonstration of high theoretical yields, cell-free technologies are already showing tremendous value for pathway construction and prototyping for metabolic engineering. However, there are still many challenges to commercializing cell-free synthetic systems for biomanufacturing. Catalyst and cofactor costs remain high relative to *in vivo* counterparts. Moreover, catalyst stability is an open question particularly for synthetic enzyme pathways. In addition, the scalability on *in vitro* systems remains a concern. In recent years, much activity in the field has begun to address these challenges, which we describe below.

4.4.1
Purification

While crude lysate systems do not require individual enzyme purification [37], synthetic enzymatic pathways do. As a result, a key challenge facing those using purified enzymes is facile strategies to produce large numbers of enzymes at a low cost. The most common robust strategies used leverage affinity tags and heat purification of thermostable enzymes.

Modifying proteins can make addressing purification challenges easier and in turn controlling a cell-free system easier. Dating back to the 1980s, chemical purification such as the use of ammonium sulfate for precipitation has been used as a common technique to purify proteins and has been optimized in recent years for cell-free systems [84, 85]. However, the most common strategy by far for purifying proteins today is by fusing the desired protein sequence with unique peptide sequence handles that can serve as an affinity tag (e.g., FLAG-, HIS_6-, etc.). This can either be done on a plasmid, or as more recently shown by Wang *et al.*, one can use genome engineering strategies to introduce affinity tags onto the genome for purification of many enzymes at once [86]. As an added benefit, protein tags can also be used to scaffold enzymes together. Scaffolding involves binding of tagged proteins to a centralized location for temporary immobilization that can be used for purification [87]. These methods by themselves or in combination are powerful and make purification of proteins of multi-step enzymatic pathways much less daunting.

Alternatively, the use of thermostability to purify enzymes has been used to speed up the purification process. The key idea is to use enzymes from thermophilic organisms to create thermotolerant biocatalytic modules. This is done by expressing these enzymes in *E. coli* followed by heat denaturation of all endogenous enzymes (one-step preparation). While thermophilic enzymes have been purified with this method for many years, long pathways have not been achieved until recently. Using six thermophilic enzymes, scientists were able to produce 2-deoxyriobse 5-phosphate, a nucleoside intermediate, from fructose [88]. Moreover, by combining traditional ammonium sulfate purification with thermostable enzymes, optimal purification that takes advantage of both ionic strength and temperature to increase purity can be achieved. In another example, four enzymes in the glucose-3-phosphate to glucose-6-phosphate pathway were studied using heat purification for 20 min then a gradient of ammonium sulfate precipitation [84]. The use of heat to modulate the system has simplified and accelerated the process of purification.

Thermostable enzymes are not always best suited for desired industrial enzymatic activities and specific activities. A key challenge for synthetic enzymatic pathways, for example, is ensuring that optimal enzyme activity for all enzymes is in the same temperature window to avoid kinetic limitations. Moreover, product inhibition has been observed. To address these issues, thermostable enzymes have been engineered. A prime example includes the characterization of two enzymes, glyceraldehyde dehydrogenase and alcohol dehydrogenase, which are thermostable [89, 90]. Sieber and colleagues used directed evolution to create a thermostable glyceraldehyde dehydrogenase and alcohol dehydrogenase that were more soluble, more tolerant to substrate and products, and utilized cofactors more efficiently. These enzymes were heat purified and isolated through chromatography, providing evidence that heat purification does not have to limit the activity, stability, or strength of the enzyme.

Unfortunately, heat purification alone removes the ability of cell-free systems to use native metabolism as in a cell lysate system. To address this, chimeric pathways for natural metabolism with thermostable enzymes could be created. Ohtake and colleagues have created a thermostable reconstitution of glycolysis to make stoichiometric amounts of lactate and malate from glucose [58, 91]. As seen here, a modular approach to engineering cell-free systems has created many tools to adapt *in vivo* attributes and allowed us to push the boundaries of the CFME landscape.

An example of using heat purification to develop long enzymatic pathways is the production of *n*-butanol from glucose [57]. Here, the Ohtake lab used 16 thermostable, oxygen-insensitive enzymes. Heat purification allowed them to quickly isolate their biosynthetic pathways for production of *n*-butanol. However, stability of cofactors when using the high reaction temperatures necessary for enzyme activity is a trade-off for simple purification. Higher temperatures (>50 °C) are known to degrade typical biological cofactors, which can cause issues with the use of cofactors in metabolic pathways. Some have looked at enzyme evolution to increase an enzyme's ability to use cofactors [59]. In this study, the Ohtake lab evolved malate enzymes, glyceraldehyde-3-phosphate dehydrogenase (GAPDH), and an ATP-generating phosphoglycerate kinase (PGK), to use NADH instead of NADPH. This was beneficial because NADPH is more thermosensitive compared to NADH. This further shows the power to tune enzymes within an *in vitro* system.

While there are many ways that purification strategies have developed over the last few years, there is still a lot to be done in this area to increase enzyme stability while using relatively simple and cheap purification methods.

4.4.2 Spatial Organization

Cell-free systems have adapted many *in vivo* properties such as organelle-like organization and molecular crowding effects to have the benefits of *in vitro* and *in vivo* synthesis [92]. An early example of this is the conversion of CO_2 to methanol in a three-enzyme pathway [93]. In this study, Obert and Dave show that enzyme stability could be improved by running their cell-free system in an alginate–silica matrix that mimics the viscous nature of an *in vivo* cell. By achieving cellular properties in the cell-free environment, they were able to enhance production. Furthermore, compartmentalization has been utilized *in vitro* to take advantage of native benefits of localization to minimize diffusion limitations for both protein and chemical production [92]. Many *in vivo* processes utilize compartmentalization and there is a large focus in synthetic biology on engineering new ways of using this [94]. Localization of proteins during protein production and compartmentalization are two ways cellular systems take advantage of the spatial organization of biological pathways.

In cell-free systems, spatially organizing metabolic pathways can increase productivities of biosynthetic pathways by potentially sequestering pathway intermediates for diffusion-limited processes by creating high local concentrations of substrates or stabilizing enzymes. By tethering metabolic enzymes together either by fusion proteins [95] or DNA-scaffolds [96], scientists have shown the ability to increase flux through desired metabolic paths; this type of multienzymatic systems are further discussed in another review [97]. In an alternative DNA-scaffold approach, enzymes are colocalized by the addition of site-specific zinc-finger binding domains to pathway enzymes of interest. By doing so, the enzymes can be stabilized to be active for longer periods of time or also minimize diffusion limitations that reduce product formation if this is an issue [22, 98, 99]. In a third example, scaffolding-dockerin modules have been used, mimicking *in vivo*, cellulosomes. For example, Zhang used a cellulosome scaffold for feedstock degradation, offering both colocalization and confinement in space of enzymatic pathways for a specific design objective, improving activity from two- to several-fold higher than noncomplexed enzymes [100]. Broadly, the technologies that address spatial organization and colocalization in cell-free systems are expanding rapidly. This provides many avenues to pursue pathways that are diffusion limited, pathways dependent upon a large number of reactions, parallel reaction cascades, or reactions in which the pathway enzymes can be stabilized by localization.

4.4.3
Cell-Free System Stability

Cofactor stability remains a central challenge for CFME. Cell-free crude-cell lysates partially address this concern by mimicking the environment of the intact cell, in one case having achieved more than 1000 turnover events of NAD^+ (which was synthesized by cells and was not added exogenously) [62]. However, there are still relatively few examples of high-level cofactor regeneration (>100 turnover events) for *in vitro* bioconversions. Without cofactor regeneration, cell-free technologies become increasingly expensive due to costly additions of new cofactors. Thus, improving turnover numbers or engineering enzymes to utilize alternate cofactors is a critical hurdle for cost-effective cell-free efforts.

Cofactors, typically electron carriers and donors, are required for almost every biosynthetic pathway that utilizes energy. Cells regenerate cofactors and energy as part of the complex regulation of homeostasis, but it is challenging (and not necessarily desirable) to reconstitute this system in cell-free systems. One approach is to modulate the amount of additional enzymes that regenerate cofactors to increase metabolic activity [101]. By introducing NADH oxidase, NADH dehydrogenase, and GPD1, the authors were able to improve the amount of NAD^+ regeneration from excess NADH. While this was performed *in vivo*, the prospect of this being done cell-free is good. A recent study developed a robust, molecular valve to balance the production and consumption of NADPH and NADH *in vitro* [102]. Earlier it was discussed that engineering enzymes to utilize different cofactors could increase thermostability [59]. Inherent in this is the idea that enzymes can change the cofactor that is used and ultimately change the cofactor constraints as a whole. By doing this one can adjust pathway flux for higher productivities *in vitro*. Indeed, there are a variety of works centered on exploring the ability to engineer biomimetic cofactors, which is described in several recent reviews [22, 103, 104].

Introducing regeneration cycles can increase the longevity of cofactors. Alternatively, using other forms of energy can improve pathway flux. By using light as an energy source, it is possible to induce NADPH regeneration. This is achieved by converting light energy through enzymes that (i) take in light, (ii) transfer electrons, and (iii) mediate electron use. One illustrative example of this type of system uses eosin Y as a photosensitizing dye, triethanolamine as an electron donor, and [Cp*Rh(bpy)H_2O] as an electron mediator [105]. Here the system was used to improve cytochrome P450 activity. A newer technology involves cofactor regeneration by directing cofactors toward enzymes that regenerate them via improved mixing [106]. This form of cofactor organization utilizes magnetic nanoparticles to create a Brownian motion effect and in turn a denser cell-free environment. The magnetic particles allow for cofactors, after their use, to stay near enzymes. Enzymes colocalized to the enzymes that used the cofactors regenerate these cofactors at a faster rate than enzymes not localized in this fashion. While the breadth of technologies that have been developed to stabilize cofactor use, there is a lot of work to be done in cofactor engineering to increase reaction lengths and final product titers.

4.4.4 Modeling

With the development of technologies for purification, spatial organization, and cofactor regeneration comes the need for more powerful computational tools. Computational approaches in evaluating metabolic engineering efforts have often benefited the design of *in vivo* systems. The immediate challenge in adapting *in silico* methods to cell-free systems requires movement away from a cell growth-focused approach. Despite this challenge, new computational techniques have been developed to model cell-free systems [48]. In this study, Bujara and Panke develop network topology analysis in cell-free systems based on genome-scale metabolic models (similar to flux analyses performed for *in vivo* systems). These methods span the production of proteins, the kinetics of biosynthetic pathways, and non-growth-related analysis of reactions. Optimization of cell extracts and inputs is sometimes the limiting factor in efficient cell-free reactions. An example of a cell-free optimization approach is the *in silico* evaluation of a complex multienzymatic system producing H_2 from cellobiose [21]. In this study, Ardao and Zeng were able to model improved productivities from approximately 2 to

355 mmol l^{-1} h^{-1}. This involved multiobjective optimizations to understand how differences in the cell-free environment interact with multiple enzymes together. Moreover, a typical modeling approach with metabolic engineering is to map metabolic pathways and their respective kinetics. Computer design of metabolic pathways *in vivo* and *in vitro* are improving production through the use of a variety of new tools that include "Metabolic Tinker" for breadboarding pathways [107]. Computational tools to design DNA for pathway enzymes have also been developed [108]. Additional effort to discover, predict, and design novel conversion pathways is embodied by programs such as (BNICE) [109, 110], PathPred [111], UM-PPS [112], and Retropath [113]. The development of new tools unique for advancing CFME will grow as our understanding of the cell-free environment is better understood.

4.5 Recent Advances

Cell-free systems for metabolic engineering expand our ability to study metabolism and access biochemicals. There have been many recent advances in purified enzyme and crude lysate systems. Purified enzyme systems recently explored the production of 1,3-propanediol, hydrogen, and farnesene. The Zeng lab produced 1,3-propanediol from glycerol showing that cell-free systems can avoid loss of product to by-product pathways [114]. The Zhang lab has continued to develop the cell-free hydrogen production platform to use additional sugars, such as sucrose, and to coutilize sugars [115, 116]. Additionally, Zhu and colleagues produced farnesene from acetyl-CoA with nine purified enzymes utilizing NADPH and ATP [117]. The purified enzyme system was used to vary and test enzymes to improve *in vivo* titers. These works demonstrate that purified enzyme systems continue to expand the cell-free palate of biochemicals and perform robust cell-free analyses of metabolic pathways.

Crude lysate systems have expanded both in the realm of metabolic engineering and CFPS metabolism. Ninh and colleagues explored one-pot conversion of glucose to pyruvate and lactate using crude *E. coli* extracts expressing over nine thermotolerant glycolytic enzymes [118]. The Park lab has continued to develop yeast cell-free bioethanol production by incorporating encapsulation of cell-free systems as well as testing different simultaneous saccharification and fermentation processes across varying pH, temperature, and physiochemical environments [119, 120]. Additionally, Kay and Jewett expressed a full heterologous pathway to 2,3-butanediol (2,3-BDO) [121]. Here, a single strain of *E. coli* was engineered to express three pathway enzymes necessary to make 2,3-BDO. With no strain optimization, Kay and Jewett observed a maximal synthesis rate of 2,3-BD of approximately 10 g l^{-1} h^{-1}, titer of approximately 80 g l^{-1}, and theoretical yield of 71%. This work led to the development of an extract-mixing approach to CFME where individual lysates containing individually overexpressed heterologous enzymes are mixed together to construct a biosynthetic pathway activated by the addition of simple substrates [124] and use CFPS to drive metabolic conversions [125]. Furthermore, energy metabolism in crude lysates has been shown to improve CFPS. Caschera and Noireaux activated a polysugar metabolism to efficiently fuel CFPS improving on the state-of-the-art CFPS platform [122]. Anderson *et al.* also explored a different energy source for CFPS [123]. This work showed that glucose could be used to fuel yeast CFPS. Crude lysate systems could transform the way we prototype biosynthesis.

4.6
Summary

Cell-free systems can be composed of purified enzymes, crude-cell lysates, or variations thereof. They provide a complement to traditional cellular systems used in biotechnology. By removing the constraints of cellular growth and decoupling catalyst synthesis from catalyst utilization, biosynthetic pathways can direct the entirety of a given substrate toward the production of a single product. This potentially provides interesting advantages for prototyping pathways and biomanufacturing. While large-scale production of multistep enzymatic systems have yet to be demonstrated commercially, the cell-free approach is already being used for biomolecular breadboarding and pathway debugging. This is clearly evident from the examples provided in the text and described in Table 4.2. Such advances show the potential power of CFME. That said, CFME is still in its early stages, and there are several key challenges to be addressed in the areas of purification, cofactor regeneration systems, and reconstitution of beneficial cellular mechanisms such as molecular crowding and compartmentalization. Underlying these needs is the overarching goal to lower catalyst and cofactor costs. Looking forward, recent advances in CFME suggest new opportunities. For example, cell-free systems may provide advantages for making products that cannot be made *in vivo* owing to toxicity or product purification limitations. They might also offer exciting new directions in the synthesis of hybrid biochemicals comprised of parts derived from organic syntheses and parts derived from biological syntheses. The ability to not only use enzymes from multiple organisms but unique metabolisms from across the phylogenic spectra are also on the horizon. In tandem with biomanufacturing prospects, cell-free systems show potential in prototyping biosynthetic pathways to guide enzyme discovery and *in vivo* production.

Acknowledgments

We gratefully acknowledge the National Science Foundation (MCB-0943393), the Office of Naval Research (N00014-11-1-0363), the DARPA YFA Program (N66001-11-1-4137), the Army Research Office (W911NF-11-1-044), the NSF Materials Network Grant (DMR-1108350), the DARPA Living Foundries Program (N66001-12-C-4211), the David and Lucile Packard Foundation (2011-37152),

ARPA-E (DE-AR0000435), and the Chicago Biomedical Consortium with support from the Searle Funds at the Chicago Community Trust for support. ASK is an NSF Graduate Fellow. QMD is funded, in part, by the Northwestern Molecular Biophysics Training Program supported by NIH via NIGMS (5T32 GM008382).

References

1. Sheldon, R.A. (2007) The E Factor: fifteen years on. *Green Chem.*, **9**, 1273.
2. Werpy, T., Petersen, G.R., Aden, A., Bozell, J.J., Holladay, J., White, J., Manheim, A., Eliot, D., Lasure, L., and Jones, S. (2004) Top Value Added Chemicals from Biomass. Volume 1 – Results of Screening for Potential Candidates from Sugars and Synthesis Gas. DTIC Document 1.
3. Bozell, J.J. and Petersen, G.R. (2010) Technology development for the production of biobased products from biorefinery carbohydrates—the US Department of Energy's "Top 10" revisited. *Green Chem.*, **12**, 539.
4. Nielsen, J., Fussenegger, M., Keasling, J., Lee, S.Y. *et al.* (2014) Engineering synergy in biotechnology. *Nat. Chem. Biol.*, **10**, 319–322.
5. Keasling, J.D. (2010) Manufacturing molecules through metabolic engineering. *Science*, **330**, 1355–1358.
6. Lee, J.W., Kim, H.U., Choi, S., Yi, J. *et al.* (2011) Microbial production of building block chemicals and polymers. *Curr. Opin. Biotechnol.*, **22**, 758–767.
7. Curran, K.A. and Alper, H.S. (2012) Expanding the chemical palate of cells by combining systems biology and metabolic engineering. *Metab. Eng.*, **14**, 289–297.
8. Nielsen, J., Larsson, C., van Maris, A., and Pronk, J. (2013) Metabolic engineering of yeast for production of fuels and chemicals. *Curr. Opin. Biotechnol.*, **24**, 398–404.
9. Galanie, S., Siddiqui, M.S., and Smolke, C.D. (2013) Molecular tools for chemical biotechnology. *Curr. Opin. Biotechnol.*, **24**, 1000–1009.
10. Keasling, J.D. (2012) Synthetic biology and the development of tools for metabolic engineering. *Metab. Eng.*, **14**, 189–195.
11. Nakamura, C.E. and Whited, G.M. (2003) Metabolic engineering for the microbial production of 1,3-propanediol. *Curr. Opin. Biotechnol.*, **14**, 454–459.
12. Renninger, N. and McPhee, D. (2008) Fuel compositions comprising farnesane and farnesane derivatives and method of making and using same. US Patent 7399323.
13. Hodgman, C.E. and Jewett, M.C. (2012) Cell-free synthetic biology: thinking outside the cell. *Metab. Eng.*, **14**, 261–269.
14. Woolston, B.M., Edgar, S., and Stephanopoulos, G. (2013) in *Annual Review of Chemical and Biomolecular Engineering*, vol. **4** (ed. J.M. Prausnitz), Annual Reviews, Palo Alto, CA, pp. 259–288.
15. Clomburg, J.M. and Gonzalez, R. (2010) Biofuel production in *Escherichia coli*: the role of metabolic engineering and synthetic biology. *Appl. Microbiol. Biotechnol.*, **86**, 419–434.
16. Kwok, R. (2010) Five hard truths for synthetic biology. *Nature*, **463**, 288–290.
17. Yadav, V.G., De Mey, M., Giaw Lim, C., Kumaran Ajikumar, P. *et al.* (2012) The future of metabolic engineering and synthetic biology: towards a systematic practice. *Metab. Eng.*, **14**, 233–241.
18. Guterl, J.-K. and Sieber, V. (2013) Biosynthesis "debugged": novel bioproduction strategies. *Eng. Life Sci.*, **13**, 4–18.
19. Swartz, J.R. (2012) Transforming biochemical engineering with cell-free biology. *AIChE J.*, **58**, 5–13.
20. Zhang, Y.H.P. (2011) What is vital (and not vital) to advance economically-competitive biofuels production. *Process Biochem.*, **46**, 2091–2110.

21. Ardao, I. and Zeng, A.-P. (2013) In silico evaluation of a complex multi-enzymatic system using one-pot and modular approaches: application to the high-yield production of hydrogen from a synthetic metabolic pathway. *Chem. Eng. Sci.*, **87**, 183–193.
22. Rollin, J.A., Tam, T.K., and Zhang, Y.H.P. (2013) New biotechnology paradigm: cell-free biosystems for biomanufacturing. *Green Chem.*, **15**, 1708.
23. Guterl, J.K., Garbe, D., Carsten, J., Steffler, F. *et al.* (2012) Cell-free metabolic engineering: production of chemicals by minimized reaction cascades. *ChemSusChem*, **5**, 2165–2172.
24. Murray, C.J. and Baliga, R. (2013) Cell-free translation of peptides and proteins: from high throughput screening to clinical production. *Curr. Opin. Chem. Biol.*, **17**, 420–426.
25. Catherine, C., Lee, K.-H., Oh, S.-J., and Kim, D.-M. (2013) Cell-free platforms for flexible expression and screening of enzymes. *Biotechnol. Adv.*, **31**, 797–803.
26. Chappell, J., Jensen, K., and Freemont, P.S. (2013) Validation of an entirely *in vitro* approach for rapid prototyping of DNA regulatory elements for synthetic biology. *Nucleic Acids Res.*, **41**, 3471–3481.
27. Shin, J. and Noireaux, V. (2012) An *E. coli* cell-free expression toolbox: application to synthetic gene circuits and artificial cells. *ACS Synth. Biol.*, **1**, 29–41.
28. Jewett, M.C., Fritz, B.R., Timmerman, L.E., and Church, G.M. (2013) *In vitro* integration of ribosomal RNA synthesis, ribosome assembly, and translation. *Mol. Syst. Biol.*, **9**, 678.
29. Goerke, A.R. and Swartz, J.R. (2009) High-level cell-free synthesis yields of proteins containing site-specific non-natural amino acids. *Biotechnol. Bioeng.*, **102**, 400–416.
30. Bundy, B.C. and Swartz, J.R. (2010) Site-specific incorporation of *p*-propargyloxyphenylalanine in a cell-free environment for direct protein–protein click conjugation. *Bioconjugate Chem.*, **21**, 255–263.
31. Albayrak, C. and Swartz, J.R. (2013) Cell-free co-production of an orthogonal transfer RNA activates efficient site-specific non-natural amino acid incorporation. *Nucleic Acids Res.*, **41**, 5949–5963.
32. Hong, S.H., Ntai, I., Haimovich, A.D., Kelleher, N.L. *et al.* (2014) Cell-free protein synthesis from a release factor 1 deficient *Escherichia coli* activates efficient and multiple site-specific nonstandard amino acid incorporation. *ACS Synth. Biol.*, **3**, 398–409.
33. Ugwumba, I.N., Ozawa, K., de la Cruz, L., Xu, Z.-Q. *et al.* (2011) Using a genetically encoded fluorescent amino acid as a site-specific probe to detect binding of low-molecular-weight compounds. *Assay Drug Dev. Technol.*, **9**, 50–57.
34. Ugwumba, I.N., Ozawa, K., Xu, Z.-Q., Ely, F. *et al.* (2010) Improving a natural enzyme activity through incorporation of unnatural amino acids. *J. Am. Chem. Soc.*, **133**, 326–333.
35. Mukai, T., Yanagisawa, T., Ohtake, K., Wakamori, M. *et al.* (2011) Genetic-code evolution for protein synthesis with non-natural amino acids. *Biochem. Biophys. Res. Commun.*, **411**, 757–761.
36. Loscha, K.V., Herlt, A.J., Qi, R., Huber, T. *et al.* (2012) Multiple-site labeling of proteins with unnatural amino acids. *Angew. Chem. Int. Ed.*, **51**, 2243–2246.
37. Swartz, J. (2006) Developing cell-free biology for industrial applications. *J. Ind. Microbiol. Biotechnol.*, **33**, 476–485.
38. Rupp, S. (2013) Next-generation bioproduction systems: cell-free conversion concepts for industrial biotechnology. *Eng. Life Sci.*, **13**, 19–25.
39. Billerbeck, S., Harle, J., and Panke, S. (2013) The good of two worlds: increasing complexity in cell-free systems. *Curr. Opin. Biotechnol.*, **24**, 1037–1043.
40. Kohler, R.E. (1972) The reception of Eduard Buchner's discovery of cell-free fermentation. *J. Hist. Biol.*, **5**, 327–353.
41. Welch, P. and Scopes, R.K. (1985) Studies on cell-free metabolism: ethanol production by a yeast glycolytic system

reconstituted from purified enzymes. *J. Biotechnol.*, **2**, 257–273.

42. Nirenberg, M.W. and Matthaei, J.H. (1961) The dependence of cell-free protein synthesis in *E. coli* upon naturally occurring or synthetic polyribonucleotides. *Proc. Natl. Acad. Sci. U.S.A.*, **47**, 1588–1602.
43. Zhu, Z., Kin Tam, T., Sun, F., You, C. *et al.* (2014) A high-energy-density sugar biobattery based on a synthetic enzymatic pathway. *Nat. Commun.*, **5**, 3026.
44. Zaks, A. (2001) Industrial biocatalysis. *Curr. Opin. Chem. Biol.*, **5**, 130–136.
45. Bogorad, I.W., Lin, T.S., and Liao, J.C. (2013) Synthetic non-oxidative glycolysis enables complete carbon conservation. *Nature*, **502**, 693–697.
46. Jensen, V.J. and Rugh, S. (1987) [33] Industrial-scale production and application of immobilized glucose isomerase. *Methods Enzymol.*, **136**, 356–370.
47. Bruggink, A., Roos, E.C., and de Vroom, E. (1998) Penicillin acylase in the industrial production of β-lactam antibiotics. *Org. Process Res. Dev.*, **2**, 128–133.
48. Bujara, M. and Panke, S. (2012) In silico assessment of cell-free systems. *Biotechnol. Bioeng.*, **109**, 2620–2629.
49. Rosenthaler, L. (1908) Durch enzyme bewirkte asymmetrische synthesen. *Biochem. Z.*, **14**, 238–253.
50. Swartz, J.R. (2011) Transforming biochemical engineering with cell-free biology. *AIChE J.*, **58**, 5–13.
51. Jewett, M.C., Calhoun, K.A., Voloshin, A., Wuu, J.J. *et al.* (2008) An integrated cell-free metabolic platform for protein production and synthetic biology. *Mol. Syst. Biol.*, **4**, 220.
52. Chen, X., Zhang, C., Zou, R., Zhou, K. *et al.* (2013) Statistical experimental design guided optimization of a one-pot biphasic multienzyme total synthesis of amorpha-4,11-diene. *PLoS One*, **8**, e79650.
53. Zhang, Y.H., Evans, B.R., Mielenz, J.R., Hopkins, R.C. *et al.* (2007) High-yield hydrogen production from starch and water by a synthetic enzymatic pathway. *PLoS One*, **2**, e456.
54. Ye, X., Wang, Y., Hopkins, R.C., Adams, M.W. *et al.* (2009) Spontaneous high-yield production of hydrogen from cellulosic materials and water catalyzed by enzyme cocktails. *ChemSusChem*, **2**, 149–152.
55. Martin del Campo, J.S., Rollin, J., Myung, S., Chun, Y. *et al.* (2013) High-yield production of dihydrogen from xylose by using a synthetic enzyme cascade in a cell-free system. *Angew. Chem. Int. Ed.*, **52**, 4587–4590.
56. Wang, Y., Huang, W., Sathitsuksanoh, N., Zhu, Z. *et al.* (2011) Biohydrogenation from biomass sugar mediated by *in vitro* synthetic enzymatic pathways. *Chem. Biol.*, **18**, 372–380.
57. Krutsakorn, B., Honda, K., Ye, X., Imagawa, T. *et al.* (2013) *In vitro* production of *n*-butanol from glucose. *Metab. Eng.*, **20**, 84–91.
58. Ye, X., Honda, K., Sakai, T., Okano, K. *et al.* (2012) Synthetic metabolic engineering-a novel, simple technology for designing a chimeric metabolic pathway. *Microb. Cell Fact.*, **11**, 120.
59. Morimoto, Y., Honda, K., Ye, X., Okano, K. *et al.* (2013) Directed evolution of thermotolerant malic enzyme for improved malate production. *J. Biosci. Bioeng.*, **117**, 147–152.
60. Korman, T.P., Sahachartsiri, B., Li, D., Vinokur, J.M. *et al.* (2014) A synthetic biochemistry system for the *in vitro* production of isoprene from glycolysis intermediates. *Protein Sci.*, **23**, 576–585.
61. Bujara, M., Schumperli, M., Billerbeck, S., Heinemann, M. *et al.* (2010) Exploiting cell-free systems: implementation and debugging of a system of biotransformations. *Biotechnol. Bioeng.*, **106**, 376–389.
62. Khattak, W.A., Ul-Islam, M., Ullah, M.W., Yu, B. *et al.* (2014) Yeast cell-free enzyme system for bio-ethanol production at elevated temperatures. *Process Biochem.*, **49**, 357–364.
63. Fessner, W.-D. and Walter, C. (1992) "Artificial metabolisms" for the asymmetric one-pot synthesis of branched-chain saccharides. *Angew. Chem. Int. Ed. Engl.*, **31**, 614–616.

64. Siegal-Gaskins, D., Tuza, Z.A., Kim, J., Noireaux, V. *et al.* (2014) Gene circuit performance characterization and resource usage in a cell-free "breadboard". *ACS Synth. Biol.*, **3**, 416–425.
65. Sun, Z.Z., Yeung, E., Hayes, C.A., Noireaux, V. *et al.* (2014) Linear DNA for rapid prototyping of synthetic biological circuits in an *Escherichia coli* based TX-TL cell-free system. *ACS Synth. Biol.*, **3**, 387–397.
66. Zhang, Y.H.P. (2011) Simpler is better: high-yield and potential low-cost biofuels production through cell-free synthetic pathway biotransformation (SyPaB). *ACS Catal.*, **1**, 998–1009.
67. You, C., Chen, H., Myung, S., Sathitsuksanoh, N. *et al.* (2013) Enzymatic transformation of nonfood biomass to starch. *Proc. Natl. Acad. Sci. U.S.A.*, **110**, 7182–7187.
68. Savile, C.K., Janey, J.M., Mundorff, E.C., Moore, J.C. *et al.* (2010) Biocatalytic asymmetric synthesis of chiral amines from ketones applied to sitagliptin manufacture. *Science*, **329**, 305–309.
69. Bechtold, M., Brenna, E., Femmer, C., Gatti, F.G. *et al.* (2012) Biotechnological development of a practical synthesis of ethyl (*S*)-2-ethoxy-3-(*p*-methoxyphenyl)propanoate (EEHP): over 100-fold productivity increase from yeast whole cells to recombinant isolated enzymes. *Org. Process Res. Dev.*, **16**, 269–276.
70. Babich, L., van Hemert, L.J.C., Bury, A., Hartog, A.F. *et al.* (2011) Synthesis of non-natural carbohydrates from glycerol and aldehydes in a one-pot four-enzyme cascade reaction. *Green Chem.*, **13**, 2895.
71. Carlson, E.D., Gan, R., Hodgman, C.E., and Jewett, M.C. (2012) Cell-free protein synthesis: applications come of age. *Biotechnol. Adv.*, **30**, 1185–1194.
72. Bujara, M., Schumperli, M., Pellaux, R., Heinemann, M. *et al.* (2011) Optimization of a blueprint for *in vitro* glycolysis by metabolic real-time analysis. *Nat. Chem. Biol.*, **7**, 271–277.
73. Harper, A.D., Bailey, C.B., Edwards, A.D., Detelich, J.F. *et al.* (2012) Preparative, *in vitro* biocatalysis of triketide lactone chiral building blocks. *ChemBioChem*, **13**, 2200–2203.
74. Keller, M.W., Schut, G.J., Lipscomb, G.L., Menon, A.L. *et al.* (2013) Exploiting microbial hyperthermophilicity to produce an industrial chemical, using hydrogen and carbon dioxide. *Proc. Natl. Acad. Sci. U.S.A.*, **110**, 5840–5845.
75. Krauser, S., Weyler, C., Blass, L.K., and Heinzle, E. (2013) Directed multistep biocatalysis using tailored permeabilized cells. *Adv. Biochem. Eng./Biotechnol.*, **137**, 185–234.
76. Krauser, S., Kiefer, P., and Heinzle, E. (2012) Multienzyme whole-cell in situ biocatalysis for the production of flaviolin in permeabilized cells of *Escherichia coli*. *ChemCatChem*, **4**, 786–788.
77. Ninh, P.H., Honda, K., Yokohigashi, Y., Okano, K. *et al.* (2013) Development of a continuous bioconversion system using a thermophilic whole-cell biocatalyst. *Appl. Environ. Microbiol.*, **79**, 1996–2001.
78. Palmore, G.T.R., Bertschy, H., Bergens, S.H., and Whitesides, G.M. (1998) A methanol/dioxygen biofuel cell that uses NAD^+–dependent dehydrogenases as catalysts: application of an electro-enzymatic method to regenerate nicotinamide adenine dinucleotide at low overpotentials. *J. Electroanal. Chem.*, **443**, 155–161.
79. Sokic-Lazic, D. and Minteer, S.D. (2009) Pyruvate/Air enzymatic biofuel cell capable of complete oxidation. *Electrochem. Solid-State Lett.*, **12**, F26.
80. Moehlenbrock, M.J., Toby, T.K., Waheed, A., and Minteer, S.D. (2010) Metabolon catalyzed pyruvate/air biofuel cell. *J. Am. Chem. Soc.*, **132**, 6288–6289.
81. Minteer, S.D., Liaw, B.Y., and Cooney, M.J. (2007) Enzyme-based biofuel cells. *Curr. Opin. Biotechnol.*, **18**, 228–234.
82. Zhu, Z., Sun, F., Zhang, X., and Zhang, Y.H. (2012) Deep oxidation of glucose in enzymatic fuel cells through a synthetic enzymatic pathway containing a cascade of two thermostable dehydrogenases. *Biosens. Bioelectron.*, **36**, 110–115.

83. Xu, S. and Minteer, S.D. (2012) Enzymatic biofuel cell for oxidation of glucose to CO_2. *ACS Catal.*, **2**, 91–94.
84. Myung, S. and Zhang, Y.H. (2013) Non-complexed four cascade enzyme mixture: simple purification and synergetic co-stabilization. *PLoS One*, **8**, e61500.
85. Yang, W.C., Patel, K.G., Wong, H.E., and Swartz, J.R. (2012) Simplifying and streamlining *Escherichia coli*-based cell-free protein synthesis. *Biotechnol. Progr.*, **28**, 413–420.
86. Wang, H.H., Huang, P.Y., Xu, G., Haas, W. *et al.* (2012) Multiplexed *in vivo* His-tagging of enzyme pathways for *in vitro* single-pot multienzyme catalysis. *ACS Synth. Biol.*, **1**, 43–52.
87. Conrado, R.J., Wu, G.C., Boock, J.T., Xu, H. *et al.* (2012) DNA-guided assembly of biosynthetic pathways promotes improved catalytic efficiency. *Nucleic Acids Res.*, **40**, 1879–1889.
88. Honda, K., Maya, S., Omasa, T., Hirota, R. *et al.* (2010) Production of 2-deoxyribose 5-phosphate from fructose to demonstrate a potential of artificial bio-synthetic pathway using thermophilic enzymes. *J. Biotechnol.*, **148**, 204–207.
89. Steffler, F. and Sieber, V. (2013) Refolding of a thermostable glyceraldehyde dehydrogenase for application in synthetic cascade biomanufacturing. *PLoS One*, **8**, e70592.
90. Steffler, F., Guterl, J.K., and Sieber, V. (2013) Improvement of thermostable aldehyde dehydrogenase by directed evolution for application in synthetic cascade biomanufacturing. *Enzyme Microb. Technol.*, **53**, 307–314.
91. Ye, X., Honda, K., Morimoto, Y., Okano, K. *et al.* (2013) Direct conversion of glucose to malate by synthetic metabolic engineering. *J. Biotechnol.*, **164**, 34–40.
92. Jandt, U., You, C., Zhang, Y.H., and Zeng, A.P. (2013) Compartmentalization and metabolic channeling for multienzymatic biosynthesis: practical strategies and modeling approaches. *Adv. Biochem. Eng./Biotechnol.*, **137**, 41–65.
93. Obert, R. and Dave, B.C. (1999) Enzymatic conversion of carbon dioxide to methanol: enhanced methanol production in silica sol–gel matrices. *J. Am. Chem. Soc.*, **121**, 12192–12193.
94. Chen, A.H. and Silver, P.A. (2012) Designing biological compartmentalization. *Trends Cell Biol.*, **22**, 662–670.
95. Iturrate, L., Sanchez-Moreno, I., Doyaguez, E.G., and Garcia-Junceda, E. (2009) Substrate channelling in an engineered bifunctional aldolase/kinase enzyme confers catalytic advantage for C–C bond formation. *Chem. Commun. (Camb.)*, 1721–1723.
96. Wilner, O.I., Weizmann, Y., Gill, R., Lioubashevski, O. *et al.* (2009) Enzyme cascades activated on topologically programmed DNA scaffolds. *Nat. Nanotechnol.*, **4**, 249–254.
97. Lopez-Gallego, F. and Schmidt-Dannert, C. (2010) Multi-enzymatic synthesis. *Curr. Opin. Chem. Biol.*, **14**, 174–183.
98. Dueber, J.E., Wu, G.C., Malmirchegini, G.R., Moon, T.S. *et al.* (2009) Synthetic protein scaffolds provide modular control over metabolic flux. *Nat. Biotechnol.*, **27**, 753–759.
99. You, C., Myung, S., and Zhang, Y.H. (2012) Facilitated substrate channeling in a self-assembled trifunctional enzyme complex. *Angew. Chem. Int. Ed.*, **51**, 8787–8790.
100. Zhang, Y.H. (2011) Substrate channeling and enzyme complexes for biotechnological applications. *Biotechnol. Adv.*, **29**, 715–725.
101. Li, Y., Su, M., Ge, X., and Tian, P. (2013) Enhanced aldehyde dehydrogenase activity by regenerating NAD^+ in *Klebsiella pneumoniae* and implications for the glycerol dissimilation pathways. *Biotechnol. Lett.*, **35**, 1609–1615.
102. Opgenorth, P.H., Korman, T.P., and Bowie, J.U. (2014) A synthetic biochemistry molecular purge valve module that maintains redox balance. *Nat. Commun.*, **5**, 4113.

103. Wang, Y., San, K.Y., and Bennett, G.N. (2013) Cofactor engineering for advancing chemical biotechnology. *Curr. Opin. Biotechnol.*, **24**, 994–999.
104. Li, Y. and Cirino, P.C. (2014) Recent advances in engineering proteins for biocatalysis. *Biotechnol. Bioeng.*, **111**, 1273–1287.
105. Lee, S.H., Kwon, Y.C., Kim, D.M., and Park, C.B. (2013) Cytochrome P450-catalyzed O-dealkylation coupled with photochemical NADPH regeneration. *Biotechnol. Bioeng.*, **110**, 383–390.
106. Zheng, M., Su, Z., Ji, X., Ma, G. *et al.* (2013) Magnetic field intensified bi-enzyme system with in situ cofactor regeneration supported by magnetic nanoparticles. *J. Biotechnol.*, **168**, 212–217.
107. McClymont, K. and Soyer, O.S. (2013) Metabolic tinker: an online tool for guiding the design of synthetic metabolic pathways. *Nucleic Acids Res.*, **41**, e113.
108. Tools BSB, https://synbiotools.bbn.com.
109. Henry, C.S., Broadbelt, L.J., and Hatzimanikatis, V. (2010) Discovery and analysis of novel metabolic pathways for the biosynthesis of industrial chemicals: 3-hydroxypropanoate. *Biotechnol. Bioeng.*, **106**, 462–473.
110. Gonzalez-Lergier, J., Broadbelt, L.J., and Hatzimanikatis, V. (2005) Theoretical considerations and computational analysis of the complexity in polyketide synthesis pathways. *J. Am. Chem. Soc.*, **127**, 9930–9938.
111. Moriya, Y., Shigemizu, D., Hattori, M., Tokimatsu, T. *et al.* (2010) PathPred: an enzyme-catalyzed metabolic pathway prediction server. *Nucleic Acids Res.*, **38**, W138–W143.
112. Gao, J., Ellis, L.B., and Wackett, L.P. (2011) The University of Minnesota Pathway Prediction System: multi-level prediction and visualization. *Nucleic Acids Res.*, **39**, W406–W411.
113. Carbonell, P., Parutto, P., Baudier, C., Junot, C. *et al.* (2013) Retropath: automated pipeline for embedded metabolic circuits. *ACS Synth. Biol.*, **3** (8), 565–577.
114. Rieckenberg, F., Ardao, I., Rujananon, R., and Zeng, A.-P. (2014) Cell-free synthesis of 1,3-propanediol from glycerol with a high yield. *Eng. Life Sci.*, **14**, 380–386.
115. Myung, S., Rollin, J., You, C., Sun, F. *et al.* (2014) *In vitro* metabolic engineering of hydrogen production at theoretical yield from sucrose. *Metab. Eng.*, **24**, 70–77.
116. Rollin, J.A., Martin del Campo, J., Myung, S., Sun, F. *et al.* (2015) High-yield hydrogen production from biomass by *in vitro* metabolic engineering: mixed sugars coutilization and kinetic modeling. *Proc. Natl. Acad. Sci. U.S.A.*, **112**, 4964–4969.
117. Zhu, F., Zhong, X., Hu, M., Lu, L. *et al.* (2014) *In vitro* reconstitution of mevalonate pathway and targeted engineering of farnesene overproduction in *Escherichia coli*. *Biotechnol. Bioeng.*, **111**, 1396–1405.
118. Ninh, P.H., Honda, K., Sakai, T., Okano, K. *et al.* (2015) Assembly and multiple gene expression of thermophilic enzymes in *Escherichia coli* for *in vitro* metabolic engineering. *Biotechnol. Bioeng.*, **112**, 189–196.
119. Ullah, M.W., Khattak, W.A., Ul-Islam, M., Khan, S. *et al.* (2014) Bio-ethanol production through simultaneous saccharification and fermentation using an encapsulated reconstituted cell-free enzyme system. *Biochem. Eng. J.*, **91**, 110–119.
120. Ullah, M.W., Khattak, W.A., Ul-Islam, M., Khan, S. *et al.* (2015) Encapsulated yeast cell-free system: a strategy for cost-effective and sustainable production of bio-ethanol in consecutive batches. *Biotechnol. Bioprocess Eng.*, **20**, 561–575.
121. Kay, J.E. and Jewett, M.C. (2015) Lysate of engineered *Escherichia coli* supports high-level conversion of glucose to 2,3-butanediol. *Metab. Eng.*, **32**, 133–142.
122. Caschera, F. and Noireaux, V. (2015) A cost-effective polyphosphate-based metabolism fuels an all *E. coli* cell-free expression system. *Metab. Eng.*, **27**, 29–37.

123. Anderson, M.J., Stark, J.C., Hodgman, C.E., and Jewett, M.C. (2015) Energizing eukaryotic cell-free protein synthesis with glucose metabolism. *FEBS Lett.*, **589**, 1723–1727.

124. Dudley, Anderson, and Jewett, A.C.S. (2016) *Synth. Biol.*, doi: 10.1021/acssynbio.6b00154.

125. Karim and Jewett (2016) *Metab. Eng.*, **36**, 116–126, http://dx.doi.org/10.1016/j.ymben.2016.03.002.

Part II
Multipurpose Bacterial Cell Factories

Industrial Biotechnology: Microorganisms, First Edition. Edited by Christoph Wittmann and James C. Liao.

5
Industrial Biotechnology: *Escherichia coli* as a Host

Matthew Theisen and James C. Liao

5.1
Introduction

5.1.1
Background

Escherichia coli has many advantages as a microbial platform for industrial applications. First, it is very well studied. Its genome was first sequenced in 1997 [1]. A large number of *E. coli* enzymes and regulatory proteins have been characterized at the level of reaction mechanism, and transcriptional and substrate-level regulation. Databases have been developed, for example, EcoCyc [2], which allow easy searches for the sequence, function, and regulation of nearly all *E. coli* genes and proteins. System-level understanding, including global regulation and adaptation are well documented. Second, a wide variety of efficient genetic tools are available to genetically transform *E. coli* or edit *E. coli* genome with high precision. Sophisticated computational tools [3] have also been developed to aid such modifications. Third, *E. coli* grows very quickly and is a facultative anaerobe, that is, it can grow with or without oxygen. Fourth, it adapts well to metabolic stress. As a result, it has been used to produce an almost unbelievable variety of compounds.

Despite its many advantages *E. coli* has some shortcomings as well. A key disadvantage of *E. coli* as an industrial organism, in comparison to eukaryotes like yeast, seems to be its sensitivity to bacteriophage infection [4, 5], although several industrial processes have become successful despite this challenge. Since it is a gram-negative bacterium, the *E. coli* outer membrane contains lipopolysaccharide, also known as endotoxin, which elicits strong immune responses in animals. *E. coli* production of compounds intended for human or animal consumption requires special clearance to be classified as "Generally Recognized as Safe" (GRAS), for example, in an *E. coli* K12 methionine production process intended for animal feed supplements [6]. Further, *E. coli* is a metabolic generalist, that is, *E. coli* is able to produce many compounds, but without optimization it is not the best at producing any given compound. For example, *E. coli* can produce ethanol, but *Saccharomyces cerevisiae* natively produces faster [7]; *E. coli* can

Industrial Biotechnology: Microorganisms, First Edition. Edited by Christoph Wittmann and James C. Liao.

produce citrate, but *Aspergillus niger* is preferred [8]; *E. coli* can produce lysine, but *Corynebacterium glutamicum* is often used [9]. However, this situation has begun to turn around, as the advanced genetic manipulation tools for *E. coli* have enabled rapid progress in optimizing *E. coli* production of many products. *E. coli* is also the host to 1,3-propanediol (PDO) in industry. Additionally, *E. coli* production of 1-butanol [10, 11] has reached a level comparable to or better than that of the native producer *Clostridium acetobutylicum*. For this reason, *E. coli* is useful not only as a demonstration-scale platform but has increasingly been used in industrial-scale production.

5.1.2 Historical Context

E. coli has been an important industrial host for many processes. Perhaps the most widely known, historically, is the use of *E. coli* as a producer of insulin. Previously, insulin for diabetic patients was produced by extraction from pig pancreases [12]. Approximately 4 ton of pig pancreases were required per pound of insulin. In 1982, Genentech and Eli Lilly introduced a much more cost-effective and humane method, producing the first microbially derived protein therapeutic version of insulin called *Humulin*. Since then, dozens more protein therapeutics have been produced in *E. coli* and other hosts [13]. Some production of therapeutic proteins has moved to more sophisticated animal cells to take advantage of protein glycosylation that is not available in bacteria.

However, *E. coli* has gained significance in the production of other non-protein metabolites. Many of these projects, industrial and academic, are successful on the research scale. However, only those published in the literature and those successfully commercialized are citable (see Table 5.1). Many more *E. coli* processes were developed successfully in industrial laboratories, but were not commercialized because of market reasons, and thus are not widely known. Industrial strain development processes are impressively successful, but details are not always disclosed. Here we summarize a few examples of *E. coli* metabolic engineering efforts reported in the literature. These examples will provide an idea of the versatility of *E. coli* metabolism. In sum, *E. coli* can serve as a host to produce nearly any compound after metabolic engineering.

5.2 *E. coli* Products

5.2.1 Amino Acids

Amino acids are building blocks of proteins but are also used extensively as nutritional supplements and in livestock feed [20, 21]. Amino acids are produced normally by all organisms, but native production typically has feedback

Table 5.1 Some of the economically successful industrial *E. coli* bioprocesses.

Product	Company	Remarks	References
1,3-Propanediol	DuPont	Introduction of heterologous production pathway allowed titer reported up to $130\,g\,l^{-1}$	[14, 15]
Various L-amino acids	Various	Some amino acids are produced fermentatively by *E. coli* or *C. glutamicum*	[16]
1,4-Butanediol	Genomatica and BASF	Process currently under commercialization	[17, 18]
Protein therapeutics	Biotech firms including Amgen, Genetech/Roche, Novartis, and others	First *E. coli* protein therapeutic was insulin in the 1980s	[13]
Cephalexin	Novo-Nordisk	Immobilized *E. coli* cells are used to catalyze some reaction steps	[19]

inhibition to prevent accumulation. For example, threonine and lysine inhibit the aspartokinases of *E. coli* which are an early step in their production from aspartate [22]. Isoleucine inhibits threonine deaminase [23]. Overproduction of amino acids must overcome such inhibition. Historically, *C. glutamicum* has been used as the main organism for production of various amino acids. However, with efficient metabolic engineering techniques, *E. coli* has surpassed *C. glutamicum* in the production of many important amino acids that are traditionally produced by *C. glutamicum*, for example, L-lysine. Unfortunately, the most crucial industrial developments are trade secrets and are not available in the open literature.

The industrial production of L-threonine or L-lysine from glucose or sucrose at titers much greater than $100\,g\,l^{-1}$ and yields greater than 80% of the theoretical maxima has been routinely accomplished by metabolic engineering of *E. coli*. The overproduction of threonine from glucose at titers of $58\,g\,l^{-1}$ has been accomplished by the removal of threonine degradation pathways [24]. Random mutants that could not utilize threonine as a nitrogen source were used [24]. After identifying mutants that were also tolerant to higher concentrations of salt, production was pushed to $74\,g\,l^{-1}$ [25]. By the use of systems metabolic engineering strategies, titers were pushed to $82\,g\,l^{-1}$ and yield was $0.393\,g\,(g\,glc)^{-1}$. Competing pathways toward lysine, methionine, and glycine were knocked out and acetyl-CoA synthetase (acs) was overexpressed to limit acetate production [26]. These efforts provide some ideas regarding how industrial strains were developed (Figure 5.1).

Valine overproduction in *E. coli* K-12 using random mutant strains is unknown. However, systems metabolic engineering approaches allowed titers of $7.3\,g\,l^{-1}$ by several methods [27]. Acetohydroxy acid synthase (AHAS), encoded by *ilvI*, was mutated to remove feedback inhibition. Competing pathways toward leucine,

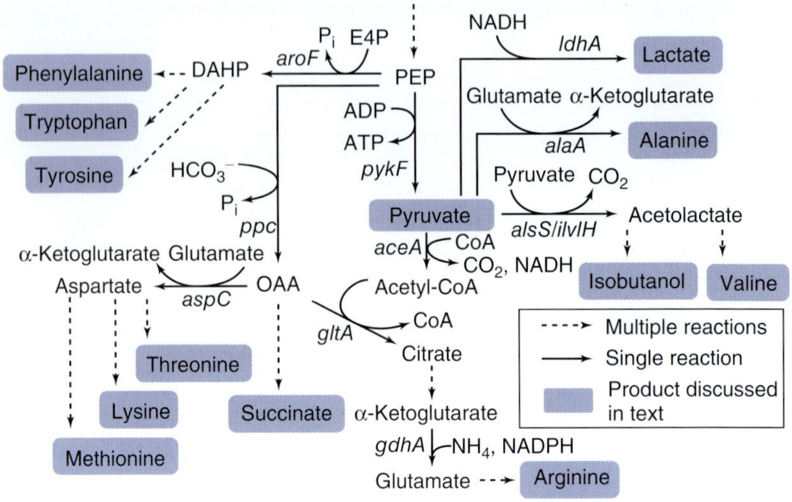

Figure 5.1 Production pathways of selected products discussed in the text. Gene/protein symbols: *aroF*, DAHP synthase; *pykF*, pyruvate kinase; *ldhA*, lactate dehydrogenase; *alaA*, alanine transaminase; *ilvIH*, acetolactate synthase (*alsS* = heterologous enzyme from *Bacillus subtilis*); *aceA*, pyruvate dehydrogenase; *ppc*, phosphoenolpyruvate carboxylase; *aspC*, aspartate transaminase; *gltA*, citrate synthase; and *gdhA*, glutamate dehydrogenase.

isoleucine, and pantothenate were removed. Valine transporter ygaZH and valine regulator Lrp were overexpressed to maximize valine export from the cell and minimize uptake of valine. *E. coli* K12 is known to be valine sensitive owing to its nonfunctional AHAS II [28]. *E. coli* W is more tolerant to valine because its functional AHAS II is resistant to valine inhibition. *E. coli* W was used as a production platform for valine. With knockouts in *ilvA* to reduce isoleucine production and overexpression of valine synthesis genes, a strain was able to produce 60 g l^{-1} valine [29].

Aromatic amino acids require the precursor 3-deoxy-D-arabino (DAHP). Strategies for increasing DAHP production focused on increasing the availability of phosphoenolpyruvate (PEP) and erythrulose-4-phosphate (E4P) which react together to form DAHP. It was found that overexpressing transketolase (tktA) and phosphoenolpyruvate synthase (pps) increased DAHP production from glucose to near theoretical yields [30]. From xylose, only overexpression of DAHP synthase itself is required to achieve theoretical yields [31]. Further, the PEP-related glucose phosphotransferase system (PTS) was replaced with a non-PTS system and this was also shown to increase DAHP production from glucose in *E. coli* [32].

Aromatic amino acid L-phenylalanine is a component of the artificial sweetener aspartame. It has been produced at 28 g l^{-1} in *E. coli* K12 by overexpressing feedback-resistant DAHP-synthase (*aroF*^{fbr}). However, it was found that overexpressing wild-type DAHP-synthase (*aroF*^{wt}), which is tyrosine sensitive, and controlling tyrosine levels allowed production of up to 34 g l^{-1} [33, 34]. Proteomic analysis of a pyruvate kinase (Pyk-F) knockout showed upregulation of the shikimate pathway and aromatic amino acid production enzymes, anthranilate synthase (AS), prephenate dehydrogenase (PRDH), chorismate synthase (CM), and shikimate dehydrogenase (SKDH) [35]. A report of 47 g l^{-1} production of phenylalanine from glucose in *E. coli* was accomplished using temperature-controlled promoters to turn off production of phenylalanine during the lag phase. Additionally, phenylalanine efflux was increased by the overexpression of transporter yddG [36]. Although this represents a promising advance in phenylalanine production, the methods used in the paper were not fully specified, including mobile phase and protocol for HPLC analysis of phenylalanine concentration. Given the interest in using alternative carbon sources other than glucose, a strain producing 13 g l^{-1} phenylalanine from glycerol, lactic acid, and ammonia was produced [37].

L-Tyrosine and L-tryptophan have also been overproduced in *E. coli*. Their production follows the same shikimate pathway as phenylalanine through chorismate where the pathway branches [38]. In a bioreactor *E. coli*, 3 g l^{-1} of L-tyrosine was produced by overexpression of feedback inhibition–insensitive cyclohexadienyl dehydrogenase (TyrC) from *Zymomonas mobilis* and native chorismate mutase-prephenate dehydratase ($PheA_{CM}$) [39]. Combinatorial and mutagenic approaches were able to increase tyrosine titers by more than 100% over a rationally designed strain. Transcriptional data indicated that the *E. coli* had mutated to impose a stringent response similar to that observed under amino acid starvation. The effect was recreated by overexpression of *evgA* and *relA* regulators and a final titer of 13.8 g l^{-1} was observed [40]. Industrial strain optimization at scales up to 200-l has shown titers up to 55 g l^{-1} in 48 h [41]. Tryptophan production was accomplished by site-directed mutagenesis of feedback-inhibited enzymes DAHP synthase (*AroF*) and anthranilate synthase (*TrpED*). Additionally, the tryptophan degradation pathway gene *tnaA* and competing pathways encoded by *pheA* and *tyrA* were removed, allowing for a final titer of 13.3 g l^{-1} [42]. Overexpression of tryptophan pathway genes *trpEDCBA* and transporter *yddG* increased the titer to 36.3 g l^{-1} in a 30-l fermenter.

More complex shikimate derivatives have also been produced. One example is the antioxidant, anti-inflammatory, and antiproliferative hydroxycinnamoyl anthranilates (also known as *avenanthramides* or Avns). The production was first shown from feeding of cinnamates and anthranilates, but the entire pathway from glucose was also demonstrated by using feedback-resistant AroG and TyrA as well as exogenous 4-coumarate: CoA ligase from *Arabidopsis thaliana* and hydroxycinnamoyl/benzoyl-CoA/anthranilate *N*-hydroxycinnamoyl/benzoyltransferase from *Dianthus caryophyllus* (carnation) [43].

Alanine was overproduced by knockout of lactate dehydrogenase (ldhA) and replacement with alanine dehydrogenase. Additionally, methylglyoxal synthase (mgsA) was deleted to reduce lactate, and alanine racemase (*dadX*) was removed to prevent conversion between L- and D-alanine. The final titer was 120 g l^{-1} with yield 0.95 g (g glc)$^{-1}$ [44].

Arginine synthesis by other organisms such as *C. glutamicum* is known up to $92.5\,g\,l^{-1}$ [45]. In *E. coli*, a production strategy overexpressing feedback-resistant alleles of the arginine synthesis pathway was used. Using the overexpression of these pathways, combined with knockout of arginine decomposition pathways, the titer of arginine production by *E. coli* was pushed to $11.6\,g\,l^{-1}$ [46].

5.2.2
Organic Acids

Lactic acid is a natural fermentative product and is a precursor for polylactide, a biodegradable polymer. Synthesis of both isomers of lactic acid by non-acetogenic *E. coli* has been shown. Strains which had the *Pta* (acetogenesis pathway) and *Ppc* (succinate production pathway) genes inactivated produced $62.2\,g\,l^{-1}$ D-lactate from glucose. With a non-native *ldhA* from *Lactobacillus casei*, $45\,g\,l^{-1}$ of L-lactate was produced [47]. Production of $52.7\,g\,l^{-1}$ of L-lactic acid was achieved by *E. coli* with mutations in catabolite repression systems which co-utilized glucose and xylose [48]. Titer of the D-isomer was pushed to $125\,g\,l^{-1}$ by elimination of several competing pathways which eliminated or greatly reduced the production of succinate, formate, ethanol, and acetate [49].

The TCA cycle intermediate succinic acid has been identified by the US Department of Energy as a top value-added chemical from biomass. It has potential to be converted to compounds of interest such as tetrahydrofuran (THF) and 1,4-butanediol (1,4-BDO) [50]. Production of succinate in *E. coli* has been accomplished by eliminating pyruvate drains (*pfl* and *ldhA*) and PEP-related glucose transporter (ptsG) combined with overexpression of pyruvate carboxylase (pyc) from *Rhizobium etli*. Employing a two-phase aerobic/anaerobic production scheme allowed production of up to $99\,g\,l^{-1}$ succinic acid [51]. Other efforts have sought to generate succinate overproducers without recombinant enzymes. This has been accomplished by knockout of anaerobic fermentation genes followed by growth-based selection. Succinate was produced up to $86\,g\,l^{-1}$ after 2000 generations [52]. Analysis indicated that the ATP-conserving phosphoenol-pyruvate carboxykinase (pck) had high activity in the producing mutants. Also, glucose phosphotransferase and transporter ptsG, which normally consumes PEP, were found to be inactivated. Instead, a permease (*galP*) and glucokinase (glk) were found to act as the glucose uptake mechanism [53].

Pyruvate has been used in nutritional supplements and a derivative, 3-bromopyruvic acid, has been approved for clinical trials by the FDA as an anticancer agent [54]. Production of pyruvate in *E. coli* (up to $66\,g\,l^{-1}$) was accomplished by knocking out pyruvate-consuming pathways pyruvate formate lyase (pflB), *ldhA*, and fumarate reductase (frdBC), and additionally by the disruption of ATP synthase (*atpFH*) and acetate kinase (ackA) [55].

Acetic acid is an important chemical precursor for vinyl acetate monomers with a small amount being used for vinegar and salad dressing [56]. Acetic acid was produced in *E. coli* by the knockout of native fermentation pathway genes, *pflB*, *frdBC*, *ldhA*, and *AdhE* (the gene for alcohol dehydrogenase). A final titer of 52 g l^{-1} was achieved [57].

Hydroxycarboxylic acids are precursors to lactones. For example, 3-hydroxy-γ-butyrolactone (HBL) can be produced by the dehydration of 3,4-dihydroxybutyric acid (DHBA) and is used in the production of a range of pharmaceutical products [58]. DHBA was produced in *E. coli* by the use of thiolase BktB which condenses glycolyl-CoA and acetyl-CoA. DHBA, its isomer 2,3-dihydroxybutyric acid and its dehydration product HBL were produced in a total titer over 1 g l^{-1} [59] from glucose and glycolic acid. By adding the pathway for production of glycolate, total production of about 1 g l^{-1} was achieved from glucose as a sole carbon source [60].

3-Aminopropionic acid is noncanonical amino acid and chemical precursor. Also known as beta-alanine because of its similarity to the canonical amino acid L-lysine, it has been overproduced in *E. coli.* A strain was prepared, which included an aspartate decarboxylase gene (*panD*) from *C. glutamicum*, overexpression of aspartase (aspA) and phosphoenolpyruvate carboxylase (ppc), and acetyl-CoA synthase (acs) and it resulted in total titers of 32 g l^{-1} after 39 h from rich media supplemented by glucose and ammonium sulfate [61].

5.2.3 Alcohols

Production of 1-butanol has been accomplished using a pathway based on the coenzyme A (CoA)-dependent pathway utilized by the native producer, *C. acetobutyricum.* However, transferring this 1-butanol pathway to *E. coli* resulted in only about 1 g l^{-1} of production [10]. The low production was attributed to the poor expression and function of the butyryl-coA dehydrogenase (Bcd)-electron transferring flavoprotein (Etf) complex. Production of 1-butanol was found to be much improved on replacing butyryl-CoA dehydrogenase/electron transfer complex (Bcd-Etf) with *trans*-enoyl-CoA reductase (ter) [11] which expresses well in *E. coli* and does not perform an electron bifurcation reaction. Furthermore, thermodynamic driving forces were created by increasing the NADH pool after deleting all other fermentative pathways, by increasing the acetyl-CoA pool after deleting phosphate acetyltransferase (Pta). The latter is particularly important as the first reaction in the 1-butanol pathway is the condensation of 2 acetyl-CoA to acetoacetyl-CoA, which is thermodynamically unfavorable. Thus, it requires a large acetyl-CoA pool to drive the reaction forward. With these strategies, the engineered *E. coli* produced 15 g l^{-1} of 1-butanol in flasks and up to 30 g l^{-1} in a bioreactor with *in situ* product stripping [11].

Another method for alcohol production is through the keto-acid pathways [62] (Figure 5.2). Multiple keto-acids exist in *E. coli* metabolism as intermediates to various products, particularly amino acids. The most common of these is the 3-carbon keto-acid pyruvate. Keto-acids can be decarboxylated to aldehydes, which in turn can be reduced to alcohols. In fact, decarboxylation of pyruvate to acetaldehyde followed by reduction to alcohol is the basis for bioethanol production. Extending this concept, any keto-acid can be converted to the corresponding

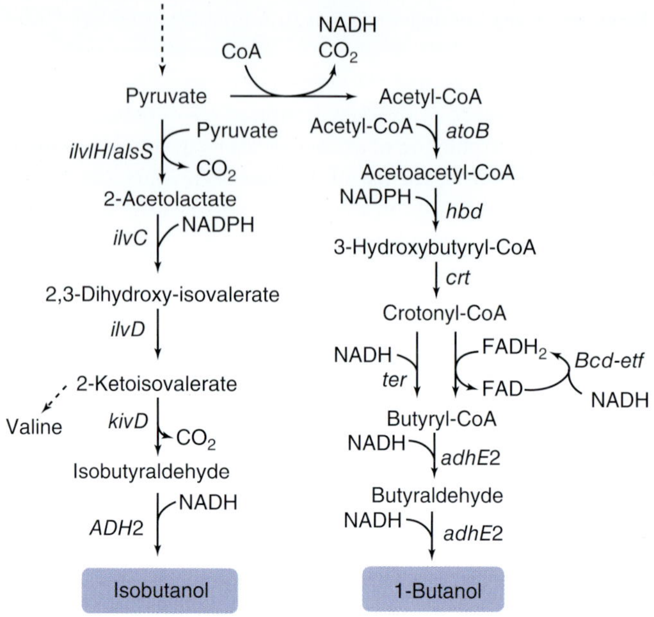

Figure 5.2 Isobutanol production via valine pathway and 1-butanol production via CoA-dependent pathway implemented. Common enzyme abbreviations or gene symbols: *ilvIH*, acetolactate synthase (*E. coli*); *alsS*, acetolactate synthase (*Bacillus subtilis*); *ilvC*, acetohydroxy acid isomeroreductase (*E. coli*); *ilvD*, dihydroxy acid dehydratase (*E. coli*); *kivd*, ketoisovalerate decarboxylase (*Lactoccus lactis*); *ADH2*, alcohol dehydrogenase (*Saccharomyces cerivisiae*); *atoB*, thiolase (*E. coli*); *hbd*, hydroxybutyryl-CoA dehydrogenase (*Clostridium acetobutylicum*); crt, crotonase (*C. acetobutylicum*); *ter*, *trans*-enoyl-CoA reductase (*Treponema denticola*); and *Bcd-etf*, butyryl-CoA dehydrogenase/electron transfer complex (*C. acetobutylicum*) *AdhE2*, alcohol dehydrogenase (bifunctional, *C. acetobutylicum*).

alcohols by using a keto-acid decarboxylase and an alcohol dehydrogense. This pathway was identified as the Ehrlich pathway [63] more than a century ago to explain the small amounts of higher alcohols produced in various fermentation broths. However, production of keto-acids longer than three carbons is highly regulated, and conventional wisdom in the past did not favor using a biosynthetic pathway for high-flux alcohol production. This concept was changed by overexpression of a catabolic acetolactate synthase (AlsS) from *Bacillus subtilis* and a ketoisovalerate decarboxylase (Kivd) from *Lactococcus lactis*, along with other native *E. coli* genes (ilvCD, yqhD) in *E. coli* [62]. This engineered *E. coli* strain produced $22\,g\,l^{-1}$ of isobutanol, with much higher toxicity level than isobutanol which was shown to be $6-8\,g\,l^{-1}$ for *E. coli*. It continued to produce isobutanol without growth and after losing viability, much like yeast production of ethanol in the stationary phase. Up to $50\,g\,l^{-1}$ of isobutanol was produced in a bioreactor with *in situ* product stripping [64]. Strains with similar productivities were also created by random mutagenesis followed by section of norvaline-resistant colonies [65]. These strains showed increased valine flux, which was directed to isobutanol production. Interestingly, AlsS also possessed a Kivd activity, such that Kivd expression was not required to achieve isobutanol production [66]. Fully anaerobic production of isobutanol was accomplished by changing the cofactor utilization for pathway enzymes from NADPH to NADH [67]. *E. coli* isobutanol tolerance was improved by passaging into increasing concentrations. The resistant strain was isolated and sequenced and the resistance was found to be mostly attributable to mutations in five genes, including the efflux pump gene *AcrA* [68]. The response network to the isobutanol challenge [69] was analyzed by using Network Component Analysis [70]. These alcohols are produced by the same pathway in yeast [71].

The advantages of the keto-acid pathways are its compatibility with many organisms and its versatility to produce many products. Isobutanol production has been demonstrated in many organisms, including *C. glutamicum* [72], *Ralstonia eutropha* [73], *Synechococcus elongatus* [74], *Geobacillus thermoglucosidasius* [75], *B. subtilis* [76], and *Clostridium thermocellum* [77]. The keto-acid pathways can also be used to produce 2-methyl-1-butanol [78], 3-methyl-1-butanol [79], and 1-butanol [80]. Longer chain alcohols have been made using a recursive pathway in *E. coli.* Starting with the threonine/1-butanol pathway in *E. coli,* the substrates can be extended, one carbon at a time [81]. Alcohols up to octanol were observed. Isopropanol, another potential fuel alcohol, has been produced in *E. coli* by a pathway through acetyl-CoA which uses thiolase to generate acetoacetyl-CoA. This is followed by decarboxylation of acetoacetate and dehydration [82]. Optimization of culture pH and gas-stripping of isopropanol pushed total titers to 143 $g\,l^{-1}$ [83].

5.2.4 Diols

PDOs are useful for synthesis of polymers, aircraft de-icing, and as food additives [84]. The synthesis of 1,3-PDO was reported at a titer of up to 130 $g\,l^{-1}$ reported in 2003 [14] as the result of an effort by DuPont and Genencor. 1,3-PDO was produced in *E. coli* through the glycolytic triose dihydroxyacetone phosphate (DHAP), which is reduced and dephosphorylated to glycerol, followed by dehydration and reduction to 1,3-PDO. The strategy involved the use of ATP-related glucose transport rather than the native PEP-related mechanism. Additionally, the previously uncharacterized native reductase YqhD was used for the final reduction from 3-hydroxypropanal to 1,3-PDO. The process has become industrially successful, supported by patents for the production [85, 86] and purification [87] of 1,3-PDO.

1,2-PDO has also been produced in *E. coli* from glucose. The production pathway used methylglyoxal and it was shown that overexpression of both *mgsA* and glycerol dehydrogenase (gldA) produced 0.7 $g\,l^{-1}$ 1,2-PDO. This was a greater titer than was accomplished by either enzyme alone [88]. *ldhA* and glyoxalase I were inactivated to prevent production of lactate side product and the final strain expressing msg, gldA, and 1,2-PDO oxidoreductase fucO generated 4.5 g l^{-1} D-1,2,-PDO [89].

Butanediols are used in the formation of polyurethanes and THF [90]. The production of 1,4-BDO in *E. coli* from glucose, sucrose, xylose, and other sugars was reported at a titer up to 18 g l^{-1}. Previously, no synthetic pathway for 1,4-BDO had been known, but one was developed through succinate or α-ketoglutarate [91]. A combination of knockouts, varying expression systems, finding exogenous enzymes, and rerouting carbon through the oxidative rather than reductive TCA cycle resulted in the final titer. Development of the process is being conducted industrially by BASF and Genomatica [17, 18]. 2,3-Butanediol (2,3-BDO) production has also been demonstrated. The *meso* isomer was produced from glucose using three enzymes from *Klebsella pneumonia*, acetolactate synthase (ALS), acetolactate decarboxylase (ALDC), and *meso*-butanediol dehydrogenase (BDH). These enzymes catalyze the conversion of two molecules of pyruvate to *meso*-2,3-BDO through acetolactate. They were found to produce titers up to 17.7 g l^{-1} from glucose [92]. L-2,3-BDO was produced from diacetyl with 73% molar yield using an enantioselective secondary dehydrogenases from *Brevibacterium saccharolyticum* [93]. *R,R*-2,3-BDO was produced from glucose using another enantioselective enzyme from *Thermanaerobacter brockii* with a maximum titer of 6 g l^{-1} [94].

5.2.5 Hydrogen and Alkanes

In addition to being used to produce fuel alcohols, *E. coli* has also been used to produce hydrogen from a variety of substrates including glucose and formate [95, 96]. Formate can be produced electrochemically from carbon dioxide and is highly soluble, making it advantageous as a carrier of reducing power for use by microbes [97, 98]. Hydrogen detection has been demonstrated using a simple colorimetric assay, potentially simplifying the process for screening the g strains produced [99]. The hydrogen production of *E. coli* was improved by inactivating competing pathways for pyruvate flux [96]. Hydrogen is produced through pyruvate and pflB from formate hydrogen lyase (fhl). Hydrogen production and uptake can also be catalyzed by the native hydrogenases of *E. coli* [100, 101].

Alkanes are a desirable fuel compound as they have fuel properties nearly identical to that of gasoline. Production of a range of alkanes including nonane and decane has been shown in *E. coli* [102]. Fatty acyl-ACPs (ACP = acyl carrier protein) were produced through the fatty acid biosynthetic pathway. They were then reduced to aldehydes and decarbonylated to alkanes by an exogenous enzyme, CER1 from *A. thaliana*.

5.2.6 Fatty Acids

Fatty acids and other oily compounds are desirable for their uses in many areas, including fuel, chemical, nutrition, and personal care [103–105]. Production in

E. coli is receiving attention due to its potential for lower land use than plants and ease of genetic modification [106, 107]. Toxicity of fatty acids in culture medium toward *E. coli* has limited titers, but productivity is known up to 4.5 g l^{-1} day^{-1} and *in situ* product removal has allowed total titers up to about 9 g l^{-1} [108, 109]. Biodiesel is typically produced by chemical synthesis from triacylglycerols (TAGs) to produce fatty acid alkyl esters (FAAE). However, FAAEs and fatty alcohols have also been produced directly from sugars in *E. coli* up to 700 mg l^{-1} [110].

Shorter chain alkyl esters have been produced in *E. coli* at much higher titer. The pathway involves the esterification of mid-chain alcohols with acetyl-CoA via an alcohol acyltransferase (ATF1) from *Saccharomyces cerivisiae.* Isobutyl acetate was initially produced by overexpression of the ATF1 enzyme and it resulted in a maximum titer of 17 g l^{-1} including a hexadecane layer [111]. On overexpressing ATF1 and the pathway for the production of isobutanol and knocking out competing pathways, 36 g l^{-1} of isobutyl acetate was produced in a bioreactor from glucose without an organic layer [112].

5.2.7
Sugar Alcohols

Sugar alcohols mannitol and xylitol are used as sweeteners [113]. Mannitol also has some medical uses [114]. Mannitol is produced in a single step from D-fructose by mannitol dehydrogenase. In one *E. coli* system, reducing power from formate was used and D-fructose was converted to mannitol with 84% molar yield and final titer about 66 g l^{-1} [115].

The production of xylitol from xylose requires only a single step, xylose reductase (XR). The XR gene *xyrA* from *Candida tropicalis* was expressed in *E. coli* and 13.3 g l^{-1} of xylitol was produced from xylose [116]. A strain that co-consumes xylose and glucose was used to generate a 56 g l^{-1} xylitol [117]. Another study used a transhydrogenase (PntAB) to achieve theoretical yield of xylitol from glucose in the presence of excess xylose [118].

5.2.8
Isoprenoids

Isoprenoids are a diverse class of compounds synthesized from the 5-carbon precursors isopentenyl pyrophosphate (IPP) and dimethylallyl pyrophosphate (DMAPP). Several isoprenoids have been synthesized in *E. coli.* Lycopene is an antioxidant that has been associated with reduced UV skin damage and some anticancer effects [119–121]. A metabolic control strategy was employed to increase production of lycopene in *E. coli.* The Ntr regulon sensed the presence of acetyl phosphate, which was an intermediate toward waste acetate signifying high glycolytic flux. In the presence of acetyl phosphate, the limiting enzymes (Pps and idi) for lycopene synthesis would be overexpressed. This strategy was shown to increase lycopene production to 170 mg l^{-1} while reducing acetate and pyruvate production [122]. Further, overexpression of Pps and knockout of PykF were also shown to improve production, even without Ntr regulon control [123]. These genetic changes balance the availability of G3P and pyruvate, which react to form a precursor of IPP.

Other isoprenoid derivatives have also been produced. Coenzyme Q10 consists of a quinone bonded to a 50-carbon isoprenoid chain. The enzyme decaprenyl diphosphate synthase (Rsdds) from *Rhodobacter sphaeroides* was found to have high specificity for the 10-isoprenoid CoQ_{10} product compared to Atdds from *Agrobacterium tumefaciens*, which yielded a higher fraction of the CoQ_9 product [124]. The entire pathway for CoQ_{10} synthesis in *E. coli* from glucose was found to generate up to 2500 μg CoQ_{10}/g CDW [125]. Limonene is a 10-carbon isoprenoid used in cosmetics and cleaning. It has been produced up to 400 $mg\,l^{-1}$ in *E. coli* from glucose [126]. 8-Hydroxycadinene and artemisinic acid were generated by engineering plant cytochrome P450 oxidases for proper folding and membrane insertion. 8-Hydroxycadinene was produced at over 100 $mg\,l^{-1}$. Artemisinic acid and the related aldehyde and alcohol were produced to a total titer around 500 $mg\,l^{-1}$ [127].

5.2.9 Polymers

Poly(lactic acid) and copolymer poly(3-hudroxybutyrate-*co*-lactic acid) rather than their monomers were synthesized in *E. coli*. Propionate CoA-transferase from *Clostridium propionicum* [128] was engineered to efficiently transfer CoA to D-lactate from acetyl-CoA forming lactyl-CoA [129]. This lactyl-CoA was incorporated into polymers via another engineered enzyme, polyhydroxyalkanoate (PHA) synthase from *Pseodomonas* sp. accumulating up to 56 wt% of the copolymer in *E. coli* [130]. PHAs are a class of compounds which are produced by the esterification of hydroxyalkyl acids. Production of poly(3-hydroxybutyrate) (PHB) in *E. coli* has been accomplished with productivity up to 2 $g\,l^{-1}\,h^{-1}$ and titer 80 $g\,l^{-1}$ [131]. It follows a similar pathway as reverse-beta oxidation of fatty acids, except that the 3-hydroxy group is not removed but esterified to an acyl-CoA. A systematic analysis of genetic perturbations on PHB production in *E. coli* was carried out. It was found that high overexpression of the PHB pathway resulted in higher flux to PHB at the expense of growth rate [132]. Poly(3-hydroxypropionate) (P3HP) is the strongest of the microbial PHAs and was produced from glucose with a P3HP-PHB copolymer at a total titer of 3 $g\,l^{-1}$ through a pathway which proceeds from glycolytic trioses. Previous reports had only produced P3HP from glycerol or 1,3-PDO [133–135]. A PHA production process using *E. coli* as an industrial host was under development by Metabolix in association with Archer Daniels Midland from 2004 until 2012 [136–138].

Hyaluronic acid is a glycosaminoglycan, a carbohydrate augmented with carboxylic acid and amide functional groups. It is a component of human tissue and used in various medical applications. It has been produced by *E. coli* at

190 mg l^{-1} by using a codon-optimized enzyme from *Streptococcus equisimilis* [139]. Another glycosaminoglycan called *chondroitin* was produced up to a total of 2.4 g l^{-1} in a bioreactor by expressing enzymes from the *E. coli* capsular polysaccharide pathway [140]. However, a majority of the product was intracellular.

Most microbially produced polymers are known to be stiff and not very flexible. However, one potential microbially derived polymer that is more supple is a lactic acid-β-methyl-δ-valerolactone(βMδVL)-lactic acid triblock copolymer. This was demonstrated by the overproduction of mevalonate in *E. coli* via enzymes from *L. casei* which generate mevalonate, a precursor of the isoprenoid pathway, from acetyl-CoA. Mevalonate is produced at over 80 g l^{-1} from glucose, dehydrated, and then reduced to form βMδVL. PLA-PβMδVL-PLA triblock copolymers were created to demonstrate physical properties such as elasticity and toughness [141].

Spider silk is composed of large polypeptides, 250–320 kDa. A 285 kDa protein from the spider *Nephila clavipes* was generated in *E. coli* and purified. Its structural properties were superior to smaller-molecular-weight proteins. The glycine-rich (44% of residues) protein was generated by increasing the pool of glycyl-tRNA [142].

5.2.10
Natural Products

Natural products refer to a general category of secondary metabolic compounds which are often generated by eukaryotes such as plants or fungi [143]. Natural products are a major source of new drug products [144, 145]. Purifying these compounds from their natural hosts may be complicated by difficulty of culturing and low concentrations. Thus, when possible, production via a model organism such as *E. coli* is preferable.

A precursor to taxol, an anticancer drug, has also been produced from glucose in *E. coli.* IPP and DMAPP react together to form geranylgeranyl diphosphate which is transformed to taxadiene by taxadiene synthase. Enzymes for some of these reactions were obtained from species in the *Taxus* genus of yews. A combinatorial approach was employed, in which expression of pathway enzymes was varied to achieve final concentrations of taxadiene over 1 g l^{-1}, a 15 000-fold improvement over the control strain [40].

Valinomycin is a circular depsipeptide isolated from *Streptomyces fulvissimus* [146]. It is noted for its antibacterial and insecticidal properties. Valinomycin was produced by *E. coli* by expressing two large non-ribosomal polypeptide synthases (NRPSs), Vlm1 (370 kDa) and Vlm2 (284 kDa). These complexes were reconstituted in *E. coli* by the coexpression of a phosphopantetheinyl transferase (PPTase) which covalently modifies the thiolase domains of the NRPSs [147]. Subsequently, it was found that by optimizing culture conditions including a nutrient boost, up to 10 mg ml^{-1} could be produced [148], while from glucose, only 2 mg ml^{-1} was achieved [149].

Coumarin is a class of compounds found notably in the tonka bean, *Dipterix odorata*, which is known for its acaricidal extracts [150]. The biosynthesis of coumarins proceeds through tyrosine by a series of reactions. The basic structure of coumarins includes a fused aromatic ring which is formed spontaneously in the final reaction of the pathway, in which a CoA-thioester bond is broken to form the ring. Overexpression of pathway enzymes from *A. thaliana* and *Ipomoea batatas* (sweet potato) resulted in total coumarin concentration of over 100 mg l^{-1} [151]. A broader class of compounds called *phenylpropanoids*, derived from phenylalanine, was demonstrated as well. Genes from *A. thaliana*, *Oryza sativa* (rice), and *Arachis hypogaea* (peanut) were overexpressed in *E. coli*, resulting in a smattering of phenylpropanoids including cinnamic acid, coumaric acid, caffeic acid, ferulic acid, reservatrol, piceatannol, pinosylvin, curcumin, and more [152].

5.2.11 Protein as Product

E. coli has been used extensively as an industrial host for protein expression and a wealth of knowledge has been accumulated [13, 153, 154]. About 30% of therapeutic proteins are produced by *E. coli*, including treatments for macular degeneration, multiple sclerosis, hepatitis C, and other conditions [13].

Protein expression requires effective tools for initiation of transcription and translation. There are numerous *E. coli* transcription systems known. IPTG-inducible promoters are frequently used. However, other transcription initiation systems such as phosphate depletion/*PhoA* and temperature shift/pL are also used because of the relatively high cost and possible toxicity of IPTG [13, 155–157]. Ribosomal binding site (RBS) design is also a concern, but progress in this area has been made by the development of automated RBS design software capable of both forward and reverse engineering RBS strength [3].

Expressed proteins may be directed toward periplasm, extracellular media, or intracellularly, sometimes as an inclusion body. Periplasmic expression is advantageous because proteins localized there are subject to decreased degradation and an oxidizing environment which facilitates folding and disulfide bond formation [158–160]. Direction toward periplasm requires a signal sequence [161]. Methods for secretion to extracellular medium are also known [162]. The hemolysin secretion system can be exploited to cause secretion through the transenvelope complex containing the proteins HlyB and HlyD [163]. Secretion through that transporter is triggered by a C-terminal amino acid sequence and has been used for heterologous proteins [164, 165]. Formation of inclusion bodies is caused by low solubility and lack of appropriate machinery in prokaryotes, resulting in nonfunctional aggregates [166]. However, renaturation of proteins from *E. coli* inclusion bodies has been shown with multiple methods [167, 168].

Glycosylation of proteins is not widely known in prokaryotes. However, a system for N-linked glycosylation was identified in *Campylobacter jejuni*, a spiral-shaped pathogenic bacteria. Further, this glycosylation system was imported to *E. coli* [169]. Using this strategy, detectable amounts of glycosylated antibody fragments were produced in *E. coli* [170].

5.3 Rewiring Central Metabolism

5.3.1 NOG and Carbon Efficiency

Introducing heterologous metabolism is sometimes as simple as extending a biosynthetic pathway directly from primary metabolism intermediates. However, in some cases, it is necessary to alter primary metabolism itself. For example, alterations such as avoiding decarboxylating reactions, incorporating one carbon molecule into a product, or combining metabolites from different metabolic pathways may be required to promote carbon efficiency or other metabolic engineering goals.

An alternative version of glycolysis, termed nonoxidative glycolysis (NOG), which has as its product three acetyl phosphate molecules, rather than two pyruvate molecules, was proposed and demonstrated [171]. This pathway route is carbon efficient as it avoids the decarboxylating pyruvate dehydrogenase reaction, and thus allows for complete carbon conservation in sugar conversion. The pathway was demonstrated in both *in vitro* purified enzyme assays and in *E. coli*, by showing xylose was transformed to acetate or acetyl phosphate in a 1 : 2.5 ratio. The pathway relies on an enzyme called *phosphoketolase* [172], which generates acetyl phosphate by cleaving a sugar phosphate, either fructose-6-phosphate or xylose-5-phosphate. Pentose phosphate pathway enzymes then reshuffle the sugar molecules to continue the cycle.

5.3.2 Methanol and CO_2 Utilization

Methanol is a feedstock which has significant potential for bioconversion [173]. Native methylotrophic hosts are known [174, 175], but *E. coli* has also generated interest as a potential methanol conversion host. Significant challenges exist, including the identification of appropriate heterologous methanol dehydrogenase enzymes, the management of formaldehyde toxicity, and the proper channeling of carbon flux [176]. While a fully methylotrophic *E. coli* has yet to be demonstrated, significant methanol assimilation has been shown, using an NAD-dependent methanol dehydrogenase and other enzymes of the ribulose monophosphate (RuMP) pathway from *Bacillus methanolicus* MGA3 [177]. These enzymes were demonstrated to have the highest *in vivo* concentration in a comparison among many different enzyme combinations. Expression of these enzymes in *E. coli* allowed for up to 40% of carbons composing intracellular hexose 6-phosphates to become labeled when cells were cultured in the presence of ^{13}C methanol. Demonstration of *E. coli* methanol consumption paves the way for further utilization of methanol by *E. coli* by implementation of natural and engineered pathways for growth and production [178, 179].

Carbon dioxide fixation by *E. coli* remains a captivating, but elusive goal. Functional expression of carbon-fixing enzymes such as RuBisCO is known [180]. More recently, expression of RuBisCO and phosphoribulokinase (prk) was shown to reduce CO_2 evolution in glucose metabolism and improve ethanol yield per CO_2 generated [181], although incorporation of external CO_2 was not tested. A proposed reverse glyoxylate shunt (rGS) was demonstrated in *E. coli.* Use of the exogenous genes malate-thiokinase and malyl-CoA lyase demonstrated the production of 2 acetyl-CoA molecules from 4-carbon carboxylates, theoretically allowing for increased carbon conservation in glucose metabolism, and growth solely on CO_2 with the addition of pyruvate:ferredoxin oxidoreductase [182]. Another approach showed that each section of the 3-hydroxypropionate (3HP) pathway for carbon fixation was active in *E. coli,* although it could not be demonstrated as a whole [183].

5.3.3
Advancing Analytical Techniques

Other challenges arise when using metabolites from different branches of metabolism. For example, the production of shikimates requires phosphoenolpyruvate, an intermediate of glycolysis, and erythrulose 4-phosphate, an intermediate. It is uncertain whether the fluxes of these would be properly balanced if the attempt were made to overexpress the shikimate pathway. In glucose medium, the production of DAHP was significantly improved by the expression of Pps as it increases the availability of PEP [30, 31, 184].

The resolution of the significant challenges faced by attempts to significantly rewire *E. coli* central metabolism may be aided by advances in techniques for proteomics, fluxomics, and metabolomics. Proteomic techniques are now capable of measuring with single-molecule sensitivity by using fluorescent imaging techniques [185]. Fluxomic techniques such as metabolic flux analysis (MFA) have been hampered by large error bars and difficulty with choosing an appropriate labeling scheme, but with the combination of multiple, parallel labeling experiments, much more accurate values can be obtained [186]. When measuring internal metabolites, a difficulty has been obtaining absolute concentrations, as opposed to relative concentrations. Absolute concentrations may be found if standards are prepared for each metabobilte to be measured, and a metabolome-wide measurement of over 100 metabolites was conducted in this way [187]. Altogether, these techniques will enhance the ability of metabolic engineers to push the boundaries of what is possible in altering metabolism.

5.4 Alternative Carbon Sources

In addition to having great flexibility for anabolic metabolism, *E. coli* has also shown some flexibility to adopt non-native catabolic metabolism as well. This is an important property in view of uncertainty about feedstock prices. It is unknown whether glucose, glycerol, or other starches will be more inexpensive or available over time. Thus it is important to develop capabilities to utilize all of these compounds (Figure 5.3).

5.4.1 Complex Carbohydrates

Cellulosic biomass has received extensive attention as a feedstock as it is not edible and therefore does not compete with food supply. Cellobiose is two glucose molecules connected by a $\beta(1 \rightarrow 4)$ glycosidic bond instead of $\alpha(1 \rightarrow 4)$ in more easily digestible maltose. Hemicellulose is a branched polysaccharide including different monomers and both α- and β-glycosidic bonds. Using β-glucosidases, production of fuels from cellobiose has been shown in *E. coli*. Ethanol was

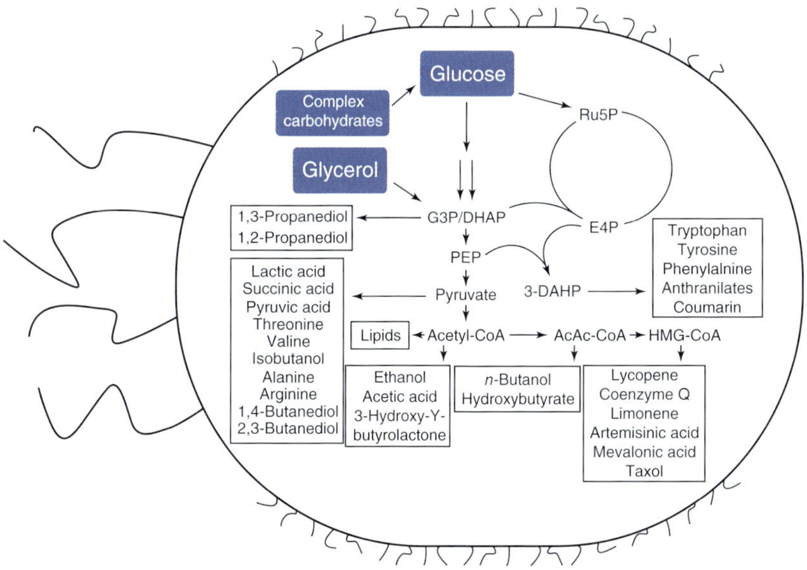

Figure 5.3 Overview of some *E. coli* feedstocks (white text on gray background) and products (black text in black boxes) discussed in this text.

produced up to 45 g l^{-1} from cellobiose and about 35 g l^{-1} from office paper, while isobutanol has been produced up to 7 g l^{-1} from cellobiose [188, 189]. Fusion of *E. coli* secretion partner OsmY to xylanases was found to enable growth on hemicellulose, a more complex carbohydrate [110, 190]. Further, *E. coli* expressing cellulases and xylanases were found to grow and produce fuel compounds on ionic-liquid pretreated plant waste [191].

Cellobiose must be cleaved into its constituent monosaccharides in order to be metabolized by *E. coli.* There are two main ways in which this reaction can occur, hydrolysis and phosphorolysis. β-Glucosidase and cellobiose phosphorylase from *Saccharophagus degradans* were expressed in *E. coli.* The results showed that phosphorolysis cells tolerate common inhibitors (sodium acetate) more effectively and produce recombinant proteins more effectively than hydrolysis cells. However, hydrolysis cells utilize xylose more effectively in combination with cellobiose [192].

Simultaneous use of different sugars such as hexoses and pentoses is an important goal in lignocellulose utilization, as lignocellulose contains xylose and glucose [193], and carbon catabolite repression can prevent some the sugars from being fully utilized [194]. Xylose and glucose were consumed simultaneously by *E. coli* by the introduction of constitutive promoters for sugar transporters and subsequent adaptations by serial passage for 50 days. The result was a strain which completely consumed glucose and xylose at almost equal rates, and grew more quickly than the wild-type parent strain in sugar mixtures [195].

5.4.2
Glycerol

Glycerol is a by-product in the esterification of TAGs for biodiesel production comprising 10 wt% of the overall product stream. This by-product has substantial volume, with 1.3 billion gallons of biodiesel produced in the United States in 2013 [196]. Additionally, crude glycerol has decreased in price from $0.25 to $0.05 per pound [197]. Use of glycerol for production of many compounds has been demonstrated in *E. coli.*

3HP and its polymer have received attention recently as compounds of interest for production from glycerol. P3HP was produced from glyerol to about 2 g l^{-1} and 12 wt% of CDW. This was accomplished using glycerol dehydratase (dhaB1) of *Clostridium butyricum*, propionaldehyde dehydrogenase (PduP) of *Salmonella enterica*, and PHA synthase of *R. eutropha* [198]. By optimizing fermentation conditions, production was increased to 10 g l^{-1} and 46 wt% CDW [135]. Production of the monomer, 3HP, has also been accomplished. By deregulation of glycerol metabolism, 3HP was produced up to 42 g l^{-1} [133].

Other compounds have also been produced. For example, hydrogen was produced by a fermentative strategy using hydrogenase (Hyd) and fhl enzymes [199]. 1,2-PDO has also been produced from glycerol in *E. coli* by a pathway through DHAP and methylglyoxal [200]. Ethanol and other fermentative products have also been produced [201]. Limonene has been produced by an *E. coli* strain from glycerol in minimal media at a concentration higher than was produced from glucose [202].

5.4.3 Protein to Fuel

A concern in the use of biofuels is that the nitrogen cycle is not closed. Ammonia fertilizers are used to grow plants and the resulting biomass is converted to carbon. Nitrogen-rich industrial and agricultural wastes are not converted back into ammonia for fertilizers [203, 204]. Additionally, *E. coli* is not naturally able to completely catabolize all natural amino acids. Enzymes catalyzing irreversible deamination reactions were introduced into *E. coli*. This allowed for the production of over $4\,g\,l^{-1}$ of biofuels from biomass containing amino acids. Further work in this area is being pursued in other organisms, notably *B. subtilis* [76].

5.5 *E. coli* Techniques and Concerns

5.5.1 Genetic Manipulation

To accomplish the many production schemes described in this chapter, genetic transformation of *E. coli* is necessary. Genetic transformation falls broadly into two categories, either increase of expression for a given polypeptide (overexpression) or decrease (knockdown or knockout), which are used to direct metabolite flow for metabolic engineering purposes. There are many effective strategies for each type of transformation. Bacterial transformation has been covered extensively elsewhere [205], so it is not covered at length here.

E. coli genetic transformation is well known and readily accomplished [206]. Plasmid-based expression is the simplest and fastest method and transformation is achievable by chemical or electroshock competency [207, 208]. Commercial competent cell kits are also available (e.g., NEB Turbo Competent *E. coli*). Chromosomal integration and gene knockout are also possible by a number of methods including Tn*7* transposon and lambda Red phage [209, 210]. Commercial kits are also available (e.g., Sigma-Aldrich TargeTron® Gene Knockout System). While transformation of *E. coli* is well known, techniques with increased efficiency or throughput become available occasionally [211].

5.5.2 Phage Contamination

Bacteriophages (phages) are viruses which inject genetic material into bacteria, potentially leading to the destruction of the bacterial cell and generation of more phage particles. A touted application of these viruses is their potential to treat bacterial infections in lieu of antibiotics [212, 213]. However, phage contamination poses a serious threat to microbial bioprocessing.

Phages have been studied extensively, pertaining to their physiology, life cycle, and genetics, with at least an entire textbook and academic journal devoted to them [214, 215]. However, the practical problem of dealing with phage contamination in microbial bioprocessing is a relatively less studied area, with far fewer sources available in the literature.

There are many concerns and confounding details of phage contamination. The infections can spread rapidly, causing problems for bioprocessing at lab and production scale, but pose the most difficulty in large, hard-to-clean bioreactors. Phages are small in size, much smaller than the bacterial cells they infect, so they can pass through some filters and may be present in upstream ingredients of bioprocesses including media and buffers [216]. A study of the effect of media conditions showed that bacteria in starvation conditions propagated phages most slowly, indicating that carbon starvation may be a prudent course of action upon detection of phage contamination [217].

The simplest method for detection of phages is the plaque assay [218]. A lawn of bacteria is grown on a plate of a suitable solid medium. Then, liquid from the suspected culture is dropped onto the bacterial lawn. After 24–48 h, a plaque may form on the bacterial lawn, indicating the infection and lysis of the bacteria by a phage. Serial dilutions may be performed to determine a more quantitative phage titer, although phages are known to be present in infected cultures at 10^{13} ml^{-1} [216]. More advanced phage detection methods were demonstrated by electric biochip and by propidium iodide labeling, allowing for detection of phage concentrations between 10^2 and 10^7 ml^{-1} in 25 min – 1 h depending on the exact method and concentrations used [219, 220].

5.6 Conclusions

The overarching theme that emerges from the explosion of research into the use of *E. coli* over the past decade or so is that there is almost no limit to what can be produced by *E. coli* in laboratory conditions [221]. However, it is becoming clear that the economical case for bioprocessing writ large is much narrower than the scientific one. Indeed several factors are keys in the economic viability of bioproduction of chemicals broadly, including titer, productivity, product value, raw material cost, and underlying economic forces such as oil price [222]. This is reflected in the bioproduction processes which have been successfully brought to market. The product is either high value (protein therapeutics) [13], or produced in very high yields, titers, and productivities (PDO, amino acids, ethanol, and lactic acid) [7, 9, 84, 223].

E. coli is metabolically flexible and easily genetically manipulated, and this allows it to be competitive for some bioprocesses. The economical production of PDO by an engineered pathway in *E. coli* remains a considerable achievement in metabolic engineering and bioprocessing, and certain amino acids are also produced economically in *E. coli* [15, 224]. Other high-value products such as protein therapeutics and industrial enzymes are also produced by *E. coli* owing to its genetic flexibility (Table 5.1). The long-term motivation for microbial biocatalysis remains strong despite variability in oil price [225]. Finite fossil fuels and carbon dioxide emissions will continue to drive the market toward biocatalysis and *E. coli* will remain an advantageous host for many applications.

References

1. Blattner, F.R. *et al* (1997) The complete genome sequence of *Escherichia coli* K-12. *Science*, **277**, 1453–1462.
2. Keseler, I.M. (2004) EcoCyc: a comprehensive database resource for *Escherichia coli*. *Nucleic Acids Res.*, **33**, D334–D337.
3. Salis, H.M., Mirsky, E.A., and Voigt, C.A. (2009) Automated design of synthetic ribosome binding sites to control protein expression. *Nat. Biotechnol.*, **27**, 946–950.
4. Mandel, M. and Higa, A. (1970) Calcium-dependent bacteriophage DNA infection. *J. Mol. Biol.*, **53**, 159–162.
5. Łos, M. *et al* (2004) Bacteriophage contamination: is there a simple method to reduce its deleterious effects in laboratory cultures and biotechnological factories? *J. Appl. Genet.*, **45**, 111–120.
6. FDA (2014) Generally Recognized as Safe (GRAS) Notification for L-Methionine 85% From a Modified *Escherichia coli* K-12.
7. Kosaric, N. *et al* (2005) Ethanol. *Ullmann's Encycl. Ind. Chem.*, **13**, 8676–8694.
8. Verhoff, F.H. and Bauweleers, H. (2012) Citric acid. *Ullmann's Encycl. Ind. Chem.*, **57**, 657–659.
9. Drauz, K. *et al* (2012) Amino acids. *Ullmann's Encycl. Ind. Chem.*, **3**, 1–58, doi: 10.1002/14356007.a02.
10. Atsumi, S. *et al* (2008) Metabolic engineering of *Escherichia coli* for 1-butanol production. *Metab. Eng.*, **10**, 305–311.
11. Shen, C.R. *et al* (2011) Driving forces enable high-titer anaerobic 1-butanol synthesis in *Escherichia coli*. *Appl. Environ. Microbiol.*, **77**, 2905–2915.
12. Gebel, E. (2013) Making Insulin, *Diabetes Forecast*, http://www.diabetesforecast.org/2013/jul/making-insulin.html (accessed 31 May 2016).
13. Huang, C.-J., Lin, H., and Yang, X. (2012) Industrial production of recombinant therapeutics in *Escherichia coli* and its recent advancements. *J. Ind. Microbiol. Biotechnol.*, **39**, 383–399.
14. Nakamura, C.E. and Whited, G.M. (2003) Metabolic engineering for the microbial production of 1,3-propanediol. *Curr. Opin. Biotechnol.*, **14**, 454–459.
15. Zeng, A.-P. and Biebl, H. (2002) Bulk chemicals from biotechnology: the case of 1,3-propanediol production and the new trends. *Adv. Biochem. Eng./Biotechnol.*, **74**, 239–259.
16. Scheper, T., Faurie, R., and Thommel, J. (2003) *Microbial Production of L-Amino Acids*, Springer.
17. Van Dien, S. *et al* (2012) Microorganisms for the production of 1,4-butanediol and related methods, US Patent: 8129169.
18. Genomatica Inc. (2013) BASF plans production of butanediol from renewable feedstock using genomatica technology. Press Release: May 10, 2013.
19. Clausen, K. (1993) An improved method for the preparation of certain beta-lactam antibiotics, US Patent: 5470717.
20. Pond, W., Church, D., and Pond, K. (1995) *Basic Animal Nutrition and Feeding*, Wiley.

21. Bucci, L.R. and Unlu, L. (2000) *Energy-Yielding Macronutrients and Energy Metabolism in Sports Nutrition*, (eds Judy A. Driskell, Ira Wolinsky) CRC Press, pp. 191–210.
22. Stadtman, E.R., Cohen, G.N., Lebras, G., and De Robichon-szulmajster, H. (1961) Feed-back inhibition and repression of aspartokinase activity in *Escherichia coli* and *Saccharomyces cerevisiae*. *J. Biol. Chem.*, **236**, 2033–2038.
23. Umbarger, H.E., Brown, B., and Edwin, H. (1958) Isoleucine and valine metabolism in *Escherichia coli* VII. *J. Biol. Chem.*, **233**, 415–420.
24. Furukawa, S., Ozaki, A., Kotani, Y., and Nakanishi, T. (1988) Breeding of L-threonine hyper-producer of *Escherichia coli* W. *Appl. Microbiol. Biotechnol.*, **29**, 253–257.
25. Song, K.H., Lee, H.H., and Hyun, H.H. (2000) Characterization of salt-tolerant mutant for enhancement of L-threonine production in *Escherichia coli*. *Appl. Microbiol. Biotechnol.*, **54**, 647–651.
26. Lee, K.H., Park, J.H., Kim, T.Y., Kim, H.U., and Lee, S.Y. (2007) Systems metabolic engineering of *Escherichia coli* for L-threonine production. *Mol. Syst. Biol.*, **3**, 149.
27. Park, J.H., Lee, K.H., Kim, T.Y., and Lee, S.Y. (2007) Metabolic engineering of *Escherichia coli* for the production of L-valine based on transcriptome analysis and in silico gene knockout simulation. *Proc. Natl. Acad. Sci. U.S.A.*, **104**, 7797–7802.
28. Lawther, R.P. *et al* (1987) The complete nucleotide sequence of the ilvGMEDA operon of *Escherichia coli* K-12. *Nucleic Acids Res.*, **15**, 2137–2155.
29. Park, J.H., Jang, Y.-S., Lee, J.W., and Lee, S.Y. (2011) *Escherichia coli* W as a new platform strain for the enhanced production of L-valine by systems metabolic engineering. *Biotechnol. Bioeng.*, **108**, 1140–1147.
30. Patnaik, R. and Liao, J.C. (1994) Engineering of *Escherichia coli* central metabolism for aromatic metabolite production with near theoretical yield. *Appl. Environ. Microbiol.*, **60**, 3903–3908.
31. Patnaik, R., Spitzer, R.G., and Liao, J.C. (1995) Pathway engineering for production of aromatics in *Escherichia coli*: comfirmation of stoichiometric analysis by independent modulation of AroG, TktA, and Pps activities. *Biotechnol. Bioeng.*, **46**, 361–370.
32. Flores, N., Xiao, J., Berry, A., Bolivar, F., and Valle, F. (1996) Pathway engineering for the production of aromatic compounds in *Escherichia coli*. *Nat. Biotechnol.*, **14**, 620–623.
33. Gerigk, M. *et al* (2002) Process control for enhanced L-phenylalanine production using different recombinant *Escherichia coli* strains. *Biotechnol. Bioeng.*, **80**, 746–754.
34. Schoner, R. and Herrmann, K.M. (1976) 3-deoxy-D-arabino-heptulosonate 7-phosphate synthase. Purification, properties, and kinetis of the tyrosine-sensitive isoenzyme from *Escherichia coli*. *J. Biol. Chem.*, **251**, 5440–5447.
35. Kedar, P., Colah, R., and Shimizu, K. (2007) Proteomic investigation on the pyk-F gene knockout *Escherichia coli* for aromatic amino acid production. *Enzyme Microb. Technol.*, **41**, 455–465.
36. Liu, S.P. *et al* (2014) A systems level engineered *E. coli* capable of efficiently producing L-phenylalanine. *Process Biochem.*, **49**, 751–757.
37. Weiner, M., Albermann, C., Gottlieb, K., Sprenger, G.A., and Weuster-Botz, D. (2014) Fed-batch production of l-phenylalanine from glycerol and ammonia with recombinant *Escherichia coli*. *Biochem. Eng. J.*, **83**, 62–69.
38. Sprenger, G.A. (2007) From scratch to value: engineering *Escherichia coli* wild type cells to the production of L-phenylalanine and other fine chemicals derived from chorismate. *Appl. Microbiol. Biotechnol.*, **75**, 739–749.
39. Chávez-Béjar, M.I. *et al* (2008) Metabolic engineering of *Escherichia coli* for L-tyrosine production by expression of genes coding for the chorismate mutase domain of the native chorismate mutase-prephenate dehydratase and a cyclohexadienyl dehydrogenase from *Zymomonas*

mobilis. *Appl. Environ. Microbiol.*, **74**, 3284–3290.
40. Santos, C.N.S., Xiao, W., and Stephanopoulos, G. (2012) Rational, combinatorial and genomic approaches for engineering L-tyrosine production in *Escherichia coli*. *Proc. Natl. Acad. Sci. U.S.A.*, **109**, 13538–13543.
41. Patnaik, R., Zolandz, R.R., Green, D.A., and Kraynie, D.F. (2008) L-Tyrosine production by recombinant *Escherichia coli*: fermentation optimization and recovery. *Biotechnol. Bioeng.*, **99**, 741–752.
42. Zhao, Z.-J. *et al* (2011) Development of L-tryptophan production strains by defined genetic modification in *Escherichia coli*. *J. Ind. Microbiol. Biotechnol.*, **38**, 1921–1929.
43. Eudes, A. *et al* (2013) Production of hydroxycinnamoyl anthranilates from glucose in *Escherichia coli*. *Microb. Cell Fact.*, **12**, 62.
44. Zhang, X., Jantama, K., Moore, J.C., Shanmugam, K.T., and Ingram, L.O. (2007) Production of L -alanine by metabolically engineered *Escherichia coli*. *Appl. Microbiol. Biotechnol.*, **77**, 355–366.
45. Park, S.H. *et al* (2014) Metabolic engineering of *Corynebacterium glutamicum* for L-arginine production. *Nat. Commun.*, **5**, 1–9.
46. M. Ginesy, J. Belotserkovsky, J. Enman, L. Isaksson, U. Rova, Metabolic engineering of *Escherichia coli* for enhanced arginine biosynthesis, *Microb. Cell Fact.* **14**, 1–11 (2015).
47. Chang, D., Jung, H., Rhee, J., and Pan, J. (1999) Homofermentative production of D- or L-lactate in metabolically engineered *Escherichia coli* RR1. *Appl. Environ. Microbiol.*, **65**, 1384–1389.
48. Dien, B.S., Nichols, N.N., and Bothast, R.J. (2002) Fermentation of sugar mixtures using *Escherichia coli* catabolite repression mutants engineered for production of L-lactic acid. *J. Ind. Microbiol. Biotechnol.*, **29**, 221–227.
49. Zhou, L. *et al* (2011) Evaluation of genetic manipulation strategies on D-lactate production by *Escherichia coli*. *Curr. Microbiol.*, **62**, 981–989.
50. US Department of Energy (2004) Top Value Added Chemicals from Biomass.
51. Vemuri, G.N., Eiteman, M.A., and Altman, E. (2002) Succinate production in dual-phase *Escherichia coli* fermentations depends on the time of transition from aerobic to anaerobic conditions. *J. Ind. Microbiol. Biotechnol.*, **28**, 325–332.
52. Jantama, K. *et al* (2008) Combining metabolic engineering and metabolic evolution to develop nonrecombinant strains of *Escherichia coli* C that produce succinate and malate. *Biotechnol. Bioeng.*, **99**, 1140–1153.
53. Zhang, X. *et al* (2009) Metabolic evolution of energy-conserving pathways for succinate production in *Escherichia coli*. *Proc. Natl. Acad. Sci. U.S.A.*, **106**, 20180–20185.
54. Reuters (2014) PreScience Labs Announced that the FDA Accepts IND Application for Novel Oncology Drug, http://www.reuters.com/article/2013/07/24/md-prescience-labs-idUSnBw246023a+100+BSW20130724 (accessed 31 May 2016).
55. Causey, T.B., Shanmugam, K.T., Yomano, L.P., and Ingram, L.O. (2004) Engineering *Escherichia coli* for efficient conversion of glucose to pyruvate. *Proc. Natl. Acad. Sci. U.S.A.*, **101**, 2235–2240.
56. Cheung, H., Tanke, R.S., and Torrence, G.P. (2012) Acetic acid. *Ullman's Encycl. Ind. Chem.*, **1**, doi: 10.1002/14356007.a01
57. Causey, T.B., Zhou, S., Shanmugam, K.T., and Ingram, L.O. (2003) Engineering the metabolism of *Escherichia coli* W3110 for the conversion of sugar to redox-neutral and oxidized products: homoacetate production. *Proc. Natl. Acad. Sci. U.S.A.*, **100**, 825–832.
58. Lee, S.-H. and Park, O.-J. (2009) Uses and production of chiral 3-hydroxy-gamma-butyrolactones and structurally related chemicals. *Appl. Microbiol. Biotechnol.*, **84**, 817–828.
59. Martin, C.H. *et al* (2013) A platform pathway for production of 3-hydroxyacids provides a biosynthetic route to 3-hydroxy-γ-butyrolactone. *Nat. Commun.*, **4**, 1414.

60. Dhamankar, H., Tarasova, Y., Martin, C.H., and Prather, K.L.J. (2014) Engineering *E. coli* for the biosynthesis of 3-hydroxy-γ-butyrolactone (3HBL) and 3,4-dihydroxybutyric acid (3,4-DHBA) as value-added chemicals from glucose as a sole carbon source. *Metab. Eng.*, **25**, 72–81.
61. Song, C.W., Lee, J., Ko, Y.-S., and Yup Lee, S. (2015) Metabolic engineering of *Escherichia coli* for the production of 3-aminopropionic acid. *Metab. Eng.*, **30**, 121–129.
62. Atsumi, S., Hanai, T., and Liao, J.C. (2008) Non-fermentative pathways for synthesis of branched-chain higher alcohols as biofuels. *Nature*, **451**, 86–89.
63. Hazelwood, L.A., Daran, J.-M., van Maris, A.J.A., Pronk, J.T., and Dickinson, J.R. (2008) The Ehrlich pathway for fusel alcohol production: a century of research on *Saccharomyces cerevisiae* metabolism. *Appl. Environ. Microbiol.*, **74**, 2259–2266.
64. Baez, A., Cho, K.-M., and Liao, J.C. (2011) High-flux isobutanol production using engineered *Escherichia coli*: a bioreactor study with in situ product removal. *Appl. Microbiol. Biotechnol.*, **90**, 1681–1690.
65. Smith, K.M. and Liao, J.C. (2011) An evolutionary strategy for isobutanol production strain development in *Escherichia coli*. *Metab. Eng.*, **13**, 674–681.
66. Atsumi, S., Li, Z., and Liao, J.C. (2009) Acetolactate synthase from *Bacillus subtilis* serves as a 2-ketoisovalerate decarboxylase for isobutanol biosynthesis in *Escherichia coli*. *Appl. Environ. Microbiol.*, **75**, 6306–6311.
67. Bastian, S. *et al* (2011) Engineered ketol-acid reductoisomerase and alcohol dehydrogenase enable anaerobic 2-methylpropan-1-ol production at theoretical yield in *Escherichia coli*. *Metab. Eng.*, **13**, 345–352.
68. Atsumi, S. *et al* (2010) Evolution, genomic analysis, and reconstruction of isobutanol tolerance in *Escherichia coli*. *Mol. Syst. Biol.*, **6**, 449.
69. Brynildsen, M.P. and Liao, J.C. (2009) An integrated network approach identifies the isobutanol response network of *Escherichia coli*. *Mol. Syst. Biol.*, **5**, 277.
70. Liao, J.C. *et al* (2003) Network component analysis: reconstruction of regulatory signals in biological systems. *Proc. Natl. Acad. Sci. U.S.A.*, **100**, 15522–15527.
71. Derrick, S. and Large, P.J. (1993) Activities of the enzymes of the Ehrlich pathway and formation of branched-chain alcohols in *Saccharomyces cerevisiae* and *Candida utilis* grown in continuous culture on valine or ammonium as sole nitrogen source. *J. Gen. Microbiol.*, **139**, 2783–2792.
72. Smith, K.M., Cho, K.-M., and Liao, J.C. (2010) Engineering *Corynebacterium glutamicum* for isobutanol production. *Appl. Microbiol. Biotechnol.*, **87**, 1045–1055.
73. Li, H. *et al* (2012) Integrated electromicrobial conversion of CO_2 to higher alcohols. *Science*, **335**, 1596.
74. Atsumi, S., Higashide, W., and Liao, J.C. (2009) Direct photosynthetic recycling of carbon dioxide to isobutyraldehyde. *Nat. Biotechnol.*, **27**, 1177–1180.
75. Lin, P.P. *et al* (2014) Isobutanol production at elevated temperatures in thermophilic *Geobacillus thermoglucosidasius*. *Metab. Eng.*, **24**, 1–8.
76. Choi, K.-Y., Wernick, D.G., Tat, C.A., and Liao, J.C. (2014) Consolidated conversion of protein waste into biofuels and ammonia using *Bacillus subtilis*. *Metab. Eng.*, **23**, 53–61.
77. Lin, P.P. *et al* (2015) Consolidated bioprocessing of cellulose to isobutanol using *Clostridium thermocellum*. *Metab. Eng.*, **31**, 44–52.
78. Cann, A.F. and Liao, J.C. (2008) Production of 2-methyl-1-butanol in engineered *Escherichia coli*. *Appl. Microbiol. Biotechnol.*, **81**, 89–98.
79. Connor, M.R., Cann, A.F., and Liao, J.C. (2010) 3-Methyl-1-butanol production in *Escherichia coli*: random mutagenesis and two-phase fermentation. *Appl. Microbiol. Biotechnol.*, **86**, 1155–1164.
80. Shen, C.R. and Liao, J.C. (2008) Metabolic engineering of *Escherichia*

coli for 1-butanol and 1-propanol production via the keto-acid pathways. *Metab. Eng.*, **10**, 312–320.
81. Marcheschi, R.J. *et al* (2012) A synthetic recursive "+1" pathway for carbon chain elongation. *ACS Chem. Biol.*, 7, 689–697.
82. Hanai, T., Atsumi, S., and Liao, J.C. (2007) Engineered synthetic pathway for isopropanol production in *Escherichia coli*. *Appl. Environ. Microbiol.*, **73**, 7814–7818.
83. Inokuma, K., Liao, J.C., Okamoto, M., and Hanai, T. (2010) Improvement of isopropanol production by metabolically engineered *Escherichia coli* using gas stripping. *J. Biosci. Bioeng.*, **110**, 696–701.
84. Sullivan, C.J. (2000) *Ullman's Encyclopedia of Industrial Chemistry*, Wiley-VCH Verlag GmbH.
85. Laffend, L.A., Nagarjan, V., and Nakamura, C.E. (1997) Bioconversion of a fermentable carbon source to 1,3-propanediol by a single microorganism, US Patent: 5686276.
86. Diaz-Torres, M., Dunn-Coleman, N.S., Chase, M.W., and Trimbur, D. (2000) Method for recombinant production of 1,3-propanediol, US Patent: 6013494.
87. Adkesson, D.M. *et al* (2011) Purification of biologically-produced 1,3-propanediol, US Patent: 7919658.
88. Altaras, N.E. and Cameron, D.C. (1999) Metabolic engineering of a 1, 2-propanediol pathway in *Escherichia coli*. *Appl. Environ. Microbiol.*, **65**, 1180–1185.
89. Altaras, N.E. and Cameron, D.C. (2000) Enhanced production of (R)-1,2-propanediol by metabolically engineered *Escherichia coli*. *Biotechnol. Progr.*, **16**, 940–946.
90. Gräfje, H., Körnig, W., Weitz, H.-M., Reiss, W., and Steffan, G. (2000) *Ullman's Encyclopedia of Industrial Chemistry*, Butanediols, Butenediol, and Butynediol, Wiley.
91. Yim, H. *et al* (2011) Metabolic engineering of *Escherichia coli* for direct production of 1,4-butanediol. *Nat. Chem. Biol.*, 7, 445–452.
92. Ui, S. *et al* (2004) Production of L-2,3-butanediol by a new pathway constructed in *Escherichia coli*. *Lett. Appl. Microbiol.*, **39**, 533–537.
93. Ui Sadaharu, *et al* (1997) Molecular Generation of an Escherichia coli Strain Producing only the meso-Isomer of 2,3-Butanediol. *J. Rementation Bioeng.*, **83** (3), 185–189.
94. Yan, Y., Lee, C.-C., and Liao, J.C. (2009) Enantioselective synthesis of pure (R,R)-2,3-butanediol in *Escherichia coli* with stereospecific secondary alcohol dehydrogenases. *Org. Biomol. Chem.*, 7, 3914–3917.
95. Seol, E. *et al* (2011) Sustained hydrogen production from formate using immobilized recombinant *Escherichia coli* SH5. *Int. J. Hydrogen Energy*, **36**, 8681–8686.
96. Yoshida, A., Nishimura, T., Kawaguchi, H., Inui, M., and Yukawa, H. (2006) Enhanced hydrogen production from glucose using ldh- and frd-inactivated *Escherichia coli* strains. *Appl. Microbiol. Biotechnol.*, **73**, 67–72.
97. Udupa, K., Subramanian, G., and Udupa, H. (1971) The electrolytic reduction of carbon dioxide to formic acid. *Electrochim. Acta*, **16**, 1593–1598.
98. Bar-Even, A., Noor, E., Flamholz, A., and Milo, R. (2013) Design and analysis of metabolic pathways supporting formatotrophic growth for electricity-dependent cultivation of microbes. *Biochim. Biophys. Acta*, **1827**, 1039–1047.
99. Hershlag (Portman), N., Hurley, I., and Woodward, J. (1998) A simple method to demonstrate the enzymatic production of hydrogen from sugar. *J. Chem. Educ.*, **75**, 1270.
100. Maeda, T., Sanchez-Torres, V., and Wood, T.K. (2007) *Escherichia coli* hydrogenase 3 is a reversible enzyme possessing hydrogen uptake and synthesis activities. *Appl. Microbiol. Biotechnol.*, **76**, 1035–1042.
101. Sawers, G. (1994) The hydrogenases and formate dehydrogenases of *Escherichia coli*. *Antonie Van Leeuwenhoek*, **66**, 57–88.
102. Choi, Y.J. and Lee, S.Y. (2013) Microbial production of short-chain alkanes. *Nature*, **502**, 571–574.

103. Kinney, A.J. (1996) Designer oils for better nutrition. *Nat. Biotechnol.*, **14**, 946.
104. Gunstone, F.D. and Padley, F.B. (1997) *Lipid Technologies and Applications*, Marcel Dekker, New York.
105. Kinney, A.J. and Knowlton, S. (1998) in *Genetic Modification in the Food Industry* (eds S. Roller and S. Harlander), Blackie, pp. 193–213.
106. Janßen, H.J. and Steinbüchel, A. (2014) Fatty acid synthesis in *Escherichia coli* and its applications towards the production of fatty acid based biofuels. *Biotechnol. Biofuels*, **7**, 7.
107. Handke, P., Lynch, S.A., and Gill, R.T. (2011) Application and engineering of fatty acid biosynthesis in *Escherichia coli* for advanced fuels and chemicals. *Metab. Eng.*, **13**, 28–37.
108. Liu, T., Vora, H., and Khosla, C. (2010) Quantitative analysis and engineering of fatty acid biosynthesis in *E. coli*. *Metab. Eng.*, **12**, 378–386.
109. Liu, H. *et al* (2012) Production of extracellular fatty acid using engineered *Escherichia coli*. *Microb. Cell Fact.*, **11**, 41.
110. Steen, E.J. *et al* (2010) Microbial production of fatty-acid-derived fuels and chemicals from plant biomass. *Nature*, **463**, 559–562.
111. Rodriguez, G.M., Tashiro, Y., and Atsumi, S. (2014) Expanding ester biosynthesis in *Escherichia coli*. *Nat. Chem. Biol.*, **10**, 259–265.
112. Tai, Y., Xiong, M., and Zhang, K. (2015) Engineered biosynthesis of medium-chain esters in *Escherichia coli*. *Metab. Eng.*, **27**, 20–28.
113. Mathieu, M. *et al* (1997) The 1.8 A crystal structure of the dimeric peroxisomal 3-ketoacyl-CoA thiolase of *Saccharomyces cerevisiae*: implications for substrate binding and reaction mechanism. *J. Mol. Biol.*, **273**, 714–728.
114. Schwartz, M. *et al* (1984) The University of Toronto head injury treatment study: a prospective, randomized comparison of phenobarbitol and mannitol. *Can. J. Neurol. Sci.*, **11**, 434–440.
115. Kaup, B., Bringer-Meyer, S., and Sahm, H. (2004) Metabolic engineering of *Escherichia coli*: construction of an efficient biocatalyst for D-mannitol formation in a whole-cell biotransformation. *Appl. Microbiol. Biotechnol.*, **64**, 333–339.
116. Suzuki, T. *et al* (1999) Expression of xyrA gene encoding for D-xylose reductase of *Candida tropicalis* and production of xylitol in *Escherichia coli*. *J. Biosci. Bioeng.*, **87**, 280–284.
117. Khankal, R., Chin, J.W., and Cirino, P.C. (2008) Role of xylose transporters in xylitol production from engineered *Escherichia coli*. *J. Biotechnol.*, **134**, 246–252.
118. Akinterinwa, O. and Cirino, P.C. (2011) Anaerobic obligatory xylitol production in *Escherichia coli* strains devoid of native fermentation pathways. *Appl. Environ. Microbiol.*, **77**, 706–709.
119. Sies, H. and Stahl, W. (1998) Lycopene: antioxidant and biological effects and its bioavailability in humans. *Proc. Soc. Exp. Biol. Med.*, **218**, 121–124.
120. Giovannucci, E. (1999) Tomatoes, tomato-based products, lycopene, and cancer: review of the epidemiologic literature. *J. Natl. Cancer Inst.*, **91**, 317–331.
121. Rizwan, M. *et al* (2011) Tomato paste rich in lycopene protects against cutaneous photodamage in humans in vivo: a randomized controlled trial. *Br. J. Dermatol.*, **164**, 154–162.
122. Farmer, W.R. and Liao, J.C. (2000) Improving lycopene production in *Escherichia coli* by engineering metabolic control. *Nat. Biotechnol.*, **18**, 533–537.
123. Farmer, W.R. and Liao, J.C. (2001) Precursor balancing for metabolic engineering of lycopene production in *Escherichia coli*. *Biotechnol. Progr.*, **17**, 57–61.
124. Zahiri, H.S., Noghabi, K.A., and Shin, Y.C. (2006) Biochemical characterization of the decaprenyl diphosphate synthase of *Rhodobacter sphaeroides* for coenzyme Q10 production. *Appl. Microbiol. Biotechnol.*, **73**, 796–806.
125. Zahiri, H.S. *et al* (2006) Coenzyme Q10 production in recombinant *Escherichia coli* strains engineered with a heterologous decaprenyl diphosphate synthase

gene and foreign mevalonate pathway. *Metab. Eng.*, **8**, 406–416.
126. Alonso-Gutierrez, J. *et al* (2013) Metabolic engineering of *Escherichia coli* for limonene and perillyl alcohol production. *Metab. Eng.*, **19**, 33–41.
127. Chang, M.C.Y., Eachus, R.A., Trieu, W., Ro, D.-K., and Keasling, J.D. (2007) Engineering *Escherichia coli* for production of functionalized terpenoids using plant P450s. *Nat. Chem. Biol.*, **3**, 274–277.
128. Selmer, T., Willanzheimer, A., and Hetzel, M. (2002) Propionate CoA-transferase from *Clostridium propionicum* cloning of the gene and identification of glutamate 324 at the active site. *Eur. J. Biochem.*, **269**, 372–380.
129. Yang, T.H. *et al* (2010) Biosynthesis of polylactic acid and its copolymers using evolved propionate CoA transferase and PHA synthase. *Biotechnol. Bioeng.*, **105**, 150–160.
130. Jung, Y.K., Kim, T.Y., Park, S.J., and Lee, S.Y. (2010) Metabolic engineering of *Escherichia coli* for the production of polylactic acid and its copolymers. *Biotechnol. Bioeng.*, **105**, 161–171.
131. Lee, S.Y. (1996) Bacterial polydroxyalkanoates. *Biotechnol. Bioeng.*, **49**, 1–14.
132. Tyo, K.E.J., Fischer, C.R., Simeon, F., and Stephanopoulos, G. (2010) Analysis of polyhydroxybutyrate flux limitations by systematic genetic and metabolic perturbations. *Metab. Eng.*, **12**, 187–195.
133. Jung, W.S., Kang, J.H., Chu, H.S., Choi, I.S., and Cho, K.M. (2014) Elevated production of 3-hydroxypropionic acid by metabolic engineering of the glycerol metabolism in *Escherichia coli*. *Metab. Eng.*, **23**, 116–122.
134. Andreessen, B., Lange, A.B., Robenek, H., and Steinbüchel, A. (2010) Conversion of glycerol to poly(3-hydroxypropionate) in recombinant *Escherichia coli*. *Appl. Environ. Microbiol.*, **76**, 622–626.
135. Wang, Q. *et al* (2013) Bioresource technology biosynthesis of poly (3-hydroxypropionate) from glycerol by recombinant *Escherichia coli*. *Bioresour. Technol.*, **131**, 548–551.
136. Bohlmann, G.M. (2006) *Feedstocks for the Future* (eds Joseph J. Bozell, Martin K. Patel), vol. 921, American Chemical Society, pp. 253–270.
137. Metabolix Inc. (2004) Metabolix and ADM enter strategic alliance to commercialize PHA natural polymers. Press Release: November 4, 2004.
138. Archer Daniels Midland, Inc (2012) ADM to end commercial alliance with metabolix. Press Release: January 12, 2012, pp. 1–2.
139. Yu, H. and Stephanopoulos, G. (2008) Metabolic engineering of *Escherichia coli* for biosynthesis of hyaluronic acid. *Metab. Eng.*, **10**, 24–32.
140. He, W. *et al* (2015) Production of chondroitin in metabolically engineered *E. coli*. *Metab. Eng.*, **27**, 92–100.
141. Xiong, M., Schneiderman, D.K., Bates, F.S., Hillmyer, M.A., and Zhang, K. (2014) Scalable production of mechanically tunable block polymers from sugar. *Proc. Natl. Acad. Sci. U.S.A.*, **111**, 8357–8362.
142. Xia, X.-X. *et al* (2010) Native-sized recombinant spider silk protein produced in metabolically engineered *Escherichia coli* results in a strong fiber. *Proc. Natl. Acad. Sci. U.S.A.*, **107**, 14059–14063.
143. Unkles, S.E., Valiante, V., Mattern, D.J., and Brakhage, A.A. (2014) Synthetic biology tools for bioprospecting of natural products in eukaryotes. *Chem. Biol.*, **21**, 502–508.
144. Cragg, G.M., Newman, D.J., and Snader, K.M. (1997) Natural products in drug discovery and development, *J. Natural Prod.*, **3864**, 52–60.
145. Newman, D.J. and Cragg, G.M. (2007) Natural products as sources of new drugs over the last 25 years. *J. Nat. Prod.*, **70**, 461–477.
146. Heisey, R.M. *et al* (1988) Production of valinomycin, an insecticidal antibiotic, by *Streptomyces griseus* var. flexiperum var. nov. *J. Agric. Food. Chem.*, **2**, 1283–1286.
147. Jaitzig, J., Li, J., Sussmuth, R.D., and Neubauer, P. (2013) Reconstituted nonribosomal biosynthesis of the peptide

antibiotic valinomycin in *Escherichia coli* as recombinant whole-cell biocatalyst, *ACS Synth. Biol.*, **3** (7), 432–438.

148. Li, J., Jaitzig, J., Hillig, F., Süssmuth, R., and Neubauer, P. (2014) Enhanced production of the nonribosomal peptide antibiotic valinomycin in *Escherichia coli* through small-scale high cell density fed-batch cultivation. *Appl. Microbiol. Biotechnol.*, **98**, 591–601.
149. Li, J., Jaitzig, J., Lu, P., Süssmuth, R.D., and Neubauer, P. (2015) Scale-up bioprocess development for production of the antibiotic valinomycin in *Escherichia coli* based on consistent fed-batch cultivations. *Microb. Cell Fact.*, **14**, 83.
150. Gleye, C. *et al* (2003) Acaricidal activity of tonka bean extracts. Synthesis and structure-activity relationships of bioactive derivatives. *J. Nat. Prod.*, **66**, 690–692.
151. Yang, S.-M., Shim, G.Y., Kim, B.-G., and Ahn, J.-H. (2015) Biological synthesis of coumarins in *Escherichia coli. Microb. Cell Fact.*, **14**, 1–12.
152. Wang, S. *et al* (2015) Metabolic engineering of *Escherichia coli* for the biosynthesis of various phenylpropanoid derivatives. *Metabolic*, **29**, 153–159.
153. Makrides, S.C. (1996) Strategies for achieving high-level expression of genes in *Escherichia coli. Microbiol. Rev.*, **60**, 512–538.
154. Sahdev, S., Khattar, S.K., and Saini, K.S. (2008) Production of active eukaryotic proteins through bacterial expression systems: a review of the existing biotechnology strategies. *Mol. Cell. Biochem.*, **307**, 249–264.
155. Shin, P.K. and Seo, J.-H. (1990) Analysis of *E. coli* phoA-lacZ fusion gene expression inserted into a multicopy plasmid and host cell's chromosome. *Biotechnol. Bioeng.*, **36**, 1097–1104.
156. Swartz, J.R. (2001) Advances in *Escherichia coli* production of therapeutic proteins. *Curr. Opin. Biotechnol.*, **12**, 195–201.
157. Valdez-Cruz, N.A., Caspeta, L., Pérez, N.O., Ramírez, O.T., and Trujillo-Roldán, M.A. (2010) Production of recombinant proteins in *E. coli* by the heat inducible expression system based on the phage lambda pL and/or pR promoters. *Microb. Cell Fact.*, **9**, 18.
158. Talmadge, K. and Gilbert, W. (1982) Cellular location affects protein stability in *Escherichia coli. Proc. Natl. Acad. Sci. U.S.A.*, **79**, 1830–1833.
159. Bessette, P.H., Aslund, F., Beckwith, J., and Georgiou, G. (1999) Efficient folding of proteins with multiple disulfide bonds in the *Escherichia coli* cytoplasm. *Proc. Natl. Acad. Sci. U.S.A.*, **96**, 13703–13708.
160. Missiakas, D., Georgopoulos, C., and Raina, S. (1994) The *Escherichia coli* dsbC (xprA) gene encodes a periplasmic protein involved in disulfide bond formation. *EMBO J.*, **13**, 2013–2020.
161. Michaelis, S. and Beckwith, J. (1982) Mechanism of incorporation of cell envelope proteins in *Escherichia coli. Annu. Rev. Microbiol.*, **36**, 435–465.
162. Blight, M.A., Chervaux, C., and Holland, I.B. (1994) Protein secretion pathways in *Escherichia coli. Curr. Opin. Biotechnol.*, **5**, 468–474.
163. Gray, L., Mackman, N., Nicaud, J.-M., and Holland, I. (1986) The carboxy-terminal region of haemolysin 2001 is required for secretion of the toxin *Escherichia coli. Mol. Gen. Genet.*, **205**, 127–133.
164. Kenny, B., Chervaux, C., and Holland, I.B. (1994) Evidence that residues -15 to -46 of the haemolysin secretion signal are involved in early steps in secretion, leading to recognition of the translocator. *Mol. Microbiol.*, **11**, 99–109.
165. Blight, M.A. and Holland, I.B. (1994) Heterologous protein secretion and the versatile *Escherichia coli* haemolysin translocator. *Trends Biotechnol.*, **12**, 450–455.
166. Graumann, K. and Premstaller, A. (2006) Manufacturing of recombinant therapeutic proteins in microbial systems. *Biotechnol. J.*, **1**, 164–186.
167. Lu, H. *et al* (2007) Expression and purification of bioactive high-purity mouse monokine induced by IFN-gamma in *Escherichia coli. Protein Expression Purif.*, **55**, 132–138.

168. Li, M. and Huang, D. (2007) On-column refolding purification and characterization of recombinant human interferon-lambda1 produced in *Escherichia coli. Protein Expression Purif.*, **53**, 119–123.
169. Wacker, M. *et al* (2002) N-linked glycosylation in *Campylobacter jejuni* and its functional transfer into *E. coli. Science*, **298**, 1790–1793.
170. Lizak, C., Fan, Y.-Y., Weber, T.C., and Aebi, M. (2011) N-linked glycosylation of antibody fragments in *Escherichia coli. Bioconjugate Chem.*, **22**, 488–496.
171. Bogorad, I.W., Lin, T.-S., and Liao, J.C. (2013) Synthetic non-oxidative glycolysis enables complete carbon conservation. *Nature*, **502**, 1–6.
172. Meile, L., Rohr, L.M., Geissmann, T.A., Herensperger, M., and Teuber, M. (2001) Characterization of the D-xylulose 5-phosphate/D-fructose 6-phosphate phosphoketolase gene (xfp) from bifidobacterium lactis. *J. Bacteriol.*, **183**, 2929–2936.
173. Conrado, R.J. and Gonzalez, R. (2014) Envisioning the bioconversion of methane to liquid fuels. *Science*, **343**, 621–623.
174. Brautaset, T., Jakobsen, Ø.M., Josefsen, K.D., Flickinger, M.C., and Ellingsen, T.E. (2007) *Bacillus methanolicus*: a candidate for industrial production of amino acids from methanol at 50 °C. *Appl. Microbiol. Biotechnol.*, **74**, 22–34.
175. Gould, S.J., McCollum, D., Spong, A.P., Heyman, J.A., and Subramani, S. (1992) Development of the yeast *Pichia pastoris* as a model organism for a genetic and molecular analysis of peroxisome assembly. *Yeast*, **8**, 613–628.
176. Whitaker, W.B., Sandoval, N.R., Bennett, R.K., Fast, A.G., and Papoutsakis, E.T. (2015) Synthetic methylotrophy: engineering the production of biofuels and chemicals based on the biology of aerobic methanol utilization. *Curr. Opin. Biotechnol.*, **33**, 165–175.
177. Müller, J.E.N. *et al* (2015) Engineering *Escherichia coli* for methanol conversion. *Metab. Eng.*, **28**, 190–201.
178. Anthony, C. (1983) *The Biochemistry of Methylotrophs*, Academic Press, New York, http://linkinghub.elsevier.com/retrieve/pii/0300962983901160 (accessed 31 May 2016).
179. Bogorad, I.W. *et al* (2014) Building carbon–carbon bonds using a biocatalytic methanol condensation cycle. *Proc. Natl. Acad. Sci. U.S.A.*, **111** (45), 15928–15933, doi: 10.1073/pnas.1413470111.
180. Somerville, C.R. and Somerville, S.C. (1984) Cloning and expression of the *Rhodospirillum rubrum* ribulosebisphosphate carboxylase gene in *E. coli. MGG Mol. Gen. Genet.*, **193**, 214–219.
181. Li, Y.-H., Ou-Yang, F.-Y., Yang, C.-H., and Li, S.-Y. (2015) The coupling of glycolysis and the Rubisco-based pathway through the non-oxidative pentose phosphate pathway to achieve low carbon dioxide emission fermentation. *Bioresour. Technol.*, **187**, 189–197.
182. Mainguet, S.E., Gronenberg, L.S., Wong, S.S., and Liao, J.C. (2013) A reverse glyoxylate shunt to build a non-native route from C4 to C2 in *Escherichia coli. Metab. Eng.*, **19**, 116–127.
183. Mattozzi, M.D., Ziesack, M., Voges, M.J., Silver, P.A., and Way, J.C. (2013) Expression of the sub-pathways of the *Chloroflexus aurantiacus* 3-hydroxypropionate carbon fixation bicycle in *E. coli*: toward horizontal transfer of autotrophic growth. *Metab. Eng.*, **16**, 130–139.
184. Liao, J.C., Hou, S.Y., and Chao, Y.P. (1996) Pathway analysis, engineering, and physiological considerations for redirecting central metabolism. *Biotechnol. Bioeng.*, **52**, 129–140.
185. Taniguchi, Y. *et al* (2010) Quantifying E. coli Proteome and Transcriptome with Single-Molecule Sensitivity in Single Cells *Science*, **329** (5991), 533–538.
186. Crown, S.B., Long, C.P., and Antoniewicz, M.R. (2015) Integrated 13C-metabolic flux analysis of 14 parallel labeling experiments in *Escherichia coli. Metab. Eng.*, **28**, 151–158.
187. Bennett, B.D. *et al* (2009) Absolute metabolite concentrations and implied enzyme active site occupancy in *Escherichia coli. Nat. Chem. Biol.*, **5**, 593–599.

188. Moniruzzaman, M. *et al* (1997) Isolation and molecular characterization of spontaneous mutants of ethanologenic *Escherichia coli* KO11 containing the Klebsiella oxytoca casAB operon. *Appl. Environ. Microbiol.*, **63**, 4633.
189. Desai, S.H., Rabinovitch-Deere, C.A., Tashiro, Y., and Atsumi, S. (2014) Isobutanol production from cellobiose in *Escherichia coli*. *Appl. Microbiol. Biotechnol.*, **98**, 3727–3736.
190. Qian, Z.-G., Xia, X.-X., Choi, J.H., and Lee, S.Y. (2008) Proteome-based identification of fusion partner for high-level extracellular production of recombinant proteins in *Escherichia coli*. *Biotechnol. Bioeng.*, **101**, 587–601.
191. Bokinsky, G. *et al* (2011) Synthesis of three advanced biofuels from ionic liquid-pretreated switchgrass using engineered *Escherichia coli*. *Proc. Natl. Acad. Sci. U.S.A.*, **108**, 19949–19954.
192. Shin, H.D., Wu, J., and Chen, R. (2014) Comparative engineering of *Escherichia coli* for cellobiose utilization: hydrolysis versus phosphorolysis. *Metab. Eng.*, **24**, 9–17.
193. Zaldivar, J., Nielsen, J., and Olsson, L. (2001) Fuel ethanol production from lignocellulose: a challenge for metabolic engineering and process integration. *Appl. Microbiol. Biotechnol.*, **56**, 17–34.
194. Stulke, J. and Hillen, W. (1999) Carbon catabolite repression in bacteria. *Curr. Opin. Microbiol.*, **2**, 195–201.
195. Kim, S.M. *et al* (2015) Simultaneous utilization of glucose and xylose via novel mechanisms in engineered *Escherichia coli*. *Metab. Eng.*, **30**, 141–148.
196. US Energy Information Administration (2013) Monthly Biodiesel Production Report, p. 100.
197. Yang, F., Hanna, M.A., and Sun, R. (2012) Value-added uses for crude glycerol – a byproduct of biodiesel production. *Biotechnol. Biofuels*, **5**, 13.
198. Fleige, C., Kroll, J., and Steinbüchel, A. (2011) Establishment of an alternative phosphoketolase-dependent pathway for fructose catabolism in *Ralstonia eutropha* H16. *Appl. Microbiol. Biotechnol.*, **91**, 769–776.
199. Trchounian, K., Sanchez-Torres, V., Wood, T.K., and Trchounian, A. (2011) *Escherichia coli* hydrogenase activity and H2 production under glycerol fermentation at a low pH. *Int. J. Hydrogen Energy*, **36**, 4323–4331.
200. Clomburg, J.M. and Gonzalez, R. (2010) Biofuel production in *Escherichia coli*: the role of metabolic engineering and synthetic biology. *Appl. Microbiol. Biotechnol.*, **86**, 419–434.
201. Shams Yazdani, S. and Gonzalez, R. (2008) Engineering *Escherichia coli* for the efficient conversion of glycerol to ethanol and co-products. *Metab. Eng.*, **10**, 340–351.
202. Willrodt, C. *et al* (2014) Engineering the productivity of recombinant of *Escherichia coli* for limonene formation from glycerol in minimal media. *Biotechnol. J.*, **9** (8), 1000–1012.
203. Wernick, D.G. and Liao, J.C. (2013) Protein-based biorefining: metabolic engineering for production of chemicals and fuel with regeneration of nitrogen fertilizers. *Appl. Microbiol. Biotechnol.*, **97**, 1397–1406.
204. Huo, Y.-X., Wernick, D.G., and Liao, J.C. (2012) Toward nitrogen neutral biofuel production. *Curr. Opin. Biotechnol.*, **23**, 406–413.
205. Snyder, L., Peters, J.E., Henkin, T.M., and Champness, W. (2014) *Molecular Genetics of Bacteria* (ASM Press), American Society for Microbiology Press, pp. 247–263.
206. Sambrook, J., Fritsch, E., and Maniatis, T. (1989) *Molecular Cloning: a Laboratory Manual*, Cold Springs Harbor Laboratory Press.
207. Dower, W.J., Miller, J.F., and Ragsdale, C.W. (1988) High efficiency transformation of *E. coli* by high voltage electroporation. *Nucleic Acids Res.*, **16**, 6127–6145.
208. Chung, C.T., Niemela, S.L., and Miller, R.H. (1989) One-step preparation of competent *Escherichia coli*: transformation and storage of bacterial cells in the same solution. *Proc. Natl. Acad. Sci. U.S.A.*, **86**, 2172–2175.
209. Lichtenstein, C. and Brenner, S. (1982) Unique insertion site of Tn7 in the

E. coli chromosome. *Nature*, **297**, 601–603.

210. Murphy, K.C. and Campellone, K.G. (2003) Lambda Red-mediated recombinogenic engineering of enterohemorrhagic and enteropathogenic *E. coli*. *BMC Mol. Biol.*, **4**, 11.
211. Wang, H.H. *et al* (2009) Programming cells by multiplex genome engineering and accelerated evolution. *Nature*, **460**, 894–898.
212. Summers, W.C. (2012) The strange history of phage therapy. *Bacteriophage*, **2**, 130–133.
213. Loc-Carrillo, C. and Abedon, S.T. (2011) Pros and cons of phage therapy. *Bacteriophage*, **1**, 111–114.
214. Calendar, R. (1988) *The Bacteriophages*, Plenum Press, New York.
215. Sulakvelidze, A. (2011) Bacteriophage. *Bacteriophage*, **1**, 1–2.
216. Potera, C. (2012) Keeping cell cultures contamination free. *Genet. Eng. Biotechnol. News*, **32**, 22–23.
217. Wegrzyn, G., Los, M., and Neubauer, P. (2006) The 4th Recombinant Protein Production Meeting: a Comparative View on Host Physiology, vol. 2, pp. 4–5.
218. Szczepankowska, A.K., Górecki, R.K., Kołakowski, P., and Bardowski, J.K. (2013) Lactic acid bacteria resistance to bacteriophage and prevention techniques to lower phage contamination in dairy fermentation, Lactic Acid Bacteria – R & D for Food, Health and Livestock Purposes (ed Marcelino Kongo), pp. 23–72, doi: 10.5772/51541
219. Wang, M.S. and Nitin, N. (2014) Rapid detection of bacteriophages in starter culture using water-in-oil-in-water emulsion microdroplets. *Appl. Microbiol. Biotechnol.*, **98**, 8347–8355.
220. Los, M., Los, J., and Wegrzyn, G. (2006) Rapid detection of bacteriophage infection and prophage induction using electric biochips. *Microb. Cell Fact.*, **5**, S38.
221. Chen, X. *et al* (2013) Metabolic engineering of *Escherichia coli*: a sustainable industrial platform for bio-based chemical production. *Biotechnol. Adv.*, **31**, 1200–1223.
222. Papoutsakis, E.T. (2015) Re-assessing the progress in the production of advanced biofuels in the current competitive environment and beyond: what are the successes and where progress eludes us and why. *Ind. Eng. Chem. Res.*, **54**, 10170–10182.
223. Starr, J.N. and Westhoff, G. (2003) Lactic acid. *Ullmann's Encycl. Ind. Chem.*, **20**, 4–6.
224. Leuchtenberger, W., Huthmacher, K., and Drauz, K. (2005) Biotechnological production of amino acids and derivatives: current status and prospects. *Appl. Microbiol. Biotechnol.*, **69**, 1–8.
225. Erickson, B., Nelson, J.E. and Winters, P. (2012) Perspective on opportunities in industrial biotechnology in renewable chemicals. *Biotechnol. J.*, **7**, 176–185.

6 Industrial Microorganisms: *Corynebacterium glutamicum*

Judith Becker and Christoph Wittmann

6.1 Introduction

"The highest level of L-glutamate production has been obtained by a new species […] yielding as much as 0.25 mol of it from one mole of glucose." With this discovery in the 1960s, Kinoshita and co-workers initiated the era of industrial amino acid fermentation [1], which has meanwhile evolved into a billion dollar market. The new species, isolated by the researchers and named *Micrococcus glutamicus*, was later classified as *Corynebacterium glutamicum*. Subsequently, numerous L-glutamate producers were isolated and classified in different genera, mainly *Brevibacterium*, *Corynebacterium*, *Micrococcus*, and *Arthrobacter* [2]. However, most of these classifications were later revised in a large taxonomic study investigating a total of 208 L-glutamate-producing strains [3]. Among others *Brevibacterium chang-fua*, *B. divaricatum*, *B. lactofermentum*, *B. flavum*, *B. glutamigenes*, *B. roseum*, *B. seonmiso*, *B. taipei*, *B. thiogenitalis*, *C. lilium*, *C. herculis*, *Microbacterium ammoniaphilum*, *Microbacterium* sp., *Micrococcus maripuniceus* were found to group into *C. glutamicum* clusters [2]. In short, *C. glutamicum* is a nonmotile, facultative anaerobic, Gram-positive biotin-auxotrophic soil bacterium, which forms rod-shaped, straight, or slightly curved cells. The genome of the wild-type strain *C. glutamicum* ATCC 13032 contains about 3000 genes located on a circular chromosome. *C. glutamicum* can assimilate carbon from many different sources, such as sugars, sugar alcohols, organic acids, and aromatic compounds. Several characteristics of *C. glutamicum* make it particularly useful for biotechnology. It is not pathogenic, has the GRAS status (generally regarded as safe), does not form spores, and grows quickly at 30 °C and neutral pH with relatively few growth requirements. Mineral salt media usually contain only some vitamins (biotin, thiamine, pantothenic acid), some salts, and trace elements (Na^+, Ca^{2+}, Fe^{2+}, Mg^{2+}, Mn^{2+}, Zn^{2+}, and Cu^{2+}) in addition to the carbon and nitrogen source. In recent years, intensive research has provided detailed insights into its biochemistry and metabolism. Combined with recombinant DNA technology, this has enabled the rational creation of efficient producer strains with superior characteristics that even outcompete

Industrial Biotechnology: Microorganisms, First Edition. Edited by Christoph Wittmann and James C. Liao.

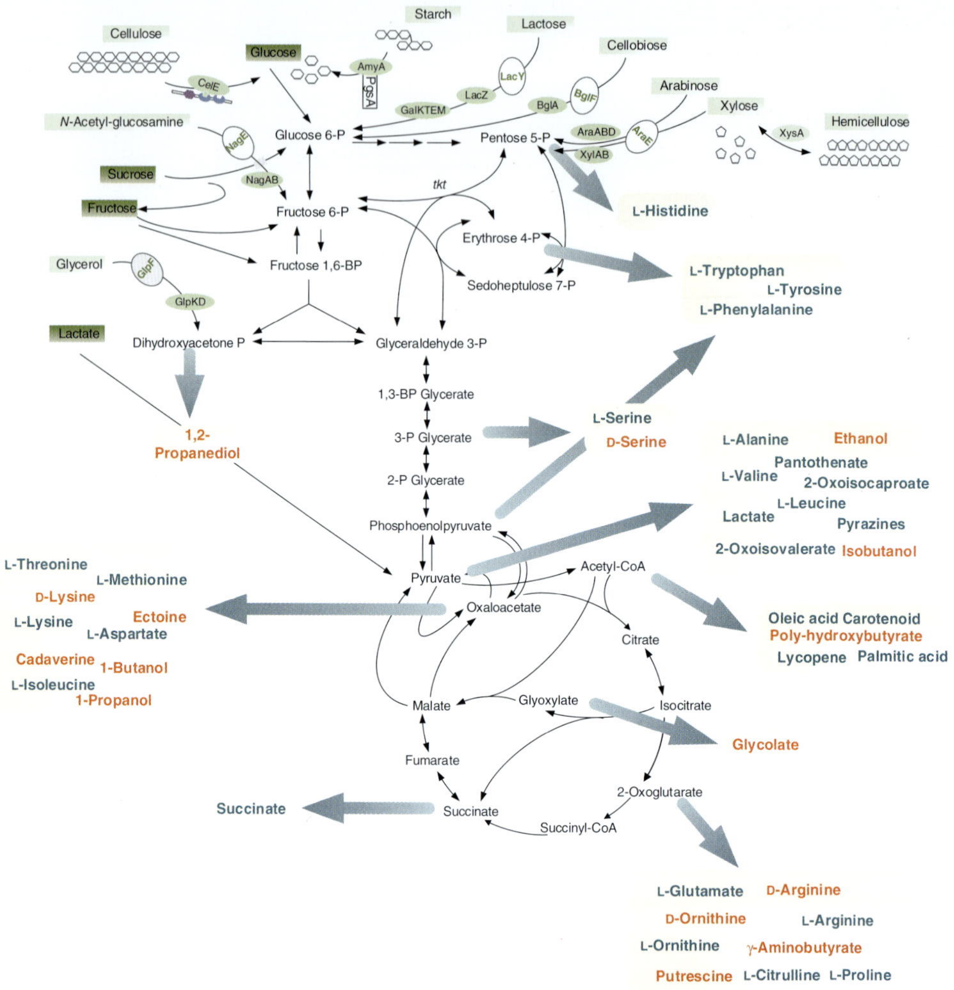

Figure 6.1 Carbon core metabolism of *Corynebacterium glutamicum* comprising the major catabolic routes of pentose phosphate pathway and Embden-Meyerhof-Parnas pathway, tricarboxylic acid cycle, glyoxylate shunt, and anaplerotic reactions. The relevance of the individual pathways and carbon building blocks for biosynthesis of the broad product portfolio of *C. glutamicum* is indicated. Additionally, the extended substrate spectrum and the heterologous assimilation routes are illustrated. Natural products and substrates are highlighted in turquoise and dark green, non-natural products and substrates in yellow and light green.

the performance of classical producers. Today, *C. glutamicum* is used to produce more than 4 million tons of amino acids per year, such as L-glutamate, L-lysine, L-threonine, and L-isoleucine; meanwhile, the product portfolio has been expanded to a wide range of other natural and non-natural products (Figure 6.1).

6.2 Physiology and Metabolism

Elucidation of the metabolic pathway repertoire of *C. glutamicum* was initiated soon after its discovery [4–6], related to the high importance of the central metabolism for amino acid fermentation. Evidence on the presence of the major catabolic routes, such as the Embden-Meyerhof-Parnas (EMP) pathway, the pentose phosphate (PP) pathway, the tricarboxylic acid (TCA) cycle, and the glyoxylate shunt had already been provided by the end of the 1950s [4, 5, 7] (Figure 6.1). However, it took more than 30 years for a more detailed resolution of the central metabolic network mainly related to the complex structure of the phosphoenolpyruvate/pyruvate – oxaloacetate/malate node [8–11]. Altogether, *C. glutamicum* possesses at least five enzymes, which interconvert carbon three and carbon four metabolites at this node. Additionally involving pyruvate kinase (PYK) and pyruvate dehydrogenase (PDH) – also present – *C. glutamicum* is characterized by a surprisingly diverse set of anaplerotic reactions [12]. The network of carbon metabolism is directly linked to the biosynthesis of industrially interesting products through the supply of building blocks, redox power, and energy (Figure 6.2).

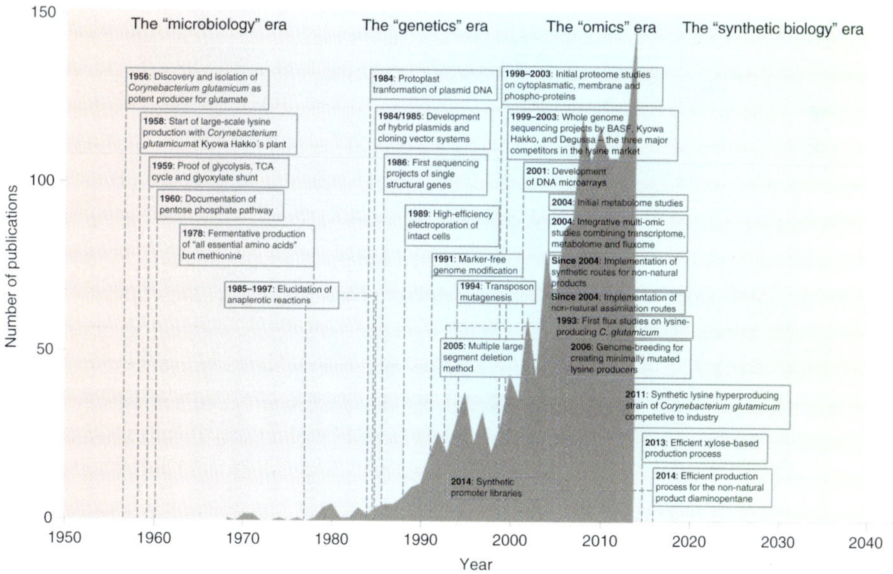

Figure 6.2 Timeline of research and development of the industrial, Gram-positive soil bacterium *Corynebacterium glutamicum*, highlighting major discovery breakthroughs in the microbiology, genetic, omics, and synthetic biology era. The gray background indicates the number of publications that appeared as deduced from PubMed (http://www.ncbi.nlm.nih.gov/pubmed).

6.2.1 Substrate Assimilation, Catabolic Pathways, and Gluconeogenesis

C. glutamicum is able to grow on different substrates such as carbohydrates, organic acids, and alcohols as well as on the sugar alcohol arabitol [13–15]. Sugar uptake in *C. glutamicum* typically involves phosphotransferase systems (PTSs) [16]. At least three different PTS transporters are present in *C. glutamicum*. Early findings of the transporter set [17] were later confirmed by whole-genome sequencing, which discovered the genes of the general components *ptsI* and *ptsH*, and of the sugar-specific components *ptsG* (previously referred to as *ptsM*), *ptsF*, and *ptsS*, respectively, specific for the uptake of glucose, fructose, and sucrose [18]. In contrast to many other bacteria [13, 19, 20], PTS expression and sugar uptake in *C. glutamicum* are not necessarily linked to carbon catabolite repression (CCR), and different strains are able to co-consume sugar mixtures and sugar/acid mixtures [21–23]. This is, however, not a general feature as ethanol assimilation is subjected to CCR [24] and there is evidence that recombinant xylose-assimilating *C. glutamicum* strains preferably consume glucose over xylose [25]. As the native xylose uptake system has not been identified so far [13], information on a potential CCR-like regulation is still missing. PTS-independent glucose uptake recruits the inositol pathway including the transporters IolT1 and IolT2 and subsequent phosphorylation of glucose by a glucokinase [26].

For catabolic breakdown of sugars, *C. glutamicum* possesses two glycolytic routes: the EMP pathway and the PP pathway. The relative contribution of these alternative routes depends on the carbon source. Metabolic flux analysis revealed that during growth on glucose both routes are almost equally active [27, 28]. On fructose, however, the PP pathway only marginally contributes to catabolism, related to the entry point of fructose into central metabolism [29, 30]. In addition to the supply of building blocks for anabolism, both routes are involved in redox and energy metabolism. The PP pathway is of high relevance as a source for NADPH. Glucose 6-phosphate dehydrogenase (G6PDH) and 6-phosphogluconate dehydrogenase belong to the first enzymes discovered in *C. glutamicum* [4, 6]. G6PDH is controlled by the NADPH/NADP ratio [31] and by signal metabolites of the energy metabolism (fructose 1,6-bisphosphate, phosphoenolpyruvate, oxaloacetate, PRPP, ATP) [32, 33] (Table 6.1) thus playing a major role in carbon flux control through this pathway. Most genes of the PP pathway – *tkt*, *tal*, *zwf*, *opcA*, and *pgl* encoding transketolase, transaldolase, G6PDH (*zwf* and *opcA*), and phosphogluconolactonase – are organized in an operon [59, 60]. Gene expression can substantially change in response to the carbon feedstock (Table 6.1). When grown on acetate, the PP pathway is significantly downregulated as compared to growth on glucose [35, 61]. The transcriptional changes further translate to the proteome level as observed from decreased enzyme activities [33]. For citrate and xylose, however, no or only marginal changes of PP pathway gene expression have been reported [38, 62]. The high relevance of the PP pathway for NADPH and precursor supply for amino acid biosynthesis entailed substantial engineering of this pathway for improving the production performance of *C. glutamicum* strains [32, 60, 63, 64].

Enzymes of the EMP pathway – the second major catabolic route in *C. glutamicum* – were identified in the 1950s [4]. Related to their metabolic regulation (Table 6.1), glyceraldehyde 3-phosphate dehydrogenase [30, 65] and PYK [8, 40] appear most relevant for metabolic flux control of the EMP pathway [59]. Most interestingly, 6-phosphofructokinase does not show the typical regulatory behavior, that is inhibition by "high-energy-metabolites" (ATP, PEP, citrate) and activation by "low-energy-metabolites" (ADP, AMP) [59], but is rather inhibited by ADP [36]. Control on the level of expression was found for *pfk*, *tpi*, *gapA*, *eno*, and *pyk*, encoding for phosphofructokinase (PFK), triosephosphate isomerase, glyceraldehyde 3-phosphate dehydrogenase, enolase, and PYK, respectively, which are downregulated during growth on acetate as compared to glucose [35] (Table 6.2). However, the enzyme activity of PFK remains unaffected [34]. For gluconeogenesis – the antagonist pathway of the EMP pathway – *C. glutamicum* possesses an essential class II fructose 1,6-bisphosphatase enzyme [34, 37]. Regulation of this pathway occurs on the level of transcription [35] and by metabolic control [34, 69]. Expressional differences in response to the supplied carbon source are also reflected in protein abundance [37] as well as in enzyme activity [34, 37, 70].

6.2.2
Tricarboxylic Acid Cycle and Glyoxylate Shunt

The TCA cycle – or at least parts of it – is invariably active during growth of *C. glutamicum* on sugars, alcohols, organic acids, or mixtures thereof [12]. It is the major energy source for the cell by (i) directly supplying ATP (or GTP) via substrate chain phosphorylation and (ii) indirectly supplying energy through NADH for subsequent respiratory chain phosphorylation. During growth on acetate or ethanol the glyoxylate shunt emerges as metabolic bypass of the TCA cycle, thereby circumventing the decarboxylating reactions of isocitrate dehydrogenase and 2-oxoglutarate dehydrogenase [15, 61]. Citrate synthase and to some extent PDH are considered rate controlling for the entry step into the TCA cycle (Table 6.1) [12, 42]. Further operation of the TCA cycle or flux through the glyoxylate shunt is controlled on the level of isocitrate dehydrogenase, mainly through metabolic inhibition by glyoxylate plus oxaloacetate [47], and on the level of expressional and metabolic control of isocitrate lyase and malate synthase [15, 35] (Table 6.1). Additional flux control is taken by the 2-oxoglutarate dehydrogenase complex (OGDHC) [71, 72]. OGDHC is strongly inhibited by aconitate, succinyl-CoA, NADPH, NADH, pyruvate, and oxaloacetate, and activated by acetyl-CoA [50, 51] Moreover, it is subject to the inhibitor protein OdhI [73, 74]. Activity of OGDHC thereby most importantly determines the flux at the 2-oxoglutarate node and is thus at least partially responsible for the glutamate production efficiency in *C. glutamicum* [75–78].

Table 6.1 Enzymes and corresponding genes of the central metabolism of *Corynebacterium glutamicum* including known effectors and transcriptional regulation.

Enzyme	Gene	Effector	Transcriptional control/regulators	References
EMPP/gluconeogenesis				
Glucose 6-phosphate isomerase	*pgi*	E4P (−)		[34]
6-Phosphofructokinase	*pfkB*	ADP (−)	Xyl (+), Ace (−)	[35, 36]
Fructose 1,6-bisphosphatase	*fbp*	AMP (−), PEP (−), S7P (−), F6P (−), FBP (−), UDP (−)	Cit (+), Lac (+),	[34, 37]
Fructose 1,6-bisphosphate aldolase	*fda*			—
Triosephosphate isomerase	*tpi*		Xyl (+), Ace (−)	[35, 38]
Glyceraldehyde 3-phosphate dehydrogenase	*gapA*	ATP (−), NADH/NAD (−)	Xyl (+), Ace (−)	[30, 35, 39]
	gapB	ATP (+)	Ace (+)	[35, 39]
3-Phosphoglycerate kinase	*pgk*			—
Phosphoglycerate mutase	*pgm*			—
Enolase	*eno*		Ace (−)	[35]
Pyruvate kinase	*pyk*	AMP (+), ATP (−)	Ace (−)	[8, 35, 40, 41]
Phosphoenolpyruvate synthetase	*pps*	ATP (+), AMP (−), PEP (−)		[41]
Pyruvate dehydrogenase complex	*aceE aceF lpd*		RamB (+)	[42]
PPP				
Glucose 6-phosphate dehydrogenase	*zwf opcA*	NADPH (−), ATP (−), PEP (−), FBP (−), PRPP (−)		[31–33]
6-Phosphogluconolactonase	*pgl*		Ace (−)	[35]
6-Phosphogluconate dehydrogenase	*gnd*	NADPH (−), ATP (−), FBP (−), G3P (−), E4P (−), Ru5P (−), OAA (−), PRPP (−)	Xyl (+), Ace (−)	[31, 35, 38, 43]
Ribulose 5-phosphate epimerase	*rpe*			—
Ribose 5-phosphate isomerase	*rpi*			—
Transketolase	*tkt*	E4P (+)		[34]
Transaldolase	*tal*			—
TCA cycle/glyoxylate shunt				
Citrate synthase	*cis*	ATP (−), ACO (−), ICI (−)	RamA (+), RamB (−), GlxR (−), Ace (+)	[35, 41, 44–46]
Aconitase	*acn*			—
Isocitrate dehydrogenase	*icd*	OAA (−), Glx (−)		[47–49]

Table 6.1 *(Continued)*

Enzyme	Gene	Effector	Transcriptional control/regulators	References
2-Oxoglutarate dehydrogenase complex	*odhA sucB lpd*	ACO (−), SucCoA (−), NADPH (−), NADH (−), PYR (−), OAA (−), AcCoA (−)		[50, 51]
Succinyl-CoA synthetase	*sucCD*		Xyl (−)	[38]
Succinate: menaquinone oxidoreductase	*sdhCAB*	OAA (−)	Ace (+)	[35, 52]
Fumarase	*fum*		Ace (+)	[35]
Malate: quinone oxidoreductase	*mqo*	FAD (+)		[52, 53]
Malate dehydrogenase	*mdh*		Ace (+), Xyl (+)	[35, 38, 53]
Isocitrate lyase	*aceA*	3PG (−), 6PG (−), PEP (−), FBP (−), Suc (−), Glx (−)	Ace (+), EtOH (+)	[15, 35, 48, 54, 55]
Malate synthase	*aceB*	ATP (−), Oxa (−), Glyc (−)	Ace (+), EtOH (+)	[15, 35, 55, 56]
Pyruvate metabolism				
Phosphoenolpyruvate carboxylase	*ppc*	AcCoA (+), FBP (+), Asp (−), Mal (−), Suc (−), 2OG (−)	Xyl (+), Ace (−)	[8, 35, 38, 41, 57]
Phosphoenolpyruvate carboxykinase	*pck*	Ace (+)	Ace (+)	[35]
Pyruvate carboxylase	*pyc*	ADP (−), AMP (−), AcCoA (−), Asp (−)	Ace (−)	[35]
Malic enzyme	*malE*	OAA (−), Glu (−)	Ace (−)	[35]
Oxaloacetate decarboxylase	*odx*	ADP (−), CoA (−), Suc (−)		[58]

6.2.3
Reactions of the Pyruvate Node

C. glutamicum possesses a highly complex network around the pyruvate node, comprising PYK, PDH, phosphoenolpyruvate synthase (PEPS), phosphoenolpyruvate carboxykinase (PEPCk), phosphoenolpyruvate carboxylase (PEPCx), pyruvate carboxylase (PCx), malic enzyme (MalE), and oxaloacetate decarboxylase (ODx) [12, 79], of which the latter five interconvert carbon-three metabolites of the EMP pathway and carbon-four metabolites of the TCA cycle. Most interestingly, fueling reactions for TCA cycle replenishment (PEPCx, PCx) act in parallel to enzymes with decarboxylating activity (PEPCk, MalE, ODx) [80]. In sum, this establishes an energetic futile cycle. As a basic advantage, simultaneous activity of the reactions might allow fine-tuning of anaplerotic and catabolic reactions, energy and redox balancing, and a rapid response to altering

Table 6.2 Precursor demand for biomass synthesis in *C. glutamicum* with additional consideration of the ATP demand for polymerization and assembly [66–68].

	Demand (μmol g^{-1})	G6P	F6P	R5P	E4P	GAP	PGA	PEP	Pyr	AcCoA	OAA	AKG	NADPH	ATP	NADH
Ala	606								1				1		
Ala_{intra}	9.8								1				1		
Arg	189											1	4	7	1
Arg_{intra}	0.73											1	4	7	1
Asx	399										1		1	1.5	
Asx_{intra}	13.2										1		1	1.5	
Cys	87						1						5	5	1
Glu	360											1	1		
Glu_{intra}	375.2											1	1		
Gln	147											1	1	1	
Gln_{intra}	91.5											1	1	1	
Gly	361						1						1		1
Gly_{intra}	5.3						1						1		1
His	71			1									1	6	3
His_{intra}	0.93			1									1	6	3
Ile	202								1		1		5	2	
Ile_{intra}	1.49								1		1		5	2	
Leu	440								2	1			2		1
Leu_{intra}	2.28								2	1			2		1
Lys	202								1		1		4	2	
Lys_{intra}	27.3								1		1		4	2	
Met	146										1		8	7	
Met_{intra}	0.62										1		8	7	
Phe	133				1			2					2	1	
Phe_{intra}	1.25				1			2					2	1	

Pro	170											1	3	1	
Ser	225						1						1		1
Ser_{intra}	3.76						1						1		1
Thr	275										1		3	2	
Thr_{intra}	1.5										1		3	2	
Trp	54			1	1			1					2	5	2
Trp_{intra}	9.3			1	1			1					2	5	2
Tyr	81				1			2					2	1	1
Tyr_{intra}	0.81				1			2					2	1	1
Val	284								2				2		
Val_{intra}	17								2				2		
Dap	149								1		1		4	2	
Sum protein+iAA		**0**	**0**	**135**	**279**	**0**	**682**	**495**	**2 684**	**442**	**1 417**	**1 333**	**10 951**	**6 497**	**1 738**
RNA				630			368				262		1152	6 540	1 104
DNA				100			50				50		272	1 089	0
Lipids						129	129			2116			3612	1 858	0
LPS		51	16	24			24	24		329			470	470	24
Peptidoglycan			292					146		292			146	1 314	0
Glycogen		154												154	
C1-units							49						49	49	49
Poly-amines												59	180	119	
Polymerization and assembly														50 270	
Total		**205**	**308**	**889**	**279**	**129**	**1 302**	**665**	**2 684**	**3 179**	**1 729**	**1 392**	**16 832**	**68 359**	**2 951**

For amino acid (AA) biosynthesis, both protein-bound (e.g., Ala) and free (iAA, e.g., Ala_{intra}) amino acids were considered. The withdrawal of G6P, F6P, R5P, E4P, GAP, PEP, Pyr, AcCoA, OAA, AKG, NADPH, ATP, and NADH is calculated from the respective AA demand ($\mu mol\ g^{-1}$) and the pathway stoichiometry, indicated by the given numbers. The total demand gives the micromolar amount of carbon (G6P, F6P, R5P, E4P, GAP, PEP, Pyr, AcCoA, OAA, AKG), redox (NADPH, NADH), and energy (ATP) required to build 1 g of cell dry mass.

environmental conditions [12]. Metabolic regulation (Table 6.1) of the different enzymes *in vivo* appears significant as PCx contributes to 90% of anaplerotic flux although the *in vitro* activity of PEPCx is substantially higher [80]. As an important switch-point in metabolism, metabolic engineering of the pyruvate metabolism of *C. glutamicum* has improved the production of different amino acids [60, 81–88].

6.2.4
Anabolism

Biomass of *C. glutamicum* is composed of five major macro-polymers – namely, protein (0.52 $g\,g^{-1}$ cell dry mass), DNA (0.01 $g\,g^{-1}$ cell dry mass), RNA (0.05 $g\,g^{-1}$ cell dry mass), lipids (0.13 $g\,g^{-1}$ cell dry mass), and cell wall carbohydrates (0.19 $g\,g^{-1}$ cell dry mass) – glycogen, C_1-units, and polyamines [66, 67]. All are synthesized from 11 carbon building blocks stemming from the central metabolism. Glucose 6-phosphate, fructose 6-phosphate, ribose 5-phosphate, erythrose 4-phosphate, glyceraldehyde 3-phosphate, 3-phosphoglycerate, phosphoenolpyruvate, pyruvate, acetyl-CoA, oxaloacetate, and 2-oxoglutarate are withdrawn to serve different biosynthetic pathways. In addition, ATP and NADPH are consumed, whereas NADH and CO_2 are produced during biosynthesis. The highest energy demand is thereby allotted to polymerization of the individual monomers and cell assembly [66, 89, 90]. The cellular demand for biomass synthesis of *C. glutamicum*, that is, the micromolar amount of the above listed building blocks required to form 1 g of cell dry mass, is given in Table 6.2.

6.3
Genetic Manipulation of *Corynebacterium glutamicum*

Driven by the huge commercial interest and the quest for improved producer strains, a broad set of genetic tools has been developed for genetic manipulation of *C. glutamicum* [91]. In 1984, the isolation of natural plasmids [92, 93] and the development of the first DNA transfer strategies [94] initiated the *genetics era* of *C. glutamicum* (Figure 6.2). Since then, the methods have been refined for advanced applications covering analysis of gene functions, gene replacement or deletion, expression systems for homologous and heterologous genes, or genome rearrangement [91, 95]. Molecular biology thereby not only leveraged the performance of existing production processes, but also enabled extension of the production portfolio to further natural compounds and beyond [96].

6.3.1
Plasmids

For the development of recombinant DNA techniques, the presence of endogenous plasmids is an important prerequisite [97]. The discovery of the native set

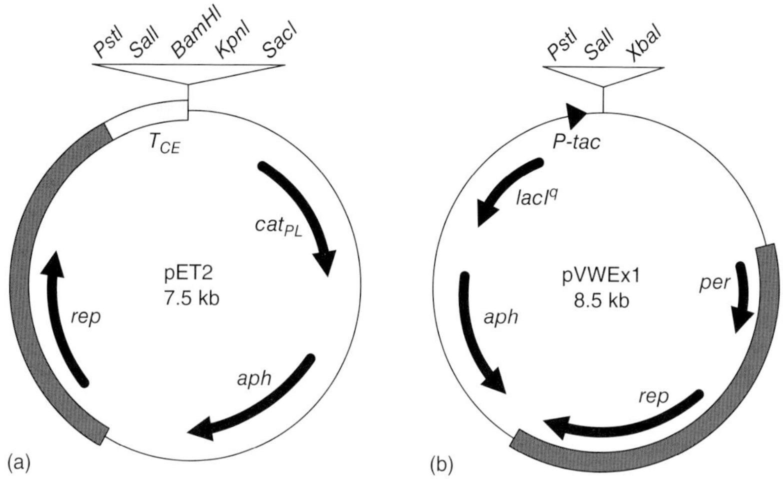

Figure 6.3 Maps of plasmid vectors for *C. glutamicum* (adapted from [100]) (a) *C. glutamicum/E. coli* shuttle promoter-probe vector pET2 and (b) *C. glutamicum/E. coli* shuttle expression vector pVWEx1. Dark areas indicate regions coming from plasmids of Corynebacteria (pBL1, pCG1, and pGA1, respectively). Arrows inside the map indicate genes. *aph*, Kanamycin resistance determinant; cat_{PL}, promoterless reporter gene coding for chloramphenicol acetyltransferase; T_{CE}, tandems of transcriptional terminators; *rep*, initiator of RC replication; *per*, positive effector of plasmid replication; $lacI^q$, lactose repressor; and P-*tac*, IPTG inducible promoter.

of corynebacteria plasmids was thus a major breakthrough for establishment of cloning vectors and genetic manipulation of *C. glutamicum* [95, 97]. The autonomously replicating vectors applied today are largely based on the initially isolated small cryptic plasmids pBL1 [92], pCG1 [98], and pGA1 [99], having copy numbers of from about 10 to 30 [100]. They basically serve as cloning vectors [100–102], promoter-probe vectors [100, 102], and expression vectors [100, 102] and are typically designed as *C. glutamicum* – *Escherichia coli* shuttle vectors [102–105], predominantly equipped with antibiotic resistance determinants as selection markers [95, 100, 106] (Figure 6.3). For evaluating the promoter activity, promoter-probe vectors with various promoterless reporter genes have been developed [100]. These in general carry genes, encoding easily detectable and quantifiable proteins such as β-galactosidase [107], melanin [108], and green or red fluorescent protein [109, 110]. Vice versa, vectors with isopropyl-β-D-thiogalactopyranoside (IPTG)-inducible or constitutive promoters serve as expression vectors for the overexpression of target genes. Although the IPTG-inducible system is of value for research, it is of limited interest for industrial-scale application. Here, temperature-sensitive systems appear more suitable and the first temperature-dependent expression system for *C. glutamicum* was already developed in 1988 on the basis of the λ phage P_RP_L promoter combined with the temperature-sensitive repressor cI857 [111]. However, repressible and inducible gene expression systems still need to be tailored for industrial applications

[91]. Recent studies have borne propionate-inducible [112] and synthetic *E. coli lac*-promoter-based [107] systems for wide-spanning expression strengths.

Stable chromosomal gene manipulations in *C. glutamicum* became possible with the development of transformation vectors lacking autonomous replication. Deletion of replicon-elements from native plasmids [113] and the use of mobilizable vectors from *E. coli* [114] were established as the first suitable systems. Subsequent to DNA transfer, homologous recombination permitted chromosomal integration. On a molecular basis, the Campbell-like mechanism (98%) is predominantly used over double-crossover events (2%) in *C. glutamicum* [91]. Other integration strategies recruit insertion sequences, predominantly IS31831 and its isoform isolates [91, 106, 115, 116], and site-specific integration functions of corynephages [117]. Genome-based manipulation in *C. glutamicum* was greatly facilitated by application of the conditionally lethal levansucrase (*sacB*) counter-selectable marker from *Bacillus subtilis* [118]. Non-self-replicating vectors equipped with the *sacB* gene have been largely used over the past decades for systems-wide genome modification of *C. glutamicum* [70, 119–123].

6.3.2
Transformation Methods

Lacking any apparent natural or chemically inducible DNA uptake, traditionally applied techniques, known from *E. coli* or *B. subtilis* transformation, were not transferable to *C. glutamicum* [91]. Hence, new experimental procedures had to be developed for fast, easy, and reproducible transformation of *C. glutamicum*. In sum, four different techniques – protoplast transformation, transduction, conjugation, and electrotransformation – crystallized as suitable methods for efficient DNA transfer. The first successful DNA-transfer method recruited protoplasts, generated by means of lysozyme treatment of cells grown in the presence of penicillin. DNA uptake was then promoted by adding polyethylene glycol [92, 94]. Protoplast formation was optimized mainly by the addition of chemicals and heat treatment [124, 125]. Despite the high transformation efficiency of up to 10^6 transformants per microgram of vector DNA, protoplast transformation did not establish itself as a state-of-the-art method related to high strainwise variation of transformation efficiency and elaborate protoplast regeneration [106]. Transduction of *C. glutamicum* was realized by infection with phage f1A transferring the cosmid vector pAJ667 [126]. However, limited knowledge on corynephages and the lack of efficient *in vitro* packaging systems clearly hampered development and application of transduction techniques [106]. Establishing conjugation and electrotransformation for DNA transfer then leveraged recombinant gene manipulation in *C. glutamicum*, as the high efficiency of these methods increased the frequency of rare integration events thus allowing gene disruption and replacement techniques [100, 106]. Electroporation is today established as most convenient technique for transforming *C. glutamicum*. Here, the cellular barrier is temporarily ruptured by high-voltage electric pulses allowing the DNA to enter the cell [127, 128]. Electric field strength, pulse length, DNA, and cell concentration affect the transformation efficiency [129]. Furthermore, known effectors of cell wall composition for optimized protoplast formation proved also valuable for improving the electrotransformation competence of *C. glutamicum* [130–132]. However, efficient transformation did not strictly rely on the use of cell wall-weakening compounds during the preceding cultivation [128, 129]. All studies commonly agree that the growth phase of the culture is an important factor for the preparation of electrocompetent cells. During evaluation of conjugation and electrotransformation, it additionally turned out to be crucial to circumvent the naturally high defense system of *C. glutamicum*, fending foreign DNA by enzymatic restriction. This was achieved by exposing the cells to exogenous stress by means of heat shock, organic solvents, pH shifts, or detergents, presumably resulting in inactivation of the restriction system [133–135], or by directly applying restriction-deficient mutants of *C. glutamicum* [134]. Meanwhile, different strategies have been used and approved to deceive the restriction system including plasmid passaging through *C. glutamicum* compatible hosts [113], using synthetic [136, 137] or nonmethylated DNA [129, 138], or by coexpression of the *C. glutamicum* specific DNA-methyltransferase in the intermediate cloning host [139].

6.3.3
Engineering Gene Expression

Transcription initiation is controlled by promoters as major regulatory signals determining the binding of RNA polymerases to the DNA. In *C. glutamicum*, seven different types of sigma factors have been described that recognize different classes of promoters [100, 140, 141]. Consensus sequences related to different regulatory patterns have been intensively studied and different review articles are available on this topic [100, 142–145]. The right promoter choice plays a significant role in genetic modification to obtain a purposeful level and timing in gene expression. Consequently, application-dependent inducible and constitutive promoters have been established for genetic manipulation.

Inducible promoter systems are valuable and convenient tools for controlled gene expression. They are broadly applicable in research and for lab-scale studies but up to now are of limited interest for industry. Most inducible systems of *C. glutamicum* expression vectors recruit heterologous promoters mainly derived from *E. coli*. IPTG-inducible promoters (P*tac*, P*lac*, P*trc*) have been broadly used for the construction of expression vectors (Figure 6.3) although the expression efficiency might vary related to the strain-specific cell permeability for IPTG. P*lac* expression was thereby improved by single-base point mutation, resulting in a 20-fold increased strength [146]. Most recently, a randomized *lac*-based promoter library was generated, allowing expression levels ranging from 284 to 1665 Miller units [107]. The high costs of the inducer, however, hamper industrial usefulness. Alternative options are the temperature-induced P_RP_L promoter of λ phage and the *araBAD* promoter of *E. coli* [111, 147]. The arabinose-dependent system especially has proven valuable as it allows a modulated gene expression by variation of the inducer concentration [147]. However, similarly to IPTG, the applicability of arabinose as inducer is limited by the low arabinose uptake of *C. glutamicum*. This issue was addressed recently by simultaneous overexpression of the *E. coli* arabinose transporter *araE* [148]. In addition, native *C. glutamicum* promoters – inducible by different organic carbon compounds including acetate [61], propionate [112], gluconate [109, 149], and maltose [149] – have been described and applied for industrial-relevant gene expression changes [142].

In the long tradition of *C. glutamicum* as industrial workhorse, the use of native constitutive promoters for genetic engineering is a rather young research field. An initial study provided a mutagenized promoter set on the basis of the *dapA* promoter, which controls the expression of dihydrodipicolinate synthase, an enzyme of lysine biosynthesis [150]. Subsequent studies predominantly recruited housekeeping promoters of genes with more global cellular functions for altered gene expression [95, 142]. Most frequently used promoters comprise the *sod* [23, 32, 70, 151–153] and the *tuf* promoters [60, 70, 120, 139, 152, 154, 155], which are naturally responsible for expression control of superoxide dismutase and the translational elongation factor EF-TU. Successful overexpression was also achieved by using promoters of the *cspB* gene [156], encoding the secreted surface-layer protein PS2, the *gapA* gene [157], encoding glyceraldehyde 3-phosphate dehydrogenase, or the *groEL* gene, encoding chaperonin [25]. Regardless of the choice of the promoter, higher expression cannot necessarily be equalized with improvement, but context-dependent fine-tuning of gene expression appears optimal. In this regard, recent development in creating promoters with variable expression strength [107, 158] and the increasing knowledge on transcriptional regulation by the different types of sigma factors [144] are promising for future development of advanced tools for modulation of gene expression.

6.4 Systems Biology of *Corynebacterium glutamicum*

Exploration of *C. glutamicum* on a more global level became possible with the development of the different omics technologies. Within just one decade, methods for analyzing the different cellular layers on the systems level – genome, transcriptome, proteome, metabolome, and fluxome – were established and applied to *C. glutamicum* providing fascinating insights into this prominent organism (Figure 6.2).

6.4.1 Genome Analysis

Whole genome sequencing projects of *C. glutamicum* were simultaneously initiated in the late 1990s by the three major competitors in the L-lysine market – BASF SE, Degussa (Evonik), and Kyowa Hakko [159]. The publicly available sequences were reported to comprise around 3.3 Mbp, harboring approximately 3000 potential genes [18, 160]. Slight but distinct differences in the reported

Table 6.3 General features of the publicly available genome sequences of *C. glutamicum* ATCC 13032 obtained in different whole genome sequencing project by Kalinowski and co-workers (Accession No. BX927147) [18], and Ikeda and co-workers (Accession No. BA000036/NC003450) [160], respectively.

Strain	*C. glutamicum* ATCC 13032	*C. glutamicum* ATCC 13032
Accession No.	BX927147	BA000036/NC003450
Size	3 282 708	3 309 401
Average G + C content	53.8	53.8
No. of ORFs	3 002	3 099/2 993
No. of rRNA operons	6	6
No. of tRNAs	60	60
Coding regions (%)	88.3	86.8
Start codon usage (%)		
AUG	66.5	62.5
GUG	23.1	24.3
UUG	10.3	13.2

Source: Data are taken from [159].

sequences are related to mobile elements obviously leading to genome changes in rather short time periods [159]. Beyond the wild type, a prophage-free variant of *C. glutamicum* ATCC 13032 with a 6% reduced genome was recently constructed and sequenced [161]. The general features of the genome sequences are summarized in Table 6.3. Unraveling of the genome strongly supported the understanding of *C. glutamicum* and had substantial industrial impact through revolutionizing genetic and metabolic engineering. It, for instance, helped to identify missing candidates to close biosynthetic pathways [18], and was the cornerstone for developing new techniques including transcriptome and proteome analysis. Comparative sequence analysis additionally gave rise to "genome-breeding" as one of the first methods for creating minimally mutated *C. glutamicum* strains for efficient L-lysine production [63, 83, 162]. Moreover, it enabled genome-scale modeling approaches for *in silico* analysis of the natural network boundaries and theoretical flux spaces. The genome-scale model comprised 446 reactions and 411 metabolites [66]. For studies on L-methionine and L-lysine, condensed models were useful in identifying optimal production pathways, promising genetic targets, and beneficial pathway correlations [163, 164]. In this regard, comprehensive *in vivo* and *in silico* modeling approaches on basis of the genome sequence in combination with systems metabolic engineering have yielded the best L-lysine hyperproducing the *C. glutamicum* strain so far reported [60].

6.4.2
Transcriptome Analysis

Release of the genome sequence of *C. glutamicum* promoted the development of global analysis of transcription. First microarrays were designed as "metabolic arrays," comprising selected genes of the central metabolism and amino acid biosynthetic pathways [35]. These were soon extended to whole-genome DNA microarrays covering most of the annotated genes of *C. glutamicum* [165]. In the beginning, most transcriptome studies focused on diverse aspects of the *C. glutamicum* physiology including sulfur [166], nitrogen [167, 168], and phosphorus metabolism [169]. Additionally, the global transcriptional response to the use of various carbon sources comprising acetate [35], citrate [62], and xylose [38] was analyzed. Beyond the investigation of expression changes to external stimuli, transcriptome studies were intensively applied to study regulons for predicting global regulatory networks of *C. glutamicum*. Meanwhile, the regulatory pattern of various transcriptional regulators involved in regulation of carbon metabolism and stress response such as AcnR [170], AmtR [171, 172], ClgR [140], CysR [173], DtxR [140], GlxR [44], HspR [140], McbR [166], RamA [174, 175], RamB [174], RbsR [176], and UreR [140] has been investigated. Today, transcriptome profiling is a valuable and convenient method for supporting rational strain design toward optimized production processes. Examples comprise producer strains for lysine [38, 177], valine [178, 179], arginine [180], citrulline [180], cadaverin [151, 152], and glycolate [181]. Recent studies for transcriptome analysis of *C. glutamicum* recruited RNA sequencing as an advanced technique for expression analysis [182–184].

6.4.3
Proteome Analysis

Proteome analysis of *C. glutamicum* was initiated in 1998 by establishing 2-D gel electrophoresis and microsequencing for protein analysis [185]. Availability of the genome sequence facilitated the fast, sensitive, and reliable identification of proteins by mass spectrometry analysis [186, 187], and optimized protocols allowed higher resolved proteome maps for cytoplasmatic, membrane-bound, cell wall-associated, secreted, and phosphorylated proteins [187–189]. Classical 2D-PAGE protocols have been successfully applied in recent years to study the cellular response to externally exposed stressors and supplements [190–193], starvation [168], or growth on different carbon sources [62, 194, 195]. This provided some interesting insights into cellular adaptation processes and suggested some systems-wide impact on physiology, which is reflected by changed levels of proteins referring to cell wall composition, energy, carbon, and amino acid metabolism [62, 194]. Moreover, it helped in identifying proteins that are essential for assimilation of different carbon sources [195].

6.4.4
Metabolome Analysis

With regard to the timescale, metabolome analysis is the most critical "omics" level related to the high turnover within the cell. Appropriate techniques for fast quenching of the metabolic activity for immediate inactivation of further metabolite turnover, however, in general affect the cell integrity thus leading to metabolite leakage and consequently biased metabolome data [196–198]. In addition, a sensitive, high-throughput analysis platform is needed to quantify a large number of metabolites in a small amount of sample. These issues are until now great challenges for metabolome analysis and are continuously addressed in current research studies [196, 198–202]. Hence, the intracellular metabolome of *C. glutamicum* was successfully assessed on the level of amino acids [203]. Related to the comparably low turnover and high stability of amino acids, fast filtration and boiling in water were appropriate for sampling and metabolite extraction [203]. Validation of different sampling protocols such as cold methanol quenching, fast filtration, and complete culture quenching could carve out some principal recommendations for proper metabolome analysis [196]. Still, analysis of the intracellular metabolome remains a critical and demanding method which has been rather scarcely applied for analyzing the metabolism of *C. glutamicum*. However, when applied in an appropriate manner, it is most valuable for the identification of metabolic bottlenecks as indicated by metabolite accumulation or decreased metabolite pools [204, 205]. Moreover, metabolome analysis helps in identifying so far unassigned pathways or metabolic reactions [204, 206, 207], gives insights into the physiological response to genetic and environmental perturbations [32, 68, 120, 205, 208, 209], and discloses transcriptional or metabolic regulatory patterns [30, 203, 210], all of which deepen the understanding of *C. glutamicum* as an industrial platform organism.

6.4.5 Fluxome Analysis

Metabolic flux analysis allows a detailed investigation of the *in vivo* pathway activities, which in sum coin the metabolic phenotype of the cell [211]. The fluxome as the physiological output of the interaction of transcriptome, proteome, and metabolome is thus most valuable to gain metabolic insights for fundamental understanding of the metabolism, and consequently most helpful to support rational strain engineering [212]. The first pioneering studies investigated the metabolic flux distribution of lysine-producing *C. glutamicum* by stoichiometric mass balances [213]. More advanced techniques on the basis of the use of stable mass isotopomers – in general, ^{13}C-labeled – were subsequently developed to overcome the limitations and pitfalls of pure stoichiometric flux estimation [212, 214–216]. Indeed, *C. glutamicum* is today one of the best-studied organisms with regard to ^{13}C metabolic flux analysis [217]. ^{13}C fluxome analysis has been used (i) for comparative strain analysis to investigate the impact of genetic modifications [28, 32, 60, 70, 210, 218, 219] or to establish pathway correlations that indicate production-supporting metabolic reaction [27, 67], (ii) to unravel substrate-specific pathway usages [29, 38, 220], or (iii) to analyze the metabolic response to environmental changes [221] or in different cultivation phases [203, 222]. The respective findings strongly guided strain improvement in recent years and had substantial impact on the industrial production of commercial goods produced by *C. glutamicum* [212, 217, 223]. Related to its systems-wide character, fluxome analysis unraveled hitherto unknown targets for metabolic engineering including overexpression of fructose 1,6-bisphosphatase for improved lysine production [29, 70].

6.5 Application in Biotechnology

Industrial amino acid production with *C. glutamicum* was already established quite soon after its discovery. Traditionally, amino acids were produced from pure carbohydrate media [224]. However, for economic reasons, more complex raw materials such as corn molasses, beet molasses, and hydrolysates from starch or cassava became standard [224, 225]. *C. glutamicum* production strains are relatively resistant to high shear stress, tolerate high osmotic pressure, are rather insensitive to moderate changes in pH, temperature, and pO_2, and exhibit amazingly high cell-specific production rates [226]. Hence, usually the fermenter system, rather than the organism, limits the productivity. In this regard, today's production processes substantially profit from scale with fermenter dimensions of up to 750 m^3 [226].

6.5.1 Wild-Type Producers and Classical Mutants

The early years after the discovery of *C. glutamicum* were characterized by the first systematic studies of its metabolism [4, 5, 7] and the development of producers of proteinogenic and non-proteinogenic amino acids, which were obtained by random mutagenesis and selection [1, 227–229]. Within a short time, large-scale production plants for L-glutamate and L-lysine were launched [224, 225] with a steady increase meanwhile in production volume to 2.5 and 1.5 million tons for glutamate and L-lysine, respectively [230]. Additionally, production strains for other amino acids such as L-tyrosine [231], L-tryptophan [232], L-threonine [233], L-valine [86], L-arginine [234], and L-serine [235] were developed. By 1978, the fermentative production of all essential amino acids but L-methionine was already established by four classes of producers: namely wild-type strain, auxotrophic mutants, regulatory mutants, and auxotrophic regulatory mutants [236] (Figure 6.2). Today, amino acids are among the most important products of *C. glutamicum* in industrial biotechnology (Table 6.4).

6.5.2 Top-Selling Amino Acids

In relation to their industrial importance, amino acid biosynthesis and its regulation have been intensively studied in *C. glutamicum*. For commercial interest,

Table 6.4 Overview on the industrial amino acid market [96] and the production performance of *Corynebacterium glutamicum* [223, 230, 237].

Product	Titer ($g\,l^{-1}$)	Yield ($g\,g^{-1}$)	Productivity ($g\,l^{-1}\,h^{-1}$)	Market size ($ton\,yr^{-1}$)
L-Glutamate	100	0.60	3.5	2 500 000
L-Lysine[a)]	120	0.55	4.0	1 500 000
L-Tryptophan	58	0.20–0.25	—	10 000
L-Glutamine	41	0.22[b)]	0.66[b)]	2 000
L-Arginine	52	0.17	—	1 200
L-Phenylalanine[c)]	28	—	—	8 000
L-Valine[c)]	99	0.3–0.4	0.27	500
L-Histidine[c)]	23	0.15–0.2	—	400

a) Given as HCl salt.
b) Estimated from [237].
c) Not exclusively produced by *C. glutamicum* fermentation.
Data refer to classical producers and/or producers obtained by metabolic engineering approaches.

most studies focused on the top-selling amino acids L-glutamate, L-lysine, and aromatic amino acids, which are briefly described here.

6.5.2.1
Glutamate Biosynthesis

L-Glutamate is the first amino acid ever produced by microbial fermentation [1]. It is directly synthesized from 2-oxoglutarate by glutamate dehydrogenase. Even the *C. glutamicum* wild type secretes substantial amounts of this amino acid in response to different triggers such as biotin-limitation, penicillin, or surfactant treatment [77, 78, 224, 238]. Enzymatic profiling indicated that the major flux control at the 2-oxoglutarate branch point is taken by the OGDHC [71].

6.5.2.2
L-Lysine Biosynthesis

In contrast to L-glutamate, L-lysine biosynthesis is feedback-regulated to avoid overproduction and secretion. The most important control point in *C. glutamicum* is the enzyme aspartokinase, which is inhibited by concerted action of L-lysine and L-threonine [239, 240]. Even in early times, this undesired regulation was bypassed and eliminated by generating and selecting random mutants with auxotrophy for homoserine, threonine, or methionine [228], and later by mutants with resistance to the lysine analog *S*-(2-aminoethyl)-L-cysteine [241] expressing desensitized aspartokinase [240]. In addition to control of biosynthesis, lysine formation is transcriptionally regulated by induction of *lysE*, encoding the lysine exporter upon intracellular accumulation of lysine [203, 242]. Despite the still limited knowledge of the underlying regulatory principles of metabolic control and export, large-scale lysine fermentation was successfully launched in 1958 at Kyowa Hakko's plant in Japan [226].

6.5.2.3 Biosynthesis of Aromatic Amino Acids

The formation of L-phenylalanine (Phe), L-tyrosine (Tyr), and L-tryptophan (Trp) is highly demanding with regard to carbon and energy supply as well as reducing power and thus regulated by a highly complex network of metabolic inhibition and repression [96]. The key enzymes 3-deoxy-D-arabino-heptulosonate 7-phosphate synthase (DS) and chorismate mutase (CM) of the Phe–Tyr-pathway are metabolically inhibited by the corresponding end products [243]. L-Tryptophan regulates its own biosynthesis at the metabolic switch point to the Phe–Tyr-pathway by inhibition of anthranilate synthetase [244], and by inhibition of anthranilate phosphoribosyltransferase [96]. Additional regulation is imposed by repression of the *trp* operon by L-tryptophan [245].

6.5.3 Tailored Strains by Recombinant DNA Technology

Throughout the years, the product portfolio of *C. glutamicum* was substantially expanded by homologous and heterologous genetic engineering [96, 223, 246]. Today, specialized cell factories are described for the production of various natural and non-natural compounds such as L- and D-amino acids [60, 180, 247, 248], vitamins [249], pyrazines [250], diamines [139, 251, 252], organic acids [181, 253–256], alcohols [257–259], pyrimidines [120], terpenes [260], terpenoids [261, 262], and polymers [263–266]. Admittedly, not all of them are yet under commercial production, but at least they provide a valuable proof of concept. Moreover, genetic engineering of *C. glutamicum* enabled the extension of its substrate spectrum to (i) sugar monomers such as xylose [25, 267] and arabinose [268] as building blocks of hemicellulose [13], (ii) amino sugars [269, 270], (iii) sugar oligomers such as cellobiose [271, 272] and lactose [146], (iv) sugar polymers including starch [273, 274], cellulose [275–277], and hemicellulose [275, 278], and (v) waste streams from biomass valorization processes [69, 153]. This allowed complementing naturally utilizable carbon sources by alternative industrial feedstocks, a milestone for integration of *C. glutamicum* biotechnology into the rising era of bio-based economy.

6.6 Conclusions and Perspectives

Isolation of *C. glutamicum* and establishment of amino acid fermentation in the late 1950s gave birth to a new era of biotechnology. Over the past decades, with the rapid development of industrial processes, analytical, and genetic engineering techniques, and the continuously increasing knowledge on metabolism, cell physiology, and regulation we have reached the dawn of synthetic biology. Recent advances in creating synthetic promoters for expressional regulation [107], the integration of *in silico* and *in vivo* modeling [60], and the prior design of new hyperproducing cell factories for traditional [60] and innovative products [38, 152] promise even faster developments for future industrial strain design and product diversity. Despite substantial studies on the core metabolism of *C. glutamicum*, challenges and knowledge gaps remain. Whole-genome sequencing, for instance, has unraveled numerous genes with hypothetical or unknown function, which may potentially be involved in regulatory programs for stress response. Indeed, the global interplay of regulatory units with the core metabolism to promote the required complex physiological changes to fend off environmentally exposed stress such as temperature, pH, salinity, or toxic substances is rather unexplored. Beyond the pure observation of changes in transcriptome or proteome, a holistic integration of these data with metabolites and metabolic fluxes is needed to decipher global regulatory mechanisms in depth. Initial multi-omics studies for *C. glutamicum* in this direction appear promising [168, 172, 203]. Most recently, a comprehensive approach of transcriptome, proteome, metabolome, and fluxome elucidated the osmo-stress response of *Bacillus subtilits* and gave new insights into physiological strategies [279]. Further progress will not only deepen our understanding of *C. glutamium* but will provide a fundament for subsequent knowledge-based design and engineering of strains with improved stress tolerance as advanced industrial cell factories, better resisting the harsh production conditions or the inhibitory effects of new and inhibitory products.

References

1. Kinoshita, S., Udaka, S., and Shimono, M. (1957) Studies on the amino acid fermentation. Part 1. Production of L-glutamic acid by various microorganisms. *J. Gen. Appl. Microbiol.*, **3** (3), 193–205.
2. Liebl, W. (2005) *Corynebacterium* taxonomy in *Handbook of Corynebacterium glutamicum* (eds L. Eggeling and M. Bott), CRC Press, Boca Raton, FL, pp. 9–34.
3. Abe, S., Takayama, K., and Kinoshita, S. (1967) Taxonomical studies on glutamic acid-producing bacteria. *J. Gen. Appl. Microbiol.*, **13** (3), 279–301.
4. Shiio, I., Otsuka, S., and Tsunoda, T. (1959) Glutamic acid formation from glucose by bacteria. I. Enzymes of the Embden-Meyerhof-Parnas pathway, the Krebs cycle, and the glyoxylate bypass in cell extracts of *Brevibacterium flavum* No. 2247. *J. Biochem.*, **46**, 1303–1311.
5. Shiio, I., Otsuka, S., and Tsunoda, T. (1960) Glutamic acid formation from glucose by bacteria. III. On the pathway of pyruvate formation in *Brevibacterium flavum* No. 2247. *J. Biochem.*, **47** (3), 414–421.
6. Otsuka, S., Miyajima, R., and Shiio, I. (1965) Comparative studies on the mechanism of microbial glutamate formation. I. Pathway of glutamate formation from glucose in *Brevibacterium flavum* and in *Micrococcus glutamicus*. *J. Gen. Appl. Microbiol.*, **11** (4), 285–294.
7. Shiio, I., Otsuka, S., and Tsunoda, T. (1959) Glutamic acid formation from glucose by bacteria. II. Glutamic acid and α-ketoglutaric acid formation by *Brevibacterium flavum* No. 2247. *J. Biochem.*, **46** (12), 1597–1605.
8. Ozaki, H. and Shiio, I. (1969) Regulation of the TCA and glyoxylate cycles in *Brevibacterium flavum*. II. Regulation of phosphoenolpyruvate carboxylase and pyruvate kinase. *J. Biochem.*, **66** (3), 297–311.
9. Mori, M. and Shiio, I. (1985) Purification and some properties of phosphoenolpyruvate carboxylase

from *Brevibacterium flavum* and its aspartate-overproducing mutant. *J. Biochem.*, **97** (4), 1119–1128.
10. Peters-Wendisch, P.G., Wendisch, V.F., de Graaf, A.A., Eikmanns, B.J., and Sahm, H. (1996) C3-carboxylation as an anaplerotic reaction in phosphoenolpyruvate carboxylase-deficient *Corynebacterium glutamicum*. *Arch. Microbiol.*, **165** (6), 387–396.
11. Cocaign-Bousquet, M., Guyonvarch, A., Lindley, N.D. (1996) Growth rate-dependent modulation of carbon flux through central metabolism and the kinetic consequences for glucose-limited chemostat cultures of *Corynebacterium glutamicum*. *Appl. Environ. Microbiol.* **62**(2), 429–436.
12. Eikmanns, B.J. (2005) Central metabolism: tricarboxylic acid cycle and anaplerotic reactions in *Handbook of Corynebacterium glutamicum* (eds L. Eggeling and M. Bott), CRC Press, Boca Raton, FL, pp. 241–276.
13. Buschke, N., Schäfer, R., Becker, J., and Wittmann, C. (2013) Metabolic engineering of industrial platform microorganisms for biorefinery applications – optimization of substrate spectrum and process robustness by rational and evolutive strategies. *Bioresour. Technol.*, **135**, 544–554.
14. Laslo, T., von Zaluskowski, P., Gabris, C., Lodd, E., Rückert, C., Dangel, P., Kalinowski, J., Auchter, M., Seibold, G., and Eikmanns, B.J. (2012) Arabitol metabolism of *Corynebacterium glutamicum* and its regulation by AtlR. *J. Bacteriol.*, **194** (5), 941–955.
15. Arndt, A., Auchter, M., Ishige, T., Wendisch, V.F., and Eikmanns, B.J. (2008) Ethanol catabolism in *Corynebacterium glutamicum*. *J. Mol. Microbiol. Biotechnol.*, **15** (4), 222–233.
16. Parche, S., Burkovski, A., Sprenger, G.A., Weil, B., Krämer, R., and Titgemeyer, F. (2001) *Corynebacterium glutamicum*: a dissection of the PTS. *J. Mol. Microbiol. Biotechnol.*, **3** (3), 423–428.
17. Mori, M. and Shiio, I. (1987) Phosphoenolpyruvate: sugar phosphotransferase systems and sugar metabolism in *Brevibacterium flavum*. *Agric. Biol. Chem.*, **51** (10), 2671–2678.
18. Kalinowski, J., Bathe, B., Bartels, D., Bischoff, N., Bott, M., Burkovski, A., Dusch, N., Eggeling, L., Eikmanns, B.J., Gaigalat, L., Goesmann, A., Hartmann, M., Huthmacher, K., Krämer, R., Linke, B., McHardy, A.C., Meyer, F., Möckel, B., Pfefferle, W., Pühler, A., Rey, D.A., Rückert, C., Rupp, O., Sahm, H., Wendisch, V.F., Wiegrabe, I., and Tauch, A. (2003) The complete *Corynebacterium glutamicum* ATCC 13032 genome sequence and its impact on the production of L-aspartate-derived amino acids and vitamins. *J. Biotechnol.*, **104** (1–3), 5–25.
19. Nichols, N.N., Dien, B.S., and Bothast, R.J. (2001) Use of catabolite repression mutants for fermentation of sugar mixtures to ethanol. *Appl. Microbiol. Biotechnol.*, **56** (1–2), 120–125.
20. Vinuselvi, P., Kim, M.K., Lee, S.K., and Ghim, C.M. (2012) Rewiring carbon catabolite repression for microbial cell factory. *BMB Rep.*, **45** (2), 59–70.
21. Dominguez, H., Cogaign-Bousquet, M., and Lindley, N.D. (1997) Simultaneous consumption of glucose and fructose from sugar mixtures during batch growth of *Corynebacterium glutamicum*. *Appl. Microbiol. Biotechnol.*, **47** (5), 600–603.
22. Wendisch, V.F., de Graaf, A.A., Sahm, H., and Eikmanns, B.J. (2000) Quantitative determination of metabolic fluxes during coutilization of two carbon sources: comparative analyses with *Corynebacterium glutamicum* during growth on acetate and/or glucose. *J. Bacteriol.*, **182** (11), 3088–3096.
23. Neuner, A. and Heinzle, E. (2011) Mixed glucose and lactate uptake by *Corynebacterium glutamicum* through metabolic engineering. *Biotechnol. J.*, **6** (3), 318–329.
24. Arndt, A. and Eikmanns, B.J. (2007) The alcohol dehydrogenase gene *adhA* in *Corynebacterium glutamicum* is subject to carbon catabolite repression. *J. Bacteriol.*, **189** (20), 7408–7416.
25. Buschke, N., Schröder, H., and Wittmann, C. (2011) Metabolic engineering of *Corynebacterium*

glutamicum for production of 1,5-diaminopentane from hemicellulose. *Biotechnol. J.*, **6** (3), 306–317.

26. Lindner, S.N., Seibold, G.M., Henrich, A., Krämer, R., and Wendisch, V.F. (2011) Phosphotransferase system-independent glucose utilization in *Corynebacterium glutamicum* by inositol permeases and glucokinases. *Appl. Environ. Microbiol.*, **77** (11), 3571–3581.
27. Wittmann, C. and Heinzle, E. (2002) Genealogy profiling through strain improvement by using metabolic network analysis: metabolic flux genealogy of several generations of lysine-producing Corynebacteria. *Appl. Environ. Microbiol.*, **68** (12), 5843–5859.
28. Becker, J., Klopprogge, C., and Wittmann, C. (2008) Metabolic responses to pyruvate kinase deletion in lysine producing *Corynebacterium glutamicum*. *Microb. Cell Fact.*, **7** (8), 8.
29. Kiefer, P., Heinzle, E., Zelder, O., and Wittmann, C. (2004) Comparative metabolic flux analysis of lysine-producing *Corynebacterium glutamicum* cultured on glucose or fructose. *Appl. Environ. Microbiol.*, **70** (1), 229–239.
30. Dominguez, H., Rollin, C., Guyonvarch, A., Guerquin-Kern, J.L., Cocaign-Bousquet, M., and Lindley, N.D. (1998) Carbon-flux distribution in the central metabolic pathways of *Corynebacterium glutamicum* during growth on fructose. *Eur. J. Biochem.*, **254** (1), 96–102.
31. Moritz, B., Striegel, K., De Graaf, A.A., and Sahm, H. (2000) Kinetic properties of the glucose-6-phosphate and 6-phosphogluconate dehydrogenases from *Corynebacterium glutamicum* and their application for predicting pentose phosphate pathway flux *in vivo*. *Eur. J. Biochem.*, **267** (12), 3442–3452.
32. Becker, J., Klopprogge, C., Herold, A., Zelder, O., Bolten, C.J., and Wittmann, C. (2007) Metabolic flux engineering of L-lysine production in *Corynebacterium glutamicum* – over expression and modification of G6P dehydrogenase. *J. Biotechnol.*, **132** (2), 99–109.
33. Sugimoto, S. and Shiio, I. (1987) Regulation of glucose-6-phosphate dehydrogenase in *Brevibacterium flavum*. *Agric. Biol. Chem.*, **51** (1), 101–108.
34. Sugimoto, S. and Shiio, I. (1989) Regulation of enzymes for erythrose 4-phosphate synthesis in *Brevibacterium flavum*. *Agric. Biol. Chem.*, **53** (8), 2081–2087.
35. Hayashi, M., Mizoguchi, H., Shiraishi, N., Obayashi, M., Nakagawa, S., Imai, J., Watanabe, S., Ota, T., and Ikeda, M. (2002) Transcriptome analysis of acetate metabolism in *Corynebacterium glutamicum* using a newly developed metabolic array. *Biosci. Biotechnol., Biochem.*, **66** (6), 1337–1344.
36. Sugimoto, S. and Shiio, I. (1989) Fructose metabolism and regulation of 1-phosphofructokinase and 6-phosphofructokinase in *Brevibacterium flavum*. *Agric. Biol. Chem.*, **53** (5), 1261–1268.
37. Rittmann, D., Schaffer, S., Wendisch, V.F., and Sahm, H. (2003) Fructose-1,6-bisphosphatase from *Corynebacterium glutamicum*: expression and deletion of the fbp gene and biochemical characterization of the enzyme. *Arch. Microbiol.*, **180** (4), 285–292.
38. Buschke, N., Becker, J., Schäfer, R., Kiefer, P., Biedendieck, R., and Wittmann, C. (2013) Systems metabolic engineering of xylose-utilizing *Corynebacterium glutamicum* for production of 1,5-diaminopentane. *Biotechnol. J.*, **8** (5), 557–570.
39. Omumasaba, C.A., Okai, N., Inui, M., and Yukawa, H. (2004) *Corynebacterium glutamicum* glyceraldehyde-3-phosphate dehydrogenase isoforms with opposite, ATP-dependent regulation. *J. Mol. Microbiol. Biotechnol.*, **8** (2), 91–103.
40. Jetten, M.S., Gubler, M.E., Lee, S.H., and Sinskey, A.J. (1994) Structural and functional analysis of pyruvate kinase from *Corynebacterium glutamicum*. *Appl. Environ. Microbiol.*, **60** (7), 2501–2507.
41. Jetten, M., Pitoc, G.A., Follettie, M.T., and Sinskey, A.J. (1994) Regulation of phospho(enol)-pyruvate and

oxaloacetate-converting enzymes in *Corynebacterium glutamicum*. *Appl. Microbiol. Biotechnol.*, **41** (1), 47–52.

42. Eikmanns, B.J. and Blombach, B. (2014) The pyruvate dehydrogenase complex of *Corynebacterium glutamicum*: an attractive target for metabolic engineering. *J. Biotechnol.*, **192**, 339–345.
43. Sugimoto, S. and Shiio, I. (1987) Regulation of 6-phosphogluconate dehydrogenase in *Brevibacterium flavum*. *Agric. Biol. Chem.*, **51** (5), 1257–1263.
44. van Ooyen, J., Emer, D., Bussmann, M., Bott, M., Eikmanns, B.J., and Eggeling, L. (2011) Citrate synthase in *Corynebacterium glutamicum* is encoded by two *gltA* transcripts which are controlled by RamA, RamB, and GlxR. *J. Biotechnol.*, **154** (2–3), 140–148.
45. Shiio, I., Ozaki, H., and Ujigawa, K. (1977) Regulation of citrate synthase in *Brevibacterium flavum*, a glutamate-producing bacterium. *J. Biochem.*, **82** (2), 395–405.
46. Eikmanns, B.J., Thum-Schmitz, N., Eggeling, L., Lüdtke, K.U., and Sahm, H. (1994) Nucleotide sequence, expression and transcriptional analysis of the *Corynebacterium glutamicum gltA* gene encoding citrate synthase. *Microbiology*, **140** (Pt. 8), 1817–1828.
47. Eikmanns, B.J., Rittmann, D., and Sahm, H. (1995) Cloning, sequence analysis, expression, and inactivation of the *Corynebacterium glutamicum icd* gene encoding isocitrate dehydrogenase and biochemical characterization of the enzyme. *J. Bacteriol.*, **177** (3), 774–782.
48. Ozaki, H. and Shiio, I. (1968) Regulation of the TCA and glyoxylate cycles in *Brevibacterium flavum*. I. Inhibition of isocitrate lyase and isocitrate dehydrogenase by organic acids related to the TCA and glyoxylate cycles. *J. Biochem.*, **64** (3), 355–363.
49. Shiio, I. and Ozaki, H. (1968) Concerted inhibition of isocitrate dehydrogenase by glyoxylate plus oxalacetate. *J. Biochem.*, **64** (1), 45–53.
50. Shiio, I. and Ujigawa-Takeda, K. (1980) Presence and regulation of α-ketoglutarate dehydrogenase complex in a glutamate-producing bacterium, *Brevibacterium flavum*. *Agric. Biol. Chem.*, **44** (8), 1897–1904.
51. Schwinde, J.W., Hertz, P.F., Sahm, H., Eikmanns, B.J., and Guyonvarch, A. (2001) Lipoamide dehydrogenase from *Corynebacterium glutamicum*: molecular and physiological analysis of the *lpd* gene and characterization of the enzyme. *Microbiology*, **147** (Pt. 8), 2223–2231.
52. Molenaar, D., van der Rest, M.E., and Petrovic, S. (1998) Biochemical and genetic characterization of the membrane-associated malate dehydrogenase (acceptor) from *Corynebacterium glutamicum*. *Eur. J. Biochem.*, **254** (2), 395–403.
53. Molenaar, D., van der Rest, M.E., Drysch, A., and Yucel, R. (2000) Functions of the membrane-associated and cytoplasmic malate dehydrogenases in the citric acid cycle of *Corynebacterium glutamicum*. *J. Bacteriol.*, **182** (24), 6884–6891.
54. Reinscheid, D.J., Eikmanns, B.J., and Sahm, H. (1994) Characterization of the isocitrate lyase gene from *Corynebacterium glutamicum* and biochemical analysis of the enzyme. *J. Bacteriol.*, **176** (12), 3474–3483.
55. Wendisch, V.F., Spies, M., Reinscheid, D.J., Schnicke, S., Sahm, H., and Eikmanns, B.J. (1997) Regulation of acetate metabolism in *Corynebacterium glutamicum*: transcriptional control of the isocitrate lyase and malate synthase genes. *Arch. Microbiol.*, **168** (4), 262–269.
56. Reinscheid, D.J., Eikmanns, B.J., and Sahm, H. (1994) Malate synthase from *Corynebacterium glutamicum*: sequence analysis of the gene and biochemical characterization of the enzyme. *Microbiology*, **140** (Pt. 11), 3099–3108.
57. Mori, M. and Shiio, I. (1985) Synergistic inhibition of phosphoenolpyruvate carboxylase by aspartate and 2-oxoglutarate in *Brevibacterium flavum*. *J. Biochem.*, **98** (6), 1621–1630.

58. Jetten, M.S. and Sinskey, A.J. (1995) Purification and properties of oxaloacetate decarboxylase from *Corynebacterium glutamicum*. *Antonie Van Leeuwenhoek*, **67** (2), 221–227.
59. Yokota, A. and Lindley, N.D. (2005) in *Handbook of Corynebacterium glutamicum* (eds L. Eggeling and M. Bott), CRC Press, Boca Raton, FL, pp. 215–240.
60. Becker, J., Zelder, O., Haefner, S., Schröder, H., and Wittmann, C. (2011) From zero to hero – design-based systems metabolic engineering of *Corynebacterium glutamicum* for L-lysine production. *Metab. Eng.*, **13** (2), 159–168.
61. Gerstmeir, R., Wendisch, V.F., Schnicke, S., Ruan, H., Farwick, M., Reinscheid, D., and Eikmanns, B.J. (2003) Acetate metabolism and its regulation in *Corynebacterium glutamicum*. *J. Biotechnol.*, **104** (1–3), 99–122.
62. Polen, T., Schluesener, D., Poetsch, A., Bott, M., and Wendisch, V.F. (2007) Characterization of citrate utilization in *Corynebacterium glutamicum* by transcriptome and proteome analysis. *FEMS Microbiol. Lett.*, **273** (1), 109–119.
63. Ohnishi, J., Katahira, R., Mitsuhashi, S., Kakita, S., and Ikeda, M. (2005) A novel *gnd* mutation leading to increased L-lysine production in *Corynebacterium glutamicum*. *FEMS Microbiol. Lett.*, **242** (2), 265–274.
64. Ikeda, M., Okamoto, K., and Katsumata, R. (1999) Cloning of the transketolase gene and the effect of its dosage on aromatic amino acid production in *Corynebacterium glutamicum*. *Appl. Microbiol. Biotechnol.*, **51** (2), 201–206.
65. Dominguez, H. and Lindley, N.D. (1996) Complete sucrose metabolism requires fructose phosphotransferase activity in *Corynebacterium glutamicum* to ensure phosphorylation of liberated fructose. *Appl. Environ. Microbiol.*, **62** (10), 3878–3880.
66. Kjeldsen, K.R. and Nielsen, J. (2009) *In silico* genome-scale reconstruction and validation of the *Corynebacterium glutamicum* metabolic network. *Biotechnol. Bioeng.*, **102** (2), 583–597.
67. Wittmann, C. and De Graaf, A.A. (2005) Metabolic flux analysis in *Corynebacterium glutamicum* in *Handbook of Corynebacterium glutamicum* (eds L. Eggeling and M. Bott), CRC Press, Boca Raton, FL, pp. 277–304.
68. Kind, S., Becker, J., and Wittmann, C. (2013) Increased lysine production by flux coupling of the tricarboxylic acid cycle and the lysine biosynthetic pathway – metabolic engineering of the availability of succinyl-CoA in *Corynebacterium glutamicum*. *Metab. Eng.*, **15**, 184–195.
69. Rittmann, D., Lindner, S.N., and Wendisch, V.F. (2008) Engineering of a glycerol utilization pathway for amino acid production by *Corynebacterium glutamicum*. *Appl. Environ. Microbiol.*, **74** (20), 6216–6222.
70. Becker, J., Klopprogge, C., Zelder, O., Heinzle, E., and Wittmann, C. (2005) Amplified expression of fructose 1,6-bisphosphatase in *Corynebacterium glutamicum* increases *in vivo* flux through the pentose phosphate pathway and lysine production on different carbon sources. *Appl. Environ. Microbiol.*, **71** (12), 8587–8596.
71. Shimizu, H., Tanaka, H., Nakato, A., Nagahisa, K., Kimura, E., and Shioya, S. (2003) Effects of the changes in enzyme activities on metabolic flux redistribution around the 2-oxoglutarate branch in glutamate production by *Corynebacterium glutamicum*. *Bioprocess Biosyst. Eng.*, **25** (5), 291–298.
72. Bott, M. (2007) Offering surprises: TCA cycle regulation in *Corynebacterium glutamicum*. *Trends Microbiol.*, **15** (9), 417–425.
73. Niebisch, A., Kabus, A., Schultz, C., Weil, B., and Bott, M. (2006) Corynebacterial protein kinase G controls 2-oxoglutarate dehydrogenase activity via the phosphorylation status of the OdhI protein. *J. Biol. Chem.*, **281** (18), 12300–12307.
74. Schultz, C., Niebisch, A., Gebel, L., and Bott, M. (2007) Glutamate production by *Corynebacterium glutamicum*:

dependence on the oxoglutarate dehydrogenase inhibitor protein OdhI and protein kinase PknG. *Appl. Microbiol. Biotechnol.*, **76** (3), 691–700.
75. Hasegawa, T., Hashimoto, K., Kawasaki, H., and Nakamatsu, T. (2008) Changes in enzyme activities at the pyruvate node in glutamate-overproducing *Corynebacterium glutamicum. J. Biosci. Bioeng.*, **105** (1), 12–19.
76. Asakura, Y., Kimura, E., Usuda, Y., Kawahara, Y., Matsui, K., Osumi, T., and Nakamatsu, T. (2007) Altered metabolic flux due to deletion of *odhA* causes L-glutamate overproduction in *Corynebacterium glutamicum. Appl. Environ. Microbiol.*, **73** (4), 1308–1319.
77. Kim, J., Fukuda, H., Hirasawa, T., Nagahisa, K., Nagai, K., Wachi, M., and Shimizu, H. (2010) Requirement of de novo synthesis of the OdhI protein in penicillin-induced glutamate production by *Corynebacterium glutamicum. Appl. Microbiol. Biotechnol.*, **86** (3), 911–920.
78. Kim, J., Hirasawa, T., Sato, Y., Nagahisa, K., Furusawa, C., and Shimizu, H. (2009) Effect of *odhA* overexpression and odhA antisense RNA expression on Tween-40-triggered glutamate production by *Corynebacterium glutamicum. Appl. Microbiol. Biotechnol.*, **81** (6), 1097–1106.
79. Sauer, U. and Eikmanns, B.J. (2005) The PEP-pyruvate-oxaloacetate node as the switch point for carbon flux distribution in bacteria. *FEMS Microbiol. Rev.*, **29** (4), 765–794.
80. Petersen, S., de Graaf, A.A., Eggeling, L., Mollney, M., Wiechert, W., and Sahm, H. (2000) *In vivo* quantification of parallel and bidirectional fluxes in the anaplerosis of *Corynebacterium glutamicum. J. Biol. Chem.*, **275** (46), 35932–35941.
81. Peters-Wendisch, P.G., Schiel, B., Wendisch, V.F., Katsoulidis, E., Möckel, B., Sahm, H., and Eikmanns, B.J. (2001) Pyruvate carboxylase is a major bottleneck for glutamate and lysine production by *Corynebacterium glutamicum. J. Mol. Microbiol. Biotechnol.*, **3** (2), 295–300.
82. Riedel, C., Rittmann, D., Dangel, P., Möckel, B., Petersen, S., Sahm, H., and Eikmanns, B.J. (2001) Characterization of the phosphoenolpyruvate carboxykinase gene from *Corynebacterium glutamicum* and significance of the enzyme for growth and amino acid production. *J. Mol. Microbiol. Biotechnol.*, **3** (4), 573–583.
83. Ohnishi, J., Mitsuhashi, S., Hayashi, M., Ando, S., Yokoi, H., Ochiai, K., and Ikeda, M. (2002) A novel methodology employing *Corynebacterium glutamicum* genome information to generate a new L-lysine-producing mutant. *Appl. Microbiol. Biotechnol.*, **58** (2), 217–223.
84. Sano, K., Ito, K., Miwa, K., and Nakamori, S. (1987) Amplification of the phosphoenol pyruvate carboxylase gene of *Brevibacterium lactofermentum* to improve amino acid production. *Agric. Biol. Chem.*, **51** (2), 597–599.
85. Sawada, K., Zen-in, S., Wada, M., and Yokota, A. (2010) Metabolic changes in a pyruvate kinase gene deletion mutant of *Corynebacterium glutamicum* ATCC 13032. *Metab. Eng.*, **12** (4), 401–407.
86. Blombach, B., Schreiner, M.E., Holatko, J., Bartek, T., Oldiges, M., and Eikmanns, B.J. (2007) L-valine production with pyruvate dehydrogenase complex-deficient *Corynebacterium glutamicum. Appl. Environ. Microbiol.*, **73** (7), 2079–2084.
87. Yao, W., Deng, X., Zhong, H., Liu, M., Zheng, P., Sun, Z., and Zhang, Y. (2009) Double deletion of *dtsR1* and *pyc* induce efficient L: –glutamate overproduction in *Corynebacterium glutamicum. J. Ind. Microbiol. Biotechnol.*, **36** (7), 911–921.
88. Peters-Wendisch, P., Stansen, K.C., Gotker, S., and Wendisch, V.F. (2012) Biotin protein ligase from *Corynebacterium glutamicum*: role for growth and L: –lysine production. *Appl. Microbiol. Biotechnol.*, **93** (6), 2493–2502.
89. Kim, T.Y., Kim, H.U., Park, J.M., Song, H., Kim, J.S., and Lee, S.Y. (2007) Genome-scale analysis of *Mannheimia succiniciproducens* metabolism. *Biotechnol. Bioeng.*, **97** (4), 657–671.
90. Reed, J.L., Vo, T.D., Schilling, C.H., and Palsson, B.O. (2003) An expanded

genome-scale model of *Escherichia coli* K-12 (iJR904 GSM/GPR). *Genome Biol.*, **4** (9), R54.
91. Vertes, A.A., Inui, M., and Yukawa, H. (2005) Manipulating corynebacteria, from individual genes to chromosomes. *Appl. Environ. Microbiol.*, **71** (12), 7633–7642.
92. Santamaria, R., Gil, J., Mesas, J., and Martin, J. (1984) Characterization of an endogenous plasmid and development of cloning vectors and a transformation system in *Brevibacterium lactofermentum*. *J. Gen. Microbiol.*, **130**, 2237–2246.
93. Miwa, K., Matsui, H., Terabe, M., Nakamori, S., Sano, K., and Momose, H. (1984) Cryptic plasmids in glutamic acid producing bacteria. *Agric. Biol. Chem.*, **48** (11), 2901–2903.
94. Katsumata, R., Ozaki, A., Oka, T., and Furuya, A. (1984) Protoplast transformation of glutamate-producing bacteria with plasmid DNA. *J. Bacteriol.*, **159** (1), 306–311.
95. Nesvera, J. and Patek, M. (2011) Tools for genetic manipulations in *Corynebacterium glutamicum* and their applications. *Appl. Microbiol. Biotechnol.*, **90** (5), 1641–1654.
96. Becker, J., Kind, S., and Wittmann, C. (2012) Systems metabolic engineering of *Corynebacterium glutamicum* for biobased production of chemicals, materials and fuels in *Systems Metabolic Engineering* (eds C. Wittmann and S.Y. Lee), Springer-Verlag, Dordrecht, Heidelberg, New York, London, pp. 152–191.
97. Tauch, A. (2005) Native plasmides of amino acid-producing Corynebacteria in *Handbook of Corynebacterium glutamicum* (eds L. Eggeling and M. Bott), CRC Press, Boca Raton, FL, pp. 57–80.
98. Ozaki, A., Katsumata, R., Oka, T., and Furuya, A. (1984) Functional expression of the genes of *Escherichia coli* in gram-positive *Corynebacterium glutamicum*. *Mol. Gen. Genet.*, **196** (1), 175–178.
99. Sonnen, H., Thierbach, G., Kautz, S., Kalinowski, J., Schneider, J., Pühler, A., and Kutzner, H.J. (1991) Characterization of pGA1, a new plasmid from *Corynebacterium glutamicum* LP-6. *Gene*, **107** (1), 69–74.
100. Patek, M. and Nesvera, J. (2013) Promoters and plasmid vectors of *Corynebacterium glutamicum* in *Corynebacterium glutamicum – Biology and Biotechnology* (eds H. Yukawa and M. Inui), Springer-Verlag, Berlin, Heidelberg, pp. 51–88.
101. Yoshihama, M., Higashiro, K., Rao, E.A., Akedo, M., Shanabruch, W.G., Follettie, M.T., Walker, G.C., and Sinskey, A.J. (1985) Cloning vector system for *Corynebacterium glutamicum*. *J. Bacteriol.*, **162** (2), 591–597.
102. Eikmanns, B.J., Kleinertz, E., Liebl, W., and Sahm, H. (1991) A family of *Corynebacterium glutamicum/Escherichia coli* shuttle vectors for cloning, controlled gene expression, and promoter probing. *Gene*, **102** (1), 93–98.
103. Yeh, P., Oreglia, J., Prevots, F., and Sicard, A.M. (1986) A shuttle vector system for *Brevibacterium lactofermentum*. *Gene*, **47** (2–3), 301–306.
104. Xu, D., Tan, Y., Shi, F., and Wang, X. (2010) An improved shuttle vector constructed for metabolic engineering research in *Corynebacterium glutamicum*. *Plasmid*, **64** (2), 85–91.
105. Barrett, E., Stanton, C., Zelder, O., Fitzgerald, G., and Ross, R.P. (2004) Heterologous expression of lactose- and galactose-utilizing pathways from lactic acid bacteria in *Corynebacterium glutamicum* for production of lysine in whey. *Appl. Environ. Microbiol.*, **70** (5), 2861–2866.
106. Kirchner, O. and Tauch, A. (2003) Tools for genetic engineering in the amino acid-producing bacterium *Corynebacterium glutamicum*. *J. Biotechnol.*, **104** (1–3), 287–299.
107. Rytter, J.V., Helmark, S., Chen, J., Lezyk, M.J., Solem, C., and Jensen, P.R. (2014) Synthetic promoter libraries for *Corynebacterium glutamicum*. *Appl. Microbiol. Biotechnol.*, **98** (6), 2617–2623.
108. Adham, S.A., Rodriguez, S., Ramos, A., Santamaria, R.I., and Gil, J.A. (2003) Improved vectors for transcriptional/translational signal screening

in corynebacteria using the *melC* operon from *Streptomyces glaucescens* as reporter. *Arch. Microbiol.*, **180** (1), 53–59.

109. Letek, M., Valbuena, N., Ramos, A., Ordonez, E., Gil, J.A., and Mateos, L.M. (2006) Characterization and use of catabolite-repressed promoters from gluconate genes in *Corynebacterium glutamicum*. *J. Bacteriol.*, **188** (2), 409–423.
110. Knoppova, M., Phensaijai, M., Vesely, M., Zemanova, M., Nesvera, J., and Patek, M. (2007) Plasmid vectors for testing *in vivo* promoter activities in *Corynebacterium glutamicum* and *Rhodococcus erythropolis*. *Curr. Microbiol.*, **55** (3), 234–239.
111. Tsuchiya, M. and Morinaga, Y. (1988) Genetic control systems of *Escherichia coli* can confer inducible expression of cloned genes in coryneform bacteria. *Bio/Technology*, **6** (4), 428–430.
112. Plassmeier, J.K., Busche, T., Molck, S., Persicke, M., Pühler, A., Rückert, C., and Kalinowski, J. (2013) A propionate-inducible expression system based on the *Corynebacterium glutamicum prpD2* promoter and PrpR activator and its application for the redirection of amino acid biosynthesis pathways. *J. Biotechnol.*, **163** (2), 225–232.
113. Reyes, O., Guyonvarch, A., Bonamy, C., Salti, V., David, F., and Leblon, G. (1991) 'Integron'-bearing vectors: a method suitable for stable chromosomal integration in highly restrictive corynebacteria. *Gene*, **107** (1), 61–68.
114. Schwarzer, A. and Pühler, A. (1991) Manipulation of *Corynebacterium glutamicum* by gene disruption and replacement. *Biotechnology (N.Y.)*, **9** (1), 84–87.
115. Bonamy, C., Labarre, J., Cazaubon, L., Jacob, C., Le Bohec, F., Reyes, O., and Leblon, G. (2003) The mobile element IS1207 of *Brevibacterium lactofermentum* ATCC21086: isolation and use in the construction of Tn5531, a versatile transposon for insertional mutagenesis of *Corynebacterium glutamicum*. *J. Biotechnol.*, **104** (1–3), 301–309.
116. Vertes, A.A., Asai, Y., Inui, M., Kobayashi, M., Kurusu, Y., and Yukawa, H. (1994) Transposon mutagenesis of coryneform bacteria. *Mol. Gen. Genet.*, **245** (4), 397–405.
117. Moreau, S., Blanco, C., and Trautwetter, A. (1999) Site-specific integration of corynephage phi16: construction of an integration vector. *Microbiology*, **145** (Pt. 3), 539–548.
118. Jäger, W., Schäfer, A., Pühler, A., Labes, G., and Wohlleben, W. (1992) Expression of the *Bacillus subtilis sacB* gene leads to sucrose sensitivity in the gram-positive bacterium *Corynebacterium glutamicum* but not in *Streptomyces lividans*. *J. Bacteriol.*, **174** (16), 5462–5465.
119. Schäfer, A., Tauch, A., Jäger, W., Kalinowski, J., Thierbach, G., and Pühler, A. (1994) Small mobilizable multi-purpose cloning vectors derived from the *Escherichia coli* plasmids pK18 and pK19: selection of defined deletions in the chromosome of *Corynebacterium glutamicum*. *Gene*, **145** (1), 69–73.
120. Becker, J., Schäfer, R., Kohlstedt, M., Harder, B.J., Borchert, N.S., Stöveken, N., Bremer, E., and Wittmann, C. (2013) Systems metabolic engineering of *Corynebacterium glutamicum* for production of the chemical chaperone ectoine. *Microb. Cell Fact.*, **12** (1), 110.
121. Tan, Y., Xu, D., Li, Y., and Wang, X. (2012) Construction of a novel *sacB*-based system for marker-free gene deletion in *Corynebacterium glutamicum*. *Plasmid*, **67** (1), 44–52.
122. Okibe, N., Suzuki, N., Inui, M., and Yukawa, H. (2011) Efficient markerless gene replacement in *Corynebacterium glutamicum* using a new temperature-sensitive plasmid. *J. Microbiol. Methods*, **85** (2), 155–163.
123. Blombach, B., Riester, T., Wieschalka, S., Ziert, C., Youn, J.W., Wendisch, V.F., and Eikmanns, B.J. (2011) *Corynebacterium glutamicum* tailored for efficient isobutanol production. *Appl. Environ. Microbiol.*, **77** (10), 3300–3310.
124. Jang, K.H., Chambers, P.J., Chun, M.H., and Britz, M.L. (2001) Characterization of the cell-surface barriers to plasmid transformation in *Corynebacterium*

glutamicum. *J. Microbiol. Biotechnol.*, **11**, 294–301.
125. Thierbach, G., Schwarzer, A., and Pühler, A. (1988) Transformation of spheroplasts and protoplasts of *Corynebacterium glutamicum*. *Appl. Microbiol. Biotechnol.*, **29** (4), 356–362.
126. Miwa, K., Matsui, K., Terabe, M., Ito, K., Ishida, M., Takagi, H., Nakamori, S., and Sano, K. (1985) Construction of novel shuttle vectors and a cosmid vector for the glutamic acid-producing bacteria *Brevibacterium lactofermentum* and *Corynebacterium glutamicum*. *Gene*, **39** (2–3), 281–286.
127. Dunican, L.K. and Shivnan, E. (1989) High frequency transformation of whole cells of amino acid producing Coryneform bacteria using high voltage electroporation. *Nat. Biotechnol.*, **7**, 1067–1070.
128. Liebl, W., Bayerl, A., Schein, B., Stillner, U., and Schleifer, K.H. (1989) High efficiency electroporation of intact *Corynebacterium glutamicum* cells. *FEMS Microbiol. Lett.*, **65** (3), 299–303.
129. Bonamy, C., Guyonvarch, A., Reyes, O., David, F., and Leblon, G. (1990) Interspecies electro-transformation in Corynebacteria. *FEMS Microbiol. Lett.*, **54** (1–3), 263–269.
130. Satoh, Y., Hatakeyama, K., Kohama, K., Kobayashi, M., Kurusu, Y., and Yukawa, H. (1990) Electrotransformation of intact cells of *Brevibacterium flavum* MJ-233. *J. Ind. Microbiol.*, **5** (2–3), 159–165.
131. Haynes, J.A. and Britz, M.L. (1989) Electrotransformation of *Brevibacterium lactofermentum* and *Corynebacterium glutamicum*: growth in tween 80 increases transformation frequencies. *FEMS Microbiol. Lett.*, **61** (3), 329–334.
132. Haynes, J.A. and Britz, M.L. (1990) The effect of growth conditions of *Corynebacterium glutamicum* on the transformation frequency obtained by electroporation. *J. Gen. Microbiol.*, **136** (2), 255–263.
133. Schäfer, A., Kalinowski, J., and Pühler, A. (1994) Increased fertility of *Corynebacterium glutamicum* recipients in intergeneric matings with *Escherichia coli* after stress exposure. *Appl. Environ. Microbiol.*, **60** (2), 756–759.
134. Schäfer, A., Kalinowski, J., Simon, R., Seep-Feldhaus, A.H., and Pühler, A. (1990) High-frequency conjugal plasmid transfer from gram-negative *Escherichia coli* to various gram-positive coryneform bacteria. *J. Bacteriol.*, **172** (3), 1663–1666.
135. van der Rest, M.E., Lange, C., and Molenaar, D. (1999) A heat shock following electroporation induces highly efficient transformation of *Corynebacterium glutamicum* with xenogeneic plasmid DNA. *Appl. Microbiol. Biotechnol.*, **52** (4), 541–545.
136. Horton, R.M. (1995) PCR-mediated recombination and mutagenesis. SOEing together tailor-made genes. *Mol. Biotechnol.*, **3** (2), 93–99.
137. Ankri, S., Reyes, O., and Leblon, G. (1996) Electrotransformation of highly DNA-restrictive corynebacteria with synthetic DNA. *Plasmid*, **35** (1), 62–66.
138. Vertes, A., Hatakeyama, K., Inui, M., Kobayashi, M., Kurusu, Y., and Yukawa, H. (1993) Replacement recombination in coryneform bacteria: high efficiency integration requirement for non-methylated plasmid DNA. *Biosci. Biotechnol., Biochem.*, **57** (12), 2036–2038.
139. Kind, S., Jeong, W.K., Schröder, H., and Wittmann, C. (2010) Systems-wide metabolic pathway engineering in *Corynebacterium glutamicum* for bio-based production of diaminopentane. *Metab. Eng.*, **12** (4), 341–351.
140. Brinkrolf, K., Brune, I., and Tauch, A. (2007) The transcriptional regulatory network of the amino acid producer *Corynebacterium glutamicum*. *J. Biotechnol.*, **129** (2), 191–211.
141. Nesvera, J., Holatko, J., and Patek, M. (2012) Analysis of *Corynebacterium glutamicum* promoters and their applications. *Subcell. Biochem.*, **64**, 203–221.
142. Patek, M., Holatko, J., Busche, T., Kalinowski, J., and Nesvera, J. (2013) *Corynebacterium glutamicum* promoters: a practical approach. *Microb. Biotechnol.*, **6** (2), 103–117.
143. Patek, M., Eikmanns, B.J., Patek, J., and Sahm, H. (1996) Promoters

from *Corynebacterium glutamicum*: cloning, molecular analysis and search for a consensus motif. *Microbiology*, **142** (Pt. 5), 1297–1309.

144. Patek, M. and Nesvera, J. (2011) Sigma factors and promoters in *Corynebacterium glutamicum*. *J. Biotechnol.*, **154** (2–3), 101–113.

145. Patek, M., Nesvera, J., Guyonvarch, A., Reyes, O., and Leblon, G. (2003) Promoters of *Corynebacterium glutamicum*. *J. Biotechnol.*, **104** (1–3), 311–323.

146. Brabetz, W., Liebl, W., and Schleifer, K.H. (1991) Studies on the utilization of lactose by *Corynebacterium glutamicum*, bearing the lactose operon of *Escherichia coli*. *Arch. Microbiol.*, **155** (6), 607–612.

147. Ben-Samoun, K., Leblon, G., and Reyes, O. (1999) Positively regulated expression of the *Escherichia coli araBAD* promoter in *Corynebacterium glutamicum*. *FEMS Microbiol. Lett.*, **174** (1), 125–130.

148. Zhang, Y., Shang, X., Lai, S., Zhang, G., Liang, Y., and Wen, T. (2012) Development and application of an arabinose-inducible expression system by facilitating inducer uptake in *Corynebacterium glutamicum*. *Appl. Environ. Microbiol.*, **78** (16), 5831–5838.

149. Okibe, N., Suzuki, N., Inui, M., and Yukawa, H. (2010) Isolation, evaluation and use of two strong, carbon source-inducible promoters from *Corynebacterium glutamicum*. *Lett. Appl. Microbiol.*, **50** (2), 173–180.

150. Vasicova, P., Patek, M., Nesvera, J., Sahm, H., and Eikmanns, B. (1999) Analysis of the *Corynebacterium glutamicum dapA* promoter. *J. Bacteriol.*, **181** (19), 6188–6191.

151. Kind, S., Kreye, S., and Wittmann, C. (2011) Metabolic engineering of cellular transport for overproduction of the platform chemical 1,5-diaminopentane in *Corynebacterium glutamicum*. *Metab. Eng.*, **13** (5), 617–627.

152. Kind, S., Neubauer, S., Becker, J., Yamamoto, M., Völkert, M., Abendroth, G.V., Zelder, O., and Wittmann, C. (2014) From zero to hero – production of bio-based nylon from renewable resources using engineered *Corynebacterium glutamicum*. *Metab. Eng.*, **25**, 113–123.

153. Neuner, A., Wagner, I., Sieker, T., Ulber, R., Schneider, K., Peifer, S., and Heinzle, E. (2013) Production of L-lysine on different silage juices using genetically engineered *Corynebacterium glutamicum*. *J. Biotechnol.*, **163** (2), 217–224.

154. Fukui, K., Koseki, C., Yamamoto, Y., Nakamura, J., Sasahara, A., Yuji, R., Hashiguchi, K., Usuda, Y., Matsui, K., Kojima, H., and Abe, K. (2011) Identification of succinate exporter in *Corynebacterium glutamicum* and its physiological roles under anaerobic conditions. *J. Biotechnol.*, **154** (1), 25–34.

155. Litsanov, B., Brocker, M., and Bott, M. (2012) Toward homosuccinate fermentation: metabolic engineering of *Corynebacterium glutamicum* for anaerobic production of succinate from glucose and formate. *Appl. Environ. Microbiol.*, **78** (9), 3325–3337.

156. Tateno, T., Okada, Y., Tsuchidate, T., Tanaka, T., Fukuda, H., and Kondo, A. (2009) Direct production of cadaverine from soluble starch using *Corynebacterium glutamicum* coexpressing alpha-amylase and lysine decarboxylase. *Appl. Microbiol. Biotechnol.*, **82** (1), 115–121.

157. Ikeda, M., Mizuno, Y., Awane, S., Hayashi, M., Mitsuhashi, S., and Takeno, S. (2011) Identification and application of a different glucose uptake system that functions as an alternative to the phosphotransferase system in *Corynebacterium glutamicum*. *Appl. Microbiol. Biotechnol.*, **90** (4), 1443–1451.

158. Yim, S.S., An, S.J., Kang, M., Lee, J., and Jeong, K.J. (2013) Isolation of fully synthetic promoters for high-level gene expression in *Corynebacterium glutamicum*. *Biotechnol. Bioeng.*, **110** (11), 2959–2969.

159. Kalinowski, J. (2005) in *Handbook of Corynebacterium glutamicum* (eds L. Eggeling and M. Bott), CRC Press, Boca Raton, FL, pp. 215–240.

160. Ikeda, M. and Nakagawa, S. (2003) The *Corynebacterium glutamicum* genome: features and impacts on biotechnological processes. *Appl. Microbiol. Biotechnol.*, **62** (2–3), 99–109.
161. Baumgart, M., Unthan, S., Ruckert, C., Sivalingam, J., Grunberger, A., Kalinowski, J., Bott, M., Noack, S., and Frunzke, J. (2013) Construction of a prophage-free variant of *Corynebacterium glutamicum* ATCC 13032 for use as a platform strain for basic research and industrial biotechnology. *Appl. Environ. Microbiol.*, **79** (19), 6006–6015.
162. Ikeda, M., Ohnishi, J., Hayashi, M., and Mitsuhashi, S. (2006) A genome-based approach to create a minimally mutated *Corynebacterium glutamicum* strain for efficient L-lysine production. *J. Ind. Microbiol. Biotechnol.*, **33** (7), 610–615.
163. Krömer, J.O., Wittmann, C., Schröder, H., and Heinzle, E. (2006) Metabolic pathway analysis for rational design of L-methionine production by *Escherichia coli* and *Corynebacterium glutamicum*. *Metab. Eng.*, **8** (4), 353–369.
164. Melzer, G., Esfandabadi, M.E., Franco-Lara, E., and Wittmann, C. (2009) Flux design: *In silico* design of cell factories based on correlation of pathway fluxes to desired properties. *BMC Syst. Biol.*, **3** (1), 120.
165. Wendisch, V.F. (2003) Genome-wide expression analysis in *Corynebacterium glutamicum* using DNA microarrays. *J. Biotechnol.*, **104** (1–3), 273–285.
166. Rey, D.A., Nentwich, S.S., Koch, D.J., Rückert, C., Pühler, A., Tauch, A., and Kalinowski, J. (2005) The McbR repressor modulated by the effector substance S-adenosylhomocysteine controls directly the transcription of a regulon involved in sulphur metabolism of *Corynebacterium glutamicum* ATCC 13032. *Mol. Microbiol.*, **56** (4), 871–887.
167. Silberbach, M., Hüser, A., Kalinowski, J., Pühler, A., Walter, B., Krämer, R., and Burkovski, A. (2005) DNA microarray analysis of the nitrogen starvation response of *Corynebacterium glutamicum*. *J. Biotechnol.*, **119** (4), 357–367.
168. Silberbach, M., Schäfer, M., Hüser, A.T., Kalinowski, J., Pühler, A., Krämer, R., and Burkovski, A. (2005) Adaptation of *Corynebacterium glutamicum* to ammonium limitation: a global analysis using transcriptome and proteome techniques. *Appl. Environ. Microbiol.*, **71** (5), 2391–2402.
169. Ishige, T., Krause, M., Bott, M., Wendisch, V.F., and Sahm, H. (2003) The phosphate starvation stimulon of *Corynebacterium glutamicum* determined by DNA microarray analyses. *J. Bacteriol.*, **185** (15), 4519–4529.
170. Krug, A., Wendisch, V.F., and Bott, M. (2005) Identification of AcnR, a TetR-type repressor of the aconitase gene *acn* in *Corynebacterium glutamicum*. *J. Biol. Chem.*, **280** (1), 585–595.
171. Beckers, G., Strosser, J., Hildebrandt, U., Kalinowski, J., Farwick, M., Krämer, R., and Burkovski, A. (2005) Regulation of AmtR-controlled gene expression in *Corynebacterium glutamicum*: mechanism and characterization of the AmtR regulon. *Mol. Microbiol.*, **58** (2), 580–595.
172. Buchinger, S., Strosser, J., Rehm, N., Hanssler, E., Hans, S., Bathe, B., Schomburg, D., Krämer, R., and Burkovski, A. (2009) A combination of metabolome and transcriptome analyses reveals new targets of the *Corynebacterium glutamicum* nitrogen regulator AmtR. *J. Biotechnol.*, **140** (1–2), 68–74.
173. Rückert, C., Milse, J., Albersmeier, A., Koch, D.J., Pühler, A., and Kalinowski, J. (2008) The dual transcriptional regulator CysR in *Corynebacterium glutamicum* ATCC 13032 controls a subset of genes of the McbR regulon in response to the availability of sulphide acceptor molecules. *BMC Genomics*, **9** (1), 483.
174. Auchter, M., Cramer, A., Hüser, A., Rückert, C., Emer, D., Schwarz, P., Arndt, A., Lange, C., Kalinowski, J., Wendisch, V.F., and Eikmanns, B.J. (2011) RamA and RamB are global transcriptional regulators in *Corynebacterium glutamicum* and control genes

for enzymes of the central metabolism. *J. Biotechnol.*, **154** (2–3), 126–139.

175. Emer, D., Krug, A., Eikmanns, B.J., and Bott, M. (2009) Complex expression control of the *Corynebacterium glutamicum* aconitase gene: identification of RamA as a third transcriptional regulator besides AcnR and RipA. *J. Biotechnol.*, **140** (1–2), 92–98.

176. Nentwich, S.S., Brinkrolf, K., Gaigalat, L., Hüser, A.T., Rey, D.A., Mohrbach, T., Marin, K., Pühler, A., Tauch, A., and Kalinowski, J. (2009) Characterization of the LacI-type transcriptional repressor RbsR controlling ribose transport in *Corynebacterium glutamicum* ATCC 13032. *Microbiology*, **155** (Pt. 1), 150–164.

177. Hayashi, M., Ohnishi, J., Mitsuhashi, S., Yonetani, Y., Hashimoto, S., and Ikeda, M. (2006) Transcriptome analysis reveals global expression changes in an industrial L-lysine producer of *Corynebacterium glutamicum. Biosci. Biotechnol., Biochem.*, **70** (2), 546–550.

178. Lange, C., Rittmann, D., Wendisch, V.F., Bott, M., and Sahm, H. (2003) Global expression profiling and physiological characterization of *Corynebacterium glutamicum* grown in the presence of L-valine. *Appl. Environ. Microbiol.*, **69** (5), 2521–2532.

179. Holatko, J., Elisakova, V., Prouza, M., Sobotka, M., Nesvera, J., and Patek, M. (2009) Metabolic engineering of the L-valine biosynthesis pathway in *Corynebacterium glutamicum* using promoter activity modulation. *J. Biotechnol.*, **139** (3), 203–210.

180. Ikeda, M., Mitsuhashi, S., Tanaka, K., and Hayashi, M. (2009) Reengineering of a *Corynebacterium glutamicum* L-arginine and L-citrulline producer. *Appl. Environ. Microbiol.*, **75** (6), 1635–1641.

181. Zahoor, A., Otten, A., and Wendisch, V.F. (2014) Metabolic engineering of *Corynebacterium glutamicum* for glycolate production. *J. Biotechnol.*, **192**, 366–375.

182. Pfeifer-Sancar, K., Mentz, A., Rückert, C., and Kalinowski, J. (2013) Comprehensive analysis of the *Corynebacterium glutamicum* transcriptome using an improved RNAseq technique. *BMC Genomics*, **14** (1), 888.

183. Mentz, A., Neshat, A., Pfeifer-Sancar, K., Pühler, A., Rückert, C., and Kalinowski, J. (2013) Comprehensive discovery and characterization of small RNAs in *Corynebacterium glutamicum* ATCC 13032. *BMC Genomics*, **14** (1), 714.

184. Kim, H.I., Nam, J.Y., Cho, J.Y., Lee, C.S., and Park, Y.J. (2013) Next-generation sequencing-based transcriptome analysis of L-lysine-producing *Corynebacterium glutamicum* ATCC 21300 strain. *J. Microbiol.*, **51** (6), 877–880.

185. Hermann, T., Wersch, G., Uhlemann, E.M., Schmid, R., and Burkovski, A. (1998) Mapping and identification of *Corynebacterium glutamicum* proteins by two-dimensional gel electrophoresis and microsequencing. *Electrophoresis*, **19** (18), 3217–3221.

186. Wendisch, V.F. and Polen, T. (2013) in *Corynebacterium glutamicum – Biology and Biotechnology* (eds H. Yukawa and M. Inui), Springer-Verlag, Berlin, Heidelberg, pp. 173–216.

187. Schaffer, S., Weil, B., Nguyen, V.D., Dongmann, G., Gunther, K., Nickolaus, M., Hermann, T., and Bott, M. (2001) A high-resolution reference map for cytoplasmic and membrane-associated proteins of *Corynebacterium glutamicum. Electrophoresis*, **22** (20), 4404–4422.

188. Hermann, T., Pfefferle, W., Baumann, C., Busker, E., Schaffer, S., Bott, M., Sahm, H., Dusch, N., Kalinowski, J., Pühler, A., Bendt, A.K., Krämer, R., and Burkovski, A. (2001) Proteome analysis of *Corynebacterium glutamicum. Electrophoresis*, **22** (9), 1712–1723.

189. Bendt, A.K., Burkovski, A., Schaffer, S., Bott, M., Farwick, M., and Hermann, T. (2003) Towards a phosphoproteome map of *Corynebacterium glutamicum. Proteomics*, **3** (8), 1637–1646.

190. Barreiro, C., Gonzalez-Lavado, E., Patek, M., and Martin, J.F. (2004) Transcriptional analysis of the *groES-groEL*1, *groEL2*, and *dnaK* genes in *Corynebacterium glutamicum*: characterization

of heat shock-induced promoters. *J. Bacteriol.*, **186** (14), 4813–4817.

191. Fanous, A., Hecker, M., Gorg, A., Parlar, H., and Jacob, F. (2010) *Corynebacterium glutamicum* as an indicator for environmental cobalt and silver stress--a proteome analysis. *J. Environ. Sci. Health, Part B*, **45** (7), 666–675.

192. Fanous, A., Weiss, W., Gorg, A., Jacob, F., and Parlar, H. (2008) A proteome analysis of the cadmium and mercury response in *Corynebacterium glutamicum*. *Proteomics*, **8** (23–24), 4976–4986.

193. Vasco-Cardenas, M.F., Banos, S., Ramos, A., Martin, J.F., and Barreiro, C. (2013) Proteome response of *Corynebacterium glutamicum* to high concentration of industrially relevant C_4 and C_5 dicarboxylic acids. *J. Proteomics*, **85**, 65–88.

194. Haussmann, U. and Poetsch, A. (2012) Global proteome survey of protocatechuate- and glucose-grown *Corynebacterium glutamicum* reveals multiple physiological differences. *J. Proteomics*, **75** (9), 2649–2659.

195. Claes, W.A., Pühler, A., and Kalinowski, J. (2002) Identification of two *prpDBC* gene clusters in *Corynebacterium glutamicum* and their involvement in propionate degradation via the 2-methylcitrate cycle. *J. Bacteriol.*, **184** (10), 2728–2739.

196. Bolten, C.J., Kiefer, P., Letisse, F., Portais, J.C., and Wittmann, C. (2007) Sampling for metabolome analysis of microorganisms. *Anal. Chem.*, **79** (10), 3843–3849.

197. Meyer, H., Weidmann, H., and Lalk, M. (2013) Methodological approaches to help unravel the intracellular metabolome of *Bacillus subtilis*. *Microb. Cell Fact.*, **12** (1), 69.

198. Van Gulik, W.M., Canelas, A.B., Taymaz-Nikerel, H., Douma, R.D., de Jonge, L.P., and Heijnen, J.J. (2012) Fast sampling of the cellular metabolome. *Methods Mol. Biol.*, **881**, 279–306.

199. Taymaz-Nikerel, H., de Mey, M., Ras, C., ten Pierick, A., Seifar, R.M., van Dam, J.C., Heijnen, J.J., and van Gulik, W.M. (2009) Development and application of a differential method for reliable metabolome analysis in *Escherichia coli*. *Anal. Biochem.*, **386** (1), 9–19.

200. Bolten, C.J. and Wittmann, C. (2008) Appropriate sampling for intracellular amino acid analysis in five phylogenetically different yeasts. *Biotechnol. Lett.*, **30** (11), 1993–2000.

201. Peifer, S., Schneider, K., Nurenberg, G., Volmer, D.A., and Heinzle, E. (2012) Quantitation of intracellular purine intermediates in different Corynebacteria using electrospray LC-MS/MS. *Anal. Bioanal. Chem.*, **404** (8), 2295–2305.

202. Luo, B., Groenke, K., Takors, R., Wandrey, C., and Oldiges, M. (2007) Simultaneous determination of multiple intracellular metabolites in glycolysis, pentose phosphate pathway and tricarboxylic acid cycle by liquid chromatography-mass spectrometry. *J. Chromatogr. A*, **1147** (2), 153–164.

203. Krömer, J.O., Sorgenfrei, O., Klopprogge, K., Heinzle, E., and Wittmann, C. (2004) In-depth profiling of lysine-producing *Corynebacterium glutamicum* by combined analysis of the transcriptome, metabolome, and fluxome. *J. Bacteriol.*, **186** (6), 1769–1784.

204. Krömer, J.O., Heinzle, E., Schröder, H., and Wittmann, C. (2006) Accumulation of homolanthionine and activation of a novel pathway for isoleucine biosynthesis in *Corynebacterium glutamicum* McbR deletion strains. *J. Bacteriol.*, **188** (2), 609–618.

205. Bolten, C.J., Schröder, H., Dickschat, J., and Wittmann, C. (2010) Towards methionine overproduction in *Corynebacterium glutamicum* – methanethiol and dimethyldisulfide as reduced sulfur sources. *J. Microbiol. Biotechnol.*, **20** (8), 1196–1203.

206. Kind, S., Jeong, W.K., Schröder, H., Zelder, O., and Wittmann, C. (2010) Identification and elimination of the competing *N*-acetyldiaminopentane pathway for improved production of diaminopentane by *Corynebacterium*

glutamicum. Appl. Environ. Microbiol., **76** (15), 5175–5180.

207. Petri, K., Walter, F., Persicke, M., Ruckert, C., and Kalinowski, J. (2013) A novel type of N-acetylglutamate synthase is involved in the first step of arginine biosynthesis in *Corynebacterium glutamicum. BMC Genomics*, **14** (1), 713.
208. Rehm, N., Buchinger, S., Strosser, J., Dotzauer, A., Walter, B., Hans, S., Bathe, B., Schomburg, D., Krämer, R., and Burkovski, A. (2010) Impact of adenylyltransferase GlnE on nitrogen starvation response in *Corynebacterium glutamicum. J. Biotechnol.*, **145** (3), 244–252.
209. Käss, F., Hariskos, I., Michel, A., Brandt, H.J., Spann, R., Junne, S., Wiechert, W., Neubauer, P., and Oldiges, M. (2014) Assessment of robustness against dissolved oxygen/substrate oscillations for *C. glutamicum* DM1933 in two-compartment bioreactor. *Bioprocess Biosyst. Eng.*, **37** (6), 1151–1162.
210. van Ooyen, J., Noack, S., Bott, M., Reth, A., and Eggeling, L. (2012) Improved L-lysine production with *Corynebacterium glutamicum* and systemic insight into citrate synthase flux and activity. *Biotechnol. Bioeng.*, **109** (8), 2070–2081.
211. Buescher, J.M., Antoniewicz, M.R., Boros, L.G., Burgess, S.C., Brunengraber, H., Clish, C.B., DeBerardinis, R.J., Feron, O., Frezza, C., Ghesquiere, B., Gottlieb, E., Hiller, K., Jones, R.G., Kamphorst, J.J., Kibbey, R.G., Kimmelman, A.C., Locasale, J.W., Lunt, S.Y., Maddocks, O.D., Malloy, C., Metallo, C.M., Meuillet, E.J., Munger, J., Noh, K., Rabinowitz, J.D., Ralser, M., Sauer, U., Stephanopoulos, G., St-Pierre, J., Tennant, D.A., Wittmann, C., Vander Heiden, M.G., Vazquez, A., Vousden, K., Young, J.D., Zamboni, N., and Fendt, S.M. (2015) A roadmap for interpreting ^{13}C metabolite labeling patterns from cells. *Curr. Opin. Biotechnol.*, **34C**, 189–201.
212. Becker, J. and Wittmann, C. (2013) in *Corynebacterium Glutamicum – Biology and Biotechnology* (eds H. Yukawa and M. Inui), Springer-Verlag, Berlin, Heidelberg, pp. 217–237.
213. Vallino, J.J. and Stephanopoulos, G. (1993) Metabolic flux distributions in *Corynebacterium glutamicum* during growth and lysine overproduction. *Biotechnol. Bioeng.*, **41** (6), 633–646.
214. Wittmann, C. (2007) Fluxome analysis using GC-MS. *Microb. Cell Fact.*, **6** (1), 6.
215. Christensen, B., Gombert, A.K., and Nielsen, J. (2002) Analysis of flux estimates based on ^{13}C-labelling experiments. *Eur. J. Biochem.*, **269** (11), 2795–2800.
216. Dauner, M. and Sauer, U. (2000) GC-MS analysis of amino acids rapidly provides rich information for isotopomer balancing. *Biotechnol. Progr.*, **16** (4), 642–649.
217. Wittmann, C. (2010) Analysis and engineering of metabolic pathway fluxes in *Corynebacterium glutamicum. Adv. Biochem. Eng./Biotechnol.*, **120**, 21–49.
218. Bartek, T., Blombach, B., Lang, S., Eikmanns, B.J., Wiechert, W., Oldiges, M., Nöh, K., and Noack, S. (2011) Comparative ^{13}C metabolic flux analysis of pyruvate dehydrogenase complex-deficient, L-valine-producing *Corynebacterium glutamicum. Appl. Environ. Microbiol.*, **77** (18), 6644–6652.
219. Bommareddy, R.R., Chen, Z., Rappert, S., and Zeng, A.P. (2014) A de novo NADPH generation pathway for improving lysine production of *Corynebacterium glutamicum* by rational design of the coenzyme specificity of glyceraldehyde 3-phosphate dehydrogenase. *Metab. Eng.*, **25**, 30–37.
220. Wittmann, C., Kiefer, P., and Zelder, O. (2004) Metabolic fluxes in *Corynebacterium glutamicum* during lysine production with sucrose as carbon source. *Appl. Environ. Microbiol.*, **70** (12), 7277–7287.
221. Rados, D., Turner, D.L., Fonseca, L.L., Carvalho, A.L., Blombach, B., Eikmanns, B.J., Neves, A.R., and Santos, H. (2014) Carbon flux analysis by ^{13}C nuclear magnetic resonance to determine the effect of CO_2 on anaerobic

succinate production by *Corynebacterium glutamicum*. *Appl. Environ. Microbiol.*, **80** (10), 3015–3024.
222. Shirai, T., Fujimura, K., Furusawa, C., Nagahisa, K., Shioya, S., and Shimizu, H. (2007) Study on roles of anaplerotic pathways in glutamate overproduction of *Corynebacterium glutamicum* by metabolic flux analysis. *Microb. Cell Fact.*, **6** (1), 19.
223. Becker, J. and Wittmann, C. (2012) Bio-based production of chemicals, materials and fuels -*Corynebacterium glutamicum* as versatile cell factory. *Curr. Opin. Biotechnol.*, **23** (4), 631–640.
224. Kimura, E. (2005) in *Handbook of Corynebacterium glutamicum* (eds L. Eggeling and M. Bott), CRC Press, Boca Raton, FL, pp. 439–463.
225. Wittmann, C. and Becker, J. (2007) in *Amino Acid Biosynthesis – Pathways, Regulation and Metabolic Engineering* (ed. V.F. Wendisch), Springer-Verlag, Berlin, Heidelberg, pp. 40–68.
226. Kelle, R., Hermann, T., and Bathe, B. (2005) in *Handbook of Corynebacterium glutamicum* (eds L. Eggeling and M. Bott), CRC Press, Boca Raton, FL, pp. 465–488.
227. Kinoshita, S. (1972) Amino-acid production by the fermentation process. *Nature*, **240** (5378), 211.
228. Kinoshita, S., Nakayama, K., and Kitada, S. (1958) L-lysine production using microbial auxotrophs. *J. Gen. Appl. Microbiol.*, **4** (2), 128–129.
229. Udaka, S. and Kinoshita, S. (1958) Studies on L-ornithine fermentation I. – The biosynthetic pathway of L-ornithine in *Micrococcus glutamicus*. *J. Gen. Appl. Microbiol.*, **4** (4), 272–275.
230. Becker, J. and Wittmann, C. (2012) Systems and synthetic metabolic engineering for amino acid production – the heartbeat of industrial strain development. *Curr. Opin. Biotechnol.*, **23** (5), 718–726.
231. Sugimoto, S., Nakagawa, M., Tsuchida, T., and Shiio, I. (1973) Regulation of aromatic amino acid biosynthesis and production of tyrosine and phenylalanine in *Brevibacterium flavum*. *Agric. Biol. Chem.*, **37** (10), 2327–2336.
232. Sugimoto, S. and Shiio, I. (1977) Enzymes of the tryptophan synthetic pathway in *Brevibacterium flavum*. *J. Biochem.*, **81** (4), 823–833.
233. Shiio, I. (1990) Threonine production by dihydrodipicolinate synthase-defective mutants of *Brevibacterium flavum*. *Biotechnol. Adv.*, **8** (1), 97–103.
234. Kubota, K., Onoda, T., Kamijo, H., Yoshinaga, F., and Okumura, S. (1973) Production of L-arginine by mutants of glutamic acid-producing bacteria. *Gen. Appl. Microbiol.*, **19**, 339–352.
235. Peters-Wendisch, P., Stolz, M., Etterich, H., Kennerknecht, N., Sahm, H., and Eggeling, L. (2005) Metabolic engineering of *Corynebacterium glutamicum* for L-serine production. *Appl. Environ. Microbiol.*, **71** (11), 7139–7144.
236. Nakayama, K., Araki, K., and Kase, H. (1978) Microbial production of essential amino acid with *Corynebacterium glutamicum* mutants. *Adv. Exp. Med. Biol.*, **105**, 649–661.
237. Liu, Q., Zhang, J., Wei, X.X., Ouyang, S.P., Wu, Q., and Chen, G.Q. (2008) Microbial production of L -glutamate and L -glutamine by recombinant *Corynebacterium glutamicum* harboring V*itreoscilla hemoglobin* gene *vgb*. *Appl. Microbiol. Biotechnol.*, **77** (6), 1297–1304.
238. Sato, H., Orishimo, K., Shirai, T., Hirasawa, T., Nagahisa, K., Shimizu, H., and Wachi, M. (2008) Distinct roles of two anaplerotic pathways in glutamate production induced by biotin limitation in *Corynebacterium glutamicum*. *J. Biosci. Bioeng.*, **106** (1), 51–58.
239. Kalinowski, J., Cremer, J., Bachmann, B., Eggeling, L., Sahm, H., and Pühler, A. (1991) Genetic and biochemical analysis of the aspartokinase from *Corynebacterium glutamicum*. *Mol. Microbiol.*, **5** (5), 1197–1204.
240. Shiio, I., Miyajima, R., and Sano, K. (1970) Genetically desensitized aspartate kinase to the concerted feedback inhibition in *Brevibacterium flavum*. *J. Biochem.*, **68** (5), 701–710.
241. Sano, K. and Shiio, I. (1970) Microbial production of L-lysine III: production by mutants resistant to

S-(2-aminoethyl)-L-cysteine. *J. Gen. Appl. Microbiol.*, **16**, 373–391.
242. Vrljic, M., Sahm, H., and Eggeling, L. (1996) A new type of transporter with a new type of cellular function: L-lysine export from *Corynebacterium glutamicum*. *Mol. Microbiol.*, **22** (5), 815–826.
243. Ikeda, M. (2006) Towards bacterial strains overproducing L-tryptophan and other aromatics by metabolic engineering. *Appl. Microbiol. Biotechnol.*, **69** (6), 615–626.
244. Shiio, I., Miyajima, R., and Nakagawa, M. (1972) Regulation of aromatic amino acid biosynthesis in *Brevibacterium flavum*. I. Regulation of anthranilate synthetase. *J. Biochem.*, **72** (6), 1447–1455.
245. Ikeda, M. (2005) in *Handbook of Corynebacterium glutamicum* (eds L. Eggeling and M. Bott), CRC Press, Boca Raton, FL, pp. 489–509.
246. Becker, J. and Wittmann, C. (2015) Advanced biotechnology: metabolically engineered cells for the bio-based production of chemicals and fuels, materials, and health-care products. *Angew. Chem. Int. Ed.*, **54** (11), 3328–3350.
247. Schneider, J., Niermann, K., and Wendisch, V.F. (2011) Production of the amino acids L-glutamate, L-lysine, L-ornithine and L-arginine from arabinose by recombinant *Corynebacterium glutamicum*. *J. Biotechnol.*, **154** (2–3), 191–198.
248. Stäbler, N., Oikawa, T., Bott, M., and Eggeling, L. (2011) *Corynebacterium glutamicum* as a host for synthesis and export of D-amino acids. *J. Bacteriol.*, **193** (7), 1702–1709.
249. Hüser, A.T., Chassagnole, C., Lindley, N.D., Merkamm, M., Guyonvarch, A., Elisakova, V., Patek, M., Kalinowski, J., Brune, I., Pühler, A., and Tauch, A. (2005) Rational design of a *Corynebacterium glutamicum* pantothenate production strain and its characterization by metabolic flux analysis and genome-wide transcriptional profiling. *Appl. Environ. Microbiol.*, **71** (6), 3255–3268.
250. Dickschat, J., Wickel, S., Bolten, C.J., Nawrath, T., Schulz, S., and Wittmann, C. (2010) Pyrazine biosynthesis in *Corynebacterium glutamicum*. *Eur. J. Org. Chem.*, **2010** (14), 2687–2695.
251. Mimitsuka, T., Sawai, H., Hatsu, M., and Yamada, K. (2007) Metabolic engineering of *Corynebacterium glutamicum* for cadaverine fermentation. *Biosci. Biotechnol., Biochem.*, **71** (9), 2130–2135.
252. Schneider, J. and Wendisch, V.F. (2010) Putrescine production by engineered *Corynebacterium glutamicum*. *Appl. Microbiol. Biotechnol.*, **88** (4), 859–868.
253. Inui, M., Murakami, S., Okino, S., Kawaguchi, H., Vertes, A.A., and Yukawa, H. (2004) Metabolic analysis of *Corynebacterium glutamicum* during lactate and succinate productions under oxygen deprivation conditions. *J. Mol. Microbiol. Biotechnol.*, 7 (4), 182–196.
254. Okino, S., Noburyu, R., Suda, M., Jojima, T., Inui, M., and Yukawa, H. (2008) An efficient succinic acid production process in a metabolically engineered *Corynebacterium glutamicum* strain. *Appl. Microbiol. Biotechnol.*, **81** (3), 459–464.
255. Okino, S., Suda, M., Fujikura, K., Inui, M., and Yukawa, H. (2008) Production of D-lactic acid by *Corynebacterium glutamicum* under oxygen deprivation. *Appl. Microbiol. Biotechnol.*, **78** (3), 449–454.
256. Becker, J., Lange, A., Fabarius, J., and Wittmann, C. (2015) Top value platform chemicals: bio-based production of organic acids. *Curr. Opin. Biotechnol.*, **36**, 168–175.
257. Inui, M., Kawaguchi, H., Murakami, S., Vertes, A.A., and Yukawa, H. (2004) Metabolic engineering of *Corynebacterium glutamicum* for fuel ethanol production under oxygen-deprivation conditions. *J. Mol. Microbiol. Biotechnol.*, **8** (4), 243–254.
258. Smith, K.M., Cho, K.M., and Liao, J.C. (2010) Engineering *Corynebacterium glutamicum* for isobutanol production. *Appl. Microbiol. Biotechnol.*, **87** (3), 1045–1055.

259. Niimi, S., Suzuki, N., Inui, M., and Yukawa, H. (2011) Metabolic engineering of 1,2-propanediol pathways in *Corynebacterium glutamicum*. *Appl. Microbiol. Biotechnol.*, **90** (5), 1721–1729.
260. Heider, S.A., Peters-Wendisch, P., Netzer, R., Stafnes, M., Brautaset, T., and Wendisch, V.F. (2014) Production and glucosylation of C50 and C 40 carotenoids by metabolically engineered *Corynebacterium glutamicum*. *Appl. Microbiol. Biotechnol.*, **98** (3), 1223–1235.
261. Frohwitter, J., Heider, S.A., Peters-Wendisch, P., Beekwilder, J., and Wendisch, V.F. (2014) Production of the sesquiterpene (+)-valencene by metabolically engineered *Corynebacterium glutamicum*. *J. Biotechnol.*, **191**, 205–213.
262. Kang, M.K., Eom, J.H., Kim, Y., Um, Y., and Woo, H.M. (2014) Biosynthesis of pinene from glucose using metabolically-engineered *Corynebacterium glutamicum*. *Biotechnol. Lett.*, **36** (10), 2069–2077.
263. Song, Y., Matsumoto, K., Yamada, M., Gohda, A., Brigham, C.J., Sinskey, A.J., and Taguchi, S. (2012) Engineered *Corynebacterium glutamicum* as an endotoxin-free platform strain for lactate-based polyester production. *Appl. Microbiol. Biotechnol.*, **93** (5), 1917–1925.
264. Song, Y., Matsumoto, K., Tanaka, T., Kondo, A., and Taguchi, S. (2013) Single-step production of polyhydroxybutyrate from starch by using alpha-amylase cell-surface displaying system of *Corynebacterium glutamicum*. *J. Biosci. Bioeng.*, **115** (1), 12–14.
265. Jo, S.J., Leong, C.R., Matsumoto, K., and Taguchi, S. (2009) Dual production of poly(3-hydroxybutyrate) and glutamate using variable biotin concentrations in *Corynebacterium glutamicum*. *J. Biosci. Bioeng.*, **107** (4), 409–411.
266. Wu, H., Li, Q., Lu, R., Wang, Y., Zhuang, X., and He, N. (2010) Fed-batch production of a bioflocculant from *Corynebacterium glutamicum*. *J. Ind. Microbiol. Biotechnol.*, **37** (11), 1203–1209.
267. Kawaguchi, H., Vertes, A.A., Okino, S., Inui, M., and Yukawa, H. (2006) Engineering of a xylose metabolic pathway in *Corynebacterium glutamicum*. *Appl. Environ. Microbiol.*, **72** (5), 3418–3428.
268. Kawaguchi, H., Sasaki, M., Vertes, A.A., Inui, M., and Yukawa, H. (2008) Engineering of an L-arabinose metabolic pathway in *Corynebacterium glutamicum*. *Appl. Microbiol. Biotechnol.*, **77** (5), 1053–1062.
269. Matano, C., Uhde, A., Youn, J.-W., Maeda, T., Clermont, L., Marin, K., Kramer, R., Wendisch, V.F., and Seibold, G.M. (2014) Engineering of *Corynebacterium glutamicum* for growth and L-lysine and lycopene production from *N*-acetyl-glucosamine. *Appl. Microbiol. Biotechnol.*, **98** (12), 5633–5643.
270. Uhde, A., Youn, J.W., Maeda, T., Clermont, L., Matano, C., Krämer, R., Wendisch, V.F., Seibold, G.M., and Marin, K. (2013) Glucosamine as carbon source for amino acid-producing *Corynebacterium glutamicum*. *Appl. Microbiol. Biotechnol.*, **97** (4), 1679–1687.
271. Adachi, N., Takahashi, C., Ono-Murota, N., Yamaguchi, R., Tanaka, T., and Kondo, A. (2013) Direct L-lysine production from cellobiose by *Corynebacterium glutamicum* displaying beta-glucosidase on its cell surface. *Appl. Microbiol. Biotechnol.*, **97** (16), 7165–7172.
272. Kotrba, P., Inui, M., and Yukawa, H. (2003) A single V317A or V317M substitution in Enzyme II of a newly identified beta-glucoside phosphotransferase and utilization system of *Corynebacterium glutamicum* R extends its specificity towards cellobiose. *Microbiology*, **149** (Pt. 6), 1569–1580.
273. Seibold, G., Auchter, M., Berens, S., Kalinowski, J., and Eikmanns, B.J. (2006) Utilization of soluble starch by a recombinant *Corynebacterium glutamicum* strain: growth and lysine production. *J. Biotechnol.*, **124** (2), 381–391.

274. Tateno, T., Fukuda, H., and Kondo, A. (2007) Production of L-Lysine from starch by *Corynebacterium glutamicum* displaying alpha-amylase on its cell surface. *Appl. Microbiol. Biotechnol.*, **74** (6), 1213–1220.

275. Adham, S.A., Honrubia, P., Diaz, M., Fernandez-Abalos, J.M., Santamaria, R.I., and Gil, J.A. (2001) Expression of the genes coding for the xylanase Xys1 and the cellulase Cel1 from the straw-decomposing *Streptomyces halstedii* JM8 cloned into the amino-acid producer *Brevibacterium lactofermentum* ATCC13869. *Arch. Microbiol.*, **177** (1), 91–97.

276. Tsuchidate, T., Tateno, T., Okai, N., Tanaka, T., Ogino, C., and Kondo, A. (2011) Glutamate production from beta-glucan using endoglucanase-secreting *Corynebacterium glutamicum*. *Appl. Microbiol. Biotechnol.*, **90** (3), 895–901.

277. Hyeon, J.E., Jeon, W.J., Whang, S.Y., and Han, S.O. (2011) Production of minicellulosomes for the enhanced hydrolysis of cellulosic substrates by recombinant *Corynebacterium glutamicum*. *Enzyme Microb. Technol.*, **48** (4–5), 371–377.

278. Adham, S.A., Campelo, A.B., Ramos, A., and Gil, J.A. (2001) Construction of a xylanase-producing strain of *Brevibacterium lactofermentum* by stable integration of an engineered *xysA* gene from *Streptomyces halstedii* JM8. *Appl. Environ. Microbiol.*, **67** (12), 5425–5430.

279. Kohlstedt, M., Sappa, P.K., Meyer, H., Maass, S., Zaprasis, A., Hoffmann, T., Becker, J., Steil, L., Hecker, M., van Dijl, J.M., Lalk, M., Mäder, U., Stülke, J., Bremer, E., Völker, U., and Wittmann, C. (2014) Adaptation of *Bacillus subtilis* carbon core metabolism to simultaneous nutrient limitation and osmotic challenge: a multi-omics perspective. *Environ. Microbiol.*, **16** (6), 1898–1917.

7
Host Organisms: *Bacillus subtilis*

Hans-Peter Hohmann, Jan M. van Dijl, Laxmi Krishnappa, and Zoltán Prágai

7.1
Introduction and Scope

Bacillus subtilis is a gram-positive bacterium that inhabits the upper layers of the soil and the plant rhizosphere [1]. In these ecological niches, *B. subtilis* is frequently exposed to harsh environmental situations, such as drought, extreme temperature changes, fluctuating osmolarity, and nutrient limitation. *B. subtilis* employs several adaptive strategies to cope with harsh environmental challenges, or to avoid them altogether. For example, synthesis of a flagellum enables *B. subtilis* to migrate toward favorable environments [2–4]. Antibiotic production and degradative enzyme secretion by *B. subtilis* ensure a competitive advantage in relation to other organisms in the soil [5–7]. Furthermore, *B. subtilis* can easily become competent to take up foreign genetic material as a nutrient or to increase its genetic diversity, which can be helpful for survival of the species [8, 9]. The uptake of potassium ions, increased synthesis of proline, and uptake of glycine betaine ensure protection during changes in environmental osmolarity [10–15]. If moderate adaptations are insufficient for cell survival, particular cells will produce highly resistant endospores that ensure the viability of *B. subtilis* in a dormant state until favorable situations are encountered [16–18].

Today, *B. subtilis* is the best-studied gram-positive bacterium in the scientific world owing to the groundbreaking work on genetic transformation by Spizizen, Anagnostopoulos, Schaeffer, and others in the late 1950s and 1960s. The complete genome sequence of *B. subtilis* became available in late 1997, just a few month after publication of the *Escherichia coli* genome sequence [19, 20]. *B. subtilis* is an ideal model organism for studies on chromosomal replication, gene regulation, metabolism, protein secretion, and bacterial cell differentiation (e.g., motility, competence development, and sporulation). Systems biology approaches in the areas of genomewide gene expression and central carbon metabolism have revealed the highly versatile abilities of *B. subtilis* [15, 21, 22].

The early availability of the genome sequence has not only enhanced our understanding of fundamental cellular processes in *B. subtilis* and other gram-positive bacteria, but it also represented a knowledge push for the application

Industrial Biotechnology: Microorganisms, First Edition. Edited by Christoph Wittmann and James C. Liao.

of *B. subtilis* in industrial settings. The latter relates to the extraordinary ability of *Bacillus* species to secrete large quantities (20–25 $g\,l^{-1}$) of protein into the culture medium, which makes it an ideal workhorse for the bulk production of degradative enzymes and other proteins. In addition, *B. subtilis* has become the workhorse for the production of fine chemicals such as nucleotides, riboflavin (vitamin B2), and pantothenic acid (vitamin B5). Here, we review advances in the application of *B. subtilis* for the production of proteins and fine chemicals with special emphasis on the methodologies that can be applied to improve this versatile cell factory.

7.2
Identification of Genetic Traits Pertinent to Enhanced Biosynthesis of a Value Product

Robust microorganisms constitute the basis of every biotechnological fermentation process to transform, at a large scale, raw materials such as simple carbohydrates or vegetable oils as efficiently as possible into useful organic compounds. In most cases, the main objectives to be reached in a fermentation process development program are to (i) recover with high yield the carbon atoms of the consumed feedstock in the value product and (ii) maximize the space-time yield, that is, the speed of the biochemical transformation process in the fermenter. Microorganisms that are suitable for industrial fermentation processes are rarely isolated from the environment and can only seldom be ordered from microbial strain depositories. Elaborate and intricate research efforts commonly referred to as *strain development* are required to adapt and optimize a microorganism to suit a biotechnological process.

To turn microbial strains into production hosts for value compounds and to enhance their performance further, the microorganisms have to be provided with pertinent stably inheritable traits. To this end, changes in the DNA sequence of their genomes are introduced. Epigenetic approaches in production strain development have not been reported so far. For a microorganism like *B. subtilis* with a highly developed genetic toolbox, metabolic engineering is the method of choice to alter the genetic makeup of the bacterium in a targeted and purposeful manner. In a first step, the metabolic capacity of the specific final pathway(s) toward a value metabolite has to be increased over the wild-type ancestor, which is frequently achieved by applying recombinant DNA methods for genetic engineering. This simple reductionist approach following a local concept of strain development [23] is based on the assumption that key nodes of the metabolic network are flexible enough to cope with the sometimes drastic changes in flux distribution that accompany the overproduction of a value compound. For example, to satisfy the GTP needs for optimal production of riboflavin by *B. subtilis* the metabolic flux through the highly regulated purine pathway has to be doubled at elevated growth rates and at low growth rates even quadrupled. Although *in vivo* flux analysis suggested a surprising rigidity of the *B. subtilis* metabolic network, at least as pertains to its central metabolic pathways [24], the network provides the required flexibility to match the GTP demand of artificial high riboflavin production without deliberate interventions.

Once a transformant with the intended genomic structure has been obtained through genetic engineering, it takes 50 or more generations to multiply the clonal founder cell to obtain a culture that can be tested for a desired phenotype and from which inocula for large-scale fermentations can be obtained. It is conceivable that during this multiplication, cells with adaptive mutations will be enriched if the engineered genetic modification has a detrimental impact on cell physiology. Such adaptive mutations will provide the necessary network flexibility and contribute to increased fitness of the cells carrying the mutant allele.

Nevertheless, when aiming for strains that convert their carbon substrate to a desired product at carbon yields close to the stoichiometrically possible yield, it will become necessary to take a global or systems view on the bacterial metabolism [25]. That includes the supply pathways for the precursors from which the value compound is biosynthesized and the central metabolism providing the handful of the basic metabolic building blocks from which all cellular components are derived. Furthermore, competing metabolic pathways, redox balance, energy charge, the ratio of phosphorylated to nonphosphorylated form of the nucleotide redox carrier NAD(P)H/H^+, and product storage within the cell or product excretion/secretion are to be considered.

Following a local concept of strain development, the genetic targets to be addressed are usually limited and fairly obvious. The genes and the corresponding expression elements to be modified are known from the targeted gene cloning efforts of the 1980s and 1990s, but latest since whole microbial genome sequencing and functional gene annotation became standard laboratory practice. In sharp contrast, pursuing metabolic engineering at a systems level, dozens of genes have to be considered as potential targets. Even worse, still about 25% [21] of the *B. subtilis*-encoded genes are not or only tentatively annotated providing no clue whether their products might influence the metabolic flux to be optimized. The post-translational regulatory network of *B. subtilis*, which is only incompletely understood, appears to be conservative in keeping the metabolic flux resilient to changes. Nevertheless, *a priori* reasoning on the basis of the vast physiological knowledge gathered over decades of *B. subtilis* research, intuition, and simply trial and error is not a bad way of proceeding and this has elicited some good results. In the later sections of this chapter, a few success stories and also failed efforts of metabolic pathway engineering in *B. subtilis* are reviewed in some detail.

Hypothesis-driven gene target identification beyond the obvious local targets resembles the search for the proverbial needle in a haystack. More systematic approaches building on the genome-scale reconstruction of the metabolic network in *B. subtilis* [26] and making use of the various "omics" technologies for in-depth profiling of the bacterial inventory of transcripts, proteins, and arguably most importantly, small molecule metabolites were developed to render target identification more predictable. These technologies together with genome-scale modeling for both experimental and *in silico* analysis of pathway fluxes facilitated the *ab initio* construction of a *Corynebacterium glutamicum* lysine producer strain with a carbon conversion yield of 55% [27], which is close to the optimum predicted yield for zero growth and matches the performance of traditionally derived lysine production strains.

A recently established class of systems-level models applies constraints-based methods to predict phenotypes from genotypic characteristics [28], taking into account the presence or absence of metabolic functions [29]. Thermodynamic analyses are available to discover limited catalytic activities [30]. Furthermore, dynamic simulations of kinetic models provide a classical method to predict the phenotype of cells with genotype perturbations, including perturbations of expression levels, that is, overexpression and expression attenuation of homologous genes. Especially for strains that have already been engineered to show a good performance, further significant improvements frequently require multiple concurrent alterations [31]. What all these methods have in common is that they rely on verified predictive models. The experimental basis for verification differs in the same manner as the effects that can be taken into account. The benefit, for example, that thermodynamic or kinetic effects can be considered is paid for by the need to verify such models by measurements of intracellular metabolites. Such a verification procedure has been exemplified by Chassagnole *et al.* [32].

The advent of affordable high-throughput sequencing platforms facilitated a rather different approach to identify performance-critical mutations. Instead of anticipating on the basis of limited knowledge and highly sophisticated systemic methods which genome alteration could possibly advance the productivity of a strain, an *ex post facto* approach can be applied searching for those mutations which in retrospect turned out as conductive to improved performance. Commercial production strains including those that were derived from hypothesis-driven genetic engineering programs frequently underwent classical strain improvement (CSI) campaigns consisting of random infliction of mutations in the genomes of a host strain population and screening for individuals with improved productivity. Since iterative rounds of harsh mutagen treatment are rather the rule in such CSI campaigns one can expect that many, say several hundreds, mutations accumulate in the genomes of the survivors of these procedures, from which the next production strain generation is subsequently selected. Only a minority of the accumulated mutations positively contribute to the superior performance of the selected strain. Most of them might either elicit no phenotype with regard to productivity and some might be even adverse. A special transformation protocol designated as lysed protoplast transformation (see Section 7.4.7), which so far has been reported only for *B. subtilis*, leads to genetic exchange between a donor and a recipient strain via numerous crossing-over events providing transformants with thoroughly mixed up genetic traits from both parent strains. Recombining *B. subtilis* production strain and wild-type genomes by lysed protoplast transformation followed by correlation analysis between the mutation repertoire of the resulting hybrids and their production performance can track down the relevant mutations. From the metabolic function of the gene products that are affected, and from the manner in which they are affected, the physiological conditions pertinent to superior product formation can be deduced. This knowledge can be used to improve the accuracy of engineering target prediction.

7.3 Traits to Be Engineered for Enhanced Synthesis and Secretion of Proteinaceous Products

Many applications of *B. subtilis* involve the secretion of protein products into the fermentation broth, particularly because this allows easy downstream processing. This has focused attention on the secretome, which includes both the secreted proteins and the protein secretion machinery [33]. *B. subtilis* naturally secretes many different proteins for a wide range of purposes, including nutrient acquisition, competition with other organisms, and adaptation to changes in the environment. Most of these secreted proteins contain N-terminal signal peptides that direct their export from the cytoplasm. Accordingly, efficient signal peptides are also needed for the secretion of proteinaceous products in industrial fermentations. In addition to signal peptides, an intricate cellular machinery is needed for the export of proteins from the cytoplasm, their translocation across the cytoplasmic membrane, and their subsequent release into the growth medium in a fully folded, biologically active, and protease-resistant state. This implies that both the properties of secreted proteins and the secretion machinery can be engineered to optimize secretory production of high-value proteins.

7.3.1 Signal Peptides and Signal Peptidases

Signal peptides are composed of three characteristic domains generally known as the *N-, H-, and C-regions* [34–37]. Arginine or lysine residues in the N-region interact with the translocation machinery and with negatively charged phospholipids in the cell membrane [38, 39]. The H-region comprises hydrophobic residues. It adopts an alpha-helical conformation in the membrane, which is necessary for the initiation of protein translocation across the membrane [40]. The C-region contains the recognition and cleavage sites for signal peptidase (SPase). These SPases cleave off the signal peptide to release the mature protein from the membrane during or shortly after translocation via a dedicated secretory apparatus. The cleaved signal peptides are then targeted for degradation by the "site-2" membrane protease RasP, and possibly other membrane-associated proteases [41].

Secreted proteins can contain one of five different types of signal peptides identified in *B. subtilis*. These are (i) secretory (Sec-type) signal peptides, (ii) twin-arginine (RR/KR) signal peptides, (iii) lipoprotein signal peptides, (iv) pseudopilin-like signal peptides, and (v) bacteriocin and pheromone signal peptides [33]. The latter signal peptides are mostly referred to as *leader peptides* or *leader sequences*. The specific properties of *B. subtilis* signal peptides have previously been extensively reviewed and are therefore not detailed here [33]. Notably, only the Sec-type signal peptides have so far been used in bioproduction, although twin-arginine signal peptide could potentially also be useful [42]. Proteins with a Sec-type signal peptide are exported via the general protein secretion (Sec) pathway, whereas proteins with a twin-arginine signal peptide are exported across the membrane via the twin-arginine translocation (Tat) pathway. There are five so-called *type I SPases* (named SipS-W) in *B. subtilis*, which commonly cleave both Sec-type and twin-arginine signal peptides [43–46].

7.3.2 Protein Transport Pathways

The most relevant and best-studied translocation pathways for secretory proteins in *B. subtilis* are the Sec pathway and the Tat pathway [33].

7.3.2.1 The Sec Pathway

The majority of secreted proteins are translocated from the cytoplasm to the exterior of *B. subtilis* via the highly conserved Sec pathway (Figure 7.1). These proteins are synthesized as precursors with N-terminal Sec-type signal peptides.

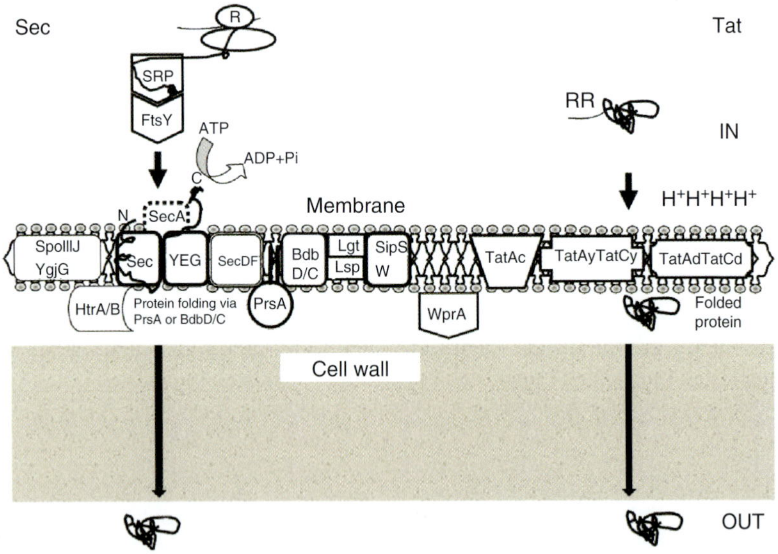

Figure 7.1 The Sec and Tat protein secretion pathways of *B. subtilis*. The cartoon depicts the components involved in Sec- and Tat-dependent export of proteins from the cytoplasm (IN) to the membrane, cell wall, and extracellular milieu (OUT) of the bacterium. A nascent precursor protein is schematically presented as emerging from the ribosome (R) and being bound by the SRP-FtsY complex for targeting to the membrane. Another precursor protein is shown as being translocated via the SecYEG channel in an unfolded state. On the right side, a folded precursor protein with twin-arginine signal peptide (RR) is targeted to the Tat translocase.

The precursor proteins are maintained in a translocation-competent state by the folding-antagonizing activity of their signal peptides and cytoplasmic chaperones, which potentially include the signal recognition particle (SRP) and the FtsY protein [47–49]. This enables translocation across the membrane through the Sec apparatus, which can only handle unfolded or loosely folded proteins. The Sec apparatus consists of SecA, the translocation motor, and the integral membrane proteins SecY, SecE, SecG, and SecDF. The energy for translocation of proteins is derived from ATP binding by SecA and the proton-motive force (PMF). The ATPase activity of SecA is triggered by the interaction of the translocation-competent protein with SecA, the SecYEG translocation channel, and the phospholipids of the cell membrane [50]. SecA then drives the protein through the SecYEG channel through cycles of ATP binding and hydrolysis [51]. The exact function of SecDF of *B. subtilis*, a "molecular Siamese twin protein" is as yet unknown, but it is indeed required for efficient export, especially under conditions of secretory protein overproduction [52, 53]. On the basis of studies in *E. coli* and *B. subtilis*, it seems likely that the SpoIIIJ and YqjG proteins can operate in concert with the Sec machinery in protein translocation. However, this seems to apply mainly to the translocation of membrane proteins. In addition, certain other membrane proteins may be inserted into the membrane in a non-Sec-dependent manner with the help of SpoIIIJ/YqjG [54–56].

The N-terminal signal peptides of precursor proteins are cleaved by SPases during or shortly after translocation via the Sec apparatus. Only after translocation through the Sec apparatus has been completed are the secreted proteins folded into a biologically active conformation by chaperones, such as the lipoprotein PrsA [57] or the thiol-disulfide oxidoreductases BdbB, BdbC, and BdbD [58, 59]. Other factors that are important for post-translocational protein folding are propeptides and metal ions [60]. For example, the serine proteases of *B. subtilis* are guided by their propeptides to achieve their biologically active conformation. Propeptides also shield the serine proteases against proteolytic activities. Metal ions, such as Ca^{2+}, Fe^{3+}, and Mg^{2+} affect the folding and stability of many secreted proteins in *B. subtilis* [60]. Specifically, the requirement of Ca^{2+} for the stability of *B. subtilis* in-like serine proteases has been demonstrated *in vitro*. Furthermore, the post-translocational folding and secretion of the model protein levansucrase was shown to require both Ca^{2+} and Fe^{3+} [60]. Lastly, the folded proteins can fulfill their biological function at the membrane-cell wall interface, in the cell wall or in the external milieu. Whether the protein remains attached to the membrane will depend on the presence of a retention signal, such as a diacyl-glyceryl modification in the case of lipoproteins. Cell wall attachment will mainly depend on the presence of cell wall-binding domains, the covalent attachment to peptidoglycan, or overall charge properties of the translocated protein. If a translocated protein lacks retention signals for the membrane or cell wall, it is usually effectively secreted into the extracellular milieu [61], which is the desired scenario for secretory protein production.

Genetic engineering of the Sec machinery was shown to increase the secretion of some heterologous proteins. For example, the yield of human interferon-α2b was 2.2-fold higher in a strain with a deletion in the C-terminal linker (CTL) domain of SecA than in the parental strain. Thus, the interactions between the engineered SecA and preproteins were apparently optimized for the secretion of hINF-α2b [62].

Recently, the *E. coli* SecB was coexpressed in a *B. subtilis* strain with a modified SecA protein. Specifically, the 32 C-terminal residues of SecA of *B. subtilis* were replaced by the corresponding amino acids of *E. coli* SecA to optimize the SecA–SecB interaction. This engineering of the protein secretion machinery resulted in the increased secretion of a mutant maltose-binding protein (MalE11) and the alkaline phosphatase PhoA of *E. coli*, which were used as model heterologous proteins [63]. It should be noted here that early studies by Collier also provided evidence that the heterologous expression of *E. coli* SecB in *B. subtilis* could be beneficial for the secretion of particular proteins [64].

The essential lipoprotein PrsA, which localizes to the outer surface of the cytoplasmic membrane (Figure 7.1), can also be used to enhance secretion of particular proteins [65]. PrsA belongs to the parvulin subfamily of peptidyl-prolyl cis/trans isomerases (PPIases) [66]. The PPIase domain of PrsA exhibits PPIase activity and a possible chaperone activity *in vivo* at the membrane-cell wall interface [65]. Overexpression of PrsA was shown to enhance the secretion of several extracellular proteins, including α-amylases [57, 65]. Furthermore, PrsA overproduction in *B. subtilis* resulted in 1.5-fold increased secretion and activity of the pharmacologically relevant human interferon-β (hINF-β) with the AmyE propeptide [67].

7.3.2.2
The Tat Pathway

The Tat pathway is conserved in prokaryotes, chloroplasts of green plants, and also in some archaea such as *Haloferax volcanii* and *Haloarcula hispanica* [68]. Substrates with a highly conserved twin-arginine motif (RR-motif) in the N-terminal domain of the signal peptide are selectively translocated via the Tat pathway. Potential RR-motifs have been predicted by *in silico* analyses in the signal peptides of 69 *B. subtilis* proteins [69]. Proteomics and biochemical analyses have so far detected only EfeB (YwbN), PhoD, QcrA, and YkuE as major substrates of the *B. subtilis* Tat system [45, 70, 71]. Furthermore, the esterase LipA, which is normally secreted via Sec, can be secreted also via the Tat translocation pathway upon severe overproduction [69, 72]. The proteins destined for translocation via the Tat pathway are usually folded prior to translocation [73, 74]. Thus, globular proteins and folded proteins with bound cofactors can be translocated via the Tat pathway across the cell membrane in a process that is driven by the PMF. These unique characteristics distinguish the Tat pathway from the Sec pathway [75]. Notably, with the exception of streptomycetes, the gram-positive bacteria have a Tat machinery that differs from the Tat machinery of gram-negative bacteria with respect to the subunit composition. While gram-negative bacteria have a three-component TatABC machinery, most gram-positive bacteria have a two-component "minimal" TatAC machinery [73, 76] (Figure 7.1).

In contrast to *E. coli*, *B. subtilis* contains two minimal TatAC translocases, namely, TatAyCy and TatAdCd, that can function independently of each other. The *tatAy-tatCy* and *tatAd-tatCd* genes encoding the minimal translocases are situated as operons at different genomic locations [46, 69, 70, 76]. The third TatA subunit of *B. subtilis*, TatAc, was shown to form active complexes with both TatCd and TatCy when expressed in *E. coli* [77]. Intriguingly, a recent study showed that in *B. subtilis* TatAc has mainly a supporting role in enhancing protein translocation via the TatAyCy translocase [78].

The ability of the Tat machinery to secrete folded proteins makes it an attractive system for use in the biotechnology industry. Unfortunately however, the use of Tat signal peptides has, in many cases, not been very successful in achieving increased levels of secreted heterologous proteins. On the other hand, it has been shown that Sec-type proteins such as subtilisin can be rerouted into the Tat pathway [42]. Furthermore, a heterologous methyl parathion hydrolase (MPH) with the signal peptide of *E. coli* TorA was successfully translocated via the Tat machinery in *B. subtilis* [79]. Also, coexpression of TatAdCd or TatAyCy with hINF-α containing the LipA signal peptide in a *B. subtilis* strain deficient in eight extracellular proteases (Eprs) resulted in elevated levels of secreted hINF-α [67].

7.3.2.3
Extracellular and Membrane-Bound Proteases of *B. subtilis*

B. subtilis produces at least eight Eprs at the end of the exponential phase of growth in liquid culture [80]. These have been identified as the alkaline serine protease subtilisin (AprE) [81, 82], the neutral protease (NprE) [83], the minor Epr [84, 85], the bacillopeptidase F (Bpr) [86], the Vpr protease [87], the metalloprotease (Mpr) [86], the NprB [83], and the cell wall-associated extracellular protease WprA [88] (Figure 7.2). Proteomics and transcriptomics studies have shown the involvement of the DegS-DegU two-component regulatory system in the expression of several of these exoproteases [89, 90].

The Eprs are synthesized with Sec-type signal peptides and exported from the cytoplasm via the Sec pathway. Although a potential RR-motif (KR) is present in the signal peptide of WprA, there is currently no published evidence for Tat-dependent WprA export under standard laboratory growth conditions. AprE and NprE are known to contribute 95% of the Epr activity, while the remaining activity is fulfilled by Epr, NprB, Bpr, Vpr, WprA, and Mpr [91]. Eprs in *B. subtilis* are generally synthesized as inactive "preproenzymes" in the cytoplasm. The prepeptide (i.e., the signal peptide) directs export from the cytoplasm. The propeptide of the exported proenzyme is essential for the proper post-translocational folding of the proteases. Upon membrane translocation, the proenzymes are processed into their proteolytically active form by the removal of the propeptide either by autoprocessing or heteroprocessing [60, 92–95]. For example, the inactive form of subtilisin (AprE) is converted into the active form by intramolecular autoprocessing in the extracytoplasmic space [93, 95], and Bpr is involved in the processing of pro-Mpr [96]. The main biological function of secreted proteases is to supply the *B. subtilis* cells with peptides and amino acids by the degradation of extracellular

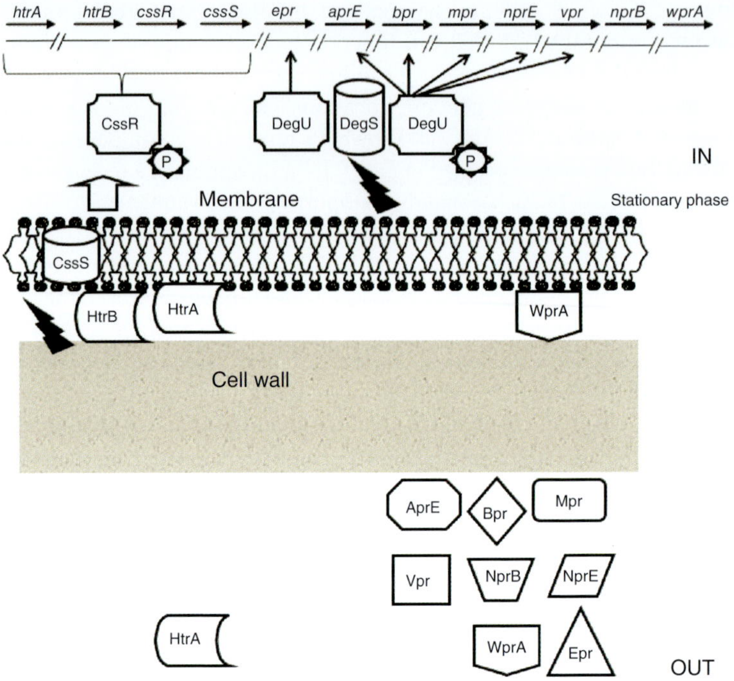

Figure 7.2 Schematic representation of the subcellular localization of the 10 extracytoplasmic proteases of *B. subtilis*. The key regulatory two-component systems CssRS and DegSU for the expression of the respective protease genes are also shown.

proteins from dead organic matter in the soil. Interestingly, these enzymes also impact on cell wall biogenesis as the stability of the cell wall lytic enzymes LytE and LytF is controlled by the cell wall-bound protease WprA and the Epr [97]. Also, the cell wall protein of unknown function WapA is a target of Eprs [98]. Epr provides signals for the swarming motility of *B. subtilis* [99]. Lastly, Eprs impact on the maturation of bacteriocins and signaling molecules. Thus, the processing of the lantibiotic subtilin of *B. subtilis* ATCC 6633 is achieved by AprE, WprA, and Vpr [100]. Likewise, the minor Eprs and Vpr of *B. subtilis* are involved in processing of the Phr pentapeptide proCSF into the mature CSF that is active in extracellular signaling [101].

The Eprs of *B. subtilis* and related bacilli are also very relevant for commercial applications. In particular, these enzymes are employed in the manufacture of detergents, tanning of leather, management of industrial and household wastes, bioprocessing of X-ray or photographic films for the recovery of silver, protein hydrolysate preparation in the food industry, synthesis of aspartame, and other applications [102]. Recently, the fibrinolytic activity of Vpr was discovered. Accordingly, Vpr has the potential to work as a thrombolytic agent in medical applications [103, 104].

Notably, Eprs are not only beneficial for industrial applications. Their downside is that they represent major bottlenecks during the production of secreted heterologous proteins in *B. subtilis*. This relates to the fact that such heterologous proteins often fold more slowly upon membrane translocation than native proteins of *B. subtilis*, or that the heterologous proteins expose cleavage sites that are recognized by particular secreted proteases. To overcome these bottlenecks, various single or multiple protease mutants were constructed, including the well-known strains WB600, lacking six protease genes, and WB800, lacking eight protease genes. Such strains can be employed for the enhanced production of α-amylase, artificial fusion lipase, streptavidin, and many more proteins of commercial importance [105, 106]. Completely markerless protease mutants were recently also reported [107].

The HtrA and HtrB proteases, which belong to the CssRS regulon, are also known to affect the secretion of heterologous proteins [108]. These proteases are both integral membrane proteins, but potentially active degradation products are detectable in the growth medium [109, 110]. Importantly, HtrA and HtrB are involved in the quality control of secreted proteins by their protease and chaperone functions [109, 111]. Overproduction of a heterologous protein, such as the α-amylase AmyQ of *Bacillus amyloliquefaciens*, is sufficient to induce so-called secretion stress. This secretion stress is most likely caused by the accumulation of misfolded AmyQ in the cell wall and at the cell wall-membrane interface, which is sensed by the CssRS two-component regulatory system [108] (Figure 7.2). The CssRS two-component system then induces the expression of the *htrA* and *htrB* genes resulting in AmyQ degradation by HtrA and HtrB. Both proteases are known to be potential bottlenecks also in the (over)production of certain membrane proteins [108, 112, 113]. Intriguingly, recent studies have shown that the lipoprotein PrsA, HtrA, and HtrB are the substrates of multiple extracytoplasmic proteases. Thus, improved protein secretion by multiple-protease-mutant strains may be related to both reduced proteolysis and improved post-translocational protein folding and quality control [114].

Several additional membrane-bound proteases for membrane protein turnover have been identified in *B. subtilis* [115]. Similarly to HtrA and HtrB, PrsW and RasP also have been implicated in the degradation of overproduced membrane proteins [113].

7.4 Engineering of Genetic Traits in *Bacillus subtilis*

Once identified as a target gene for strain improvement, the activity of the encoded gene product has to be modulated in the production strain. Commonly, a DNA sequence comprising the target gene or a fragment of it and DNA elements, which affect the intracellular steady concentration of the gene product at various levels, is assembled *in vitro* and introduced into the chromosome of the recipient cell. For further or alternative activity modulation, a suited mutant allele of the target gene can be synthesized *in vitro* and provided to the recipient host. The ability of *B. subtilis* to develop natural competence is frequently used to firmly embed the *in vitro* generated DNA constructs by double crossover recombination in the recipient chromosome. In this chapter, methods that are unique (or nearly so) to *B. subtilis* are discussed. For description of general molecular biological methods for *B. subtilis*, the reader is referred to relevant reviews and books, for example, [116].

7.4.1 Natural Competence and Double Crossover Integration

B. subtilis and many other Bacilli have the intrinsic potential to differentiate into distinct physiological states. One such state is natural competence, which allows the bacteria to take up DNA from the surrounding environment and to incorporate it into their chromosome by homologous recombination. This process involves over 50 different genes with many different functions ranging from regulated expression of the DNA uptake machinery to incorporation of the ingested DNA into the chromosome [117–119]. For a *B. subtilis* cell to enter into a state of competence, several criteria must be met and this means that only approximately 10% of the population will ever develop full competence. Nowadays, the most frequently used protocol for transformation of *B. subtilis* via competence is the one published by Kunst and Rapoport [120].

In vitro synthesized or recombined DNA sequences selected as potential performance-enhancing elements have to become inheritable traits of a production strain. Preferably the exogenous DNA should be chromosomally integrated by double crossover recombination to ensure sufficient genetic stability over many replication cycles in the absence of selection pressure during inoculum preparation and the production fermentation. If the exogenous DNA encodes a recessive trait, for example, a protein with attenuated enzyme activity or a less active expression element, the endogenous wild-type allele has to be replaced by site-specific integration of the exogenous DNA at the target locus.

For ectopic integration and expression, a number of well-characterized, nonessential regions of the chromosome are available, such as the *amyE, gltA, lacA, pyrD, sacA, sacB,* and *thrC* genes [121–124] and the minor Epr genes *bpr, epr,* and *vpr* [125]. Successful integration at the *amyE, sacA,* and *sacB* genes can be easily confirmed by testing colonies with the simple color assays for α-amylase and levansucrase [124].

To introduce the overexpression cassettes with the target genes into the ectopic integration sites of the *B. subtilis* production strains, antibiotic resistance genes were the preferred option owing to their rapid, efficient, and easy use. The most frequently used antibiotic resistance genes are the *spc, erm, tet, neo,* and *cat* genes conferring resistance to spectinomycin, erythromycin, tetracycline, neomycin/kanamycin, and chloramphenicol, respectively [126–131]. However, the number of modifications around the genome is limited by the number of available resistance genes, and use of multiple antibiotics for selection could affect the physiology of the engineered strain [125]. In addition, the regulatory agencies FDA in the US and EFSA in the EU are more critical with commercial enzyme and metabolite production strains which carry antibiotic resistance genes [132].

Accordingly, the removal of antibiotic resistance genes from the overexpression constructs after integration or simply construction of genetically engineered strains without any antibiotic resistance gene have increasingly gained favor (see Section 7.4.5).

7.4.2 Tools to Modify Protein Expression Levels in *B. subtilis* Production Strains

A well-equipped toolbox is available to the *B. subtilis* metabolic engineer to modify the intracellular steady-state concentration(s) of the target gene product(s) at the transcriptional or post-transcriptional level.

7.4.2.1 Promoters of Various Strength for Customized mRNA Synthesis

In general, the activities in production strains of final pathway enzymes should be as high as possible for maximal pull toward the target molecule. To increase gene transcription, the very strong, constitutive phage promoters SPO1-15 and SPO1-26 [133] are the promoters of choice [124, 134, 135]. These promoters are obtained from the lytic *Bacillus* SPO1 bacteriophage, which after infection takes over the transcription and translation machinery of the host to its own advantage; the genes encoded by the phage genome are then expressed from their own promoters in addition to, and at the later stages of the infection, at the expense of, the host genes [136]. During the infection, which is a naturally occurring process, the promoters SPO1-15 and SPO1-26 become part of the gene pool of *B. subtilis*. Consequently, production strains provided with *in vitro* recombined DNA sequences encoding these promoters linked to endogenous *B. subtilis* genes do not contain foreign DNA sequences and are considered self-clone strains.

Some of the native *B. subtilis* chromosomal genes have also strong constitutive promoters such as the promoter of the *rpsD* gene, the P43 promoter [137], the *lepA* promoter [123], and $P_{veg\mathrm{I}}$, the first RNA polymerase binding site of the tandem promoter complex of the *veg* gene [138]. The strength of the *veg*I promoter is comparable with that of the SPO1 promoters. Interestingly, the transcription initiated from $P_{veg\mathrm{I}}$ and its native partner, the second *veg* promoter, collide with each other and produce much less mRNA [139].

In many cases, particularly when aiming for increased expression of genes encoding proteins with functions in pathways common to several metabolites or in the central metabolic pathways, the usage of very strong constitutive promoters can lead to undesirable effects on cellular metabolism [140], the translational machinery of the host cell [141], and also induce proteolytic activities mediated by stress responses [142] which could lower production titers. For best results, several promoters of different strength should be tested, for example, the weak constitutive *rplU* and *yvfV* promoters, the medium *citZ*, *ybxF*, and *infC* constitutive promoters together with the tandem promoters of *veg*, the strong constitutive promoters of *cspD* and *veg*I, and the very strong SPO1-15, PE3, and SPO1-26 promoters.

The inducible, weak *B. subtilis aprE, nprE,* and *amyE* promoters are used in high-copy-number plasmid expression systems [143, 144]. These promoters are induced in the transition growth phase when the maximum biomass is reached in the fermentation culture and therefore, minimizing the possibility of mutations and plasmid segregation. The growth-phase-dependent, natural induction avoids the necessity of artificial inducers such as isopropyl-beta-D-thiogalactopyranoside (IPTG), xylose, or heat. By replacing the −35 sequences with consensus sigA sequences, the strength of the *aprE* promoter improved drastically and the mutated still inducible *aprE* promoter produced up to 100-fold more protein in *B. subtilis,* allowing single copy chromosomal expression [143].

7.4.2.2
Gene Amplification and Plasmid Vectors

In addition to using strong promoters to enhance gene expression, the copy number of the target gene(s) can be increased via gene amplification or using multicopy plasmid expression systems. The copy number of the target gene can be increased up to about 10 copies per bacterial chromosome if the target gene is linked to a selectable antibiotic resistance gene, such that direct DNA sequence repeats are generated (e.g., target gene–resistance gene–target gene) during a single crossover transformation with circular DNA (nonreplicative plasmids or circularized DNA fragments). Stepwise cultivation at increasing concentrations of the antibiotic against which the marker provides resistance will enrich for individuals with increased copy numbers of the resistance marker and, hence, the target gene. In *B. subtilis,* such fundamentally instable genetic arrangements are maintained in the absence of any selection pressure, at least through a standard 48 h production fermentation [125].

Most plasmid expression systems for *B. subtilis* are based on *Staphylococcus aureus* plasmids (e.g., pUB110, pC194, and pE194) which replicate and express their antibiotic resistance in *B. subtilis* [145, 146]. Some of these vectors and their derivatives are available from the Bacillus Genetic Stock Center (BGSC) at The Ohio State University (http://www.bgsc.org). The major problems of these vectors are the low efficiency of cloning and high instability of the recombinant plasmid mostly due to the rolling-circle replication mechanism that uses a single-stranded DNA intermediate [145]. Although significant progresses were made for improved *B. subtilis* plasmid vectors (e.g., application of cryptic *Bacillus* plasmids, plasmids with theta-type replication, and donor-helper plasmids) [145], the *S. aureus* plasmids and their derivatives are dominant in industrial enzyme production [144, 147–149]. For metabolite production, only a handful of commercial production strains contain expression plasmids with entire biosynthetic operons (e.g., some riboflavin production strains [150, 151].

7.4.2.3
mRNA-Stabilizing Elements and Transcriptional Terminators

Microorganisms have developed mechanisms by which the stability of mRNA transcripts is enhanced. To achieve this, the 5′-end of the stabilized transcripts contains strong ribosome or protein binding sites, or nucleotide sequences capable of forming secondary structures which impose an impediment to mRNA-degrading enzymes in *B. subtilis* [152, 153].

Introducing mRNA stabilization elements (RSEs) between the native promoters and target genes was shown to be a promising method for enabling the right levels of overproduction of biosynthetic enzymes without the unwanted side effects of very strong promoters [154]. Using the strong SPO1-26 promoter, deregulation of the *ilvBNC-leuABCD* operon (important to supply the key intermediate, 2-ketoisovaleric acid, see Section 7.9.3) resulted in 3.5-fold lower pantothenate production and a higher rate of cell lysis during prolonged growth. However, using the native promoter and the RSE of the *gapA* gene [155] for the stabilization of the *ilvBNC-leuABCD* mRNA in the pantothenate overproducing *B. subtilis* strain resulted in 1.4-fold increase in the pantothenate production compared to the wild-type expression [154].

The application of RSEs in combination with strong non-native promoters will result in further enhanced overexpression of target genes and improved performance, if tolerated by the host strain. This method is particularly useful if the replacement of the native promoter by a strong constitutive promoter does not result in the desired expression level of the gene firmly integrated in single copy into the chromosome and gene amplification or plasmid expression is not sufficiently stable. The RSE elements of the genes *aprE* [156], *grpE* [157], and *gsiB* [158] in combination with the strong promoter SPO1-15 to drive the riboflavin genes *ribDEAHT* resulted in *B. subtilis* strains of very high riboflavin productivity [159]. Placing the SPO1-15 promoter directly in front of *ribDEAHT* generated highly instable transcripts. Instead of full length mRNA only heavily degraded species were detected in a Northern blot analysis. Accordingly, even after amplification of the *rib* expression cassette, only a modest riboflavin production strain was obtained [135].

Insertion of rho-independent transcription terminators at the 3′-end of the overexpression cassettes should prevent the RNA polymerase from transcribing downstream chromosomal or plasmid genes that may cause instability and the loss of the cassette. The T_1T_2 terminators from the *rrnB* operon of *E. coli* together with the lambda T_0 terminator are known to be active and used widely in *B. subtilis* [160, 161]. Transcription terminators from the *Bacillus thuringiensis* crystal protein (*cry*) gene (T_{cry}) and from the *B. subtilis* SPO1 phage such as the TH1 terminator are also popular in *B. subtilis* [124, 136]. Interestingly, the 381-bp DNA fragment containing the T_{cry} terminator increased the mRNA half-life and therefore, enhanced protein synthesis of the target upstream gene(s) [162].

7.4.2.4
RBS Elements

The activity of a target gene product can also be modulated at the post-transcriptional level by adjusting the efficiency at which mRNA is translated into a polypeptide chain. Translation is conceptually divided into four phases: initiation, elongation, termination, and ribosome recycling [163]. Regulation of translation occurs primarily in the rate-limiting initiation phase, in which the large and small subunits of ribosome are assembled on the ribosomal binding site (RBS) of the mRNA. The efficiency of translation initiation is determined mainly by (i) the conservation of the Shine-Dalgarno (SD) sequence, that is, the core part of the RBS that interacts with the anti-SD sequence of the 16S rRNA of the ribosome, (ii) the distance between the SD sequence and the translational start codon, and (iii) the sequence immediately surrounding the translational start codon [164–166]. The accessibility of the RBS is regulated by temperature and binding of small metabolites, proteins, or antisense RNAs. Secondary structures at the mRNA inhibit translation initiation, while an exposed, single-stranded translational start codon correlates well with high protein expression rates in *E. coli* [167, 168]. The SD sequence is an A/G-rich stretch which can vary in sequence and length. The consensus SD motif for *B. subtilis* is eight nucleotides [AAAGGAGG] [164]. Genes with higher codon adaptation index (CAI), a frequently used parameter for characterization of codon usage bias, showed a trend toward more conserved SD sequences than genes with lower CAI [169].

The RBSs of the SPO1 bacteriophage host-takeover proteins are highly efficient in translating early phage proteins soon after infection [136] and might be well suited to forcefully initiate translation of target genes. Application of the strong RBSs of the *B. subtilis gsiB* gene and the most actively expressed SPO1 early genes (e.g., gene 41) are a good option to achieve very efficient translation of heterologous biosynthetic genes in *B. subtilis* [124].

The usage of non-native or mutant RBSs provides a means to up- or down-modulate the expression level of a target gene in a production strain while keeping it subjected to the natural regulatory control, which is in most cases exerted at the mRNA level. This feature of expression modulation by RBSs could become important during an industrial fed-batch fermentation run with its steady succession of various physiological conditions and the corresponding gene expression pattern from the state of overflow metabolism to severe carbon source limitation. Constitutive promoters will inevitably override the expression control mechanisms and it might also not be expedient for regulated promoters to faithfully follow the natural gene expression pattern. Selection of a less-efficient RBS and engineering it in front of the open reading frame to be translated at reduced efficiency should be straightforward. To increase the translation efficiency requires a more extended screening for a suited RBS but should also be possible.

7.4.2.5
Signal Peptides

The signal peptides of *B. subtilis* can be easily manipulated for exploitation in the biotechnology industry. Signal peptides of various proteins of *B. subtilis* could be potential candidates for the secretion of heterologous proteins. In this regard, all natural signal peptides of *B. subtilis* have been screened with the final conclusion that not every *B. subtilis* signal peptide is suited for the efficient secretion of a particular heterologous protein. Instead, there is a need to optimize the selection of the best-suited signal peptide for each particular heterologous protein [170].

7.4.3
Modification of the Amino Acid Sequence of an Enzyme

Randomly inflicted mutations within the open reading frame of an enzyme that result in amino acid exchanges in the primary sequence typically lead to attenuation of the activity of the enzyme, the release from repressive feedback, or allosteric regulation. The former can be instrumental to prevent the drainage of a value product precursor metabolite into competing pathways, while ensuring a minimal flux to serve the basic metabolic needs of the organism. An example is the presence of an attenuated transketolase in riboflavin production strains (see Section 7.9.2), which prevents drainage of ribulose 5-phosphate and ribose 5-phosphate back into the central metabolism. PanB, an enzyme of the (*R*)-pantoate branch of pantothenic acid biosynthesis, is feedback repressed by pantoate. PanB mutant alleles might be resistant to pantoate repression (see Section 7.9.3). Provided that a high-throughput screening or selection assay is available, mutant enzymes with increased turnover numbers can be isolated out of a random mutant library generated, for example, by error prone PCR. For instance, a mutant cyclohydrolase II, the entrance enzyme into the riboflavin pathway, carrying seven point mutations showed a twofold activity increase compared to the wild-type ($k_{cat} = 4.3$ and $2.1\ min^{-1}$, respectively) [171].

7.4.4
Transforming DNA Constructs Obtained by Template-Free Synthesis or by PCR

The ability of *B. subtilis* to develop natural competence allows double crossover transformation with DNA in the form of PCR fragments, synthetic DNA, or linearized plasmid vectors encoding the trait(s) to be provided to the recipient strain, the pertinent expression elements, and, if required, a selectable marker (SM) flanked at the 3′ and 5′ ends by DNA sequences, which are homolog to the desired site of integration. The size of the flanking sequences should be at least 400–500 bp [172, 173]. If a linearized plasmid vector assembled *in vitro* and cloned in *E. coli* is used as transforming agent, it obviously must not contain a *B. subtilis* replication origin to prevent subsequent recircularization in the host strain plasmid inheritance.

The transforming agent can be conveniently generated in multiple PCR reactions. A DNA fragment with the antibiotic resistance marker cassette, engineered regulatory elements, or the modified gene could be fused to the 5′- and 3′-flanking regions of the genomic integration site by long flanking homology PCR (LFH-PCR) reaction [174]. In a first step, two approximately 400–500 bp long DNA fragments homolog to the 5′- and 3′-ends of the chromosomal target region are amplified in such way that short extensions (e.g., 26 bp) homologous to the marker or the modified gene are added to one of their ends. In a second step, one strand of each of these two flanks then serves as a long primer in a PCR reaction using the DNA fragment carrying the marker, the regulatory elements, or the modified gene as template. The final LFH-PCR product contains all the three DNA fragments as one molecule in the order of 5′-flank, target DNA, followed by the 3′-flank [159].

To generate a DNA construct from multiple DNA fragments of different origins, for example, an artificial operon with biosynthetic genes from different microorganisms, template-free DNA synthesis is very convenient. The method is fast and allows codon optimization for efficient expression in *B. subtilis* [124]. The synthesis of such DNA constructs is offered by several companies such as DNA 2.0 (Menlo Park, USA, https://www.dna20.com/), GeneArt now Life Technologies (Regensburg, Germany, http://www.lifetechnologies.com/), and SGI-DNA (La Jolla, USA, https://www.sgidna.com/). An alternative is the Gibson Assembly method that combines multiple double-stranded DNA fragments carrying 16–40 bp overlaps with the adjacent fragments in a guided combinatorial fashion in a one-step, isothermal approach without the need of restriction enzymes [175]. The double-stranded, sequence-verified DNA fragments are provided by Integrated DNA Technologies (Coralville, USA, http://eu.idtdna.com/site) as gBlocks® Gene Fragments (up to 2 kb in length) and Life Technologies as GeneArt® Strings™ DNA Fragments (up to 3 kb in length).

7.4.5 Production Strains Devoid of Antibiotic Resistance Markers

Antibiotic resistance genes are convenient dominant selection markers frequently used in *Bacillus* research. However, the presence of these genes in production strains, even if only traces of production strain biomass are left in the final product, are not appreciated by the registration authorities, who are concerned with the spread of antibiotic resistance by horizontal gene transfer. Marker replacement by auxotrophic/prototrophic growth switches or marker removal by Cre/*loxP* site-specific recombination are convenient procedures to provide production strains devoid of exogenous antibiotic resistance genes. Both methods together with the recombineering technique to avoid dominant selection markers for genetic modification are addressed in the following.

7.4.5.1 Chromosomal Modification with Transiently Employed Selection Markers

Typically transient selection markers are employed when the endogenous expression controlling DNA sequences of a gene are to be replaced by non-native sequences. The usage of transient selection markers is particularly easy if the product of the gene to be modified is involved in an essential metabolic pathway, such that deletion of (or parts of) the gene by introduction of a selection marker, conveniently an antibiotic resistance cassette, results in an auxotrophic growth phenotype. The auxotrophic, antibiotic-resistant mutant is, in a second step, transformed with a DNA fragment containing the modified gene (e.g., a gene with a strong constitutive promoter and RBS or a mutant gene encoding an enzyme with altered kinetic characteristics) flanked with the wild-type chromosomal sequences surrounding the target engineering region. During the transformation, the antibiotic resistance marker is replaced by the modified gene restoring prototrophic growth, the selection criterion for successful reversion. The resulting transformants are antibiotic sensitive, which is a convenient screening criterion for correct chromosomal integration. The method has been used to replace native, regulated promoters with strong constitutive promoters for genes and operons involved in the biosynthesis of riboflavin [135], biotin [176], pantothenate [134, 177], folate, and thiamin [178]. This method can also be used to introduce point mutations, as long as sufficient enzymatic activity remains to allow for selection of prototrophic colonies [179, 180].

In a similar approach, a nonselectable genetic trait, such as an acquired mutation or an engineered DNA sequence, can be transferred from a donor into a recipient strain, if the locus of the trait is in transformation or transduction linkage to an antibiotic resistance marker, that is placed in a biosynthetic pathway gene and, thus, causes an auxotrophic growth phenotype. In a first step, the prototrophic recipient strain is provided with the antibiotic resistance marker resulting in auxotrophic transformants or transductants. Chromosomal DNA or a phage lysate from the prototrophic donor strain or a fusion PCR product is then used in a second step to revert the auxotrophic mutant from the first step to prototrophic growth. The resulting transformants or transductants are screened phenotypically or by PCR for transfer via linkage of the desired nonselectable genetic trait.

Under congression or cotransformation conditions the genetic exchange based on prototrophic/auxotrophic growth switches also works when the antibiotic resistance marker is not linked to the nonselectable genetic trait. Since the congression frequency (frequency of cotransfer of an unlinked locus during natural competence transformation) is 0.2–3% in *B. subtilis* [181], screening of 100–1000 prototrophic transformants of the second step for congression of the desired nonselectable genetic trait is required. To expedite the screening work, the locus to be replaced by congression in the recipient strain should contain a marker, conveniently an antibiotic resistance gene, the replacement of which is easy to monitor.

7.4.5.2
Counterselection Strategies to Force the Deletion of Dominant Markers

In recent years, several counterselection systems were developed allowing positive selection for the loss of dominant antibiotic resistance markers and thereby marker-free manipulation of the *B. subtilis* genome even at multiple sites. These systems usually allow the use of antibiotic resistance markers for mutant selection, followed by efficient removal of the dominant markers, leaving unmarked mutations in the genome. The counterselectable markers (CSMs) are (i) the *upp* gene in combination with 5-fluorouracil selection [182], (ii) the BlaI cassette coupled with conditional lysine auxotrophic phenotype [183], (iii) the *araR* repressor gene in combination with neomycin-resistance gene under the control of the arabinose operon promoter (*Para-neo*) [184], and (iv) the heat-inducible hen egg white lysozyme (hewl) gene [185]. The methods using these markers require the use of specific genetic backgrounds, such as inactivation of the native *upp* gene for 5-FU selection, or replacement of the *araR* gene with the *Para-neo* construct. However, the utilization of the *E. coli* toxin gene *mazF* as a CSM under the control of an IPTG-inducible expression system could be applied in any genetic background for markerless engineering of the *B. subtilis* chromosome [186]. The *mazF* gene encodes the toxin (an endoribonuclease) of the *E. coli* toxin-antitoxin system *mazEF* that specifically cleaves free mRNAs at ACA sequences and is also highly toxic in *B. subtilis* [187]. The drawbacks of this MazF cassette are the dependency on cloning and the non-strictly-controlled *mazF* expression leading to spontaneous *mazF*-resistance in high frequency. Therefore, the mini-*mazF*-cassette was developed with zeocin resistance gene as positive selection marker and *mazF* was placed under the strict control of the *xyl* promoter [188]. DNA constructs containing the mini-*mazF*-cassette were generated by PCR (no cloning step was required) and used successfully to delete a gene, a larger gene cluster, and to insert a green fluorescent protein (GFP) expression cassette. The frequency of spontaneous *mazF*-resistant mutants was shown to be lower, and thereby, the frequency of generating desired clones was nearly 100% [188].

To the authors of this chapter, it is unclear to what extent the elaborate marker replacement strategies outlined above have found their way into the reality of industrial production strain development.

7.4.5.3
Removal of the Selection Marker by Cre/*loxP* Site-Specific Recombination

After successful genetic recombination involving a selection marker, frequently an antibiotic resistance gene, the marker can be removed by site-specific recombination such as the bacteriophage P1 Cre/*loxP* site-specific recombination system [189]. The Cre recombinase catalyzes the reciprocal site-specific recombination between two *loxP* sites without requiring any host factor or accessory protein. During the recombination event, the DNA fragment flanked by convergently oriented *loxP* sites is excised, leaving one intact *loxP* site behind in the genomic integration site. Accordingly, the use of native *loxP* sites for multiple consecutive genomic manipulations would result in multiple *loxP* sites around the chromosome that could still be recognized by Cre leading to genetic instability in further modification. To minimize this instability, a pair of mutant *lox* sites, a right element mutant *lox* site (*lox66*) and a left element mutant *lox* site (*lox71*), were developed [190]. Recombination of *lox71* and *lox66* results in a double-mutant *lox72* site that has a strongly reduced binding affinity for Cre and therefore does not interfere during the next rounds of chromosomal manipulation involving Cre.

The Cre/*lox* technique was developed for genome engineering in *B. subtilis* allowing rapid introduction of mutations and subsequent removal of the selection markers [191]. During a PCR fusion reaction, the antibiotic resistance gene bordered by two mutant *lox* sites (*lox71* and *lox66*) was fused with the 5′ and 3′-end flanking regions and the PCR product was directly transformed into *B. subtilis*. After integration of the *lox71-zeo-lox66* (zeocin resistance (*zeo*) gene, only recently used in *B. subtilis* research) or the *lox71-spc-lox66* (spectinomycin resistance gene) cassettes into the target chromosomal locus, the Cre recombinase was transiently expressed from a temperature-sensitive vector and the *lox71* and *lox66* sites recombined into a double-mutant *lox72* site, thereby excising the marker gene and the wild-type *loxP* site from the genome. The *lox72* site left behind on the chromosome is a 34-bp scar that did not interfere with subsequent rounds of Cre/*lox* mutagenesis. In the end, the temperature-sensitive vector carrying the *cre* gene has to be eliminated by growing the bacteria at elevated temperature to get the marker-free target strain. The method was successfully used to introduce or delete genes and remove chromosomal regions in three consecutive genomic manipulations.

7.4.5.4
Recombineering

Recombineering (recombination-mediated genetic engineering) has become a convenient method to modify large DNA molecules and bacterial chromosomes during the last decade, circumventing the need for most standard *in vitro* cloning techniques. With more than 10^7 recombinants within 10^8 viable *E. coli* cells that had been subjected to a recombineering protocol, the method is efficient enough to generate *E. coli* transformants without employing a dominant selection marker [192–195].

For recombineering, bacteriophage-encoded recombination systems are applied, such as the lambda Red or *E. coli* RecET systems [193, 196]. The lambda Red system consisting of three proteins, Exo, Beta, and Gam, recombines both double-stranded DNA (dsDNA) and single-stranded DNA (ssDNA). Gam prevents the degradation of linear dsDNA, while Exo (lambda exonuclease) degrades the exogenously supplied dsDNA in a 5′ to 3′ manner generating ssDNA. Beta protein is a single-strand annealing protein that is the only lambda function required for efficient recombination with exogenous ssDNA and even oligonucleotides [192]. Beta protein anneals the oligonucleotide to the complementary strand of the replication fork. An oligonucleotide corresponding in sequence to the lagging strand is more efficient than an oligonucleotide with the reverse sequence annealing to the leading strand [192, 193]. In the host cell, the methyl-directed mismatch repair (MMR) system can affect the ssDNA recombination events by eliminating the recombinant allele and restoring the original sequence. In this way, MMR reduces the ssDNA recombination frequency by approximately 100-fold in *E. coli*. Nevertheless, for the sake of high-fidelity DNA replication, it is advisable to keep the MMR function intact and rather to keep MMR away from the transforming oligonucleotides through appropriate oligonucleotide design such as (i) a C–C mismatch 6 bp from that target base change, (ii) four or more adjacent mismatches, or (iii) mismatches at four or more consecutive wobble positions [194].

Recently, a lambda Red beta protein-based genome editing method was developed in *B. subtilis* to inactivate target genes using ssDNA fragments obtained by asymmetric PCR [185]. The ssDNA comprised a 1081-nt disruption cassette and two 70-nt regions homolog to the intended chromosomal integration sites. The cassette encoded, in this order, an upstream *lox71* site, the bleomycin (*ble*) resistance gene, the *hewl* CSM, and the downstream *lox66* site (see Sections 7.4.5.2 and 7.4.5.3). The genes of the lambda beta protein and the Cre recombinase were expressed transiently under the control of the lambda *cI857*-P_{RM}-P_R promoter system on a low-copy number, temperature-sensitive plasmid. Accordingly, beta protein was produced at 30 °C for the beta-mediated integration of the disruption cassette at the chromosomal target site. After primary selection against bleomycin, a temperature shift to 42 °C was applied to produce Cre for the deletion of the disruption cassette. Using this genome editing tool, insertional gene inactivation was generated in *B. subtilis* with an efficiency of 3×10^3 transformants per microgram ssDNA targeting the lagging strand of the replication fork. Upon usage of phosphorothioate modified ssDNA to confer exonuclease resistance, a fivefold improvement of the recombination efficiency was obtained.

The above results demonstrated that beta protein-mediated recombination worked in *B. subtilis*, but owing to low efficiency, the method required a dominant selection marker to isolate the desired transformants from the bulk of nontransformed cells. With the recent improvements of *B. subtilis* electroporation protocols using osmoprotective agents and high electric fields that can generate up to 10^7 transformants per microgram replicative plasmid DNA (see Section 7.4.6.3), we expect to see in the close future reports on efficient ssDNA recombineering techniques using mutant oligonucleotides in *B. subtilis.*

7.4.6 Introduction of Exogenous DNA into *Bacillus* Strains Incapable of Developing Natural Competence

Transformation of *B. subtilis* via natural competence is the preferred method to introduce exogenous DNA and integrate it in the chromosome of the recipient cells. However, certain strains of *B. subtilis*, particularly highly developed production strains, do not become readily competent. In this case, it may help to ectopically express the master regulator for competence, ComK [197]. Alternatively, an easy to handle laboratory strain is the primary recipient of the exogenous DNA, whose modified genome is then recombined with the genome of the production strain by phage transduction or protoplast fusion. *B. subtilis* protoplasts can also be used as recipients for replicative plasmids (protoplast transformation). Electroporation provides another means to introduce replicative plasmids into *B. subtilis* and, according to recent literature, even linear DNA fragments, such as PCR products or linearized integrative plasmids (electrotransformation).

7.4.6.1 Phage Transduction

Different bacteriophages have been used to transduce noncompetent *B. subtilis* strains. Detailed protocols for transduction with the PBS1 and SPP1 phages are presented in the book entitled "Molecular Biological Methods for *Bacillus*" [116].

7.4.6.2
Protoplast Transformation and Protoplast Fusion

A classical way to transform noncompetent *Bacilli* with replicative plasmids is protoplast transformation. A detailed protocol for protoplast transformation has been documented in the handbook "Molecular biological methods for *Bacillus*" by Harwood and Cutting [116].

By traditional protoplast fusion mediated through polyethylene glycol, genetic material can be transferred from a donor to a recipient strain that is recalcitrant to phage infection and incapable to develop natural competence [198]. Only a small number of the protoplasts, which revert to the bacillary form after polyethylene glycol treatment, are haploid recombinants. These haploids are stable progenies of the diploids generated immediately after cell fusion. For convenient isolation of the recombinants from the bulk of regenerated parent cells, selection and counterselection markers (CSMs) are required.

7.4.6.3
Electrotransformation

Electroporation is an alternative to natural competence transformation to bring exogenous DNA into *B. subtilis* [199]. It is a less tedious and rapid technique using high-voltage electric pulses for inducing a membrane potential. The potential renders the cells temporarily permeable to allow the entry of DNA into the cells [200]. One of the most important parameters for electrotransformation is the applied voltage and accordingly, the published protocols can be assigned to two groups: (i) the conventional iso-osmolarity electrotransformation protocols with lower optimal electric field strengths ranging from 5 to 12.5 $kV\,cm^{-1}$ [199, 201, 202] and (ii) the recent high osmolarity electroporation protocols using high voltage, 20–25 $kV\,cm^{-1}$ [203–206].

The transformation efficiencies of the conventional iso-osmolarity electrotransformation protocols were $10^3 - 10^5$ transformants per microgram replicative plasmid DNA, depending on the *B. subtilis* strain and method applied [145, 202, 207, 208]. Using the high-osmolarity protocols, the transformation efficiencies improved two to three orders of magnitude compared to the conventional protocols and were $10^6 - 10^7$ transformants per microgram replicative plasmid DNA [185, 203]. This drastic improvement is partly due to the much better cell survival rate as the result of applying osmoprotecants (such as trehalose or glycine betaine, in addition to sorbitol, mannitol, and glycerol) in the electroporation medium [204–206]; and partly due to the weaken cell wall and elevated membrane fluidity as the results of growing *B. subtilis* in the presence of glycine, DL-threonine, and Tween 80 during the preparation of electrocompetent cells [185]. In addition, the transformation efficiency also improved owing to the increased electric field strength (e.g., 23-fold higher efficiency at 23 $kV\,cm^{-1}$ than at 12 $kV\,cm^{-1}$) [203].

With the use of highly efficient high osmolarity protocols, these days, circular and linearized integrative plasmid DNA could be efficiently integrated into the chromosome of *B. subtilis* via single and double crossover recombination events. While the efficiency of the conventional electrotransformation was only about 5 transformants per microgram DNA with circular integrative plasmid DNA (single crossover recombination into the biotin operon) or linearized delivery plasmid DNA (double crossover recombination into the biotin operon) [209], the efficiency of the high osmolarity protocols improved drastically: 700 transformants per microgram linearized delivery plasmid DNA (double crossover recombination to delete *phrC* gene by insertion of a *cat* gene) [204]; and 5×10^5 transformants per microgram linearized integrative plasmid DNA (double crossover recombination into the *sacA* locus) [205].

Applying the high-osmolarity method, insertional gene inactivation was easily achieved also with ssDNA construct in *B. subtilis* cells expressing lambda red Beta protein [185] (see Section 7.4.5.4). The transformation efficiencies were $10^3 - 10^4$ transformants per microgram lagging-strand annealing ssDNA construct that was protected with phosphorothioate modification at the 5′-end. For guiding to the chromosomal integration sites, the ssDNA disruption cassette was flanked by very short (70 nucleotide) homology regions on both sides. Recently, it was shown that employing a 5′ phosphorothioated lagging-strand targeting full-length ssDNA construct and the high osmolarity electrotransformation, the heterologous lipase expression construct was efficiently introduced into the *amyE* locus through the *B. subtilis*' own homologous recombination machinery. The recombination frequency was $3 \times 10^{-4}\ \mu g^{-1}$ DNA per viable cell [210].

7.4.7
Lysed Protoplast Transformation

A special transformation protocol designated as the lysed protoplast transformation has been reported for *B. subtilis* [211, 212]. Adding *B. subtilis* protoplasts as transforming agent to natural competent *B. subtilis* cells in a hypotonic buffer results in transformants at a frequency of 1 transformant per 10 genomes of donor strain. Thus, the transformation frequency with lysed protoplasts is up to 1000-fold higher compared to transformation protocols with isolated *B. subtilis* DNA. Lysed protoplast transformation requires competent cells and is *comK* dependent similarly to conventional transformation, suggesting that both employ similar mechanisms. The cotransformation frequency of markers located at opposite regions of the chromosome (*purB* 59.8° and *trpC* 202.8°) is 9%. In the labs of the authors of the present chapter (Z.P., H.P.H.), even 30% cotransformation frequency of distant markers was observed (S. Potot, unpublished). Furthermore, high cotransformation frequency was observed even if the number of donor chromosomes is 60-fold lower than the number of competent cells in the transformation mixture [212] indicating the uptake of whole-genome donor DNA into the competent cell. While mutants in *ecsB* encoding an ABC transporter are nontransformable using purified DNA as transforming agent, lysed protoplast transformation was essentially possible with these mutants [213]. It is proposed that proteins covering chromosomal donor DNA and interacting with DNA receptors on the receiving cell facilitate high-frequency transformation. The chromosomal DNA released after the protoplasts burst in the hypotonic transformation buffer keeps the protein/DNA interaction intact, whereas conventional DNA preparation protocols will remove the proteins from the DNA and inflict DNA strand breaks and lesions [181].

By applying lysed protoplast transformation large genomic regions in the recipient strain can be replaced. However, the replacement occurs discontinuously with multiple crossover events between the recipient chromosome and a single donor fragment. If the genomic region to be replaced is inverted [214] in the recipient chromosome, the continuous replacement of several hundred kilobases long genome regions is possible [215].

The multiple crossover events between a large donor DNA fragment and the complementary chromosomal region of the recipient cell during lysed protoplast transformation result in transformants with thoroughly mixed up genetic traits of the recipient and the donor strain. For instance, a heavily mutagenized *B. subtilis* 168 strain with about 200 mutated loci randomly distributed over the entire chromosome and provided with a SM as donor strain (for illustration see Figure 7.3) and wild-type *B. subtilis* 168 encoding a CSM as recipient strain were subjected to lysed protoplast transformation (S. Potot, unpublished). The chromosomes of the resulting SM/CSM plus transformants contained the mutant alleles of the donor strain with an average probability of 30%, independent of whether the locus was a few thousand base pairs away from the SM or localized on the opposite region of the chromosome. Recombining the chromosomes of a production strain comprising hundreds of mutations inflicted during multiple random mutagenesis rounds and a wild-type strain by lysed protoplast transformation followed by correlation analysis between the mutation repertoire of the transformants and their production performance should allow to separate performance-critical mutations from those which are irrelevant or even detrimental.

7.5 Genome Reduction

Several publications have reported on the reduction of the *B. subtilis* genome with the aim of generating improved *B. subtilis* platforms for protein production [216–218]. The genome of the *B. subtilis* strain 168 is 4 215 606 bp in length with 4175 protein-encoding genes out of a total of 4421 identified genes (GenBank: AL009126.3) [20]. After systematic inactivation of individual genes, only 271 (i.e., <7%) of the *B. subtilis* genes were found to be essential for growth in the nutrient-rich LB medium at 37 °C [219]. Accordingly, the majority of the other genes are likely to be unnecessary under controlled industrial fermentation conditions as many of them are activated only under specific conditions or in response to environmental stresses (e.g., sporulation or motility). Furthermore, some of

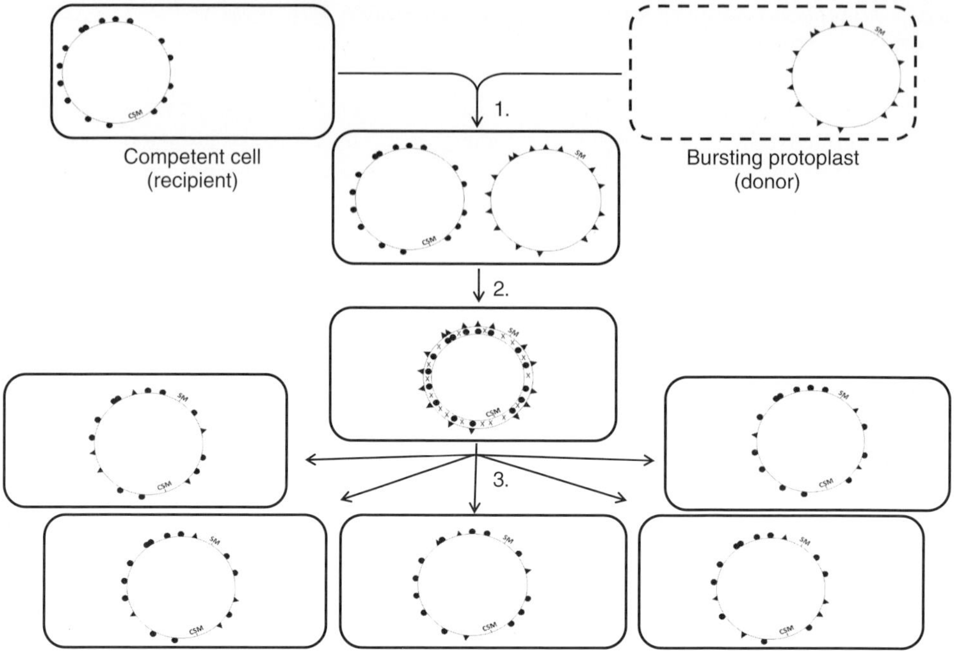

Figure 7.3 "Crossbreeding" by lysed protoplast transformation. (1) A competent *B. subtilis* cell with wild-type alleles (black dots) is mixed in a hypotonic buffer with a *B. subtilis* protoplast comprising a multitude of mutant alleles (black triangles). The competent cell takes up the chromosome of the burst protoplast. CSM is a counterselection marker on the chromosome of the competent wild-type *B. subtilis* cell and SM is a selection marker on the chromosome of the mutant *B. subtilis* protoplast. (2) Intensive crossing over between the chromosomes of the donor and recipient bacteria. (3) Selection and counterselection provide transformants with thoroughly mixed-up genetic traits of the recipient and the donor strains.

the nonessential genes are likely to be detrimental for the production of particular enzymes or metabolites by consuming energy, building blocks, and/or important cofactors. Deletion of genomic regions with unnecessary and/or detrimental genes is therefore potentially beneficial for the production of particular target compounds.

To generate *B. subtilis* strains with significantly reduced genomes, dispensable chromosomal regions had to be identified systematically by evaluating mutants carrying single deletions of individual genes [216] or larger genomic regions flanked by indispensable genes [217, 220]. Apart from the 271 essential genes [219], other genes are also desirable to be retained in the final strains, such as genes coding for rRNA and tRNA, genes involved in primary and DNA metabolism for maintaining growth and avoiding genome instability, and genes involved in the natural competence development for allowing further genetic manipulations [217]. Next, the selected individual dispensable genomic regions had to be sequentially deleted to build the final strains. To date, three "minimal"

B. subtilis strains have been generated and characterized [216–218]. The combination of sequential deletions leading to these three strains started with the removal of prophages, prophage-like sequences, the polyketide synthase (*pks*) genes, and the plipastatin peptide antibiotic synthetase (*pps*) operons. The latter two operons encode genes for the production of secondary metabolites and encompass large parts (2.8%) of the *B. subtilis* genome. The first strain *B. subtilis* Δ6 with 7.7% (0.32 Mb; 332 genes) of the genome deleted showed no beneficial effects on high-level heterologous protein production, and also the growth and key physiological and developmental processes were not affected under laboratory conditions [218]. Strain MG1M with almost a quarter (23.5%; 0.991 Mb; ~1000 genes) of the genome deleted was unstable for biomass and recombinant protein production, making this strain inappropriate for industrial protein production [216, 217]. The third strain MGB874 lacked 20.7% of the genome (0.874 Mb; 865 genes). Interestingly, the production of two enzymes, an extracellular alkaline cellulase and a subtilisin-like alkaline protease was significantly (~50%) enhanced in strain MGB874 [217]. Notably, strains MG1M and MGB874 have a common ancestor, strain MGB469, which lacks 11.1% (0.469 Mb) of the genome. Strain MGB469 showed normal cell growth and had no beneficial effect on protein production. Starting from strain MGB469, sequential deletion of an additional 6 genomic regions led to strain MG1M with the unstable phenotype, while sequential deletion of an additional 11 nonessential genomic regions resulted in MGB874 with beneficial effects on protein production. This underscores the challenges and difficulties of the rational and stepwise genome reduction approach for the development of microbial production platforms. The success of such a genome reduction program depends on (i) the accurate identification of the dispensable genomic regions that are not required for essential cellular functions or fermentative production of proteins or metabolites; (ii) the systematic generation and phenotypic characterization of mutant strains carrying single deletions of the dispensable genomic regions; and (iii) stepwise combination of the individual dispensable genomic regions to build the final strain displaying improved characteristics relevant for industrial production of proteins or metabolites. In strain MGB874, the enhanced enzyme production was found to be due to the synergistic effects of multiple deletions: (i) deletion of the *rocDEF-rocR* region improved the growth and biomass formation to wild-type levels and (ii) deletion of yet unidentified gene(s) enhanced specific enzyme productivity due to the higher expression levels of the target genes resulting from an increased plasmid copy number and higher promoter activity [221].

7.6 Significance of Classical Strain Improvement in Times of Synthetic Biology

Companies active in the fermentation business tend to treat the genetic details of the production strains they actually are using in their large-scale processes as a trade secret. However, it is fair to assume that still today many of these strains are derived from CSI campaigns. CSI aims at phenotypic improvements pertinent to increased strain performance by modifying the genome in a random fashion – contrary to genetic engineering aiming for targeted genetic modifications – and screening of the offspring for a relevant phenotype. CSI strains are considered as nongenetically modified organisms (non-GMOs) – in scientific and technical terms an obvious hogwash – that eases registration and customer acceptance. In addition to complying with political constraints and consumer behavior, CSI is still an indispensable tool for the development of high-performance strains, also for microorganisms with well-developed genetic tools such as *B. subtilis*. This is because of the randomness by which CSI afflicts mutations, neither driven by biochemical hypotheses nor relying on model predictions. A prerequisite for a successful CSI campaign is the availability of a relevant assay to screen at high or ultra-high throughput mode for phenotypically improved individuals within a mutagenized population. CSI typically delivers good results in cases where single mutations can lead to strain improvements, for example, removing rigid metabolic nodes caused by repressive regulatory mechanisms, or limiting otherwise the metabolic flux toward the target molecule. CSI finds its limits when it comes to more complex genetic modifications, such as the exchange of the natural, weak promoter in front of a gene of interest by a stronger one or adding a novel gene to the genetic repertoire of a production strain for further performance improvement.

7.6.1
Generation of Genetic Diversity in Random Fashion

The availability of highly diverse, template-free synthesized oligonucleotides at low costs should facilitate the generation of *B. subtilis* mutant populations *a la carte* from individuals with point mutations dispersed discretionarily all over the bacterial chromosomes. Unfortunately, efficient transformation of the oligonucleotide libraries into the *B. subtilis* genome – the method of choice would be recombineering (see Section 7.4.5.4) – without relying on a dominant selection marker is still not efficient enough to generate the desired mutant diversity.

B. subtilis random mutant libraries can be obtained if DNA replication proceeds under low fidelity conditions during the expansion of a bacterial culture. PolC, the major DNA polymerase of *B. subtilis*, comprises within its N-terminal half a proof-reading exonuclease activity. The enzyme copies DNA with a fidelity of 10^{-9} mismatches per base [222, 223]. Point mutations in the proof-reading domain, for example, mut-1a G430D, reduces the replication fidelity of PolC about 1000-fold [224]. The error-prone replication by this PolC mutant should be well suited to generate populations almost completely devoid of wild-type bacteria, instead consisting of individuals varying greatly with regard to the number of randomly introduced mutations. The switch from high-fidelity to mutagenic replication and back should be possible by the conditional expression of polC-G430D. Despite the obvious advantages, the application of low-fidelity replication procedures to obtain mutant libraries – the usage of a *polC* mutant is just one example, other interferences with the DNA repair machinery of *B. subtilis* are conceivable – has not been reported so far.

For a number of decades, the method of choice to obtain *B. subtilis* mutant libraries is the treatment of an exponentially growing culture with a mutagen, frequently ethyl methanesulfonate (EMS) or nitrosomethylguanidine (NTG). NTG treatment leads to the methylation of O^6-guanine and to a minor extent of O^4-thymidine at random positions within the DNA sequence. During DNA replication, O^6-methylguanine pairs with thymine rather than with cytosine, thereby inflicting GC to AT pair transitions. The course of the mutagenesis can be monitored with readily measurable parameters such as the killing rate and occurrence of auxotrophic growth phenotypes, or the frequency of rifampicin-resistant mutants among the survivors. Today's high capacity and deep sequencing capabilities provide the means to scrutinize directly the quality of the library with regard to randomness, mutation frequency, and the sort of mutation inflicted by the mutagen.

7.6.2
Screening Platforms

Lipophilic metabolites tend to accumulate within the production host. If in addition the metabolite is colored or fluorescing, a carotenoid for instance, flow cytometry-based cell sorting protocols are suited to isolate individuals with improved production traits out of a mutagenized population, even if they are present at very low frequency. Value compounds typically produced with *B. subtilis* production hosts, such as water-soluble vitamins, purine nucleosides, ribose, or proteins, are secreted and accumulate in the culture medium. Screening of improved production strains for such metabolites requires clonal cultivation of the members of the mutagenized population in compartmentalized reactors and individual analyses of the culture supernatants in each of the reactors.

The frequency of improved individuals in a mutant library derived from an advanced production strain depends on the mutation rate aimed for during the generation of the library. The trait sought after might be present in 1 out of 1000 members in the library and contribute not more than 5% performance improvement – to expect more would be unrealistic. High-throughput cultivation and analytical procedures, which have to be applied, are notoriously inaccurate with standard deviations of 10% or above. For having a good chance to capture descendants of an improved individual within the selected population, a sufficient number of clones must be screened and the selection threshold should not be too high. Assuming a standard deviation of 10% in the performance read out signal and 5% performance increase afforded by a novel trait, statistics suggest to cultivate and screen 20 000 clones of the mutagenized library and collect all those that feature an apparent performance of at least 120% of the median of the library. Following this procedure, an enrichment factor for improved mutants of three to four will thereby be achieved. Depending on the available fermentation infrastructure, four to five rounds of recultivation and screening will be necessary to enrich the population for improved individuals to a level such that they can be identified and characterized in more detail applying low-throughput devices, for example, small or medium-sized pilot fermenters. A CSI campaign of this size can be realized with commercially available, although costly, highly automated multititer plate screening platforms.

If improved individuals are even scarcer and library members carrying a performance improving trait are present at or below a frequency of 10^{-4}, campaign sizes readily have to exceed 10^6 samples. High plastics and reagents costs as well as prohibitively long analyses times disqualify multititer plates as primary reaction or growth arrays but solutions are provided by microconfinement approaches [225] employing homogeneously sized aqueous microdroplets as reaction or growth compartments. The droplets are charged with the cells and then incubated in hydrophobic carrier solvents supplemented by surfactants serving as sealants in order to avoid crosstalk (exchange of nutrients, educts, or products), prevent contamination of the confined cultures or cells, and increase the stability of the emulsion. If required, droplets can be transformed into hydrogels [226] thereby further increasing their operational stability.

Analysis of the droplet interior for quantitative or qualitative product analysis typically requires a fluorescence signal which can be recorded at the kilohertz range with commercially available flow cytometers [227] or customized microchips [228]. Naturally, most industrially relevant products are not fluorescent, which requires that an appropriate chemical or biological assays is integrated into the microdroplets too. Yet, once all components are in place, libraries containing 10^6 entries can be readily screened.

7.6.3
High-Throughput, Low-Intensity Fed-Batch Cultivation

Industrial fermentation processes with *B. subtilis* host strains are frequently carried out under carbon limited fed-batch conditions. After a short initial batch phase of unlimited nutrient supply, a concentrated solution of the carbon source, starch hydrolysate for instance, is pumped into the production fermenter at a constant rate to ensure that both the availability of the carbon source becomes the growth limiting factor and the capabilities of the fermenter with regard to oxygen transfer and heat removal are fully used as early as possible to maximize space-time yield. Under these conditions, the specific carbon uptake rate of the biomass accumulating in the fermenter gradually decreases and the growth rate falls below $0.05\,h^{-1}$ early in the feeding phase, approaching zero toward the end of the fermentation. The metabolic conditions during strong carbon limitation are very different from carbon excess conditions [229]. There is obviously no carbon catabolite repression when the carbon source is scarce and, instead, the culture might be close to the induction of the stringent response. Under these conditions, the number of ribosomes per cell is drastically reduced [230]. The energy demand for maintenance purposes becomes dominant over the resource requirements for metabolite synthesis and biomass formation [231]. To ensure that the traits selected during a screening campaign are relevant for industrial production runs, the cultivation conditions during the screening should reflect the production conditions, particularly the limited availability of the carbon source.

Several protocols have been developed to mimic fed-batch conditions in high-throughput cultivations during the screening for improved traits. A liquid screening system for *B. subtilis* that is based on the slow metabolism of the tri-hexose saccharide raffinose is easy to perform [231]. Although the carbon uptake rate during *B. subtilis* cultivation with raffinose as the carbon source is sufficiently reduced to prevent overflow metabolism, the growth rate achieved under these conditions is reported as 0.17 h^{-1} and, thus, is well above the growth rates realized in industrial fed-batch fermentations, which decrease continuously from about 0.1 h^{-1} during the early feeding phase to 0.01 h^{-1} or lower toward the end of the fermentation.

Glucose-limited cultivation conditions in high-throughput mode can be realized if a deposit of metabolically inactive starch is added to a cultivation medium devoid of any other carbon source. A starch-degrading enzyme, for example, glucoamylase, is provided to reach such an activity that the rate of starch depolymerization and, consequently, glucose release stays below the maximal glucose uptake rate of the microbe inoculated into the medium [232]. The lower the activity of glucoamylase in the medium, the lower will be the growth rate. The cultivation can be readily carried out in milliliter scale. Obviously, to apply this protocol, the strain to be cultivated has to be deficient for starch-degrading enzymes.

A controlled-release system made of glucose-containing silicone elastomer disks can be applied to realize fed-batch cultivation conditions in 96-deep-well microtiter plates. Screening of *Hansenula polymorpha* clones, transformed with GFP expression plasmids for GFP secretion, demonstrated superior screening results using the controlled-release system as compared to glucose- or glycerol-batch screening [233]. A deposit of a glucose-containing silicone elastomer fixed at the bottom of U-shaped 96-deep-well microtiter plates was successfully used to screen for superior *B. subtilis* riboflavin production strains (C. Bähr and J. Büchs, unpublished).

7.6.4 Medium-Throughput, High-Intensity Fed-Batch Cultivation

To maximize space-time yield industrial fermentation runs in stirred tank reactors are frequently performed in fed-batch mode at high feeding intensities, requiring oxygen uptake rates of 200 $mmol\,l^{-1}\,h^{-1}$ and above. Furthermore, high-intensity fermentations require the continuous supply of a nitrogen source for biomass- and product formation. If gaseous ammonia or a concentrated ammonia solution is used, the nitrogen supply can be coupled with online pH control. High feeding intensity, high oxygen demand, the required continuous nitrogen feeding, and pH adjustments preclude microtiter plate-based high-intensity cultivation of a microorganism. For screening under high-intensity fed-batch conditions, a block with 48 stirred-tank 10-ml bioreactors was developed [234, 235]. High oxygen transfer capabilities of up to $k_L a = 0.4\,s^{-1}$ to support the required oxygen uptake rates are achieved through gas-inducing stirrers propelled by a magnetic inductive drive. Individual dissolved oxygen and pH monitoring and control is possible by intermittent substrate and base additions through a standard liquid handler. However, owing to the limited throughput, the screening of random mutant libraries with the 10-ml stirred-tank bioreactor block might be precluded. Nonetheless, this device is useful for the screening of transformants obtained from targeted gene engineering or to enable multivariable bioprocess development by statistical or stochastic design.

7.7 Resource-Efficient *B. subtilis* Fermentation Processes

In general, the economics of a microbial production process first and foremost depends on the fermentation raw material consumption to produce the target molecule. Carbon sources, such as starch syrups, sucrose, vegetable oils, or carbohydrates derived from lignocellulosic biomass are usually the most important cost drivers. In an aerobic fermentation, the standard regime for cultivation of many microorganisms including *B. subtilis*, the carbon source delivers the carbon atoms for building up organic matter and, while being oxidized preferably to carbon dioxide, serves as energy source for growth of the microorganism and product formation in the fermenter. If the fermentation substrate availability is not or only slightly restricted and favorable physicochemical conditions (temperature, pH) are met, the biomass will grow at its maximal rate or slightly below that. At high growth rates, microorganisms tend to incompletely oxidize their carbon source. Instead of forming carbon dioxide, which is removed from the fermenter via the off gas, organic breakdown products are synthesized and released into the fermentation broth. In case of *B. subtilis*, these molecules, or overflow metabolites as they are called, are *acetate*, *lactate*, *butandiol*, and *acetoin*. The physiological condition of carbon catabolite repression leading to overflow metabolism is well studied in *B. subtilis* (for a more recent review, see [236]). Production of overflow metabolites is a waste of resources and has to be avoided (if an overflow metabolite is not the intended product). At reduced growth rates ($\mu < 0.2\,h^{-1}$) resulting from a limited carbon source supply, overflow metabolite production ceases and almost complete oxidation to carbon dioxide of the portion of the carbon, that is not recovered in the biomass or in the product, is achieved. The respiratory coefficient, a parameter that can be determined online as the quotient of the molar consumption and production of oxygen and carbon dioxide, respectively, is close to one under conditions of complete oxidation of the carbon source.

High growth rates in the fermenter thwart good product yield also for another reason. In highly developed production strains, the metabolic carbon flux toward the product of interest is deregulated and largely released from repressive mechanisms. In consequence, carbon source limitation affects product formation less severely than biomass formation and relatively more carbon will end up in the product than in the concomitantly produced biomass. At less restrictive carbon source supply conditions and high growth rates, the competitive advantage of product over biomass formation is lost, resulting in reduced product yields.

Low growth rates are technically realized in fed-batch fermentation runs. After an initial, short batch phase of unlimited substrate supply, the carbon source is pumped into the fermenter at a rate that is lower than the maximal specific carbon source uptake rate of the biomass growing in the fermenter. Furthermore, the feeding rate has to be aligned with the aeration capacity of the fermenter to prevent oxygen-limiting conditions. Thus, sooner or later the feed has to be provided at a constant rate and the carbon substrate for an individual bacterium in the fermenter becomes increasingly scarce. As a consequence of the decreasing specific carbon uptake rate, the growth rate gradually decreases approaching $\mu = 0\,h^{-1}$ toward the end of the fermentation run. The specific productivity is also reduced but, in general, to a lesser extent than the specific biomass formation rate (i.e., the growth rate), leading to a higher product yield.

Biomass consumes the carbon substrate not only for reproduction but also for keeping itself in a viable state. The maintenance demand of a bacterial culture is defined as the carbon source requirements for purposes other than growth and product formation. The maintenance parameter can be determined by plotting the carbon source uptake rates during continuous fermentation runs at different dilution rates as a function of the growth rates and linear extrapolation to growth rate $\mu = 0\,h^{-1}$. The y-intercept will then provide the maintenance parameter.

At low and very low growth rates ($\mu < 0.05\,h^{-1}$) prevailing, for instance, during the latter phases of fed-batch fermentation runs, the maintenance demand can become a significant sink for the carbon source provided [231]. This is particularly true for strains with high maintenance parameters. For glucose-cultivated *C. glutamicum* [237] and *Pseudomonas putida* [238], two bacterial species of industrial importance, maintenance parameters m_s below 10 mg glucose g^{-1} DCM^{-1} (DCM dry cell mass) and between 4 and 14 mg glucose g^{-1} DCM^{-1}, respectively, were reported. However, a maintenance parameter $m_s = 37$ mg glucose g^{-1} DCM^{-1} for *P. putida* KT2440 can also be found in the literature [239]. With reported 81 mg glucose g^{-1} DCM^{-1} the maintenance parameter for *B. subtilis* is much higher than for *C. glutamicum* or *P. putida* [231, 240]. A *B. subtilis*-based riboflavin production strain showed even higher maintenance requirements with 119 mg glucose g^{-1} DCM^{-1}. Nevertheless, the growth yield, that is, the biomass yield on the carbon source adjusted for non-growth-associated carbon source consumption [241], was within the range determined for other bacterial species (0.46 g DCM g^{-1} glucose^{-1}). In contrast to *B. subtilis,* the maintenance requirements (25 mg glucose g^{-1} DCM^{-1}) for *Bacillus licheniformis* [242] were found to be closer to those of *P. putida.*

To reduce the very high maintenance demand of the riboflavin-overproducing *B. subtilis* strain 1012 pRF93 a *cydBC* knockout mutant was constructed [243]. The gene product of *cydB* is the cytochrome bd oxidase subunit II, whereas *cydC* together with *cydD* encodes an ABC transporter required for expression of cytochrome bd. From the various terminal oxidases of the branched *Bacillus* respiratory chains, the bd oxidase is the least energy efficient terminal oxidase, translocating only one proton over the plasma membrane per electron transferred to oxygen [244, 245]. The maintenance parameter of the mutant was still very high (70 mg glucose g^{-1} DCM^{-1}), but clearly lower than the *cyd* wild-type demand (121 mg glucose g^{-1} DCM^{-1}). Importantly, riboflavin productivity was positively affected by the mutation. The reduced maintenance demand was explained by an electron flux redirection toward the more energy efficient cytochrome aa3 oxidase, translocating two protons per electron. Improved riboflavin production performance in a *cyd* deletion mutant was also reported in a publication from 2006 [246]. Despite the positive effects on product formation obtained through redistribution of the electron flux through the respiratory chain, a real understanding of the biochemical mechanisms causing the unusually high carbon consumption and respiratory rate at very slow growth is a prerequisite for a further significant reduction of the *B. subtilis* maintenance metabolism.

Besides its high maintenance demand, *B. subtilis*, similar to the other *Bacillus* species of industrial relevance, is well suited for cultivation in large aerobic stirred-tank reactors. The bacteria are of fairly uniform size with some tendency to form chains of a few individual cells generated from incomplete cell wall separation during the growth cycle. There is no growth on the solid surfaces of the fermentation equipment. *B. subtilis* cultures are of low viscosity even at elevated cell density and therefore easy to mix and to aerate. The biomass separation after completion of the fermentation run by centrifugation, decantation, or microfiltration usually does not pose a serious problem. Downstream processing comprising value product isolation and purification depends on the chemical characteristics of the product and is very product specific.

7.8
Safety of *Bacillus subtilis*

For centuries, *B. subtilis* have been deliberately introduced into the food chain, as exemplified by the traditional Japanese food "natto" prepared from soybeans fermented with *B. subtilis* natto strains, without any adverse effect on human health [247]. In addition, *B. subtilis* is dominant among several *Bacillus* species that are involved in the preparation of traditional fermented soybean (Kinema, Hawaijar), and locust bean (Soumbala) dishes in Asia and Africa [248–250].

B. subtilis is the organism of choice for many commercial products, such as fine chemicals (B vitamins and nucleotides) and food and feed enzymes [251]. On the basis of either a recognized history of safe use in foods before 1958, or on scientific procedures similar to those required for a food additive regulation, many compounds and enzymes such as B-vitamins, carbohydrase, and protease enzymes produced by *B. subtilis* have GRAS (generally recognized as safe) status from the US Food and Drug Administration. A status similar to GRAS, called QPS (qualified presumption of safety), is used by the European Food Safety Authority (EFSA) [252]; however, the GRAS status is granted for the product, while the

QPS status is for the production microorganisms themselves. QPS is a concept for a harmonized, generic, premarket safety assessment system of microorganisms deliberately introduced into the food chain as live organisms and microorganisms used as a source of food and feed additives and enzymes. QPS is made on the basis of four pillars: (i) taxonomical unit identification at least on the basis of rRNA sequences and other housekeeping genes, (ii) body of knowledge, (iii) possible pathogenicity, and (iv) end use. Microorganisms not considered suitable for QPS would remain subject to a full safety assessment. Microbial strains with unambiguously established identity and which are assigned to a QPS group are freed from the need for complete safety assessment. For some QPS taxonomic units, safety concerns are identified and designated as "qualifications" in connection with a recommendation for a QPS status. The generic qualification for all bacterial taxonomic units on the QPS recommended list is that the strains should not harbor any acquired antimicrobial resistance genes to clinically relevant antibiotics [253]. Bacterial strains such as the *Bacillus* strains carrying acquired resistances should not be intentionally introduced into the food and feed chain (see Section 7.8.1). Several *Bacillus* species including the industrially relevant *B. subtilis, B. amyloliquefaciens*, and *B. licheniformis* are on the list of QPS granted taxonomic units due to the substantial body of knowledge (e.g., annotated genome sequence) available about these bacteria and therefore decreasing the probability of unforeseen hazards. However, these *Bacillus* QPS species are with the qualification "absence of toxigenic activity." This is based on the observation that some rare strains among these *Bacillus* species have caused foodborne intoxication in the past, and that these intoxications have been attributed to the production of compounds with toxigenic potential by these strains [253]. Therefore, the absence of toxigenic activity (see Sections 7.8.2 and 7.8.3) for these *Bacillus* strains needs to be verified for QPS qualification.

7.8.1
Absence of Acquired Added Antimicrobial Resistance Genes

The development of bacterial resistance to antimicrobials remains a serious concern for human health and well-being. Therefore, acquired added resistance genes (genes acquired by the bacteria via horizontal exogenous DNA transfer) to antimicrobials of clinical importance must be absent from bacterial strains deliberately introduced into the food chain [132]. The QPS approach to the safety assessment of the commercially relevant Bacillus species requires the unambiguous identification of the strain, a demonstration of susceptibility to clinically relevant antibiotics, and evidence that the strain lacks a capacity for toxin production [252, 253]. After the taxonomical characterization of the production strains with molecular methods (sequences of the 16S rRNA, gyrA, and gyrB genes) to verify that they belong unambiguously to the *B. subtilis* species, the susceptibility to relevant antibiotics has to be determined. According to the EFSA guidance [132], the minimum inhibitory concentration (MIC) of the antimicrobials of human and veterinary importance should be determined as a basic requirement.

In the cases of *Bacillus* species, cutoff values for various antimicrobial substances are $4\,\mathrm{mg\,l^{-1}}$ for vancomycin, gentamicin, erythromycin, or clindamycin, and $8\,\mathrm{mg\,l^{-1}}$ for kanamycin, streptomycin, tetracycline, or chloramphenicol. The EFSA Panel on Additives and Products or Substances used in Animal Feed (FEEDAP) does not require the determination of a microbiological cut-off value for ampicillin in the cases of *Bacillus* species. This is mostly due to the presence of the native ampicillin resistance (β-lactamase) genes on the chromosomes of many *Bacillus* species [254]. When a *B. subtilis* strain demonstrates higher resistance to a specific antimicrobial than indicated above, the genetic basis of the antimicrobial resistance needs to be determined. Such a *B. subtilis* strain may be used as a feed additive if the antimicrobial resistance is due to intrinsic resistance (inherent to *B. subtilis*, all *B. subtilis* strains carry it) or acquired resistance due to chromosomal mutation(s). These antimicrobial resistance genes present a minimal or low potential for horizontal spread. In contrast, a *B. subtilis* strain carrying an acquired added antimicrobial resistance gene due to the acquisition of genetic determinant(s) presents the greatest potential for horizontal spread and should not be used as a feed additive [132].

7.8.2
Safety of Surfactin

Bacillus species not belonging to the *B. cereus* group are a rare cause of foodborne diseases [255]. *B. subtilis* and the closely related species *Bacillus pumilus* and *B. licheniformis* have periodically been suggested to play a role in rare food-poisoning incidents despite the fact that the organisms do not possess the *B. cereus*-like toxin-producing genes [256]. Merely the production of cyclic lipopeptides, such as surfactins with hemolytic activities, is of some concern among these Bacilli [257–259].

Surfactins are produced by the surfactin synthetases through a thiotemplate mechanism independent of ribosomes. Surfactin synthetases are composed of modularly arranged enzymatic domains, which catalyze the necessary steps in the surfactin biosynthesis [260]. The surfactin synthetases of *B. subtilis* are encoded by the surfactin operon with four open reading frames: *srfAA*, *srfAB*, *srfAC*, and *srfAD* [261]. Three modules are present in the surfactin synthetases SrfAA and SrfAB, one module in SrfAC, and a thioesterase/acyltransferase (Te/At) domain in the fourth enzyme SrfAD [247, 260]. Despite having the complete and functional surfactin biosynthetic (*srfA*) operon on the genome [19], *B. subtilis* 168 is unable to produce surfactin owing to the presence of a frame-shifted sfp gene leading to a truncated Sfp enzyme (165 of the 224 amino acids) [253, 255]. The Sfp enzyme 4′ phosphopantetheinyl transferase is responsible for the transformation of apo surfactin synthetases into the holoenzymes that is essential for surfactin synthesis [262]. Therefore, the *B. subtilis* 168 strain is designated genotypically as sfp^0 (surfactin nonproducing) [263]. The role of surfactin and the importance of the presence of sfp was demonstrated in *Bacillus mojavensis*, an organism closely related to *B. subtilis*. The wild-type B. mojavensis produces a complex of three different surfactin analogs with toxigenic activities in a series of cytotoxicity assays (lactate dehydrogenase release, hemolysis, inhibition of both protein synthesis in Vero cells, and motility in boar sperm). The enterotoxicity was entirely due to the production of surfactin cyclic lipopeptides as deletion of sfp abolished the cytotoxicity in all assays [256]. Accordingly, industrial *B. subtilis* strains carrying the *sfp*^{0} mutation are expected to be nontoxigenic in the in vitro cytotoxicity assays described in the EFSA guidance [264].

Altogether, the concerns with the *B. subtilis* group appear to be associated with the production of surfactin like-lipopeptides, although the relation between the presence of these compounds and/or other toxic factors and the risk of illness in humans has not yet been established [264]. Therefore, it is important to note that some *B. subtilis* strains, such as the *B. subtilis* subsp. natto strains are capable of producing surfactin as the *sfp* gene is intact. Accordingly, surfactin is found in relatively large amounts (~0.3 mg g^{-1}) in the Japanese traditional food "natto" that is fermented using *B. subtilis* subsp. natto strains [260]. Notably, surfactin was suggested to be safe on the basis of experiments for acute toxicity in mice (LD_{50} at >100 mg kg^{-1}, i.v.) and from long-term Japanese experience that the consumption of up to 10 mg of surfactin (30 g of natto) daily does not disturb health [247]. In addition, surfactin may even have anticancer properties as it inhibited growth of human breast cancer cells [262] and strongly blocked the proliferation of LoVo cells, a human colon carcinoma cell line, by inducing proapoptotic activity and arresting the cell cycle [261].

7.8.3 Absence of Toxin Production

Most food poisoning incidents attributed to *Bacillus* species are associated with *B. cereus* owing to its ability to produce enterotoxins and emetic toxin [265]. The three different heat-labile enterotoxins cause the diarrheal type of foodborne illness when produced during vegetative growth of *B. cereus* in the small intestine after ingestion. On the other hand, the heat-stable emetic toxin, cereulide, is produced during vegetative growth of *B. cereus* in foods before ingestion [266].

Other *Bacillus* species have generally been considered of little significance in food poisoning incidents, but their ability to produce both enterotoxins and emetic toxin has been increasingly recognized [257, 267]. Mainly certain strains of *B. licheniformis*, *B. mojavensis*, *B. pumilus*, and *Bacillus megaterium* and rarely a few strains of *B. subtilis* have raised suspicion of toxin production in incidents of foodborne illness [268–272]. It is believed that the enterotoxins produced by *Bacillus* spp. other than *B. cereus* are proteins transcribed from genes that are similar to those of *B. cereus* enterotoxins [257, 273, 274]. However, with the exception of an emetic toxin from *B. licheniformis*, the toxins from *Bacillus* species other than *B. cereus* have never been purified and characterized [257]. The "majority of strains 'from the *B. subtilis* group' evidently lack the gene(s) encoding toxin(s) or, if present, the capacity for toxin production or fail to produce detectable levels of toxin under the conditions employed and these could be considered safe for use" [255]. The view of EFSA is that the very few reports of *B. cereus*-like enterotoxins occurring in other species of *Bacillus* are likely to have resulted from a misidentification of the strains involved [257, 264].

7.8.4 Absence of Sporulation

From environmental and governmental regulatory viewpoints, it is important that genetically modified microorganisms do not have the ability to live in the natural environment if they escape from the manufacturing plant during the fermentation process or during disposal of the biomass. For this reason, sporulation-deficient strains of *B. subtilis* are used in fermentative processes for riboflavin [135] and biotin [176]. Typically, the *spo0A* mutation is used to arrest sporulation. The *spo0A* gene encodes a protein that regulates the initiation of sporulation. Inactivation of *spo0A* arrests sporulation at its earliest stage, resulting in strains that are no longer resistant to chemical solvents, radiation, and/or heat. However, in the absence of Spo0A, pantothenate production decreased drastically in the *B. subtilis* pantothenate overproducing strains [134]. In this case, sporulation-deficient pantothenate overproducing strains can be constructed by incorporation of mutations in later-stage transcription factors such as SigE and SigG, which prevent sporulation yet maintain high pantothenate production. An alternative way to enhance pantothenate production and keep sporulation deficiency is the introduction of both *spo0A* and *abrB*-null mutations into the vitamin B5-producing strains [134]. To produce commercially important enzymes, a method that uses a mutation in the *B. licheniformis spoIIAC* gene, which encodes the sporulation-specific transcription sigma factor F (σ^F), was described to obtain bacteria of the genus *Bacillus* other than *B. subtilis* that are incapable of sporulating [275]. In another patent, *B. licheniformis*- or *B. subtilis*-producing keratinase are described carrying an inactive endogenous *spoIIAC* gene that renders sporulation deficiency [276].

7.9 *Bacillus* Production Strains on the Factory Floor: Some Examples

7.9.1 Nucleotides

Together with monosodium L-glutamate, 5′-ribonucleotides evoke a strong, rich taste sensation known as umami taste, which is mediated by the G-protein-coupled receptor heterodimer T1R1/T1R3 of the human tongue. The nucleotides enhance the responsiveness of T1R1/T1R3 to L-glutamate explaining the synergistic effect of the amino acid and the nucleotides on the taste buds [277]. Nucleotides, mainly inosine monophosphate (IMP) and guanosine monophosphate (GMP), in combination with monosodium L-glutamate are therefore used in a wide range of processed foods such as soups and flavored seasonings. In addition to being used as food additives, nucleosides serve as starting material for the production of antiviral drugs such as ribavirin and acyclovir.

The global demand for nucleotides in 2009 was approximately 23 000 metric tons according to Ajinomoto Co. of Japan, the nucleotide market leader with a market share of about 40%. Other competitors in the field are CJ Group and Daesang Group of Korea, Kirin Kyowa Foods Company of Japan, and Star Lake Bioscience of China. The demand for nucleotides grows in parallel with the market for processed foods (Source: FLEXNEWS 06/10/2010).

The purine nucleotides IMP and GMP are produced in a chemical process step by phosphorylation of the corresponding nucleosides, which are obtained by fermentation [278–280]. The phosphorylation agent is phosphoryl chloride [281]. The traditional purine nucleoside production strains are mutants of *B. amyloliquefaciens*, in the older literature designated as *B. subtilis K*, but *B. subtilis* strains might also be used at industrial scale.

Purine biosynthesis (Figure 7.4) starts from phosphoribosylpyrophosphate (PRPP), which is derived from the pentose phosphate pathway intermediate ribose 5-phosphate. The biochemical reaction is catalyzed by PRPP synthetase encoded by the *prs* gene. Purine biosynthesis commences with glutamine-dependent amidotransfer to PRPP, affording phosphoribosylamine. IMP is derived in nine additional consecutive steps and converted to adenine monophosphate (AMP) or GMP, each in two separate additional steps. The genes encoding the enzymes converting PRPP to IMP are localized at map position 55° of the *B. subtilis* chromosome forming the pur operon [282, 283]. Expression of the pur operon is twofold negatively regulated, at the transcription initiation level by PurR [284] and at the transcription attenuation level by a riboswitch located in the leader sequence of the *pur* operon [285, 286]. PRPP acts as effector of the PurR protein preventing the binding of the repressor to the *pur*-Boxes upstream of the *pur* core promoter [287, 288]. ADP and, to a lesser extent, AMP and GDP indirectly control the *pur* transcription initiation by allosterically inhibiting PRPP synthase [289]. The sensor molecule of the riboswitch is guanine, which upon binding to the riboswitch results in transcription termination before the first structural gene of the operon. Although not experimentally confirmed, the DNA sequence of the *pur* locus and the presence of a PurR homolog suggests that *pur* gene transcription regulation in *B. amyloliquefaciens* is similar to the regulation in *B. subtilis*.

In the purine degradation pathway IMP and GMP are dephosphorylated by nonspecified phosphatases to the corresponding nucleosides, which can leave the bacteria either spontaneously or facilitated by an exporter, for example, PbuE [290, 291]. Further degradation of the nucleosides to the nucleobases is catalyzed by the nucleotide phosphorylases PunA and DeoD [292]. The bases can be salvaged by Hpt-catalyzed reactions with PRPP to IMP and GMP.

A common property of all *Bacillus* purine production strains is their auxotrophy for adenine caused by a dysfunctional adenylosuccinate synthase gene (*purA*) of the AMP-specific branch of the purine pathway. Furthermore, the

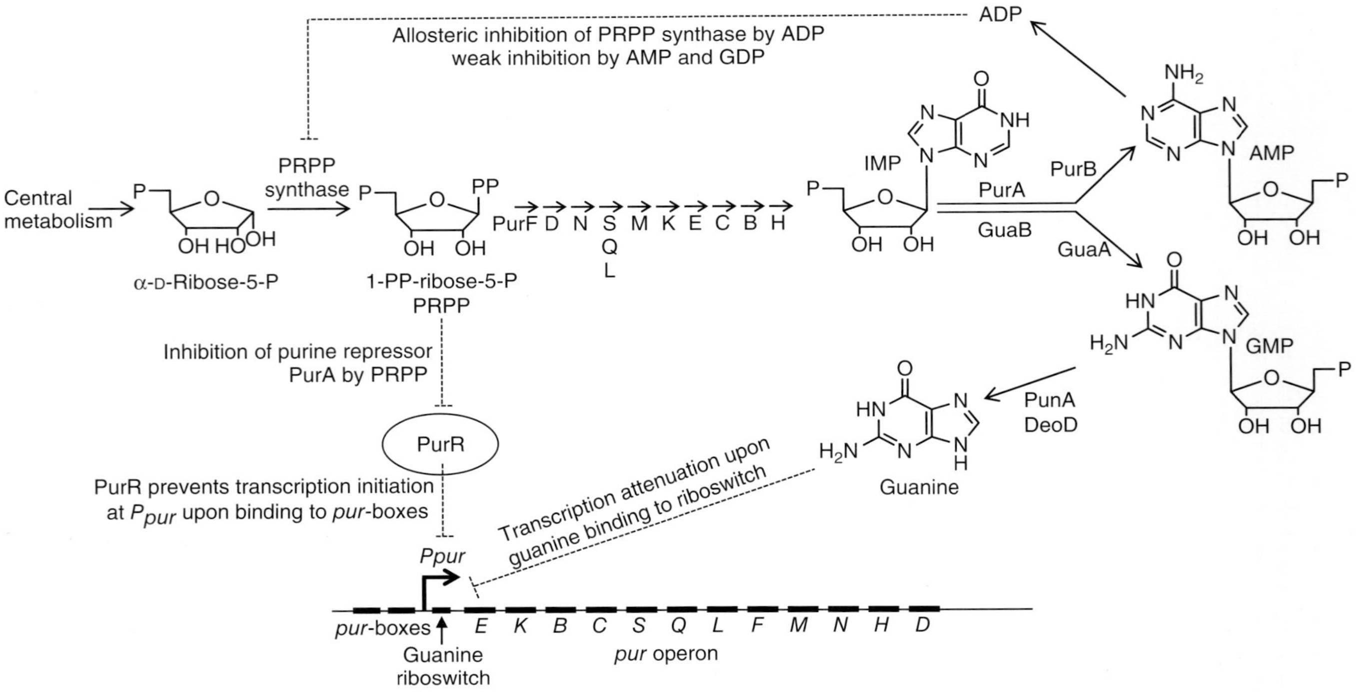

Figure 7.4 Regulation of *pur* gene expression in *B. subtilis* (simplified). In addition to the *pur* operon PurR also interferes with *purAB* transcription.

strains are phosphorylase defective, preventing nucleoside degradation to the nucleobases. Literature from the 1960s and 1970s, mainly from Ajinomoto researchers, report on *B. amyloliquefaciens* strains provided with these two mutations, which produced 10 g l^{-1} inosine from 80 g l^{-1} glucose upon cultivation under adenine-limiting conditions. The product accumulated in the culture supernatant [279, 293]. Under adenine, but not guanosine surplus conditions, inosine secretion ceased almost completely. A double mutant with defects in the genes encoding the first enzymes of the AMP and the GMP branch of the purine pathway (PurA and GuaB) produced 15 g l^{-1} inosine from 80 g l^{-1} glucose upon cultivation with growth-limiting amounts of adenine and guanosine. According to the current understanding of the regulation of *pur* gene expression, under adenine- and guanosine-restrictive cultivation conditions, inhibition of the PRPP synthase is prevented. The resulting elevated intracellular PRPP levels ensure that (i) the amidophosphoribosyltransferase PurF, the entrance enzyme into the purine pathway, is well provided with substrate and (ii) the *pur* gene repressor PurR is kept inactive, allowing unhindered initiation of *pur* gene transcription. The low repressing activity of guanosine is surprising as guanine (but not guanosine) is a powerful inducer of *pur* transcription termination [286]. Probably, the impaired phosphorylase activity in the inosine-producing mutants prevents the accumulation of guanine to intracellular levels that interact with the guanine riboswitch and induce transcription termination.

B. amyloliquefaciens inosine production strains can be converted into guanosine production strains by deregulation of the GMP branch of the purine pathway [294]. Deregulation is achieved in mutants resistant to the nucleotide analogs decoyinine and psicofuranine. In one of these mutants, repression by guanine of the *guaB* encoding IMP dehydrogenase and inhibition by adenosine of the GMP synthase is greatly reduced. This strain delivered 16 g l^{-1} guanosine from 80 g l^{-1} glucose. Neither inosine nor xanthosine secretion was observed.

The genome sequences of two *B. amyloliquefaciens* inosine production strains designated as XH7 [295a] and TA208 [295b] were reported recently. The report on XH7 was coauthored by researchers from the Chinese nucleotide producer Star Lake Bioscience. Both genomes encode a defective *purA* gene in line with the presumed adenine auxotrophy of the strains. Besides a nonsense mutation in the GMP reductase gene *guaC*, which interferes with the conversion of GMP back to IMP, key genes involved in purine biosynthesis, degradation, and salvaging, such as *purF*, *purR*, and *deoD*, are unchanged in the genome of XH7. No information regarding the nucleoside production capabilities of XH7 was provided.

In the genome of strain TA208, mutations in *purL* (D624N), *purM* (Q322K, H334Q), and *purC* (N6S, I89V, H166L) were identified. These mutations might modify the enzymatic characteristics of the encoded biosynthetic enzymes of the de novo purine pathway proteins, for example, release from putative feedback inhibition, and thus could contribute to enhanced inosine productivity. A nonsense mutation in the *pbuX* gene encoding a guanine permease might contribute to the 8-azaguanine resistance of the strain. Mutations that could affect the regulation of the expression of the *pur* genes or the nucleoside efflux pumps were not observed. As for XH7, no information regarding the nucleoside production capabilities of TA208 was provided.

Sequencing of the DNA up- and downstream of the pur transcription initiation site of a *B. subtilis* inosine production strain revealed mutations in and around the PurR binding sites (*pur*-Box1 and *pur*-Box2) upstream of the pur core promoter and a highly mutagenized untranslated leader region [296]. These alterations should interfere with both the PurR-dependent inhibition of the transcription initiation of the *pur* operon and the guanine riboswitch-mediated premature transcription termination of the operon. The inosine production capabilities of the studied strain as well as the steps by which the strain was derived from wild-type *B. subtilis* have not been reported.

An early attempt to breed *B. subtilis* guanosine production strains involving metabolic pathway engineering was reported by scientists from Takeda of Japan. A 6.4 kb *Pst*I fragment comprising the *guaB* locus (in the publication designated as *guaA*) from the mainly guanosine-overproducing strain *B. subtilis* NA7821 was isolated by complementation cloning in a xanthine-dependent E. coli mutant. The 6.4 kb fragment was then introduced into the medium copy number vector pC194, and the resulting plasmid pBX121 was transformed into the mainly inosine-overproducing strain *B. subtilis* NA6128 (19 $g\,l^{-1}$ inosine from 100 $g\,l^{-1}$ glucose). Compared to the parent, a pBX121-containing transformant showed a 10-fold increased IMP dehydrogenase activity in a cell free lysate (1.5 and 15.8 $nmol\,mg^{-1}\,min^{-1}$ in parent and transformant, respectively) and produced mainly guanosine (25 $g\,l^{-1}$ from 100 $g\,l^{-1}$ glucose) [297].

While the strains discussed so far were derived from CSI campaigns, recent literature reports on the *ab initio* rational design of inosine production strains. In an attempt by Chinese researchers [372] to construct inosine production strains, *deoD* and *purA* encoding a nucleoside phosphorylase and the adenylosuccinate synthetase, respectively, were deleted from the wild-type *B. subtilis* chromosome. The double mutant accumulated 7.6 $g\,l^{-1}$ inosine with a 4.7% (w/w) yield on glucose in shake flask cultivation, which, according to the glucose consumption and growth kinetic, should be adenine limited despite the highly complex composition of the cultivation medium.

A more comprehensive approach to construct *de novo* inosine production strains was reported by Ajinomoto researchers [292]. From the genome of *B. subtilis* wild-type Marburg 168, the genes encoding the nucleoside phosphorylases DeoD and PunA, the adenylosuccinate synthetase encoding *purA* gene, and the gene encoding the PurR repressor were deleted. Furthermore, the strain was provided with a shortened *pur* leader sequence upstream of *purE* lacking the riboswitch triggered transcription terminator sequence. Finally, wild-type *guaB* was replaced by a mutant allele encoding an enzyme with attenuated IMP dehydrogenase activity. Cultivated under adenine-limited conditions, the strain designated as KMBS375 delivered 6 $g\,l^{-1}$ inosine and traces of guanosine from 30 $g\,l^{-1}$ initial glucose that was completely consumed (20% w/w yield) in tube fermentation. In a follow-up article [298], KMBS375 was reported to have reached a yield (w/w) of 17%. Upon replacement of the wild-type PRPP synthase encoding prs gene by mutant alleles encoding Prs(N120S) and Prs(L135I), which were shown to be resistant to ADP inhibition, the inosine yield rose to 22.2% and 23.1%, respectively.

The remarkable metabolic flux redistribution toward inosine in the classic *Bacillus* production strains as well as those derived by metabolic engineering resulting in product yields above 20% was achieved by the endogenous biosynthetic capabilities released from (physiologically meaningful) repressive mechanisms of *B. amyloliquefaciens* and *B. subtilis*. This is conceivable considering that purine nucleotides are also, in terms of quantity, important biomass constituents which have to be rapidly produced to sustain the purine building block supply of fast-growing bacterial cultures. The situation is different when it comes to the construction of production strains for compounds that play quantitatively a minor role in the bacterial metabolism, for example, vitamins. Mere deregulation of the metabolic potential of wild-type strains and interfering with degradative pathways are inadequate to breed commercially viable production strains for these compounds.

7.9.2 Riboflavin

In the late 1920s, it became evident that the so-called vitamin B complex, which was discovered 10 years earlier in aqueous extracts of brewer's yeast as a growth-promoting additive in the diet of young rats [299], consisted of several micronutritional components [300]. One component was designated as vitamin B1 or the antineuritic factor, the other as vitamin B2 or the rat antipellagra factor. While vitamin B1 turned out to be a unique chemical entity, that is thiamine, originally identified in and isolated from rice bran, vitamin B2 was recognized to consist of several different components including a yellow, intensively fluorescing molecule designated riboflavin. Riboflavin was the first vitamin isolated from the vitamin B2 complex in 1933 by Kuhn, György, and Wagner. Thereafter pantothenic acid (vitamin B5) and vitamin B6 (pyridoxine, pyridoxal) were identified as the other main vitamins in the vitamin B2 complex. Today the term vitamin B2 refers to riboflavin only. Riboflavin had already been isolated 50 years earlier by Blyth from whey without recognizing its nutritional function. Kuhn, Weygand, and Karrer determined the structure of the yellow pigment and proved it by chemical synthesis in 1933 and 1934. Riboflavin is biosynthesized in plants and in many microorganisms. Vegetables and milk are major sources of the vitamin in human nutrition.

By phosphorylation, riboflavin is converted to flavin mononucleotide (FMN), from which flavin adenine dinucleotide (FAD) is derived by adenylation. FMN and FAD are the cofactors of the broad class of flavoenzymes coupling the two-electron oxidation of many organic substrates to the one-electron transfers of the respiratory chain.

At present, over 4000 ton of riboflavin are industrially produced each year, nowadays exclusively by fermentation. About 70% of this material is used as feed additive in the form of free-flowing, spray-dried granules, or microgranules. The remaining 30% are required for the fortification of foods like breakfast cereals, pastas, sauces, processed cheese, fruit drinks, vitamin-enriched milk products, baby formulas, and clinical infusions.

Owing to fierce price competition, the riboflavin market is consolidated to a high degree with only a few producers remaining. The market leader is DSM Nutritional Products from Switzerland producing the vitamin in a plant in Southern Germany with a production strain derived from *B. subtilis* Marburg 168. BASF from Germany moved their riboflavin production facilities to South Korea. BASF's process is based on the yeast *Ashbya gossypii*. Hubei Guangji from Hubei Province, the main Chinese riboflavin producer, uses a *B. subtilis* strain whose key traits for high riboflavin productivity were developed by the Russian Institute for Genetics and Selection of Industrial Microorganisms, Moscow.

Riboflavin biosynthesis [301] in prokaryotes starts from guanosine triphosphate and ribulose-5-phosphate in a 1–2 M ratio, respectively (Figure 7.5). The hydrolytic opening of the imidazole ring of GTP (RibA: cyclohydrolase II reaction) is followed by (i) deamination of the resulting pyrimidinone to afford

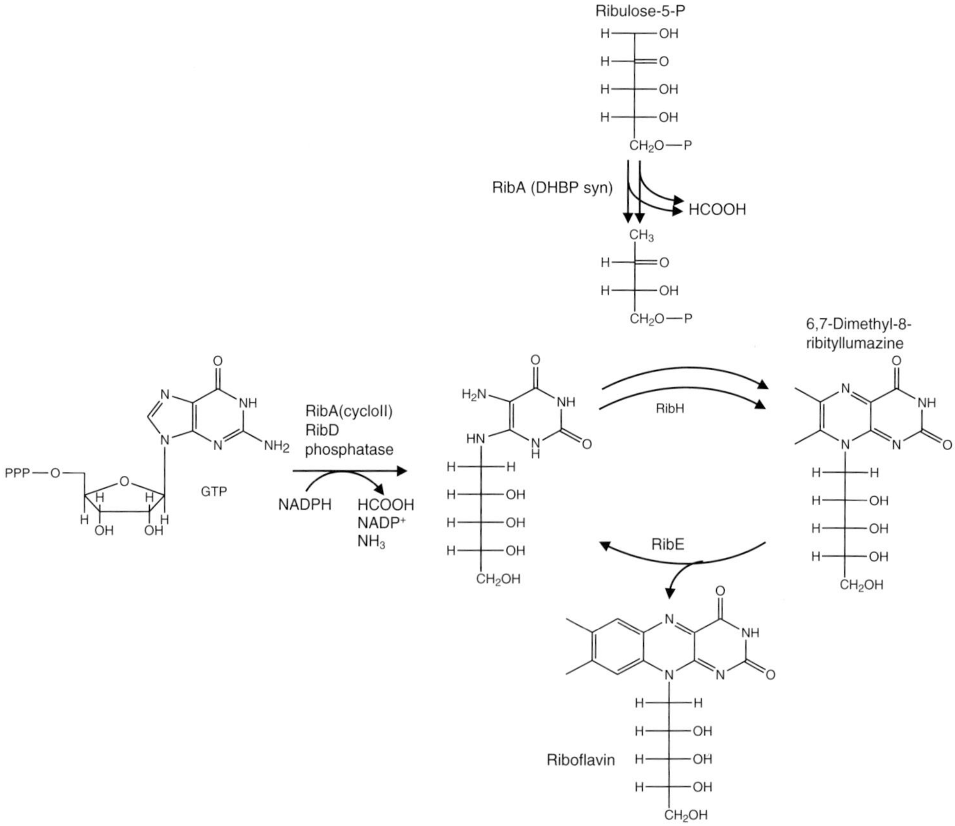

Figure 7.5 Riboflavin biosynthesis in *B. subtilis*.

a pyrimidinedione (RibD: deaminase reaction), (ii) reduction of the ribosyl side chain (RibD: reductase reaction), and (iii) dephosphorylation of the resulting ribityl side chain (phosphatase reaction). 6,7-Dimethyl-8-ribityl lumazine (DMRL), the direct biosynthetic precursor of riboflavin, is synthesized by adding the 4-carbon compound 3,4-dihydroxy-2-butanone-4-phosphate (DHBP) derived from ribulose-5-phosphate to the pyrimidinedione intermediate (RibA: DHBP synthase reaction and RibH: DMRL synthase reaction, respectively). Finally, two molecules of DMRL are converted in a dismutation reaction to riboflavin and the pyrimidinedione intermediate of the pathway (RibE: riboflavin synthase reaction). *B. subtilis* DMRL synthase forms a remarkable structure consisting of 60 monomers arranged into 12 pentamers, which assemble into a capsid of icosahedral 532 symmetry (similar to capsids of viruses). The inner core of the capsid contains a riboflavin synthase homotrimer.

The genes encoding the riboflavin biosynthetic enzymes are clustered together with a fifth gene of unknown function (*ribT*) as an operon at map position 207.50° of the *B. subtilis* chromosome [302, 303]. Dephosphorylation of the pyrimidinedione phosphate, which is a prerequisite for the pyrimidinedione to become a substrate for DMRL synthase, is presumably achieved by the action of one or several phosphatases with broad substrate specificity [304].

Riboflavin gene expression in *B. subtilis* is subjected to tight control involving a cis-acting element [305], that later turned out to be an FMN sensing riboswitch [306]. Deregulated, riboflavin-secreting mutants can be isolated with a frequency of 10^{-6} upon cultivating a *B. subtilis* population (not subjected to forced mutagenesis) on a solid medium in the presence of the riboflavin analog roseoflavin. Such mutants either acquired a mutation in the cis-acting control element [305] or encode a mutant riboflavin kinase [307, 308] with drastically reduced FMN forming activity leading to low FMN levels in the cell [309]. Since FMN, but not riboflavin acts as the effector molecule triggering the riboswitch-based riboflavin repression mechanism in *B. subtilis*, riboflavin kinase mutants overproduce and secrete riboflavin.

At Roche Vitamins, now DSM Nutritional Products, the development of a microbial riboflavin production strain started from *B. subtilis* Marburg 168 [310]. The roseoflavin-resistant mutant RB50, a progenitor of the current production strains, encodes an attenuated riboflavin kinase in addition to mutations affecting the regulation of purine biosynthesis. Genetic engineering was applied to further enhance *rib* gene expression by making use of strong, constitutive promoters and by increasing the *rib* gene dosage [310]. Several rounds of forced mutagenesis and selection for deep yellow colonies completed the development of the first-generation DSM Nutritional Products production strain.

Further improvement in production performance was achieved with a strain containing a *tkt* mutant allele (instead of the wild type) with drastically reduced enzymatic transketolase activity [180]. Transketolase is a key enzyme in the pentose phosphate pathway delivering ribulose-5-P and ribose-5-P, which are important building blocks for riboflavin biosynthesis. Presumably, the mutation impedes drainage of the pentose phosphate pathway intermediates back toward glycolysis resulting in higher levels of ribulose-5-P and ribose-5-P for riboflavin biosynthesis. The residual activity of the mutant enzyme allows some minor carbon flux through the pentose phosphate pathway facilitating the biosynthesis of erythrose-4-P from which the aromatic amino acids are derived. Hence, the improved riboflavin production mutant is prototrophic with regard to these amino acids.

The intracellular C5 carbon sugar pools of *Bacillus* strains comprising transketolase knockout mutations, which are auxotroph for aromatic amino acids, can rise to concentrations by far exceeding the physiological requirements of the bacteria. After dephosphorylation, excess ribose is excreted into the fermentation broth. Highly efficient microbial processes based on *Bacillus* transketolase knockout mutants were used in the past to obtain D-ribose for chemical riboflavin production at industrial level. For a review on Bacillus ribose production strains, see [311].

After a detailed study of the *rib* leader sequence comprising regulatory and mRNA-stabilizing elements, a new generation of highly effective riboflavin production strains became available with a single copy of a carefully deregulated *rib* operon driven by an SPO1 promoter element at the 5′ end of the operon [171]. These strains are free of antibiotic markers and fulfill the legal requirements for self-cloning strains (EC Council directive 98/81), which is of some significance to market fermentative riboflavin in Europe.

A 48 h fed-batch fermentation protocol was developed with glucose as the growth-limiting substrate delivering significantly above $15\,g\,l^{-1}$ riboflavin [310]. During the fermentation, riboflavin is excreted from the production strain and crystallizes in the fermentation broth. It is still unclear by which mechanism the rather efficient translocation of the vitamin through the plasma membrane is accomplished. The long needle-shaped crystals can be easily recovered and separated from the biomass by centrifugation. An acid treatment of the recovered riboflavin crystals at elevated temperatures followed by intensive washing resulted in a 96% pure product. Food/pharma-grade riboflavin of over 99% purity is obtained after recrystallization of the 96% product [312]. Since the year 2000, DSM Nutritional Products has been producing riboflavin exclusively by the *B. subtilis*-based microbial process in Southern Germany.

Riboflavin overexpressing and excreting *B. subtilis* strains were also developed at the Russian Institute for Genetics and Selection of Industrial Microorganisms, Moscow. In fact, *B. subtilis* VNIIGenetika 304 containing the plasmid pMX45 was the first riboflavin production strain, maybe even the first production strain for a small organic molecule at all, that was obtained by a genetic engineering program. The Russian patent application SU3599355 claiming the strain and a fermentation process for riboflavin was filed with priority date June 2, 1983. The host strain was deregulated with regard to the purine and the riboflavin biosynthetic pathway. The plasmid pMX45 contained a 10 kb *EcoRI* fragment comprising the entire *rib* operon of *B. subtilis* driven from its natural promoter [313]. With VNIIGenetika 304/pMX45, $4.5\,g\,l^{-1}$ riboflavin was produced from a total of $100\,g\,l^{-1}$ saccharose supplied during a 25 h fermentation run [150].

The co-occurrence of chromosomal and episomal copies of the *rib* operon in *B. subtilis* production strains provided with pMX45 gave rise to plasmid instability. Genetically stable riboflavin production strains were obtained using an integrative vector construct which targeted the entire *rib* operon of *B. amyloliquefaciens* in close proximity to the *pur* operon of a *B. subtilis* production strain [151]. The resulting strain Y32 produced $3\,g\,l^{-1}$ riboflavin during a 72 h shake flask fermentation, but should have the potential to produce considerably more riboflavin under industrial fed-batch fermentation conditions. It is presumed that riboflavin-producing companies from China employ *B. subtilis* production strains with a genetic makeup similar to that of Y32.

A research group from the Tianjin University, Tianjin, China, is actively publishing on genetic traits positively affecting the metabolic flux toward riboflavin in *B. subtilis*. A DNA fragment encoding the *B. subtilis* riboflavin operon lacking its natural upstream promoter and regulatory sequences was linked to either a nondisclosed promoter element or the *cdd* promoter (p43) and integrated into the *rib* locus of a roseoflavin and purine analog-resistant *B. subtilis* host strain resulting in strains RH33 [314] and PK [315], respectively. With RH33, small-scale aerobic fed-batch fermentation runs with molasses sucrose as the main carbon source afforded up to $12\,g\,l^{-1}$ riboflavin at a yield of 5.8% (w/w) on sucrose [314].

Engineering of the metabolic flux in *B. subtilis* RH33 or PK through the purine pathway [315] or through the oxidative branch of the pentose phosphate pathway [316, 317], or of the electron flux through the respiratory chain [246], or deregulation of gluconeogenesis [318] and other measures were all reported to improve riboflavin productivity. Additive or synergistic beneficial effects on riboflavin production by combining the individual traits in a single production strain were not reported. It is unclear to what extent these traits have found their application in industrially used production strains in China.

Lastly, riboflavin is the paradigm vitamin whose industrial production process switched completely from chemistry to biotechnology. The latter is superior with regard to economic efficiency, but in addition, the advantages for the environment have been demonstrated in several ecological footprint studies. For more details on microbial riboflavin production, the reader is referred to a review by Hohmann and Stahmann [319].

7.9.3 (R)-Pantothenic Acid

(R)-pantothenic acid and its pantothenate salts (vitamin B5) belong to the group of water-soluble B-vitamins originally identified as one of the vitamins in the vitamin B2 complex (see Section 7.9.2). Its name originates from the Greek word "pantos," meaning "everywhere," as it can be found throughout all living cells. The vitamin is present in virtually all plant and animal tissues and in microbial cells and, thus, it is widely distributed in foods. (*R*)-Pantothenic acid is an essential precursor for the biosynthesis of coenzyme A (CoA) and the acyl carrier protein (ACP). CoA and

ACP play critical roles in many cellular processes and essential reactions, such as energy metabolism and fatty acid synthesis.

Currently, (*R*)-pantothenic acid is industrially produced by conventional chemical synthesis and marketed as the calcium salt (calpan). The global market of calpan was valued at approximately 75 million US dollars in sales in 2005 with a world output over 10 000 ton. The key starting material for large-scale chemical processes is (*R*)-pantolactone. This molecule can be coupled with either calcium β-alanine to form (*R*)-pantothenate or with 3-aminopropanol to form panthenol, an important ingredient in skin and hair care consumer products. Purification of (*R*)-pantolactone from racemic mixtures occurs by chemical or enzymatic racemic resolution [320, 321].

Biosynthesis of (*R*)-pantothenate (Figure 7.6) occurs in microorganisms, plants, and fungi. In *B. subtilis,* as in other organisms, the (*R*)-pantothenic acid biosynthetic pathway consists of two separate branches providing (*R*)-pantoic acid and β-alanine (β-Ala), which are condensed by PanC to (*R*)-pantothenic acid. (*R*)-pantoic acid is derived from 2-ketoisovaleric acid (*α*-KIV) by the PanB and PanE reactions [322–324]. During the latter, ketopantoate is reduced to (*R*)-pantoate in a NADPH-dependent manner. PanB transfers the methylene group of the 5,10-methylenetetrahydrofolate cofactor (MTHF) to α-KIV to form ketopantoate. For MTHF cofactor regeneration, the methylene group is derived from GlyA-catalyzed conversion of serine to glycine. In addition, the glycine cleavage cycle from the gene products of *gcvPA*, *gcvPB*, and *gcvT* degrades glycine to CO_2, NH_3, and water and adds a methylene moiety to tetrahydrofolate (Figure 7.6) [325, 326].

α-KIV is formed from two molecules of pyruvate by the enzymes IlvBH, IlvC, and IlvD (Figure 7.6). Since α-KIV is an intermediate of the valine and leucine biosynthetic pathways as well, genetic manipulations to improve α-KIV supply for increased (*R*)-pantothenate production could affect the synthesis of these branch-chain amino acids too.

The second key intermediate in the vitamin B5 pathway is β-Ala, which is formed from aspartic acid in an α-decarboxylation reaction catalyzed by PanD (Figure 7.6). PanD is a unique pyruvoyl-containing enzyme that is translated as a preprotein and undergoes processing for activity.

Several microbial fermentation processes were developed aiming to replace the existing chemical processes. The most efficient of them is the *B. subtilis* process described in a series of OmniGene Bioproducts/BASF patents [177, 322]. The wild-type strains of *B. subtilis* excrete very low amounts of (*R*)-pantothenate ($<1\ mg\ l^{-1}$) into the culture medium. After substantial genetic engineering, the best *B. subtilis* vitamin B5 strains produced approximately five orders higher (R)-pantothenate titers (82 and $86\ g\ l^{-1}$ in 48 h fed-batch fermentations) [177]. Overproduction of (*R*)-pantothenate was achieved by increasing the transcription levels of (i) the *ilv* genes (*ilvB*, *ilvH*, *ilvC*, and *ilvD*), which are required for the formation of the precursor α-KIV and (ii) the (*R*)-pantothenate biosynthetic genes (*panB*, *panC*, *panD*, and *panE*). Overexpression of these genes was achieved by (i) replacing the native promoters and ribosome binding sites (RBSs) with

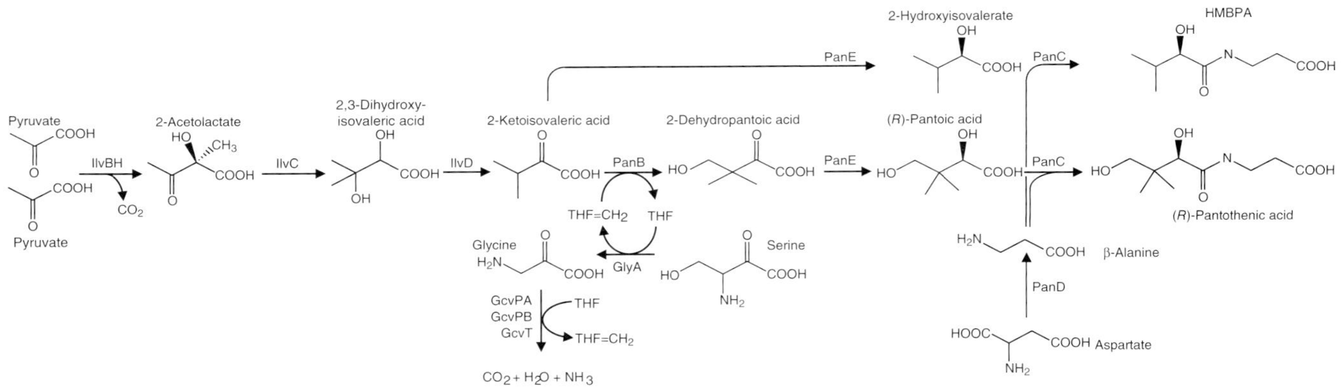

Figure 7.6 Biosynthesis of (*R*)-pantothenic acid and the side metabolite 3-(2-hydroxy-3-methyl-butyrylamino)-propionic acid (HMBPA) in *B. subtilis*. Regeneration of methylene tetrahydrofolate (THF=CH_2) in the GlyA-catalyzed reaction and in the glycine cleavage cycle (GcvPA, GcvPB, and GcvT) is shown as well.

stronger constitutive promoters and RBSs upstream of the native genes and operons and (ii) increasing the copy numbers of some of the genes using gene amplification.

A side product, 3-(2-hydroxy-3-methyl-butyrylamino)-propionic acid (HMBPA), can accumulate to high concentrations in the cultivation broth of (*R*)-pantothenate-overproducing strains [177]. HMPBA formation is caused by premature α-KIV reduction to 2-hydroxyisovalerate (α-HIV) most probably by PanE before the molecule is hydroxymethylated (Figure 7.6). PanC-catalyzed condensation of α-HIV and β-alanine delivers HMBPA.

Significant HMBPA accumulation indicates a metabolic bottleneck at the PanB-catalyzed hydroxymethylation step. Engineering the supply chain for methylene tetrahydrofolate, the cosubstrate of the PanB reaction, by overexpression of SerA and GlyA or of the enzymes of the glycine cleavage cycle drastically reduced HMBPA accumulation in favor of further increased (*R*)-pantothenic acid production [177]. This result indicates that insufficient availability of the PanB cosubstrate is a major cause of unwanted side-product formation.

The *in vitro* activities of the PanB enzymes from *B. subtilis* Marburg 168 and *B. subtilis* W23, which differ at four amino acid positions, were determined at 1.15 and 0.38 $\mu mol\ mg^{-1}\ min^{-1}$, respectively [327]. The activities, particularly the former, are quite high indicating that the activity of PanB *per se* should be sufficiently high to facilitate a smooth flux toward 2-ketopantoate. However, PanB is feedback inhibited by (*R*)-pantoate ($K_i = 100\ \mu mol$). A library of *B. subtilis* Marburg 168 *panB* mutants was generated by error-prone PCR and provided to a low-level pantothenate overproducing *B. subtilis* recipient strain [179]. Upon screening for pantothenate productivity and excretion, several PanB mutant alleles were identified that stimulated superior productivity compared to the wild-type PanB. Although no further characterization was reported, the mutated enzymes might be less sensitive to inhibition, for example, by pantoate.

In accordance with environmental and governmental regulatory requirements, sporulation-deficient strains of *B. subtilis* are used in fermentative processes. Therefore, sporulation mutations (*spo0A*, *sigH*, *sigF*, *sigE*, and *sigG*) that affected sporulation at different stages were introduced into *B. subtilis* (*R*)-pantothenate-overproducing strains [134]. The results indicated that sporulation-deficient strains for commercial production can be constructed by incorporation of mutations in later-stage transcription factors such as SigE and SigG, which prevent sporulation yet maintain high (*R*)-pantothenate production. (*R*)-pantothenate production levels responded to the levels of Spo0A, the master regulator of sporulation: (i) in the absence of Spo0A, (*R*)-pantothenate production decreased drastically (sevenfold) and (ii) overproduction of Spo0A increased Pan production and spore formation. To avoid sporulation in the strains overexpressing Spo0A, the sporulation was disrupted at a later stage by introducing a *sigE*-null mutation. An alternative way to enhance pantothenate production and retain sporulation deficiency is the introduction of both *spo0A* and *abrB*-null mutations into the vitamin B5-producing strains [134].

In conclusion, the engineering of the (*R*)-pantothenate biosynthetic genes provided strains yielding rather high concentrations of vitamin B5 in fed-batch fermentation runs. With the performance levels reported, a superior microbial process for pantothenic acid production is coming up on the horizon.

7.9.4
Production Strains for Metabolites with Unusual Biochemistry

Conventional metabolic pathway engineering concepts and tools were appropriate to develop *B. subtilis* Marburg 168 into efficient production strains for the vitamins riboflavin and pantothenic acid. For both compounds, the biochemical reactions from common metabolites to the final products were known to a great detail. Pantothenic acid biosynthesis involves efficient enzymes with high turnover numbers that catalyze standard biochemical reactions. The enzymes of the riboflavin pathway catalyzing more complex reactions are relatively slow, but strains overexpressing these enzymes could be furnished with sufficient biocatalytic activity.

The biosynthesis pathways of the vitamins thiamine (B1), pyridoxine (B6), and biotin (B7) have been elucidated during the last 10 years to some detail. It became clear that in all cases enzymes catalyzing unusual or complex biochemical reaction mechanisms are involved, which perform at least *in vitro* with very low catalytic efficiency. Previous attempts to breed *B. subtilis* production strains for these vitamins as a base for the development of superior processes to supersede the decades old chemical processes are reviewed here briefly. These efforts followed the blueprints that were successful for other metabolites like amino acids, nucleotides, the vitamins mentioned above, and others. However, the metabolic fluxes toward the vitamins B1, B6, and B7 have proven to be particularly adamant to engineering. The strains that were obtained overproduced these value compounds only at marginal levels.

7.9.4.1
Biotin (Vitamin B7)

Industrial biotin production is carried out by a lengthy, multistep chemical process including an optical resolution step, which was devised in the late 1940s and further improved over time [328]. Thus, biotin seemed a promising compound to be produced by a microbial process.

Biosynthesis of biotin in *B. subtilis* (Figure 7.7) starts from a long-chain fatty acid, probably myristic acid, which is cleaved by the cytochrome P450-depending BioI via an in-chain cleavage reaction to afford the seven-carbon dicarbonic pimelic acid. Pimeloyl-CoA is derived by the BioW-catalyzed reaction. Conversion of pimeloyl-CoA to dethiobiotin via BioF, BioA, and BioD is quite effective. Abundant substrates and conventional biosynthetic reactions are involved in these steps. The final step, however, the BioB-catalyzed sulfur recruitment and thiophane ring formation becomes a significant bottleneck upon attempts to enhance the metabolic flux through the biotin pathway [329]. The intriguing

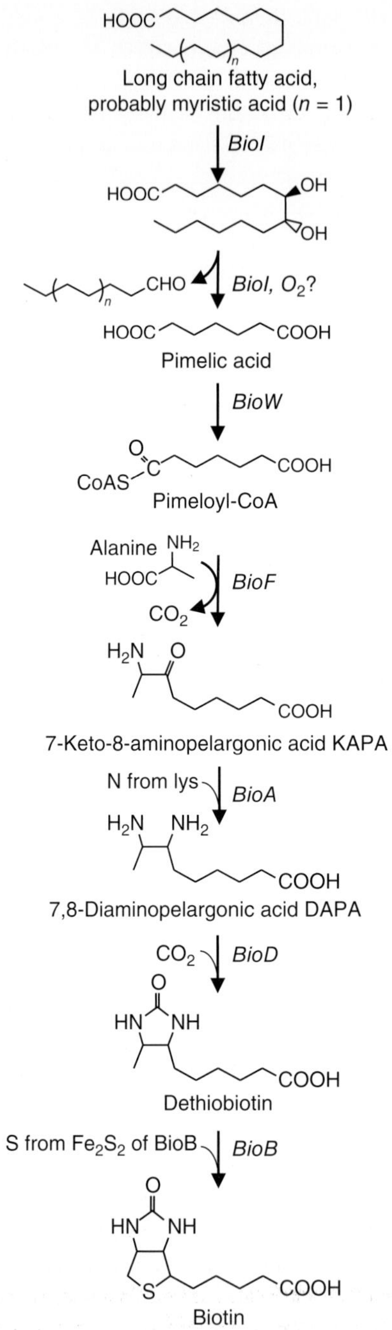

Figure 7.7 Biotin biosynthesis in *B. subtilis*.

BioB reaction has been elucidated in some detail recently (for a review, see [330]). It became clear that a 5′-desoxyadenosyl radical mechanism activates the nonfunctionalized C6 and C9 positions of dethiobiotin involving the Fe4S4 iron sulfur cluster of BioB. Sulfur is recruited from a Fe2S2 iron sulfur cluster that is present in BioB as well. It was speculated that BioB is used as a reactant rather than a catalyst. *In vitro* experiments, however, demonstrated that biotin synthase is catalytic, but that catalysis puts the protein at risk of proteolytic destruction [331]. Furthermore, 5′-deoxyadenosine, a coproduct of the BioB reaction, is a potent biotin synthase inhibitor *in vitro* and *in vivo*.

B. subtilis strains containing multiple copies of the biotin biosynthetic genes under the control of a strong phage promoter were constructed at OmniGene Inc. and Roche Vitamins Ltd. It was recognized that in *B. subtilis* the amino donor required for the conversion of 7-keto-8-aminopelargonic acid to 7,8-diaminopelargonic acid is not methionine as in *E. coli*, but lysine. Upon lysine cofeeding, the engineered *B. subtilis* strains converted pimelic acid quite effectively to dethiobiotin, but further conversion to biotin was rather poor [332].

Realizing that the conventional biotin biosynthesis route via BioB was resistant to pathway engineering, metagenomic libraries from various habitats were screened for alternative routes using an *E. coli* screening host that was devoid of all *bio* genes. Furthermore, the screening host was deleted for *pfs* encoding a 5′-deoxyadenosine nucleosidase to remove hits that encoded biotin pathways relying on 5′-deoxyadenosine-inhibited BioB. Indeed, 5′-deoxyadenosine-resistant BioB homologs were isolated, but their biocatalytic efficiency was not improved over the known BioB enzymes. Biotin pathways employing an alternative biochemistry and circumventing the intriguing sulfur recruitment were not identified (BRAIN AG, unpublished).

The highest biotin titer reached in a microbial process so far is $500\,mg\,l^{-1}$ in addition to $100\,mg\,l^{-1}$ dethiobiotin in a 10-day fed-batch fermentation. This process, based on a genetically engineered *Serratia marcescens* mutant, was developed in the mid-1990s at Tanabe Seiyaku, Japan [333].

7.9.4.2
Thiamine (Vitamin B1)

Several chemical routes are used for commercial production of thiamine chloride and thiamine nitrate, the thiamine sales forms introduced in the market. Despite the high functionalization of the thiamine molecule, the current industrial processes starting from chemical commodities deliver thiamine at reasonable yield and are, thus, quite cost effective [334]. Only few attempts to develop microbial production processes have been reported in the scientific or patent literature [334].

Following a convergent biosynthetic pathway (Figure 7.8) in *B. subtilis* [335, 336], as in other microorganisms, the assembly of two building blocks, a pyrimidine moiety (4-amino-5-hydroxymethyl-2-methylpyrimidine pyrophosphate, HMP-PP), and a thiazole moiety (5-(2-hydroxyethyl)-4-methylthiazole phosphate, HET-P) affords thiamine monophosphate (TMP). The condensation

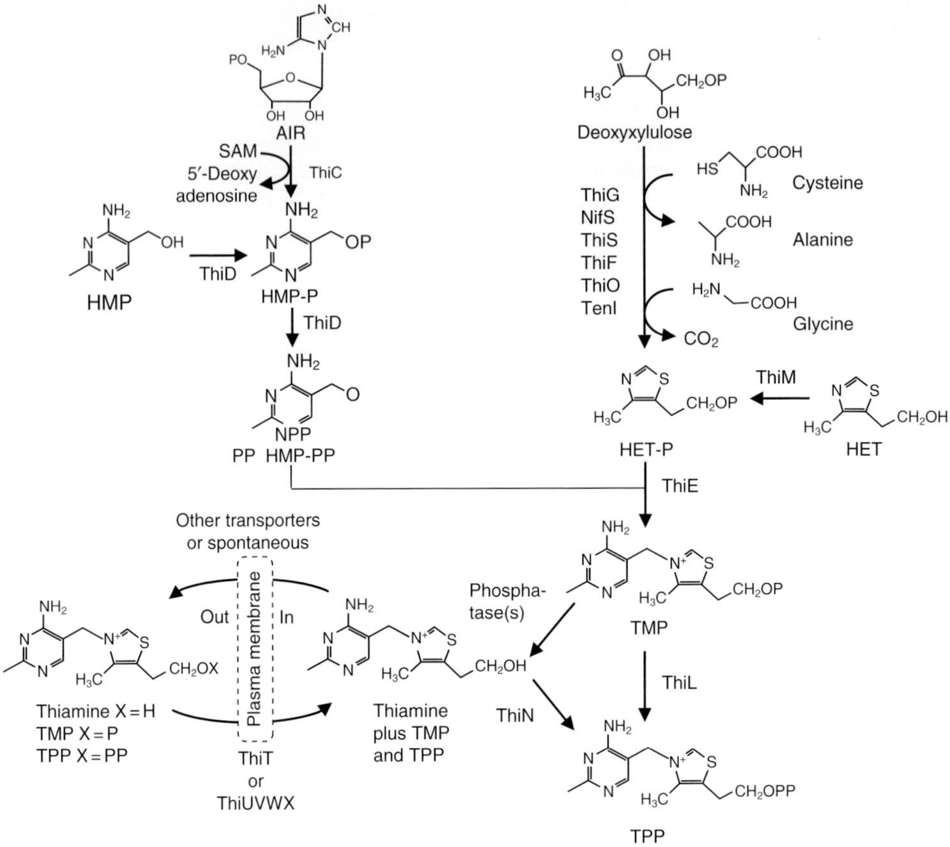

Figure 7.8 Thiamine pyrophosphate (TPP) biosynthesis in *B. subtilis*. In addition to the *de novo* route, salvage reactions to recruit nonphosphorylated 4-amino-5-hydroxymethyl-2-methylpyrimidine (HMP), 5-(2-hydroxyethyl)-4-methylthiazole (HET), and thiamine are shown.

reaction is catalyzed by ThiE. HMP-PP is derived from HMP monophosphate (HMP-P). The phosphorylation reaction is catalyzed by ThiD [337]. HMP-P is the product of the remarkable ThiC-catalyzed conversion of the purine pathway intermediate 5-aminoimidazole ribotide (AIR). HET-P is derived from 1-deoxy-D-xylulose 5-phosphate, cysteine, and glycine in a complex oxidative cyclization reaction requiring the products of at least six different genes: thiG, nifS, thiS, thiF, thiO, and tenI. HET-P and HMP-P can also be derived from the corresponding nonphosphorylated compounds through kinase reactions catalyzed by ThiM [338] and ThiD [337], respectively. Additional phosphorylation of TMP by ThiL [335] yields thiamine pyrophosphate (TPP), the biologically active form of the cofactor. Expression of the *thi* genes is regulated by a TPP-specific riboswitch mechanism [339].

B. subtilis mutants in *thiL* are dependent on ThiN, a thiamine pyrophosphokinase [178, 340], indicating the presence of phosphatase(s) converting TMP to the

ThiN substrate thiamine. *B. subtilis* mutants defective for TMP biosynthesis can be rescued by exogenous supply of thiamine, TMP, or TPP. The uptake of the thiamine compounds is dependent on the thiamine permease ThiT or the thiamine ABC transporter ThiUVWX [178]. Mutants defective for TMP kinase ThiL and both of the importers show elevated intracellular and extracellular levels of thiamine compounds possibly because of excretion of the metabolites, which cannot be recaptured as functional importers are missing in the mutant. If, in addition, an attenuated thiamine pyrophosphokinase ThiN is present, further increased thiamine compound levels, mainly thiamine, slightly above 1 mg l^{-1} are observed. Presumably, because of the reduced thiamine pyrophosphokinase activity and the absence of TMP kinase activity, the intracellular TPP levels are low, resulting in deregulated expression of the TPP riboswitch-controlled *thi* genes and increased metabolic flux through the thiamine pathway until TMP, which is excreted as such or after dephosphorylation.

A *B. subtilis* strain deregulated for *thi* gene expression and cultivated to cell densities around 40 g l^{-1} DCM in glucose-limited fed-batch fermentation runs with excess supply of the precursor compounds HET and HMP accumulated up to 120 mg l^{-1} thiamine TMP. Upon considerable overexpression of ThiE and ThiM, the genes of which form an operon, in the deregulated *B. subtilis* strain background up to 300 mg l^{-1} thiamine TMP was measured in the cofeeding cultivation regime. A *thi* gene-deregulated strain considerably overexpressing ThiC afforded maximal 40 mg l^{-1} in the glucose-limited cultivation regime that provided only HET as a cofeed (J. Perkins and G. Schyns, unpublished). Obviously, the reported productivities of the thiamine-secreting strains are far below any commercial value.

In search of alternative thiamine pathways that might be more susceptible to metabolic pathway engineering, metagenomic libraries from various habitats were screened using a thiamine auxotrophic, *thiC*-deleted *E. coli* screening host. In addition, the host contained a deletion in an early *pur* gene to prevent the supply of AIR, the ThiC substrate, to remove those clones which are complemented by the conventional ThiC dependent pathway. None of the hits encoded a genuinely novel thiamine pathway (BRAIN AG, unpublished).

7.9.4.3 Pyridoxine (Vitamin B6)

The major sales form of vitamin B6 is the hydrochlorid salt of the primary alcohol pyridoxine. Another vitamin B6 form introduced in the market is the dihydrochlorid salt of pyridoxamine. Both vitamin B6 forms are commercially produced via various straightforward chemical synthesis routes. The biologically active cofactor is the aldehyde pyridoxal-5′-phosphate, which is derived in human or animals from the vitamin B6 forms by oxidation or transamination before or after 5′ phosphorylation by pyridoxal kinases.

Two different pathways (Figure 7.9) toward pyridoxal 5′-phosphate, the DXP-dependent pathway and the R5P (DXP-independent) pathway, are known. For a review, see [341]. The R5P pathway is more prevalent and is found, for example, in

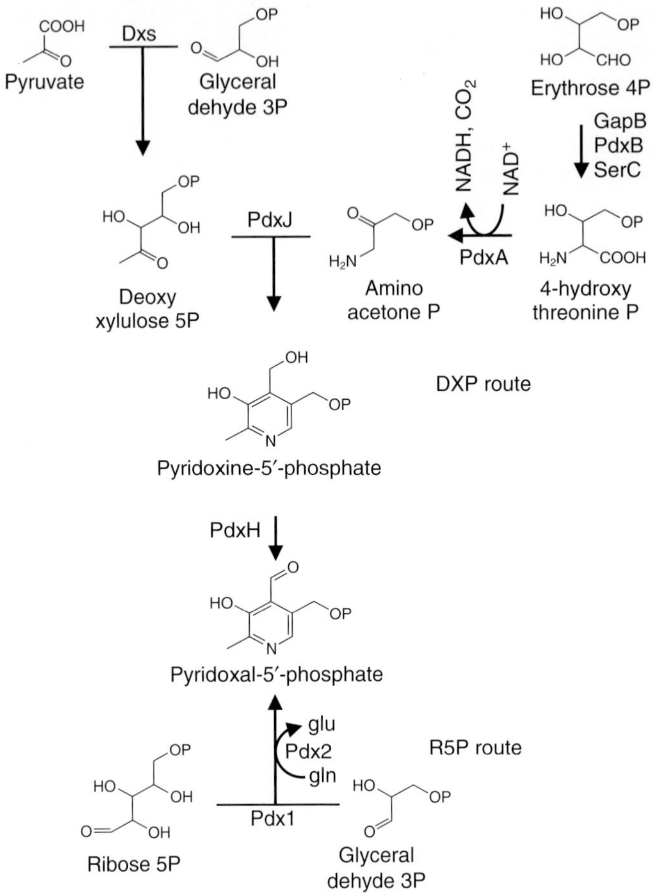

Figure 7.9 Pyridoxal-5′-phosphate biosynthesis. In the upper part of the figure the DXP route is depicted as realized, for example, in *E. coli*. The lower part shows the R5P (DXP-independent) route of, for example, in *B. subtilis*.

archaebacteria, fungi, plants, plasmodia, some metazoa, and a subset of eubacteria including *B. subtilis*. In this pathway, Pdx1 and Pdx2 assembled into an ornate multisubunit complex are the only enzymes required to convert three substrates from the central metabolism, that is, ribose 5-phosphate (R5P), glyceraldehyde 3-phosphate (GAP), and glutamine into pyridoxal 5′-phosphate (PLP). Ammonia derives from glutamine by the glutaminase activity of Pdx2. Pdx1 catalyzes the cyclization reaction of R5P and GAP via a series of tautomerization, transimination, and dehydration steps to directly afford PLP [342, 343]. It was suggested [344] that PLP does not spontaneously dissociate from Pdx1, but that an apoenzyme in quest of PLP receives its cofactor directly from Pdx1. This mechanism ensures that the reactive aldehyde product is not released into the cellular milieu, where it could cause damage to exposed lysine residues of the cytosolic proteins. If indeed PLP synthase is primed for donating its reaction product only when required for PLP holoenzyme formation, this would generate a strong metabolic flux bottleneck toward extracellular pyridoxal accumulation that cannot be overcome by simple Pdx1 overexpression.

In accordance with the idea of a strongly repressed enzyme, the *in vitro* activity of *B. subtilis* Pdx1 is extremely low ($k_{cat} = 0.018\ min^{-1}$). A rough estimate based on this turnover number suggests that a *B. subtilis* strain considerably overexpressing Pdx1 should not produce above $100\ mg\ l^{-1}$ pyridoxal upon cultivation for 40 h at 10 g DCM per liter. In fact, attempts to produce vitamin B6 compounds in *B. subtilis* overexpressing Pdx1 and Pdx2 [345] resulted in a maximal product titer of $61\ mg\ l^{-1}$ total B6 vitamers, that is, the sum of pyridoxamine, pyridoxamine phosphate, and pyridoxine, in 48-h shake-flask culture. The aldehyde pyridoxal or its phosphate, although being the direct products of the DXP pathway, were hardly overproduced.

As in the R5P pathway, in the alternative deoxyxylulose 5-phosphate (DXP) pathway toward PLP as present in *E. coli* and other gram-negative bacteria, a 5-carbon sugar, DXP, and a three-carbon moiety, amino hydroxyacetone phosphate, are linked to afford the pyridine ring of the vitamin B6 compounds. DXP resulting from the condensation of glyceraldehyde phosphate and pyruvate catalyzed by DXP synthase can be considered as a compound of the central metabolism, as it is also recruited for isoprene and thiamine biosynthesis. Amino-hydroxyacetone phosphate is a dedicated metabolite synthesized from erythrose 4-phosphate involving GapB (Epd), PdxB (or PdxR), SerC, and PdxA [346–348]. The cyclization reaction catalyzed by PdxJ consists similar to the Pdx1 reaction of a series of tautomerization, transimination, and dehydration steps, but affording instead of the aldehyde PLP, the primary alcohol pyridoxine 5-phosphate (PNP) [349]. PLP is derived from PNP oxidation catalyzed by PdxH [350].

The catalytic activity of PdxJ ($k_{cat} = 4\ min^{-1}$) [351] is much higher than that of Pdx1 (see above) and similar to the activity of cyclohydrolase II ($k_{cat} = 2.1\ min^{-1}$) [352], the entrance enzyme into the riboflavin pathway. Thus, PdxJ overexpression should provide the cyclization activity that a competitive pyridoxine production strain would require. The catalytic efficiency of PdxJ and the fact that the enzyme provides PNP, which after dephosphorylation is the molecule present in the most important vitamin B6 sales forms, encouraged the development of a vitamin B6 production strain based on the DXP pathway. Reasoning that product accumulation could be kept below the radar screen of the host's regulation and repression mechanisms presumably sensing PLP, but not PNP, a *B. subtilis* pyridoxine production strain was constructed by overexpressing the genes of the non-native DXP pathway except pdxH. The genes were synthesized according to the *E. coli* and *Sinorhizobium meliloti* sequences [124]. Although significant expression of the heterologous genes was achieved, only $14\ mg\ l^{-1}$ pyridoxine in a small-scale production assay was obtained. Upon optimizing the growth conditions and cofeeding of 4-hydroxy-threonine, the dephosphorylated form of the PdxA substrate, and deoxyxylulose, the productivity was increased to $54\ mg\ l^{-1}$ [124]. Optimization of the condensing reaction by expressing different PdxA (4-hydroxy-L-threonine-phosphate dehydrogenase) and PdxJ (PNP synthase) alleles in a *B. subtilis* strain carrying adaptive mutations increased further the pyridoxine (PN) production (78 mg l^{-1}); however, this is still far below any commercial relevance [353].

The best pyridoxine production process reported so far is based on the α-proteobacterium *S. meliloti* IFO14782, a natural pyridoxine overproducer. A derivative of IFO14782 obtained by classical mutagenesis and overexpressing PdxJ and Epd provided 1.3 g l^{-1} pyridoxine [348], which is still too low to beat the chemical processes.

7.9.5 Enzymes

The use of enzymes for a wide range of applications has become a major success story where *Bacillus* species have played pivotal roles both as the source of enzymes and as cell factories. Essentially, the field of industrial enzyme production was opened up in the early 1900s by Otto Röhm, who started to commercialize pancreatic enzymes (trypsin) from slaughtered animals as leather mordants under the product name OROPON®. Soon thereafter, enzymes started to be used in laundry detergents, wound and body-care products, silk production, and hide glue. From the 1930s onward, enzymes were also used for food processing, especially in the clarification of fruit (apple) juices and in baking products. An important milestone was the market introduction of Biotex by Kortman & Schulte (now Kortman Intradal) in 1963, which contained a *Bacillus* protease active in the alkaline pH range. The latter meant a substantial improvement over earlier detergent enzymes, which were mostly active at neutral pH while detergents are usually alkaline. Today, enzymes have an immense impact on our daily lives. They are standard components of detergents for the removal of stains, color care, bleaching, or the removal of surfactants and bleach. In the production of textiles, enzymes are used for desizing and finishing, thereby replacing chemical bleaches and peroxides. Enzymes are present in many products for personal care, including solutions for contact lens cleaning, skin care, and oral care. Many of our food products are prepared with the help of enzymes, including for example, bread, cheese, and whey products, as well as low-calorie beer. Enzymes are added to animal feeds for improved digestion and the reduction of waste. Lastly, in industrial processes, enzymes are crucial for starch processing. Since vast amounts of enzymes are needed on a daily basis to satisfy the ever-growing market and consumer needs, enzymes must be produced in bulk and at low costs. Altogether, the global enzyme market in 2013 was estimated at $4.8 billion and is forecast to grow to $7.1 billion in 2018 (BCC Research, Global Markets for Enzymes in Industrial Applications, Market Research Report June 2014). With selling prices of less than \$500 kg^{-1} active protein, depending on the enzyme, it can be assumed that on a global scale up to 8000 metric tons of enzymes are currently produced per year (Eugenio Ferrari, personal communication).

Many of today's industrial enzymes are produced by *Bacillus* species, especially *B. subtilis*, *B. amyloliquefaciens*, and *B. licheniformis*. These include amylases, amylomaltase, asparaginase, cellulases, glucanases, glycerophospholipid cholesterol acyltransferase, 1,4-α-glucan branching enzymes, glucose isomerases, maltotetraohydrolase, proteases, pectate lyases, pectinases, phytases, pullulanases, and xylanases, several of which have the GRAS status. The specific production processes for these enzymes are not reviewed here, as many excellent reviews covering the respective scientific and patent literature have been published on this topic [251, 354–357]. Important future markets for enzymes relate to the exploitation of biomass as there are strong societal demands for more sustainable industrial processes and renewable alternatives to fossil fuels without interfering with food supply. Cellulose and hemicellulose, especially, are the major polymers of plant biomass, which can be transformed by cellulases, xylanases, and accessory enzymes into biosourced value products (carbohydrates and peptides) and new-generation biofuels. A major barrier in this process is the difficulty to degrade cellulose in combination with related polysaccharides and lignin in a cost-efficient manner. It will therefore be important to develop enzymes or enzyme cocktails that effectively degrade lignocellulose. Such cocktails would combine cellulases, hemicellulases, xylanases, pectinases, and enzymes with other activities (e.g., oxidases) [355, 358, 359].

With the steadily growing demand for a wide range of different enzymes, there is an ever-increasing need for new-generation *Bacillus* cell factories that can deliver these enzymes in greater yields, with higher quality, and at lower costs. The engineering of *Bacillus* species into new-generation cell factories will most likely require combined systems and synthetic biology approaches [360]. In such an effort, the protein secretion machinery should be optimized in a product-specific manner and, at the same time, this engineering effort has to take into account the balanced use of cellular resources. In recent years, *B. subtilis* was shown to be a highly tractable model for systems biological analyses [15, 21, 22]. These studies were paralleled by detailed proteome analyses that allowed the identification and quantification of the majority of expressed *B. subtilis* proteins, providing deep insights into this organism's cell physiology [361–365]. Notably, proteomics was crucial for defining the extracellular proteome of *B. subtilis*, which includes all secreted proteins [61, 109, 364]. This allowed the identification of all signal peptides of *B. subtilis* [33, 61, 366] and has already facilitated synthetic biology approaches where the best signal peptides for secretion of particular proteins were selected following a so-called *bio-brick* approach [170]. What needs to be achieved in the coming years is a systems-level understanding of the *Bacillus* secretion machinery and its interactions with secretory precursor proteins from single molecules to complexes and cellular physiology. At the single-molecule level, it will be crucial to better understand the roles of individual secretion machinery components and their interactions with each other and with secretory precursor proteins in order to achieve optimal precursor targeting, translocation and post-translocational folding. Fine-tuning of protein secretion processes at the physiological level may even be more beneficial for increased protein production. To this end, the systems biology approaches recently explored in the analysis of cellular responses to changing nutritional conditions [21] can probably be applied to minimize any counterproductive stress responses triggered by high-level protein secretion. This can, for example, be achieved through physiological proteomics where particular proteomic stress and starvation signatures pinpoint the nature of the respective stresses that can then be addressed by genetic engineering [367, 368]. Lastly, to optimally tune the cells for high-level protein secretion, it will be necessary to take into account the management of cellular resources and growth rate. Although this may appear highly ambitious, it is an achievable goal as the theoretical and mathematical tools to predict optimal protein expression levels with minimal side effects on cell physiology have already been developed for *B. subtilis* [369–371].

Acknowledgments

ZP and JMvD were supported by the CEU BaSynthec project 244093. LK and JMvD were supported by the transnational SysMO initiative through projects BACELL SysMO1 and 2 with funding from the Research Council for Earth and Life Sciences of the Netherlands Organization for Scientific Research (NWO-ALW).

References

1. Earl, A.M., Losick, R., and Kolter, R. (2008) Ecology and genomics of *Bacillus subtilis*. *Trends Microbiol.*, **16** (6), 269–275.
2. Rao, C.V., Glekas, G.D., and Ordal, G.W. (2008) The three adaptation systems of *Bacillus subtilis* chemotaxis. *Trends Microbiol.*, **16** (10), 480–487.
3. Kearns, D.B. and Losick, R. (2003) Swarming motility in undomesticated *Bacillus subtilis*. *Mol. Microbiol.*, **49** (3), 581–590.
4. Kearns, D.B. *et al.* (2004) Genes governing swarming in *Bacillus subtilis* and evidence for a phase variation mechanism controlling surface motility. *Mol. Microbiol.*, **52** (2), 357–369.
5. Stein, T. (2005) *Bacillus subtilis* antibiotics: structures, syntheses and specific functions. *Mol. Microbiol.*, **56** (4), 845–857.
6. Dubois, J.Y. *et al.* (2009) Immunity to the bacteriocin sublancin 168 is determined by the SunI (YolF) protein of *Bacillus subtilis*. *Antimicrob. Agents Chemother.*, **53** (2), 651–661.
7. Paik, S.H., Chakicherla, A., and Hansen, J.N. (1998) Identification and characterization of the structural and transporter genes for, and the chemical and biological properties of, sublancin 168, a novel lantibiotic produced by *Bacillus subtilis* 168. *J. Biol. Chem.*, **273** (36), 23134–23142.
8. Chen, I., Christie, P.J., and Dubnau, D. (2005) The ins and outs of DNA transfer in bacteria. *Science*, **310** (5753), 1456–1460.
9. Hamoen, L.W., Venema, G., and Kuipers, O.P. (2003) Controlling competence in *Bacillus subtilis*: shared use of regulators. *Microbiology*, **149** (1), 9–17.
10. Whatmore, A.M., Chudek, J.A., and Reed, R.H. (1990) The effects of osmotic upshock on the intracellular solute pools of *Bacillus subtilis*. *J. Gen. Microbiol.*, **136** (12), 2527–2535.
11. Whatmore, A.M. and Reed, R.H. (1990) Determination of turgor pressure in *Bacillus subtilis*: a possible role for K+ in turgor regulation. *J. Gen. Microbiol.*, **136** (12), 2521–2526.

12. Kappes, R.M., Kempf, B., and Bremer, E. (1996) Three transport systems for the osmoprotectant glycine betaine operate in *Bacillus subtilis*: characterization of OpuD. *J. Bacteriol.*, **178** (17), 5071–5079.
13. Kempf, B. and Bremer, E. (1995) OpuA, an osmotically regulated binding protein-dependent transport system for the osmoprotectant glycine betaine in *Bacillus subtilis*. *J. Biol. Chem.*, **270** (28), 16701–16713.
14. Lin, Y. and Hansen, J.N. (1995) Characterization of a chimeric proU operon in a subtilin-producing mutant of *Bacillus subtilis* 168. *J. Bacteriol.*, **177** (23), 6874–6880.
15. Kohlstedt, M. *et al.* (2014) Adaptation of *Bacillus subtilis* carbon core metabolism to simultaneous nutrient limitation and osmotic challenge: a multi-omics perspective. *Environ. Microbiol.*, **16** (6), 1898–1917.
16. Higgins, D. and Dworkin, J. (2012) Recent progress in *Bacillus subtilis* sporulation. *FEMS Microbiol. Rev.*, **36** (1), 131–148.
17. Hecker, M., Schumann, W., and Volker, U. (1996) Heat-shock and general stress response in *Bacillus subtilis*. *Mol. Microbiol.*, **19** (3), 417–428.
18. Losick, R., Youngman, P., and Piggot, P.J. (1986) Genetics of endospore formation in *Bacillus subtilis*. *Annu. Rev. Genet.*, **20**, 625–669.
19. Kunst, F. *et al.* (1997) The complete genome sequence of the gram-positive bacterium *Bacillus subtilis*. *Nature*, **390** (6657), 249–256.
20. Barbe, V. *et al.* (2009) From a consortium sequence to a unified sequence: the *Bacillus subtilis* 168 reference genome a decade later. *Microbiology*, **155** (6), 1758–1775.
21. Buescher, J.M. *et al.* (2012) Global network reorganization during dynamic adaptations of *Bacillus subtilis* metabolism. *Science*, **335** (6072), 1099–1103.
22. Nicolas, P. *et al.* (2012) Condition-dependent transcriptome reveals high-level regulatory architecture in *Bacillus subtilis*. *Science*, **335** (6072), 1103–1106.
23. Becker, J. and Wittmann, C. (2012) Systems and synthetic metabolic engineering for amino acid production – the heartbeat of industrial strain development. *Curr. Opin. Biotechnol.*, **23** (5), 718–726.
24. Fischer, E. and Sauer, U. (2005) Large-scale in vivo flux analysis shows rigidity and suboptimal performance of *Bacillus subtilis* metabolism. *Nat. Genet.*, **37** (6), 636–640.
25. Lee, J.W. *et al.* (2012) Systems metabolic engineering of microorganisms for natural and non-natural chemicals. *Nat. Chem. Biol.*, **8** (6), 536–546.
26. Oh, Y.-K. *et al.* (2007) Genome-scale reconstruction of metabolic network in *Bacillus subtilis* based on high-throughput phenotyping and gene essentiality data. *J. Biol. Chem.*, **282** (39), 28791–28799.
27. Becker, J. *et al.* (2011) From zero to hero-design-based systems metabolic engineering of *Corynebacterium glutamicum* for L-lysine production. *Metab. Eng.*, **13** (2), 159–168.
28. Lewis, N.E., Nagarajan, H., and Palsson, B.O. (2012) Constraining the metabolic genotype–phenotype relationship using a phylogeny of *in silico* methods. *Nat. Rev. Microbiol.*, **10** (4), 291–305.
29. Pharkya, P., Burgard, A.P., and Maranas, C.D. (2004) OptStrain: a computational framework for redesign of microbial production systems. *Genome Res.*, **14** (11), 2367–2376.
30. Kümmel, A., Panke, S., and Heinemann, M. (2006) Putative regulatory sites unraveled by network-embedded thermodynamic analysis of metabolome data. *Mol. Syst. Biol.*, **2** (1), 2006–2024.
31. Halpin, C. and Ryan, M. (2004) in *Metabolic Engineering in the Post Genomic Era* (eds B.N. Kholodenko and H.V. Westerhof), Horizon Bioscience, Norfolk, pp. 377–408.
32. Chassagnole, C. *et al.* (2002) Dynamic modeling of the central carbon metabolism of *Escherichia coli*. *Biotechnol. Bioeng.*, **79** (1), 53–73.
33. Tjalsma, H. *et al.* (2000) Signal peptide-dependent protein transport in *Bacillus*

subtilis: a genome-based survey of the secretome. *Microbiol. Mol. Biol. Rev.*, **64** (3), 515–547.
34. von Heijne, G. (1984) How signal sequences maintain cleavage specificity. *J. Mol. Biol.*, **173** (2), 243–251.
35. von Heijne, G. (1989) The structure of signal peptides from bacterial lipoproteins. *Protein Eng.*, **2** (7), 531–534.
36. von Heijne, G. (1990) Protein targeting signals. *Curr. Opin. Cell Biol.*, **2** (4), 604–608.
37. von Heijne, G. (1990) The signal peptide. *J. Membr. Biol.*, **115** (3), 195–201.
38. Akita, M. *et al.* (1990) SecA interacts with secretory proteins by recognizing the positive charge at the amino terminus of the signal peptide in *Escherichia coli*. *J. Biol. Chem.*, **265** (14), 8164–8169.
39. de Vrije, G.J. *et al.* (1990) Lipid involvement in protein translocation in *Escherichia coli*. *Mol. Microbiol.*, **4** (1), 143–150.
40. Briggs, M.S. *et al.* (1986) Conformations of signal peptides induced by lipids suggest initial steps in protein export. *Science*, **233** (4760), 206–208.
41. Saito, A. *et al.* (2011) Post-liberation cleavage of signal peptides is catalyzed by the site-2 protease (S2P) in bacteria. *Proc. Natl. Acad. Sci. U.S.A.*, **108** (33), 13740–13745.
42. Kolkman, M.A. *et al.* (2008) The twin-arginine signal peptide of *Bacillus subtilis* YwbN can direct either Tat- or Sec-dependent secretion of different cargo proteins: secretion of active subtilisin via the *B. subtilis* Tat pathway. *Appl. Environ. Microbiol.*, **74** (24), 7507–7513.
43. van Roosmalen, M.L. *et al.* (2004) Type I signal peptidases of Gram-positive bacteria. *Biochim. Biophys. Acta*, **1694** (1-3), 279–297.
44. Tjalsma, H. *et al.* (1998) Functional analysis of the secretory precursor processing machinery of *Bacillus subtilis*: identification of a eubacterial homolog of archaeal and eukaryotic signal peptidases. *Genes Dev.*, **12** (15), 2318–2331.
45. Goosens, V.J. *et al.* (2013) Novel twin-arginine translocation pathway-dependent phenotypes of *Bacillus subtilis* unveiled by quantitative proteomics. *J. Proteome Res.*, **12** (2), 796–807.
46. Jongbloed, J.D. *et al.* (2004) Two minimal Tat translocases in *Bacillus*. *Mol. Microbiol.*, **54** (5), 1319–1325.
47. Zanen, G. *et al.* (2006) Proteomic dissection of potential signal recognition particle dependence in protein secretion by *Bacillus subtilis*. *Proteomics*, **6** (12), 3636–3648.
48. Zanen, G. *et al.* (2005) Signal peptide hydrophobicity is critical for early stages in protein export by *Bacillus subtilis*. *FEBS J.*, **272** (18), 4617–4630.
49. Honda, K. *et al.* (1993) Cloning and characterization of a *Bacillus subtilis* gene encoding a homolog of the 54-kilodalton subunit of mammalian signal recognition particle and *Escherichia coli* Ffh. *J. Bacteriol.*, **175** (15), 4885–4894.
50. Economou, A. (2002) Bacterial secretome: the assembly manual and operating instructions (review). *Mol. Membr. Biol.*, **19** (3), 159–169.
51. Lycklama, A.N.J.A. and Driessen, A.J. (2012) The bacterial Sec-translocase: structure and mechanism. *Philos. Trans. R. Soc. London, Ser. B*, **367** (1592), 1016–1028.
52. Bolhuis, A. *et al.* (1998) SecDF of *Bacillus subtilis*, a molecular Siamese twin required for the efficient secretion of proteins. *J. Biol. Chem.*, **273** (33), 21217–21224.
53. van Wely, K.H. *et al.* (2001) Translocation of proteins across the cell envelope of Gram-positive bacteria. *FEMS Microbiol. Rev.*, **25** (4), 437–454.
54. Tjalsma, H., Bron, S., and van Dijl, J.M. (2003) Complementary impact of paralogous Oxa1-like proteins of *Bacillus subtilis* on post-translocational stages in protein secretion. *J. Biol. Chem.*, **278** (18), 15622–15632.
55. Saller, M.J., Fusetti, F., and Driessen, A.J. (2009) *Bacillus subtilis* SpoIIIJ and YqjG function in membrane protein biogenesis. *J. Bacteriol.*, **191** (21), 6749–6757.

56. Chiba, S., Lamsa, A., and Pogliano, K. (2009) A ribosome-nascent chain sensor of membrane protein biogenesis in *Bacillus subtilis*. *EMBO J.*, **28** (22), 3461–3475.
57. Kontinen, V.P. and Sarvas, M. (1993) The PrsA lipoprotein is essential for protein secretion in *Bacillus subtilis* and sets a limit for high-level secretion. *Mol. Microbiol.*, **8** (4), 727–737.
58. Kouwen, T.R. and van Dijl, J.M. (2009) Interchangeable modules in bacterial thiol-disulfide exchange pathways. *Trends Microbiol.*, **17** (1), 6–12.
59. Bolhuis, A. *et al.* (1999) Evaluation of bottlenecks in the late stages of protein secretion in *Bacillus subtilis*. *Appl. Environ. Microbiol.*, **65** (7), 2934–2941.
60. Sarvas, M. *et al.* (2004) Post-translocational folding of secretory proteins in Gram-positive bacteria. *Biochim. Biophys. Acta*, **1694** (1-3), 311–327.
61. Tjalsma, H. *et al.* (2004) Proteomics of protein secretion by *Bacillus subtilis*: separating the "secrets" of the secretome. *Microbiol. Mol. Biol. Rev.*, **68** (2), 207–233.
62. Kakeshita, H. *et al.* (2010) Enhanced extracellular production of heterologous proteins in *Bacillus subtilis* by deleting the C-terminal region of the SecA secretory machinery. *Mol. Biotechnol.*, **46** (3), 250–257.
63. Diao, L. *et al.* (2012) Functional implementation of the posttranslational SecB-SecA protein-targeting pathway in *Bacillus subtilis*. *Appl. Environ. Microbiol.*, **78** (3), 651–659.
64. Collier, D.N. (1994) Expression of Escherichia coli SecB in *Bacillus subtilis* facilitates secretion of the SecB-dependent maltose-binding protein of *E. coli*. *J. Bacteriol.*, **176** (16), 4937–4940.
65. Vitikainen, M. *et al.* (2001) Quantitation of the capacity of the secretion apparatus and requirement for PrsA in growth and secretion of alpha-amylase in *Bacillus subtilis*. *J. Bacteriol.*, **183** (6), 1881–1890.
66. Vitikainen, M. *et al.* (2004) Structure-function analysis of PrsA reveals roles for the parvulin-like and flanking N- and C-terminal domains in protein folding and secretion in *Bacillus subtilis*. *J. Biol. Chem.*, **279** (18), 19302–19314.
67. Kakeshita, H. *et al.* (2011) Secretion of biologically-active human interferon-beta by *Bacillus subtilis*. *Biotechnol. Lett.*, **33** (9), 1847–1852.
68. Yuan, J. *et al.* (2010) Protein transport across and into cell membranes in bacteria and archaea. *Cell. Mol. Life Sci.*, **67** (2), 179–199.
69. Jongbloed, J.D. *et al.* (2002) Selective contribution of the twin-arginine translocation pathway to protein secretion in *Bacillus subtilis*. *J. Biol. Chem.*, **277** (46), 44068–44078.
70. Jongbloed, J.D. *et al.* (2000) TatC is a specificity determinant for protein secretion via the twin-arginine translocation pathway. *J. Biol. Chem.*, **275** (52), 41350–41357.
71. Monteferrante, C.G. *et al.* (2012) Specific targeting of the metallophosphoesterase YkuE to the bacillus cell wall requires the twin-arginine translocation system. *J. Biol. Chem.*, **287** (35), 29789–29800.
72. Kouwen, T.R. *et al.* (2009) Overflow of a hyper-produced secretory protein from the Bacillus Sec pathway into the Tat pathway for protein secretion as revealed by proteogenomics. *Proteomics*, **9** (4), 1018–1032.
73. Goosens, V.J., Monteferrante, C.G., and van Dijl, J.M. (2014) The Tat system of Gram-positive bacteria. *Biochim. Biophys. Acta*, **1843** (8), 1698–1706.
74. Goosens, V.J., Monteferrante, C.G., and van Dijl, J.M. (2014) Co-factor insertion and disulfide bond requirements for twin-arginine translocase-dependent export of the *Bacillus subtilis* Rieske protein QcrA. *J. Biol. Chem.*, **289** (19), 13124–13131.
75. Robinson, C. and Bolhuis, A. (2004) Tat-dependent protein targeting in prokaryotes and chloroplasts. *Biochim. Biophys. Acta*, **1694** (1-3), 135–147.
76. Jongbloed, J.D., van der Ploeg, R., and van Dijl, J.M. (2006) Bifunctional TatA subunits in minimal Tat protein

translocases. *Trends Microbiol.*, **14** (1), 2–4.

77. Monteferrante, C.G. *et al.* (2012) TatAc, the third TatA subunit of *Bacillus subtilis*, can form active twin-arginine translocases with the TatCd and TatCy subunits. *Appl. Environ. Microbiol.*, **78** (14), 4999–5001.
78. Goosens, V.J., De-San-Eustaquio-Campillo, A., Carballido-López, R., and van Dijl, J.M. (2015) A Tat menage a trois-- The role of *Bacillus subtilis* TatAc in twin-arginine protein translocation. *Biochim. Biophys. Acta*, **1853** (10, Pt. A), 2745–2753.
79. Yang, C. *et al.* (2010) Twin-arginine translocation of methyl parathion hydrolase in *Bacillus subtilis*. *Environ. Sci. Technol.*, **44** (19), 7607–7612.
80. Priest, F.G. (1977) Extracellular enzyme synthesis in the genus *Bacillus*. *Bacteriol. Rev.*, **41** (3), 711–753.
81. Stahl, M.L. and Ferrari, E. (1984) Replacement of the *Bacillus subtilis* subtilisin structural gene with an in vitro-derived deletion mutation. *J. Bacteriol.*, **158** (2), 411–418.
82. Wong, S.L. *et al.* (1984) The subtilisin E gene of *Bacillus subtilis* is transcribed from a sigma 37 promoter *in vivo*. *Proc. Natl. Acad. Sci. U.S.A.*, **81** (4), 1184–1188.
83. Tran, L., Wu, X.C., and Wong, S.L. (1991) Cloning and expression of a novel protease gene encoding an extracellular neutral protease from *Bacillus subtilis*. *J. Bacteriol.*, **173** (20), 6364–6372.
84. Sloma, A. *et al.* (1988) Gene encoding a minor extracellular protease in *Bacillus subtilis*. *J. Bacteriol.*, **170** (12), 5557–5563.
85. Bruckner, R., Shoseyov, O., and Doi, R.H. (1990) Multiple active forms of a novel serine protease from *Bacillus subtilis*. *Mol. Gen. Genet.*, **221** (3), 486–490.
86. Rufo, G.A. Jr., *et al.* (1990) Isolation and characterization of a novel extracellular metalloprotease from *Bacillus subtilis*. *J. Bacteriol.*, **172** (2), 1019–1023.
87. Sloma, A. *et al.* (1991) Cloning and characterization of the gene for an additional extracellular serine protease of *Bacillus subtilis*. *J. Bacteriol.*, **173** (21), 6889–6895.
88. Margot, P. and Karamata, D. (1996) The wprA gene of *Bacillus subtilis* 168, expressed during exponential growth, encodes a cell-wall-associated protease. *Microbiology*, **142** (12), 3437–3444.
89. Veening, J.W. *et al.* (2008) Transient heterogeneity in extracellular protease production by *Bacillus subtilis*. *Mol. Syst. Biol.*, **4**, 184.
90. Davidson, F.A., Seon-Yi, C., and Stanley-Wall, N.R. (2012) Selective heterogeneity in exoprotease production by *Bacillus subtilis*. *PLoS One*, **7** (6), e38574.
91. Pero, J.G. and Sloma, A. (1993) in Bacillus subtilis *and Other Gram-Positive Bacteria: Biochemistry, Physiology, and Molecular Genetics* (eds A.L. Sonenshein, J.A. Hoch, and R. Losick), American Society for Microbiology, Washington, DC, pp. 939–952.
92. Khan, A.R. and James, M.N. (1998) Molecular mechanisms for the conversion of zymogens to active proteolytic enzymes. *Protein Sci.*, **7** (4), 815–836.
93. Ohta, Y. and Inouye, M. (1990) Pro-subtilisin E: purification and characterization of its autoprocessing to active subtilisin E *in vitro*. *Mol. Microbiol.*, **4** (2), 295–304.
94. Inouye, M. (1991) Intramolecular chaperone: the role of the pro-peptide in protein folding. *Enzyme*, **45** (5-6), 314–321.
95. Strausberg, S. *et al.* (1993) Catalysis of a protein folding reaction: thermodynamic and kinetic analysis of subtilisin BPN' interactions with its propeptide fragment. *Biochemistry*, **32** (32), 8112–8119.
96. Park, C.H. *et al.* (2004) Hetero- and autoprocessing of the extracellular metalloprotease (Mpr) in *Bacillus subtilis*. *J. Bacteriol.*, **186** (19), 6457–6464.
97. Yamamoto, H., Kurosawa, S., and Sekiguchi, J. (2003) Localization of the vegetative cell wall hydrolases LytC, LytE, and LytF on the *Bacillus subtilis* cell surface and stability of these

enzymes to cell wall-bound or extracellular proteases. *J. Bacteriol.*, **185** (22), 6666–6677.
98. Antelmann, H. *et al.* (2002) Stabilization of cell wall proteins in *Bacillus subtilis*: a proteomic approach. *Proteomics*, **2** (5), 591–602.
99. Dixit, M., Murudkar, C.S., and Rao, K.K. (2002) Epr is transcribed from a final sigma(D) promoter and is involved in swarming of *Bacillus subtilis*. *J. Bacteriol.*, **184** (2), 596–599.
100. Corvey, C. *et al.* (2003) Activation of subtilin precursors by *Bacillus subtilis* extracellular serine proteases subtilisin (AprE), WprA, and Vpr. *Biochem. Biophys. Res. Commun.*, **304** (1), 48–54.
101. Lanigan-Gerdes, S. *et al.* (2007) Identification of subtilisin, Epr and Vpr as enzymes that produce CSF, an extracellular signalling peptide of *Bacillus subtilis*. *Mol. Microbiol.*, **65** (5), 1321–1333.
102. Gupta, R., Beg, Q.K., and Lorenz, P. (2002) Bacterial alkaline proteases: molecular approaches and industrial applications. *Appl. Microbiol. Biotechnol.*, **59** (1), 15–32.
103. Park, S.G. *et al.* (2002) A functional proteomic analysis of secreted fibrinolytic enzymes from *Bacillus subtilis* 168 using a combined method of two-dimensional gel electrophoresis and zymography. *Proteomics*, **2** (2), 206–211.
104. Kho, C.W. *et al.* (2005) Confirmation of Vpr as a fibrinolytic enzyme present in extracellular proteins of *Bacillus subtilis*. *Protein Expression Purif.*, **39** (1), 1–7.
105. Wu, X.C. *et al.* (1991) Engineering a *Bacillus subtilis* expression-secretion system with a strain deficient in six extracellular proteases. *J. Bacteriol.*, **173** (16), 4952–4958.
106. Wu, S.C. *et al.* (2002) Functional production and characterization of a fibrin-specific single-chain antibody fragment from *Bacillus subtilis*: effects of molecular chaperones and a wall-bound protease on antibody fragment production. *Appl. Environ. Microbiol.*, **68** (7), 3261–3269.
107. Pohl, S. *et al.* (2013) Proteomic analysis of *Bacillus subtilis* strains engineered for improved production of heterologous proteins. *Proteomics*, **13** (22), 3298–3308.
108. Hyyrylainen, H.L. *et al.* (2001) A novel two-component regulatory system in *Bacillus subtilis* for the survival of severe secretion stress. *Mol. Microbiol.*, **41** (5), 1159–1172.
109. Antelmann, H. *et al.* (2003) The extracellular proteome of *Bacillus subtilis* under secretion stress conditions. *Mol. Microbiol.*, **49** (1), 143–156.
110. Krishnappa, L. *et al.* (2013) Extracytoplasmic proteases determining the cleavage and release of secreted proteins, lipoproteins, and membrane proteins in *Bacillus subtilis*. *J. Proteome Res.*, **12** (9), 4101–4110.
111. Noone, D. *et al.* (2001) YkdA and YvtA, HtrA-like serine proteases in *Bacillus subtilis*, engage in negative autoregulation and reciprocal cross-regulation of ykdA and yvtA gene expression. *J. Bacteriol.*, **183** (2), 654–663.
112. Darmon, E. *et al.* (2002) A novel class of heat and secretion stress-responsive genes is controlled by the autoregulated CssRS two-component system of *Bacillus subtilis*. *J. Bacteriol.*, **184** (20), 5661–5671.
113. Zweers, J.C., Wiegert, T., and van Dijl, J.M. (2009) Stress-responsive systems set specific limits to the overproduction of membrane proteins in *Bacillus subtilis*. *Appl. Environ. Microbiol.*, **75** (23), 7356–7364.
114. Krishnappa, L. *et al.* (2014) Degradation of extracytoplasmic catalysts for protein folding in *Bacillus subtilis*. *Appl. Environ. Microbiol.*, **80** (4), 1463–1468.
115. Dalbey, R.E., Wang, P., and van Dijl, J.M. (2012) Membrane proteases in the bacterial protein secretion and quality control pathway. *Microbiol. Mol. Biol. Rev.*, **76** (2), 311–330.
116. Harwood, C.R. and Cutting, S.M. (1990) in *Molecular Biological Methods for Bacillus*, Modern Microbiological Methods (ed. M. Goodfellow), John Wiley & Sons, Ltd., Chichester.

117. Chen, I. and Dubnau, D. (2004) DNA uptake during bacterial transformation. *Nat. Rev. Microbiol.*, **2** (3), 241–249.

118. Leisner, M. *et al.* (2008) Stochastic switching to competence. *Curr. Opin. Microbiol.*, **11** (6), 553–559.

119. Lopez, D., Vlamakis, H., and Kolter, R. (2009) Generation of multiple cell types in *Bacillus subtilis*. *FEMS Microbiol. Rev.*, **33** (1), 152–163.

120. Kunst, F. and Rapoport, G. (1995) Salt stress is an environmental signal affecting degradative enzyme synthesis in *Bacillus subtilis*. *J. Bacteriol.*, **177** (9), 2403–2407.

121. Middleton, R. and Hofmeister, A. (2004) New shuttle vectors for ectopic insertion of genes into *Bacillus subtilis*. *Plasmid*, **51** (3), 238–245.

122. Kiel, J.A. *et al.* (1995) A general method for the consecutive integration of single copies of a heterologous gene at multiple locations in the *Bacillus subtilis* chromosome by replacement recombination. *Appl. Environ. Microbiol.*, **61** (12), 4244–4250.

123. Radeck, J. *et al.* (2013) The Bacillus BioBrick Box: generation and evaluation of essential genetic building blocks for standardized work with *Bacillus subtilis*. *J. Biol. Eng.*, **7** (1), 29.

124. Commichau, F.M. *et al.* (2014) Overexpression of a non-native deoxyxylulose-dependent vitamin B6 pathway in *Bacillus subtilis* for the production of pyridoxine. *Metab. Eng.*, **25**, 38–49.

125. Perkins, J.B. *et al.* (2010) in *The Metabolic Pathway Engineering Handbook: Fundamentals* (ed. C.D. Smolke), CRC Press, Taylor & Francis Group, Boca Raton, FL, pp. 1–36.

126. Chary, V.K., Amaya, E.I., and Piggot, P.J. (1997) Neomycin- and spectinomycin-resistance replacement vectors for *Bacillus subtilis*. *FEMS Microbiol. Lett.*, **153** (1), 135–139.

127. Palva, A. *et al.* (1990) Nucleotide sequence of the tetracycline resistance gene of pBC16 from *Bacillus cereus*. *Nucleic Acids Res.*, **18** (6), 1635.

128. Bernhard, K., Schrempf, H., and Goebel, W. (1978) Bacteriocin and antibiotic resistance plasmids in *Bacillus cereus* and *Bacillus subtilis*. *J. Bacteriol.*, **133** (2), 897–903.

129. Gryczan, T.J. and Dubnau, D. (1978) Construction and properties of chimeric plasmids in *Bacillus subtilis*. *Proc. Natl. Acad. Sci. U.S.A.*, **75** (3), 1428–1432.

130. Itaya, M., Kondo, K., and Tanaka, T. (1989) A neomycin resistance gene cassette selectable in a single copy state in the *Bacillus subtilis* chromosome. *Nucleic Acids Res.*, **17** (11), 4410.

131. Horinouchi, S. and Weisblum, B. (1982) Nucleotide sequence and functional map of pC194, a plasmid that specifies inducible chloramphenicol resistance. *J. Bacteriol.*, **150** (2), 815–825.

132. EFSA_FEEDAP_Panel (2012) Guidance on the assessment of bacterial susceptibility to antimicrobials of human and veterinary importance. *EFSA J.*, **10** (6), 2740.

133. Lee, G. and Pero, J. (1981) Conserved nucleotide sequences in temporally controlled bacteriophage promoters. *J. Mol. Biol.*, **152** (2), 247–265.

134. Perkins, J.B. and Prágai, Z. (2003) Production of pantothenate using microorganisms incapable of sporulation. Patent WO2004113510 A2, DSM.

135. Perkins, J.B., Sloma, A., and Pero, J.G. (1998) Bacterial strains which overproduce riboflavin. Patent US5837528 A, Roche.

136. Stewart, C.R. *et al.* (1998) Genes and regulatory sites of the "host-takeover module" in the terminal redundancy of *Bacillus subtilis* bacteriophage SPO1. *Virology*, **246** (2), 329–340.

137. Wang, L.F. and Doi, R.H. (1987) Promoter switching during development and the termination site of the sigma 43 operon of *Bacillus subtilis*. *Mol. Gen. Genet.*, **207** (1), 114–119.

138. Lam, K.H., Chow, K.C., and Wong, W.K. (1998) Construction of an efficient *Bacillus subtilis* system for extracellular production of heterologous proteins. *J. Biotechnol.*, **63** (3), 167–177.

139. Le Grice, S.F. *et al.* (1986) Separation and analysis of the RNA polymerase binding sites of a complex *Bacillus* *subtilis* promoter. *Mol. Gen. Genet.*, **204** (2), 229–236.

140. Jensen, P.R. and Hammer, K. (1996) Artificial promoter libraries for selected organisms and promoters derived from such libraries. Patent EP0934406 B1, Jensen.
141. Dong, H., Nilsson, L., and Kurland, C.G. (1995) Gratuitous overexpression of genes in *Escherichia coli* leads to growth inhibition and ribosome destruction. *J. Bacteriol.*, **177** (6), 1497–1504.
142. Ramirez, D.M. and Bentley, W.E. (1995) Fed-batch feeding and induction policies that improve foreign protein synthesis and stability by avoiding stress responses. *Biotechnol. Bioeng.*, **47** (5), 596–608.
143. Ferrari, E. and Valle, F. (2000) Mutant aprE promoter. Patent EP1244794 B1, Genencor.
144. Ye, R. *et al.* (1999) High-level secretory production of intact, biologically active staphylokinase from *Bacillus subtilis*. *Biotechnol. Bioeng.*, **62** (1), 87–96.
145. Bron, S. (1990) in *Molecular Biological Methods for Bacillus* (eds C.R. Harwood and S.M. Cutting), John Wiley & Sons, Ltd., Chichester, pp. 75–174.
146. Jannière, L., Gruss, A., and Ehrlich, S.D. (1993) in Bacillus subtilis *and Other Gram-Positive Bacteria: Biochemistry, Physiology, and Molecular Genetics* (eds A.L. Sonenshein, J.A. Hoch, and R. Losick), American Society for Microbiology, Washington, DC, pp. 625–644.
147. Wu, S.C. and Wong, S.L. (1999) Development of improved pUB110-based vectors for expression and secretion studies in *Bacillus subtilis*. *J. Biotechnol.*, **72** (1-2), 185–195.
148. Lu, F. et al. (2009) Alkaline pectinase genetic engineering bacteria and construction method thereof. Patent CN101531988 B, Tianjin University.
149. Zhang, P. and Huang, Y.M. (2010) Shuttle plasmid and method for efficiently expressing cecropin and lysozyme genes. Patent CN101899469 B, Shanghai E K M Biotechnology.
150. Stepanov, A.I. et al. (1984) Riboflavin preparation. Patent FR2546907, Inst Genetiki.
151. Stoinova, N.V. *et al.* (1996) The construction of recombinant strains that produce riboflavin. II. Development of methods of integration of foreign genes into the *Bacillus subtilis* chromosome on a model of the riboflavin operon of *Bacillus amyloliquefaciens*. *Biotekhnologiya*, **11**, 7–10.
152. Condon, C. (2003) RNA processing and degradation in *Bacillus subtilis*. *Microbiol. Mol. Biol. Rev.*, **67** (2), 157–174.
153. Sharp, J.S. and Bechhofer, D.H. (2005) Effect of 5′-proximal elements on decay of a model mRNA in *Bacillus subtilis*. *Mol. Microbiol.*, **57** (2), 484–495.
154. Ferrandez, A., Perkins, J.B., and Schaber, M. (2005) Method for mRNA stabilization. Patent EP1957651 B1, DSM.
155. Meinken, C. *et al.* (2003) Expression of the glycolytic gapA operon in *Bacillus subtilis*: differential syntheses of proteins encoded by the operon. *Microbiology*, **149** (3), 751–761.
156. Hambraeus, G., von Wachenfeldt, C., and Hederstedt, L. (2003) Genome-wide survey of mRNA half-lives in *Bacillus subtilis* identifies extremely stable mRNAs. *Mol. Genet. Genomics*, **269** (5), 706–714.
157. Homuth, G., Mogk, A., and Schumann, W. (1999) Post-transcriptional regulation of the *Bacillus subtilis* dnaK operon. *Mol. Microbiol.*, **32** (6), 1183–1197.
158. Jurgen, B., Schweder, T., and Hecker, M. (1998) The stability of mRNA from the gsiB gene of *Bacillus subtilis* is dependent on the presence of a strong ribosome binding site. *Mol. Gen. Genet.*, **258** (5), 538–545.
159. Lehmann, M., Schraber, M., and Prágai, Z. (2007) Increased production of a target product via stabilization of mRNA. Patent US8551732 B2, DSM.
160. Peschke, U. *et al.* (1985) Efficient utilization of *Escherichia coli* transcriptional signals in *Bacillus subtilis*. *J. Mol. Biol.*, **186** (3), 547–555.

161. Vagner, V., Dervyn, E., and Ehrlich, S.D. (1998) A vector for systematic gene inactivation in *Bacillus subtilis*. *Microbiology*, **144** (11), 3097–3104.

162. Wong, H.C. and Chang, S. (1986) Identification of a positive retroregulator that stabilizes mRNAs in bacteria. *Proc. Natl. Acad. Sci. U.S.A.*, **83** (10), 3233–3237.

163. Laursen, B.S. *et al.* (2005) Initiation of protein synthesis in bacteria. *Microbiol. Mol. Biol. Rev.*, **69** (1), 101–123.

164. Rocha, E.P., Danchin, A., and Viari, A. (1999) Translation in *Bacillus subtilis*: roles and trends of initiation and termination, insights from a genome analysis. *Nucleic Acids Res.*, **27** (17), 3567–3576.

165. Tats, A., Remm, M., and Tenson, T. (2006) Highly expressed proteins have an increased frequency of alanine in the second amino acid position. *BMC Genomics*, 7, 28.

166. Vellanoweth, R.L. and Rabinowitz, J.C. (1992) The influence of ribosome-binding-site elements on translational efficiency in *Bacillus subtilis* and *Escherichia coli in vivo*. *Mol. Microbiol.*, **6** (9), 1105–1114.

167. Zhang, W. *et al.* (2006) mRNA secondary structure at start AUG codon is a key limiting factor for human protein expression in *Escherichia coli*. *Biochem. Biophys. Res. Commun.*, **349** (1), 69–78.

168. Nishikubo, T. *et al.* (2005) Improved heterologous gene expression in Escherichia coli by optimization of the AT-content of codons immediately downstream of the initiation codon. *J. Biotechnol.*, **120** (4), 341–346.

169. Sakai, H. *et al.* (2001) Correlation between Shine–Dalgarno sequence conservation and codon usage of bacterial genes. *J. Mol. Evol.*, **52** (2), 164–170.

170. Brockmeier, U. *et al.* (2006) Systematic screening of all signal peptides from *Bacillus subtilis*: a powerful strategy in optimizing heterologous protein secretion in Gram-positive bacteria. *J. Mol. Biol.*, **362** (3), 393–402.

171. Lehmann, M. et al. (2008) Improved production of riboflavin. Patent EP2356211 A1, DSM.

172. Dubnau, D. (1993) in Bacillus subtilis *and Other Gram-Positive Bacteria: Biochemistry, Physiology, and Molecular Genetics* (eds A.L. Sonenshein, J.A. Hoch, and R. Losick), American Society for Microbiology, Washington, DC, pp. 555–584.

173. Melnikov, A. and Youngman, P.J. (1999) Random mutagenesis by recombinational capture of PCR products in *Bacillus subtilis* and *Acinetobacter calcoaceticus*. *Nucleic Acids Res.*, **27** (4), 1056–1062.

174. Wach, A. (1996) PCR-synthesis of marker cassettes with long flanking homology regions for gene disruptions in S. cerevisiae. *Yeast*, **12** (3), 259–265.

175. Gibson, D.G. *et al.* (2008) Complete chemical synthesis, assembly, and cloning of a Mycoplasma genitalium genome. *Science*, **319** (5867), 1215–1220.

176. Bower, S.G. et al. (1993) Biotin biosynthesis in *Bacillus subtilis*. Patent US6057136 A, Roche Vitamins.

177. Hermann, T. et al. (2001) Process for enhanced production of pantothenate. Patent WO02057474 (A2), OmniGene.

178. Schyns, G. *et al.* (2005) Isolation and characterization of new thiamine-deregulated mutants of *Bacillus subtilis*. *J. Bacteriol.*, **187** (23), 8127–8136.

179. Perkins, J.B. and Prágai, Z. (2008) Increased production of pantothenate (vitamin B5). Patent EP2157174 A1, DSM.

180. Lehmann, M. et al. (2005) Modified transketolase and use thereof. Patent EP1957640 B1, DSM.

181. Saito, Y., Taguchi, H., and Akamatsu, T. (2006) Fate of transforming bacterial genome following incorporation into competent cells of *Bacillus subtilis*: a continuous length of incorporated DNA. *J. Biosci. Bioeng.*, **101** (3), 257–262.

182. Fabret, C., Ehrlich, S.D., and Noirot, P. (2002) A new mutation delivery system for genome-scale approaches in *Bacillus subtilis*. *Mol. Microbiol.*, **46** (1), 25–36.

183. Brans, A. *et al.* (2004) New integrative method to generate *Bacillus subtilis* recombinant strains free of selection markers. *Appl. Environ. Microbiol.*, **70** (12), 7241–7250.

184. Liu, S. *et al.* (2008) Introduction of marker-free deletions in *Bacillus subtilis* using the AraR repressor and the ara promoter. *Microbiology*, **154** (9), 2562–2570.

185. Wang, Y. *et al.* (2012) *Bacillus subtilis* genome editing using ssDNA with short homology regions. *Nucleic Acids Res.*, **40** (12), e91.

186. Zhang, X.Z. *et al.* (2006) mazF, a novel counter-selectable marker for unmarked chromosomal manipulation in *Bacillus subtilis*. *Nucleic Acids Res.*, **34** (9), e71.

187. Morimoto, T. *et al.* (2009) A new simple method to introduce marker-free deletions in the *Bacillus subtilis* genome. *Genes Genet. Syst.*, **84** (4), 315–318.

188. Lin, Z. *et al.* (2013) A versatile mini-mazF-cassette for marker-free targeted genetic modification in *Bacillus subtilis*. *J. Microbiol. Methods*, **95** (2), 207–214.

189. Abremski, K., Hoess, R., and Sternberg, N. (1983) Studies on the properties of P1 site-specific recombination: evidence for topologically unlinked products following recombination. *Cell*, **32** (4), 1301–1311.

190. Albert, H. *et al.* (1995) Site-specific integration of DNA into wild-type and mutant lox sites placed in the plant genome. *Plant J.*, **7** (4), 649–659.

191. Yan, X. *et al.* (2008) Cre/lox system and PCR-based genome engineering in *Bacillus subtilis*. *Appl. Environ. Microbiol.*, **74** (17), 5556–5562.

192. Ellis, H.M. *et al.* (2001) High efficiency mutagenesis, repair, and engineering of chromosomal DNA using single-stranded oligonucleotides. *Proc. Natl. Acad. Sci. U.S.A.*, **98** (12), 6742–6746.

193. Costantino, N. and Court, D.L. (2003) Enhanced levels of lambda Red-mediated recombinants in mismatch repair mutants. *Proc. Natl. Acad. Sci. U.S.A.*, **100** (26), 15748–15753.

194. Sawitzke, J.A. *et al.* (2011) Probing cellular processes with oligo-mediated recombination and using the knowledge gained to optimize recombineering. *J. Mol. Biol.*, **407** (1), 45–59.

195. Wang, H.H. *et al.* (2009) Programming cells by multiplex genome engineering and accelerated evolution. *Nature*, **460** (7257), 894–898.

196. Zhang, Y. *et al.* (1998) A new logic for DNA engineering using recombination in *Escherichia coli*. *Nat. Genet.*, **20** (2), 123–128.

197. Duitman, E.H. *et al.* (2007) Novel methods for genetic transformation of natural *Bacillus subtilis* isolates used to study the regulation of the mycosubtilin and surfactin synthetases. *Appl. Environ. Microbiol.*, **73** (11), 3490–3496.

198. Schaeffer, P., Cami, B., and Hotchkiss, R.D. (1976) Fusion of bacterial protoplasts. *Proc. Natl. Acad. Sci. U.S.A.*, **73** (6), 2151–2155.

199. Brigidi, P., Rossi, M., and Matteuzzi, D. (2000) in *Electrotransformation of Bacteria* (eds N. Eynard and J. Teissie), Springer-Verlag, Berlin, pp. 42–46.

200. Chassy, B.M., Saunders, J.A., and Sowers, A.E. (1992) in *Guide to Electroporation and Electrofusion* (eds D.C. Chang *et al.*), Academic Press, San Diego, CA, pp. 2555–2569.

201. Ohse, M. *et al.* (1995) Effects of plasmid DNA sizes and several other factors on transformation of *Bacillus subtilis* ISW1214 with plasmid DNA by electroporation. *Biosci. Biotechnol., Biochem.*, **59** (8), 1433–1437.

202. Vehmaanpera, J. (1989) Transformation of *Bacillus amyloliquefaciens* by electroporation. *FEMS Microbiol. Lett.*, **52** (1–2), 165–169.

203. Xue, G.-P., Johnson, J.S., and Dalrymple, B.P. (1999) High osmolarity improves the electro-transformation efficiency of the gram-positive bacteria *Bacillus subtilis* and *Bacillus licheniformis*. *J. Microbiol. Methods*, **34** (3), 183–191.

204. Cao, G. *et al.* (2011) A modified electro-transformation method for *Bacillus subtilis* and its application in the production of antimicrobial lipopeptides. *Biotechnol. Lett.*, **33** (5), 1047–1051.

205. Meddeb-Mouelhi, F., Dulcey, C., and Beauregard, M. (2012) High transformation efficiency of *Bacillus subtilis*

with integrative DNA using glycine betaine as osmoprotectant. *Anal. Biochem.*, **424** (2), 127–129.

206. Lu, Y.P. *et al.* (2012) Study on the electro-transformation conditions of improving transformation efficiency for *Bacillus subtilis. Lett. Appl. Microbiol.*, **55** (1), 9–14.
207. Brigidi, P. *et al.* (1990) Genetic transformation of intact cells of *Bacillus subtilis* by electroporation. *FEMS Microbiol. Lett.*, **67** (1-2), 135–138.
208. Ohse, M., Kawade, K., and Kusaoke, H. (1997) Effects of DNA topology on transformation efficiency of *Bacillus subtilis* ISW1214 by electroporation. *Biosci. Biotechnol., Biochem.*, **61** (6), 1019–1021.
209. Yang, M.M. *et al.* (2010) Electroporation is a feasible method to introduce circularized or linearized DNA into *B. subtilis* chromosome. *Mol. Biol. Rep.*, **37** (5), 2207–2213.
210. Wen, S., Yang, J., and Tan, T. (2013) Full-length single-stranded PCR product mediated chromosomal integration in intact *Bacillus subtilis. J. Microbiol. Methods*, **92** (3), 273–277.
211. Akamatsu, T. and Sekiguchi, J. (1987) Characterization of chromosome and plasmid transformation in *Bacillus subtilis* using gently lysed protoplasts. *Arch. Microbiol.*, **146** (4), 353–357.
212. Akamatsu, T. and Taguchi, H. (2001) Incorporation of the whole chromosomal DNA in protoplast lysates into competent cells of *Bacillus subtilis. Biosci. Biotechnol., Biochem.*, **65** (4), 823–829.
213. Takeno, M., Taguchi, H., and Akamatsu, T. (2011) Essential involvement of the *Bacillus subtilis* ABC transporter, EcsB, in genetic transformation of purified DNA but not native DNA from protoplast lysates. *J. Biosci. Bioeng.*, **112** (3), 209–214.
214. Toda, T., Tanaka, T., and Itaya, M. (1996) A method to invert DNA segments of the *Bacillus subtilis* 168 genome by recombination between two homologous sequences. *Biosci. Biotechnol., Biochem.*, **60** (5), 773–778.
215. Liu, S. *et al.* (2007) The accurate replacement of long genome region more than several hundreds kilobases in *Bacillus subtilis. Genes Genet. Syst.*, **82** (1), 9–19.
216. Ara, K. *et al.* (2007) *Bacillus* minimum genome factory: effective utilization of microbial genome information. *Biotechnol. Appl. Biochem.*, **46** (Pt. 3), 169–178.
217. Morimoto, T. *et al.* (2008) Enhanced recombinant protein productivity by genome reduction in *Bacillus subtilis. DNA Res.*, **15** (2), 73–81.
218. Westers, H. *et al.* (2003) Genome engineering reveals large dispensable regions in *Bacillus subtilis. Mol. Biol. Evol.*, **20** (12), 2076–2090.
219. Kobayashi, K. *et al.* (2003) Essential *Bacillus subtilis* genes. *Proc. Natl. Acad. Sci. U.S.A.*, **100** (8), 4678–4683.
220. Tanaka, K. *et al.* (2013) Building the repertoire of dispensable chromosome regions in *Bacillus subtilis* entails major refinement of cognate large-scale metabolic model. *Nucleic Acids Res.*, **41** (1), 687–699.
221. Manabe, K. *et al.* (2011) Combined effect of improved cell yield and increased specific productivity enhances recombinant enzyme production in genome-reduced *Bacillus subtilis* strain MGB874. *Appl. Environ. Microbiol.*, **77** (23), 8370–8381.
222. Drake, J.W. *et al.* (1998) Rates of spontaneous mutation. *Genetics*, **148** (4), 1667–1686.
223. Fijalkowska, I.J., Schaaper, R.M., and Jonczyk, P. (2012) DNA replication fidelity in *Escherichia coli*: a multi-DNA polymerase affair. *FEMS Microbiol. Rev.*, **36** (6), 1105–1121.
224. Sanjanwala, B. and Ganesan, A. (1991) Genetic structure and domains of DNA polymerase III of *Bacillus subtilis. Mol. Gen. Genet.*, **226** (3), 467–472.
225. Wessel, A.K. *et al.* (2013) Going local: technologies for exploring bacterial microenvironments. *Nat. Rev. Microbiol.*, **11** (5), 337–348.
226. Eun, Y.-J. *et al.* (2010) Encapsulating bacteria in agarose microparticles using microfluidics for high-throughput cell analysis and isolation. *ACS Chem. Biol.*, **6** (3), 260–266.

227. Meyer, A. *et al.* (2015) Optimization of a whole-cell biocatalyst by employing genetically encoded product sensors inside nanolitre reactors. *Nat. Chem.*, **7** (8), 673–678. doi: 10.1038/nchem.2301, Epub 2015 Jul 13.
228. Wang, B.L. *et al.* (2014) Microfluidic high-throughput culturing of single cells for selection based on extracellular metabolite production or consumption. *Nat. Biotechnol.*, **32** (5), 473–478.
229. Dauner, M., Storni, T., and Sauer, U. (2001) *Bacillus subtilis* metabolism and energetics in carbon-limited and excess-carbon chemostat culture. *J. Bacteriol.*, **183** (24), 7308–7317.
230. van Dijk-Salkinoja, M.S. and Planta, R.J. (1971) Rate of ribosome production in *Bacillus licheniformis*. *J. Bacteriol.*, **105** (1), 20–27.
231. Tännler, S. *et al.* (2008) Screening of *Bacillus subtilis* transposon mutants with altered riboflavin production. *Metab. Eng.*, **10** (5), 216–226.
232. Panula-Perala, J. *et al.* (2008) Enzyme controlled glucose auto-delivery for high cell density cultivations in microplates and shake flasks. *Microb. Cell Fact.*, **7** (1), 31.
233. Scheidle, M. *et al.* (2010) High-throughput screening of Hansenula polymorpha clones in the batch compared with the controlled-release fed-batch mode on a small scale. *FEMS Yeast Res.*, **10** (1), 83–92.
234. Knorr, B. *et al.* (2007) Scale-down and parallel operation of the riboflavin production process with *Bacillus subtilis*. *Biochem. Eng. J.*, **33** (3), 263–274.
235. Vester, A. *et al.* (2009) Discrimination of riboflavin producing *Bacillus subtilis* strains based on their fed-batch process performances on a millilitre scale. *Appl. Microbiol. Biotechnol.*, **84** (1), 71–76.
236. Fujita, Y. (2009) Carbon catabolite control of the metabolic network in *Bacillus subtilis*. *Biosci. Biotechnol., Biochem.*, **73** (2), 245–259.
237. Varela, C.A., Baez, M.E., and Agosin, E. (2004) Osmotic stress response: quantification of cell maintenance and metabolic fluxes in a lysine-overproducing strain of *Corynebacterium glutamicum*. *Appl. Environ. Microbiol.*, **70** (7), 4222–4229.
238. Ebert, B.E. *et al.* (2011) Response of *Pseudomonas putida* KT2440 to increased NADH and ATP demand. *Appl. Environ. Microbiol.*, **77** (18), 6597–6605.
239. van Duuren, J. *et al.* (2013) Reconciling *in vivo* and *in silico* key biological parameters of *Pseudomonas putida* KT2440 during growth on glucose under carbon-limited condition. *BMC Biotech.*, **13** (1), 93.
240. Sauer, U. *et al.* (1996) Physiology and metabolic fluxes of wild-type and riboflavin-producing *Bacillus subtilis*. *Appl. Environ. Microbiol.*, **62** (10), 3687–3696.
241. Pirt, S.J. (1965) The maintenance energy of bacteria in growing cultures. *Proc. R. Soc. London, Ser. B*, **163** (991), 224–231.
242. Frankena, J., Verseveld, H.W., and Stouthamer, A.H. (1985) A continuous culture study of the bioenergetic aspects of growth and production of exocellular protease in Bacillus licheniformis. *Appl. Microbiol. Biotechnol.*, **22** (3), 169–176.
243. Zamboni, N. *et al.* (2003) Reducing maintenance metabolism by metabolic engineering of respiration improves riboflavin production by *Bacillus subtilis*. *Metab. Eng.*, **5** (1), 49–55.
244. Sone, N., Tsukita, S., and Sakamoto, J. (1999) Direct correlationship between proton translocation and growth yield: an analysis of the respiratory chain of *Bacillus stearothermophilus*. *J. Biosci. Bioeng.*, **87** (4), 495–499.
245. Winstedt, L. and von Wachenfeldt, C. (2000) Terminal oxidases of *Bacillus subtilis* strain 168: one quinol oxidase, cytochrome aa(3) or cytochrome bd, is required for aerobic growth. *J. Bacteriol.*, **182** (23), 6557–6564.
246. Li, X.-J. *et al.* (2006) Redirection electron flow to high coupling efficiency of terminal oxidase to enhance riboflavin biosynthesis. *Appl. Microbiol. Biotechnol.*, **73** (2), 374–383.

247. Kikuchi, T. and Hasumi, K. (2002) Enhancement of plasminogen activation by surfactin C: augmentation of fibrinolysis *in vitro* and *in vivo*. *Biochim. Biophys. Acta*, **1596** (2), 234–245.
248. Ouoba, L.I. *et al.* (2004) Genotyping of starter cultures of *Bacillus subtilis* and *Bacillus pumilus* for fermentation of African locust bean (Parkia biglobosa) to produce Soumbala. *Int. J. Food Microbiol.*, **90** (2), 197–205.
249. Sarkar, P.K., Hasenack, B., and Nout, M.J. (2002) Diversity and functionality of *Bacillus* and related genera isolated from spontaneously fermented soybeans (Indian Kinema) and locust beans (African Soumbala). *Int. J. Food Microbiol.*, 77 (3), 175–186.
250. Jeyaram, K. *et al.* (2008) Molecular identification of dominant microflora associated with 'Hawaijar' - a traditional fermented soybean (Glycine max (L.)) food of Manipur, India. *Int. J. Food Microbiol.*, **122** (3), 259–268.
251. Schallmey, M., Singh, A., and Ward, O.P. (2004) Developments in the use of *Bacillus* species for industrial production. *Can. J. Microbiol.*, **50** (1), 1–17.
252. EFSA (2007) Introduction of a Qualified Presumption of Safety (QPS) approach for assessment of selected microorganisms referred to EFSA. Opinion of the Scientific Committee. *EFSA J.*, **587**, 1–16.
253. EFSA_BIOHAZ_Panel (2013) Scientific opinion on the maintenance of the list of QPS biological agents intentionally added to food and feed. *EFSA J.*, **11** (11), 3449.
254. Collins, J.F. (1979) in *Beta-Lactamases* (eds J.M.T. Hamilton-Miller and J.T. Smith), Academic Press, New York, pp. 351–368.
255. European_Commission (2000) Opinion of the Scientific Committee on Animal Nutrition on the safety of use of Bacillus species in animal nutrition. EC Directorate B – Scientific Health Opinions.
256. From, C. *et al.* (2007) Cytotoxicity in Bacillus mojavensis is abolished following loss of surfactin synthesis: implications for assessment of toxicity and food poisoning potential. *Int. J. Food Microbiol.*, **117** (1), 43–49.
257. From, C. *et al.* (2005) Toxin-producing ability among Bacillus spp. outside the Bacillus cereus group. *Appl. Environ. Microbiol.*, **71** (3), 1178–1183.
258. Taylor, J.M. *et al.* (2005) Heat-stable toxin production by strains of *Bacillus cereus, Bacillus firmus, Bacillus megaterium, Bacillus simplex* and *Bacillus licheniformis. FEMS Microbiol. Lett.*, **242** (2), 313–317.
259. EFSA_FEEDAP_Panel (2011) Technical guidance on the assessment of the toxigenic potential of Bacillus species used in animal nutrition. *EFSA J.*, **9** (11), 2445.
260. Oka, K. *et al.* (1993) Satisfactory separation and MS–MS spectrometry of six surfactins isolated from *Bacillus subtilis* natto. *Chem. Pharm. Bull. (Tokyo)*, **41** (5), 1000–1002.
261. Kim, S.Y. *et al.* (2007) Surfactin from *Bacillus subtilis* displays antiproliferative effect via apoptosis induction, cell cycle arrest and survival signaling suppression. *FEBS Lett.*, **581** (5), 865–871.
262. Lee, J.H. *et al.* (2012) The production of surfactin during the fermentation of cheonggukjang by potential probiotic *Bacillus subtilis* CSY191 and the resultant growth suppression of MCF-7 human breast cancer cells. *Food Chem.*, **131** (4), 1347–1354.
263. Marahiel, M.A., Nakano, M.M., and Zuber, P. (1993) Regulation of peptide antibiotic production in *Bacillus. Mol. Microbiol.*, 7 (5), 631–636.
264. EFSA_FEEDAP_Panel (2014) Guidance on the assessment of the toxigenic potential of *Bacillus* species used in animal nutrition. *EFSA J.*, **12** (5), 3665.
265. Granum, P.E. (1997) in *Food Microbiology: Fundamentals and Frontiers* (eds M.P. Doyle, L.R. Beuchat, and T.J. Montville), American Society for Microbiology, Washington, DC, pp. 327–336.
266. Granum, P.E. (2001) in *Food Microbiology: Fundamentals and Frontiers* (eds M.P. Doyle, L.R. Beuchat, and T.J. Montville), American Society

for Microbiology, Washington, DC, pp. 373–381.

267. Logan, N.A. (2012) Bacillus and relatives in foodborne illness. *J. Appl. Microbiol.*, **112** (3), 417–429.

268. Turnbull, P.C.B. and Kramer, J.M. (1995) in *Manual of Clinical Microbiology*, 6th edn (eds P.R. Murray *et al.*), American Society for Microbiology, Washington, DC, pp. 349–356.

269. Rowan, N.J. *et al.* (2001) Putative virulence factor expression by clinical and food isolates of *Bacillus* spp. after growth in reconstituted infant milk formulae. *Appl. Environ. Microbiol.*, **67** (9), 3873–3881.

270. Kramer, J.M. and Gilbert, R.J. (1989) in *Foodborne Bacterial Pathogens* (ed. M.P. Doyle), Marcel Dekker Inc., New York, pp. 21–70.

271. Pedersen, P.B. *et al.* (2002) Cytotoxic potential of industrial strains of *Bacillus* sp. *Regul. Toxicol. Pharmacol.*, **36** (2), 155–161.

272. Salkinoja-Salonen, M.S. *et al.* (1999) Toxigenic strains of *Bacillus licheniformis* related to food poisoning. *Appl. Environ. Microbiol.*, **65** (10), 4637–4645.

273. Rowan, N.J. *et al.* (2003) Production of diarrheal enterotoxins and other potential virulence factors by veterinary isolates of *bacillus* species associated with nongastrointestinal infections. *Appl. Environ. Microbiol.*, **69** (4), 2372–2376.

274. Ouoba, L.I., Thorsen, L., and Varnam, A.H. (2008) Enterotoxins and emetic toxins production by *Bacillus cereus* and other species of *Bacillus* isolated from Soumbala and Bikalga, African alkaline fermented food condiments. *Int. J. Food Microbiol.*, **135** (3), 224–230.

275. Priest, F.G., Fleming, A.B., and Tangney, M. (1995) Production of proteins using *Bacillus* incapable of sporulation. Patent WO9703185 A1, Novozyme.

276. Shih, J.C.H. (2007) Production of an asporogenic strain of Bacillus licheniformis and its use for the production of keratinase. Patent WO2007011410 A1, UNIV NC.

277. Li, X. *et al.* (2002) Human receptors for sweet and umami taste. *Proc. Natl. Acad. Sci. U.S.A.*, **99** (7), 4692–4696.

278. Ishii, K. and Shiio, I. (1972) Improved inosine production and derepression of purine nucleotide biosynthetic enzymes in 8-azaguanine resistant mutants of *Bacillus subtilis. Agric. Biol. Chem.*, **36** (9), 1511–1522.

279. Shiio, I. (1989) in Bacillus subtilis*: Molecular Biology and Industrial Application* (eds B. Maruo and H. Yoshikawa), Elsevier Science Publishers, Amsterdam, pp. 191–211.

280. Ledesma-Amaro, R. *et al.* (2013) Biotechnological production of feed nucleotides by microbial strain improvement. *Process Biochem.*, **48** (9), 1263–1270.

281. Yoshikawa, M., Kato, T., and Takenishi, T. (1969) Studies of phosphorylation III. Selective phosphorylation of unprotected nucleosides. *Bull. Chem. Soc. Jpn.*, **42** (12), 3505–3508.

282. Zalkin, H. (1993) in Bacillus subtilis *and Other Gram-Positive Bacteria: Biochemistry, Physiology, and Molecular Genetics* (eds A.L. Sonenshein, J.A. Hoch, and R. Losick), American Society for Microbiology, Washington, DC, pp. 335–341.

283. Ebbole, D.J. and Zalkin, H. (1987) Cloning and characterization of a 12-gene cluster from *Bacillus subtilis* encoding nine enzymes for de novo purine nucleotide synthesis. *J. Biol. Chem.*, **262** (17), 8274–8287.

284. Weng, M., Nagy, P., and Zalkin, H. (1995) Identification of the *Bacillus subtilis* pur operon repressor. *Proc. Natl. Acad. Sci. U.S.A.*, **92** (16), 7455–7459.

285. Ebbole, D.J. and Zalkin, H. (1988) Detection of pur operon-attenuated mRNA and accumulated degradation intermediates in *Bacillus subtilis. J. Biol. Chem.*, **263** (22), 10894–10902.

286. Mandal, M. *et al.* (2003) Riboswitches control fundamental biochemical pathways in *Bacillus subtilis* and other bacteria. *Cell*, **113** (5), 577–586.

287. Bera, A.K. *et al.* (2003) Functional dissection of the *Bacillus subtilis* pur

operator site. *J. Bacteriol.*, **185** (14), 4099–4109.

288. Ebbole, D.J. and Zalkin, H. (1989) Interaction of a putative repressor protein with an extended control region of the *Bacillus subtilis* pur operon. *J. Biol. Chem.*, **264** (6), 3553–3561.

289. Arnvig, K., Hove-Jensen, B., and Switzer, R.L. (1990) Purification and properties of phosphoribosyl-diphosphate synthetase from *Bacillus subtilis. Eur. J. Biochem.*, **192** (1), 195–200.

290. Nygaard, P. and Saxild, H.H. (2005) The purine efflux pump PbuE in *Bacillus subtilis* modulates expression of the PurR and G-box (XptR) regulons by adjusting the purine base pool size. *J. Bacteriol.*, **187** (2), 791–794.

291. Zakataeva, N.P. *et al.* (2007) A new function for the *Bacillus* PbuE purine base efflux pump: efflux of purine nucleosides. *Res. Microbiol.*, **158** (8–9), 659–665.

292. Asahara, T. *et al.* (2010) Accumulation of gene-targeted *Bacillus subtilis* mutations that enhance fermentative inosine production. *Appl. Microbiol. Biotechnol.*, **87** (6), 2195–2207.

293. Ishii, K. and Shiio, I. (1973) Regulation of purine nucleotide biosynthesis in *Bacillus subtilis. Agric. Biol. Chem.*, **37** (2), 287–300.

294. Matsui, H. *et al.* (1979) Production of guanosine by psicofuranine and decoyinine resistant mutants of *Bacillus subtilis. Agric. Biol. Chem.*, **43** (8), 1739–1744.

295. a) Yang, H., *et al.* (2011) Complete genome sequence of Bacillus amyloliquefaciens XH7, which exhibits production of purine nucleosides. *J. Bacteriol.*, **193** (19), 5593–5594; b) Zhang, G. *et al.* (2011) Complete genome sequence of *Bacillus amyloliquefaciens* TA208, a strain for industrial production of guanosine and ribavirin. *J. Bacteriol.*, **193** (12), 3142–3143.

296. Qian, J. *et al.* (2006) Nucleotide mutations in purA gene and pur operon promoter discovered in guanosine- and inosine-producing *Bacillus subtilis* strains. *Biotechnol. Lett.*, **28** (12), 937–941.

297. Miyagawa, K. *et al.* (1986) Cloning of the *Bacillus subtilis* IMP dehydrogenase gene and its application to increased production of guanosine. *Nat. Biotechnol.*, **4** (3), 225–228.

298. Zakataeva, N.P. *et al.* (2012) Wild-type and feedback-resistant phosphoribosyl pyrophosphate synthetases from *Bacillus amyloliquefaciens*: purification, characterization, and application to increase purine nucleoside production. *Appl. Microbiol. Biotechnol.*, **93** (5), 2023–2033.

299. Funk, C. and Poklop, J. (1916) The study of certain dietary conditions bearing on the problem of growth in rats. *J. Biol. Chem.*, **27**, 1–14.

300. György, P. (1976) Reminiscences on the discovery and significance of some of the B vitamins. *Nutr. Rev.*, **34** (5), 141–144.

301. Fischer, M. and Bacher, A. (2005) Biosynthesis of flavocoenzymes. *Nat. Prod. Rep.*, **22** (3), 324–350.

302. Chikindas, M.L. *et al.* (1987) Determining the riboflavin operon boundaries in *Bacillus subtilis. Mol. Gen. Mikrobiol. Virusol. (Russian)*, **4**, 22–26.

303. Perkins, J.B., Pero, J.G., and Sloma, A. (1989) Riboflavin overproducing strains of bacteria. Patent EP0405370 B1, Roche.

304. Haase, I. *et al.* (2013) Enzymes from the haloacid dehalogenase (HAD) superfamily catalyse the elusive dephosphorylation step of riboflavin biosynthesis. *ChemBioChem*, **14** (17), 2272–2275.

305. Kil, Y.V. *et al.* (1992) Riboflavin operon of *Bacillus subtilis*: unusual symmetric arrangement of the regulatory region. *Mol. Gen. Genet.*, **233** (3), 483–486.

306. Mironov, A.S. *et al.* (2002) Sensing small molecules by nascent RNA: a mechanism to control transcription in bacteria. *Cell*, **111** (5), 747–756.

307. Gusarov, I.I. *et al.* (1997) Primary structure and functional activity of the *Bacillus subtilis* ribC gene. *Mol. Biol. (Mosk)*, **31** (5), 820–825.

308. Coquard, D. *et al.* (1997) Molecular cloning and characterisation of the ribC gene from *Bacillus subtilis*: a point mutation in ribC results in riboflavin

overproduction. *Mol. Gen. Genet.*, **254** (1), 81–84.

309. Mack, M., van Loon, A.P.G.M., and Hohmann, H.-P. (1998) Regulation of riboflavin biosynthesis in *Bacillus subtilis* is affected by the activity of the flavokinase/flavin adenine dinucleotide synthetase encoded by ribC. *J. Bacteriol.*, **180** (4), 950–955.
310. Perkins, J.B. *et al.* (1999) Genetic engineering of *Bacillus subtilis* for the commercial production of riboflavin. *J. Ind. Microbiol. Biotechnol.*, **22** (1), 8–18.
311. de Wulf, P. and Vandamme, E.J. (1997) Production of D-ribose by fermentation. *Appl. Microbiol. Biotechnol.*, **48** (2), 141–148.
312. Bretzel, W. *et al.* (1999) Commercial riboflavin production by recombinant *Bacillus subtilis*: down-stream processing and comparison of the composition of riboflavin produced by fermentation or chemical synthesis. *J. Ind. Microbiol. Biotechnol.*, **22** (1), 19–26.
313. Panina, L.I. *et al.* (1983) Cloning the operon genes of riboflavin biosynthesis in *Bacillus subtilis* on plasmid vector pBR322 in Escherichia coli. *Genetika*, **19** (1), 174–176.
314. Zhao, X.M. et al. (2006) Riboflavin-produced engineering strain and its method for producing riboflavin. Patent CN1891814 A, Tianjin University.
315. Shi, S. *et al.* (2009) Increased production of riboflavin by metabolic engineering of the purine pathway in *Bacillus subtilis*. *Biochem. Eng. J.*, **46** (1), 28–33.
316. Wang, Z. *et al.* (2011) Enhancement of riboflavin production with *Bacillus subtilis* by expression and site-directed mutagenesis of zwf and gnd gene from *Corynebacterium glutamicum*. *Bioresour. Technol.*, **102** (4), 3934–3940.
317. Duan, Y.X. *et al.* (2010) Overexpression of glucose-6-phosphate dehydrogenase enhances riboflavin production in *Bacillus subtilis*. *Appl. Microbiol. Biotechnol.*, **85** (6), 1907–1914.
318. Wang, G. *et al.* (2014) Enhancement of riboflavin production by deregulating gluconeogenesis in *Bacillus subtilis*. *World J. Microbiol. Biotechnol.*, **30**, 1893–1900.
319. Hohmann, H.-P. and Stahmann, K.-P. (2010) in *Comprehensive Natural Products II: Chemistry and Biology* (eds L. Mander and H.-W. Liu), Elsevier, Oxford, pp. 115–139.
320. Blum, R. (1996) *Ullmann's Encyclopedia of Industrial Chemistry*, VCH Verlagsgesellschaft, Weinheim, pp. 559–572.
321. Sakamoto, K., Yamada, H., and Shimizu, S. (1989) Method of producing D-pantolactone. Patent EP0436730 B1, Fuji.
322. Yocum, R.R. et al. (1999) Methods and microorganisms for production of panto-compounds. Patent WO0121772 A2, OmniGene.
323. Jackowski, S. (1996) in *Escherichia Coli and Salmonella: Cellular and Molecular Biology* (ed. F.C. Neidhardt), ASM Press, Washington, DC, pp. 687–694.
324. Elischewski, F., Pühler, A., and Kalinowski, J. (1999) Pantothenate production in *Escherichia coli* K12 by enhanced expression of the panE gene encoding ketopantoate reductase. *J. Biotechnol.*, **75** (2–3), 135–146.
325. Matthews, R.G. (1996) in *Escherichia Coli and Salmonella: Cellular and Molecular Biology* (ed. F.C. Neidhardt), ASM Press, Washington, DC, pp. 600–611.
326. Mizuno, M. *et al.* (1996) Systematic sequencing of the 283 kb 210°–232° region of the *Bacillus subtilis* genome containing the skin element and many sporulation genes. *Microbiology*, **142** (11), 3103–3111.
327. Zeigler, D.R. *et al.* (2008) The origins of 168, W23, and other *Bacillus subtilis* legacy strains. *J. Bacteriol.*, **190** (21), 6983–6995.
328. Casutt, M., Koppe, T., and Schwarz, E. (2000) *Ullmann's Encyclopedia of Industrial Chemistry*, Wiley-VCH Verlag GmbH, Weinheim, pp. 130–139.
329. Bower, S.G. *et al.* (1996) Cloning, sequencing, and characterization of the *Bacillus subtilis* biotin biosynthetic operon. *J. Bacteriol.*, **178** (14), 4122–4130.

330. Fugate, C.J. and Jarrett, J.T. (2012) Biotin synthase: insights into radical-mediated carbon-sulfur bond formation. *Biochim. Biophys. Acta*, **1824** (11), 1213–1222.

331. Reyda, M.R. *et al.* (2008) Loss of iron-sulfur clusters from biotin synthase as a result of catalysis promotes unfolding and degradation. *Arch. Biochem. Biophys.*, **471** (1), 32–41.

332. Bower, S.G. et al. (1993) Biotin biosynthesis in *Bacillus subtilis*. Patent US6841366 B1, DSM.

333. Sakurai, N. *et al.* (1994) Improvement of a d-biotin-hyperproducing recombinant strain of *Serratia marcescens*. *J. Biotechnol.*, **36** (1), 63–73.

334. Moine, G. and Hohmann, H.-P. (2000) *Ullmann's Encyclopedia of Industrial Chemistry*, Wiley-VCH Verlag GmbH, Weinheim, pp. 67–82.

335. Begley, T.P. *et al.* (1999) Thiamin biosynthesis in prokaryotes. *Arch. Microbiol.*, **171** (5), 293–300.

336. Begley, T.P., Ealick, S.E., and McLafferty, F.W. (2012) Thiamin biosynthesis: still yielding fascinating biological chemistry. *Biochem. Soc. Trans.*, **40** (3), 555–560.

337. Park, J.H. *et al.* (2004) Characterization of two kinases involved in thiamine pyrophosphate and pyridoxal phosphate biosynthesis in *Bacillus subtilis*: 4-amino-5-hydroxymethyl-2-methylpyrimidine kinase and pyridoxal kinase. *J. Bacteriol.*, **186** (5), 1571–1573.

338. Zhang, Y. *et al.* (1997) Characterization of the *Bacillus subtilis* thiC operon involved in thiamine biosynthesis. *J. Bacteriol.*, **179** (9), 3030–3035.

339. Winkler, W., Nahvi, A., and Breaker, R.R. (2002) Thiamine derivatives bind messenger RNAs directly to regulate bacterial gene expression. *Nature*, **419** (6910), 952–956.

340. Melnick, J. *et al.* (2004) Identification of the two missing bacterial genes involved in thiamine salvage: thiamine pyrophosphokinase and thiamine kinase. *J. Bacteriol.*, **186** (11), 3660–3662.

341. Mukherjee, T. *et al.* (2011) Pyridoxal phosphate: biosynthesis and catabolism. *Biochim. Biophys. Acta*, **1814** (11), 1585–1596.

342. Raschle, T. *et al.* (2007) Reaction mechanism of pyridoxal 5′-phosphate synthase. Detection of an enzyme-bound chromophoric intermediate. *J. Biol. Chem.*, **282** (9), 6098–6105.

343. Burns, K.E. *et al.* (2005) Reconstitution and biochemical characterization of a new pyridoxal-5′-phosphate biosynthetic pathway. *J. Am. Chem. Soc.*, **127** (11), 3682–3683.

344. Moccand, C., Kaufmann, M., and Fitzpatrick, T.B. (2011) It takes two to tango: defining an essential second active site in pyridoxal 5′-phosphate synthase. *PLoS One*, **6** (1), e16042.

345. Yocum, R.R., Williams, M.K., and Pero, J.G. (2002) Methods and organisms for production of B6 vitamers. Patent WO2004035010 A2, BASF AG.

346. Yang, Y. *et al.* (1998) Involvement of the gapA- and epd (gapB)-encoded dehydrogenases in pyridoxal 5′-phosphate coenzyme biosynthesis in *Escherichia coli* K-12. *J. Bacteriol.*, **180** (16), 4294–4299.

347. Banks, J. and Cane, D.E. (2004) Biosynthesis of vitamin B6: direct identification of the product of the PdxA-catalyzed oxidation of 4-hydroxy-L-threonine-4-phosphate using electrospray ionization mass spectrometry. *Bioorg. Med. Chem. Lett.*, **14** (7), 1633–1636.

348. Hoshino, T. et al. (2002) Microorganism and process for preparing vitamin B6. Patent EP1543104 B1, DSM.

349. Yeh, J.I. *et al.* (2002) Multistate binding in pyridoxine 5′-phosphate synthase: 1.96 A crystal structure in complex with 1-deoxy-D-xylulose phosphate. *Biochemistry*, **41** (39), 11649–11657.

350. di Salvo, M. *et al.* (1998) Expression, purification, and characterization of recombinant *Escherichia coli* pyridoxine 5′-phosphate oxidase. *Protein Expression Purif.*, **13** (3), 349–356.

351. Cane, D.E. *et al.* (1999) Biosynthesis of vitamin B6: enzymatic conversion of 1-deoxy-D-xylulose-5-phosphate to pyridoxol phosphate. *J. Am. Chem. Soc.*, **121** (33), 7722–7723.

352. Lehmann, M. *et al.* (2009) Biosynthesis of riboflavin. Screening for an improved GTP cyclohydrolase II mutant. *FEBS J.*, **276** (15), 4119–4129.
353. Commichau, F.M. *et al.* (2015) Engineering *Bacillus subtilis* for the conversion of the antimetabolite 4-hydroxy-l-threonine to pyridoxine. *Metab. Eng.*, **29**, 196–207. doi: 10.1016/j.ymben.2015.03.007, Epub 2015 Mar 14.
354. Harwood, C.R. (1992) *Bacillus subtilis* and its relatives: molecular biological and industrial workhorses. *Trends Biotechnol.*, **10** (7), 247–256.
355. Liu, G. *et al.* (2013) Development of highly efficient, low-cost lignocellulolytic enzyme systems in the post-genomic era. *Biotechnol. Adv.*, **31** (6), 962–975.
356. Nijland, R. and Kuipers, O.P. (2008) Optimization of protein secretion by *Bacillus subtilis. Recent Pat. Biotechnol.*, **2** (2), 79–87.
357. Jarnagin, A.S. and Ferrari, E. (1992) Extracellular enzymes: gene regulation and structure function relationship studies. *Biotechnology*, **22**, 189–217.
358. Dimarogona, M., Topakas, E., and Christakopoulos, P. (2013) Recalcitrant polysaccharide degradation by novel oxidative biocatalysts. *Appl. Microbiol. Biotechnol.*, **97** (19), 8455–8465.
359. Blumer-Schuette, S.E. *et al.* (2014) Thermophilic lignocellulose deconstruction. *FEMS Microbiol. Rev.*, **38** (3), 393–448.
360. van Dijl, J.M. and Hecker, M. (2013) *Bacillus subtilis*: from soil bacterium to super-secreting cell factory. *Microb. Cell Fact.*, **12**, 3.
361. Otto, A. *et al.* (2010) Systems-wide temporal proteomic profiling in glucose-starved *Bacillus subtilis. Nat. Commun.*, **1**, 137.
362. Becher, D. *et al.* (2011) From the genome sequence to the protein inventory of *Bacillus subtilis. Proteomics*, **11** (15), 2971–2980.
363. Maass, S. *et al.* (2011) Efficient, global-scale quantification of absolute protein amounts by integration of targeted mass spectrometry and two-dimensional gel-based proteomics. *Anal. Chem.*, **83** (7), 2677–2684.
364. Antelmann, H. *et al.* (2001) A proteomic view on genome-based signal peptide predictions. *Genome Res.*, **11** (9), 1484–1502.
365. Maass, S. *et al.* (2014) Highly precise quantification of protein molecules per cell during stress and starvation responses in *Bacillus subtilis. Mol. Cell. Proteomics*, **13** (9), 2260–2276.
366. Tjalsma, H. and van Dijl, J.M. (2005) Proteomics-based consensus prediction of protein retention in a bacterial membrane. *Proteomics*, **5** (17), 4472–4482.
367. Tam, L.T. *et al.* (2006) Proteome signatures for stress and starvation in *Bacillus subtilis* as revealed by a 2-D gel image color coding approach. *Proteomics*, **6** (16), 4565–4585.
368. Kabisch, J. *et al.* (2013) Characterization and optimization of *Bacillus subtilis* ATCC 6051 as an expression host. *J. Biotechnol.*, **163** (2), 97–104.
369. Bansal, A.K. (2005) Bioinformatics in microbial biotechnology – a mini review. *Microb. Cell Fact.*, **4**, 19.
370. Goelzer, A. and Fromion, V. (2011) Bacterial growth rate reflects a bottleneck in resource allocation. *Biochim. Biophys. Acta*, **1810** (10), 978–988.
371. Goelzer, A., Fromion, V., and Scorletti, G. (2011) Cell design in bacteria as a convex optimization problem. *Automatica*, **47** (6), 1210–1218.
372. Li, H., *et al.* (2011). De novo engineering and metabolic flux analysis of inosine biosynthesis in Bacillus subtilis. *Biotechnol. Lett.*, **33** (8), 1575–1580.

8
Host Organism: *Pseudomonas putida*

Ignacio Poblete-Castro, José M. Borrero-de Acuña, Pablo I. Nikel, Michael Kohlstedt, and Christoph Wittmann

8.1
Introduction

Fast growth, a low nutrient demand, an extremely versatile metabolism, and a high capacity to supply redox power are only a few advantages that render *Pseudomonas putida* an interesting host for biotechnological applications. The species furthermore tolerates a wide operational process window, including extreme pH values, high temperature, as well as high levels of toxic substances and organic solvents [1–4]. *P. putida* strains are typically found in soil and water and members of this species have a broad metabolic versatility, which allows them to adapt to different habitats and nutritional environments [5]. Among others, strains of industrial relevance comprise *P. putida* KT2440, *P. putida* BIRD-1, *P. putida* UW4, and *P. putida* S11, isolated from rhizosphere soil and developed as plant-growth-promoting rhizobacteria; *P. putida* F1, originating from polluted soil and developed as a bioremediation agent; nicotine-degrading *P. putida* S16; and naphthalene-degrading *P. putida* ND6 [6]. In addition, solvent-tolerant strains such as *P. putida* S12, *P. putida* DOT-T1E, or *P. putida* VLB120 have been isolated, which exhibit effective defense mechanisms and are even more robust against organic solvents [7–9]. In particular, the biosafety strain *P. putida* KT2440, designated for safe recombinant DNA experimental procedures and applications [10], has been considered for biotechnological application [11]. It is a plasmid-free derivative of a toluene-degrading bacterium previously classified as *Pseudomonas arvilla* strain mt-2 and later reclassified as *P. putida* mt-2 [12]. The genome of the strain is 6.18 Mb in size and indicates an exceptional repertoire of enzymes needed to thrive in diverse environments [13]. Although *P. putida* KT2440 partly shares coding regions with pathogenic *Pseudomonas* strains such as *P. aeruginosa*, it lacks virulence factors, such as Type III secretion system components or exotoxin A [12]. Taken together, these features render *P. putida* a workhorse for white biotechnology, involving the production of biopolymers such as polyhydroxyalkanoates (PHAs), industrial enzymes, and fine chemicals such as epoxides, substituted catechols, enantiopure alcohols, and heterocyclic

Industrial Biotechnology: Microorganisms, First Edition. Edited by Christoph Wittmann and James C. Liao.

compounds. In addition, *P. putida* is well known for its capacity to degrade xenobiotics and resist heavy metals, opening plentiful applications in environmental biotechnology [2, 14]. In addition to *P. putida*, other related *Pseudomonas* species, such as *P. fluorescens*, *P. aeruginosa*, *P. chlororaphis*, *P. denitrificans*, and *P. oleovorans*, also have interesting properties and traits for biotechnological application. These are not covered here, but are described in an excellent manner in different reviews [11, 15–17].

8.2 Physiology and Metabolism

8.2.1 Substrate Uptake

P. putida is capable of metabolizing hexoses such as glucose and fructose. However, glucose does not play the same central role in *Pseudomonas* as it does in *Escherichia coli*, *Bacillus subtilis*, or lactic acid bacteria [18, 19]. In fact, the preferred carbon sources for *Pseudomonas* are some organic acids or amino acids, rather than glucose [20]. In the presence of succinate and citrate, glucose metabolism is inhibited until these compounds are completely consumed [21]. In addition, catabolic enzymes of phenol degradation are repressed in the presence of organic acids [22]. Unlike many other bacteria, such as *E. coli*, the uptake of glucose in *P. putida* is not mediated by a phosphotransferase transport system (PTS). In fact, glucose enters the periplasmic space through the OprB-1 porin and can then be directly transported into the cell, or converted to gluconate or to 2-ketogluconate in the periplasmic space, compounds that are internalized using specific transporters [19]. In contrast, a PTS transporter takes up fructose [23]. All other sugars metabolized by pseudomonads are transported through PTS-independent systems [19]. Most *Pseudomonas* strains cannot grow on carbon five sugars such as D-xylose or L-arabinose [11]. Only selected isolates such as *Pseudomonas taiwanensis* can assimilate D-xylose via the Weimberg pathway [24]. Glycerol, organic acids, fatty acids, and amino acids, preferred substrates of *P. putida*, are assimilated by the cell via specific transporters. For example, glycerol, a side product from the biodiesel industry and an interesting industrial substrate, is incorporated by an outer-membrane porin (OprB-1, PP_1019) and is then further transported into the cytoplasm by a facilitator (GlpF, PP_1076) [25, 26].

8.2.2 Major Catabolic Pathways

Similarly to other rhizosphere symbiotic bacteria, *P. putida* KT2440 lacks the enzyme 6-phophofructokinase (Pfk) [18] and therefore it exhibits a nonfunctional glycolytic Embden–Meyerhof–Parnas (EMP) pathway (Figure 8.1). Accordingly,

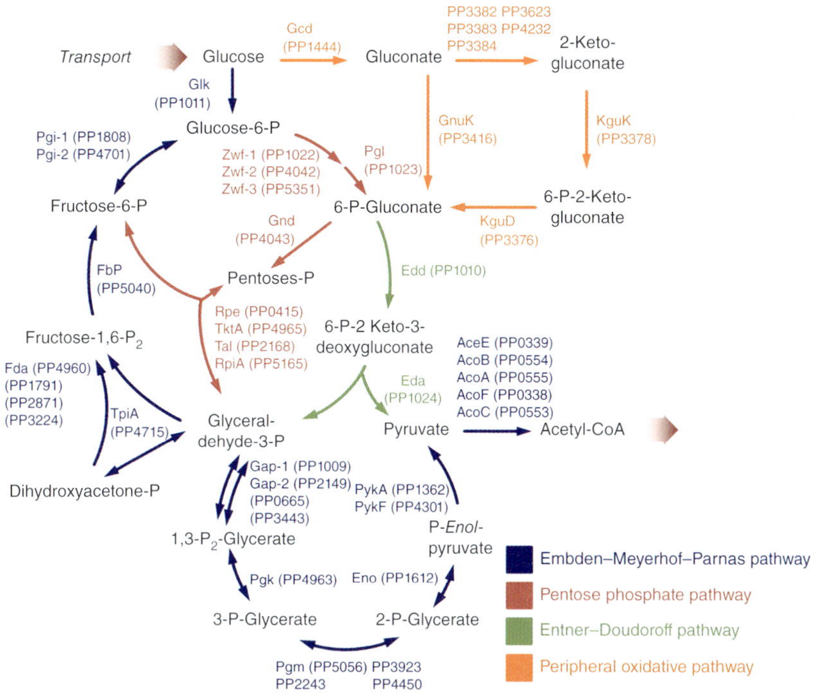

Figure 8.1 Biochemical pathways involved in glucose catabolism in *Pseudomonas putida* KT2440. The metabolic network depicted is sketched around four main metabolic blocks, identified with different colors: (i) the peripheral oxidative pathways, that encompass the oxidative transformation of glucose into gluconate and 2-ketogluconate (and the corresponding phosphorylated derivatives of these metabolites); (ii) the Embden–Meyerhof–Parnas (EMP) pathway (nonfunctional, due to the absence of a 6-phosphofructokinase activity); (iii) the pentose phosphate (PP) pathway; and (iv) the Entner–Doudoroff (ED) pathway. Note that further catabolism downward acetyl-CoA (i.e., the tricarboxylic acid cycle) is represented by a wide shaded arrow. Some reactions and pools of metabolites in this scheme have been lumped to simplify the diagram. The main enzymes involved in this biochemical network are identified; in the instances in which no gene name has been assigned, the PP number is given for each open reading frame encoding the corresponding activity.

carbon compounds are processed mainly through the Entner–Doudoroff (ED) pathway and partially via the pentose phosphate (PP) pathway [27]. While the ED pathway yields only half of the ATP generated by the EMP route, it is superior with regard to the supply of reducing power. Elevated amounts of NADPH provided by the ED route serve anabolic pathways and oxidative stress response metabolism [28, 29]. As mentioned earlier, periplasmic glucose can be transported into the cytoplasm, followed by phosphorylation into glucose-6 phosphate and further metabolized. In addition, glucose can be oxidized to gluconate and 2-ketogluconate (through a number of sequential reactions termed peripheral oxidative pathway), which are either secreted into the medium or transported into the cytoplasm. Excreted gluconate and 2-ketogluconate can be reimported upon glucose depletion. After phosphorylation, they feed the ED route, yielding the central pathway intermediates pyruvate and glyceraldehyde 3-phosphate, respectively. Interestingly, a fraction of these trioses was found to be recycled back to hexoses-phosphate through the concerted action of enzymes belonging to the ED, EMP, and PP pathways. The cyclic operation of central carbon metabolism in strain KT2440 can further boost the formation of NADPH [30].

The PTS-mediated uptake of fructose involves phosphorylation to fructose 1-phosphate (F1P) and then to fructose 1,6-bis-phosphate (Figure 8.1). This entry point into the metabolism results from the fact that fructose-derived carbon is channeled through the ED pathway, the PP pathway, and the reactions of the lower EMP pathway, simultaneously. The tricarboxylic acid (TCA) cycle is an essential pathway for energy generation in aerobic pseudomonads and this cycle is operational during growth of *P. putida* on most carbon sources [25]. The TCA cycle is fueled by anaplerotic reactions. Strains of *P. putida* use the pyruvate carboxylase instead of the PEP carboxylase to form the TCA cycle intermediate oxaloacetate, as demonstrated by ^{13}C metabolic flux analysis [29]. In addition, ^{13}C metabolic flux analysis revealed that *P. putida* KT2440 has a more active pyruvate shunt when growing on fructose as compared to growth on glucose [31]. Remarkably, *P. putida* KT2440 has a huge capacity to provide acetyl-CoA. Even under carbon limitation, it can simultaneously fuel the synthesis of medium-chain-length polyhydroxyalkanoates (mcl-PHAs) from this building block and the anabolic pathways originating from acetyl-CoA [32]. With regard to the catabolic breakdown of glycerol, a kinase (GlpK) phosphorylates glycerol into glycerol 3-phosphate, which is then oxidized to dihydroxyacetone phosphate (DHAP) by a membrane-bound dehydrogenase (GlpD). Intriguingly, glycerol provokes a long lag phase in various *P. putida* strains [26, 33]. Deletion of GlpR in *P. putida* KT2440 alleviates the repression of several genes involved in glycerol metabolism, thus reducing the length of the lag phase [34]. It has also been shown that glycerol metabolism activates glycolytic and gluconeogenic pathways [26].

8.2.3
Degradation of Xenobiotics

Certain pseudomonads differentiate themselves from other bacteria like *E. coli* and *B. subtilis* by their ability to degrade aromatic compounds. In bioremediation applications, specialized *P. putida* strains and enzymes serve as biocatalysts for the degradation of xenobiotics. They make use of the powerful arsenal of monooxygenases, aromatic ring-cleaving dioxygenases, and hydroxylases found in *P. putida*. For example, biodegradation of toluene and xylenes largely relies on activities encoded by the TOL plasmid [35]. Two operons located on the TOL plasmid rule catabolism and subsequent conversion of toluene and xylenes into benzoate and alkyl-benzoates, respectively, and further into TCA cycle intermediates, while two regulator genes, namely, *xylS* and *xylR*, encode the two main transcriptional regulators [36]. The two operons *xylUWCAMBN* and *xylXYZLTEGFJQKIH* encode the corresponding enzyme cascades of the upper and the lower (*meta*) pathways, respectively [37]. Transcription of genes of the TOL plasmid integrates a wide range of environmental signals such as different carbon sources, temperature, and the metabolic status of the cell. For example, XylR responds to the level of toluene and *m*-xylene by activating transcription of genes of the upper pathway together with the regulator gene *xylS*. As a consequence, XylS triggers the expression of the *meta*-operon for expression of the entire catabolic route. Likewise, the introduction of the TOL plasmid into other pseudomonads enables the degradation of toluene, *m*-xylene, and *p*-xylene [3]. More recently, the available genomic sequence of *P. putida* KT2440 stimulated the elucidation of other aromatic degradation pathways. It turned out that the bacterium possesses five major catabolic routes to degrade a variety of xenobiotics into metabolites such as pyruvate, succinyl-CoA, and acetyl-CoA: the β-ketoadipate pathway, the phenylacetyl-CoA pathway, the homogentisate pathway, the gentisate pathway, and the homo-proto-catechuate pathway [1, 38]. On the basis of this repertoire, *P. putida* KT2440 has been successfully applied to degrade recalcitrant aromatic hydrocarbons [1, 39, 40], chloro- and heterocyclic aromatics [41], alkyl branched aromatic naphthenic acids [42], aromatic amines [1, 43], and organic solvents [44, 45]. Using heterologous gene expression, pollutants such as polycyclic aromatic hydrocarbons (PAHs) [46], organophosphates [47–50], pyrethroids [50], organohalides [51], and aromatic sulfonates [52] can be also be mineralized.

8.2.4
Regulation of Core Carbon and Energy Metabolism

Bacteria such as *P. putida*, which can catabolize diverse carbon sources, possess an intricate regulatory control to fish out the most efficient substrates from mixtures of different compounds. It is interesting to note that pseudomonads do not prefer glucose as a carbon source, unlike most other bacteria. For example, *P. putida* exhibits sequential use of succinate and glucose, when both substrates are provided simultaneously [20]. The preferential use of specific carbon sources is termed *carbon catabolite repression* (CCR). In pseudomonads, the two key regulators Crc and Hfq cooperate in catabolite repression and optimization of metabolism [53]. They form a stable cocomplex with mRNAs containing catabolite activity motifs (AAnAAnAA, where n represents any nucleotide) resulting in the inhibition of translation initiation and therefore repression. Upon depletion of preferred carbon sources, cells switch off the CCR system and metabolize remaining substrates in parallel. Actually, succinate exerts an intense CCR response because of which this substrate is preferred over glucose in *P. aeruginosa* [19, 21]. In contrast, *P. putida* seems to generate a weak Crc-mediated CCR response upon succinate presence [53]. Glucose uptake appears strongly inhibited by the Crc/Hfq regulatory system, when cells are cultivated in complex medium.

8.3 Genetic Manipulation

8.3.1 Transformation

Owing to the high relevance of *P. putida*, a rich set of plasmids has been developed as workhorses for molecular biology (Table 8.1). These can be used in a straightforward manner to transform *P. putida* strains. Efficient transformation is based on electroporation, that is, transitory cell envelope damage (generation of pores) to allow plasmid transfer into the cell [63, 64]. Alternatively, plasmids can be introduced into *P. putida* by bi- and triparental mating, respectively, whereby a donor strain such as *E. coli* S17-1 (lambda *pir*) along with a helper strain such as *E. coli* HB101 carrying pRK600, delivers the plasmid of interest through conjugation. The donor strain(s) and recipient *P. putida* are grown to mid-log phase and are subsequently combined to equal cell concentrations and plated on a mating plate. After incubation for a couple of hours, appropriate suspensions prepared from grown cells are plated onto selective medium to select exconjugant target cells and eliminate the donor strains [65].

8.3.2 Engineering Gene Expression

Genetic modification of *P. putida* is highly relevant for metabolic engineering of producing strains. Modifications of interest consist of deletions and insertions attributing new features to the genome [66]. Such manipulations can be obtained by (i) homologous recombination (mediated by the RecA catalytic activity), (ii) site-specific recombination (subordinated to either the resolvase–invertase family or the FLP-*FRT*, Cre-*loxP* family, also referred to as the Int family), and

Table 8.1 Available plasmids for genetic engineering of *P. putida*.

Plasmid	Application	References
pSEVAs	Gene cloning, overexpression, transcriptional and translational fusions, gene deletion	[54, 55]
pEMG, pSEVAx12S	Gene deletion	[56]
pSW-I	I-*Sce*I-mediated gene deletion	[56]
Mini-Tn*5* and Tn*10* derivatives	Gene insertion mutagenesis, promoter probing, chromosomal insertion of cloned DNA	[57–60]
Mini-Tn*7* derivatives	Gene insertion mutagenesis, promoter probing, chromosomal insertion of cloned DNA	[61, 62]

(iii) transposon-mediated gene integration (involving phage integration, excision through site-specific recombination between the phage and bacteria attachment sites *attP* and *attB*, respectively). A breakthrough approach to eliminate genes from *P. putida* relies on the generation of double-strand breaks directed by the action of a conditionally expressed I-*Sce*I homing endonuclease, not found in any microbial genome known [56]. In short, I-*Sce*I target sites are inserted in the genome upon cointegration of a suicide vector (pEMG) by homologous recombination (Figure 8.2). The endowed antibiotic resistance can be used

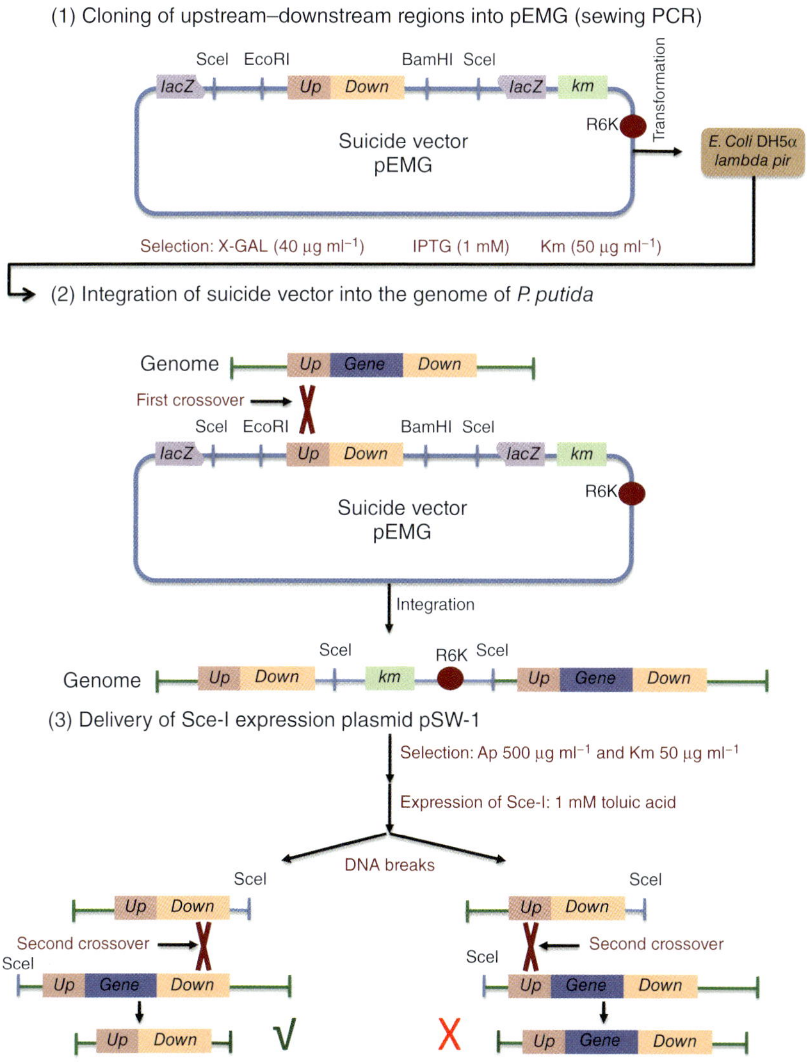

Figure 8.2 Molecular basis for the generation of knockout mutants in *P. putida* [56].

to select cells harboring the integrated vector. After this, a second plasmid-encoding I-*Sce*I endonuclease is introduced into *P. putida*. Upon induction, the I-*Sce*I-mediated DNA cleavage provides four base pair staggered fragments, which are reassembled to a large extent in a way that leads to the elimination of the gene of interest. The system provides a much higher efficiency to knockout genes as compared to other approaches (such as the *sacB*-based counterselection method).

In addition, *P. putida* can be genetically modified by means of transposons, for example, by mini-Tn*5*, mini-Tn*7*, and mini-Tn*10* transposons (Table 8.1), respectively [57, 67]. Such mini-transposon vectors have been available for about 20 years and have proved valuable for the stable engineering of diverse Gram-negative bacteria [68]. Transpositions can be random (most transposons, *mariner* and Mu) or site specific (conjugative transposons, AT-rich specific and Tn*7*). Transposable elements, such as mini-Tn*5* and mini-Tn*10*, favor insertion into AT-rich sequences, whereas mini-Tn*7* targets are site specific and they integrate at the downstream region of the *glmS* gene. Therefore, these transposons are site-specific transposable elements, which have been widely exploited for genetic modification of *P. putida* [66]. Tn*5*-derived constructs are attractive because they can be maintained in cells without antibiotic selection and can be reused for different insertions within the same cells [68]. Originating from the pUT plasmid series [58], these vectors can be delivered to *P. putida* by triparental mating or electroporation. The Tn*5* transposition system requires only a transposase (*tnpA*), which is encoded within the plasmid, and the terminal ends of the transposon so that this system functions independently from the host offering a broad application range beyond *P. putida*. This genetic tool is not only suitable for genomic mutagenesis, but, to name a few applications, for successful integration of functional cargos such as transcriptional reporters, which have a large potential for determination of transcriptional regulation scenarios. In addition, Tn*7* plasmids have been designed over the last decades and applications spanning genetic footprinting, transcriptional and translational fusions, signature-tagged mutagenesis, DNA or cDNA sequencing, transposon site hybridization, and scanner-linker mutagenesis have been developed [69].

Besides gene disruption or deletion, cloning and overexpression of genes has been widely investigated. In recent years, a plasmid collection termed SEVA (*Standard European Vector Architecture*), has been created and displays an invaluable source for construction of recombinant *P. putida* strains. A rich set of cloning and expression vectors (some of them activated by orthogonal inducers such as toluic acid or cyclohexanone), are publically available upon request (http://wwwuser.cnb.csic.es/~seva/) [54]. The architecture of SEVA vectors is based on three modules: a cargo (blue), a replication origin (green), and an antibiotic marker (brown) (Figure 8.3). The different modules are exchangeable upon restriction, so that the inventory of available vectors spans a broad application range (Table 8.2).

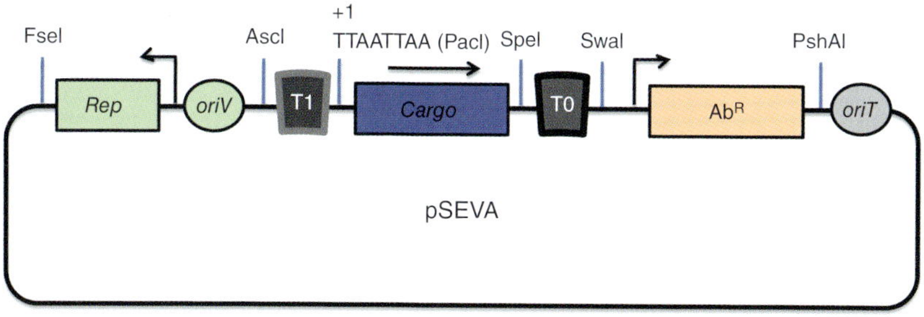

Figure 8.3 Architecture of pSEVA vector modules for genetic engineering of *P. putida*. The different restriction enzymes available for exchange of these modules are shown. Rep, replication origin and Ab^R, antibiotic resistance. The three modules are separated by three permanent regions (shared by all vectors) comprising the T0 and T1 transcriptional terminators and the *oriT* origin, which are intended for broad host range transfer to diverse Gram-negative recipients mediated by the RP4 conjugation machinery.

Table 8.2 Modules available among the pSEVA vectors (http://wwwuser.cnb.csic.es/~seva/).

Antibiotic markers	Replication origins	Cargo types
Ampicillin	R6K	Cloning (MCS, *lacZ*-alpha and *lacZ*-alpha-I-SceI)
Kanamycin	RK2	Expression systems (*lacIq*-Ptrc and *xylS*-Pm)
Chloramphenicol	pBBR1	Promoter probe (*lacZ*, *luxCDABE*, YFP, CFP, DsRed2, and mCherry)
Streptomycin/spectinomycin	pRO1600/ColE1	—
Tetracycline	RFS1010	—
Gentamicin	p15A	—
—	pSC101	—
—	pUC	—
—	pBR322/ROP	—

8.4 Systems Biology

As shown in the section above, a sophisticated toolbox allows targeted genetic modification of *P. putida*. Toward industrial application, the inherent complexity of its metabolism and the superimposed regulation displays a huge challenge to identify the right genetic targets for improved performance. A much better understanding of cellular physiology is needed to reshape the intermediary metabolism toward desired functions. The rise of powerful technologies of the *omics* era over the past decades and their application to *P. putida* have significantly contributed to our current understanding of this bacterium and enabled knowledge-based metabolic engineering of *P. putida* strains [70].

8.4.1
Genomics and Genome Scale-Modeling

Around 20 years ago, the pioneering sequencing of the 6.18 Mb genome of *P. putida* KT2440 revealed the existence of diverse transport and metabolic systems hitherto unknown in pseudomonads. Most important for industrial purposes was the obviously nonpathogenic nature of the strain, which became immediately clear from the lack of pathogenic determinants. The presence of approximately 600 transcriptional factors and more than 20 alternative σ-subunits of RNA polymerase indicates a complex architecture of transcriptional control to modulate gene expression in *P. putida*. Driven by the desire to exploit the valuable genome sequence as much as possible, extensive online data bases on strain taxonomy, gene and protein annotation, and metabolic pathways were soon developed [71]. Today, we have detailed knowledge of the genetic elements of the core genome and the pangenome of strain KT2440 [72]. The core genome comprises 3386 genes with a vast collection of nutrient transporters. Other conserved genes relate to central carbon metabolism, such as ED and PP pathway, arginine and proline metabolism, and pathways for the degradation of aromatics. The related strains *P. putida* W619, F1, GB-1, and KT2440 share specific features in their genome [72]. For instance, all harbor a circular genome of higher GC content and their chromosomal replication (*oriC*) origin is distinct from enteric-type origins.

Most important for strain engineering, several genome-scale metabolic models were constructed on the basis of sequence information, enriched by biochemical and physiological data [73–75]. As outlined below in more detail, they serve as an invaluable resource *in silico* flux predictions and strain engineering [76]. To date, the most complete metabolic model spans 1071 reactions and 1044 metabolites and is related to 900 fully annotated genes. All models are continuously updated with novel findings, involving experimental data on stoichiometric demand for anabolism and cellular maintenance, to increase their predictive power. In addition, the genome sequence has strongly boosted other systems biology efforts with *P. putida*. For example, a comprehensive interaction database of *P. putida* KT2440 (PutidaNET) has been generated from three protein–protein interaction methods [77]. Mutant and promoter libraries have also been generated [67, 78]. In addition, massive genome reduction provided strains with enhanced energy supply and redox power for challenging biosynthesis [79, 80].

8.4.2
Transcriptomics

Genome-wide transcriptomics have been used to quantify gene expression under conditions that are relevant for industrial production. For example, studies focused on growth on different nutrients [27, 81, 82], different types of abiotic stresses [3, 83–85], low temperature [86–88], elevated pressure [89], nitrogen deficiency [90, 91], and accumulation of PHAs [92, 93]. A remarkable feature of *P. putida* as revealed by means of high-resolution RNA sequencing is the differential expression of more than 20% of the *P. putida* genome, when cells shift from an entirely glycolytic to a gluconeogenic metabolism, depending on the carbon source [94]. In a similar study, the growth of *P. putida* mt-2 on *m*-xylene and related aromatic compounds was studied at the transcriptional level using tiling arrays [95]. This allowed to identify all genes activated by certain aromatic substrates and to reannotate genes that had been overlooked previously. An even more detailed inspection of the regulatory inventory of *P. putida* recently revealed 36 novel small, noncoding RNAs, 143 novel open reading frames in 106 intergenic regions, 42 unclassified genes and 8 highly expressed, leaderless mRNA transcripts [96].

8.4.3 Proteomics

In a pioneering attempt, a two-dimensional gel protein reference map was set up for the study of *P. putida* KT2440, describing the pattern of protein abundance for cells growing with glucose as the carbon source [97]. This revealed about 200 individual proteins, reflecting most prominent pathways for hexose degradation. More recently, the transition from exponential growth to stationary conditions was investigated using label-free quantification of proteins in *P. putida* F1 [98]. It turned out that the bacterium adapts to stationary metabolism by gradually increasing the capacity to use amino acids as carbon, nitrogen, and energy sources. These metabolic and physiological adaptations take place in a continuous fashion, instead of a sudden physiological switch. Another interesting study considered culture and environmental conditions similar to the type of *P. putida* normally encounters in nature [99]. In a continuous culture system with cell retention, proliferating *P. putida* cells attained a steady state, in which slow growth created two populations: growing cells and viable but nonculturable cells. Obviously, near-zero growth requires a controlled partial self-digestion and deep reconfiguration of the metabolic machinery that results in the biosynthesis of new products and development of broad stress resistance. Several other studies explored the proteomic response of *P. putida* to stressful conditions, such as presence of toluene [100–102], naphthalene [103], styrene [104], phenol [4, 105], *n*-butanol [45], heavy metals (Cd^{2+}, Ni^{2+}, and Co^{2+}) [106, 107], and shift to low temperature [87]. The picture emerging from these studies – in combination with the evidence gathered through detailed transcriptome analysis – indicates a rather *general* response of *P. putida* to stress conditions: the formation of proteins involved in heat shock stress and the universal stress-response protein, proteins related to oxidative stress detoxification (e.g., superoxide dismutase and catalase), and a profound rearrangement of the pattern of transporters for metabolites and solvents.

8.4.4
Metabolomics and Fluxomics

The metabolic network of *P. putida* was, most importantly, extensively studied on the level of metabolic fluxes, that is, *in vivo* reactions rates through individual enzymes and metabolic pathways [27, 30, 108–110]. As for other microorganisms, comprehensive approaches for flux analysis combined ^{13}C tracer studies with mass spectrometric labeling analysis and isotopomer and metabolite balancing for flux calculation, the latter using sophisticated software [111]. This provided important insights into the metabolism of *P. putida*. Until now, fluxome studies of different *P. putida* strains have focused on the elucidation of network operation during catabolism of glucose [27, 30, 108, 112] or other carbon sources [31, 113]. Furthermore, growth under severe stress conditions, such as the presence of a solvent [44, 109, 112], an impaired energy and redox cofactor metabolism [110], or iron limitation [114], was studied. In *P. putida* grown on glucose, the ED pathway is the main catabolic route yielding pyruvate and glyceraldehyde 3-phosphate (Figure 8.1). Thereby, the PP pathway serves only anabolic demands. The previously proposed cyclic operation of the ED pathway in pseudomonads [115] was recently proved using well-defined tracer experiments followed by LC-MS analysis of intracellular intermediates of central metabolism [30]. This study also revealed that instead of direct glucose phosphorylation via glucose kinase (Glk), most of the imported glucose is directly oxidized in the periplasm into gluconate and, to a lesser extent, to 2-ketogluconate. These are transiently secreted and eventually enter metabolism at the level of 6-phosphogluconate. Additionally, glucose-grown cells exhibit a high gluconeogenic flux through phosphoenolpyruvate carboxykinase (PEPCK) and the anaplerotic pyruvate shunt, thus bypassing malate dehydrogenase [31, 108]. A naturally high NAD(P)H regeneration rate and the presence of two transhydrogenases – one is soluble and non-energy-dependent, and the other is membrane-bound, which translocates protons – ensure a sufficient redox cofactor supply, for example, during whole-cell redox biocatalysis [109, 110]. Several stress studies show that *P. putida* is generally highly resistant against energetic stress owing to its flexible and robust metabolism. Substrate uptake, flux rewiring as well as anabolic demand can be fine-tuned to face challenging situations, which does not necessarily correlate with changes on the transcript level [113]. ^{13}C metabolic flux analysis has also been performed in other pseudomonads, [116, 117]. The findings are in line with previous studies on *P. putida* and confirm the metabolic plasticity and capability to reorganize carbon catabolism upon genetic modifications and environmental changes. Although highly informative to understand cellular physiology and identify metabolic bottlenecks for strain engineering, only a few selected studies report on the level of intracellular metabolites in *P. putida*. This is likely due to the fact that accurate microbial metabolomics are challenging and huge efforts are needed to develop and validate appropriate protocols for sampling and analysis [118]. However, first studies highlight the advance that has been made recently in this field as well. For example, nontargeted analysis of *P. putida* S12 provided prominent "snapshot" metabolomes, which could be used to derive signature metabolic intermediates for growth on fructose, glucose, gluconate, and succinate, respectively [119]. More recently, thorough quantification of metabolic intermediates and cofactors allowed for the *in vitro* investigation of major catabolic dehydrogenases of *P. putida* KT2440 reproducing quasi *in vivo* conditions [30, 120]. This revealed a promiscuity of the dehydrogenases in terms of cofactor use.

8.4.5
Multiomics Studies

Without doubt, the integration of different omics technologies provides extensive additional information toward the complete understanding of bacterial physiology [121–123]. Such studies are most straightforward for systems metabolic engineering [124, 125]. For example, the integration of metabolic modeling on the level of elementary flux modes with transcriptomics allowed the tailored design of *P. putida* KT2440, overproducing PHAs [76]. A large-scale model of carbon core metabolism was used for *in silico* strain design. The implementation of even single genetic targets, predicted by the modeling, almost doubled the accumulation of this valuable biopolymer, which underlines the power of this approach. Transcriptomics proved helpful for analysis of the constructed mutants toward understanding of the molecular basis of their improved performance. Another prominent multiomics explored PHA biosynthesis further [126]. The combination of proteomics, transcriptomics, and metabolomics demonstrated that PHA formation and hydrolysis affect a number of cellular processes in *P. putida*, and not only, as it may seem obvious, those strictly related to carbon and nitrogen metabolism. In addition, the cellular response of *P. putida* KT2440 to *n*-butanol was evaluated at the transcript, the proteomic, and the metabolic levels [127]. Chemostat cultivation, using *n*-butanol either as sole carbon source or together with glucose, was implemented for this purpose. Metabolic flux analysis revealed that the central metabolism was split into a glucose-fueled combination of the ED and the PP pathway and an *n*-butanol-fueled TCA cycle, when both substrates were coconsumed. The hitherto unknown degradation pathway of *n*-butanol of *P. putida* KT2440 was unraveled, thus setting the base to upgrade *P. putida* KT2440 from an *n*-butanol consumer to a biofuel producer. Hopefully, we will see more studies like these in the future.

8.5
Application in Biotechnology

Initially, *P. putida* strains were evaluated as degrader for aromatic xenobiotics in bioremediation processes [38, 128, 129], as plant-growth promoting microorganism [130], as well as biocontrol agent [131]. Today, *P. putida* is regarded as an efficient producer for biotechnological production of a broad portfolio of biopolymers, chemicals, and pharmaceuticals [11]. However, most biotechnology-related studies of *P. putida* provide feasibility at a small scale and large-scale industrial application is still rare. Nevertheless, the different achievements demonstrate successful proofs of concept and thus provide promising starting points that can mature to economic applicability [124]. *P. putida* KT2440, a certified biosafety strain that is generally regarded as safe (GRAS), is most promising toward industrial application. As shown, the well-established organism is also amenable as host to heterologous genes, thereby enabling the recombinant production of valuable natural compounds [132, 133].

8.5.1 Biopolymers and Advanced Materials

Plastics are part of our daily life. Traditionally, they are based on petrochemical production from fossil fuels. Owing to massive environmental pollution, global warming, and increasing prices linked to petrochemical polymers, biobased polymers from renewable feedstocks are now being regarded as an excellent alternative, provided that they offer competitive material properties in terms of thermostability, elasticity, and durability [134]. Most interestingly, *P. putida* naturally incorporates PHAs for carbon and energy storage, when cell growth is otherwise limited [126, 135–140]. PHAs are biodegradable and biocompatible and thus ideal for environment-friendly packaging and medical drug delivery carriers or implants [141, 142]. As a result, industrial PHA production has now reached several ten thousand tons per year [141]. PHAs show structural variation in terms of monomer and macromolecule composition, which can be precisely adjusted by feeding of precursors, cultivation conditions, and strain engineering, respectively [136–138]. PHAs can be produced from various carbon sources, especially sugars and organic and fatty acids [33, 92, 93, 135, 143]. In *P. putida* KT2440, PHA biosynthesis from carbohydrates is carried out via *de novo* fatty acid biosynthesis, whereas the β-oxidation pathway delivers the building block from fatty acids (Figure 8.4) [144]. Renewable feedstocks such as crude glycerol from biodiesel production, plant-derived fatty acids, and aromatics have become of interest as inexpensive raw materials for PHA production [145–148]. Upon heterologous expression of xylose isomerase and xylulokinase, *P. putida* produces PHAs also from the hemicellulose monomer xylose [149]. The recent use of non-degradable plastic waste and its biotransformation into degradable PHA has attracted attention [150, 151]. Indeed, certain strains of *P. putida* are able to degrade phthalate and ethylene glycol, both PET monomers [152–155]. Next to PHAs, other polymer precursors have been successfully produced in *P. putida* KT2440. A promising compound is *cis,cis*-muconic acid, a precursor to adipic acid, which in turn serves as a building block for nylon 6,6. It can be obtained at high yield from aromatics via the catechol branch of the *β*-ketoadipate pathway (Figure 8.4) [156–158]. Finally, recombinant *P. putida* strains have been successfully applied for *de novo* synthesis of ethylene, monomeric hydroxyvalerate, and 4-valerolactone as further biobased building blocks [159–161].

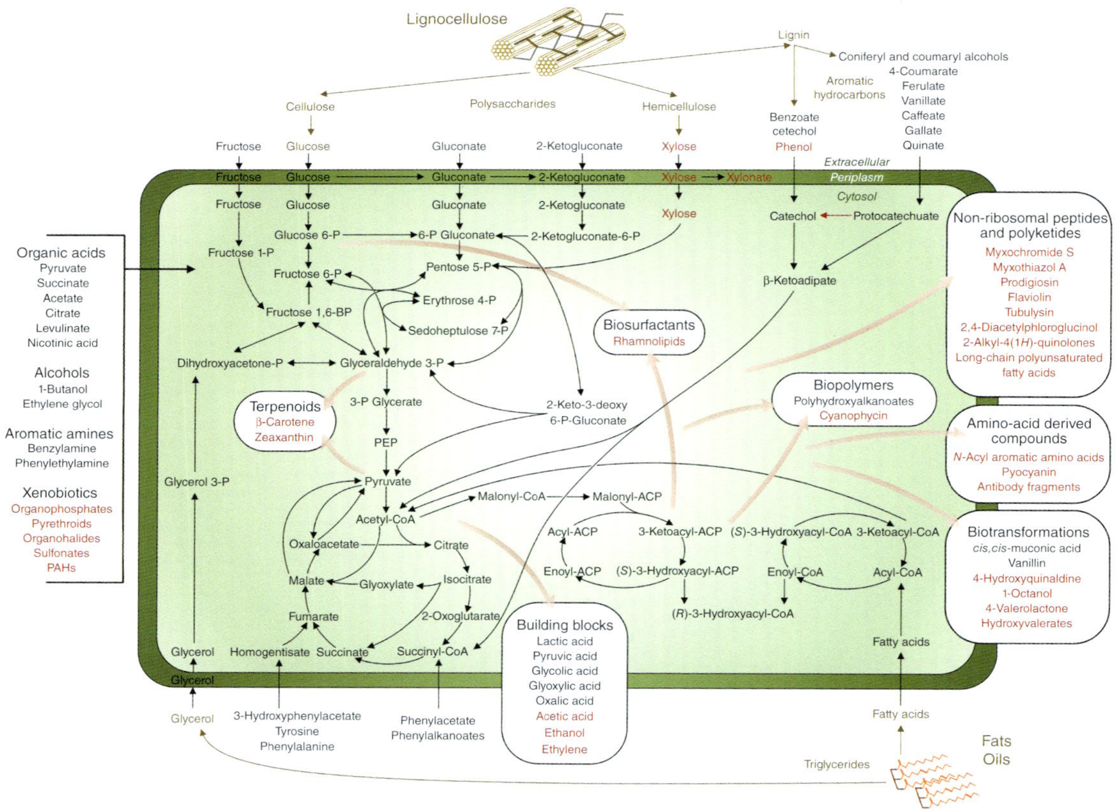

Figure 8.4 Biosynthetic potential of *Pseudomonas putida*. Extended carbon core metabolism of *Pseudomonas putida* KT2440 including the major catabolic routes of Entner–Doudoroff pathway, Embden–Meyerhof–Parnas pathway, pentose phosphate pathway, tricarboxylic acid cycle, glyoxylate shunt, anaplerotic reactions, fatty acid *de novo* biosynthesis, β-oxidation of fatty acids, as well as the convergent *β*-ketoadipate pathway for catabolism of aromatics. Known pathways for respective precursor supply for the broad product spectrum of *P. putida* KT2440 are indicated by light red arrows. Natural products and substrates are highlighted in black, heterologous products and substrates in red.

8.5.2
High-Value Chemicals

Various examples describe the successful application of *P. putida* KT2440 for whole-cell biocatalysis. In its genomic repertoire, *P. putida* KT2440 possesses 18 dioxygenases, 15 monooxygenases, and 80 oxidoreductases [12], an interesting arsenal for chemical conversions. Such bioconversions typically involve one or a few reaction steps to transform a preformed substrate into a product of interest. Among others, selective oxyfunctionalization capacities were exploited for conversion of octane to 1-octanol [162]. Using resting cells of *P. putida*, ferulic acid was efficiently converted into the flavoring agent vanillin, which was supported by deletion of vanillin-degrading pathways [163]. Other industrial biocatalysis applications based on *P. putida* spp. or isolated enzymes thereof comprise the synthesis of nonproteinogenic amino acids, semisynthetic penicillins, and chiral compounds [11, 16]. Besides its use as a whole-cell catalyst, *P. putida* KT2440 served as donor for several oxidoreductases that were successfully cloned and expressed in other microorganisms [164–166].

In addition, *P. putida* KT2440 has emerged as a reliable host for heterologous expression of biosynthetic pathways for natural products of relevance as pharmaceuticals and specialty chemicals. Many microorganisms that produce promising secondary metabolites, such as actino- or myxobacteria, as well as other sensitive species, are either pathogens, unculturable, or genetically difficult to manipulate. To make use of their biosynthetic capabilities, the corresponding gene clusters of these organisms can be isolated and cloned into suitable vectors and heterologously expressed in easy-to-handle and harmless pseudomonads [132, 167]. In this manner, a large number of bioactive compounds comprising rhamnolipids, carotenoids, polyketides, nonribosomally synthesized peptides, and amino acid-derived molecules [167] could be obtained. Quite recently, the first synthetic polyketide synthase-like pathway was successfully expressed in *P. putida* for the production of long-chain polyunsaturated fatty acids [168]. With *P. putida* KT2440 as heterologous host, microbial production of rhamnolipids, a class of biodegradable, and sustainable surfactants, has been established [15, 169–171]. Expressing the relevant *rhlAB* genes from the opportunistic pathogen *P. aeruginosa* and using glucose as sole carbon source, mono-rhamnolipid *de novo* synthesis in *P. putida* KT2440 has been described. The overall performance was comparable to that reported in previous studies employing the natural producer, thus offering a promising alternative for industrial application [169]. In other studies, the carotenoids β-carotene and zeaxanthin, relevant for feed and food applications, were produced [172, 173]. Myxobacterial metabolites such as the antibiotics myxochromide S, myxothiazol A, or tubulysin [133, 174, 175], as well as the UV-protective pigment flaviolin [176], were successfully synthesized. Other interesting products are prodigiosin [177], a red-pigmented tripyrrole with antibiotic and antitumoral potential and 2,4-diacetylphloroglucinol [178], exhibiting biological activity against plant pathogens. Generally, *P. putida* has shown great potential in metagenome-based detection and production of natural products due to its high tolerance against the metabolites of interest [179, 180]. Using environmental DNA libraries, secretion of antibacterial *N*-acyl aromatic amino acids and pigments of different colors could be shown, giving rise to more discoveries by metagenomic approaches [179].

8.6 Future Outlook

Today, most industrial processes based on *Pseudomonas* strains are enzymatic biotransformations and production of biopolymers. Beyond these applications, first steps are currently being taken to evaluate *P. putida* KT2440 as a cell factory strain for biobased production of fuels [45, 181] and organic acids [182]. Moreover, *P. putida* KT2440 has been proposed as a reliable expression host for recombinant production of functional proteins [80, 183] and for potential applications in microbial electrosynthesis for the production of chemicals [184]. In order to enable anaerobic growth and biosynthesis under oxygen-limiting conditions in the otherwise obligate aerobe *P. putida*, the direct electron transfer to an anode as electron acceptor via a heterologously expressed phenazine-type redox mediator from *P. aeruginosa* was evaluated [184]. This novel bioelectrochemical process allowed production of the pigment phenazine 1-carboxylic acid with subsequent conversion into pyocynin. Another emerging feature of *P. putida* is the ability of this bacterium to execute harsh biotransformations (e.g., dehalogenations) in programmable catalytic biofilms [185]. On the basis of current developments, the already broad product and application spectrum of different *P. putida* strains will be further expanded. Especially in the scope of a sustainable bioeconomy based on renewable feedstocks displaying complex mixtures with partly toxic impurities, the versatile and robust *P. putida* might be perfectly integrated into novel biorefinery concepts.

References

1. Jiménez, J.I., Miñambres, B., García, J.L., and Díaz, E. (2002) Genomic analysis of the aromatic catabolic pathways from *Pseudomonas putida* KT2440. *Environ. Microbiol.*, **4** (12), 824–841.
2. Martins dos Santos, V.A.P., Heim, S., Moore, E.R.B., Strätz, M., and Timmis, K.N. (2004) Insights into the genomic basis of niche specificity of *Pseudomonas putida* KT2440. *Environ. Microbiol.*, **6** (12), 1264–1286.
3. Domínguez-Cuevas, P., González-Pastor, J.-E., Marqués, S., Ramos, J.-L., and de Lorenzo, V. (2006) Transcriptional tradeoff between metabolic and stress-response programs in *Pseudomonas putida* KT2440 cells exposed to toluene. *J. Biol. Chem.*, **281** (17), 11981–11991.
4. Roma-Rodrigues, C., Santos, P.M., Benndorf, D., Rapp, E., and Sá-Correia, I. (2010) Response of *Pseudomonas putida* KT2440 to phenol at the level of membrane proteome. *J. Proteomics*, **73** (8), 1461–1478.
5. Molina, L., Udaondo, Z., Duque, E., Fernández, M., Molina-Santiago, C., Roca, A., Porcel, M., de la Torre, J., Segura, A., Plesiat, P., Jeannot, K., and Ramos, J.L. (2014) Antibiotic resistance

determinants in a *Pseudomonas putida* strain isolated from a hospital. *PLoS One*, **9** (1), e81604.
6. Sharma, P.K., Fu, J., Zhang, X., Fristensky, B., Sparling, R., and Levin, D.B. (2014) Genome features of *Pseudomonas putida* LS46, a novel polyhydroxyalkanoate producer and its comparison with other *P. putida* strains. *AMB Express*, **4**, 37.
7. Ramos, J.L., Duque, E., Gallegos, M.T., Godoy, P., Ramos-González, M.I., Rojas, A., Terán, W., and Segura, A. (2002) Mechanisms of solvent tolerance in gram-negative bacteria. *Annu. Rev. Microbiol.*, **56**, 743–768.
8. Heipieper, H.J., Neumann, G., Cornelissen, S., and Meinhardt, F. (2007) Solvent-tolerant bacteria for biotransformations in two-phase fermentation systems. *Appl. Microbiol. Biotechnol.*, **74** (5), 961–973.
9. Segura, A., Molina, L., Fillet, S., Krell, T., Bernal, P., Muñoz-Rojas, J., and Ramos, J.L. (2012) Solvent tolerance in Gram-negative bacteria. *Curr. Opin. Biotechnol.*, **23** (3), 415–421.
10. Regenhardt, D., Heuer, H., Heim, S., Fernandez, D.U., Strompl, C., Moore, E.R., and Timmis, K.N. (2002) Pedigree and taxonomic credentials of *Pseudomonas putida* strain KT2440. *Environ. Microbiol.*, **4** (12), 912–915.
11. Poblete-Castro, I., Becker, J., Dohnt, K., Martins Dos Santos, V.A.P., and Wittmann, C. (2012) Industrial biotechnology of *Pseudomonas putida* and related species. *Appl. Microbiol. Biotechnol.*, **93** (6), 2279–2290.
12. Nelson, K.E., Weinel, C., Paulsen, I.T., Dodson, R.J., Hilbert, H., Martins dos Santos, V.A.P., Fouts, D.E., Gill, S.R., Pop, M., Holmes, M., Brinkac, L., Beanan, M., DeBoy, R.T., Daugherty, S., Kolonay, J., Madupu, R., Nelson, W., White, O., Peterson, J., Khouri, H., Hance, I., Lee, P.C., Holtzapple, E., Scanlan, D., Tran, K., Moazzez, A., Utterback, T., Rizzo, M., Lee, K., Kosack, D., Moestl, D., Wedler, H., Lauber, J., Stjepandic, D., Hoheisel, J., Straetz, M., Heim, S., Kiewitz, C., Eisen, J., Timmis, K.N., Dusterhoft, A., Tummler, B., and Fraser, C.M. (2002) Complete genome sequence and comparative analysis of the metabolically versatile *Pseudomonas putida* KT2440. *Environ. Microbiol.*, **4** (12), 799–808.
13. Wu, X., Monchy, S., Taghavi, S., Zhu, W., Ramos, J., and van der Lelie, D. (2011) Comparative genomics and functional analysis of niche-specific adaptation in *Pseudomonas putida*. *FEMS Microbiol. Rev.*, **35** (2), 299–323.
14. Timmis, K.N. (2002) *Pseudomonas putida*: a cosmopolitan opportunist par excellence. *Environ. Microbiol.*, **4** (12), 779–781.
15. Rehm, B.H.A. (2009) in *Encyclopedia of Industrial Biotechnology* (ed. M.C. Flickinger), John Wiley & Sons, Inc., Hoboken, NJ, pp. 1–15.
16. Schmid, A., Dordick, J.S., Hauer, B., Kiener, A., Wubbolts, M., and Witholt, B. (2001) Industrial biocatalysis today and tomorrow. *Nature*, **409** (6817), 258–268.
17. Tiso, T., Wierckx, N., and Blank, L.M. (2014) in *Industrial Biocatalysis* (ed. P. Grunwald), CRC Press, Boca Raton, FL, pp. 323–372.
18. Vicente, M. and Cánovas, J.L. (1973) Glucolysis in *Pseudomonas putida*: physiological role of alternative routes from the analysis of defective mutants. *J. Bacteriol.*, **116** (2), 908–914.
19. Rojo, F. (2010) Carbon catabolite repression in *Pseudomonas*: optimizing metabolic versatility and interactions with the environment. *FEMS Microbiol. Rev.*, **34** (5), 658–684.
20. La Rosa, R., Behrends, V., Williams, H.D., Bundy, J.G., and Rojo, F. (2015) Influence of the Crc regulator on the hierarchical use of carbon sources from a complete medium in *Pseudomonas*. *Environ. Microbiol.*, **18** (3), 807–818.
21. Collier, D.N., Hager, P.W., and Phibbs, P.V. Jr., (1996) Catabolite repression control in the *Pseudomonads*. *Res. Microbiol.*, **147** (6–7), 551–561.
22. Müller, C., Petruschka, L., Cuypers, H., Burchhardt, G., and Herrmann, H. (1996) Carbon catabolite repression of phenol degradation in *Pseudomonas putida* is mediated by the inhibition of the activator protein PhlR. *J. Bacteriol.*, **178** (7), 2030–2036.

23. Velázquez, F., Pflüger, K., Cases, I., De Eugenio, L.I., and de Lorenzo, V. (2007) The phosphotransferase system formed by PtsP, PtsO, and PtsN proteins controls production of polyhydroxyalkanoates in *Pseudomonas putida*. *J. Bacteriol.*, **189** (12), 4529–4533.
24. Köhler, K.A.K., Blank, L.M., Frick, O., and Schmid, A. (2015) D-Xylose assimilation via the Weimberg pathway by solvent-tolerant *Pseudomonas taiwanensis* VLB120. *Environ. Microbiol.*, **17** (1), 156–170.
25. Clarke, P.H. (1982) The metabolic versatility of pseudomonads. *Antonie van Leeuwenhoek*, **48** (2), 105–130.
26. Nikel, P.I., Kim, J., and de Lorenzo, V. (2014) Metabolic and regulatory rearrangements underlying glycerol metabolism in *Pseudomonas putida* KT2440. *Environ. Microbiol.*, **16** (1), 239–254.
27. del Castillo, T., Ramos, J.L., Rodríguez-Herva, J.J., Fuhrer, T., Sauer, U., and Duque, E. (2007) Convergent peripheral pathways catalyze initial glucose catabolism in *Pseudomonas putida*: genomic and flux analysis. *J. Bacteriol.*, **189** (14), 5142–5152.
28. Chavarría, M., Nikel, P.I., Pérez-Pantoja, D., and de Lorenzo, V. (2013) The Entner–Doudoroff pathway empowers *Pseudomonas putida* KT2440 with a high tolerance to oxidative stress. *Environ. Microbiol.*, **15** (6), 1772–1785.
29. Klingner, A., Bartsch, A., Dogs, M., Wagner-Dobler, I., Jahn, D., Simon, M., Brinkhoff, T., Becker, J., and Wittmann, C. (2015) Large-scale ^{13}C flux profiling reveals conservation of the Entner–Doudoroff pathway as a glycolytic strategy among marine bacteria that use glucose. *Appl. Environ. Microbiol.*, **81** (7), 2408–2422.
30. Nikel, P.I., Chavarría, M., Fuhrer, T., Sauer, U., and de Lorenzo, V. (2015) *Pseudomonas putida* KT2440 metabolizes glucose through a cycle formed by enzymes of the Entner–Doudoroff, Embden–Meyerhof–Parnas, and pentose phosphate pathways. *J. Biol. Chem.*, **290** (43), 25920–25932.
31. Chavarría, M., Kleijn, R.J., Sauer, U., Pflüger-Grau, K., and de Lorenzo, V. (2012) Regulatory tasks of the phosphoenolpyruvate-phosphotransferase system of *Pseudomonas putida* in central carbon metabolism. *mBio*, **3** (2), e00028-12.
32. Escapa, I.F., García, J.L., Bühler, B., Blank, L.M., and Prieto, M.A. (2012) The polyhydroxyalkanoate metabolism controls carbon and energy spillage in *Pseudomonas putida*. *Environ. Microbiol.*, **14** (4), 1049–1063.
33. Escapa, I.F., del Cerro, C., García, J.L., and Prieto, M.A. (2013) The role of GlpR repressor in *Pseudomonas putida* KT2440 growth and PHA production from glycerol. *Environ. Microbiol.*, **15** (1), 93–110.
34. Nikel, P.I., Romero-Campero, F.J., Zeidman, J.A., Goñi-Moreno, A., and de Lorenzo, V. (2015) The glycerol-dependent metabolic persistence of *Pseudomonas putida* KT2440 reflects the regulatory logic of the GlpR repressor. *mBio*, **6** (2), e00340-15.
35. Assinder, S.J. and Williams, P.A. (1990) The TOL plasmids: determinants of the catabolism of toluene and the xylenes. *Adv. Microb. Physiol.*, **31**, 1–69.
36. Ramos, J.L., Marqués, S., and Timmis, K.N. (1997) Transcriptional control of the *Pseudomonas* TOL plasmid catabolic operons is achieved through an interplay of host factors and plasmid-encoded regulators. *Annu. Rev. Microbiol.*, **51**, 341–373.
37. Nikel, P.I., Silva-Rocha, R., Benedetti, I., and de Lorenzo, V. (2014) The private life of environmental bacteria: pollutant biodegradation at the single cell level. *Environ. Microbiol.*, **16** (3), 628–642.
38. Jiménez, J.I., Miñambres, B., García, J.L., and Díaz, E. (2004) in *Pseudomonas* (ed. J.-L. Ramos), Springer, New York, pp. 425–462.
39. Kim, Y.H., Cho, K., Yun, S.H., Kim, J.Y., Kwon, K.H., Yoo, J.S., and Kim, S.I. (2006) Analysis of aromatic catabolic pathways in *Pseudomonas putida* KT 2440 using a combined proteomic approach: 2-DE/MS and cleavable isotope-coded affinity tag analysis. *Proteomics*, **6** (4), 1301–1318.

40. Loh, K.-C. and Cao, B. (2008) Paradigm in biodegradation using *Pseudomonas putida*—A review of proteomics studies. *Enzyme Microb. Technol.*, **43** (1), 1–12.
41. Jiménez, J.I., Canales, A., Jiménez-Barbero, J., Ginalski, K., Rychlewski, L., García, J.L., and Díaz, E. (2008) Deciphering the genetic determinants for aerobic nicotinic acid degradation: the *nic* cluster from *Pseudomonas putida* KT2440. *Proc. Natl. Acad. Sci. U.S.A.*, **105** (32), 11329–11334.
42. Johnson, R.J., Smith, B.E., Rowland, S.J., and Whitby, C. (2013) Biodegradation of alkyl branched aromatic alkanoic naphthenic acids by *Pseudomonas putida* KT2440. *Int. Biodeterior. Biodegrad.*, **81**, 3–8.
43. Mallavarapu, M., Möhler, I., Krüger, M., Hosseini, M.M., Bartels, F., Timmis, K.N., and Holtel, A. (1998) Genetic requirements for the expression of benzylamine dehydrogenase activity in *Pseudomonas putida*. *FEMS Microbiol. Lett.*, **166** (1), 109–114.
44. Rühl, J., Schmid, A., and Blank, L.M. (2009) Selected *Pseudomonas putida* strains able to grow in the presence of high butanol concentrations. *Appl. Environ. Microbiol.*, **75** (13), 4653–4656.
45. Simon, O., Klebensberger, J., Mükschel, B., Klaiber, I., Graf, N., Altenbuchner, J., Huber, A., Hauer, B., and Pfannstiel, J. (2015) Analysis of the molecular response of *Pseudomonas putida* KT2440 to the next-generation biofuel *n*-butanol. *J. Proteomics*, **122**, 11–25.
46. Fernández, M., Niqui-Arroyo, J.L., Conde, S., Ramos, J.L., and Duque, E. (2012) Enhanced tolerance to naphthalene and enhanced rhizoremediation performance for *Pseudomonas putida* KT2440 via the NAH7 catabolic plasmid. *Appl. Environ. Microbiol.*, **78** (15), 5104–5110.
47. Walker, A.W. and Keasling, J.D. (2002) Metabolic engineering of *Pseudomonas putida* for the utilization of parathion as a carbon and energy source. *Biotechnol. Bioeng.*, **78** (7), 715–721.
48. Gilbert, E.S., Walker, A.W., and Keasling, J.D. (2003) A constructed microbial consortium for biodegradation of the organophosphorus insecticide parathion. *Appl. Microbiol. Biotechnol.*, **61** (1), 77–81.
49. de la Pena Mattozzi, M., Tehara, S.K., Hong, T., and Keasling, J.D. (2006) Mineralization of paraoxon and its use as a sole C and P source by a rationally designed catabolic pathway in *Pseudomonas putida*. *Appl. Environ. Microbiol.*, **72** (10), 6699–6706.
50. Zuo, Z., Gong, T., Che, Y., Liu, R., Xu, P., Jiang, H., Qiao, C., Song, C., and Yang, C. (2015) Engineering *Pseudomonas putida* KT2440 for simultaneous degradation of organophosphates and pyrethroids and its application in bioremediation of soil. *Biodegradation*, **26** (3), 223–233.
51. Nikel, P.I. and de Lorenzo, V. (2013) Engineering an anaerobic metabolic regime in *Pseudomonas putida* KT2440 for the anoxic biodegradation of 1,3-dichloroprop-1-ene. *Metab. Eng.*, **15**, 98–112.
52. Kertesz, M.A. and Wietek, C. (2001) Desulfurization and desulfonation: applications of sulfur-controlled gene expression in bacteria. *Appl. Microbiol. Biotechnol.*, **57** (4), 460–466.
53. Moreno, R., Hernández-Arranz, S., La Rosa, R., Yuste, L., Madhushani, A., Shingler, V., and Rojo, F. (2015) The Crc and Hfq proteins of *Pseudomonas putida* cooperate in catabolite repression and formation of ribonucleic acid complexes with specific target motifs. *Environ. Microbiol.*, **17** (1), 105–118.
54. Martínez-García, E., Aparicio, T., Goñi-Moreno, A., Fraile, S., and de Lorenzo, V. (2015) SEVA 2.0: an update of the Standard European Vector Architecture for de-/re-construction of bacterial functionalities. *Nucleic Acids Res.*, **43**, D1183–D1189.
55. Silva-Rocha, R., Martínez-García, E., Calles, B., Chavarría, M., Arce-Rodríguez, A., de Las Heras, A., Páez-Espino, A.D., Durante-Rodríguez, G., Kim, J., Nikel, P.I., Platero, R., and de Lorenzo, V. (2013) The Standard European Vector Architecture (SEVA): a coherent platform for the analysis

and deployment of complex prokaryotic phenotypes. *Nucleic Acids Res.*, **41** (Database issue), D666–D675.

56. Martínez-García, E. and de Lorenzo, V. (2011) Engineering multiple genomic deletions in gram-negative bacteria: analysis of the multi-resistant antibiotic profile of *Pseudomonas putida* KT2440. *Environ. Microbiol.*, **13** (10), 2702–2716.
57. Martínez-García, E., Aparicio, T., de Lorenzo, V., and Nikel, P.I. (2014) New transposon tools tailored for metabolic engineering of gram-negative microbial cell factories. *Front. Bioeng. Biotechnol.*, **2**, 46.
58. de Lorenzo, V., Herrero, M., Jakubzik, U., and Timmis, K.N. (1990) Mini-Tn*5* transposon derivatives for insertion mutagenesis, promoter probing, and chromosomal insertion of cloned DNA in gram-negative eubacteria. *J. Bacteriol.*, **172** (11), 6568–6572.
59. Herrero, M., de Lorenzo, V., and Timmis, K.N. (1990) Transposon vectors containing non-antibiotic resistance selection markers for cloning and stable chromosomal insertion of foreign genes in gram-negative bacteria. *J. Bacteriol.*, **172** (11), 6557–6567.
60. Martínez-García, E., Calles, B., Arévalo-Rodríguez, M., and de Lorenzo, V. (2011) pBAM1: an all-synthetic genetic tool for analysis and construction of complex bacterial phenotypes. *BMC Microbiol.*, **11**, 38.
61. Silva-Rocha, R. and de Lorenzo, V. (2014) Chromosomal integration of transcriptional fusions. *Methods Mol. Biol.*, **1149**, 479–489.
62. Lambertsen, L., Sternberg, C., and Molin, S. (2004) Mini-Tn*7* transposons for site-specific tagging of bacteria with fluorescent proteins. *Environ. Microbiol.*, **6** (7), 726–732.
63. Iwasaki, K., Uchiyama, H., Yagi, O., Kurabayashi, T., Ishizuka, K., and Takamura, Y. (1994) Transformation of *Pseudomonas putida* by electroporation. *Biosci. Biotechnol., Biochem.*, **58** (5), 851–854.
64. Itoh, N., Kouzai, T., and Koide, Y. (1994) Efficient transformation of *Pseudomonas* strains with pNI vectors by electroporation. *Biosci. Biotechnol., Biochem.*, **58** (7), 1306–1308.
65. Dennis, J.J. and Zylstra, G.J. (1998) Plasposons: modular self-cloning minitransposon derivatives for rapid genetic analysis of gram-negative bacterial genomes. *Appl. Environ. Microbiol.*, **64** (7), 2710–2715.
66. Choi, K.H. and Kim, K.J. (2009) Applications of transposon-based gene delivery system in bacteria. *J. Microbiol. Biotechnol.*, **19** (3), 217–228.
67. Zobel, S., Benedetti, I., Eisenbach, L., de Lorenzo, V., Wierckx, N., and Blank, L.M. (2015) A Tn*7*-based device for calibrated heterologous gene expression in *Pseudomonas putida*. *ACS Synth. Biol.*, **4** (12), 1341–1351.
68. Martínez-García, E. and de Lorenzo, V. (2012) Transposon-based and plasmid-based genetic tools for editing genomes of gram-negative bacteria. *Methods Mol. Biol.*, **813**, 267–283.
69. Peters, J.E. and Craig, N.L. (2001) Tn*7*: smarter than we thought. *Nat. Rev. Mol. Cell Biol.*, **2** (11), 806–814.
70. Nikel, P.I., Martínez-García, E., and de Lorenzo, V. (2014) Biotechnological domestication of pseudomonads using synthetic biology. *Nat. Rev. Microbiol.*, **12**, 368–379.
71. Wackett, L.P. (2015) *Pseudomonas*. *Environ. Microbiol.*, **17** (1), 253–254.
72. Udaondo, Z., Molina, L., Segura, A., Duque, E., and Ramos, J.L. (2015) Analysis of the core genome and pangenome of *Pseudomonas putida*. *Environ. Microbiol.*, doi: 10.1111/1462-2920.13015.
73. Nogales, J., Palsson, B.Ø., and Thiele, I. (2008) A genome-scale metabolic reconstruction of *Pseudomonas putida* KT2440: *i*JN746 as a cell factory. *BMC Syst. Biol.*, **2**, 79.
74. Puchałka, J., Oberhardt, M.A., Godinho, M., Bielecka, A., Regenhardt, D., Timmis, K.N., Papin, J.A., and Martins dos Santos, V.A.P. (2008) Genome-scale reconstruction and analysis of the *Pseudomonas putida* KT2440 metabolic network facilitates applications in biotechnology. *PLoS Comput. Biol.*, **4** (10), e1000210.

75. Sohn, S.B., Kim, T.Y., Park, J.M., and Lee, S.Y. (2010) In silico genome-scale metabolic analysis of *Pseudomonas putida* KT2440 for polyhydroxyalkanoate synthesis, degradation of aromatics and anaerobic survival. *Biotechnol. J.*, **5**, 739–750.
76. Poblete-Castro, I., Binger, D., Rodrigues, A.L., Becker, J., Martins dos Santos, V.A.P., and Wittmann, C. (2013) In-silico-driven metabolic engineering of *Pseudomonas putida* for enhanced production of poly-hydroxyalkanoates. *Metab. Eng.*, **15**, 113–123.
77. Park, S.J., Choi, J.S., Kim, B.C., Jho, S.W., Ryu, J.W., Park, D., Lee, K.A., Bhak, J., and Kim, S.I. (2009) PutidaNET: interactome database service and network analysis of *Pseudomonas putida* KT2440. *BMC Genomics*, **10** (Suppl. 3), S18.
78. Duque, E., Molina-Henares, A.J., Torre, J.D.L., Molina-Henares, M.A., Castillo, T.D., Lam, J., and Ramos, J.L. (2007) in *Pseudomonas* (eds J.-L. Ramos and A. Filloux), Springer, Dordrecht, pp. 227–251.
79. Martínez-García, E., Nikel, P.I., Aparicio, T., and de Lorenzo, V. (2014) *Pseudomonas* 2.0: genetic upgrading of *P. putida* KT2440 as an enhanced host for heterologous gene expression. *Microb. Cell Fact.*, **13**, 159.
80. Lieder, S., Nikel, P.I., de Lorenzo, V., and Takors, R. (2015) Genome reduction boosts heterologous gene expression in *Pseudomonas putida*. *Microb. Cell Fact.*, **14**, 23.
81. Yuste, L., Hervás, A.B., Canosa, I., Tobes, R., Jiménez, J.I., Nogales, J., Pérez-Pérez, M.M., Santero, E., Díaz, E., Ramos, J.L., de Lorenzo, V., and Rojo, F. (2006) Growth phase-dependent expression of the *Pseudomonas putida* KT2440 transcriptional machinery analysed with a genome-wide DNA microarray. *Environ. Microbiol.*, **8** (1), 165–177.
82. Miyakoshi, M., Shintani, M., Terabayashi, T., Kai, S., Yamane, H., and Nojiri, H. (2007) Transcriptome analysis of *Pseudomonas putida* KT2440 harboring the completely sequenced IncP-7 plasmid pCAR1. *J. Bacteriol.*, **189** (19), 6849–6860.
83. Gülez, G., Dechesne, A., Workman, C.T., and Smets, B.F. (2012) Transcriptome dynamics of *Pseudomonas putida* KT2440 under water stress. *Appl. Environ. Microbiol.*, **78** (3), 676–683.
84. Reva, O.N., Weinel, C., Weinel, M., Böhm, K., Stjepandic, D., Hoheisel, J.D., and Tümmler, B. (2006) Functional genomics of stress response in *Pseudomonas putida* KT2440. *J. Bacteriol.*, **188** (11), 4079–4092.
85. Cánovas, D., Cases, I., and de Lorenzo, V. (2003) Heavy metal tolerance and metal homeostasis in *Pseudomonas putida* as revealed by complete genome analysis. *Environ. Microbiol.*, **5** (12), 1242–1256.
86. Fonseca, P., Moreno, R., and Rojo, F. (2008) Genomic analysis of the role of RNase R in the turnover of *Pseudomonas putida* mRNAs. *J. Bacteriol.*, **190** (18), 6258–6263.
87. Fonseca, P., Moreno, R., and Rojo, F. (2011) Growth of *Pseudomonas putida* at low temperature: global transcriptomic and proteomic analyses. *Environ. Microbiol. Rep.*, **3** (3), 329–339.
88. Frank, S., Schmidt, F., Klockgether, J., Davenport, C.F., Gesell Salazar, M., Völker, U., and Tümmler, B. (2011) Functional genomics of the initial phase of cold adaptation of *Pseudomonas putida* KT2440. *FEMS Microbiol. Lett.*, **318** (1), 47–54.
89. Follonier, S., Escapa, I.F., Fonseca, P.M., Henes, B., Panke, S., Zinn, M., and Prieto, M.A. (2013) New insights on the reorganization of gene transcription in *Pseudomonas putida* KT2440 at elevated pressure. *Microb. Cell Fact.*, **12**, 30.
90. Hervás, A.B., Canosa, I., and Santero, E. (2008) Transcriptome analysis of *Pseudomonas putida* in response to nitrogen availability. *J. Bacteriol.*, **190** (1), 416–420.
91. Yeom, S., Yeom, J., and Park, W. (2010) NtrC-sensed nitrogen availability is important for oxidative stress defense in *Pseudomonas putida* KT2440. *J. Microbiol.*, **48** (2), 153–159.

92. Wang, Q. and Nomura, C.T. (2010) Monitoring differences in gene expression levels and polyhydroxyalkanoate (PHA) production in *Pseudomonas putida* KT2440 grown on different carbon sources. *J. Biosci. Bioeng.*, **110** (6), 653–659.
93. Borrero-de Acuña, J.M., Bielecka, A., Häussler, S., Schobert, M., Jahn, M., Wittmann, C., Jahn, D., and Poblete-Castro, I. (2014) Production of medium chain length polyhydroxyalkanoate in metabolic flux optimized *Pseudomonas putida*. *Microb. Cell Fact.*, **13**, 88.
94. Kim, J., Oliveros, J.C., Nikel, P.I., de Lorenzo, V., and Silva-Rocha, R. (2013) Transcriptomic fingerprinting of *Pseudomonas putida* under alternative physiological regimes. *Environ. Microbiol. Rep.*, **5** (6), 883–891.
95. Kim, J., Pérez-Pantoja, D., Silva-Rocha, R., Oliveros, J.C., and de Lorenzo, V. (2015) High-resolution analysis of the *m*-xylene/toluene biodegradation subtranscriptome of *Pseudomonas putida* mt-2. *Environ. Microbiol.*, in press.
96. Frank, S., Klockgether, J., Hagendorf, P., Geffers, R., Schöck, U., Pohl, T., Davenport, C.F., and Tümmler, B. (2011) *Pseudomonas putida* KT2440 genome update by cDNA sequencing and microarray transcriptomics. *Environ. Microbiol.*, **13** (5), 1309–1326.
97. Heim, S., Ferrer, M., Heuer, H., Regenhardt, D., Nimtz, M., and Timmis, K.N. (2003) Proteome reference map of *Pseudomonas putida* strain KT2440 for genome expression profiling: distinct responses of KT2440 and *Pseudomonas aeruginosa* strain PAO1 to iron deprivation and a new form of superoxide dismutase. *Environ. Microbiol.*, **5** (12), 1257–1269.
98. Herbst, F.A., Danielsen, H.N., Wimmer, R., Nielsen, P.H., and Dueholm, M.S. (2015) Label-free quantification reveals major proteomic changes in *Pseudomonas putida* F1 during the exponential growth phase. *Proteomics*, **15** (18), 3244–3252.
99. Panikov, N.S., Mandalakis, M., Dai, S., Mulcahy, L.R., Fowle, W., Garrett, W.S., and Karger, B.L. (2015) Near-zero growth kinetics of *Pseudomonas putida* deduced from proteomic analysis. *Environ. Microbiol.*, **17** (1), 215–228.
100. Segura, A., Godoy, P., van Dillewijn, P., Hurtado, A., Arroyo, N., Santacruz, S., and Ramos, J.L. (2005) Proteomic analysis reveals the participation of energy- and stress-related proteins in the response of *Pseudomonas putida* DOT-T1E to toluene. *J. Bacteriol.*, **187** (17), 5937–5945.
101. Volkers, R.J., de Jong, A.L., Hulst, A.G., van Baar, B.L., de Bont, J.A., and Wery, J. (2006) Chemostat-based proteomic analysis of toluene-affected *Pseudomonas putida* S12. *Environ. Microbiol.*, **8** (9), 1674–1679.
102. Wijte, D., van Baar, B.L., Heck, A.J., and Altelaar, A.F. (2011) Probing the proteome response to toluene exposure in the solvent tolerant *Pseudomonas putida* S12. *J. Proteome Res.*, **10** (2), 394–403.
103. Li, S.S., Hu, X., Zhao, H., Li, Y.X., Zhang, L., Gong, L.J., Guo, J., and Zhao, H.B. (2015) Quantitative analysis of cellular proteome alterations of *Pseudomonas putida* to naphthalene-induced stress. *Biotechnol. Lett.*, **37** (8), 1645–1654.
104. Nikodinovic-Runic, J., Flanagan, M., Hume, A.R., Cagney, G., and O'Connor, K.E. (2009) Analysis of the *Pseudomonas putida* CA-3 proteome during growth on styrene under nitrogen-limiting and non-limiting conditions. *Microbiology*, **155** (10), 3348–3361.
105. Santos, P.M., Benndorf, D., and Sa-Correia, I. (2004) Insights into *Pseudomonas putida* KT2440 response to phenol-induced stress by quantitative proteomics. *Proteomics*, **4** (9), 2640–2652.
106. Manara, A., DalCorso, G., Baliardini, C., Farinati, S., Cecconi, D., and Furini, A. (2012) *Pseudomonas putida* response to cadmium: changes in membrane and cytosolic proteomes. *J. Proteome Res.*, **11** (8), 4169–4179.
107. Ray, P., Girard, V., Gault, M., Job, C., Bonneu, M., Mandrand-Berthelot, M.A., Singh, S.S., Job, D., and Rodrigue, A. (2013) *Pseudomonas putida* KT2440 response to nickel or cobalt induced

stress by quantitative proteomics. *Metallomics*, **5** (1), 68–79.

108. Fuhrer, T., Fischer, E., and Sauer, U. (2005) Experimental identification and quantification of glucose metabolism in seven bacterial species. *J. Bacteriol.*, **187**, 1581–1590.
109. Blank, L.M., Ionidis, G., Ebert, B.E., Bühler, B., and Schmid, A. (2008) Metabolic response of *Pseudomonas putida* during redox biocatalysis in the presence of a second octanol phase. *FEBS J.*, **275** (20), 5173–5190.
110. Ebert, B.E., Kurth, F., Grund, M., Blank, L.M., and Schmid, A. (2011) Response of *Pseudomonas putida* KT2440 to increased NADH and ATP demand. *Appl. Enivron. Microbiol.*, **77**, 6597–6605.
111. Wittmann, C. (2007) Fluxome analysis using GC-MS. *Microb. Cell Fact.*, **6**, 6.
112. Wierckx, N., Ruijssenaars, H.J., de Winde, J.H., Schmid, A., and Blank, L.M. (2009) Metabolic flux analysis of a phenol producing mutant of *Pseudomonas putida* S12: verification and complementation of hypotheses derived from transcriptomics. *J. Biotechnol.*, **143**, 124–129.
113. Sudarsan, S., Dethlefsen, S., Blank, L.M., Siemann-Herzberg, M., and Schmid, A. (2014) The functional structure of central carbon metabolism in *Pseudomonas putida* KT2440. *Appl. Environ. Microbiol.*, **80** (17), 5292–5303.
114. Sasnow, S.S., Wei, H., and Aristilde, L. (2015) Bypasses in intracellular glucose metabolism in iron-limited *Pseudomonas putida*. *Microbiol. Open*, **5** (1), 3–20.
115. Conway, T. (1992) The Entner–Doudoroff pathway: history, physiology and molecular biology. *FEMS Microbiol. Rev.*, **103**, 1–28.
116. Berger, A., Dohnt, K., Tielen, P., Jahn, D., Becker, J., and Wittmann, C. (2014) Robustness and plasticity of metabolic pathway flux among uropathogenic isolates of *Pseudomonas aeruginosa*. *PLoS One*, **9** (4), e88368.
117. Lien, S.K., Niedenführ, S., Sletta, H., Nöh, K., and Bruheim, P. (2015) Fluxome study of *Pseudomonas fluorescens* reveals major reorganisation of carbon flux through central metabolic pathways in response to inactivation of the anti-sigma factor MucA. *BMC Syst. Biol.*, **9**, 6.
118. Bolten, C.J., Kiefer, P., Letisse, F., Portais, J.-C., and Wittmann, C. (2007) Sampling for metabolome analysis of microorganisms. *Anal. Chem.*, **79**, 3843–3849.
119. van der Werf, M.J., Overkamp, K.M., Muilwijk, B., Koek, M.M., van der Werff-van der Vat, B.J.C., Jellema, R.H., Coulier, L., and Hankemeier, T. (2008) Comprehensive analysis of the metabolome of *Pseudomonas putida* S12 grown on different carbon sources. *Mol. Biosyst.*, **4**, 315–327.
120. Olavarria, K., Marone, M.P., da Costa Oliveira, H., Roncallo, J.C., da Costa Vasconcelos, F.N., da Silva, L.F., and Gomez, J.G. (2015) Quantifying NAD(P)H production in the upper Entner–Doudoroff pathway from *Pseudomonas putida* KT2440. *FEBS Open Bio.*, **5**, 908–915.
121. Krömer, J.O., Sorgenfrei, O., Klopprogge, K., Heinzle, E., and Wittmann, C. (2004) In-depth profiling of lysine-producing *Corynebacterium glutamicum* by combined analysis of the transcriptome, metabolome, and fluxome. *J. Bacteriol.*, **186** (6), 1769–1784.
122. Kohlstedt, M., Becker, J., and Wittmann, C. (2010) Metabolic fluxes and beyond – systems biology understanding and engineering of microbial metabolism. *Appl. Microbiol. Biotechnol.*, **88** (5), 1065–1075.
123. Kohlstedt, M., Sappa, P.K., Meyer, H., Maaß, S., Zaprasis, A., Hoffmann, T., Becker, J., Steil, L., Hecker, M., van Dijl, J.M., Lalk, M., Mäder, U., Stülke, J., Bremer, E., Völker, U., and Wittmann, C. (2014) Adaptation of *Bacillus subtilis* carbon core metabolism to simultaneous nutrient limitation and osmotic challenge: a multi-omics perspective. *Environ. Microbiol.*, **16** (6), 1898–1917.
124. Becker, J. and Wittmann, C. (2012) Bio-based production of chemicals, materials and fuels – *Corynebacterium*

glutamicum as versatile cell factory. *Curr. Opin. Biotechnol.*, **23** (4), 631–640.

125. Becker, J. and Wittmann, C. (2015) Advanced biotechnology: metabolically engineered cells for the bio-based production of chemicals and fuels, materials, and health-care products. *Angew. Chem. Int. Ed.*, **54** (11), 3328–3350.
126. Poblete-Castro, I., Escapa, I.F., Jäger, C., Puchałka, J., Lam, C.M.C., Schomburg, D., Prieto, M.A., and Martins dos Santos, V.A.P. (2012) The metabolic response of *P. putida* KT2442 producing high levels of polyhydroxyalkanoate under single- and multiple-nutrient-limited growth: highlights from a multi-level omics approach. *Microb. Cell Fact.*, **11**, 34.
127. Vallon, T., Simon, O., Rendgen-Heugle, B., Frana, S., Mückschel, B., Broicher, A., Siemann-Herzberg, M., Pfannenstiel, J., Hauer, B., Huber, A., Breuer, M., and Takors, R. (2015) Applying systems biology tools to study *n*-butanol degradation in *Pseudomonas putida* KT2440. *Eng. Life Sci.*, **15** (8), 760–771.
128. Pieper, D.H. and Reineke, W. (2004) in *Pseudomonas* (ed. J.-L. Ramos), Springer, New York, pp. 509–574.
129. de Lorenzo, V. (2008) Systems biology approaches to bioremediation. *Curr. Opin. Biotechnol.*, **19** (6), 579–589.
130. Fernández, M., Conde, S., Duque, E., and Ramos, J.L. (2013) *In vivo* gene expression of *Pseudomonas putida* KT2440 in the rhizosphere of different plants. *Microb. Biotechnol.*, **6** (3), 307–313.
131. Molina, L., Ramos, C., Duque, E., Ronchel, M.C., García, J.M., Wyke, L., and Ramos, J.L. (2000) Survival of *Pseudomonas putida* KT2440 in soil and in the rhizosphere of plants under greenhouse and environmental conditions. *Soil Biol. Biochem.*, **32** (3), 315–321.
132. Wenzel, S.C., Gross, F., Zhang, Y., Fu, J., Stewart, A.F., and Müller, R. (2005) Heterologous expression of a myxobacterial natural products assembly line in pseudomonads via red/ET recombineering. *Chem. Biol.*, **12** (3), 349–356.
133. Stephan, S., Heinzle, E., Wenzel, S.C., Krug, D., Müller, R., and Wittmann, C. (2006) Metabolic physiology of *Pseudomonas putida* for heterologous production of myxochromide. *Process Biochem.*, **41** (10), 2146–2152.
134. Becker, J., Lange, A., Fabarius, J., and Wittmann, C. (2015) Top value platform chemicals: bio-based production of organic acids. *Curr. Opin. Biotechnol.*, **36**, 168–175.
135. Follonier, S., Panke, S., and Zinn, M. (2011) A reduction in growth rate of *Pseudomonas putida* KT2442 counteracts productivity advances in medium-chain-length polyhydroxyalkanoate production from gluconate. *Microb. Cell Fact.*, **10**, 25.
136. Ouyang, S.P., Luo, R.C., Chen, S.S., Liu, Q., Chung, A., Wu, Q., and Chen, G.Q. (2007) Production of polyhydroxyalkanoates with high 3-hydroxydodecanoate monomer content by *fadB* and *fadA* knockout mutant of *Pseudomonas putida* KT2442. *Biomacromolecules*, **8** (8), 2504–2511.
137. Wang, H.H., Zhou, X.R., Liu, Q., and Chen, G.Q. (2011) Biosynthesis of polyhydroxyalkanoate homopolymers by *Pseudomonas putida*. *Appl. Microbiol. Biotechnol.*, **89** (5), 1497–1507.
138. Liu, W. and Chen, G.Q. (2007) Production and characterization of medium-chain-length polyhydroxyalkanoate with high 3-hydroxytetradecanoate monomer content by *fadB* and *fadA* knockout mutant of *Pseudomonas putida* KT2442. *Appl. Microbiol. Biotechnol.*, **76** (5), 1153–1159.
139. Huijberts, G.N., Eggink, G., de Waard, P., Huisman, G.W., and Witholt, B. (1992) *Pseudomonas putida* KT2442 cultivated on glucose accumulates poly(3-hydroxyalkanoates) consisting of saturated and unsaturated monomers. *Appl. Environ. Microbiol.*, **58** (2), 536–544.
140. Huijberts, G.N.M. and Eggink, G. (1996) Production of poly(3-hydroxyalkanoates) by *Pseudomonas putida* KT2442 in continuous cultures.

Appl. Microbiol. Biotechnol., **46** (3), 233–239.

141. Berezina, N. and Martelli, S.M. (2014) in *Renewable Resources for Biorefineries* (eds C.L. Lin and R. Luque), The Royal Society of Chemistry, London, pp. 1–28.
142. López, N.I., Pettinari, M.J., Nikel, P.I., and Méndez, B.S. (2015) in *Advances in Applied Microbiology* (eds S. Sima and M.G. Geoffrey), Academic Press, Cambridge, pp. 73–106.
143. Sun, Z., Ramsay, J.A., Guay, M., and Ramsay, B.A. (2007) Carbon-limited fed-batch production of medium-chain-length polyhydroxyalkanoates from nonanoic acid by *Pseudomonas putida* KT2440. *Appl. Microbiol. Biotechnol.*, **74**, 69–77.
144. Prieto, M.A., Eugenio, L.D., Galán, B., Luengo, J.M., and Witholt, B. (2007) in *Pseudomonas* (eds J.-L. Ramos and A. Filloux), Springer, Dordrecht, pp. 397–428.
145. Poblete-Castro, I., Binger, D., Oehlert, R., and Rohde, M. (2014) Comparison of mcl-Poly(3-hydroxyalkanoates) synthesis by different *Pseudomonas putida* strains from crude glycerol: citrate accumulates at high titer under PHA-producing conditions. *BMC Biotechnol.*, **14**, 110.
146. Davis, R., Kataria, R., Cerrone, F., Woods, T., Kenny, S., O'Donovan, A., Guzik, M., Shaikh, H., Duane, G., Gupta, V.K., Tuohy, M.G., Padamatti, R.B., Casey, E., and O'Connor, K.E. (2013) Conversion of grass biomass into fermentable sugars and its utilization for medium chain length polyhydroxyalkanoate (mcl-PHA) production by *Pseudomonas* strains. *Bioresour. Technol.*, **150**, 202–209.
147. Cerrone, F., Choudhari, S.K., Davis, R., Cysneiros, D., O'Flaherty, V., Duane, G., Casey, E., Guzik, M.W., Kenny, S.T., Babu, R.P., and O'Connor, K. (2014) Medium chain length polyhydroxyalkanoate (mcl-PHA) production from volatile fatty acids derived from the anaerobic digestion of grass. *Appl. Microbiol. Biotechnol.*, **98** (2), 611–620.
148. Linger, J.G., Vardon, D.R., Guarnieri, M.T., Karp, E.M., Hunsinger, G.B., Franden, M.A., Johnson, C.W., Chupka, G., Strathmann, T.J., Pienkos, P.T., and Beckham, G.T. (2014) Lignin valorization through integrated biological funneling and chemical catalysis. *Proc. Natl. Acad. Sci. U.S.A.*, **111** (33), 12013–12018.
149. Le Meur, S., Zinn, M., Egli, T., Thony-Meyer, L., and Ren, Q. (2012) Production of medium-chain-length polyhydroxyalkanoates by sequential feeding of xylose and octanoic acid in engineered *Pseudomonas putida* KT2440. *BMC Biotechnol.*, **12**, 53.
150. Kenny, S.T., Runic, J.N., Kaminsky, W., Woods, T., Babu, R.P., Keely, C.M., Blau, W., and O'Connor, K.E. (2008) Up-cycling of PET (polyethylene terephthalate) to the biodegradable plastic PHA (polyhydroxyalkanoate). *Environ. Sci. Technol.*, **42** (20), 7696–7701.
151. Wierckx, N., Prieto, M.A., Pomposiello, P., de Lorenzo, V., O'Connor, K., and Blank, L.M. (2015) Plastic waste as a novel substrate for industrial biotechnology. *Microb. Biotechnol.*, **8** (6), 900–903.
152. Nomura, Y., Nakagawa, M., Ogawa, N., Harashima, S., and Oshima, Y. (1992) Genes in *pht* plasmid encoding the initial degradation pathway of phthalate in *Pseudomonas putida*. *J. Ferment. Bioeng.*, **74** (6), 333–344.
153. Liang, D.W., Zhang, T., Fang, H.H., and He, J. (2008) Phthalates biodegradation in the environment. *Appl. Microbiol. Biotechnol.*, **80** (2), 183–198.
154. Kenny, S.T., Runic, J.N., Kaminsky, W., Woods, T., Babu, R.P., and O'Connor, K.E. (2012) Development of a bioprocess to convert PET derived terephthalic acid and biodiesel derived glycerol to medium chain length polyhydroxyalkanoate. *Appl. Microbiol. Biotechnol.*, **95** (3), 623–633.
155. Mückschel, B., Simon, O., Klebensberger, J., Graf, N., Rosche, B., Altenbuchner, J., Pfannstiel, J., Huber, A., and Hauer, B. (2012) Ethylene glycol metabolism by *Pseudomonas putida*. *Appl. Environ. Microbiol.*, **78** (24), 8531–8539.

156. van Duuren, J.B.J.H., Wijte, D., Leprince, A., Karge, B., Puchałka, J., Wery, J., Martins dos Santos, V.A.P., Eggink, G., and Mars, A.E. (2011) Generation of a *catR* deficient mutant of *P. putida* KT2440 that produces *cis,cis*-muconate from benzoate at high rate and yield. *J. Biotechnol.*, **156**, 163–172.
157. van Duuren, J.B.J.H., Wijte, D., Karge, B., Martins dos Santos, V.A.P., Yang, Y., Mars, A.E., and Eggink, G. (2012) pH-stat fed-batch process to enhance the production of *cis,cis*-muconate from benzoate by *Pseudomonas putida* KT2440-JD1. *Biotechnol. Progr.*, **28**, 85–92.
158. Vardon, D.R., Franden, M.A., Johnson, C.W., Karp, E.M., Guarnieri, M.T., Linger, J.G., Salm, M.J., Strathmann, T.J., and Beckham, G.T. (2014) Adipic acid production from lignin. *Energy Environ. Sci.*, **8**, 617–628.
159. Wang, J.P., Wu, L.X., Xu, F., Lv, J., Jin, H.J., and Chen, S.F. (2010) Metabolic engineering for ethylene production by inserting the ethylene-forming enzyme gene (*efe*) at the 16S rDNA sites of *Pseudomonas putida* KT2440. *Bioresour. Technol.*, **101** (16), 6404–6409.
160. Martin, C.H. and Prather, K.L. (2009) High-titer production of monomeric hydroxyvalerates from levulinic acid in *Pseudomonas putida*. *J. Biotechnol.*, **139** (1), 61–67.
161. Martin, C.H., Wu, D., and Prather, K.L. (2010) Integrated bioprocessing for the pH-dependent production of 4-valerolactone from levulinate in *Pseudomonas putida* KT2440. *Appl. Environ. Microbiol.*, **76** (2), 417–424.
162. Vallon, T., Glemser, M., Malca, S.H., Scheps, D., Schmid, J., Siemann-Herzberg, M., Hauer, B., and Takors, R. (2013) Production of 1-octanol from *n*-octane by *Pseudomonas putida* KT2440. *Chem. Ing. Tech.*, **85** (6), 841–848.
163. Graf, N. and Altenbuchner, J. (2014) Genetic engineering of *Pseudomonas putida* KT2440 for rapid and high-yield production of vanillin from ferulic acid. *Appl. Microbiol. Biotechnol.*, **98** (1), 137–149.
164. Kumar, C., Yadav, K., Archana, G., and Naresh Kumar, G. (2013) 2-Ketogluconic acid secretion by incorporation of *Pseudomonas putida* KT 2440 gluconate dehydrogenase (*gad*) operon in *Enterobacter asburiae* PSI3 improves mineral phosphate solubilization. *Curr. Microbiol.*, **67** (3), 388–394.
165. Rehdorf, J., Kirschner, A., and Bornscheuer, U.T. (2007) Cloning, expression and characterization of a Baeyer–Villiger monooxygenase from *Pseudomonas putida* KT2440. *Biotechnol. Lett.*, **29** (9), 1393–1398.
166. Bell, S.G., French, L., Rees, N.H., Cheng, S.S., Preston, G., and Wong, L.L. (2013) A phthalate family oxygenase reductase supports terpene alcohol oxidation by CYP238A1 from *Pseudomonas putida* KT2440. *Biotechnol. Appl. Biochem.*, **60** (1), 9–17.
167. Loeschcke, A. and Thies, S. (2015) *Pseudomonas putida* – a versatile host for the production of natural products. *Appl. Microbiol. Biotechnol.*, **99** (15), 6197–6214.
168. Gemperlein, K., Zipf, G., Bernauer, H.S., Müller, R., and Wenzel, S.C. (2016) Metabolic engineering of *Pseudomonas putida* for production of docosahexaenoic acid based on a myxobacterial PUFA synthase. *Metab. Eng.*, **33**, 98–108.
169. Wittgens, A., Tiso, T., Arndt, T.T., Wenk, P., Hemmerich, J., Müller, C., Wichmann, R., Küpper, B., Zwick, M., Wilhelm, S., Hausmann, R., Syldatk, C., Rosenau, F., and Blank, L.M. (2011) Growth independent rhamnolipid production from glucose using the non-pathogenic *Pseudomonas putida* KT2440. *Microb. Cell Fact.*, **10**, 80.
170. Nitschke, M., Costa, S.G.V.A.O., and Contiero, J. (2011) Rhamnolipids and PHAs: recent reports on *Pseudomonas*-derived molecules of increasing industrial interest. *Process Biochem.*, **46** (3), 621–630.
171. Müller, M.M., Kügler, J.H., Henkel, M., Gerlitzki, M., Hörmann, B., Pöhnlein, M., Syldatk, C., and Hausmann, R.

(2012) Rhamnolipids – next generation surfactants? *J. Biotechnol.*, **162** (4), 366–380.

172. Beuttler, H., Hoffmann, J., Jeske, M., Hauer, B., Schmid, R.D., Altenbuchner, J., and Urlacher, V.B. (2011) Biosynthesis of zeaxanthin in recombinant *Pseudomonas putida*. *Appl. Microbiol. Biotechnol.*, **89** (4), 1137–1147.

173. Loeschcke, A., Markert, A., Wilhelm, S., Wirtz, A., Rosenau, F., Jaeger, K.E., and Drepper, T. (2013) TREX: a universal tool for the transfer and expression of biosynthetic pathways in bacteria. *ACS Synth. Biol.*, **2** (1), 22–33.

174. Gross, F., Ring, M.W., Perlova, O., Fu, J., Schneider, S., Gerth, K., Kuhlmann, S., Stewart, A.F., Zhang, Y., and Müller, R. (2006) Metabolic engineering of *Pseudomonas putida* for methylmalonyl-CoA biosynthesis to enable complex heterologous secondary metabolite formation. *Chem. Biol.*, **13** (12), 1253–1264.

175. Perlova, O., Fu, J., Kuhlmann, S., Krug, D., Stewart, A.F., Zhang, Y., and Müller, R. (2006) Reconstitution of the myxothiazol biosynthetic gene cluster by Red/ET recombination and heterologous expression in *Myxococcus xanthus*. *Appl. Environ. Microbiol.*, **72** (12), 7485–7494.

176. Gross, F., Luniak, N., Perlova, O., Gaitatzis, N., Jenke-Kodama, H., Gerth, K., Gottschalk, D., Dittmann, E., and Müller, R. (2006) Bacterial type III polyketide synthases: phylogenetic analysis and potential for the production of novel secondary metabolites by heterologous expression in pseudomonads. *Arch. Microbiol.*, **185** (1), 28–38.

177. Domröse, A., Klein, A.S., Hage-Hülsmann, J., Thies, S., Svensson, V., Classen, T., Pietruszka, J., Jaeger, K.E., Drepper, T., and Loeschcke, A. (2015) Efficient recombinant production of prodigiosin in *Pseudomonas putida*. *Front. Microbiol.*, **6**, 972.

178. Martinez, A., Kolvek, S.J., Yip, C.L., Hopke, J., Brown, K.A., MacNeil, I.A., and Osburne, M.S. (2004) Genetically modified bacterial strains and novel bacterial artificial chromosome shuttle vectors for constructing environmental libraries and detecting heterologous natural products in multiple expression hosts. *Appl. Environ. Microbiol.*, **70** (4), 2452–2463.

179. Craig, J.W., Chang, F.-Y., Kim, J.H., Obiajulu, S.C., and Brady, S.F. (2010) Expanding small-molecule functional metagenomics through parallel screening of broad-host-range cosmid environmental DNA libraries in diverse *Proteobacteria*. *Appl. Enivron. Microbiol.*, **76** (5), 1633–1641.

180. Troeschel, S.C., Thies, S., Link, O., Real, C.I., Knops, K., Wilhelm, S., Rosenau, F., and Jaeger, K.E. (2012) Novel broad host range shuttle vectors for expression in *Escherichia coli, Bacillus subtilis* and *Pseudomonas putida*. *J. Biotechnol.*, **161** (2), 71–79.

181. Nikel, P.I. and de Lorenzo, V. (2014) Robustness of *Pseudomonas putida* KT2440 as a host for ethanol biosynthesis. *New Biotechnol.*, **31** (6), 562–571.

182. Johnson, C.W. and Beckham, G.T. (2015) Aromatic catabolic pathway selection for optimal production of pyruvate and lactate from lignin. *Metab. Eng.*, **28**, 240–247.

183. Dammeyer, T., Steinwand, M., Krüger, S.C., Dübel, S., Hust, M., and Timmis, K.N. (2011) Efficient production of soluble recombinant single chain Fv fragments by a *Pseudomonas putida* strain KT2440 cell factory. *Microb. Cell Fact.*, **10**, 11.

184. Schmitz, S., Nies, S., Wierckx, N., Blank, L.M., and Rosenbaum, M.A. (2015) Engineering mediator-based electroactivity in the obligate aerobic bacterium *Pseudomonas putida* KT2440. *Front. Microbiol.*, **6**, 284.

185. Benedetti, I., de Lorenzo, V., and Nikel, P.I. (2016) Genetic programming of catalytic *Pseudomonas putida* biofilms for boosting biodegradation of haloalkanes. *Metab. Eng.*, **33**, 109–118.

Part III
Exploiting Anaerobic Biosynthetic Power

Industrial Biotechnology: Microorganisms, First Edition. Edited by Christoph Wittmann and James C. Liao.

9
Host Organisms: *Clostridium acetobutylicum/Clostridium beijerinckii* and Related Organisms

Frank R. Bengelsdorf, Anja Poehlein, Stefanie K. Flitsch, Sonja Linder, Bettina Schiel-Bengelsdorf, Benjamin A. Stegmann, Preben Krabben, Edward Green, Ying Zhang, Nigel Minton, and Peter Dürre

9.1
Introduction

In 1861, the famous microbiologist Louis Pasteur demonstrated that microorganisms are able to grow without oxygen [1]. Subsequently, he created the term anaerobic and named the anaerobic butyrate-producing organism he was working with *Vibrion butyrique.* Most likely, Pasteur was not working with a pure culture, but his reports allow to assume that the culture contained a butyric clostridial species such as *Clostridium butyricum* or *C. acetobutylicum* as their fermentation products were mainly butyrate and butanol [1]. In the early years, the term Clostridium was only used to describe the morphology of bacteria, irrespective of any metabolic properties [2]. The name *Clostridium* originates from the Greek word "klostridion" (small spindle) which was latinized to "clostridium" [2].

About 40 years after the ground-breaking studies of Louis Pasteur, anaerobic acetone production was described using a microorganism called at that time *Paenibacillus macerans* [3, 4]. At that time, isopropanol also was identified as a by-product of butanol fermentation by Pringsheim [5, 6].

British scientists started in 1912 to develop the fermentation process for *n*-butanol [7]. *n*-Butanol was to be used as a starting material for synthetic rubber. The First World War dramatically changed the market for acetone, which was used for cordite production. Acetone production was a by-product of the butanol process developed by Synthetic Products Company Ltd, and they gained a government contract to supply acetone. The Synthetic Products process involved prefermentation with a proteolytic microorganism followed by fermentation of the hydrolysates by the butylic bacteria of the type Fitz [8]. The absence of significant proteolytic activity of the butanol-producing bacteria indicates that it was probably a *Clostridium beijerinckii* or a *Clostridium saccharobutylicum* strain.

In parallel to the Synthetic Products development of the acetone process, Weizmann isolated a microorganism originally named *Bacillus granulobacter*

Industrial Biotechnology: Microorganisms, First Edition. Edited by Christoph Wittmann and James C. Liao.

pectinovorum [9]. This bacterium had plenty of proteolytic activity and could grow on pure maize meal or potato meal. The bacterium was later renamed *C. acetobutylicum* and the protease activity was identified as that of acidolysin [10, 11], a member of the peptidase M4 family. None of the sequenced *C. beijerinckii* or *C. saccharobutylicum* strains contains a peptidase M4 gene. The Weizmann process was scaled up to production scale at the British Acetones Toronto, transferred to Terre Haute, and after the war converted into a commercial company called Commercial Solvents Corporation (CSC).

The Weizmann process for producing acetone is characterized by the formation of twice the amount of butanol as acetone. Toward the end of the 1910s there was limited need for butanol and Northrup developed *P. macerans*, at that time called *Bacillus acetoethylicum*, as an acetone/ethanol producer [12, 13]. The Northrop process was never scaled up to production scale, presumably because *n*-butanol became a valuable product in the lacquer business.

Around 1930, high corn prices coincided with excess of cheap molasses. As a consequence, CSC isolated acetone/butanol-producing strains which were better than *C. acetobutylicum* on molasses medium [PK1] [PK2]. This included the famous "strain 8" [14], which was initially named *C. saccharo-acetobutylicum* alpha [15, 16] and deposited in NRRL as B-591. Genome sequencing has shown that this strain is identical to *C. beijerinckii* NCIMB 8052. *C. saccharo-acetobutylicum* beta and gamma strains were subsequently isolated and a representative was deposited in NRRL as B-597, and has subsequently also been identified as a *C. beijerinckii* strain.

The next improvement came from the isolation of *C. saccharobutyl-acetonicum-liquifaciens* delta and gamma in 1938 [17, 18]. These strains later became known as *C. saccharobutylicum* type I and II, and had faster growth on molasses concentrations above 5% sugars. The next improvement came by the isolation of *C. granulobacter acetobutylicum*, which is now the ancestor strain for all the National Chemical Products (NCP) *C. beijerinckii* strains [19].

The last installment of classical acetone/butanol producers is *Clostridium saccharoperbutylacetonicum*, which was isolated in 1959 by Hongo [20]. This strain never made it into a commercial plant owing to phage issues in the developing labs and a weaker economic basis owing to low oil prices.

9.2 Microorganisms

9.2.1 *C. acetobutylicum*

The strain type *C. acetobutylicum* ATCC 824 was isolated from garden soil in Connecticut in 1924 [21]. Its genome sequence including comparative analysis was published in 2001 [22]. It is a rod-shaped ($1-1.6\,\mu m \times 0.8\,\mu m$), gram-positive, obligately anaerobe, mesophil, endospore-forming bacterium and it possesses peritrichous flagella. It has a 4 Mb chromosome and harbors the megaplasmid pSOL1, which carries the genes necessary for solvent formation. *C. acetobutylicum* ATCC 824 is able to ferment glucose, lactose, sucrose, levulose, galactose, mannose, arabinose, xylose, starch, pectin, inulin, mannitol, glycogen, dextrin, melezitose [21], amygdalin, and raffinose and it produces the solvents acetone, butanol, and ethanol [23].

Etymology: N.L. n. *acetonum*, acetone; N.L. adj. *butylicum*, butylic; N.L. neut. adj. *acetobutylicum*, referring to production of acetone and butyl alcohol.

9.2.2
C. beijerinckii

C. beijerinckii is a member of the solvent-producing clostridia and has a biphasic metabolism. *C. beijerinckii* produces solvents in approximately the same ratio as *C. acetobutylicum*, but isopropanol is produced in place of acetone [24]. *C. beijerinckii* (formerly *C. acetobutylicum*) NCIMB 8052 prefers glucose as substrate but also ferments glucitol, galactose, and the disaccharides cellobiose, lactose, maltose, and sucrose [25]. *C. beijerinckii* NCIMB 8052 is a gram-positive, rod-shaped, obligately anaerobic, mesophylic bacterium and possesses peritrichous flagella.

C. beijerinckii is also a producer of clostrubin, an exceptional polyphenolic polyketide antibiotic [26]. Small doses (0.12–0.97 μM minimal inhibitory concentration (MIC)) of clostrubin inhibit growth of various pathogenic bacteria, such as methicillin-resistant *Staphylococcus aureus*, vancomycin-resistant *Enterococcus*, and mycobacteria [26].

Etymology: N.L. gen. masc. n. *beijerinckii*, of Beijerinck, named for M. W. Beijerinck, Dutch bacteriologist.

9.2.3
C. saccharobutylicum

The CSC firstly isolated and patented industrial strains of the *C. saccharobutylicum* species and designated them *Clostridium saccharo-butyl-acetonicum-liquefaciens* [27]. Arzberger (1938) and Carnarius and McCutchan [17] submitted the US patents delineating those *Clostridium* species. The industrial strains were subsequently named *C. acetobutylicum*; however, Keis and coworkers suggested the name *C. saccharobutylicum* [23].

C. saccharobutylicum is a gram-positive, rod-shaped, mesophilic (optimum temperature for growth is between 30 and 34 °C) endospore-forming bacterium and has peritrichous flagella. It ferments arabinose, xylose, glucose, mannose, cellobiose, lactose, maltose, saccharose, inositol, melibiose, methyl-glucopyranoside, raffinose, salicin, trehalose, turanose, amygdalin, starch, glycogen, and dextrin. The G-C content of the DNA amounts to about 28 ± 32 mol%. It also shows a metabolic shift from acidogenic to solventogenic phase in which the solvents acetone, butanol, and ethanol are formed [23].

Etymology: Sac.cha.ro.bu.tyli. cum. Gr. n. *saccharon*, sugar juice; N.L. n. *butylum*, butanol; N.L. neut. adj. *saccharobutylicum*, denoting the production of butanol from sugar [23].

9.2.4 *C. saccharoperbutylacetonicum*

Hongo was the first who named and described *C. saccharoperbutylacetonicum* in 1960 in the US patent no. 2945786 [20] and Hongo and Murata deposited the strain N1-4 (ATCC 13564) and its derivatives N1-4 (HMT) (ATCC 27021T) and N1-504 (ATCC 27022) in the American Type Culture Collection. *C. saccharoperbutylacetonicum* is a gram-positive, rod-shaped, mesophilic, (optimum temperature is between 25 and 35 °C), endospore-forming bacterium and has peritrichous flagella. It is able to ferment arabinose, xylose, glucose, mannose, cellobiose, lactose, maltose, saccharose, D-arabitol, L-arabitol, mannitol, melibiose, methyl-glucopyranoside, raffinose, salicin, trehalose, turanose, amygdalin, starch, glycogen, dextrin, pectin, melezitose, and inulin. In the patent by Hongo of 1960, melezitose and inulin were not listed as possible substrates. It performs a metabolic shift from acid production to solvent production, during which it forms acetone, butanol, and ethanol [23].

Etymology: Sac.cha.ro. per.bu.tyl.a.ce.toni.cum. Gr. n. *saccharon*, sugar juice; Gr. pron. per throughout; N.L. n. *butylum*, butanol; N.L. adj. *acetonicus*, acetonic; N.L. adj. *saccharoperbutylacetonicum*, denoting the production of a large amount of butanol and acetone from sugar [23].

9.3 Bacteriophages

9.3.1 History of Bacteriophages Infecting Solventogenic Clostridia

The acetone–butanol–ethanol (ABE) fermentation process, in keeping with other large-scale microbial fermentations, suffered occasionally from bacteriophage contaminations that resulted in reduced productivity or at the worst necessitated the fermentation being aborted. Accordingly, abnormal and sluggish fermentations have been reported as a major issue in ABE fermentation plants, although documented cases of phage infection are scarce [28–31].

One of the earliest unconfirmed examples of an ABE fermentation phage infection concerned an abnormal fermentation at the CSC plant in Terre Haute, USA, using maize mash as substrate for a *C. acetobutylicum* strain [32]. In this instance, the sluggish fermentation lasted for about 1 year and halved the solvent yield. CSC expended considerable efforts into the investigation of the cause of this abnormal fermentation performance and developed a number of countermeasures for preventing its recurrence. These included the use of phage-immunized clostridial strains. Despite these measures, phage infection subsequently occurred in at least one of the immunized strains, that is, *C. saccharoacetobutylicum* NRRL B591 [33], subsequently renamed *C. beijerinckii* NRRL B591 [34].

The first confirmed bacteriophage that affected the ABE fermentation process was isolated in 1943 from a culture of *C. beijerinckii* 214, at that time designated *Clostridium madisonii* 214 [35–37]. A fermentation plant located in Puerto Rico used this strain to convert molasses to solvents, and the abnormal fermentation that occurred was shown to be linked to four different infections with the bacteriophages A, B, C, and D. In further studies, it was possible to relate phages A and B to a serological phage group that was distinct from phages C and D. In the course of creating an immunized *C. beijerinckii* 214 that was no longer susceptible to all four phages (A, B, C, and D), the lysogenic strain *C. beijerinckii* 4J9 was created. This strain contained a temperate phage, designated CMX, and was capable of infecting and lysogenizing a number of industrial strains that were phylogenetically closely related to *C. beijerinckii* 214 and *C. beijerinckii* NCIMB 8052 [37].

The most well-documented phage infections are reported for a Japanese ABE fermentation process that was established the 1940s and operated until the late 1960s. The process organisms used were *C. acetobutylicum* and *C. saccharoperbutylacetonicum* [30]. In literature referred to as sleeping sickness of fermentation, several ABE fermentation plants operated during 1943, and based on *C. acetobutylicum* strains utilizing starch or sugar as the feedstock, exhibited abnormal process performance. It was not before 1951 that it could be shown that the cause of this irregular process were bacteriophage infections. Three phages (designated a, b, and c) were identified [38]. After the Second World War, the Japanese fermentation plants switched from starch or sugar to molasses as substrates, but were never able to exclude phage infections entirely, as reports throughout the 1950s indicated. A new strain for the Japanese butanol ABE process was isolated in 1959 from soil samples near Fukuoko and patented as *C. saccharoperbutylacetonicum* N1-4 [20]. Soon after the fermentation process was established with this strain, an infection with the phage HM 1 occurred. After isolation of a phage-resistant mutant of *C. saccharoperbutylacetonicum* N1-4, a resumed fermentation with the phage-resistant variant led to an infection with another bacteriophage. The isolation of phage-resistant mutant, followed by an infection with a new bacteriophage, became a cyclic phenomenon such that in just 1 year 12 different bacteriophages were reported. These phages, named HM 1 to HM 12 in accordance to their chronological appearance, could be classified morphologically and serologically into three distinct types [39, 40]: group I (HM 1, HM 2, HM 8, HM 9, HM 10, HM 11, and HM 12) contained short-tailed *Podoviridae*. Group II (HM 3, HM 4, HM 5, and HM 6) belonged to the *Myoviridae* and infected both strains of the *C. saccharoperbutylacetonicum* N1-4 subgroup. Group III contained only the phage HM 7, which belonged to the *Siphoviridae* and only infected strains belonging to the *C. saccharoperbutylacetonicum* N1-504 subgroup. Alongside these 12 lysogenic phages, further investigation of the original infected fermentation, in which the HM 1 phage was discovered, identified a bacteriophage capable of lysogenizing *C. saccharoperbutylacetonicum*

N1-4. The lysogen created was designated *C. saccharoperbutylacetonicum* N1-4 (HMT). The identified phage HM T was released spontaneously from its host and showed morphological and serological similarities to HM 1 and other members of group I and was, therefore, presumed to be the parental phage of this group [30, 41]. After suffering these setbacks, the use of *C. saccharoperbutylacetonicum* N1-4 and its derivatives in the ABE fermentation was discontinued.

Other examples of phage contamination were reported in an ABE fermentation plant in Germiston, South Africa, operated by the NCP. Between 1936 and 1982, a total of four phage infections were confirmed in their batch fermentation process. Thus, in 1943, a maize-based fermentation using a derived *C. acetobutylicum* Weizmann strain (the Melle strain), showed an abnormal fermentation. This led NCP to switch their fermentation process in the following year (1944) to molasses-fermenting strains. The process subsequently ran without phage contamination until 1947, when solvent yields exhibited a dramatic decrease as a result of "sluggish" fermentation. They were, however, unable to confirm that the observed downturn in productivity was caused by a bacteriophage. In 1960, after one of the fermenters burst apart, most of the remaining fermenters were infected with a long-tailed phage belonging to the *Myoviridae* that led to sluggish and irregular fermentation. The next documented phage infection in the NCP fermentation plant was in 1976 due to contaminations with two different phages, one belonging to the *Siphoviridae*, the other to the *Myoviridae*. Protoplasts, sluggish fermentations, and poor solvent yield were the result of this dual infection. The two phages most likely infiltrated the fermenters when the process was switched from carbon dioxide to unfiltered nitrogen as gas atmosphere. Another phage infection was assumed to have taken place in 1977 in a fermentation run that used a subculture of *C. beijerinckii* P261 that was supposed to be lysogenic, but neither the infection nor the lysogeny were confirmed. The problems with the fermentation carried on over a year, despite the use of a different production strain. The last confirmed phage contamination at the NCP fermentation factory and the last documented phage outbreak in an ABE fermentation plant was reported in 1980 [42]. The effect of this phage, designated CA1 and a member of the *Podoviridae*, had a minor impact on process productivity. Consequently, the fermentation process was not aborted and the phage was not further investigated.

Apart from the phage infections in Japan, the United States, and South Africa in the fermentation plants discussed above, other countries (i.e., Republic of Cuba, Soviet Union, and the United Arab Republic) faced similar problems, and sporadic reports documenting these cases can be found in the literature [30, 43].

Since the well-documented phage infection of the NCP plant in 1980, there have been no further reports in the literature of phage infections affecting the ABE process. Nonetheless, phage typing has been used by Keis and co-workers [37], in combination with other typing methods, to help in the reclassification of solventogenic clostridia. In another study, Illumina sequencing was applied to sequence the genome of the temperate bacteriophage HM T, conserved in the strain *C. saccharoperbutylacetonicum* N1-4 (HMT), and represents the first fully sequenced bacteriophage of the solventogenic clostridia [44].

9.3.2 Phage Specificity

The specificity of the bacteriophages infecting solventogenic clostridia is very high [28, 30, 37, 45, 46]. Individual phage exhibit a very limited host range, infecting often only closely related strains and never crossing the species barrier [37, 47–50]. For instance, the HM phages described in the 1960s [45] only infect strains of *C. saccharoperbutylacetonicum* N1-4, and never more distantly related strains, including pathogens or butyric acid producers.

9.3.3 Phage Effect

Although measures were taken by companies to prevent an infection of their production strains with bacteriophages, reports of such incidences have constantly appeared throughout the history of the commercial ABE fermentation process. The effects of a phage infection were quite similar for all reported incidents, irrespective of the substrate or production strain used. An infected fermentation exhibit a delayed and incomplete consumption of sugars and as a consequence a slow or sluggish fermentation accompanied by an extended fermentation time. The infection influenced the solvent production negatively, as the uptake of organic acids and their conversion to solvents was disturbed. The pH of the culture typically dropped rapidly, resulting in high acid levels and low solvent yields. At a macroscopic scale, a bacteriophage infection is indicated by decreased gas formation by the culture and retention of a dark coloration to the media, indicative of a failure to utilize the substrate. In the final stages of the infection, the number of viable cells dramatically decreases, culminating in complete cell lysis. In some cases, a continued incubation for 24–48 h of a dying culture led to a restart of bacterial growth, giving rise to clostridial cells that were no longer lysed by the bacteriophages. However, these phage-resistant variants invariably no longer resembled the original producer strains, either in productivity or in morphology. The solvent yields of the fermentation diminished to moderate and even poor amounts, whereas the overall cell number was reduced. The remaining cells had lost their motility and the cultures were characterized by the presence of "etched" or "moth-eaten" cells, as well as elongated cells or protoplasts.

The overall effect of bacteriophage contaminations is influenced by several factors related to the characteristics of the phage itself as well as the timing of infection relative to the stage of the fermentation process. Consequently, independent infections even with the same phage can result in different outcomes. Thus, minor contaminations can only result in extended fermentation times and decreased solvent yield, whereas more severe infections can lead to complete cessation of the fermentation and/or nearly complete cell lysis within hours.

Bacteriophages with a short latent period and large burst size tend to have the most serious consequences for the fermentation process. Such characteristics were reported for the bacteriophage HM 2, infecting *C. saccharoperbutylaceton-icum* N1-4. In contrast, bacteriophages that have a long latent period and are released in small amounts from their host cells can be present in a fermentation without serious consequences for the process, as described for the phage HM 7, infecting again *C. saccharoperbutylacetonicum* [30].

Phage contamination of bacterial cells growing in minimal medium suffer more dramatic effects than cultures using complex medium, characterized by a faster decimation of viable cells [51]. Widespread infections in several fermenters are often associated with an initial contamination of the seed or starter cultures that were prepared in the preaccession phase of the fermentation. Infections that occur later in the fermentation process might have limited effects and are therefore less obvious. Bacteria that contain temperate phages, either in a lysogenic or pseudolysogenic state, display less obvious symptoms if cultivated and the effects occur permanently. The elimination of such bacteriophages can be exceedingly problematic and should be avoided in advance by avoiding phage-harboring hosts as process organisms. Known representatives of such lysogenic clostridia are *C. beijerinckii* 4J9, formerly known as *C. madisonii* or *C. saccharoperbutylacetonicum* N1-4 (HMT), harboring the bacteriophages CMX and HM T, respectively [37, 41, 46, 52].

An infection with bacteriophages in large-scale fermentations resulted in most cases in the abandonment of the fermentation process. The ensuing cleanup operations necessitated the complete shutdown of the fermentation plant. The operators were therefore confronted with the considerable expenses associated with production losses and the additional costs incurred in removing the phages from the process equipment and environment.

One exception of an infected fermentation process that was continued after the infection was detected was reported by the NCP factory in Germiston, South Africa, in 1980 [31]. After the operators became aware of an infection with CA1 phage, a decision was made to continue the fermentation process, although solvent concentrations and yields were decreasing. As the effects of this contamination were not as severe as previous infections, the eradication of the bacteriophage was carried out during the fermentation process. The application of improved culture practices and plant hygiene, eradication of corrupted precultures and their replacement by phage-free production strains, and steam cleaning of the contaminated fermentation system led to the reestablishment of normal fermentation conditions within a couple of weeks.

9.4
ABE Fermentation of Solvent-Producing *Clostridium* Strains

C. acetobutylicum is the model microorganism to study clostridial ABE fermentation. The respective biphasic metabolism is similar to other solventogenic clostridial species (Table 9.1, see Section 9.5). The fermentation starts with uptake of the substrate (e.g., hexoses and pentoses). Sugars are metabolized to

Table 9.1 General GenBank features of genomes of solventogenic clostridia.

Status	Strain	Accession	BioProject	BioSample	IMG Taxon-ID
F	*C. acetobutylicum* ATCC 824	AE001437 AE001438	77	2603243	637000076
F	*C. acetobutylicum* DSM 1731	CP002660 CP002661 CP002662	63241	2603396	650716024
F	*C. acetobutylicum* EA 2018	CP002118 CP002119	50455	2603410	651053017
PD	*C. acetobutylicum* GXAS18-1	JRWL01000000	263987	3112708	2630968745
F	*C. beijerinckii* NCIMB 14988	CP010086	260606	3024437	2608642293
PD	*C. beijerinckii* AAU1_buffalo	AUPH01000000	168249	2469634	2563367072
PD	*C. beijerinckii* G117	AKWA01000000	168606	2470653	2516493007
PD	*C. beijerinckii* HUN142	JHXK01000000	223437	2744012	2558860976
F	*C. beijerinckii* NCIMB 8052	CP000721	12637	2598276	640753016
PD	*C. beijerinckii* SA-1, ATCC 35702	CP006777	207367	3120854	2574179756
F	*C. saccharobutylicum* DSM 13864	CP006721	217481	2603278	2534682367
F	*C. saccharoperbutylacetonicum* N1-4 (HMT)	CP004121 CP004122	60605	2603265	2608642274

pyruvate through glycolysis (hexoses) or pentose phosphate pathway (pentoses). The fermentation metabolism of solventogenic clostridia is split into two separate phases: first acidogenesis, then solventogenesis. Solvent production in uncontrolled batch fermentations starts shortly after bacteria switch from the exponential to the stationary growth phase. Therefore, exponentially growing cells produce the acids acetate and butyrate (molar ratio 1 : 2) and to some extent CO_2, H_2, and ethanol. Production of organic acids leads to an acidification of the environment and the external pH decreases. In consequence, the internal pH of the cell drops in concert with the external pH, the internal pH being continuously one pH unit more alkaline [53]. At an external pH of 4.5, significant amounts of undissociated acetic ($pK_a = 4.75$) and butyric ($pK_a = 4.82$) acids are present and cross the cytoplasmic membrane by diffusion. Once the internal pH is at approximately 5, undissociated acids dissociate inside the cell. Thus, collapse of the proton gradient of the cytoplasmic membrane is caused by the release of respective protons. In order to neutralize the acidification of the environment, cells switch to solvent production. In the subsequent solventogenic phase, stationary cells take up the organic acids and transform them into butanol and acetone (molar ratio butanol:acetone:ethanol = 6 : 3 : 1). Although the solvents are neutral, their production is not an appropriate way to prevent the death of cells, because high butanol concentrations (up to 2%) are toxic as well. When solvent production starts, the sporulation process is initiated simultaneously. Thus, solvent production is a process that buys time to finalize the sporulation process (for reviews see [54, 55]).

Enzymes involved in the biphasic acid and solvent production are shown in Figure 9.1. The respective enzymes for acid production (starting from acetyl-CoA) are phosphotransacetylase (Pta) and acetate kinase (Ack) for acetate, thiolase (ThlA), 3-hydroxybutyryl-CoA dehydrogenase (Hbd), crotonase (Crt), butyryl-CoA dehydrogenase (Bcd) coupled to a pair of electron-transferring flavoproteins (EtfA, EtfB), phosphotransbutyrulase (Ptb), and butyrate kinase

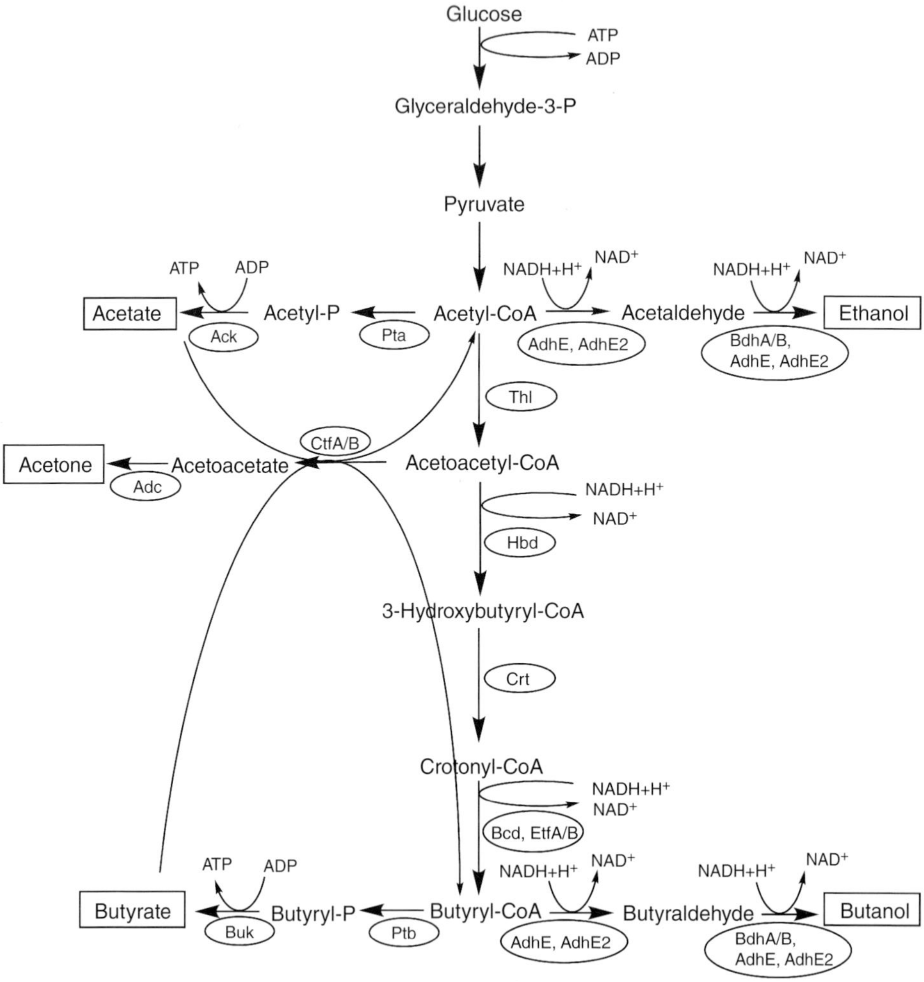

Figure 9.1 Metabolism and products of solventogenic bacteria. Ack, acetate kinase; Adc, acetoacetate decarboxylase; AdhE, AdhE2, aldehyde/alcohol dehydrogenase; BdhA, BdhB, alcohol dehydrogenase; Bcd, butyryl-CoA dehydrogenase; Buk, butyrate kinase; Crt, crotonase; CtfA/B, CoA transferase; EtfA/B, electron-transfer flavoproteins A and B; Hbd, 3-hydroxybutyryl-CoA dehydrogenase; NADH, reduced nicotinamide adenine dinucleotide (reduced form); Pta, phosphotransacetylase; Ptb, phosphotransbutyrylase; Thl, thiolase. The single reactions do not represent stoichiometric fermentation balances.

(Buk) for butyrate. The proteins EtfA and EtfB are crucial for the activity of the Bcd protein [56]. The required enzymes for solvent production are coenzyme A transferase (CtfA, CtfB), acetoacetate decarboxylase (Adc) as well as different aldehyde and alcohol dehydrogenases (BdhA, BdhB, AdhE, AdhE2). In *C. beijerinckii* [57], an additional secondary alcohol dehydrogenase allows the reduction of acetone to isopropanol. All genes and enzymes were discussed in detail by Köpke and Dürre [58].

Strains of *C. acetobutylicum* and *C. beijerinckii* were studied mostly with respect to ABE fermentation. Although strains of both species show a similar phenotype, they are phylogenetically different [37]. As of now, *C. acetobutylicum* is the only solventogenic strain that harbors a megaplasmid (pSOL1). The genome of *C. beijerinckii* G117 represents so far the largest known clostridial genome with a size of 8.8 Mbp. Genome features of *C. acetobutylicum, C. beijerinckii, C. saccharobutylicum,* and *C. saccharoperbutylacetonicum* are displayed in Table 9.2 (see Section 9.5).

In solventogenic clostridia, the genes encoding enzymes required for acid production are clustered, with the exception of the *thlA* gene (encoding the thiolase), in the chromosomes of respective strains. The clustered genes *crt, bcd, etfA,* and *etfB* as well as *hbd* form the *bcs* operon (Figure 9.2a) [59]. Likewise, genes encoding enzymes necessary for solvent production are also clustered in the so-called *sol* operon. However, the gene encoding the acetoacetate decarboxylase (*adc* gene) is directly located downstream of the *sol* operon, but in convergent direction (Figure 9.2b). In *C. acetobutylicum,* the *sol* operon is located on the megaplasmid pSOL1 and consists of the genes *adhE, ctfA,* and *ctfB* (CAC0162-CAC0164) (Figure 9.2b). Furthermore, a small peptide (gene *orfL*) of still unknown function [60, 61] is also encoded in the *sol* operon. In *C. beijerinckii, C. saccharobutylicum,* and *C. saccharoperbutylacetonicum* the respective genes

Table 9.2 General features of genomes of solventogenic clostridia.

Strain	Size	Genes	CRISPRs	Scaffolds	16S rRNAs	23S rRNAs	5S rRNAs	tRNAs
C. acetobutylicum ATCC 824	4 132	4 022	0	2	11	11	11	73
C. acetobutylicum DSM 1731	4 145	4 029	0	3	11	11	11	73
C. acetobutylicum EA 2018	4 132	4 024	0	2	11	11	11	75
C. acetobutylicum GXAS18-1	3 796	3 650	15	49	2	2	1	43
C. beijerinckii NCIMB 14988	6 455	5 699	1	1	3	5	7	69
C. beijerinckii AAU1_buffalo	6 954	10 315	0	6 093	4	5	7	74
C. beijerinckii G117	8 748	7 799	0	151	1	2	10	63
C. beijerinckii HUN142	6 106	5 589	0	53	12	7	15	56
C. beijerinckii NCIMB 8052	6 001	5 290	0	1	14	14	15	94
C. beijerinckii SA-1, ATCC 35702	5 880	5 330	1	191	7	5	14	51
C. saccharobutylicum DSM 13864	5 108	4 532	2	1	12	12	13	85
C. saccharoperbutylacetonicum N1-4 (HMT)	6 666	5 937	2	2	11	11	13	70

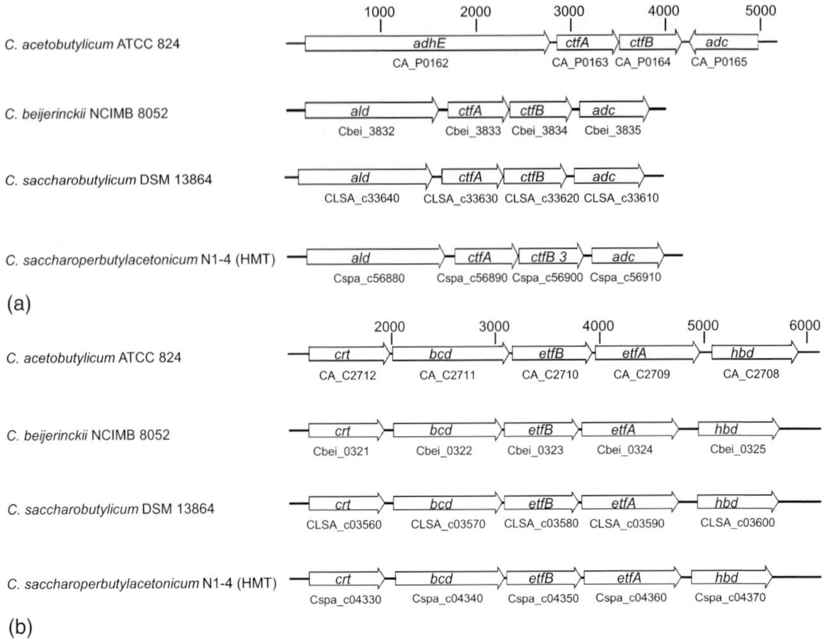

Figure 9.2 Genes involved in butanol production. (a) *sol* operon and (b) *bcs* operon. *adhE*, aldehyde/alcohol dehydrogenases; *ctfA/B*, CoA transferase; *adc*, acetoacetate decarboxylase; *ald*, aldehyde dehydrogenase; *crt*, crotonase; *bcd*, butyryl-CoA dehydrogenase; *etfA/B*, electron-transferring flavoproteins; *hbd*, 3-hydroxybutyryl-CoA dehydrogenase. Numbers indicate length in base pairs.

of the *sol* operon are located on the chromosomes. However, the arrangement of genes differs and comprise *ald* (encoding an aldehyde dehydrogenase), *ctfA*, *ctfB*, and *adc* in the same orientation [62, 63].

Wiesenborn *et al.* [64] purified the CoA transferase of *C. acetobutylicum* ATCC 824 and determined K_m values for acetate (1200 mM) and butyrate (660 mM). In *C. acetobutylicum*, intracellular concentrations of acetate and butyrate are generally much lower than the apparent K_m values and do not exceed 300 and 700 mM, respectively. Enzyme assays also showed that acetate and butyrate conversion was suppressed by physiological levels of acetone (200 mM), butanol (200 mM), acetoacetate (20 mM), and acetyl-CoA (1 mM). The activity of Adc is equally dependent on concentration of acetate in the medium [65]. Conversion of organic acids by CoA transferase was not observed until induction of Adc. Overall, the reaction of Adc is thermodynamically advantageous (ΔG_0, $\sim -27.2\,\text{kJ mol}^{-1}$) and forces the less favorable CoA transferase reaction (ΔG_0, -9.6 to $-7.1\,\text{kJ mol}^{-1}$) toward formation of acetoacetate [66]. Nevertheless, the inactivation of the *adc* gene in *C. acetobutylicum* almost eliminated acetone production and improved the butanol:ethanol ratio [67–69].

The bifunctional aldehyde/alcohol dehydrogenase AdhE is crucial to trigger the solvent production in uncontrolled batch fermentations using glucose as substrate [69, 70]. The inactivation of the respective gene (*adhE*) in a *C. acetobutylicum* mutant strain resulted in a limited solvent and enhanced acid production [69]. In addition to the *adhE* gene, there are three other genes that code for alcohol dehydrogenases present in the genome of *C. acetobutylicum*. One gene (*adhE2*) is located on the pSOL1 megaplasmid and the other genes are located on the chromosome (*bdhA* and *bdhB*, encoding butanol dehydrogenase). The enzyme BdhA apparently eliminates surplus of reducing equivalents by the formation of ethanol, whereas BdhB is involved in butanol production [70, 71]. A so-called "alcohologenic" fermentation can be induced by providing reduced substrate mixtures such as glucose/glycerol. Under these conditions, *C. acetobutylicum* expresses the *adhE2* gene (functional gene product AdhE2) and acetone production is blocked [72–74]. "Alcohologenic" fermentation can also be stimulated by addition of methyl viologen to the medium [74–76].

In 2008, the first genome-scale model was presented for *C. acetobutylicum* [77, 78]. The first part describes reconstruction of the metabolic network for *C. acetobutylicum* (ATCC 824) applying a new semiautomated reverse-engineering algorithm. Thereby, a set of genes encoding clostridial enzymes were *in silico* identified by systematic gene knockout simulations that are essential for growth. In the second part, specific proton fluxes were modulated for an accurate prediction of the extracellular medium pH during vegetative growth of batch cultures. The specific proton flux refers to the influx or efflux of free protons (per unit biomass) across the cell membrane. In early stages of acidogenesis, proton efflux through cation channels was found to be significant and 3.5 mol of protons were secreted per mole of butyric and acetic acids. Millat and coworkers [79] provided an integrative model of pH-dependent enzyme activities and transcriptomic regulation of *C. acetobutylicum* during ABE fermentation in continuous culture. Seven steady-state conditions between pH 5.7 and pH 4.5 were compared and revealed that bacteria switch from acidogenesis to solventogenesis between pH 5.3 and pH 5.0 with an intermediate state at pH 5.1. A further modeling approach showed that the pH-induced metabolic shift under experimental conditions is caused by a phenotypic change of predominantly acidogenic subpopulation to a predominantly solventogenic subpopulation [80].

Recently, Yoo *et al.* [81] contributed a complete transcriptomic, proteomic, and fluxomic analysis of different metabolic states of *C. acetobutylicum*. Accordingly, during acidogenesis, the total butyraldehyde dehydrogenase flux is contributed to the activity of AdhE2, while the butanol dehydrogenase flux is primarily due to the activity of BdhB and BdhA. During solventogenesis, the butyraldehyde dehydrogenase flux mainly refers to AdhE, and the butanol dehydrogenase flux primarily to BdhB and to a lesser extent to BdhA, Lastly, during "alcohologenic" fermentation, the entire flux of butyraldehyde and large amounts of butanol are generated because of the activity of AdhE2.

Liao *et al.* [82] have recently presented a system-level computational framework for ABE fermentation by *C. acetobutylicum* and combined metabolic reactions, gene regulation, and environmental cues. ABE fermentation was used as an example to model mathematically the complex microbial physiology to provide an integrated and quantitative understanding of biological systems (systems biology). The applied model successfully mirrored ABE fermentations of *C. acetobutylicum* (ATCC 824), and distinct mutant strains. Thus, there is a quantitative tool that permits the development of systems engineering strategies for the production of advanced biofuels.

In terms of energy conservation, *C. beijerinckii*, *C. saccharobutylicum*, and *C. saccharoperbutylacetonicum* possess genes encoding an *Rnf* (*Rhodobacter nitrogen fixation*) complex, which is not present in and *C. acetobutylicum*. An energy-converting hydrogenase (Ech) complex is not present in any solventogenic bacterium.

9.5 Genome-Based Comparison of Solvent-Producing *Clostridium* Strains

To date, the genome sequences of 12 solvent-producing strains belonging to four different species can be found in public databases such as National Center for Biotechnology Information (NCBI) or Integrated Microbial Genomes (IMG). These are *C. acetobutylicum* strains ATCC 824 [22], DSM 1731 [83], EA 2018 [84], and GXAS18-1 [85]; and *C. beijerinckii* strains NCIMB 8052 [63], AAU1_buffalo, G117 [86], HUN142, NCIMB 14988 [87], ATCC 35702 [88], and DSM 525 [89]; *C. saccharobutylicum* DSM 13864 [90], and *C. saccharoperbutylacetonicum* N1-4 (HMT) [91]. General GenBank features of these genomes are listed in Table 9.1.

Closed genomes are available for seven of these strains, namely, *C. acetobutylicum* strains ATCC 824, DSM 1731, and EA 2018; *C. beijerinckii* strains NCIMB 14988 and NCIMB 8052; *C. saccharobutylicum* DSM 13864, and *C. saccharoperbutylacetonicum* N1-4 (HMT).The general features of all publicly available genomes are listed in Table 9.2. The genome sizes range from 8.75 Mb for *C. beijerinckii* G117, representing the largest clostridial genome sequenced so far, to 3.8 Mb for *C. beijerinckii* GXAS18-1. A plasmid was found in two of the *C. acetobutylicum* strains (ATCC 824, EA 2018) and in *C. saccharoperbutylacetonicum* N1-4 (HMT), whereas *C. acetobutylicum* DSM 1731 is the only strain harboring two plasmids. The two draft genomes of *C. beijerinckii* HUN142 and AAU1_buffalo seem to contain at least one plasmid as genes encoding a protein involved in plasmid replication are present. However, the latter genome seems to have a low-quality assembly (6093 contigs) and a bad annotation, as there is an extraordinary high number of genes annotated (10 315) compared to a genome size of 6.5 Mb. The lowest number of genes are present in *C. acetobutylicum* GXAS18-1: approximately 375 genes are missing compared to the other *C. acetobutylicum* strains, which harbor between 4022 and 4029 genes. Inspection of the genome revealed that the contigs representing the megaplasmid, which is typical for *C. acetobutylicum* [60], is missing in strain GXAS18-1. In *C. acetobutylicum* strains, genes encoding key enzymes for solvent production (coenzyme A transferase (*ctfA* and *ctfB*), aldehyde/alcohol dehydrogenase (*adhE*), and acetoacetate decarboxylase (*adc*)) are usually located on the megaplasmid, but in the draft genome of strain GXAS18-1 these genes are completely missing, although Mo *et al.* [85] described this strain as a novel biobutanol-producing strain [85].

The numbers of genes encoded in the *C. beijerinckii* genomes vary between 5290 for *C. beijerinckii* NCIMB 8052 and 7799 for strain G117. The genomes of *C. saccharobutylicum* DSM 13684 and *C. saccharoperbutylacetonicum* N1-4 (HMT) encode 4532 and 5937 genes, respectively. Solventogenic clostridia encode a relatively high number of rRNA clusters; *C. beijerinckii* NCIMB 8052 encodes with 14 rRNA clusters the highest number, while 12 rRNA clusters are encoded by *C. saccharobutylicum* DSM 13864 and *C. beijerinckii* HUN142 and 11 by all *C. acetobutylicum* strains and *C. saccharoperbutylacetonicum.* For genomes of which only draft assemblies are available, the exact number of rRNA clusters cannot be counted, as gaps often represent this kind of repeats. We could identify one or two CRISPR (clustered regularly interspaced short palindromic repeats) clusters in 10 of the analyzed genomes (see Table 9.2). CRISPR's are widespread in many bacterial and almost all archaeal genomes [92]. The CRISPR/Cas system is a prokaryotic defense mechanism and provides immunity against invading mobile genetic elements such as phages and plasmids in an RNA interference-like manner. CRISPR loci typically consist of different numbers of noncontiguous repeats with lengths ranging from 20 to 47 bp [93] and unique spacers of different length and sequence between the repeats. Spacers arise from the integration of invader sequences that are integrated into CRISPR loci and act as immunity memory of past invasive elements [94]. The *C. beijerinckii* strains AAU1_buffalo, HUN142, G117, and NCIMB 8052 do not harbor any CRISPR loci. There are also no CRISPR repeats present in the genomes of the *C. acetobutylicum* strains except of strain GXAS18_1, where 15 CRISPR loci could be identified [95]. This comparatively high number might be misleading as CRISPR loci are highly repetitive regions and some loci might be interrupted by a gap.

Phylogenetic analysis revealed that the 16S rRNA genes of solventogenic clostridia group into two distinct clusters (Figure 9.3). One cluster is formed by all *C. acetobutylicum* strains sequenced so far. All strains have identical 16S rRNA sequences except for one substitution in strain GXAS18-1.

The second cluster is rather diverse, as *C. beijerinckii* AAU1_buffalo, NCIMB 14988 form a subcluster and a second subcluster is formed by *C. beijerinckii* strains NCIMB 8052 and ATCC 35702 and *C. saachroperbutylacetonicum* DSM 14923 (N1-4 (HMT)). Strains *C. beijerinckii* G117 and HUN142 were excluded from this analysis, as there are no or only very short 16S rRNA sequences available for these strains.

Whole-genome comparison of all solventogenic clostridia with sequenced genomes revealed a core genome of 861 ortholog groups (OGs) and a pan genome of 13 614 OGs (Figure 9.4). Core genome implies that the OGs are shared by all organisms in an analyzed set of genomes. Pan genome includes the core genome, OGs shared by at least two genomes and the genome-specific OGs [98]. The genome of *C. beijerinckii* AAU1_buffalo was excluded from this analysis owing

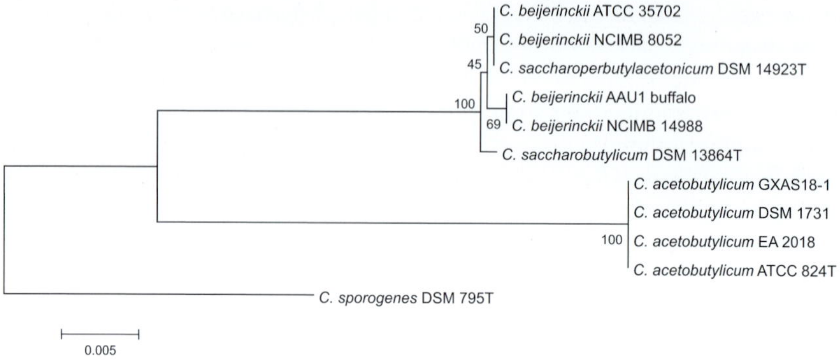

Figure 9.3 Phylogenetic tree of solventogenic clostridia based on 16S rRNA gene sequences. Sequences were aligned using ClustalW 1.6 [96]. The phylogenetic tree was obtained by using the neighbor-joining method within MEGA 5.05 software [97]. Numbers at nodes are bootstrap values calculated from 1000 resamplings to generate a majority consensus tree. *C. sporogenes* DSM 795T was used as outgroup. The scale bar indicates the nucleotide sequence divergence. The length of the tree branches was scaled according to the number of substitutions per site (see size bar).

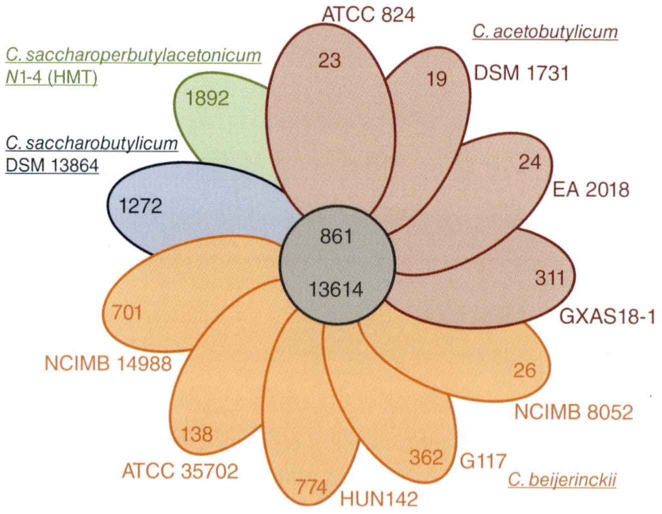

Figure 9.4 Simplified Venn diagram representing the pan and core genomes and the number of genome-specific orthologous groups of solventogenic clostridia. Ortholog detection was done with the Proteinortho software (blastp) with an identity cutoff of 50% and an *E*-value of 1e−10 [98]. The genome of *C. beijerinckii* AAU1_buffalo was excluded from this analysis owing to an extraordinarily high number of genes annotated (10 315) compared to a genome size of 6.5 Mb.

to an extraordinarily high number of genes annotated (10 315) compared to a genome size of 6.5 Mb.

The number of singletons (genes only present in one of the analyzed genomes) is rather different and ranges from 1892 found in *C. saccharoperbutylacetonicum* N1-4 (HMT) to only 19 identified in *C. acetobutylicum* DSM 1731. A relatively high number of singletons could also be identified in *C. saccharobutylicum* DSM 13864 (1272) and *C. beijerinckii* HUN142 (774). The number of singletons detected in the genomes is mostly correlated to the size of the respective genome, as *C. saccharoperbutylacetonicum* N1-4 (HMT), *C. saccharobutylicum* DSM 13864, and *C. beijerinckii* HUN142, respectively are the largest ones in this set of genomes. The lowest number of singletons could be detected in the *C. acetobutylicum* strains and with sizes of approximately 4.1 Mb, these genomes are 1–2 Mb smaller than the other genomes of solventogenic clostridia.

9.6 Regulation of Solvent Formation in *C. acetobutylicum*

Regulation of solvent formation is complex in *C. acetobutylicum* and the precise signal(s) initiating solventogenesis are still unknown. Besides concentration of undissociated acids, pH value, temperature, metabolic intermediates, cofactors, salts, and DNA supercoiling [55, 61, 99, 100], regulation includes several transcription factors: Spo0A [101], Rex [102, 103], CcpA [104, 105], CodY [106], and AdcR/S [107]. In addition, solvent production is regulated by RNA processing (*adhE* transcript [108], possibly *adhE2* transcript [101]), protein modification (Adc protein, [101]), and a small noncoding regulatory RNA (sRNA) (SolB) [109].

The transcription regulator Spo0A links the physiological processes of sporulation and solventogenesis in *C. acetobutylicum* [101, 110, 111], allowing the organism survival of otherwise lethal conditions (for reviews, see [54, 55, 99, 112]). Phosphorylated Spo0A (Spo0A∼P) regulates transcription of genes by DNA binding to so-called 0A boxes [113]. Respective binding sites are verified upstream of key operons of solvent production (*adc* and *sol* operon), clearly demonstrating the role of Spo0A in solventogenesis [108, 113, 114]. In contrast to *Bacillus subtilis*, *C. acetobutylicum* Spo0A is phosphorylated directly by histidine kinases [115, 116]. Spo0A is present in all solventogenic *Clostridium* species (BLAST analysis using JGI/IMG website) and all of them are missing a phosphorelay (BLAST analysis using JGI/IMG website).

The transcriptional repressor Rex is a redox-sensing protein playing a role in the shift from acidogenesis to solventogenesis in *C. acetobutylicum* [102]. Rex controls the expression of butanol biosynthetic genes in response to NADH/NAD^+ ratio in the cell [102, 103]. Rex is also present in *C. beijerinckii* [103] and in *C. saccharobutylicum* as well as *C. saccharoperbutylacetonicum* (BLAST analysis using JGI/IMG website).

The catabolite control protein A (CcpA) is a pleiotropic regulator involved in the regulation of many different cellular processes in *C. acetobutylicum* [104, 105].

In *C. acetobutylicum*, CcpA activates genes responsible for D-glucose-specific phosphotransferase system (PTS), enhances D-glucose utilization, mediates carbon catabolite repression on nonpreferred carbon sources (D-xylose and L-arabinose), is involved in regulation of acidogenesis and solventogenesis, and seems to be necessary for efficient sporulation [105]. CcpA regulates transcription of genes by DNA binding to so-called *cre* sequences [117, 118] which could also be identified in *C. acetobutylicum* [105]. CcpA is also present in *C. beijerinckii* [119], *C. saccharobutylicum* [120], and in *C. saccharoperbutylacetonicum* (BLAST analysis using JGI/IMG website).

The transcription factor CodY regulates metabolism, sporulation, and virulence in gram-positive bacteria [121]. In *C. acetobutylicum*, CodY seems to be a repressor of solvent formation [106]. CodY regulates transcription of genes by DNA binding [122] and a binding site was identified upstream of the *sol* operon [106]. CodY is also present in *C. beijerinckii*, *C. saccharobutylicum*, and *C. saccharoperbutylacetonicum* (BLAST analysis using JGI/IMG website).

The transcription factor AdcR binds upstream of the *adc* and *sol* operons and seems to play a role in solvent production [107, 123]. Its detailed function is yet to be characterized.

Overproduction of the sRNA SolB [109, 124] of *C. acetobutylicum* inhibits production of acetone and butanol as well as sporulation, and the genes responsible for solvent formation are downregulated (*adhE*, *ctfA*, *ctfB*, *adc*, *adcR*, and *adcS*) [125]. Increasing the protein concentration of Spo0A reverses these physiological effects [125]. Comparison of a *C. acetobutylicum solB* overexpression mutant and a *C. acetobutylicum spo0A* deletion mutant at the transcriptomic level revealed that the master regulator Spo0A and the small noncoding RNA SolB act divergently on the same targets, especially genes involved in solvent production and sporulation. These results clearly emphasize the important role of SolB as antagonist of Spo0A in *C. acetobutylicum*. The RNA-binding protein Hfq binds SolB. SolB is not present in *C. beijerinckii*, *C. saccharobutylicum*, and *C. saccharoperbutylacetonicum*.

9.7 Genetic Tools for Clostridial Species

Interest in exploiting the metabolic diversity of clostridial species for industrial purposes has led to concerted efforts to develop the necessary genetic tools for generating mutant derivatives (Table 9.3). Initial strategies sought to use traditional recombination-based methods to inactivate genes of *C. acetobutylicum* [126–129]. In the absence of counterselection markers, these approaches relied on the integration of the entire plasmid carrying an internal segment of the targeted gene [126–129]. The vectors used carried a marker gene that encoded resistance to an antibiotic, which allowed the desired single crossover integrant to be selected on the basis of acquisition of resistance to that antibiotic. Accordingly, the mutants selected on agar media that incorporated that antibiotic carried

Table 9.3 Main directed genome editing systems relevant to *Clostridium acetobutylicum* and *C. beijerinckii*.

Clostridium spp. used for exemplification	Wild type or mutant	Negative selection marker	Positive selection marker	Selection criteria[a)]	Plasmid example	Principle genes inactivated/added	References
C. acetobutylicum ATCC 824 *C. difficile* CD630 *C. botulinum* ATCC 3502 *C. sporogenes* NCIMB 10696 *C. beijerinckii* NCIMB 8052	Wild type	NAP	*ermB*	Em^R	pMTL007, pMTL007C-E2, pMTL007S-E2	*spo0A, pyrF, agrA, adc, ptb, ack, adhE, pta, adhE2, bdhA, bdhB, ctfA, ctfB, adc*	[7, 8, 10]
C. acetobutylicum ATCC 824	Wild type	NAP	NAP	PCR screening	pSY6	*solR, buk*	[6]
C. thermocellum DSM 1313	Δ*pyrF*	*pyrF*	NAP	FOA^R	pMU482	*Pta*	[18]
C. perfringens strain 13	Δ*galKT*	*galK*	NAP	DOG^R	pCM-GALK	α-, θ-, and k-toxin genes, *virRS*	[17]
C. thermocellum DSM 1313	Δ*hpt*	*hpt/tdk*	NAP	FUD^R/AZH^R	pMU1647, pMU1657	*ldh, pta*	[22]
C. difficile R20291 *C. difficile* CD630	Wild type	*codA*	NAP	FC^R	pMTL-SC7215, pMTL-SC7315	*tcdC*	[20]
C. acetobutylicum ATCC 824 *C. acetobutylicum* M5	Wild type	*mazF*	NAP	Cell survival	pKORD_mazF, pKOD_mazF	*CA_P0167, sigK, CA_P0167, sigF*	[21]
C. acetobutylicum ATCC 824	Wild type	NAP	*ermB*	Em^R	pMTL-JN12, pMTL-JN14	Addition of *adh*	[24]
C. acetobutylicum ATCC 824	Δ*pyrE*		*pyrE*	Uracil prototrophy	pMTL-JN15, pMTL-JN16, pMTL-JN30	Phage lambda genome	[24]
C. difficile R20291 *C. difficile* CD630 Δ*erm*	Δ*pyrE*	*pyrE*	NAP	FOA^R	pMTL-YN3, pMTL-YN4	*spo0A, cwp84, mtlD*	[25]

(*continued overleaf*)

Table 9.3 (*continued*)

Clostridium spp. used for exemplification	Wild type or mutant	Negative selection marker	Positive selection marker	Selection criteria[a]	Plasmid example	Principle genes inactivated/added	References
C. beijerinckii NCIMB 8052	Wild type	NAP	NAP	PCR screening	pJIR750ai	*pta, buk*	[13]
C. thermocellum DSM 1313	Wild type	NAP	NAP	PCR screening	pHK-TT1A	*cipA, hfat, hyd, ldh, pta, pyrF*	[16]
C. beijerinckii NCIMB 8052	Wild type	CRISPR	NAP	Cell survival	pYWCas-derivatives	*spo0A*	[36]
C. beijerinckii NCIMB 8052	Wild type	*sceC*	NAP	Cell survival	pXY1/pIMP1-derivatives	*adc, xylR*	[23]
C. acetobutylicum ATCC 824	Wild type	*codA*	NAP	FC^R	pMTL-SC7515	*spo0A, cac824I*	[26]
C. acetobutylicum ATCC 824	Δ*pyrE*	*pyrE*	NAP	FOA^R	pMTL-ME3	*spo0A, cac824I, glgA, amyP*	[26]

a) The abbreviated counterselection agents, used in conjunction with the listed counter selection markers, are as follows: 5-fluorocytosine (FC), 5-fluoroorotic acid (FOA), 8-aza-hypoxanthine (AZH), 5-fluorodeoxyuridine (FUDR), D-galactose analog 2-deoxy-D-galactose (DOG). In other cases, no selective agent is added, rather cells have to survive the consequences of the expression of the *mazF* toxin, or of cleavage of the genome by I-SceI activity (*sceC*) or Cas9 (CRISPR).

duplicated DNA regions that acted as sites at which homologous recombination led to excision of the integrated mutagenic plasmid. Such mutants were, therefore, intrinsically unstable. Subsequently, researchers explored the use of group II intron-retargeting systems in which an intron-encoding segment of DNA could be deliberately inserted into the genomic target. Provided the cell loses the retargeted plasmid, and in that process the encoded LtrA protein responsible for intron retro-homing, the inserted intron remains "fixed" in the genome, and does not get excised [130–132]. The mutants generated were, therefore, extremely stable. Initial intron insertional mutants were generated using the "TargeTron®" system, commercialized and sold by Sigma Aldrich, in which clones carrying the inserted intron-encoding DNA are detected by appropriate polymerase chain reaction (PCR) screening. This was rapidly superseded by the "ClosTron system." In this case, intron insertions are selected on the basis of acquisition of resistance to erythromycin due to the incorporation of an inactivated *ermB* gene, or RAM (Retrotransposition-Activated Marker), that became functional as a consequence of the successful insertion of the group II intron-encoding DNA into the genomic target [133, 134].

The efficiency, reproducibility, speed, and ease of use have led ClosTron technology to become one of the most widely used clostridial mutagens [135, 136]. In order to facilitate the repeated use of the ClosTron, marker recycling was introduced in which the *ermB* gene was removed from the created mutant using flippase (FLP) recombinase, encoded by a subsequently introduced plasmid. The enzyme acted on deliberately introduced flippase recognition target (FRT) sites that flank the *ermB* gene. Those cells carrying derivatives in which the antibiotic gene had been successfully "flipped" are detected by screening for loss of resistance to erythromycin [133]. Once a cell that is sensitive to antibiotic has been obtained, it may be used to generate a second ClosTron mutant through selection for cells that are erythromycin resistant. Inexplicably, flipping of *ermB* has only ever been demonstrated in *C. acetobutylicum* [133], and never in *Clostridium difficile, C. botulinum,* or *C. sporogenes,* for instance (Y. Zhang and N. Minton, unpublished data). Alternatively, it is possible to repeatedly use RAM-less intron retargeting to obtain multiple mutants [137, 138]. It should be realized, however, that if the intron is inserted into the positive DNA strand, and the intron delivery plasmid carrying the *ltrA* gene remains in the cell, then the clones carrying the intron will not be null mutants. This is because the LtrA protein will splice the intron-encoding region from the mRNA, enabling translation of an encoded wild-type protein. The mutants described in certain studies are, therefore, merely knockdown mutants, not knockouts [139]. The retargeting system of the ClosTron is based on the Group II intron and intron-encoded protein (responsible for RNA splicing and insertion) of *Lactococcus lactis*. The system is temperature sensitive, and does not function above 41 °C and so is unsuitable for use in a thermophilic clostridial species [140]. Accordingly, a mobile group II intron from the thermophilic cyanobacterium *Thermosynechococcus elongatus* was used to disrupt six different chromosomal genes (*cipA, hfat, hyd, ldh, pta,* and *pyrF*) in *Clostridium thermocellum* [141].

As with any insertional tool, however, retargeted introns are disadvantaged by the possible occurrence of polar effects on downstream genes. This can complicate the subsequent analysis of mutant phenotypes. Accordingly, markerless, in-frame deletion mutants generated by homologous recombination are preferred. In recent years, a number of groups have developed such methods through the successful deployment of a range of counterselection markers that had previously been exemplified in other bacteria. These included a number of markers that could not be used in a standard wild-type host, but required a specific mutant background. Examples are the genes *galK* (galactokinase), *pyrF* (5-phosphate decarboxylase), and *upp* (uracil phosphor-ribosyltransferase) which were first exemplified in Clostridium *perfringens* [142], *C. thermocellum* [143], and *C. acetobutylicum* [144], respectively. Subsequently, a number of alternative counterselection markers were shown to be useable in a wild-type background, most notably the *Escherichia coli* genes *codA* (cytosine deaminase) and *mazF* (mRNA interferase) in *C. difficile* [145] and *C. acetobutylicum* [146], respectively, and the *Thermoanaerobacterium saccharolyticum tdk* (thymidylate synthetase) and *C. thermocellum hpt* (hypoxanthine phosphoribosyl transferase) genes, which were used to make knockouts in *C. thermocellum* [147]. More recently, a recombineering approach developed for use in *E.coli*, and based on the enzyme I-SceI, was exemplified in both *C. acetobutylicum* and *C. beijerinckii* [148]. Restriction recognition sites (a 10 nt palindrome) for this enzyme do not occur in either genome. Thus, the integration by homologous recombination of plasmids carrying appropriate homology arms and this recognition sequence into the genome by single crossover result in cells which contain a single I-SceI restriction site. The introduction of a second plasmid into such cells carrying a gene (*sceC*) encoding this restriction enzyme leads to the enrichment of cells in which the integrated plasmid excises resulting a portion of the population that are double crossover mutants. Application of the method was demonstrated by deleting two genes (*adc* and *glcG*) from *C. acetobutylicum* ATCC 824 and one gene (*adc*) from *C. beijerinckii* NCIMB 8052, and by introducing point mutations into *xylR* of *C. beijerinckii* NCIMB 8052 [148].

A further method of particular utility is Allele-Coupled Exchange (ACE), which allows the rapid insertion of heterologous DNA of any size of complexity into the genome [149]. The system is designed such that, following integration of the plasmid by single-crossover recombination, the desired second recombination event results in a plasmid-borne allele becoming "coupled" to a genome-located allele. This leads to the creation of a new allele that is selectable, and the isolation of double-crossover cells. The use of asymmetric homology arms dictates the order of recombination events. Thus, a long homology arm directs the first recombination event (plasmid integration) and a much shorter homology arm directs the second recombination event (plasmid excision). A number of different genetic loci may be used to insert heterologous DNA via ACE. Of particular utility is the native *pyrE* gene, which is inactivated by replacement of the wild-type allele with a mutant allele lacking approximately 300 bp from the 3′ end of the structural gene. The *pyrE* gene encodes orotate phosphoribosyltransferase [E.C.2.4.2.10], an enzyme involved in *de novo* pyrimidine biosynthesis. It may be used as a positive/negative selection marker as it is essential in the absence of exogenous pyrimidines and renders cells sensitive to 5-fluoroorotate (FOA). As a consequence, a heterologous *pyrE* gene can be used as a counterselection marker in such a strain, in an equivalent manner to that described for the *pyrF* gene [143]. This approach was initially demonstrated in two different strains of *C. difficile*, using a heterologous *pyrE* allele from *C. sporogenes* [150], and more recently in *C. acetobutylicum* ATCC 824 [151]. Crucially, the design of the created *pyrE* mutant strain is such that its *pyrE* allele can be rapidly (2 days) restored to wild type using an appropriate ACE correction vector allowing the specific in-frame deletion mutant to be characterized in a clean, wild-type background. More importantly, the system provides the in-parallel opportunity to complement the mutant at an appropriate gene dosage, through insertion of a wild-type copy of the gene, either under the control of its native promoter or the strong P_{fdx} promoter (derived from the ferredoxin gene of *C. sporogenes*), concomitant with restoration of the *pyrE* allele back to wild type [150]. The value of this approach was recently demonstrated in *C. acetobutylicum* ATCC 824 through deletion and complementation of *spo0A* and a putative glycogen synthase (CA_C2239, *glgA*) and the pSOL1 amylase gene (CA_P0168, *amyP*), leading to lack of production of spores, granulose, and amylase, respectively, in the derived mutants, and restoration of these phenotypes to wild type in the complemented clones. Interestingly, during this study, genome resequencing revealed the presence of a surprising number of errors in the published ATCC 824 genome sequence, equating to 175 SNVs (single nucleotide variations) and 48 Indels (Insertions–deletions) [151].

In recent years, genome engineering has been revolutionized by the discovery of the CRISPR and CRISPR-associated (Cas) system [152, 153], an immune system in bacteria and archaea that can efficiently cleave foreign DNA such as phage or plasmids entering the cells. Short nucleic acid sequences from invading pathogens can be captured and integrated in the CRISPR loci. Small RNAs, produced by transcription of the CRISPR loci, can then guide a set of bacterial endonucleases to cleave the foreign DNA. This system has since been adapted into a powerful tool in genome engineering where editing is achieved by introducing a cell with the Cas protein along with a specially designed guide RNA (gRNA) that directs the cut through hybridization with its matching genomic sequence. When the cell repairs that break, errors can occur to generate a knockout of that gene or additional modifications can be introduced. CRISPR is particularly good for the efficient and simultaneous generation of multiple knockouts in more than one gene. Thus, following the initial demonstration in 2012 that Cas9 could be programmed to cut various DNA sites *in vitro* [154], the CRISPR/Cas9 system has been widely adopted in eukaryotes for genome editing *in vivo* [155–158]. Recently, the CRISPR/Cas9 system has been applied for genome engineering in both gram-positive [159–161] and gram-negative [159, 162, 163] bacteria, including *C. beijerinckii* NCIMB 8052 [161]. In the case of *C. beijerinckii*, a single clean deletion of *spo0A* gene was made using the native CRISPR/Cas9 system from *Streptococcus pyogenes*. This is effectively using the system as a counterselection marker. In the future, however, one can expect the full potential of CRISPR/Cas9 to be reached with its use as a true genome-editing tool for multiple, simultaneous knockouts in clostridia.

Some progress has also been made in devising tools that can be used in forward genetics, where random mutants with the desired phenotype are isolated, and the nature of the genotype responsible is determined. The most useful tool in this context is a transposon. Two conjugative transposons Tn*916* and Tn*1514* have been studied in *C. acetobutylicum* ATCC 824, *C. saccharobutylicum* P262, and *C. beijerinckii* NCIMB 8052 [164–166]. These were used to isolate a number of interesting mutants, including one deficient in "degeneration" (the loss in prolonged culture of the ability to produce solvents) [167] and others exhibiting increased butanol tolerance [168]. However, a number of factors mean these tools are far from ideal. Principal among these factors are their large size and low efficiency of transposition, and in the case of Tn*916*, the existence of "hot spots" in the chromosome where the majority of the insertions locate, while in the case of Tn*1545*, the predilection to insert in multiple copies (probably attributed to a high rate of vector retention). This latter property significantly complicates the association of genotype with phenotype and hence, requires further experimentation to understand the basis of the observed phenotypes.

The first random nonconjugative transposon mutagenesis systems in clostridia are preassembled transposomes, which were developed for *C. perfringens* owing to its routinely achieved high electroporation efficiencies [169, 170]. Although a few "hot spots" of rRNA genes were preferred by both the Mu bacteriophage and Tn*5* transposon-based system and the EZ-Tn*5* random mutagenesis system (Epicentre®). Compared to conjugative transposons, both of the aforementioned transposome systems represented a considerable improvement on the use of conjugative transposons. They are, however, unlikely to secure wide applicability in industrial clostridia as they are entirely reliant on attaining high frequencies of transformation in the target organism. Although *C. perfringens*, and in particular strain 13, are relatively highly transformable by electroporation [171], such frequencies are relatively rare in other *Clostridium* sp.

In the search for better forward genetic tools, the *mariner* transposable element *Himar1*, previously exemplified in *C. perfringens* [172] and *C. difficile* [173], has recently been adapted for use in *C. acetobutylicum* ATCC 824 and *C. sporogenes*. This was achieved by equipping the host cell with the ability to produce the foreign sigma factor, TcdR, responsible for expression of the transposase through the introduction of the encoding gene at the *pyrE* locus of the cell using ACE [174]. *Himar1* appears particularly suited to low G + C microbes such as clostridia as its target recognition sequence is "TA." Thus in all cases, the transposon has been shown to insert itself entirely at random around the various clostridial genomes and at high frequency. Its effectiveness opens up the possibility of using high-throughput screening methods to identify essential genes using next-generation DNA sequencing (NGS) technology to sequence the site of insertion of every transposon in a mutant pool, so-called transposon-directed insertion site sequencing or TraDIS. At present, such an approach has only been applied to *C. difficile* where a core set of 404 essential genes, and 798 genes that are likely to have an impact on spore production, were identified [175]. The application of such technology to the saccharolytic clostridial species is likely and would be expected to be accomplished in the near future.

9.8
Industrial Application of ABE Fermentation

The ABE fermentation is an important industrial biotechnological process in which carbohydrate-based feedstocks are readily converted into acetone and butanol by solventogenic clostridia under anaerobic conditions.

Today, acetone and butanol are used as solvents and chemical building blocks for a diverse range of applications from paints, adhesives, and inks to food ingredients, cosmetic and personal care applications, to pharmaceuticals, plastics, and polymers. According to ICIS market data, the annual global markets for butanol is 3.4 million tonnes (worth over \$5 billion) and for acetone it is 6 million tonnes (worth $\sim$ \$6 billion). Today, the solvent markets are supplied petroleum-derived acetone and butanol. There is increasing demand for renewable chemicals derived from sustainable feedstocks as replacements for chemicals derived from oil. At present, renewable chemicals comprise about 10% of the chemical market and this is forecast to grow at 18% per year to \$98.5 billion by 2020 [176].

Biobutanol is also an attractive biofuel with superior fuel properties but this application has not yet been established in the market. Biobutanol fits the existing fuel infrastructure; it has a higher energy density (similar to petrol/gasoline) and better performance than bioethanol. In addition, butanol can be dehydrated to 1-butene and catalyzed into longer-chain oligomers for jet-fuel applications. Biobutanol can substitute for both ethanol and biodiesel in the biofuel market estimated to be \$247 billion by 2020 [177].

The ABE fermentation process was first developed by C. Weizmann at Manchester University in 1912. Commercial production quickly spread to the United States and then worldwide during the First and Second World Wars; first to produce acetone for ammunitions and then later to produce butanol for paint lacquers. The fermentation process fell out of favor in the United States and Europe in the 1950s when renewable solvents could no longer compete with their synthetic equivalents on price. Some production, via fermentation, continued in China, Russia, and South Africa until the early 1980s [54].

For example, an ABE plant was established at Germinston, South Africa in 1937 and ran successfully until 1983, first producing solvent from starch but switching to molasses. The fermentation and distillation recovery process operated in batch mode. The fermentation produced approximately 20 $g\,l^{-1}$ of mixed solvents from 55 to 60 $g\,l^{-1}$ of substrate with solvent yields of about 0.35 $g\,g^{-1}$ sugar. The butanol: acetone molar ratio is typically 2 : 1 [178].

Commercial solvent production was established in Russia during 1929–1935 using starch-based feedstocks [43]. The Russian plants produced approximately

15–25 kT of solvent per year and were more advanced than those plants operating in South Africa. The plant at Dokshukino operated continuously incorporating both continuous feeding and continuous distillation for solvent recovery. This mode of operation improved volumetric productivities by 31%. In addition, the plant switched from starch-based feedstocks toward sugars derived from agricultural wastes and forestry byproducts. The Russian process produced, alongside acetone and butanol, a range of by-products: ethanol, H_2, liquid CO_2, vitamin B_{12}, animal feed, and biogas.

Around 2006, six new ABE fermentation plants were established in China, costing over $200 million and creating more than 200 kT of solvent capacity [43]. The plants shared a common design to produce about 30 kT butanol per year from corn and cassava starch in a semicontinuous mode of operation lasting up to 21 days. The fermentation process consisted of several trains of eight fermentation tanks (300–400 m^3 volumes) linked together in a cascade. Continuous distillation was used to recover solvents from multiple trains. Most plants are located next to ethanol plants to reduce utility and operating costs with integrated effluent treatment based on anaerobic digestion. Biogas is used to generate combined heat and power.

Unfortunately, by 2009, most plants had become uneconomic because of a significant drop in butanol prices (linked to a corresponding fall in oil prices) and a dependency on corn starch which was relatively expensive. One plant located at Songyuan (Lignicell Refining Biotechnologies), in partnership with the UK company Green Biologics, continued to operate by switching to cheaper hexose and pentose sugars derived from agricultural residues, corn stover and rice hulls. The partners successfully demonstrated the production of acetone and butanol from a cellulosic feedstock using advances in pretreatment, cellulose hydrolysis, and fermentation. Several hundred tonnes of purified solvent were shipped to customers in North America for chemical applications (information supplied by Green Biologics).

More recently, Green Biologics has demonstrated an advanced, high-productivity fermentation process to produce acetone and butanol using a novel *Clostridium* strain. This process offers a step change in process economics. In addition, the company has identified an opportunity to convert existing ethanol assets to produce butanol, thus not only saving significant capital costs but also shortening the time required to get operational. In December 2014, Green Biologics acquired the assets of Central MN Ethanol Co-op LLC in Little Falls, MN including a 65 kT per year ethanol plant (http://www.greenbiologics.com/news.php). The plant is being converted to produce approximately 30 kT of acetone and butanol with an anticipated start-up in mid-2016. The plant will initially use corn starch but has the flexibility to utilize wood-based lignocellulosic feedstocks. Once proved, the ethanol plant retrofit model will be rolled out to other underperforming ethanol assets in North America.

The ABE fermentation is an established industrial process but has, over the last century, struggled to compete with fluctuating commodity prices that affect both the input feedstock cost and the product pricing. These factors have a huge impact on process economics and this challenge is particularly acute at times with unusually low oil pricing. However, on the plus side, there is strong market demand (and incentives) for renewable chemicals. Also, the flexibility of the modern fermentation provides some insulation against changes in commodity pricing. For example, the fermentation can utilize a broad range of feedstocks including sugars derived from lignocellulosic waste products. The fermentation can also produce a broad range of chemical products in addition to acetone and butanol and is, therefore, less reliant on the solvent market. The fermentation process has also been improved through advances in process design and control which together with improvements in microbial strain development using genetic engineering offer a step change in both performance and process economics. In summary, solventogenic clostridia are robust and versatile bacteria that can make a significant contribution toward renewable chemical production for industrial applications, even in a low oil price world.

Acknowledgments

This work was supported by the ERA-IB project REACTIF (Rational Engineering of Advanced Clostridia for Transformational Improvements in Fermentation), EIB 12.050.

References

1. Pasteur, L. (1861) Animalcules infusoires vivant sans gaz oxygène libre et déterminant des fermentations. *C.R. Hebd. Seances Acad. Sci.*, **52**, 344–347.
2. Dürre, P. (2001) in *Clostridia. Biotechnology and Medical Applications* (eds H. Bahl and P. Dürre), Wiley-VCH Verlag GmbH, Weinheim, pp. 1–17.
3. Schardinger, F. (1904) Acetongärung. *Wien. Klin. Wochenschr.*, **17**, 207–209.
4. Schardinger, F. (1905) *Bacillus macerans*, ein Aceton bildender Rottebacillus. *Zentralbl. Bakteriol. Parasitenkd. Infektionskr. Hyg. Abt. II*, **14**, 772–781.
5. Pringsheim, H. (1906) Über den Ursprung des Fuselsöls und eine Alkohole bildende Bakterienform. *Zentralbl. Bakteriol. Parasitenkd. Infektionskr. Hyg. Abt. II*, **15**, 300–321.
6. Weimer, P.J. (1984) Fermentation of 6-deoxyhexoses by *Bacillus macerans*. *Appl. Environ. Microbiol.*, **47**, 263–267.
7. Krabben, P. (2014) Acetone production during the First World War. *Microbiol. Today*, **41**, 74–77.
8. Fernbach, A. and Strange, E.H. (1911) Improvements in the manufacture of higher alcohols. British Patent 15204, filed June 29, 1911 and issued Sep. 30, 1912.
9. Weizmann, C. (1915). Improvement in the bacterial fermentation of carbohydrates and in bacterial cultures for the same. British Patent 4845, filed Feb. 29, 1915 and issued Mar. 6, 1919.
10. Croux, C., Paquet, V., Goma, G., and Soucaille, P. (1990) Purification and characterization of acidolysin, an acidic metalloprotease produced by *Clostridium acetobutylicum* ATCC 824. *Appl. Environ. Microbiol.*, **56**, 3634–3642.
11. Doi, S., Kaneko, Y., and Uchino, F. (1958) Proteolytic enzymes of butyl bacteria part II. The effect of calcium carbonate on the proteolytic activity in the acetone–butanol fermentation and some properties of the enzyme. *J. Agric. Chem. Soc. Jpn.*, **32**, 558–562.
12. Northrop, J.H. (1918) Process for the production of acetone. US Patent

1,293,172, filed July 8, 1918 and issued Feb. 4, 1919.

13. Northrop, J.H., Ashe, L.H., and Morgan, R.R. (1919) A fermentation process for the production of acetone and ethyl alcohol. *Ind. Eng. Chem.*, **11**, 723–727.

14. Kelly, F.C. (1936) *One Thing Leads to Another: the Growth of an Industry*, Houghton Mifflin Company, Boston, MA and New York.

15. Woodruff, J.C., Stiles, H.R., and Legg, D.A. (1933) Process of producing butyl alcohol. US Patent 2,089,522, filed June 12, 1933 and issued Aug. 10, 1937.

16. Jones, D.T. (2001) in *Clostridia: Biotechnology and Medical Applications* (eds H. Bahl and P. Dürre), Wiley-VCH Verlag GmbH, Weinheim, pp. 125–168.

17. Carnarius, E.H. and McCutchan, W.N. (1937) Process for production of butyl alcohol by fermentation. US Patent 2,139,111, filed Feb. 8, 1937 and issued Dec. 6, 1938.

18. Carnarius, E.H. (1938) Butyl alcohol fermentation process. US Patent 2,198,104, filed Dec. 15, 1938 and issued Apr. 23, 1940.

19. Müller, J. (1937) Butyl-acetone fermentation process and inoculant. US Patent 2,195,629, filed May 1, 1937 and issued Apr. 2, 1940.

20. Hongo, M. (1959) Process for producing butanol by fermentation. US Patent 2,945,786, filed Oct. 15, 1959 and issued July 19, 1960.

21. Weyer, E.R. and Rettger, L.F. (1927) A comparative study of six different strains of the organism commonly concerned in large-scale production of butyl alcohol and acetone by the biological process. *J. Bacteriol.*, **14**, 399–424.

22. Nölling, J., Breton, G., Omelchenko, M.V., Makarova, K.S. *et al.* (2001) Genome sequence and comparative analysis of the solvent-producing bacterium *Clostridium acetobutylicum*. *J. Bacteriol.*, **183**, 4823–4838.

23. Keis, S., Shaheen, R., and Jones, D.T. (2001) Emended descriptions of *Clostridium acetobutylicum* and *Clostridium beijerinckii*, and descriptions of *Clostridium saccharoperbutylacetonicum* sp. nov. and *Clostridium saccharobutylicum* sp. nov. *Int. J. Syst. Evol. Bacteriol.*, **51**, 2095–2103.

24. George, H.A. and Chen, J.-S. (1983) Acidic conditions are not obligatory for onset of butanol formation by *Clostridium beijerinckii* (synonym, *C. butylicum*). *Appl. Environ. Microbiol.*, **46**, 321–327.

25. Tangney, M., Brehm, J.K., Minton, N.P., and Mitchell, W.J. (1998) A gene system for glucitol transport and metabolism in *Clostridium beijerinckii* NCIMB 8052. *Appl. Environ. Microbiol.*, **64**, 1612–1619.

26. Pidot, S., Ishida, K., Cyrulies, M., and Hertweck, C. (2014) Discovery of clostrubin, an exceptional polyphenolic polyketide antibiotic from a strictly anaerobic bacterium. *Angew. Chem. Int. Ed.*, **126**, 7990–7993.

27. Jones, D.T. and Keis, S. (1995) Origins and relationships of industrial solvent-producing clostridial strains. *FEMS Microbiol. Rev.*, **17**, 223–232.

28. Beesch, S.C. (1952) Acetone–butanol fermentation of sugars. *J. Ind. Eng. Chem.*, **44**, 1677–1682.

29. Hastings, J.J. (1971) in *Advances in Applied Microbiology*, vol. **14** (ed. D. Perlman), Academic Press, New York, pp. 1–45.

30. Ogata, S. and Hongo, M. (1979) Bacteriophages of the genus *Clostridium*. *Adv. Appl. Microbiol.*, **25**, 241–273.

31. Jones, D.T., Shirley, M., Wu, X., and Keis, S. (2000) Bacteriophage infections in the industrial acetone butanol (AB) fermentation process. *J. Mol. Microbiol. Biotechnol.*, **2**, 21–26.

32. Gabriel, C.L. (1928) Butanol fermentation process. *J. Ind. Eng. Chem.*, **20**, 1063–1067.

33. McCutchan, W.N. and Hickey, R.J. (1954) in *Industrial Fermentations*, vol. **1** (eds L.A. Underkofler and R.J. Hickey), Chemical Publishing, New York, pp. 347–388.

34. Johnson, J.L., Toth, J., Santiwatanakul, S., and Chen, J.-S. (1997) Cultures of "*Clostridium acetobutylicum*" from various collections comprise *Clostridium*

acetobutylicum, *Clostridium beijerinckii*, and two other distinct types based on DNA-DNA reassociation. *Int. J. Syst. Evol. Bacteriol.*, **47**, 420–424.

35. McCoy, E., McDaniel, L.E., and Sylvester, J.C. (1944) Bacteriophage in a butyl fermentation plant. *J. Bacteriol.*, **47**, 443.
36. McCoy, E.F. (1940) Production of butyl alcohol and acetone by fermentation. US Patent 2,398,837, filed July 30, 1940 and issued Apr. 23, 1946.
37. Keis, S., Bennett, C.F., Ward, V.K., and Jones, D.T. (1995) Taxonomy and phylogeny of industrial solvent-producing clostridia. *Int. J. Syst. Bacteriol.*, **45**, 693–705.
38. Kinoshita, S., Itagaki, S., Okumura, T., Terada, M. *et al.* (1952) Studies on bacteriophages for *Cl. acetobutylicum* part 1. Electron microscopic observation. *Nippon Nōgeikagaku Kaishi*, **26**, 104–107.
39. Hongo, M. and Murata, A. (1966) Bacteriophages of *Clostridium saccharoperbutylacetonicum* part III. Morphology of the twelve HM-phages. *Agric. Biol. Chem.*, **30**, 292–295.
40. Hongo, M., Aono, T., and Murata, A. (1966) Bacteriophages of *Clostridium saccharoperbutylacetonicum* part IV. Serological characteristics of the twelve HM-phages. *Agric. Biol. Chem.*, **30**, 399–405.
41. Hongo, M., Murata, A., and Ogata, S. (1969) Bacteriophages of *Clostridium saccharoperbutylacetonicum* part XVI. Isolation and some characters of a temperate phage. *Agric. Biol. Chem.*, **33**, 331–342.
42. Reid, S.J., Allock, E.R., Jones, D.T., and Woods, D.R. (1983) Transformation of *Clostridium acetobutylicum* protoplasts with bacteriophage DNA. *Appl. Environ. Microbiol.*, **45**, 305–307.
43. Zverlov, V.V., Berezina, O., Velikodvorskaya, G.A., and Schwarz, W.H. (2006) Bacterial acetone and butanol production by industrial fermentation in the Soviet Union: use of hydrolyzed agricultural waste for biorefinery. *Appl. Microbiol. Biotechnol.*, **71**, 587–597.
44. Stegmann, B.A. (2015) Molecular procedures for improvement of genetic accessibility of *Clostridium acetobutylicum* with regard to recombineering. PhD thesis. Universität Ulm.
45. Hongo, M. and Murata, A. (1965) Bacteriophages of *Clostridium saccharoperbutylacetonicum* part I. Some characteristics of the twelve phages obtained from the abnormally fermented broths. *Agric. Biol. Chem.*, **29**, 1135–1145.
46. Keis, S. (1992) Characterization of two bacteriophages infecting *Clostridium acetobutylicum*. M.Sc. thesis. University of Otago.
47. Sylvester, J.C. (1943) Studies on bacteriophage affecting an acetone–butanol anaerobe. PhD thesis. University of Wisconsin.
48. Gold, W. and Watson, D.W. (1950) Studies on the bacteriophage infection cycle I.: a lytic activity assay for bacteriophages of *Clostridium madisonii*. *J. Bacteriol.*, **59**, 13–16.
49. Kinoshita, S. and Teramoto, K. (1955) Studies on the bacteriophage for *Cl. acetobutylicum* part 6. (1) Classification of strains by phage (2) On the relation between sluggish fermentation and phage action. *Nippon Nōgeikagaku Kaishi*, **29**, 955–957.
50. Maeda, A., Ishii, K., Tanaka, M., Mikami, Y. *et al.* (1986) KM1, a bacteriophage of *Clostridium butyricum*. *J. Gen. Microbiol.*, **132**, 2271–2275.
51. Wu, X. (1998) Characterisation of a large temperate bacteriophage of *Clostridium beijerinckii*. M.Sc. thesis. University of Otago.
52. Peri, B.A. and Watson, D.W. (1948) Studies on host-virus relationships: lysogenesis in a strain of *Clostridium madisonii*. Proceedings of the 48th Annual Meeting of the Society of American Bacteriologists, vol. **1**, p. 26.
53. Dürre, P., Bahl, H., and Gottschalk, G. (1988) in *Handbook of Anaerobic Fermentation* (eds L.E. Erickson and D.Y.C. Fung), Marcel Dekker, New York, pp. 187–206.

54. Jones, D.T. and Woods, D.R. (1986) Acetone–butanol fermentation revisited. *Microbiol. Mol. Biol. Rev.*, **50**, 484–524.
55. Dürre, P. (1998) New insights and novel developments in clostridial acetone/butanol/isopropanol fermentation. *Appl. Microbiol. Biotechnol.*, **49**, 639–648.
56. Inui, M., Suda, M., Kimura, S., Yasuda, K. *et al.* (2007) Expression of *Clostridium acetobutylicum* butanol synthetic genes in *Escherichia coli*. *Appl. Microbiol. Biotechnol.*, **77**, 1305–1316.
57. Ismaiel, A.A., Zhu, C.-X., Colby, G.D., and Chen, J.-S. (1993) Purification and characterization of a primary–secondary alcohol dehydrogenase from two strains of *Clostridium beijerinckii*. *J. Bacteriol.*, **175**, 5097–5105.
58. Köpke, M. and Dürre, P. (2011) in *Handbook of Biofuels Production: Processes and Technologies*, Woodhead Publishing Series in Energy, vol. **15** (eds R. Luque, J. Campelo, and J. Clark), Woodhead, Cambridge, pp. 221–257.
59. Boynton, Z.L., Bennett, G.N., and Rudolph, F.B. (1996) Cloning, sequencing, and expression of clustered genes encoding β-hydroxybutyryl-coenzyme A (CoA) dehydrogenase, crotonase, and butyryl-CoA dehydrogenase from *Clostridium acetobutylicum* ATCC 824. *J. Bacteriol.*, **178**, 3015–3024.
60. Fischer, R.-J., Helms, J., and Dürre, P. (1993) Cloning, sequencing, and molecular analysis of the *sol* operon of *Clostridium acetobutylicum*, a chromosomal locus involved in solventogenesis. *J. Bacteriol.*, **175**, 6959–6969.
61. Dürre, P., Böhringer, M., Nakotte, S., Schaffer, S. *et al.* (2002) Transcriptional regulation of solventogenesis in *Clostridium acetobutylicum*. *J. Mol. Microbiol. Biotechnol.*, **4**, 295–300.
62. Wang, Y., Li, X., Mao, Y., and Blaschek, H.P. (2012) Genome-wide dynamic transcriptional profiling in *Clostridium beijerinckii* NCIMB 8052 using single-nucleotide resolution RNA-Seq. *BMC Genomics*, **13**, 102.
63. Wang, Y., Li, X., Mao, Y., and Blaschek, H.P. (2011) Single-nucleotide resolution analysis of the transcriptome structure of *Clostridium beijerinckii* NCIMB 8052 using RNA-Seq. *BMC Genomics*, **12**, 479.
64. Wiesenborn, D.P., Rudolph, F.B., and Papoutsakis, E.T. (1989) Coenzyme A transferase from *Clostridium acetobutylicum* ATCC 824 and its role in the uptake of acids. *Appl. Environ. Microbiol.*, **55**, 323–329.
65. Ballongue, J., Amine, J., Masion, E., Petitdemange, H. *et al.* (1985) Induction of acetoacetate decarboxylase in *Clostridium acetobutylicum*. *FEMS Microbiol. Lett.*, **29**, 273–277.
66. Petersen, D.J. and Bennett, G.N. (1990) Purification of acetoacetate decarboxylase from *Clostridium acetobutylicum* ATCC 824 and cloning of the acetoacetate decarboxylase gene in *Escherichia coli*. *Appl. Environ. Microbiol.*, **56**, 3491–3498.
67. Jiang, Y., Xu, C., Dong, F., Yang, Y. *et al.* (2009) Disruption of the acetoacetate decarboxylase gene in solvent-producing *Clostridium acetobutylicum* increases the butanol ratio. *Metab. Eng.*, **11**, 284–291.
68. Lehmann, D., Hönicke, D., Ehrenreich, A., Schmidt, M. *et al.* (2012) Modifying the product pattern of *Clostridium acetobutylicum*: physiological effects of disrupting the acetate and acetone formation pathways. *Appl. Microbiol. Biotechnol.*, **94**, 743–754.
69. Cooksley, C.M., Zhang, Y., Wang, H., Redl, S. *et al.* (2012) Targeted mutagenesis of the *Clostridium acetobutylicum* acetone–butanol–ethanol fermentation pathway. *Metab. Eng.*, **14**, 630–641.
70. Feustel, L., Nakotte, S., and Dürre, P. (2004) Characterization and development of two reporter gene systems for *Clostridium acetobutylicum*. *Appl. Environ. Microbiol.*, **70**, 798–803.
71. Sauer, U. and Dürre, P. (1995) Differential induction of genes related to solvent formation during the shift from acidogenesis to solventogenesis in continuous culture of *Clostridium acetobutylicum*. *FEMS Microbiol. Lett.*, **125**, 115–120.

72. Fontaine, L., Meynial-Salles, I., Girbal, L., Yang, X. *et al.* (2002) Molecular characterization and transcriptional analysis of *adhE2*, the gene encoding the NADH-dependent aldehyde/alcohol dehydrogenase responsible for butanol production in alcohologenic cultures of *Clostridium acetobutylicum* ATCC 824. *J. Bacteriol.*, **184**, 821–830.
73. Girbal, L., Croux, C., Vasconcelos, I., and Soucaille, P. (1995) Regulation of metabolic shifts in *Clostridium acetobutylicum* ATCC 824. *FEMS Microbiol. Rev.*, **17**, 287–297.
74. Hönicke, D., Janssen, H., Grimmler, C., Ehrenreich, A. *et al.* (2012) Global transcriptional changes of *Clostridium acetobutylicum* cultures with increased butanol:acetone ratios. *New Biotechnol.*, **29**, 485–493.
75. Rao, G. and Mutharasan, R. (1986) Alcohol production by *Clostridium acetobutylicum* induced by methyl viologen. *Biotechnol. Lett.*, **8**, 893–896.
76. Rao, G. and Mutharasan, R. (1987) Altered electron flow in continuous cultures of *Clostridium acetobutylicum* induced by viologen dyes. *Appl. Environ. Microbiol.*, **53**, 1232–1235.
77. Senger, R.S. and Papoutsakis, E.T. (2008) Genome-scale model for *Clostridium acetobutylicum*: part I. Metabolic network resolution and analysis. *Biotechnol. Bioeng.*, **101**, 1036–1052.
78. Senger, R.S. and Papoutsakis, E.T. (2008) Genome-scale model for *Clostridium acetobutylicum*: part II. Development of specific proton flux states and numerically determined sub-systems. *Biotechnol. Bioeng.*, **101**, 1053–1071.
79. Millat, T., Janssen, H., Bahl, H., Fischer, R.J. *et al.* (2013) Integrative modelling of pH-dependent enzyme activity and transcriptomic regulation of the acetone–butanol–ethanol fermentation of *Clostridium acetobutylicum* in continuous culture. *Microb. Biotechnol.*, **6**, 526–539.
80. Millat, T., Janssen, H., Thorn, G.J., King, J.R. *et al.* (2013) A shift in the dominant phenotype governs the pH-induced metabolic switch of *Clostridium acetobutylicum* in phosphate-limited continuous cultures. *Appl. Microbiol. Biotechnol.*, **97**, 6451–6466.
81. Yoo, M., Bestel-Corre, G., Croux, C., Riviere, A. *et al.* (2015) A quantitative system-scale characterization of the metabolism of *Clostridium acetobutylicum*. *mBio*, **6**, e01808–e01815.
82. Liao, C., Seo, S.-O., Celik, V., Liu, H. *et al.* (2015) Integrated, systems metabolic picture of acetone–butanol–ethanol fermentation by *Clostridium acetobutylicum*. *Proc. Natl. Acad. Sci. U.S.A.*, **112**, 8505–8510.
83. Bao, G., Wang, R., Zhu, Y., Dong, H. *et al.* (2011) Complete genome sequence of *Clostridium acetobutylicum* DSM 1731, a solvent-producing strain with multireplicon genome architecture. *J. Bacteriol.*, **193**, 5007–5008.
84. Hu, S., Zheng, H., Gu, Y., Zhao, J. *et al.* (2011) Comparative genomic and transcriptomic analysis revealed genetic characteristics related to solvent formation and xylose utilization in *Clostridium acetobutylicum* EA 2018. *BMC Genomics*, **12**, 93.
85. Mo, X., Pei, J., Guo, Y., Lin, L. *et al.* (2015) Genome sequence of *Clostridium acetobutylicum* GXAS18-1, a novel biobutanol production strain. *Genome Announc.*, **3**, e00033-15.
86. Wu, Y.R., Li, Y., Yang, K.L., and He, J. (2012) Draft genome sequence of butanol–acetone- producing *Clostridium beijerinckii* strain G117. *J. Bacteriol.*, **194**, 5470–5471.
87. Little, G.T., Winzer, K., and Minton, N.P. (2015) Genome sequence of the solvent-producing *Clostridium beijerinckii* strain 59B, isolated from Staffordshire garden soil. *Genome Announc.*, **3**, e00108-15.
88. Noar, J., Makwana, S.T., and Bruno-Bárcena, J.M. (2014) Complete genome sequence of solvent-tolerant *Clostridium beijerinckii* strain SA-1. *Genome Announc.*, **2**, e01310-14.
89. Poehlein, A., Grosse-Honebrink, A., Zhang, Y., Minton, N.P. *et al.* (2015) Complete genome sequence of the nitrogen-fixing and solvent-producing

Clostridium pasteurianum DSM 525. *Genome Announc.*, **3**, e01591-14.

90. Poehlein, A., Hartwich, K., Krabben, P., Ehrenreich, A. *et al.* (2013) Complete genome sequence of the solvent producer *Clostridium saccharobutylicum* NCP262 (DSM 13864). *Genome Announc.*, **1**, e00997-13.
91. Poehlein, A., Krabben, P., Dürre, P., and Daniel, R. (2014) Complete genome sequence of the solvent producer *Clostridium saccharoperbutylacetonicum* strain DSM 14923. *Genome Announc.*, **2**, e01056-14.
92. Horvath, P. and Barrangou, R. (2010) CRISPR/Cas, the immune system of bacteria and archaea. *Science*, **327**, 167–170.
93. Haft, D.H., Selengut, J., Mongodin, E.F., and Nelson, K.E. (2005) A guild of 45 CRISPR-associated (Cas) protein families and multiple CRISPR/Cas subtypes exist in prokaryotic genomes. *PLoS Comput. Biol.*, **1**, e60.
94. Westra, E.R. and Brouns, S.J. (2012) The rise and fall of CRISPRs – dynamics of spacer acquisition and loss. *Mol. Microbiol.*, **85**, 1021–1025.
95. Bland, C., Ramsey, T.L., Sabree, F., Lowe, M. *et al.* (2007) CRISPR Recognition Tool (CRT): a tool for automatic detection of clustered regularly interspaced palindromic repeats. *BMC Bioinf.*, **8**, 209.
96. Thompson, J.D., Higgins, D.G., and Gibson, T.J. (1994) CLUSTAL W: improving the sensitivity of progressive multiple sequence alignment through sequence weighting, position-specific gap penalties and weight matrix choice. *Nucleic Acids Res.*, **22**, 4673–4680.
97. Kumar, S., Nei, M., Dudley, J., and Tamura, K. (2008) MEGA: a biologist-centric software for evolutionary analysis of DNA and protein sequences. *Briefings Bioinf.*, **9**, 299–306.
98. Tettelin, H., Masignani, V., Cieslewicz, M.J., Donati, C. *et al.* (2005) Genome analysis of multiple pathogenic isolates of *Streptococcus agalactiae*: implications for the microbial "pan-genome". *Proc. Natl. Acad. Sci. U.S.A.*, **102**, 13950–13955.
99. Dürre, P. and Bahl, H. (1996) in *Biotechnology: Products of Primary Metabolism*, 2nd edn, vol. **6** (ed. M. Roehr), Wiley-VCH Verlag GmbH, Weinheim, pp. 229–268.
100. Girbal, L. and Soucaille, P. (1998) Regulation of solvent production in *Clostridium acetobutylicum. Trends Biotechnol.*, **16**, 11–16.
101. Dürre, P. (2005) in *Handbook on Clostridia* (ed. P. Dürre), CRC Press-Taylor and Francis Group, Boca Raton, FL, pp. 846–874.
102. Wietzke, M. and Bahl, H. (2012) The redox-sensing protein Rex, a transcriptional regulator of solventogenesis in *Clostridium acetobutylicum. Appl. Microbiol. Biotechnol.*, **96**, 749–761.
103. Zhang, L., Nie, X., Ravcheev, D.A., Rodionov, D.A. *et al.* (2014) Redox-responsive repressor Rex modulates alcohol production and oxidative stress tolerance in *Clostridium acetobutylicum. J. Bacteriol.*, **196**, 3949–3963.
104. Ren, C., Gu, Y., Hu, S., Wu, Y. *et al.* (2010) Identification and inactivation of pleiotropic regulator CcpA to eliminate glucose repression of xylose utilization in *Clostridium acetobutylicum. Metab. Eng.*, **12**, 446–454.
105. Ren, C., Gu, Y., Wu, Y., Zhang, W. *et al.* (2012) Pleiotropic functions of catabolite control protein CcpA in butanol-producing *Clostridium acetobutylicum. BMC Genomics*, **13**, 349.
106. Standfest, T., Nold, N., Feustel, L., Schiel, B. *et al.* (2009) CodY, a potential repressor of the butanol formation in *Clostridium acetobutylicum. Biospektrum Spec. Issue*, **PS21**, 174.
107. Schiel, B., Böhringer, M., Schaffer, S., and Dürre, P. (2003) Identification and characterization of a potential transcriptional regulator of the acetoacetate decarboxylase gene of *Clostridium acetobutylicum. Biospektrum Spec. Issue*, **KB003**, 39.
108. Thormann, K., Feustel, L., Lorenz, K., Nakotte, S. *et al.* (2002) Control of butanol formation in *Clostridium acetobutylicum* by transcriptional activation. *J. Bacteriol.*, **184**, 1966–1973.

109. Schiel, B., Nold, N., and Dürre, P. (2010) Identification of a small noncoding RNA in *Clostridium acetobutylicum*. *Biospektrum Spec. Issue*, **GRV09**, 135.
110. Paredes, C.J., Alsaker, K.V., and Papoutsakis, E.T. (2005) A comparative genomic view of clostridial sporulation and physiology. *Nat. Rev. Microbiol.*, **3**, 969–978.
111. Harris, L.M., Welker, N.E., and Papoutsakis, E.T. (2002) Northern, morphological, and fermentation analysis of *spo0A* inactivation and overexpression in *Clostridium acetobutylicum* ATCC 824. *J. Bacteriol.*, **184**, 3586–3597.
112. Dürre, P. (2014) Physiology and sporulation in *Clostridium*. *Microbiol. Spectr.*, **2** (4), TBS-0010-2012.
113. Ravagnani, A., Jennert, K.C., Steiner, E., Grünberg, R. *et al.* (2000) Spo0A directly controls the switch from acid to solvent production in solvent-forming clostridia. *Mol. Microbiol.*, **37**, 1172–1185.
114. Dürre, P. and Hollergschwandner, C. (2004) Initiation of endospore formation in *Clostridium acetobutylicum*. *Anaerobe*, **10**, 69–74.
115. Steiner, E., Dago, A.E., Young, D.I., Heap, J.T. *et al.* (2011) Multiple orphan histidine kinases interact directly with Spo0A to control the initiation of endospore formation in *Clostridium acetobutylicum*. *Mol. Microbiol.*, **80**, 641–654.
116. Dürre, P. (2011) Ancestral sporulation initiation. *Mol. Microbiol.*, **80**, 584–587.
117. Görke, B. and Stülke, J. (2008) Carbon catabolite repression in bacteria: many ways to make the most out of nutrients. *Nat. Rev. Microbiol.*, **6**, 613–624.
118. Fujita, Y. (2009) Carbon catabolite control of the metabolic network in *Bacillus subtilis*. *Biosci. Biotechnol., Biochem.*, **73**, 245–259.
119. Rafudeen, M.S. (2002) Investigation of carbon catabolite repression in *Clostridium beijerinckii* NCIMB 8052. PhD thesis. University of Cape Town.
120. Davison, S.P., Santangelo, J.D., Reid, S.J., and Woods, D.R. (1995) A *Clostridium acetobutylicum* regulator gene (*regA*) affecting amylase production in *Bacillus subtilis*. *Microbiology*, **141**, 989–996.
121. Sonenshein, A.L. (2005) CodY, a global regulator of stationary phase and virulence in Gram-positive bacteria. *Curr. Opin. Microbiol.*, **8**, 203–207.
122. Joseph, P., Ratnayake-Lecamwasam, M., and Sonenshein, A.L. (2005) A region of *Bacillus subtilis* CodY protein required for interaction with DNA. *J. Bacteriol.*, **187**, 4127–4139.
123. Janssen, H., Döring, C., Ehrenreich, A., Voigt, B. *et al.* (2010) A proteomic and transcriptional view of acidogenic and solventogenic steady-state cells of *Clostridium acetobutylicum* in a chemostat culture. *Appl. Microbiol. Biotechnol.*, **87**, 2209–2226.
124. Chen, Y., Indurthi, D.C., Jones, S.W., and Papoutsakis, E.T. (2011) Small RNAs in the genus *Clostridium*. *mBio*, **2**, e00340-10.
125. Riester, E., Zimmermann, T., Thormann, K., Schiel-Bengelsdorf, B. *et al.* (2014) SolB, a small noncoding RNA of *Clostridium acetobutylicum*. *Biospektrum Spec. Issue*, **BTP19**, 176–177.
126. Green, E.M. and Bennett, G.N. (1996) Inactivation of an aldehyde/alcohol dehydrogenase gene from *Clostridium acetobutylicum* ATCC 824. *Appl. Biochem. Biotechnol.*, **57**, 213–221.
127. Green, E.M., Boynton, Z.L., Harris, L.M., Rudolph, F.B. *et al.* (1996) Genetic manipulation of acid formation pathways by gene inactivation in *Clostridium acetobutylicum* ATCC 824. *Microbiology*, **142**, 2079–2086.
128. Nair, R.V., Green, E.M., Watson, D.E., Bennett, G.N. *et al.* (1999) Regulation of the *sol* locus genes for butanol and acetone formation in *Clostridium acetobutylicum* ATCC 824 by a putative transcriptional repressor. *J. Bacteriol.*, **181**, 319–330.
129. Wilkinson, S.R. and Young, M. (1994) Targeted integration of genes into the *Clostridium acetobutylicum* chromosome. *Microbiology*, **140**, 89–95.
130. Chen, Y., Caruso, L., McClane, B., Fisher, D. *et al.* (2007) Disruption of a toxin gene by introduction of a foreign gene into the chromosome of

Clostridium perfringens using targetron-induced mutagenesis. *Plasmid*, **58**, 182–189.

131. Shao, L., Hu, S., Yang, Y., Gu, Y. *et al.* (2007) Targeted gene disruption by use of a group II intron (targetron) vector in *Clostridium acetobutylicum*. *Cell Res.*, **17**, 963–965.
132. Heap, J.T., Pennington, O.J., Cartman, S.T., Carter, G.P. *et al.* (2007) The ClosTron: a universal gene knock-out system for the genus *Clostridium*. *J. Microbiol. Methods*, **70**, 452–464.
133. Heap, J.T., Kuehne, S.A., Ehsaan, M., Cartman, S.T. *et al.* (2010) The ClosTron: mutagenesis in *Clostridium*: refined and streamlined. *J. Microbiol. Methods*, **80**, 49–55.
134. Kuehne, S.A., Heap, J.T., Cooksley, C.M., Cartman, S.T. *et al.* (2011) ClosTron-mediated engineering of *Clostridium*. *Methods Mol. Biol.*, **765**, 389–407.
135. Kuehne, S.A. and Minton, N.P. (2012) ClosTron-mediated engineering of *Clostridium*. *Bioengineered*, **3**, 247–254.
136. Liu, Y.J., Zhang, J., Cui, G.Z., and Cui, Q. (2015) Current progress of targetron technology: development, improvement and application in metabolic engineering. *Biotechnol. J.*, **10**, 855–865.
137. Li, Y., Tschaplinski, T.J., Engle, N.L., Hamilton, C.Y. *et al* (2012) Combined inactivation of the *Clostridium cellulolyticum* lactate and malate dehydrogenase genes substantially increases ethanol yield from cellulose and switchgrass fermentations. *Biotechnol. Biofuels*, **5**, 2.
138. Wang, Y., Li, X.Z., Milne, C.B., Janssen, H. *et al.* (2013) Development of a gene knockout system using mobile group II introns (targetron) and genetic disruption of acid production pathways in *Clostridium beijerinckii*. *Appl. Environ. Microbiol.*, **79**, 5853–5863.
139. Jang, Y.-S., Im, J.A., Choi, S.Y., Lee, J.I. *et al.* (2014) Metabolic engineering of *Clostridium acetobutylicum* for butyric acid production with high butyric acid selectivity. *Metab. Eng.*, **23**, 165–174.
140. Yao, J., Zhong, J., Fang, Y., Geisinger, E. *et al.* (2006) Use of targetrons to disrupt essential and nonessential genes in *Staphylococcus aureus* reveals temperature sensitivity of Ll.LtrB group II intron splicing. *RNA*, **12**, 1271–1281.
141. Mohr, G., Hong, W., Zhang, J., Cui, G.Z. *et al.* (2013) A targetron system for gene targeting in thermophiles and its application in *Clostridium thermocellum*. *PLoS One*, **8**, e69032.
142. Nariya, H., Miyata, S., Suzuki, M., Tamai, E. *et al.* (2011) Development and application of a method for counterselectable in-frame deletion in *Clostridium perfringens*. *Appl. Environ. Microbiol.*, **77**, 1375–1382.
143. Tripathi, S.A., Olson, D.G., Argyros, D.A., Miller, B.B. *et al.* (2010) Development of *pyrF*-based genetic system for targeted gene deletion in *Clostridium thermocellum* and creation of a *pta* mutant. *Appl. Environ. Microbiol.*, **76**, 6591–6599.
144. Dusseaux, S., Croux, C., Soucaille, P., and Meynial-Salles, I. (2013) Metabolic engineering of *Clostridium acetobutylicum* ATCC 824 for the high-yield production of a biofuel composed of an isopropanol/butanol/ethanol mixture. *Metab. Eng.*, **18**, 1–8.
145. Cartman, S.T., Kelly, M.L., Heeg, D., Heap, J.T. *et al.* (2012) Precise manipulation of the *Clostridium difficile* chromosome reveals a lack of association between the *tcdC* genotype and toxin production. *Appl. Environ. Microbiol.*, **78**, 4683–4690.
146. Al-Hinai, M.A., Fast, A.G., and Papoutsakis, E.T. (2012) Novel system for efficient isolation of *Clostridium* double-crossover allelic exchange mutants enabling markerless chromosomal gene deletions and DNA integration. *Appl. Environ. Microbiol.*, **78**, 8112–8121.
147. Argyros, D.A., Tripathi, S.A., Barrett, T.F., Rogers, S.R. *et al* (2011) High ethanol titers from cellulose by using metabolically engineered thermophilic, anaerobic microbes. *Appl. Environ. Microbiol.*, **77**, 8288–8294.
148. Zhang, N., Shao, L.J., Jiang, Y., Gu, Y. *et al.* (2015) I-SceI-mediated scarless gene modification via allelic exchange in *Clostridium*. *J. Microbiol. Methods*, **108**, 49–60.

149. Heap, J.T., Ehsaan, M., Cooksley, C.M., Ng, Y.-K. *et al* (2012) Integration of DNA into bacterial chromosomes from plasmids without a counter-selection marker. *Nucleic Acids Res.*, **40**, e59.
150. Ng, Y.K., Ehsaan, M., Philip, S., Collery, M.M. *et al.* (2013) Expanding the repertoire of gene tools for precise manipulation of the *Clostridium difficile* genome: allelic exchange using *pyrE* alleles. *PLoS One*, **8**, e56051.
151. Ehsaan, M., Kuit, W., Zhang, Y., Cartman, S.T. *et al.* (2016) Mutant generation by allelic exchange and genome resequencing of the biobutanol organism *Clostridium acetobutylicum* ATCC 824. *Biotechnol. Biofuels*, **9**, 4.
152. Ishino, M., Sawada, Y., Yaegashi, T., Demura, M. *et al.* (1987) Nucleotide sequence of the adenovirus type 40 inverted terminal repeat: close relation to that of adenovirus type 5. *Virology*, **15**, 414–416.
153. Barrangou, R., Fremaux, C., Deveau, H., Richards, M. *et al.* (2007) CRISPR provides acquired resistance against viruses in prokaryotes. *Science*, **315**, 1709–1712.
154. Jinek, M., Chylinski, K., Fonfara, I., Hauer, M. *et al.* (2012) A programmable dual-RNA-guided DNA endonuclease in adaptive bacterial immunity. *Science*, **337**, 816–821.
155. Mali, P., Yang, L., Esvelt, K.M., Aach, J. *et al.* (2013) RNA-guided human genome engineering via Cas9. *Science*, **339**, 823–826.
156. DiCarlo, J.E., Norville, J.E., Mali, P., Rios, X. *et al.* (2013) Genome engineering in *Saccharomyces cerevisiae* using CRISPR-Cas systems. *Nucleic Acids Res.*, **41**, 4336–4343.
157. Feng, Z., Zhang, B., Ding, W., Liu, X. *et al* (2013) Efficient genome editing in plants using a CRISPR/Cas system. *Cell Res.*, **23**, 1229–1232.
158. Hai, T., Teng, F., Guo, R., Li, W. *et al.* (2014) One-step generation of knockout pigs by zygote injection of CRISPR/Cas system. *Cell Res.*, **24**, 372–375.
159. Jiang, W., Bikard, D., Cox, D., Zhang, F. *et al.* (2013) RNA-guided editing of bacterial genomes using CRISPR-Cas systems. *Nat. Biotechnol.*, **31**, 233–239.
160. Oh, J.H. and van Pijkeren, J.P. (2014) CRISPR-Cas9-assisted recombineering in *Lactobacillus reuteri*. *Nucleic Acids Res.*, **42**, e131.
161. Wang, Y., Zhang, Z.T., Seo, S.O., Choi, K. *et al.* (2015) Markerless chromosomal gene deletion in *Clostridium beijerinckii* using CRISPR/Cas9 system. *J. Biotechnol.*, **200**, 1–5.
162. Jiang, Y., Chen, B., Duan, C., Sun, B. *et al.* (2015) Multigene editing in the *Escherichia coli* genome via the CRISPR-Cas9 system. *Appl. Environ. Microbiol.*, **81**, 2506–2514.
163. Li, Y., Lin, Z., Huang, C., Zhang, Y. *et al.* (2015) Metabolic engineering of *Escherichia coli* using CRISPR-Cas9 meditated genome editing. *Metab. Eng.*, **31**, 13–21.
164. Woolley, R.C., Pennock, A., Ashton, R.J., Davies, A. *et al.* (1989) Transfer of Tn*1545* and Tn*916* to *Clostridium acetobutylicum*. *Plasmid*, **22**, 169–174.
165. Bertram, J., Kuhn, A., and Dürre, P. (1990) Tn*916*-induced mutants of *Clostridium acetobutylicum* defective in regulation of solvent formation. *Arch. Microbiol.*, **153**, 373–377.
166. Mattsson, D.M. and Rogers, P. (1994) Analysis of Tn*916*-induced mutants of *Clostridium acetobutylicum* altered in solventogenesis and sporulation. *J. Ind. Microbiol. Biotechnol.*, **13**, 258–268.
167. Kashket, E.R. and Cao, Z.Y. (1993) Isolation of a degeneration-resistant mutant of *Clostridium acetobutylicum* NCIMB 8052. *Appl. Environ. Microbiol.*, **59**, 4198–4202.
168. Liyanage, H., Young, M., and Kashket, E.R. (2000) Butanol tolerance of *Clostridium beijerinckii* NCIMB 8052 associated with down-regulation of *gldA* by antisense RNA. *J. Mol. Microbiol. Biotechnol.*, **2**, 87–93.
169. Lanckriet, A., Timbermont, L., Happonen, L.J., Pajunen, M.I. *et al.* (2009) Generation of single-copy transposon insertions in *Clostridium perfringens* by electroporation of phage mu DNA transposition complexes. *Appl. Environ. Microbiol.*, **75**, 2638–2642.
170. Vidal, J.E., Chen, J., Li, J., and McClane, B.A. (2009) Use of an EZ-Tn5-based

random mutagenesis system to identify a novel toxin regulatory locus in *Clostridium perfringens* strain 13. *PLoS One*, **4**, e6232.

171. Scott, P.T. and Rood, J.I. (1989) Electroporation-mediated transformation of lysostaphin-treated *Clostridium perfringens*. *Gene*, **82**, 327–333.
172. Liu, H., Bouillaut, L., Sonenshein, A.L., and Melville, S.B. (2013) Use of a mariner-based transposon mutagenesis system to isolate *Clostridium perfringens* mutants deficient in gliding motility. *J. Bacteriol.*, **195**, 629–636.
173. Cartman, S.T. and Minton, N.P. (2010) A mariner-based transposon system for *in vivo* random mutagenesis of *Clostridium difficile*. *Appl. Environ. Microbiol.*, **76**, 1103–1109.
174. Zhang, Y., Grosse-Honebrink, A., and Minton, N.P. (2015) A universal mariner transposon system for forward genetic studies in the genus *Clostridium*. *PLoS One*, **10**, e0122411.
175. Dembek, M., Barquist, L., Boinett, C.J., Cain, A.K. *et al.* (2015) High-throughput analysis of gene essentiality and sporulation in *Clostridium difficile*. *mBio*, **6**, 02383-14.
176. Golden, J.S. and Handfield, R.B. (2014) *Why Biobased? Opportunities in the Emerging Bioeconomy*, United States Department of Agriculture, Washington, DC.
177. Cascone, R. (2008) Biobutanol – A replacement for bioethanol? *Chem. Eng. Prog.*, **104**, S4–S9.
178. Green, E.M. (2011) Fermentative production of butanol-the industrial perspective. *Curr. Opin. Biotechnol.*, **22**, 337–343.

10
Advances in Consolidated Bioprocessing Using *Clostridium thermocellum* and *Thermoanaerobacter saccharolyticum*

Lee R. Lynd, Adam M. Guss*, Michael E. Himmel, Dhananjay Beri, Chris Herring, Evert K. Holwerda, Sean J. Murphy, Daniel G. Olson, Julie Paye, Thomas Rydzak, Xiongjun Shao, Liang Tian, and Robert Worthen*

10.1
Introduction

Biological conversion of lignocellulose (plant cell walls) to fuels and chemicals has been a major research focus since the 1970s. To be deployed industrially, an economically viable process is needed for the deconstruction of cellulosic biomass and fermentation of biomass-derived sugars. The current state of the art involves thermochemical pretreatment, followed by the addition of fungal cellulase to hydrolyze cellulose and hemicellulose to fermentable sugars, with fermentation of the liberated sugars by a noncellulolytic microorganism, such as *Saccharomyces cerevisiae*. Some naturally occurring, saccharolytic anaerobic bacteria – including but not limited to thermophiles – are able to solubilize lignocellulosic biomass without added enzymes [1]. Converting lignocellulose to biofuels in one process step without added enzymes, consolidated bioprocessing or "CBP" [2] is conceptually attractive as a means to avoid the substantial cost of added enzymes [3, 4]. Cost-competitive processing via CBP requires organism development and represents a challenge for metabolic engineering that is at once formidable and impactful. Recent reviews address thermophilic lignocellulose deconstruction [5] and ethanol production using engineered thermophilic bacteria [6]. The multicomponent, self-assembling cellulosome complex of *C. thermocellum* is addressed in detail elsewhere [7] and contrasted to cellulase systems consisting of a single catalytic domain with discretely acting cellulase components characteristic of commercial cellulases [8, 9].

Here, we draw upon recently published and unpublished works, to provide an aggregated perspective on the status and prospects for CBP using thermophilic bacteria. We focus on two microbes that have received particular attention in this context: the cellulose-fermenting *Clostridium thermocellum* and the hemicellulose-fermenting *Thermoanaerobacterium saccharolyticum*. We focus

* These authors contributed equally.

Industrial Biotechnology: Microorganisms, First Edition. Edited by Christoph Wittmann and James C. Liao.

on ethanol because it has received the most attention in the context of organism development for CBP, and because cellulosic ethanol production is the logical point of entry and proving ground for CBP.

In Section 10.2, we comment on CBP organism development strategies. Section 10.3 addresses plant cell wall solubilization by *C. thermocellum*, with an emphasis on comparison to industry-standard fungal cellulase. Bioenergetics of *C. thermocellum* cellulose fermentation is addressed in Section 10.4. Metabolic engineering is addressed in Section 10.5, with topics addressed including development of genetic tools, the high-functioning ethanol pathway in *T. saccharolyticum*, and the current state of strain development. We conclude by offering perspectives on future directions in Section 10.6.

10.2 CBP Organism Development Strategies

No known microorganism is capable of both efficient lignocellulose deconstruction in high yield and titer biofuel production. As previously reviewed [1, 4], development of CBP-enabling microorganisms must proceed via metabolic engineering aimed at either (i) improving properties related to product formation in naturally occurring plant cell wall solubilizing microbes (known as the native cellulolytic strategy) or (ii) conferring the ability to solubilize plant cell walls to a noncellulolytic host that is good at product formation (known as the recombinant cellulolytic strategy). To date, thermophilic anaerobes have received the most attention as hosts pursuant to the native cellulolytic strategy. Yeasts have most often been considered as hosts for the recombinant cellulolytic strategy.

We regard the native cellulolytic strategy as more promising for industrial deployment of CBP. As detailed in Section 10.3, thermophilic, cellulolytic microbes such as *C. thermocellum* solubilize lignocellulosic biomass particularly well. In our view, it is simpler and likely easier to replicate high-functioning product formation pathways in native plant cell wall solubilizing microbes than to replicate plant cell wall solubilization in noncellulolytic hosts. We observe that the strong ability of thermophiles to solubilize plant cell walls will likely be difficult to fully reproduce in a noncellulolytic host microbe, while the high product tolerance of yeast will likely be difficult to fully reproduce in thermophiles. Considering the case of ethanol, strong solubilizing capability has a greater impact on process economics than high ethanol titer [10]. This is especially the case for lignocellulose processing, where the mechanical difficulty of processing high solids concentration constrains titers of ethanol independent of the tolerance of the fermenting microbe (Section 10.5.2). Forecasting whether the native or recombinant strategy will prove more successful is, of course, in the realm of speculation and cannot be definitively settled until one of these strategies is fully developed and deployed. Both reconstructing a plant cell wall solubilizing system one heterologous component at a time, central to the recombinant strategy, and understanding and manipulating metabolic flux in cellulolytic microbes, central to the native strategy, are interesting and informative fundamental exercises.

10.3 Plant Cell Wall Solubilization by *C. thermocellum*

10.3.1 Understanding and Describing Solubilization

As reviewed elsewhere [11–14], factors implicated in biomass recalcitrance and strategies to overcome it include physical inaccessibility from plant tissue and cell wall porosity, biopolymer features such as crystallinity and degree of polymerization, down to molecular-level shielding, and the diversity of chemical bonds that must be cleaved. South *et al.* [15] proposed a model for simultaneous saccharification and fermentation (SSF; a process in which cellulolytic enzymes are added simultaneously with fermentation by a noncellulolytic organism, such as *S. cerevisiae*) featuring an adsorption equation capable of exhibiting saturation with either enzyme or biocatalyst and a conversion-dependent rate constant exhibiting declining activity as solubilization proceeds. This model fit data for the microcrystalline cellulose, Avicel®, and pretreated hardwood; it was successfully used in conjunction with a particle population model to predict results in continuous culture and extended in subsequent studies [16, 17]. More detailed models have been created that incorporate multiple solubilizing activities and/or substrate state variables in addition to concentration and conversion [18], but these models are not sufficiently advanced to be valuable for predictive purposes. High solubilization becomes more challenging to achieve at increasing solids loading due to known factors, including mass transfer, mixing, water interactions, and inhibitors generated by pretreatment [11, 19–23].

For lignocellulose fermentation by cellulolytic microbes, the concentration of saccharolytic enzymes is an output variable and changes over time, whereas this concentration is an input variable and is often approximately constant for SSF. Kinetics and reactor design for microbial fermentation of particulate feedstocks are substantially more complicated than for soluble substrates for several reasons, including that particle reactivity is determined by extent of reaction and concentration [1]. Both description and understanding of such kinetics are a work in progress. Rates of microbial mediated solubilization have been described as being proportional to cell concentration and substrate concentration, and proceeding according to the Monod equation. These simplified descriptions may be applicable over limited operating conditions, but have conceptual problems and inconsistencies for general application [1]. Holwerda and Lynd [24] proposed a second-order model that describes well Avicel hydrolysis over a batch culture in which first cells, and then substrate, are the primary rate-limiting factor. Descriptive models at this level of complexity have yet to be developed for fermentation of plant cell walls by pure cellulolytic cultures, for which substrate features are more complex and experimental measurements and validation more difficult.

Accurate modeling of microbial solubilization of lignocellulose will be dependent on knowledge of the dynamics of microbial cell concentration over the course of bioconversion. While measurement of cell concentrations distinct from the concentration of substrate is trivial for soluble substrates, it is a substantial and not-yet-resolved challenge for fermentation of particulate substrates based on plant cell walls. Cell measurement has been approached on the basis of elemental composition (pellet nitrogen, [25]), concentration of cellular macromolecules (total protein [26] or DNA via quantitative PCR [27]), and estimated by indirect methods, such as off-gas analysis [25] and detection of enzymes (ELISA assays [28]). Future efforts using quantitative proteomics approaches also hold promise.

10.3.2
Comparative Solubilization Effectiveness

The yield of fermentable sugars ($g\,g^{-1}$ dry biomass) from lignocellulose determines the maximum amount of fuel that can be produced per unit biomass and thus is a major cost driver. The literature contains a few studies comparing initial rates of hydrolysis of model substrates by *C. thermocellum* and fungal cellulase on model substrates. As reviewed elsewhere [1], it appears on the basis of these studies that the specific activity of the complexed cellulosome of *C. thermocellum* is about an order of magnitude greater than that of the noncomplexed cellulase system of the industry-standard cellulase system produced by *Trichoderma reesei.* After replicating solubilization data in the literature and/or from other labs for six different biocatalysts, Paye *et al.* [29] compared final solubilization yields achieved by these biocatalysts on a common lignocellulosic substrate, mid-season-harvested switchgrass. Total carbohydrate solubilization ranged from 24% for *Caldicellulosiruptor bescii* to 65% for *C. thermocellum,* with intermediate values observed for a thermophilic horse manure enrichment, *Clostridium clariflavum, Clostridium cellulolyticum,* and SSF featuring fungal cellulases and yeast. Feedstock initial particle size over the range 0.08–6 mm was found to have little impact on solubilization of mid-season and senescent switchgrass by *C. thermocellum. C. thermocellum* was found to give substantially higher solubilization compared to SSF on all tested substrates and particle sizes.

Comprehensive data comparing carbohydrate solubilization by *C. thermocellum* cultures and a commercial fungal cellulase (Cellic® Ctec2 supplemented with Htec2) is presented in Table 10.1. Given the small inoculum used for *C. thermocellum,* cell and enzyme levels are initially diluted 20-fold or more and increase during the course of the experiment as a result of cell growth. By contrast, cellulase concentrations remain at the initial levels throughout the incubation period for the added fungal cellulase during SSF. Notwithstanding this difference, solubilization yields are consistently several-fold higher for *C. thermocellum* even when compared to likely unaffordable levels of added fungal cellulase (more than ~20 mg cellulase per gram solids[1]). We note, in particular, the much stronger performance of *C. thermocellum* on as-received industrial by-products (paper sludge and corn fiber) and an agricultural residue (corn stover). In initial experiments of *C. thermocellum* at solids concentrations on the order of $100\,g\,l^{-1}$, solubilization yields after the same incubation time are only slightly lower than at low concentrations and still much higher than for fungal cellulase. Additional diagnostic and remedial work aimed at obtaining high solubilization at high concentrations is underway.

Further comparative insights are provided in Figure 10.1. Here, we plot with open symbols and reference to the left y-axis the solubilization yield realized by *C. thermocellum* cultures (b) as a function of the solubilization yield using SSF with fungal cellulase and *S. cerevisiae* (a) on the x-axis. Data are without pretreatment other than autoclaving taken from conversion of mid-season or senescent switchgrass at various particle sizes from Paye *et al.* [29] at a loading of 5 mg protein g^{-1} dry solids, and from conversion of winter rye at various harvest times as reported by Shao *et al.* [32] at a loading of 15 mg protein g^{-1} dry solids. Solubilization yields are seen to be higher for *C. thermocellum* under all conditions, with every open symbol being above the $a = b$ dashed line. On the right axis with closed symbols, we plot the ratio of solubilization by *C. thermocellum* relative to fungal enzymes (b/a); all values are ≥ 1.7. Notably, the lower the solubilization achieved by SSF, the greater the relative advantage of *C. thermocellum* – for example, threefold higher solubilization yields for senescent switchgrass compared to SSF.

The data in Table 10.1 and Figure 10.1, believed to be the most comprehensive such comparison to date, provide strong evidence that *C. thermocellum* is more effective at mediating plant cell wall solubilization than a representative commercial fungal cellulase preparation. We note that these data are under controlled conditions for which intrinsic biocatalyst capability is, by design, the limiting factor. Comparison under commercially viable process conditions awaits further work as these conditions are defined.

Although the methodological capability to do comparative studies such as those in Table 10.1 and Figure 10.1 has been in hand for decades, few studies of this type have been reported. Reasons for this likely include the following:

- Performing fair and informative comparisons of different biocatalysts, although doable, is a significant undertaking, involving both nuance and careful attention to controls.
- Comparative studies have seldom if ever been the top priorities of funding agencies and reviewers.
- Investigators in the field were initially focused on finding microbes that could secrete high levels of saccharolytic enzymes that could be recovered and used to generate soluble sugars for fermentation by noncellulolytic microbes such as yeast (see discussion in [1]).

1) Readers wishing to convert from cellulase loading in milligram protein per gram solids to filter paper units per gram solids may note that the specific activity of *T. reesei* cellulase preparations is about 0.6 filter paper units mg^{-1} protein.

Table 10.1 Comparative solubilization by *C. thermocellum* and fungal cellulase.

Substrate, biocatalyst[a)]	Vessel[b)]	Initial particle size (mm)[c)]	Initial catalyst loading[d)]	Solids loading ($g\,l^{-1}$)	Duration (d)	Solubilization (%)[e)]	References
Avicel							
C. therm	B	0.02	2%	5	5	99.6 (0.1)^G	[29]
FC + yeast	B	0.02	5 kg/Mg	5	5	55 (2.0)^G	
C. therm	F	0.02	0.5%	72.7	6.5	99.0^G	[30, 31]
C. therm	F	0.02	0.5%	103	6.5	93.5^G	
C. therm	F	0.02	0.5%	123	9	89.2^G	Holwerda[f)]
Paper sludge							
C. therm	F	ND	3%	50	8	98^G	Holwerda[f)]
C. therm	F	ND	3%	100	8	96^G	
FC	B	ND	5 kg/Mg	10	5	30.1^T	Herring[f)]
FC	B	ND	20 kg/Mg	10	5	80.1^T	
Corn fiber							
C. therm	F	ND	5%	10	5	95.2^T	Beri and Herring[f)]
FC	B	ND	5 kg/Mg	10	5	43.1^T	
FC	B	ND	20 kg/Mg	10	5	56.2^T	
Corn stover							
C. therm	F	5	2%	15.8	5	66.9 (1.5)^T	Worthen and Holwerda[f)]
C. therm	F	5	2.5%	79.0	10	60.3 (1.5)^T	
FC + yeast, 37°	F	5	5 kg/Mg	14.6	5	20.9 (5.4)^T	
Switchgrass, midseason, washed							
C. therm	B	Ave of six	2%	14	5	66.4 (2.3)^G	[29]
C. therm PC	B	4	Ave offour	14	5	61.0 (4.3)^G	
FC + yeast, 37°	B	Ave of six	5 kg/Mg	14	5	35.3 (4.4)^G	
FC + yeast, 37°	B	4	Ave offour	14	5	33.4 (2.3)^G	
FC, 37°	B	4	20 kg/Mg	14	5	37.5 (3.9)^G	
FC, 50°	B	4	5 kg/Mg	14	5	33.7 (7.7)^G	
C. therm	F	4	0.5%	9.2	6	75.2^T	Murphy and Shao[f)]
C. therm	F	4	5%	91.7	6	60.0^T	
Switchgrass, midseason, unwashed							
C. therm	F	0.08	2.5%	16.1	6	80^G	Holwerda[f)]
C. therm	F	0.08	1%	32.2	6	75.5^G	
Switchgrass, senescent, washed (4 mm initial particle size)							
C. therm	B	Ave of 6	2%	13	5	34.7 (4.3)^G	[29]
FC + yeast	B	Ave of 6	5 kg/Mg	13	5	12.4 (2.3)^G	

Experimental methods, to be reported in more detail in archival manuscripts, are in general as presented by Paye *et al.* [29]. Solubilization values for *C. thermocellum* in gray, SSF in black and SSF with ≥20 mg cellulase g^{-1} solids is denoted by light gray (this is deemed unaffordable : >\$3 gal^{-1} EtOH at \$15 kg^{-1} protein).

a) *FC* refers to fungal cellulase, a 9 : 1 mixture of Ctec2 and Htec2. *PC* refers to purified cellulase. "*C. therm*" refers to cultures.
b) B refers to 158 ml serum bottles, 50 ml working volume, F refers to 1–1.5 l fermenter.
c) Values listed are mesh sizes. Ave of six, average of six particle sizes from 0.08 to 6 mm.
d) % refers to volume % inoculum, kg/Mg is per gram of dry solids. Ave offour, enzyme loading from 5 to 20 kg/Mg.
e) Superscripts G and T indicate values based on glucan and total carbohydrate, respectively. Values in parentheses denote standard deviations.
f) Data not yet published.

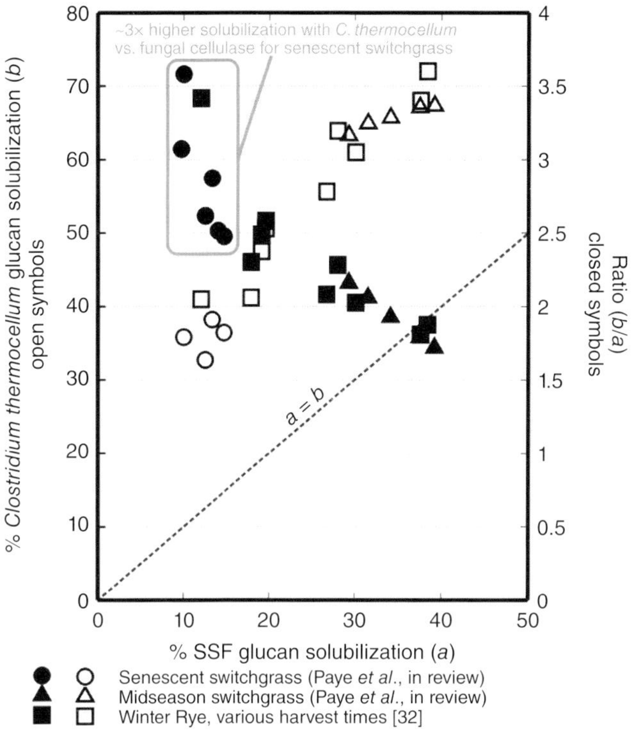

Figure 10.1 Comparison of biomass solubilization by *C. thermocellum* relative to fungal cellulases. Open symbols, percent glucan solubilization by *C. thermocellum* (left axis) as a function of the percent glucan solubilization by fungal enzymes during SSF. The dashed line represents values where solubilization would be equal between *C. thermocellum* and fungal cellulases. Closed symbols, the same data plotted as the ratio between *C. thermocellum* and fungal enzyme solubilization (right axis).

The recent realization that *C. thermocellum* has greater intrinsic capability to deconstruct plant cell walls relative to a commercial cellulase after extensive development invites mechanistically oriented studies that aim to understand this observation – particularly in light of the observation that the *C. thermocellum* cellulosome is minimally an order of magnitude larger than fungal systems. Large size was once thought to be a liability for cellulase enzyme systems due to exclusion from small pores in the feedstock [33]. However, recent studies provide evidence that *T. reesei* cellulases sharpen the ends of cellulose fibers via ablative action, whereas *C. thermocellum* cellulosomes unravel the ends of fibers [34]. This unraveling functionality was shown to be dependent on the presence of the CipA scaffoldin protein responsible for formation of the multimolecular cellulosome complex. We therefore speculate that deconstruction of cell walls may be enabled by a larger molecular machine tethered to the microbial cells when compared to simple, unimolecular fungal cellulases. Moreover, the presence of known modalities of cellulase action: cell-associated enzyme complexes, cell-free enzyme complexes, and noncomplexed enzymes have also been implicated as factors contributing to the effectiveness of *C. thermocellum* at solubilizing cellulosic biomass [35]. In addition to providing impetus for fundamental studies, the strong biomass-solubilizing capability of *C. thermocellum* – and we think likely other thermophilic anaerobes – adds impetus to addressing the current limitations of these microbes for CBP.

10.4 Bioenergetics of *C. thermocellum* Cellulose Fermentation

Bioenergetics of anaerobic cellulose fermentation is important because cellulase synthesis is energy intensive, and because functioning in industrial conditions can be expected to place added stress on the cell. *C. thermocellum* has evolved to secrete large amounts of biomass-solubilizing enzymes; thus, it may not be surprising that *C. thermocellum* has also evolved to conserve more ATP than many fermenting organisms. Zhang and Lynd presented an analysis of bioenergetics of *C. thermocellum* showing that this organism allocates 0.26–0.32 mol ATP/mol glucosyl monomer to cellulase synthesis, but that energy conservation specific to growth on cellulose more than offsets this burden [36]. Understanding of energy conservation and metabolic flux in *C. thermocellum* has since progressed considerably. We endeavor to review and extend this new understanding in this section.

Organisms such as *S. cerevisiae* and *Escherichia coli* conserve 2–3 ATP per glucose. However, at standard conditions, the thermodynamics of cellodextrin conversion to ethanol, acetate, lactate, H_2, and formate is quite favorable, with the $\Delta G^{\circ\prime}$ ranging from −246 to −258 kJ mol^{-1} glucose [37–39]. In many well-studied organisms, *in vivo* ATP formation requires approximately −45 to −50 kJ mol^{-1} of energy [38, 39]; thus, at standard condition, a maximum of approximately 5 ATP could be formed per glucose. Under some physiological/environmental conditions, the thermodynamics can be even more favorable, for instance when other organisms keep the concentration of acetate and H_2 low.

10.4.1 Membrane Energetics

The exact amount of ATP conserved per fermented glucose in *C. thermocellum* is uncertain, both because of the multitude of potential enzymatic pathways for carbon and electron flux encoded in the genome [40], as well as the unknown coupling numbers when enzymatic reactions are coupled to membrane gradients. However, a range can be estimated on the basis of known pathways that are present and the most likely coupling numbers.

Membrane-bound pyrophosphatases have been reported to translocate 1–2 protons (H^+) per pyrophosphatase (PP_i) [41, 42]. While the coupling number in *C. thermocellum* is not known, a value of 2 H^+ per PP_i will be used here (Figure 10.2b) because this value would result in a higher intracellular concentration of PP_i,

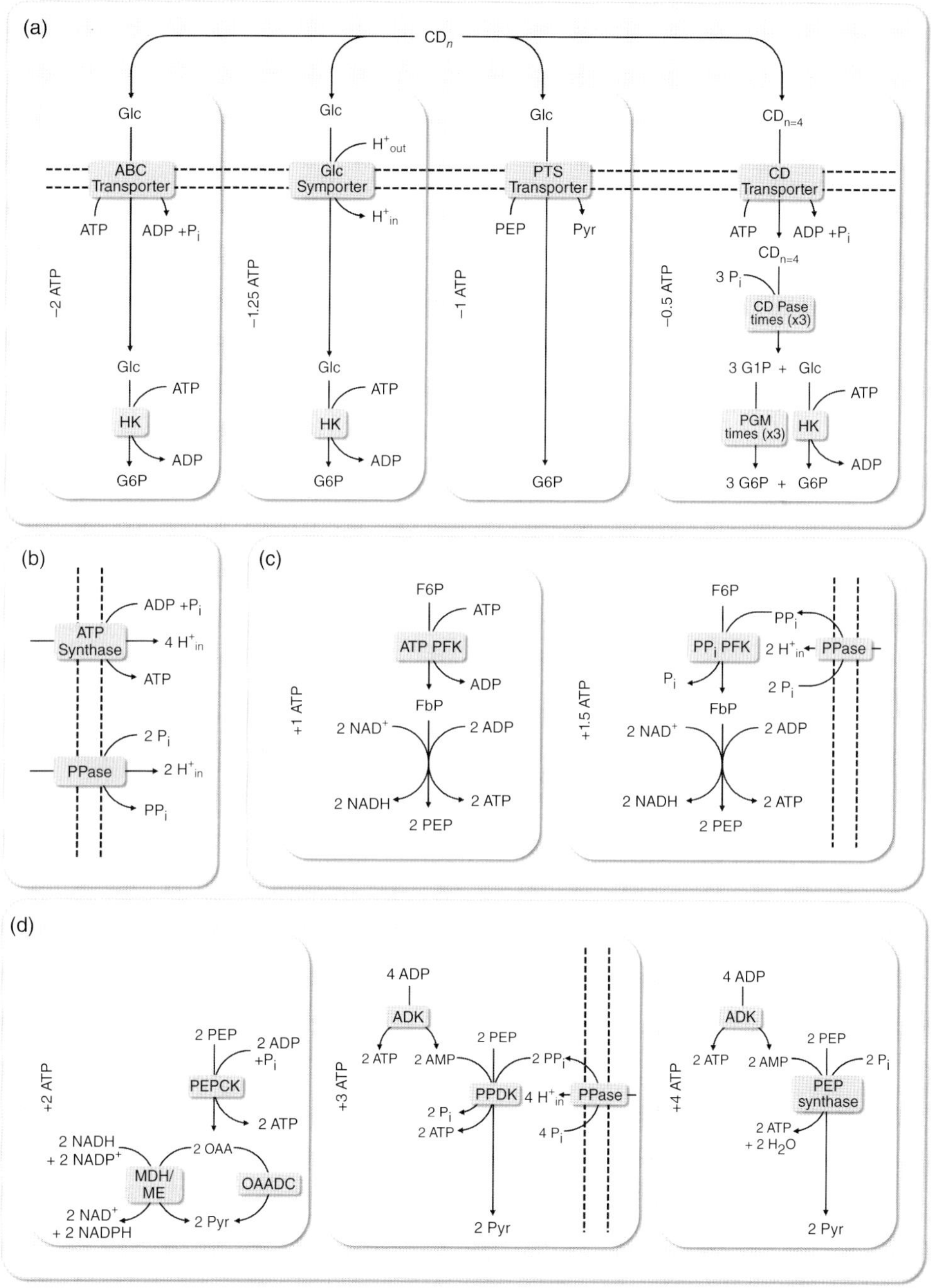
(a)
CD_n
Glc
ABC Transporter
ATP
ADP +Pi
−2 ATP
Glc
HK
ATP
ADP
G6P
Glc
H+out
Glc Symporter
H+in
−1.25 ATP
Glc
HK
ATP
ADP
G6P
Glc
PTS Transporter
PEP
Pyr
−1 ATP
G6P
CDn=4
CD Transporter
ATP
ADP +Pi
CDn=4
3 Pi
CD Pase times (x3)
−0.5 ATP
3 G1P + Glc
PGM times (x3)
HK
ATP
ADP
3 G6P + G6P
(b)
ADP +Pi
ATP Synthase
4 H+in
ATP
2 Pi
PPase
2 H+in
PPi
(c)
F6P
ATP
ATP PFK
ADP
FbP
+1 ATP
2 NAD+
2 ADP
2 NADH
2 ATP
2 PEP
F6P
PPi
PPi PFK
2 H+in
PPase
Pi
2 Pi
FbP
+1.5 ATP
2 NAD+
2 ADP
2 NADH
2 ATP
2 PEP
(d)
+2 ATP
2 PEP
2 ADP +Pi
PEPCK
2 ATP
2 NADH + 2 NADP+
2 OAA
MDH/ ME
OAADC
2 NAD+ + 2 NADPH
2 Pyr
4 ADP
ADK
2 ATP
2 AMP
2 PEP
2 PPi
+3 ATP
PPDK
4 H+in
PPase
2 Pi
2 ATP
4 Pi
2 Pyr
4 ADP
ADK
2 ATP
2 AMP
2 PEP
2 Pi
+4 ATP
PEP synthase
2 ATP + 2 H2O
2 Pyr

Figure 10.2 Typical pathways that can convert cellodextrins (CDs) to pyruvate in *C. thermocellum*. (a) Transport and conversion of CD to glucose-6-phosphate (G6P), (b) coupling of the membrane gradient to ATP and pyrophosphate (PP_i) formation, (c) catabolism of fructose-6-phosphate (F6P) to phosphoenolpyruvate (PEP), and (d) conversion of PEP to pyruvate (Pyr). ATP yields per hexose equivalent are provided. Glc, glucose; FbP, fructose-1,6-bisphosphate; OAA, oxaloacetate; HK, hexokinase; CD Pase, cellodextrin phosphorylase; PGM, phosphoglucomutase; PPase, pyrophosphatase; PFK, phosphofructokinase; PEPCK, phosphoenolpyruvate carboxykinase; MDH/ME, malate dehydrogenase/malic enzyme; OAADC, oxaloacetate decarboxylase; and PPDK, pyruvate phosphate dikinase.

which would help drive glycolytic reactions. Similarly, ATP synthase reversibly couples the membrane gradient to ATP synthesis, and the coupling number can vary. The ATP synthase complex makes one complete rotation per 3 ATP synthesized, and the number of c-subunits dictates the number of H^+ pumped per 3 ATP [43–46]. The number of c-subunits in studied ATP synthases varies from organism to organism and ranges from 10 to 14 subunits, giving a coupling number of 10/3 (3.33 H^+ per ATP) to 14/3 (4.66 H^+ per ATP) [47]. For the sake of simplicity, we use a coupling number of 12/3 (4 H^+ per ATP) (Figure 10.2b). Combining our estimates for the ATP synthase and pyrophosphatase, hydrolysis of 1 ATP would pump 4 H^+ outside the cell, which could be used to synthesize 2 PP_i. Thus, using PP_i is the equivalent of using 0.5 ATP. A coupling number of 10 or 14 would change the ATP yield by approximately 0.1 ATP per PP_i (see above).

Much less is known about the coupling numbers of the ion-pumping Ferredoxin (Fd):NAD^+ oxidoreductase (Rnf) and Energy converting hydrogenase (Ech). Isozymes of Rnf and Ech can translocate an unknown number of either Na^+ or H^+ during the transfer of electrons between Fd and NAD(H) or Fd and H^+/H_2, respectively. In the present analysis, we will presume 2 H^+ each, similar to previous reports [48, 49].

Another strategy used by many anaerobes including *C. thermocellum* is flavin-based electron bifurcation, in which exergonic electron-transfer reactions are coupled to endergonic electron-transfer reactions to provide a thermodynamic driving force for a reaction or to capture excess free energy available in a reaction. Relevant examples of these types of reactions for *C. thermocellum* include the bifurcating transhydrogenase NfnAB (NADH-dependent reduced ferredoxin:$NADP^+$ oxidoreductase) (reduced ferredoxin + NADH + 2 $NADP^+ \rightarrow$ oxidized ferredoxin + NAD^+ + 2 NADPH) [50] and a bifurcating hydrogenase (reduced ferredoxin + NADH + 4 $H^+ \rightarrow$ oxidized ferredoxin + NAD^+ + 2 H_2) [51, 52].

10.4.2
Sugar Conversion to Pyruvate

The first point at which *C. thermocellum* differs in ATP yield from organisms such as *E. coli* or *Zymomonas mobilis* is the combined import of glucose and conversion to glucose-6-phosphate (G6P) [36]. Many organisms use an ABC transporter, PTS transporter, or symporter to import glucose into the cell (Figure 10.2a). Depending on the method of import and subsequent phosphorylation, this can result in the expenditure of 1–2 ATP to make one G6P. *C. thermocellum*, on the other hand, uses ABC-type cellodextrin transporters [53] that take up glucose oligomers of average length $n = 4$ (cellotetraose) [36]. Once inside the cell, a cellodextrin phosphorylase (CD Pase) performs phosphorolytic cleavage to make a free cellodextrin of length $n - 1$ and a molecule of glucose-1-phosphate (G1P) without expending ATP. CD Pase can then act repeatedly on the shortening cellodextrin until finally cellobiose is cleaved into one glucose and one G1P. These G1P are then isomerized to G6P, and the free glucose is phosphorylated to G6P using 1 ATP. Thus, when starting with cellotetraose, 2 ATP are expended per 4 G6P produced, for an average of only 0.5 ATP used per glucose. This represents a 0.5–1.5 ATP savings per glucose relative to transport and phosphorylation via common mechanisms.

Most model organisms use ATP to phosphorylate fructose-6-phosphate (F6P) to fructose-1,6-bisphosphate (FbP) before gaining 2 ATP from the conversion of FbP to phosphoenolpyruvate (PEP), resulting in a net of +1 ATP during the conversion of F6P to 2 PEP molecules. *C. thermocellum* encodes both an ATP-dependent and PP_i-dependent phosphofructokinase (Pfk) (Figure 10.2c). Gene expression analysis [54, 55] and enzymology [56] revealed that the PP_i-dependent enzyme is likely the main Pfk used by *C. thermocellum*. Because PP_i synthesis costs only 0.5 ATP, the net ATP production in *C. thermocellum* during conversion of F6P to 2 PEP is 1.5 ATP.

Unlike most organisms, *C. thermocellum* does not encode a pyruvate kinase, which would generate 2 ATP during the conversion of 2 PEP to pyruvate. Instead, the reaction(s) responsible for conversion of PEP to pyruvate are uncertain (Figure 10.2d). One possibility is the so-called malate shunt [55, 57], in which PEP is first carboxylated to oxaloacetate (OAA) with the concomitant synthesis of GTP, which can be used directly to make ATP via nucleoside-diphosphate kinase. Then, OAA can either be directly decarboxylated to pyruvate or reduced to malate and then oxidatively decarboxylated to pyruvate. The net result of these pathways is the synthesis of 2 ATP equiv. per glucose, the same as using pyruvate kinase, with the possibility of electron transfer from NADH to make NADPH.

Alternatively, pyruvate phosphate dikinase (PPDK) can be used to directly convert PEP to pyruvate using AMP and PP_i as substrates (Figure 10.2d). The pyrophosphatase and ATP synthase together would spend 1 ATP to produce the 2 PP_i, while adenylate kinase used to synthesize the AMP would produce 2 ATP, and PPDK would produce an additional 2 ATP. Thus, this pathway would form a net of 3 ATP. A very similar possibility would be the use of PEP synthase, which also uses AMP but uses free phosphate instead of PP_i, such that the conversion of PEP to pyruvate would net 4 ATP. The genes encoding PPDK and the malate shunt are all highly expressed in *C. thermocellum* while PEP synthase expression is barely detectable, suggesting that PPDK and the malate shunt are the primary paths used in *C. thermocellum* for PEP conversion to pyruvate [55]. If flux were evenly distributed between PPDK and the malate shunt, then an average of 2.5 ATP per mole of glucose would be generated in this process.

10.4.3 End-Product Formation

Pyruvate is a major branch point in fermentation, with most flux distributing between lactate, acetate, ethanol, formate, and H_2. Lactate is a minor fermentation product, especially during growth on cellulose. It also yields no additional ATP and is not considered here.

During growth in a mixed community with H_2- and acetate-consuming organisms, *C. thermocellum* has been shown to synthesize predominantly acetate and H_2 as fermentation products (Figure 10.3a) [58]. The *in situ* removal of these products by other microbes makes the pathway more thermodynamically favorable than during growth in pure culture. For acetate formation from pyruvate, 2 pyruvate would be converted to 2 acetyl-CoA with the concomitant reduction of ferredoxin using pyruvate:Fd oxidoreductase (PFOR). The 2 acetyl-CoA would then be converted to 2 acetate, generating 2 ATP. This leaves four pairs of electrons (2 NADH and 2 reduced ferredoxin) that would be used in H_2 production. If a bifurcating hydrogenase is used, then H_2 is formed with no additional ATP production. However, the combined use of Rnf and Ech could result in a net translocation of 4 H^+ outside the cell, resulting in an additional 1 ATP formed per glucose.

In pure culture, wild-type *C. thermocellum* makes a mixture of acetate and ethanol, with varying levels of formate and H_2 (Figure 10.3b). If formate is produced during the conversion of 2 pyruvate to 2 acetyl-CoA, then 1 acetyl-CoA can be converted to acetate, generating 1 ATP, while the electrons from glycolysis would be used to reduce the other acetyl-CoA to ethanol for a net production of 1 ATP. If instead, pyruvate were converted to acetyl-CoA via PFOR, the carbon would have the same fate, but Ech would pump additional 4 H^+ outside the cell, resulting in the formation of additional 1 ATP for a total of 2 ATP per glucose.

As strains are engineered to make a single biofuel compound such as ethanol, with the understanding that potential ATP yields from a homoethanologenic strain is necessary (Figure 10.3c). The most straightforward pathway would involve conversion of pyruvate to acetyl-CoA via PFOR, followed by reduction of acetyl-CoA to ethanol using NADH as the electron donor. Two NADH would be generated by glycolysis, and the reduced ferredoxin generated by PFOR would be used by Rnf to generate additional 2 NADH while also pumping out 4 H^+, resulting in the production of additional 1 ATP. Alternative pathways to ethanol yield no additional ATP. One would rely on a potential pyruvate decarboxylase (PDC) side activity of PFOR; some thermophilic PFORs are able to catalyze the PDC reaction [59, 60], although this has not been demonstrated in *C. thermocellum*. Alternatively, NfnAB could be used to take electrons from ferredoxin and NADH to make NADPH [50]. Then, if the ethanol production pathway is capable of utilizing NADPH as the electron donor, this would provide an additional possible path to ethanol formation without the additional production of ATP.

While the absolute number of ATP formed by *C. thermocellum* is not certain, and the number likely varies with growth conditions, we are able to put bounds on the likely number of ATP formed during fermentation of sugars derived from

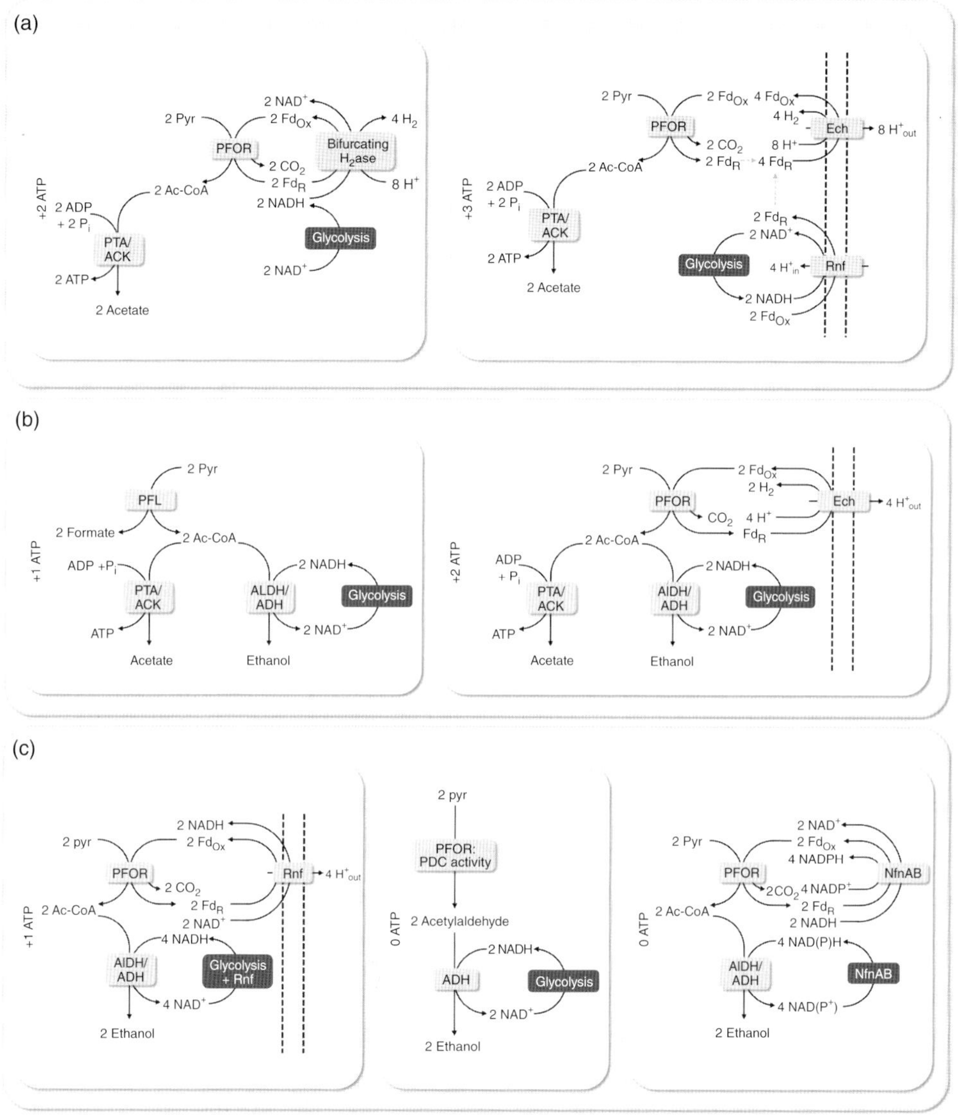

Figure 10.3 Possible pathways involved in conversion of pyruvate to end products. (a) Homoacetate fementation, (b) mixed acetate/ethanol fermentation, and (c) homoethanol fermentation *C. thermocellum*. ATP yields per hexose equivalent are provided. Pyr, pyruvate; Ac-CoA, acetyl-Coenzyme A; Fd, ferredoxin; PFL, pyruvate:formate lyase; PTA/ACK, phosphotransacetylase/acetate kinase; ALDH/ADH, aldehyde dehydrogenase/alcohol dehydrogenase; PFOR, pyruvate:ferredoxin oxidoreductase; Ech, energy converting hydrogenase; Rnf, ferredoxin:NAD^+ oxidoreductase; and NfnAB, NADH-dependent reduced ferredoxin:$NADP^+$ oxidoreductase.

cellulose. All pathways utilize a cellodextrin transporter and CD Pase, as well as likely utilizing a PP_i-dependent Pfk. The conversion of PEP to pyruvate likely utilizes both the malate shunt and PPDK. Therefore, the conversion of cellodextrins to pyruvate likely results in the production of 3.5 ATP per glucose, with a theoretical range of 2.5–5 ATP per glucose. On the basis of current understanding, conversion of pyruvate to ethanol or acetate as sole fermentation products would result in an additional 1 or 2 ATP per glucose, respectively. Thus, the expected total ATP yield on cellulose by *C. thermocellum* is 4.5 ATP per glucosyl monomer for conversion to ethanol and 5.5 ATP for conversion to acetate.

10.5 Metabolic Engineering

10.5.1 Transformation and Genetic Tool Development

While the primary focus of this section is application of metabolic engineering tools, the development of such tools is a critical undertaking for nonstandard host microbes such as *C. thermocellum* and *T. saccharolyticum*. Transformation of *T. saccharolyticum* was first reported by Mai and Wiegel [61] using kanamycin resistance as a selection. A second selectable marker conferring erythromycin resistance has also been used [62], and two counterselectable markers, *pyrF* and *pta-ack*, have also been developed [63]. Subsequently, it was found that this organism and many other members of the *Thermoanaerobacter* and *Thermoanaerobacterium* genera are naturally competent [64].

Reliable transformation of *C. thermocellum* was only realized in the 2007 timeframe, and was – and remains – more difficult than *T. saccharolyticum* transformation. In *C. thermocellum*, the development of selectable [65, 66] and counterselectable [66, 67] markers allowed for the first demonstrations of gene deletion and rational strain engineering. More recent work has focused on expanding the available tools and making them easier to implement. Increases in transformation efficiency have made transformation more consistent and increased the number of plasmid origins available for use [68]. Temperature-sensitive origins of replication have been developed [69], as have retrotransposon-based gene disruption technologies [70]. The tools needed for heterologous expression are also improving, with a variety of newly characterized promoters [71]. Together, these tools are allowing rapid advances in *C. thermocellum* metabolic engineering.

10.5.2 Ethanol Tolerance and Titer

For ethanol, or any other fermentation product, there are limiting concentrations at which metabolic functions cease to be operative. Because the limiting concentration for catabolism is often greater than for anabolism, it is possible – and indeed common in industrial practice – for titers to be achieved that exceed the upper limit for growth. The upper limits for both growth and product formation are different for different biological taxa and strains, and are often highly amenable to being increased by both evolutionary and targeted approaches. The maximum product concentration a microbe is able to produce is often less than the maximum it can tolerate (for growth or fermentation), a situation referred to as a titer gap [4]. Closing the titer gap, which is a common component of industrial microorganism development, has often been achieved although typically with substantial effort (see, e.g., the case of propanediol [72]), and generally has to be approached on a case-by-case basis.

Wild type *C. thermocellum* can initiate growth in the presence of approximately $15\,g\,l^{-1}$ ethanol [73]. However, *C. thermocellum* strains able to grow in the presence of $\geq 50\,g\,l^{-1}$ ethanol, including on cellulose, have been repeatedly obtained by serial transfer in the presence of increasing concentrations [74–76]. Sequencing of three such strains revealed mutations in the bifunctional alcohol and aldehyde dehydrogenase (ALDH) gene, *adhE* [73, 75]. Replacement of wild-type *adhE* with the mutant *adhE** allele was sufficient to triple the maximum concentration of ethanol tolerated in *C. thermocellum*. *Clostridium phytofermentans* also acquired mutations in its *adhE* gene after adaptation for ethanol tolerance [77]. Wild-type strains of thermophilic anaerobic bacteria typically exhibit ethanol inhibition at concentrations several-fold below the maximum amount of added ethanol that selected strains are able to tolerate. Available evidence suggests that such inhibition is due to metabolic factors rather than biophysical factors such as the loss of membrane integrity. If and when metabolic limitations are removed, we expect that biophysical factors will become more important limiting factors as has been observed for high ethanol tolerant microbes in the literature [78, 79].

Thermoanaerobacter species have been shown to tolerate up to $68\,g\,l^{-1}$ ethanol at temperatures of up to 70 °C [80, 81]. To date, the highest reported ethanol titer produced by a thermophilic bacterium is $70\,g\,l^{-1}$ for an engineered and selected strain of *T. saccharolyticum* [82]. By contrast, the highest concentrations of ethanol produced by *C. thermocellum* in our lab or other labs is, to our knowledge is $<30\,g\,l^{-1}$, although higher concentrations have been produced from cellulose in co-cultures that include *C. thermocellum* [67].

An ethanol titer $\geq 40\,g\,l^{-1}$ is generally thought to be necessary in order to avoid high costs and distillation energy requirements [83, 84]. As discussed in more detail elsewhere [2], it is difficult to make more than 5 or 6 wt% ethanol owing to the mechanical properties of lignocellulose-water mixtures. For example, converting corn stover at an initial concentration of 20 wt% and relatively high solubilization and fermentation yields results in about 5 wt% ethanol. Solids loadings higher than 20 wt% entail progressively greater materials handling and mixing challenges, the cost of which may not be fully captured in models for processes that have not yet been run at scale.

On the basis of these considerations, the feasible window for stand-alone facilities processing lignocellulose to ethanol is bounded on the low side by distillation energy requirements and on the high side by the difficulty of handling biomass slurries at high concentrations. The ethanol tolerance of evolved strains of *C. thermocellum* would appear to be compatible with this window. As presented in more detail below (Section 10.5.3), the maximum reported concentrations of produced ethanol are at the upper bound of this window for *T. saccharolyticum* but have not yet reached the lower bound for *C. thermocellum*. While not rigorously documented, it appears to us that the titer gap is largely closed for *T. saccharolyticum* but not for *C. thermocellum*. Thermophilic ethanol production in a "bolt-on" configuration, for example, colocated with a facility producing ethanol from sugarcane or corn, may be feasible with titers of produced ethanol that are lower than required for stand-alone production.

Notwithstanding recent progress, it remains to be demonstrated that thermophiles can produce economically viable titers under industrial conditions – for example, an affordable lignocellulosic feedstock, inexpensive growth media, and in the presence of potential inhibitors. It has been shown that thermophiles can be adapted to manifest increased resistance to inhibitors other than ethanol, in particular in the case of by-products resulting from *Populus* pretreatment [85]. We speculate that it may be difficult and perhaps not feasible to develop strains of thermophiles that are as resistant to chemical inhibition as yeast, and that the merits of thermophiles may be best realized in the context of processes that minimize formation of inhibitors in the first place.

10.5.3
Metabolic Engineering for High Ethanol Yield

There are theoretical and practical considerations that limit the yield of ethanol or any product. Since carbohydrate has four available electrons per carbon and ethanol has six, conservation of available electrons places an upper limit on the ethanol yield of 0.67 mol of ethanol carbon per mole of carbohydrate carbon, corresponding to 2 mol of ethanol per mole of hexose or 5/3 mol of ethanol per mole of pentose. Yields are expressed as a percentage of this limit, which is independent of the pathway used. Exceeding the stoichiometric limit based on conservation of available electrons can only occur if a source of reducing power is metabolized in addition to carbohydrate [86, 87]. Actual product yields are generally less than the theoretical limit due to carbon and electrons originating from the feedstock being diverted to synthesis of cells and secondary metabolic products. Yields of ethanol based on fermented sugar ≥90% of the theoretical yield are realized in industrial production based on yeast, and this is a reasonable standard against which to compare performance of other microbes.

A prominent early success in engineering thermophiles for high ethanol yield was realized with *T. saccharolyticum*, for which ethanol was produced at 90% of the maximum theoretical yield at a titer of 33 $g\,l^{-1}$ [62] with further improvement to 54 $g\,l^{-1}$ [88] and eventually 70 $g\,l^{-1}$ [82]. Metabolic engineering of this strain was achieved by first eliminating lactate production by deletion of *ldh* and then eliminating acetate production by deleting *pta-ack* resulting in strain ALK2, as previously reviewed in detail [6]. Two separate attempts to apply a similar strategy to the metabolic engineering of *C. thermocellum* followed by strain evolution both met with limited success [67, 89]. While the *ldh* and *pta* deletions resulted in an ethanol yield of 90% of the theoretical maximum in *T. saccharolyticum*, the same mutations increased ethanol production to only 60% of the theoretical maximum in *C. thermocellum* [67]. Contrasting the ease of metabolic engineering of *T. saccharolyticum* for high ethanol yield with the difficulty of engineering *C. thermocellum* for the same purpose has inspired recent investigations into the metabolism of both organisms.

10.5.3.1
Metabolic Engineering of *T. saccharolyticum*

In order to allow for more-advanced engineering of *T. saccharolyticum*, the *ldh-pta*-ALK2 strain from Shaw *et al.* [62] was reconstructed using a maker-removal strategy to create an *ldh-pta* strain that does not contain antibiotic resistance genes in the chromosome [63]. Here the order of genetic modifications was different than that of the ALK2 strain. First the *pta-ack* operon was deleted to eliminate acetate production, and then the *ldh* gene was deleted to eliminate lactate production. The deletion of *pta-ack* (to create strain M0350) resulted in an ethanol yield of 56%, which is not substantially different from the wild-type ethanol yield of 58%. Interestingly, the subsequent reintroduction of the *pta-ack* operon (to use as a counterselectable marker for the purpose of deleting the *ldh* gene) *increased* ethanol yield from 56% to 70%. A final redeletion of the *pta-ack* operon, resulting in strain M0355 (Δ*pta-ack* Δ*ldh*), increased ethanol yield from 70% to 77%. Subsequent reintroduction of the *pta-ack* operon did not result in increased acetate production [90]. The varied effect of *pta-ack* deletions between strains ALK2, M0350, and M0355 suggests that the observed changes in ethanol production are not due to *pta-ack* itself, but rather secondary mutation(s).

This possibility was investigated by resequencing the genomes of six strains derived from M0355 via evolution targeting improved fermentation fitness [90]. In all cases, mutations were repeatedly observed in the subunits of the *hfs* hydrogenase operon and the bifunctional aldehyde and alcohol dehydrogenase (ADH), *adhE*. The *hfs* and *adhE* mutations from one strain, which had a point mutation in *adhE*, a frameshift in *hfsB*, and a point mutation in *hfsD*, was selected for further analysis. Reintroduction of the *adhE* mutation into wild-type *T. saccharolyticum* did not change the ethanol yield. Reintroduction of these *hfs* mutations, on the other hand, increased ethanol yield from 60% to 77% of the theoretical maximum. This would seem to suggest that the *hfs* mutation is important for achieving high ethanol yields while the *adhE* mutation is not; however, other studies have revealed a more complicated picture (see below). All of the strains derived from M0355 showed ethanol yield of >90% of the theoretical maximum. This result appears to be due to a combination of targeted genetic modifications and spontaneous modifications. We are of the opinion that mutations in *hfs* and *adhE* are some of the key modifications, but have not yet confirmed this definitively.

10.5.3.2
Hydrogenases

T. saccharolyticum encodes three hydrogenases, including a membrane-bound Ech hydrogenase (*ech*), the four subunit Hfs hydrogenase, and a cluster of eight genes that encode five hydrogenase subunits (*hyd*). The roles of these hydrogenases in hydrogen formation and ethanol production were analyzed by gene-deletion experiments, both individually and in combination in an otherwise wild-type genetic background [91]. Deletion of *ech* or *hyd* resulted in no change in H_2 production, but deletion of *hfs* resulted in a 25-fold decrease in H_2 production, with concomitant increase in carbon flux to lactate, suggesting that *hfs* is the primary H_2-producing hydrogenase. Interestingly, none of the single mutations had a large effect on ethanol production, and in fact, the *hfs* deletion strain decreased ethanol production slightly, in contrast to the phenotype of the *hfsB* and *hfsD* mutations above. To eliminate lactate production, *ldh* was deleted in the *hfs* deletion strain, resulting in a 44% increase in ethanol yield but incomplete utilization of the growth substrate. None of the strains made more than 67% of the theoretical maximum ethanol yield. One hypothesis to explain the different results obtained from different *hfs* modifications is that perhaps the *hfs* point mutations are still partially active, and small amounts of H_2 need to be produced to balance the excess reducing equivalents generated during production of microbial biomass [92]. This phenomenon is observed when *S. cerevisiae* ferments glucose to ethanol, where glycerol (which could serve the same metabolic purpose as H_2) is also produced [93]. Further studies will be needed to determine the role of *hfs* mutations in *T. saccharolyticum*.

10.5.3.3
The Pyruvate to Ethanol Pathway in *T. saccharolyticum*

Similar to *C. thermocellum*, the general reactions that convert pyruvate to ethanol in *T. saccharolyticum* are known. Generally, pyruvate is reductively decarboxylated to acetyl-CoA, followed by reduction to acetaldehyde and then reduction to ethanol. However, the specific genes responsible for each reaction and their cofactor specificity were in general not known until recently, and are still not completely described. Further, the pathway(s) used by different engineered strains may not always be the same. To transfer the ethanol production capabilities from *T. saccharolyticum* to *C. thermocellum*, it will be necessary to understand the gene-enzyme-activity relationships in the *T. saccharolyticum* pathway.

In *T. saccharolyticum*, there are two possible pathways from pyruvate to acetyl-CoA, pyruvate formate lyase (PFL: pyruvate + CoA → acetyl-CoA + formate) and pyruvate ferredoxin oxidoreductase (PFOR: pyruvate + Fd_{ox} + CoA → acetyl-CoA + CO_2 + Fd_{red}). The Tsac_0628 and Tsac_0629 genes encode *pfl* and its activating enzyme [94]. The Tsac_0046 gene encodes the primary *pfor*. Wild-type *T. saccharolyticum* does not produce formate as a fermentation end product, so the PFOR reaction is the primary route for acetyl-CoA generation. Increased flux through the PFL reaction can compensate to a degree in strains where the PFOR reaction has been eliminated. Inactivation of both pathways resulted in a strain that produced only lactate, confirming that those are the only routes for acetyl-CoA generation in *T. saccharolyticum* and that ethanol production is dependent on this pathway [94].

Conversion of acetyl-CoA to ethanol is a two-step process: first acetyl-CoA is reduced to acetaldehyde with NADH or NADPH via the ALDH reaction, then acetaldehyde is further reduced to ethanol with NADH or NADPH via the ADH reaction. Both of these reactions can be catalyzed by the bifunctional AdhE protein. The AdhE protein from *T. saccharolyticum* was expressed in *E. coli*, purified, and characterized and found to have predominantly NADH-linked activity for each reaction. Interestingly, in two high ethanol-yielding strains of *T. saccharolyticum* (ALK2 and M1442), the cofactor specificity for both reactions had switched from NADH to NADPH [95]. Deletion of the *adhE* gene completely eliminated ALDH activity, but a substantial amount of NADPH-linked ADH activity remained, suggesting that multiple genes encode ADH activity [96].

In order for ethanol to be produced at high yield, electrons from the Fd_{red} generated by the PFOR reaction must be transferred to NAD^+ or $NADP^+$, depending on the cofactor specificity of the ALDH and ADH reactions. The general transfer of electrons from ferredoxin to nicotinamide cofactors is known as ferredoxin-NAD(P) oxidoreductase (FNOR) activity. In some organisms, this conversion is coupled with transhydrogenation of NADH to NADPH via the NFN reaction[2] ($NADH + Fd_{red} + 2\ NADP^+ \rightarrow NAD^+ + Fd_{ox} + 2\ NADPH$) [50]. The Tsac_2086 and Tsac_2085 genes code for proteins with >60% amino acid identity to known NfnAB proteins. To explore their role in metabolism, the *nfnAB* genes were deleted and overexpressed in several strains of *T. saccharolyticum* [97]. In the wild-type strain, deletion of *nfnAB* had no effect on ethanol production. By contrast, in the high ethanol-yielding strain M1442, deletion of *nfnAB* reduced ethanol yield from 84% to 28%, and complementation of *nfnAB* restored ethanol yield to 69%, thus demonstrating its role in ethanol production in strain M1442.

In another high ethanol-yielding strain, deletion of *nfnAB increased* ethanol yield from 76% to 86%. The disparate results of *nfnAB* deletion (M0353 vs M1442) can be explained by coordination of cofactor specificity between FNOR, ALDH, and ADH reactions (Table 10.2). In strain M0353, the FNOR, ALDH, and ADH reactions are all predominantly NADH linked. In strain M1442, the same reactions are all NADPH linked. This suggests that wild-type *T. saccharolyticum* has the potential for high yield ethanol production via two distinct pathways, one that is NADPH linked (Figure 10.4a) and the other that is NADH linked (Figure 10.4b) [97].

10.5.3.4
Engineering *C. thermocellum* and Comparison with *T. saccharolyticum*

Although there are a number of notable differences between the metabolic networks of *C. thermocellum* and *T. saccharolyticum* [56], the potential pathways for conversion of pyruvate to ethanol is similar in both organisms. Both have PFOR

2) In our nomenclature, NFN is a specific type of FNOR reaction.

Table 10.2 Genes associated with reactions in the *T. saccharolyticum* pyruvate-to-ethanol pathway.

Reaction	Gene	References
NADPH-linked pathway		
PFOR	Tsac_0046	[94]
NFN	Tsac_2086, Tsac_2085	[97]
NADPH-ALDH	Tsac_0416 (with G544D mutation)	[95]
NADPH-ADH	Tsac_0416 (with G544D mutation)	[95]
NADH-linked pathway		
PFOR	Tsac_0046	[94]
NADH-FNOR	(Unknown)	—
NADH-ALDH	Tsac_0416 (wild type)	[97]
NADH-ADH	Tsac_0416 (wild type)	[97]

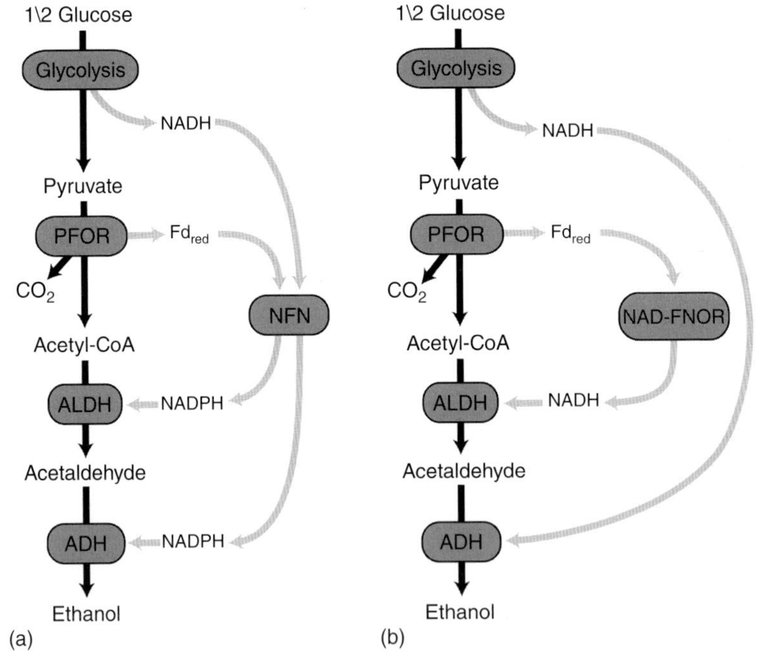

Figure 10.4 Ethanol production pathways in *T. saccharolyticum*. (a) The NADPH-linked pathway. (b) The NADH-linked pathway. Carbon fluxes are shown in black, electron fluxes are shown in gray. For simplicity, only the reduced form of electron carriers are shown (i.e., Fd_{red}, NADH, and NADPH are shown, but Fd_{ox}, NAD^+, and $NADP^+$ are not shown). (This figure is adapted from Lo *et al.* [75].)

(Tsac_0046 and Clo1313_0673), NFN (Tsac_2086, Tsac_2085, Clo1313_1848, and Clo1313_1849), and a bifunctional AdhE protein that performs the ALDH and ADH reactions (Tsac_0416, Clo1313_1798). Despite these similarities, these two organisms responded very differently to knockout of genes responsible for acetate and lactate formation, with higher ethanol yields achieved for *T. saccharolyticum* as described earlier.

One possible explanation for this different response involves cofactor balancing. In order to use the NFN reaction for ethanol production, the organism needs a supply of NADH and Fd_{red}, as well as ALDH and ADH reactions that can use the NADPH that is generated. In *C. thermocellum*, there are potential problems with both the supply and demand sides. The "malate shunt" that *C. thermocellum* uses to convert PEP to pyruvate also results in the net conversion of NADH to NADPH [55, 57]. Thus when *C. thermocellum* uses the "malate shunt" to convert PEP to pyruvate and then uses the PFOR reaction to convert pyruvate to acetyl-CoA, it generates NADPH and Fd_{red} instead of the NADH and Fd_{red} needed by the NFN reaction. Even if the "malate shunt" were eliminated (which has been done [57]), *C. thermocellum* would not be able to use the resulting NADPH generated by the NFN reaction because it does not have an NADPH-linked ALDH reaction [95]. Therefore, transferring the *T. saccharolyticum* ethanol production pathway to *C. thermocellum* may require engineering both the NADH supply (i.e., eliminating the malate shunt) and the NADPH demand (i.e., engineering the ALDH reaction to use NADPH).

Another possible explanation is the hydrogenase system. In *T. saccharolyticum*, adaptation for improved growth after elimination of acetate production resulted in mutations in both *adhE* and the *hfs* hydrogenases. In *C. thermocellum*, the same treatment resulted in *adhE* mutations, but no mutations in the hydrogenases (Holwerda *et al.*, unpublished). It is possible that there are properties of the *hfs* hydrogenase system in *T. saccharolyticum* that make it uniquely suited for high-yield ethanol production. If this is true, part or all of this system may need to be transferred to *C. thermocellum*.

A further complication is that while wild-type *T. saccharolyticum* produces ethanol at yields >50% of the theoretical maximum, wild-type *C. thermocellum* typically produces ethanol at yields closer to 25–30% of the theoretical maximum, with the excess carbon and electrons going to nontraditional fermentation products such as secreted amino acids; organic acids such as pyruvate, fumarate, and malate, and alcohols such as isobutanol and butanediol [30, 31]. This may reflect a fundamental difference in either central metabolism or metabolic control between these two organisms.

Despite the uncertainties in *C. thermocellum* carbon and electron flux detailed above and differences compared to *T. saccharolyticum*, pathway engineering has further increased ethanol yield in *C. thermocellum*. Early approaches described above included deletion of *pta* [66], deletion of *ldh* in combination with *pta* [67], and heterologous expression of a pyruvate kinase and deletion of malic enzyme [57]. More recently, deletion of *pfl* eliminated formate production [98], while deletion of the [FeFe] hydrogenase maturase HydG along with [NiFe] hydrogenase

Table 10.3 Current state of strain development of *C. thermocellum* and *T. saccharolyticum*.

Metric, strain	Values	Conditions	Source
Ethanol yields (engineered strains), % theoretical			
T. saccharolyticum M1442	90	Several feedstocks, ethanol titers $\geq 50\,g\,l^{-1}$, inhibitors present	[73]
C. thermocellum LL1210	75	Titer $25\,g\,l^{-1}$, no inhibitors	a)
Ethanol titer (selected strains) ($g\,l^{-1}$)			
T. saccharolyticum M1442	70	Produced from maltodextrins and cellobiose in 90 h	[73]
C. thermocellum LL1210	25	Produced from Avicel in 122 h	a)
C. thermocellum E50C and E50A	50	Added, tolerated for growth on Avicel	[75]

a) [102]

Ech eliminated H_2 production and greatly increased ethanol yield [99]. Stacking deletions of *hydG, ldh, pfl,* and *pta-ack* eliminated organic acid synthesis and greatly reduced H_2 production, which resulted in an increase in ethanol yield to approximately 70% of the theoretical maximum [100]. More recently, this strain has been evolved via serial transfer, resulting in strain LL1210 that can produce $22.4 \pm 1.4\,g\,l^{-1}$ ethanol at a yield of 75% of the theoretical maximum (L. Tian *et al.*, in preparation). Prospects for further strain improvement are also bright. In addition to the heterologous expression of the *T. saccharolyticum* pathways described above, approaches such as further strain evolution, modification of nitrogen assimilation and amino acid synthesis, and prevention of overflow metabolism are attractive targets for further improving ethanol yield.

10.5.3.5
Current State of Strain Development

The current state of strain development for ethanol production via *C. thermocellum* and *T. saccharolyticum* is summarized in Table 10.3 with reference to key performance metrics. It may be noted that solubilization data and fermentation of high substrate concentrations have been summarized in Table 10.1. Volumetric productivities (g ethanol $l^{-1}\,h^{-1}$) calculated from the data in Table 10.3 include $0.78\,g\,l^{-1}\,h^{-1}$ for *T. saccharolyticum* fermentation of mixed cellodextrins and $0.20\,g\,l^{-1}\,h^{-1}$ for *C. thermocellum* fermentation of Avicel. Similarly to the wild type, *C. thermocellum* mutant strains still secrete amino acids into the culture medium, providing a target for further increasing ethanol yields and titers.

10.6
Summary and Future Directions

To date, the overwhelming balance of effort and industrial application in the metabolic engineering field has been based on "standard" microbial hosts, that is, hosts such as *E. coli* or certain yeasts from the *Saccharomyces* or *Pichia* genera for which metabolic understanding and genetic tools are highly developed. Metabolic engineering of nonstandard hosts involves challenges that are distinctive and formidable compared to the "chassis organism" approach that dominates the synthetic biology field today. We speculate in Section 10.2 that CBP organism development is more promising starting with naturally occurring cellulolytic microbes as compared to starting with noncellulolytic hosts. If this proves to be correct, CBP will provide an important early example of a phenotype that is difficult to transfer and fully replicate in a "chassis" organism – in this case, lignocellulose solubilization – providing sufficient motivation to address the added complexities of engineering nonstandard hosts. Time will tell.

The key aspects of the status of thermophilic bacteria relevant to CBP, addressed in detail in the preceding sections, include the following:

- *Solubilization* (Section 10.3). An extensive recent body of work spanning a broad range of feedstocks and conditions supports the proposition that *C. thermocellum* is considerably more effective at achieving high solubilization yields from cellulosic biomass than industry-standard fungal cellulase. The relative effectiveness of *C. thermocellum* compared to fungal cellulase increases with increasing feedstock recalcitrance. A modest decrease in fractional solubilization yield is observed in work to date involving high (e.g., $100\,g\,l^{-1}$) initial feedstock concentrations.
- *Bioenergetics* (Section 10.4). The primary metabolism of *C. thermocellum* has many distinctive features compared to other fermentative anaerobes. Energy conservation mechanisms in addition to those known 10 years ago are operative in *C. thermocellum*. Although quantitative understanding of the ATP yield in this organism, and other cellulolytic anaerobes, is not complete, it is clear that cellulose-specific bioenergetics benefits are larger than the bioenergetic cost of cellulase synthesis. Thus there are no evident bioenergetic barriers to anaerobic conversion of cellulose to ethanol and potentially other fuels without exogenous enzymes.
- *Metabolic engineering* (Section 10.5). Over the last 5 years, metabolic engineering tools have matured to the point that targeted manipulation of *C. thermocellum* and *T. saccharolyticum* is possible, although more time and effort are required than would be required for the same manipulations in standard hosts. Near-theoretical ethanol yields and commercially viable titers of produced ethanol have been achieved in *T. saccharolyticum*. Ethanol yield and titers in *C. thermocellum* have been substantially increased of late, but are not yet equal to those of *T. saccharolyticum*. The genes responsible for the PFOR-dependent ethanol production pathway of *T. saccharolyticum* have recently been characterized, and work to transfer this pathway to *C. thermocellum* is underway. Available data indicate that ethanol inhibition is metabolic rather than biophysical.

Notwithstanding the substantial progress made, there are still important challenges to be overcome in order for thermophiles to be applied commercially for ethanol production via CBP. The solution of some of these challenges will benefit greatly from new fundamental understanding and application of systems biology, while others will rely more on empiricism. While *C. thermocellum* is capable of converting some industrially interesting feedstocks – for example, paper sludge and corn fiber – without pretreatment, it appears that most lignocellulosic feedstocks will require some form of nonbiological processing in order for *C. thermocellum* to achieve high fractional solubilization yields. Thermochemical pretreatment has been the standard answer to this dilemma. However, we regard mechanical disruption during fermentation (cotreatment) as a promising alternative to be investigated in the future [29], particularly in light of the resistance of gram positive bacteria to mechanical disruption as well as the example provided by the ruminating cow [101]. Continuous processing offers many advantages over batch processing, and is a promising direction for future work. Industrial growth media and culture management protocols will also need to be established.

Further improvements in the yield and titer produced by *C. thermocellum* are desirable and likely necessary for many applications. Progress will be fostered by further improvements in genetic tools, both for strains that are genetically tractable as well as strains that are currently intractable. Understanding and development of strains pursuant to robustness under industrial conditions is an important area for future work that is not widely addressed in the literature and has great opportunities for application of genomic science.

Initial commercial application of thermophilic ethanol production via CBP is most likely and most economically advantageous as a bolt-on to an existing ethanol mill – for example, processing corn, sugar cane, or lignocellulose. Such application would be a significant milestone for thermophiles and use of nonstandard host microbes, and appears to be a possibility in the near future.

Acknowledgments

This work was supported by the BioEnergy Science Center (BESC) and Enchi Corporation. BESC is a US DOE Bioenergy Research Center supported by the Office of Biological and Environmental Research in the DOE Office of Science. Oak Ridge National Laboratory is managed by UT-Battelle, LLC, for the US DOE under contract DE-AC05-00OR22725.

References

1. Lynd, L.R., Weimer, P.J., van Zyl, W.H., and Pretorius, I.S. (2002) Microbial cellulose utilization: fundamentals and biotechnology. *Microbiol. Mol. Biol. Rev.*, **66** (3), 506–577.
2. Lynd, L.R. (1996) Overview and evaluation of fuel ethanol from cellulosic biomass: technology, economics, the environment, and policy. *Annu. Rev. Energy Env.*, **21** (1), 403–465.
3. Klein-Marcuschamer, D., Oleskowicz-Popiel, P., Simmons, B.A., and Blanch, H.W. (2012) The challenge of enzyme cost in the production of lignocellulosic biofuels. *Biotechnol. Bioeng.*, **109** (4), 1083–1087.

4. Olson, D.G., McBride, J.E., Shaw, A.J., and Lynd, L.R. (2012) Recent progress in consolidated bioprocessing. *Curr. Opin. Biotechnol.*, **23** (3), 396–405.
5. Blumer-Schuette, S.E., Brown, S.D., Sander, K.B., Bayer, E.A., Kataeva, I., Zurawski, J.V., Conway, J.M., Adams, M.W.W., and Kelly, R.M. (2014) Thermophilic lignocellulose deconstruction. *FEMS Microbiol. Rev.*, **38** (3), 393–448.
6. Olson, D.G., Sparling, R., and Lynd, L.R. (2015) Ethanol production by engineered thermophiles. *Curr. Opin. Biotechnol.*, **33**, 130–141.
7. Bayer, E.A., Belaich, J.P., Shoham, Y., and Lamed, R. (2004) The cellulosomes: multienzyme machines for degradation of plant cell wall polysaccharides. *Annu. Rev. Microbiol.*, **58**, 521–554.
8. Decker, S.R., Sheehan, J., Dayton, D.C., Bozell, J.J., Adney, W.S., Hames, B., Thomas, S.R., Bain, R.L., Czernik, S., Zhang, M. *et al* (2007) in *Kent and Riegel's Handbook of Industrial Chemistry and Biotechnology* (ed. J.A. Kent), Springer, Boston, MA, pp. 1449–1548.
9. Singh, A., Taylor, L.E. II,, Vander Wall, T.A., Linger, J., Himmel, M.E., Podkaminer, K., Adney, W.S., and Decker, S.R. (2015) Heterologous protein expression in *Hypocrea jecorina*: a historical perspective and new developments. *Biotechnol. Adv.*, **33** (1), 142–154.
10. Lynd, L.R., Laser, M.S., Bransby, D., Dale, B.E., Davison, B., Hamilton, R., Himmel, M., Keller, M., McMillan, J.D., Sheehan, J. *et al* (2008) How biotech can transform biofuels. *Nat. Biotechnol.*, **26** (2), 169–172.
11. Hendriks, A.T. and Zeeman, G. (2009) Pretreatments to enhance the digestibility of lignocellulosic biomass. *Bioresour. Technol.*, **100** (1), 10–18.
12. Himmel, M.E., Ding, S.Y., Johnson, D.K., Adney, W.S., Nimlos, M.R., Brady, J.W., and Foust, T.D. (2007) Biomass recalcitrance: engineering plants and enzymes for biofuels production. *Science*, **315** (5813), 804–807.
13. Meng, X. and Ragauskas, A.J. (2014) Recent advances in understanding the role of cellulose accessibility in enzymatic hydrolysis of lignocellulosic substrates. *Curr. Opin. Biotechnol.*, **27**, 150–158.
14. Kumar, R. and Wyman, C.E. (2013) *Aqueous Pretreatment of Plant Biomass for Biological and Chemical Conversion to Fuels and Chemicals*, John Wiley & Sons, Ltd, pp. 281–310.
15. South, C.R., Hogsett, D.A.L., and Lynd, L.R. (1995) Modeling simultaneous saccharification and fermentation of lignocellulose to ethanol in batch and continuous reactors. *Enzyme Microb. Technol.*, **17** (9), 797–803.
16. Shao, X., Lynd, L., Wyman, C., and Bakker, A. (2009) Kinetic modeling of cellulosic biomass to ethanol via simultaneous saccharification and fermentation: part I. Accommodation of intermittent feeding and analysis of staged reactors. *Biotechnol. Bioeng.*, **102** (1), 59–65.
17. Zhang, J., Shao, X., Townsend, O.V., and Lynd, L.R. (2009) Simultaneous saccharification and co-fermentation of paper sludge to ethanol by *Saccharomyces cerevisiae* RWB222—Part I: kinetic modeling and parameters. *Biotechnol. Bioeng.*, **104** (5), 920–931.
18. Zhang, Y.H. and Lynd, L.R. (2004) Kinetics and relative importance of phosphorolytic and hydrolytic cleavage of cellodextrins and cellobiose in cell extracts of *Clostridium thermocellum*. *Appl. Environ. Microbiol.*, **70** (3), 1563–1569.
19. Kristensen, J.B., Felby, C., and Jørgensen, H. (2009) Yield-determining factors in high-solids enzymatic hydrolysis of lignocellulose. *Biotechnol. Biofuels*, **2** (1), 1–10.
20. Lavenson, D.M., Tozzi, E.J., Karuna, N., Jeoh, T., Powell, R.L., and McCarthy, M.J. (2012) The effect of mixing on the liquefaction and saccharification of cellulosic fibers. *Bioresour. Technol.*, **111**, 240–247.
21. Palmqvist, E. and Hahn-Hägerdal, B. (2000) Fermentation of lignocellulosic hydrolysates. II: inhibitors and mechanisms of inhibition. *Bioresour. Technol.*, **74** (1), 25–33.
22. Roberts, K., Lavenson, D., Tozzi, E., McCarthy, M., and Jeoh, T. (2011) The

effects of water interactions in cellulose suspensions on mass transfer and saccharification efficiency at high solids loadings. *Cellulose*, **18** (3), 759–773.
23. Thygesen, L.G., Thybring, E.E., Johansen, K.S., and Felby, C. (2014) The mechanisms of plant cell wall deconstruction during enzymatic hydrolysis. *PLoS One*, **9** (9), e108313.
24. Holwerda, E.K. and Lynd, L.R. (2013) Testing alternative kinetic models for utilization of crystalline cellulose (Avicel) by batch cultures of *Clostridium thermocellum*. *Biotechnol. Bioeng.*, **110** (9), 2389–2394.
25. Holwerda, E.K., Ellis, L.D., and Lynd, L.R. (2013) Development and evaluation of methods to infer biosynthesis and substrate consumption in cultures of cellulolytic microorganisms. *Biotechnol. Bioeng.*, **110** (9), 2380–2388.
26. Jensen, P.D., Hardin, M.T., and Clarke, W.P. (2008) Measurement and quantification of sessile and planktonic microbial populations during the anaerobic digestion of cellulose. *Water Sci. Technol.*, **57** (4), 465–469.
27. Tang, H., Ou, J.F., and Zhu, M.J. (2015) Development of a quantitative real-time PCR assay for direct detection of growth of cellulose-degrading bacterium *Clostridium thermocellum* in lignocellulosic degradation. *J. Appl. Microbiol.*, **118** (6), 1333–1344.
28. Zhang, Y. and Lynd, L.R. (2003) Quantification of cell and cellulase mass concentrations during anaerobic cellulose fermentation: development of an enzyme-linked immunosorbent assay-based method with application to *Clostridium thermocellum* batch cultures. *Anal. Chem.*, **75** (2), 219–227.
29. Paye, J.M., Guseva, A., Hammer, S.K., Gjersing, E., Davis, M.F., Davison, B.H., Olstad, J., Donohoe, B.S., Nguyen, T.Y., Wyman, C.E., Pattathil, S., Hahn, M.G., Lynd, L.R. (2016) Biological lignocellulose solubilization: comparative evaluation of biocatalysts and enhancement via cotreatment. *Biotechnol Biofuels.*, **12**, 2907–2915. doi: 10.1186/s13068-015-0412-y.
30. Ellis, L.D., Holwerda, E.K., Hogsett, D., Rogers, S., Shao, X., Tschaplinski, T., Thorne, P., and Lynd, L.R. (2012) Closing the carbon balance for fermentation by *Clostridium thermocellum* (ATCC 27405). *Bioresour. Technol.*, **103** (1), 293–299.
31. Holwerda, E.K., Thorne, P.G., Olson, D.G., Amador-Noguez, D., Engle, N.L., Tschaplinski, T.J., van Dijken, J.P., and Lynd, L.R. (2014) The exometabolome of *Clostridium thermocellum* reveals overflow metabolism at high cellulose loading. *Biotechnol. Biofuels*, **7** (1), 155.
32. Shao, X., DiMarco, K., Richard, T.L., and Lynd, L.R. (2015) Winter rye as a bioenergy feedstock: impact of crop maturity on composition, biological solubilization and potential revenue. *Biotechnol. Biofuels*, **8** (1), 1–10.
33. Himmel, M.E., Adney, W.S., Baker, J.O., Elander, R., McMillan, J.D., Nieves, R.A., Sheehan, J.J., Thomas, S.R., Vinzant, T.B., and Zhang, M. (1997) *Fuels and Chemicals from Biomass*, vol. **666**, American Chemical Society, pp. 2–45.
34. Donohoe, B.S. and Resch, M.G. (2015) Mechanisms employed by cellulase systems to gain access through the complex architecture of lignocellulosic substrates. *Curr. Opin. Chem. Biol.*, **29**, 100–107.
35. Xu, Q., Resch, M.G., Podkaminer, K., Yang, S., Baker, J.O., Donohoe, B.S., Wilson, C., Klingeman, D.M., Olson, D.G., Decker, S.R., Giannone, R.J., Hettch, R.L., Brown, S.D., Lynd, L.R., Bayer, E., Himmel, M.E., Bomble, Y. (2016) Dramatic performance of Clostridium thermocellum explained by its wide range of cellulase modalities Sci. Adv. Vol. 2 e1501254 DOI: 10.1126/sciadv.1501254.
36. Zhang, Y.H. and Lynd, L.R. (2005) Cellulose utilization by *Clostridium thermocellum*: bioenergetics and hydrolysis product assimilation. *Proc. Natl. Acad. Sci. U.S.A.*, **102** (20), 7321–7325.
37. Sparling, R., Carere, C., Rydzak, T., Schellenberg, J., and Levin, D.B. (2012) Thermodynamic and biochemical aspect of hydrogen production by dark fermentation, in *State of the Art and*

Progress in Production of Biohydrogen (eds N. Azbar and D.B. Levin), Bentham Science Publishers.

38. Erecinska, M. and Wilson, D.F. (1982) Regulation of cellular energy metabolism. *J. Membr. Biol.*, **70** (1), 1–14.
39. Thauer, R.K., Jungermann, K., and Decker, K. (1977) Energy conservation in chemotrophic anaerobic bacteria. *Bacteriol. Rev.*, **41** (1), 100–180.
40. Feinberg, L., Foden, J., Barrett, T., Davenport, K.W., Bruce, D., Detter, C., Tapia, R., Han, C., Lapidus, A., Lucas, S. *et al* (2011) Complete genome sequence of the cellulolytic thermophile *Clostridium thermocellum* DSM1313. *J. Bacteriol.*, **193** (11), 2906–2907.
41. Schmidt, A.L. and Briskin, D.P. (1993) Energy transduction in tonoplast vesicles from red beet (*Beta vulgaris* L.) storage tissue: H+/substrate stoichiometries for the H(+)-ATPase and H(+)-PPase. *Arch. Biochem. Biophys.*, **301** (1), 165–173.
42. Sosa, A. and Celis, H. (1995) H^+/PPi stoichiometry of membrane-bound pyrophosphatase of *Rhodospirillum rubrum*. *Arch. Biochem. Biophys.*, **316** (1), 421–427.
43. Abrahams, J.P., Leslie, A.G., Lutter, R., and Walker, J.E. (1994) Structure at 2.8 A resolution of F1-ATPase from bovine heart mitochondria. *Nature*, **370** (6491), 621–628.
44. Boyer, P.D. (1993) The binding change mechanism for ATP synthase – some probabilities and possibilities. *Biochim. Biophys. Acta*, **1140** (3), 215–250.
45. Boyer, P.D. (1997) The ATP synthase – a splendid molecular machine. *Annu. Rev. Biochem.*, **66**, 717–749.
46. Nakamoto, R.K., Baylis Scanlon, J.A., and Al-Shawi, M.K. (2008) The rotary mechanism of the ATP synthase. *Arch. Biochem. Biophys.*, **476** (1), 43–50.
47. Cross, R.L. and Muller, V. (2004) The evolution of A-, F-, and V-type ATP synthases and ATPascs: reversals in function and changes in the H+/ATP coupling ratio. *FEBS Lett.*, **576** (1-2), 1–4.
48. Buckel, W. and Thauer, R.K. (2013) Energy conservation via electron bifurcating ferredoxin reduction and proton/Na + translocating ferredoxin oxidation. *Biochim. Biophys. Acta, Bioenerg.*, **1827** (2), 94–113.
49. Hackmann, T.J. and Firkins, J.L. (2015) Electron transport phosphorylation in rumen butyrivibrios: unprecedented ATP yield for glucose fermentation to butyrate. *Front. Microbiol.*, **6**.
50. Wang, S., Huang, H., Moll, J., and Thauer, R.K. (2010) NADP+ reduction with reduced ferredoxin and NADP+ reduction with NADH are coupled via an electron-bifurcating enzyme complex in *Clostridium kluyveri*. *J. Bacteriol.*, **192** (19), 5115–5123.
51. Schut, G.J. and Adams, M.W.W. (2009) The iron-hydrogenase of *Thermotoga maritima* utilizes ferredoxin and NADH synergistically: a new perspective on anaerobic hydrogen production. *J. Bacteriol.*, **191** (13), 4451–4457.
52. Wang, S., Huang, H., Kahnt, J., and Thauer, R.K. (2013) A reversible electron-bifurcating ferredoxin- and NAD-dependent [FeFe]-hydrogenase (HydABC) in *Moorella thermoacetica*. *J. Bacteriol.*, **195** (6), 1267–1275.
53. Nataf, Y., Yaron, S., Stahl, F., Lamed, R., Bayer, E.A., Scheper, T.H., Sonenshein, A.L., and Shoham, Y. (2009) Cellodextrin and laminaribiose ABC transporters in *Clostridium thermocellum*. *J. Bacteriol.*, **191** (1), 203–209.
54. Raman, B., McKeown, C.K., Rodriguez, M. Jr.,, Brown, S.D., and Mielenz, J.R. (2011) Transcriptomic analysis of *Clostridium thermocellum* ATCC 27405 cellulose fermentation. *BMC Microbiol.*, **11**, 134.
55. Rydzak, T., McQueen, P.D., Krokhin, O.V., Spicer, V., Ezzati, P., Dwivedi, R.C., Shamshurin, D., Levin, D.B., Wilkins, J.A., and Sparling, R. (2012) Proteomic analysis of *Clostridium thermocellum* core metabolism: relative protein expression profiles and growth phase-dependent changes in protein expression. *BMC Microbiol.*, **12**, 214.
56. Zhou, J., Olson, D.G., Argyros, D.A., Deng, Y., van Gulik, W.M., van Dijken, J.P., and Lynd, L.R. (2013) Atypical

glycolysis in *Clostridium thermocellum. Appl. Environ. Microbiol.*, **79** (9), 3000–3008.

57. Deng, Y., Olson, D.G., Zhou, J., Herring, C.D., Joe Shaw, A., and Lynd, L.R. (2013) Redirecting carbon flux through exogenous pyruvate kinase to achieve high ethanol yields in *Clostridium thermocellum. Metab. Eng.*, **15**, 151–158.
58. Weimer, P.J. and Zeikus, J.G. (1977) Fermentation of cellulose and cellobiose by *Clostridium thermocellum* in the absence of *Methanobacterium thermoautotrophicum. Appl. Environ. Microbiol.*, **33** (2), 289–297.
59. Eram, M.S., Wong, A., Oduaran, E., and Ma, K. (2015) Molecular and biochemical characterization of bifunctional pyruvate decarboxylases and pyruvate ferredoxin oxidoreductases from *Thermotoga maritima* and *Thermotoga hypogea. J. Biochem.*, mvv058.
60. Ma, K., Hutchins, A., Sung, S.J., and Adams, M.W. (1997) Pyruvate ferredoxin oxidoreductase from the hyperthermophilic archaeon, *Pyrococcus furiosus*, functions as a CoA-dependent pyruvate decarboxylase. *Proc. Natl. Acad. Sci. U.S.A.*, **94** (18), 9608–9613.
61. Mai, V., Lorenz, W.W., and Wiegel, J. (1997) Transformation of *Thermoanaerobacterium* sp. strain JW/SL-YS485 with plasmid pIKM1 conferring kanamycin resistance. *FEMS Microbiol. Lett.*, **148** (2), 163–167.
62. Shaw, A.J., Podkaminer, K.K., Desai, S.G., Bardsley, J.S., Rogers, S.R., Thorne, P.G., Hogsett, D.A., and Lynd, L.R. (2008) Metabolic engineering of a thermophilic bacterium to produce ethanol at high yield. *Proc. Natl. Acad. Sci. U.S.A.*, **105** (37), 13769–13774.
63. Shaw, A.J., Covalla, S.F., Hogsett, D.A., and Herring, C.D. (2011) Marker removal system for *Thermoanaerobacterium saccharolyticum* and development of a markerless ethanologen. *Appl. Environ. Microbiol.*, **77** (7), 2534–2536.
64. Shaw, A.J., Hogsett, D.A., and Lynd, L.R. (2010) Natural competence in *Thermoanaerobacter* and *Thermoanaerobacterium* species. *Appl. Environ. Microbiol.*, **76** (14), 4713–4719.
65. Olson, D.G., Tripathi, S.A., Giannone, R.J., Lo, J., Caiazza, N.C., Hogsett, D.A., Hettich, R.L., Guss, A.M., Dubrovsky, G., and Lynd, L.R. (2010) Deletion of the Cel48S cellulase from *Clostridium thermocellum. Proc. Natl. Acad. Sci. U.S.A.*, **107** (41), 17727–17732.
66. Tripathi, S.A., Olson, D.G., Argyros, D.A., Miller, B.B., Barrett, T.F., Murphy, D.M., McCool, J.D., Warner, A.K., Rajgarhia, V.B., Lynd, L.R. *et al* (2010) Development of *pyrF*-based genetic system for targeted gene deletion in *Clostridium thermocellum* and creation of a *pta* mutant. *Appl. Environ. Microbiol.*, **76** (19), 6591–6599.
67. Argyros, D.A., Tripathi, S.A., Barrett, T.F., Rogers, S.R., Feinberg, L.F., Olson, D.G., Foden, J.M., Miller, B.B., Lynd, L.R., Hogsett, D.A. *et al* (2011) High ethanol titers from cellulose using metabolically engineered thermophilic, anaerobic microbes. *Appl. Environ. Microbiol.*, **77** (23), 8288–8294.
68. Guss, A.M., Olson, D.G., Caiazza, N.C., and Lynd, L.R. (2012) Dcm methylation is detrimental to plasmid transformation in *Clostridium thermocellum. Biotechnol. Biofuels*, **5** (1), 30.
69. Olson, D.G. and Lynd, L.R. (2012) Computational design and characterization of a temperature-sensitive plasmid replicon for gram positive thermophiles. *J. Biol. Eng.*, **6** (1), 5.
70. Mohr, G., Hong, W., Zhang, J., Cui, G.Z., Yang, Y., Cui, Q., Liu, Y.J., and Lambowitz, A.M. (2013) A targetron system for gene targeting in thermophiles and its application in *Clostridium thermocellum. PLoS One*, **8** (7), e69032.
71. Olson, D.G., Maloney, M., Lanahan, A.A., Hon, S., Hauser, L.J., and Lynd, L.R. (2015) Identifying promoters for gene expression in *Clostridium thermocellum. Metab. Eng. Commun.*, **2**, 23–29.
72. Nakamura, C.E. and Whited, G.M. (2003) Metabolic engineering for

the microbial production of 1,3-propanediol. *Curr. Opin. Biotechnol.*, **14** (5), 454–459.

73. Brown, S.D., Guss, A.M., Karpinets, T.V., Parks, J.M., Smolin, N., Yang, S., Land, M.L., Klingeman, D.M., Bhandiwad, A., Rodriguez, M. Jr., *et al* (2011) Mutant alcohol dehydrogenase leads to improved ethanol tolerance in *Clostridium thermocellum*. *Proc. Natl. Acad. Sci. U.S.A.*, **108** (33), 13752–13757.
74. Rani, K.S. and Seenayya, G. (1999) High ethanol tolerance of new isolates of *Clostridium thermocellum* strains SS21 and SS22. *World J. Microbiol. Biotechnol.*, **15** (2), 173–178.
75. Shao, X., Raman, B., Zhu, M., Mielenz, J.R., Brown, S.D., Guss, A.M., and Lynd, L.R. (2011) Mutant selection and phenotypic and genetic characterization of ethanol-tolerant strains of *Clostridium thermocellum*. *Appl. Microbiol. Biotechnol.*, **92** (3), 641–652.
76. Williams, T.I., Combs, J.C., Lynn, B.C., and Strobel, H.J. (2007) Proteomic profile changes in membranes of ethanol-tolerant *Clostridium thermocellum*. *Appl. Microbiol. Biotechnol.*, **74** (2), 422–432.
77. Tolonen, A.C., Zuroff, T.R., Ramya, M., Boutard, M., Cerisy, T., and Curtis, W.R. (2015) Physiology, genomics, and pathway engineering of an ethanol-tolerant strain of *Clostridium phytofermentans*. *Appl. Environ. Microbiol.*, **81** (16), 5440–5448.
78. Henderson, C.M. and Block, D.E. (2014) Examining the role of membrane lipid composition in determining the ethanol tolerance of *Saccharomyces cerevisiae*. *Appl. Environ. Microbiol.*, **80** (10), 2966–2972.
79. Osman, Y.A. and Ingram, L.O. (1985) Mechanism of ethanol inhibition of fermentation in *Zymomonas mobilis* CP4. *J. Bacteriol.*, **164** (1), 173–180.
80. Burdette, D.S., Jung, S.-H., Shen, G.-J., Hollingsworth, R.I., and Zeikus, J.G. (2002) Physiological function of alcohol dehydrogenases and long-chain (C30) fatty acids in alcohol tolerance of *Thermoanaerobacter ethanolicus*. *Appl. Environ. Microbiol.*, **68** (4), 1914–1918.
81. Georgieva, T., Mikkelsen, M., and Ahring, B. (2007) High ethanol tolerance of the thermophilic anaerobic ethanol producer *Thermoanaerobacter* BG1L1. *Cent. Eur. J. Biol.*, **2** (3), 364–377.
82. Herring, C.D., Kenealy, W.R., Shaw, A.J., Raman, B., Tschaplinski, T.J., Brown, S.D., Davison, B.H., Covalla, S.F., Sillers, W.R., Xu, H. *et al.* (2012) Development of *Thermoanaerobacterium saccharolyticum* for the conversion of lignocellulose to ethanol. Final Report.
83. Zacchi, G. and Axelsson, A. (1989) Economic evaluation of preconcentration in production of ethanol from dilute sugar solutions. *Biotechnol. Bioeng.*, **34** (2), 223–233.
84. Dien, B., Cotta, M., and Jeffries, T. (2003) Bacteria engineered for fuel ethanol production: current status. *Appl. Microbiol. Biotechnol.*, **63** (3), 258–266.
85. Linville, J.L., Rodriguez, M., Land, M., Syed, M.H., Engle, N.L., Tschaplinski, T.J., Mielenz, J.R., and Cox, C.D. (2013) Industrial robustness: understanding the mechanism of tolerance for the *Populus* hydrolysate-tolerant mutant strain of *Clostridium thermocellum*. *PLoS One*, **8** (10), e78829.
86. Bogorad, I.W., Lin, T.-S., and Liao, J.C. (2013) Synthetic non-oxidative glycolysis enables complete carbon conservation. *Nature*, **502** (7473), 693–697.
87. Schramm, M., Klybas, V., and Racker, E. (1958) Phosphorolytic cleavage of fructose-6-phosphate by fructose-6-phosphate phosphoketolase from *Acetobacter xylinum*. *J. Biol. Chem.*, **233** (6), 1283–1288.
88. Shaw, A.J., Covalla, S.F., Miller, B.B., Firliet, B.T., Hogsett, D.A., and Herring, C.D. (2012) Urease expression in a *Thermoanaerobacterium saccharolyticum* ethanologen allows high titer ethanol production. *Metab. Eng.*, **14** (5), 528–532.
89. van der Veen, D., Lo, J., Brown, S.D., Johnson, C.M., Tschaplinski, T.J., Martin, M., Engle, N.L., van den Berg, R.A., Argyros, A.D., Caiazza, N.C. *et al*

(2013) Characterization of *Clostridium thermocellum* strains with disrupted fermentation end-product pathways. *J. Ind. Microbiol. Biotechnol.*, **40** (7), 725–734.

90. Shaw, A.J., Miller, B.B., Rogers, S.R., Kenealy, W.R., Meola, A., Bhandiwad, A., Sillers, W.R., Shikhare, I., Hogsett, D.A., and Herring, C.D. (2015) Anaerobic detoxification of acetic acid in a thermophilic ethanologen. *Biotechnol. Biofuels*, **8** (1), 1–12.
91. Shaw, A.J., Hogsett, D.A., and Lynd, L.R. (2009) Identification of the [FeFe]-hydrogenase responsible for hydrogen generation in *Thermoanaerobacterium saccharolyticum* and demonstration of increased ethanol yield via hydrogenase knockout. *J. Bacteriol.*, **191**, 6457–6464.
92. Fuhrer, T. and Sauer, U. (2009) Different biochemical mechanisms ensure network-wide balancing of reducing equivalents in microbial metabolism. *J. Bacteriol.*, **191** (7), 2112–2121.
93. Ansell, R., Granath, K., Hohmann, S., Thevelein, J.M., and Adler, L. (1997) The two isoenzymes for yeast NAD(+)-dependent glycerol 3-phosphate dehydrogenase encoded by GPD1 and GPD2 have distinct roles in osmoadaptation and redox regulation. *EMBO J.*, **16** (9), 2179–2187.
94. Zhou, J., Olson, D.G., Lanahan, A.A., Tian, L., Murphy, S.J., Lo, J., and Lynd, L.R. (2015) Physiological roles of pyruvate ferredoxin oxidoreductase and pyruvate formate-lyase in *Thermoanaerobacterium saccharolyticum* JW/SL-YS485. *Biotechnol. Biofuels*, **8**, 138.
95. Zheng, T., Olson, D.G., Tian, L., Bomble, Y.J., Himmel, M.E., Lo, J., Hon, S., Shaw, A.J., van Dijken, J.P., and Lynd, L.R. (2015) Cofactor specificity of the bifunctional alcohol and aldehyde dehydrogenase (AdhE) in wild-type and mutant *Clostridium thermocellum* and *Thermoanaerobacterium saccharolyticum*. *J. Bacteriol.*, **197** (15), 2610–2619.
96. Lo, J., Zheng, T., Hon, S., Olson, D.G., and Lynd, L.R. (2015) The bifunctional alcohol and aldehyde dehydrogenase gene, *adhE*, is necessary for ethanol production in *Clostridium thermocellum* and *Thermoanaerobacterium saccharolyticum*. *J. Bacteriol.*, **197** (8), 1386–1393.
97. Lo, J., Zheng, T., Olson, D.G., Ruppertsberger, N., Tripathi, S.A., Guss, A.M., and Lynd, L.R. (2015) Deletion of *nfnAB* in *Thermoanaerobacterium saccharolyticum* and its effect on metabolism. *J. Bacteriol.*, **197**, 2920–2929.
98. Rydzak, T., Lynd, L., and Guss, A. (2015) Elimination of formate production in *Clostridium thermocellum*. *J. Ind. Microbiol. Biotechnol.*, **42**, 1263–1272.
99. Biswas, R., Zheng, T., Olson, D.G., Lynd, L.R., and Guss, A.M. (2015) Elimination of hydrogenase active site assembly blocks H2 production and increases ethanol yield in *Clostridium thermocellum*. *Biotechnol. Biofuels*, **8**, 20.
100. Papanek, B., Biswas, R., Rydzak, T., and Guss, A.M. (2015) Elimination of metabolic pathways to all traditional fermentation products increases ethanol yields in *Clostridium thermocellum*. *Metab. Eng.*, **32**, 49–54.
101. Weimer, P.J., Russell, J.B., and Muck, R.E. (2009) Lessons from the cow: what the ruminant animal can teach us about consolidated bioprocessing of cellulosic biomass. *Bioresour. Technol.*, **100** (21), 5323–5331.
102. Tian, L., Papanek, B., Olson, D. G., Rydzak, T., Holwerda, E. K., Zheng, T., Zhou, J., Maloney, M., Jiang, N., Giannone, R. J., *et al.* (2016) Simultaneous achievement of high ethanol yield and titer in *Clostridium Thermocellum*. *Biotechnol. Biofuels*, **9** (1), 116.

11
Lactic Acid Bacteria

Luciana Ruiz-Rodríguez[*], *Juliana Bleckwedel*[*], *Maria Eugenia Ortiz, Micaela Pescuma, and Fernanda Mozzi*

11.1
Introduction

Lactic acid bacteria (LAB) are a very diverse group of fermenting, acid-tolerant bacteria characterized for the production of lactic acid as the major or sole final product. LAB are gram-positive cocci or rods, catalase-negative, generally nonsporulating, nonmotile with low guanine cytosine (GC) content [1]. Although the majority of LAB have nonaerobic habit, these microorganisms are aerotolerant. Interestingly, when heme or heme plus menaquinone are present in the medium, some species can respire aerobically using oxygen as final electron acceptor. Respiration has a positive effect on cell growth as bacterial biomass increases together with oxygen resistance and survival [2].

LAB belong to the Firmicutes phylum, Bacilli class, and Lactobacillales order, which includes six families (Aerococcaceae, Carnobacteriaceae, Enterococcaceae, Lactobacillaceae, Leuconostocaceae, Streptococcaceae); to date, 43 genera (*Abiotrophia, Aerococcus, Agitococcus, Alkalibacterium, Allofustis, Alloiococcus, Atopobacter, Atopococcus, Atopostipes, Bavariicoccus, Carnobacterium, Catellicoccus, Desemzia, Dolosicoccus, Dolosigranulum, Enterococcus, Eremococcus, Facklamia, Fructobacillus, Globicatella, Granulicatella, Ignavigranum, Isobaculum, Lacticigenium, Lactobacillus, Lactococcus, Lactosphaera, Lactovum, Leuconostoc, Marinilactibacillus, Melissococcus, Oenococcus, Paralactobacillus, Pediococcus, Pilibacter, Pisciglobus, Sharpea, Streptococcus, Symbiobacterium, Tetragenococcus, Trichococcus, Vagococcus, Weissella*) have been described [3]. From all these genera, the genus *Lactobacillus* is the most diverse genera of LAB for which 214 species were described in October 2015.

Although LAB are ubiquitous microorganisms widespread in a plethora of niches, they have complex nutritional requirements for amino acids, vitamins, minerals, fatty acids, peptides, nucleotide bases, and carbohydrates. These microorganisms can be found in plants, fermented foods, the gastrointestinal

[*] These authors contributed equally.

Industrial Biotechnology: Microorganisms, First Edition. Edited by Christoph Wittmann and James C. Liao.

tract and vagina of humans and animals, and in soil and water. They can be classified as homofermentative or heterofermentative according to the manner in which they use carbohydrates as energy source under standard growth conditions (nonlimited supply of glucose, growth factors such as amino acids, vitamins, and nucleic acid precursors, and limited oxygen availability). Thus, while homofermentative bacteria produce only lactic acid as end product through the Embden–Meyerhof–Parnas (EMP) pathway (Figure 11.1), heterofermentative bacteria produce equimolar quantities of lactic acid, carbon dioxide, and ethanol or acetate through the phospoketolase pathway (Figure 11.2) [1–3, 6].

In addition to the synthesis of organic acids, many LAB can also produce other interesting metabolites such as antibacterial compounds (e.g., bacteriocins) [7], aroma compounds (diacetyl, acetoin, etc.), vitamins, exopolysaccharides (EPS), low-calorie sugars (e.g., mannitol), short-chain fatty acids, and γ-amino butyric acid (GABA) [4, 8]. Owing to this metabolic versatility, these bacteria can be used as microbial cell factories in the production of chemicals, pharmaceuticals, or other industrially relevant products [1, 9].

As the majority of the LAB species possess food-grade status (considered GRAS, generally recognized as safe), LAB are the most important microorganisms used in the food industry. Generally, the carbon source available in raw foods is fermented by the LAB strains producing lactic acid with concomitant reduction in pH, which improves the safety and storage quality of foods by preventing the growth of pathogenic microorganisms with consequent food spoilage [8]. In addition, LAB contribute to the palatability, nutritional value, texture, and flavor of fermented foods such as dairy products (e.g., yogurts and fermented milks, cheeses, and buttermilk), sourdough breads, fermented meats and vegetables, and fermented beverages. For these reasons, LAB have been used as starter cultures, which are represented by a large number of at least one-microorganism cells added to a raw material in order to produce a fermented food. The use of starter cultures allows the standardization of the obtained end product by controlling the fermentation [9].

In addition to the technological properties mentioned above, certain LAB species can be used as probiotics as they are components of the normal gastrointestinal human microbiota and can survive low pH and presence of bile salts during passage through the gastrointestinal tract. Probiotic strains are those that exert beneficial effects to the host such as direct antagonist effect against specific microbial groups or stimulation of the immune system. Probiotics can prevent the colonization of pathogens by producing antimicrobial compounds, competing for the adhesion to the gut epithelial surface, and for the environmental carbon sources. In addition, they can reduce the intolerance to lactose by means of their galactosidase enzyme [10].

This chapter deals with the biotechnological application of LAB as starter cultures, although briefly, for the elaboration of fermented foods, and the production of industrially interesting metabolites such as food ingredients, nutraceutics, and high added-value compounds.

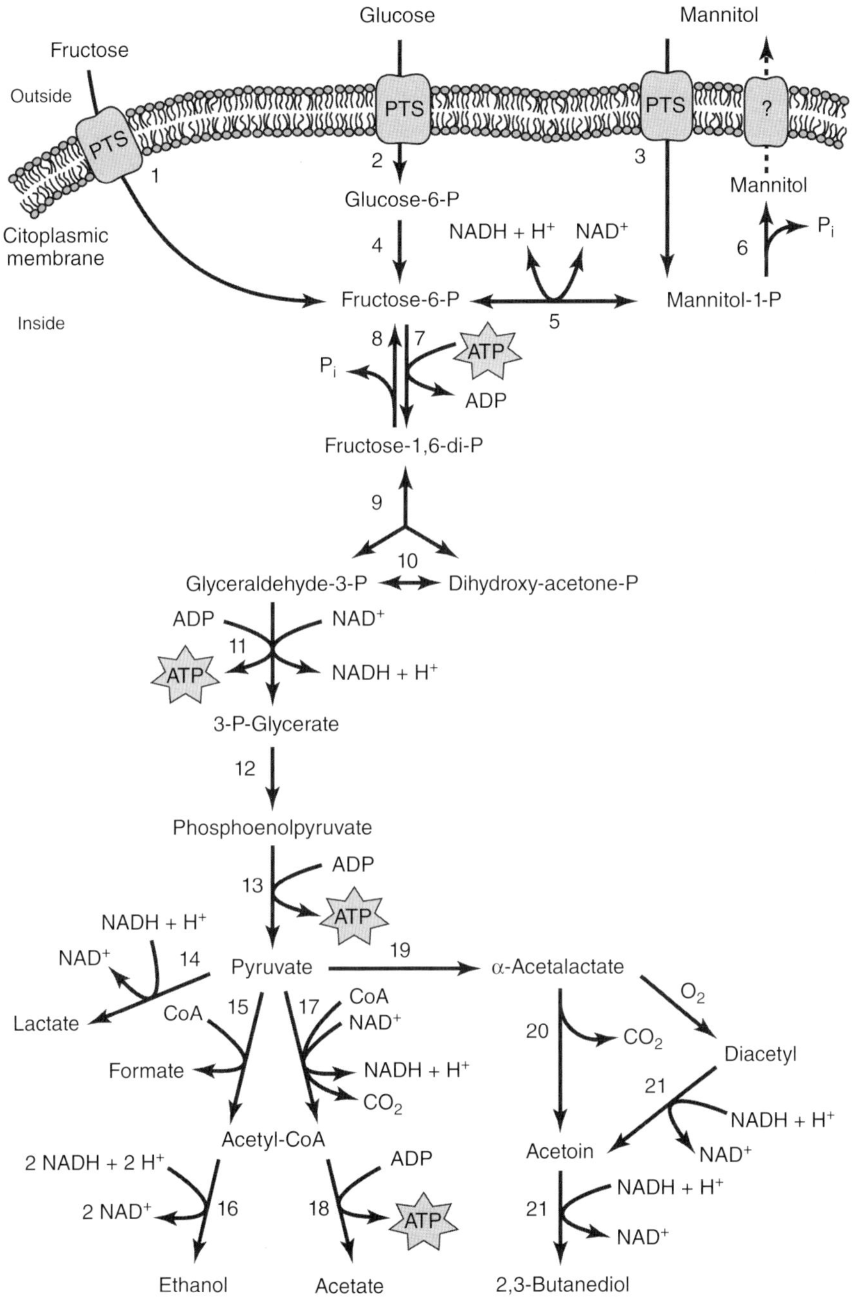
Glucose
Mannitol
Fructose
Outside
PTS
PTS
PTS
?
1
2
3
Glucose-6-P
Mannitol
Citoplasmic membrane
4
NADH + H+
NAD+
6
Pi
Fructose-6-P
Mannitol-1-P
Inside
5
8
7
ATP
Pi
ADP
Fructose-1,6-di-P
9
10
Glyceraldehyde-3-P
Dihydroxy-acetone-P
ADP
NAD+
11
ATP
NADH + H+
3-P-Glycerate
12
Phosphoenolpyruvate
ADP
13
ATP
NADH + H+
NAD+
14
Pyruvate
19
α-Acetalactate
O2
Lactate
CoA
15
17
CoA
NAD+
20
CO2
Diacetyl
Formate
NADH + H+
CO2
21
NADH + H+
Acetyl-CoA
Acetoin
NAD+
2 NADH + 2 H+
ADP
NADH + H+
2 NAD+
16
18
ATP
21
NAD+
Ethanol
Acetate
2,3-Butanediol

Figure 11.1 Proposed pathway for hexose metabolism of homofermentative LAB (1) and (2) phosphoenolpyruvate (PEP)-dependent sugar phosphotransferase system (PTS); (3) mannitol-specific PTS; (4) phosphoglucose isomerase; (5) mannitol-1-phosphate dehydrogenase; (6) mannitol-1-phosphatase; (7) 6-phosphofructokinase; (8) fructose-diphosphatase; (9) fructose-1,6-diphosphate aldolase; (10) triosephosphate isomerase; (11) glyceraldehyde-3-phosphate dehydrogenase and phosphoglycerate kinase; (12) phosphoglyceromutase and enolase; (13) pyruvate kinase; (14) lactate dehydrogenase; (15) pyruvate-formate lyase; (16) acetaldehyde dehydrogenase and alcohol dehydrogenase; (17) pyruvate dehydrogenase; (18) acetate kinase; (19) α-acetolactate synthase; (20) α-acetolactate decarboxylase; and (21) 2,3-butanediol dehydrogenase. (Modified from Wisselink *et al.* [4].)

11.2 Fermented Foods

Foods normally contain simple or complex sugars from plant carbohydrates that cannot be catabolyzed by the human digestive system. In addition, foods contain vitamins, amino acids, and lipids, making them suitable for microbial growth. Several processes have been developed to prevent food spoilage and enhance their digestibility. Fermentation is one of the oldest methods applied for food preservation during which the growth of selected microorganisms prevents development of deteriorating or harmful bacteria and yeasts by production of organic acids, bacteriocins, and nutrient depletion by competition. These starter cultures are most frequently LAB, yeasts, and certain molds. LAB evolution was directed by human food consumption, the first evidence of their use being in China in 7000 BC [11].

Nowadays, food industrialization continuously searches for stable and novel products which can be achieved by controlled fermentation using defined starter cultures with specific properties. The most desirable characteristics of a starter culture are robustness during manufacturing, fast growth, high biomass and product yields, and specific organoleptic properties. Initially, starter cultures with specific characteristics were sought in nature; later in the 1950s, random mutagenesis and selection programs were applied to improve the use of starter cultures for fermentation. In this sense, engineered bacteria had poor acceptance and classical microbial evolution is applied even now. However, the advantage of new DNA sequencing methods enables to rapidly search for beneficial random mutations in the genome [12]. These techniques need new tools to analyze and compile the obtained data; bioinformatics playing a key role on predicting desired and undesired characteristics of microorganisms during their growth in different food matrices that have to be verified by laboratory analyses [13].

LAB are industrially used as starter cultures and as probiotics to perform controlled fermentations for the production of dairy, meat, vegetable, and bakery-fermented foods. LAB with specific characteristics have been isolated from a variety of traditional fermented products or from diverse raw materials used as starter cultures to obtain controlled and uniform fermented foods [14–16].

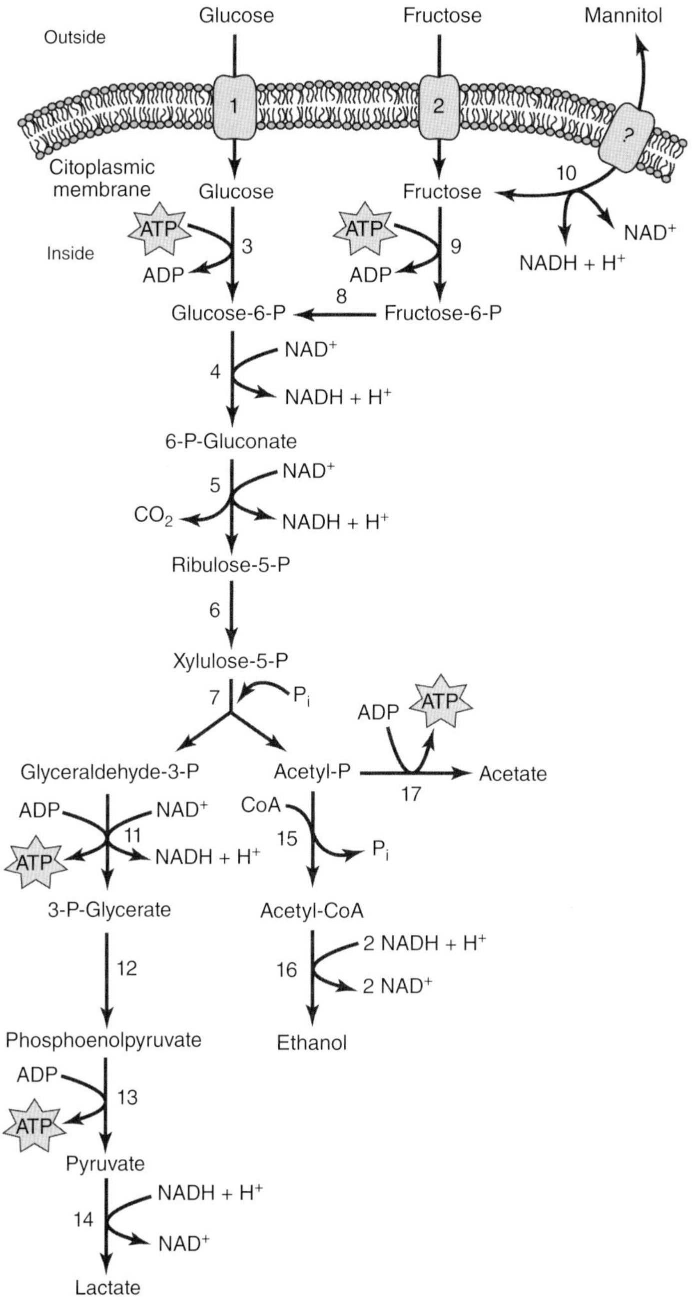
Glucose
Fructose
Mannitol
Outside
1
2
?
Citoplasmic membrane
Glucose
Fructose
10
NAD+
NADH + H+
ATP
ADP
3
Inside
ATP
ADP
9
8
Glucose-6-P
Fructose-6-P
NAD+
4
NADH + H+
6-P-Gluconate
NAD+
5
CO2
NADH + H+
Ribulose-5-P
6
Xylulose-5-P
7
Pi
ADP
ATP
Glyceraldehyde-3-P
Acetyl-P
Acetate
17
ADP
NAD+
11
ATP
NADH + H+
CoA
15
Pi
3-P-Glycerate
Acetyl-CoA
12
16
2 NADH + H+
2 NAD+
Phosphoenolpyruvate
Ethanol
ADP
13
ATP
Pyruvate
NADH + H+
14
NAD+
Lactate

Figure 11.2 Proposed pathway for hexose metabolism of heterofermentative LAB. (1) glucose permease; (2) fructose permease; (3) glucokinase; (4) glucose 6-phosphate dehydrogenase; (5) 6-phosphogluconate dehydrogenase; (6) epimerase; (7) phosphoketolase; (8) glucose phosphate isomerase; (9) fructokinase; (10) mannitol 2-dehydrogenase; (11) glyceraldehyde 3-phosphate dehydrogenase and phosphoglycerate kinase; (12) phosphoglyceromutase and enolase; (13) pyruvate kinase; (14) lactate dehydrogenase; (15) phosphate acetyltransferase; (16) acetaldehyde dehydrogenase and alcohol dehydrogenase; (17) acetate kinase; and ? Unknown mannitol transport system. (Modified from Ortiz *et al.* [5].)

Different species of LAB are used depending on the raw material needed to produce a fermented product (Table 11.1). Dairy products such as cheese normally contain *Lactococcus* (*Lc.*) *lactis* subsp. *lactis*, *Lc. lactis* subsp. *cremoris*, *Lc. lactis* subsp. *lactis* var. *diacetylactis*, *Leuconostoc* (*Leuc.*) *mesenteroides* subsp. *cremoris*, *Lactobacillus* (*Lb.*) *delbrueckii* subsp. *lactis*, *Lb. helveticus*, and *Lb. casei*; the species being dependent on the type of cheese to be manufactured. On the other hand, fermented milks are produced mainly using strains of *Lb. delbrueckii* subsp. *bulgaricus* and *Streptococcus* (*Strep.*) *thermophilus* (for yogurt production), *Lb. casei*, *Lb. acidophilus*, *Lb. rhamnosus*, and *Lb. johnsonii* while other fermented milks such as Kefir contain *Lb. kefir*, *Lb. kefiranofaciens*, and *Lb. brevis* together with yeasts [9]. In fermented meat products (e.g., sausages), the LAB species commonly applied are *Lb. sakei*, *Lb. curvatus*, *Pediococcus* (*P.*) *acidilactici*, and *P. pentosaceus*, while in fermented fish products the species used are *Lb. plantarum*, *Lb. alimentarius*, and *Carnobacterium* (*C.*) *piscicola* [11–16, 22]. In fermented vegetables such as pickles, sauerkraut, and olives, species of *Leuc. mesenteroides*, *Lb. plantarum*, *P. acidilactici*, *P. cerevisiae*, *Lb. brevis*, *Lb. pentosus*, *P. pentosaceus*, *Lb. fermentum*, *Tetragenococcus* (*T.*) *halophilus*, and *Lb. buchneri* [9, 24, 25] are used, while in cereal fermentation *Lb. sanfransiscensis*, *Lb. farciminis*, *Lb. fermentum*, *Lb. brevis*, *Lb. plantarum*, *Lb. amylovorus*, *Lb. reuteri*, *Lb. pontis*, *Lb. panis*, *Lb. alimentarius*, and *Weissella* (*W.*) *cibaria* are employed. Finally, for alcoholic beverages such as wine and sake *Oenococcus* (*O.*) *oeni* and *Lb. sakei*, respectively, are normally used as starter or adjunct cultures during fermentation [9]. Although the use of LAB in dairy, meat, sauerkraut, pickles, olives, and sourdough has been extensively revised [15, 39–42], general aspects of the major commercial fermented food products are briefly discussed in this chapter.

11.2.1
Fermented Milks

The dairy market has shown the highest economic growth within the food market in recent years. In this respect, the production of yogurt alone reached values of 2.14 million tons in the United States in 2013 [43]. Fermented milks are largely consumed because of their claimed functional properties that are related with probiotics or the generation of functional molecules such as organic acids and bioactive components [9]. Yogurt is one of the most consumed fermented

Table 11.1 Fermented foods and beverages, their applications, and the lactic acid bacteria involved in each fermentative process.

Fermented foods	LAB involved	Applications/functions	References
Classical			
Fermented milks			
Yogurt	*Strep. thermophilus, Lb. delbrueckii* subsp. *bulgaricus*	Technological and functional properties related with probiotics. Production of functional molecules	[9]
Fermented milks	*Lb. casei, Lb. acidophilus, Lb. rhamnosus, Lb. johnsonii*		
Kefir	*Lb. kefir, Lb. kefiranofaciens, Lb. brevis*		
Cheeses			
Hard cheeses without eyes	*Lc. lactis* subsp. *lactis, Lc. lactis* subsp. *cremoris*		[9]
Hard cheeses with small eyes (e.g., emmental)	*Lc. lactis* subsp. *lactis* var. *diacetylactis, Leuc. mesenteroides* subsp. *cremoris, Lb. helveticus, Strep. salivarius* subsp. *thermophilus*		[9, 17]
Semihard cheeses (Swiss and Italian-type)	*Lb. delbrueckii* subsp. *lactis, Lb. helveticus, Lb. casei*		[9, 18, 19]
Fermented meat products			
Sausages (e.g., *Wieners, Bratwurst,* and *Currywurst* from Germany; *Saucisson* from France; *Salamino Cacciatore, Salsiccia passita, Salame Romagnolo, Salame Napoli, Milano type,* and *Lombardo type sausages* from Italy)	*Lb. sakei, Lb. curvatus, P. acidilactici, P. pentosaceus*	Prevention of food spoilage	[20, 21]
Fermented fish products	*Lb. alimentarius, C. piscicola, Lb. sakei, Leuc. mesenteroide, Lb. plantarum, Lc. garvieae, Strep. bovis, W. cibaria, P. pentosaceus*	Prevention of food spoilage	[9, 22, 23]

(*continued overleaf*)

Table 11.1 (*continued*)

Fermented foods	LAB involved	Applications/functions	References
Fermented vegetables: pickles, sauerkraut, olives, and so on	*Leuc. mesenteroides, Lb. plantarum, P. acidilactici, P. cerevisiae, Lb. brevis, Lb. pentosus, Lb. plantarum, P. acidilactici, P. pentosaceus, Lb. plantarum, Lb. fermentum, T. halophilu, Lb. buchneri*	Prevention of food spoilage	[9, 24, 25]
Cereal fermentation (Sourdough)	*Lb. sanfransiscensis, Lb. farciminis, Lb. fermentum, Lb. brevis, Lb. plantarum, Lb. amylovorus, Lb. reuteri, Lb. pontis, Lb. panis, Lb. alimentarius, W. cibaria*	Artisanal bread manufacture	[9]
Non-classical fermented foods and beverages			
Yacon-based products			
Yogurt	*Strep. thermophilus, Lb. delbrueckii* subsp. *bulgaricus*		[26–28]
Yacon	*Lb. acidophilus*	Probiotic properties	[29, 30]
Yacon and soy-based	*Ent. faecium, Lb. helveticus* subsp. *jugurti*	Colesterol-lowering effects	[31]
Fruits and legumes-based products			
Mango juice	*Lb. plantarum*	Prolongation of fruit shelf life and maintainance of the fruit nutritional quality	[32]
Coconut drink	*Lb. plantarum*		[33]
Passion fruit yogurt-like product	*Lb. fermentum, Lb. plantarum* (amylolytic LAB) *Lb. acidophilus, Lb. casei* (probiotic LAB)		[34]
Soy-peanut beverage	*Lb. acidophilus, P. acidilactici*	Technological properties for production of a fermented vegetable non-alcoholic beverage	[35]
Fermented melon	Riboflavin-producing *Lb. plantarum, Lb. fermentum*	Increase in the riboflavin content	[36]
Gluten-free foods			
Quinoa, teff, or buckwheat breads	*W. cibaria*	Improved baking, sensory and nutritional properties of breads	[37]
Quinoa bread	*Lb. amylovorus*	Increased bread shelf life. Enhanced bread volume, crumb hardness, and nutritional value	[38]

milks, which is defined by the symbiosis of the two LAB strains belonging to *Strep. thermophilus* and *Lb. delbrueckii* subsp. *bulgaricus.* Fermentation by these microorganisms yields a characteristic acidic flavor and a product with an improved appearance, taste, consistency, and digestibility. Nowadays, several types of fermented milks are produced and many of them contain probiotic strains or LAB displaying specific characteristics with the final product giving health benefits to the host [26].

11.2.2 Cheeses

Cheese is produced worldwide at industrial and artisanal scales. The main cheese producers are the United States, which produced 4.12 million tons in 2013, followed by Germany (2.4 million tons), and France (2.1 million tons). LAB are the dominant microbiota in cheese; these microorganisms are added as starter and/or adjuvant cultures and can be present in nonpasteurized milk. At the beginning of the manufacture process, the LAB cell count canexceed 1×10^9 CFU ml^{-1} but the unfavorable conditions of cheese ripening (low pH, high salt concentrations, and low temperature) result in a decrease in bacterial viability [44]. LAB contribute to cheese flavor ripening by casein proteolysis and release of peptides by cell wall-associated proteinases; further, peptide hydrolysis is carried out by intracellular peptidases which are released from cell lysates. The most proteolytic LAB species are *Lb. helveticus* and *Lb. delbrueckii* subsp. *bulgaricus* [17, 18]. The most important aroma compounds in cheese are produced through amino acid catabolism where aldehydes, alcohols, and carboxylic acids are released; in this catabolic process, species such as *Lb. paracasei*, and *Lc. lactis* are involved [19].

11.2.3 Fermented Meats

Fermented meat products include different types of fermented sausages that vary according to climatic conditions, local ingredients, and regional customs [45]. These products are widespread around the world, the European countries being the major producers and consumers. For example, Italy produced 1 180 000 tons of cured meat products in 2013 with an economic value of €7900 million [20]. The microorganisms involved in sausage fermentation are mainly LAB, gram positive–catalase-positive cocci, certain molds, and yeasts. During fermented sausage preparations, facultative homofermentative lactobacilli such as *Lb. sakei* and *Lb. curvatus* dominate the fermentation and contribute to the typical aroma profile [21]. LAB contribute to flavor by generating large amounts of lactic acid and some volatile compounds from sugar fermentation and by the release of amino acids by LAB exopeptidases, which degrade peptides released by meat endogenous proteases [46]. In addition to fermented meat products, several LAB species are involved in different kinds of fermented fish foods. *Plaa-som* is a Thai fermented fish product where, during the early stage of the fermentation,

Lc. garvieae, Strep. bovis, and *W. cibaria* are the dominant bacteria, while after 48 h, *W. cibaria, P. pentosaceus*, and *Lb. plantarum* are prevalent, the last species becoming dominant at the end of the fermentation [22]. *Gajami-sikhae* is a traditional Korean food made with fermented flat fish where *Lb. sakei* and *Leuc. mesenteroides* are the dominant species among the bacterial community [23].

Apart from the conventional fermented foods above mentioned, some nondairy fermented foods are gaining interest, in particular, those from soy, cereal, and of vegetable origin.

11.2.4
Nondairy Drinks and Foods

Although dairy fermented and probiotic foods are extensively commercialized worldwide, new healthy nondairy alternative foods are being considered mainly owing to allergy to animal proteins and lactose intolerance [47]. In addition, the food industry continuously seeks novel nondairy fermented and probiotic drinks owing to the increasing demand from vegetarian and vegan consumers. In this regards, novel formulations of nonalcoholic fermented beverages using roots, legumes, and fruits are being explored.

11.2.4.1
Roots (Yacon)

Yacon (*Smallanthus sonchifolius*), an Andean root growing at high altitudes (1000–3200 m.s.l.) from Colombia to Northeastern Argentina has been recently introduced in the United States, Europe, New Zealand, Japan, Korea, and Brazil [27]. Yacon is considered a functional food due to its high content of fructooligosacharides (FOS), inulin, and polyphenols. FOS are nondigestible sugars that promote growth of bifidobacteria and other beneficial bacteria in the colon; these compounds being considered prebiotics. While nowadays this sweet root is consumed boiled, baked, or as juice, new products such as flour, jam, chips, and pickles are being developed [28].

LAB and yeasts are commonly present during spontaneous yacon fermentation and LAB growth contributes to yacon preservation by reducing pH and adding to the food flavor and texture [48]. Parra [27] developed a fermented milk with the addition of 3.5% of yacon juice; the product had good sensorial properties and consumer acceptance, although the final pH was lower than in a conventional yogurt after storage. On the other hand, Vasconcelos *et al.* [48] formulated a low-calorie fermented milk with yacon flour added in different concentrations (1.6–3.9% w/v). The obtained product had low-fat and low-caloric content and higher dietary fiber than the minimum value required by Brazilian regulations. In another study, Dias *et al.* [29] analyzed the behavior of the probiotic strains *Lb. acidophilus* La-5E and *Bifidobacterium bifidum* BB12 in a mixture of pasteurized whey (50%), sucrose (10%), rehydrated whole milk powder (25%), and yacon pulp (15%). The obtained beverage had satisfactory sensory evaluation and high probiotic cell count for storage of up to 21 days. In addition, the health-promoting effects of fermented yacon with some probiotic LAB strains was analyzed taking advantage of the prebiotic characteristics of the Andean root. The effect of yacon and *Lb. acidophilus* CRL 1014 as a symbiotic preparation on early phases of colon carcinogenesis was studied [30] and the results indicated that the fermented product could prevent carcinogenesis in early stage of colon tumors in male Wistar rats. Roselino *et al.* [31] evaluated the effect of a product composed of soy and yacon fermented with the strains *Ent. faecium* CRL 183 and *Lb. helveticus* subsp. *jugurti* 416 on diabetes development; the symbiotic product had a positive effect on triglyceride level and High-density lipoprotein (HDL) cholesterol although it was not able to reduce blood glucose.

11.2.4.2
Fruits and Legumes

Generally, fruit consumption is accepted as healthy due to its high mineral, vitamin, antioxidant, and dietary fiber content. Fruits are usually consumed fresh although new outgrowths of pathogenic bacteria are frequent. One way to prolong fruit shelf life is by heating it, but this process diminishes its nutritional quality. This fact and the increasing search for healthier foods led to formulate novel fruit-based foods and new carriers for probiotics. Although fruit fermentation has not been extensively exploited yet, there exist a few studies involving LAB strains and fermented fruit beverages. In this respect, Reddy *et al.* [32] formulated a healthy mango beverage by fermenting mango juice with a probiotic *Lb. plantarum* strain. The resulting beverage had low pH (3.2) and a cell count of 10^7 CFU ml^{-1}, which remained constant throughout the storage period. Russo *et al.* [36] formulated a functional food of fresh-cut melon inoculated with two riboflavin-producing *Lactobacillus* strains. The technological parameters of melon were unaffected, while riboflavin content increased twofold during the 11 day storage period. The probiotic strains used were able to reduce levels of *Listeria* (*L.*) *monocytogenes* in artificial contaminated fruits. In another study, strawberry preservation and its antioxidant capacity were enhanced by addition of a lysate of a *Lb. pentosus* strain [49], which was increased 45% after 3 days of fermentation, while proliferation of yeasts and molds was inhibited.

do Amaral Santos *et al.* [35] formulated a potential nonalcoholic probiotic drink by fermenting a soy-peanut mixture with the probiotic strains *Lb. acidophilus* LACA 4, *P. acidilactici* UFLA BFFCX 27.1, and the yeast *Saccharomyces* (*S.*) *cerevisiae* UFLA YFFBM 18.03. The beverage was considered nonalcoholic as the ethanol content was below 0.03% (v/v). Prado *et al.* [33] developed a nondairy fermented functional drink using coconut water, soy protein hydrolysate, yeast extract, and sucrose; a *Lb. plantarum* strain isolated from the coconut fruit was used to ferment the nondairy mixture for 8 h at 37 °C. The fermented product, supplemented with coconut flavor and sucrose, had a successful sensorial evaluation and the advantage of being cheaper than the dairy counterparts, especially in countries were soy is extensively produced. In another study, do Espirito-Santo *et al.* [34] formulated a gruel containing soy, rice, and passion fruit fiber fermented by amylolytic and probiotic LAB strains. Amylolytic bacteria were necessary to reduce the pH of the gruel and to produce EPS, increasing the viscosity of the yogurtlike product.

11.2.4.3 Gluten-Free Foods

The rising incidence of celiac disease has increased the interest in using gluten-free crops, which usually have lower nutritional value than those containing gluten. Moreover, gluten-free breads have low volume and texture due to the lack of foaming proteins and use of pure starches. It has been observed that bread production using sourdough (composed of LAB and baker's yeast) improves the quality of wheat and other gluten-free breads.

Quinoa (*Chenopodium quinoa*) is an Andean grainlike crop with high nutritional value. It is rich in essential amino acids, vitamins, and large amounts of polyunsaturated fatty acids compared with other grains [38]. Quinoa starch has high viscosity and water-binding capacity; these characteristics making this pseudocereal interesting for formulating gluten-free breads [50]. Wolter *et al.* [51] studied the effect of an EPS-producing *W. cibaria* strain on gluten-free bread texture and observed that fermentation improved the baking characteristics and the sensorial and nutritional properties of bread made with quinoa, teff, or buckwheat flours. Moreover, the highest EPS production which was correlated with the highly fermentable sugars (glucose, maltose, fructose), protein, and mineral content was detected in quinoa bread. On the other hand, Axel *et al.* [38] studied the effect of the antifungal strain *Lb. amylovorus* DSM 19280 on quinoa fermentation and bread spoilage during storage under pilot-plant conditions. High concentrations of the antifungal compounds 4-hydroxyphenillactic acid, 3-phenyllactic acid, phloretic acid, and hydroferulic acid were produced by the *Lactobacillus* strain increasing the shelf life of the bread. Moreover, an improvement in the bread specific volume, crumb hardness, and the nutritional values of these gluten-free breads was observed.

11.3 Industrially Relevant Compounds

11.3.1 Food Ingredients

11.3.1.1 Antimicrobial Compounds

LAB produce several compounds with antimicrobial properties as a result of nutrient competition with other microorganisms growing in the same niche. These antimicrobial active metabolites consist of nonspecific antimicrobial substances such as short-chain fatty acids (formic, acetic, propionic, butyric, and lactic acids), ethanol and hydrogen peroxide, and toxins including bacteriocins (nisin, reuterin, reutericyclin, pediocin, lacticin, enterocin, and others), bacteriocin-like inhibitory substances (BLIS), and antifungal compounds (propionate, phenyl-lactate, hydroxyphenyl-lactate, cyclic dipeptides, and 3-hydroxy fatty acids). Owing to this functional property, interest in the use of LAB has increased as an alternative to the use of synthetic chemicals and additives in food biopreservation [37, 52]. Among the antimicrobial molecules produced by LAB, bacteriocins are the most promising.

Bacteriocins *Bacteriocins* are a heterogeneous group of antimicrobial proteinaceus compounds (small proteins or peptides) of low molecular weight (usually below 10 kDa), ribosomally synthesized during the primary phase of growth and generally active against a narrow range of species closely related to the bacteriocin-producer strain [52–55]. Klaenhammer [225] claims that 99% of all bacteria produce at least one bacteriocin; LAB genera which were reported to produce bacteriocins are *Lactococcus, Streptococcus, Lactobacillus, Pediococcus, Leuconostoc, Enterococcus, Carnobacterium, Aerococcus, Oenococcus, Tetragenococcus, Vagococcus,* and *Weisella* [52]. Bacteriocins are considered suitable as food preservatives for several reasons: (i) they are safe for human consumption. (ii) They are selective toward prokaryotes. (iii) They can be inactivated by digestive proteases having little effect on the gut microbiota. (iv) Most of them are thermal resistant and are active after pasteurization and sterilization. (v) Finally, they act against many foodborne pathogens and spoilage bacteria [52, 55]. These active molecules can show bactericidal or bacteriostatic effect depending on the dose and degree of purification as well as the physiological state of the susceptible strain and other factors such as temperature, presence of agents able to disrupt cell wall, and the pH [52]. When acting as a bactericidal compound, their blank of action is usually the bacterial cytoplasmic membrane [55].

Bacteriocin classification, based on the structure, mode, and spectrum of action, is in continuous revision. Bacteriocins are commonly divided into three classes: (i) *Class I* (lantibiotics): they are small (<5 kDa) heat-stable peptides with post-translationally modified amino acids (lanthionine, β-methyl lanthionine, dehydroalanine, and dehydrobutyrine) in their primary structure. They are mainly active against gram-positive bacteria through mechanisms such as pore formation and enzyme inhibition. (ii) *Class II* (non-lantibiotics): they are thermoresistant, small (<10 kDa) peptides subdivided into two subclasses: IIa (pediocin-like) with a strong antilisterial effect and IIb that requires the complementary action of two peptides in order to display activity. (iii) *Class III* (bacteriolysins): they are large (>30 kDa), heat-labile proteins that catalyze cell wall hydrolysis causing the lysis of sensitive cells [37, 52, 54, 55].

Bacteriocins have been used to control spoilage and pathogenic bacteria in foods offering several benefits such as prolongation of the product shelf life, reduction of the risk of foodborne disease transmission, and decrease in the use of chemical compounds as preservatives and in the intensity of heat treatments to achieve high-quality foods [37, 52].

Bacteriocins can be *ex situ* added to foods as food ingredients (purified or semipurified bacteriocins, or as a previously fermented product with a bacteriocin-producing LAB) or through a bacteriocinogenic starter culture to produce it *in situ*. One drawback of using bacteriocin-producing strains is that they may lack compatibility with other cultures needed for fermentation, so in these cases, the incorporation of bacteriocin preparations as food ingredients is preferable [52, 54]. However, as bacteriocins can be degraded by proteases and peptidases often present in the food matrix, it is more advantageous to purify and immobilize them into the food packaging than to directly add them into the food [55, 56]. There exist several bioactive packing films such as a polyethylene film with the immobilized bacteriocin 32Y from *Lb. curvatus* active against *Listeria monocytogenes* during storage in packaged ground beef, pork steaks, and frankfurters [57, 58]. Other examples are nisin methylcellulose coatings for polyethylene films used on poultry meat, with the adsorption of nisin on polyethylene, ethylene, vinyl acetate, polypropylene, polyamide, polyester acrylics, and polyvinyl chloride [59]. On the other hand, *in situ* bacteriocin production is advantageous from legal and economic aspects when compared to *ex situ* production. The lower process cost and fewer legal obstacles to biotechnological production may be beneficial and attractive, especially for developing economies [52]. The most important bacteriocins from LAB from an industrial standpoint are presented here (Tables 11.2 and 11.3).

Nisin *Nisin*, produced by *Lc. lactis* subsp. *lactis*, was the first bacteriocin to be described in detail. In 1969, it was recognized as a food preservative (E234) by the Food and Agriculture Organization/World Health Organization (FAO/WHO), and in 1988, the US Food and Drug Administration (FDA) approved its use as an additive in canned products to inhibit the growth of *Clostridium botulinum*. Currently, this compound is the only bacteriocin authorized in the food industry in 45 countries worldwide [53, 138]. Nisin is a Class I bacteriocin, and presents two variants A and Z with a structural modification that does not affect the activity but gives nisin Z solubility and diffusion characteristics, important for food application [52]. Nisin is active against gram-positive pathogens and spoilage bacteria including *Staphylococcus aureus* and *L. monocytogernes*, even at nanomolar concentrations, depending on the target strain. The mechanism of action is the disruption of membrane function by pore formation in the bacterial cell membrane [52]. Nisin is commercially produced by several companies and one of the best known products is Nisaplin® (Danisco) that is produced by controlled fermentation of *Lc. lactis* in a milk-based medium. Other nisin manufacturers worldwide can be checked in the report of Jones *et al.* [60] while production values are presented in Table 11.3. Several researchers have aimed to optimize nisin production and to develop strategies combining this bacteriocin with other antimicrobial compounds. Jozala *et al.* [61] studied the optimization of nisin production from skimmed milk aiming to reduce the process cost. Gao and Ju [139] achieved the reduction in *C. botulinum* 33A spores by the combined effect of nisin (0–333 IU ml^{-1}), pressure (300.0–700.0 MPa), pressure holding times (7.5–17.5 min), and temperature (30–70 °C) in milk processing. Al-Holy *et al.* [140] completely eliminated *L. innocua* from

Table 11.2 Industrially relevant compounds produced by lactic acid bacteria and their applications.

Interesting metabolites	LAB involved	Applications/ functions	Commercial names	References
Food ingredients				
Bacteriocins				
Nisin	*Lc. lactis*	Food preservative in salami production	Nisaplin® (Danisco)	[53, 60, 61]
Pediocin	*P. acidilactici*, *P. pentosaceus*, *P. parvulus*, *Lb. plantarum*		ALTA® 2351 (Kerry Biosciences), Bactoflavor® Flora Italia (Chr. Hansen), SafePro® B-LC-20 (Chr. Hansen)	[53, 54, 62, 63]
Sakacin	*Lb. sakei*			
Aroma compounds				
Diacetyl	*Lc. lactis*, *Strep. thermophilus*, *Leuc. mesenteroides*	Improvement of aroma (buttery) to dairy products		[64, 65]
Acetaldehyde	*Lb. delbrueckii* subsp. *bulgaricus*, *Strep. thermophilus*	Main aroma compound of yogurt		[66]
Esters	*Lc. lactis*, *Lb. rhamnosus*, *Lb. casei*, *Lb. plantarum*, *O. oeni*	Confer fruity aroma when are added in food products; generally used in beverages, candies, dairy products, and wines		[64, 65, 67, 68]
Enzymes				
Starch-modifying enzymes: amylases and amylopullulanases	*Lb. amylophilus*, *Lb. amylovorus*, *Lb. acidophilus*, *Lb. fermentum*, *Lb. plantarum*, *Lb. paracasei*, *Strep. bovis*, *Strep. macedonicus*, *P. pentosaceus*, *P. acidilactici*, *Leuc. mesenteroides* subsp. *dextranicum*, *W. confusa*	Lactic acid production from starchy products. Starter cultures for the production of probiotics, cereal-based fermented foods (sourdough fermentation), and hypoallergenic foods for children		[69–72]

(*continued overleaf*)

Table 11.2 (*continued*)

Interesting metabolites	LAB involved	Applications/ functions	Commercial names	References
Proteases	*Lc. lactis*, *Lb. helveticus*	Cheese manufacture	Acelerase® (Danisco)	[73]
		Enzyme for modified cheese manufacture	Savorase® (Danisco)	[74]
		Casein and whey hydrolysate production	Debitrase® (Medipharm) Enzobact® (Medipharm)	[74]
Winemaking-related enzymes				
1. *Malolactic enzymes*	*O. oeni*	Malolactic fermentation		[75, 76]
2. *Proteolytic enzymes*	*Lactobacillus*, *Pediococcus*, and *Leuconostoc* genera	Removal of haze-forming proteins		[77]
3. *Glycosidases*	*O. oeni*	Improvement of wine sensorial profile		[78]
Nutraceutics				
Low-calorie sugars				
Mannitol	*Lb. fermentum*, *Lb. reuteri*, *Leuc. mesenteroides*, *Lb. intermedius*, *Leuc. fructosum*	Sweeteners in "sugar-free" and diabetic food products (food industry); osmotic diuretic (medicine); excipient (pharmaceutical industry)		[79–83]
Erythritol	*O. oeni*; *Lb. sanfranciscensis*, *Leuc. oenos*			[84–86]
Vitamins				
Folate (B_9)	*Lc. lactis*, *Strep. thermophilus*, *Lb. plantarum*, *Lb. delbrueckii* subsp. *bulgaricus*, *Lb. helveticus*, *Lb. rhamnosus*	DNA and RNA biosynthesis; amino acid repair, methylation, and interconversions; antioxidant competence		[87–104]
Riboflavin (B_2)	*Lc. lactis*, *Lb. plantarum*, *Leuc. mesenteroides*	Additive for animal feed; component of vitamin mixtures (medicine)		[98–101]
Cobalamin (B_{12})	*Propionibacterium freudenreichii*, *Lb. reuteri*	Required for the metabolism of amino acids, fatty acids, nucleic acids, and carbohydrates		[103–105]
Menaquinone (K_2)	*Lactococcus*, *Lactobacillus*, *Enterococcus*, *Leuconostoc* and *Streptococcus* genera	Involved in blood clotting, bone and kidney function, and tissue calcification		[106, 107]

Table 11.2 (*continued*)

Interesting metabolites	LAB involved	Applications/ functions	Commercial names	References
Other nutraceutics				
CLA	*Streptococcus, Lactobacillus, Leuconostoc, Enterococcus, Pediococcus, Propionibacterium* genera	Dietary supplement to provide beneficial effects (antidiabetic, antiatherogenic, anticarcinogenic, antiobesity)		[70, 108–111]
GABA	*Lb. brevis, Lc. lactis, Lb. paracasei, Lb. delbrueckii* subsp. *bulgaricus, Lb. buchneri,*	Starter cultures for production of GABA-enriched functional foods and beverages		[112–114]
	Lb. plantarum, Lb. helveticus, Lb. rhamnosus, Lb. hilgardii, P. pentosaceus, Strep. thermophilus, Strep. salivarius subsp. *thermophilus, Leuc. mesenteroides, Leuc. lactis, W. viridesce*	Food ingredient	PharmaGABA™ (natural factors)	[115, 116]
Seleno-proteins	*Strep. thermophilus, Lb. delbrueckii* subsp. *bulgaricus, Lb. acidophilus, Lb. paracasei, Bifidobacterium lactis*	Beneficial health effects (antioxidant activity, redox signaling, thyroid hormone, metabolism) when added to yogurt		[117, 118]
EPS	*Leuconostoc, Lactobacillus, Streptococcus, Pediococcus, Weissella* and *Oenococcus* genera	Food-grade additives for improving smoothness, rheological properties, texture, and stability of dairy products	Dextran from *Leuconostoc* spp. (Sigma-Aldrich Chemical CO)	[119, 122]

(*continued overleaf*)

Table 11.2 (*continued*)

Interesting metabolites	LAB involved	Applications/ functions	Commercial names	References
Bulk compounds				
Lactic acid	*Lactobacillus* genus, *Ent. faecalis, Ent. mundtii, Lc. Lactis*	Preservative (acidifier); flavor-enhancing agent (food industry); emulsifying agent (cosmetic industry). Substrate for production of polylactic acid		[55, 123–126]
Biofuels				
Ethanol	*Lb. pentosus, P. acidilactici, P. pentosaceus, Lb. plantarum*	Partial replacement of fossil-derived fuels		[127–129]
Butanol	*Lb. brevis, Lb. buchneri*			[130, 131]

brined white cheese stored at 4 and 10 °C by associating mild heating with nisin (1000 and 1500 IU ml^{-1}).

Pediocin *Pediocin* is a cationic membrane-permeabilizing antimicrobial peptide of Class IIa. The best-known pediocin is pediocin PA-1/AcH, which is a small antilisterial plasmid-encoded peptide produced by *P. acidilactici.* Pediocins are commercially traded as bacteriocin-containing fermentative powder by companies such as Kerry Bio-Sciences, Chr. Hansen, and Handary [52, 53] (Table 11.2). Other species such as *P. pentosaceus* and *P. parvulus* and strains belonging to *Lb. plantarum* are also capable of producing this compound [62, 63]. Although pediocin is mainly used in meat products, its production in heterologous hosts is considered an attractive alternative in dairy products [52, 141] (Table 11.3).

Sakacin *Sakacin* is produced by several strains of *Lb. sakei* used in fermented meat products. Sakacin is a Class IIa bacteriocin active against *Listeria* species; the most important type is sakacin A produced by *Lb. sakei* Lb706. Some commercial formulations of sakacin A manufactured by Danisco and Chr. Hansen are available in the market [52, 53]. More detailed information about other bacteriocins can be found in the review of Reis *et al.* [52].

Large-scale application of LAB bacteriocins requires yield optimization, determination of the activity spectrum, and improvement of their protease resistance through genetic modifications [55]. Several companies have been developing new methods to enhance the heterologous expression of bacteriocins using currently well-established industrial LAB to apply these antimicrobial substances in foods, for clinical applications, and in other areas [53]. Benmechernene *et al.* [53] have reviewed the most relevant patents regarding the innovations about bacteriocin applications both at food and clinical levels, in probiotics, for the inhibition of

Table 11.3 Examples of production values of industrially relevant metabolites by lactic acid bacteria.

Industrially relevant metabolites	LAB involved	Production	References
Food ingredients			
Bacteriocins			
Nisin	*Lc. lactis* UQ2	$20\,g\,l^{-1}$	[60, 61]
	Lc. lactis [YP-Nisin 1000 (YP-Bio-tech Co., Ltd)]	$49\,mg\,ml^{-1}$; $117\,mg\,ml^{-1}$	
	Lc. lactis ATCC 11454	$31.38\,mg\,ml^{-1}$; $82.80\,mg\,ml^{-1}$	
Aroma compounds			
Diacetyl	*Lc. lactis*[a)]	1.07–4.16 mM	[132]
Acetaldehyde	*Lc. lactis* NZ9000[a)]	21.3 mM	[66]
Esters (ethyl octanoate)	*O. oeni* AWRI B551	$33\,\mu g\,l^{-1}$; $175\,\mu g\,l^{-1}$	[68]
Nutraceutics			
Low-calorie sugars			
Mannitol	*Leuc. mesenteroides* ATCC 9135, *Lb. fermentum* NRRL	$16–26.2\,g\,l^{-1}$	[5, 79–83, 133–136]
	Lb. intermedius NRRL	$228\,g\,l^{-1}$	
	Leuc. Fructosum	$44\,g\,l^{-1}$	
	Leuc. Mesenteroides	$18\,g\,l^{-1}$	
	Lb. fermentum CRL 573	$56\,g\,l^{-1}$	
	Lb. reuteri, CRL 1101	$22\,g\,l^{-1}$; $38\,g\,l^{-1}$; $41.5\,g\,l^{-1}$	
Erythritol *Leuc. Oenos*		16–37 μmol	[84–86]
Vitamins			
Folate (B_9)	*Lb. bulgaricus* CRL 871- *Strep. thermophilus* CRL 803- *Strep. thermophilus* CRL 415	$180\,\mu g\,l^{-1}$ (yogurt)	[94, 95]
	Lc. lactis NZ9000[a)]	$80\,ng\,ml^{-1}$	
Riboflavin (B_2)	*Lb. plantarum* CRL 725	$700\,ng\,ml^{-1}$ (soy milk)	[101, 102]
	Lb. plantarum CRL 725[b)]	$1860\,ng\,ml^{-1}$ (soy milk)	
	Lc. lactis CB010[a)]	$1200\,\mu g\,l^{-1}$	
Cobalamin (B_{12})	*Lb. reuteri* CRL 1098	$50–500\,\mu g\,l^{-1}$	[103]
Other nutraceutics			
CLA	*Lb. plantarum* AKU1009a	$20\,mg\,ml^{-1}$; $40\,g\,l^{-1}$	[110, 111]
	Lb. plantarum JCM 1551	$2.7\,mg\,ml^{-1}$; $3.4\,mg\,ml^{-1}$	
	Lb. acidophilus-Lb. plantarum co-culture	$316.5\,\mu g\,ml^{-1}$	
GABA	*Lb. plantarum* DSM 19463	$59\,\mu M\,h^{-1}$	[112–116]
	Lb. brevis	$27.6\,mg\,ml^{-1}$	
	Lb. paracasei NFRI 7415	302 mM	
	Strep. thermophilus	$7984\,mg\,l^{-1}$	
EPS	*Weissella* subsp.	$16\,g$ EPS kg^{-1} dough	[119–121, 137]
	Lb. sanfranciscensis	$5\,g$ EPS kg^{-1} dough	

(*continued overleaf*)

Table 11.3 (*continued*)

Industrially relevant metabolites	LAB involved	Production	References
Bulk compounds			
Lactic acid	*Lb. helveticus* milano	$9.7\,g\,l^{-1}\,h^{-1}$	[55, 123–126]
	Lb. delbruekii	$57\,g\,l^{-1}$	
	Lb. delbruekii IFO 3202	$13.92\,g\,l^{-1}\,h^{-1}$	
	Lb. brevis 53 F4	$39.1\,g\,l^{-1}$	
	Lc. lactis IO-1	$10.85\,g\,l^{-1}$	
Biofuels			
Butanol	*Lb. brevis*	$0.8\,g\,l^{-1}$	[130, 131]

a) Genetically modified bacteria.
b) Spontaneous mutant riboflavin-overproducing strain.

biofilms, and also as coadjuvants of other antimicrobial agents in novel therapeutic strategies. To learn more about the patents, readers are invited to check the review mentioned above.

11.3.1.2
Aroma Compounds

Flavor is generally the result of the presence of many volatile and non-volatile components. The non-volatile compounds are mainly related to taste and the volatile ones contribute to both taste and aroma. Several chemicals, such as aldehydes, esters, dicarbonyl groups, alcohols, short to medium-chain free fatty acids, phenolic and sulfur compounds, methyl ketones, and lactones, can be used as flavoring compounds [64]. Flavors and fragrances are widely used in beverage, feed, cosmetic, food, detergent, chemical, and pharmaceutical formulations. Currently, they are mainly produced through artificial means particularly by chemical synthesis or by extraction from plant or animal sources. Flavor chemical synthesis is not often preferred by consumers and sometimes it can be quite expensive, time consuming, and non-eco-friendly. Synthesis of aroma compounds based on fermentation (*de novo* microbial processes) or biocatalysis (bioconversions of precursors using microbial cells or enzymes) such as aldehydes and ketones (acetaldehyde, diacetyl), lactones (gamma-decalactone), acids (acetic, butyric, caproic, caprylic, isobutyric, isovaleric, 2-methyl butyric), and esters (ethyl and butyl acetates, ethyl butyrate, caproate, isobutyrate, isovalerate, 2-methyl butyrate, menthyl acetate) are attractive alternative methods. Some advantages attributed to microbial production over chemical synthesis are that microbes are more eco-friendly and natural, and they can provide an abundant supply of uniform-quality products independently of the environmental conditions [65].

As food additives, flavor compounds may be incorporated into products from an external source or be naturally synthesized *in situ* as secondary metabolites by some starter cultures used in food manufacture. As an example of the latter case, certain LAB species involved in cocoa fermentation, such as *Lb. fermentum*, are responsible for the production of relevant flavor compounds [142, 143]. In this regard, LAB are able to produce interesting aroma compounds such as diacetyl, acetaldehyde, and esters (Tables 11.2 and 11.3).

Diacetyl *Diacetyl* (butanedione, $CH_3COCOCH_3$) can be industrially obtained by oxidation of 2-butanone using a copper catalyst at 300 °C, by dehydrogenation of 2,3-butanediol over a copper or silver catalyst in the presence of air, or with 3-hydroxy-2-butanone (acetoin) as a by-product [144]. On the other hand, this compound is naturally produced by LAB conferring a strong buttery aroma to many fermented dairy products (butter, buttermilk, and cheese). Diacetyl is synthesized by oxidative decarboxylation of the intermediate product α-acetolactate [55]. The most important diacetyl-producing LAB species have been shown to be *Lc. lactis*, *Lactobacillus* spp., *Strep. thermophilus*, and *Leuc. mesenteroides* [65]; *Lc. lactis* biovar. *diacetylactis* being widely known for producing large amounts of this compound [64]. Numerous metabolic engineering strategies have been designed to improve diacetyl production. Hugenholtz *et al.* [145] reported that an engineered strain *Lc. lactis* strain efficiently converted sugar into diacetyl by combining NADH-oxidase overproduction and alpha-acetolactate decarboxylase inactivation. Papagianni [132] has described other genetic modifications in LAB for diacetyl production; for more details, readers may check this review.

Acetaldehyde *Acetaldehyde* is also an aroma compound naturally produced by LAB in dairy products. It is one of the most important aroma compounds present in yogurt, where it can be synthesized by strains of *Lb. delbrueckii* subsp. *bulgaricus* and *Strep. thermophilus* [63]. There are at least two pathways for acetaldehyde production in LAB: (i) pyruvate decarboxylation by the pyruvate decarboxylase enzyme and (ii) threonine conversion into acetaldehyde and glycine by the threonine aldolase enzyme [55]. These enzymes have been submitted to metabolic engineering to increase acetaldehyde production by different LAB strains. Bongers *et al.* [66] overexpressed the *Zymomonas mobilis* pyruvate decarboxylase gene under the NICE system in *Lc. lactis* achieving efficient acetaldehyde production from glucose.

Esters *Esters* belong to other important aromatic chemical group that confers fruity aroma when they are added in food products. These compounds are generally used in some beverages, candies, baked goods, dairy products, wines, and so on. Ester derivatives (i.e., ethyl acetate, isoamyl acetate, hexyl acetate, and 2-phenylethyl acetate) may be responsible for flavor in grape-derived alcoholic beverages such us wine. In addition, certain LAB species such as *Lc. lactis* can produce ethyl esters and thioesters during cheese production [64, 65]. Abeijón Mukdsi *et al.* [67] studied short-chain fatty acid ester synthesis in several LAB strains isolated from goat's and ewe's milk and cheeses. The authors found that the strains mainly produced ethyl butanoate and ethyl hexanoate, which are desirable key flavor compounds in this kind of cheeses. Esterification was the prevalent mechanism for ester synthesis in lactobacilli strains such as *Lb. rhamnosus* ETC14, *Lb. casei* ETC19, and *Lb. plantarum* Ov156 while among enterococci isolates, the alcoholysis (transferase activity) mechanism was also involved. Costello *et al.* [68] reported that the strain *O. oeni* AWRI B551 produced important levels of ethyl hexanoate and ethyl octanoate, which contributed to the fruity sensory characteristics in wine. The best culture conditions for high production of ethyl esters by this strain was a combination of pH 4.5, anaerobiosis, and precursor supplementation. Two different enzyme activities involved in ethyl ester synthesis were found in strains of *O. oeni* and *Lb. plantarum* strains, acyl coenzyme A: alcohol acyltransferase and reverse esterase.

11.3.1.3 Enzymes

Although LAB have high nutritional requirements, these microorganisms have developed numerous strategies for nutrient scavenging from diverse matrices and environments, including several enzymatic activities. Thus, these bacteria may be considered as a potential source of a wide battery of different types of enzymes (Table 11.2). Various enzymes are released by LAB into the gastrointestinal tract, exerting potential synergistic effects on digestion and alleviating symptoms of intestinal mal-absorption [146]. From an industrial perspective, and referring to the topic related to this chapter, LAB are capable of synthesizing enzymes, which can affect the composition, processing, quality, and organoleptic characteristics of foods and feeds [37]. These bacteria can produce the enzymes directly into the food matrix when LAB are used as a starter or associated cultures or they can be used as a source of enzymatic extracts to be applied under the fermentation conditions [147]. Owing to their industrial interest, the following LAB enzymes have been extensively studied.

Starch-Modifying Enzymes Starch is one of the most abundant carbohydrates in terrestrial plants and the most important polysaccharide used by humankind. This polymer is normally processed and used in a variety of products such as starch hydrolysates, starch or maltodextrin derivatives, fructose, glucose syrups, and cyclodextrins [148, 149]. In addition, starch is widely used as a raw material in the paper industry, in polyol production, and as economic substrate for many microbial fermentations [149]. Starch consists of a large number of glucose units that can be linearly attached as helical amylose [~99% α-(1–4) and ~1% α-(1–6) bonds] or branched as amylopectin [~95% α-(1–4) and ~5% α-(1–6) bonds] [69]. In nature, four types of starch-converting enzymes exist: (i) endoamylases; (ii) exoamylases; (iii) debranching enzymes; and (iv) transferases.

Though LAB strains with starch-degrading activities are very rare, many amylolytic LAB (ALAB) have been isolated and several biochemical and genetic studies on starch hydrolysis have been performed. Fermented meals based on sorghum, maize, rice, millet, cassava, or taro [70, 71, 150–152]; fermented cereal beverages such as maize *pozol* [72], the non-alcoholic drink *bushera* [153], sorghum beer [154], and other beverages [155–157]; sorghum, maize, wheat, or rye sourdough [158–162]; and other habitats such as animal digestive tracts and plant and vegetal wastes have been sources of ALAB strains [72, 163–165].

Despite the fact that information from available genome sequencing and annotations revealed the presence of amylase genes in the majority of LAB, only strains belonging to the genera *Lactobacillus*, *Lactococcus*, and *Streptococcus* have been capable of producing lactic acid directly from starch as a carbon source. In general, the starch-modifying enzymes in ALAB belong to α-amylases, oligo-1,6-glucosidases, maltogenic amylases, pullulanases, amylopullulanases, neopullulanases, and glycogen phosphorylases, some of which have been reported as extracellular while others as intracellular enzymes [149].

The most efficient ALAB strains to utilize starch belongs to the *Lactobacillus* genus. The first ALAB species reported was *Lb. amylophilus* in 1979 [166]; this species together with *Lb. amylovorus* [163] possesses high amylase activity [149, 167, 168]. Also, many starch-degrading *Lb. acidophilus* strains showing intracellular [164], cell-bond, and extracellular activities were isolated [169]. *Lb. fermentum* has been described to produce extracellular or cell-bond amylases [149]. *Lb. plantarum* A6 was reported to synthesize an extracellular amylase enzyme [150], while the strain L137 presented an amylopullulanase able to hydrolyze starch, glycogen, pullulan, amylopectin, and, to a small extent, amylose, but with no activity on dextran and cyclodextrins [170]. More recently, *Lb. paracasei* B41 was identified as producing extracellular amylopullulanase [157]. Other starch-degrading *Lactobacillus* species reported in the literature are *Lb. amylolyticus* [155], *Lb. amylotrophicus* [171], and *Lb. manihotivorans* [151]. Among *Streptococcus* species, the strains *Strep. bovis* JB1 and 148 were able to degrade starch by producing intra- and extracellular α-amylases [172–174], while strains of *Strep. bovis* and *Strep. macedonicus* have reported amylolytic activity [72]. ALAB strains belonging to species such as *P. pentosaceus* and *P. acidilactici* [71]; *Leuc. mesenteroides* [71, 153, 175] and *Leuc. mesenteroides* subsp. *dextranicum*; and *W. confusa* [153] have also been reported.

The activity of the starch-modifying enzymes are dependent on the temperature and pH; the highest activity values reported being found with temperatures between 40 and 50 °C and a pH range of 4.5–6.0 with a few exceptions [149, 150, 167].

One of the main applications of these enzymes or the ALAB themselves is production of lactic acid from starchy products [149, 176, 177]. The current conversion of cellulosic and starchy materials into lactic acid is an expensive process considering that hydrolysis steps are required before fermentation. On the other hand, as ALAB can directly convert starch into lactic acid, unifying saccharification and fermentation steps in a single process, their use is economically attractive [149]. However, only a few ALAB strains convert starch in high substrate concentration; for instance, *Lb. amylophilus* GV6 [178], *Lb. amylovorus* ATCC 33620 [179], and *Lb. plantarum* A6 [180] produced interesting amounts (8–10% w/v) of lactic acid when fermenting soluble starch, although the yields were low (<4% w/v) when raw starchy substrates were used. In addition, ALAB strains could be applied in the production of cereal-based fermented foods and beverages [181] as well as in the elaboration of probiotics [71, 156, 182] or hypoallergenic children's foods [183, 184]. ALAB-producing extracellular amylases are also used as starters in sourdough technology for improving bread texture and shelf life of the product [73, 149, 185–187].

Proteases A LAB proteolytic system comprises three components [74, 188, 189]: (i) *Proteases* (cell wall-bound extracellular proteinases-CEP) able to degrade proteins (e.g., casein) producing oligopeptides, which can be classified into three groups according to their caseinolytic activity: (a) PI-type that predominantly degrade β-casein (e.g., from strains *Lc. lactis* subsp. *cremoris* HP and WG2); (b) PIII-type that degrade αs_1-, β-, and k-casein (e.g., from *Lc. lactis* subsp. *cremoris* AM1 and SK11); and (c) PI/PIII-type, which display substrate activity that is a combination of the two previous types. (ii) *A specific transport system* that transports oligopeptides generated by CEP into the cytoplasm. (iii) *Intracellular peptidases* that can fully degrade oligopeptides into shorter peptides and amino acids. According to their cleavage site specificity, these peptidases can be classified as: (i) aminopeptidases; (ii) endopeptidases; (iii) di- and tripeptidases; and (iv) proline-specific peptidases.

Proteases and peptidases are involved in two fundamental steps during cheese manufacture: coagulation and ripening, these enzymes playing a key role in the flavor, aroma, and texture of the cheese. LAB proteases and peptidases have been applied in ripening processes where cheeses acquire their differential sensorial characteristics through proteolysis, lipolysis, and glycolysis reactions. Starter LAB strains are source of proteolytic and lipolytic enzymes during the ripening process. These LAB hydrolyze intermediate-size and short peptides derived from casein after the coagulation process by CEP. Some studies reported that certain peptidases produced by *Lc. lactis* subsp. *cremoris* improved the sensory properties of cheese [37, 73]. On the other hand, noncoagulant proteases can be used in the dairy industry for casein and whey hydrolysate production as source of proteins with technological properties and reduced allergenicity, and in enzyme-modified cheese (EMC) manufacture [74]. Some LAB proteases and peptidases are now commercially available: (i) Acelerase®, an exopeptidase used for accelerated cheese ripening and Savorase® for EMC production, both synthesized by *Lc. lactis* are produced by Danisco. (ii) Debitrase® manufactured by Medipharm using a *Lc. lactis* strain is useful in whey processing and debittering, and (iii) Enzobact®, an aminopeptidase produced by *Lb. helveticus*, which is used for cheese debittering and flavor development [74]. Another developing approach in cheese maturation is the use of genetically modified bacterial strains. Cheddar cheese with increased amount of free amino acids has been produced using the recombinant *Lc. lactis* subsp. *cremoris* SK11 strain expressing the aminopeptidase PepN from *Lb. helveticus* CNRZ32 [190]; the same type of cheese with enhanced flavor and increased level of specific amino acids was developed by overexpressing either *pepN* or *pepC* in *Lc. lactis* subsp. *cremoris* NM1 [191]. A process with accelerated proteolysis and, hence, cheese ripening, was achieved with the recombinant strains of *Lc. lactis* MG1363 and *Strep. thermophilus* carrying a *Lb. bulgaricus* NCDO1489 protease (PrtP) gene (Patent EP0810289) [192]. The fact that the use of genetically modified foodstuffs is subjected to different laws worldwide and is not yet accepted by many consumers is the main drawback faced by the food industry.

Winemaking-Related Enzymes Winemaking involves two key groups of organisms: yeasts and LAB. Among yeasts, *S. cerevisiae* is the main species used for the primary or alcoholic fermentation (sugars are converted to ethanol and CO_2), while LAB, typically *O. oeni*, carry out the secondary or malolactic fermentation (MLF, decarboxylation of L-malic acid to L-lactic acid and CO_2) [193]. Apart from these two main reactions, several other processes take place during grape vinification where many flavor-, taste-, and quality-related compounds are synthesized, degraded, or modified. Certain enzymes are the main responsible for these changes; the source of these biocatalysts can be the grapes themselves, the grape indigenous microbiota, the inoculated microorganisms, microbes associated with wineries, or enzyme addition. A current strategy in winemaking is the addition of specific microorganisms or enzyme preparation to confer a particular group of enzymatic activities [77]. Numerous studies concerning the potential of LAB as source of enzymes or enzymatic activities have been reported; however, only few enzymes showed good activity under wine preparation conditions that include low pH (pH 3.0–4.0) and temperatures (<15 °C), presence of ethanol (15% (v/v) or higher) and SO_2, and high sugar (glucose and fructose) content. The most interesting activities are those provided by a single enzyme, ideally with an extracellular localization [77]. Winemaking-relevant LAB enzymes include the malolactic enzyme, proteases and peptidases, glycosidases, polysaccharide-degrading enzymes, esterases, ureases, phenoloxidases, and lipases [73, 77].

1) *Malolactic enzyme.* This enzyme is responsible for decarboxylating L-malic acid to L-lactic acid and carbon dioxide during MLF, which is an unpredictable step difficult to control or manipulate [75]. For this reason, research focusing on the optimization of L-malic acid to L-lactic acid conversion in grape juice and wine in order to separate this reaction from the often-difficult growth of LAB in wine has been carried out. For this purpose, some bioreactor systems comprising LAB cells immobilized alone [194–199], coimmobilized with yeasts [76], free *O. oeni* cells [200], or enzymes and cofactors [197, 201, 202] have been developed. In another approach, the *mleS* gene encoding the malolactic enzyme from *Lc. lactis* was cloned in a *S. cerevisiae* wine strain to perform the MLF and alcoholic fermentations simultaneously [203].
2) *Proteolytic and peptidolytic enzymes.* During winemaking, some undesirable haze is produced due to the precipitation of unstable grape proteins. Typically, bentonite is used to remove haze-forming proteins; however, this compound also removes assimilable nitrogen, modifying the characteristics of the fermentation [77]. The use of proteases has been thought as an alternative to bentonite treatment. Nevertheless, utilization of commercial enzyme

preparations or proteases from non-winemaking-related organisms such as *Aspergillus niger* has been unsuccessful under winemaking conditions [204–208]. Thus, the use of proteolytic and peptidolytic LAB belonging to the genera *Lactobacillus, Pediococcus,* and *Leuconostoc,* including several wine-related species is being evaluated [77], although further studies are needed to develop an outstanding application of their proteolytic enzymes.

3) *Glycosidases.* Several compounds closely related to the sensorial characteristics of wine can be present as their glycoside form linked to sugars such as monoterpenes, C13-norisoprenoids, and benzene derivatives (related to wine aroma and color), and aliphatic compounds (related to flavor). The characteristics and properties of glycosides and their corresponding aglycones are quite different; important wine-quality-related molecules are not sensorially detected when they are in the glycoside form. Consequently, glycosidase enzymes that cleave the sugar moiety from glycosides are relevant in the wine sensorial profile [77]. In this respect, some commercial wine isolates of *O. oeni* showed interesting glycosidase activities such as nitrophenyl glycosidase, β-D-glucosidase, β-D-xylopyranosidase, and α-L-arabinopyranosidase activities [78].

Other enzymatic activities that may be of relevance in winemaking are still in the early stages of studies; for more details, the readers may see the review of Matthews *et al.* [77].

11.3.2 Nutraceutics

DeFelice (2002) coined the term "Nutraceutical" from "Nutrition" and "Pharmaceutical," and defined it as "a food or part of food (components) that provides medical or health benefits, including the prevention and/or treatment of a disease." Although nutraceuticals are commonly used in marketing, the term has no regulatory definition in the majority of the countries worldwide. The production of nutraceutics (polyols, conjugated linoleic acids, GABA, vitamins, etc.) and other interesting metabolites such as EPS by LAB is presented here [209].

11.3.2.1 Low-Calorie Sugars

Sugar alcohols are part of the low-calorie sweeteners that can replace sucrose, lactose, glucose, or fructose in foods. These compounds are widely used in the pharmaceutical, medical, chemical, and food industries. Sugar alcohols or polyols are noncyclic hydrogenated carbohydrates where the carbonyl group (aldehyde or ketone) of the precursor sugar is reduced to the corresponding primary or secondary alcohol [210]. Thus, sorbitol comes from reduction of glucose, mannitol from fructose, xylitol from xylose, erythritol from erythrose, and so on.

Polyols have a sweet taste and low energy; they are nonmetabolizable, so they do not affect the blood sugar (glycemia) level. These compounds are either partially absorbed or not absorbed in the intestine where they can be metabolized by bacteria with the concomitant production of short-chain fatty acids thus conferring health benefits to the host [210, 211]. For this reason, these sweeteners are used in diabetic foods and in sugar-free and tooth-friendly products such as chewing gums. Furthermore, they promote the hydration of the colonic content acting as laxative or as a potent osmotic diuretic [212, 213]. Polyols are also used as texturizing agents, softeners, color stabilizers, and humectants not taking part in the Maillard reaction [214]. Because of their capacity to mask undesirable flavors and their low hygroscopicity, they are used in pharmaceutical formulations [215, 216]. Mannitol, sorbitol, and erythritol have all received GRAS status by the US FDA [217].

Sugar alcohols can be naturally found in fruits and vegetables and can be produced by certain bacteria, yeast, fungi, and algae. Only mannitol and erythritol are naturally produced by LAB (Tables 11.2 and 11.3), while sorbitol and xylitol can be synthesized through metabolic engineering processes. Nowadays, industrial production of sorbitol and mannitol are performed by catalytic reduction of a mixture of sugars (glucose/fructose or fructose syrup) with hydrogen gas and a nickel catalyst using high pressures and temperatures [218]. In addition to the high cost of this process using pure substrates and extreme conditions, separation and purification methods are required. In an attempt to solve this problem, investigations on the biotechnological production of low-calorie sugars by LAB have been carried out in the last decades.

Mannitol *Mannitol* is a six-carbon polyol obtained from reduction of fructose; as it is half as sweet as sucrose, mannitol is widely used in the food industry as a natural sweetener in "sugar-free" and diabetic food products [210]. Owing to its noncariogenic properties and the positive enthalpy of its solution in water imparting a "breath-freshening," it is used in chewing gums and candies. In the pharmaceutical industry, it is used as an excipient because it is chemically inert, has low hygroscopicity, and masks undesirable flavors [219, 220]. It is also used in medicine as a potent osmotic diuretic to prevent or treat acute renal failure, to remove toxic substances, to prevent kidney failure in surgeries, and to reduce brain edema [220].

Some heterofermentative LAB are capable to produce mannitol when fructose is present in the medium; these bacteria have the ability to use D-fructose as an alternative electron acceptor and convert it into D-mannitol using the mannitol 2-dehydrogenase enzyme (MDH) regenerating NAD(P)H, thus making it advantageous for microbial growth. von Weymarn *et al.* [79, 80] evaluated mannitol production by the strains *Lb. fermentum* NRRL B-1932 and *Leuc. mesenteroides* ATCC 9135, which showed a maximum yield of 93.6% and 96.6% with a volumetric productivity of 16 and 26.2 $g\,l^{-1}\,h^{-1}$, respectively. Saha *et al.* [81–83, 221, 222] studied mannitol production by *Lb. intermedius* NRRL B-3693 in different growth media; the authors reported the highest mannitol production (228 $g\,l^{-1}$) in an inulin-based medium. A patent on continuous fed-batch and recycled fermentative production processes for mannitol synthesis by this strain was published by Saha *et al.* [223]. Moniz *et al.* [224] and Carvalheiro *et al.* [133] evaluated mannitol production by heterofermentative LAB strains in hydrolyzed carob syrup-supplemented medium, the highest mannitol production ($44\,g\,l^{-1}$) being obtained with a *Leuc. fructosum* strain. Fontes *et al.* [134] reported a production of $18\,g\,l^{-1}$ with a strain of *Leuc. mesenteroides* and cashew apple juice as carbon source. More recently, Rodríguez *et al.* [135] studied the production of mannitol by the strains *Lb. mucosae* CRL 573 and *Lb. reuteri* CRL 1101 in MRS to a fructose/glucose (6.5 : 1.0) mixture as carbon source at different pH values; the highest production (56 and $22\,g\,l^{-1}$, respectively) being attained at pH 5.0. Ortiz *et al.* [136] evaluated mannitol synthesis by *Lb. reuteri* CRL 1101 using sugarcane molasses as carbon source in the culture medium; a mannitol concentration of $38.4\,g\,l^{-1}$ was obtained with 7.5% (w/w) of sugars from molasses at 37 °C under agitation conditions after 24 h. More recently, the same authors [225] improved mannitol production ($41.5\,g\,l^{-1}$) by this *Lactobacillus* strain using a simplified culture medium containing sugarcane molasses as carbon source, yeast extract, and beef extract at constant pH of 5.0. Mannitol production by BAL is summarized in Table 11.3.

Erythritol *Erythritol* is a four-carbon sugar alcohol derived from erythrose that has properties similar to those of other polyols. As this compound possesses the lowest caloric value among all polyols, it is used in "sugar-free" and "tooth-friendly" foods. Erythritol is rapidly absorbed in the small intestine, making it unviable for colonic fermentation, generating a glycemic response of zero. These nutritional properties are due to its small molecular size [226]. In addition, it is a caries-reducing agent because it can decrease the presence of *Strep. mutans* in the mouth [227]. Veiga-da-Cunha *et al.* [84] described the production and the biosynthetic route of erythritol by LAB. They reported that *O. oeni* produced erythritol anaerobically from glucose. In these conditions, the ratio NADPH/NADP is high, inhibiting the enzymes glucose 6-phosphate dehydrogenase and 6-phosphogluconate dehydrogenase producing the accumulation of glucose 6-phosphate and fructose 6-phosphate. Then, a nonspecific phosphoketolase cleaves fructose-6-phosphate, leading to the formation of erythrose 4-phosphate, and finally of erythritol. Richter *et al.* [84] showed that under pantothenic acid limitation (CoA deficiency), *O. oeni* shifted the metabolism from heterolactic to erythritol/acetate fermentation to regenerate $NAD(P)^+$. Stolz *et al.* [86] studied the fermentation of maltose present in sourdough by lactobacilli. They demonstrated that resting cells of *Lb. sanfranciscensis* LTH 1729 and LTH 2581 produced 20% of erythritol from maltose in the presence of citrate under stress and anaerobic conditions.

Metabolic engineering techniques have been applied to modify bacterial metabolism to direct it to the synthesis of compounds of interest. Ferain *et al.* [228] and Neves *et al.* [229] reported the production of mannitol by *Lb. plantarum* and *Lc. lactis* (homolactic bacteria) strains when the lactate dehydrogenase gen was disrupted (Δldh). In these cases, mannitol was produced by reduction of fructose-6-phosphate to mannitol-1-phosphate and further dephosphorylated. Gaspar *et al.* [230] generated a *L. lactis* (Δ*ldh*) deficient in mannitol transport system (*mtlF* or *mtlA* genes), to prevent mannitol consumption, a common feature in homofermentative LAB. These mutants were able to produce 2.4 g l^{-1} of mannitol from glucose. Aarnikunnas *et al.* [231] observed that in heterofermentative LAB (*Lb. fermentum*), the disruption of D- and L-*ldh* genes lead to the production of mannitol and pyruvate instead of lactic acid. Furthermore, *Leuc. pseudomesenteroides* containing a mutation in the fructokinase gene showed normal fructose uptake and improved its conversion into mannitol [232].

Sorbitol is not naturally produced by LAB; in this regard, Nissen *et al.* [233] used the lactose operon of a *Lb. casei* strain to control the expression of D-sorbitol 6-phosphate dehydrogenase (*lac::gutF*). These engineered strains were able to synthesize sorbitol from glucose. In addition, the authors inactivated the *ldh* gen increasing 2.5-fold the sorbitol production. De Boeck *et al.* [234] continued working with the modified *Lb. casei* (*lac::gutF* Δ*ldh*) strain inactivating the gene-encoding mannitol 1-phosphate dehydrogenase and sorbitol phosphotransferase system (PTSGutB) (responsible for sorbitol utilization). This strain showed an increase sorbitol production, whereas no mannitol synthesis was detected. Similar strategies were used by Ladero *et al.* [235] in a strain of *Lb. plantarum* deficient in both L- and D-*ldh*. The authors expressed the two enzymes of sorbitol-6-phosphate dehydrogenase (*srlD1* and *srlD2*) under the control of *ldh*L gene of *Lb. plantarum*, achieving a high efficiency conversion (61–65%) of glucose into sorbitol.

As sorbitol, xylitol is not naturally synthesized by LAB, its production was obtained by heterologous expression [236] of the D-xylose reductase gene from the yeast *Pichia stipitis* CBS 5773 and the xylose transporter gene from *Lb. brevis* ATCC 8287 in the strain *Lc. lactis* NZ9800. D-xylose was quantitatively converted to D-xylitol, reaching an amount of 54.4 g l^{-1} similar to that produced by yeast.

More detailed information on polyol production by LAB or on the metabolic engineering strategies applied for the synthesis of these compounds by LAB can be found in the reviews of Saha and Racine, Moon *et al.*, and Ortiz *et al.* [220, 226, 237].

11.3.2.2
Vitamins

Vitamins are micronutrients that are essential for the metabolism of all living organisms. They are involved in essential functions such as cell metabolism and antioxidant activities, synthesis of nucleic acids, and are precursors of intracellular coenzymes necessary to regulate vital biochemical reactions in the cell [73, 238]. Human and animals cannot synthesize vitamins or they are formed in insufficient amounts, so they must be regularly consumed with the diet or exogenously. Some yeasts, fungi, and bacteria can produce vitamins; in this regard, certain gastrointestinal LAB species are able to synthesize B-group vitamins (i.e., folate or vitamin B_9, riboflavin or vitamin B_2, vitamin B_{12} or cobalamin, vitamin K_2 or menaquinone) (Tables 11.2 and 11.3). In addition, the presence of vitamins has been found in foods fermented by LAB such as cheese, yogurt, cultured buttermilk [11, 105].

Folates *Folates* are folic acid derivatives that are naturally present in foods. Folic acid is the chemically synthesized form of folate. Folates are involved in many metabolic pathways such as DNA and RNA biosynthesis, repair and methylation, and amino acid interconversions. These compounds possess antioxidant competence that protects the genome by preventing free radical attack [238]. In humans, folate deficiency is associated with a variety of disorders such as coronary heart disease, osteoporosis, Alzheimer's disease, and increased risk of breast and colorectal cancer [239].

Sybesma *et al.* [87] reported for the first time that LAB species such as *Lc. lactis, Strep. thermophilus,* and *Leuconostoc* spp. were capable of producing folates while the majority of the *Lactobacillus* spp. with the exception of *Lb. plantarum* and *Lb. delbrueckii* subsp. *bulgaricus* were not. LAB are capable of producing different forms of folate. *Lc. lactis* produces folate which remains in the cell as 5,10-methenyl-tetrahydrofolate (THF) and 10-methyl-THF, both with four, five, or six glutamate residues. In *Strep. thermophilus,* folate is also retained inside the cell although as 5-formyl-THF and 5,10-methenyl-THF, both with three glutamate residues [238]. Genes encoding folate synthesis were identified in *Lb. plantarum* [88], *Lc. lactis* [89], and in *Lb. delbrueckii* subsp. *bulgaricus* although incompletely in this last species [90].

In situ production of folates in fermented foods was studied by several authors. Crittenden *et al.* [240] demonstrated that the folate content increased in skim milk fermented by *Bifidobacterium* with *Strep. thermophilus* and/or *Lb. delbrueckii* subsp. *bulgaricus* strains. Goswami [91] reported that the strains *Lb. helveticus* MTCC 5463 and *Lb. rhamnosus* MTCC 5462 increased the folate content in fermented milks. Holasova *et al.* [92] studied the folate content in yogurt fermented with strains of *Strep. thermophilus*; after 12 h incubation, a sixfold higher amount of 5-metyl-THF as compared to the control was achieved. Murdock and Fields [93] fermented corn flour with LAB and observed a threefold folate increase after 4 days of fermentation. More recently, Laiño *et al.* [94] studied 15 different yogurts fermented with combined folate producing strains of *Lb. delbrueckii* subsp *bulgaricus* and *Strep. thermophilus*; the authors found a 2.5-fold higher folate amount in the yogurt than in the nonfermented milk.

Interestingly, it is possible to increase folate biosynthesis through metabolic engineering strategies. For example, a 10-fold increase of extracellular folate production in *Lc. lactis* was achieved when the 6-hydroxymethyl-dihydropterinpyrophosphokinase (*folK*) and GTP cyclohydrolase (*folE*) genes were overexpressed [95]. Wegkamp *et al.* [96] converted the strain *Lb. gasseri* ATCC 33323 from a folate consumer into a folate-producing strain by introducing a plasmid harboring the complete folate gene cluster (*folA, folB, folKE, folP, ylgG,* and *folC*) from *Lc. lactis* MG1363.

Riboflavin *Riboflavin*, also named vitamin B_2, is very important for cellular metabolism as it is precursor of the coenzymes flavin mononucleotide and flavin adenine dinucleotide acting as hydrogen carriers in biological redox reactions. A deficiency of these vitamins can produce damages in skin, liver, changes in the brain glucose metabolism with symptoms like hyperemia, sore throat, edema of oral and mucous membranes, cheilosis, and glossitis [97, 238]. Currently, the majority of the produced riboflavin is used in agriculture as an additive for animal feed, in the food industry as a yellow colorant, and in medicine as a component of vitamin mixtures.

Most of riboflavin-producing LAB were obtained by selecting roseoflavin-resistant mutants. Burgess *et al.* [98, 99] obtained by this means riboflavin-producing strains of *Lc. lactis, Lb. plantarum*, and *Leuc. mesenteroides*. Capozzi *et al.* [173] screened roseoflavin-resistant LAB strains from durum wheat flour and isolated two strains of *Lb. plantarum* that overproduced riboflavin. These strains were used to bioenrich bread and pasta, achieving a two- and threefold more riboflavin content in these foods, respectively. Juarez del Valle *et al.* [101] evaluated the ability to produce riboflavin in 179 LAB isolates from food products; the strain *Lb. plantarum* CRL 725 was able to increase the riboflavin concentration up to 700 ng ml^{-1} in soymilk after 12 h fermentation. When this strain was exposed to roseoflavin to obtain resistant variants, one variant was able to increase six times the initial riboflavin level in soymilk.

Metabolic engineering approaches were used to convert *Lc. lactis* subsp. *cremoris* NZ9000 from a riboflavin consumer into a riboflavin producer by overexpressing the biosynthetic genes (*ribG, ribH, ribB*, and *ribA*) [89]. In another study, a *Lc. lactis* strain was modified through directed mutagenesis followed by selection and metabolic engineering to produce folate and riboflavin [102].

Cobalamin *Cobalamin* or vitamin B_{12} is another vitamin required for the metabolism of amino acids, fatty acids, nucleic acids, and carbohydrates. Cobalamin deficiency can cause severe and irreversible damage in the brain, nervous, cardiovascular, and hematopoietic (pernicious anemia) systems [4]. Animals, plants, and fungi do not synthesize vitamin B_{12}, which can only be produced by some microorganisms. To date, the best cobalamin producer is *Propionibacterium freudenreichii*, which is currently applied for industrial production.

LAB are traditionally known to be auxotrophic for cobalamin. However, Taranto *et al.* [103] described for the first time the production of a cobalamin-like compound by the strain *Lb. reuteri* CRL 1098. Cobalamin concentrations of 50–500 μg l^{-1} were detected in cell extracts of this strain using different bioassays; its cobalamin synthesis was confirmed by genetic analysis finding at least 30 genes involved in the *novo* synthesis [104]. Other *Lb. reuteri* strains (DCM 20016, JCM1112, and CRL 1324 and 1327) were reported also to synthesize this vitamin [105]. Cobalamin production by other species such as *Lb. coryniformis* has also been studied [241].

Vitamin K *Vitamin K* is an essential cofactor for the formation of γ-carboxyglutamic acid (Gla) residues in proteins, which can bind calcium ions and influence blood clotting, bones, and kidney function, tissue calcification, and atherosclerotic plaque. Deficiency in vitamin K is associated with clinical disorders such as intracranial hemorrhage in newborn infants and possible bone fracture resulting from osteoporosis [73]. Vitamin K is present in green plants as vitamin K_1 and as menaquinone, which is produced by some intestinal LAB genera such as *Lactococcus, Lactobacillus, Enterococcus, Leuconostoc,* and *Streptococcus* [106]. In addition, Cooke *et al.* [107] described the production of this vitamin in *Enterobacter agglomerans, Serratia marcescens,* and *Ent. faecium.*

More information about vitamin production by LAB and their applications in foods is reported in the works of Hugenholtz *et al.* [7], LeBlanc *et al.* [167], Papagianni *et al.* [132], Patel *et al.* [73], and Panagiota Florou-Paneri *et al.* [37].

11.3.2.3
Conjugated Linoleic Acids

Conjugated linoleic acids (CLAs) are a group of at least 28 positional and geometric isomeric forms of linoleic acid (octadecadienoic acid) with conjugated double bonds. CLA isomers can be classified as cis–cis, trans–trans, cis–trans, and trans–cis isomers of 7, 9; 8, 10; 9, 11; 10, 12; and 11, 13 C18 diene acids. These compounds are naturally present in meat, milk, and dairy foods derived from ruminants [108]. However, the CLA levels in dairy products are insufficient to cause a physiological response [242]; for this reason, CLA are used as dietary supplements to enhance their positive effects on human health (Table 11.2). These lipids have shown antidiabetic, antiatherogenic, anticarcinogenic, antiobesity, ant-inflammatory, and antioxidant properties among other beneficial effects [55, 109, 243, 244]. Nevertheless, only some isomers such as the *cis*-9, *trans*-11 (c9, t11 CLA; rumenic acid), and the t10, c12 CLAs have been reported as biologically active compounds [245–247]. Interestingly, biological production of CLA has the advantage of being more isomer selective than chemical processes; in addition, it is possible to modify the isomer ratio by changing the bacterial culture conditions [248].

LAB are able to produce biologically active CLA isomers by biohydrogenation and oxidation processes and also by hydration and dehydration reactions [55], probably as key step for fatty acid detoxification in bacteria [249, 250]. Several LAB belonging to the genera *Streptococcus, Lactobacillus,* and *Leuconostoc* are capable of synthesizing CLA when LA is present in MRS medium [172, 173]. Kishino *et al.* [109] reported that some LAB strains belonging to the *Enterococcus, Pediococcus, Propionibacterium,* and *Lactobacillus* genera were able to produce significant amounts of CLA. Among them, the strain *Lb. plantarum* AKU 1009 was selected as a potential CLA-producer strain synthesizing 20 mg CLA ml^{-1} reaction mixture after 96 h of incubation using 2.6% (wt/v) LA as substrate (Table 11.3).

As some experimental results suggest that different CLA isomers are correlated with dangerous health effects, it is important to select the beneficial isomer-producing strains [55]. Aiming for a nutraceutical application, Kishino *et al.* [109] developed an isomer-selective method for CLA production by *Lb. plantarum* AKU1009a (40 g l^{-1}) controlling the ratio produced between c9, t11 (biologically active), and t9, t11 (less biologically active) CLA. In addition, several studies on the production of CLA by LAB using different substrates were carried out. Ando *et al.* [190] used castor oil pretreated with lipases and ricinoleic acid to evaluate CLA synthesis by a *Lb. plantarum* strain which produced 2.7 and 2.4 mg ml^{-1} CLA after 99 and 90 h of incubation, respectively. Ye *et al.* [191] used response surface methodology to establish the optimal fermentation parameters for CLA production in skim milk supplemented with hydrolyzed safflower oil as substrate. A maximum CLA production (316.5 μg ml^{-1}) was achieved with a *Lb. acidophilus–Lb. plantarum* coculture. Kim and Liu [251] analyzed sunflower oil (70% of LA)-supplemented-milk fermentation by 14 LAB strains and the results showed that the strain *Lc. lactis* I-01 was the best CLA producer. In general, the achieved CLA production was lower than that obtained when free LA was used as substrate [248]. Although the industrial production of CLA by LAB is not yet established, the study of these biological systems may provide novel strategies for developing functional foods bioenriched with CLA isomers.

11.3.2.4
γ-Aminobutyric Acid

γ-Aminobutyric acid (GABA) is a nonprotein amino acid widely produced by different organisms displaying a variety of interesting functions [252]. In vertebrates, GABA is one of the major inhibitory neurotransmitters in the central nervous system. This compound has different physiological functions including induction of hypotension, controlling lipid levels in serum, diabetes prevention, immune cells stimulation, inhibition of cancer cell proliferation, and diuretic and tranquilizer effects [112, 253–260]. GABA is involved in regulation of important processes such as sleeplessness, depression, autonomic disorders [261], stress, pain, and anxiety [55, 255]. Owing to these valuable benefits on human health, there is an increasing industrial interest in this compound. Biosynthetic methods of GABA are preferable over chemical synthesis [252]. GABA is biosynthesized in an irreversible one-step reaction where glutamic acid is decarboxylated by glutamate decarboxylase (GAD), a pyridoxal-5′-phosphate-dependent enzyme. Among microorganisms, the main GABA producers are LAB, where production is related with pH tolerance and ATP synthesis [113, 114].

The GABA-producing ability is species- and strain-dependent in LAB and has been described in several species including *Lb. brevis* [115, 262–268], *Lc. lactis* [115, 264, 269–274], *Lb. paracasei* [264, 272, 275], *Lb. delbrueckii* subsp. *bulgaricus* [264], *Lb. buchneri* [276–278], *Lb. plantarum* [115, 264, 272, 279], *Lb. helveticus* [280], *Lb. rhamnosus* [272], *P. pentosaceus* [272], *Strep. thermophilus* [272], *Leuc. mesenteroides* [115], *Leuc. lactis* [115], and *W. viridesce* [115] (Tables 11.2 and 11.3). The best GABA-producer strains belong to the genus *Lactobacillus* [112]. GABA production can be improved by optimizing different factors and fermentation parameters such as incubation and fermentation time, temperature, pH, and the presence of culture additives such as glutamate. *Lb. plantarum* DSM 19463 produced the highest amount of GABA (59 $\mu M\,h^{-1}$) when the incubation temperature was between 30 and 35 °C, while 37 °C was the optimum for the strain *Lb. brevis* CGMCC 1306. During black raspberry juice fermentation, maximum GABA production (27.6 mg ml^{-1}) by *Lb. brevis* GABA 100 was achieved at pH 3.5 and 30 °C after 12 days. In another study, the highest GABA amount (302 mM) synthesized by *Lb. paracasei* NFRI 7415 was found at 37 °C, decreasing drastically at 43 °C [252, 266, 275, 281]. The optimum pH value for GABA production is species dependent and varies in the range 4.5–8.0. Thus, maximum GABA synthesis by *Lb. plantarum* DSM 19463 (59 μM h^{-1}) was attained at pH 6.0, while for *Lb. paracasei* NFRI 7415, the optimum production (210 mM) was attained at pH 5.0. *Strep. thermophilus* produced the maximum GABA concentration (7984 mg l^{-1}) at pH 4.5, while in *Lb. lactis*, production was highest (7.2 g l^{-1}) at higher pH values (7.5 and 8.0) [252, 282]. Regarding the fermentation time, the highest GABA synthesis 26.5 mg ml^{-1} (37 °C, pH 5.5) was reached after 15 days of black raspberry juice fermentation by *Lb. brevis* GABA100. *Lb. plantarum* DSM 19463 produced the maximum GABA amount (4.83 mM) after 72 h, while *Lb. paracasei* NFRI 7415 required 144 h to produce 60 mM of GABA [113, 252, 281]. One of the major factors affecting GABA production is the presence of glutamate and pyridoxal-5′-phosphate (PLP). *Lb. paracasei* NFRI 7415 produced 161 mM GABA in a glutamate-containing medium after 144 h. In contrast, *Strep. thermophilus* Y2 did not considerably improve GABA production when 10–20 g l^{-1} of glutamate was added to the medium, while the presence of PLP markedly increased GABA synthesis by the strains *Strep. salivarius* subsp. *thermophilus* Y2, *Lb. paracasei* FRI 74150, and *Lb. plantrum* C48 [252, 275, 282].

Diverse strategies have been used to enhance GABA biosynthesis by LAB: (i) *optimization based on GAD properties* [112, 275, 279, 282] as reported in the study using response surface methodology based on the GAD activity from *Lb. plantarum* Taj-Apis362 [279]. (ii) *Grading-controlling fermentation*, as high cell density is necessary for effective synthesis of GABA although optimal cell growth and GABA-synthesis conditions do not always fit [176, 283]. (iii) *Immobilized cell technology* is a promising technique for efficient GABA production [112, 268, 284]. In this respect, Choi *et al.* [268] optimized a semicontinuous cell entrapment system to produce GABA by *Lb. brevis* GABA057; glutamate conversion increasing from 2% to 12% (w/v).

GABA-producing LAB may be applied as *starter cultures* to produce GABA-enriched functional foods and beverages, in dairy products [221, 258, 260], black raspberry juice [281], cheese [273], kimchi [285], and soymilk [286]. In addition, they can be used to produce GABA as a food ingredient; a commercial natural GABA formulation known as PharmaGABA™ produced by using the kimchi strain *Lb. hilgardii* K-3 is licensed by the US FDA [115, 116].

11.3.2.5
Seleno-Proteins

One interesting property of LAB is their capacity to bind, uptake, and biotransform metal ions present in the environment [287]; in this regard, LAB can transform selenite into elemental Se_0 and organic selenium (Se) [288] (Table 11.2). Se is an essential element for almost every organism. Se is present in nature in the inorganic chemical forms Se_2^{-2}, SeO_4, and $SeO_3^{2,-}$ which are toxic, while Se_0 is nontoxic and biologically available [289]. Se can be covalently bound to proteins in their active site as seleno-cysteine (SeCys); seleno-proteins are essential for several functions within our cells as these proteins have antioxidant activity, are involved in redox signaling, and in thyroid hormone metabolism as well as in Se transport and storage [290]. Se deficiency affects 1 billion persons in the world; the symptoms associated with its low intake being poor immune function, oxidative stress, cancer, and increased risk of mortality [291].

Certain LAB can insert Se into cysteine, which can further be incorporated into proteins. It has been observed that LAB can grow in low Se concentrations and that this mineral can affect its protein expression. Some authors [117, 118] showed that yogurt strains could release seleno-proteins and produce SeCys and SeMetCys during their growth in milk; the resulting fermented milk had adequate sensorial characteristics and seleno-peptides could be found in yogurt samples after a simulated gastrointestinal digestion.

11.3.2.6 Exopolysaccharides

Several LAB synthesize exocellular polysaccharides, which include both capsular polysaccharides (CPSs) and slime EPS. The latter can be loosely bound to the cell wall or released into the extracellular environments [119]. The reason for EPS being produced by LAB remains still unclear; however, it has been claimed that these biopolymers may protect the cells from adverse environmental conditions [55].

EPS from LAB have become industrially attractive compounds due to their great potential economic applications in the food industry as natural and safe food additives for improving smoothness, rheological properties, texture, mouthfeel, stability, and water retention of the final (mainly dairy) food products (Table 11.2). EPS are also important in the clinical and pharmaceutical industries. More recently, it has been suggested that these biopolymers display health-promoting properties such as antiulcer, antitumor, cholesterol-lowering, or immunomodulating activities, and that they could act as prebiotics [292, 293].

EPS are commonly classified according to their monomer composition into two different groups: homopolysaccharides (HoPSs) and heteropolysaccharides (HePSs). HoPS are composed of one monosaccharide type: glucose (α-glucans or β-glucans), fructose (β-fructans, levan-type or inulin-type), or galactose (polygalactan). These polymers are mainly synthesized extracellularly using sucrose as substrate, this reaction being catalyzed by a single type of enzyme transglycosylase or glycansucrase (glucansucrases and fructansucrases). In addition, this enzyme can also catalyze the synthesis of low-molecular-mass oligosaccharides (fructooligosaccharides and glucooligosaccharides), with potential importance as prebiotics [55, 119]. LAB capable of synthesizing α-glucans belong to the genera *Leuconostoc*, *Lactobacillus*, *Streptococcus*, *Pediococcus*, and *Weissella* while β-glucans are synthesized intracellularly without sucrose as substrate mainly by LAB strains isolated from alcoholic fermented beverages (*Pediococcus, Lactobacillus,* and *Oenococcus*) [120]. Fructan production has been reported for *W. confusa, Lb. reuteri, Lb. pontis,* and some *Leuconostoc* and *Streptococcus* strains [119]. Polygalactans are rare and were only reported to be produced by strains of *Lc. lactis* subsp. *lactis* [283] and *Lb. delbrueckii* subsp. *bulgaricus* [294]. LAB produce HoPS in lower amounts compared to other bacteria such as *Xanthomonas campestris,* which produces xanthan a polymer already applied in the food industry as a viscosifying agent. Production of HoPS from LAB has been intensively studied during sourdough fermentation; *Weisella* spp. and *Lb. sanfranciscenis* strains produced 16 and 5 g EPS kg^{-1} dough, respectively [137] while *Lb. reuteri* Lb121 was able to produce up to 10 g l^{-1} growing on sucrose [295] (Table 11.3). The presence of EPS had a positive effect on the viscosity, volume, firmness, and mouthfeel in cereal and pseudocereal doughs [137] Currently, the only commercially exploited EPS produced by LAB is the HoPS dextran synthesized by *Leuc. mesenteroides* NRRL B512F, which produces a linear polysaccharide containing 95% α-1,6 linkages. This polymer is employed in the biochemical and pharmaceutical industries [121].

In contrast to HoPS, HePS are complex polymers composed of repeating subunits (consisting of three to eight monosaccharides) usually composed of glucose, galactose, xylose, mannose, arabinose, and rhamnose while amino-sugars, polyols, or glucuronic acid can be occasionally present [55]. HePS are intracellularly synthesized, and exported and polymerized outside the cell. Production levels for these polymers are lower than those for HoPS with reported yield values between 25 and 600 mg l^{-1} [119]. Although most EPS are not industrially produced, the *in situ* food production of biologically active and/or texturing EPS is consider as a promising alternative due to the GRAS status of these compounds [119].

A broad diversity of HePS, regarding their monomer composition, molecular mass, and yields, from LAB has been reported by Mozzi *et al.* [294] and Vaningelgem *et al.* [122]. Interestingly, these latter authors observed that small quantities of EPS remarkably increase the texture of fermented milks.

Different metabolic engineering approaches have been developed for improving production yields or for structural engineering of EPS produced by LAB [66]. However, to date, only a modest increase in EPS production has been achieved [55].

11.3.3 Bulk Compounds

11.3.3.1 Lactic Acid

Lactic acid (LA) or 2-hydroxypropanoic acid, is a naturally occurring hydroxycarboxylic acid and the sole or main final fermentation product produced by LAB. LA is widely used in the food, pharmaceutical, cosmetic, textile, and chemical industries and it is considered GRAS and recognized as harmless by the US FDA. In the food industry, LA is used as a preservative (acidifier) and flavor-enhancing agent, while it is used as an emulsifying and moisturizing agent in the cosmetic industry [123]. Recently, the demand for LA increased because of its use as feedstock for the production of poly-lactic acid (PLA), an environmentally friendly biodegradable plastic used in medical applications that can replace synthetic plastics derived from petroleum resources [124] (Table 11.2).

LA can be produced by humans, animals, plants, and microorganisms. In nature, it can occur as two optical isomers D(−) and L(+); L(+)-lactic acid is preferred in the food and drug industries as high levels of the D(−) isomer are harmful to humans [296]. Industrial production of LA can be synthesized either chemically or biotechnologically through fermentation. Chemical synthesis is based on hydrolysis of lactonitrile, derived from acetaldehyde and hydrogen cyanide, but this synthesis results in a racemic mixture of D- and L-lactic acid, which is difficult and expensive to separate. The biotechnological process may yield either D or L alone or a mixture of both in different ratios depending on the microorganism, substrate, and conditions used. Currently, almost 90% of LA is produced by fermentation using *Lactobacillus* spp. or *Ent. mundtii* as the main producer microorganisms. Glucose, sucrose, lactose, and starch/maltose derived from feedstocks such as beet sugar, molasses, whey, and barley malt are commonly used as carbon source [123]. Microorganisms used for the synthesis of LA are homolactic LAB such as *Lactobacillus* strains; these bacteria produce almost 90% of LA as end product from sugar fermentation through the EMP pathway and the pentose phosphate (PP)/glycolic pathway, when hexose and pentose are used as carbon source, respectively. The theoretical yield of LA from glucose is $1.0\,g\,g^{-1}$ ($2.0\,mol\,mol^{-1}$) via the EMP pathway, while pentose exhibits a theoretical yield of $1.0\,g\,g^{-1}$ ($1.67\,mol\,mol^{-1}$) of LA via the PP/glycolic pathway [186]. The species *Lb. amylophilus, Lb. brevis, Lb. buchneri, Lb. casei, Lb. delbrueckii, Lb. rhamnosus, Lc. lactis,* and *Streptococcus* spp. can produce L(+) LA; the D(−) isomer can be synthesized by *Lb. coryniformis,* while a mixture of isomers is produced by *Lb. helveticus, Lb. plantarum,* and *Lb. pentosus* [123]. The main fermentation processes and parameters affecting LA production were recently reviewed by Ghaffar *et al.* [297] and Wang *et al.* [123]. In batch fermentations with LAB, using glucose as carbon source, temperatures of 35–45 °C, pH values of 5.0–6.5% and 5–10% inoculum are the usual conditions applied. Other methods used are fed-batch, repeated-batch, and continuous-batch fermentations; although higher LA productivity is obtained using continuous fermentations, batch- and fed-batch cultures enable obtaining higher LA concentration.

As the raw material for LA fermentative synthesis is almost 34% of the total manufacturing cost, the use of cheap alternative sources has been extensively studied [123, 297]. In this respect, employment of paper sludge, lignocellulose, and starch material from agricultural residues and forestry resources are used. The main inconvenience is that the majority of these sources need pretreatment or supplementation with additional nutrients, thus increasing the production cost. LA production using whey as growth substrate by strains of *Lb. helveticus, Lb. delbrueckii* subsp. *bulgaricus, Lb. acidophilus,* and *Lb. casei* has been investigated [55]. The main drawback of using whey as substrate is that owing to the low proteolytic activity of many LAB, they cannot use whey proteins and supplementation of the medium with yeast extract is required [126]. *Lb. helveticus* strain milano, the preferred LAB species for LA production, reached a productivity of 9.7 $g\,l^{-1}\,h^{-1}$ during continuous fermentation of cheese whey-yeast extract permeate medium [298]. On the other hand, and as an alternative to yeast extract supplementation, Vasala *et al.* [299] treated *in situ* fermentation media with proteolytic microorganisms achieving an efficient conversion of lactose to lactic acid by *Lb. salivarius.* Another inexpensive raw material for lactic acid fermentation substrate is molasses; in this regard, Monteagudo *et al.* [300] studied kinetics of beet molasses fermentation by *Lb. delbrueckii* and found that at concentrations higher than 57 $g\,l^{-1}$, LA production stopped. In addition, using this raw substrate Göksungur and Güvenç [301] have performed a comparative study of LA production yield by batch, repeated batch, and continuous fermentation using *Lb. delbrueckii* IFO 3202 immobilized in calcium alginate gel beads. LA yield of 90% and productivity of 4.74 $g\,l^{-1}\,h^{-1}$ were achieved with batch fermentation, while in continuous fermentation a productivity value of 13.92 $g\,l^{-1}\,h^{-1}$ and 4.22% yield were obtained. In another approach, different amylolytic LAB have been used aiming to directly produce LA from starchy materials. Altaf *et al.* [302] screened several types of inexpensive organic nitrogen supplements to be used in starchy substrates fermentation by the strain *Lb. amylophilus* GV6. Red lentil and baker's yeast cells were the best alternative supplements for LA production with a 92% product yield. Good LA yields using synthetic mixed sugar and starch medium was obtained with *Lb. plantarum* A6 [303]. Other starchy agricultural resources such as corn, wheat, and barley have been fermented into LA with the strain *Ent. faecalis* RKY1 in a combination procedure of a previous degradation treatment with commercial amylolytic enzymes. LA productivity obtained from wheat and barley without nutrient supplementation was higher than 0.8 $g\,l^{-1}\,h^{-1}$ [304]. Among agricultural and forestry residues, lignocellulosic materials are considered as attractive feedstocks for LA production. *Lb. brevis* S3F4 was able to use xylose and glucose from lignocellulosic hydrolysates producing LA in concentrations of 39.1 $g\,l^{-1}$ [305]. *Ent. mundtii* QU25 efficiently metabolized lignocellulose-derived sugars synthesizing optically pure homo L(+)-LA with a theoretical yield close to 100% [188]. In another attempt at LA production, detoxified hydrolysate of sugarcane bagasse supplemented with yeast extract was fermented by *Lc. lactis* IO-1; this strain synthesized 10.85 $g\,l^{-1}$ from xylose (89%) as main fermentable sugar [306]. Marques *et al.* [307] evaluated LA synthesis by fermenting recycled paper sludge with *Lb. rhamnosus* ATCC 7469; LA productivity was maximum (2.9 $g\,l^{-1}\,h^{-1}$) when the enzymatic hydrolysis and fermentation steps occurred simultaneously. LA production by BAL is summarized in Table 11.3.

Several strategies, including metabolic engineering, have been applied to improve the optical purity of LA optimizing enantioselective biosynthesis. For instance, a *Lc. lactis* mutant strain obtained by UV mutagenesis was able to produce D-LA from molasses and hydrolyzed sugarcane with 73% yield [308], while a recombinant strain of *Lb. plantarum* NCIMB 8826 produced homo-D-LA from xylose and from arabinose [309, 310]. All these attempts to increase LA production result from the increasing demand of this compound to be used as substrate for PLA synthesis. This plastic material is obtained by chemical condensation of LA [311] and presents interesting characteristics such as thermoplasticity, high tensile strength, biocompatibility, and biodegradability making it promising for several purposes. Currently, several PLA-based products are available in the market, replacing petroleum-based consumables such as bottles, films, extrusion-thermoformed containers, fiberfill products, carpets, and coverings [312]. However, the amount of PLA produced (450 million kg per year) is still lower than the total amount of plastics produced (200 billion kg per year) per year [123]. Nevertheless, it is expected that worldwide annual PLA production capacity will up to 216 000 metric tons in 2015 [55] with the concomitant LA production.

11.3.3.2
Biofuels

Biofuels are defined as fuels of biological origin obtained renewably from organic residues. They can partly replace fossil-derived fuels (i.e., oil) because they can be used in current internal combustion engines and are compatible with the infrastructure that exists now. Biofuels have positive carbon balance compared to fossil fuels; the CO_2 released to the atmosphere is the same, but in the case of biofuels this is offset by the CO_2 captured by the plants from which biodiesel is produced occurring in a closed cycle; this reduces the contamination of emissions and improves the ambient air quality [313] (Table 11.2).

Biofuels can be classified as "first generation" when they are produced from fermentation of sugar-based raw materials, "second generation" if they are produced from feedstocks such as lignocellulose and municipal solid wastes, and "third generation" when they are formed from algal biomass [314, 315]. Currently, the most used biofuels are ethanol and butanol.

Ethanol *Ethanol* is industrially produced by fermentation from sugarcane, corn, or wheat by strains of *S. cerevisiae* [316, 317]. The commercial starter Superstart™ (Lallemand Biofuels and Distilled Spirits, Montreal, Canada) is an active dry yeast used in fuel ethanol and beverage alcohol fermentations. It contains a selected strain of *S. cerevisiae* distillers yeast in a highly concentrated and stable form. Also, *S. cerevisiae* has been metabolically engineered to produce other several biofuels such as 1-butanol and isobutanol, the sesquiterpenoids farnesene, bisabolene, and amorphadiene, and fatty acid ethyl esters (biodiesel) [316]. During ethanol production by *S. cerevisiae*, LAB are contaminant microorganisms usually present in raw feedstocks and/or in production lines competing with the yeast for the sugar source and producing compounds that may inhibit the yeast growth and thus reducing the overall biofuel yield. Several studies using bacteriophage-encoded lytic enzymes to control the growth of contaminating lactobacilli have been reported [317–319]. Conversely, certain LAB species have been evaluated for the production of ethanol and butanol. One interesting feature of these bacterial groups is that they display high ethanol tolerance and different catabolic activities, thus becoming interesting candidates for biofuel production [127, 128, 320]. Recently, Boguta *et al.* [127] studied the potential of some LAB strains to produce ethanol from lignocellulose feedstocks; the authors demonstrated that strains of *Lactobacillus* and *Pediococcus* were capable of using xylose and arabinose as they are highly resistant to the key inhibitors from chemically pretreated lignocellulosic biomass. In this regard, Moraïs *et al.* [127, 128, 320] proposed a strain of *Lb. plantarum* as candidate for ethanol production through metabolic engineering due to its high ethanol tolerance and ability to metabolize pentoses and hexoses from lignocellulolytic biomass. The potential of the *Lb. plantarum* strain to express different enzymes to degrade the lignocellulosic feedstocks for ethanol production has been demonstrated.

Butanol *Butanol* and its isomers (1-butanol, 2-butanol, iso-butanol) are emerging biofuels. Since butanol is less hygroscopic and has a longer chain length than ethanol, it can be blended up to 85% with gasoline compared to only 5–25% of ethanol mixed with gasoline [316]. Because some lactobacilli are capable of producing 2-butanol [130, 321], Ghiaci *et al.* [131] selected those LAB strains able to consume *meso*-2,3-butanediol. In *Lb. brevis*, *meso*-2,3-butanediol is converted to 2-butanone by a diol dehydratase enzyme and then converted to 2-butanol by a secondary alcohol dehydrogenase [130]. From nine LAB species evaluated, only *Lb. brevis* and *Lb. buchneri* were able to produce 2-butanol; from these species, two isolates of *Lb. brevis* were the best 2-butanol producers synthesizing $0.8\,g\,l^{-1}$ after 7 days (Table 11.3).

Although LAB have not been chosen yet to produce biofuels at industrial level, their high ethanol tolerance and ability to metabolize different sugars give the possibility of using renewable sources such as lignocellulosic feedstock or *meso*-2,3-butanediol make them interesting candidates.

11.4 Conclusions

In this chapter, an array of multiple applications of LAB for the elaboration of fermented foods and production of industrially interesting metabolites such as food ingredients, nutraceutics, and commodity chemicals was presented. Additionally, new biotechnological approaches for the production of these compounds were described. The metabolic versatility of this bacterial group demonstrates their potential for their use as cell factories beyond their classical use as food starter cultures and preservatives or as probiotics. These bacteria have been systematically assayed for novel specific traits or applications due to the availability of new molecular techniques and the consumers' demands for healthier foods. LAB, considered the horsepower of the food industry, have been the focus of numerous studies to increase their product yields and repertoire.

Conflict of Interest

The authors do not have potential conflict of interest to disclose.

References

1. Halász, A. (2009) Lactic acid bacteria. *Food Qual. Stand.*, **3**, 70–82.
2. Duwat, P., Sourice, S., Cesselin, B., Lamberet, G. *et al.* (2001) Respiration capacity of the fermenting bacterium *Lactococcus lactis* and its positive effects on growth and survival. *J. Bacteriol.*, **183**, 4509–4516.
3. LPSN http://www.bacterio.net/-taxonomy.html (accessed 27 August 2015).
4. Wisselink, H, Weusthuis, R, Eggink, G, Hugenholtz, J, et al. 2002. Mannitol production by lactic acid bacteria: a review. *International Dairy Journal*, **12**: 151–61.
5. Ortiz, M.E., Raya, R.R., and Mozzi, F. (2015) Efficient mannitol production by wild-type *Lactobacillus reuteri* CRL 1101 is attained at constant pH using a simplified culture medium. *Appl. Microbiol. Biotechnol.*, 1–13.
6. Mozzi, F. (2016) in *Encyclopedia of Food and Health* (eds B. Caballero, P.M. Finglas, and F. Toldrá), Academic Press, Oxford, pp. 501–508.
7. Hugenholtz, J. and Smid, E.J. (2002) Nutraceutical production with food-grade microorganisms. *Curr. Opin. Biotechnol.*, **13**, 497–507.
8. Stiles, M.E. (1996) Biopreservation by lactic acid bacteria. *Antonie van Leeuwenhoek*, **70**, 331–345.
9. Leroy, F. and De Vuyst, L. (2004) Lactic acid bacteria as functional starter cultures for the food fermentation industry. *Trends Food Sci. Technol.*, **15**, 67–78.
10. Fuller, R. (1989) A review: probiotics in man and animals. *J. Appl. Bacteriol.*, **66**, 365–378.
11. Papadimitriou, K., Pot, B., and Tsakalidou, E. (2015) How microbes adapt to a diversity of food niches. *Curr. Opin. Food Sci.*, **2**, 29–35.
12. Bachmann, H., Pronk, J.T., Kleerebezem, M., and Teusink, B. (2015) Evolutionary engineering to enhance starter culture performance in food fermentations. *Curr. Opin. Biotechnol.*, **32**, 1–7.
13. Alkema, W., Boekhorst, J., Wels, M., and van Hijum, S.A. (2015) Microbial bioinformatics for food safety and production. *Briefings Bioinf.*, **17**, 283–292.
14. Yépex, L. and Tenea, G.N. (2015) Genetic diversity of lactic acid bacteria strains towards their potential probiotic application. *Rom. Biotechnol. Lett.*, **20**, 10191.
15. Montel, M.-C., Buchin, S., Mallet, A., Delbes-Paus, C. *et al.* (2014) Traditional cheeses: rich and diverse microbiota with associated benefits. *Int. J. Food Microbiol.*, **177**, 136–154.
16. Mohammadi, R., Sohrabvandi, S., and Mohammad Mortazavian, A. (2012) The starter culture characteristics of probiotic microorganisms in fermented milks. *Eng. Life Sci.*, **12**, 399–409.
17. Gagnaire, V., Piot, M., Camier, B., Vissers, J.P. *et al.* (2004) Survey of bacterial proteins released in cheese: a proteomic approach. *Int. J. Food Microbiol.*, **94**, 185–201.
18. Griffiths, M.W. and Tellez, A.M. (2013) *Lactobacillus helveticus*: the proteolytic system. *Front. Microbiol.*, **4**, 30.
19. Ardö, Y. (2006) Flavour formation by amino acid catabolism. *Biotechnol. Adv.*, **24**, 238–242.
20. Montanari, C., Bargossi, E., Gardini, A., Lanciotti, R. *et al.* (2016) Correlation between volatile profiles of Italian fermented sausages and their size and starter culture. *Food Chem.*, **192**, 736–744.
21. Leroy, F., Verluyten, J., and De Vuyst, L. (2006) Functional meat starter cultures

for improved sausage fermentation. *Int. J. Food Microbiol.*, **106**, 270–285.
22. Kopermsub, P. and Yunchalard, S. (2010) Identification of lactic acid bacteria associated with the production of plaa-som, a traditional fermented fish product of Thailand. *Int. J. Food Microbiol.*, **138**, 200–204.
23. Kim, H.J., Kim, M.-J., Turner, T.L., Kim, B.-S. *et al.* (2014) Pyrosequencing analysis of microbiota reveals that lactic acid bacteria are dominant in Korean flat fish fermented food, gajami-sikhae. *Biosci. Biotechnol., Biochem.*, **78**, 1611–1618.
24. Elmacı, S.B., Tokatlı, M., Dursun, D., Özçelik, F. *et al.* (2015) Phenotypic and genotypic identification of lactic acid bacteria isolated from traditional pickles of the Çubuk region in Turkey. *Folia Microbiol.*, **60**, 241–251.
25. Beganović, J., Kos, B., Pavunc, A.L., Uroić, K. *et al.* (2014) Traditionally produced sauerkraut as source of autochthonous functional starter cultures. *Microbiol. Res.*, **169**, 623–632.
26. Fisberg, M. and Machado, R. (2015) History of yogurt and current patterns of consumption. *Nutr. Rev.*, **73**, 4–7.
27. Parra, R.A.H. (2014) Efecto de la adición de yacon (*Smallanthus sonchifolius*) en las características fisicoquímicas, microbiológicas, proximales y sensoriales de yogur durante el almacenamiento bajo refrigeración. @ Limentech, Ciencia Y Tecnología Alimentaria, vol. **12**.
28. Vasconcelos, C.M., Minim, V.P.R., and Chaves, J.B.P. (2012) Low-calorie yogurt added with yacon flour: development and physicochemical evaluation. *Rev. Chil. Nutr.*, **39**, 65–71.
29. Dias, M.L.L.A., Salgado, S.M., Guerra, N.B., Livera, A.V.S. *et al.* (2013) Physicochemical, sensory, and microbiological evaluation and development of symbiotic fermented drink. *Food Sci. Technol. (Campinas)*, **33**, 805–811.
30. da Silva Almeida, A.P., Avi, C.M., Barbisan, L.F., de Moura, N.A. *et al.* (2015) Yacon (*Smallanthus sonchifolius*) and *Lactobacillus acidophilus* CRL 1014 reduce the early phases of colon carcinogenesis in male Wistar rats. *Food Res. Int.*, **74**, 48–54.
31. Roselino, M.N., Pauly-Silveira, N.D., Cavallini, D.C., Celiberto, L.S. *et al.* (2012) A potential synbiotic product improves the lipid profile of diabetic rats. *Lipids Health Dis.*, **11**, 1–9.
32. Reddy, L.V., Min, J.-H., and Wee, Y.-J. (2015) Production of probiotic mango juice by fermentation of lactic acid bacteria, **106**, 270–285.
33. Prado, F.C., Lindner, J.D.D., Inaba, J., Thomaz-Soccol, V. *et al.* (2015) Development and evaluation of a fermented coconut water beverage with potential health benefits. *J. Funct. Foods*, **12**, 489–497.
34. do Espirito-Santo, A.P., Mouquet-Rivier, C., Humblot, C., Cazevieille, C. *et al.* (2014) Influence of cofermentation by amylolytic *Lactobacillus* strains and probiotic bacteria on the fermentation process, viscosity and microstructure of gruels made of rice, soy milk and passion fruit fiber. *Food Res. Int.*, **57**, 104–113.
35. do Amaral Santos, C.C.A., da Silva Libeck, B., and Schwan, R.F. (2014) Co-culture fermentation of peanut-soy milk for the development of a novel functional beverage. *Int. J. Food Microbiol.*, **186**, 32–41.
36. Russo, P., Peña, N., de Chiara, M.L.V., Amodio, M.L. *et al.* (2015) Probiotic lactic acid bacteria for the production of multifunctional fresh-cut cantaloupe. *Food Res. Int.*, **77**, 762–772.
37. Florou-Paneri, P., Christaki, E., and Bonos, E. (2013) Lactic acid bacteria as source of functional ingredients, in *Lactic Acid Bacteria – R & D for Food, Health and Livestock Purposes* (ed. J.M. Kongo), InTech.
38. Axel, C., Röcker, B., Brosnan, B., Zannini, E. *et al.* (2015) Application of *Lactobacillus amylovorus* DSM19280 in gluten-free sourdough bread to improve the microbial shelf life. *Food Microbiol.*, **47**, 36–44.
39. Fernández, M., Hudson, J.A., Korpela, R., and de los Reyes-Gavilán, C.G. (2015) Impact on human health of microorganisms present in fermented

dairy products: an overview. *Biomed Res. Int.*, **2015**, 1–13.
40. Gatti, M., Bottari, B., Lazzi, C., Neviani, E. *et al.* (2014) Invited review: microbial evolution in raw-milk, long-ripened cheeses produced using undefined natural whey starters. *J. Dairy Sci.*, **97**, 573–591.
41. Nielsen, B., Gürakan, G.C., and Ünlü, G. (2014) Kefir: a multifaceted fermented dairy product. *Probiot. Antimicrob. Proteins*, **6**, 123–135.
42. Cocolin, L., Dolci, P., and Rantsiou, K. (2011) Biodiversity and dynamics of meat fermentations: the contribution of molecular methods for a better comprehension of a complex ecosystem. *Meat Sci.*, **89**, 296–302.
43. http://future.aae.wisc.edu/data/annual_values/by_area/3469?tab=production (accessed 1 October 2015).
44. Steele, J., Broadbent, J., and Kok, J. (2013) Perspectives on the contribution of lactic acid bacteria to cheese flavor development. *Curr. Opin. Biotechnol.*, **24**, 135–141.
45. Leroy, F., Scholliers, P., and Amilien, V. (2014) Elements of innovation and tradition in meat fermentation: conflicts and synergies. *Int. J. Food Microbiol.*, **212**, 2–8.
46. Demeyer, D., Raemaekers, M., Rizzo, A., Holck, A. *et al.* (2000) Control of bioflavour and safety in fermented sausages: first results of a European project. *Food Res. Int.*, **33**, 171–180.
47. Kumar, B.V., Vijayendra, S.V.N., and Reddy, O.V.S. (2015) Trends in dairy and non-dairy probiotic products-a review. *J. Food Sci. Technol.*, 1–13.
48. Reina, L., Pérez-Díaz, I., Breidt, F., Azcarate-Peril, M.A. *et al.* (2015) Characterization of the microbial diversity in yacon spontaneous fermentation at 20 °C. *Int. J. Food Microbiol.*, **203**, 35–40.
49. Kraiem, M., Kachouri, F., Ghoul, M., and Hamdi, M. (2015) Antioxidative and bioprotective effect of lactic acid bacteria on postharvest strawberry: intact and cell lysates. *J. Food Sci. Technol.*, 1–8.
50. Elgeti, D., Nordlohne, S.D., Föste, M., Besl, M. *et al.* (2014) Volume and texture improvement of gluten-free bread using quinoa white flour. *J. Cereal Sci.*, **59**, 41–47.
51. Wolter, A., Hager, A.-S., Zannini, E., Galle, S. *et al.* (2014) Evaluation of exopolysaccharide producing *Weissella cibaria* MG1 strain for the production of sourdough from various flours. *Food Microbiol.*, **37**, 44–50.
52. Reis, J., Paula, A., Casarotti, S., and Penna, A. (2012) Lactic acid bacteria antimicrobial compounds: characteristics and applications. *Food Eng. Rev.*, **4**, 124–140.
53. Benmechernene, Z., Fernandez-No, I., Kihal, M., Bohme, K. *et al.* (2013) Recent patents on bacteriocins: food and biomedical applications. *Recent Pat. DNA Gene Sequences*, **7**, 66–73.
54. Zacharof, M. and Lovitt, R. (2012) Bacteriocins produced by lactic acid bacteria a review article. *APCBEE Procedia*, **2**, 50–56.
55. Mazzoli, R., Bosco, F., Mizrahi, I., Bayer, E.A. *et al.* (2014) Towards lactic acid bacteria-based biorefineries. *Biotechnol. Adv.*, **32**, 1216–1236.
56. Jin, T., Zhang, H., and Boyd, G. (2010) Incorporation of preservatives in polylactic acid films for inactivating *Escherichia coli* O157: H7 and extending microbiological shelf life of strawberry puree. *J. Food Prot.*, **73**, 812–818.
57. Ercolini, D., Storia, A., Villani, F., and Mauriello, G. (2006) Effect of a bacteriocin-activated polythene film on *Listeria monocytogenes* as evaluated by viable staining and epifluorescence microscopy. *J. Appl. Microbiol.*, **100**, 765–772.
58. Mauriello, G., Ercolini, D., La Storia, A., Casaburi, A. *et al.* (2004) Development of polythene films for food packaging activated with an antilisterial bacteriocin from *Lactobacillus curvatus* 32Y. *J. Appl. Microbiol.*, **97**, 314–322.
59. Deegan, L.H., Cotter, P.D., Hill, C., and Ross, P. (2006) Bacteriocins: biological tools for bio-preservation and shelf-life extension. *Int. Dairy J.*, **16**, 1058–1071.

60. Jones, E., Salin, V., and Williams, G.W. (2005) Nisin and the Market for Commercial Bacteriocins. Consumer and Product Research CP-01-05, Texas Agribusiness Market Research Center, Texas A&M University, College Station, TX.
61. Jozala, A.F., De Andrade, M.S., De Arauz, L.J., Pessoa, A. *et al.* (2007) Nisin production utilizing skimmed milk aiming to reduce process cost. *Appl. Biochem. Biotechnol.*, **137**, 515–528.
62. Devi, S.M. and Halami, P.M. (2011) Detection and characterization of pediocin PA-1/AcH like bacteriocin producing lactic acid bacteria. *Curr. Microbiol.*, **63**, 181–185.
63. Nes, I.F., Diep, D.B., Håvarstein, L.S., Brurberg, M.B. *et al.* (1996) Biosynthesis of bacteriocins in lactic acid bacteria. *Antonie van Leeuwenhoek*, **70**, 113–128.
64. Longo, M.A. and Sanromán, M.A. (2006) Production of food aroma compounds: microbial and enzymatic methodologies. *Food Technol. Biotechnol.*, **44**, 335–353.
65. Gupta, C., Prakash, D., and Gupta, S. (2015) A biotechnological approach to microbial based perfumes and flavours. *J. Microbiol. Exp.*, **2** (1).
66. Bongers, R.S., Hoefnagel, M.H.N., and Kleerebezem, M. (2005) High-level acetaldehyde production in *Lactococcus lactis* by metabolic engineering. *Appl. Environ. Microbiol.*, **71**, 1109–1113.
67. Mukdsi, M.C.A., Medina, R.B., Alvarez, M.F., and González, S.N. (2009) Ester synthesis by lactic acid bacteria isolated from goat's and ewe's milk and cheeses. *Food Chem.*, **117**, 241–247.
68. Costello, P., Siebert, T., Solomon, M., and Bartowsky, E. (2013) Synthesis of fruity ethyl esters by acyl coenzyme A: alcohol acyltransferase and reverse esterase activities in *Oenococcus oeni* and *Lactobacillus plantarum*. *J. Appl. Microbiol.*, **114**, 797–806.
69. Stam, M.R., Danchin, E.G., Rancurel, C., Coutinho, P.M. *et al.* (2006) Dividing the large glycoside hydrolase family 13 into subfamilies: towards improved functional annotations of α-amylase-related proteins. *Protein Eng. Des. Sel.*, **19**, 555–562.
70. Sanni, A., Morlon-Guyot, J., and Guyot, J. (2002) New efficient amylase-producing strains of *Lactobacillus plantarum* and *L. fermentum* isolated from different Nigerian traditional fermented foods. *Int. J. Food Microbiol.*, **72**, 53–62.
71. Turpin, W., Humblot, C., and Guyot, J.-P. (2011) Genetic screening of functional properties of lactic acid bacteria in a fermented pearl millet slurry and in the metagenome of fermented starchy foods. *Appl. Environ. Microbiol.*, **77** (24), 8722–8734.
72. Diaz-Ruiz, G., Guyot, J.-P., Ruiz-Teran, F., Morlon-Guyot, J. *et al.* (2003) Microbial and physiological characterization of weakly amylolytic but fast-growing lactic acid bacteria: a functional role in supporting microbial diversity in pozol, a Mexican fermented maize beverage. *Appl. Environ. Microbiol.*, **69**, 4367–4374.
73. Patel, A., Shah, N., and Prajapati, J. (2013) Biosynthesis of vitamins and enzymes in fermented foods by lactic acid bacteria and related genera-A promising approach. *Croat. J. Food Sci. Technol.*, **5**, 85–91.
74. Feijoo-Siota, L., Blasco, L., Rodriguez-Rama, J., Barros-Velazquez, J. *et al.* (2013) Recent patents on microbial proteases for the dairy industry. *Recent Adv. DNA Gene Sequences*, **8**, 44–55.
75. Henschke, P. (1993) An overview of malolactic fermentation research. *Aust. N. Z. Wine Ind. J.*, **8**, 69.
76. Nedovic, V., Durieuxb, A., Van Nedervelde, L., Rosseels, P. *et al.* (2000) Continuous cider fermentation with co-immobilized yeast and *Leuconostoc oenos* cells. *Enzyme Microb. Technol.*, **26**, 834–839.
77. Matthews, A., Grimaldi, A., Walker, M., Bartowsky, E. *et al.* (2004) Lactic acid bacteria as a potential source of enzymes for use in vinification. *Appl. Environ. Microbiol.*, **70**, 5715–5731.
78. Grimaldi, A., McLean, H., and Jiranek, V. (2000) Identification and partial

characterization of glycosidic activities of commercial strains of the lactic acid bacterium, *Oenococcus oeni*. *Am. J. Enol. Vitic.*, **51**, 362–369.

79. von Weymarn, N., Hujanen, M., and Leisola, M. (2002) Production of D-mannitol by heterofermentative lactic acid bacteria. *Process Biochem.*, **37**, 1207–1213.

80. von Weymarn, N., Kiviharju, K., and Leisola, M. (2002) High-level production of d-mannitol with membrane cell-recycle bioreactor. *J. Ind. Microbiol. Biotechnol.*, **29**, 44–49.

81. Saha, B.C. (2006) A low-cost medium for mannitol production by *Lactobacillus intermedius* NRRL B-3693. *Appl. Microbiol. Biotechnol.*, **72**, 676–680.

82. Saha, B.C. and Nakamura, L.K. (2003) Production of mannitol and lactic acid by fermentation with *Lactobacillus intermedius* NRRL B-3693. *Biotechnol. Bioeng.*, **82**, 864–871.

83. Saha, B.C. and Racine, F.M. (2010) Effects of pH and corn steep liquor variability on mannitol production by *Lactobacillus intermedius* NRRL B-3693. *Appl. Microbiol. Biotechnol.*, **87**, 553–560.

84. Veiga-da-Cunha, M., Santos, H., and Van Schaftingen, E. (1993) Pathway and regulation of erythritol formation in *Leuconostoc oenos*. *J. Bacteriol.*, **175**, 3941–3948.

85. Richter, H., Vlad, D., and Unden, G. (2001) Significance of pantothenate for glucose fermentation by *Oenococcus oeni* and for suppression of the erythritol and acetate production. *Arch. Microbiol.*, **175**, 26–31.

86. Stolz, P., Vogel, R.F., and Hammes, W.P. (1995) Utilization of electron acceptors by lactobacilli isolated from sourdough. *Z. Lebensm. Unters. Forsch.*, **201**, 402–410.

87. Sybesma, W., Starrenburg, M., Tijsseling, L., Hoefnagel, M.H.N. *et al.* (2003) Effects of cultivation conditions on folate production by lactic acid bacteria. *Appl. Environ. Microbiol.*, **69**, 4542–4548.

88. Kleerebezem, M., Boekhorst, J., van Kranenburg, R., Molenaar, D. *et al.* (2003) Complete genome sequence of *Lactobacillus plantarum* WCFS1. *Proc. Natl. Acad. Sci. U.S.A.*, **100**, 1990–1995.

89. Hugenholtz, J., Kleerebezem, M., Starrenburg, M., Delcour, J. *et al.* (2000) *Lactococcus lactis* as a cell factory for high-level diacetyl production. *Appl. Environ. Microbiol.*, **66**, 4112–4114.

90. Van de Guchte, M., Penaud, S., Grimaldi, C., Barbe, V. *et al.* (2006) The complete genome sequence of *Lactobacillus bulgaricus* reveals extensive and ongoing reductive evolution. *Proc. Natl. Acad. Sci. U.S.A.*, **103**, 9274–9279.

91. Goswami, R. (2012) Biosynthesis of folic acid and biotin from probiotic strain *L. helveticus* MTCC 5463. M. Sc. thesis submitted to Dairy Microbiology Department, SMC College of Dairy Science, Anand Agricultural University, Anand, Gujarat, India.

92. Holasova, M., Fiedlerova, V., Roubal, P., and Pechacova, M. (2004) Biosynthesis of folates by lactic acid bacteria and propionibacteria in fermented milk. *Czech. J. Food Sci.*, **22**, 175–181.

93. Murdock, F.A. and Fields, M.L. (1984) B-Vitamin content of natural lactic acid fermented cornmeal. *J. Food Sci.*, **49**, 373–375.

94. Laiño, J.E., del Valle, M.J., de Giori, G.S., and LeBlanc, J.G. (2013) Development of a high folate concentration yogurt naturally bio-enriched using selected lactic acid bacteria. *LWT Food Sci. Technol.*, **54**, 1–5.

95. Sybesma, W., Starrenburg, M., Kleerebezem, M., Mierau, I. *et al.* (2003) Increased production of folate by metabolic engineering of *Lactococcus lactis*. *Appl. Environ. Microbiol.*, **69**, 3069–3076.

96. Wegkamp, A., Starrenburg, M., de Vos, W.M., Hugenholtz, J. *et al.* (2004) Transformation of folate-consuming *Lactobacillus gasseri* into a folate producer. *Appl. Environ. Microbiol.*, **70**, 3146–3148.

97. Wilson, J.A. (1983) in *Harrison's Principles of Internal Medicine* (eds R.G. Petersdorf and T.R. Harrison),

McGraw-Hill Book Co, New York, pp. 461–470.
98. Burgess, C., O'Connell-Motherway, M., Sybesma, W., Hugenholtz, J. *et al.* (2004) Riboflavin production in *Lactococcus lactis*: potential for in situ production of vitamin-enriched foods. *Appl. Environ. Microbiol.*, **70**, 5769–5777.
99. Burgess, C.M., Smid, E.J., Rutten, G., and Van Sinderen, D. (2006) A general method for selection of riboflavin-overproducing food grade micro-organisms. *Microb. Cell Fact.*, **5**, 24.
100. Capozzi, V., Menga, V., Digesu, A.M., De Vita, P. *et al.* (2011) Biotechnological production of vitamin B2-enriched bread and pasta. *J. Agric. Food Chem.*, **59**, 8013–8020.
101. Juarez del Valle, M., Laiño, J.E., de Giori, G.S., and LeBlanc, J.G. (2014) Riboflavin producing lactic acid bacteria as a biotechnological strategy to obtain bio-enriched soymilk. *Food Res. Int.*, **62**, 1015–1019.
102. Sybesma, W., Burgess, C., Starrenburg, M., van Sinderen, D. *et al.* (2004) Multivitamin production in *Lactococcus lactis* using metabolic engineering. *Metab. Eng.*, **6**, 109–115.
103. Taranto, M.P., Vera, J.L., Hugenholtz, J., De Valdez, G.F. *et al.* (2003) *Lactobacillus reuteri* CRL1098 produces cobalamin. *J. Bacteriol.*, **185**, 5643–5647.
104. Saulnier, D.M., Santos, F., Roos, S., Mistretta, T.-A. *et al.* (2011) Exploring metabolic pathway reconstruction and genome-wide expression profiling in *Lactobacillus reuteri* to define functional probiotic features. *PLoS One*, **6**, e18783.
105. LeBlanc, J.G., Milani, C., de Giori, G.S., Sesma, F. *et al.* (2013) Bacteria as vitamin suppliers to their host: a gut microbiota perspective. *Curr. Opin. Biotechnol.*, **24**, 160–168.
106. O'Connor, E.B., Barrett, E., Fitzgerald, G., Hill, C. *et al.* (2005) in *Probiotic Dairy Products* (ed. A.Y. Tamime), Blackwell Publishing Ltd, pp. 167–194.
107. Cooke, G., Behan, J., and Costello, M. (2006) Newly identified vitamin K-producing bacteria isolated from the neonatal faecal flora. *Microb. Ecol. Health Dis.*, **18**, 133–138.
108. Al-Hindi, R.R. and El Ghani, S.A. (2015) Production of free conjugated linoleic acid by fermentation performed using *Lactobacillus casei* and *Bifidobacterium bifidum*. *Global Vet.*, **14** (5), 720–728.
109. Kishino, S., Ogawa, J., Omura, Y., Matsumura, K. *et al.* (2002) Conjugated linoleic acid production from linoleic acid by lactic acid bacteria. *J. Am. Oil Chem. Soc.*, **79**, 159–163.
110. Ando, A., Ogawa, J., Kishino, S., and Shimizu, S. (2004) Conjugated linoleic acid production from castor oil by *Lactobacillus plantarum* JCM 1551. *Enzyme Microb. Technol.*, **35**, 40–45.
111. Ye, S., Yu, T., Yang, H., Li, L. *et al.* (2013) Optimal culture conditions for producing conjugated linoleic acid in skim-milk by co-culture of different *Lactobacillus* strains. *Ann. Microbiol.*, **63**, 707–717.
112. Li, H. and Cao, Y. (2010) Lactic acid bacterial cell factories for gamma-aminobutyric acid. *Amino Acids*, **39**, 1107–1116.
113. Higuchi, T., Hayashi, H., and Abe, K. (1997) Exchange of glutamate and gamma-aminobutyrate in a *Lactobacillus* strain. *J. Bacteriol.*, **179**, 3362–3364.
114. Small, P.L. and Waterman, S.R. (1998) Acid stress, anaerobiosis and gadCB: lessons from *Lactococcus lactis* and *Escherichia coli*. *Trends Microbiol.*, **6**, 214–216.
115. Kim, M.-J. and Kim, K.-S. (2012) Isolation and Identification of γ-Aminobutyric acid (GABA)-producing lactic acid bacteria from Kimchi. *J. Korean Soc. Appl. Biol. Chem.*, **55**, 777–785.
116. Takeshima, K., Yamatsu, A., Yamashita, Y., Watabe, K. *et al.* (2014) Subchronic toxicity evaluation of γ-aminobutyric acid (GABA) in rats. *Food Chem. Toxicol.*, **68**, 128–134.
117. Alzate, A., Pérez-Conde, M., Gutiérrez, A., and Cámara, C. (2010) Selenium-enriched fermented milk: a suitable dairy product to improve selenium

intake in humans. *Int. Dairy J.*, **20**, 761–769.
118. Palomo, M., Gutiérrez, A.M., Pérez-Conde, M.C., Cámara, C. *et al.* (2014) Se metallomics during lactic fermentation of Se-enriched yogurt. *Food Chem.*, **164**, 371–379.
119. Torino, M.I., Font de Valdez, G., and Mozzi, F. (2015) Biopolymers from lactic acid bacteria. Novel applications in foods and beverages. *Front. Microbiol.*, **6**, 834.
120. Werning, M.L., Ibarburu, I., Dueñas, M.T., Irastorza, A. *et al.* (2006) *Pediococcus parvulus* gtf gene encoding the GTF glycosyltransferase and its application for specific PCR detection of *β*-d-glucan–producing bacteria in foods and beverages. *J. Food Prot.*, **69**, 161–169.
121. Monsan, P., Bozonnet, S., Albenne, C., Joucla, G. *et al.* (2001) Homopolysaccharides from lactic acid bacteria. *Int. Dairy J.*, **11**, 675–685.
122. Vaningelgem, F., Zamfir, M., Mozzi, F., Adriany, T. *et al.* (2004) Biodiversity of exopolysaccharides produced by *Streptococcus thermophilus* strains is reflected in their production and their molecular and functional characteristics. *Appl. Environ. Microbiol.*, **70**, 900–912.
123. Wang, Y., Tashiro, Y., and Sonomoto, K. (2015) Fermentative production of lactic acid from renewable materials: recent achievements, prospects, and limits. *J. Biosci. Bioeng.*, **119**, 10–18.
124. Nampoothiri, K.M., Nair, N.R., and John, R.P. (2010) An overview of the recent developments in polylactide (PLA) research. *Bioresour. Technol.*, **101**, 8493–8501.
125. Martinez, F.A.C., Balciunas, E.M., Salgado, J.M., González, J.M.D. *et al.* (2013) Lactic acid properties, applications and production: a review. *Trends Food Sci. Technol.*, **30**, 70–83.
126. Abdel-Rahman, M.A., Tashiro, Y., and Sonomoto, K. (2013) Recent advances in lactic acid production by microbial fermentation processes. *Biotechnol. Adv.*, **31**, 877–902.
127. Boguta, A., Bringel, F., Martinussen, J., and Jensen, P. (2014) Screening of lactic acid bacteria for their potential as microbial cell factories for bioconversion of lignocellulosic feedstocks. *Microb. Cell Fact.*, **13**, 113.
128. Moraïs, S., Shterzer, N., Lamed, R., Bayer, E.A. *et al.* (2014) A combined cell-consortium approach for lignocellulose degradation by specialized *Lactobacillus plantarum* cells. *Biotechnol. Biofuels*, **7**, 112.
129. Moraïs, S., Shterzer, N., Grinberg, I.R., Mathiesen, G. *et al.* (2013) Establishment of a simple Lactobacillus plantarum cell consortium for cellulase–xylanase synergistic interactions. *Appl. Environ. Microbiol.*, **79**, 5242–5249.
130. Radler, F. and Zorg, J. (1986) Characterization of the enzyme involved in formation of 2-butanol from meso-2, 3-butanediol by lactic acid bacteria. *Am. J. Enol. Vitic.*, **37**, 206–210.
131. Ghiaci, P., Lameiras, F., Norbeck, J., and Larsson, C. (2014) Production of 2-butanol through *meso*-2,3-butanediol consumption in lactic acid bacteria. *FEMS Microbiol. Lett.*, **360**, 70–75.
132. Papagianni, M. (2012) Metabolic engineering of lactic acid bacteria for the production of industrially important compounds. *Comput. Struct. Biotechnol. J.*, **3**, 1–8.
133. Carvalheiro, F., Moniz, P., Duarte, L.C., Esteves, M.P. *et al.* (2011) Mannitol production by lactic acid bacteria grown in supplemented carob syrup. *J. Ind. Microbiol. Biotechnol.*, **38**, 221–227.
134. Fontes, C.P.M.L., Honorato, T.L., Rabelo, M.C., and Rodrigues, S. (2009) Kinetic study of mannitol production using cashew apple juice as substrate. *Bioprocess. Biosyst. Eng.*, **32**, 493–499.
135. Rodríguez, C., Rimaux, T., Fornaguera, M.J., Vrancken, G. *et al.* (2012) Mannitol production by heterofermentative *Lactobacillus reuteri* CRL 1101 and *Lactobacillus fermentum* CRL 573 in free and controlled pH batch fermentations. *Appl. Microbiol. Biotechnol.*, **93**, 2519–2527.
136. Ortiz, M.E., Fornaguera, M.J., Raya, R.R., and Mozzi, F. (2012) *Lactobacillus reuteri* CRL 1101 highly produces

mannitol from sugarcane molasses as carbon source. *Appl. Microbiol. Biotechnol.*, **95**, 991–999.

137. Galle, S. and Arendt, E.K. (2014) Exopolysaccharides from sourdough lactic acid bacteria. *Crit. Rev. Food Sci. Nutr.*, **54**, 891–901.
138. Riley, M.A. and Wertz, J.E. (2002) Bacteriocins: evolution, ecology, and application. *Annu. Rev. Microbiol.*, **56**, 117–137.
139. Gao, Y.-L. and Ju, X.-R. (2008) Exploiting the combined effects of high pressure and moderate heat with nisin on inactivation of *Clostridium botulinum* spores. *J. Microbiol. Methods*, **72**, 20–28.
140. Al-Holy, M.A., Al-Nabulsi, A., Osaili, T.M., Ayyash, M.M. *et al.* (2012) Inactivation of *Listeria innocua* in brined white cheese by a combination of nisin and heat. *Food Control*, **23**, 48–53.
141. Sobrino-López, A. and Martín-Belloso, O. (2008) Use of nisin and other bacteriocins for preservation of dairy products. *Int. Dairy J.*, **18**, 329–343.
142. De Vuyst, L. and Weckx, S. (2015) *Biotechnology of Lactic Acid Bacteria* (eds F. Mozzi, R. Raya, G. Vignolo), John Wiley & Sons, Ltd., pp. 248–278.
143. Adler, P., Bolten, C.J., Dohnt, K., Hansen, C.E. *et al.* (2013) Core fluxome and metafluxome of lactic acid bacteria under simulated cocoa pulp fermentation conditions. *Appl. Environ. Microbiol.*, **79**, 5670–5681.
144. Siegel, H. and Eggersdorfer, M. (2000) Ketones, in *Ullmann's Encyclopedia of Industrial Chemistry* (eds M. Placke, N. Fischer, M. Colditz, E. Kunkel, K.-H. Bohne), Wiley-VCH Verlag GmbH & Co. KGaA.
145. Hugenholtz, J., Sybesma, W., Groot, M.N., Wisselink, W. *et al.* (2002) *Lactic Acid Bacteria: Genetics, Metabolism and Applications* (eds W. Konings, O.P. Kuipers, J.H.J. Huis in 't Veld), Springer, pp. 217–235.
146. O'May, G. and Macfarlane, G. (2005) in *Probiotic Dairy Products* (ed. A.Y. Tamime), Blackwell Publishing, pp. 138–166.
147. Tamang, J.P. (2011) Prospects of Asian fermented foods in global markets. The 12th Asian Food Conference.
148. Van Der Maarel, M.J., Van Der Veen, B., Uitdehaag, J.C., Leemhuis, H. *et al.* (2002) Properties and applications of starch-converting enzymes of the α-amylase family. *J. Biotechnol.*, **94**, 137–155.
149. Petrova, P., Petrov, K., and Stoyancheva, G. (2013) Starch-modifying enzymes of lactic acid bacteria–structures, properties, and applications. *Starch – Stärke*, **65**, 34–47.
150. Giraud, E., Brauman, A., Keleke, S., Lelong, B. *et al.* (1991) Isolation and physiological study of an amylolytic strain of *Lactobacillus plantarum*. *Appl. Microbiol. Biotechnol.*, **36**, 379–383.
151. Morlon-Guyot, J., Guyot, J., Pot, B., De Haut, I.J. *et al.* (1998) *Lactobacillus manihotivorans* sp. nov., a new starch-hydrolysing lactic acid bacterium isolated during cassava sour starch fermentation. *Int. J. Syst. Bacteriol.*, **48**, 1101–1109.
152. Yousif, N.M., Huch, M., Schuster, T., Cho, G.-S. *et al.* (2010) Diversity of lactic acid bacteria from Hussuwa, a traditional African fermented sorghum food. *Food Microbiol.*, **27**, 757–768.
153. Muyanja, C., Narvhus, J., Treimo, J., and Langsrud, T. (2003) Isolation, characterisation and identification of lactic acid bacteria from bushera: a Ugandan traditional fermented beverage. *Int. J. Food Microbiol.*, **80**, 201–210.
154. Sawadogo-Lingani, H., Lei, V., Diawara, B., Nielsen, D.S. *et al.* (2007) The biodiversity of predominant lactic acid bacteria in dolo and pito wort for the production of sorghum beer. *J. Appl. Microbiol.*, **103**, 765–777.
155. Bohak, I., Back, W., Richter, L., Ehrmann, M. *et al.* (1998) *Lactobacillus amylolyticus* sp. nov., isolated from beer malt and beer wort. *Syst. Appl. Microbiol.*, **21**, 360–364.
156. Petrova, P., Emanuilova, M., and Petrov, K. (2010) Amylolytic *Lactobacillus* strains from Bulgarian fermented beverage boza. *Z. Naturforsch., C*, **65**, 218–224.

157. Petrova, P. and Petrov, K. (2012) Direct starch conversion into l-(+)-lactic acid by a novel amylolytic strain of *Lactobacillus paracasei* B41. *Starch – Stärke*, **64**, 10–17.
158. Agati, V., Guyot, J.-P., Morlon-Guyot, J., Talamond, P. *et al.* (1998) Isolation and characterization of new amylolytic strains of *Lactobacillus fermentum* from fermented maize doughs (mawe and ogi) from Benin. *J. Appl. Microbiol.*, **85**, 512–520.
159. Hamad, S., Dieng, M., Ehrmann, M., and Vogel, R. (1997) Characterization of the bacterial flora of Sudanese sorghum flour and sorghum sourdough. *J. Appl. Microbiol.*, **83**, 764–770.
160. Rocha, J.M. and Malcata, F.X. (1999) On the microbiological profile of traditional Portuguese sourdough. *J. Food Prot.*, **62**, 1416–1429.
161. Corsetti, A., Lavermicocca, P., Morea, M., Baruzzi, F. *et al.* (2001) Phenotypic and molecular identification and clustering of lactic acid bacteria and yeasts from wheat (species *Triticum durum* and *Triticum aestivum*) sourdoughs of Southern Italy. *Int. J. Food Microbiol.*, **64**, 95–104.
162. Petrov, K., Urshev, Z., and Petrova, P. (2008) l(+)-Lactic acid production from starch by a novel amylolytic *Lactococcus lactis* subsp. *lactis* B84. *Food Microbiol.*, **25**, 550–557.
163. Nakamura, L. (1981) *Lactobacillus amylovorus*, a new starch-hydrolyzing species from cattle waste-corn fermentations. *Int. J. Syst. Bacteriol.*, **31**, 56–63.
164. Champ, M., Szylit, O., Raibaud, P., and Aïut-Abdelkader, N. (1983) Amylase production by three *Lactobacillus* strains isolated from chicken crop. *J. Appl. Bacteriol.*, **55**, 487–493.
165. Sen, S. and Chakrabarty, S. (1986) Amylase from *Lactobacillus cellobiosus* D-39 isolated from vegetable wastes: purification and characterization. *J. Appl. Bacteriol.*, **60**, 419–423.
166. Nakamra, L. and Crowell, C. (1979) *Lactobacillus amylophilus*, a new starchhydrolyzing species from swine waste-corn fermentation. *Dev. Ind. Microbiol.*, **20**, 531–540.
167. Pompeyo, C.C., Gómez, M.S., Gasparian, S., and Morlon-Guyot, J. (1993) Comparison of amylolytic properties of *Lactobacillus amylovorus* and of *Lactobacillus amylophilus*. *Appl. Microbiol. Biotechnol.*, **40**, 266–269.
168. Giraud, E. and Cuny, G. (1997) Molecular characterization of the α-amylase genes of *Lactobacillus plantarum* A6 and *Lactobacillus amylovorus* reveals an unusual 3′ end structure with direct tandem repeats and suggests a common evolutionary origin. *Gene*, **198**, 149–157.
169. Lee, H., Gilliland, S., and Carter, S. (2001) Amylolytic cultures of *Lactobacillus acidophilus*: potential probiotics to improve dietary starch utilization. *J. Food Sci.*, **66**, 338–344.
170. Olympia, M., Fukuda, H., Ono, H., Kaneko, Y. *et al.* (1995) Characterization of starch-hydrolyzing lactic acid bacteria isolated from a fermented fish and rice food,"Burong Isda", and its amylolytic enzyme. *J. Ferment. Bioeng.*, **80**, 124–130.
171. Naser, S.M., Vancanneyt, M., Snauwaert, C., Vrancken, G. *et al.* (2006) Reclassification of *Lactobacillus amylophilus* LMG 11400 and NRRL B-4435 as *Lactobacillus amylotrophicus* sp. nov. *Int. J. Syst. Evol. Microbiol.*, **56**, 2523–2527.
172. Satoh, E., Niimura, Y., Uchimura, T., Kozaki, M. *et al.* (1993) Molecular cloning and expression of two alpha-amylase genes from *Streptococcus bovis* 148 in *Escherichia coli*. *Appl. Environ. Microbiol.*, **59**, 3669–3673.
173. Freer, S.N. (1993) Purification and characterization of the extracellular alpha-amylase from *Streptococcus bovis* JB1. *Appl. Environ. Microbiol.*, **59**, 1398–1402.
174. Whitehead, T.R. and Cotta, M.A. (1995) Identification of intracellular amylase activity in *Streptococcus bovis* and *Streptococcus salivarius*. *Curr. Microbiol.*, **30**, 143–148.
175. Oguntoyinbo, F. (2007) Identification and functional properties of dominant lactic acid bacteria isolated at different stages of solid state fermentation of

cassava during traditional gari production. *World J. Microbiol. Biotechnol.*, **23**, 1425–1432.

176. Xiaodong, W., Xuan, G., and Rakshit, S. (1997) Direct fermentative production of lactic acid on cassava and other starch substrates. *Biotechnol. Lett.*, **19**, 841–843.
177. Chatterjee, M., Chakrabarty, S., Chattopadhyay, B., and Mandal, R. (1997) Production of lactic acid by direct fermentation of starchy wastes by an amylase-producing *Lactobacillus*. *Biotechnol. Lett.*, **19**, 873–874.
178. Vishnu, C., Seenayya, G., and Reddy, G. (2002) Direct fermentation of various pure and crude starchy substrates to l(+)-lactic acid using *Lactobacillus amylophilus* GV6. *World J. Microbiol. Biotechnol.*, **18**, 429–433.
179. Zhang, D. and Cheryan, M. (1991) Direct fermentation of starch to lactic acid by *Lactobacillus amylovorus*. *Biotechnol. Lett.*, **13**, 733–738.
180. Giraud, E., Champailler, A., and Raimbault, M. (1994) Degradation of raw starch by a wild amylolytic strain of *Lactobacillus plantarum*. *Appl. Environ. Microbiol.*, **60**, 4319–4323.
181. Blandino, A., Al-Aseeri, M., Pandiella, S., Cantero, D. *et al.* (2003) Cereal-based fermented foods and beverages. *Food Res. Int.*, **36**, 527–543.
182. Petrova, P.M. and Petrov, K.K. (2011) Antimicrobial activity of starch-degrading *Lactobacillus* strains isolated from boza. *Biotechnol. Biotechnol. Equip.*, **25**, 114–116.
183. Brown, A.C. and Valiere, A. (2004) The medicinal uses of poi. *Nutr. Clin. Care*, **7**, 69.
184. Nguyen, T.T.T., Loiseau, G., Icard-Vernière, C., Rochette, I. *et al.* (2007) Effect of fermentation by amylolytic lactic acid bacteria, in process combinations, on characteristics of rice/soybean slurries: a new method for preparing high energy density complementary foods for young children. *Food Chem.*, **100**, 623–631.
185. Novik, G., Astapovich, N., and Ryabaya, N. (2007) Production of hydrolases by lactic acid bacteria and bifidobacteria and their antibiotic resistance. *Appl. Biochem. Microbiol.*, **43**, 164–172.
186. Mogensen, G. (1993) *Technology of Reduced-Additive Foods* (ed J. Smith), Springer, pp. 1–25.
187. Reddy, G., Altaf, M., Naveena, B., Venkateshwar, M. *et al.* (2008) Amylolytic bacterial lactic acid fermentation—a review. *Biotechnol. Adv.*, **26**, 22–34.
188. Pederson, J.A., Steele, J.L., Christensen, J.E., and Dudley, E.G. (1999) *Lactic Acid Bacteria: Genetics, Metabolism and Applications* (eds W. Konings, O.P. Kuipers, J.H.J. Huis in 't Veld), Springer, pp. 217–246.
189. Savijoki, K., Ingmer, H., and Varmanen, P. (2006) Proteolytic systems of lactic acid bacteria. *Appl. Microbiol. Biotechnol.*, **71**, 394–406.
190. Christensen, J.E., Johnson, M.E., and Steele, J.L. (1995) Production of Cheddar cheese using a *Lactococcus lactis* ssp. *cremoris* SK11 derivative with enhanced aminopeptidase activity. *Int. Dairy J.*, **5**, 367–379.
191. Guldfeldt, L.U., Sørensen, K.I., Strøman, P., Behrndt, H. *et al.* (2001) Effect of starter cultures with a genetically modified peptidolytic or lytic system on Cheddar cheese ripening. *Int. Dairy J.*, **11**, 373–382.
192. Germond, J.E., Lapierre, L., and Mollet, B. (2004) Starter strains expressing a protease of *Lactobacillus bulgaricus*. Google Patents EP0810289 A1.
193. Wibowo, D., Eschenbruch, R., Davis, C., Fleet, G. *et al.* (1985) Occurrence and growth of lactic acid bacteria in wine: a review. *Am. J. Enol. Vitic.*, **36**, 302–313.
194. Crapisi, A., Nuti, M.P., Zamorani, A., and Spettoli, P. (1987) Improved stability of immobilized *Lactobacillus* sp. cells for the control of malolactic fermentation in wine. *Am. J. Enol. Vitic.*, **38**, 310–312.
195. Crapisi, A., Spettoli, P., Nuti, M., and Zamorani, A. (1987) Comparative traits of *Lactobacillus brevis, Lact. fructivorans* and *Leuconostoc oenos* immobilized cells for the control of malo-lactic fermentation in wine. *J. Appl. Bacteriol.*, **63**, 513–521.

196. Divies, C. (1989) *Biotechnology Applications in Beverage Production* (ed C. Cantarelli), Springer, pp. 153–167.
197. McCord, J. and Ryu, D.D. (1985) Development of malolactic fermentation process using immobilized whole cells and enzymes. *Am. J. Enol. Vitic.*, **36**, 214–218.
198. Naouri, P., Chagnaud, P., Arnaud, A., Galzy, P. *et al.* (1991) A new technology for malolactic bioconversion in wine. *J. Wine Res.*, **2**, 5–20.
199. Spettoli, P., Bottacin, A., Nuti, M.P., and Zamorani, A. (1982) Immobilization of *Leuconostoc oenos* ML 34 in calcium alginate gels and its application to wine technology. *Am. J. Enol. Vitic.*, **33**, 1–5.
200. Gao, C. and Fleet, G. (1995) Cell-recycle membrane bioreactor for conducting continuous malolactic fermentation. *Aust. J. Grape Wine Res.*, **1**, 32–38.
201. Formisyn, P., Vaillant, H., Lantreibecq, F., and Bourgois, J. (1997) Development of an enzymatic reactor for initiating malolactic fermentation in wine. *Am. J. Enol. Vitic.*, **48**, 345–351.
202. Vaillant, H. and Formisyn, P. (1996) Purification of the malolactic enzyme from a *Leuconostoc oenos* strain and use in a membrane reactor for achieving the malolactic fermentation of wine. *Biotechnol. Appl. Biochem.*, **24**, 217–223.
203. Volschenk, H., Viljoen, M., Grobler, J., Bauer, F. *et al.* (1997) Malolactic fermentation in grape musts by a genetically engineered strain of *Saccharomyces cerevisiae. Am. J. Enol. Vitic.*, **48**, 193–197.
204. Bakalinsky, A.T. and Boulton, R. (1985) The study of an immobilized acid protease for the treatment of wine proteins. *Am. J. Enol. Vitic.*, **36**, 23–29.
205. Modra, E., Williams, P., Lee, T., and Wallace, W. (1990) Effects of commercial peptidase on must and wine. *Actualities oenologiques*, **89**, 217–221.
206. Woiwodov, K., Galunsky, B., Djankov, S., Gorinova, N. *et al.* (1982) Immobilisierte saure protease zur eiweissstabilisierung von weinen. *Mitt. Klosterneuburg*, **32**, 117–121.
207. Colagrande, O., Silva, A., and Fumi, M.D. (1994) Recent applications of biotechnology in wine production. *Biotechnol. Progr.*, **10**, 2–18.
208. Heatherbell, D., Ngaba, P., Fombin, J., Watson, B. *et al.* (1984) Recent developments in the application of ultrafiltration and protease enzymes to grape juice and wine processing. Proceedings of the International Symposium in Cool Climate Viticulture and Enology, June 1984, pp. 25–28.
209. DeFelice, S. (2002) FIM Rationale and Proposed Guidelines for the Nutraceutical Research & Education Act-NREA, November 10, 2002, Foundation for Innovation in Medicine.
210. Livesey, G. (2003) Health potential of polyols as sugar replacers, with emphasis on low glycaemic properties. *Nutr. Res. Rev.*, **16**, 163–191.
211. Liong, M.T. and Shah, N.P. (2005) Production of organic acids from fermentation of mannitol, fructooligosaccharide and inulin by a cholesterol removing *Lactobacillus acidophilus* strain. *J. Appl. Microbiol.*, **99**, 783–793.
212. Marteau, P. and Flourié, B. (2001) Tolerance to low-digestible carbohydrates: symptomatology and methods. *Br. J. Nutr.*, **85**, S17–S21.
213. Livesey, G. (2001) Tolerance of low-digestible carbohydrates: a general view. *Br. J. Nutr.*, **85**, S7–S16.
214. Monedero, V., Pérez-Martínez, G., and Yebra, M.J. (2010) Perspectives of engineering lactic acid bacteria for biotechnological polyol production. *Appl. Microbiol. Biotechnol.*, **86**, 1003–1015.
215. Lam, X.M., Oeswein, J.Q., Ongpipattanakul, B., Shahrokh, Z. *et al.* (2006) Antibody formulation. Google Patents US6171586 B1.
216. Masten, L.W. (1993) Pharmaceutical formulations employing esterified alkoxylated polyols as vehicles. Google Patents US5213802 A.
217. Salminen, S., Halikainen, A. (1990) in *Sweeteners, Food Additives* (eds A.L. Branen, P.M. Davidson, S. Salminen, J.H. Thorngate, Marcel Dekker Inc., New York, pp. 297–321.

218. Moreland, M., Geigert, J., and Neidleman, S.L. (1987) Process for the production of mannitol and sorbitol. Google Patents.
219. Debord, B., Lefebvre, C., Guyot-Hermann, A.M., Hubert, J. *et al.* (1987) Study of different crystalline forms of mannitol: comparative behaviour under compression. *Drug Dev. Ind. Pharm.*, **13**, 1533–1546.
220. Saha, B.C. and Racine, F.M. (2011) Biotechnological production of mannitol and its applications. *Appl. Microbiol. Biotechnol.*, **89**, 879–891.
221. Hayakawa, K., Kimura, M., Kasaha, K., Matsumoto, K. *et al.* (2004) Effect of a γ-aminobutyric acid-enriched dairy product on the blood pressure of spontaneously hypertensive and normotensive Wistar–Kyoto rats. *Br. J. Nutr.*, **92**, 411–417.
222. Saha, B.C. (2006) Production of mannitol from inulin by simultaneous enzymatic saccharification and fermentation with *Lactobacillus intermedius* NRRL B-3693. *Enzyme Microb. Technol.*, **39**, 991–995.
223. Saha, B., Racine, F.M., and Terentieva, E. (2008) Fermentative production of mannitol. Google Patents US7358072 B2.
224. Moniz, P., Carvalheiro, F., Moura, P., Pereira, J. *et al.* (2009) Screening and characterization of lactic acid bacteria for the production of mannitol in carob based syrups. BioMicroWorld2009, Lisbon.
225. Klaenhammer, T.R. (1988) Bacteriocins of lactic acid bacteria. *Biochimie*, **70**, 337–349.
226. Moon, H.-J., Jeya, M., Kim, I.-W., and Lee, J.-K. (2010) Biotechnological production of erythritol and its applications. *Appl. Microbiol. Biotechnol.*, **86**, 1017–1025.
227. Mäkinen, K.K., Isotupa, K.P., Kivilompolo, T., Mäkinen, P.L. *et al.* (2001) Comparison of erythritol and xylitol saliva stimulants in the control of dental plaque and mutans streptococci. *Caries Res.*, **35**, 129–135.
228. Ferain, T., Schanck, A.N., and Delcour, J. (1996) 13C nuclear magnetic resonance analysis of glucose and citrate end products in an ldhL-ldhD double-knockout strain of *Lactobacillus plantarum*. *J. Bacteriol.*, **178**, 7311–7315.
229. Neves, A.R., Ramos, A., Shearman, C., Gasson, M.J. *et al.* (2000) Metabolic characterization of *Lactococcus lactis* deficient in lactate dehydrogenase using in vivo ^{13}C-NMR. *Eur. J. Biochem.*, **267**, 3859–3868.
230. Gaspar, P., Neves, A.R., Ramos, A., Gasson, M.J. *et al.* (2004) Engineering *Lactococcus lactis* for production of mannitol: high yields from food-grade strains deficient in lactate dehydrogenase and the mannitol transport system. *Appl. Environ. Microbiol.*, **70**, 1466–1474.
231. Aarnikunnas, J., von Weymarn, N., Rönnholm, K., Leisola, M. *et al.* (2003) Metabolic engineering of *Lactobacillus fermentum* for production of mannitol and pure l-lactic acid or pyruvate. *Biotechnol. Bioeng.*, **82**, 653–663.
232. Helanto, M., Aarnikunnas, J., von Weymarn, N., Airaksinen, U. *et al.* (2005) Improved mannitol production by a random mutant of *Leuconostoc pseudomesenteroides*. *J. Biotechnol.*, **116**, 283–294.
233. Nissen, L., Pérez-Martínez, G., and Yebra, M.J. (2005) Sorbitol synthesis by an engineered *Lactobacillus casei* strain expressing a sorbitol-6-phosphate dehydrogenase gene within the lactose operon. *FEMS Microbiol. Lett.*, **249**, 177–183.
234. De Boeck, R., Sarmiento-Rubiano, L.A., Nadal, I., Monedero, V. *et al.* (2010) Sorbitol production from lactose by engineered *Lactobacillus casei* deficient in sorbitol transport system and mannitol-1-phosphate dehydrogenase. *Appl. Microbiol. Biotechnol.*, **85**, 1915–1922.
235. Ladero, V., Ramos, A., Wiersma, A., Goffin, P. *et al.* (2007) High-level production of the low-calorie sugar sorbitol by *Lactobacillus plantarum* through metabolic engineering. *Appl. Environ. Microbiol.*, **73**, 1864–1872.
236. Nyyssölä, A., Pihlajaniemi, A., Palva, A., Von Weymarn, N. *et al.* (2005) Production of xylitol from d-xylose

by recombinant *Lactococcus lactis*. *J. Biotechnol.*, **118**, 55–66.

237. Ortiz, M.E., Bleckwedel, J., Raya, R.R., and Mozzi, F. (2013) Biotechnological and in situ food production of polyols by lactic acid bacteria. *Appl. Microbiol. Biotechnol.*, **97**, 4713–4726.

238. LeBlanc, J.G., Laino, J.E., del Valle, M.J., Vannini, V. *et al.* (2011) B-Group vitamin production by lactic acid bacteria – current knowledge and potential applications. *J. Appl. Microbiol.*, **111**, 1297–1309.

239. Rossi, M., Amaretti, A., and Raimondi, S. (2011) Folate production by probiotic bacteria. *Nutrients*, **3**, 118–134.

240. Crittenden, R.G., Martinez, N.R., and Playne, M.J. (2003) Synthesis and utilisation of folate by yoghurt starter cultures and probiotic bacteria. *Int. J. Food Microbiol.*, **80**, 217–222.

241. Martín, R., Olivares, M., Marín, M.L., Xaus, J. *et al.* (2005) Characterization of a reuterin-producing Lactobacillus coryniformis strain isolated from a goat's milk cheese. *Int. J. Food Microbiol.*, **104**, 267–277.

242. Chung, S.H., Kim, I.H., Park, H.G., Kang, H.S. *et al.* (2008) Synthesis of conjugated linoleic acid by human-derived *Bifidobacterium breve* LMC 017: utilization as a functional starter culture for milk fermentation. *J. Agric. Food Chem.*, **56**, 3311–3316.

243. Wahle, K.W., Heys, S.D., and Rotondo, D. (2004) Conjugated linoleic acids: are they beneficial or detrimental to health? *Prog. Lipid Res.*, **43**, 553–587.

244. Yang, B., Chen, H., Gu, Z., Tian, F. *et al.* (2014) Synthesis of conjugated linoleic acid by the linoleate isomerase complex in food-derived lactobacilli. *J. Appl. Microbiol.*, **117**, 430–439.

245. Andrade, J.C., Ascencao, K., Gullon, P., Henriques, S. *et al.* (2012) Production of conjugated linoleic acid by food-grade bacteria: a review. *Int. J. Dairy Technol.*, **65**, 467–481.

246. Pariza, M.W., Park, Y., and Cook, M.E. (2001) The biologically active isomers of conjugated linoleic acid. *Prog. Lipid Res.*, **40**, 283–298.

247. Park, Y. (2009) Conjugated linoleic acid (CLA): good or bad trans fat? *J. Food Compos. Anal.*, **22**, S4–S12.

248. Ogawa, J., Kishino, S., Ando, A., Sugimoto, S. *et al.* (2005) Production of conjugated fatty acids by lactic acid bacteria. *J. Biosci. Bioeng.*, **100**, 355–364.

249. Maia, M.R., Chaudhary, L.C., Figueres, L., and Wallace, R.J. (2007) Metabolism of polyunsaturated fatty acids and their toxicity to the microflora of the rumen. *Antonie van Leeuwenhoek*, **91**, 303–314.

250. Maia, M.R., Chaudhary, L.C., Bestwick, C.S., Richardson, A.J. *et al.* (2010) Toxicity of unsaturated fatty acids to the biohydrogenating ruminal bacterium, *Butyrivibrio fibrisolvens*. *BMC Microbiol.*, **10**, 52.

251. Kim, Y. and Liu, R. (2002) Increase of conjugated linoleic acid content in milk by fermentation with lactic acid bacteria. *J. Food Sci.*, **67**, 1731–1737.

252. Dhakal, R., Bajpai, V.K., and Baek, K.-H. (2012) Production of gaba (γ-aminobutyric acid) by microorganisms: a review. *Braz. J. Microbiol.*, **43**, 1230–1241.

253. Jakobs, C., Jaeken, J., and Gibson, K. (1993) Inherited disorders of GABA metabolism. *J. Inherit. Metab. Dis.*, **16**, 704–715.

254. Wong, T., Guin, C., Bottiglieri, T., and Snead, O.C. (2003) Gaba, γ-hydroxybutyric acid, and neurological disease. *Ann. Neurol.*, **54**, S3–S12.

255. Miura, D., Ito, Y., Mizukuchi, A., Kise, M. *et al.* (2006) Hypocholesterolemic action of pre-germinated brown rice in hepatoma-bearing rats. *Life Sci.*, **79**, 259–264.

256. Adeghate, E. and Ponery, A. (2002) GABA in the endocrine pancreas: cellular localization and function in normal and diabetic rats. *Tissue Cell*, **34**, 1–6.

257. Hagiwara, H., Seki, T., and Ariga, T. (2004) The effect of pre-germinated brown rice intake on blood glucose and PAI-1 levels in streptozotocin-induced diabetic rats. *Biosci. Biotechnol., Biochem.*, **68**, 444–447.

258. Inoue, K., Shirai, T., Ochiai, H., Kasao, M. *et al.* (2003) Blood-pressure-lowering effect of a novel fermented milk containing γ-aminobutyric acid (GABA) in mild hypertensives. *Eur. J. Clin. Nutr.*, **57**, 490–495.
259. Oh, S.-H. and Oh, C.-H. (2003) Brown rice extracts with enhanced levels of GABA stimulate immune cells. *Food Sci. Biotechnol.*, **12**, 248–252.
260. Park, K.-B. and Oh, S.-H. (2007) Production of yogurt with enhanced levels of gamma-aminobutyric acid and valuable nutrients using lactic acid bacteria and germinated soybean extract. *Bioresour. Technol.*, **98**, 1675–1679.
261. Okada, T., Sugishita, T., Murakami, T., Murai, H. *et al.* (2000) Effect of the defatted rice germ enriched with GABA for sleeplessness, depression, autonomic disorder by oral administration. *Nippon Shokuhin Kagaku Kogaku Kaishi= J. Jpn. Soc. Food Sci. Technol.*, **47**, 596–603.
262. Kim, S.-H., Shin, B.-H., Kim, Y.-H., Nam, S.-W. *et al.* (2007) Cloning and expression of a full-length glutamate decarboxylase gene from *Lactobacillus brevis* BH2. *Biotechnol. Bioprocess Eng.*, **12**, 707–712.
263. Li, H., Gao, D., Cao, Y., and Xu, H. (2008) A high γ-aminobutyric acid-producing *Lactobacillus brevis* isolated from Chinese traditional paocai. *Ann. Microbiol.*, **58**, 649–653.
264. Siragusa, S., De Angelis, M., Di Cagno, R., Rizzello, C. *et al.* (2007) Synthesis of γ-aminobutyric acid by lactic acid bacteria isolated from a variety of Italian cheeses. *Appl. Environ. Microbiol.*, **73**, 7283–7290.
265. Yokoyama, S., J-i, H., and Hayakawa, K. (2002) Production of γ-aminobutyric acid from alcohol distillery lees by *Lactobacillus brevis* IFO-12005. *J. Biosci. Bioeng.*, **93**, 95–97.
266. Ueno, Y., Hayakawa, K., Takahashi, S., and Oda, K. (1997) Purification and characterization of glutamate decarboxylase from *Lactobadllus brevis* IFO 12005. *Biosci. Biotechnol., Biochem.*, **61**, 1168–1171.
267. Park, K.-B. and Oh, S.-H. (2007) Cloning, sequencing and expression of a novel glutamate decarboxylase gene from a newly isolated lactic acid bacterium, *Lactobacillus brevis* OPK-3. *Bioresour. Technol.*, **98**, 312–319.
268. Choi, S.-I., Lee, J.-W., Park, S.-M., Lee, M.-Y. *et al.* (2006) Improvement of γ-aminobutyric acid (GABA) production using cell entrapment of *Lactobacillus brevis* GABA 057. *J. Microbiol. Biotechnol.*, **16**, 562–568.
269. Lu, X., Xie, C., and Gu, Z. (2009) Optimisation of fermentative parameters for GABA. *Czech. J. Food Sci.*, **27**, 433–442.
270. Nomura, M., Kimoto, H., Someya, Y., and Suzuki, I. (1999) Novel characteristic for distinguishing *Lactococcus lactis* subsp. *lactis* from subsp. *cremoris*. *Int. J. Syst. Bacteriol.*, **49**, 163–166.
271. Nomura, M., Nakajima, I., Fujita, Y., Kobayashi, M. *et al.* (1999) *Lactococcus lactis* contains only one glutamate decarboxylase gene. *Microbiology*, **145**, 1375–1380.
272. Franciosi, E., Carafa, I., Nardin, T., Schiavon, S. *et al.* (2015) Biodiversity and γ-aminobutyric acid production by lactic acid bacteria isolated from traditional Alpine raw cow's milk cheeses. *Biomed Res. Int.*, **2015**, 625740.
273. Nomura, M., Kimoto, H., Someya, Y., Furukawa, S. *et al.* (1998) Production of γ-aminobutyric acid by cheese starters during cheese ripening. *J. Dairy Sci.*, **81**, 1486–1491.
274. Lu, X., Chen, Z., Gu, Z., and Han, Y. (2008) Isolation of γ-aminobutyric acid-producing bacteria and optimization of fermentative medium. *Biochem. Eng. J.*, **41**, 48–52.
275. Komatsuzaki, N., Shima, J., Kawamoto, S., Momose, H. *et al.* (2005) Production of γ-aminobutyric acid (GABA) by *Lactobacillus paracasei* isolated from traditional fermented foods. *Food Microbiol.*, **22**, 497–504.
276. Cho, S.Y., Park, M.J., Kim, K.M., Ryu, J.-H. *et al.* (2011) Production of high γ-aminobutyric acid (GABA) sour kimchi using lactic acid bacteria isolated from mukeunjee kimchi. *Food Sci. Biotechnol.*, **20**, 403–408.

277. Cho, Y.R., Chang, J.Y., and Chang, H.C. (2007) Production of gamma-aminobutyric acid (GABA) by *Lactobacillus buchneri* isolated from kimchi and its neuroprotective effect on neuronal cells. *J. Microbiol. Biotechnol.*, **17**, 104–109.
278. Park, K.-B. and Oh, S.-H. (2006) Isolation and characterization of *Lactobacillus buchneri* strains with high γ-aminobutyric acid producing capacity from naturally aged cheese. *Food Sci. Biotechnol.*, **15**, 86–90.
279. Tajabadi, N., Ebrahimpour, A., Baradaran, A., Rahim, R.A. *et al.* (2015) Optimization of γ-aminobutyric acid production by *Lactobacillus plantarum* Taj-Apis362 from honeybees. *Molecules*, **20**, 6654–6669.
280. Sun, T., Zhao, S., Wang, H., Cai, C. *et al.* (2009) ACE-inhibitory activity and gamma-aminobutyric acid content of fermented skim milk by *Lactobacillus helveticus* isolated from Xinjiang koumiss in China. *Eur. Food Res. Technol.*, **228**, 607–612.
281. Kim, J.Y., Lee, M.Y., Ji, G.E., Lee, Y.S. *et al.* (2009) Production of γ-aminobutyric acid in black raspberry juice during fermentation by *Lactobacillus brevis* GABA100. *Int. J. Food Microbiol.*, **130**, 12–16.
282. Yang, S.-Y., Lü, F.-X., Lu, Z.-X., Bie, X.-M. *et al.* (2008) Production of γ-aminobutyric acid by *Streptococcus salivarius* subsp. *thermophilus* Y2 under submerged fermentation. *Amino Acids*, **34**, 473–478.
283. Van Kranenburg, R., Vos, H.R., van Swam, I.I., Kleerebezem, M. *et al.* (1999) Functional analysis of glycosyltransferase genes from *Lactococcus lactis* and other gram-positive cocci: complementation, expression, and diversity. *J. Bacteriol.*, **181**, 6347–6353.
284. Huang, J., Mei, L.-H., Wu, H., and Lin, D.-Q. (2007) Biosynthesis of γ-aminobutyric acid (GABA) using immobilized whole cells of *Lactobacillus brevis*. *World J. Microbiol. Biotechnol.*, **23**, 865–871.
285. Seok, J.-H., Park, K.-B., Kim, Y.-H., Bae, M.-O. *et al.* (2008) Production and characterization of kimchi with enhanced levels of γ-aminobutyric acid. *Food Sci. Biotechnol.*, **17**, 940–946.
286. Tsai, J., Lin, Y., Pan, B., and Chen, T. (2006) Antihypertensive peptides and γ-aminobutyric acid from prozyme 6 facilitated lactic acid bacteria fermentation of soymilk. *Process Biochem.*, **41**, 1282–1288.
287. Mrvčić, J., Stanzer, D., Šolić, E., and Stehlik-Tomas, V. (2012) Interaction of lactic acid bacteria with metal ions: opportunities for improving food safety and quality. *World J. Microbiol. Biotechnol.*, **28**, 2771–2782.
288. Pophaly, S.D., Singh, P., Kumar, H., Tomar, S.K. *et al.* (2014) Selenium enrichment of lactic acid bacteria and bifidobacteria: a functional food perspective. *Trends Food Sci. Technol.*, **39**, 135–145.
289. Deng, Y., Man, C., Fan, Y., Wang, Z. *et al.* (2015) Preparation of elemental selenium-enriched fermented milk by newly isolated *Lactobacillus brevis* from kefir grains. *Int. Dairy J.*, **44**, 31–36.
290. Mangiapane, E., Lamberti, C., Pessione, A., Galano, E. *et al.* (2014) Selenium effects on the metabolism of a Se-metabolizing *Lactobacillus reuteri*: analysis of envelope-enriched and extracellular proteomes. *Mol. Biosyst.*, **10**, 1272–1280.
291. Pusztahelyi, T., Kovács, S., Pócsi, I., and Prokisch, J. (2015) Selenite-stress selected mutant strains of probiotic bacteria for Se source production. *J. Trace Elem. Med. Biol.*, **30**, 96–101.
292. Ryan, P., Ross, R., Fitzgerald, G., Caplice, N. *et al.* (2015) Sugar-coated: exopolysaccharide producing lactic acid bacteria for food and human health applications. *Food Funct.*, **6**, 679–693.
293. Salazar, N., Gueimonde, M., De Los Reyes-Gavilán, C.G., and Ruas-Madiedo, P. (2015) Exopolysaccharides produced by lactic acid bacteria and bifidobacteria as fermentable substrates by the intestinal microbiota. *Crit. Rev. Food Sci. Nutr.*, **56**, 1440–1453.
294. Mozzi, F., Vaningelgem, F., Hébert, E.M., Van der Meulen, R. *et al.* (2006) Diversity of heteropolysaccharide-producing lactic acid bacterium strains

and their biopolymers. *Appl. Environ. Microbiol.*, **72**, 4431–4435.

295. Van Geel-Schutten, G., Faber, E., Smit, E., Bonting, K. *et al.* (1999) Biochemical and structural characterization of the glucan and fructan exopolysaccharides synthesized by the *Lactobacillus reuteri* wild-type strain and by mutant strains. *Appl. Environ. Microbiol.*, **65**, 3008–3014.
296. Vijayakumar, J., Aravindan, R., and Viruthagiri, T. (2008) Recent trends in the production, purification and application of lactic acid. *Chem. Biochem. Eng. Q.*, **22**, 245–264.
297. Ghaffar, T., Irshad, M., Anwar, Z., Aqil, T. *et al.* (2014) Recent trends in lactic acid biotechnology: a brief review on production to purification. *J. Radiat. Res. Appl. Sci.*, 7, 222–229.
298. Roy, D., Goulet, J., and Le Duy, A. (1987) Continuous production of lactic acid from whey perméate by free and calcium alginate entrapped *Lactobacillus helveticus*. *J. Dairy Sci.*, **70**, 506–513.
299. Vasala, A., Panula, J., and Neubauer, P. (2005) Efficient lactic acid production from high salt containing dairy by-products by *Lactobacillus salivarius* ssp. *salicinius* with pre-treatment by proteolytic microorganisms. *J. Biotechnol.*, **117**, 421–431.
300. Monteagudo, J.M., Rodríguez, L., Rincón, J., and Fuertes, J. (1997) Kinetics of lactic acid fermentation by *Lactobacillus delbrueckii* grown on beet molasses. *J. Chem. Technol. Biotechnol.*, **68**, 271–276.
301. Göksungur, Y. and Güvenç, U. (1999) Production of lactic acid from beet molasses by calcium alginate immobilized *Lactobacillus delbrueckii* IFO 3202. *J. Chem. Technol. Biotechnol.*, **74**, 131–136.
302. Altaf, M., Naveena, B.J., and Reddy, G. (2005) Screening of inexpensive nitrogen sources for production of L (+) lactic acid from starch by amylolytic *Lactobacillus amylophilus* GV6 in single step fermentation. *Food Technol. Biotechnol.*, **43**, 235–239.
303. Thomsen, M., Guyot, J., and Kiel, P. (2007) Batch fermentations on synthetic mixed sugar and starch medium with amylolytic lactic acid bacteria. *Appl. Microbiol. Biotechnol.*, **74**, 540–546.
304. Oh, H., Wee, Y.-J., Yun, J.-S., Ho Han, S. *et al.* (2005) Lactic acid production from agricultural resources as cheap raw materials. *Bioresour. Technol.*, **96**, 1492–1498.
305. Guo, W., Jia, W., Li, Y., and Chen, S. (2010) Performances of *Lactobacillus brevis* for producing lactic acid from hydrolysate of lignocellulosics. *Appl. Biochem. Biotechnol.*, **161**, 124–136.
306. Laopaiboon, P., Thani, A., Leelavatcharamas, V., and Laopaiboon, L. (2010) Acid hydrolysis of sugarcane bagasse for lactic acid production. *Bioresour. Technol.*, **101**, 1036–1043.
307. Marques, S., Santos, J.A.L., Gírio, F.M., and Roseiro, J.C. (2008) Lactic acid production from recycled paper sludge by simultaneous saccharification and fermentation. *Biochem. Eng. J.*, **41**, 210–216.
308. Joshi, D.S., Singhvi, M.S., Khire, J.M., and Gokhale, D.V. (2010) Strain improvement of *Lactobacillus lactis* for d-lactic acid production. *Biotechnol. Lett.*, **32**, 517–520.
309. Okano, K., Yoshida, S., Tanaka, T., Ogino, C. *et al.* (2009) Homo-d-lactic acid fermentation from arabinose by redirection of the phosphoketolase pathway to the pentose phosphate pathway in l-lactate dehydrogenase gene-deficient *Lactobacillus plantarum*. *Appl. Environ. Microbiol.*, **75**, 5175–5178.
310. Okano, K., Yoshida, S., Yamada, R., Tanaka, T. *et al.* (2009) Improved production of homo-d-lactic acid via xylose fermentation by introduction of xylose assimilation genes and redirection of the phosphoketolase pathway to the pentose phosphate pathway in l-lactate dehydrogenase gene-deficient *Lactobacillus plantarum*. *Appl. Environ. Microbiol.*, **75**, 7858–7861.

311. Cheng, Y., Deng, S., Chen, P., and Ruan, R. (2009) Polylactic acid (PLA) synthesis and modifications: a review. *Front. Chem. China*, **4**, 259–264.
312. Henton, D.E., Gruber, P., Lunt, J., and Randall, J. (2005) *Natural Fibers, Biopolymers, and Biocomposites*, (eds A.K. Mohanty, M. Misra, L.T. Drzal) Taylor & Francis, Boca Raton, FL, pp. 527–577.
313. Hill, J., Nelson, E., Tilman, D., Polasky, S. *et al.* (2006) Environmental, economic, and energetic costs and benefits of biodiesel and ethanol biofuels. *Proc. Natl. Acad. Sci. U.S.A.*, **103**, 11206–11210.
314. Kang, Q., Appels, L., Tan, T., and Dewil, R. (2014) Bioethanol from lignocellulosic biomass: current findings determine research priorities. *Sci. World J.*, **2014**, 1–13.
315. Lee, R.A. and Lavoie, J.-M. (2013) From first-to third-generation biofuels: challenges of producing a commodity from a biomass of increasing complexity. *Anim. Front.*, **3**, 6–11.
316. Buijs, N.A., Siewers, V., and Nielsen, J. (2013) Advanced biofuel production by the yeast *Saccharomyces cerevisiae. Curr. Opin. Chem. Biol.*, **17**, 480–488.
317. Silva, J.B. and Sauvageau, D. (2013) Bacteriophages as antimicrobial agents against bacterial contaminants in yeast fermentation processes. Doctoral dissertation. University of Alberta.
318. Albers, E., Johansson, E., Franzén, C.J., and Larsson, C. (2011) Selective suppression of bacterial contaminants by process conditions during lignocellulose based yeast fermentations. *Biotechnol. Biofuels*, **4**, 1–8.
319. Roach, D.R., Khatibi, P.A., Bischoff, K.M., Hughes, S.R. *et al.* (2013) Bacteriophage-encoded lytic enzymes control growth of contaminating *Lactobacillus* found in fuel ethanol fermentations. *Biotechnol. Biofuels*, **6**, 20.
320. Dreschke, G., Probst, M., Walter, A., Pümpel, T. *et al.* (2015) Lactic acid and methane: Improved exploitation of biowaste potential. *Bioresour. Technol.*, **176**, 47–55.
321. Speranza, G., Manitto, P., Fontana, G., Monti, D. *et al.* (1996) Evidence for enantiomorphic–enantiotopic group discrimination in diol dehydratase-catalyzed dehydration of meso-2, 3-butanediol. *Tetrahedron Lett.*, **37**, 4247–4250.
322. Waters, D.M., Mauch, A., Coffey, A., Arendt, E. K., & Zannini, E. (2015). Lactic acid bacteria as a cell factory for the delivery of functional biomolecules and ingredients in cereal-based beverages: a review. *Crit. Rev. Food Sci. Nutr.*, **55**(4), 503–520.

Edited by
Christoph Wittmann and James C. Liao

Industrial Biotechnology

Related Titles

Mozzi, F., Raya, R.R., Vignolo, G.M. (eds.)

Biotechnology of Lactic Acid Bacteria – Novel Applications 2e

2nd Edition
2015
Print ISBN: 978-1-118-86840-9

Gupta, V.K., Mach, R., Sreenivasaprasad, S. (eds.)

Fungal Biomolecules - Sources, Applications and Recent Developments

2015
Print ISBN: 978-1-118-95829-2

Sharma, N.K., Stal, L.J., Rai, A.K. (eds.)

Cyanobacteria - An Economic Perspective

2013
Print ISBN: 978-1-119-94127-9

(Further Volumes of the "Advanved Biotechnology" Series:)

Published:

Villadsen, J. (ed.)

Fundamental Bioengineering

2016
Print ISBN: 978-3-527-33674-6

Love, J. Ch. (ed.)

Micro- and Nanosystems for Biotechnology

2016
Print ISBN: 978-3-527-33281-6

Coming soon:

Yoshida, T. (ed.)

Applied Bioengineering

2017
Print ISBN: 978-3-527-34075-0

Wittmann, Ch., and Liao, J.C. (eds.)

Industrial Biotechnology

Products and Processes

2017
Print ISBN: 978-3-527-34181-8

Planned:

Systems Biology

J. Nielsen & S. Hohmann (Chalmers University, Sweden)

2017
Print ISBN: 9783527335589

Emerging Areas in Bioengineering

H. N. Chang (KAIST, South Korea)

2017
Print ISBN: 9783527340880

Edited by Christoph Wittmann and James C. Liao

Industrial Biotechnology

Microorganisms

Volume 2

WILEY-VCH
Verlag GmbH & Co. KGaA

Volume Editors

Christoph Wittmann
Saarland University
Institute of Systems Biotechnology
Campus A1.5
66123 Saarbrücken
Germany

James C. Liao
University of California
Chemistry & Biochemistry
5531 Boelter Hall
420 Westwood Plaza
Los Angeles, CA 90095
United States

Series Editors

Sang Yup Lee
KAIST
373-1 Guseong-Dong
291 Daehak-ro, Yuseong-gu
305-701 Daejon
South Korea

Jens Nielsen
Chalmers University
Department of Chemical and Biological Engineering
Kemivägen 10
412 96 Göteborg
Sweden

Gregory Stephanopoulos
Massachusetts Institute of Technology
Department of Chemical Engineering
77 Massachusetts Avenue
Cambridge, MA 02139
United States

Library of Congress Card No.: applied for

British Library Cataloguing-in-Publication Data
A catalogue record for this book is available from the British Library.

Bibliographic information published by the Deutsche Nationalbibliothek
The Deutsche Nationalbibliothek lists this publication in the Deutsche Nationalbibliografie; detailed bibliographic data are available on the Internet at <http://dnb.d-nb.de>.

Print ISBN: 978-3-527-34179-5
ePDF ISBN: 978-3-527-80778-9
ePub ISBN: 978-3-527-80780-2
Mobi ISBN: 978-3-527-80781-9
oBook ISBN: 978-3-527-80779-6

Cover Design Adam-Design, Weinheim, Germany
Typesetting SPi Global, Chennai, India
Printing and Binding Markono Print Media Pte Ltd, Singapore
Printed on acid-free paper

To Heike, Isabelle, Felix, and Florian, with thanks from Christoph for their love, support, and inspiration.

To Kelly, Carol, and Clara, with thanks from Jim.

Contents to Volume 1

Contents to Volume 2

List of Contributors

Michael W.W. Adams
University of Georgia
Department of Biochemistry and Molecular Biology
Life Sciences Bldg.
Athens, GA 30602–7229
USA

Hal Alper
Department of Chemical Engineering
The University of Texas at Austin
200 E Dean Keeton Street
Stop C0400
Austin, TX 78712
USA

Andriy Luzhetskyy
Helmholtz Institute for Pharmaceutical Research, Actinobacteria Metabolic Engineering Group
Universitätscampus E8
66123 Saarbrücken
Germany

and

University of Saarland
Department of Pharmaceutical Biotechnology
UdS Campus C2.3
66123 Saarbrücken
Germany

William S. Ansari
University of California
California Center for Algae Biotechnology, Division of Biological Sciences
9500 Gilman Drive
San Diego, La Jolla, CA 92093
USA

Judith Becker
Saarland University
Institute of Systems Biotechnology
Campus A 15
66123 Saarbrücken
Germany

Frank R. Bengelsdorf
Universität Ulm
Institut für Mikrobiologie und Biotechnologie
Albert-Einstein-Allee 11
89081 Ulm
Germany

Dhananjay Beri
Dartmouth College
Thayer School of Engineering
14 Engineering Drive
Hanover, NH 03755
USA

and

BioEnergy Science Center
Oak Ridge, TN
USA

Oksana Bilyk
Helmholtz Institute for
Pharmaceutical Research,
Actinobacteria Metabolic
Engineering Group
Universitätscampus E8
66123 Saarbrücken
Germany

Juliana Bleckwedel
Centro de Referencia para
Lactobacilos (CERELA)-
CONICET
Chacabuco 145
San Miguel de Tucumán 4000
Argentina

José M. Borrero-de Acuña
Universidad Andrés Bello
Center for Bioinformatics and
Integrative Biology
Biosystems Engineering
Laboratory
Faculty of Biological Sciences
Av. República 239
8340176 Santiago de Chile
Chile

Klaus Buchholz
Technical University
Braunschweig
Institute of Chemical
Engineering
Hans-Sommer-Str. 10
38106 Braunschweig
Germany

Derrick S.W. Chuang
University of California
Department of Chemical and
Biomolecular Engineering
420 Westwood Plaza
5531 Boelter Hall
Los Angeles, CA 90095
USA

John Collins
Science historian
Leipziger Straße 82A
38124 Braunschweig
Germany

Jonathan M. Conway
North Carolina State University
Department of Chemical and
Biomolecular Engineering
EB-1, 911 Partners Way
Raleigh, NC 27695-7905
USA

James A. Counts
North Carolina State University
Department of Chemical and
Biomolecular Engineering
EB-1, 911 Partners Way
Raleigh, NC 27695-7905
USA

Arnold L. Demain
Drew University
Charles A. Dana Research
Institute for Scientists Emeriti
(R.I.S.E.)
36, Madison Ave
Madison, NJ 07940
USA

Fabienne Duchoud
University of California
Department of Chemical and
Biomolecular Engineering
420 Westwood Plaza
5531 Boelter Hall
Los Angeles, CA 90095
USA

Quentin M. Dudley
Northwestern University
Department of Chemical and
Biological Engineering
2145 Sheridan Road
Evanston, IL 60208
USA

and

Northwestern University
Chemistry of Life Processes
Institute
2170 Campus Drive
Evanston, IL 60208
USA

Peter Dürre
Universität Ulm
Institut für Mikrobiologie und
Biotechnologie
Albert-Einstein-Allee 11
89081 Ulm
Germany

Michael Egermeier
BOKU – University of Natural
Resources and Life Sciences
Department of Biotechnology
Muthgasse 18
1190 Vienna
Austria

and

BOKU – University of Natural
Resources and Life Sciences
CD-Laboratory for Biotechnology
of Glycerol
Muthgasse 18
1190 Vienna
Austria

Maria Eugenia Ortiz
Centro de Referencia para
Lactobacilos
(CERELA)-CONICET
Chacabuco 145
San Miguel de Tucumán 4000
Argentina

Stefanie K. Flitsch
Universität Ulm
Institut für Mikrobiologie und
Biotechnologie
Albert-Einstein-Allee 11
89081 Ulm
Germany

Brigitte Gasser
BOKU – University of Natural Resources and Life Sciences
Department of Biotechnology
Muthgasse 18
1190 Vienna
Austria

and

Austrian Centre of Industrial Biotechnology (ACIB GmbH)
Muthgasse 18
1190 Vienna
Austria

Matthias P. Gerstl
Austrian Centre of Industrial Biotechnology (ACIB)
Muthgasse 11
1190 Vienna
Austria

Javier A. Gimpel
Centre for Biotechnology and Bioengineering
Department of Chemical Engineering and Biotechnology
Universidad de Chile
851 Beaucheff, Santiago
Chile

Edward Green
CHAIN Biotechnology Limited
Imperial College Incubator
Imperial College London
Level 1 Bessemer Building
London SW7 2AZ
UK

Adam M. Guss
BioEnergy Science Center
Oak Ridge, TN
USA

and

Oak Ridge National Laboratory
Biosciences Division
1 Bethel Valley Road
Oak Ridge, TN 37831
USA

Michael Hanscho
Austrian Centre of Industrial Biotechnology (ACIB)
Muthgasse 11
1190 Vienna
Austria

Chris Herring
Enchi Corporation
Hanover, NH 03755
USA

Michael E. Himmel
BioEnergy Science Center
Oak Ridge, TN
USA

and

Biosciences Center, National Renewable Energy Laboratory
15013 Denver West Parkway
Golden, CO 80401
USA

Hans-Peter Hohmann
Nutrition Innovation Center
R&D Biotechnology
DSM Nutritional Products Ltd
Wurmisweg 576
4303 Kaiseraugst
Switzerland

Evert K. Holwerda
Dartmouth College
Thayer School of Engineering
14 Engineering Drive
Hanover, NH 03755
USA

and

BioEnergy Science Center
Oak Ridge, TN
USA

Michael C. Jewett
Northwestern University
Department of Chemical and
Biological Engineering
2145 Sheridan Road
Evanston, IL 60208
USA

and

Northwestern University
Chemistry of Life Processes
Institute
2170 Campus Drive
Evanston, IL 60208
USA

and

Northwestern University
Robert H. Lurie Comprehensive
Cancer Center
676 North St. Clair
Chicago, IL 60611
USA

and

Northwestern University
Simpson Querrey Institute for
Bionanotechnology
303 E. Superior
Chicago, IL 60611
USA

Christian Jungreuthmayer
TGM - Technologisches
Gewerbemuseum
Wexstraße 19-23
1200 Vienna
Austria

and

Austrian Centre of Industrial
Biotechnology (ACIB)
Muthgasse 11
1190 Vienna
Austria

Ashty S. Karim
Northwestern University
Department of Chemical and
Biological Engineering
2145 Sheridan Road
Evanston, IL 60208
USA

and

Northwestern University
Chemistry of Life Processes
Institute
2170 Campus Drive
Evanston, IL 60208
USA

Prema S. Karunanithi
University of California
California Center for Algae
Biotechnology, Division of
Biological Sciences
9500 Gilman Drive
San Diego, La Jolla, CA 92093
USA

Matthew W. Keller
University of Georgia
Department of Biochemistry and
Molecular Biology
Life Sciences Bldg.
Athens, GA 30602–7229
USA

Robert M. Kelly
North Carolina State University
Department of Chemical and
Biomolecular Engineering
EB-1, 911 Partners Way
Raleigh, NC 27695-7905
USA

Piyum A. Khatibi
North Carolina State University
Department of Chemical and
Biomolecular Engineering
EB-1, 911 Partners Way
Raleigh, NC 27695-7905
USA

Michael Kohlstedt
Saarland University
Institute of Systems Biology
Biosciences Campus A1.5
66123 Saarbrücken
Germany

Preben Krabben
Green Biologics Limited
45A Western Avenue
Milton Park
Abingdon
Oxfordshire OX14 4RU
UK

Laxmi Krishnappa
University of Groningen,
University Medical Center
Groningen
Department of Medical
Microbiology
Hanzeplein 1
9700 RB Groningen
The Netherlands

Laura L. Lee
North Carolina State University
Department of Chemical and
Biomolecular Engineering
EB-1, 911 Partners Way
Raleigh, NC 27695-7905
USA

James C. Liao
University of California
Department of Chemical
and Biomolecular Engineering
& Departmant of Bioengineering
420 Westwood Plaza
5531 Boelter Hall
Los Angeles, CA, 90095
USA

Sonja Linder
Universität Ulm
Institut für Mikrobiologie und
Biotechnologie
Albert-Einstein-Allee 11
89081 Ulm
Germany

Gina L. Lipscomb
University of Georgia
Department of Biochemistry and Molecular Biology
Life Sciences Bldg.
Athens, GA 30602–7229
USA

Andrew J. Loder
North Carolina State University
Department of Chemical and Biomolecular Engineering
EB-1, 911 Partners Way
Raleigh, NC 27695-7905
USA

Lee R. Lynd
Dartmouth College
Thayer School of Engineering
14 Engineering Drive
Hanover, NH 03755
USA

and

Dartmouth College
Department of Biological Sciences
Hanover, NH
USA

and

Enchi Corporation
Hanover, NH 03755
USA

and

BioEnergy Science Center
Oak Ridge, TN
USA

Hans Marx
BOKU – University of Natural Resources and Life Sciences
Department of Biotechnology
Muthgasse 18
1190 Vienna
Austria

and

BOKU – University of Natural Resources and Life Sciences
CD-Laboratory for Biotechnology of Glycerol
Muthgasse 18
1190 Vienna
Austria

Diethard Mattanovich
BOKU – University of Natural Resources and Life Sciences
Department of Biotechnology
Muthgasse 18
1190 Vienna
Austria

and

Austrian Centre of Industrial Biotechnology (ACIB GmbH)
Muthgasse 18
1190 Vienna
Austria

Stephen P. Mayfield
University of California
California Center for Algae Biotechnology, Division of Biological Sciences
9500 Gilman Drive
San Diego, La Jolla, CA 92093
USA

Nigel Minton
University of Nottingham
BBSRC/EPSRC Synthetic Biology
Research Centre (SBRC)
School of Life Sciences
University Park
Nottingham NG7 2RD
UK

Fernanda Mozzi
Centro de Referencia para
Lactobacilos (CERELA)-
CONICET
Chacabuco 145
San Miguel de Tucumán 4000
Argentina

Rolf Müller
Saarland University, Helmholtz
Centre for Infection Research
and Pharmaceutical
Biotechnology
Department of Microbial Natural
Products, Helmholtz-Institute
for Pharmaceutical Research
Saarland, Saarland University
Campus, Building E8.1,
66123 Saarbrücken
Germany

Stefan Müller
Austrian Academy of Sciences
Johann Radon Institute for
Computational and Applied
Mathematics
Altenberger Straße 69
4040 Linz
Austria

Sean J. Murphy
Dartmouth College
Thayer School of Engineering
14 Engineering Drive
Hanover, NH 03755
USA

and

BioEnergy Science Center
Oak Ridge, TN
USA

Govind Nair
Department of Biotechnology
University of Natural Resources
and Life Sciences
Vienna, Muthgasse 18
1190 Vienna
Austria

and

Austrian Centre of Industrial
Biotechnology (ACIB)
Muthgasse 11
1190 Vienna
Austria

Pablo I. Nikel
Systems and Synthetic Biology
Program
National Spanish Center for
Biotechnology (CNB-CSIC)
Calle Darwin, 3
28049 Madrid
Spain

Daniel G. Olson
Dartmouth College
Thayer School of Engineering
14 Engineering Drive
Hanover, NH 03755
USA

and

BioEnergy Science Center
Oak Ridge, TN
USA

Julie Paye
Dartmouth College
Thayer School of Engineering
14 Engineering Drive
Hanover, NH 03755
USA

and

BioEnergy Science Center
Oak Ridge, TN
USA

and

Novo Nordisk
West Lebanon, NH
USA

Micaela Pescuma
Centro de Referencia para
Lactobacilos (CERELA)-
CONICET
Chacabuco 145
San Miguel de Tucumán 4000
Argentina

Jennifer Pfizenmaier
University of Stuttgart
Institute of Biochemical
Engineering
Allmandring 31
70569 Stuttgart
Germany

Ignacio Poblete-Castro
Universidad Andrés Bello
Center for Bioinformatics and
Integrative Biology
Biosystems Engineering
Laboratory
Faculty of Biological Sciences
Av. República 239
8340176 Santiago de Chile
Chile

Anja Poehlein
Georg-August University
Genomic and Applied
Microbiology and Göttingen
Genomics Laboratory
Grisebachstr. 8
37077 Göttingen
Germany

Zoltán Prágai
Nutrition Innovation Center
R&D Biotechnology
DSM Nutritional Products Ltd
Wurmisweg 576
4303 Kaiseraugst
Switzerland

Georg Regensburger
Johannes Kepler University Linz
Institute for Algebra
Altenberger Straße 69
4040 Linz
Austria

Amanda M. Rhaesa
University of Georgia
Department of Biochemistry
and Molecular Biology
Life Sciences Bldg.
Athens, GA 30602–7229
USA

Gabe M. Rubinstein
University of Georgia
Department of Biochemistry
and Molecular Biology
Life Sciences Bldg.
Athens, GA 30602–7229
USA

Luciana Ruiz-Rodríguez
Centro de Referencia para
Lactobacilos (CERELA)-
CONICET
Chacabuco 145
San Miguel de Tucumán 4000
Argentina

Thomas Rydzak
BioEnergy Science Center
Oak Ridge, TN
USA

and

Oak Ridge National Laboratory
Biosciences Division
Oak Ridge, TN
USA

Michael Sauer
BOKU – University of Natural
Resources and Life Sciences
Department of Biotechnology
Muthgasse 18
1190 Vienna
Austria

and

Austrian Centre of Industrial
Biotechnology (ACIB GmbH)
Muthgasse 18
1190 Vienna
Austria

and

BOKU – University of Natural
Resources and Life Sciences,
CD-Laboratory for
Biotechnology of Glycerol
Muthgasse 18
1190 Vienna
Austria

Bettina Schiel-Bengelsdorf
Universität Ulm
Institut für Mikrobiologie und
Biotechnologie
Albert-Einstein-Allee 11
89081 Ulm
Germany

Israel M. Scott
University of Georgia
Department of Biochemistry
and Molecular Biology
Life Sciences Bldg.
Athens, GA 30602–7229
USA

Xiongjun Shao
Dartmouth College
Thayer School of Engineering
14 Engineering Drive
Hanover, NH 03755
USA

and

BioEnergy Science Center
Oak Ridge, TN
USA

Elizabeth A. Specht
University of California
California Center for Algae Biotechnology, Division of Biological Sciences
9500 Gilman Drive
San Diego, La Jolla, CA 92093
USA

Benjamin A. Stegmann
Universität Ulm
Institut für Mikrobiologie und Biotechnologie
Albert-Einstein-Allee 11
89081 Ulm
Germany

Christopher T. Straub
North Carolina State University
Department of Chemical and Biomolecular Engineering
EB-1, 911 Partners Way
Raleigh, NC 27695-7905
USA

Jie Sun
Department of Chemical Engineering
The University of Texas at Austin
200 E Dean Keeton Street
Stop C0400
Austin, TX 78712
USA

Ralf Takors
University of Stuttgart
Institute of Biochemical Engineering
Allmandring 31
70569 Stuttgart
Germany

Matthew Theisen
University of California
Department of Chemical and Biomolecular Engineering
& Department of Bioengineering
420 Westwood Plaza
5531 Boelter Hall,
Los Angeles, CA, 90095
USA

Liang Tian
Dartmouth College
Thayer School of Engineering
14 Engineering Drive
Hanover, NH 03755
USA

and

BioEnergy Science Center
Oak Ridge, TN
USA

Erick J. Vandamme
Ghent University
Department of Biochemical and Microbial Technology
Block B, 2nd floor Coupure links
6539000 Ghent
Belgium

Jan M. van Dijl
University of Groningen
University Medical Center Groningen
Department of Medical Microbiology
Hanzeplein 1
9700 RB Groningen
The Netherlands

Nicholas P. Vitko
North Carolina State University
Department of Chemical and Biomolecular Engineering
EB-1, 911 Partners Way
Raleigh, NC 27695-7905
USA

Silke C. Wenzel
Saarland University
Helmholtz Centre for Infection Research and Pharmaceutical Biotechnology
Department of Microbial Natural Products
Helmholtz-Institute for Pharmaceutical Research
Saarland, Saarland University Campus, Building E8.1,
66123 Saarbrücken
Germany

Christoph Wittmann
Saarland University
Institute of Systems Biology, Biosciences, Campus A1.5
66123 Saarbrücken
Germany

Robert Worthen
Dartmouth College
Thayer School of Engineering
14 Engineering Drive
Hanover, NH 03755
USA

and

BioEnergy Science Center
Oak Ridge, TN
USA

Ying Zhang
University of Nottingham
BBSRC/EPSRC Synthetic Biology Research Centre (SBRC)
School of Life Sciences
University Park
Nottingham NG7 2RD
UK

Jürgen Zanghellini
Department of Biotechnology
University of Natural Resources and Life Sciences
Vienna, Muthgasse 18
1190 Vienna
Austria

and

Austrian Centre of Industrial Biotechnology (ACIB)
Muthgasse 11
1190 Vienna
Austria

Benjamin M. Zeldes
North Carolina State University
Department of Chemical and Biomolecular Engineering
EB-1, 911 Partners Way
Raleigh, NC 27695-7905
USA

About the Series Editors

Sang Yup Lee is Distinguished Professor at the Department of Chemical and Biomolecular Engineering at the Korea Advanced Institute of Science and Technology. At present, Prof. Lee is the Director of the Center for Systems and Synthetic Biotechnology, Director of the BioProcess Engineering Research Center, and Director of the Bioinformatics Research Center. He has published more than 500 journal papers, 64 books, and book chapters, and has more than 580 patents (either registered or applied) to his credit. He has received numerous awards, including the National Order of Merit, the Merck Metabolic Engineering Award, the ACS Marvin Johnson Award. Charles Thom Award, Amgen Biochemical Engineering Award, Elmer Gaden Award, POSCO TJ Park Prize, and HoAm Prize. He is Fellow of American Association for the Advancement of Science, the American Academy of Microbiology, American Institute of Chemical Engineers, Society for Industrial Microbiology and Biotechnology, American Institute of Medical and Biological Engineering, the World Academy of Science, the Korean Academy of Science and Technology, and the National Academy of Engineering of Korea. He is also Foreign Member of National Academy of Engineering, USA. In addition, he is honorary professor of the University of Queensland (Australia), honorary professor of the Chinese Academy of Sciences, honorary professor of Wuhan University (China), honorary professor of Hubei University of Technology (China), honorary professor of Beijing University of Chemical Technology (China), and advisory professor of the Shanghai Jiaotong University (China). Apart from his academic associations, Prof. Lee is the editor-in-chief of the *Biotechnology Journal* and is also contributing to numerous other journals as associate editor and board member. Prof. Lee is serving as a member of Presidential Advisory Committee on Science and Technology (South Korea).

Jens Nielsen is Professor and Director to Chalmers University of Technology (Sweden) since 2008. He obtained an MSc degree in chemical engineering and a PhD degree (1989) in biochemical engineering from the Technical University of Denmark (DTU) and after that established his independent research group and was appointed full professor there in 1998. He was Fulbright visiting professor at MIT in 1995–1996. At DTU, he founded and directed the Center for Microbial Biotechnology. Prof. Nielsen has published more than 350 research papers and coauthored more than 40 books, and he is inventor of more than 50 patents. He has founded several companies that have raised more than 20 million in venture capital. He has received numerous Danish and international awards and is member of the Academy of Technical Sciences (Denmark), the National Academy of Engineering (USA), the Royal Danish Academy of Science and Letters, the American Institute for Medical and Biological Engineering and the Royal Swedish Academy of Engineering Sciences.

Gregory Stephanopoulos is the W.H. Dow Professor of Chemical Engineering at the Massachusetts Institute of Technology (MIT, USA) and Director of the MIT Metabolic Engineering Laboratory. He is also Instructor of Bioengineering at Harvard Medical School (since 1997). He received his BS degree from the National Technical University of Athens and his PhD from the University of Minnesota (USA). He has coauthored about 400 research papers and 50 patents, along with the first textbook on metabolic engineering. He has been recognized by numerous awards from the American Institute of Chemical Engineers (AIChE) (Wilhelm, Walker and Founders awards), American Chemical Society (ACS), Society of Industrial Microbiology (SIM), BIO (Washington Carver Award), the John Fritz Medal of the American Association of Engineering Societies, and others. In 2003, he was elected member of the National Academy of Engineering (USA) and in 2014 President of AIChE.

Preface

Over the past 100 years, industrial biotechnology has grown into a multibillion dollar market, which now has even begun to include large parts of chemical, material, and fuel production in a rapidly growing bioeconomy. Through decades of research and discovery, industrial biotechnology offers a cornucopia of possibilities. This volume is focused on the microorganisms, which are at the very heart of industrial production and determine its success through their biocatalytic efficiency. Optimized and streamlined through billion years of evolution, microbial cells exhibit networks of hundreds to thousands of biochemical conversions. Embedded into a remarkable architecture of fine-tuned control and regulation, their cellular networks can operate at highest efficiency, versatility, selectivity, vitality, and robustness. This allows top efficiency and quality production processes, provided we are able tame and orchestrate this huge natural power. It is therefore more than worthwhile to collect and compile our current knowledge on the most relevant industrial cell factories, including state-of-the-art strategies from systems biology, systems metabolic engineering, and synthetic biology to design, improve, and upgrade their properties, as has been done in this volume.

Organized in six thematic parts, this volume comprises 19 well-elaborated chapters by leading experts in the field and provides a most comprehensive view on important aspects of industrial microorganisms, including the latest trends in research and development: the level of global analysis, design, and engineering of biological systems.

Part A "From Pioneers to Visionary" covers the period from the early days of industrial biotechnology (Chapter 1) till the time of development of novel concepts of strain design and production, including synthetic biology, genome-scale modeling, and cell-free production (Chapters 2–4) and bridges 100 years of discovery and innovation.

Part B "Multipurpose Bacterial Cell Factories" offers four chapters on some of the most widely used industrial microorganisms. *Escherichia coli* (Chapter 5), *Corynebacterium glutamicum* (Chapter 6), *Bacillus subtilis* (Chapter 7), and *Pseudomonas putida* (Chapter 8) have evolved into synthetic platforms with a broad range of applications. Their product portfolios include fine chemicals, bulk chemicals, drugs, flavors and fragrances, materials, fuels, therapeutic and diagnostic proteins, and enzymes, among others. This progress has taken place largely as a result of several decades of intensive research so that these bacteria belong to the best-characterized biological systems today – a vast knowledge base for further exploration.

Part C "Exploiting Anaerobic Synthetic Power" deals with microorganisms that live without oxygen. As the aeration of the production vessels is expensive and challenging at large scale, this lifestyle offers interesting applications. Chapters 9 and 10 focus on different *Clostridium* strains that are specifically suited to produce solvents such as acetone and butanol, but also discussing the use of thermostable cellulolytic enzymes toward consolidated bioprocessing: an elegant way to couple the decomposition of polymeric raw materials with conversion of the formed sugar into the product of interest in one operation and thereby streamlining production. Chapter 11 introduces lactic acid bacteria, which are well-accepted strains applied in human nutrition and other industrial areas.

Part D "Microbial Treasure Chests for High-Value Molecules" touches the world of natural products – complex molecule structures with biologically unique properties and high value as therapeutics to fight infections, cancer, and other threatening diseases, but also for many other applications, including herbicides, insecticides, or fungicides in agriculture. Chapters 12 and 13 highlight two of the most important bacterial groups that supply these high-value products: *Myxobacteria* and *Streptomyces.*

Part E discusses the use of microorganisms as novel sustainable feedstocks in industrial biotechnology. Chapter 14 sheds light on thermophilic bacteria, whereas Chapters 15 and 16 deal with autotrophic systems that enable production simply from sunlight and carbon dioxide: cyanobacteria and algae.

Part F discusses eukaryotic cell factories of high industrial relevance. Mammalian cells offer great potential to synthesize complex proteins with high therapeutic value, but are far more complex and more difficult to grow, hence requiring specific handling (Chapter 17). Two interesting chapters (Chapter 18 and 19) introduce yeasts, including *Saccharomyces cerevisiae* and *Pichia pastoris.*

Thanks to the leading experts and their excellent contributions, which are greatly appreciated, this volume – together with its sister volume "Industrial Biotechnology Products and Processes" – hopefully sets a milestone of perpetual value. Finally, we would like to thank Claudia Ley and Waltraud Wuest at Wiley for their assistance in the production.

Saarbrücken, July 2016 — *Christoph Wittmann*
Los Angeles, July 2016 — *James C. Liao*

Part IV
Microbial Treasure Chests for High-Value Natural Compounds

Industrial Biotechnology: Microorganisms, First Edition. Edited by Christoph Wittmann and James C. Liao.

12
Host Organisms: Myxobacterium

Silke C. Wenzel and Rolf Müller

12.1
Introduction into the Myxobacteria

Myxobacteria are rod-shaped, gram-negative bacteria which are found widely distributed in nature [1, 2]. They are frequently isolated from the dung of herbivorous animals or from decaying plant material. Myxobacteria are often described as some of the most unique prokaryotes due to a number of unusual features including their extraordinarily large genomes and their complex life cycle (see Figure 12.1): During vegetative growth, these prokaryotes spread over surfaces as multicellular swarms by gliding, which is also referred to as *the wolf-pack-strategy* as they hunt and feed on prey microorganisms cooperatively via the secretion of enzymes and specific natural products termed *secondary metabolites*. Facing nutrient limitation, hundreds of thousands of myxobacteria cooperatively begin a swarming process leading to initial aggregation of cells. Subsequently, the bacteria form small (up to 2 mm in size) knob- or treelike structures called *fruiting bodies* displaying bright colors and thus enabling identification with the naked eye. Depending on the myxobacterial species, various fruiting body types exhibiting more or less elaborate structural features are constructed. Upon completion of the developmental program, dormant heat- and desiccation-resistant myxospores are built within sporangioles and await improved nutrient conditions to germinate and form a swarm colony again.

Owing to their ability to degrade bio-macromolecules and to form bioactive natural products, these bacteria play an important role in various ecological niches. Isolated from samples collected worldwide, including diverse biotopes such as rain forests, arctic tundra; in steppes and deserts; and at sea level as well as at high altitudes, currently most myxobacteria appear to be primarily soil organisms. However, recently, a widely distributed clade of exclusively marine myxobacteria (MMC) was identified by metagenomic sequencing although not a single representative has to date been isolated in the laboratory [3]. As most strains can be characterized as "micropredators," [4] they are attracted by habitats enriched in microbial communities.

Industrial Biotechnology: Microorganisms, First Edition. Edited by Christoph Wittmann and James C. Liao.

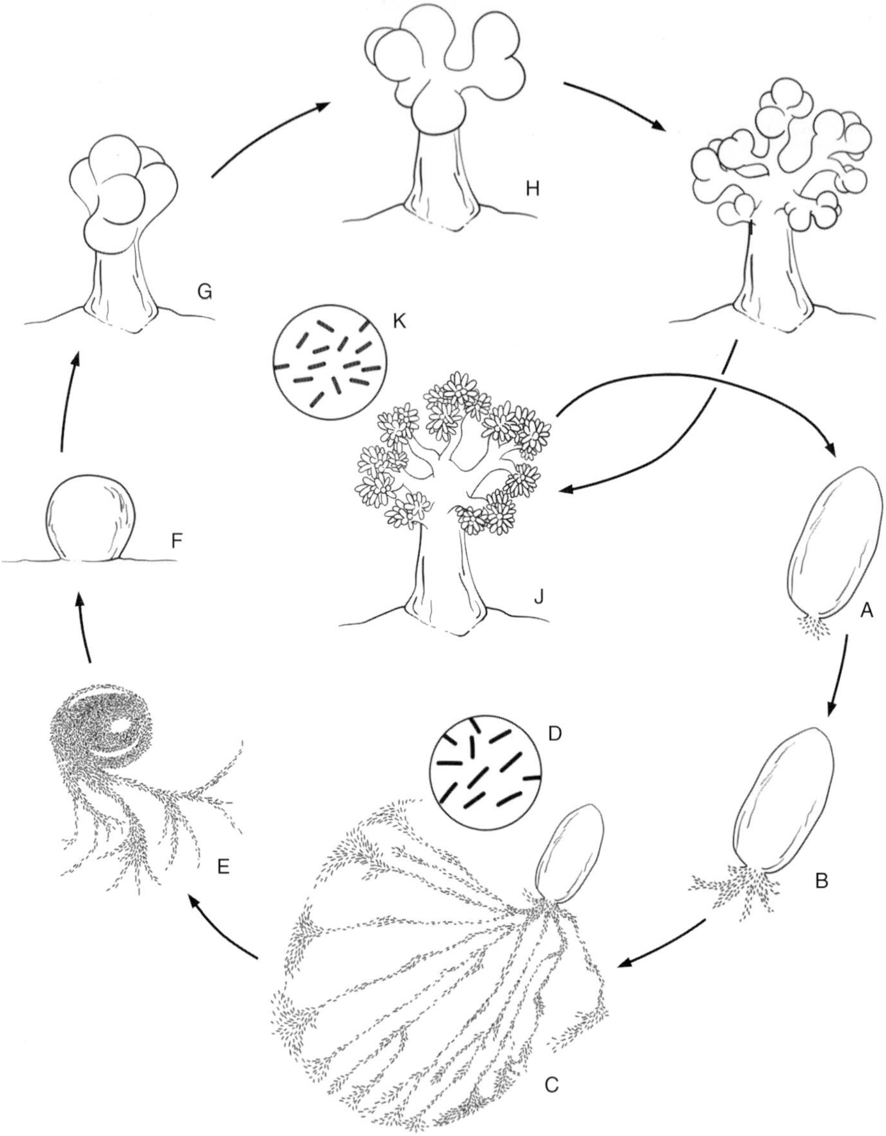

Figure 12.1 Cooperative morphogenesis in myxobacteria. The organism shown in the diagram is *Chondromyces crocatus*. Symbols: A, myxospores released from a sporangiole; B, myxospores' germination and initial stage of swarming; C, advancing and radiating swarm; D, vegetative-type cells in the swarm; E, aggregation of vegetative cells; F–J, fruiting body development; F, knob-/mound-stage; G, development of the slime stalk and cell mass differentiation; H, formation of slime branches; I, clustering of undifferentiated sporangioles; J, matured fruiting body borne with a stalk and bearing multiple branches with clusters of dense ovoid sporangioles; and K, myxospores inside the sporangiole.

Our current knowledge about myxobacteria has been summarized repeatedly in recent years. We would like to draw the reader's attention to two excellent and comprehensive book chapters by Shimkets *et al.* [5] and Reichenbach [6], which should be consulted for additional details. These references were also used here as information source for summarizing aspects of myxobacterial phylogeny, physiology, and growth/nutrient requirements (see Sections 12.2–12.4.).

12.2 Phylogeny and Classification

Phylogenetically, myxobacteria are united in the single order *Myxococcales* belonging to the Deltaproteobacteria. Their relationship has been elucidated by comparing 16S rRNA gene sequences [7–11]. Currently, 64 species of myxobacteria have been validly assigned to 29 different genera, 11 families, and 3 suborders: *Cystobacterineae, Sorangineae,* and *Nannocystineae* [6] (see Figure 12.2). Their classification has traditionally been performed according to morphological features and chemotaxonomic markers such as fatty acid content. The most important morphological recognition elements to assign suborders, families, and genera are shape and size of fruiting bodies, vegetative cells, and myxospores as well as the morphology of the swarm (e.g., see Figure 12.2). Fruiting bodies exhibit various complexities: they can be very simple, such as the soft knobs produced by *Myxococcus* species, or very elaborate, including branched slime stalks found for *Chondromyces crocatus* (see Figure 12.1). Structure and color are species specific but may vary significantly within one species. Importantly, the ability to form fruiting bodies may be lost during cultivation under laboratory conditions, leading to the appearance of degenerated structures.

The shape of vegetative cells, which measure up to 1 μm in diameter and up to 20 μm in length, represent another important criterion to distinguish taxonomic groups of myxobacteria [6]. Long and slender cells with tapering ends are characteristic for the suborder *Cystobacterineae,* whereas cylindrical cells with rounded ends seem to be typical for the suborder *Sorangineae.* Another cell type that is characteristic of the third suborder, *Nannocystineae,* is described as short, often almost cube-shaped cylindrical cells with rounded or truncated ends. Additional variation within each of the three cell types, for example, cell length, helps to differentiate at the family or genus level.

A very unusual representative of the myxobacteria is *Anaeromyxobacter dehalogenans,* which was described as an anaerobe using acetate as an electron donor [21]. This species is not known to form fruiting bodies and was placed within the *Myxococcales* as a separate genus. Additional novel genera and families are constantly validly described [12–19], illustrating that the taxonomic and phylogenetic boundaries of the myxobacteria have not yet been reached. Especially the study of previously untapped habitats and ecological niches representing biodiverse microbial populations seem to allow for the continuous isolation of novel species, genera, and families of the myxobacteria.

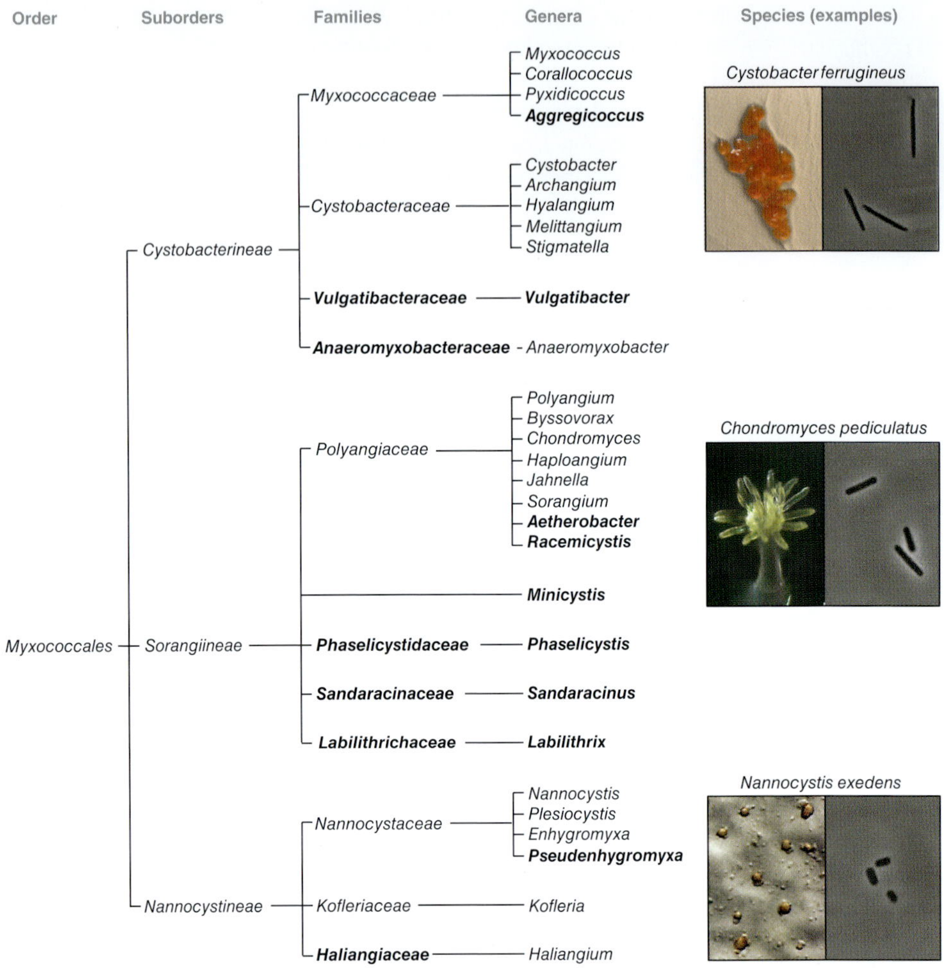

Figure 12.2 Classification of myxobacteria. Recently identified novel families/genera are highlighted in bold: *Aggregicoccus* [12], *Vulgatibacteraceae/Vulgatibacter* [13], *Aetherobacter* [14], *Racemicystis* [15], *Minicystis* [16], *Phasilicystidaceae/Phasilicystis* [17], *Sandaracinaceae/Sandaracinus* [18], *Labilithrichaceae/Labilithrix* [13], *Pseudenhygromyxa* [19]. In addition, the family of *Haliangiaceae* was recently proposed [20] and the genus *Anaeromyxobacter* was reclassified into *Anaeromyxobacteraceae* family [13]. The three suborders are distinct phenotypically by growth-stage characteristics. For each suborder, pictures of a fruiting body (left) and vegetative cells (right) are shown for a selected species. (Pictures: Ronald Garcia, ©HZI.) *Cystobacter* fruiting bodies are composed of clusters of sporangioles held by transparent slime with typical long and needle-shaped rod cells. The fruiting stage of *Chondromyces* shows a complex treelike structure with cylindrical rod vegetative cells. In case of *Nannocystis*, several fruiting bodies are visible in agar plate as simple, spherical, and tiny sporangioles with almost cuboidal vegetative rod cells.

12.3 Physiology

Myxobacteria are generally regarded as mesophilic and strictly aerobic soil microorganisms exhibiting a temperature optimum between 26 and 34 °C [2]. *A. dehalogenans* is currently the only facultative anaerobe classified within the order *Myxococcales* [21] (see Section 12.2). In antarctic soil samples, psychrophilic species were found growing at low temperatures of 4–9 °C only [22]. In contrast, some strains discovered from hot spring or semiarid and warm climate zones grow at temperatures above 40 °C [23, 24]. The optimal pH for growth is usually found to be between 6.6 and 8.0, while some species (cellulose degraders) may grow at pH 5.8–6.4 [6]. However, some alkalophilic myxobacteria isolated from the shore of an alkaline lake in East Africa grow at pH 9.5, but not at 7.2 [2]. Genuine acidophilic myxobacteria are currently not known, but there are indications for acid-tolerant species isolated from peat bogs (pH 3.2–3.3) [2]. Salt tolerance of myxobacteria is normally low, but some species were found to allow for growth at elevated NaCl concentrations. Truly halophilic species growing best at NaCl concentrations of 2–3% have been obtained from samples collected at the Japanese coast [25].

Primary metabolism in myxobacteria has been studied only to a limited extent and seems to be similar to other gram-negative bacteria. However, carbohydrate metabolism is, at least in *Myxococcus xanthus*, unusual because the Emden–Meyerhof pathway is deficient in hexokinase and pyruvate kinase activity explaining why this species is unable to utilize glucose [26]. The tricarboxylic acid (TCA) cycle is present and is speculated to be supplied with acetate from amino acid degradation. However, other members of the *Cystobacterineae* such as *Stigmatella aurantiaca* and *Cystobacter* strains utilize glucose concomitant with acid production [27–29]. For *Sorangium cellulosum* species, the presence of carbohydrates is essential for growth and some strains have even been reported to grow on glucose and nitrate-based minimal media [6] (see Section 12.4). On the basis of the annotated genome sequence of strain *S. cellulosum* So ce56 [30], a metabolic network comprising catabolic and anabolic pathways of the primary carbon metabolism was reconstituted [31]. Flux analysis carried out for growth in glucose as carbon source and asparagine as nitrogen source revealed that glycolysis and the pentose phosphate pathway are employed as major catabolic routes, whereas the Entner–Doudoroff pathway and the glycolate shunt are not active. A high flux through the TCA cycle was observed, resulting in a strong formation of ATP from which only 10% was utilized for growth and the remaining 90% was required for maintenance of basic functions. These first insights into fluxes of primary metabolism in myxobacteria provide valuable information for future studies on energy and cofactor supply as well as availability of precursors for the biotechnological production of target compounds, for example, bioactive secondary metabolites.

A prominent metabolic feature of myxobacteria besides their extensive secondary metabolism (see Section 12.6) is their ability to specifically decompose a variety of biomacromolecules [32] including casein, gelatin, other proteins, dextrin, β-glucans, starch, chitin, as well as DNA and RNA. The degradation of cellulose seems to be restricted to the genera *Sorangium* and *Byssovorax* and connected to the ability for utilization of inorganic nitrogen sources. The nature of degrading enzymes acting mostly as diffusing exoenzymes is only partly known, for example, protease pattern varies from strain to strain. The best-studied myxobacterial species, *M. xanthus*, is described to have a large potential for protein secretion [33] and was indeed shown to secrete protease(s), lysozyme(s), amidase(s), and glucoamidase(s) [34, 35]. The composition of the *M. xanthus* secretome has not been experimentally analyzed in detail yet. However, *in silico* analyses of genome data revealed that approximately 10% among the potentially secreted proteins are assumed to be hydrolases, whereas for 58% of all proteins predicted to be secreted, no functions can currently be assigned [33]. These numbers indicate a large potential of myxobacterial enzymes for industrial and biotechnological applications. Production of exoenzymes as well as bioactive secondary metabolites are also discussed to play an important role in the socio-biology of myxobacteria [36] including the formation of fruiting bodies and the wolf-pack strategy involved in hunting, killing, and degrading other microbial species as prey (see Section 12.1). The ability of myxobacteria to lyse and consume various microorganisms is exploited for isolation of novel myxobacterial species from crude samples by employing microbes as food organisms [5]. Typically *Micrococcus luteus*, *Escherichia coli*, or *Saccharomyces cerevisiae* are used, which works better if the prey microbes are alive. Myxobacteria were also found to be capable of forming holes in the cell wall of fungi and green alga and they were described to enter the cells and form fruiting bodies [37, 38].

It is worth mentioning that besides their fascinating sociobiology, myxobacteria exhibit several features which were previously thought to be characteristic of eukaryotes only [39]. They produce true steroids, possess an aromatic L-amino acid decarboxylase (DOPA decarboxylase), and employ regulatory mechanisms heavily relying on protein kinases [40]. Overall, it is not surprising to find enormous physiological discrepancies between the various myxobacteria: In light of the unexpected size differences of myxobacterial genomes and the low synteny found, for example, between representatives of the *Cystobacterineae* and the *Sorangiineae* [30, 41] (see Section 12.5), it appears that the *Myxococcales* represent a genetically enormously diverse order of bacteria.

12.4
Growth and Nutritional Requirements

Myxobacteria have a reputation of being particularly fastidious organisms, but in fact many species can be grown in pure culture. As with all other microorganisms, myxobacteria exhibit peculiarities, of course, but they do not have unusual or particularly complex nutrient and growth requirements. In general, there are two nutritional types among the myxobacteria: Most species are characterized by their proteolytic and bacteriolytic properties, whereas the genera *Sorangium* and *Byssovorax* represent cellulose decomposers [6]. All myxobacteria, including the cellulose degraders, seem to grow well on agar plates containing autoclaved baker's yeast [5]. The usage of VY/2 agar containing 0.5% baker's yeast (commercial yeast cake), 0.1% $CaCl_2 \times 2\ H_2O$, 0.5 mg ml^{-1} cyanocobalamin, and 1.5% agar (pH adjusted to 7.2) is described as a standard method for preparing stock cultures [42]. As myxobacterial plate cultures are often incubated for extended periods (1–4 weeks) during which dehydration may occur, plates have to be poured thicker than for standard microbiological work [5]. Liquid and agar media for myxobacterial cultivation are mostly of the same composition. However, requirements show more pronounced effects in liquid media where the organisms respond very fast to nutritional limitations [6].

Proteolytic species can grow well on peptone-based media at ordinary concentrations between 0.3% and 1%, but the choice of the peptone type is critical and its suitability has to be tested for each strain. Pancreatic and tryptic digests of casein or tryptone typically work well and soybean peptones can be also used, whereas meat peptones are usually not suitable owing to their high phosphate content. Many myxobacteria are sensitive to high phosphate concentrations above 0.5 mM which is already reached in 0.2% casitone solution, also explaining why there is no need to add additional phosphate to the medium. Addition of glucose or starch to the peptone medium may have positive effects on growth as many myxobacteria, for example, *Stigmatella* and *Cystobacter* species, efficiently metabolize sugars. It seems that only members of the genera *Myxococcus*, *Nannocystis*, and perhaps some *Corallococcus* strains are not capable to utilize glucose, mono-, and disaccharides. However, in some cases, polysaccharides such as starch can be consumed [5]. Cellulose decomposers will usually not grow in the absence of a suitable carbohydrate such as glucose or starch as carbon (C) source. They can utilize nitrate or ammonia as sole nitrogen (N) source and respond favorably to the addition of organic N sources, such as peptone (0.1–0.2%) [5]. Cellulose decomposers can even be grown in simple synthetic minimal media, which turned out quite difficult to define for proteolytic myxobacteria as the composition and concentration of each amino acid need to be carefully adjusted [5]. Branched-chain and aromatic amino acids as well as glycine and methionine are essential for many species. As the cost of the medium has to be taken into account when large-scale culture is necessary, many myxobacteria have been cultivated in various media supplemented with technical substrates such as soy meal, skim milk, corn steep powder, or single-cell protein that can be used for large-scale fermentations [43].

Homogeneous growth in liquid culture can be achieved for most strains of *Myxococcus*. Most other myxobacteria, especially members of the suborder *Sorangiineae*, tend to form cell clumps. For small-scale cultivations, standard Erlenmeyer flasks without baffles are suitable for myxobacterial growth as the oxygen demand is moderate. In large-scale fermenters, optimal conditions depend on the fermenter used and oxygen levels between 90% and 100% saturation give satisfactory results [44–53]. Ideal growth conditions result in doubling times between 3 and 15 h reaching cell densities of up to $10^{10}\ ml^{-1}$. Cell mass yields of 10 $g\,l^{-1}$ and more can be achieved. Optimal pH for growth depends on the strain and is usually found at levels of 6.6–8.0. Stabilization of pH can be achieved by addition of various buffers such as HEPES, TRIS, or MOPS which are well tolerated by myxobacteria. Vitamins are generally not required, but in some instances, thiamine or biotin dependencies were observed. Mineral requirements are very similar to other bacteria with the exception of magnesium, which is often needed at concentrations up to 10 mM.

Quite a number of myxobacterial natural products have been produced by cultivation of the respective species in large-scale fermenters, which is often achieved on technical media. This includes processes for generating potential agrochemicals and pharmaceuticals such as the clinically used anticancer compound epothilone (see Section 12.8.3).

12.5 Genetics and Genomics

Genetic tools to study and also engineer myxobacteria have become increasingly available over the years. Most developments in this area were directly connected to *M. xanthus* and later applied and/or adapted for other species. All tools currently in use have recently and excellently been reviewed [54] and include methods for mutagenesis, transduction, conjugation, and transformation. Within the field of natural products ("secondary metabolites"), research developments toward heterologous expression of complex biosynthetic pathways have gained a lot of interest [55–57] (see also Sections 12.7.3 and 12.7.4), including methods to transfer complete biosynthetic gene clusters within a transposable element [58]. Regulated promoters as well as strong constitutive ones (e.g., [59, 60]) and even an autonomously replicating plasmid [61] have become available. In addition to *M. xanthus*, a number of tools have been applied to the species *S. cellulosum*, *Myxococcus fulvus*, *Myxococcus virescens*, *S. aurantiaca*, *C. crocatus*, *Cystobacter fuscus*, and *Angiococcus disciformis*.

These genetic tools have become even more important with the advent of the genomics era as researchers intend to study at the molecular level the overwhelming variety of fascinating features connected to myxobacteria. Functional genomic studies of myxobacteria, such as transcriptome and proteome analyses, have become possible with the complete sequencing of the *M. xanthus* [62] and *S. cellulosum* genomes [30]. Myxobacterial genomes are typically circular chromosomes of very large size with an average GC content of around 70% (see Table 12.1). At the time of manuscript preparation, complete genome sequence information was available for representatives of all suborders of the *Myxococcales* (GenBank accession numbers in brackets): *A. dehalogenans* (CP000251 [63] and CP001359) plus two unassigned *Anaeromyxobacter* species (CP000769 and CP001131), *Archangium gephyra* (CP011509), *C. crocatus* (CP012159 [64]), *Corallococcus coralloides* (CP003389 [65]), *Haliangium ochraceum* (CP001804 [66]), *Labilithrix luteola* (CP012333), *M. fulvus* (CP002830 [67]),

Table 12.1 Characteristics of 13 myxobacteria representatives with completely sequenced genomes.

Species[a)]	Native environment	O_2 requirement	Genome size (bp)	GC content (%)	CDS[b)]	SM pathways[c)]
A. dehalogenans	Terrestrial	Facultative anaerobic	513 482	74.9	4287	4 (2)
A. gephyra	Terrestrial	Aerobic	12 489 432	69.4	10 121	42 (16)
C. crocatus	Terrestrial	Aerobic	11 388 132	68.7	8339	36 (19)
C. coralloides	Terrestrial	Aerobic	10 080 619	69.9	8033	36 (17)
H. ochraceum	Marine	Aerobic	9 446 314	69.5	6898	25 (8)
L. luteola	Terrestrial	Aerobic	12 191 466	66.1	11 518	15 (4)
M. fulvus	Marine	Aerobic	9 003 593	70.6	7284	25 (13)
M. stipitatus	Terrestrial	Aerobic	10 350 586	69.2	8043	29 (17)
M. xanthus	Terrestrial	Aerobic	9 139 763	68.9	7316	24 (16)[d)]
S. amylolyticus	Terrestrial	Aerobic	10 327 335	72.0	8961	20 (6)
S. cellulosum	Terrestrial	Aerobic	13 033 779	71.4	9367	31 (11)
S. aurantiaca	Terrestrial	Aerobic	10 260 755	67.5	8352	36 (19)[e)]
V. incomptus	Terrestrial	Aerobic	4 350 553	68.9	3723	4 (2)

a) Genome sequences of the following strains were analyzed (GenBank accession numbers in brackets): *A. dehalogenans* 2CP-C (CP000251 [63]), *A. gephyra* DSM 2261 (CP011509), *C. crocatus* Cm c5 (CP012159 [64]), *C. corraloides* DSM 2259 (CP003389 [65]), *H. ochraceum* DSM 14365 (CP001804 [66]), *L. luteola* DSM 27648 (CP012333), *M. fulvus* HW-1 (CP002830 [67]), *M. stipitatus* DSM 14675 (CP004025 [68]), *M. xanthus* DK1622 (CP000113 [62]), *S. amylolyticus* DSM 53668 (CP011125), *S. cellulosum* So ce56 (AM746676 [30]), *S. aurantiaca* DW4/3-1 (CP002271 [69]), and *V. incomptus* strain DSM 27710 (CP012332).

b) CDS (coding sequence) indicates the total number of predicted protein-coding genes in a genome.

c) SM pathways indicates the total number of predicted secondary metabolite (SM) biosynthetic pathways after subjecting whole genome sequences to antiSMASH v3.0.4 analysis [70]; number in brackets indicates gene clusters encoding polyketide synthases (PKS) and/or nonribosomal peptide synthetases (NRPS).

d) In a previous *in silico* analysis, 18 (putative) PKS and/or NRPS pathways have been reported [71], which, however, did not include "T2pks," "T3pks," and one of the "T1pks" pathways, which might be involved in fatty acid biosynthesis (see antiSMASH classification [70]). Considering the revised pathway definition in this analysis [71] (partition of pathways), the number of SM pathways would increase to 29 (21).

e) In a previous *in silico* analysis, 18 (putative) PKS and/or NRPS pathways have been reported [72], which, however, did not include "T3pks" and "Otherks" pathways (see antiSMASH classification [70]). Considering the revised pathway definition in this analysis [72] (partition of one pathway), the number of SM pathways would increase to 37 (20).

Myxococcus stipitatus (CP004025 [68]), *M. xanthus* (CP000113 [62]), *Sandaracinus amylolyticus* (CP011125), *S. cellulosum* (CP003969 [73], and AM746676 [30]), *S. aurantiaca* (CP002271 [69]), and *Vulgatibacter incomptus* (CP012332). Several draft genomes sequences were available too, including *Anaeromyxobacter* sp. (BAZG01000000 [74]), *Chondromyces apiculatus* (NZ_ASRX01000000), *C. fuscus* (NZ_ANAH02000000), *Cystobacter violaceus* (NZ_JPMI01000000), *Enhygromyxa salina* (JMCC02000000), *Hyalangium minutum* (NZ_JMCB01000000), *M. xanthus* (NZ_AKYI02000000 [75], and AOBT01000000 [76]) as well as *Plesiocystis pacifica* (NZ_ABCS01000000).

Bioinformatic analyses of these genomes has already revealed quite a number of significant and unexpected findings which have been excellently reviewed with focus on genome evolution and content recently [41]. One of the most striking features of myxobacterial genomes is the enormous number of proteins encoded with functions in signal transduction and transcriptional regulation. Myxobacterial regulation seems to be especially strongly governed by post-translational modification via protein kinases ("eukaryotic-like kinases," ELKs) other than typical bacterial two-component systems [40]. Although overall synteny between the different myxobacteria appears unexpectedly low [30, 41], comparative genomics allowed to connect different forms of fruiting bodies to genomic information [69]. *Anaeromyxobacter* sp., the first facultative anaerobes that group with the order *Myxococcales* (see Section 12.2) and the recently described *Vulgaribacter incomptus* gen. nov., sp. nov. [13], turned out to be very different from the other myxobacteria. The genomes from *A. dehalogenans* (5.0 Mb) and *V. incomptus* (4.4 Mb) are only about half the size of typical myxobacterial genomes (see Table 12.1) and more similar in size to the genomes of other Deltaproteobacteria. Intriguingly, for both species to date, fruiting body formation has not been detected under any of the culture conditions examined [13, 21, 63]. Another abnormality is the relatively low number of putative natural product biosynthetic pathways identified in the *A. dehalogenans* and *V. incomptus* genome sequence. In contrast, currently available genome data indicate that other myxobacteria have an enormous genomic potential for secondary metabolite production [77]. Sequence analyses reveal the presence of numerous natural product biosynthetic pathways (15–42 pathways per strain, see Table 12.1) the majority of which are currently of unknown function. This genomic potential represents a large yet underexploited resource for the discovery of novel bioactive compounds from myxobacteria as discussed in Sections 12.6 and 12.7.2.

12.6 Secondary Metabolism

In addition to their unusual life-cycle, one additional major reason to study myxobacteria is their metabolic physiology, including the production of secondary metabolites exhibiting promising biological activities [78, 79]. Research aiming to find, produce, optimize, and study the molecular mechanism of action of these natural products is highly promising in terms of development of such evolutionary optimized small molecules to application as pharmaceuticals and agrochemicals. With this respect, it is worth mentioning that a high percentage of today's pharmaceuticals is either derived from natural products or represents derivatives of such [80] and thus natural product research continues to be a major driving force of drug development. Currently, already one drug derived from a myxobacterial compound, epothilone, is used in the clinic [81] (see Section 12.8.3) and several other metabolites are being evaluated in preclinical studies owing to their promising mechanisms of action [82–85]. It is particularly remarkable that myxobacterial compounds exhibit modes of action that are found extremely rarely in other organisms [78, 79, 86]. Although the number of myxobacterial species known today is still limited [5, 6, 87], the number of compounds is constantly growing and has already reached several hundreds. Starting from the chemical description of the antifungal compounds ambruticin [88] and myxothiazol [89] more than three decades ago, numerous highly bioactive compounds from myxobacteria have been described.

As mentioned in Section 12.5, each strain harbors quite a number of different secondary metabolite pathways. Although most of these are currently of unknown function, each strain usually produces several compound families which is why myxobacteria are considered as multiproducers of bioactive compounds. One impressive example is *S. cellulosum* So ce12 (see Table 12.2). From this species, five compound families have so far been identified: the highly effective disorazoles as tubulin destabilizers currently under preclinical development

Table 12.2 Myxobacteria as multi-producers of bioactive compounds exemplified with two species.

Species	Natural products[a)]	Biological activity; mode of action[b)]
Chondromyces crocatus Cmc5	Ajudazoles	Antifungal, antibacterial; not reported
	Chondramides	Cytotoxic, antifungal; inhibits actin polymerization
	Chondrochlorens	Antibacterial, antifungal; not reported
	Crocacins	Antifungal, cytotoxic; inhibits respiration
	Crocapeptins	Antiinflammatory; kallikrein inhibitors
	Thuggacins	Antibacterial; inhibits respiration
Sorangium cellulosum So ce12	Chivosazoles	Cytotoxic; inhibits actin polymerization
	Disorazoles	Cytotoxic, antifungal; inhibits tubulin polymerization
	Sorangicins	Antibacterial; inhibits RNA polymerase
	Sorangolids	Antibacterial; disrupts membrane integrity
	Sulfangolids	Antibacterial (weak); not reported

a) References for production of compounds by *C. crocatus* Cmc5 and *S. cellulosum* So ce12: Ajudazoles [90], Chondramides [91], Chondrochlorens [92], Crocacins [93], Crocapeptin [94], Thuggacins [95], Chivosazoles [96], Disorazoles [97], Sorangicins [50], Sorangolids [98], and Sulfangolids [99].

b) For references on biological activity and mode of action, please see [79], for Crocapeptins [94, 100] and for Sulfangolids [99].

(see Section 12.8.4) and sorangicins as potent eubacterial RNA polymerase inhibitors. The same strain also generates bactericidal sorangiolides, antifungal chivosazoles as well as the sulfangolides. A second and equally intriguing example is *C. crocatus* Cm c5 (see Table 12.2), which produces the highly cytotoxic actin polymerizing agents chondramides as well as two classes of electron transport inhibitors: ajudazols interfere with complex I of the cytochrome bc_1, whereas the crocacins interact with complex III, resulting in an efficient inhibition of growth of competing fungi and yeast. Bacteria are also targeted by secondary metabolites from *C. crocatus* Cm c5 as chondrochlorens and thuggacins exhibit antibacterial activities. Crocapeptins, the sixth class of natural products from this species, are currently studied because of their activity as kallikrein inhibitors.

Such molecules are biosynthetically derived from relatively simple building blocks from primary metabolism (e.g., acetate, malonate, amino acids, chorismate, isopentenyl pyrophosphate, fatty acids), which are modified, condensed, and further derivatized in various ways giving rise to the final compound. Two of the most common bacterial biosynthetic systems for secondary metabolites are the polyketide synthases (PKSs) [101, 102] and the nonribosomal peptide synthetases (NRPSs) [103, 104]. These represent highly complex megasynthetases characterized by a modular architecture in which each module exhibits a size of approximately 200 kDa thus leading to biosynthetic machineries rivaling the size of the ribosome. Biosynthetic intermediates are channeled through such megasynthetase while they are tethered to way stations called *carrier proteins* (see Figure 12.3), which themselves require post-translational activation via the modifying action of phosphopantetheinyl transferases [108]. PKS and NRPS can also form hybrids with each other and combine with other biosynthetic logic such as isoprenoid or alkaloid biosynthesis giving rise to the enormous structural diversity found in nature. In addition to this diversity, each single secondary metabolite assembly line usually generates a number of derivatives of one class of compounds which is due to the enzyme's substrate promiscuity and the fact that some biosynthetic steps are only optional. It has been described that 37 variants of the epothilones are made in the strain *S. cellulosum* So ce90 (see Figure 12.3) [107]. Intriguingly, such derivatives may even exhibit different biological activities: in case of the eight argyrin variants isolated from *A. gephyra* cultures (argyrins A–H [109]), the compounds are thought to inhibit the proteasome as well as elongation factor G [110–112] and thus exhibit cytotoxic and antibacterial activities, respectively.

Figure 12.3 Stepwise biosynthesis of epothilones by a PKS/NRPS megasynthetase consisting of 6 protein subunits (EpoA, P, B–E) and 50 catalytic domains (shown as circles), which are arranged into 10 modules, each incorporating one precursor molecule into the growing linear epothilone chain [105, 106]. Besides the main products epothilones A–D, 33 additional epothilone derivatives could be identified in culture extracts of the epothilone producer *S. cellulosum* So ce90 [107] (not all derivatives shown).

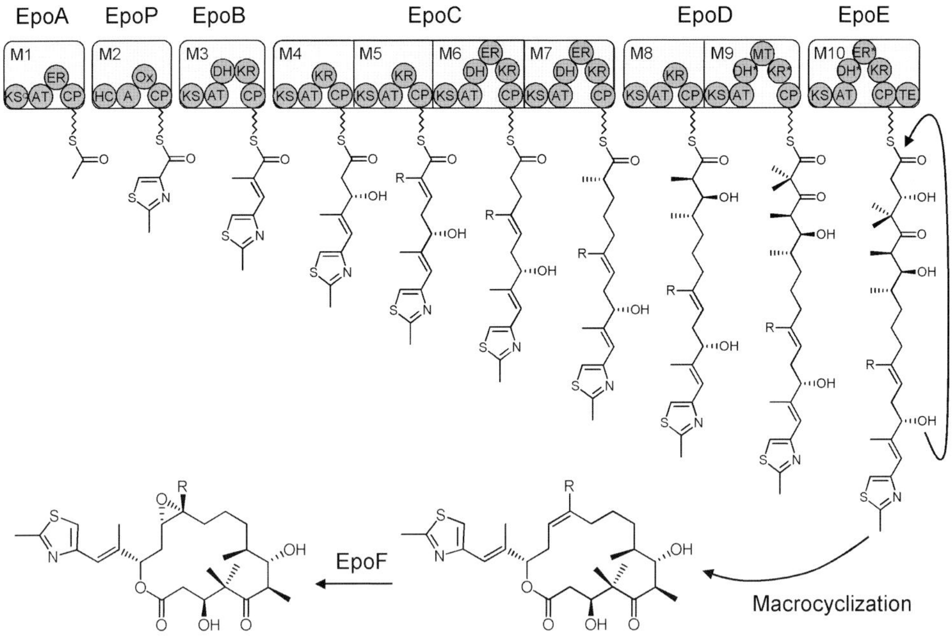

Epothilone A (R = H)
B (R = Me)

Epothilone C (R = H)
D (R = Me)

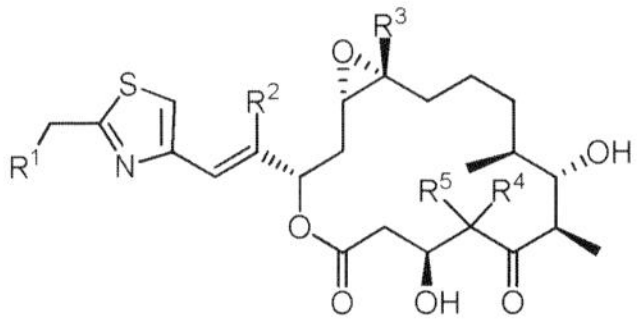

Epothilone E (R^3 = H; $R^{2,4,5}$ = Me; R^1 = OH)
F ($R^{2,3,4,5}$ = Me; R^1 = OH)
A_1 ($R^{1,3}$ = H; $R^{2,4,5}$ = Me)
A_2 ($R^{1,3,4}$ = H; $R^{2,5}$ = Me)
A_8 ($R^{1,2,3}$ = H; $R^{4,5}$ = Me)
A_9 ($R^{1,3}$ = H; $R^{4,5}$ = Me; R^2 = CH_2OH)
B_{10} ($R^{1,2,3,4,5}$ = Me)

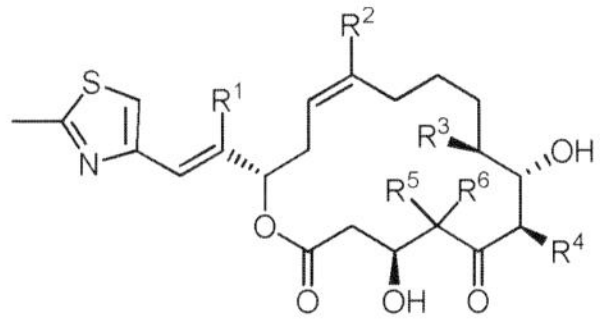

Epothilone C_1 ($R^{2,5}$ = H; $R^{1,3,4,6}$ = Me)
D_1 (R^6 = H; $R^{1,2,3,4,5}$ = Me)
C_2 ($R^{2,6}$ = H; $R^{1,3,4,5}$ = Me)
D_2 (R^5 = H; $R^{1,2,3,4,6}$ = Me)
C_3 ($R^{2,4}$ = H; $R^{1,3,5,6}$ = Me)
C_4 ($R^{2,3}$ = H; $R^{1,4,5,6}$ = Me)

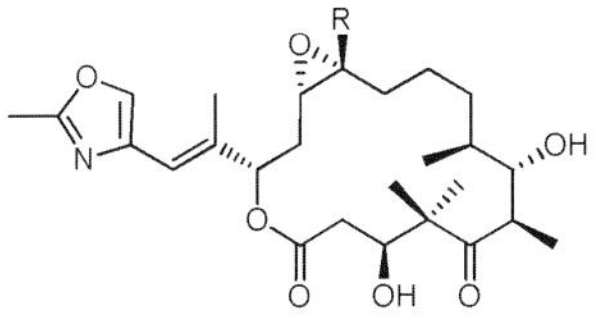

Epothilone G_1 (R = H)
G_2 (R = Me)

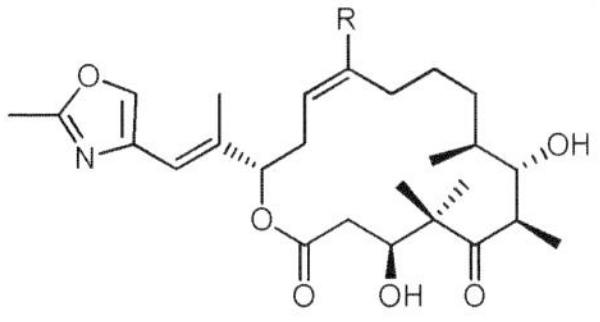

Epothilone H_1 (R = H)
H_2 (R = Me)

In most cases studied, all genes required for the biosynthesis of a specific metabolite class are found in one genomic region, a so-called biosynthetic gene cluster. On the basis of the growing body of genomic data from myxobacteria (see Section 12.5) and the fact that sequence homology search can be used to identify PKS and NRPS genes and even in part predict their substrates, it has become possible to describe the genomic potential of sequenced species for secondary metabolite production. According to these analyses, myxobacteria do harbor a large number of PKS, NRPS, and PKS/NRPS biosynthetic gene clusters [77] (see Table 12.1) which are expected to encode a class of secondary metabolite each (basic structure) such as the epothilones or argyrins. Most of these gene clusters are currently of unknown function and their numbers thus shows the hidden potential in the genome. Despite the large quantity of such "silent" pathways identified in myxobacterial genomes, quite a number of secondary metabolites are known already. These resemble the secondary metabolome of analyzed species under laboratory conditions and represent different basic classes of molecules (each encoded in one gene cluster).

Taken together, it is clear that myxobacteria continue to be a most promising source for novel bioactive secondary metabolites with potential use in therapy. In the following sections, two myxobacterial genera most prominent for their secondary metabolism are described together with some of their intriguing natural products.

12.7 Myxococcus

12.7.1 Introduction

In comparison to most other myxobacteria, the genus *Myxococcus* is characterized by relative ease of isolation and cultivation and has thus been intensively studied, including searching for secondary metabolites and deciphering genomics as exemplified by the availability of five genome sequences belonging to the three species of *M. xanthus*, *M. stipitatus*, and *M. fulvus*. Genome sizes range between 9 and 10.4 Mb encoding a large variety of secondary metabolite gene clusters. *Myxococcus* species were indeed shown to produce novel compounds such as the antibiotics myxovirescin [46], myxovalargin [48], and myxopyronin, [49] and the cytotoxins althiomycin [113], bengamide [114, 115], and saframycin Mx1 [116].

For several decades, *M. xanthus* has served as the model organism to study the various features unique to myxobacteria. Strains DK1622 and DZF1, especially, are being intensively used to analyze and decipher aspects of prokaryotic development such as cooperative behavior required for swarming, predation, and fruiting body formation just to name a few [117–120]. Consequently, the sequence of DK1622 was the first completely deciphered myxobacterial genome described in a publication. It revealed numerous unusual traits including the large size (see

Table 12.1) and its enormous genetic capacity for secondary metabolism which was found to occupy more than 8.6% of the total genome equaling 786 kbp [62].

12.7.2
Secondary Metabolism of *Myxococcus xanthus*

Until the genome sequence became available, almost nothing was known about secondary metabolism in *M. xanthus* DK1622. Some early reports on *M. xanthus* DK1622 speculated that it may produce myxovirescin (antibiotic TA) which was known from other species. Comparison of the genomic biosynthetic gene clusters with such clusters from published data of other myxobacteria, in conjunction with chemical analyses, enabled the identification of myxochelins as iron chelators, and myxochromides, myxovirescins, and myxalamids as secondary metabolites produced by the strain [121–124]. An additional novel class of secondary metabolites was later described and found to be essential for the formation of viable spores thus representing an internal signaling molecule called *DKxanthene* [125, 126]. Consequently, 5 out of 18 PKS, NRPS, and PKS/NRPS biosynthetic gene clusters could be correlated to their products. Efforts aiming at elucidating the function of all other such gene loci were subsequently performed by analyzing secondary metabolite production profiles of wild-type and mutant strains in each of these gene clusters using statistical tools. Following this approach, three further gene loci could be correlated to potential metabolites using LC-MS analysis revealing compounds c329, c506, and c844 [71] (see Figure 12.4). Searching for an improved producer in a database of myxobacterial extracts and subsequent genetic modification by promotor insertion in the strain identified led to the elucidation of the structure of myxoprincomides (c506 and derivatives thereof), a sixth class of natural products from DK1622 [71], while work on c329 and c844 is still ongoing.

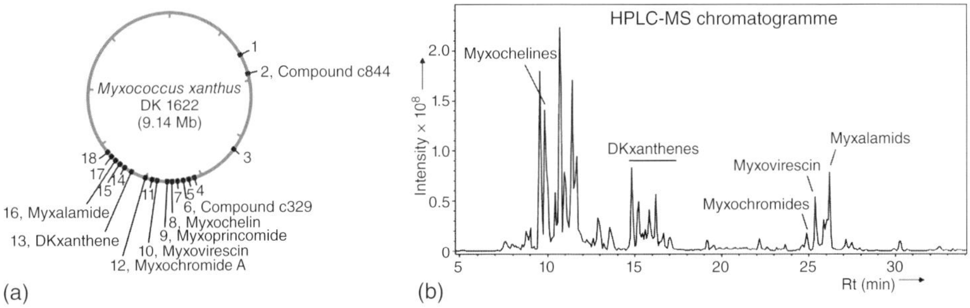

Figure 12.4 Biosynthetic potential of the myxobacterial model strain *Myxococcus xanthus* DK1622. (a) To date, 8 out of 18 identified biosynthetic gene clusters encoding PKS, NRPS, and PKS/NRPS hybrid systems could be assigned to products as indicated in the genome circle. (b) HPLC-MS chromatogram of a *M. xanthus* DK1622 culture extract showing the presence/detection of five natural product families among various other metabolites and media components (myxoprincomides and compounds c329 and c844 could only be detected in trace amounts, which is not indicated here).

Additionally, transcriptomic and proteomic analyses suggest that the remaining 10 unassigned biosynthetic gene clusters are also active under standard cultivation conditions and should thus produce additional secondary metabolites awaiting their discovery [127, 128]. Thus, *M. xanthus* DK1622 can be regarded as a terrific producer of natural products and not least owing to the availability of numerous genetic tools (see Section 12.5), it has become a logical choice to heterologously express biosynthetic pathways from other myxobacteria which turn out to be difficult to handle genetically. Relatively fast and also suspended growth as well as already available fermentation know-how for this strain are further arguments to employ *M. xanthus* DK1622 as host for heterologous expression of myxobacterial biosynthetic gene clusters. An additional advantage of this species for heterologous expression is the presence of three different phosphopantetheinyl transferases, two of which exhibit broad spectrum substrate acceptance [129]. Such proteins are required for post-translational activation of PKS and NRPS biosynthetic proteins.

12.7.3
Myxococcus xanthus as Expression Host

Since its first use as host for heterologous expression of secondary metabolites, *M. xanthus* has been widely used for this purpose (see Table 12.3). As it is thought that phylogenetic relatedness is of utmost importance for successful heterologous expression due to issues such as promoter recognition, precursor supply, and self-resistance [55–57], most examples are related to gene clusters from other myxobacteria, and we are currently only aware of one example of the usage of this strain for the expression of a streptomycete-derived secondary metabolite (oxytetracycline; see Table 12.3). Large gene constructs are usually reconstituted

Table 12.3 Heterologous production of secondary metabolites in *Myxococcus xanthus*.

Compound	Native producer	Pathway-type (size)	Yield	References
Bengamide	*Myxococcus virescens*	PKS/NRPS (25 kb)	$>10\ mg\ l^{-1}$	[115]
Dawenol	*Stigmatella aurantiaca*	PKS (21 kb)	n.d.	[72]
Epothilone	*Sorangium cellulosum*	PKS/NRPS (56 kb)	$0.1-23\ mg\ l^{-1}$	[130–132]
Myxochromide S	*Stigmatella aurantiaca*	PKS/NRPS (29 kb)	$>1\ g\ l^{-1}$	[58]
Myxothiazol	*Stigmatella aurantiaca*	PKS/NRPS (57 kb)	$20\ mg\ l^{-1}$	[133]
Oxytetracycline	*Streptomyces rimosus*	PKS (32 kb)	$10\ mg\ l^{-1}$	[134]
Pretubulysin	*Cystobacter species*	PKS/NRPS (40 kb)	$0.2\ mg\ l^{-1}$ [a]	[135]
PUFAs	*Aetherobacter fasciculatus*	PKS/FAS	~1 mg/CDW[b]	[136]

n.d. – Not determined.

a) $0.1\ mg\ l^{-1}$ supplemental of racemic D,L-pipecolic acid.

b) Mixture of <0.1 mg/CDW Arachidonic acid (AA, 20 : 4, $n-6$), 0.2 mg/CDW Eicosapentaenoic acid (EPA, 20 E: 5, $n-3$), <0.1 mg/CDW docosapentaenoic acid (DPA, 22 : 5, $n-6$), <0.1 mg/CDW docosapentaenoic acid (DPA, 22 : 5, $n-3$), and 0.8 mg/CDW docosapentaenoic acid (DPA, 22 : 6, $n-3$); CDW: cell dry weight.

in *E. coli* and then moved into *M. xanthus* by electroporation or conjugation. Subsequent insertion into the chromosome is achieved via homologous recombination, integration into phage attachment sites, or by transposition of complete gene clusters (see Table 12.3 for references and [54]). Yields achieved vary from hardly detectable small quantities of the product under study to superb production titers. In some cases, simple promotor insertion in front of the PKS/NRPS operon has tremendous effects on production titers. In general, according to our experience with the heterologous production of myxobacterial compounds, there seems to be a positive correlation between the relatedness of the producer strain to the host and the yields achieved.

12.7.4 Heterologous Expression of the Epothilone Gene Cluster from *Sorangium cellulosum* in *Myxococcus xanthus*

One of the most prominent examples of heterologous expression in *M. xanthus* relates to the epothilones, *S. cellulosum* secondary metabolites lactame derivatives of which were approved for the treatment of breast cancer by the Food and Drug Administration (FDA) in 2007 [81]. Identification of the biosynthetic gene cluster and initial heterologous expression in the host bacterium *Streptomyces coelicolor* yielded only very low quantities (0.05–0.1 $\mu g\,l^{-1}$) of the compound [137], leading to the assumption that a myxobacterium should provide better chances for heterologous expression. After the transfer of the whole pathway in two pieces into two different phage attachment sites of *M. xanthus*, production of epothilone A and B was established and measured at initial levels of 0.18 mg^{-1} [130]. This successful chassis for epothilone production was subsequently employed to generate a number of analogs of epothilone by biosynthetic pathway engineering in *E. coli* and transfer of the modified pathway into *M. xanthus* [138, 139]. Epothilone production profiles could be also modulated through media design and dissolved oxygen supply [140, 141]. The issue of yield improvement was addressed by the incorporation of an adsorber resin (XAD-16), addition of trace elements to the medium, identification of a suitable carbon source (methyl oleate), and implementation of a continuous fed-batch fermentation process generating higher cell density and yielding an average production titer of 23 $mg\,l^{-1}$ [131]. In parallel, a purification process was developed to recover epothilone from 1000 l fermentations from *M. xanthus* cultures at high yield and purity [142]. In the meantime, a synthetic biotechnology approach was also used to synthesize the total epothilone biosynthetic gene cluster for expression in *M. xanthus* [132]. The codon composition was adapted to a modified codon table for *M. xanthus* and the applied constructional pathway design allows for flexible engineering approaches. Although the initially achieved production yield (0.1 $mg\,l^{-1}$) has to be improved in upcoming work, the established production system sets the stage for detailed studies to decipher bottlenecks of production and also to generate derivatives of epothilone.

12.7.5 Production of Myxovirescin by Fermentation of *Myxococcus virescens*

Species of the genus *Myxococcus* are also well established as production hosts for their naturally occurring secondary metabolites, for example, *M. virescens* Mx v48 was studied intensely for production of myxovirescin [143] which can provide clues for how to optimize heterologous expression in *M. xanthus.* In general, *Myxococcus* species cannot metabolize simple sugars but can use amino acids as sole source of carbon, nitrogen, and energy. During growth in a peptone-based medium, amino nitrogen from amino acids is metabolized and large quantities of ammonia accumulate within the cells and in the culture supernatant (intracellular concentrations up to 140 mM were reported [144]), eventually leading to a decline in growth and productivity. A pH-dependent gaseous ammonium extraction device applying hydrophobic porous membranes was developed, allowing for the continuous extraction of ammonium from fermentations resulting in a decrease of ammonium concentration from 10–140 to 3–7.5 mM. In parallel, an elevation in myxovirescin yield from 8 to 100 $mg\,l^{-1}$ and an increase of cell mass by 233% was observed. Interestingly, it turned out that at a certain production level, myxovirescin inhibits growth of the natural host. This problem could be circumvented by the addition of an adsorber resin (XAD-16) for continuous extraction of the metabolite in a bypass to the fermenter [145].

12.8 Sorangium

12.8.1 Introduction

Members of the myxobacterial genus *Sorangium* are saprophytes capable of deriving their nutrition by cellulose degradation aerobically. They appear to be extraordinarily gifted secondary metabolite producers. Many species of *S. cellulosum* were analyzed in the past, resulting in a bias of the total known myxobacterial secondary metabolome, which currently indicates almost 50% of the myxobacterial natural products to be derived from this species [143]. Recently, novel genera and families within the *Sorangiineae* suborder were discovered showing even more potential for the isolation of new metabolites. Among them, *Aetherobacter* species [14] have been shown to be prolific producers of polyunsaturated fatty acids (PUFAs) [136, 146] and to produce aetheramides [147], a novel natural product scaffold inhibiting HIV-1 infection

12.8.2 *Sorangium cellulosum* So ce56

The capacity of the genus *Sorangium* has raised considerable attention and motivated the corresponding myxobacterial genome project of the representative model species *S. cellulosum* So ce56, aiming at establishing the genomic basis for secondary metabolite formation. The project was completed in 2007, resulting in the publication of the highquality sequence of the largest bacterial genome at that time comprising 13.0 Mb of closed circular DNA sequence [30]. Since then some additional species of *S. cellulosum* were sequenced in our laboratory revealing genome sizes of up to 16.2 Mb (GenBank accession CP012671). Indeed, analyses of these genomes predict an even larger genomic capacity for the formation of natural products than previously anticipated: As outlined in Sections 12.6 and 12.7.2, the number of typical secondary metabolite biosynthetic gene clusters found is much higher than the number of compounds known from each species. This finding underpins the future potential of the genus *Sorangium* and myxobacteria in general for the isolation of new bioactive compounds [77]. For *S. cellulosum* So ce56, the genome sequence enabled the identification of the biosynthetic gene clusters for the etnangien antibacterials and the highly cytotoxic actin binders chivosazols [148, 149].

The availability of the genome sequence also made possible for the first time to search for direct or indirect regulators of secondary metabolite biosynthesis. It is worth mentioning that for myxobacteria, such regulators are usually not found in close proximity to the corresponding biosynthetic gene cluster. In a proof-of-concept study, a direct positive regulator of chivosazol biosynthesis in strain *S. cellulosum* So ce56 (ChiR) was identified using pulldown assays employing the presumed promotor region of the gene cluster as a bait [150]. The encoding gene *chiR* was located several megabases apart from the chivosazol biosynthesis genes in the genome. While inactivation of *chiR* resulted in hardly detectable chivosazol production, overexpression of the ChiR protein led to a significant increase of productivity in a time course study of the fermentation process from 40 mg l^{-1} to more than 200 mg l^{-1}. This example represents one of the few cases where knowledge gained from a functional genome project was transferred into a directed approach to increase the yield of a secondary metabolite via genetic engineering.

Although such directed genetic and metabolic engineering has nowadays become feasible, traditional strain improvement has industrially remained the method of choice for elevating production titers (e.g., see Sections 12.8.3 and 12.8.4). This is largely caused by our limited understanding of the factors contributing to efficient production of secondary metabolites in myxobacteria due to the diversity of myxobacterial physiology and the resulting lack of ability to generalize species-specific findings. Additionally, understanding the complexity of regulation in myxobacteria is challenging as exemplified by the presence of almost 350 protein kinases of unknown function in *S. cellulosum* So ce56. Especially because the growth requirements for proficient secondary metabolite production in fermenters usually do not allow the use of minimal media, it is currently hardly possible to perform in-depth studies and predict which genetic changes would result in optimal productivity for most microbial producers. However, initial attempts have been made in this direction to define the metabolic network and quantify pathway fluxes in *S. cellulosum* So ce56 [31] (see Section 12.3).

12.8.3
The Anticancer Drug Ixabepilone Is a Derivative of Epothilone from *Sorangium cellulosum* So ce90

Epothilones are the most famous secondary metabolites from *S. cellulosum* and from myxobacteria in general, because their semisynthetic derivative ixabepilone was approved for cancer therapy by the American FDA in 2007 [81]. Epothilones were discovered as early as in 1987, but it took until 1995 to identify their mode of action which is highly similar to the anticancer drug taxol [151]: tubulin binding and subsequent destruction of the cellular tubulin network (see Figure 12.5). This finding led to an intensive drug developmental program in search for a derivative suitable for pharmaceutical application by several pharmaceutical companies. Eventually, ixabepilone, a lactam derivative of epothilone generated by few chemical steps from the natural product, was FDA approved as an anticancer agent, more than 10 years later.

Obviously, this process required multi-gram amounts of epothilone which could only be supplied after significant improvement of initial production titers found for epothilone in *S. cellulosum* So ce90. Total synthesis did not turn out as a viable alternative due to the complexity of the molecule. In general and despite the fact that directed metabolic engineering is gaining increasing popularity, classical strain improvement by random mutagenesis and subsequent screening for improved features of the strain remains an essential tool, not only when working with myxobacteria. This approach was successfully used in several instances with the genus *Sorangium* and enabled production of epothilone derivatives for generating ixabepilone by semisynthesis. In comparison to the heterologous expression of epothilone in *M. xanthus* this approach yielded a significantly more effective *S. cellulosum* production strain, reaching yields of more than 500 mg l^{-1} starting with titers of less than 1 mg l^{-1} [152].

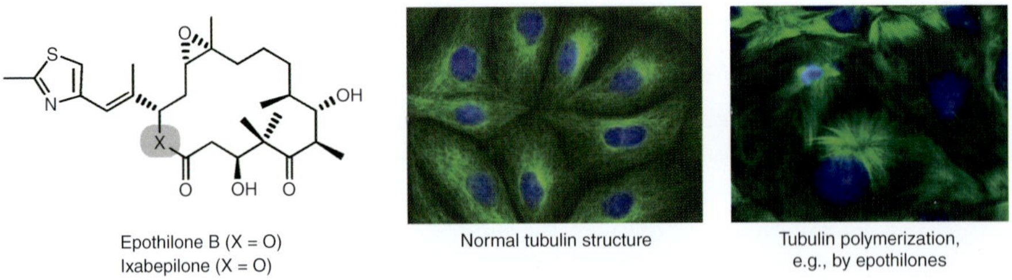

Figure 12.5 Structures of epothilone B in comparison to its semisynthetic derivative ixabepilone and the effect of both compounds on eukaryotic cells (nuclei were stained blue, microtubules were labeled with green).

12.8.4 Other *Sorangium* Secondary Metabolites Holding Promise for Future Application

Soraphen represents another prominent *Sorangium* compound (see Figure 12.6) and was identified from *S. cellulosum* So ce26 as a potent antibiotic active against a broad spectrum of fungi [153]. When employed in field trials, soraphen, for example, showed very good activity against the disease apple scab and Riesling–Silvaner infection with *Botrytis cinerea*. To allow for these tests, soraphen had to be produced in kilogram scale, which became possible owing to classical strain optimization and fermentation process improvement, resulting in a final production titer of $1.5\,g\,l^{-1}$ starting from approximately $3\,mg\,l^{-1}$ after 4 years of intensive work [143]. Soraphen, unfortunately, was not further developed as an agro-fungicide because of some potential toxicity issues. Nevertheless, it continued to serve as an important tool compound due to its potent activity as acetyl-CoA carboxylase inhibitor which is under study for various applications including the recent findings of exciting immunoregulatory activity as demonstrated in mice studies [82, 154] and very promising antiviral activities [155, 156].

Several additional compounds derived from *S. cellulosum* species are currently under study for pharmaceutical development (see Figure 12.6): Chlorotonil was recently shown to exhibit impressive antimalarial activity in mouse models [83] and the tubulin binder disorazole has demonstrated imposing activity in cancer models in mice when administered as peptide conjugate targeting

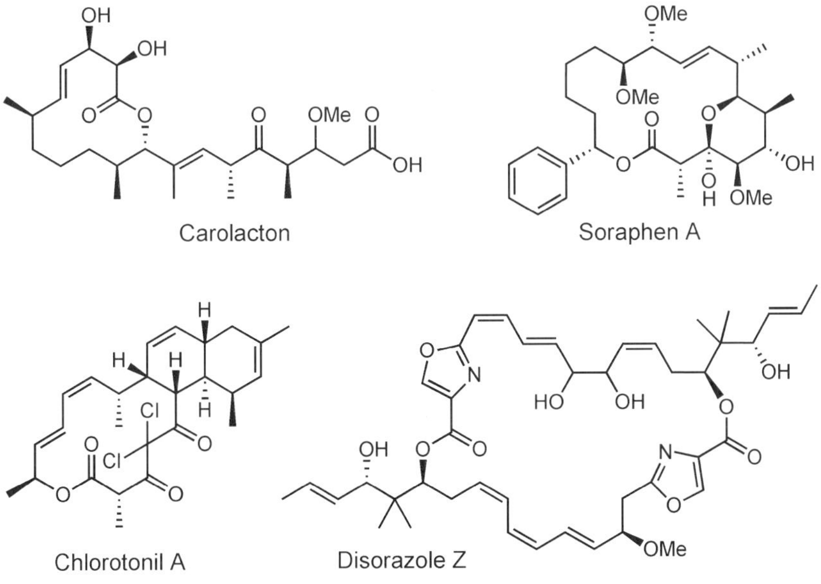

Figure 12.6 Examples for bioactive compounds from *Sorangium cellulosum* species.

cancer cells\nobreak [84]. Additionally, carolacton was described as an almost selective agent active against caries-forming streptococci via an unprecedented mechanism of action [157–159]. Overall, it can thus be expected that natural products derived from the genus *Sorangium* will continue to be of special interest due to their potential for pharmaceutical application.

12.9 Outlook

During the last decades, myxobacteria have been largely accepted as a prominent resource for novel and exciting natural products exhibiting various promising biological activities. In contrast to the more established bacterial producers for bioactive compounds, such as the streptomycetes, it is worth mentioning that the biodiversity of myxobacteria is mostly untapped as can be seen by the continuous discovery of novel species, genera, and even families belonging to this class of bacteria. Although only few genomes of myxobacteria have been analyzed to date, it is clear already that these species encode an enormous genomic potential for new natural products. Taken together, these facts make myxobacteria a most promising resource for the novel natural products of the future. Although a lot needs to be done to identify more and biodiverse myxobacteria, to develop genetic tools for such species, to study their potential to generate promising natural products, and to develop such products for the market, it can be expected that additional myxobacterial compounds will make their way through pharmaceutical and agrochemical development and can hopefully advance to application.

References

1. Reichenbach, H. (1999) The ecology of the myxobacteria. *Environ. Microbiol.*, **1** (1), 15–21.
2. Dawid, W. (2000) Biology and global distribution of myxobacteria in soils. *FEMS Microbiol. Rev.*, **24** (4), 403–427.
3. Brinkhoff, T., Fischer, D., Vollmers, J., Voget, S. *et al.* (2012) Biogeography and phylogenetic diversity of a cluster of exclusively marine myxobacteria. *ISME J.*, **6** (6), 1260–1272.
4. Singh, B.N. (1947) Myxobacteria in soils and composts; their distribution, number and lytic action on bacteria. *J. Gen. Microbiol.*, **1** (1), 1–10.
5. Shimkets, L., Dworkin, M., and Reichenbach, H. (2006) in *The Prokaryotes*, 3rd edn (eds M. Dworkin, S. Falkow, E. Rosenberg, K. Schleifer, and E. Stackebrandt), Springer-Verlag, New York, pp. 31–115.
6. Reichenbach, H. (2005) in *Bergey's Manual of Systematic Bacteriology*, vol. 2, Part C, 2nd edn (eds D. Brenner, N. Krieg, and J. Staley), Springer-Verlag, pp. 1059–1144.
7. Ludwig, W., Schleifer, K.H., Reichenbach, H., and Stackebrandt, E. (1983) A phylogenetic analysis of the myxobacteria *Myxococcus fulvus, Stigmatella aurantiaca, Cystobacter fuscus, Sorangium cellulosum* and *Nannocystis exedens*. *Arch. Microbiol.*, **135** (1), 58–62.
8. Shimkets, L. and Woese, C.R. (1992) A phylogenetic analysis of the myxobacteria: basis for their classification.

Proc. Natl. Acad. Sci. U.S.A., **89** (20), 9459–9463.
9. Spröer, C., Reichenbach, H., and Stackebrandt, E. (1999) The correlation between morphological and phylogenetic classification of myxobacteria. *Int. J. Syst. Bacteriol.*, **49** (3), 1255–1262.
10. Oyaizu, H. and Woese, C.R. (1985) Phylogenetic-relationships among the sulfate respiring bacteria, myxobacteria and purple bacteria. *Syst. Appl. Microbiol.*, **6** (3), 257–263.
11. Garcia, R., Gerth, K., Stadler, M., Dogma, I.J. Jr., *et al.* (2010) Expanded phylogeny of myxobacteria and evidence for cultivation of the 'unculturables'. *Mol. Phylogenet. Evol.*, **57** (2), 878–887.
12. Sood, S., Awal, R.P., Wink, J., Mohr, K.I. *et al.* (2015) *Aggregicoccus edonensis* gen. nov., sp. nov., an unusually aggregating myxobacterium isolated from a soil sample. *Int. J. Syst. Evol. Microbiol.*, **65** (Pt. 3), 745–753.
13. Yamamoto, E., Muramatsu, H., and Nagai, K. (2014) *Vulgatibacter incomptus* gen. nov., sp. nov. and *Labilithrix luteola* gen. nov., sp. nov., two myxobacteria isolated from soil in Yakushima Island, and the description of *Vulgatibacteraceae* fam. nov., *Labilitrichaceae* fam. nov. and *Anaeromyxobacteraceae* fam. nov. *Int. J. Syst. Evol. Microbiol.*, **64** (Pt. 10), 3360–3368.
14. Garcia, R., Stadler, M., Gemperlein, K., and Müller, R. (2016) *Aetherobacter fasciculatus* gen. nov., sp. nov. and *Aetherobacter rufus* sp. nov., novel myxobacteria with promising biotechnological applications. *Int. J. Syst. Evol. Microbiol.*, **66**, 928–938.
15. Awal, R.P., Garcia, R.O., and Müller, R. (2016) *Racemicystis crocea* gen. nov., sp. nov., a soil myxobacterium in *Polyangiaceae. Int. J. Syst. Evol. Microbiol.*, **66**, 2389–2395.
16. Garcia, R., Gemperlein, K., and Müller, R. (2014) *Minicystis rosea* gen. nov., sp. nov., a polyunsaturated fatty acid-rich and steroid-producing soil myxobacterium. *Int. J. Syst. Evol. Microbiol.*, **64** (Pt. 11), 3733–3742.
17. Garcia, R.O., Reichenbach, H., Ring, M.W., and Müller, R. (2009) *Phaselicystis flava* gen. nov., sp. nov., an arachidonic acid-containing soil myxobacterium, and the description of *Phaselicystidaceae* fam. nov. *Int. J. Syst. Evol. Microbiol.*, **59** (6), 1524–1530.
18. Mohr, K.I., Garcia, R.O., Gerth, K., Irschik, H. *et al.* (2012) *Sandaracinus amylolyticus* gen. nov., sp. nov., a starch-degrading soil myxobacterium, and description of *Sandaracinaceae* fam. nov. *Int. J. Syst. Evol. Microbiol.*, **62** (Pt. 5), 1191–1198.
19. Iizuka, T., Jojima, Y., Hayakawa, A., Fujii, T. *et al.* (2013) *Pseudenhygromyxa salsuginis* gen. nov., sp. nov., a myxobacterium isolated from an estuarine marsh. *Int. J. Syst. Evol. Microbiol.*, **63** (Pt. 4), 1360–1369.
20. Garcia, R. and Müller, R. (2014) in *The Prokaryotes*, 4th edn (eds E. Rosenberg, E. DeLong, S. Lory, E. Stackebrandt, and F. Thompson), Springer-Verlag, Berlin, pp. 173–181.
21. Sanford, R.A., Cole, J.R., and Tiedje, J.M. (2002) Characterization and description of *Anaeromyxobacter dehalogenans* gen. nov., sp. nov., an aryl-halorespiring facultative anaerobic myxobacterium. *Appl. Environ. Microbiol.*, **68** (2), 893–900.
22. Dawid, W., Gallikowski, C.A., and Hirsch, P. (1988) Psychrophilic myxobacteria from Antarctic soils. *Polarforschung*, **58** (2/3), 271–278.
23. Gerth, K. and Müller, R. (2005) Moderately thermophilic myxobacteria: novel potential for the production of natural products isolation and characterization. *Environ. Microbiol.*, **7** (6), 874–880.
24. Iizuka, T., Tokura, M., Jojima, Y., Hiraishi, A. *et al.* (2006) Enrichment and phylogenetic analysis of moderately thermophilic myxobacteria from hot springs in Japan. *Microbes Environ.*, **21** (3), 189–199.
25. Iizuka, T., Jojima, Y., Fudou, R., and Yamanaka, S. (1998) Isolation of

myxobacteria from the marine environment. *FEMS Microbiol. Lett.*, **169** (2), 317–322.
26. Watson, B.F. and Dworkin, M. (1968) Comparative intermediary metabolism of vegetative cells and microcysts of *Myxococcus xanthus*. *J. Bacteriol.*, **96** (5), 1465–1473.
27. Reichenbach, H. and Dworkin, M. (1969) Studies on *Stigmatella aurantiaca* (Myxobacterales). *J. Gen. Microbiol.*, **58**, 3–14.
28. Gerth, K. and Reichenbach, H. (1978) Induction of myxospore formation in *Stigmatella aurantiaca* (Myxobacterales). *Arch. Microbiol.*, **177** (2), 173–182.
29. Reichenbach, H. and Dworkin, M. (1992) in *The Prokaryotes*, 2nd edn (eds A. Balows, H. Trüper, M. Dworkin, W. Harder, and K. Schleifer), Springer-Verlag, New York, pp. 3416–3487.
30. Schneiker, S., Perlova, O., Kaiser, O., Gerth, K. *et al.* (2007) Complete genome sequence of the myxobacterium *Sorangium cellulosum*. *Nat. Biotechnol.*, **25** (11), 1281–1289.
31. Bolten, C.J., Heinzle, E., Müller, R., and Wittmann, C. (2009) Investigation of the central carbon metabolism of *Sorangium cellulosum*: metabolic network reconstruction and quantification of pathway fluxes. *J. Microbiol. Biotechnol.*, **19** (1), 23–36.
32. Dworkin, M. (1966) Biology of the myxobacteria. *Annu. Rev. Microbiol.*, **20**, 75–106.
33. Konovalova, A., Petters, T., and Sogaard-Andersen, L. (2010) Extracellular biology of *Myxococcus xanthus*. *FEMS Microbiol. Rev.*, **34** (2), 89–106.
34. Hart, B.A. and Zahler, S.A. (1966) Lytic enzyme produced by *Myxococcus xanthus*. *J. Bacteriol.*, **92** (6), 1632–1637.
35. Sudo, S. and Dworkin, M. (1972) Bacteriolytic enzymes produced by *Myxococcus xanthus*. *J. Bacteriol.*, **110** (1), 236–245.
36. Velicer, G.J. and Vos, M. (2009) Sociobiology of the myxobacteria. *Annu. Rev. Microbiol.*, **63**, 599–623.
37. Homma, Y. (1984) Perforation and lysis of hyphae of rhizoctonia-solani and conidia of cochliobolus-miyabeanus by soil myxobacteria. *Phytopathology*, **74** (10), 1234–1239.
38. Geitler, L. (1925) Über *Polyangium parasiticum* n. sp., eine submerse, parasitische *Myxobacteriacee*. *Zeit. Bot.*, **17**, 600–603.
39. Bode, H.B. and Müller, R. (2003) Possibility of bacterial recruitment of plant genes associated with the biosynthesis of secondary metabolites. *Plant Physiol.*, **132** (3), 1153–1161.
40. Perez, J., Castaneda-Garcia, A., Jenke-Kodama, H., Müller, R. *et al.* (2008) Eukaryotic-like protein kinases in the prokaryotes and the myxobacterial kinome. *Proc. Natl. Acad. Sci. U.S.A.*, **105** (41), 15950–15955.
41. Huntley, S., Wuichet, K., and Søgaard-Andersen, L. (2014) in *Myxobacteria – Genomics, Cellular and Molecular Biology* (eds Z. Yang and P.I. Higgs), Caister Academic Press, Norfolk, pp. 31–50.
42. Reichenbach, R. and Dworkin, M. (1981) *The Order Myxobacterales*, Springer-Verlag, Berlin.
43. Gerth, K., Trowitzsch, W., Piehl, G., Schultze, R. *et al.* (1984) Inexpensive media for mass cultivation of myxobacteria. *Appl. Microbiol. Biotechnol.*, **19** (1), 23–28.
44. Mizrahi, A., Arnan, J., Miller, G., Liron, Z. *et al.* (1976) Scaling-up of process for production of antibiotic of *Myxococcus xanthus* TA. *J. Appl. Chem. Biotechnol.*, **26**, 160–166.
45. Gerth, K., Irschik, H., Reichenbach, H., and Trowitzsch, W. (1980) Myxothiazol, an antibiotic from *Myxococcus fulvus* (myxobacterales). I. Cultivation, isolation, physico-chemical and biological properties. *J. Antibiot.*, **33** (12), 1474–1479.
46. Gerth, K., Irschik, H., Reichenbach, H., and Trowitsch, W. (1982) The myxovirescins, a family of antibiotics from *Myxococcus virescens* (Myxobacterales). *J. Antibiot.*, **35** (11), 1454–1459.
47. Gerth, K., Jansen, R., Reifenstahl, G., Höfle, G. *et al.* (1983) The myxalamids, new antibiotics from *Myxococcus xanthus* (Myxobacterales). I. Production,

physico-chemical and biological properties, and mechanism of action. *J. Antibiot.*, **36** (9), 1150–1156.
48. Irschik, H., Gerth, K., Kemmer, T., Steinmetz, H. *et al.* (1983) The Myxovalargins, new peptide antibiotics from *Myxococcus fulvus* (Myxobacterales) I. Cultivation, isolation, and some chemical and biological properties. *J. Antibiot.*, **36** (1), 6–12.
49. Irschik, H., Gerth, K., Höfle, G., Kohl, W. *et al.* (1983) The myxopyronins, new inhibitors of bacterial RNA synthesis from *Myxococcus fulvus* (Myxobacterales). *J. Antibiot.*, **36** (12), 1651–1658.
50. Irschik, H., Jansen, R., Gerth, K., Höfle, G. *et al.* (1987) The sorangicins, novel and powerful inhibitors of eubacterial RNA polymerase isolated from myxobacteria. *J. Antibiot.*, **40** (1), 7–13.
51. Kunze, B., Kemmer, T., Höfle, G., and Reichenbach, H. (1984) Stigmatellin, a new antibiotic from *Stigmatella aurantiaca* (Myxobacterales). I. Production, physico-chemical and biological properties. *J. Antibiot.*, **37** (5), 454–461.
52. Kunze, B., Kohl, W., Höfle, G., and Reichenbach, H. (1985) Production, isolation, physico-chemical and biological properties of angiolam A, a new antibiotic from *Angiococcus disciformis* (Myxobacterales). *J. Antibiot.*, **38** (12), 1649–1654.
53. Kunze, B., Höfle, G., and Reichenbach, H. (1987) The aurachins, new quinoline antibiotics from myxobacteria: production, physico-chemical and biological properties. *J. Antibiot.*, **40** (3), 258–265.
54. Black, W.P., Julien, B., Rodriguez, E., and Yang, Z. (2010) in *Manual of Industrial Microbiology and Biotechnology*, 3rd edn (eds R.H. Baltz, J.E. Davies, and A.L. Demain), ASM Press, Washington, DC, pp. 262–272.
55. Wenzel, S.C. and Müller, R. (2005) Recent developments towards the heterologous expression of complex bacterial natural product biosynthetic pathways. *Curr. Opin. Biotechnol.*, **16** (6), 594–606.
56. Ongley, S., Bian, X., Neilan, B.A., and Müller, R. (2013) Recent advances in the heterologous expression of microbial natural product biosynthetic pathways. *Nat. Prod. Rep.*, **30** (8), 1121–1138.
57. Zhang, H., Boghigian, B.A., Armando, J., and Pfeifer, B.A. (2011) Methods and options for the heterologous production of complex natural products. *Nat. Prod. Rep.*, **28** (1), 125–151.
58. Fu, J., Wenzel, S.C., Perlova, O., Wang, J. *et al.* (2008) Efficient transfer of two large secondary metabolite pathway gene clusters into heterologous hosts by transposition. *Nucleic Acids Res.*, **36** (17), e113.
59. Letouvet-Pawlak, B., Monnier, C., Barray, S., Hodgson, D.A. *et al.* (1990) Comparison of beta-galactosidase production by two inducible promoters in *Myxococcus xanthus*. *Res. Microbiol.*, **141** (4), 425–435.
60. Iniesta, A.A., García-Heras, F., Abellón-Ruiz, J., Gallego-García, A. *et al.* (2012) Two systems for conditional gene expression in *Myxococcus xanthus* inducible by isopropyl-ß-D-thiogalactopyranoside or vanillate. *J. Bacteriol.*, **194** (21), 5875–5885.
61. Zhao, J.-Y., Zhong, L., Shen, M.-J., Xia, Z.-J. *et al.* (2008) Discovery of the autonomously replicating plasmid pMF1 from *Myxococcus fulvus* and development of a gene cloning system in *Myxococcus xanthus*. *Appl. Environ. Microbiol.*, **74** (7), 1980–1987.
62. Goldman, B.S., Nierman, W.C., Kaiser, D., Slater, S.C. *et al.* (2006) Evolution of sensory complexity recorded in a myxobacterial genome. *Proc. Natl. Acad. Sci. U.S.A.*, **103** (41), 15200–15205.
63. Thomas, S.H., Wagner, R.D., Arakaki, A.K., Skolnick, J. *et al.* (2008) The mosaic genome of *Anaeromyxobacter dehalogenans* strain 2CP-C suggests an aerobic common ancestor to the delta-proteobacteria. *PLoS One*, **3** (5), e2103.
64. Zaburannyi, N., Bunk, B., Maier, J., Overmann, J. *et al.* (2016) Genome analysis of the fruiting body forming myxobacterium *Chondromyces crocatus* reveals a high potential for natural product biosynthesis. *Appl. Environ. Microbiol.*, **82**, 1945–1957.

65. Huntley, S., Zhang, Y., Treuner-Lange, A., Kneip, S. *et al.* (2012) Complete genome sequence of the fruiting myxobacterium *Corallococcus coralloides* DSM 2259. *J. Bacteriol.*, **194** (11), 3012–3013.
66. Ivanova, N., Daum, C., Lang, E., Abt, B. *et al.* (2010) Complete genome sequence of *Haliangium ochraceum* type strain (SMP-2). *Stand. Genomic Sci.*, **2** (1), 96–106.
67. Li, Z.F., Li, X., Liu, H., Liu, X. *et al.* (2011) Genome sequence of the halotolerant marine bacterium *Myxococcus fulvus* HW-1. *J. Bacteriol.*, **193** (18), 5015–5016.
68. Huntley, S., Kneip, S., Treuner-Lange, A., and Sogaard-Andersen, L. (2013) Complete genome sequence of *Myxococcus stipitatus* strain DSM 14675, a fruiting myxobacterium. *Genome Announc.*, **1** (2), e0010013.
69. Huntley, S., Hamann, N., Wegener-Feldbrügge, S., Treuner-Lange, A. *et al.* (2011) Comparative genomic analysis of fruiting body formation in myxococcales. *Mol. Biol. Evol.*, **28** (2), 1083–1097.
70. Weber, T., Blin, K., Duddela, S., Krug, D. *et al.* (2015) antiSMASH 3.0-a comprehensive resource for the genome mining of biosynthetic gene clusters. *Nucleic Acids Res.*, **43** (W1), W237–W243.
71. Cortina, N.S., Krug, D., Plaza, A., Revermann, O. *et al.* (2012) Myxoprincomide: a natural product from *Myxococcus xanthus* discovered by comprehensive analysis of the secondary metabolome. *Angew. Chem. Int. Ed.*, **51** (3), 811–816.
72. Oßwald, C., Zaburannyi, N., Burgard, C., Hoffmann, T. *et al.* (2014) A highly unusual polyketide synthase directs dawenol polyene biosynthesis in *Stigmatella aurantiaca*. *J. Biotechnol.*, **191**, 54–63.
73. Han, K., Li, Z.F., Peng, R., Zhu, L.P. *et al.* (2013) Extraordinary expansion of a *Sorangium cellulosum* genome from an alkaline milieu. *Sci. Rep.*, **3** (2101), 1–7.
74. Tonomura, M., Ehara, A., Suzuki, H., and Amachi, S. (2015) Draft genome sequence of *Anaeromyxobacter* sp. Strain PSR-1, an arsenate-respiring bacterium isolated from arsenic-contaminated soil. *Genome Announc.*, **3** (3), e00472-15.
75. Müller, S., Willett, J.W., Bahr, S.M., Darnell, C.L. *et al.* (2013) Draft genome sequence of *Myxococcus xanthus* wild-type strain DZ2, a model organism for predation and development. *Genome Announc.*, **1** (3), e00217-13.
76. Müller, S., Willett, J.W., Bahr, S.M., Scott, J.C. *et al.* (2013) Draft genome of a type 4 pilus defective *Myxococcus xanthus* strain, DZF1. *Genome Announc.*, **1** (3), e00392-13.
77. Wenzel, S.C. and Müller, R. (2009) The impact of genomics on the exploitation of the myxobacterial secondary metabolome. *Nat. Prod. Rep.*, **26** (11), 1385–1407.
78. Wenzel, S.C. and Müller, R. (2009) The biosynthetic potential of myxobacteria and their impact on drug discovery. *Curr. Opin. Drug Discovery Dev.*, **12** (2), 220–230.
79. Weissman, K.J. and Müller, R. (2010) Myxobacterial secondary metabolites: bioactivities and modes-of-action. *Nat. Prod. Rep.*, **27** (9), 1276–1295.
80. Newman, D.J. and Cragg, G.M. (2012) Natural products as sources of new drugs over the 30 years from 1981 to 2010. *J. Nat. Prod.*, **75** (3), 311–335.
81. Altmann, K.H., Höfle, G., Müller, R., Mulzer, J. *et al.* (2009) *The Epothilones: An Outstanding Family of Anti-Tumor Agents*, Springer-Verlag, Wien, New York.
82. Berod, L., Friedrich, C., Nandan, A., Freitag, J. *et al.* (2014) De novo fatty acid synthesis controls the fate between regulatory T and T helper 17 cells. *Nat. Med.*, **20** (11), 1327–1333.
83. Held, J., Gebru, T., Kalesse, M., Jansen, R. *et al.* (2014) Antimalarial activity of the myxobacterial macrolide chlorotonil A. *Antimicrob. Agents Chemother.*, **58** (11), 6378–6384.
84. Guenther, E.G., Paulini, K.W., Aicher, B., Baasner, S. *et al.* (2008) Targeting Disorazol Z to LHRH-receptor positive tumors by the cytotoxic conjugate AEZS-125. American Association

for Cancer Research (AACR) Annual Meeting.

85. Baumann, S., Herrmann, J., Raju, R., Steinmetz, H. *et al.* (2014) Cystobactamids: myxobacterial topoisomerase inhibitors exhibiting potent antibacterial activity. *Angew. Chem. Int. Ed.*, **53** (52), 14605–14609.
86. Reichenbach, H. (2001) Myxobacteria, producers of novel bioactive substances. *J. Ind. Microbiol. Biotechnol.*, **27** (3), 149–156.
87. Garcia, R.O., Krug, D., and Müller, R. (2009) Discovering natural products from myxobacteria with emphasis on rare producer strains in combination with improved analytical methods. *Methods Enzymol.*, **458**, 59–91.
88. Ringel, S.M., Greenough, R.C., Roemer, S., Connor, D. *et al.* (1977) Ambruticin (W7783), a new antifungal antibiotic. *J. Antibiot.*, **30** (5), 371–375.
89. Trowitzsch, W., Reifenstahl, G., Wray, V., and Höfle, G. (1980) Myxothiazol, an antibiotic from *Myxococcus fulfus* (Myxobacteraoles) II. Structure elucidation. *J. Antibiot.*, **33** (12), 1480–1490.
90. Kunze, B., Jansen, R., Höfle, G., and Reichenbach, H. (2004) Ajudazols, new inhibitors of the mitochondrial electron transport from *Chondromyces crocatus*. Production, antimicrobial activity and mechanism of action. *J. Antibiot.*, **57** (2), 151–155.
91. Kunze, B., Jansen, R., Reichenbach, H., and Höfle, G. (1996) Chondramides A-D, new cytostatic and antifungal cyclodepsipeptides from *Chondromyces crocatus* (myxobacteria): Isolation and structure elucidation. *Liebigs Ann. Chem.*, **1996** (2), 285–290.
92. Jansen, R., Kunze, B., Reichenbach, H., and Höfle, G. (2003) Chondrochloren A and B, new beta-amino styrenes from *Chondromyces crocatus* (myxobacteria). *Eur. J. Org. Chem.*, **2003** (14), 2684–2689.
93. Kunze, B., Jansen, R., Höfle, G., and Reichenbach, H. (1994) Crocacin, a new electron transport inhibitor from *Chondromyces crocatus* (myxobacteria). Production, isolation, physico-chemical and biological properties. *J. Antibiot.*, **47** (8), 881–886.
94. Viehrig, K., Surup, F., Harmrolfs, K., Jansen, R. *et al.* (2013) Concerted action of P450 plus helper protein to form the amino-hydroxy-piperidone moiety of the potent protease inhibitor crocapeptin. *J. Am. Chem. Soc.*, **135** (45), 16885–16894.
95. Steinmetz, H., Irschik, H., Kunze, B., Reichenbach, H. *et al.* (2007) Thuggacins, macrolide antibiotics active against *Mycobacterium tuberculosis*: isolation from myxobacteria, structure elucidation, conformation analysis and biosynthesis. *Chem. Eur. J.*, **13** (20), 5822–5832.
96. Jansen, R., Irschik, H., Reichenbach, H., and Höfle, G. (1997) Antibiotics from gliding bacteria, LXXX. Chivosazoles A-F: novel antifungal and cytotoxic macrolides from *Sorangium cellulosum* (Myxobacteria). *Liebigs Ann. Chem.*, **1997** (8), 1725–1732.
97. Jansen, R., Irschik, H., Reichenbach, H., Wray, V. *et al.* (1994) Disorazoles, highly cytotoxic metabolites from the Sorangicin-producing bacterium *Sorangium cellulosum*, strain So ce12. *Liebigs Ann. Chem.*, **1994** (8), 759–773.
98. Irschik, H., Jansen, R., Gerth, K., Höfle, G. *et al.* (1995) Sorangiolid A, a new antibiotic isolated from the myxobacterium *Sorangium cellulosum* So ce 12. *J. Antibiot.*, **48** (8), 886–887.
99. Zander, W., Irschik, H., Augustiniak, H., Herrmann, M. *et al.* (2012) Sulfangolids, macrolide sulfate esters from *Sorangium cellulosum*. *Chem. Eur. J.*, **18** (20), 6264–6271.
100. Krastel, P., Liechty, B.M., Schmitt, E., and Schreiner, E.P. (2011) Use of cyclic depsipeptides to inhibit kallikrein 7. US Patent 2011/0092437 A1.
101. Staunton, J. and Weissman, K.J. (2001) Polyketide biosynthesis: a millennium review. *Nat. Prod. Rep.*, **18** (4), 380–416.
102. Hertweck, C. (2009) The biosynthetic logic of polyketide diversity. *Angew. Chem. Int. Ed.*, **48** (26), 4688–4716.
103. Marahiel, M.A., Stachelhaus, T., and Mootz, H.D. (1997) Modular peptide synthetases involved in nonribosomal

peptide synthesis. *Chem. Rev.*, **97** (7), 2651–2674.

104. Marahiel, M.A. (2009) Working outside the protein-synthesis rules: insights into non-ribosomal peptide synthesis. *J. Pept. Sci.*, **15** (12), 799–807.
105. Molnar, I., Schupp, T., Ono, M., Zirkle, R. *et al.* (2000) The biosynthetic gene cluster for the microtubule-stabilizing agents epothilones A and B from *Sorangium cellulosum* So ce90. *Chem. Biol.*, 7 (2), 97–109.
106. Julien, B., Shah, S., Ziermann, R., Goldman, R. *et al.* (2000) Isolation and characterization of the epothilone biosynthetic gene cluster from *Sorangium cellulosum. Gene*, **249** (1–2), 153–160.
107. Hardt, I.H., Steinmetz, H., Gerth, K., Sasse, F. *et al.* (2001) New natural epothilones from *Sorangium cellulosum*, strains So ce90/B2 and So ce90/D13: isolation, structure elucidation, and SAR studies. *J. Nat. Prod.*, **64** (7), 847–856.
108. Walsh, C.T., Gehring, A.M., Weinreb, P.H., Quadri, L.E. *et al.* (1997) Post-translational modification of polyketide and nonribosomal peptide synthases. *Curr. Opin. Chem. Biol.*, **1** (3), 309–315.
109. Sasse, F., Steinmetz, H., Schupp, T., Petersen, F. *et al.* (2002) Argyrins, immunosuppressive cyclic peptides from myxobacteria. I. Production, isolation, physico-chemical and biological properties. *J. Antibiot.*, **55** (6), 543–551.
110. Nickeleit, I., Zender, S., Sasse, F., Geffers, R. *et al.* (2008) Argyrin A reveals a critical role for the tumor suppressor protein p27kip1 in mediating antitumor activities in response to proteasome inhibition. *Cancer Res.*, **14**, 23–35.
111. Bielecki, P., Lukat, P., Hüsecken, K., Dötsch, A. *et al.* (2012) Mutation in elongation factor G confers resistance to the antibiotic argyrin in the opportunistic pathogen *Pseudomonas aeruginosa. ChemBioChem*, **13** (16), 2339–2345.
112. Nyfeler, B., Hoepfner, D., Palestrant, D., Kirby, C.A. *et al.* (2012) Identification of elongation factor G as the conserved cellular target of argyrin B. *PLoS One*, **7** (9), e42657.
113. Kunze, B., Reichenbach, H., Augustiniak, H., and Höfle, G. (1982) Isolation and identification of althiomycin from *Cystobacter fuscus* (myxobacterales). *J. Antibiot.*, **35** (5), 635–636.
114. Johnson, T.A., Sohn, J., Vaske, Y.M., White, K.N. *et al.* (2012) Myxobacteria versus sponge-derived alkaloids: the bengamide family identified as potent immune modulating agents by scrutiny of LC-MS/ELSD libraries. *Bioorg. Med. Chem.*, **20** (14), 4348–4355.
115. Wenzel, S.C., Hoffmann, H., Zhang, J., Debussche, L. *et al.* (2015) Production of the bengamide class of marine natural products in Myxobacteria: biosynthesis and structure-activity relationships. *Angew. Chem. Int. Ed.*, **54** (51), 15560–15564.
116. Irschik, H., Trowitzsch-Kienast, W., Gerth, K., Höfle, G. *et al.* (1988) Saframycin Mx1, a new natural saframycin isolated from a myxobacterium. *J. Antibiot.*, **41** (8), 993–998.
117. Rosenberg, E. (1984) *Myxobacteria, Development and Cell Interactions*, Springer-Verlag, New York.
118. Dworkin, M. and Kaiser, D. (1993) *Myxobacteria II*, American Society for Microbiology, Washington, DC.
119. Whitworth, D. (ed.) (2007) *Myxobacteria: Multicellularity and Differentiation*, ASM Press, Chicago, IL.
120. Yang, Z. and Higgs, P.I. (eds) (2014) *Myxobacteria – Genomics, Cellular and Molecular Biology*, Caister Academic Press, Norfolk.
121. Bode, H.B. and Müller, R. (2007) in *Myxobacteria: Multicellularity and Differentiation* (ed. D. Whitworth), ASM Press, Chicago, IL, pp. 259–282.
122. Wenzel, S.C., Meiser, P., Binz, T., Mahmud, T. *et al.* (2006) Nonribosomal peptide biosynthesis: point mutations and module skipping lead to chemical diversity. *Angew. Chem. Int. Ed.*, **45** (14), 2296–2301.
123. Simunovic, V., Zapp, J., Rachid, S., Krug, D. *et al.* (2006) Myxovirescin biosynthesis is directed by hybrid polyketide synthases/nonribosomal

peptide synthetase, 3-hydroxy-3-methylglutaryl CoA synthases and trans-acting acyltransferases. *ChemBioChem*, **7** (8), 1206–1220.

124. Bode, H.B., Meiser, P., Klefisch, T., Cortina, N.S. *et al.* (2007) Mutasynthesis-derived myxalamids and origin of the isobutyryl-CoA starter unit of myxalamid B. *ChemBioChem*, **8** (17), 2139–2144.
125. Meiser, P., Bode, H.B., and Müller, R. (2006) The unique DKxanthene secondary metabolite family from the myxobacterium *Myxococcus xanthus* is required for developmental sporulation. *Proc. Natl. Acad. Sci. U.S.A.*, **103** (50), 19128–19133.
126. Meiser, P., Weissman, K.J., Bode, H.B., Krug, D. *et al.* (2008) DKxanthene biosynthesis - understanding the basis for diversity-oriented synthesis in myxobacterial secondary metabolism. *Chem. Biol.*, **15** (8), 771–781.
127. Bode, H.B., Ring, M.W., Schwär, G., Altmeyer, M.O. *et al.* (2009) Identification of additional players in the alternative biosynthesis pathway to isovaleryl-CoA in the myxobacterium *Myxococcus xanthus*. *ChemBioChem*, **10** (1), 128–140.
128. Schley, C., Altmeyer, M.O., Swart, R., Müller, R. *et al.* (2006) Proteome analysis of *Myxococcus xanthus* by off-line two-dimensional chromatographic separation using monolithic poly-(styrene-divinylbenzene) columns combined with ion-trap tandem mass spectrometry. *J. Proteome Res.*, **5** (10), 2760–2768.
129. Meiser, P. and Müller, R. (2008) Two functionally redundant Sfp-type 4'-phosphopantetheinyl transferases differentially activate biosynthetic pathways in *Myxococcus xanthus*. *ChemBioChem*, **9** (10), 1549–1553.
130. Julien, B. and Shah, S. (2002) Heterologous expression of epothilone biosynthetic genes in *Myxococcus xanthus*. *Antimicrob. Agents Chemother.*, **46** (9), 2772–2778.
131. Lau, J., Frykman, S., Regentin, R., Ou, S. *et al.* (2002) Optimizing the heterologous production of epothilone D in *Myxococcus xanthus*. *Biotechnol. Bioeng.*, **78** (3), 280–288.
132. Oßwald, C., Zipf, G., Schmidt, G., Maier, J. *et al.* (2014) Modular construction of a functional artificial epothilone polyketide pathway. *ACS Synth. Biol.*, **3** (10), 759–772.
133. Perlova, O., Fu, J., Kuhlmann, S., Krug, D. *et al.* (2006) Reconstitution of myxothiazol biosynthetic gene cluster by Red/ET recombination and heterologous expression in *Myxococcus xanthus*. *Appl. Environ. Microbiol.*, **72** (12), 7485–7494.
134. Stevens, D.C., Henry, M.R., Murphy, K.A., and Boddy, C.N. (2010) Heterologous expression of the oxytetracycline biosynthetic pathway in *Myxococcus xanthus*. *Appl. Environ. Microbiol.*, **76** (8), 2681–2683.
135. Chai, Y., Shan, S., Weissman, K.J., Hu, S. *et al.* (2012) Heterologous expression and genetic engineering of the tubulysin biosynthetic gene cluster using Red/ET recombineering and inactivation mutagenesis. *Chem. Biol.*, **19** (3), 361–371.
136. Gemperlein, K., Rachid, S., Garcia, R., Wenzel, S.C. *et al.* (2014) Polyunsaturated fatty acid biosynthesis in myxobacteria: Different PUFA synthases and their product diversity. *Chem. Sci.*, **5** (5), 1733–1741.
137. Tang, L., Shah, S., Chung, L., Carney, J. *et al.* (2000) Cloning and heterologous expression of the epothilone gene cluster. *Science*, **287** (5453), 640–642.
138. Starks, C.M., Zhou, Y.Q., Liu, F.H., and Licari, P.J. (2003) Isolation and characterization of new epothilone analogues from recombinant *Myxococcus xanthus* fermentations. *J. Nat. Prod.*, **66** (10), 1313–1317.
139. Tang, L., Chung, L., Carney, J.R., Starks, C.M. *et al.* (2005) Generation of new epothilones by genetic engineering of a polyketide synthase in *Myxococcus xanthus*. *J. Antibiot.*, **58** (3), 178–184.
140. Frykman, S., Tsuruta, H., Lau, J., Regentin, R. *et al.* (2002) Modulation of epothilone analog production through media design. *J. Ind. Microbiol. Biotechnol.*, **28** (1), 17–20.

141. Frykman, S.A., Tsuruta, H., Starks, C.M., Regentin, R. *et al.* (2002) Control of secondary metabolite congener distributions via modulation of the dissolved oxygen tension. *Biotechnol. Progr.*, **18** (5), 913–920.
142. Arslanian, R.L., Parker, C.D., Wang, P.K., McIntire, J.R. *et al.* (2002) Large-scale isolation and crystallization of epothilone D from *Myxococcus xanthus* cultures. *J. Nat. Prod.*, **65** (4), 570–572.
143. Gerth, K., Pradella, S., Perlova, O., Beyer, S. *et al.* (2003) Myxobacteria: proficient producers of novel natural products with various biological activities – past and future biotechnological aspects with the focus on the genus *Sorangium*. *J. Biotechnol.*, **106**, 233–253.
144. Gerth, K. and Reichenbach, H. (1986) Determination of bacterial ammonia pools using *Myxococcus virescens* as an example. *Anal. Biochem.*, **152** (1), 78–82.
145. Hecht, V., Vorlop, J., Kalbitz, H., Gerth, K. *et al.* (1987) Vortex chamber for in situ recovery of the antibiotic Myxovirescin A in continuous cultivation. *Biotechnol. Bioeng.*, **29**, 222–227.
146. Garcia, R., Pistorius, D., Stadler, M., and Müller, R. (2011) Fatty acid-related phylogeny of myxobacteria as an approach to discover polyunsaturated omega-3/6 fatty acids. *J. Bacteriol.*, **193** (8), 1930–1942.
147. Plaza, A., Garcia, R., Bifulco, G., Martinez, J.P. *et al.* (2012) Aetheramides A and B, potent HIV-inhibitory depsipeptides from a myxobacterium of the new genus "*Aetherobacter*". *Org. Lett.*, **14** (11), 2854–2857.
148. Menche, D., Arikan, F., Perlova, O., Horstmann, N. *et al.* (2008) Stereochemical determination and complex biosynthetic assembly of etnangien, a highly potent RNA polymerase inhibitor from the myxobacterium *Sorangium cellulosum*. *J. Am. Chem. Soc.*, **130** (43), 14234–14243.
149. Perlova, O., Gerth, K., Hans, A., Kaiser, O. *et al.* (2006) Identification and analysis of the chivosazol biosynthetic gene cluster from the myxobacterial model strain *Sorangium cellulosum* So ce56. *J. Biotechnol.*, **121** (2), 174–191.
150. Rachid, S., Gerth, K., Kochems, I., and Müller, R. (2007) Deciphering regulatory mechanisms for secondary metabolite production in the myxobacterium *Sorangium cellulosum* So ce56. *Mol. Microbiol.*, **63** (6), 1783–1796.
151. Bollag, D.M., McQueney, P.A., Zhu, J., Hensens, O. *et al.* (1995) Epothilones, a new class of microtubule-stabilizing agents with taxol-like mechanism of action. *Cancer Res.*, **55** (11), 2325–2333.
152. Müller, R. (2008) Biosynthesis and heterologous production of epothilones, in *The Epothilones – An Outstanding Family of Anti-Tumor Agents* (ed. J. Mulzer), Springer-Verlag.
153. Reichenbach, H. and Höfle, G. (1994) *Scientific Annual Report*, Gesellschaft für Biotechnologische Forschung GmbH, Braunschweig, pp. 5–22.
154. Beckers, A., Organe, S., Tinunermans, L., Scheys, K. *et al.* (2007) Chemical inhibition of acetyl-CoA carboxylase induces growth arrest and cytotoxicity selectively in cancer cells. *Cancer Res.*, **67** (17), 8180–8187.
155. Singaravelu, R., Desrochers, G.F., Srinivasan, P., O'Hara, S. *et al.* (2015) Soraphen A: a probe for investigating the role of de novo lipogenesis during viral infection. *ACS Infect. Dis.*, **1** (3), 130–134.
156. Koutsoudakis, G., Romero-Brey, I., Berger, C., Pérez-Vilaró, G. *et al.* (2015) Soraphen A: a broad-spectrum antiviral natural product with potent anti-hepatitis C virus activity. *J. Hepatol.*, **63** (4), 813–821.
157. Jansen, R., Irschik, H., Huch, V., Schummer, D. *et al.* (2010) Carolacton – a macrolide ketocarbonic acid reducing biofilm formation by the caries- and endocarditis-associated bacterium *Streptococcus mutans*. *Eur. J. Org. Chem.*, **2010** (7), 1284–1289.

158. Kunze, B., Reck, M., Dötsch, A., Lemme, A. *et al.* (2010) Damage of *Streptococcus mutans* biofilms by carolacton, a secondary metabolite from the myxobacterium *Sorangium cellulosum*. *BMC Microbiol.*, **10** (1), 199.

159. Reck, M., Rutz, K., Kunze, B., Tomasch, J. *et al.* (2011) The biofilm inhibitor carolacton disturbs membrane integrity and cell division of *Streptococcus mutans* through the serine/threonine protein kinase PknB. *J. Bacteriol.*, **193**, 5692–5706.

13
Host Organism: *Streptomyces*

Oksana Bilyk and Andriy Luzhetskyy

13.1
Introduction

Streptomycetes are gram-positive saprophytic bacteria with characteristic high GC DNA content. They are most abundant in various soils; however, they can be found also in other ecological environments such as fresh and salt water, air, and in the association with plants and animals [1–3]. The majority of streptomycetes are chemoheterotrophic and can utilize a wide variety of complex organic substrates. There are, however, few examples of *Streptomyces* species able to use carbon monoxide as an energy source and carbon dioxide as a sole carbon source [4]. Streptomycetes undergo a complex cycle of morphological differentiation involving formation of a layer of hyphae that often differentiate into a chain of spores, which makes them an interesting, but also challenging laboratory organism [5]. Streptomycetes are the most productive bacteria with respect to the synthesis of bioactive metabolites. These bacteria therefore have a long-standing successful history of production of secondary metabolites: streptomycetes have been in the focus of industrial application for many decades, for example, in the production of antibiotics, anthelmintics, immunosuppressant, anticancer drugs, pesticides, or insecticides [6, 7]. In addition, several industrially relevant enzymes (e.g., Mouse Tumor Necrosis Factor-α (mTNF-α)) could only be efficiently produced in *Streptomyces* [8–10]. However, actinomycetes are particularly known as the most prolific producers of antibiotics, providing approximately two-thirds of all known antibiotics, the majority of which are produced by streptomycetes. The market volume of antibiotics is expected to reach USD 44.68 billion by 2016. Considering the urgent need for new antibiotics, streptomycetes were of intensive screening programs in academic and industrial sectors. This led to the high rate of the rediscovery of known compounds and therefore streptomycetes have been considered as an exhausted source of bioactive natural products [6]. However, the whole genome sequencing programs have revealed that the biosynthetic potential of *Streptomyces* has been greatly underexplored relying on traditional approaches. A common feature of all streptomycete genomes is a

Industrial Biotechnology: Microorganisms, First Edition. Edited by Christoph Wittmann and James C. Liao.

length of about 8–11 Mb and the presence of approximately 20–40 biosynthetic gene clusters encoding the synthesis of secondary metabolites [11]. Moreover, the majority of these biosynthetic gene clusters are unique and potentially encode for novel chemical entity. Experimental data strongly show that the expression of secondary metabolite pathways is tightly regulated and is triggered by complex environmental factors [12–14]. Therefore, only a minority of potential chemical structures is produced, as the laboratory cultivation conditions have little in common with the complexity of environmental factors that bacteria face in their natural habitats [15]. Such silent genetic loci are referred to as cryptic or orphan pathways. The mathematical model of Watve *et al.* predicts that over 150 000 bioactive metabolites from streptomycetes alone could be discovered [16]. However, only approximately 17 000 natural products have been isolated from the streptomycetes species so far [6]. Considering the high demand for new antibiotics required to combat antibiotic-resistant pathogens, an exploitation of the biosynthetic potential of streptomycetes became a focus of many academic and industrial institutions. Various technologies to access this untapped reservoir of potentially bioactive structures have emerged over the last years. However, they all can be classified into two major strategies: (i) activation of the biosynthetic gene clusters *in situ* in the organism carrying it originally and (ii) activation of the biosynthetic gene cluster in the heterologous host [7, 17].

The first approach is based on the reconstitution of environmental conditions in a laboratory by mimicking the signals and stress factors that microorganisms undergo in nature or genetic manipulation of the native organism in order to trigger the expression of the gene cluster [12–14]. This strategy relies on our understanding of the complex regulatory cascades controlling the expression of biosynthetic gene clusters, and our ability to grow and genetically manipulate the native producers. Many different producers of bioactive natural products are not genetically accessible and this prevents the high throughput application of the first approach [5]. The second approach is to bypass natural regulatory networks that control the expression of cryptic gene clusters via their expression in a heterologous host [12, 18]. In this case, biosynthetic gene clusters could be expressed in the native form or modified with various genetic controlling elements (e.g., promoters, ribosome binding sites, terminators).

The successful implementation of the second approach requires three major components: (i) biosynthetic gene clusters; (ii) library of synthetic controlling elements (promoters, RBSs, terminators, etc.); and (iii) the chassis or host for the heterologous expression. There are several straightforward techniques allowing access to the biosynthetic gene clusters starting from the conventional BAC/PAC/cosmid/fosmid libraries generation and ending with the direct cloning of gene clusters from the chromosome using TAR and/or redET methods [19, 20] (Figure 13.1). In this chapter, we focus on different *Streptomyces* species used as a platform for heterologous expression of small bioactive molecules and proteins, and also give a brief overview of existing genome engineering tools and genetic controlling elements driving the gene expression in streptomycetes.

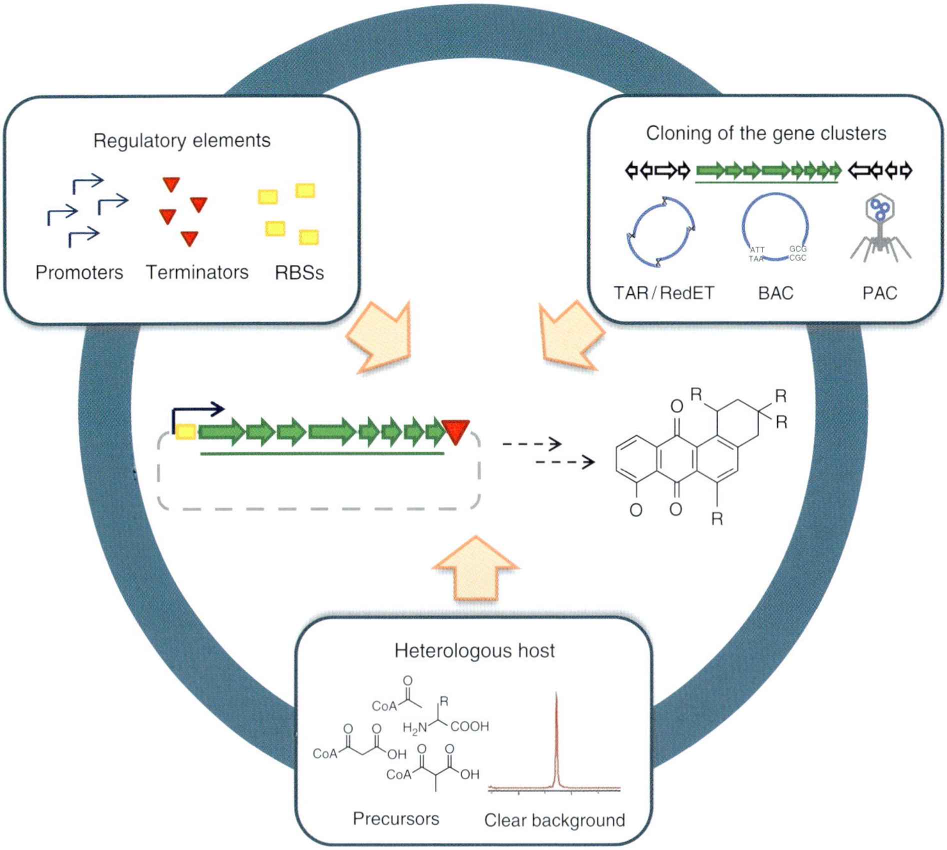

Figure 13.1 Schematic approach of activation of cryptic biosynthetic gene clusters.

13.2 *Streptomyces* Genome Manipulation Toolkits

Although the streptomycetes are the first choice for the heterologous expression of gene clusters, very often their growth characteristics suboptimal, exhibiting a long exponential phase, production of unwanted secondary metabolites, and reaching relatively low cell density in fermentations [21, 22]. There are number of genome engineering techniques developed during the last decade serving to improve the host characteristics of streptomycetes. In this chapter, we review the major actinomycetes, genome engineering techniques exploiting site-specific recombinases, transposons, meganucleases, CRISPR Cas9 system, and β-glucuronidase reporter [23–33]. The utilization of the above-mentioned system for genome manipulation of actinomycetes helps to overcome many technical difficulties and significantly facilitates the host development process.

13.2.1 Site-Specific Recombinases

Site-specific recombination (SSR) systems are powerful and efficient tools for high-throughput genome manipulation in the post genomic era [34]. Various SSRs have been described for a number of bacteria and higher organisms [35, 36]. Usually, SSRs are used for the generation of marker-free mutations [37], integration of heterologous DNA into the chromosome [38], marker-free heterologous expression of genes [39], transient and timed expression of genes [40], deletion of large genomic fragments [25], and *in vivo* cloning of genomic DNA regions, among others [39]. Several SSRs have been successfully employed in actinomycetes for various genome engineering experiments. The Cre, Dre, and Flp tyrosine recombinases catalyzing reciprocal SSR of DNA at 34 bp target sites (loxP, rox, and FRT, respectively), have been used for construction of unmarked mutants in various actinomycetes [23–25]. The first two works with an excellent efficiency in most of the actinobacteria, where the Flp recombinase is significantly less active. SSR-mediated excision of the antibiotic resistance cassette leaves a "scar" in the chromosome, that is, a single target site (e.g., rox site in the case of Dre). The length of this scar is usually less than 50 bp and does not exert polar transcriptional effects [25]. Marker rescue and reuse offers a simple and efficient way to introduce multiple gene deletions. However, consecutive deletions of additional genes in the same genetic background may be limited because of undesired DNA rearrangements generated by the multiple target sites left after each excision. This issue has been solved with the introduction of an iterative marker excision system (IMES). On the basis of serine recombinase of the phage phiC31 and its mutant att sites, IMES can be used for highly effective deletion of DNA fragments between inversely oriented B-CC and P-GG sites. The B-CC and P-GG sites are derived from attB and attP by substitution of the central core TT dinucleotide with CC and GG, respectively. After the recombination, an unnatural RR site that resides in the chromosome is the joining product of the right shoulders of B-CC and P-GG. The RR site does not interfere with the other similar site in the chromosome and thus, IMES can be used for multistep genomic engineering without risking unwanted DNA recombination [26, 41].

However, serine recombinases such as from phages phiC31 and φBT1 are usually unidirectional, mediating the integration of a circular DNA molecule into the genome [42, 43]. Most of the biosynthetic gene clusters are integrated into the chromosome of *Streptomyces* heterologous host via phage φC31 integration system [26, 44]. It provides a stable and nonreversible integration. Usually, such integration leaves behind a selection marker as well as additional unnecessary delivery vehicle sequences, such as an *int* gene and an *Escherichia coli* origin of replication creating major limitations for the iterative use of such systems. Recently, two integrative vectors (pTES and pTOS) were constructed, which allow the removal of unwanted delivery vehicle DNA such as the phage integrase gene and the antibiotic selection marker after the successful integration of the genes of interest into the chromosome. pTES is based on the integrative system of φC31 and its bacterial attachment site (attB) is located in a gene encoding a chromosome condensation protein. The pTOS vector is based on the VWB phage integration system having its attB site within the arginine tRNA. An excision of the integrase genes together with an apramycin resistance marker from the pTES and pTOS vectors is performed by Cre and Dre recombinase, respectively. Both vectors enable the engineering of stable recombinant streptomycete strains without any selection marker. The availability of the marker-free expression system allows the generation of multiple, stable, unmarked integrations, and the expression of several genes simultaneously [25].

One of the most important features of the site-specific recombinases is their ability to perform an excision of large DNA stretches within the chromosome. Two recombination target sites should be introduced on both ends of a DNA fragment to be deleted as direct repeats for the subsequent recombination. This approach has been utilized for generating three deletions spanning the α-lipomycin, phenalinolactone, and monensin biosynthetic gene clusters in *Streptomyces aureofaciens* Tü117, *Streptomyces* sp. Tü6071, and *Streptomyces cinnamonensis* A519, respectively [25]. Strikingly, the Cre recombinase was able to excise a fragment of more than 1.5 Mb in one step in order to construct the improved *Streptomyces avermitilis* host strains for the heterologous production of natural products [45]. The same recombinase has been employed to excise an approximately 200 kb region from the chromosome of Streptomyces *albus* J1074 [46].

Site-specific recombinases enabled the application of various genome engineering methods including the construction of unmarked multiple mutations, marker-free expression of target genes, large-fragment deletions, and chromosomal integration of biosynthetic gene clusters to be applied for *Streptomyces*. Obviously, these techniques will facilitate the generation of a *Streptomyces* host panel for heterologous expression of biosynthetic gene clusters.

13.2.2 Transposons

In vivo transposon mutagenesis has been successfully used for genetic manipulation of bacteria and higher organisms. It has a great impact on studying gene functions and generating mutants with desired phenotype traits [47]. Several transposon mutagenesis systems have been established in streptomycetes, including Himar1 and Tn5 derivatives, natural *Streptomyces* transposons, and different IS-elements [48–51]. Baltz and colleagues have developed many transposons derived from IS493 and show their usefulness in a couple of streptomycetes. Transposon mutagenesis based on IS elements from related organisms may interfere with the indigenous transposons, or transpose in response to the introduction of related elements leading to unmarked mutations. Therefore, the orthogonal systems based on Tn*5* and Himar1 transposon mutagenesis in streptomycetes have been established [28, 52]. Both transposon systems are based on synthetic transposase genes optimized for expression in streptomycetes. These transposons also contain an apramycin resistance marker and the R6Kgamma origin of replication, allowing quick and easy cloning of the insertion site directly from the chromosome. Corresponding plasmids can be introduced into *Streptomyces* cells via conjugation from *E. coli* on the vector carrying temperature-sensitive replicons. In addition, Himar 1 transposon is equipped with further features such as Cre and Dre recombination target sites and the use of three different antibiotic resistance cassettes. Both transposons are fairly random and provide an insertion frequency over 90% [53]. These two transposon systems are widely applied for genome engineering of *Streptomyces coelicolor*, *Streptomyces lividans*, and *S. albus* J1074 and have a potential to be utilized for other streptomycetes.

13.2.3
I-SceI Endonuclease and CRISPR/Cas9

The rare-cutting homing endonucleases are efficient tools for targeted mutagenesis in many bacteria, for example, *Bacillus anthracis*, *E. coli*, and *Burkhodelia* [54, 55]. These enzymes introduce a double strand break into the DNA and thus facilitate homologous recombination at a certain locus on the chromosome. The endonuclease I-SceI-based system was used to facilitate gene targeting in streptomycetes [27, 56]. Briefly, the mutant allele is cloned into the suicide plasmid which contains an antibiotic selection marker and a recognition site for I-SceI. Such construct is afterward introduced into streptomycetes and integrates into the chromosome via homologous recombination. Then, the I-SceI endonuclease introduces the double strand break and thus stimulates intramolecular recombination between parental and mutant alleles. An additional advantage of the method is that I-SceI serves not only as a stimulator of homologous recombination at the target locus, but also as a counterselection tool against cointegrates.

A more elegant and precise solution is provided by the Cas9 endonuclease [57], which is the part of CRISPRs (Clustered Regularly Interspaced Short Palindromic Repeats) system. The advantage of the CRISPR/Cas9 system is that it allows introduction of a double strand break in a desired target genomic sequence. This is achieved by the *cas9* gene coexpression with the customized single guide RNA [58]. It makes the CRISPR/Cas9 system very versatile and provides unprecedented flexibility in terms of genome editing. Recently, several papers described the adaptation of the system to streptomycetes. A couple of genes have been targeted with the help of CRISPR/Cas9 system and large genomic deletion also could be performed [29–32].

13.2.4
Reporter Genes for Streptomycetes

Reporter genes are applied for genetic manipulations in bacterial genetic systems. They are used to generate the synthetic biobrick libraries, to study gene regulation and expression, translation and transcription of operons, to perform gene deletion, and many others [59]. Several reporter systems are available for streptomycetes including neo and cat resistance genes [60], catechol 2,3-dioxygenase [61], green fluorescent protein [62], luciferase [63], and *gusA* [33]. The GFP and luciferase are the most widely used reporters in many bacteria. Usually, GFP is used as a tool to monitor the trafficking and subcellular localization of proteins. Since this protein does not provide enzymatic signal amplification, its use for gene expression studies is limited due to low sensitivity [64]. Moreover, many streptomycetes have a high level of autofluorescence, which interferes with GFP and complicates the analysis, because of a low signal-to-noise ratio [65]. The luciferase assays are more sensitive and rapid. The luciferases are especially useful to monitor the dynamics of gene transcription because of their short half-life time [66, 67]. The disadvantage of the luciferase-based system is the complexity of enzymatic reactions. For example, bacterial luciferase utilizes $FMNH_2$, O_2, and a 10-carbon aldehyde, decanal, whereas eukaryotic luciferases use ATP, $Mg_2 + O_2$, and the substrate luciferin [68]. A few years ago, the gusA-based reporter system was applied to streptomycetes and provided a number of advantages: high sensitivity, simple assays, enzyme stability, and the possibility of visual, spectrophotometric, and fluorogenic assays. Most of its substrates, including X-gluc, MUG, and *p*-nitrophenyl-β-D-glucuronide, are commercially available [69]. The majority of the described streptomycetes strains lack the indigenous β-glucuronidase activity. Several vectors have been constructed for functional gene studies based on the transcriptional and translational fusion with the gusA reporter gene [33]. However, one of the most useful applications for which the gusA reporter was used is the generation of gene inactivation vectors for rapid visual screening of clones with double crossovers. For this purpose, the gusA gene was introduced into the backbone of the suicidal vectors and a replicative vector carrying temperature-sensitive replicon. Such an approach facilitates the selection of clones with double crossovers by visual screening without picking thousands of colonies. Clones containing the vector will turn blue after overlaying with the substrate solution, and only clones that have lost the plasmid will stay white [33]. The combination of the gusA gene and an antibiotic resistance marker for gene inactivation provides nearly 100% frequency of clones with the mutated allele. Successful utilization of gusA-containing vectors for gene inactivation significantly simplifies the genome engineering procedure. Recently, this approach has been used to generate large chromosomal deletion s in *S. albus* J1074 [46].

13.2.5 Genetic Controlling Elements for Streptomycetes

Well-characterized genetic controlling elements are important tools for the heterologous expression of biosynthetic gene clusters and various metabolic engineering approaches in which the fine-tuning of gene expression is necessary [70, 71]. Promoters are the most important regulators of gene expression in bacteria [72]. There are number of well-characterized constitutive and inducible promoters described for streptomycetes. Recently, several comprehensive synthetic promoter libraries have been constructed for gene expression in streptomycetes. The popular approach based on the randomization of nucleotides surrounding the −10 and −35 consensus sequences within the ermEp1 and ermEp2 promoter [73, 74] was applied to generate first two libraries of synthetic promoters. The sequences upstream, in between and downstream of the −10 and −35 consensus regions were randomized using degenerate primers and the strength of promoters was evaluated with the help of reporter genes (*gusA* and *aphII* conferring neomycin resistance). Both libraries comprise the wide range of the promoter's strengths from 2% to 319% when compared with ermEp1 and ermEp2. The strongest synthetic promoter was successfully applied to overexpress the *rppA* gene encoding for type III polyketide synthase. For the successful manipulation or overexpression of biosynthetic gene clusters consisting of several operons, it is imperative to consider the appropriate balance between all genes involved. Therefore, the availability of well-characterized promoters of different strengths is of key importance. Recently, two sets of native constitutive promoters have been identified in *S. coelicolor* A3(2) and *S. albus* J1074 using microarray and RNA sequencing, respectively [18, 75]. Several selected promoters from these libraries have been successfully used to activate silent gene clusters in *S. griseus* and *S. venezuelae*.

A number of inducible promoter systems have been developed for *Streptomyces*. The most widely used tipAp promoter was cloned from *S. lividans* and can be induced in the presence of the antibiotic thiostrepton [76]. tipAp is a strong promoter providing up to 200-fold induction upon addition of thiostrepton [77]. tipAp is a component of many vectors successfully utilized for gene expression in streptomycetes [5]. However, the tipAL gene encoding the positive regulator of tipAp is not present in all streptomycetes [5]. To avoid this problem, the tipAL gene was placed together with tipAp on plasmid vectors [78]. The drawbacks to the tipAp promoter system are that the presence of the thiostrepton resistance gene is necessary in many cases to protect host cells against toxicity from the inducer compound and the high basal level of transcription initiation even without the inducer [65]. Despite these disadvantages, tipAp remains the most used inducible promoter in *Streptomyces* genetics. Another widely distributed set of promoters can be induced by tetracycline. These promoters have been developed via combination of ermEp1 and tetO1 and tetO2 operator regions from the *E. coli* transposon Tn*10*, which are responsible for binding of the TetR repressor protein [79]. One promoter, tcp830p, showed the largest change in transcription levels between the uninduced and induced states and was among the strongest *Streptomyces* promoters. This system is functional in several commonly used streptomycetes strains and was successfully used to control the expression of the novobiocin biosynthetic gene cluster in *S. coelicolor* [80]. However, both promoters, tcp and tipA, are not fully repressed, and some level of transcription still can be observed even without an inducer. Recently, two new inducible gene switches have been adapted for use in *Streptomyces*. The first system originates from *Pseudomonas putida* and represents the combination of the CymR regulator, the operator sequence (cmt) from the cumate degradation operon, and a P21 synthetic promoter. This promoter is induced by cumate (*p*-isopropylbenzoic acid) up to 45-fold in comparison with the uninduced state. The second system is composed of the RolR regulator and the PA3 promoter fused with the operator (rolO) from the *Corynebacterium glutamicum* resorcinol catabolic operon. This promoter has the low basal level of transcription and can be induced by resorcinol. In summary, there are significant numbers of well-characterized synthetic and inducible promoters to control the expression of gene and gene clusters in streptomycetes [81].

Further control of gene expression is performed on the translational level. A synthetic ribosomal binding site (RBS) with the sequence AAGGAGG has been successfully used to initiate the translation of mRNA in streptomycetes [82]. However, the change in RBS flanking regions as well as within the first 60 nucleotides of the coding region can have a dramatic impact on the translation initiation rate. Therefore, the RBS flanking regions should be optimized to every individual open reading frame, which is very laborious. Recently, the group of Lixin Zhang developed an insulator (RiboJ) eliminating the interference between promoters and RBSs. Such modular genetic controlling elements were applied to overproduce lycopene in the heterologous host *S. avermitilis* [70].

13.3 Hosts for Heterologous Production of Natural Products

E. coli is the most widely used host for heterologous gene expression in order to obtain a single protein for further studies. However, for the expression of complex biosynthetic gene clusters from actinobacteria several issues need to be considered. The high GC content of actinomycetes DNA causes translational rate of corresponding mRNA that is not optimal owing to significant differences in the tRNA pool. Another challenge is usually insufficient concentrations of precursors for the formation of natural products. In addition, the genetic controlling elements from actinomycetes very often do not function in the *E. coli* cells. Despite all these challenges, the successful heterologous production of several natural products, such as 6-methylsalicylic acid and 6-deoxyerythronolide B, have been reported in *E. coli* [83]. However, their yields are still to be optimized. There are successful examples of yield optimization of the natural products in *E. coli*, such as amorpha-4,11-diene, which could be increased from microgram quantities to more than $27\,g\,l^{-1}$ [84]. Industrially relevant titers of some natural products may enrich the multigram per liter scale (e.g., $\geq 10\,g\,l^{-1}$ erythromycin; $\geq 30\,g\,l^{-1}$ tetracycline; $\geq 80\,g\,l^{-1}$ salinomycin; personal communication with Prof. Hrvoje Petkovich, Ljubljana University). However, the optimization of the production level in a distant heterologous host and is very time consuming and laborious [84]. In this context, the utility of the *Streptomyces* strains as hosts for heterologous production of natural products seems to be more feasible. Such strains should meet the requirements of a sufficient precursor supply besides providing an environment that is amenable to the expression of introduced heterologous pathways. Three different approaches have been utilized to develop a chassis strain: (i) deletion of redundant genetic information, such as IS elements and secondary metabolite clusters, thus increasing genetic stability and reducing the metabolic burden; (ii) development and introduction of host-independent regulatory mechanisms for gene expression [85]; and (iii) inactivation, silencing, and activation of a number of host metabolic steps as well as introduction and balancing of foreign metabolic pathways, which together, should direct the flux to the production of higher yields of biosynthetic precursors [86]. In addition, the fermentation conditions should be optimized to turn a *Streptomyces* strain into the industrially relevant producing factory. Owing to the filamentous nature of streptomycetes, it is quite challenging to optimize their growth and production capacities in a submerged culture. The typical problem to be solved is the formation of unfavorable pellets or clumps, resulting in uneven distribution of oxygen and nutrients in the bacterial culture. Downstream process optimization approaches such as increasing culture broth mixing or a more intensive shearing of the culture are usually applied to solve this problem. In addition, an elegant solution based on strain engineering has been applied by the group of van Wezel. They overexpressed the SsgA protein, required for the activation of cell division in streptomycetes, and thus enhanced the fragmentation of mycelial clumps in *S. coelicolor* and *S. lividans* [87]. Hopefully, the comprehensive *Streptomyces* genetic toolbox described above will enable the application of genome engineering approaches for the production optimization of highly valuable molecules derived from streptomycetes.

At present, several *Streptomyces* strains have been utilized as hosts for the production of natural products. Baltz performed an excellent comparative analysis of different *Streptomyces* hosts for heterologous production [7, 88]. The most widely used strains are *S. coelicolor, S. avermitilis,* and *S. albus* J1074 [89].

13.3.1
S. coelicolor

S. coelicolor A3(2) is the model strain of actinomycetes genetics and is one of the most popular hosts for the heterologous expression of small-molecule biosynthetic gene clusters originating from actinobacteria. The first attempt to engineer the host strain-based *S. coelicolor* was made in 1993 [90]. Authors successfully deleted the entire actinorhodin biosynthetic gene cluster from the chromosome of the strain lacking endogenous plasmids and carrying a mutation that blocks biosynthesis of red-pigmented undecylprodigiosin. The expression of 6-deoxyerythronolide B polyketide synthase gene in this engineered strain yielded more than $40\,mg\,l^{-1}$ of 6-deoxyerythronolide. Gomez-Escribano and Bibb [91] further developed *S. coelicolor* M1154 deleting the Act, Red, CDA, and Cpk biosynthetic gene clusters, and introducing rpsL (K88E) and rpoB (S433L) mutations that usually increase the yield of natural products. The engineered host strain was used for the heterologous production of a broad range of compound classes, including: aminocoumarins, ribosomally synthesized peptides, polyketides, nucleosides, and indolocarbazoles [91]. Another *S. coelicolor*-based host strain has been developed by Zhou *et al.* [92]. The authors have deleted 10 PKS and NRPS gene clusters from *S. coelicolor.* In addition, a telomeric region of around 900 kb has been excised as well, resulting in the genome-minimized strain ZM12. The expression of the native actinorhodin gene cluster in ZM12 yielded larger amounts of actinorhodin than the wild-type strain. However, further experiments need to be performed to determine whether this strain offers advantages over the wild-type strain of *S. coelicolor.*

13.3.2
S. avermitillis

S. avermitilis is known for its industrial production of avermectin [93]. Komatsu *et al.* [94] developed genome-minimized *S. avermitilis* hosts (SUKA17 and SUKA22), which harbor a deletion corresponding to a 1.4-Mb subtelomeric region along with several secondary metabolite gene clusters and transposase-encoding genes [94]. The authors compared the ability of the genome-engineered strains with the wild type to act as a host for the heterologous production of various small molecules. The *S. avermitilis* engineered host is able to produce polyketides, aminoglycosides, peptides, shikimate-derived compounds, and terpenes. The yield of heterologously produced natural compounds was between 1 and 360 mg l^{-1}. The authors also could demonstrate the advantages of genome-minimized strains in terms of production levels of several small molecules, making SUKA17/22 strains a valid platform for heterologous expression of biosynthetic gene clusters [94].

13.3.3
S. albus

S. albus J1074, a derivative of the *S. albus* G1 strain, defective in SalG1 restriction-modification system, is perhaps the most versatile expression host among many streptomycetes. Various natural products scaffolds (e.g., steffimycin, fredericamycin, isomigrastatin, napyradiomycin, cyclooctatin, thiocoraline, moenomycin) could be heterologously produced in this strain, highlighting the availability of broad precursor variety [7, 89]. Several examples of natural product classes synthesized by *S. albus* J1074 are presented in Figure 13.2. The superior production capacity of *S. albus* J1074 in comparison with the other streptomycetes hosts was nicely demonstrated on the examples of isomigrastatin. The compound yield in *S. albus* J1074 was twofold higher in comparison to *S. coelicolor* and *S. lividans* and approximately 10-fold higher when compared with the *S. avermitilis* SUKA strains [95]. Interestingly, we have recently identified a second active attB site within the chromosome of *S. albus* J1074 at which integration of phiC31-based vectors occurs with a high frequency. The availability of the two highly active attB sites in the *S. albus* genome at least partially explains the ability of this strain to heterologously produce higher

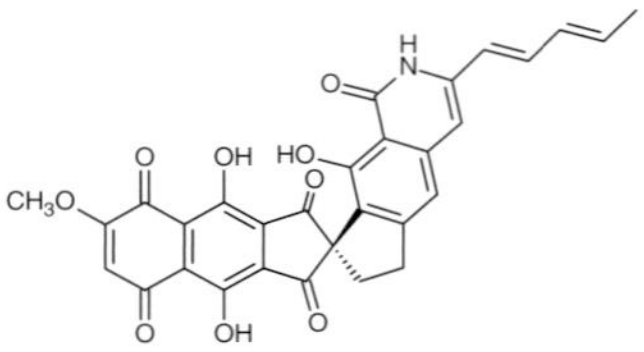

Fredericamycin
type II PKS

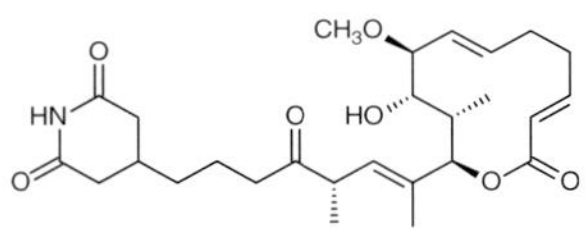

iso-Migrastatin
trans-AT type I PKS

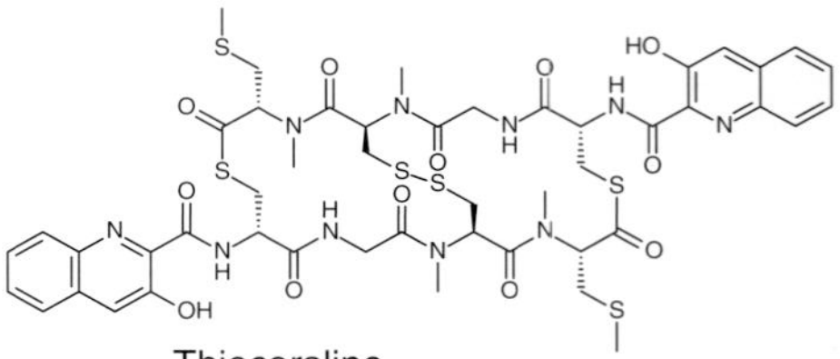

Thiocoraline
NRPS

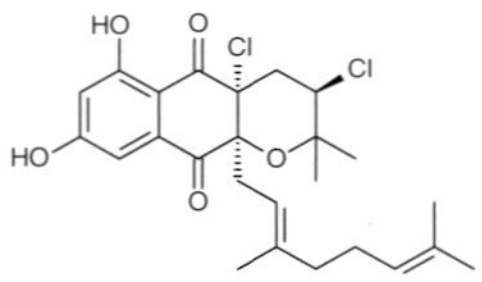

Napyradiomycin A1
type III PKS

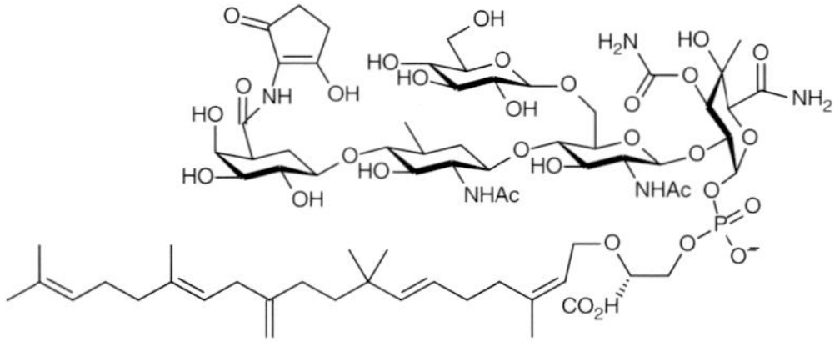

Moenomycin A
phosphoglycolipid

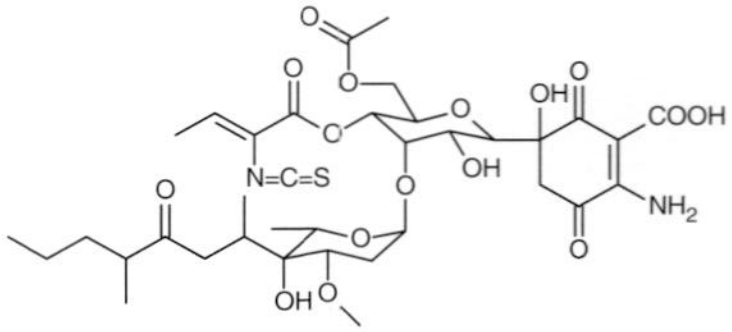

Paulomycin A
Glycoside?

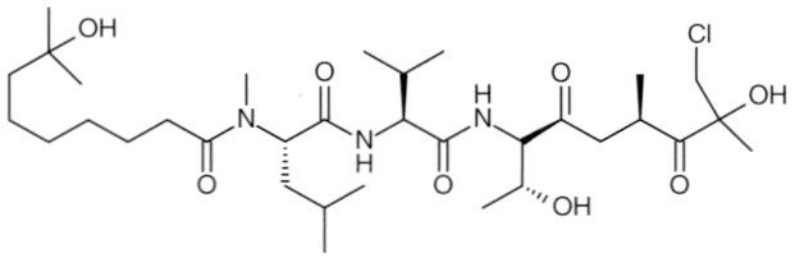

Clarepoxcin E
NRPS-PKS

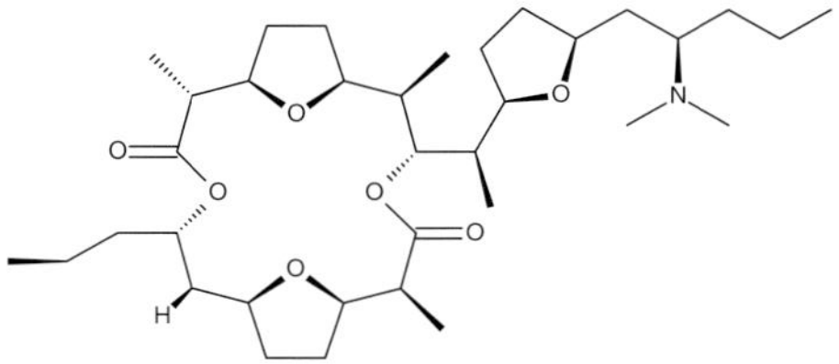

Pamamycin 607
macrodiolide

Figure 13.2 Natural products produced by *Streptomyces albus* J1074.

yields of natural products in comparison to other hosts [96]. Furthermore, when cryptic biosynthetic gene clusters from metagenome were integrated into the chromosomes of *S. viridochromogenes*, *S. roseosporus*, *S. coelicolor* M145, *S. coelicolor* M1152, and *S. albus* J1074, only the last named yielded novel natural products landepoxcins A and B, highlighting its extraordinary capacity as a heterologous host [97]. The genome of *S. albus* J1074 is naturally minimized and comprises 6.8 Mb. It was further reduced by deleting 204 kb fragments containing two large hybrid PKS–NRPS clusters. However, the genome-reduced mutant of *S. albus* J1074 was not characterized for its ability to host secondary metabolite gene clusters [46].

Although, several hosts for the heterologous small-molecule production are already existing, their productivity is typically in the milligram range and therefore suboptimal for industrial use. Obviously, that new generation of *Streptomyces*-based chassis with reduced and fine-tuned genomes need to be developed using state-of-the art technologies, including systems and synthetic biology, to channel more metabolic potential into the biosynthesis of the heterologous biomolecules.

Acknowledgments

This work was supported by the European Commission under the Seventh Framework Program through the "Collaborative Project" action "STREPSYNTH" grant No. 613877 and through the ERC starting grant EXPLOGEN No. 281623 to AL. The authors declare no conflict of interest.

References

1. Gohain, A., Gogoi, A., Debnath, R., Yadav, A. *et al.* (2015) Antimicrobial biosynthetic potential and genetic diversity of endophytic actinomycetes associated with medicinal plants. *FEMS Microbiol. Lett.*, **362** (19).
2. Hanshew, A.S., McDonald, B.R., Díaz Díaz, C., Djiéto-Lordon, C. *et al.* (2015) Characterization of actinobacteria associated with three ant-plant mutualisms. *Microb. Ecol.*, **69** (1), 192–203.
3. Wardecki, T., Brötz, E., de Ford, C., von Loewenich, F.D. *et al.* (2015) Endophytic *Streptomyces* in the traditional medicinal plant *Arnica montana* L.: secondary metabolites and biological activity. *Antonie Van Leeuwenhoek*, **108** (2), 391–402.
4. Kim, S.B. and Goodfellow, M. (2002) *Streptomyces thermospinisporus* sp. nov., a moderately thermophilic carboxydotrophic streptomycete isolated from soil. *Int. J. Syst. Evol. Microbiol.*, **52** (Pt. 4), 1225–1228.
5. Kieser, T., Bibb, M.J., Buttner, M.J., Chater, K.F., and Hopwood, D.A. (2000) *An Introduction to Streptomyces. Practical Streptomyces Genetics*, John Innes Foundation.
6. Sánchez, S. and Demain, A.L. (eds) (2015) *Antibiotics: Current Innovations and Future Trends*, Caister Academic Press.
7. Baltz, R.H. (2010) *Streptomyces* and *Saccharopolyspora* hosts for heterologous expression of secondary metabolite gene clusters. *J. Ind. Microbiol. Biotechnol.*, **37** (8), 759–772.
8. Pozidis, C., Lammertyn, E., Politou, A.S., Anné, J. *et al.* (2001) Protein

secretion biotechnology using *Streptomyces lividans*: large-scale production of functional trimeric tumor necrosis factor alpha. *Biotechnol. Bioeng.*, **72** (6), 611–619.

9. Anné, J., Vrancken, K., van Mellaert, L., van Impe, J. *et al.* (2014) Protein secretion biotechnology in Gram-positive bacteria with special emphasis on *Streptomyces lividans*. *Biochim. Biophys. Acta*, **1843** (8), 1750–1761.
10. Vrancken, K. and Anné, J. (2009) Secretory production of recombinant proteins by *Streptomyces*. *Future Microbiol.*, **4** (2), 181–188.
11. Rebets, Y., Brötz, E., Tokovenko, B., and Luzhetskyy, A. (2014) Actinomycetes biosynthetic potential: how to bridge *in silico* and *in vivo*? *J. Ind. Microbiol. Biotechnol.*, **41** (2), 387–402.
12. Brakhage, A.A. and Schroeckh, V. (2011) Fungal secondary metabolites – strategies to activate silent gene clusters. *Fungal Genet. Biol.*, **48** (1), 15–22.
13. Onaka, H., Mori, Y., Igarashi, Y., and Furumai, T. (2011) Mycolic acid-containing bacteria induce natural-product biosynthesis in *Streptomyces* species. *Appl. Environ. Microbiol.*, **77** (2), 400–406.
14. Tanaka, Y., Hosaka, T., and Ochi, K. (2010) Rare earth elements activate the secondary metabolite-biosynthetic gene clusters in *Streptomyces coelicolor* A3(2). *J. Antibiot.*, **63** (8), 477–481.
15. van Pham, H.T. and Kim, J. (2012) Cultivation of unculturable soil bacteria. *Trends Biotechnol.*, **30** (9), 475–484.
16. Watve, M.G., Tickoo, R., Jog, M.M., and Bhole, B.D. (2001) How many antibiotics are produced by the genus *Streptomyces*? *Arch. Microbiol.*, **176** (5), 386–390.
17. Lewis, K. (2013) Platforms for antibiotic discovery. *Nat. Rev. Drug Discovery*, **12** (5), 371–387.
18. Luo, Y., Huang, H., Liang, J., Wang, M. *et al.* (2013) Activation and characterization of a cryptic polycyclic tetramate macrolactam biosynthetic gene cluster. *Nat. Commun.*, **4**, 2894.
19. Fu, J., Bian, X., Hu, S., Wang, H. *et al.* (2012) Full-length RecE enhances linear–linear homologous recombination and facilitates direct cloning for bioprospecting. *Nat. Biotechnol.*, **30** (5), 440–446.
20. Yamanaka, K., Reynolds, K.A., Kersten, R.D., Ryan, K.S. *et al.* (2014) Direct cloning and refactoring of a silent lipopeptide biosynthetic gene cluster yields the antibiotic taromycin A. *Proc. Natl. Acad. Sci. U.S.A.*, **111** (5), 1957–1962.
21. van Dissel, D., Claessen, D., and van Wezel, G.P. (2014) Morphogenesis of *Streptomyces* in submerged cultures. *Adv. Appl. Microbiol.*, **89**, 1–45.
22. Barka, E.A., Vatsa, P., Sanchez, L., Gaveau-Vaillant, N. *et al.* (2016) Taxonomy, physiology, and natural products of actinobacteria. *Microbiol. Mol. Biol. Rev.*, **80** (1), 1–43.
23. Fedoryshyn, M., Petzke, L., Welle, E., Bechthold, A. *et al.* (2008) Marker removal from actinomycetes genome using Flp recombinase. *Gene*, **419** (1–2), 43–47.
24. Fedoryshyn, M., Welle, E., Bechthold, A., and Luzhetskyy, A. (2008) Functional expression of the Cre recombinase in actinomycetes. *Appl. Microbiol. Biotechnol.*, **78** (6), 1065–1070.
25. Herrmann, S., Siegl, T., Luzhetska, M., Petzke, L. *et al.* (2012) Site-specific recombination strategies for engineering actinomycete genomes. *Appl. Environ. Microbiol.*, **78** (6), 1804–1812.
26. Myronovskyi, M. and Luzhetskyy, A. (2013) Genome engineering in actinomycetes using site-specific recombinases. *Appl. Microbiol. Biotechnol.*, **97** (11), 4701–4712.
27. Siegl, T., Petzke, L., Welle, E., and Luzhetskyy, A. (2010) I-SceI endonuclease: a new tool for DNA repair studies and genetic manipulations in *Streptomyces*. *Appl. Microbiol. Biotechnol.*, **87** (4), 1525–1532.
28. Bilyk, B. and Luzhetskyy, A. (2014) Transposon mutagenesis in streptomycetes. Saarbrücken, Universität des Saarlandes, Diss., 2014. Saarländische Universitäts- und Landesbibliothek.
29. Cobb, R.E., Wang, Y., and Zhao, H. (2015) High-efficiency multiplex genome editing of *Streptomyces* species using an

engineered CRISPR/Cas system. *ACS Synth. Biol.*, **4** (6), 723–728.

30. Huang, H., Zheng, G., Jiang, W., Hu, H. *et al.* (2015) One-step high-efficiency CRISPR/Cas9-mediated genome editing in *Streptomyces. Acta Biochim. Biophys. Sin. (Shanghai)*, **47** (4), 231–243.
31. Tong, Y., Charusanti, P., Zhang, L., Weber, T. *et al.* (2015) CRISPR-Cas9 based engineering of actinomycetal genomes. *ACS Synth. Biol.*, **4** (9), 1020–1029.
32. Zeng, H., Wen, S., Xu, W., He, Z. *et al.* (2015) Highly efficient editing of the actinorhodin polyketide chain length factor gene in *Streptomyces coelicolor* M145 using CRISPR/Cas9-CodA(sm) combined system. *Appl. Microbiol. Biotechnol.*, **99** (24), 10575–10585.
33. Myronovskyi, M., Welle, E., Fedorenko, V., and Luzhetskyy, A. (2011) Beta-glucuronidase as a sensitive and versatile reporter in actinomycetes. *Appl. Environ. Microbiol.*, **77** (15), 5370–5383.
34. Branda, C.S. and Dymecki, S.M. (2004) Talking about a revolution: the impact of site-specific recombinases on genetic analyses in mice. *Dev. Cell*, **6** (1), 7–28.
35. Kühn, R. and Torres, R.M. (2002) Cre/loxP recombination system and gene targeting. *Methods Mol. Biol.*, **180**, 175–204.
36. Leibig, M., Krismer, B., Kolb, M., Friede, A. *et al.* (2008) Marker removal in staphylococci via Cre recombinase and different lox sites. *Appl. Environ. Microbiol.*, **74** (5), 1316–1323.
37. Suzuki, N., Nonaka, H., Tsuge, Y., Inui, M. *et al.* (2005) New multiple-deletion method for the *Corynebacterium glutamicum* genome, using a mutant lox sequence. *Appl. Environ. Microbiol.*, **71** (12), 8472–8480.
38. Kuhstoss, S. and Rao, R.N. (1991) A thiostrepton-inducible expression vector for use in *Streptomyces* spp. *Gene*, **103** (1), 97–99.
39. Schweizer, H.P. (2003) Applications of the *Saccharomyces cerevisiae* Flp-FRT system in bacterial genetics. *J. Mol. Microbiol. Biotechnol.*, **5** (2), 67–77.
40. Slauch, J.M. and Camilli, A. (2000) IVET and RIVET: use of gene fusions to identify bacterial virulence factors specifically induced in host tissues. *Methods Enzymol.*, **326**, 73–96.
41. Myronovskyi, M., Rosenkränzer, B., and Luzhetskyy, A. (2014) Iterative marker excision system. *Appl. Microbiol. Biotechnol.*, **98** (10), 4557–4570.
42. Thorpe, H.M. and Smith, M.C. (1998) In vitro site-specific integration of bacteriophage DNA catalyzed by a recombinase of the resolvase/invertase family. *Proc. Natl. Acad. Sci. U.S.A.*, **95** (10), 5505–5510.
43. Baltz, R.H. (2012) *Streptomyces* temperate bacteriophage integration systems for stable genetic engineering of actinomycetes (and other organisms). *J. Ind. Microbiol. Biotechnol.*, **39** (5), 661–672.
44. Huff, J., Czyz, A., Landick, R., and Niederweis, M. (2010) Taking phage integration to the next level as a genetic tool for mycobacteria. *Gene*, **468** (1–2), 8–19.
45. Komatsu, M., Uchiyama, T., Omura, S., Cane, D.E. *et al.* (2010) Genome-minimized *Streptomyces* host for the heterologous expression of secondary metabolism. *Proc. Natl. Acad. Sci. U.S.A.*, **107** (6), 2646–2651.
46. Myronovskyi, M., Tokovenko, B., Brötz, E., Rückert, C. *et al.* (2014) Genome rearrangements of *Streptomyces albus* J1074 lead to the carotenoid gene cluster activation. *Appl. Microbiol. Biotechnol.*, **98** (2), 795–806.
47. Goryshin, I.Y., Jendrisak, J., Hoffman, L.M., Meis, R. *et al.* (2000) Insertional transposon mutagenesis by electroporation of released Tn5 transposition complexes. *Nat. Biotechnol.*, **18** (1), 97–100.
48. Baltz, R.H., McHenney, M.A., Cantwell, C.A., Queener, S.W. *et al.* (1997) Applications of transposition mutagenesis in antibiotic producing streptomycetes. *Antonie Van Leeuwenhoek*, **71** (1–2), 179–187.
49. Solenberg, P.J. and Baltz, R.H. (1991) Transposition of Tn5096 and other IS493 derivatives in *Streptomyces griseofuscus. J. Bacteriol.*, **173** (3), 1096–1104.
50. Gehring, A.M., Nodwell, J.R., Beverley, S.M., and Losick, R. (2000) Genomewide insertional mutagenesis in *Streptomyces coelicolor* reveals additional genes

involved in morphological differentiation. *Proc. Natl. Acad. Sci. U.S.A.*, **97** (17), 9642–9647.

51. Volff, J.N. and Altenbuchner, J. (1997) High frequency transposition of the Tn5 derivative Tn5493 in *Streptomyces lividans*. *Gene*, **194** (1), 81–86.
52. Petzke, L. and Luzhetskyy, A. (2009) In vivo Tn5-based transposon mutagenesis of Streptomycetes. *Appl. Microbiol. Biotechnol.*, **83** (5), 979–986.
53. Bilyk, B., Weber, S., Myronovskyi, M., Bilyk, O. *et al.* (2013) In vivo random mutagenesis of streptomycetes using mariner-based transposon Himar1. *Appl. Microbiol. Biotechnol.*, **97** (1), 351–359.
54. Pósfai, G., Kolisnychenko, V., Bereczki, Z., and Blattner, F.R. (1999) Markerless gene replacement in *Escherichia coli* stimulated by a double-strand break in the chromosome. *Nucleic Acids Res.*, **27** (22), 4409–4415.
55. Janes, B.K. and Stibitz, S. (2006) Routine markerless gene replacement in *Bacillus anthracis*. *Infect. Immun.*, **74** (3), 1949–1953.
56. Fernández-Martínez, L.T. and Bibb, M.J. (2014) Use of the meganuclease I-SceI of *Saccharomyces cerevisiae* to select for gene deletions in actinomycetes. *Sci. Rep.*, **4**, 7100.
57. van der Oost, J., Jore, M.M., Westra, E.R., Lundgren, M. *et al.* (2009) CRISPR-based adaptive and heritable immunity in prokaryotes. *Trends Biochem. Sci*, **34** (8), 401–407.
58. Jinek, M., Chylinski, K., Fonfara, I., Hauer, M. *et al.* (2012) A programmable dual-RNA-guided DNA endonuclease in adaptive bacterial immunity. *Science*, **337** (6096), 816–821.
59. Shuman, H.A. and Silhavy, T.J. (2003) Microbial genetics: the art and design of genetic screens: *Escherichia coli*. *Nat. Rev. Genet.*, **4** (6), 419–431.
60. Shaw, W.V. and Hopwood, D.A. (1976) Chloramphenicol acetylation in *Streptomyces*. *J. Gen. Microbiol.*, **94** (1), 159–166.
61. Sekurova, O.N., Brautaset, T., Sletta, H., Borgos, S.E.F. *et al.* (2004) In vivo analysis of the regulatory genes in the nystatin biosynthetic gene cluster of *Streptomyces noursei* ATCC 11455 reveals their differential control over antibiotic biosynthesis. *J. Bacteriol.*, **186** (5), 1345–1354.
62. Sun, J., Kelemen, G.H., Fernández-Abalos, J.M., and Bibb, M.J. (1999) Green fluorescent protein as a reporter for spatial and temporal gene expression in *Streptomyces coelicolor* A3(2). *Microbiology (Reading, Engl.)*, **145** (Pt. 9), 2221–2227.
63. Craney, A., Hohenauer, T., Xu, Y., Navani, N.K. *et al.* (2007) A synthetic luxCDABE gene cluster optimized for expression in high-GC bacteria. *Nucleic Acids Res.*, **35** (6), e46.
64. Hutter, H. (2006) Fluorescent reporter methods. *Methods Mol. Biol.*, **351**, 155–173.
65. Kieser, T. (2000) *Practical Streptomyces Genetics*, Innes.
66. Weiser, J., Buriánková, K., Kalachová, L., Branny, P. *et al.* (2006) Construction and testing of a bacterial luciferase reporter gene system for in vivo measurement of nonsense suppression in *Streptomyces*. *Folia Microbiol. (Praha)*, **51** (1), 62–64.
67. Sohaskey, C.D., Im, H., Nelson, A.D., and Schauer, A.T. (1992) Tn4556 and luciferase: synergistic tools for visualizing transcription in *Streptomyces*. *Gene*, **115** (1–2), 67–71.
68. Wilson, T. and Hastings, J.W. (1998) Bioluminescence. *Annu. Rev. Cell Dev. Biol.*, **14**, 197–230.
69. Jefferson, R.A., Kavanagh, T.A., and Bevan, M.W. (1987) GUS fusions: beta-glucuronidase as a sensitive and versatile gene fusion marker in higher plants. *EMBO J.*, **6** (13), 3901–3907.
70. Bai, C., Zhang, Y., Zhao, X., Hu, Y. *et al.* (2015) Exploiting a precise design of universal synthetic modular regulatory elements to unlock the microbial natural products in *Streptomyces*. *Proc. Natl. Acad. Sci. U.S.A.*, **112** (39), 12181–12186.
71. Boyle, P.M. and Silver, P.A. (2012) Parts plus pipes: synthetic biology approaches to metabolic engineering. *Metab. Eng.*, **14** (3), 223–232.
72. Hammer, K., Mijakovic, I., and Jensen, P.R. (2006) Synthetic promoter libraries--tuning of gene expression. *Trends Biotechnol.*, **24** (2), 53–55.

73. Siegl, T., Tokovenko, B., Myronovskyi, M., and Luzhetskyy, A. (2013) Design, construction and characterisation of a synthetic promoter library for fine-tuned gene expression in actinomycetes. *Metab. Eng.*, **19**, 98–106.
74. Seghezzi, N., Amar, P., Koebmann, B., Jensen, P.R. *et al.* (2011) The construction of a library of synthetic promoters revealed some specific features of strong *Streptomyces* promoters. *Appl. Microbiol. Biotechnol.*, **90** (2), 615–623.
75. Sohoni, S.V., Fazio, A., Workman, C.T., Mijakovic, I. *et al.* (2014) Synthetic promoter library for modulation of actinorhodin production in *Streptomyces coelicolor* A3(2). *PLoS One*, **9** (6), e99701.
76. Chiu, M.L., Folcher, M., Katoh, T., Puglia, A.M. *et al.* (1999) Broad spectrum thiopeptide recognition specificity of the *Streptomyces lividans* TipAL protein and its role in regulating gene expression. *J. Biol. Chem.*, **274** (29), 20578–20586.
77. Ali, N., Herron, P.R., Evans, M.C., and Dyson, P.J. (2002) Osmotic regulation of the *Streptomyces lividans* thiostrepton-inducible promoter, ptipA. *Microbiology (Reading, Engl.)*, **148** (Pt. 2), 381–390.
78. Takano, E., White, J., Thompson, C.J., and Bibb, M.J. (1995) Construction of thiostrepton-inducible, high-copy-number expression vectors for use in *Streptomyces* spp. *Gene*, **166** (1), 133–137.
79. Rodríguez-García, A., Combes, P., Pérez-Redondo, R., Smith, M.C.A. *et al.* (2005) Natural and synthetic tetracycline-inducible promoters for use in the antibiotic-producing bacteria *Streptomyces*. *Nucleic Acids Res.*, **33** (9), e87.
80. Dangel, V., Westrich, L., Smith, M.C.M., Heide, L. *et al.* (2010) Use of an inducible promoter for antibiotic production in a heterologous host. *Appl. Microbiol. Biotechnol.*, **87** (1), 261–269.
81. Horbal, L., Fedorenko, V., and Luzhetskyy, A. (2014) Novel and tightly regulated resorcinol and cumate-inducible expression systems for *Streptomyces* and other actinobacteria. *Appl. Microbiol. Biotechnol.*, **98** (20), 8641–8655.
82. Horinouchi, S., Nishiyama, M., Nakamura, A., and Beppu, T. (1987) Construction and characterization of multicopy expression-vectors in *Streptomyces* spp. *Mol. Gen. Genet.*, **210** (3), 468–475.
83. Pfeifer, B.A. and Khosla, C. (2001) Biosynthesis of polyketides in heterologous hosts. *Microbiol. Mol. Biol. Rev.*, **65** (1), 106–118.
84. Paddon, C.J. and Keasling, J.D. (2014) Semi-synthetic artemisinin: a model for the use of synthetic biology in pharmaceutical development. *Nat. Rev. Microbiol.*, **12** (5), 355–367.
85. Rudolph, M.M., Vockenhuber, M., and Suess, B. (2013) Synthetic riboswitches for the conditional control of gene expression in *Streptomyces coelicolor*. *Microbiology (Reading, Engl.)*, **159** (Pt. 7), 1416–1422.
86. Borodina, I., Krabben, P., and Nielsen, J. (2005) Genome-scale analysis of *Streptomyces coelicolor* A3(2) metabolism. *Genome Res.*, **15** (6), 820–829.
87. van Wezel, G.P., van der Meulen, J., Kawamoto, S., Luiten, R.G. *et al.* (2000) ssgA is essential for sporulation of *Streptomyces coelicolor* A3(2) and affects hyphal development by stimulating septum formation. *J. Bacteriol.*, **182** (20), 5653–5662.
88. Baltz, R.H. (2016) Genetic manipulation of secondary metabolite biosynthesis for improved production in *Streptomyces* and other actinomycetes. *J. Ind. Microbiol. Biotechnol.*, **43** (2–3), 343–370.
89. Baltz, R.H. (2015) Genetic manipulation of secondary metabolite biosynthesis for improved production in *Streptomyces* and other actinomycetes. *J. Ind. Microbiol. Biotechnol.*
90. McDaniel, R., Ebert-Khosla, S., Hopwood, D.A., and Khosla, C. (1993) Engineered biosynthesis of novel polyketides. *Science*, **262** (5139), 1546–1550.
91. Gomez-Escribano, J.P. and Bibb, M.J. (2011) Engineering *Streptomyces coelicolor* for heterologous expression of secondary metabolite gene clusters. *Microb. Biotechnol.*, **4** (2), 207–215.

92. Zhou, M., Jing, X., Xie, P., Chen, W. *et al.* (2012) Sequential deletion of all the polyketide synthase and nonribosomal peptide synthetase biosynthetic gene clusters and a 900-kb subtelomeric sequence of the linear chromosome of *Streptomyces coelicolor*. *FEMS Microbiol. Lett.*, **333** (2), 169–179.

93. Omura, S., Ikeda, H., Ishikawa, J., Hanamoto, A. *et al.* (2001) Genome sequence of an industrial microorganism *Streptomyces avermitilis*: deducing the ability of producing secondary metabolites. *Proc. Natl. Acad. Sci. U.S.A.*, **98** (21), 12215–12220.

94. Komatsu, M., Komatsu, K., Koiwai, H., Yamada, Y. *et al.* (2013) Engineered *Streptomyces avermitilis* host for heterologous expression of biosynthetic gene cluster for secondary metabolites. *ACS Synth. Biol.*, **2** (7), 384–396.

95. Yang, D., Zhu, X., Wu, X., Feng, Z. *et al.* (2011) Titer improvement of iso-migrastatin in selected heterologous *Streptomyces* hosts and related analysis of mRNA expression by quantitative RT-PCR. *Appl. Microbiol. Biotechnol.*, **89** (6), 1709–1719.

96. Bilyk, B. and Luzhetskyy, A. (2014) Unusual site-specific DNA integration into the highly active pseudo-attB of the *Streptomyces albus* J1074 genome. *Appl. Microbiol. Biotechnol.*, **98** (11), 5095–5104.

97. Owen, J.G., Charlop-Powers, Z., Smith, A.G., Ternei, M.A. *et al.* (2015) Multiplexed metagenome mining using short DNA sequence tags facilitates targeted discovery of epoxyketone proteasome inhibitors. *Proc. Natl. Acad. Sci. U.S.A.*, **112** (14), 4221–4226.

Part V
Extending the Raw Material Basis for Bioproduction

Industrial Biotechnology: Microorganisms, First Edition. Edited by Christoph Wittmann and James C. Liao.

14
Extreme Thermophiles as Metabolic Engineering Platforms: Strategies and Current Perspective

Andrew J. Loder, Benjamin M. Zeldes, Jonathan M. Conway, James A. Counts, Christopher T. Straub, Piyum A. Khatibi, Laura L. Lee, Nicholas P. Vitko, Matthew W. Keller, Amanda M. Rhaesa, Gabe M. Rubinstein, Israel M. Scott, Gina L. Lipscomb, Michael W.W. Adams, and Robert M. Kelly

14.1
Introduction

In the past decade, interest in developing biological routes to fuels and chemicals has risen dramatically, as the world attempts to find sustainable methods of replacing petroleum-derived fuels and chemicals [1–3]. In the past, fuels or solvents such as ethanol, *n*-butanol, and acetone were produced by fermentation with wild-type organisms, but these approaches were phased out as advances in petroleum-based chemical production became more cost competitive [4]. To make biological production of fuels and chemicals competitive with petroleum-based processes, the capabilities of natural producers are no longer adequate and must be improved upon. To develop organisms capable of producing fuels and chemicals with high yields and titers from the desired feed stocks, the chemical industry has turned to metabolic engineering.

Metabolic engineering involves modifying the metabolic pathways of organisms to generate a desired phenotype, such as production of a fuel or chemical. Such modifications are made through the application of genetic engineering tools, which were first developed for model mesophilic microorganisms, such as the bacterium *Escherichia coli* and the yeast *Saccharomyces cerevisiae*. As a result, the majority of metabolic engineering initiatives focus upon these organisms as hosts. Significant efforts have been directed at developing *E. coli* as a producer of fuel alcohols, terpenes, fatty acids, and pharmaceuticals [1, 2, 5–9]. Likewise, *S. cerevisiae* has been developed as a producer for many of the same products [10–13]. In fact, these organisms have also been used to produce advanced biofuels, such as isobutanol and farnesene, and commodity chemicals, such as 1,3-propanediol and the antimalarial drug artemisinin [1, 13, 14].

While genetic engineering of *E. coli* and *S. cerevisiae* is trivial, the growth and metabolic characteristics of these model organisms are not ideally suited to every metabolic engineering goal. Often, a nonmodel organism can have desirable traits that make it a more attractive starting point for certain metabolic

Industrial Biotechnology: Microorganisms, First Edition. Edited by Christoph Wittmann and James C. Liao.

engineering initiatives. As genetic tools become available for more species, nonmodel organisms are increasingly being used as hosts in metabolic engineering approaches. For example, members of the cyanobacterium genus *Synechococcus* have been engineered for photoautotrophic production of fuel alcohols, alkanes, 3-hydroxypropionate, and polyhydroxybutyrates [15–19]. The native *n*-butanol producer *Clostridium acetobutylicum* has been engineered for increased *n*-butanol production [20, 21]. Similarly, natural fatty acid accumulation in the yeast *Yarrowia lypolytica* has been enhanced through metabolic engineering [22].

One category of organisms that is also being explored as metabolic engineering hosts is the extreme thermophiles, as genetic tools are now becoming available for multiple species, making it possible to engineer them for industrial use. Extreme thermophiles are defined as organisms with optimal growth temperatures of 70 °C and above, and these belong exclusively to the domains of Bacteria and Archaea [23, 24]. Extreme thermophiles have unusual metabolic pathways and adaptations that offer intrinsic advantages for industrial bioprocessing (Box 14.1).

Box 14.1 Metabolic Classification of Organisms

The metabolism of an organism can be classified on the basis of its sources of energy and carbon. Because some Bacteria and Archaea have unusual metabolic capabilities that are not seen in eukaryotes, these classifications are defined here [23].

Energy metabolism:
- Chemoorganotroph – energy from oxidation of organic molecules
- Chemolithotroph – energy from oxidation of inorganic molecules
- Carboxydotroph – energy from oxidation of carbon monoxide (subset of chemolithotrophy)
- Phototroph – energy from light

Carbon source:
- Heterotroph – carbon from organic molecules
- Autotroph – carbon from carbon dioxide.

This chapter focuses on the current state of tools and applications of extreme thermophiles for metabolic engineering. Discussed here are the advantages and drawbacks of using extreme thermophiles as industrial hosts, as well as considerations for selecting an appropriate host from this group of microorganisms. The genetic strategies and tools available for several extreme thermophiles are discussed, including the variety of genetic elements that could be considered. Finally, applications of metabolic engineering with extreme thermophiles are discussed along with perspectives on future developments in this emerging technology.

14.2
Bioprocessing Advantages for Extremely Thermophilic Hosts

Extreme thermophiles are noteworthy for their high optimal growth temperatures, and initial research efforts to study these organisms was with regard to their being outliers to conventional temperature limits for life. However, more recently, these unique organisms are being recognized to offer opportunities for exploiting extreme thermophilicity for commercial uses. Consistent with their high optimal growth temperatures, extreme thermophiles produce enzymes that are likewise stable at elevated temperatures. The most notable and earliest example of exploiting an enzyme from an extreme thermophile was the development of the *Thermus aquaticus* (Taq) DNA polymerase for use in the Polymerase Chain Reaction (PCR) [25]. *Taq* polymerase became an essential tool in the advancement of biotechnology, and use of thermophile-sourced enzymes in PCR is now ubiquitous [26]. Now, with the advent of available genetic systems for these microorganisms, there are opportunities to use extreme thermophiles to enable economically viable commercial-scale production of fuels, chemicals, enzymes, or other products. As an added advantage, certain extreme thermophiles also possess natural tolerances for other extremes such as pH, salinity, metals, and other toxic compounds. It remains to be seen whether extreme thermophiles are more resistant to solvents, metabolic intermediates, and end-products, additional factors that could favor them over their mesophilic counterparts as metabolic engineering hosts [27].

A potential advantage of extreme thermophiles is the lower likelihood of contamination in industrial settings. Highly genetically modified organisms usually have a fitness disadvantage [28] and can therefore be easily overtaken in culture when contaminating organisms are present. The high growth temperatures of extreme thermophiles preclude growth or survival of most potential contaminating organisms. This not only reduces the likelihood of low yield or unproductive batches but can also reduce sterilization costs along with capital costs associated with maintaining a highly sterile industrial facility. It is unclear at this point whether there should be less concern with the escape of the genetically modified organism into the surrounding environment or into the final product, as extreme thermophiles are unable to grow in ambient conditions. This issue warrants further attention.

One possible concern about the use of thermophiles for industrial bioprocessing is that the elevated process temperatures will require additional energy. In fact, the opposite could be the case, and bioprocesses using thermophiles could lead to significant energy savings. All organisms generate heat as a by-product of metabolism, which must be dissipated to the environment to maintain a constant temperature. As the scale of the process increases, the surface area-to-volume ratio decreases, making it more difficult to maintain an isothermal process by heat removal. At the scales required for industrial production, metabolic heat production vastly outweighs heat loss to the environment through the bioreactor walls, and must be removed by cooling [29]. In fact, for a large process, cooling duty can be a limiting factor in scale-up. For mesophilic fermentations, the small temperature differential between the fermentor and the ambient surroundings is often not adequate for the use of cooling water, and therefore necessitates the use of refrigerated cooling fluid, which is an additional energy cost [30]. The refrigeration requirement decreases as fermentation temperature increases, and this can generate significant cost savings, even when only moderate thermophiles (40–45 °C) are used [30]. Thus, extreme thermophiles are attractive from a reactor cooling standpoint because the large temperature differential between the reactor and the environment would allow the use of nonrefrigerated cooling water. Furthermore, the low-grade heat produced by cooling the reactor offers the potential for additional energy savings via process heat integration.

An additional large energy cost associated with bioprocessing is the requirement for sterilization, which is usually achieved by contact with steam at 121 °C. Because extreme thermophiles have a lower potential for contamination and phage infection, it may not be necessary to sterilize process components at 121 °C, and this could allow additional energy savings as well as capital cost savings through the use of lower-grade materials. Finally, using extreme thermophiles to produce volatile products, such as ethanol and butanol, should allow facilitated product removal. Separation processes for purifying volatile products, including distillation, pervaporation, and gas stripping, are energy intensive because they operate at elevated temperatures. The elevated temperatures required for thermophilic fermentations are more compatible with these separation processes, making thermophilic fermentations a favorable match [31, 32]. For *in situ* separations, elevated fermentation temperatures may be amenable to distillation or stripping directly from the fermentation vessel, thereby maintaining a constant product concentration and potentially allowing for a continuous process. In contrast, if the separation strategy involves solvent extraction or phase separation, the higher temperatures may be undesirable owing to solubility increases.

14.3 Biobased Chemicals and Fuels: Targets and Opportunities

Examination of metabolic pathways in extreme thermophiles has provided interesting insights into the necessary adaptations for life at high temperatures and possible explanations for the mechanisms used by early life. These pathways include enzymes not found in mesophiles and that can perform unique reactions which often bypass steps found in conventional pathways characterized in mesophilic organisms (e.g., [33]). These shortened metabolic pathways presumably compensate for changes in equilibrium or instability of intermediates at higher temperatures. Thus, these shortcuts may be of some use in construction of synthetic metabolic pathways.

Thermophilic organisms are often found in environments that are not only characterized by high temperature but the chemical composition and nature of these environments can also be unusual. Since naturally thermal environments are a consequence of subterranean geothermal activity, these niches are often high in inorganic sulfur and biologically atypical metals, and exhibit extremes in salinity and pH. The fact that certain extreme thermophiles thrive under these otherwise biologically hostile conditions adds to their potential advantages as metabolic engineering hosts. A good example of this can be found in biomining, where engineered extremely thermoacidophilic hosts could have enhanced capacity for solubilizing base, precious, and strategic metals from low-grade ores (Figure 14.1) [34].

Current metabolic engineering efforts often focus on low-value fuels, such as ethanol and hydrogen, due to the myriad of existing known pathways in microorganisms and the potential large market for these biofuels (Figure 14.1). The highly efficient conversion of starches to ethanol by *S. cerevisiae* [35] is not likely to be surpassed by engineered extreme thermophiles; however, certain thermophiles could be utilized for processing lignocellulosic feedstocks by combining natural capacities for cellulose and hemicellulose degradation with a synthetic ethanol pathway. For example, this has been considered for species in the genus *Caldicellulosiruptor* which have shown extraordinary capacity to deconstruct complex lignocellulosic feedstocks [36]. Interest in biohydrogen production has waxed and waned over the years, depending to a large extent on natural gas prices. Many anaerobic microorganisms produce hydrogen gas as a by-product of fermentative metabolism, but the thermodynamics of this removal of excess reducing equivalents, such as NADH, becomes much more thermodynamically favorable at higher temperatures [37]. At lower temperatures, mesophilic organisms are more likely to add their reducing equivalents to organic molecules and form products such as ethanol or lactate, which at present are more desirable targets.

On the other end of the spectrum are chemicals that have a smaller market but have a much higher value compared to products such as ethanol. Many of these higher value compounds are target monomers for fibers and plastics. A current

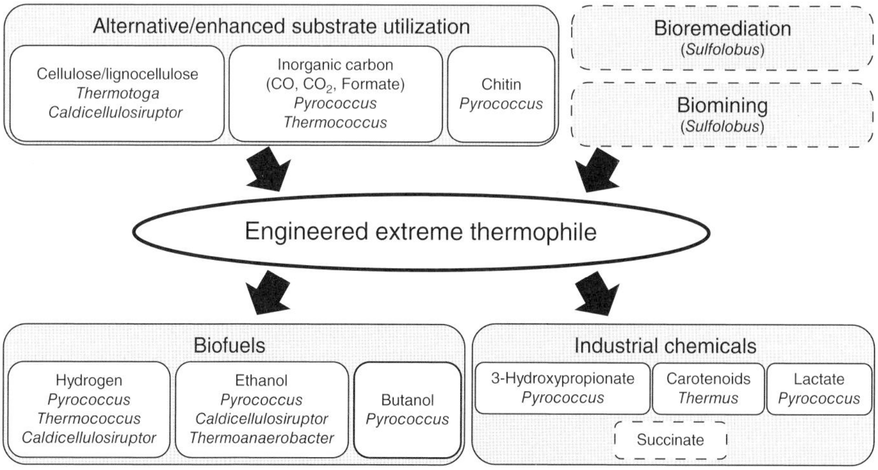

Figure 14.1 Current and future targets for metabolic engineering of extreme thermophiles. Boxes with dotted outlines represent potential targets that have not yet been demonstrated.

target at present is succinic acid, which is an excellent monomer, as well as a valuable four-carbon compound (Figure 14.1). As a component of the tricarboxylic acid (TCA) cycle, the presence of native pathways to this molecule lends optimism for near-term optimization. Compounds present in other metabolic pathways and cycles, such as 3-hydroxypropionate, are similarly attractive metabolic engineering targets. The key advantage to producing these high-value compounds by biocatalysis is their specific functional groups that are difficult to create using traditional petrochemical processing methods such as terminal diols and diacids, as well as specific alkenes and fatty acids [3].

As the chemical industry turns more and more to biotechnological solutions for chemical and fuel production, the end goal is an optimization of the process to meet given economic criteria. Thus, extreme thermophiles must find their niche on the basis of their intrinsic properties as hosts. While there are currently no known industrial products produced through the use of metabolically engineered extreme thermophiles, the attributes of these organisms make them excellent prospects for further development for future industrial use.

14.4
Considerations for Selecting an Extremely Thermophilic Host

While extreme thermophiles have many intrinsic features that make them attractive metabolic engineering hosts for industrial bioprocessing, specific aspects of the industrial target must be considered in selecting a host organism. Genetic systems are available for both aerobic and anaerobic extreme thermophiles, and both offer advantages and trade-offs as potential metabolic engineering hosts. The choice between an aerobe or anaerobe/facultative anaerobe for a host is often dependent on the degree of reduction of the product and the energetics of its production pathway. For a product with a high degree of reduction, for example, fuel alcohols and fatty acids, it is advantageous to select an anaerobic host. This allows pathway engineering strategies in which the product serves as the terminal electron acceptor for the host, providing a driving force for its production that is also coupled with growth [7, 38]. In an aerobe, electrons are typically donated to oxygen via energy-conserving mechanisms, potentially making it difficult to redirect electron flow to a reduced product that likely does not allow energy conservation. However, aerobes are frequently a better choice of hosts for products such as 3-hydroxypropionate that have a high ATP requirement for production.

The selection of a host organism can also be influenced by the desired carbon and/or energy sources to be used. For high-value products, it is useful to choose a facile, high-energy carbon source such as sugars. However, for low-value products, alternative sources must be used to control costs. For example, efforts toward cellulosic fuel ethanol production are driven in part by the reduced cost of lignocellulosic biomass relative to that of corn, potentially allowing for bioethanol prices that are cost competitive with petroleum-derived fuels [39, 40]. As a result, organisms that natively degrade lignocellulose are attractive metabolic engineering hosts for production of biofuels. Such organisms, for example, *Clostridium thermocellum* and *Caldicellulosiruptor bescii*, are currently being engineered for fuel molecule production [41, 42]. Another example comes from the concept of producing so-called electrofuels, in which host organisms are often selected for their ability to use inorganic energy (electric current or H_2) and carbon (CO_2) [43–45].

In addition, it is often advantageous to choose an organism that natively produces the desired product. This is not always possible, as only a small minority of known organisms have available genetic systems, particularly in extreme thermophiles. However, in cases where an organism with a genetic system produces the desired product, it can be easier to increase titers of the product in the native organism rather than try to engineer high-titer production into an alternate host. For instance, engineering of native thermophilic ethanol producers, such as *Thermoanaerobacter mathranii*, has led to higher yields and titers than adding ethanol production pathways to non-native hosts such as *C. bescii* [42]. This has been demonstrated in mesophilic organisms, in which much more effort was required to produce similar amounts of *n*-butanol in *E. coli* as compared to the native producer *C. acetobutylicum* [7, 20]. In a similar vein, it can be easier to use a host that natively makes a product that resembles the desired target, such as engineering an ethanologen to produce *n*-butanol [46]. However, engineering in pathways that are orthogonal to the host's natural metabolism should not be ruled out. There is some evidence that this type of heterologous pathway can improve production of recombinant products, even when some intermediates are shared with the host's endogenous metabolism.

14.5 General Strategies for Genetic Manipulation of Extreme Thermophiles

There are a number of extreme thermophiles with established genetic systems (Table 14.1). Many of these are naturally competent for DNA uptake, such as species of *Pyrococcus*, *Thermococcus*, and *Thermus*. Others require more stringent means for inducing DNA uptake, with electroporation being the most commonly used method. Transformation is typically performed with circular plasmid DNA, although some thermophiles are also readily transformed with linear DNA. Selectable markers are limited as the high growth temperatures required by these organisms preclude the use of many mesophilic markers and means of selection. Two antibiotics that are stable at high temperatures and have corresponding thermostable resistance markers are kanamycin and chloramphenicol, and these have been employed in bacterial thermophiles. Thermostable antibiotics that are functional in Archaea are statin-based compounds (simvastatin and mevinolin) that target the isoprenoid biosynthesis pathway for production of archaeal phospholipids. A number of nutritional markers are also employed using corresponding auxotrophic strains, and some of these markers have the added benefit

Table 14.1 Methods for genetic manipulation of extreme thermophiles ($T_{opt} \geq 70\,°C$).

Organism	Genetic modification	Vector	Transformation method	Selection[a)]	Counterselection[a)]	Effective transformation efficiency[a), b)] (CFU/μg DNA)	References
P. furiosus	Homologous recombination	Plasmid, BAC, or linear DNA	Natural competence	-Ura or -Agm	5FOA (2.5 mM) + Ura (20 μM) or 5FOA + Agm (4 mM)	10^4-10^5 per 10^8 cells	[47, 48]
	Replicating vector	Plasmid	Natural competence, heat shock	Sim (10 μM)	6MP (50 μM)	~10^2 per 10^8 cells	[49, 50]
P. yayanosii	Homologous recombination	Plasmid	$CaCl_2$ + heat shock	Sim (5 μM), -Ura	5FOA (0.1%)	10^2	[51]
T. kodakarensis	Homologous recombination	Plasmid or linear DNA	$CaCl_2$ + heat shock	-Ura, -Trp Sim (4 μM)	5FOA (4.3 mM) + Ura (89 mM) 6MP (10 μM)	10^0-10^1	[52–55]
	Replicating vector	Plasmid	$CaCl_2$ + heat shock	-Trp, Mev (15 uM)	NA	~10^0 per 10^8 cells	[56]
T. onnurineus	Homologous recombination	Plasmid	$CaCl_2$ + heat shock	Sim (10 μM)	NA	NR	[54, 57]
T. barophilus	Homologous recombination	Plasmid	Heat shock	Sim (6 μM)	5FOA (46 mM)	10^2	[58, 59]
S. solfataricus	Homologous recombination	Plasmid or linear DNA	Electroporation	Lactose -Ura	NR	NR	[60–63]
	Replicating vector	Plasmid	Electroporation	Lactose, butanol/benzyl alcohol, HygB	NA	10^2-10^4 per 10^8 cells	[64–66]

S. acidocaldarius	Homologous recombination	Plasmid or linear DNA	Electroporation	-Ura	5FOA (100 $\mu g\,ml^{-1}$) + Ura (10 $\mu g\,ml^{-1}$)	NR	[67, 68]
	Replicating vector	Plasmid	Electroporation	-Ura	NA	10^2-10^4 per 10^8 cells	[69]
S. islandicus	Homologous recombination	Plasmid or linear DNA	Electroporation	-Ura	5FOA (50 $\mu g\,ml^{-1}$) + Ura (20 $\mu g\,ml^{-1}$)	NR	[70]
	Homologous recombination	Plasmid (pSEF)	Electroporation	-Ura	5FOA (50 $\mu g\,ml^{-1}$) + Ura (20 $\mu g\,ml^{-1}$)	NR	[71]
	Replicating plasmid	Plasmid (pZH2)	Electroporation	-Ura	NA	10^2-10^3 per 10^8 cells	[70, 71]
	Replicating plasmid	Plasmid (pSSR)	Electroporation	Sim	NA	NR	[72]
	Homologous recombination	Plasmid or linear DNA	Electroporation	Sim	NA	NR	[73]
T. maritima	Replicating vector	Plasmid	Liposome-mediated	Kan	NA	5	[74]
	Replicating vector	Plasmid	Electroporation	Kan	NA	0.25	[74]
Thermotoga sp. RQ7	Replicating vector	Plasmid	Natural competence	Kan	NA	19	[75]
T. thermophilus	Homologous recombination	Plasmid or linear DNA	Natural competence	-Ura	5FOA (200–500 $\mu g\,ml^{-1}$) + Ura (20 $\mu g\,ml^{-1}$)	6×10^5	[76–78]
	Homologous recombination	Plasmid or linear DNA	Natural competence	-Trp, -Leu, -Pro	NR	0.0005–14.1% (frequency)	[79–84]
	Homologous recombination	Plasmid or linear DNA	Natural competence	Kan (20–40 $\mu g\,ml^{-1}$)	NR	10^4-10^5	[85–89]

(continued overleaf)

Table 14.1 (*Continued*).

Organism	Genetic modification	Vector	Transformation method	Selection[a)]	Counterselection[a)]	Effective transformation efficiency[a) b)] (CFU/μg DNA)	References
	Homologous recombination	Plasmid	Natural competence	Kan (30 μg ml^{-1})	*p*-Cl-Phe (15 mM)	NR	[90]
	Replicating vector	Plasmid	Natural competence	Kan (50–120 μg ml^{-1})	NR	10^2-10^4	[91–97]
	Homologous recombination	Linear DNA	Conjugation	Cm (20 μg ml^{-1}), nitrate(40 mM), anaerobic	NR	NR	[98, 99]
	Homologous recombination	Plasmid	Natural competence	Kan (25 μg ml^{-1})	5FC (15 μg ml^{-1})	NR	[100]
T. flavus	Homologous recombination	Plasmid or linear	Natural competence	-Ura	5FOA (500 μg ml^{-1}) + Ura (40 μg ml^{-1})	NR	[101]
	Homologous recombination	Plasmid or linear	Natural competence	Kan (20 μg ml^{-1})	NR	NR	[102]
C. bescii	Homologous recombination	Plasmid	Electroporation	-Ura, Kan (50 μg ml^{-1})	5FOA (4 mM) + Ura (40 μM)	10^3	[103, 104]
	Replicating vector	Plasmid	Electroporation	-Ura, Kan (50 μg ml^{-1})	NA	5×10^2	[104, 105]
C. hydrothermalis	Homologous recombination	Plasmid	Electroporation	-Ura	5FOA (6 mM) + Ura (40 μM)	NR	[106]
	Replicating vector	Plasmid	Electroporation	-Ura	NA	$\sim10^2$	[106]

C. subterraneus subsp. *tengcongensis*	Replicating vector	Plasmid	Natural competence	Kan (400 μg ml^{-1})	NA	7×10^4	[107]
T. ethanolicus JW200 DSM2246	Replicating vector	Plasmid	Natural competence	Kan (1000 μg ml^{-1})	NA	10^5	[108]
T. brockii ATCC 35047	Replicating vector	Plasmid	Natural competence	Kan (1000 μg ml^{-1})	NA	1.9×10^2	[108]
T. pseudoethanolicus 39E ATCC33223	Replicating vector	Plasmid	Natural competence	Kan (400 μg ml^{-1})	NA	6.3×10^3	[108]
T. thermohydrosul-furicum	Replicating vector	Plasmid	Tris-PEG6000 incubation	Kan (25 μg ml^{-1}) or Th (10 μg ml^{-1})	NA	~10^2 per 10^8 cells	[109]
T. ethanolicus JW200	Replicating vector	Plasmid	Electroporation of partial protoplasts	Cm (40–80 μg ml^{-1})	NA	50	[110]
	Homologous recombination	Plasmid	Electroporation of partial protoplasts	Cm (40–80 μg ml^{-1})	NA	30	
T. mathranii BG1	Homologous recombination	Plasmid	Electroporation	Kan (50 μg ml^{-1})	NA	NR	[111]
Thermoanaerobacter sp. X514	Replicating vector	Plasmid	Sonoporation or electroporation	Kan (250 μg ml^{-1})	NA	6×10^2	[112]

a) NA, not applicable; NR, not reported; BAC, bacterial artificial chromosome; 5FOA, 5-fluoroorotic acid; 6MP, 6-methylpurine; Agm, agmatine; Cm, chloramphenicol; Kan, kanamycin; 5FC, 5-fluorocytosine; Leu, leucine; Mev, mevinolin; *p*-Cl-Phe, *p*-chlorophenylalanine; Pro, proline; Sim, simvastatin; Th, thiamphenicol; Trp, tryptophan; and Ura, uracil.

b) Efficiency of DNA uptake for replicating vectors or combined efficiency of DNA uptake and homologous recombination for nonreplicating vectors.

of also being counterselectable. The most commonly employed nutritional marker involves genes in the uracil biosynthetic pathway. Deletions in *pyrF* and *pyrE* generate a background strain that is auxotrophic for uracil. Supplying the corresponding marker restores uracil prototrophy. Furthermore, these markers can be counterselected for their absence with the compound 5-fluoroorotic acid (5FOA), an analog to a uracil biosynthetic intermediate that can be converted to a toxic product by the enzyme encoded by *pyrF* or *pyrE*. Other counterselectable markers based on resistance to 6-methylpurine, *p*-chlorophenylalanine (*p*-Cl-Phe), and 5-fluorocytosine (5FC) have also been used to a limited extent (see Table 14.1).

Replicating shuttle vectors have been employed in a number of extreme thermophiles, and these vectors are generally based on use of the backbone of a native plasmid from the same organism or a related thermophilic organism. There are also instances for use of the host's chromosomal replication origin to construct a replicating shuttle vector. In general, the autonomously replicating plasmids employed in thermophiles have low copy numbers which do not approach the high copy numbers obtained for a variety of plasmid replication origins used in mesophiles such as *E. coli*. Replicating shuttle vectors can be useful as an initial screen for expression of metabolic pathway genes as plasmid-based expression might be higher depending on the plasmid copy number, and in general, transformation with a replicating vector is less labor intensive than constructing chromosomal mutations. However, use of a replicating plasmid is generally not a desirable long-term solution for expression of heterologous genes for genetic engineering applications as size limitations and plasmid maintenance and stability can be issues.

Genetic engineering via chromosomal integration of heterologous metabolic pathways has several advantages: higher stability in the absence of selective pressure, a greater capacity for large constructs, and the availability of multiple chromosomal locations for insertion of a variety of constructs.

All current techniques for chromosomal modification of extreme thermophiles take advantage of host homologous recombination systems to integrate foreign DNA into the chromosome, and this requires the use of homologous regions (typically 0.5–1 kb) flanking the target genome region within the transforming DNA construct. All of these chromosomal modification methods are variations on a few general strategies, as shown in Figure 14.2. Design of the transforming

Figure 14.2 General strategies for chromosomal deletion or insertion. DNA architecture and transformation steps shown for markerless insertion or deletion using circular or linear DNA (a), (b), (e), (f); and marked insertion or deletion using circular or linear DNA (c), (d), (g), (h). Note that *pyrF* is used as both a marker for selection (uracil prototrophy) and counterselection (5FOA resistance); variations on these strategies may use an alternate marker for selection and a second marker for counterselection, with appropriate chemicals replacing uracil and 5FOA. Abbreviations: 5′ HR and 3′ HR, homology regions 5′ or 3′ to the targeted chromosomal region; *pyrF*, uracil prototrophy/5FOA resistance marker; *kat*, kanamycin resistance marker; GOI, gene of interest; PO, "pop-out" region homologous to portion of 5′ HR; ura, uracil; 5FOA, 5-fluoroorotic acid; and kan, kanamycin.

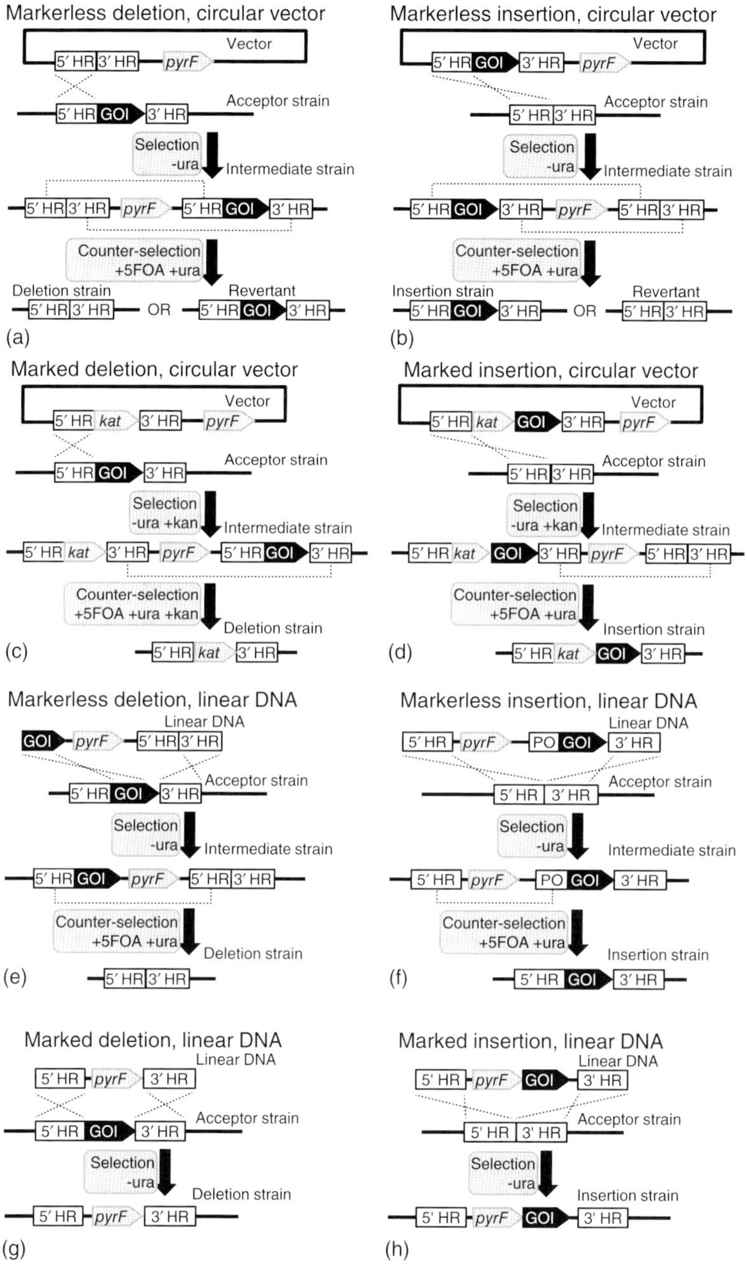
Markerless deletion, circular vector
Vector
5′ HR 3′ HR pyrF
Acceptor strain
5′ HR GOI 3′ HR
Selection -ura
Intermediate strain
5′ HR 3′ HR pyrF 5′ HR GOI 3′ HR
Counter-selection +5FOA +ura
Deletion strain
5′ HR 3′ HR
OR
Revertant
5′ HR GOI 3′ HR
(a)
Markerless insertion, circular vector
Vector
5′ HR GOI 3′ HR pyrF
Acceptor strain
5′ HR 3′ HR
Selection -ura
Intermediate strain
5′ HR GOI 3′ HR pyrF 5′ HR 3′ HR
Counter-selection +5FOA +ura
Insertion strain
5′ HR GOI 3′ HR
OR
Revertant
5′ HR 3′ HR
(b)
Marked deletion, circular vector
Vector
5′ HR kat 3′ HR pyrF
Acceptor strain
5′ HR GOI 3′ HR
Selection -ura +kan
Intermediate strain
5′ HR kat 3′ HR pyrF 5′ HR GOI 3′ HR
Counter-selection +5FOA +ura +kan
Deletion strain
5′ HR kat 3′ HR
(c)
Marked insertion, circular vector
Vector
5′ HR kat GOI 3′ HR pyrF
Acceptor strain
5′ HR 3′ HR
Selection -ura +kan
Intermediate strain
5′ HR kat GOI 3′ HR pyrF 5′ HR 3′ HR
Counter-selection +5FOA +ura
Insertion strain
5′ HR kat GOI 3′ HR
(d)
Markerless deletion, linear DNA
Linear DNA
GOI pyrF 5′ HR 3′ HR
Acceptor strain
5′ HR GOI 3′ HR
Selection -ura
Intermediate strain
5′ HR GOI pyrF 5′ HR 3′ HR
Counter-selection +5FOA +ura
Deletion strain
5′ HR 3′ HR
(e)
Markerless insertion, linear DNA
Linear DNA
5′ HR pyrF PO GOI 3′ HR
Acceptor strain
5′ HR 3′ HR
Selection -ura
Intermediate strain
5′ HR pyrF PO GOI 3′ HR
Counter-selection +5FOA +ura
Insertion strain
5′ HR GOI 3′ HR
(f)
Marked deletion, linear DNA
Linear DNA
5′ HR pyrF 3′ HR
Acceptor strain
5′ HR GOI 3′ HR
Selection -ura
Deletion strain
5′ HR pyrF 3′ HR
(g)
Marked insertion, linear DNA
Linear DNA
5' HR pyrF GOI 3' HR
Acceptor strain
5' HR 3' HR
Selection -ura
Insertion strain
5' HR pyrF GOI 3' HR
(h)

DNA and the genetic selection steps vary depending on whether circular or linear DNA is used, whether a marked or markerless mutant is desired, and whether the goal is gene deletion or insertion. For methods using circular DNA vectors, a positive selection step is first required to select for a single-crossover event that integrates the vector into the chromosome at one of the homologous flanking regions on the vector. Following selection for the presence of the marker, a counterselection step is used to select for a second crossover event that excises the marker(s) (Figure 14.2a,b). Any genes between the 3′ and 5′ homology regions in the host chromosome are deleted following counterselection (Figure 14.2a). If genes are included between the homology regions in the vector, they are inserted into the host chromosome (Figure 14.2b). In both of these cases, the second selection step will generate either the desired mutation or a reversion to the parent strain, depending on the location of the second crossover event; therefore, requiring resulting colonies to be screened for the mutation. A drawback to this strategy is that there is no selective pressure to force the mutation; therefore, in cases where a mutation creates a fitness defect, the number of colonies which contain the desired mutation will be proportional to the severity of the defect, and excessive screening may be required to obtain the desired strain. To avoid excessive screening for difficult-to-obtain mutations, a second selectable marker, such as an antibiotic resistance gene, can be placed between the homology regions, allowing a second selective pressure to be applied during the selection and counterselection steps to ensure that surviving colonies have the desired mutation (Figure 14.2c,d).

Some extreme thermophiles are capable of uptake and integration of linear DNA via double-crossover homologous recombination, allowing construction of mutants using linear DNA, generated by linearizing a plasmid or amplified as a PCR product. Similar to using circular vectors, using linear DNA to generate markerless mutants requires selection for integration of the construct and counterselection for marker removal. The transforming DNA contains the marker and 3′ homology region placed between the gene of interest and the 5′ homology region, allowing a double crossover event using the gene of interest and the 5′ homology region. A markerless deletion is generated by homologous recombination at the duplicate 3′ homology regions (Figure 14.2e). For markerless insertions, the 3′ and 5′ homology regions are used for the first crossover, with the marker, gene(s) of interest, and a third homology "pop-out" region included between them (Figure 14.2f). The length of the "pop-out" region required will depend upon the recombination efficiency of the organism and can range from 65 to 500 bp. The "pop-out" region can be a random sequence duplicated on either side of the marker or a homologous region to a portion of the marker gene, allowing the second crossover to occur, leaving a "scar" behind. A pop-out sequence that is homologous to a portion of either the 5′ or 3′ flanking regions can also be used in order to generate a final markerless, scarless strain, with the caveat that during initial selection of integration, there is the possibility of integrating the transforming DNA at only that flanking region. The advantage of using linear DNA is that only the desired mutant can be obtained following counterselection, with no possibility of reversion to the parent strain.

The advantage of generating markerless mutants, whether using circular or linear DNA, is that iterative genetic modifications can be made to the organism using the same selectable markers. This is advantageous for metabolic engineering in which multiple gene insertions or deletions are required, especially in extreme thermophiles for which a limited number of selectable markers is available. However, in cases where iterative modifications are not required, marked deletion or insertion mutants can be generated using linear DNA and only a single selection step (Figure 14.2g,h).

14.6 Limitations and Barriers to Genetic Modification of Extreme Thermophiles

The pace of metabolic engineering of extreme thermophiles is largely influenced by practical factors in generating mutant strains. The ease of use of a genetic system, the time required to generate a mutant, and the consistency with which the system works all impact the pace of strain development. Multiple factors affect how easily and quickly new strains can be generated, including transformation efficiency, selectable marker type, plating conditions and incubation time, number of selection steps required, and the transformation method. Perhaps, the most important factor for genetic engineering is transformation efficiency. Transformation efficiencies must be high enough to both generate sufficient numbers of isolates for screening and exceed any background growth that does not contain the desired modification. Extreme thermophiles with available genetic systems have a wide range of transformation efficiencies, from on the order of 1 transformant per microgram of DNA for *Thermotoga maritima*, up to 10^5 transformants per microgram of DNA for *Pyrococcus furiosus* and *Thermus thermophilus* (Table 14.1). Transformation efficiencies above 10^2 seem to be the point at which a genetic system gains traction. The type of selectable marker can also have an impact on ease of mutant construction. Nutritional markers, such as those for uracil prototrophy, often suffer from background growth as a result of trace uracil in media components, such as Gelrite [113], requiring screening of additional colonies. In contrast, antibiotic resistance markers have fewer background colonies [74].

Another consideration is the plating conditions and time required to grow colonies after transformation of the organism. Anaerobic extreme thermophiles may require anaerobic conditions during transformation and plating, which necessitates additional equipment, such as an anaerobic chamber and anaerobic jars suitable for high-temperature incubations (these can be made by modifying paint canisters) (Figure 14.3). Certain anaerobic thermophiles, such as *P. furiosus* and *Thermotoga* species, can be transformed and plated under aerobic conditions, but must still be incubated under anaerobic conditions [47, 74]. Aerobic thermophiles do not require this specialized equipment and thus do not suffer from decreased plating efficiency caused by oxygen toxicity. Plate incubation time is another challenge. Even extreme thermophiles with rapid doubling times, such as *P. furiosus*, *C. bescii*, and *T. maritima*, require 2–4 days of incubation to develop

(a) (b)

Figure 14.3 Equipment required for plating anaerobic thermophiles. Anaerobic glove box (a) for manipulating cells and paint canister (b) adapted for incubating plates anaerobically at high temperatures.

colonies [47, 74]; other organisms, such as *Sulfolobus* and *Thermoanaerobacter* species, require up to 7 days to develop colonies [67, 108, 109]. Thus, the typical time to construct a markerless mutant is several weeks, as a result of the long plating times, two selection steps, and, for some species, multiple rounds of plate purification to obtain clonal isolates. The slow turnaround time for mutant construction is a key challenge in developing extreme thermophiles as metabolic engineering platforms, and highlights the need for more predictive engineering tools to identify target genes, rather than relying on trial-and-error approaches.

An additional barrier to genetic manipulation of extreme thermophiles is the lack of knowledge on restriction modification and DNA methylation systems. Active restriction enzymes can reduce or prevent transformation of host organisms, unless the vector DNA is protected by methylation patterns similar to the host. While the REBASE database predicts many restriction enzymes and methylases from genomes, the recognition motifs are not always known [114]. As a result, biochemical characterization has been required to determine restriction enzyme targets before effective transformation of some extreme thermophiles. For example, a restriction modification system was identified in *C. bescii* that required plasmid DNA to be methylated *in vitro* prior to transformation [115, 116]. *Thermoanaerobacter* sp. X514 appears to have a restriction methylation system for GATC sites, which is protected by Dam^+/Dcm^+ *E. coli* strains, as compared to unmethylated plasmid from Dam^-/Dcm^- *E. coli* [112]. Similarly, *Sulfolobus acidocaldarius* has the *SuaI* restriction modification system, requiring vector DNA to be methylated before transformation [67, 117]. As of now, yet unknown restriction modification systems in other hosts may be responsible for low transformation efficiencies. Fortunately, with the advent of SMRT sequencing technology, which allows direct identification of DNA methylation patterns, unknown DNA restriction modification systems may become a thing of the past [114, 118].

14.7
Current Status of Metabolic Engineering Efforts and Prospects in Extreme Thermophiles

There are only a small number of extreme thermophiles for which genetic tools have been developed; however, they represent a variety of growth environments and possess an assortment of metabolic features. These species come from both the Bacteria and the Archaea, including aerobes and anaerobes, grow at a variety of temperatures, salt concentrations, and pH optima (Table 14.2). While most are chemoorganotrophs, at least one is a carboxydotroph (*Thermococcus onnurineus*). Certain strains of *Sulfolobus islandicus* may be chemolithotrophs, but no extremely thermophilic chemolithotroph currently has a genetic system. Similarly, no autotrophic extreme thermophiles currently have genetic systems. Extreme thermophiles with currently available genetic systems are discussed below, with an emphasis on how these organisms can be used in metabolic engineering efforts.

14.7.1
Pyrococcus furiosus

P. furiosus is a heterotrophic, fermentative, anaerobic marine archaeon that can grow on peptides when elemental sulfur is present, or on a variety of saccharides whether sulfur is available or not (Table 14.2) [119, 120]. It grows at temperatures from 70 to 103 °C, with an optimum near 100 °C, and at pH ranging from 5 to 9, with an optimum near pH 7 [119]. *P. furiosus* produces CO_2, H_2, acetate, and alanine as major fermentation products in the absence of sulfur, with H_2S replacing H_2 production when sulfur is present [119, 121, 157]. *P. furiosus* has a rapid doubling time of 38 min, on par with many mesophilic heterotrophs, and can reach maximum cell densities approaching 10^9 cells ml^{-1} [119, 122]. *P. furiosus* has become a model organism within the extremely thermophilic archaea, and, as such, many details of its metabolism and physiology are well understood. *P. furiosus* has several natural attributes that make it attractive as an industrial host, even compared to other thermophiles. From an industrial standpoint, its extremely high growth temperature obviates the chance of contamination. As a marine organism, seawater-based media could be used on an industrial scale, avoiding the financial and environmental costs associated with using large quantities of freshwater. In addition, *P. furiosus* can tolerate relatively high levels of oxygen for an obligate anaerobe [145], alleviating the need for strict anaerobic conditions in industrial-scale cultures.

14.7.1.1
Genetic Tools – *P. furiosus*

There are two genetic systems that have been developed for *P. furiosus*. The first system uses a heat-shock method to transform wild-type cells with a replicating shuttle vector (Table 14.1) [49]. The transformation plasmid contained the gene for HMG-CoA reductase (*hmg*) as a selectable marker for simvastatin resistance, allowing cells that take up the plasmid to grow in the presence of simvastatin. In

Table 14.2 Extremely thermophilic metabolic engineering hosts.[a)]

Organism	Growth conditions			Metabolism				Doubling time (min)	Maximum density (cells ml^{-1})	References
	Temperature range (optimum) (°C)	pH range (optimum)	Salinity	O_2 requirement	Energy metabolism	Growth substrates	Major waste products			
P. furiosus (Archaea)	70–103 (100)	5.0–9.0 (7.0)	Marine	Obligate anaerobe	Chemoorganotroph	Peptides (S^0 optional) Hexose di- and polysaccharides	H_2 (H_2S) CO_2 Acetate Alanine	38	5×10^8	[119–122]
T. kodakarensis (Archaea)	60–100 (85)	5.0–9.0 (6.5)	Marine	Obligate anaerobe	Chemoorganotroph	Peptides + S^0 Starch	H_2, H_2S CO_2 Acetate	60	1×10^8	[124, 125]
T. onnurineus (Archaea)	63–90 (80)	5.0–9.0 (8.5)	Marine	Obligate anaerobe	Chemoorganotroph Carboxydotroph	Peptides + S^0 Starch + S^0 CO Formate	H_2, H_2S CO_2 Organic acids	92	4.5×10^8	[127, 128]
T. barophilus (Archaea)	48–100 (85)	4.5–9.5 (7.0)	Marine	Obligate anaerobe	Chemoorganotroph Carboxydotroph (?)	Peptides (S^0 optional) CO (?)	H_2 (H_2S) CO_2 Organic acids	28	5×10^7	[129]
S. acidocaldarius (Archaea)	55–83 (80)	0.9–5.8 (3.0)	Terrestrial	Obligate aerobe	Chemoorganotroph	Peptides Amino acids Dextrin/starch	CO_2	180	3×10^9	[130]

S. solfataricus (Archaea)	65–87 (80)	2.0–4.0 (3.5)	Terrestrial	Obligate aerobe	Chemoorganotroph	Peptides Amino acids Pentoses Hexose di- and polysaccharides Phenol	CO_2	300	~10^9	[130, 131]
S. islandicus (Archaea)	65–85	NR	Terrestrial	Obligate aerobe	Chemoorganotroph Chemolithotroph (?)	NR	NR	NR	NR	[34]
C. bescii (Bacteria)	65–90 (75)	6.8–7.2	Terrestrial	Obligate anaerobe	Chemoorganotroph	Hexose and pentose di- and polysaccharides Lignocellulose	H_2, CO_2 Acetate Lactate	120	~5×10^8	[36, 133–137]
T. maritima (Bacteria)	65–90 (80)	6.5–7.0	Marine	Obligate anaerobe	Chemoorganotroph	Peptides (with sugars) Hexose and pentose di- and polysaccharides (S^0, $S_2O_3^{2-}$ optional)	H_2 (H_2S) CO_2 Acetate	90	1×10^9	[138–142]
T. flavus (Bacteria)	40–81 (70)	6.0–9.0 (7.0–7.5)	Terrestrial	Obligate aerobe	Chemoorganotroph	Polysaccharides Peptides	Organic acids	45	NR	[144]

(continued overleaf)

Table 14.2 (*Continued*).

Organism	Growth conditions			Metabolism				Doubling time (min)	Maximum density (cells ml^{-1})	References
	Temperature range (optimum) (°C)	pH range (optimum)	Salinity	O_2 requirement	Energy metabolism	Growth substrates	Major waste products			
T. thermophilus (Bacteria)	47–85 (62–72)	5.1–9.6 (7.0)	Terrestrial	Aerobe Facultative aerobe	Chemoorganotroph	Polysaccharides Peptides (NO_3^- for anaerobic growth)	Organic acids	18–20	8×10^8	[150, 151]
T. ethanolicus (Bacteria)	37–78 (69)	4.4–9.8 (5.8–8.5)	Terrestrial	Obligate anaerobe	Chemoorganotroph	Hexose and pentose disaccharides Starch	CO_2 Ethanol Acetate Lactate H_2	60–74	NR	[152–155]
Thermoanaerobacter mathranii (Bacteria)	50–75 (75)	4.7–8.8 (7.0–7.5)	Terrestrial	Obligate anaerobe	Chemoorganotroph			NR	NR	
T. sp. X514 (Bacteria)	45–75 (60)	6.9–8.9	Terrestrial	Obligate anaerobe	Chemoorganotroph			NR	NR	
C. subterraneus subsp. *tengcongensis* (Bacteria)	50–80 (75)	5.5–9.0 (7.0–7.5)	Terrestrial	Obligate anaerobe	Chemoorganotroph	Hexose disaccharides Starch	H_2 CO_2 Ethanol Acetate Alanine	65	NR	[156]

The "?" symbol indicates there is uncertainty or conflicting literature for the indicated metabolism or growth substrate for this organism. E.g., it is uncertain that S. islandicus is capable of chemolithotrophy.

a) NR, not reported.

this method, plasmid was mixed with a concentrated cell suspension at 4 °C and then heat-shocked at 80 °C in the presence of $CaCl_2$. After another 4 °C incubation, the cells were grown for 2 days at 85 °C in complex liquid medium, containing simvastatin to select for transformants, then plated on solid medium, also containing simvastatin, and grown for another 2 days. The transformation efficiency for this method was approximately 10^2 transformants per microgram DNA per 10^8 cells. With a plating efficiency of approximately 15%, the effective transformation efficiency was low, on the order of 10^1 colonies per microgram DNA per 10^8 cells. Cell manipulations were performed under anaerobic conditions for this method.

The heat-shock method was extended to allow chromosomal modifications via double-crossover homologous recombination [50]. In this modified genetic system, a guanosine auxotrophic strain was transformed with linear DNA containing the *hmg* marker for positive selection and xanthine-guanine phosphoribosyltransferase for counterselection with 6-methylpurine (6MP); the two markers were flanked by homology regions that target integration at a specific genome locus. Simvastatin was used to select for chromosomal integration of the plasmid, and 6MP was then used to select for cells that had undergone a second homologous recombination event for removal of the marker cassette, thereby generating a markerless mutant. This method was used to repair a chitinase gene in *P. furiosus*, and has the potential to be used as a general method to insert or delete genes [50].

The second genetic system reported for *P. furiosus* takes advantage of the naturally competent uracil auxotrophic strain COM1, which contains a targeted deletion of the *pyrF* gene [47]. This method used circular or linear DNA containing the *pyrF* gene as a selectable marker for uracil prototrophy. The DNA was mixed with a culture aliquot of *P. furiosus* strain COM1 under aerobic conditions, the mixture was plated directly on solid defined medium lacking uracil, and the plates were incubated under anaerobic conditions for 3 days at 95 °C to select for transformants [47]. Resulting colonies were then subcultured in defined medium lacking uracil and screened by PCR amplification of the target genome region. Counterselection for marker removal was performed with 5FOA. Plating on 5FOA selects for a second crossover event that removes the *pyrF* marker, resulting in a uracil auxotroph that can be used as a parent strain for additional chromosomal modifications. Transformant colonies had to be plated again two to three times to obtain pure mutant cultures prior to sequence confirmation of the mutant strains.

Genetic manipulation of the COM1 strain is significantly easier than the system established for wild-type *P. furiosus*, primarily owing to increased transformation efficiency. The COM1 strain of *P. furiosus* is highly competent for DNA uptake and recombination, with an effective transformation efficiency of 10^5 transformants per microgram DNA per 10^8 cells [47]. Additionally, it was shown that *P. furiosus* COM1 has remarkably efficient recombinogenic properties such that transformants can be obtained using DNA with homologous recombination regions as short as 40 bp [48]. This allows the use of very small homologous flanking regions (65 bp) around the *pyrF* marker cassette that enable counterselection for marker loss, thus improving the versatility of the system. This high efficiency of transformation and recombination, in combination with the high oxygen tolerance of *P. furiosus*, allows transformation and plating to be done under aerobic rather than anaerobic conditions, resulting in adequate numbers of transformant colonies after anaerobic incubation, despite variable plating efficiencies. Although the *P. furiosus* COM1 genome has been sequenced, the exact nature of the mutation(s) leading to the high competence of the COM1 strain is unknown, despite comparative analysis with the wild-type genome [48, 146]. The ease and versatility of the COM1-based system has allowed for phenotypic studies of multiple mutants, overexpression of homologous and heterologous proteins, as well as application of several genetic engineering strategies that will be discussed below.

The genetic toolbox for the COM1 strain has also been extended with the application of additional selectable markers. An agmatine auxotroph has been developed, based on the COM1 strain, which allows use of *pdaD* as a selectable marker for agmatine prototrophy [147]. The advantage of selecting for agmatine prototrophy is that it allows the use of complex media after transformation, because complex peptide sources such as yeast extract do not contain agmatine, simplifying media preparation and possibly increasing plating efficiency [147]. Additionally, a uracil and tryptophan double auxotroph strain has been developed that allows simultaneous selection for two genetic markers, allowing for possibilities such as maintaining a replicating vector with one marker and using the other for chromosomal manipulations [48].

14.7.1.2
Metabolic Engineering – *P. furiosus*

P. furiosus has been used as the host for several metabolic engineering targets, including utilization of two non-native substrates as well as synthesis of four non-native products (Table 14.3). Extensive work has been done to engineer *P. furiosus* to produce 3-hydroxypropionate (3HP), a valuable commodity chemical [3]. 3HP is an intermediate in the 3HP/4HB carbon-fixation cycle of *Metallosphaera sedula*, and is produced from acetyl-CoA and bicarbonate in three steps catalyzed by acetyl-CoA carboxylase, malonyl-CoA reductase, and malonate semialdehyde reductase [143]. To produce 3HP in *P. furiosus*, the genes for these enzymes were inserted under the control of the strong constitutive S-layer protein promoter (Table 14.4). This initial approach resulted in production of 0.6 mM 3HP [43]. To improve 3HP production, several strategies were implemented. First, genes for the enzyme acetyl-CoA synthase I (ACSI) were deleted to reduce conversion of acetyl-CoA to acetate; the knockdown of this competing pathway nearly doubled the titer of 3HP [148]. Another strategy was the addition of two "helper" enzymes, the *M. sedula* biotin-protein ligase (BPL) and carbonic anhydrase (CA). The BPL assists in the maturation of the acetyl-CoA carboxylase enzyme, a crucial enzyme in the 3HP production pathway, while the CA was included to ensure high availability of bicarbonate, the substrate for acetyl-CoA carboxylase [149]. The addition of the BPL and CA increased the 3HP titer to 3.1 mM. A strain containing both the ACSI deletion and insertion of BPL and CA would be promising, and efforts are currently under way to evaluate such a strain.

Table 14.3 Metabolic engineering applications of extreme thermophiles.

Organism	Target	Titer or productivity	Strain genotype	References
P. furiosus	3-Hydroxypropionate	$0.276\,g\,l^{-1}$	Δ*pyrF*::P_{gdh}-*pyrF* P_{slp}-Msed_0147-0148, 1375, 0709, 1993	[43, 148, 149]
	Ethanol	20 mM	Δ*pyrF*::P_{gdh}-*pyrF* P_{slp}-Teth514_0564	[158]
	CO oxidation	NA	Δ*pyrF*::P_{gdh}-*pyrF* P_{slp}-Teth514_0564 P_{mbh1}TON_1017-1031	[158]
	n-Butanol	$65\,mg\,l^{-1}$	Δ*pyrF*::P_{gdh}-*pyrF* P_{slp}-TTE0549, 0548, 0544, Stherm_c16300, Teth514_1942, 1935	[159]
	Lactate	3 mM	Δ*pyrF* Δ*pdaD*::*pdaD* P_{cipA}-Athe_1918	[160]
	H_2 from formate	45 mmol H_2 per liter	Δ*pyrF*::P_{gdh}-*pyrF* P_{mbh1}-TON1563-1580	[161]
	Chitin utilization	NA	Δ*xgprt* ΔPF1233-1234::PF1233/1234	[50]
T. kodakarensis	Increased H_2 production	28% increase	Δ*pyrF* Δ*trpE*::*pyrF* ΔTK2069-72	[162]
T. onnurineus	Increased H_2 from CO	$192\,mmol\,l^{-1}\,h^{-1}$	Δ*corR*::P_{gdh}-*corQR*	[163]
	Increased H_2 from CO	$131\,mmol\,l^{-1}\,h^{-1}$	P_{TN0510}-CODH	[164]
C. bescii	Hydrogen	~15.0 mM (~$3.4\,mol\,mol^{-1}$ of glucose)	Δ*pyrFA* Δ*ldh*	[135]
	Ethanol	12.8 mM	Δ*pyrFA* Δ*ldh*::*ISCbe4* Δ*cbe1*::P_{slp} *Cthe-adhE*[2]/(ura^-/$5FOA^R$)	[41]
		1.4 mM	Δ*pyrFA* Δ*ldh* ClS1::P_{slp}Teth39_0218 (*adhB*)/(ura^-/$5FOA^R$)	[165]
		2.3 mM	Δ*pyrFA* Δ*ldh* ClS1::P_{slp}Teth39_0206 (*adhE*)/(ura^-/$5FOA^R$)	[165]
	Lignocellulose solubilization	47% increased xylan solubilization	Δ*pyrFA* Δ*cbe1*::P_{slp}-Calkro_0402	[166]

(*continued overleaf*)

Table 14.3 (*Continued*).

Organism	Target	Titer or productivity	Strain genotype	References
Thermotoga sp. RQ2	Cellulose utilization	NA	pHX02: P_{TM1840}-Csac_1078 (*celB*) Kan^R pHX04: P_{TM0070}-Csac_1078 (*celB*) Kan^R pHX07: P_{TM0070}-Csac_1076 (*celA*) Kan^R	[167]
T. thermophilus HB27	Carotenoid	1.2 A_{460} mg^{-1} dry cell	HB27*pro4*::pCOP1 Crt^+ + *crtB*	[79, 81, 168]
T. ethanolicus JW200	Ethanol	9.4 mM	pTE16::*adhE* cm^R	[169]
T. mathranii	Ethanol	50.8 mM on glucose 51.2 mM on xylose	Δ*ldh*::*kan*^{R}	[111, 170–172]
		Δ*adhE* in strain AD4 abolishes ethanol production	AD1: Δ*adhA* *kan*^{R} AD2: Δ*adhB* *kan*^{R} AD3: Δ*bdhA* *kan*^{R} AD4: Δ*adhE* *kan*^{R}	[111]
		43.0 mM on glucose 57.0 mM on xylose	Δ*ldh*::P_{xylAB}-*adhE* *kan*^{R}	[111]
		38.5 mM on glucose 49.0 mM on xylose	Δ*ldh*::P_{xylAB}-*gldA* *kan*^{R}	[173]
		3.0 mM on glucose 14.7 mM on xylose	*gap-pgk*:: P_{xylAB}-*gldA* *kan*^{R}	[173]
Thermoanaerobacter sp. X514	β-Glucanase activity	NA	pIKM1::endo β-1,4-glucanase from *C. thermocellum* *kan*^{R}	[112]
	Ethanol	~50 mM Improved EtOH tolerance	pIKM1::*adh* Teth514_0145-0146 *kan*^{R}	[174]

Table 14.4 Genetic parts available for engineering of extreme thermophiles.

Organism	Promoters		RBS[a)]	Terminators[a)]	Selectable markers			Reporters		References
	Type	Gene[a)]			Type[b)]	Phenotype[c)]	Gene[a)]	Type	Gene[a)]	
P. furiosus	Constitutive	*slp* (PF1399)	*porγ* (PF0791)	*hpyA1* (PF1722)	S	Ura prototrophy	*pyrF* (PF1114)			[43, 47–50, 159–161]
	Cold induced	*cipA* (PF0190)	*slp* (PF1399)		S	Sim^R	*hmg* (PF1848)			
	Constitutive	*gdh* (PF1602)	*pep* (PF0043)		S	Agm prototrophy	*pdaD* (PF1623)			
	Inducible (non-sulfur conditions)	*mbh1* (PF1423)			S	Trp prototrophy	*trpAB* (PF1705-6)			
					CS	5FOA^R	Δ*pyrF*			
					CS	6MP^R	Δ*xgprt* (PF1950)			
T. kodakarensis	Constitutive	*gdh*	AGGTGG		S	Ura prototrophy	*pyrF* (TK2276)	Enzyme activity	*bglA* (TK1761 or TK1827)	[54–56, 175]
	Constitutive	*csg* (TK0895)			S	Trp prototrophy	*trpE* (TK0254)			
					S	Agm prototrophy	*pdaD* (TK0149)			
					S	Sim^R	*hmg* (TK0914 or PF1848)			
					CS	6MP^R	Δ*hgprt* (TK0664)			
T. onnurineus	Constitutive	*gdh* (TN0157)			S	Sim^R	hmg_{Pfu} (PF1848)			[57, 163, 176]
	Constitutive	*slh* (TN0413)								
	Constitutive	TN0510								

(*continued overleaf*)

Table 14.4 (*Continued*).

Organism	Promoters		RBS[a)]	Terminators[a)]	Selectable markers			Reporters		References
	Type	Gene[a)]			Type[b)]	Phenotype[c)]	Gene[a)]	Type	Gene[a)]	
T. barophilus					S	Sim^R	hmg_{Pfu} (PF1848)			[58, 59]
					CS	5FOA^R	Δ*pyrF*			
S. acidocaldarius	Constitutive	*dps* *lacS* *gdhA* *sac7d*			S	Ura prototrophy	*pyrEF* (SSO0615-0616)	Blue/white screen	*lacS* (SSO3019)	[67, 177, 178]
	Maltose inducible	*mbp*						Fluore-scence	*eCGP123*	
	Copper inducible	*copMA* + *copT*								
S. solfataricus	Heat inducible	*t5ffα*			S	Lactose catabolism	*lacS* and *lacTr*	Blue/white screen	*lacS*	[60–62, 64, 179, 180]
	Mercury inducible	*merA*			S	Alcohol^R	*adh*			
	Arabinose inducible	*araS*								
S. islandicus					S	Sim^R	*hmg* (ST1352)			[70, 72, 181]
					S	Ura prototrophy	*pyrEF* (SSO0615-0616)			
T. maritima	Constitutive	*tac* *slpA* *amyA* (TM1840) *xynB* (TM0070)			S	Kan^R				[74, 167, 182]

T. thermophilus	Constitutive	*slpA* (X57333)	*slpA* (X57333)	*slpA* (X57333)	S	Ura prototrophy	*pyrF*	Colony size	*mdh*	[81, 86, 89, 97, 183–185]
	Constitutive	J17		pheST (*T. flavus*)	S	Trp/Pro/Lys/Met/Leu prototrophy	NR	Blue/white screen	*bgl* (TTP0042)	
	Heat-shock inducible	*dnaK* (Y07826)			S	$Strep^R$	NR	Colony color	*crtB* (KC875824)	
	Arginine inducible	*arg* (Y18353)			S	Kan^R	*kat*			
	Succinate/malate inducible	*scs-mdh* (X54073)			CS	$5FOA^R$	Δ*pyrF/E*			
	Anaerobic inducible	*nar* (AJ504993)								
T. flavus					S	Ura prototrophy	*pyrE* (U27180)			[101, 102]
					CS	$5FOA^R$	Δ*pyrE*			
					S	Kan^R	*kat*			
C. bescii	Constitutive	*slp* (Cbes2303)		rho	S	Kan^R	*htk*			[41, 104, 165, 166]
	Constitutive	S30EA (Cbes2105)			CS	$5FOA^R$	Δ*pyrF*			
C. subterraneus subsp. *tengcongensis*	Constitutive	*pta* (TTE1482)			S	Kan^R	*kat*			[107, 186]
					S	His prototrophy	*hisG* (TTE2139)			

(*continued overleaf*)

Table 14.4 (*Continued*).

Organism	Promoters		RBS[a)]	Terminators[a)]	Selectable markers			Reporters		References
	Type	Gene[a)]			Type[b)]	Phenotype[c)]	Gene[a)]	Type	Gene[a)]	
T. ethanolicus JW200	Constitutive	*adhB*			S	Cm^R	*cat* (K01998)			[110]
T. mathranii BG1	Xylose inducible	*xylAB* from *T. ethanolicus*			S	Kan^R	*htk* (AB121443)			[87, 111, 187]
Thermoanaerobacter sp. X514	Undefined	β-Glucanase promoter from *C. thermocellum*			S	Kan^R	*kan*^{R} from pIMK1			[112, 188]
T. thermohydrosulfuricum					S	Kan^R	*kat* (M37273)			[109]
					S	Cm^R/Th^R	*cat* (K01998)			

a) Locus tags or accession numbers, when reported, are given in parentheses.
b) S, selection; CS, counterselection.
c) 5FOA, 5-fluoroorotic acid; 6MP, 6-methylpurine; Agm, agmatine; Cm, chloramphenicol; Kan, kanamycin; Leu, leucine; Mev, mevinolin; *p*-Cl-Phe, *p*-chlorophenylalanine; Pro, proline; Sim, simvastatin; Strep, streptomycin; Th, thiamphenicol; Trp, tryptophan; and Ura, uracil.

In addition to 3HP, *P. furiosus* has been engineered to produce lactate, ethanol, and *n*-butanol. Lactate production was achieved via expression of the lactate dehydrogenase from *C. bescii* under control of the cold-induced promoter P_{cipA} (Table 14.4), changing the fermentation end products from acetate to a mixture of lactate and acetate under low-temperature conditions [160]. Ethanol production by *P. furiosus* was likewise achieved by the insertion of a single gene, in this case an alcohol dehydrogenase (ADH) from *Thermoanaerobacter* sp. X514. The inserted ADH, together with the native *P. furiosus* aldehyde oxidoreductase (AOR), allowed conversion of acetate to ethanol [158]. Using this strain, up to 20 mM ethanol could be produced from maltose. Production of *n*-butanol was achieved by inserting a hybrid synthetic pathway assembled from several thermophilic bacteria for the conversion of the intermediate acetyl-CoA to *n*-butanol. This six-step pathway allowed production of *n*-butanol from maltose, but only in trace quantities, while greater quantities of ethanol were produced as an undesired side product of the pathway. Further examination of the pathway by *in vitro* enzyme characterization and kinetic modeling suggests that optimization of enzyme expression levels would greatly increase the selectivity for butanol production (Figure 14.4), providing further motivation for the development of tools to optimize expression levels in *P. furiosus* [132].

All of the examples using *P. furiosus* as a platform for chemical production take advantage of a temperature-controlled metabolic shift due to the difference between the optimum growth temperature of *P. furiosus* and temperature optima for the inserted enzymes. When *P. furiosus* is grown at its optimum temperature of 100 °C, the inserted pathway enzymes are not functional, as they come from organisms with growth temperatures from 60 to 75 °C. As a result, the product is not produced until the temperature drops to a point at which the enzymes are active [43, 159]. In the case of lactate production, an additional switch was employed through the use of the cold-induced protein promoter, which upregulates expression of the lactate dehydrogenase mRNA at low temperatures [160]. Repressing transcription at high temperatures prevents translation of the enzymes, decreasing production of unfolded heterologous protein that must be turned over and thus decreasing the metabolic burden on the host. A further possible advantage to using a temperature-controlled metabolic shift is that the host organism's native metabolic activity should be reduced at the suboptimal temperatures required for product formation, theoretically allowing more substrate to be directed to the product rather than used for biomass generation [189, 190]. However, transcriptional analysis of *P. furiosus* strains suggests that the high-level expression of the 3HP pathway results in increased energy metabolism compared to the parent strain, even at 72 °C. Thus, the expected state of significantly reduced metabolic activity at suboptimal temperatures may not be true for engineered strains, presumably a result of the increased burden of heterologous protein expression and metabolic activity stimulated by the enzymes corresponding to the heterologously expressed genes [149]. Balancing of heterologous pathway gene expression through the use of a

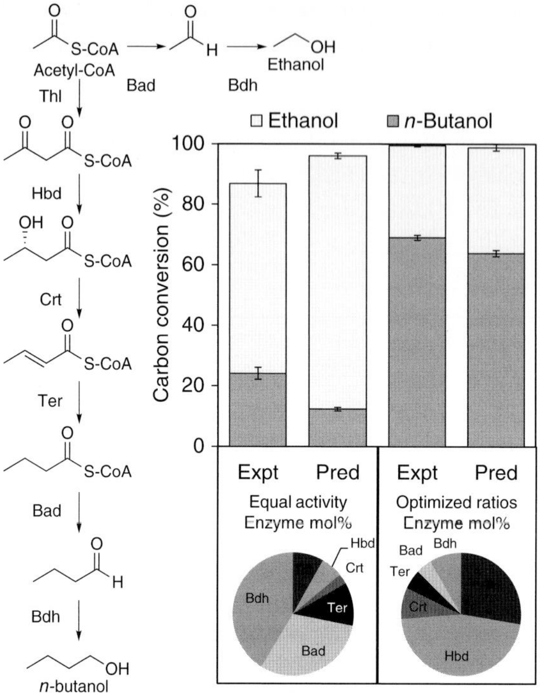

Figure 14.4 Modeling and optimization of a thermophilic *n*-butanol pathway *in vitro*. Enzymes from multiple thermophilic bacteria were assembled *in vitro* to measure conversion of carbon in acetyl-CoA to products. Reaction kinetic modeling was used to optimize the relative amount of each enzyme (as mol%) to maximize selectivity for *n*-butanol production. (Adapted from [132]; © American Society for Microbiology.)

range of promoters may better alleviate the burden on the native host metabolism at low temperatures.

P. furiosus has also been engineered to utilize non-native substrates. By repairing a disrupted chitinase, utilization of chitin as a carbon source was restored [50]. *P. furiosus* strains have also been engineered to oxidize formate and carbon monoxide by insertion of the formate hydrogen lyase (FHL) complex or carbon monoxide dehydrogenase (CODH) complex from *T. onnurineus*. Heterologous expression of the FHL required the insertion of an 18-gene cluster into the *P. furiosus* chromosome, allowing the organism to oxidize formate to CO_2 coupled with H_2 evolution [161]. However, unlike *T. onnurineus*, *P. furiosus* was unable to use FHL to conserve energy from formate oxidation. The successful insertion and expression of a large complex in *P. furiosus* demonstrates its utility as a potential metabolic engineering platform. Similarly, expression of the CODH complex in *P. furiosus* was obtained by inserting a 16-gene cluster into the chromosome, thereby allowing *P. furiosus* to utilize CO as an electron donor for the conversion of organic acids to their corresponding alcohols via the ADH/AOR pathway [158].

The genetic systems available for *P. furiosus* have been used to study several aspects of its physiology. These include hydrogen metabolism [47, 126], oxygen tolerance [145], sulfur metabolism [191], motility [122], compatible solutes [192], and acetoin production [193]. Taken together, the metabolic engineering and physiology studies indicate the relative ease of the current genetic systems for *P. furiosus* and promise significant future progress.

14.7.2
Thermococcus Species

The archaeal genus *Thermococcus* is closely related to *Pyrococcus*, and shares many of its attributes, with the main difference being lower optimum growth temperatures clustering around 85 °C as compared to 100 °C (Table 14.2). All species require seawater-based medium and are obligate anaerobes. Growth is heterotrophic, with the preferred substrate being peptides in the presence of sulfur, but some *Thermococcus* species are also capable of utilizing sugar and growing in the absence of sulfur [194]. Carboxydotrophic growth has been observed in *Thermococcus* sp. AM4 and *T. onnurineus*, although a continued requirement for organic substrates suggests that CO oxidation may provide energy, but not carbon [123, 127]. Genome sequences indicate carboxydotrophy may also be present in several other members of the genus, including the piezophilic *Thermococcus barophilus* [195]. *Thermococcus kodakarensis* (formerly *Pyrococcus* sp. KOD1 [124], then *T. kodakaraensis* [125]) is the best understood *Thermococcus* species, and it has been the subject of intense study for decades [196]. Its sugar metabolism has been thoroughly characterized [197], as have its systems of electron disposal to H_2S, H_2, and alanine, which mirror the systems found in *P. furiosus* [162].

Thermococcus species share many of the qualities that make *P. furiosus* a promising industrial host, but with the added benefit of more diverse metabolisms. CO-utilizing carboxydotrophic strains could serve as hosts for production of chemicals from syngas, or in microbial fuel cells [198]. In addition, several species, of which *T. onnurineus* is the most thoroughly examined, are capable of utilizing formate for growth [199]. Formate has been proposed as an electron source for biofuel production, overcoming the low solubility of hydrogen gas [44, 200].

14.7.2.1
Genetic Tools – *Thermococcus* Species

The genetic system for *T. kodakarensis* was among the first to be developed for a hyperthermophilic archaeon [52] and has served as a model for developing genetic systems in a variety of archaea. The first success relied on the use of a uracil auxotrophic mutant strain containing a random mutation in *pyrF*. Chromosomal DNA integration of either circular or linear DNA containing the *pyrF* marker gene was selected for through uracil prototrophy, and marker loss was counterselected with 5FOA [52] (Table 14.1). The transformation method, adapted from another archaeon [201], involved $CaCl_2$ and heat shock, although omitting $CaCl_2$ did not impede transformation. A later study reported the construction of a targeted deletion of the *pyrF* gene and also a deletion in the *trp* operon which allowed for tryptophan-based selection [53]. Since then, other selection methods have been applied in *T. kodakarensis*, including resistance to the antibiotic simvastatin [54], agmatine prototrophy, and resistance to 6-methylpurine [55].

The many developments in expanding the *T. kodakarensis* genetic toolbox have opened the door for recent successful genetic manipulations in other *Thermococcus* species, including *T. onnurineus* [57] and *T. barophilus* [58, 59]. The versatile genetic tools available for *Thermococcus* species have been used to elucidate many aspects of their physiology, most recently glycolysis [197], amino acid metabolism [202, 203], nucleotide metabolism [204], and compatible solutes [59, 205]. In addition, genetic tools have been used to study and engineer the CRISPR-Cas system of *T. kodakarensis* [206].

14.7.2.2
Metabolic Engineering – *Thermococcus* Species

The only metabolic engineering attempts with *Thermococcus* species so far have been to improve upon their natural ability to produce hydrogen (Table 14.3). *T. kodakarensis* generates H_2 during growth on sugars, and hydrogen production can be increased by deleting H_2-consuming hydrogenases [162]. Further improvements in *T. kodakarensis* hydrogen production have been realized by overexpressing the membrane-bound hydrogenase with a strong constitutive promoter [207].

The natural ability of *T. onnurineus* to produce H_2 from CO has been improved by overexpression of the operon encoding the CODH complex [57]. The application of ever stronger promoters continues to result in improvements in H_2 production [164, 176]. Overexpression of two transcriptional regulators of the CODH operon was also found to increase production [163].

14.7.3
Sulfolobus Species

The genus *Sulfolobus* is comprised of at least 12 species of thermoacidophilic archaea. These archaea are found worldwide in various sulfur-containing geothermal springs at temperatures of 55–87 °C and a pH range of 1.0–5.9 [34]. The most well studied of the *Sulfolobus* species are the ones for which genetic tools are available: Sulfolobus *solfataricus*, *S. acidocaldarius*, and *S. islandicus*. *S. solfataricus* and *S. acidocaldarius* grow as obligate aerobic organotrophs and can use both peptides and free amino acids as sole carbon and nitrogen sources. They differ in their catabolism of sugars, temperature/pH optima, and growth rates (Table 14.2) [130]. All three species are predicted to encode a branched Entner–Doudoroff pathway, a reverse Ribulose Monophosphate Pathway, and a complete oxidative TCA-cycle and are thought to generate CO_2 as the primary end product of sugar catabolism [131, 208–210]. *S. solfataricus* and *S. acidocaldarius* lack the ability to fix CO_2 and oxidize sulfur, but are predicted to encode many of the putative enzymes involved in these metabolic pathways. The fact that many other *Sulfolobus* species can fix CO_2 and oxidize sulfur suggests that this capability may have been lost in *S. solfataricus* and *S. acidocaldarius* as a result of repeated passages on laboratory medium containing organic materials [34]. Little is known about the metabolic capabilities and optimal growth parameters of *S. islandicus.*

The high temperature, low pH, and aerobic growth of the Sulfolobales make them attractive candidates for industrial use, as there is little risk of contamination and no need to purchase specialized equipment for anaerobic manipulation. All three *Sulfolobus* species have sequenced genomes making targeted genetic manipulation possible [208–210]. Despite this, only *S. acidocaldarius* is recommended as a metabolic engineering platform. This is because *S. solfataricus* and *S. islandicus* have been shown to possess numerous active mobile genetic elements [208, 210]. In fact, *S. solfataricus* P2 encodes 344 putative insertion elements, representing approximately 10% of its entire genome [211]. Furthermore, there is yet no agreed upon type strain for *S. islandicus.* However, several genetic markers, selection strategies, and plasmids have been developed for *S. solfataricus* and *S. islandicus* that may be useful for genetic manipulation of *S. acidocaldarius.*

14.7.3.1
Genetics Tools – *Sulfolobus* Species

Plasmid-based expression and homologous recombination of engineered DNA have both been achieved in *Sulfolobus* species (Table 14.1). In *S. solfataricus,* there are several selection strategies used to accomplish these goals: growth on lactose as a sole carbon source, alcohol resistance, uracil prototrophy, and hygromycin B resistance [60, 61, 64, 65, 212]. Most *E. coli/S. solfataricus* shuttle vectors are derived from either the genome of the *S. islandicus* virus SSV1 or the naturally occurring *Sulfolobus* plasmid pRN1 [213, 214]. In 2007, Berkner *et al.* made an *E. coli/S. solfataricus* shuttle vector derived from pRN1 called *pJlacS* that contains the *S. solfataricus lacS* gene [69]. Successful transformation of *S. solfataricus ΔlacS* (PBL2025) with pJ*lacS* was selected for by growth on lactose as a sole carbon source. With selection, pJ*lacS* was stably maintained at 2–3 copies per cell for over 100 generations. In 2003, Jonuscheit *et al.* developed a strategy for chromosomal insertion using a vector called *pJM03* [61]. Specifically, they cloned *S. solfataricus pyrEF* and *lacS* into an SSV1/pUC18 fusion vector and then transformed this vector (pMJ03) into a *pyrF/lacS S. solfataricus* double mutant (PH1-16) via electroporation, and used uracil prototrophy for selection. This vector was stably maintained in the *S. solfataricus* chromosome at the SSV1 integration site (at the 3′ end of the arginyl tRNA gene) for over 100 generations with selection. An additional strategy for chromosomal mutation was first employed in 2003, wherein the *S. solfataricus lacS* gene was inserted into the *S. solfataricus amyA* gene on a plasmid, and then inserted into the *S. solfataricus* genome via homologous recombination following electroporation with growth on lactose as the selective pressure [60].

In *S. acidocaldarius,* there is only one selection strategy that has been used for plasmid maintenance and homologous recombination. This strategy requires a *S. acidocaldarius pyrE* mutant background and utilizes uracil prototrophy for positive selection [67]. The first system was reported in 2007 and uses electroporation to transform a natural *S. acidocaldarius pyrE* mutant with a replicating vector [69]. The vector, pC, is an *E. coli*/*S. acidocaldarius* shuttle vector derived from pRN1. pC is stably maintained in *S. acidocaldarius* at 2–8 plasmids per cell when grown in the absence of uracil. Reported transformation efficiencies range from 10^2 to 6×10^4 cfu μg^{-1} DNA. Use of *pyrEF* for mutation of the *S. acidocaldarius* genome (knock-in and knockout) was first reported in 2012. *S. solfataricus pyrEF* was cloned into a replicating *E. coli* vector (e.g., pUC19) adjacent to the *S. acidocaldarius* sequence specific to the upstream and downstream regions of a gene targeted for insertion/deletion [67]. This plasmid was then transformed into a Δ*pyrE S. acidocaldarius* mutant (MW001) and the transformed cells were plated on defined medium in the absence of uracil to select for initial recombinants. Recombinants were subcultured in liquid medium in the presence of uracil and then plated on defined medium supplemented with uracil (10 $\mu g\, ml^{-1}$) and 5FOA (100 $\mu g\, ml^{-1}$) to select for loss of the integrated plasmid.

Important considerations for use of these genetics systems are timing for growth, background growth, and DNA methylation. Growth of *S. solfataricus* on lactose as a sole carbon source takes several days of outgrowth because the organisms do not grow well on carbohydrates [60, 113]. This increases the probability of generating secondary mutations and/or causing genomic rearrangements, an important concern in *S. solfataricus*. On the other hand, the primary issue with uracil prototrophic selection is uracil contamination [113]. Specifically, Gelrite, the solidifying agent used for making plates for *Sulfolobus* species, as well as many sources of peptides (e.g., tryptone), contains trace amounts of uracil. To prevent background growth of uracil auxotrophs in *S. acidocaldarius*, the *S. solfataricus lacS* gene has been used in combination with *pyrEF* [67]. This allows for blue-white screening of "prototrophic" *S. acidocaldarius* MW001 transformants; the background *S. acidocaldarius* LacS activity is too low to generate blue colonies, and thus there is no need to use a *lacS* mutant. Finally, *S. acidocaldarius* encodes an active *Sau*I restriction modification system targeting GGCC motifs [117, 215]. Thus, foreign DNA must be methylated before transformation into *S. acidocaldarius*. This is accomplished *in vivo*, using *E. coli* K12 ER1821 (New England Biolabs) bearing the plasmid pM.*Esa*BC4I [215]. DNA transformed into *S. solfataricus* does not require methylation.

Recently, a new selection strategy has been implemented for *S. islandicus* that may prove useful for the genetic engineering of *S. acidocaldarius* [72]. This strategy relies on the use of simvastatin, which acts as a competitive inhibitor of HMG-CoA reductase. Importantly, HMG-CoA reductase activity is required for *Sulfolobus* lipid biogenesis, which is essential for cell membrane formation. In 2012, simvastatin treatment was shown to inhibit *S. islandicus* growth. However, growth was restored via transformation of *S. islandicus* with a replicating vector containing the *Sulfolobus. tokodaii hmg* allele fused to the *S. acidocaldarius sac7d* promoter. Used to replace uracil prototrophic selection, simvastatin selection could greatly reduce the impact of background growth during genetic manipulation of *S. acidocaldarius*.

14.7.3.2
Metabolic Engineering – *Sulfolobus* Species

Metabolic engineering of *Sulfolobus* species has yet to be reported; however, there are many attributes that make *Sulfolobus* species attractive candidates for metabolic engineering. First, a genome-scale metabolic network has been published for *S. solfataricus* [131]. Second, a suite of inducible (*araS, mbp, copT, merA, tf55α*) and constitutive (*dps, lacS, slaA, gdhA, sac7d*) promoters of varying strengths have been used and characterized in these organisms (Table 14.4) [61, 62, 67, 177, 179]. Third, expression and tag-based purification of foreign metabolic genes has been accomplished in both *S. acidocaldarius* and *S. solfataricus*. Fourth, there are at least four experimentally verified secretion signals. Finally, *S. acidocaldarius* is predicted to encode no active transposable elements, and is known to be highly genetically stable. There are intriguing industrial applications for these organisms that would require metabolic engineering for either pathway completion or optimization. *S. solfataricus* encodes genes with high sequence homology to 14 of the 16 genes encoding enzymes for carbon fixation via the hydroxypropionate–hydroxybutyrate cycle, which could be completed and further engineered for the production of biofuels or biomolecules from carbon dioxide. *S. solfataricus* grows on phenol as a sole carbon source and could therefore be engineered for bioremediation and/or production of biomolecules from phenol (Figure 14.1). Lastly, other *Sulfolobus* species are known to oxidize and solubilize various metal species and could be engineered for biomining (Figure 14.1).

14.7.4
Thermotoga maritima

T. maritima is a gram-positive, anaerobic bacterium that was isolated from geothermally heated seafloors alongside Vulcano Island, Italy [138]. This bacterium grows at temperatures that range from 65 to 90 °C, optimally at 80 °C, and within a pH range of 6.5–7.0 [138, 139] (Table 14.2). Similar other members of the Thermotogales, *T. maritima* contains an outer envelope called the toga that serves as an environmental barrier [138, 216] and acts as a structure to display polysaccharide hydrolases [217, 218]. *T. maritima* is an obligate chemoorganoheterotroph, fermenting a wide range of simple sugars and complex polysaccharides including xylose, glucose, xylan, starch, barley glucan, carboxymethylcellulose, glucomannan, galactomannan, laminarin, and chitin [140–142, 219]. Absent in *T. maritima* is the ability to metabolize crystalline cellulose, owing to the lack of exoglucanases, endoglucanases, and carbohydrate binding module (CBM) domains [167, 220]. Most of the carbon *T. maritima* consumes (~85%) is directed to the Embden–Meyerhof–Parnas pathway, but it also has an active Entner–Doudoroff pathway for carbon metabolism [221, 222].

It can ferment peptides, but not as the sole carbon source [223]. Hydrogen (or hydrogen sulfide), acetate, and CO_2 are the primary fermentation products [224]. In addition to its fermentative metabolism, *T. maritima* may also respire by using Fe(III) [225], elemental sulfur [226], and thiosulfate as electron acceptors.

Growth rates for *T. maritima* can vary depending on monosaccharide- or polysaccharide-based growth, with average doubling times of 201 and 90 min, respectively, and final cell densities reaching up to 10^9 cell ml^{-1} [140]. Under respiration, growth conditions change in the presence of electron acceptors [225, 226]. For example, while ATP is generated when Fe(III) is the terminal electron acceptor [225], energy is not produced in the presence of sulfur [227]. However, the specific growth rate and final growth yield increases with added sulfur or thiosulfate under high hydrogen concentrations [226, 227], demonstrating that sulfur functions as an electron sink to aid fermentation rather than funneling electrons for the production of H_2.

Interest in *T. maritima* as a metabolic engineering candidate largely stems from its ability to produce hydrogen at high yields (3–4 mol H_2/mol glucose) [223]. Hydrogen is a promising and unique alternative energy source because it can be converted to mechanical energy with great efficiency, and the CO_2 generated during H_2 production can be captured, with no pollutants generated [228, 229]. As a marine organism, *T. maritima* uses seawater-based media, which has better economics than requiring copious amounts of freshwater for the production of fuels and chemicals.

14.7.4.1
Genetic Tools – *Thermotoga* Species

Transformation of *Thermotoga* species has been effective using liposome-mediated transformation [182], electroporation [74], and natural competence [75] (Table 14.1). While *Thermotoga.* sp. RQ7 is naturally competent [75], *T. maritima* has been successfully transformed using liposomes and electroporation [74, 182]. Initial attempts to develop a genetics strategy for *T. maritima* were moderately successful and exploited a natively occurring plasmid, called pRQ7, that was discovered in *Thermotoga* sp. strain RQ7 [230]. The plasmid (pRQ7) was modified to contain either a thermostable chloramphenicol acetyltransferase (cat) or kanamycin nucleotidyltransferase and was transformed into *Thermotoga neapolitana* and *T. maritima* cells, respectively, using liposome-mediated transformation [182]. Although resistance to the antibiotics was obtained, it was short lived in liquid media, and transformants could not be isolated from plates, indicating that the plasmid was unstable.

As demonstrated in a more recent and improved protocol by Han *et al.*, transformation was conducted under aerobic conditions at the bench [74] versus an anaerobic chamber [182] and utilized SVO medium [231] for cell culturing. For liposome-mediated transformation, spheroplasts were generated by harvesting early stationary-phase cells and suspending them in 20 mM HEPES buffer containing 300 μg ml^{-1} lysozyme at pH 7.4 [182]. Cells were incubated for 30 min at 37 °C followed by 5 min at 77 °C. Formation of spheroplasts was monitored using a microscope. To generate liposomes harboring plasmid DNA, a mixture of 1 μg DNA, HEPES buffer, and DOTAP was incubated at room temperature for 15 min. For transformation, the DNA-liposome mixture was added to the spheroplasts and incubated for 1 h at 37 °C, followed by recovery in complex medium for 3–6 h at 77 °C.

Transformation via electroporation was carried out by growing *Thermotoga* cells in SVO medium to an optical density of 0.2 at 600 nm (OD_{600}) [74]. Cells were washed in cold deionized (DI) water and then washed and suspended in a solution of 10% glycerol and 0.85 M sucrose. Plasmid DNA (4 μg) was mixed with a 50 μl cell suspension. The DNA-cell mixture was added to a prechilled cuvette, and an electronic pulse was applied. Following electroporation, cells were added to a volume of SVO liquid medium in an anaerobic N_2-containing serum bottle and incubated with slow rotation at 77 °C for a 3 h recovery. Half of the recovered culture was then mixed with hot SVO medium containing 3% Gelrite and 250 $\mu g\,ml^{-1}$ kanamycin, this mixture was poured in plates, and plates were then incubated anaerobically for 48 h, resulting in approximately 50% plating efficiency. This plating method differs from a previously described method [182], which used an overlay method where cells in a suspension of 0.3% Gelrite were poured on top of a bottom layer of medium containing 0.7% Gelrite. Once colonies were formed, they were transferred to liquid SVO medium containing 0.0075% Gelrite (soft SVO), which, by virtue of its viscous nature, restricts the mass transfer of O_2 into the medium [74]. Following 12–24 h of incubation, soft SVO cultures were transferred to liquid SVO using a syringe. The use of soft SVO cultures increased cell viability during transfer to liquid medium by minimizing cell exposure to oxygen and allowed the isolation of *Thermotoga* transformants from colonies, which was previously not possible.

Reported transformation efficiencies of liposome-mediated and electroporation methods for *T. maritima* were 5 and 0.25 transformants per microgram of DNA, respectively [74], and natural transformation efficiency for *Thermotoga.* sp. RQ7 was approximately 19 transformants per microgram of plasmid DNA [75]. These low transformation efficiencies might be improved by identifying relevant restriction-modification systems in *T. maritima* and appropriately methylating the plasmid DNA. Additionally, performing transformation steps in an anaerobic environment may improve efficiency but may also be logistically difficult. While the genetic strategies of *T. maritima* continue to evolve, development of nutritional auxotrophs should be considered as an additional selection method. Stability of transformed DNA should also be examined when planning metabolic engineering strategies. Transformed DNA in *T. maritima* was stable through 3 days after culture transfers every 12 h to fresh SVO medium containing kanamycin. Yet plasmid stability was shown to be less effective over time [74, 182]. Furthermore, antibiotic susceptibility is different for *T. maritima* (kanamycin) and *T. neapolitana* (chloramphenicol) [182], indicating antibiotic selection should be evaluated for any *Thermotoga* species under study.

14.7.4.2
Metabolic Engineering – *Thermotoga* Species

To date, there have been no published reports of metabolic engineering of *T. maritima* and only one metabolic engineering example in a *Thermotoga* species (Table 14.3). Xu *et al.* [167] heterologously expressed two *Caldicellulosiruptor saccharolyticus* cellulases, Csac_1076 (*celA*) and Csac_1078 (*celB*), in *Thermotoga.* sp. strain RQ2. These cellulases were fused to the promoter and signal peptide of TM1840 (*amyA*) and TM0070 (*xynB*) and transformed using the strain's natural competency, with the goal of providing *Thermotoga* sp. strain RQ2 the ability to degrade and utilize cellulose as a carbon and energy source. TM1840 and TM0070 are both secreted proteins in *T. maritima*, with TM1840 anchored on the "toga" and TM0070 fully secreted out of the cell and into the medium. When the promoter and signal sequences of these genes were used to control expression of Csac1076 and Csac1078, about half of the endoglucanase activity was localized to the secretome. The exoglucanase activity of CelB was functional under transcriptional control of both TM1840 and TM0070. However, no exoglucanase activity for CelA was detected, suggesting the exodomain of CelA was lost or not functional in *Thermotoga.* sp. strain RQ2. Ultimately, all transformants lost their recombinant genes after three consecutive transfers under kanamycin selection. Thus, while previous work demonstrated stability in *T. maritima and Thermotoga.* sp. strain RQ7 through 3 days, work in RQ2 was not stable, indicating the differences in genetics of the *Thermotoga* hosts. For future metabolic engineering efforts in *T. maritima*, techniques for chromosomal modifications will be needed to ensure stable expression and propagation of heterologous genes; to date, this has not been demonstrated.

14.7.5
Thermus Species

The *Thermus* genus consists of chemoorganotrophic, nonsporulating, halotolerant, gram-positive bacteria found in a variety of high-temperature environments, such as terrestrial and submarine thermal springs, compost piles, and even hot tap water [144, 232–236]. *Thermus* species can grow at temperatures ranging from 40 to 85 °C (optimum typically 62–75 °C) and pH values from 4 to 9 (optimum 7.0–8.0) (Table 14.2). These microbes are obligately aerobic, with the only exception thus far being *Thermus thermophilus* HB8 recorded to grow anaerobically on nitrate [150]. Members of the genus are capable of growth on sugars, peptides, and lipids [237], although some species oxidize sulfur [238] and even arsenic [239]. The genus also produces carotenoids, resulting in orange- or yellow-colored colonies [79, 144, 168, 240, 241].

For use in the biotechnology, *Thermus* species are already highly relevant. The first identified *Thermus* species, *T. aquaticus*, is well known for the widely commercially available Taq DNA polymerase first isolated from this organism [232, 242]. Additionally, *Thermus* species have acted as model organisms for understanding structural biology related to thermophily [243] and natural competence

[244–249]. As the insertion and expression of foreign genetic material has been well studied in *Thermus*, the concept of converting these microbes into high-temperature cell factories is becoming a reality [80, 85, 91–93, 234, 250–254].

14.7.5.1
Genetic Tools – *Thermus* Species

Extensive efforts have been directed at developing genetics tools for *Thermus* species, and have been reviewed elsewhere [234, 255, 256]. Despite the mature genetic tools available for the *Thermus* genus, very limited work to date has focused on metabolic engineering of *Thermus* strains. Here, an overview of genetic tools in *Thermus* is provided.

Since *Thermus* species are naturally competent at all stages of growth [81, 257], the main hurdle for genetic modification in these microbes was the identification and construction of selectable markers. Current genetic engineering efforts in *Thermus* species primarily focus on *T. thermophilus* strains. The natural competence of *T. thermophilus HB27* was first studied by the classical method showing that genomic DNA could be taken up to complement auxotrophic mutants [81]. Natural transformation was performed by incubating cultures with added DNA at the optimal growth temperature of 70 °C and selecting transformants by plating cells on minimal medium with incubation at 70 °C for 2–3 days. The transformation frequencies varied from 0.75% to 11.9%, depending on the mutant auxotrophic strain that was used. To initially investigate antibiotic selection, spontaneous streptomycin mutants were isolated in *T. thermophilus* HB27, as well as *T. thermophilus* HB8, *T. flavus*, *T. caldophilus* GK24, and *T. aquaticus* YTI [81]. The streptomycin resistance gene was then transformed into these species and selected for on solid medium containing 50 μg ml^{-1} streptomycin. Positive transformants were identified for each, although transformation frequencies differed for all species, with *T. aquaticus* YTI having the lowest frequency (0.064%). This study also demonstrated plasmid transformation into the *T. thermophilus* proline auxotroph strain, using a cryptic plasmid, pTT8, from *T. thermophilus* HB8. With no selective pressure, only 9 out of 1000 colonies contained the pTT8 plasmid. However, once inside the cell, the vector was stable for more than 20 generations. Other instances of plasmid transformation using only natural competence and auxotrophy are the individual deletions of the 3-isopropylmalate dehydrogenase (*leuB*) [258] and tryptophan synthetase (*trpBA*) genes [82], resulting in leucine/isoleucine and tryptophan auxotrophy, respectively.

Once a basic level of genetic insertion in *Thermus* species had been demonstrated, more complex selection methods were developed. As with many extreme thermophiles, uracil auxotrophy by deleting/disrupting *pyrE* or *pyrF* with counterselection on 5FOA has been used, first in *T. flavus* [101] and subsequently in *T. thermophilus* strains [76]. Aside from being utilized in a target gene replacement-deletion system [77], the *pyrE* deletion was also generated in a previously engineered *T. thermophilus* Δ*leuB* strain, with transformation efficiencies of the now dual-selection mechanism (Δ*leuB* Δ*pyrE*) being 4×10^5 or 6×10^5 transformants per microgram of DNA for *leuB* or *pyrE* selection, respectively [78].

No native antibiotic resistance markers have been found in *Thermus* species to date [81, 94], but a high-temperature kanamycin resistance gene (encoding kanamycin nucleotidyltransferase, *knt*), able to function at 65 °C, was generated via passage through an *E. coli* mutator strain followed by selection in *Bacillus stearothermophilus* at temperatures up to 70 °C [259]. The first demonstration of using this gene as a marker in a *Thermus* species was the insertion of *knt* into the middle of the highly expressed surface layer protein A (slpA) gene of *T. thermophilus HB8*, disrupting expression and causing a defective S-layer on the cells [86]. Kanamycin resistance has also been utilized as a chromosomal integration system in *T. flavus* [102]. Reported transformation efficiencies were 4×10^7 transformants per microgram of DNA, and the genetic modification was stable even in the absence of selective pressure. The thermostability of the *knt* gene has been further improved by directed evolution in *T. thermophilus*. This resulted in a protein having 19 amino acid substitutions and stability up to 80 °C, which was aptly renamed HTK for "Highly Thermostable Kanamycin nucleotidyltransferase" [87]. Although not used as extensively as kanamycin, streptomycin has also been utilized as a selection in *T. thermophilus*, and the resulting mutants were studied to further understand the mechanisms of bacterial antibiotic resistance [260–263].

While antibiotic resistance is the most common method of transformation selection, different methods of genetic selection continue to be investigated. A recent counterselection method that has been previously demonstrated in mesophilic systems and recently applied in *Thermus* involves a mutated phenylalanyl-tRNA synthetase gene (*pheS*) and the amino acid analog *p*-Cl-Phe [90]. The mutant version of PheS incorporates *p*-Cl-Phe into cellular proteins, resulting in unviability of the cell. Antibiotic selection with kanamycin was utilized to integrate a mutant *pheS* marker into the chromosome, and *p*-Cl-Phe was used as a counterselection to remove both markers. This method can be used to construct gene deletions, point mutations, and/or insertions resulting in a final, unmarked mutant strain. Another counterselectable deletion system was also developed in *T. thermophilus* HB27, employing a heterologous cytosine deaminase gene (*codA*) from a *Thermoanaerobacter* species. The presence of the *codA* gene confers sensitivity to the compound 5FC; therefore, kanamycin was used as the selection for chromosomal integration of the construct and 5FC was used for counterselection for loss of the markers [100]. These two counterselection methods have advantages in that they can be used in a wild-type strain without previous construction of a chromosomal deletion in order to be able to apply the counterselection (as compared with the *pyrF* or *pyrE* markers, for example).

With its high transformation frequency and well-established genetic technologies, even more sophisticated tools are being applied in *T. thermophilus*. Transposon mutagenesis was performed in *T. thermophilus* HB27 via natural transformation of genomic DNA mutagenized *in vitro* via the EZ-Tn*5* system using a transposon engineered to contain the *kat*-selectable marker [88]. Although the number of transformants was only 186, this first application of the transposon mutagenesis method in an extreme thermophile holds promise for future study in this and other organisms. Several promoters have been characterized in *Thermus* species, both inducible and noninducible [95, 183]. In addition, a variety of reporter genes have been demonstrated [184, 264, 265].

One barrier to metabolic engineering of *Thermus* species is the presence of overlapping genes, such as those present in *T. thermophilus* [90], which makes accurate genome editing nontrivial. Efforts to combat these issues have been made [96], and recent available genome sequence data should improve the current understanding of the *Thermus* genomes [237, 240, 243, 266–271].

14.7.5.2
Metabolic Engineering – *Thermus* Species

Despite the well-established genetic systems in *Thermus* species, specifically *T. thermophilus*, the number of metabolic engineering efforts for this genus has been limited. Typically, *T. thermophilus* has been used as a cell factory, designed to express proteins. Many of the initial efforts focused on expressing proteins from other *Thermus* species, as with the cloning and expression of α- and β-galactosidase genes from *Thermus* strain T2 [80], secreted aqualysin I from *T. aquaticus* [250], and his-tagged thermozymes and catalases from different *T. thermophilus* strains [91, 251]. Since then, expression of multiple heterologous genes in *T. thermophilus* has been completed [85, 252, 253], and *T. thermophilus* can serve as a host for the thermoadaptation of proteins encoded in genes from mesophiles [92, 93, 254].

A few efforts have been directed at modifying the metabolism and physiology of *T. thermophilus*. Although defined as an obligate aerobe, *T. thermophilus* HB8 had been previously shown to grown anaerobically in the presence of nitrate [150]. A different strain, *T. thermophilus* HB27, does not share this ability, and as such the nitrate reductase complex operon (*nar*) was transferred from *T. thermophilus* to HB8 via conjugation, successfully converting the strain into a facultative aerobe [98]. A very similar method was also utilized to transfer a denitrification process to *T. thermophilus* HB27 [99]. In terms of true engineering of products generated by *Thermus*, carotenoids are produced in several *Thermus* species [168, 241], and these genes were targeted for overproduction in *T. thermophilus* [79] (Table 14.3). First, a replicating plasmid encoding kanamycin resistance and a phytoene synthase gene (*crtB*) from *T. thermophilus* HB27 was transformed into *T. thermophilus* HB27; integration into the chromosome was also attempted but was unsuccessful. The presence of the vector increased carotenoid production from the host strain by over 20-fold compared to the wild-type strain. For future metabolic engineering, more information is required about the carotenoid biosynthesis pathway, as it is likely that an unknown regulatory gene is involved [79].

Aside from the cases mentioned above, *Thermus* metabolic engineering efforts have been limited. Although the native products (e.g., carotenoids) in *Thermus* species have not been an area of great interest, *Thermus* species can serve as effective hosts for expression of thermophilic proteins. Furthermore, *T. thermophilus* has the potential to be engineered with multiple genes [253] for *in vivo* construction of metabolic pathways. Expressing mesophilic proteins in thermophilic systems can be nontrivial, which is why improving the understanding of protein thermostability is crucial [92, 93, 254].

14.7.6
Caldicellulosiruptor bescii

C. bescii DSM 6725 is a gram-positive, anaerobic bacterium that was isolated from a geothermal hot spring in Kamchatka, Russia over two decades ago [133, 134]. Along with other species of the *Caldicellulosiruptor* genus, *C. bescii* is cellulolytic, fermenting simple and complex polysaccharides to predominantly H_2, CO_2, acetate, and lactate [36, 134, 135] (Table 14.2). This extremely thermophilic bacterium grows at temperatures that range from 65 to 90 °C, optimally at 75 °C [36, 134], and within a pH range of 6.8–7.2 [136, 137]. The generation time during exponential growth for *C. bescii* is highly dependent on the medium. Cellobiose, cellulose, and switchgrass are primarily used as substrates in media when studying *C. bescii*, resulting in generation times that can range from 120 min on complex medium to 960 min on switchgrass [41, 136, 272]. Other factors that affect growth of *C. bescii* are whether growth occurs in closed serum bottles or bioreactors. Closed serum bottles lead to the buildup of organic acids and hydrogen gas, which are inhibitive to fermentation and cellulose degradation [137, 273]. Cell densities in serum bottles can reach over 5.0×10^8 cells ml^{-1} on cellulose and switchgrass [272]. Growth in pH-controlled bioreactors is similar to that found in serum bottles with maximum cell densities reaching approximately 5.0×10^8 cells ml^{-1}, but yielding fermentation products (acetate and lactate) that are several times higher, indicative of a more favorable bioreactor environment [137, 272].

To date, *C. bescii* is the most thermophilic cellulolytic bacterium known and has attracted a lot of interest as a metabolic engineering host in the biofuels field due to its unique ability to degrade unpretreated plant biomass and metabolize both C_5 and C_6 sugars. These rare traits make *C. bescii* an excellent candidate for consolidated bioprocessing (CBP) in the production of biofuels, and from an industrial perspective, *C. bescii* offers several distinct economic advantages over mesophilic ethanologens. For example, high fermentation temperatures (>70 °C) eliminate any potential contamination by lactic acid bacteria, which are common contaminants of yeast-based fuel ethanol fermentations [274]. An additional cost-saving measure is that, as an anaerobe, *C. bescii* does not require aeration or vigorous agitation [36].

14.7.6.1
Genetics Tools – *Caldicellulosiruptor bescii*

Since the time of its reclassification in 2010 (formerly *Anaerocellum thermophilum*) [134], numerous advancements in *C. bescii* genetics have made genetic modification and engineering of this thermophile possible [115, 116, 136] (Table 14.1). Currently, electroporation is the only method used to successfully transform *C. bescii* for genetic modification. With that in mind, there are primarily two strains used for genetic modification: strain JWCB005 and JWCB018. JWCB005 (Δ*pyrFA*) is an auxotrophic mutant that contains a random deletion in the *pyrFA* locus for selection of transformants through complementation via the *pyrF* gene [105]. Transformation of JWCB005 requires methylation of vector DNA with the methyltransferase M.CbeI to prevent the native CbeI restriction enzyme from degrading the transformed plasmid DNA. On the other hand, JWCB018 (Δ*pyrFA* Δ*cbeI*) was developed to circumvent this step through chromosomal deletion of *cbeI*, permitting the use of recombinant vector DNA that is directly harvested from *E. coli* [103]. Transformation of *C. bescii* may be carried out using either a replicating shuttle vector or a nonreplicating/integrating vector, with both systems containing the *pyrF* gene as a selectable marker [103, 105]. Nonreplicating vectors contain homology regions specific for the 5′- and 3′-DNA flanking regions of the gene targeted for deletion, or for the genomic DNA region targeted for gene insertion.

To transform *C. bescii* via electroporation, mid-log-phase cells were harvested and washed several times in 10% sucrose. Plasmid DNA (0.5–1.0 μg) in a volume of DI water, no larger than 5 μl, was then added to a concentrated 50 μl cell suspension in 10% sucrose [115, 166, 275]. The plasmid-cell mixture was added to an electroporation cuvette and a single electric pulse at defined parameters was applied. Electroporated cells were transferred to recovery medium (LOC) at 75 °C for 1–2 h and then transferred to low osmolarity defined medium lacking uracil (LOD_{-ura}) for selection of transformants [136, 166, 275]. Selective outgrowth cultures were incubated at 70–75 °C for up to 72 h, and those with growth were transferred several times in LOD_{-ura} or plated on LOD_{-ura} to select individual colonies. Transformation efficiency for integrating plasmids was reported to be on the order of 50 transformants per microgram of plasmid DNA [103], but was later improved to 10^3 transformants per microgram [103], likely aided by the development of an optimized growth medium that was described to have increased plating efficiency by ~1000-fold [136]. Transformation efficiency of a stably replicating plasmid was 500 transformants per microgram of DNA [105]. For knockout/knock-in vectors, any confirmed successful first crossover transformants were counterselected for marker removal by plating on solid medium containing both uracil and 5FOA for 3–4 days. Colonies of strains containing gene knockouts or knock-ins were confirmed with PCR amplification and were subcultured two to three times on solid medium containing uracil to achieve strain purity [166].

Recently, the *htk* gene developed in *T. thermophilus* discussed above, has been shown to function in conferring kanamycin resistance to *C. bescii*, which is otherwise sensitive to kanamycin [104]. A codon-optimized version of the *htk* gene was used successfully in both replicating and integrating vectors for transformation of *C. bescii*. Furthermore, it was used to construct a targeted deletion of the *pyrE* gene in wild-type *C. bescii*, generating a uracil auxotroph that was resistant to 5FOA. The *pyrE* gene was then demonstrated to function as a selectable marker akin to *pyrF*, resulting in a dual marker system whereby chromosomal integration was selected for via kanamycin resistance and plasmid loss was counterselected via 5FOA resistance. Also, the kanamycin resistance marker was used as a tool to directly select for chromosomal mutations, effectively replacing a chromosomal target with the gene itself, resulting in a strain that is resistant to kanamycin. This strategy will be useful for constructing deletions that cause reduced fitness levels.

While progress on genetic tools for *C. bescii* continues, several issues remain. The lack of promoter selection (including inducible promoters), reporter genes, and additional counterselection methods presents a major difficulty for the use of *C. bescii* for metabolic engineering. These issues may be combated by adapting promoters, reporter genes, and counterselection methods demonstrated in other thermophilic bacteria, such as *Thermus* and *Thermoanaerobacter*. Low transformation efficiency makes it difficult to develop strains in a time-efficient manner. Furthermore, a practical issue that should be considered when engineering *C. bescii* is the presence of transposons (mobile DNA elements), which may insert themselves into functional genes and disable them [276]. Finally, it would be useful to have genetic tools available for other species of *Caldicellulosiruptor*, as each species possesses a different suite of enzymes for biomass deconstruction [277]. Recently, *C. bescii* genetics tools were extended for use in *C. hydrothermalis* [106].

14.7.6.2
Metabolic Engineering – *Caldicellulosiruptor bescii*

While development of genetic tools and strategies continues to progress for *C. bescii*, there are a number of examples in the literature detailing metabolic engineering efforts in this organism (Table 14.3). Cha *et al.* [135] increased H_2 production in *C. bescii* through deletion of the L-lactate dehydrogenase gene (ldh). In *C. bescii*, carbon flux is directed to lactate or acetyl-CoA with electrons routed to either lactate or H_2 and ATP produced during the production of acetate via acetate kinase. Thus, by removing *ldh* from the genome, acetate and H_2 levels increased together with the culture density, likely due to the increase in energy production.

In addition to biohydrogen, engineering *C. bescii* to produce appreciable amounts of fuel ethanol from lignocellulosic plant biomass is a critical step in the movement toward a CBP strategy of second-generation biofuels production. *C. bescii* and other *Caldicellulosiruptor* spp. are known to produce trace amounts of ethanol [133, 134, 272, 278] along with H_2, CO_2, acetate, and lactate. To increase ethanol titer, the *ldh* deletion strain (JWCB018) was further modified to contain the *C. thermocellum* bifunctional acetaldehyde/ADH under the transcriptional control of the *C. bescii* S-layer protein promoter (P_{slp}), converting acetyl-CoA to acetaldehyde followed by further reduction to ethanol [41]. In this new strain (JWCB032), acetate levels decreased by 38%, and ethanol concentrations reached 12.8 mM, well above 0.17 mM ethanol as reported for wild-type organisms [272] and well below ethanol concentrations that start to inhibit growth (400 mM) [41]. The growth rate and growth yield between wild-type, Δ*ldh*, and Δ*ldh::adhE* strains were similar, abating any concerns of poor growth due to redox imbalance from importing a new fermentative pathway.

Fermentation of switchgrass by *C. bescii* strain JWCB032 was done at 65 °C owing to the thermostability limitations of the *C. thermocellum adhE*, yet the optimum growth temperature for *C. bescii* is 75 °C. *Thermoanaerobacter pseudoethanolicus* 39E is an anaerobic thermophile that produces ethanol at 70 °C, thus the *adhB* and *adhE* genes from *T. pseudoethanolicus* 39E were introduced separately into strain JWCB018 under transcriptional control of P_{slp} [165], and substrates were fermented at 75 °C. The AdhB- and AdhE-expressing strains produced 0.4 and 1.6 mM ethanol on switchgrass, respectively, with slightly higher levels on Avicel (1.4 and 2.3 mM, respectively). These ethanol concentrations were not nearly as high as those produced with the *C. thermocellum* AdhE-expressing strain. It was hypothesized that the lower ethanol concentrations observed is due to the limitation of NADPH (the cofactor for *T. pseudoethanolicus* AdhB and AdhE) availability in *C. bescii* relative to NADH (the cofactor for *C. thermocellum* AdhE). Taken together, efforts to engineer *C. bescii* to produce biofuels at growth optimum temperatures and at economical levels are still relatively in their early stages and are continuing to improve [279, 280].

Metabolic engineering strategies in *C. bescii* will improve as further insights into its native metabolic pathways are obtained. Recent work revealed the presence of a gene cluster involved in the uptake and utilization of tungsten [275]. The ability of the gene cluster to support tungstate uptake and incorporate it into a tungstoenzyme was tested by heterologously expressing the tungsten-requiring the aldehyde-oxidizing enzyme AOR from *P. furiosus* in *C. bescii*, creating strain MACB1002 [275]. When *C. bescii* strain MACB1002 was grown in the presence of tungsten, AOR was shown to contain tungsten, thus demonstrating an active tungsten utilization pathway.

C. bescii engineering strategies for the production of fuels and chemicals on lignocellulosic plant biomass should not be limited to metabolic pathways, but must encompass a comprehensive approach that also includes improving plant biomass deconstruction. Surface layer homology (SLH) glycoside hydrolases are key enzymes involved in the degradation of plant biomass and may play an important role in the attachment of *Caldicellulosiruptor* spp. to insoluble plant substrates [281–283]. The gene encoding an SLH containing xylanase from *C. kronotskyensis* (Calkro 0402) was inserted into the *C. bescii* genome under the transcriptional control of the P_{slp} promoter [166]. The engineered strain (RKCB103) was shown to attach and degrade xylans more extensively than wild-type *C. bescii* [166].

Importing cellulose- and hemicellulose-degrading enzymes together with altering metabolic pathways demonstrates the need for a multidimensional approach when improving the ability of *C. bescii* to degrade and convert lignocellulosic substrates into fuels and chemicals. Moreover, while improvement of the genetic systems available for *C. bescii* must continue, future work may entail improving product tolerance (e.g., ethanol) as well as taking a look at phenomena of potential significance (i.e., quorum sensing signaling, cyclic dinucleotide signaling, chemotaxis, and motility) that may be important to the overall goal of developing *C. bescii* into a robust CBP organism.

14.7.7
Thermoanaerobacter Species

Thermoanaerobacter species are heterotrophic fermentative anaerobic bacteria that grow on a variety of pentose and hexose saccharides as well as starch, but are not cellulolytic (Table 14.2). The growth medium requires the addition of yeast extract, which cannot be replaced by tryptone or casein hydrolysate, but yeast extract alone without glucose does not support growth [152]. *Thermoanaerobacter* species grow at temperatures from 40 to 78 °C, with an optimum of approximately 70 °C, and at pH ranging from 4.4 to 9.8 with an optimum near pH 7 [152, 153]. The major fermentation products produced by *Thermoanaerobacter* include ethanol, CO_2, H_2, acetate, and lactate. The doubling time of these species is slightly longer than 1 h [153]. *Thermoanaerobacter* is distinct from the genus *Thermoanaerobacterium* in that *Thermoanaerobacter* species grow at slightly higher temperatures and reduce thiosulfate to hydrogen sulfide as opposed to *Thermoanaerobacterium* species, which reduce thiosulfate to elemental sulfur [284]. The most well-studied *Thermoanaerobacter* species in relation to metabolic engineering include *Thermoanaerobacter ethanolicus*, *T. mathranii*, and *Thermoanaerobacter.* sp. X514. Many *Thermoanaerobacter* species have completely sequenced genomes [285].

The primary attribute of *Thermoanaerobacter* species that makes them of interest as an industrial host is the ability to produce ethanol at high yield from sugars [154]. Various alcohol dehydrogenaseADHs have been studied from and manipulated in *T. mathranii* [111] and *T. ethanolicus* [169]. Metabolomics and transcriptomic approaches have been applied to *Thermoanaerobacter.* sp. X514 to elucidate central metabolism pathways [286] and ethanol-tolerance mechanisms [174]. While these bacteria are not cellulolytic, coculture strategies with cellulolytic organisms such as *C. thermocellum* have been proposed [287]. Strategies for using *T. mathranii* strain A3N for the production of H_2 have also been investigated [288].

14.7.7.1
Genetic Tools – *Thermoanaerobacter* Species

Various methods for transforming DNA into *Thermoanaerobacter* species have been investigated over the past 30 years (Table 14.1), all of which employ antibiotic resistance selectable markers for kanamycin, chloramphenicol, or thiamphenicol. A method of Tris-PEG6000 incubation was reported in 1985 to successfully transform *T. thermohydrosulfuricus* to confer kanamycin or thiamphenicol resistance with selection on plates for 4–6 days at 55 °C [109]. Electroporation of partial protoplasts was used to transform *T. ethanolicus* JW200 with both a replicating and integrating vector with chloramphenicol selection at 45 °C for 5–7 days or 63 °C for 2–3 days [110]. *Thermoanaerobacter.* sp. X514 has been transformed using either sonoporation or electroporation with kanamycin selection on plates at 45 °C for 5–7 days [112]. Natural competence was identified for *T. ethanolicus* JW200, *T. pseudethanolicus* 39E, and *T. brockii* ATCC35047 using a replicating vector with a kanamycin resistance marker at 55 °C [108]. In the same study of natural competence, *T. mathranii* was not transformed, but this does not rule out that this species may be naturally competent under other conditions [108]. *T. mathranii* BG1 has been transformed by electroporation with kanamycin selection at 70 °C [173].

Replicating vectors for *Thermoanaerobacter* sp. have been constructed using the replicative origins of vectors found natively in or developed for other gram-positive bacteria (*Bacillus subtilis* plasmids pUB110 and pIM13, *Thermoanaerobacterium saccharolyticum* plasmid B6A-RI, and *Clostridium perfringens–E. coli* shuttle vector pJIR715) [108–110, 112]. Homologous recombination of integrating vectors has also been used to delete or add genes to *Thermoanaerobacter* sp. [110, 111, 173, 174]. The only method of selection used to date in *Thermoanaerobacter* sp. is antibiotic resistance to kanamycin, chloramphenicol, or thiamphenicol. The relatively lower growth temperatures of *Thermoanaerobacter* species compared to other extreme thermophiles enables the use of antibiotics and antibiotic resistance genes that may be heat labile at higher temperatures. As with all extreme thermophiles, the stability of both antibiotics and antibiotic resistance proteins at elevated temperatures is of concern, and thus selection is typically performed at a temperature well below the growth optimum of extreme thermophiles. Chloramphenicol/thiamphenicol selection in *Thermoanaerobacter* spp. has used the *cat* from *B. subtilis* plasmid pC194 at temperatures ranging from 45 to 63 °C [109, 110]. In *B. stearothermophilus*, chloramphenicol resistance via the pC194 *cat* is maintained to the maximum growth temperature of *B. stearothermophilus* at 68 °C [289], suggesting it could be used at higher temperatures in *Thermoanaerobacter*. Previously used kanamycin resistance markers necessitated lower cultivation temperatures for selection (45–55 °C), but the development of the HTK resistance marker [87] and recent use in *T. mathranii* BG1 at 70 °C [111, 173] demonstrates kanamycin selection can be achieved at the growth temperature optimum of *Thermoanaerobacter*. While positive selection using antibiotics is an important selection tool, the lack of a counterselection mechanism, such as 5FOA counter-selection of *pyrF* used in other genetic systems, is a limitation of the genetic toolbox in *Thermoanaerobacter* sp. that hinders the ease of performing multiple genetic manipulations in a single strain.

Restriction methylation systems in *Thermoanaerobacter* species are largely uncharacterized, but are predicted in the genomes of many species in the REBASE database [114]. In one investigation, *Thermoanaerobacter.* sp. X514 cell extracts appear to possess a restriction enzyme, putatively identified as TspX514ORF458P, capable of cutting unmethylated GATC sites similar to DpnII, and Dam^+/Dcm^+ methylation from *E. coli* 5α protects plasmid DNA from digestion, suggesting this methylation is important for transformation of *Thermoanaerobacter.* sp. X514 [112].

14.7.7.2
Metabolic Engineering – *Thermoanaerobacter* Species

Metabolic engineering in *Thermoanaerobacter* species has been almost exclusively focused on elucidating the function of ADHs and enhancing ethanol production and tolerance (Table 14.3). The majority of metabolically engineered strains have been produced in patented strain *T. mathranii* BG1 [290] by the company BioGasol (www.biogasol.com). *T. mathranii* BG1L1 is a Δ*ldh* mutant of *T. mathranii* BG1 which has been adapted to ethanol concentrations of 6–8.3% [170] and is able to produce 0.39–0.42 g ethanol per gram glucose consumed from dilute acid corn stover hydrolysate [171] and wet-exploded wheat straw hydrolysate [172]. A set of ADH deletion mutants (AD1: Δ*adhA*, AD2: Δ*adhB*, AD3: Δ*bdhA*, and AD4: Δ*adhE*) were constructed in *T. mathranii* BG1 to elucidate that *adhE* is the primary aldehyde dehydrogenase in the ethanol formation pathway, where mutant strain AD4 (Δ*adhE*) no longer produced ethanol [111]. Another strain was developed to simultaneously delete *ldh* and insert *adhE* under the control of the *xylAB* xylose-inducible promoter. This strain, *T. mathranii* strain BG1E1, was able to produce elevated levels of ethanol (57.0 mM ethanol, 95.2% of the theoretical maximum) when xylose was used as a substrate relative to both strain BG1L1 (Δ*ldh*; 51.2 mM ethanol) or wild-type (49.4 mM ethanol) [111]. Deletion of the *ldh* in these strains eliminates an NADH oxidation pathway, disrupting redox balance; so the *T. mathranii* strain BG1G1 was constructed to heterologously express *gldA* (a NAD^+-dependent glycerol dehydrogenase) from *T. maritima* while simultaneously deleting *ldh* to facilitate redox balance through NADH regeneration [173]. Strain BG1G1 employed the *xylAB* promoter to drive expression of *gldA* in a xylose-inducible manner. Another strain, BG1G2, was also constructed to incorporate *gldA* as part of the *gap-pgk* locus to express it constitutively in a strain with *ldh* intact [173]. Fermentation from batch cultures of strains BG1G1, BG1G2, and BG1L1 compared to wild-type BG1 were conducted on glucose and xylose [173]. On glucose, BG1G1 produces elevated levels of ethanol (38.5 mM) compared to wild-type (33.4 mM), but BG1L1 produces the most ethanol (43.6 mM); however on xylose, BG1G1 (49.0 mM ethanol, 94.5% theoretical maximum) outperforms both BG1L1 (40.0 mM ethanol) and wildtype (34.3 mM ethanol). Strain BG1G2, which has an intact *ldh* and constitutively expressed *gldA*, produces severely decreased levels of ethanol (3.0 mM on glucose and 14.7 mM on xylose) and elevated levels of H_2 (up to 37.1 mM on xylose) and lactate (up to 41.0 mM on glucose).

Metabolic engineering in *Thermoanaerobacter.* sp. X514 has also been pursued to improve ethanol production and tolerance. Ethanol shock and adaptation experiments with exogenous ethanol and transcriptomic profiling of the resulting strains identified changes in *Thermoanaerobacter.* sp. X514 genes responsible for improved ethanol productivity and tolerance. Using these findings, *Thermoanaerobacter.* sp. X514 *adh* and NADH oxidase genes (Teth514_0145-0146), and separately, *Thermoanaerobacter.* sp. X514 σ^{24} factor and hypothetical protein genes (Teth514_1847-1848), were overexpressed from the replicating vector pIKM1 using the native promoter, RBS, and terminator elements from each [174]. Both strains showed improved ethanol production over the wild type (~38 mM), with the σ^{24} overexpression strain producing approximately 46 mM ethanol and the *adh* overexpression strain producing approximately 50 mM ethanol. Tolerance and growth in 1% exogenous ethanol was also improved in both strains, with the σ^{24} overexpression strain growing even more rapidly than the *adh* overexpression strain. These experiments were carried out at 45 °C, which is significantly below the growth optimum of this species, but these genetic manipulations demonstrate key genes for improving ethanol tolerance and productivity in *Thermoanaerobacter.* sp. X514.

Only one example of metabolic engineering in *T. ethanolicus* JW200 is available in which *adhE* was homologously expressed using the putative natural promoter and terminator [169]. This improved the ethanol production of the strain (9.4 mM) compared to the wild type (6.7 mM).

14.7.8 *Caldanaerobacter subterraneus* subsp. *tengcongensis*

Caldanaerobacter subterraneus subsp. *tengcongensis* was isolated from a hot spring in Tengcong, China, and initially classified as *Thermoanaerobacter tengcongensis* [156]. The genus *Caldanaerobacter* was established due to differences in the metabolic and phylogenetic characteristics of the *Caldanaerobacter* sp. and *Thermoanaerobacter* sp. [291]. *Caldanaerobacter* have slightly higher growth temperatures, ranging from 40 to 85 °C with an optimum of 75 °C for *C. subterraneus* subsp. *tengcongensis* (Table 14.2). *Caldanaerobacter* sp. grow in pH 5.5–9 with optimum of 7–7.5 [291]. *C. subterraneus* subsp. *tengcongensis* grows on various hexose sugars, disaccharides, and starch, but is not able to degrade cellulose or xylan [156]. Similar to *Thermoanaerobacter* sp., *Caldanaerobacter* sp. are reported to require yeast extract for carbohydrate metabolism, but yeast extract cannot serve as a sole carbon source [156, 291]. However, a synthetic defined medium lacking yeast extract was recently used for genetic manipulation of *C. subterraneus* subsp. *tengcongensis* [107]. *Caldanaerobacter* species produce ethanol, acetate, CO_2, H_2, and L-alanine as major fermentation products, and reduce thiosulfate to sulfide, but do not reduce sulfate [291].

Like *Thermoanaerobacter* species, *C. subterraneus* subsp. *tengcongensis* is of interest for its ability to produce ethanol as a major fermentation product [156]. However, not all *C. subterraneus* subspecies are able to produce ethanol [291]. Research on *C. subterraneus* subsp. *tengcongensis* to date has focused primarily on the *in vitro* characterization of proteins as well as proteomic [292, 293] and transcriptomic [294] studies as temperature is varied. To date, little effort has focused on the industrial application of *C. subterraneus* subsp. *tengcongensis*, despite its ability to produce 0.7 mol ethanol per mol glucose fermented [156]. *C. subterraneus* subsp. *tengcongensis* is the most studied member of the genus and has a sequenced genome [295].

14.7.8.1 Genetics Tools – *Caldanaerobacter subterraneus* subsp. *tengcongensis*

The use of natural competence for the transformation of *C. subterraneus* subsp. *tengcongensis* was reported in 2012 using both a replicating and integrating plasmid with kanamycin selection [107] (Table 14.1). No endogenous plasmids were found in *C. subterraneus* subsp. *tengcongensis*, and replicating plasmids such as pMK18 from *T. thermophilus* HB27 [186] did not function. As such, a replicating vector was created using the chromosome origin of replication *oriC* and the kanamycin selectable marker from pMK18. It appears that the plasmid with *oriC* replicates independent of the chromosome at low copy number, but the exact copy number and potential problems related to segregation of the plasmid between daughter cells were not investigated [107]. A promoter region upstream of *pta* (phosphate transacetylase, *Tte1482*) was capable of driving expression of kanamycin resistance protein production. Using homologous recombination, an integrating vector was used to disrupt *hisG* to create a mutant strain auxotrophic for histidine, which was then complemented with the *hisG* gene. This demonstrates, at a proof-of-concept level, that histidine auxotrophy and prototrophy could be used as selection methods; however, without a counter-selective mechanism, this system relies on homologous recombination to occur in both flanking regions to yield final markerless strains.

14.7.8.2 Metabolic Engineering – *Caldanaerobacter subterraneus* subsp. *tengcongensis*

Because methods for genetic manipulation in *C. subterraneus* subsp. *tengcongensis* were so recently established, metabolic engineering has not yet been applied to this host.

14.8 Metabolic Engineering of Extreme Thermophiles – Tool Kit Needs

As with any microbial hosts being considered, metabolic engineering of extreme thermophiles requires a suite of genetic parts in addition to genes targeted for deletion or insertion. Selectable marker genes are required for strain construction methods, and examples of these have been introduced above. In addition, promoters and ribosomal binding sites are used to drive transcription and translation of the gene or synthetic operon, and terminator sequences are used to halt transcription (although these are sometimes omitted). Reporter genes are occasionally used as a way to easily evaluate successful gene insertion or expression. Here, possible options of these genetic parts available for metabolic engineering of extreme thermophiles are discussed, and they are also summarized in Table 14.4.

14.8.1 Promoters

A variety of promoters have been successfully used in extremely thermophilic hosts, all of which have been derived from promoters for native genes (Table 14.4). Promoter sequences are typically obtained by taking 100–200 bases 5′ to the native gene from which they are derived [159, 161, 177]. Both constitutive and inducible promoters have been used to drive gene expression in engineered extreme thermophiles. The constitutive promoters are usually chosen on the basis of gene expression data showing high-level expression of the gene under most conditions [43]. Inducible promoters are typically identified from previous literature defining the amount of gene expression under various conditions, such as for *mbp* in *S. acidocaldarius* [177, 296] and for *mbh1* in *P. furiosus* [297, 298].

Most of the promoters used in extreme thermophiles are constitutive with a high level of expression, although a few inducible promoters have been characterized and used (Table 14.4). In *P. furiosus*, the promoter for the S-layer protein, one of the most highly expressed genes in *P. furiosus*, has been most commonly used [299]. A cold-induced promoter and a promoter activated under non-sulfur conditions have also been successfully used [160, 161]. A promoter and RBS 295 bp upstream from the *xylAB* operon in *T. ethanolicus* [187], which is repressed during growth on glucose and induced during growth on xylose, has been used to drive expression of proteins in *T. mathranii* [111, 173]. The native promoter and RBS from *adhB* (500 bp upstream) was used to drive *cat* gene expression in *T. ethanolicus* [110]. For expression of the β-glucanase from *C. thermocellum* in *Thermoanaerobacter* sp. X514, the native promoter and RBS (651 bp upstream) and terminator (142 bp downstream) were amplified along with the β-glucanase from the genomic DNA of *C. thermocellum* and shown to function in *Thermoanaerobacter.* sp. X514 [112]. In *C. bescii*, all genetic engineering achievements to date have used the strong constitutive promoters of the *slp* gene and the 30S ribosomal protein gene, and there are as yet no reports for the use of regulated or inducible promoters. However, the promoters used in *Thermoanaerobacter*, a closely related genus, may also be functional in *C. bescii.* Among the organisms considered here, by far the most characterized promoters are for *Sulfolobus*, allowing inducers such as maltose, arabinose, copper, mercury, and heat shock to be used (Table 14.4). There are also a few inducible promoters available for bacteria in the *Thermus* genus.

Although some native promoters have been characterized for several extremely thermophilic hosts, there are currently limited options from which to choose, and most of them offer high-level expression. However, high-level gene expression for metabolic engineering often does not result in optimal product formation; better results are often obtained through controlled gene expression [1, 300]. For mesophilic hosts, controlled expression of inserted genes has been accomplished through the development of advanced tools, such as promoter libraries [300, 301] and dynamic sensor-regulator systems [302]. The development of these powerful tools for extreme thermophiles would greatly increase the success of metabolic engineering efforts.

14.8.2 Ribosomal Binding Sites

When inserting multiple genes into a host organism, these are often assembled into artificial operons which are driven by a single promoter but require ribosomal binding sequences (RBSs) between genes. For cases in which the inserted genes come from closely related organisms, such as the expression of the *T. onnurineus* FHL complex in *P. furiosus*, native RBSs can be used successfully [161]. However, when using genes from distantly related organisms, modifications to RBSs may have to be made. For instance, in order to successfully express three enzymes of the 3-hydroxypropionate/4-hydroxybutyrate cycle from *M. sedula* in *P. furiosus*, the *M. sedula* RBSs did not result in effective translation [43]. Upon replacing these RBSs with those for highly expressed genes in *P. furiosus*, such as *slp* and *porγ* (Table 14.4), the heterologous genes were successfully expressed. No attempts to regulate protein expression through RBS sequences have been made in other extreme thermophiles. It should be noted that, because translation initiation is impacted by sequences both upstream and downstream of the start codon [303], simply cutting and pasting RBSs between genes may not work as expected. To combat this, libraries could be constructed to allow for testing of multiple RBSs with each gene. Alternatively, biophysical models could be developed to predict translation rates from the mRNA sequence, such as those developed for *E. coli* [304], allowing RBSs to be designed for the desired expression level of each gene. Further exploration of translational regulatory sequences for gene expression in extremely thermophilic hosts would be of great use moving forward.

14.8.3 Reporter Genes

A few reporter genes have been characterized for extreme thermophiles (Table 14.4). Beta-glycosidase has been used as a reporter gene in *T. kodakarensis* to determine the effect of promoter strength and other factors on expression. However, screening requires activity assays of cell lysate, so it cannot be used as a colorimetric plate assay to screen for positive transformants [55, 305]. Reporter genes for the popular blue/white colony screening method have been successfully demonstrated in *S. acidocaldarius*, *S. solfataricus*, and *T. thermophilus*. Another popular reporter gene, green fluorescent protein (GFP), has been successfully used in *S. acidocaldarius*. A GFP reporter was attempted in *Thermoanaerobacter.* sp. X514, but while the vector transformed, GFP was not functionally expressed, likely due to protein instability at elevated temperature or lack of molecular oxygen needed for formation of the fluorescent site [112].

14.9 Conclusions and Future Perspectives

Extreme thermophiles possess unique pathways and adaptations that offer intrinsic advantages for their use in industrial bioprocessing. With the advent of genetic tools for a variety of extreme thermophiles, these organisms are now poised to be used as hosts for metabolic engineering. Multiple barriers exist to future metabolic engineering projects using extremely thermophilic hosts. Practical limitations of the available genetic tools, such as long plate incubation times, make it difficult to quickly develop strains to test hypothesized metabolic engineering strategies. Likewise, the limited availability of genetic parts makes it difficult to express inserted pathways at the desired levels and activities. Despite these limitations, successful metabolic engineering of a few extreme thermophiles has already been demonstrated.

Some of the same tools that have been applied to overcoming challenges in metabolic engineering of mesophiles would be extremely useful in thermophiles. Advanced genetic tools and predictive modeling methods could be used to partially offset the long strain development times required for thermophiles. For example, genetic methods, such as multiplex automated genome engineering (MAGE) and global transcription machinery engineering, would allow high-throughput strain generation, reducing the time required to test metabolic engineering hypotheses [306]. To avoid the trial-and-error approach to metabolic engineering, predictive modeling methods could be applied. For example, genome-scale metabolic models have been successfully applied to identify gene targets for metabolic engineering projects [307]. Among the organisms discussed here, genome-scale metabolic models exist only for *T. maritima* and *S. solfataricus* [131, 307], and they have yet to be used for metabolic engineering. Kinetic modeling approaches also show promise as a tool for metabolic engineering [132].

One advanced gene editing technology that has seen dramatic growth in applications in recent years is CRISPR/Cas. These tools allow highly efficient, highly specific chromosomal modifications and can be used to make multiple modifications simultaneously [308]. Gene editing tools based on CRISPR/Cas most commonly use the effector protein from Type II CRISPR/Cas systems, Cas9, an RNA-guided nuclease that catalyzes targeted double-stranded DNA breaks [309]. In bacteria, gene editing is accomplished by using a synthetic guide RNA to induce double-stranded cleavage of wild-type genomic DNA by Cas9, resulting in the death of cells that do not undergo homologous recombination with donor DNA lacking the target sequence [309, 310]. The guide RNA, Cas9, and donor DNA can be introduced on the same plasmid, which is transformed into the cell [308]. In a sense, this artificial Cas9 system is used as a replacement for selective pressure, resulting in highly efficient gene editing, without the need for a selectable marker. In order to develop a Cas9-based gene editing tool for extreme thermophiles, a thermostable Cas9 is necessary. Unfortunately, Type II CRISPR/Cas systems are less common than Type I and III systems [311], and no thermophilic Cas9 enzymes have been characterized. However, at least two extreme thermophiles are predicted to have Type II systems: *Acidothermus cellulolyticus* and *Persephonella marina*, both of which grow optimally near 70 °C [311–313]. Thus, it is possible that Cas9-based gene editing tools could be developed by adaptation of a thermophilic Cas9.

While Type II CRISPR/Cas systems are most commonly adapted for gene editing due to the simplicity of the effector protein, the Type I and III systems can also be used. Type I and III systems use a protein complex for their RNA-guided nuclease, requiring expression of multiple genes to allow for targeted cleavage of DNA [310]. Type I and III systems are much more widely distributed than Type II; in fact, all of the extremely thermophilic hosts discussed here possess a Type I system, with the exception of *S. acidocaldarius* and *C. subterraneus* (which possess only Type III systems), and most also have a Type III system [311]. Recently, the endogenous Type I and III systems were used to demonstrate gene editing in *S. islandicus* by introducing plasmids containing sequences for the guide RNA and donor DNA [314], a technique that could in principle be extended to any host with a Type I or III system. Alternatively, an artificial Type I or III system could be introduced; the Type I and III systems in *P. furiosus* have been well characterized and could be adapted for such use in extreme thermophiles [315, 316]. Of particular interest is the Type III system, the only system known to cleave RNA targets, allowing it to be used for gene knockdown [309]. In addition to advances in genetic tools and methods for target identification, expanding the selection of genetic parts for extremely thermophilic hosts would be advantageous. Promoter libraries [317], predictors or libraries for ribosomal binding sites [303, 304], and synthetic biology constructs such as dynamic sensor-regulator systems [302], have been developed and used to assist metabolic engineering of mesophilic hosts. Applied to thermophilic hosts, these same tools would vastly expand the genetic parts library and allow fine-tuning of enzyme expression.

The metabolic engineering of extreme thermophiles is still in its early stages. Despite the current limitations, successful metabolic engineering of a few extreme thermophiles has already been demonstrated. As our understanding of extreme thermophiles as hosts grows and more tools become developed, these organisms have a promising future as relevant and impactful cell factories for industrial biotechnology.

Acknowledgments

This work was supported in part by the US National Science Foundation (CBET-1264052, CBET-1264053), the US Air Force Office of Scientific Research (AFOSR) (FA9550-13-1-0236), and the BioEnergy Science Center (BESC), a US Department of Energy Bioenergy Research Center supported by the Office of Biological and Environmental Research in the DOE Office of Science. AJ Loder, BM Zeldes and LL Lee acknowledge support from an NIH Biotechnology Traineeship (NIH T32GM008776-11). CT Straub and JM Conway acknowledge support from a U.S. Department of Education GAANN Fellowship (P200A100004-12).

References

1. Peralta-Yahya, P.P., Zhang, F., del Cardayre, S.B., and Keasling, J.D. (2012) Microbial engineering for the production of advanced biofuels. *Nature*, **488**, 320–328.
2. Gronenberg, L.S., Marcheschi, R.J., and Liao, J.C. (2013) Next generation biofuel engineering in prokaryotes. *Curr. Opin. Chem. Biol.*, **17**, 462–471.
3. Werpy, T. and Petersen, G. (2004) *Top Value Added Chemicals from Biomass: Volume I – Results of Screening for Potential Candidates from Sugars and Synthesis Gas*, National Renewable Energy Laboratory, Golden, CO.
4. Lee, S.Y., Park, J.H., Jang, S.H., Nielsen, L.K., Kim, J., and Jung, K.S. (2008) Fermentative butanol production by

clostridia. *Biotechnol. Bioeng.*, **101**, 209–228.

5. Ajikumar, P.K., Xiao, W.H., Tyo, K.E., Wang, Y., Simeon, F., Leonard, E., Mucha, O., Phon, T.H., Pfeifer, B., and Stephanopoulos, G. (2010) Isoprenoid pathway optimization for Taxol precursor overproduction in *Escherichia coli*. *Science*, **330**, 70–74.
6. Marcheschi, R.J., Li, H., Zhang, K., Noey, E.L., Kim, S., Chaubey, A., Houk, K.N., and Liao, J.C. (2012) A synthetic recursive "+1" pathway for carbon chain elongation. *ACS Chem. Biol.*, **7**, 689–697.
7. Shen, C.R., Lan, E.I., Dekishima, Y., Baez, A., Cho, K.M., and Liao, J.C. (2011) Driving forces enable high-titer anaerobic 1-butanol synthesis in *Escherichia coli*. *Appl. Environ. Microbiol.*, **77**, 2905–2915.
8. Kirby, J., Nishimoto, M., Chow, R.W.N., Baidoo, E.E.K., Wang, G., Martin, J., Schackwitz, W., Chan, R., Fortman, J.L., and Keasling, J.D. (2014) Enhancing terpene yield from sugars via novel routes to 1-deoxy-D-xylulose 5-phosphate. *Appl. Environ. Microbiol.*, **81**, 130–138.
9. Steen, E.J., Kang, Y.S., Bokinsky, G., Hu, Z.H., Schirmer, A., McClure, A., del Cardayre, S.B., and Keasling, J.D. (2010) Microbial production of fatty-acid-derived fuels and chemicals from plant biomass. *Nature*, **463**, 559–U182.
10. Matsuda, F., Ishii, J., Kondo, T., Ida, K., Tezuka, H., and Kondo, A. (2013) Increased isobutanol production in *Saccharomyces cerevisiae* by eliminating competing pathways and resolving cofactor imbalance. *Microb. Cell Fact.*, **12**, 119.
11. Si, T., Luo, Y., Xiao, H., and Zhao, H. (2014) Utilizing an endogenous pathway for 1-butanol production in *Saccharomyces cerevisiae*. *Metab. Eng.*, **22**, 60–68.
12. Peralta-Yahya, P.P. and Keasling, J.D. (2010) Advanced biofuel production in microbes. *Biotechnol. J.*, **5**, 147–162.
13. George, K.W., Alonso-Gutierrez, J., Keasling, J.D., and Lee, T.S. (2015) Isoprenoid drugs, biofuels, and chemicals – artemisinin, farnesene, and beyond. *Adv. Biochem. Eng./Biotechnol.*, **148**, 355–389.
14. Nakamura, C.E. and Whited, G.M. (2003) Metabolic engineering for the microbial production of 1,3-propanediol. *Curr. Opin. Biotechnol.*, **14**, 454–459.
15. Lan, E.I. and Liao, J.C. (2012) ATP drives direct photosynthetic production of 1-butanol in cyanobacteria. *Proc. Natl. Acad. Sci. U.S.A.*, **109**, 6018–6023.
16. Lan, E.I., Chuang, D.S., Shen, C.R., Lee, A.M., Ro, S.Y., and Liao, J.C. (2015) Metabolic engineering of cyanobacteria for photosynthetic 3-hydroxypropionic acid production from CO_2 using *Synechococcus elongatus* PCC 7942. *Metab. Eng.*, **31**, 163–170.
17. Li, X.Q., Shen, C.R., and Liao, J.C. (2014) Isobutanol production as an alternative metabolic sink to rescue the growth deficiency of the glycogen mutant of *Synechococcus elongatus* PCC 7942. *Photosynth. Res.*, **120**, 301–310.
18. Yoshino, T., Liang, Y., Arai, D., Maeda, Y., Honda, T., Muto, M., Kakunaka, N., and Tanaka, T. (2014) Alkane production by the marine cyanobacterium *Synechococcus* sp. NKBG15041c possessing the α-olefin biosynthesis pathway. *Appl. Microbiol. Biotechnol.*, **99**, 1521–1529.
19. Zhang, S., Liu, Y., and Bryant, D.A. (2015) Metabolic engineering of *Synechococcus* sp. PCC 7002 to produce poly-3-hydroxybutyrate and poly-3-hydroxybutyrate-co-4-hydroxybutyrate. *Metab. Eng.*, **32**, 174–183.
20. Jang, Y.S., Lee, J.Y., Lee, J., Park, J.H., Im, J.A., Eom, M.H., Lee, J., Lee, S.H., Song, H., Cho, J.H., Seung, D.Y., and Lee, S.Y. (2012) Enhanced butanol production obtained by reinforcing the direct butanol-forming route in *Clostridium acetobutylicum*. *mBio*, **3** (5), e00314-12. doi: 10.1128/mBio.00314-12
21. Lee, J., Jang, Y.S., Choi, S.J., Im, J.A., Song, H., Cho, J.H., Seung, D.Y., Papoutsakis, E.T., Bennett, G.N., and Lee, S.Y. (2012) Metabolic engineering of *Clostridium acetobutylicum* ATCC 824 for isopropanol-butanol-ethanol

fermentation. *Appl. Environ. Microbiol.*, **78**, 1416–1423.

22. Lee, S.Y., Kim, H.M., and Cheon, S. (2015) Metabolic engineering for the production of hydrocarbon fuels. *Curr. Opin. Biotechnol.*, **33**, 15–22.
23. Madigan, M.T. and Martinko, J.M. (2006) *Brock Biology of Microorganisms*, 11th edn, Pearson, Upper Saddle River, NJ.
24. Frock, A.D. and Kelly, R.M. (2012) Extreme thermophiles: moving beyond single-enzyme biocatalysis. *Curr. Opin. Chem. Eng.*, **1**, 363–372.
25. Saiki, R., Gelfand, D., Stoffel, S., Scharf, S., Higuchi, R., Horn, G., Mullis, K., and Erlich, H. (1988) Primer-directed enzymatic amplification of DNA with a thermostable DNA polymerase. *Science*, **239**, 487–491.
26. Bartlett, J. and Stirling, D. (2003) A short history of the polymerase chain reaction, in *PCR Protocols*, Methods in Molecular Biology, vol. 226 (eds J. Bartlett and D. Stirling), Humana Press Inc, Totowa, NJ.
27. Berezovsky, I.N. and Shakhnovich, E.I. (2005) Physics and evolution of thermophilic adaptation. *Proc. Natl. Acad. Sci. U.S.A.*, **102**, 12742–12747.
28. Portnoy, V.A., Bezdan, D., and Zengler, K. (2011) Adaptive laboratory evolution – harnessing the power of biology for metabolic engineering. *Curr. Opin. Biotechnol.*, **22**, 590–594.
29. Yang, X. (2010) in *Manual of Industrial Microbiology and Biotechnology*, 3rd edn (eds R.H. Baltz, J.E. Davies, and A.L. Demain), ASM Press, Washington, DC, pp. 657–742.
30. Abdel-Banat, B.M., Hoshida, H., Ano, A., Nonklang, S., and Akada, R. (2010) High-temperature fermentation: how can processes for ethanol production at high temperatures become superior to the traditional process using mesophilic yeast? *Appl. Microbiol. Biotechnol.*, **85**, 861–867.
31. Vane, L.M. (2008) Separation technologies for the recovery and dehydration of alcohols from fermentation broths. *Biofuels, Bioprod. Biorefin.*, **2**, 553–588.
32. Vane, L.M. and Alvarez, F.R. (2013) Hybrid vapor stripping-vapor permeation process for recovery and dehydration of 1-butanol and acetone/butanol/ethanol from dilute aqueous solutions. Part 1. Process simulations. *J. Chem. Technol. Biotechnol.*, **88**, 1436–1447.
33. Brasen, C., Esser, D., Rauch, B., and Siebers, B. (2014) Carbohydrate metabolism in archaea: current insights into unusual enzymes and pathways and their regulation. *Microbiol. Mol. Biol. Rev.*, **78**, 89–175.
34. Wheaton, G., Counts, J., Mukherjee, A., Kruh, J., and Kelly, R. (2015) The confluence of heavy metal biooxidation and heavy metal resistance: implications for bioleaching by extreme thermoacidophiles. *Minerals*, **5**, 397–451.
35. Pereira, F.B., Guimarães, P.M.R., Teixeira, J.A., and Domingues, L. (2010) Selection of *Saccharomyces cerevisiae* strains for efficient very high gravity bio-ethanol fermentation processes. *Biotechnol. Lett.*, **32**, 1655–1661.
36. Blumer-Schuette, S.E., Brown, S.D., Sander, K.B., Bayer, E.A., Kataeva, I., Zurawski, J.V., Conway, J.M., Adams, M.W.W., and Kelly, R.M. (2014) Thermophilic lignocellulose deconstruction. *FEMS Microbiol. Rev.*, **38**, 393–448.
37. Verhaart, M.R.A., Bielen, A.A.M., van der Oost, J., Stams, A.J.M., and Kengen, S.W.M. (2010) Hydrogen production by hyperthermophilic and extremely thermophilic bacteria and archaea: mechanisms for reductant disposal. *Environ. Technol.*, **31**, 993–1003.
38. Trinh, C.T. (2012) Elucidating and reprogramming *Escherichia coli* metabolisms for obligate anaerobic *n*-butanol and isobutanol production. *Appl. Microbiol. Biotechnol.*, **95**, 1083–1094.
39. Shao, X., Raman, B., Zhu, M., Mielenz, J.R., Brown, S.D., Guss, A.M., and Lynd, L.R. (2011) Mutant selection and phenotypic and genetic characterization of ethanol-tolerant strains of *Clostridium thermocellum*. *Appl. Microbiol. Biotechnol.*, **92**, 641–652.
40. Humbird, D., Davis, R., Tao, L., Kinchin, C., Hsu, D., and Aden, A.

(2011) *Process Design and Economics for Biochemical Conversion of Lignocellulosic Biomass to Ethanol: Dilute-Acid Pretreatment and Enzymatic Hydrolysis of Corn Stover*, National Renewable Energy Laboratory, Golden, CO.

41. Chung, D., Cha, M., Guss, A.M., and Westpheling, J. (2014) Direct conversion of plant biomass to ethanol by engineered *Caldicellulosiruptor bescii*. *Proc. Natl. Acad. Sci. U.S.A.*, **111**, 8931–8936.
42. Olson, D.G., Sparling, R., and Lynd, L.R. (2015) Ethanol production by engineered thermophiles. *Curr. Opin. Biotechnol.*, **33**, 130–141.
43. Keller, M.W., Schut, G.J., Lipscomb, G.L., Menon, A.L., Iwuchukwu, I.J., Leuko, T.T., Thorgersen, M.P., Nixon, W.J., Hawkins, A.S., Kelly, R.M., and Adams, M.W.W. (2013) Exploiting microbial hyperthermophilicity to produce an industrial chemical, using hydrogen and carbon dioxide. *Proc. Natl. Acad. Sci. U.S.A.*, **110**, 5840–5845.
44. Li, H., Opgenorth, P.H., Wernick, D.G., Rogers, S., Wu, T.Y., Higashide, W., Malati, P., Huo, Y.X., Cho, K.M., and Liao, J.C. (2012) Integrated electromicrobial conversion of CO_2 to higher alcohols. *Science*, **335**, 1596.
45. Fast, A.G. and Papoutsakis, E.T. (2012) Stoichiometric and energetic analyses of non-photosynthetic CO_2-fixation pathways to support synthetic biology strategies for production of fuels and chemicals. *Curr. Opin. Chem. Eng.*, **1**, 380–395.
46. Bhandiwad, A., Shaw, A.J., Guss, A., Guseva, A., Bahl, H., and Lynd, L.R. (2014) Metabolic engineering of *Thermoanaerobacterium saccharolyticum* for *n*-butanol production. *Metab. Eng.*, **21**, 17–25.
47. Lipscomb, G.L., Stirrett, K., Schut, G.J., Yang, F., Jenney, F.E., Scott, R.A., Adams, M.W.W., and Westpheling, J. (2011) Natural competence in the hyperthermophilic archaeon *Pyrococcus furiosus* facilitates genetic manipulation: construction of markerless deletions of genes encoding the two cytoplasmic hydrogenases. *Appl. Environ. Microbiol.*, 77, 2232–2238.
48. Farkas, J., Stirrett, K., Lipscomb, G.L., Nixon, W., Scott, R.A., Adams, M.W.W., and Westpheling, J. (2012) Recombinogenic properties of *Pyrococcus furiosus* strain COM1 enable rapid selection of targeted mutants. *Appl. Environ. Microbiol.*, **78**, 4669–4676.
49. Waege, I., Schmid, G., Thumann, S., Thomm, M., and Hausner, W. (2010) Shuttle vector-based transformation system for *Pyrococcus furiosus*. *Appl. Environ. Microbiol.*, **76**, 3308–3313.
50. Kreuzer, M., Schmutzler, K., Waege, I., Thomm, M., and Hausner, W. (2013) Genetic engineering of *Pyrococcus furiosus* to use chitin as a carbon source. *BMC Biotechnol.*, **13**, 1.
51. Li, X., Fu, L., Li, Z., Ma, X., Xiao, X., and Xu, J. (2015) Genetic tools for the piezophilic hyperthermophilic archaeon *Pyrococcus yayanosii*. *Extremophiles*, **19**, 59–67.
52. Sato, T., Fukui, T., Atomi, H., and Imanaka, T. (2003) Targeted gene disruption by homologous recombination in the hyperthermophilic archaeon *Thermococcus kodakaraensis* KOD1. *J. Bacteriol.*, **185**, 210–220.
53. Sato, T., Fukui, T., Atomi, H., and Imanaka, T. (2005) Improved and versatile transformation system allowing multiple genetic manipulations of the hyperthermophilic archaeon *Thermococcus kodakaraensis*. *Appl. Environ. Microbiol.*, **71**, 3889–3899.
54. Matsumi, R., Manabe, K., Fukui, T., Atomi, H., and Imanaka, T. (2007) Disruption of a sugar transporter gene cluster in a hyperthermophilic archaeon using a host-marker system based on antibiotic resistance. *J. Bacteriol.*, **189**, 2683–2691.
55. Santangelo, T.J., Cubonova, L., and Reeve, J.N. (2010) *Thermococcus kodakarensis* genetics: TK1827-encoded β-glycosidase, new positive-selection protocol, and targeted and repetitive deletion technology. *Appl. Environ. Microbiol.*, **76**, 1044–1052.
56. Santangelo, T.J., Cubonova, L., and Reeve, J.N. (2008) Shuttle vector expression in *Thermococcus*

kodakaraensis: contributions of *cis* elements to protein synthesis in a hyperthermophilic archaeon. *Appl. Environ. Microbiol.*, **74**, 3099–3104.

57. Kim, M.S., Bae, S.S., Kim, Y.J., Kim, T.W., Lim, J.K., Lee, S.H., Choi, A.R., Jeon, J.H., Lee, J.H., Lee, H.S., and Kang, S.G. (2013) CO-dependent H_2 production by genetically engineered *Thermococcus onnurineus* NA1. *Appl. Environ. Microbiol.*, **79**, 2048–2053.
58. Thiel, A., Michoud, G., Moalic, Y., Flament, D., and Jebbar, M. (2014) Genetic manipulations of the hyperthermophilic piezophilic archaeon *Thermococcus barophilus. Appl. Environ. Microbiol.*, **80**, 2299–2306.
59. Cario, A., Mizgier, A., Thiel, A., Jebbar, M., and Oger, P.M. (2015) Restoration of the di-myo-inositol-phosphate pathway in the piezo-hyperthermophilic archaeon *Thermococcus barophilus. Biochimie*, **118**, 286–293.
60. Worthington, P., Hoang, V., and Perez-Pomares, F. (2003) Targeted disruption of the α-amylase gene in the hyperthermophilic archaeon *Sulfolobus solfataricus. J. Bacteriol.*, **185**, 482–488.
61. Jonuscheit, M., Martusewitsch, E., Stedman, K.M., and Schleper, C. (2003) A reporter gene system for the hyperthermophilic archaeon *Sulfolobus solfataricus* based on a selectable and integrative shuttle vector. *Mol. Microbiol.*, **48**, 1241–1252.
62. Schelert, J., Dixit, V., Hoang, V., Simbahan, J., Drozda, M., and Blum, P. (2004) Occurrence and characterization of mercury resistance in the hyperthermophilic archaeon *Sulfolobus solfataricus* by use of gene disruption. *J. Bacteriol.*, **186**, 427–437.
63. Albers, S.V. and Driessen, A.J.M. (2007) Conditions for gene disruption by homologous recombination of exogenous DNA into the *Sulfolobus solfataricus* genome. *Archaea*, **2**, 145–149.
64. Aravalli, R.N. and Garrett, R.A. (1997) Shuttle vectors for hyperthermophilic archaea. *Extremophiles*, **1**, 183–191.
65. Cannio, R., Contursi, P., Rossi, M., and Bartolucci, S. (1998) An autonomously replicating transforming vector for *Sulfolobus solfataricus. J. Bacteriol.*, **180**, 3237–3240.
66. Contursi, P., Cannio, R., Prato, S., Fiorentino, G., Rossi, M., and Bartolucci, S. (2003) Development of a genetic system for hyperthermophilic Archaea: expression of a moderate thermophilic bacterial alcohol dehydrogenase gene in *Sulfolobus solfataricus. FEMS Microbiol. Lett.*, **218**, 115–120.
67. Wagner, M., van Wolferen, M., Wagner, A., Lassak, K., Meyer, B.H., Reimann, J., and Albers, S.V. (2012) Versatile genetic tool box for the crenarchaeote *Sulfolobus acidocaldarius. Front. Microbiol.*, **3**, 214.
68. Sakofsky, C.J., Runck, L.A., and Grogan, D.W. (2011) *Sulfolobus* mutants, generated via PCR products, which lack putative enzymes of UV photoproduct repair. *Archaea*, **2011**, 864015.
69. Berkner, S., Grogan, D., Albers, S.V., and Lipps, G. (2007) Small multicopy, non-integrative shuttle vectors based on the plasmid pRN1 for *Sulfolobus acidocaldarius* and *Sulfolobus solfataricus*, model organisms of the (cren-)archaea. *Nucleic Acids Res.*, **35**, e88.
70. Deng, L., Zhu, H., Chen, Z., Liang, Y.X., and She, Q. (2009) Unmarked gene deletion and host-vector system for the hyperthermophilic crenarchaeon *Sulfolobus islandicus. Extremophiles*, **13**, 735–746.
71. Jaubert, C., Danioux, C., Oberto, J., Cortez, D., Bize, A., Krupovic, M., She, Q., Forterre, P., Prangishvili, D., and Sezonov, G. (2013) Genomics and genetics of *Sulfolobus islandicus* LAL14/1, a model hyperthermophilic archaeon. *Open Biol.*, **3**, 130010.
72. Zheng, T., Huang, Q., Zhang, C., Ni, J., She, Q., and Shen, Y. (2012) Development of a simvastatin selection marker for a hyperthermophilic acidophile, *Sulfolobus islandicus. Appl. Environ. Microbiol.*, **78**, 568–574.
73. Zhang, C. and Whitaker, R.J. (2012) A broadly applicable gene knockout system for the thermoacidophilic archaeon *Sulfolobus islandicus* based on simvastatin selection. *Microbiology*, **158**, 1513–1522.

74. Han, D.M., Norris, S.M., and Xu, Z.H. (2012) Construction and transformation of a *Thermotoga-E. coli* shuttle vector. *BMC Biotechnol.*, **12**, 2.
75. Han, D.M., Xu, H., Puranik, R., and Xu, Z.H. (2014) Natural transformation of *Thermotoga* sp. strain RQ7. *BMC Biotechnol.*, **14**, 39.
76. Yamagishi, A., Tanimoto, T., Suzuki, T., and Oshima, T. (1996) Pyrimidine biosynthesis genes (*pyrE* and *pyrF*) of an extreme thermophile, *Thermus thermophilus*. *Appl. Environ. Microbiol.*, **62**, 2191–2194.
77. Tamakoshi, M., Yaoi, T., Oshima, T., and Yamagishi, A. (1999) An efficient gene replacement and deletion system for an extreme thermophile, *Thermus thermophilus*. *FEMS Microbiol. Lett.*, **173**, 431–437.
78. Tamakoshi, M., Uchida, M., Tanabe, K., Fukuyama, S., Yamagishi, A., and Oshima, T. (1997) A new *Thermus-Escherichia* coli shuttle integration vector system. *J. Bacteriol.*, **179**, 4811–4814.
79. Hoshino, T., Fujii, R., and Nakahara, T. (1994) Overproduction of carotenoids in *Thermus thermophilus*. *J. Ferment. Bioeng.*, **77**, 423–424.
80. Koyama, Y., Okamoto, S., and Furukawa, K. (1990) Cloning of α- and β-galactosidase genes from an extreme thermophile, *Thermus* strain T2, and their expression in *Thermus thermophilus* HB27. *Appl. Environ. Microbiol.*, **56**, 2251–2254.
81. Koyama, Y., Hoshino, T., Tomizuka, N., and Furukawa, K. (1986) Genetic transformation of the extreme thermophile *Thermus thermophilus* and of other *Thermus* spp. *J. Bacteriol.*, **166**, 338–340.
82. Koyama, Y., Arikawa, Y., and Furukawa, K. (1990) A plasmid vector for an extreme thermophile, *Thermus thermophilus*. *FEMS Microbiol. Lett.*, **72**, 97–101.
83. Koyama, Y. and Furukawa, K. (1990) Cloning and sequence analysis of tryptophan synthetase genes of an extreme thermophile, *Thermus thermophilus* HB27: plasmid transfer from replica-plated *Escherichia coli* recombinant colonies to competent *T. thermophilus* cells. *J. Bacteriol.*, **172**, 3490–3495.
84. Takada, T., Akanuma, S., Kotsuka, T., Tamakoshi, M., Yamagishi, A., and Oshima, T. (1993) Recombination-deficient mutants of an extreme thermophile, *Thermus thermophilus*. *Appl. Environ. Microbiol.*, **59**, 2737–2739.
85. Takagi, H., Suzumura, A., Hoshino, T., and Nakamori, S. (1999) Gene expression of *Bacillus subtilis* subtilisin E in *Thermus thermophilus*. *J. Ind. Microbiol. Biotechnol.*, **23**, 214–217.
86. Lasa, I., Castón, J.R., Fernández-Herrero, L.A., de Pedro, M.A., and Berenguer, J. (1992) Insertional mutagenesis in the extreme thermophilic eubacteria *Thermus thermophilus* HB8. *Mol. Microbiol.*, **6**, 1555–1564.
87. Hoseki, J., Yano, T., Koyama, Y., Kuramitsu, S., and Kagamiyama, H. (1999) Directed evolution of thermostable kanamycin-resistance gene: a convenient selection marker for *Thermus thermophilus*. *J. Biochem.*, **126**, 951–956.
88. Carr, J.F., Gregory, S.T., and Dahlberg, A.E. (2014) Transposon mutagenesis of the extremely thermophilic bacterium *Thermus thermophilus* HB27. *Extremophiles*, **19**, 221–228.
89. Kayser, K.J. and Kilbane, J.J. (2001) New host-vector system for *Thermus* spp. based on the malate dehydrogenase gene. *J. Bacteriol.*, **183**, 1792–1795.
90. Carr, J.F., Danziger, M.E., Huang, A.L., Dahlberg, A.E., and Gregory, S.T. (2015) Engineering the genome of *Thermus thermophilus* using a counterselectable marker. *J. Bacteriol.*, **197**, 1135–1144.
91. Hidalgo, A., Betancor, L., Moreno, R., Zafra, O., Cava, F., Fernández-Lafuente, R., Guisán, J.M., and Berenguer, J. (2004) *Thermus thermophilus* as a cell factory for the production of a thermophilic Mn-dependent catalase which fails to be synthesized in an active form in *Escherichia coli*. *Appl. Environ. Microbiol.*, **70**, 3839–3844.
92. Fridjonsson, O., Watzlawick, H., and Mattes, R. (2002) Thermoadaptation

of α-galactosidase AgaB1 in *Thermus thermophilus*. *J. Bacteriol.*, **184**, 3385–3391.

93. Nakamura, A., Takakura, Y., Kobayashi, H., and Hoshino, T. (2005) *In vivo* directed evolution for thermostabilization of *Escherichia coli* hygromycin B phosphotransferase and the use of the gene as a selection marker in the host-vector system of *Thermus thermophilus*. *J. Biosci. Bioeng.*, **100**, 158–163.
94. Mather, M.W. and Fee, J.A. (1992) Development of plasmid cloning vectors for *Thermus thermophilus* HB8: expression of a heterologous, plasmid-borne kanamycin nucleotidyltransferase gene. *Appl. Environ. Microbiol.*, **58**, 421–425.
95. Kayser, K.J., Kwak, J.H., Park, H.S., and Kilbane, J.J. (2001) Inducible and constitutive expression using new plasmid and integrative expression vectors for *Thermus* sp. *Lett. Appl. Microbiol.*, **32**, 412–418.
96. Wayne, J. and Xu, S.Y. (1997) Identification of a thermophilic plasmid origin and its cloning within a new *Thermus-E. coli* shuttle vector. *Gene*, **195**, 321–328.
97. Lasa, I., de Grado, M., de Pedro, M.A., and Berenguer, J. (1992) Development of *Thermus-Escherichia* shuttle vectors and their use for expression of the *Clostridium thermocellum celA* gene in *Thermus thermophilus*. *J. Bacteriol.*, **174**, 6424–6431.
98. Ramírez-Arcos, S., Fernández-Herrero, L.A., Marín, I., and Berenguer, J. (1998) Anaerobic growth, a property horizontally transferred by an Hfr-like mechanism among extreme thermophiles. *J. Bacteriol.*, **180**, 3137–3143.
99. Alvarez, L., Bricio, C., Gomez, M.J., and Berenguer, J. (2011) Lateral transfer of the denitrification pathway genes among *Thermus thermophilus* strains. *Appl. Environ. Microbiol.*, **77**, 1352–1358.
100. Wang, L., Hoffmann, J., Watzlawick, H., and Altenbuchner, J. (2015) Markerless gene deletion with cytosine deaminase in *Thermus thermophilus* HB27. *Appl. Environ. Microbiol.*, **82**, 1249–1255.
101. Vonstein, V., Johnson, S.P., Yu, H., Casadaban, M.J., Pagratis, N.C., Weber, J.M., and Demirjian, D.C. (1995) Molecular cloning of the *pyrE* gene from the extreme thermophile *Thermus flavus*. *J. Bacteriol.*, **177**, 4540–4543.
102. Weber, J.M., Johnson, S.P., Vonstein, V., Casadaban, M.J., and Demirjian, D.C. (1995) A Chromosome integration system for stable gene transfer into *Thermus flavus*. *Nat. Biotechnol.*, **13**, 271–275.
103. Chung, D.W., Farkas, J., and Westpheling, J. (2013) Overcoming restriction as a barrier to DNA transformation in *Caldicellulosiruptor species* results in efficient marker replacement. *Biotechnol. Biofuels*, **6**, 82.
104. Lipscomb, G.L., Conway, J.M., Blumer-Schuette, S.E., Kelly, R.M., and Adams, M.W.W. (2016) Thermostable kanamycin resistance marker expands the toolkit for genetic manipulation of *Caldicellulosiruptor bescii*. *Appl. Environ. Microbiol.* doi: 10.1128/AEM.00570-16.
105. Chung, D., Cha, M., Farkas, J., and Westpheling, J. (2013) Construction of a stable replicating shuttle vector for *Caldicellulosiruptor* species: use for extending genetic methodologies to other members of this genus. *PLoS One*, **8**, e62881.
106. Groom, J., Chung, D., Young, J., and Westpheling, J. (2014) Heterologous complementation of a *pyrF* deletion in *Caldicellulosiruptor hydrothermalis* generates a new host for the analysis of biomass deconstruction. *Biotechnol. Biofuels*, **7**, 132.
107. Liu, B., Wang, C., Yang, H.H., and Tan, H.R. (2012) Establishment of a genetic transformation system and its application in *Thermoanaerobacter tengcongensis*. *J. Genet. Genomics*, **39**, 561–570.
108. Shaw, A.J., Hogsett, D.A., and Lynd, L.R. (2010) Natural competence in *Thermoanaerobacter* and *Thermoanaerobacterium* species. *Appl. Environ. Microbiol.*, **76**, 4713–4719.

109. Soutschek-Bauer, E., Hartl, L., and Staudenbauer, W.L. (1985) Transformation of *Clostridium thermohydrosulfuricum* DSM568 with plasmid DNA. *Biotechnol. Lett.*, **7**, 705–710.
110. Peng, H., Fu, B.L., Mao, Z.G., and Shao, W.L. (2006) Electrotransformation of *Thermoanaerobacter ethanolicus* JW200. *Biotechnol. Lett.*, **28**, 1913–1917.
111. Yao, S. and Mikkelsen, M.J. (2010) Identification and overexpression of a bifunctional aldehyde/alcohol dehydrogenase responsible for ethanol production in *Thermoanaerobacter mathranii*. *J. Mol. Microbiol. Biotechnol.*, **19**, 123–133.
112. Lin, L., Song, H., Ji, Y., He, Z., Pu, Y., Zhou, J., and Xu, J. (2010) Ultrasound-mediated DNA transformation in thermophilic Gram-positive anaerobes. *PLoS One*, **5**, e12582.
113. Wagner, M., Berkner, S., Ajon, M., Driessen, A.J.M., Lipps, G., and Albers, S.V. (2009) Expanding and understanding the genetic toolbox of the hyperthermophilic genus *Sulfolobus*. *Biochem. Soc. Trans.*, **37**, 97–101.
114. Roberts, R.J., Vincze, T., Posfai, J., and Macelis, D. (2015) REBASE-a database for DNA restriction and modification: enzymes, genes and genomes. *Nucleic Acids Res.*, **43**, D298–D299.
115. Chung, D., Farkas, J., Huddleston, J.R., Olivar, E., and Westpheling, J. (2012) Methylation by a unique α-class N4-cytosine methyltransferase is required for DNA transformation of *Caldicellulosiruptor bescii* DSM6725. *PLoS One*, **7**, e43844.
116. Chung, D.H., Huddleston, J.R., Farkas, J., and Westpheling, J. (2011) Identification and characterization of *Cbe*I, a novel thermostable restriction enzyme from *Caldicellulosiruptor bescii* DSM 6725 and a member of a new subfamily of *Hae*III-like enzymes. *J. Ind. Microbiol. Biotechnol.*, **38**, 1867–1877.
117. Prangishvili, D.A., Vashakidze, R.P., Chelidze, M.G., and Gabriadze, I.Y. (1985) A restriction endonuclease *Sua*I from the thermoacidophilic archaebacterium *Sulfolobus acidocaldarius*. *FEBS Lett.*, **192**, 57–60.
118. Roberts, R.J., Carneiro, M.O., and Schatz, M.C. (2013) The advantages of SMRT sequencing. *Genome Biol.*, **14**, 405.
119. Fiala, G. and Stetter, K.O. (1986) *Pyrococcus furiosus* sp. nov. represents a novel genus of marine heterotrophic archaebacteria growing optimally at 100°C. *Arch. Microbiol.*, **145**, 56–61.
120. Lee, H.S., Shockley, K.R., Schut, G.J., Conners, S.B., Montero, C.I., Johnson, M.R., Chou, C.J., Bridger, S.L., Wigner, N., Brehm, S.D., Jenney, F.E. Jr.,, Comfort, D.A., Kelly, R.M., and Adams, M.W. (2006) Transcriptional and biochemical analysis of starch metabolism in the hyperthermophilic archaeon *Pyrococcus furiosus*. *J. Bacteriol.*, **188**, 2115–2125.
121. Chou, C.J., Shockley, K.R., Conners, S.B., Lewis, D.L., Comfort, D.A., Adams, M.W., and Kelly, R.M. (2007) Impact of substrate glycoside linkage and elemental sulfur on bioenergetics of and hydrogen production by the hyperthermophilic archaeon *Pyrococcus furiosus*. *Appl. Environ. Microbiol.*, **73**, 6842–6853.
122. Lewis, D.L., Notey, J.S., Chandrayan, S.K., Loder, A.J., Lipscomb, G.L., Adams, M.W., and Kelly, R.M. (2015) A mutant ('lab strain') of the hyperthermophilic archaeon *Pyrococcus furiosus*, lacking flagella, has unusual growth physiology. *Extremophiles*, **19**, 269–281.
123. Sokolova, T.G., Jeanthon, C., Kostrikina, N.A., Chernyh, N.A., Lebedinsky, A.V., Stackebrandt, E., and Bonch-Osmolovskaya, E.A. (2004) The first evidence of anaerobic CO oxidation coupled with H_2 production by a hyperthermophilic archaeon isolated from a deep-sea hydrothermal vent. *Extremophiles*, **8**, 317–323.
124. Morikawa, M., Izawa, Y., Rashid, N., Hoaki, T., and Imanaka, T. (1994) Purification and characterization of a thermostable thiol protease from a newly isolated hyperthermophilic *Pyrococcus* sp. *Appl. Environ. Microbiol.*, **60**, 4559–4566.
125. Atomi, H., Fukui, T., Kanai, T., Morikawa, M., and Imanaka, T. (2004) Description of *Thermococcus*

kodakaraensis sp. nov., a well studied hyperthermophilic archaeon previously reported as *Pyrococcus* sp. KOD1. *Archaea*, **1**, 263–267.

126. Schut, G.J., Nixon, W.J., Lipscomb, G.L., Scott, R.A., and Adams, M.W.W. (2012) Mutational analyses of the enzymes involved in the metabolism of hydrogen by the hyperthermophilic archaeon *Pyrococcus furiosus*. *Front. Microbiol.*, **3**, 163.
127. Bae, S.S., Kim, T.W., Lee, H.S., Kwon, K.K., Kim, Y.J., Kim, M.S., Lee, J.H., and Kang, S.G. (2012) H_2 production from CO, formate or starch using the hyperthermophilic archaeon, *Thermococcus onnurineus*. *Biotechnol. Lett.*, **34**, 75–79.
128. Bae, S.S., Kim, Y.J., Yang, S.H., Lim, J.K., Jeon, J.H., Lee, H.S., Kang, S.G., Kim, S.J., and Lee, J.H. (2006) *Thermococcus onnurineus* sp. nov., a hyperthermophilic Archaeon isolated from a deep-sea hydrothermal vent area at the PACMANUS field. *J. Ind. Microbiol. Biotechnol.*, **16**, 1826–1831.
129. Marteinsson, V.T., Birrien, J.L., Reysenbach, A.L., Vernet, M., Marie, D., Gambacorta, A., Messner, P., Sleytr, U.B., and Prieur, D. (1999) *Thermococcus barophilus* sp. nov., a new barophilic and hyperthermophilic archaeon isolated under high hydrostatic pressure from a deep-sea hydrothermal vent. *Int. J. Syst. Bacteriol.*, **49** (Pt. 2), 351–359.
130. Grogan, D.W. (1989) Phenotypic characterization of the archaebacterial genus *Sulfolobus*: comparison of five wild-type strains. *J. Bacteriol.*, **171**, 6710–6719.
131. Ulas, T., Riemer, S.A., Zaparty, M., Siebers, B., and Schomburg, D. (2012) Genome-scale reconstruction and analysis of the metabolic network in the hyperthermophilic archaeon *Sulfolobus solfataricus*. *PLoS One*, **7**, e43401.
132. Loder, A.J., Zeldes, B.M., Garrison, G.D. II,, Lipscomb, G.L., Adams, M.W., and Kelly, R.M. (2015) Alcohol selectivity in a synthetic thermophilic *n*-butanol pathway is driven by biocatalytic and thermostability characteristics of constituent enzymes. *Appl. Environ. Microbiol.*, **81**, 7187–7200.
133. Svetlichnyi, V.A., Svetlichnaya, T.P., Chernykh, N.A., and Zavarzin, G.A. (1990) *Anaerocellum thermophilum* gen. nov. sp. nov. – an extremely thermophilic cellulolytic eubacterium isolated from hot springs in the Valley of Geysers. *Microbiology*, **59**, 598–604.
134. Yang, S.J., Kataeva, I., Wiegel, J., Yin, Y.B., Dam, P., Xu, Y., Westpheling, J., and Adams, M.W.W. (2010) Classification of '*Anaerocellum thermophilum*' strain DSM 6725 as *Caldicellulosiruptor bescii* sp. nov. *Int. J. Syst. Evol. Microbiol.*, **60**, 2011–2015.
135. Cha, M., Chung, D.W., Elkins, J.G., Guss, A.M., and Westpheling, J. (2013) Metabolic engineering of *Caldicellulosiruptor bescii* yields increased hydrogen production from lignocellulosic biomass. *Biotechnol. Biofuels*, **6**, 85.
136. Farkas, J., Chung, D.W., Cha, M., Copeland, J., Grayeski, P., and Westpheling, J. (2013) Improved growth media and culture techniques for genetic analysis and assessment of biomass utilization by *Caldicellulosiruptor bescii*. *J. Ind. Microbiol. Biotechnol.*, **40**, 41–49.
137. Basen, M., Rhaesa, A.M., Kataeva, I., Prybol, C.J., Scott, I.M., Poole, F.L., and Adams, M.W.W. (2014) Degradation of high loads of crystalline cellulose and of unpretreated plant biomass by the thermophilic bacterium *Caldicellulosiruptor bescii*. *Bioresour. Technol.*, **152**, 384–392.
138. Huber, R., Langworthy, T.A., Konig, H., Thomm, M., Woese, C.R., Sleytr, U.B., and Stetter, K.O. (1986) *Thermotoga maritima* sp. nov. represents a new genus of unique extremely thermophilic eubacteria growing up to 90°C. *Arch. Microbiol.*, **144**, 324–333.
139. Nguyen, T.A.D., Kim, J.P., Kim, M.S., Oh, Y.K., and Sim, S.J. (2008) Optimization of hydrogen production by hyperthermophilic eubacteria, *Thermotoga maritima* and *Thermotoga neapolitana* in batch fermentation. *Int. J. Hydrogen Energy*, **33**, 1483–1488.

140. Chhabra, S.R., Shockley, K.R., Conners, S.B., Scott, K.L., Wolfinger, R.D., and Kelly, R.M. (2003) Carbohydrate-induced differential gene expression patterns in the hyperthermophilic bacterium *Thermotoga maritima*. *J. Biol. Chem.*, **278**, 7540–7552.
141. Bronnenmeier, K., Kern, A., Liebl, W., and Staudenbauer, W.L. (1995) Purification of *Thermotoga maritima* enzymes for the degradation of cellulosic materials. *Appl. Microbiol. Biotechnol.*, **61**, 1399–1407.
142. Conners, S.B., Montero, C.I., Comfort, D.A., Shockley, K.R., Johnson, M.R., Chhabra, S.R., and Kelly, R.M. (2005) An expression-driven approach to the prediction of carbohydrate transport and utilization regulons in the hyperthermophilic bacterium *Thermotoga maritima*. *J. Bacteriol.*, **187**, 7267–7282.
143. Berg, I.A., Kockelkorn, D., Buckel, W., and Fuchs, G. (2007) A 3-hydroxypropionate/4-hydroxybutyrate autotrophic carbon dioxide assimilation pathway in Archaea. *Science*, **318**, 1782–1786.
144. Saiki, T., Kimura, R., and Arima, K. (1972) Isolation and characterization of extremely thermophilic bacteria from hot springs. *Agric. Biol. Chem.*, **36**, 2357–2366.
145. Thorgersen, M.P., Stirrett, K., Scott, R.A., and Adams, M.W.W. (2012) Mechanism of oxygen detoxification by the surprisingly oxygen-tolerant hyperthermophilic archaeon, *Pyrococcus furiosus*. *Proc. Natl. Acad. Sci. U.S.A.*, **109**, 18547–18552.
146. Bridger, S.L., Lancaster, W.A., Poole, F.L. II, Schut, G.J., and Adams, M.W. (2012) Genome sequencing of a genetically tractable *Pyrococcus furiosus* strain reveals a highly dynamic genome. *J. Bacteriol.*, **194**, 4097–4106.
147. Hopkins, R.C., Sun, J., Jenney, F.E., Chandrayan, S.K., McTernan, P.M., and Adams, M.W.W. (2011) Homologous expression of a subcomplex of *Pyrococcus furiosus* hydrogenase that interacts with pyruvate ferredoxin oxidoreductase. *PLoS One*, **6**, e26569.
148. Thorgersen, M.P., Lipscomb, G.L., Schut, G.J., Kelly, R.M., and Adams, M.W.W. (2014) Deletion of acetyl-CoA synthetases I and II increases production of 3-hydroxypropionate by the metabolically-engineered hyperthermophile *Pyrococcus furiosus*. *Metab. Eng.*, **22**, 83–88.
149. Hawkins, A.B., Lian, H., Zeldes, B.M., Loder, A.J., Lipscomb, G.L., Schut, G.J., Keller, M.W., Adams, M.W., and Kelly, R.M. (2015) Bioprocessing analysis of *Pyrococcus furiosus* strains engineered for CO_2-based 3-hydroxypropionate production. *Biotechnol. Bioeng.*, **112**, 1533–1543.
150. Ramıćrez-Arcos, S., Fernández-Herrero, L.A., and Berenguer, J. (1998) A thermophilic nitrate reductase is responsible for the strain specific anaerobic growth of *Thermus thermophilus* HB8. *Biochim. Biophys. Acta*, **1396**, 215–227.
151. Oshima, T. and Imahori, K. (1974) Description of *Thermus thermophilus* (Yoshida and Oshima) comb. nov., a nonsporulating thermophilic bacterium from a Japanese thermal spa. *Int. J. Syst. Evol. Microbiol.*, **24**, 102–112.
152. Wiegel, J. and Ljungdahl, L.G. (1981) *Thermoanaerobacter ethanolicus* gen. nov., spec. nov., a new, extreme thermophilic, anaerobic bacterium. *Arch. Microbiol.*, **128**, 343–348.
153. Larsen, L., Nielsen, P., and Ahring, B.K. (1997) *Thermoanaerobacter mathranii* sp. nov., an ethanol-producing, extremely thermophilic anaerobic bacterium from a hot spring in Iceland. *Arch. Microbiol.*, **168**, 114–119.
154. Hemme, C.L., Fields, M.W., He, Q., Deng, Y., Lin, L., Tu, Q., Mouttaki, H., Zhou, A., Feng, X., Zuo, Z., Ramsay, B.D., He, Z., Wu, L., Van Nostrand, J., Xu, J., Tang, Y.J., Wiegel, J., Phelps, T.J., and Zhou, J. (2011) Correlation of genomic and physiological traits of T*hermoanaerobacter* species with biofuel yields. *Appl. Environ. Microbiol.*, **77**, 7998–8008.
155. Roh, Y., Liu, S.V., Li, G., Huang, H., Phelps, T.J., and Zhou, J. (2002) Isolation and characterization of metal-reducing *Thermoanaerobacter* strains

from deep subsurface environments of the Piceance Basin, Colorado. *Appl. Environ. Microbiol.*, **68**, 6013–6020.

156. Xue, Y., Xu, Y., Liu, Y., Ma, Y., and Zhou, P. (2001) *Thermoanaerobacter tengcongensis* sp. nov., a novel anaerobic, saccharolytic, thermophilic bacterium isolated from a hot spring in Tengcong, China. *Int. J. Syst. Evol. Microbiol.*, **51**, 1335–1341.
157. Kengen, S.W.M. and Stams, A.J.M. (1994) Formation of L-alanine as a reduced end product in carbohydrate fermentation by the hyperthermophilic archaeon *Pyrococcus furiosus. Arch. Microbiol.*, **161**, 168–175.
158. Basen, M., Schut, G.J., Nguyen, D.M., Lipscomb, G.L., Benn, R.A., Prybol, C.J., Vaccaro, B.J., Poole, F.L., Kelly, R.M., and Adams, M.W.W. (2014) Single gene insertion drives bioalcohol production by a thermophilic archaeon. *Proc. Natl. Acad. Sci. U.S.A.*, **111**, 17618–17623.
159. Keller, M.W., Lipscomb, G.L., Loder, A.J., Schut, G.J., Kelly, R.M., and Adams, M.W.W. (2015) A hybrid synthetic pathway for butanol production by a hyperthermophilic microbe. *Metab. Eng.*, **27**, 101–106.
160. Basen, M., Sun, J., and Adams, M.W.W. (2012) Engineering a hyperthermophilic archaeon for temperature-dependent product formation. *mBio*, **3** (2), e00053-12. doi: 10.1128/mBio.00053-12
161. Lipscomb, G.L., Schut, G.J., Thorgersen, M.P., Nixon, W.J., Kelly, R.M., and Adams, M.W. (2014) Engineering hydrogen gas production from formate in a hyperthermophile by heterologous production of an 18-subunit membrane-bound complex. *J. Biol. Chem.*, **289**, 2873–2879.
162. Santangelo, T.J., Čuboňová, L., and Reeve, J.N. (2011) Deletion of alternative pathways for reductant recycling in *Thermococcus kodakarensis* increases hydrogen production. *Mol. Microbiol.*, **81**, 897–911.
163. Kim, M.S., Choi, A.R., Lee, S.H., Jung, H.C., Bae, S.S., Yang, T.J., Jeon, J.H., Lim, J.K., Youn, H., Kim, T.W., Lee, H.S., and Kang, S.G. (2015) A novel CO-responsive transcriptional regulator and enhanced H_2 production by an engineered *Thermococcus onnurineus* NA1 strain. *Appl. Environ. Microbiol.*, **81**, 1708–1714.
164. Lee, S.H., Kim, M.S., Jung, H.C., Lee, J., Lee, J.H., Lee, H.S., and Kang, S.G. (2015) Screening of a novel strong promoter by RNA sequencing and its application to H_2 production in a hyperthermophilic archaeon. *Appl. Microbiol. Biotechnol.*, **99**, 4085–4092.
165. Chung, D., Cha, M., Snyder, E.N., Elkins, J.G., Guss, A.M., and Westpheling, J. (2015) Cellulosic ethanol production via consolidated bioprocessing at 75°C by engineered *Caldicellulosiruptor bescii. Biotechnol. Biofuels*, **8**, 163.
166. Conway, J.M., Pierce, W.S., Le, J.H., Harper, G.W., Wright, J.H., Tucker, A.L., Zurawski, J.V., Lee, L.L., Blumer-Schuette, S.E., and Kelly, R.M. (2016) Multi domain, surface layer-associated glycoside hydrolases contribute to plant polysaccharide degradation by *Caldicellulosiruptor* species. *J. Biol. Chem.*, **291**, 6732–6747.
167. Xu, H., Han, D., and Xu, Z. (2015) Expression of heterologous cellulases in *Thermotoga* sp. strain RQ2. *Biomed Res. Int.*, **2015**, 304523.
168. Hoshino, T., Yoshino, Y., Guevarra, E.D., Ishida, S., Hiruta, T., Fujii, R., and Nakahara, T. (1994) Isolation and partial characterization of carotenoid underproducing and overproducing mutants from an extremely thermophilic *Thermus thermophilus* HB27. *J. Ferment. Bioeng.*, **77**, 131–136.
169. Peng, H., Wu, G.G., and Shao, W.L. (2008) The aldehyde/alcohol dehydrogenase (AdhE) in relation to the ethanol formation in *Thermoanaerobacter ethanolicus* JW200. *Anaerobe*, **14**, 125–127.
170. Georgieva, T.I., Mikkelsen, M.J., and Ahring, B.K. (2007) High ethanol tolerance of the thermophilic anaerobic ethanol producer *Thermoanaerobacter* BG1L1. *Cent. Eur. J. Biol.*, **2**, 364–377.
171. Georgieva, T.I. and Ahring, B.K. (2007) Evaluation of continuous ethanol fermentation of dilute-acid corn stover

hydrolysate using thermophilic anaerobic bacterium *Thermoanaerobacter* BG1L1. *Appl. Microbiol. Biotechnol.*, **77**, 61–68.

172. Georgieva, T.I., Mikkelsen, M.J., and Ahring, B.K. (2008) Ethanol production from wet-exploded wheat straw hydrolysate by thermophilic anaerobic bacterium *Thermoanaerobacter* BG1L1 in a continuous immobilized reactor. *Appl. Biochem. Biotechnol.*, **145**, 99–110.

173. Yao, S. and Mikkelsen, M.J. (2010) Metabolic engineering to improve ethanol production in *Thermoanaerobacter mathranii*. *Appl. Microbiol. Biotechnol.*, **88**, 199–208.

174. Lin, L., Ji, Y., Tu, Q., Huang, R., Teng, L., Zeng, X., Song, H., Wang, K., Zhou, Q., Li, Y., Cui, Q., He, Z., Zhou, J., and Xu, J. (2013) Microevolution from shock to adaptation revealed strategies improving ethanol tolerance and production in *Thermoanaerobacter*. *Biotechnol. Biofuels*, **6**, 103.

175. Mueller, M., Takemasa, R., Schwarz, A., Atomi, H., and Nidetzky, B. (2009) "Short-chain" α-1,4-glucan phosphorylase having a truncated N-terminal domain: functional expression and characterization of the enzyme from *Sulfolobus solfataricus*. *Biochim. Biophys. Acta*, **1794**, 1709–1714.

176. Lee, S.H., Kim, M.S., Bae, S.S., Choi, A.R., Lee, J.W., Kim, T.W., Lee, J.H., Lee, H.S., and Kang, S.G. (2014) Comparison of CO-dependent H_2 production with strong promoters in *Thermococcus onnurineus* NA1. *Appl. Microbiol. Biotechnol.*, **98**, 979–986.

177. Berkner, S., Wlodkowski, A., Albers, S.V., and Lipps, G. (2010) Inducible and constitutive promoters for genetic systems in *Sulfolobus acidocaldarius*. *Extremophiles*, **14**, 249–259.

178. Henche, A.L., Koerdt, A., Ghosh, A., and Albers, S.V. (2012) Influence of cell surface structures on crenarchaeal biofilm formation using a thermostable green fluorescent protein. *Environ. Microbiol.*, **14**, 779–793.

179. Albers, S.V., Jonuscheit, M., Dinkelaker, S., Urich, T., Kletzin, A., Tampe, R., Driessen, A.J.M., and Schleper, C. (2006) Production of recombinant and tagged proteins in the hyperthermophilic archaeon *Sulfolobus solfataricus*. *Appl. Microbiol. Biotechnol.*, **72**, 102–111.

180. Elferink, M.G., Schleper, C., and Zillig, W. (1996) Transformation of the extremely thermoacidophilic archaeon *Sulfolobus solfataricus* via a self-spreading vector. *FEMS Microbiol. Lett.*, **137**, 31–35.

181. Zhang, C., Guo, L., Deng, L., Wu, Y., Liang, Y., Huang, L., and She, Q. (2010) Revealing the essentiality of multiple archaeal *pcna* genes using a mutant propagation assay based on an improved knockout method. *Microbiology*, **156**, 3386–3397.

182. Yu, J.S., Vargas, M., Mityas, C., and Noll, K.M. (2001) Liposome-mediated DNA uptake and transient expression in *Thermotoga*. *Extremophiles*, **5**, 53–60.

183. Park, H.S. and Kilbane, J.J. (2004) Gene expression studies of *Thermus thermophilus* promoters P*dnaK*, P*arg* and P*scs-mdh*. *Lett. Appl. Microbiol.*, **38**, 415–422.

184. Fujita, A., Misumi, Y., Honda, S., Sato, T., and Koyama, Y. (2013) Construction of new cloning vectors that employ the phytoene synthase encoding gene for color screening of cloned DNA inserts in *Thermus thermophilus*. *Gene*, **527**, 655–662.

185. Moreno, R., Zafra, O., Cava, F., and Berenguer, J. (2003) Development of a gene expression vector for *Thermus thermophilus* based on the promoter of the respiratory nitrate reductase. *Plasmid*, **49**, 2–8.

186. de Grado, M., Castan, P., and Berenguer, J. (1999) A high-transformation-efficiency cloning vector for *Thermus thermophilus*. *Plasmid*, **42**, 241–245.

187. Erbeznik, M., Dawson, K.A., and Strobel, H.J. (1998) Cloning and characterization of transcription of the *xylAB* operon in *Thermoanaerobacter ethanolicus*. *J. Bacteriol.*, **180**, 1103–1109.

188. Mai, V., Lorenz, W.W., and Wiegel, J. (1997) Transformation of *Thermoanaerobacterium* sp. strain JW/SL-YS485

with plasmid pIKM1 conferring kanamycin resistance. *FEMS Microbiol. Lett.*, **148**, 163–167.
189. Hawkins, A.S., Han, Y., Lian, H., Loder, A.J., Menon, A.L., Iwuchukwu, I.J., Keller, M., Leuko, T.T., Adams, M.W.W., and Kelly, R.M. (2011) Extremely thermophilic routes to microbial electrofuels. *ACS Catal.*, **1**, 1043–1050.
190. Hawkins, A.S., McTernan, P.M., Lian, H., Kelly, R.M., and Adams, M.W.W. (2013) Biological conversion of carbon dioxide and hydrogen into liquid fuels and industrial chemicals. *Curr. Opin. Biotechnol.*, **24**, 376–384.
191. Bridger, S.L., Clarkson, S.M., Stirrett, K., DeBarry, M.B., Lipscomb, G.L., Schut, G.J., Westpheling, J., Scott, R.A., and Adams, M.W.W. (2011) Deletion strains reveal metabolic roles for key elemental sulfur-responsive proteins in *Pyrococcus furiosus*. *J. Bacteriol.*, **193**, 6498–6504.
192. Esteves, A.M., Chandrayan, S.K., McTernan, P.M., Borges, N., Adams, M.W.W., and Santos, H. (2014) Mannosylglycerate and di-myo-inositol phosphate have interchangeable roles during adaptation of *Pyrococcus furiosus* to heat stress. *Appl. Microbiol. Biotechnol.*, **80**, 4226–4233.
193. Nguyen, D.M.N., Lipscomb, G.L., Schut, G.J., Vaccaro, B.J., Basen, M., Kelly, R.M., and Adams, M.W.W. (2016) Temperature-dependent acetoin production by *Pyrococcus furiosus* is catalyzed by a biosynthetic acetolactate synthase and its deletion improves ethanol production. *Metab. Eng.*, **34**, 71–79.
194. Schut, G.J., Lipscomb, G.L., Han, Y., Notey, J.S., Kelly, R.M., and Adams, M.M.W. (2014) in *The Prokaryotes: Other Major Lineages of Bacteria and The Archaea*, 4th edn (eds E. Rosenberg, E.F. DeLong, S. Lory, E. Stackebrandt, and F. Thompson), Springer, Berlin, Heidelberg, pp. 363–383.
195. Vannier, P., Marteinsson, V.T., Fridjonsson, O.H., Oger, P., and Jebbar, M. (2011) Complete genome sequence of the hyperthermophilic, piezophilic, heterotrophic, and carboxydotrophic archaeon *Thermococcus barophilus* MP. *J. Bacteriol.*, **193**, 1481–1482.
196. Imanaka, T. and Atomi, H. (2002) Catalyzing "hot" reactions: enzymes from hyperthermophilic archaea. *Chem. Rec.*, **2**, 149–163.
197. Matsubara, K., Yokooji, Y., Atomi, H., and Imanaka, T. (2011) Biochemical and genetic characterization of the three metabolic routes in *Thermococcus kodakarensis* linking glyceraldehyde 3-phosphate and 3-phosphoglycerate. *Mol. Microbiol.*, **81**, 1300–1312.
198. Tiquia-Arashiro, S.M. (2014) *Thermophilic Carboxydotrophs and Their Applications in Biotechnology*, Springer, New York.
199. Kim, Y.J., Lee, H.S., Kim, E.S., Bae, S.S., Lim, J.K., Matsumi, R., Lebedinsky, A.V., Sokolova, T.G., Kozhevnikova, D.A., Cha, S.S., Kim, S.J., Kwon, K.K., Imanaka, T., Atomi, H., Bonch-Osmolovskaya, E.A., Lee, J.H., and Kang, S.G. (2010) Formate-driven growth coupled with H_2 production. *Nature*, **467**, 352–355.
200. Lim, J.K., Bae, S.S., Kim, T.W., Lee, J.H., Lee, H.S., and Kang, S.G. (2012) Thermodynamics of formate-oxidizing metabolism and implications for H_2 production. *Appl. Environ. Microbiol.*, **78**, 7393–7397.
201. Bertani, G. and Baresi, L. (1987) Genetic transformation in the methanogen *Methanococcus voltae* PS. *J. Bacteriol.*, **169**, 2730–2738.
202. Hiyama, T., Sato, T., Imanaka, T., and Atomi, H. (2014) The tryptophan synthase β-subunit paralogs TrpB1 and TrpB2 in *Thermococcus kodakarensis* are both involved in tryptophan biosynthesis and indole salvage. *FEBS J.*, **281**, 3113–3125.
203. Tomita, H., Yokooji, Y., Ishibashi, T., Imanaka, T., and Atomi, H. (2014) An archaeal glutamate decarboxylase homolog functions as an aspartate decarboxylase and is involved in β-alanine and coenzyme A biosynthesis. *J. Bacteriol.*, **196**, 1222–1230.
204. Aono, R., Sato, T., Imanaka, T., and Atomi, H. (2015) A pentose bisphosphate pathway for nucleoside

degradation in Archaea. *Nat. Chem. Biol.*, **11**, 355–360.
205. Borges, N., Matsumi, R., Imanaka, T., Atomi, H., and Santos, H. (2009) *Thermococcus kodakarensis* mutants deficient in di-myo-inositol phosphate use aspartate to cope with heat stress. *J. Bacteriol.*, **192**, 191–197.
206. Elmore, J.R., Yokooji, Y., Sato, T., Olson, S., Glover, C.V.C. III,, Graveley, B.R., Atomi, H., Terns, R.M., and Terns, M.P. (2014) Programmable plasmid interference by the CRISPR-Cas system in *Thermococcus kodakarensis*. *RNA Biol.*, **10**, 828–840.
207. Kanai, T., Simons, J.R., Tsukamoto, R., Nakajima, A., Omori, Y., Matsuoka, R., Beppu, H., Imanaka, T., and Atomi, H. (2015) Overproduction of the membrane-bound [NiFe]-hydrogenase in *Thermococcus kodakarensis* and its effect on hydrogen production. *Front. Microbiol.*, **6**, 847.
208. She Q, Singh R.K., Confalonieri F., Zivanovic Y., Allard G, Awayez M.J., Chan-Weiher C.C., Clausen I.G., Curtis BA, De Moors A, Erauso G, Fletcher C., Gordon P.M., Heikamp-de Jong I., Jeffries AC, Kozera C.J., Medina N., Peng X., Thi-Ngoc H.P., Redder P., Schenk M.E., Theriault C., Tolstrup N., Charlebois R.L., Doolittle W.F., Duguet M., Gaasterland T., Garrett R.A., Ragan M.A., Sensen C.W., Van der Oost J. 2001. The complete genome of the crenarchaeon *Sulfolobus solfataricus* P2. *Proc. Natl. Acad. Sci. U.S.A.* **98**:7835-7840.
209. Chen, L., Brügger, K., Skovgaard, M., Redder, P., She, Q., Torarinsson, E., Greve, B., Awayez, M., Zibat, A., Klenk, H.P., and Garrett, R.A. (2005) The genome of *Sulfolobus acidocaldarius*, a model organism of the Crenarchaeota. *J. Bacteriol.*, **187**, 4992–4999.
210. Guo, L., Brugger, K., Liu, C., Shah, S.A., Zheng, H., Zhu, Y., Wang, S., Lillestol, R.K., Chen, L., Frank, J., Prangishvili, D., Paulin, L., She, Q., Huang, L., and Garrett, R.A. (2011) Genome analyses of Icelandic strains of *Sulfolobus islandicus*, model organisms for genetic and virus–host interaction studies. *J. Bacteriol.*, **193**, 1672–1680.
211. Brugger, K., Torarinsson, E., Redder, P., Chen, L., and Garrett, R.A. (2004) Shuffling of *Sulfolobus* genomes by autonomous and non-autonomous mobile elements. *Biochem. Soc. Trans.*, **32**, 179–183.
212. Zillig, W., Kletzin, A., Schleper, C., and Holz, I. (1993) Screening for Sulfolobales, their plasmids and their viruses in Icelandic solfataras. *Syst. Appl. Microbiol.*, **16**, 609–628.
213. Schleper, C., Kubo, K., and Zillig, W. (1992) The particle SSV1 from the extremely thermophilic archaeon *Sulfolobus* is a virus: demonstration of infectivity and of transfection with viral DNA. *Proc. Natl. Acad. Sci. U.S.A.*, **89**, 7645–7649.
214. Lipps, G. (2009) Molecular biology of the pRN1 plasmid from *Sulfolobus islandicus*. *Biochem. Soc. Trans.*, **37**, 42–45.
215. Grogan, D.W. (2003) Cytosine methylation by the *Sua*I restriction–modification system: implications for genetic fidelity in a hyperthermophilic archaeon. *J. Bacteriol.*, **185**, 4657–4661.
216. Petrus, A.K., Swithers, K.S., Ranjit, C., Wu, S., Brewer, H.M., Gogarten, J.P., Pasa-Tolic, L., and Noll, K.M. (2012) Genes for the major structuralcomponents of Thermotogales species' togas revealed by proteomic and evolutionary analyses of OmpA and OmpB homologs. *PLoS One*, **7**, e40236.
217. Schumann, J., Wrba, A., Jaenicke, R., and Stetter, K.O. (1991) Topographical and enzymatic characterization of amylases from the extremely thermophilic eubacterium *Thermotoga maritima*. *FEBS Lett.*, **282**, 122–126.
218. Liebl, W.G., Winterhalter, C., Baumeister, W., Armbrech, M., and Valdez, M. (2008) Xylanase attachment to the cell wall of the hyperthermophilic bacterium *Thermotoga maritima*. *J. Bacteriol.*, **190**, 1350–1358.
219. Galperin, M.Y., Noll, K.M., and Romano, A.H. (1996) The glucose transport system of the hyperthermophilic anaerobic bacterium

Thermotoga neapolitana. *Appl. Environ. Microbiol.*, **62**, 2915–2918.

220. Blumer-Schuette, S.E., Kataeva, I., Westpheling, J., Adams, M.W.W., and Kelly, R.M. (2008) Extremely thermophilic microorganisms for biomass conversion: status and prospects. *Curr. Opin. Biotechnol.*, **19**, 210–217.
221. Selig, M., Xavier, K.B., Santos, H., and Schonheit, P. (1997) Comparative analysis of Embden–Meyerhof and Entner–Doudoroff glycolytic pathways in hyperthermophilic archaea and the bacterium *Thermotoga*. *Arch. Microbiol.*, **167**, 217–232.
222. Schroder, C., Selig, M., and Schonheit, P. (1994) Glucose fermentation to acetate, CO_2 and H_2 in the anaerobic hyperthermophilic eubacterium *Thermotoga maritima* – Involvement of the Embden–Meyerhof pathway. *Arch. Microbiol.*, **161**, 460–470.
223. Chou, C.J., Jenney, F.E., Adams, M.W.W., and Kelly, R.M. (2008) Hydrogenesis in hyperthermophilic microorganisms: implications for biofuels. *Metab. Eng.*, **10**, 394–404.
224. Conners, S.B., Mongodin, E.F., Johnson, M.R., Montero, C.I., Nelson, K.E., and Kelly, R.M. (2006) Microbial biochemistry, physiology, and biotechnology of hyperthermophilic *Thermotoga* species. *FEMS Microbiol. Rev.*, **30**, 872–905.
225. Vargas, M., Kashefi, K., Blunt-Harris, E.L., and Lovley, D.R. (1998) Microbiological evidence for Fe(III) reduction on early Earth. *Nature*, **395**, 65–67.
226. Ravot, G., Ollivier, B., Magot, M., Patel, B.K.C., Crolet, J.L., Fardeau, M.L., and Garcia, J.L. (1995) Thiosulfate reduction, an important physiological feature shared by members of the order Thermotogales. *Appl. Environ. Microbiol.*, **61**, 2053–2055.
227. Janssen, P.H. and Morgan, H.W. (1992) Heterotrophic sulfur reduction by *Thermotoga* sp. strain FJSS3.B1. *FEMS Microbiol. Lett.*, **96**, 213–218.
228. Zannoni, D. and De Philippis, R. (2014) *Microbial BioEnergy: Hydrogen Production*, vol. 38, Springer, Dordrecht, Netherlands.
229. Kotay, S.M. and Das, D. (2008) Biohydrogen as a renewable energy resource – Prospects and potentials. *Int. J. Hydrogen Energy*, **33**, 258–263.
230. Harriott, O.T., Huber, R., Stetter, K.O., Betts, P.W., and Noll, K.M. (1994) A cryptic miniplasmid from the hyperthermophilic bacterium *Thermotoga* sp. strain RQ7. *J. Bacteriol.*, **176**, 2759–2762.
231. van Ooteghem, S.A., Beer, S.K., and Yue, P.C. (2002) Hydrogen production by the thermophilic bacterium, *Thermotoga neapolitana*. *Appl. Biochem. Biotechnol.*, **90**, 177–189.
232. Brock, T.D. and Freeze, H. (1969) *Thermus aquaticus* gen. n. and sp. n., a nonsporulating extreme thermophile. *J. Bacteriol.*, **98**, 289–297.
233. Kristjansson, J.K., Hreggvidsson, G.O., and Alfredsson, G.A. (1986) Isolation of halotolerant *Thermus* spp. from submarine hot springs in Iceland. *Appl. Environ. Microbiol.*, **52**, 1313–1316.
234. Cava, F., Hidalgo, A., and Berenguer, J. (2009) *Thermus thermophilus* as biological model. *Extremophiles*, **13**, 213–231.
235. Burton, N.P., Williams, T.D., and Norris, P.R. (1995) A potential anti-oxidant protein in a ferrous iron-oxidizing Sulfolobus species. *FEMS Microbiol. Lett.*, **134**, 91–95.
236. Degryse, E., Glansdorff, N., and Piérard, A. (1978) A comparative analysis of extreme thermophilic bacteria belonging to the genus *Thermus*. *Arch. Microbiol.*, **117**, 189–196.
237. Leis, B., Angelov, A., Li, H., and Liebl, W. (2014) Genetic analysis of lipolytic activities in *Thermus thermophilus* HB27. *J. Biotechnol.*, **191**, 150–157.
238. Bjornsdottir, S.H., Petursdottir, S.K., Hreggvidsson, G.O., Skirnisdottir, S., Hjorleifsdottir, S., Arnfinnsson, J., and Kristjansson, J.K. (2009) *Thermus islandicus* sp. nov., a mixotrophic sulfur-oxidizing bacterium isolated from the Torfajokull geothermal area. *Int. J. Syst. Evol. Microbiol.*, **59**, 2962–2966.
239. Gihring, T.M. and Banfield, J.F. (2001) Arsenite oxidation and arsenate respiration by a new *Thermus* isolate. *FEMS Microbiol. Lett.*, **204**, 335–340.

240. Mandelli, F., Ramires, B.O., Couger, M.B., Paixão, D.A.A., Camilo, C.M., Polikarpov, I., Prade, R., Riaño-Pachón, D.M., and Squina, F.M. (2015) Draft genome sequence of the thermophile *Thermus filiformis* ATCC 43280, producer of carotenoid-(di)glucoside-branched fatty acid (di)esters and source of hyperthermostable enzymes of biotechnological interest. *Genome Announc.*, **3**, e00475-15.
241. Hoshino, T., Fujii, R., and Nakahara, T. (1993) Molecular cloning and sequence analysis of the *crtB* gene of *Thermus thermophilus* HB27, an extreme thermophile producing carotenoid pigments. *Appl. Environ. Microbiol.*, **59**, 3150–3153.
242. Chien, A., Edgar, D.B., and Trela, J.M. (1976) Deoxyribonucleic acid polymerase from the extreme thermophile *Thermus aquaticus*. *J. Bacteriol.*, **127**, 1550–1557.
243. Henne, A., Brüggemann, H., Raasch, C., Wiezer, A., Hartsch, T., Liesegang, H., Johann, A., Lienard, T., Gohl, O., Martinez-Arias, R., Jacobi, C., Starkuviene, V., Schlenczeck, S., Dencker, S., Huber, R., Klenk, H.-P., Kramer, W., Merkl, R., Gottschalk, G., and Fritz, H.J. (2004) The genome sequence of the extreme thermophile *Thermus thermophilus*. *Nat. Biotechnol.*, **22**, 547–553.
244. Friedrich, A., Hartsch, T., and Averhoff, B. (2001) Natural transformation in mesophilic and thermophilic bacteria: identification and characterization of novel, closely related competence genes in *Acinetobacter* sp. strain BD413 and *Thermus thermophilus* HB27. *Appl. Environ. Microbiol.*, **67**, 3140–3148.
245. Averhoff, B. (2009) Shuffling genes around in hot environments: the unique DNA transporter of *Thermus thermophilus*. *FEMS Microbiol. Rev.*, **33**, 611–626.
246. Blesa, A., César, C.E., Averhoff, B., and Berenguer, J. (2015) Noncanonical cell-to-cell DNA transfer in *Thermus* spp. is insensitive to argonaute-mediated interference. *J. Bacteriol.*, **197**, 138–146.
247. Friedrich, A., Prust, C., Hartsch, T., Henne, A., and Averhoff, B. (2002) Molecular analyses of the natural transformation machinery and identification of pilus structures in the extremely thermophilic bacterium *Thermus thermophilus* strain HB27. *Appl. Environ. Microbiol.*, **68**, 745–755.
248. Schwarzenlander, C. and Averhoff, B. (2006) Characterization of DNA transport in the thermophilic bacterium *Thermus thermophilus* HB27. *FEBS J.*, **273**, 4210–4218.
249. César, C.E., Álvarez, L., Bricio, C., van Heerden, E., Littauer, D., and Berenguer, J. (2012) Unconventional lateral gene transfer in extreme thermophilic bacteria. *Int. Microbiol.*, **14**, 187–199.
250. Touhara, N., Taguchi, H., Koyama, Y., Ohta, T., and Matsuzawa, H. (1991) Production and extracellular secretion of aqualysin I (a thermophilic subtilisin-type protease) in a host-vector system for *Thermus thermophilus*. *Appl. Environ. Microbiol.*, **57**, 3385–3387.
251. Moreno, R., Haro, A., Castellanos, A., and Berenguer, J. (2005) High-level overproduction of His-tagged *Tth* DNA polymerase in *Thermus thermophilus*. *Appl. Environ. Microbiol.*, **71**, 591–593.
252. Fridjonsson, O. and Mattes, R. (2001) Production of recombinant α-galactosidases in *Thermus thermophilus*. *Appl. Environ. Microbiol.*, **67**, 4192–4198.
253. Park, H.S., Kayser, K.J., Kwak, J.H., and Ii, J.J.K. (2004) Heterologous gene expression in *Thermus thermophilus*: β-galactosidase, dibenzothiophene monooxygenase, PNB carboxy esterase, 2-aminobiphenyl-2,3-diol dioxygenase, and chloramphenicol acetyl transferase. *J. Ind. Microbiol. Biotechnol.*, **31**, 189–197.
254. Kotsuka, T., Akanuma, S., Tomuro, M., Yamagishi, A., and Oshima, T. (1996) Further stabilization of 3-isopropylmalate dehydrogenase of an extreme thermophile, *Thermus thermophilus*, by a suppressor mutation method. *J. Bacteriol.*, **178**, 723–727.
255. Tamakoshi, M. and Oshima, T. (2011) in *Extremophiles Handbook* (ed. K.

Horikoshi), Springer, Tokyo, pp. 547–566.

256. Inoue, T. and Sako, Y. (2013) in *Thermophilic Microbes in Environmental and Industrial Biotechnology: Biotechnology of Thermophiles* (eds T. Satyanarayana, J. Littlechild, and Y. Kawarabayasi), Springer, Dordrecht, pp. 351–373.

257. Hidaka, Y., Hasegawa, M., Nakahara, T., and Hoshino, T. (1994) The entire population of *Thermus thermophilus* cells is always competent at any growth phase. *Biosci. Biotechnol., Biochem.*, **58**, 1338–1339.

258. Tamakoshi, M., Yamagishi, A., and Oshima, T. (1995) Screening of stable proteins in an extreme thermophile, *Thermus thermophilus. Mol. Microbiol.*, **16**, 1031–1036.

259. Matsumura, M. and Aiba, S. (1985) Screening for thermostable mutant of kanamycin nucleotidyltransferase by the use of a transformation system for a thermophile, *Bacillus stearothermophilus. J. Biol. Chem.*, **260**, 15298–15303.

260. Blas-Galindo, E., Cava, F., López-Viñas, E., Mendieta, J., and Berenguer, J. (2007) Use of a dominant *rpsL* allele conferring streptomycin dependence for positive and negative selection in *Thermus thermophilus. Appl. Microbiol. Biotechnol.*, **73**, 5138–5145.

261. Gregory, S.T., Cate, J.H.D., and Dahlberg, A.E. (2001) Streptomycin-resistant and streptomycin-dependent mutants of the extreme thermophile *Thermus thermophilus. J. Mol. Biol.*, **309**, 333–338.

262. Carr, J.F., Lee, H.J., Jaspers, J.B., Dahlberg, A.E., Jogl, G., and Gregory, S.T. (2015) Phenotypic suppression of streptomycin resistance by mutations in multiple components of the translation apparatus. *J. Bacteriol.*, **197**, 2981–2988.

263. Cameron, D.M., Thompson, J., Gregory, S.T., March, P.E., and Dahlberg, A.E. (2004) Thiostrepton-resistant mutants of *Thermus thermophilus. Nucleic Acids Res.*, **32**, 3220–3227.

264. Fujita, A., Sato, T., Koyama, Y., and Misumi, Y. (2015) A reporter gene system for the precise measurement of promoter activity in *Thermus thermophilus* HB27. *Extremophiles*, **19**, 1193–1201.

265. Ohta, T., Tokishita, S., Imazuka, R., Mori, I., Okamura, J., and Yamagata, H. (2006) beta-Glucosidase as a reporter for the gene expression studies in *Thermus thermophilus* and constitutive expression of DNA repair genes. *Mutagenesis*, **21**, 255–260.

266. Brüggemann, H. and Chen, C. (2006) Comparative genomics of *Thermus thermophilus*: plasticity of the megaplasmid and its contribution to a thermophilic lifestyle. *J. Biotechnol.*, **124**, 654–661.

267. Brumm, P.J., Monsma, S., Keough, B., Jasinovica, S., Ferguson, E., Schoenfeld, T., Lodes, M., and Mead, D.A. (2015) Complete genome sequence of *Thermus aquaticus* Y51MC23. *PLoS One*, **10**, e0138674.

268. Dwivedi, V., Sangwan, N., Nigam, A., Garg, N., Niharika, N., Khurana, P., Khurana, J.P., and Lal, R. (2012) Draft genome sequence of *Thermus* sp. strain RL, isolated from a hot water spring located atop the Himalayan ranges at Manikaran, India. *J. Bacteriol.*, **194**, 3534.

269. Gounder, K., Brzuszkiewicz, E., Liesegang, H., Wollherr, A., Daniel, R., Gottschalk, G., Reva, O., Kumwenda, B., Srivastava, M., Bricio, C., Berenguer, J., van Heerden, E., and Litthauer, D. (2011) Sequence of the hyperplastic genome of the naturally competent *Thermus scotoductus* SA-01. *BMC Genomics*, **12**, 577.

270. Ohtani, N., Tomita, M., and Itaya, M. (2010) An extreme thermophile, *Thermus thermophilus*, is a polyploid bacterium. *J. Bacteriol.*, **192**, 5499–5505.

271. Teh, B.S., Lau, N.S., Ng, F.L., Abdul Rahman, A.Y., Wan, X., Saito, J.A., Hou, S., Teh, A.H., Najimudin, N., and Alam, M. (2015) Complete genome sequence of the thermophilic *Thermus* sp. CCB_US3_UF1 from a hot spring in Malaysia. *Stand. Genomic Sci.*, **10**, 76.

272. Zurawski, J.V., Conway, J.M., Lee, L.L., Simpson, H.J., Izquierdo, J.A.,

Blumer-Schuette, S., Nookaew, I., Adams, M.W.W., and Kelly, R.M. (2015) Comparative analysis of extremely thermophilic *Caldicellulosiruptor* species reveals common and unique cellular strategies for plant biomass utilization. *Appl. Environ. Microbiol.*, **81**, 7159–7170.
273. van Niel, E.W.J., Claassen, P.A.M., and Stams, A.J.M. (2002) Substrate and product inhibition of hydrogen production by the extreme thermophile, *Caldicellulosiruptor saccharolyticus*. *Biotechnol. Bioeng.*, **81**, 255–262.
274. Skinner, K.A. and Leathers, T.D. (2004) Bacterial contaminants of fuel ethanol production. *J. Ind. Microbiol. Biotechnol.*, **31**, 401–408.
275. Scott, I.M., Rubinstein, G.M., Lipscomb, G.L., Basen, M., Schut, G.J., Rhaesa, A.M., Lancaster, W.A., Poole, F.L., Kelly, R.M., and Adams, M.W.W. (2015) A new class of tungsten-containing oxidoreductase in *Caldicellulosiruptor*, a genus of plant biomass-degrading thermophilic bacteria. *Appl. Environ. Microbiol.*, **81**, 7339–7347.
276. Cha, M., Wang, H., Chung, D., Bennetzen, J.L., and Westpheling, J. (2013) Isolation and bioinformatic analysis of a novel transposable element, ISCbe4, from the hyperthermophilic bacterium, *Caldicellulosiruptor bescii*. *J. Ind. Microbiol. Biotechnol.*, **40**, 1443–1448.
277. Blumer-Schuette, S.E., Giannone, R.J., Zurawski, J.V., Ozdemir, I., Ma, Q., Yin, Y., Xu, Y., Kataeva, I., Poole, F.L. II,, Adams, M.W., Hamilton-Brehm, S.D., Elkins, J.G., Larimer, F.W., Land, M.L., Hauser, L.J., Cottingham, R.W., Hettich, R.L., and Kelly, R.M. (2012) *Caldicellulosiruptor* core and pangenomes reveal determinants for noncellulosomal thermophilic deconstruction of plant biomass. *J. Bacteriol.*, **194**, 4015–4028.
278. Hamilton-Brehm, S.D., Mosher, J.J., Vishnivetskaya, T., Podar, M., Carroll, S., Allman, S., Phelps, T.J., Keller, M., and Elkins, J.G. (2010) *Caldicellulosiruptor obsidiansis* sp. nov., an anaerobic, extremely thermophilic, cellulolytic bacterium isolated from Obsidian Pool, Yellowstone National Park. *Appl. Environ. Microbiol.*, **76**, 1014–1020.
279. Chung, D., Young, J., Cha, M., Brunecky, R., Bomble, Y.J., Himmel, M.E., and Westpheling, J. (2015) Expression of the *Acidothermus cellulolyticus* E1 endoglucanase in *Caldicellulosiruptor bescii* enhances its ability to deconstruct crystalline cellulose. *Biotechnol. Biofuels*, **8**, 113.
280. Cha, M., Chung, D., and Westpheling, J. (2015) Deletion of a gene cluster for [Ni-Fe] hydrogenase maturation in the anaerobic hyperthermophilic bacterium *Caldicellulosiruptor bescii* identifies its role in hydrogen metabolism. *Appl. Microbiol. Biotechnol.*, **100**, 1823–1831.
281. Ozdemir, I., Blumer-Schuette, S.E., and Kelly, R.M. (2012) S-layer homology domain proteins Csac_0678 and Csac_2722 are implicated in plant polysaccharide deconstruction by the extremely thermophilic bacterium *Caldicellulosiruptor saccharolyticus*. *Appl. Environ. Microbiol.*, **78**, 768–777.
282. Dam, P., Kataeva, I., Yang, S.J., Zhou, F.F., Yin, Y.B., Chou, W.C., Poole, F.L., Westpheling, J., Hettich, R., Giannone, R., Lewis, D.L., Kelly, R., Gilbert, H.J., Henrissat, B., Xu, Y., and Adams, M.W.W. (2011) Insights into plant biomass conversion from the genome of the anaerobic thermophilic bacterium *Caldicellulosiruptor bescii* DSM 6725. *Nucleic Acids Res.*, **39**, 3240–3254.
283. Morris, D.D., Gibbs, M.D., Ford, M., Thomas, J., and Bergquist, P.L. (1999) Family 10 and 11 xylanase genes from *Caldicellulosiruptor* sp. strain Rt69B.1. *Extremophiles*, **3**, 103–111.
284. Lee, Y.E., Jain, M.K., Lee, C., and Zeikus, J.G. (1993) Taxonomic distinction of saccharolytic thermophilic anaerobes. *Int. J. Syst. Bacteriol.*, **43**, 41–51.
285. Hemme, C.L., Mouttaki, H., Lee, Y.J., Zhang, G., Goodwin, L., Lucas, S., Copeland, A., Lapidus, A., Glavina del Rio, T., Tice, H., Saunders, E., Brettin, T., Detter, J.C., Han, C.S., Pitluck, S., Land, M.L., Hauser, L.J., Kyrpides, N., Mikhailova, N.,

He, Z., Wu, L., Van Nostrand, J.D., Henrissat, B., He, Q., Lawson, P.A., Tanner, R.S., Lynd, L.R., Wiegel, J., Fields, M.W., Arkin, A.P., Schadt, C.W., Stevenson, B.S., McInerney, M.J., Yang, Y., Dong, H., Xing, D., Ren, N., Wang, A., Huhnke, R.L., Mielenz, J.R., Ding, S.Y., Himmel, M.E., Taghavi, S., van der Lelie, D., Rubin, E.M., and Zhou, J. (2010) Sequencing of multiple clostridial genomes related to biomass conversion and biofuel production. *J. Bacteriol.*, **192**, 6494–6496.

286. Feng, X., Mouttaki, H., Lin, L., Huang, R., Wu, B., Hemme, C.L., He, Z., Zhang, B., Hicks, L.M., Xu, J., Zhou, J., and Tang, Y.J. (2009) Characterization of the central metabolic pathways in *Thermoanaerobacter* sp. strain X514 via isotopomer-assisted metabolite analysis. *Appl. Environ. Microbiol.*, **75**, 5001–5008.

287. Verbeke, T.J., Zhang, X.L., Henrissat, B., Spicer, V., Rydzak, T., Krokhin, O.V., Fristensky, B., Levin, D.B., and Sparling, R. (2013) Genomic evaluation of *Thermoanaerobacter* spp. for the construction of designer co-cultures to improve lignocellulosic biofuel production. *PLoS One*, **8**, e59362.

288. Jayasinghearachchi, H.S., Sarma, P.M., and Lal, B. (2012) Biological hydrogen production by extremely thermophilic novel bacterium *Thermoanaerobacter mathranii* A3N isolated from oil producing well. *Int. J. Hydrogen Energy*, **37**, 5569–5578.

289. Soutschek-Bauer, E., Scholz, W., Grill, E., and Staudenbauer, W.L. (1987) Thermostability and superhelicity of plasmid DNA in *Bacillus stearothermophilus. Mol. Gen. Genet.*, **209**, 575–579.

290. Mikkelsen, M.J., Ahring, B.K., and Georgieva, T.I. (2013) *Thermoanaerobacter mathranii* strain BG1. US Patent US8512714 B2.

291. Fardeau, M.L., Bonilla Salinas, M., L'Haridon, S., Jeanthon, C., Verhe, F., Cayol, J.L., Patel, B.K., Garcia, J.L., and Ollivier, B. (2004) Isolation from oil reservoirs of novel thermophilic anaerobes phylogenetically related to *Thermoanaerobacter subterraneus*: reassignment of *T. subterraneus, Thermoanaerobacter yonseiensis, Thermoanaerobacter tengcongensis* and *Carboxydibrachium pacificum* to *Caldanaerobacter subterraneus* gen. nov., sp. nov., comb. nov. as four novel subspecies. *Int. J. Syst. Evol. Microbiol.*, **54**, 467–474.

292. Wang, J.Q., Zhao, C.F., Meng, B., Xie, J.H., Zhou, C.Q., Chen, X.S., Zhao, K., Shao, J.M., Xue, Y.F., Xu, N.Z., Ma, Y.H., and Liu, S. (2007) The proteomic alterations of *Thermoanaerobacter tengcongensis* cultured at different temperatures. *Proteomics*, **7**, 1409–1419.

293. Meng, B., Qian, Z., Wei, F., Wang, W.W., Zhou, C.Q., Wang, Z.W., Wang, Q.H., Tong, W., Wang, Q., Ma, Y., Xu, N.Z., and Liu, S.Q. (2009) Proteomic analysis on the temperature-dependent complexes in *Thermoanaerobacter tengcongensis. Proteomics*, **9**, 3189–3200.

294. Wang, C., Jin, C., Zhang, J., Bao, Q., Liu, B., and Tan, H. (2015) Transcriptomic analysis of *Thermoanaerobacter tengcongensis* grown at different temperatures by RNA sequencing. *J. Genet. Genomics*, **42**, 335–338.

295. Bao, Q., Tian, Y., Li, W., Xu, Z., Xuan, Z., Hu, S., Dong, W., Yang, J., Chen, Y., Xue, Y., Xu, Y., Lai, X., Huang, L., Dong, X., Ma, Y., Ling, L., Tan, H., Chen, R., Wang, J., Yu, J., and Yang, H. (2002) A complete sequence of the *T. tengcongensis* genome. *Genome Res.*, **12**, 689–700.

296. Wagner, M., Wagner, A., Ma, X., Kort, J.C., Ghosh, A., Rauch, B., Siebers, B., and Albers, S.V. (2013) Investigation of the *malE* Promoter and MalR, a positive regulator of the maltose regulon, for an improved expression system in *Sulfolobus acidocaldarius. Appl. Environ. Microbiol.*, **80**, 1072–1081.

297. Schut, G.J., Bridger, S.L., and Adams, M.W.W. (2007) Insights into the metabolism of elemental sulfur by the hyperthermophilic archaeon *Pyrococcus furiosus*: characterization of a coenzyme A-dependent NAD(P)H sulfur oxidoreductase. *J. Bacteriol.*, **189**, 4431–4441.

298. Lipscomb, G.L., Keese, A.M., Cowart, D.M., Schut, G.J., Thomm, M., Adams, M.W.W., and Scott, R.A. (2009) SurR: a transcriptional activator and repressor controlling hydrogen and elemental sulphur metabolism in *Pyrococcus furiosus*. *Mol. Microbiol.*, **71**, 332–349.
299. Chandrayan, S.K., McTernan, P.M., Hopkins, R.C., Sun, J., Jenney, F.E., and Adams, M.W.W. (2012) Engineering hyperthermophilic archaeon *Pyrococcus furiosus* to overproduce its cytoplasmic [NiFe]-hydrogenase. *J. Biol. Chem.*, **287**, 3257–3264.
300. Alper, H., Fischer, C., Nevoigt, E., and Stephanopoulos, G. (2005) Tuning genetic control through promoter engineering. *Proc. Natl. Acad. Sci. U.S.A.*, **102**, 12678–12683.
301. De Mey, M., Maertens, J., Lequeux, G.J., Soetaert, W.K., and Vandamme, E.J. (2007) Construction and model-based analysis of a promoter library for *E. coli*: an indispensable tool for metabolic engineering. *BMC Biotechnol.*, **7**, 34.
302. Zhang, F., Carothers, J.M., and Keasling, J.D. (2012) Design of a dynamic sensor-regulator system for production of chemicals and fuels derived from fatty acids. *Nat. Biotechnol.*, **30**, 354–359.
303. Salis, H.M., Mirsky, E.A., and Voigt, C.A. (2009) Automated design of synthetic ribosome binding sites to control protein expression. *Nat. Biotechnol.*, **27**, 946–950.
304. Salis, H.M. (2011) The ribosome binding site calculator. *Methods Enzymol.*, **498**, 19–42.
305. Santangelo, T.J., Cubonova, L., Matsumi, R., Atomi, H., Imanaka, T., and Reeve, J.N. (2008) Polarity in archaeal operon transcription in *Thermococcus kodakaraensis*. *J. Bacteriol.*, **190**, 2244–2248.
306. Liu, W. and Jiang, R. (2015) Combinatorial and high-throughput screening approaches for strain engineering. *Appl. Microbiol. Biotechnol.*, **99**, 2093–2104.
307. King, Z.A., Lloyd, C.J., Feist, A.M., and Palsson, B.O. (2015) Next-generation genome-scale models for metabolic engineering. *Curr. Opin. Biotechnol.*, **35**, 23–29.
308. Jakociunas, T., Jensen, M.K., and Keasling, J.D. (2016) CRISPR/Cas9 advances engineering of microbial cell factories. *Metab. Eng.*, **34**, 44–59.
309. Terns, R.M. and Terns, M.P. (2014) CRISPR-based technologies: prokaryotic defense weapons repurposed. *Trends Genet.*, **30**, 111–118.
310. Barrangou, R. and van Pijkeren, J.P. (2016) Exploiting CRISPR-Cas immune systems for genome editing in bacteria. *Curr. Opin. Biotechnol.*, **37**, 61–68.
311. Makarova, K.S., Haft, D.H., Barrangou, R., Brouns, S.J., Charpentier, E., Horvath, P., Moineau, S., Mojica, F.J., Wolf, Y.I., Yakunin, A.F., van der Oost, J., and Koonin, E.V. (2011) Evolution and classification of the CRISPR-Cas systems. *Nat. Rev. Microbiol.*, **9**, 467–477.
312. Barabote, R.D., Xie, G., Leu, D.H., Normand, P., Necsulea, A., Daubin, V., Medigue, C., Adney, W.S., Xu, X.C., Lapidus, A., Parales, R.E., Detter, C., Pujic, P., Bruce, D., Lavire, C., Challacombe, J.F., Brettin, T.S., and Berry, A.M. (2009) Complete genome of the cellulolytic thermophile *Acidothermus cellulolyticus* 11B provides insights into its ecophysiological and evolutionary adaptations. *Genome Res.*, **19**, 1033–1043.
313. Gotz, D., Banta, A., Beveridge, T.J., Rushdi, A.I., Simoneit, B.R., and Reysenbach, A.L. (2002) *Persephonella marina* gen. nov., sp. nov. and *Persephonella guaymasensis* sp. nov., two novel, thermophilic, hydrogen-oxidizing microaerophiles from deep-sea hydrothermal vents. *Int. J. Syst. Evol. Microbiol.*, **52**, 1349–1359.
314. Li, Y., Pan, S., Zhang, Y., Ren, M., Feng, M., Peng, N., Chen, L., Liang, Y.X., and She, Q. (2016) Harnessing type I and type III CRISPR-Cas systems for genome editing. *Nucleic Acids Res.*, **44**, e34.
315. Elmore, J., Deighan, T., Westpheling, J., Terns, R.M., and Terns, M.P. (2015)

DNA targeting by the type I-G and type I-A CRISPR-Cas systems of *Pyrococcus furiosus*. *Nucleic Acids Res.*, **43**, 10353–10363.

316. Hale, C.R., Cocozaki, A., Li, H., Terns, R.M., and Terns, M.P. (2014) Target RNA capture and cleavage by the Cmr type III-B CRISPR-Cas effector complex. *Genes Dev.*, **28**, 2432–2443.

317. Blazeck, J., Liu, L., Redden, H., and Alper, H. (2011) Tuning gene expression in *Yarrowia lipolytica* by a hybrid promoter approach. *Appl. Microbiol. Biotechnol.*, **77**, 7905–7914.

15
Cyanobacteria as a Host Organism

Fabienne Duchoud, Derrick S.W. Chuang, and James C. Liao

15.1
Introduction and Relevance: Cyanobacteria as a Host Organism

Cyanobacteria, a group of prokaryotes previously known as *blue-green algae*, belong to the most ancient group of photosynthetic organisms and are the only prokaryotes capable of carrying out plantlike oxygenic photosynthesis. In fact, plants chloroplasts are considered to be originated from cyanobacteria through endosymbiosis. With a theoretical maximum photosynthetic efficiency in the range of 8–10% [1], compared to 4.6–6% for terrestrial plants [2], they convert solar energy, CO_2, and water into chemicals with only minimal nutrient requirement. Cyanobacteria exhibit a large morphological diversity and are comprised of unicellular, filamentous, colonial, and planktonic types. They can thrive in a broad range of habitats, from aquatic ecosystems, including fresh and marine water, to terrestrial ecosystems. As a result of this large biodiversity, cyanobacteria produce a wide array of molecules which have gained great importance in biotechnological and chemical industries, with compounds that may act as pesticide or have antibacterial, antiviral, and antifungal properties [3]. Moreover, because cyanobacteria are susceptible to genetic transformation, recent investment has been made for their potential use as a platform for production of biofuels and chemicals. In addition to having photosynthesis capabilities superior to that in plants, cyanobacteria cultivation does not require agricultural land, minimizing competition with food crops. Extensive research has been carried out mainly in the *Synechococcus* and *Synechocystis* genera, with some lesser effort accomplished in *Anabaena* and *Cyanothece*.

This chapter first gives a general introduction to cyanobacteria physiology as well as their genetic tools, which are important background for utilizing these organisms for industrial applications. We then review works on the development of strains as potential cell factories for the production of commercially relevant compounds, with a focus on primary metabolites.

Industrial Biotechnology: Microorganisms, First Edition. Edited by Christoph Wittmann and James C. Liao.

15.2 General Description of Cyanobacteria

15.2.1 A Diverse Bacterial Group and Its Metabolisms

Cyanobacteria form a phylum of bacteria that can perform plantlike photosynthesis. The cyanobacteria phylum includes some 2000 species from 150 genera and 5 orders, which vary greatly in shapes and sizes. All cyanobacteria are unicellular, although many grow in aggregates of cells, in more structured colonies, or in filaments. They represent the most genetically diverse group of microorganisms and survive in a large range of habitats across the globe but are most common in freshwater, marine, and terrestrial environments. They can be found in extreme ecological niches such as deserts, hot springs, and hypersaline waters. Besides their wild diversity in terms of morphology and habitats, cyanobacteria also differ in terms of physiology and metabolic capacities [4]. They have the ability, throughout different species, to synthesize a broad range of structurally and functionally diverse natural products [3].

Most cyanobacteria carry out oxygenic photosynthesis which allows them to use minimal nutrients for growth, namely, carbon dioxide, water, and some salts as their carbon and nitrogen sources, while deriving energy from light. Some of them can also utilize organic compounds for heterotrophy. Many cyanobacteria fix atmospheric nitrogen under what seems to be aerobic conditions [5]. This is surprising as nitrogenase, the enzymatic complex responsible for nitrogen fixation, is very oxygen sensitive. Cyanobacteria deal with that problem by separating, either spatially or temporarily, the two processes of oxygenic photosynthesis and nitrogen fixation. In filamentous strains, the nitrogenase is confined to heterocysts, differentiated thick-walled cells that lack an oxygenic photosystem and are the site of N_2 fixation. Other species, unicellular as well as filamentous, induce nitrogenase activity only in the dark periods of light-dark growth cycles, when photosynthesis is shut down.

Even though oxygenic photosynthesis is the principal mode of growth of these organisms, some species can also be found in anoxic environments [6]. These cyanobacteria are able to switch to sulfide-dependent anoxygenic photosynthesis where inorganic sulfur compounds act as electron donors [7, 8]. Furthermore, other species show a distinct ability to assimilate sugars or other organic compounds in the light (photoheterotrophic growth) [9, 10]. It is also known that certain cyanobacteria can survive for prolonged periods of time in complete darkness. They have indeed the ability to accumulate and store a range of different compounds that can be utilized later when nutrients are low, as it is the case during a dark phase. Among them is glycogen that serves as a carbon source and can be respired to generate energy, polyphosphate to provide a source of phosphorus, cyanophycin to supply nitrogen, and poly-β-hydroxybutyrate, although its function is uncertain [11]. Under dark anoxic condition, some species can perform fermentation for maintenance requirements [12].

15.2.2 Nitrogen Fixation

Cyanobacteria are among the few organisms, all prokaryotes, capable of fixing atmospheric nitrogen, thus decreasing their dependence on reduced nitrogen sources [13]. They play a major role in the global nitrogen cycle. Many strains of cyanobacteria, termed *diazotrophs*, are able to fix N_2. Because they simultaneously produce oxygen through photosynthesis and can fix nitrogen, they possess particularly well-developed mechanisms to protect the extremely oxygen sensitive nitrogen-fixing enzyme, nitrogenase. The two main strategies employed are spatial separation and temporal separation.

Spatial separation is mainly used by filamentous heterocysts-forming cyanobacteria [14]. When the cells are deprived of dissolved inorganic nitrogen, some of them differentiate into heterocysts in a semiregularly spaced pattern along the filament. These thick-walled cells are the site of N_2 fixation. They have decreased permeability to O_2 and high rate of respiration to protect nitrogenase from oxygen inactivation. They also only contain photosystem I for energy production in the form of ATP, but lack photosystem II to avoid oxygen production [15]. Heterocysts rely on vegetative cells for carbon sources as they cannot fix CO_2.

While all heterocystous cyanobacteria can fix nitrogen through nitrogenase, only some nonheterocystous cyanobacteria can (see [16] for a review of these strains). In addition, a gradation of tolerance to O_2 exists among those strains capable of N_2 fixation. Most do it only under micro-oxic or anoxic conditions, yet a few can fix nitrogen aerobically [16]. The main strategy employed to prevent inactivation of nitrogenase by O_2 is to temporally separate photosynthesis and its associated oxygen production from nitrogen fixation. Typically, nitrogen is fixed in the dark, while photosynthetic O_2 production occurs in the light [17–20]. During the transition from light to dark phase, cellular respiration plays a critical role, consuming the intracellular oxygen generated by photosynthesis to reach low enough O_2 concentration levels to allow for nitrogenase activity. Besides serving as a metabolic strategy for oxygen consumption, respiration also sustains the process of nitrogen fixation by generating the reducing equivalents and ATP needed by this energy-intensive process [21]. In many cases, the pattern of nitrogenase activity and respiration indeed overlap in cultures grown under alternating light and dark periods [22]. In the dark, respiration relies on the glycogen stored during the photosynthetic light period as a carbon source [23], while nitrogen is accumulated as cyanophycin that can later be used under light conditions [24].

Only a few cyanobacteria do not follow these two main nitrogen fixation strategies. Members of the genus *Trichodesmium* show nitrogenase activity peaking in the light [25], although they do not technically produce heterocyst. A lot of ongoing studies have been conducted to elucidate how *Trichodesmium* fix both CO_2 and N_2 concomitantly without any obvious spatial separation. It seems that part of the solution is achieved by restricting nitrogenase localization to a subset of cells along the trichomes where photosynthesis is downregulated [26]. Temporal separation may also play a role in metabolic partition of CO_2 and N_2 fixation in *Trichodesmium* as carbon fixation rate is higher in the morning and nitrogen fixation peaks in the afternoon [27].

15.2.3
Circadian Clock

Circadian rhythms are endogenous oscillations of about 24 h a cycle observed in many physiological activities in a wide range of organisms [28]. These rhythms are controlled by a circadian clock assumed to be a fundamental function of most living organisms to cope with the predictable environment changes that come with the daily day/night cycle [29]. Circadian timekeeping was initially assumed to be restricted to eukaryotes, until 1986 when Huang and co-workers demonstrated bacterial circadian rhythms in the cyanobacteria *Synechococcus* sp. RF-1 for nitrogen fixation first [30, 31] and later amino acid uptake [32]. These rhythms persisted for some time in absence of external cues, were reset by light signals and were temperature compensated, satisfying the three characteristics of a true circadian oscillator [33]. Since then, circadian rhythms have been identified in a number of cyanobacterial strains, with *Synechococcus elongatus* PCC7942 emerging as the main model system to further study the circadian clock mostly because of its genetic tractability and available genetic tools. To date, cyanobacteria remain the simplest organisms and only prokaryotes known to have a true circadian clock.

In *S. elongatus* the *kai* gene cluster, comprised of *kaiA*, *kaiB*, and *kaiC*, is responsible for the generation of circadian oscillations [34]. It has been shown that almost all genes in this organism are rhythmically controlled by these three genes [35]. Most of them can be divided into two groups with either high expression late in the day, such as genes involved in nitrogen fixation, or high expression in the dark phase just before the next light phase comes around, such as genes for glycogen synthesis. Additionally, the circadian clock also regulates cell division timing [36, 37] as well as chromosome compaction [38, 39]. A full review on cyanobacterial circadian clock has recently been published [40].

15.2.4
Light/Dark Regulation

The physiology of cyanobacteria as a photosynthetic organism is greatly impacted by the availability of light in its environment. Cyanobacteria are one of the few groups capable of carrying out both photosynthesis and respiration in the same cellular compartment. Light plays a role in the regulation of both pathways. Under high light intensity, photosynthesis prevails, and its electron flow capacity is greater than that of the respiratory chain. Under low light or in complete dark, the tendency is reversed [41].

Another response to changes in light quality and quantity occurs at the level of the light-harvesting apparatus of cyanobacteria. The light-harvesting complexes surround the reaction center of a photosystem, enhancing light absorption owing to their high pigment density [42]. To adapt to the different lights in their natural environment the antenna of the light-harvesting apparatus can adjust the pigment composition, pigment organization, and size [43, 44]. These strategies tend to be long-term acclimations (hours) to light conditions, while on the short term (seconds to minutes), modifications in pigment–pigment interactions or rapid chemical or structural variation of pigments help to survive when the variations of light intensity becomes too high [45].

Finally, physiological adaptations between light and dark metabolism involve a diversity of mechanisms for the activation/inactivation of enzymes controlled by the redox state of the cell [46]. Disulfide/dithiol exchange catalyzed by thioredoxins is a classic example of redox regulation already well established in higher plants [47]. In cyanobacteria, despite a presumable common origin, the functions of thioredoxins were poorly understood until recently. As in plant chloroplasts, cyanobacterial thioredoxins receive reducing equivalent from the photosynthetic electron transport chain via the ferredoxin-dependent thioredoxin reductase [48]. Screening for potential thioredoxin target proteins has led to the *in vitro* identification of 77 candidate thioredoxin target proteins in *Synechocystis* sp. PCC 6803 [49]. Only a small portion, approximately a third, has homologs in other photosynthetic organisms. These common thioredoxin targets are known in plants to control carbon fixation and storage, nitrogen and sulfur metabolism, stress response and protein translation, and folding. The other proteins identified as possible receptors of redox signals might have evolved separately and many still need to be investigated to evaluate their significance *in vivo* as true targets of thioredoxin regulation.

15.3 Genetic Tools

15.3.1 Transformation

Introduction of foreign DNA into cyanobacteria has been demonstrated in laboratory for several strains, and is now a common practice [50]. A few unicellular cyanobacteria are naturally competent for transformation, and can uptake foreign DNA from their environment in the form of plasmid or linear DNA [51]. Among naturally competent strains are the model freshwater cyanobacteria *S. elongatus* PCC 7942 and *Synechocystis* PCC 6803, as well as the marine *Synechococcus* PCC 7002 and the thermophile *Thermosynechococcus elongatus* BP-1 [52–56].

The DNA is incorporated in the cyanobacteria genome through homologous recombination when the foreign DNA is flanked by adjacent fragments from the recipient's genome. Double crossover is favored over single crossover during natural transformation, probably due to a fragmentation process of the DNA during incorporation [56]. This robust mechanism for integration of foreign

DNA has allowed the generation of mutants with gene disruption or heterologous gene overexpression when the incoming DNA is incorporated into a neutral site.

Natural transformation has not been demonstrated with any filamentous cyanobacteria. However, conjugation with Escherichia *coli* carrying the mobilization gene (*mob*) has been widely used to transfer plasmid DNA into these strains, particularly in cyanobacteria from the genus *Nostoc* and *Anabaena* [57–60]. Conjugation has also been successfully used with unicellular strains such as *Synechococcus* PCC 7942 and *Synechocystis* PCC 6803 when transfer of an intact self-replicating plasmid is preferred, but also with plasmid-containing DNA that can be integrated into the genome [61, 62]. In brief, this technique is based on the ability to mobilize DNA from one bacterium, usually *E. coli* carrying a conjugal plasmid, to another bacterium [50]. The DNA to be transferred into the cyanobacterium is cloned on a cargo plasmid that carries a *bom* site (*oriT* region). The same plasmid, or a helper plasmid provided *in trans*, carries the *mob* gene encoding a specific nickase, that recognizes the *bom* site and cuts the plasmid to be transferred. Additionally, because restriction of incoming DNA in the recipient cell may occur, the DNA is protected by methylases cloned in a helper plasmid. As mentioned above, unlike natural transformation, conjugation results in a circular plasmid in the recipient strain. It can be either a self-replicating plasmid or can be integrated in the genome if it contains a homologous flanking DNA fragment from the receiver genomic DNA. Electroporation can also be used to transform a number of cyanobacterial strains. It is generally done using plasmids developed for conjugation, but the procedure has a higher probability to be mutagenic [63].

Most cyanobacteria strains are multiploid, carrying an estimated 8–72 genome copies per cell [64], with *Synechocystis* containing up to 218 copies during exponential growth phase [65]. However, after transformation, the newly introduced DNA fragment is inserted in only one chromosome. Thus, numerous growth cycles with selection pressure are likely required to obtain complete segregation. Moreover, the choice of integration loci plays a significant role in the likeliness of achieving full segregation as disruption of an essential gene would make it impossible [66].

15.3.2
Promoters

Gene expression in cyanobacteria, whether on a self-replicating plasmid or integrated in the genome, needs to be driven by a specific promoter that can be recognized by the host. Only a limited number of promoters have been evaluated for this purpose in cyanobacteria [50]. Most known promoters are native to the host organism from genes required for photosynthesis and expressed in relatively high levels, such as P_{rbc}, P_{psbA2}, and P_{cpc}, or metal-ion induced promoters such as the P_{nrsB} promoter linked to nickel homeostasis in *Synechocystis* [67]. Even though some native promoters are well characterized with clear environmental triggers for regulation, their utilization is limited by potential cross-talk with the regulatory network of the organism. Several foreign promoters, orthogonal to cyanobacteria, have also been characterized for gene expression. Strong *E. coli* promoters P_{trc} and P_{tac} have been frequently used in cyanobacteria, with or without a regulated expression of LacI. But because the RNA polymerase (RNAP) holoenzymes in cyanobacteria are quite different from most bacteria, common promoters from *E. coli* might perform differently in cyanobacteria. For example, while P_{trc} is not inducible in *Synechocystis* PCC 6803, it can be regulated by LacI in *Synechococcus* 7942 [68]. Other IPTG-inducible promoters have been found to be more suited for use in *Synechocystis*, for example, PA1lacO-1, which exhibits an eightfold induction ratio [69]. Recently, a library of TetR-regulated promoters with varying dynamic ranges was designed in *Synechocystis*. The best-performing promoter induced a 290-fold induction ratio under red light [70]. A "super-strong" promoter, P_{cpc560}, that contains two predicted promoters and depends on 14 predicted transcription factor binding sites (TFBS) crucial for its promoter strength has been characterized [67]. Proteins expressed under its regulation were produced to up to 15% of the total soluble protein content. In *E. coli*, strong promoters usually contain only one or two TFBS. This example illustrates how important it is to understand the difference in cellular machinery between cyanobacteria and other bacteria, such as *E. coli*, to design a list of cyanobacteria promoters. The structural difference of the RNAP mentioned above, as well as the different sets of sigma factors present in cyanobacteria [71], further support the need to build a basic understanding of these organism-specific differences to engineer and identify promoters.

15.3.3
Terminators

The importance of transcriptional termination in regulating natural genetic systems is well known [72], yet when it comes to characterizing elements that control transcription and translation in bacteria, most research focuses on the initiation part [73, 74]. Similarly, in cyanobacteria, very little work has been done to characterize terminators. It is known nevertheless, that the introduction of a terminator downstream of an overexpressed gene is important for the stabilization of the transcript [75] and, in the case of chromosomal insertion, will prevent effect on the adjacent genes. In some strains, for example, *Synechococcus* PCC7942, it has been reported that out of the two signals for transcription termination identified in bacteria [76], intrinsic Rho-independent terminators play a larger role than Rho-dependent terminators [77]. In accordance with this finding, it was previously shown that most bacteria possess a homolog of the *E. coli* Rho protein, but surprisingly the cyanobacteria *S. elongatus* and *Synechocystis* PCC 6803 do not [76]. Only a few terminators, native or foreign, have been successfully used in cyanobacteria. Some examples are the bacteriophage T7 terminator, the strong terminator of the gene *rrnB* of *E. coli*, and the cyanobacteria ribulose-1,5-bisphosphate carboxylase/oxygenase (RuBisCO) terminator.

15.3.4 Ribosome Binding Sites

The ribosome binding site (RBS) affects the frequency with which the translation of downstream target genes begins. The Shine-Dalgarno (SD) sequence of the RBS is usually more or less conserved in bacteria such as *E. coli* where 57% of the genes contain this sequence. In *Synechocystis* PCC 6803, it was found that only 26% of the genes possess a conserved core SD sequence [78]. Recently, an RBS complementary to the anti-SD of *Synechocystis* PCC 6803 was tested and was found to be two- to fourfold more efficient than other RBS with core SD sequence not exactly complementary to the anti-SD [50]. The spacing between the core SD and the start codon also plays an important role in the effectiveness of an RBS. In the past few years, three independent translation rate calculators have been developed. The RBS Calculator [79], developed in 2009 and further updated in 2011, the RBS Designer [80] released in 2010, and finally the UTR Designer in 2013 [81]. Each calculator demonstrates comparable accurate predictions when compared to experimental data in *E. coli.* Unfortunately, because these three calculators were developed for *E. coli,* and because of the lack of conserved RBS sequence in cyanobacteria as mentioned above, they might not show the same accuracy if used for these organisms. It would be useful if similar models could be developed specifically for RBS prediction in cyanobacteria.

15.4 Improving Photosynthetic Efficiency

15.4.1 Improving Light Harvesting

In cyanobacteria, light is captured through light-harvesting antenna, supramolecular complexes consisting of proteins and photosynthetic pigments [82]. The absorbed light can then be converted to chemical energy by the process known as *photosynthesis.* Although solar to biomass energy conversion of cyanobacteria is two to three times higher than for crop plants [83], photosynthesis efficiency is still theoretically limited to 8–10% conversion, attainable under low-light conditions [1]. Under ambient sunlight (high-light) conditions, the energy conversion efficiency drops substantially lower to 1–2% for the best species [84]. In fact, under high light, cyanobacteria cannot absorb all incoming sunlight, as the photon absorption rate by the light-harvesting antennae is much faster than the rate at which photosynthesis can utilize them. This results in wasteful dissipation of excess light mostly via non-photochemical quenching (NPQ), which dissipates excess energy as heat [45]. Moreover, the shading effect of cells on the surface layer of high-density cultures causes limited light penetration for the cells in the lower layers [1]. To address this issue, it has been proposed to use strains with smaller antenna sizes. This approach can reduce both the problem of saturation at the surface and shading deeper in the culture. Even grown in high density, these mutants should still allow for good light penetration into the inner layers of the photobioreactor, thus resulting in more uniform illumination and increased whole-culture productivity. This concept, first demonstrated in green microalgae [85–87], is known as the *Truncated Light-harvesting Antenna* (TLA) concept [1, 88, 89]. Mutants expressing genetically truncated light-harvesting chlorophyll antenna size (*tla*) show enhanced solar energy conversion efficiency and productivity. RNA interference (RNAi) technology to downregulate the whole light-harvesting complex gene family also validated the concept independently [90]. Various strategies have been employed in cyanobacteria to create or isolate this kind of mutants. Disruption of genes encoding phycobilisome subunits first failed to produce mutants with increased biomass productivity [91]. More recently, a phycocyanin-deletion mutant of *Synechocystis* sp. PCC 6803 with a highly truncated phycobilisome antenna size allowed for a 57% improvement in the productivity as compared to the wild type when grown under high cell density and saturating light [84]. Although theoretical improvements in photosynthetic energy conversion efficiency of TLA cells could be as high as threefold that of the wild type [1], the study demonstrates the possibility of improved productivity in cyanobacteria by reducing the light-harvesting antenna size, in opposition to what was previously described [92].

15.4.2 Improving Carbon Fixation

Cyanobacteria use the Calvin–Benson–Bassham (CBB) cycle to fix CO_2, similarly to algae and higher plants. The first step is catalyzed by RuBisCO, which converts ribulose-1,5-bisphosphate (RuBP) and CO_2 to two molecules of 3-phosphoglycerate (3PG). One 3PG is recycled to RuBP, while the other goes to central metabolic pathways. Although being the first major step in carbon fixation, RuBisCO is also the primary rate-limiting factor of the CBB cycle, because of its slow catalytic turnover and low specificity. Each RuBisCO molecule can only fix 3–10 CO_2 per second, while some enzymes can carry out thousands of reactions in the same time [93]. In addition, RuBisCO can take O_2 as a substrate in a process called *photorespiration*, which reduces the overall carbon fixation efficiency. Studies have shown that improving specificity via site-directed mutagenesis comes with a loss in the catalytic rate. Thus, RuBisCO is thought to have been optimized for the best compromise between substrate specificity and maximum rate of catalytic turnover [94, 95]. Moreover, cyanobacteria and some microalgae have evolved an extremely effective CO_2-concentrating mechanism (CCM) composed of at least two elements: a C_i transport system and RuBisCO-rich polyhedral carboxysomes. This mechanism allows active uptake of inorganic carbon from the external medium in order to elevate the CO_2 concentration around the active site of RuBisCO [96]. RuBisCO can in turn operate at higher efficiency, lowering the need for high specificity [97]. In cyanobacteria, RuBisCO is encapsulated with carbonic anhydrase in unique microcompartments, the carboxysomes. Carbonic anhydrase catalyzes the conversion of HCO_3^- to CO_2, the latter staying locked in the carboxysomes in the vicinity of RuBisCO [98]. In *Synechococcus* 7942, it has been shown that cells with more carboxysomes have approximately a 50% higher carbon fixation rate than cells with fewer carboxysomes [99]. Expression of a heterologous RuBisCO from *Allochromatium vinosum*, a purple sulfur bacterium, also resulted in about a 50% increase in CO_2 assimilation [100]. More recently, overexpression of *Synechococcus* 6301 RuBisCO in *Synechococcus* 7942 led to a twofold higher isobutyraldehyde production, possibly due to a more efficient carbon fixation [101].

In addition, cyanobacteria possess a second carbon fixation mechanism through phosphoenolpyruvate carboxylase (PPC) that is responsible for close to 25% of CO_2 fixation [102]. PPC fixes HCO_3^- rather than CO_2 and combines it with phosphoenolpyruvate to form oxaloacetate and inorganic phosphate. An *in silico* modeling study has proposed to couple PPC with the core C4 plant carbon fixation cycle to increase the overall carbon fixation rate compare to the CBB cycle [103].

15.5 Direct Conversion of CO_2 into Biofuels and Chemicals

Cyanobacterial biofuel production, which captures atmospheric CO_2 as the carbon source, is an intriguing solution to relieve the excess amount of greenhouse gases released by fossil fuel combustion. Ethanol is the most common biofuel produced in various microorganisms. The first ethanol production in cyanobacteria is accomplished by introducing a pyruvate decarboxylase, which directly decarboxylates pyruvate following by a primary alcohol dehydrogenase [104]. This heterologous pathway enables to direct carbon flux from pyruvate to acetaldehyde and then ethanol. On the other hand, longer chain alcohols such as 1-butanol [105], isobutanol [101], and 2-methyl-1-butanol [106] have been produced from CO_2 using *S. elongatus* 7942. In addition, several other chemicals have been produced through heterologous pathway expression in cyanobacteria [107, 108] (Figure 15.1 and Table 15.1). In order to efficiently utilize cyanobacteria for production of biofuels and chemicals, several important strategies can be employed such as altering cofactor preference [119, 120], developing a synthetic driving force [121], managing the toxicity of the desired product [101], and reducing by-product formation [106]. These strategies improve cyanobacteria for specific and high titer production.

15.5.1 Fuels and Chemicals from Acetyl-CoA

There are many compounds that can be produced from acetyl-coenzyme A (CoA), including 1-butanol, acetone, 2-propanol, 3-hydroxypropinate, and 3-hydroxybutyrate. Ethanol production has been discussed above and has been

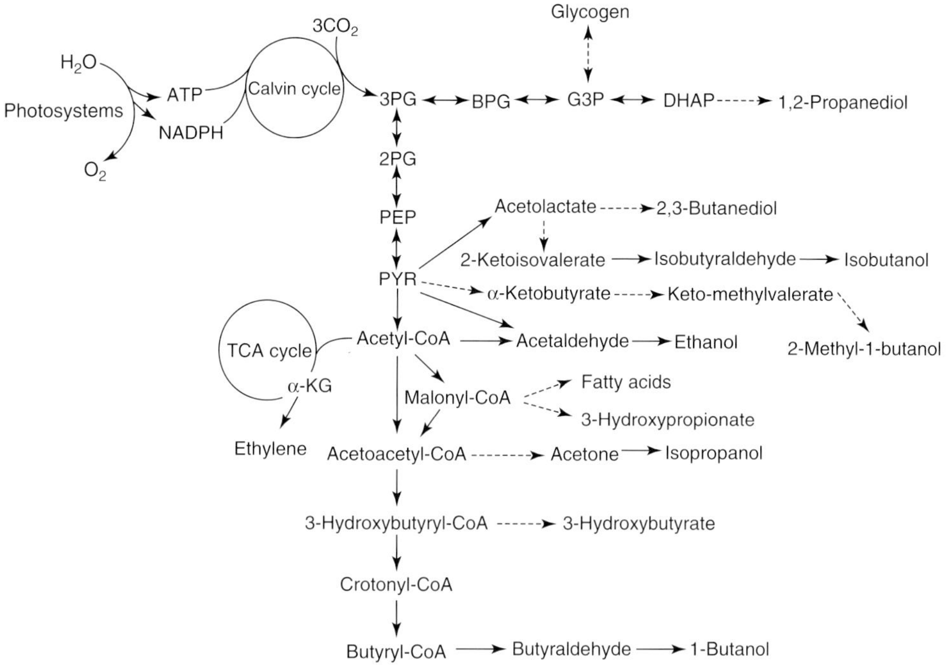

Figure 15.1 Overview of pathway for biofuels and chemicals produced from cyanobacteria.

Table 15.1 Titers for biofuel and chemical productions from cyanobacteria.

Compound	Organism	Titers	References
Acetone	*Synechocystis* sp. PCC6803	$36\,mg\,l^{-1}$	Zhou *et al.* [109]
2-Propanol	*S. elongatus* sp. PCC7942	$26.5\,mg\,l^{-1}$	Kusakabe *et al.* [110]
2,3-Butanediol	*S. elongatus* sp. PCC7942	$2.4\,g\,l^{-1}$	Oliver *et al.* [111]
1-Butanol	*S. elongatus* sp. PCC7942	$404\,mg\,l^{-1}$	Lan *et al.* [105]
Ethylene	*Synechocystis* sp. PCC6803	$718 \pm 19\,\mu l\,l^{-1}\,h^{-1}$ per A_{730nm}	Xiong *et al.* [112]
1,2-Propanediol	*S. elongatus* sp. PCC7942	$150\,mg\,l^{-1}$	Li and Liao [113]
3-Hydroxybutyrate	*Synechocystis* sp. PCC6803	$533.4\,mg\,l^{-1}$	Wang *et al.* [114]
3-Hydroxy propionate	*S. elongatus* sp. PCC7942	$665\,mg\,l^{-1}$	Lan *et al.* [115]
	Synechocystis sp. PCC6803	$837\,mg\,l^{-1}$	Wang *et al.* [116]
Isobutanol	*S. elongatus* sp. PCC7942	$450\,mg\,l^{-1}$	Atsumi *et al.* [101]
Isobutyraldehyde	*S. elongatus* sp. PCC7942	$1.1\,g\,l^{-1}$	Atsumi *et al.* [101]
2-Methyl-1-butanol	*S. elongatus* sp. PCC7942	$200\,mg\,l^{-1}$	Shen and Liao [106]
Fatty acid	*Synechocystis* sp. PCC6803	$197\,mg\,l^{-1}$	Liu *et al.* [117]
Ethanol	*Synechocystis* sp. PCC6803	$5.5\,g\,l^{-1}$ [a)]	Gao *et al.* [118]

a) By photobioreactor.

attempted in industry. Photosynthetic production of 1-butanol in oxygenic cyanobacteria has been difficult. 1-Butanol is naturally produced by *Clostridia* via the CoA-dependent pathway as part of the acetone–butanol–ethanol (ABE) fermentation. Implementation of this pathway in more user-friendly hosts was successful after rounds of improvements to build up both NADH and acetyl-CoA driving forces, enabling production of up to $15\,\text{g}\,\text{l}^{-1}$ 1-butanol anaerobically in *E. coli* [122]. In contrast, engineering a similar pathway in the oxygenic cyanobacterium *S. elongatus* PCC 7942 was even more difficult. Expression of the same modified 1-butanol pathway enzymes in this oxygenic organism failed to produce any 1-butanol owing to the oxygen sensitivity of pathway enzymes [123]. The product was detected only under anoxic conditions or with photosystem II inhibited. The problem was solved by using an oxygen-tolerant butyraldehyde dehydrogenase (PduP) [105], and by introducing a malonyl-CoA-based synthetic pathway [121] as shown in Figure 15.2. This new approach inspired by fatty acid biosynthesis utilizes ATP as a driving force, thus enabling the organism to compensate for the thermodynamically unfavorable condensation of two acetyl-CoA molecules. The combined strategy enhanced 1-butanol production to $404\,\text{mg}\,\text{l}^{-1}$ [105].

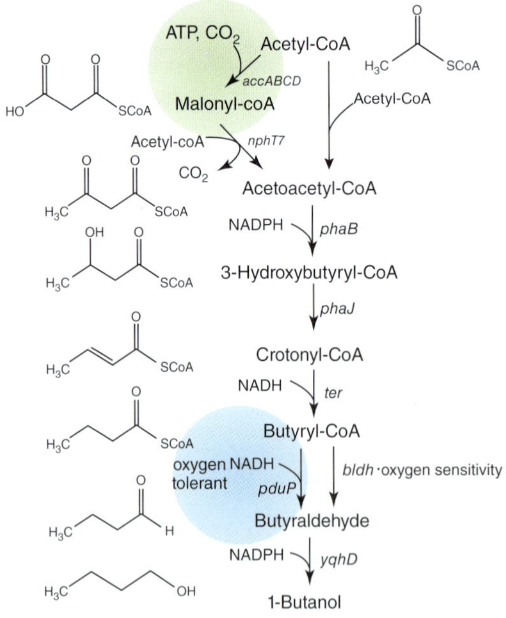

Figure 15.2 1-Butanol production via the CoA-dependent pathway implemented in *S. elongatus* PCC7942. Gene symbols are *accABCD*, acetyl-CoA carboxylase; *nphT7*, acetoacetyl-CoA synthase; *phaB*, acetoacetyl-CoA reductase; *phaJ*, R-specific enoyl-CoA hydratase; *ter*, trans enoyl-CoA reductase; *pduP*, CoA-acylating propionaldehyde dehydrogenase; *bldh*, butyraldehyde dehydrogenase; and *yqhD*, NADPH-dependent alcohol dehydrogenase.

Other fuels and chemicals produced from acetyl-CoA in cyanobacteria include acetone, 2-propanol, 3-hydroxypropionate, and 3-hydroxybutyrate. One of the difficulties in 2-propanol production is the step catalyzed by acetate/3-ketoacid CoA transferase. This enzyme needs acetate as a substrate in order to catalyze the reaction toward acetoacetate biosynthesis. However, under autotrophic condition, cyanobacteria do not produce much acetate, and the acetate production pathway remains unclear. Providing external acetate cannot increase 2-propanol titer, potentially due to regulation of acetate uptake. 3-Hydroxypropionate (3HP) is an important compound as it can be converted to several industrially important chemicals such as acrylic acid, acrylamide, and 1,3-propanediol. Furthermore, biodegradable and biocompatible poly-3HP can be made by using 3HP as the monomer. Three pathways have been proposed for 3HP production by microbial platforms (Figure 15.3). The glycerol dependent pathway [124, 125] needs a coenzyme B12-dependent glycerol dehydratase. However, providing external coenzyme B12 raises additional cost. Moreover, B12-independent glycerol dehydratase is an oxygen-sensitive enzyme which is not an ideal candidate for application in oxygenic cyanobacteria [105]. Thus, direct transfer of the glycerol-dependent pathway into cyanobacteria for 3HP production would become difficult. On the other hand, 3HP can be produced from malonyl-CoA,

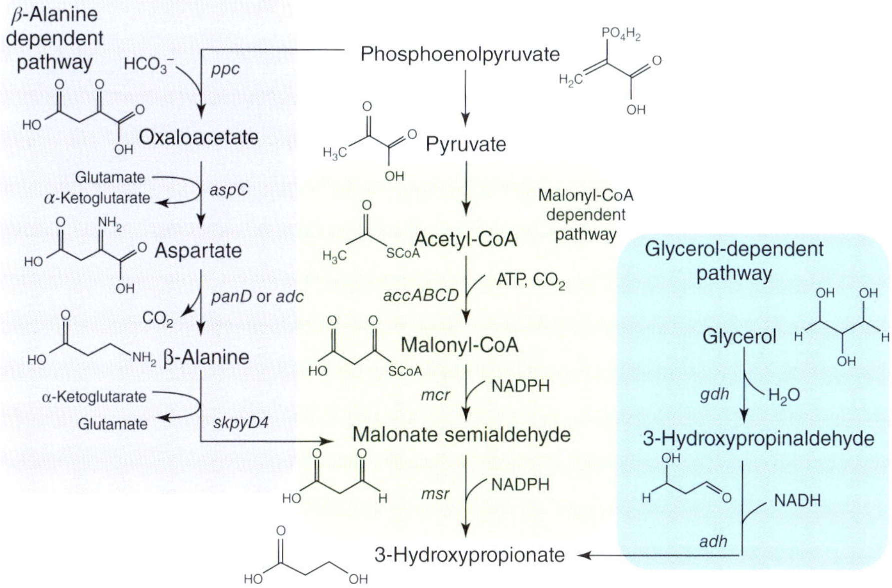

Figure 15.3 3-Hydroxypropionate (3HP) synthesis via glycerol-dependent, malonyl-CoA-dependent, and β-alanine-dependent pathways. Gene symbols are *ppc*, phosphoenolpyruvate carboxylase; *aspC*, aspartate transaminase; *panD* or *adc*: aspartate decarboxylase; *skpyD4*, β-alanine aminotransferase; *mcr*, malonyl-CoA reductase; *msr*, malonate semialdehyde reductase; *gdh*, glycerol dehydratase; and *adh*, alcohol dehydrogenase.

a precursor for fatty acid synthesis which is common to microorganisms. By introducing different malonyl-CoA reductases and malonate semialdehyde reductases, 3HP productions achieve titers between 124 and 659 $mg\,l^{-1}$ in *S. elongatus* PCC 7942. Another pathway is through β-alanine, which is able to redirect carbon fluxes from phosphoenolpyruvate to 3HP. By coupling the β-alanine-dependent pathway with the malonyl-CoA-dependent pathway, 3HP production by utilizing both pathways outperforms only the individual pathways. However, this enhancement is potentially limited by 3HP toxicity to cyanobacteria. It is also reported that knocking out acetyl-CoA synthase is able to enhance tolerance to 3HP and other organic acids in *Synechococcus* sp. PCC 7002 [126].

15.5.2
Fuels and Chemicals from Keto Acids

Keto acids can be decarboxylated to aldehydes and then reduced to the corresponding alcohols. The most common example is the conversion of pyruvate to ethanol, which has been demonstrated in multiple organisms, including cyanobacteria. The keto acid intermediates in several amino acid biosynthesis pathways can also be converted to their corresponding branched-chain higher alcohols by introducing a keto acid decarboxylase and a primary alcohol dehydrogenase [127]. The first keto acid-pathway-dependent biofuel production in cyanobacteria is through valine biosynthesis to produce isobutyraldehyde and isobutanol [101] (Figure 15.4). Isobutyraldehyde is an important precursor for industrial applications and isobutanol is a promising alternative energy as it has higher energy density and lower hygroscopicity than ethanol. In order to enhance carbon flux from pyruvate to valine biosynthesis, an *alsS* gene from *Bacillus subtilis*, *ilvC* and *ilvD* genes from *E. coli*, and *kivd* from *Lactococcus lactis* have been overexpressed. In this strain, isobutyraldehyde titer achieved 723 $mg\,l^{-1}$. Since aldehyde is a toxic compound to cells and RuBisCO is considered as one of the limiting steps in photosynthetic organisms, further improvement is accomplished by *in situ* product removal and overexpression of *rbcLS* from *S. elongatus* PCC6301. The combined strategy successfully enhanced the isobutyraldehyde titer up to 1.1 $g\,l^{-1}$. Isobutanol can be produced by introducing an alcohol dehydrogenase in the isobutyraldehyde production strains. Among the three alcohol dehydrogenases used in the test, namely, *yqhD* from *E. coli*, *adh2* from *Saccharomyces cerevisiae*, and *adhA* from *L. lactis*, *yqhD* is a NADPH-utilizing enzyme and indicates the highest isobutanol titer of 450 $mg\,l^{-1}$. This result suggests the importance of the cofactor choice in cyanobacterial biofuel production.

2-Methyl-butanol (2MB) is synthesized via the isoleucine pathway (Figure 15.4), which is similar to the valine biosynthesis pathway. One of the challenges for 2MB production is that it passes over the same cassette of genes as in isobutanol production. A recent study [106] indicated that enhancing the citramalate pathway to direct the carbon flux from pyruvate to α-ketobutyrate led 2MB production to about 200 $mg\,l^{-1}$ with minimum side products such as 1-propanol and isobutanol.

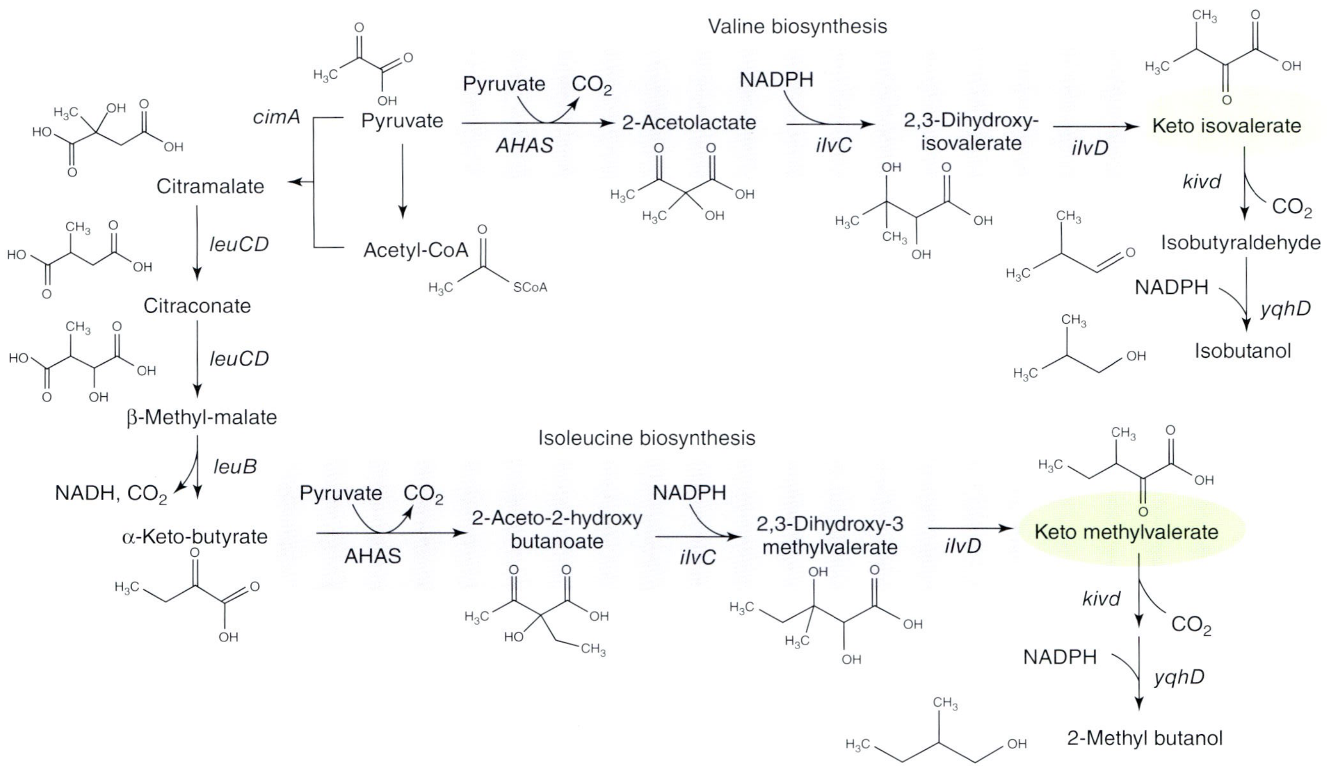

Figure 15.4 Isobutanol and 2-methyl butanol (2MB) production via keto acid pathway in cyanobacteria. Gene/protein symbols are *cimA*, citramalate synthase; *leuCD*, isopropylmalate isomerase; *leuB*, 3-isopropylmalate dehydrogenase; AHAS, acetohydroxyacid synthase; *ilvC*, acetohydroxy acid isomeroreductase; *ilvD*, dihydroxy acid dehydratase; *kivd*, ketoisovalerate decarboxylase; *yqhD*, alcohol dehydrogenase.

Kinetic analysis of native AHAS from *S. elongatus* PCC7942 demonstrated a preferred selectivity toward α-ketobutyrate.

Cyanobacteria are known to synthesize glycogen as a storage compound. Knocking out glucose-1-phosphate adenylyltransferase (*glgC*), the initial step for glycogen synthesis, has a negative effect on growth and glycogen synthesis under high light conditions (150 $\mu Es^{-1} m^{-2}$). This effect is attributed to the lack of an electron sink. However, the isobutanol pathway can serve as a metabolic sink for replacing glycogen synthesis to rescue growth in the Δ*glgC* mutant [128]. This phenotype may be exploited as a selection platform for enhancing product formation.

15.5.3 Chemicals from TCA Cycle Intermediates

The TCA cycle oxidizes acetyl-CoA derived from glycogen degradation in the dark to produce NADH and $FADH_2$ for ATP generation in the respiratory chain. Moreover, TCA cycle intermediates are important precursors for amino acid biosynthesis. In cyanobacteria, it had been generally considered that the TCA cycle was not complete, lacking the 2-oxoglutarate dehydrogenase complex (2-OGDH), until a recent study showed that their unique TCA cycle is closed by 2-oxoglutarate decarboxylase (2-OGDC) and succinic semialdehyde dehydrogenase (SSADH) [129].

It is generally accepted that cyanobacterial TCA cycle is not very active under phototrophic conditions. A recent study further proved that it operates in a bifurcated way [112]. However, the bifurcated pattern switches to a dominantly cyclic pattern by introducing an ethylene-forming enzyme from *Pseudomonas.* Ethylene can be sustainably and efficiently produced from 2-oxoglutarate while also activating the TCA cycle metabolism. The enhanced flux via the remodeled TCA is 37% of the total fixed carbon which is higher than the 13% found in wild type. This result indicates that the metabolic malleability of the cyanobacterial TCA cycle depends on what it produces.

15.5.4 Hydrogen

In recent years, hydrogen produced by living organisms has gained increased attention. Hydrogen is particularly attractive because its utilization is carbon-free, generating only water and no greenhouse gases. A number of photosynthetic organisms naturally produce small quantities of hydrogen. In cyanobacteria, several enzymes are involved in hydrogen metabolism: nitrogenase which produces hydrogen as a side product during the reduction of nitrogen to ammonia, a hydrogenase that can catalyze the reversible reaction $2H^+ + 2e^- \rightarrow H_2$, and an uptake hydrogenase with the function of recapturing the hydrogen produced by nitrogenase [130]. Several species of cyanobacteria able to make H_2 have been described and are reviewed elsewhere [131]. Studies have been focusing on selecting strains with specific hydrogen metabolism, improving conditions in bioreactors cultivation, and increasing H_2 production by genetic engineering [132]. This last point focuses on reinforcing the electron flow toward hydrogenases. In one study, hydrogen production in *S. elongatus* sp. 7942 was approximately doubled by strengthening the electron flow through overexpression of *Clostridium acetobutylicum* ferredoxin [133]. Effort has also been made to eliminate competitive pathways in term of reducing consumption of agents. In *Synechococcus* 7002, inactivation of the gene coding for lactate dehydrogenase, the main fermentative reductant sink, resulted in a fivefold higher H_2 production than wild type under dark anoxic conditions [134]. Another major problem to overcome for the development of more efficient biohydrogen production is the high oxygen sensitivity of hydrogenases [135]. In current processes, H_2 production is temporally separated from O_2-generating photosynthesis by first growing the cells normally and then transferring them in anaerobic conditions coupled with PSII inactivation [133]. Hydrogen is then produced through fermentation using internal carbohydrates. Engineering of current hydrogenases or discovery of new ones with higher O_2 tolerance remains one of the biggest challenges for a sustained continuous hydrogen production in cyanobacteria.

15.6 Conclusions

Cyanobacteria possess several advantageous features, making them promising host organisms for various sustainable biotechnological applications. They are quite easy to genetically manipulate, have only simple growth requirements, and can fix carbon dioxide to produce industrially relevant molecules using sunlight as energy. Moreover, several downstream products such as sucrose [136], 2,3-butanediol [111], and ethylene [112] have been shown to act as carbon sinks able to pull the upstream carbon flux from carbon fixation toward product formation, enhancing the photosynthetic efficiency. These results highlight the potential of cyanobacterial platforms to make progress toward carbon-neutral biofuel and chemical production.

Although it is promising, there are still a number of challenges that need to be overcome to see cyanobacteria become an industrial microorganism. The productivity of metabolites of interest is still low and needs to be increased through both metabolic engineering to create improved strains and development of more efficient bioreactors. With advances in both fields, cyanobacteria may become a valuable addition to the current industrial microbial platforms.

References

1. Melis, A. (2009) Solar energy conversion efficiencies in photosynthesis: minimizing the chlorophyll antennae to maximize efficiency. *Plant Sci.*, **177** (4), 272–280.
2. Zhu, X.-G., Long, S.P., and Ort, D.R. (2008) What is the maximum efficiency with which photosynthesis can convert solar energy into biomass? *Curr. Opin. Biotechnol.*, **19** (2), 153–159.

3. Rastogi, R.P. and Sinha, R.P. (2009) Biotechnological and industrial significance of cyanobacterial secondary metabolites. *Biotechnol. Adv.*, **27** (4), 521–539.
4. Stal, L.J. (1991) The metabolic versatility of the mat-building cyanobacteria *Microcoleus chthonoplastes* and *Oscillatoria limosa* and its ecological significance. *Algol. Stud. Hydrobiol.*, **64**, 453–467.
5. Herrero, A., Muro-Pastor, A.M., and Flores, E. (2001) Nitrogen control in cyanobacteria. *J. Bacteriol.*, **183** (2), 411–425.
6. Castenholz, R.W. (1973) in *The Biology of Blue-Green Algae*, Botanical Monographs, vol. **9** (eds N.G. Carr and B.A. Whitton), Blackwell Scientific Publishing, pp. 379–414.
7. Cohen, Y., Jørgensen, B.B., Revsbech, N.P., and Poplawski, R. (1986) Adaptation to hydrogen sulfide of oxygenic and anoxygenic photosynthesis among cyanobacteria. *Appl. Environ. Microbiol.*, **51** (2), 398–407.
8. Peschek, G.A., Löffelhardt, W., and Schmetterer, G. (2012) *The Phototrophic Prokaryotes*, Springer Science & Business Media.
9. Fay, P. (1965) Heterotrophy and nitrogen fixation in Chlorogloea fritschii. *J. Gen. Microbiol.*, **39** (1), 11–20.
10. Rippka, R. (1972) Photoheterotrophy and chemoheterotrophy among unicellular blue-green algae. *Arch. Mikrobiol.*, **87** (1), 93–98.
11. Kromkamp, J. (1987) Formation and functional significance of storage products in cyanobacteria. *N. Z. J. Mar. Freshwater Res.*, **21** (3), 457–465.
12. Stal, L.J. and Moezelaar, R. (1997) Fermentation in cyanobacteria. *FEMS Microbiol. Rev.*, **21** (2), 179–211.
13. Bergman, B., Flores, E., Böhme, H., Happe, T., Osborne, B., Sivonen, K., Ventura, S., Wilmotte, A., Moth-Wiklund, A. (1999) Cyanobacterial Nitrogen Fixation, European Science Foundation (http://www.esf.org/fileadmin/Public_documents/Publications/CIANOFIX.pdf)
14. Maldener, I. and Muro-Pastor, A.M. (2010) Cyanobacterial heterocysts, in *eLS*, John Wiley & Sons, Ltd.
15. Kumar, K., Mella-Herrera, R.A., and Golden, J.W. (2010) Cyanobacterial heterocysts. *Cold Spring Harbor Perspect. Biol.*, **2** (4), a000315.
16. Bergman, B., Gallon, J.R., Rai, A.N., and Stal, L.J. (1997) N2 fixation by non-heterocystous cyanobacteria. *FEMS Microbiol. Rev.*, **19** (3), 139–185.
17. Gallon, J.R., LaRue, T.A., and Kurz, W.G.W. (1974) Photosynthesis and nitrogenase activity in the blue-green alga Gloeocapsa. *Can. J. Microbiol.*, **20** (12), 1633–1637.
18. Mitsui, A., Kumazawa, S., Takahashi, A. *et al.* (1986) Strategy by which nitrogen-fixing unicellular cyanobacteria grow photoautotrophically. *Nature*, **323** (6090), 720–722.
19. Misra, H.S. and Tuli, R. (2000) Differential expression of photosynthesis and nitrogen fixation genes in the cyanobacterium *Plectonema boryanum*. *Plant Physiol.*, **122** (3), 731–736.
20. Compaoré, J. and Stal, L.J. (2010) Oxygen and the light–dark cycle of nitrogenase activity in two unicellular cyanobacteria. *Environ. Microbiol.*, **12** (1), 54–62.
21. Bandyopadhyay, A., Elvitigala, T., Liberton, M., and Pakrasi, H.B. (2013) Variations in the rhythms of respiration and nitrogen fixation in members of the unicellular diazotrophic cyanobacterial genus Cyanothece. *Plant Physiol.*, **161** (3), 1334–1346.
22. Colon-Lopez, M.S., Sherman, D.M., and Sherman, L.A. (1997) Transcriptional and translational regulation of nitrogenase in light-dark-and continuous-light-grown cultures of the unicellular cyanobacterium Cyanothece sp. strain ATCC 51142. *J. Bacteriol.*, **179** (13), 4319–4327.
23. Schneegurt, M.A., Sherman, D.M., Nayar, S., and Sherman, L.A. (1994) Oscillating behavior of carbohydrate granule formation and dinitrogen fixation in the cyanobacterium Cyanothece sp. strain ATCC 51142. *J. Bacteriol.*, **176** (6), 1586–1597.

24. Li, H., Sherman, D.M., Bao, S., and Sherman, L.A. (2001) Pattern of cyanophycin accumulation in nitrogen-fixing and non-nitrogen-fixing cyanobacteria. *Arch. Microbiol.*, **176** (1-2), 9–18.
25. Capone, D.G., Zehr, J.P., Paerl, H.W. *et al.* (1997) *Trichodesmium*, a globally significant marine cyanobacterium. *Science*, **276** (5316), 1221–1229.
26. Bergman, B., Sandh, G., Lin, S. *et al.* (2013) Trichodesmium–a widespread marine cyanobacterium with unusual nitrogen fixation properties. *FEMS Microbiol. Rev.*, **37** (3), 286–302.
27. Finzi-Hart, J.A., Pett-Ridge, J., Weber, P.K. *et al.* (2009) Fixation and fate of C and N in the cyanobacterium Trichodesmium using nanometer-scale secondary ion mass spectrometry. *Proc. Natl. Acad. Sci. U.S.A.*, **106** (15), 6345–6350.
28. Edmunds, L.N. (1983) Chronobiology at the cellular and molecular levels: models and mechanisms for circadian timekeeping. *Am. J. Anat.*, **168** (4), 389–431.
29. Kippert, F. (1987) Endocytobiotic coordination, intracellular calcium signaling, and the origin of endogenous rhythms. *Ann. N. Y. Acad. Sci.*, **503** (1), 476–495.
30. Grobbelaar, N., Huang, T.C., Lin, H.Y., and Chow, T.J. (1986) Dinitrogen-fixing endogenous rhythm in Synechococcus RF-1. *FEMS Microbiol. Lett.*, **37** (2), 173–177.
31. Huang, T.-C., Tu, J., Chow, T.-J., and Chen, T.-H. (1990) Circadian rhythm of the prokaryote *Synechococcus* sp. RF-1. *Plant Physiol.*, **92** (2), 531–533.
32. Chen, T.-H., Chen, T.-L., Hung, L.-M., and Huang, T.-C. (1991) Circadian rhythm in amino acid uptake by Synechococcus RF-1. *Plant Physiol.*, **97** (1), 55–59.
33. Dong, G., Kim, Y.-I., and Golden, S.S. (2010) Simplicity and complexity in the cyanobacterial circadian clock mechanism. *Curr. Opin. Genet. Dev.*, **20** (6), 619–625.
34. Ishiura, M., Kutsuna, S., Aoki, S. *et al.* (1998) Expression of a gene cluster kaiABC as a circadian feedback process in cyanobacteria. *Science*, **281** (5382), 1519–1523.
35. Ito, H., Mutsuda, M., Murayama, Y. *et al.* (2009) Cyanobacterial daily life with Kai-based circadian and diurnal genome-wide transcriptional control in *Synechococcus elongatus*. *Proc. Natl. Acad. Sci. U.S.A.*, **106** (33), 14168–14173.
36. Dong, G., Yang, Q., Wang, Q. *et al.* (2010) Elevated ATPase activity of KaiC applies a circadian checkpoint on cell division in *Synechococcus elongatus*. *Cell*, **140** (4), 529–539.
37. Mori, T., Binder, B., and Johnson, C.H. (1996) Circadian gating of cell division in cyanobacteria growing with average doubling times of less than 24 hours. *Proc. Natl. Acad. Sci. U.S.A.*, **93** (19), 10183–10188.
38. Smith, R.M. and Williams, S.B. (2006) Circadian rhythms in gene transcription imparted by chromosome compaction in the cyanobacterium *Synechococcus elongatus*. *Proc. Natl. Acad. Sci. U.S.A.*, **103** (22), 8564–8569.
39. Woelfle, M.A., Xu, Y., Qin, X., and Johnson, C.H. (2007) Circadian rhythms of superhelical status of DNA in cyanobacteria. *Proc. Natl. Acad. Sci. U.S.A.*, **104** (47), 18819–18824.
40. Cohen, S.E. and Golden, S.S. (2015) Circadian rhythms in cyanobacteria. *Microbiol. Mol. Biol. Rev.*, **79** (4), 373–385.
41. Vermass, W.F. (2001) Photosynthesis and respiration in cyanobacteria, in *eLS*, Nature Publishing Group, pp. 1–7.
42. Croce, R. and van Amerongen, H. (2014) Natural strategies for photosynthetic light harvesting. *Nat. Chem. Biol.*, **10** (7), 492–501.
43. Green, B. and Parson, W.W. (2003) *Light-Harvesting Antennas in Photosynthesis*, Springer Science & Business Media.
44. Gutu, A. and Kehoe, D.M. (2012) Emerging perspectives on the mechanisms, regulation, and distribution of light color acclimation in cyanobacteria. *Mol. Plant*, **5** (1), 1–13.
45. Niyogi, K.K. and Truong, T.B. (2013) Evolution of flexible non-photochemical quenching mechanisms that regulate

light harvesting in oxygenic photosynthesis. *Curr. Opin. Plant Biol.*, **16** (3), 307–314.

46. Ansong, C., Sadler, N.C., Hill, E.A. *et al.* (2014) Characterization of protein redox dynamics induced during light-to-dark transitions and nutrient limitation in cyanobacteria. *Front. Microbiol.*, **5**, 325.

47. Buchanan, B.B. (1991) Regulation of CO_2 assimilation in oxygenic photosynthesis: the ferredoxin/thioredoxin system: perspective on its discovery, present status, and future development. *Arch. Biochem. Biophys.*, **288** (1), 1–9.

48. Florencio, F.J., Pérez-Pérez, M.E., López-Maury, L. *et al.* (2006) The diversity and complexity of the cyanobacterial thioredoxin systems. *Photosynth. Res.*, **89** (2-3), 157–171.

49. Lindahl, M. and Kieselbach, T. (2009) Disulphide proteomes and interactions with thioredoxin on the track towards understanding redox regulation in chloroplasts and cyanobacteria. *J. Proteomics*, **72** (3), 416–438.

50. Heidorn, T., Camsund, D., Huang, H.-H. *et al.* (2011) Synthetic biology in cyanobacteria engineering and analyzing novel functions. *Methods Enzymol.*, **497**, 539–579.

51. Kufryk, G.I., Sachet, M., Schmetterer, G., and Vermaas, W.F.J. (2002) Transformation of the cyanobacterium Synechocystis sp. PCC 6803 as a tool for genetic mapping: optimization of efficiency. *FEMS Microbiol. Lett.*, **206** (2), 215–219.

52. Grigorieva, G. and Shestakov, S. (1982) Transformation in the cyanobacterium Synechocystis sp. 6803. *FEMS Microbiol. Lett.*, **13** (4), 367–370.

53. Shestakov, S.V. and Khyen, N.T. (1970) Evidence for genetic transformation in blue-green alga Anacystis nidulans. *Mol. Gen. Genet.*, **107** (4), 372–375.

54. Stevens, S.E. and Porter, R.D. (1980) Transformation in *Agmenellum quadruplicatum*. *Proc. Natl. Acad. Sci. U.S.A.*, **77** (10), 6052–6056.

55. Onai, K., Morishita, M., Kaneko, T. *et al.* (2004) Natural transformation of the thermophilic cyanobacterium *Thermosynechococcus elongatus* BP-1: a simple and efficient method for gene transfer. *Mol. Genet. Genomics*, **271** (1), 50–59.

56. Iwai, M., Katoh, H., Katayama, M., and Ikeuchi, M. (2004) Improved genetic transformation of the thermophilic cyanobacterium, *Thermosynechococcus elongatus* BP-1. *Plant Cell Physiol.*, **45** (2), 171–175.

57. Wolk, C.P., Vonshak, A., Kehoe, P., and Elhai, J. (1984) Construction of shuttle vectors capable of conjugative transfer from *Escherichia coli* to nitrogen-fixing filamentous cyanobacteria. *Proc. Natl. Acad. Sci. U.S.A.*, **81** (5), 1561–1565.

58. Cohen, M.F., Wallis, J.G., Campbell, E.L., and Meeks, J.C. (1994) Transposon mutagenesis of Nostoc sp. strain ATCC 29133, a filamentous cyanobacterium with multiple cellular differentiation alternatives. *Microbiology*, **140** (12), 3233–3240.

59. Flores, E. and Wolk, C.P. (1985) Identification of facultatively heterotrophic, N2-fixing cyanobacteria able to receive plasmid vectors from *Escherichia coli* by conjugation. *J. Bacteriol.*, **162** (3), 1339–1341.

60. Murry, M.A. and Wolk, C.P. (1991) Identification and initial utilization of a portion of the smaller plasmid of *Anabaena variabilis* ATCC 29413 capable of replication in Anabaena sp. strain M-131. *Mol. Gen. Genet.*, **227** (1), 113–119.

61. Tsinoremas, N.F., Kutach, A.K., Strayer, C.A., and Golden, S.S. (1994) Efficient gene transfer in *Synechococcus* sp. strains PCC 7942 and PCC 6301 by interspecies conjugation and chromosomal recombination. *J. Bacteriol.*, **176** (21), 6764–6768.

62. Marraccini, P., Bulteau, S., Cassier-Chauvat, C. *et al.* (1993) A conjugative plasmid vector for promoter analysis in several cyanobacteria of the genera *Synechococcus and Synechocystis*. *Plant Mol. Biol.*, **23** (4), 905–909.

63. Vioque, A. (2007) *Transgenic Microalgae as Green Cell Factories* (eds R. León, A. Galván, E. Fernández), Springer, pp. 12–22.

64. Herdman, M., Janvier, M., Rippka, R., and Stanier, R.Y. (1979) Genome size of cyanobacteria. *Microbiology*, **111** (1), 73–85.
65. Griese, M., Lange, C., and Soppa, J. (2011) Ploidy in cyanobacteria. *FEMS Microbiol. Lett.*, **323** (2), 124–131.
66. Vermaas, W. (1996) Molecular genetics of the cyanobacterium Synechocystis sp. PCC 6803: principles and possible biotechnology applications. *J. Appl. Phycol.*, **8** (4-5), 263–273.
67. Zhou, J., Zhang, H., Meng, H. *et al.* (2014) Discovery of a super-strong promoter enables efficient production of heterologous proteins in cyanobacteria. *Sci. Rep.*, **4**, 4500.
68. Camsund, D. and Lindblad, P. (2014) Engineered transcriptional systems for cyanobacterial biotechnology. *Front. Bioeng. Biotechnol.*, **2**, 40.
69. Guerrero, F., Carbonell, V., Cossu, M. *et al.* (2012) Ethylene synthesis and regulated expression of recombinant protein in Synechocystis sp. PCC 6803. *PLoS One*, **7** (11), e50470.
70. Huang, H.-H. and Lindblad, P. (2013) Wide-dynamic-range promoters engineered for cyanobacteria. *J. Biol. Eng.*, **7**, 10.
71. Imamura, S. and Asayama, M. (2009) Sigma factors for cyanobacterial transcription. *Gene Regul. Syst. Biol.*, **3**, 65.
72. Peters, J.M., Vangeloff, A.D., and Landick, R. (2011) Bacterial transcription terminators: the RNA 3′-end chronicles. *J. Mol. Biol.*, **412** (5), 793–813.
73. Mutalik, V.K., Guimaraes, J.C., Cambray, G. *et al.* (2013) Precise and reliable gene expression via standard transcription and translation initiation elements. *Nat. Methods*, **10** (4), 354–360.
74. Cambray, G., Guimaraes, J.C., Mutalik, V.K. *et al.* (2013) Measurement and modeling of intrinsic transcription terminators. *Nucleic Acids Res.*, **41** (9), 5139–5148.
75. Sharma, N.K., Rai, A.K., and Stal, L.J. (2013) *Cyanobacteria: an Economic Perspective*, John Wiley & Sons, Ltd.
76. De Hoon, M.J., Makita, Y., Nakai, K., and Miyano, S. (2005) Prediction of transcriptional terminators in *Bacillus subtilis* and related species. *PLoS Comput. Biol.*, **1** (3), e25.
77. Vijayan, V., Jain, I.H., and O'Shea, E.K. (2011) A high resolution map of a cyanobacterial transcriptome. *Genome Biol.*, **12** (5), R47.
78. Ma, J., Campbell, A., and Karlin, S. (2002) Correlations between Shine–Dalgarno sequences and gene features such as predicted expression levels and operon structures. *J. Bacteriol.*, **184** (20), 5733–5745.
79. Salis, H.M., Mirsky, E.A., and Voigt, C.A. (2009) Automated design of synthetic ribosome binding sites to control protein expression. *Nat. Biotechnol.*, **27** (10), 946–950.
80. Na, D. and Lee, D. (2010) RBSDesigner: software for designing synthetic ribosome binding sites that yields a desired level of protein expression. *Bioinformatics*, **26** (20), 2633–2634.
81. Seo, S.W., Yang, J.-S., Kim, I. *et al.* (2013) Predictive design of mRNA translation initiation region to control prokaryotic translation efficiency. *Metab. Eng.*, **15**, 67–74.
82. Ting, C.S., Rocap, G., King, J., and Chisholm, S.W. (2002) Cyanobacterial photosynthesis in the oceans: the origins and significance of divergent light-harvesting strategies. *Trends Microbiol.*, **10** (3), 134–142.
83. Blankenship, R.E., Tiede, D.M., Barber, J. *et al.* (2011) Comparing photosynthetic and photovoltaic efficiencies and recognizing the potential for improvement. *Science*, **332** (6031), 805–809.
84. Kirst, H., Formighieri, C., and Melis, A. (2014) Maximizing photosynthetic efficiency and culture productivity in cyanobacteria upon minimizing the phycobilisome light-harvesting antenna size. *Biochim. Biophys. Acta, Bioenerg.*, **1837** (10), 1653–1664.
85. Kirst, H., García-Cerdán, J.G., Zurbriggen, A., and Melis, A. (2012) Assembly of the light-harvesting chlorophyll antenna in the green alga *Chlamydomonas reinhardtii* requires

expression of the TLA2-CpFTSY gene. *Plant Physiol.*, **158** (2), 930–945.
86. Kirst, H., Garcia-Cerdan, J.G., Zurbriggen, A. *et al.* (2012) Truncated photosystem chlorophyll antenna size in the green microalga *Chlamydomonas reinhardtii* upon deletion of the TLA3-CpSRP43 gene. *Plant Physiol.*, **160** (4), 2251–2260.
87. Polle, J.E., Kanakagiri, S.-D., and Melis, A. (2003) tla1, a DNA insertional transformant of the green alga Chlamydomonas reinhardtii with a truncated light-harvesting chlorophyll antenna size. *Planta*, **217** (1), 49–59.
88. Melis, A. (2012) Photosynthesis-to-fuels: from sunlight to hydrogen, isoprene, and botryococcene production. *Energy Environ. Sci.*, **5** (2), 5531–5539.
89. Kirst, H. and Melis, A. (2014) The chloroplast signal recognition particle (CpSRP) pathway as a tool to minimize chlorophyll antenna size and maximize photosynthetic productivity. *Biotechnol. Adv.*, **32** (1), 66–72.
90. Mussgnug, J.H., Thomas-Hall, S., Rupprecht, J. *et al.* (2007) Engineering photosynthetic light capture: impacts on improved solar energy to biomass conversion. *Plant Biotechnol. J.*, **5** (6), 802–814.
91. Page, L.E., Liberton, M., and Pakrasi, H.B. (2012) Reduction of photoautotrophic productivity in the cyanobacterium *Synechocystis* sp. strain PCC 6803 by phycobilisome antenna truncation. *Appl. Environ. Microbiol.*, **78** (17), 6349–6351.
92. Lea-Smith, D.J., Bombelli, P., Dennis, J.S. *et al.* (2014) Phycobilisome-deficient strains of *Synechocystis* sp. PCC 6803 have reduced size and require carbon-limiting conditions to exhibit enhanced productivity. *Plant Physiol.*, **165** (2), 705–714.
93. Ellis, R.J. (2010) Biochemistry: tackling unintelligent design. *Nature*, **463** (7278), 164–165.
94. Tcherkez, G.G., Farquhar, G.D., and Andrews, T.J. (2006) Despite slow catalysis and confused substrate specificity, all ribulose bisphosphate carboxylases may be nearly perfectly optimized. *Proc. Natl. Acad. Sci. U.S.A.*, **103** (19), 7246–7251.
95. Ninomiya, N., Ashida, H., and Yokota, A. (2008) *Photosynthesis. Energy from the Sun* (eds J.F. Allen, E. Gantt, J.H. Golbeck, B. Osmond), Springer, pp. 867–870.
96. Price, G.D., Badger, M.R., Woodger, F.J., and Long, B.M. (2008) Advances in understanding the cyanobacterial CO_2-concentrating-mechanism (CCM): functional components, Ci transporters, diversity, genetic regulation and prospects for engineering into plants. *J. Exp. Bot.*, **59** (7), 1441–1461.
97. Price, G.D. (2011) Inorganic carbon transporters of the cyanobacterial CO_2 concentrating mechanism. *Photosynth. Res.*, **109** (1-3), 47–57.
98. Espie, G.S. and Kimber, M.S. (2011) Carboxysomes: cyanobacterial RubisCO comes in small packages. *Photosynth. Res.*, **109** (1-3), 7–20.
99. Savage, D.F., Afonso, B., Chen, A.H., and Silver, P.A. (2010) Spatially ordered dynamics of the bacterial carbon fixation machinery. *Science*, **327** (5970), 1258–1261.
100. Iwaki, T., Haranoh, K., Inoue, N. *et al.* (2006) Expression of foreign type I ribulose-1, 5-bisphosphate carboxylase/oxygenase (EC 4.1. 1.39) stimulates photosynthesis in cyanobacterium *Synechococcus* PCC7942 cells. *Photosynth. Res.*, **88** (3), 287–297.
101. Atsumi, S., Higashide, W., and Liao, J.C. (2009) Direct photosynthetic recycling of carbon dioxide to isobutyraldehyde. *Nat. Biotechnol.*, **27** (12), 1177–1180.
102. Yang, C., Hua, Q., and Shimizu, K. (2002) Metabolic flux analysis in Synechocystis using isotope distribution from 13 C-labeled glucose. *Metab. Eng.*, **4** (3), 202–216.
103. Bar-Even, A., Noor, E., Lewis, N.E., and Milo, R. (2010) Design and analysis of synthetic carbon fixation pathways. *Proc. Natl. Acad. Sci. U.S.A.*, **107** (19), 8889–8894.
104. Deng, M.D. and Coleman, J.R. (1999) Ethanol synthesis by genetic engineering in cyanobacteria. *Appl. Environ. Microbiol.*, **65** (2), 523–528.

105. Lan, E.I., Ro, S.Y., and Liao, J.C. (2013) Oxygen-tolerant coenzyme A-acylating aldehyde dehydrogenase facilitates efficient photosynthetic n-butanol biosynthesis in cyanobacteria. *Energy Environ. Sci.*, **6** (9), 2672–2681.
106. Shen, C.R. and Liao, J.C. (2012) Photosynthetic production of 2-methyl-1-butanol from CO2 in cyanobacterium *Synechococcus elongatus* PCC7942 and characterization of the native acetohydroxyacid synthase. *Energy Environ. Sci.*, **5** (11), 9574–9583.
107. Nozzi, N.E., Oliver, J.W.K., and Atsumi, S. (2013) Cyanobacteria as a platform for biofuel production. *Front. Bioeng. Biotechnol.*, **1**, 7.
108. Angermayr, S.A., Gorchs Rovira, A., and Hellingwerf, K.J. (2015) Metabolic engineering of cyanobacteria for the synthesis of commodity products. *Trends Biotechnol.*, **33** (6), 352–361.
109. Zhou, J., Zhang, H., Zhang, Y. *et al.* (2012) Designing and creating a modularized synthetic pathway in cyanobacterium Synechocystis enables production of acetone from carbon dioxide. *Metab. Eng.*, **14** (4), 394–400.
110. Kusakabe, T., Tatsuke, T., Tsuruno, K. *et al.* (2013) Engineering a synthetic pathway in cyanobacteria for isopropanol production directly from carbon dioxide and light. *Metab. Eng.*, **20**, 101–108.
111. Oliver, J.W.K., Machado, I.M.P., Yoneda, H., and Atsumi, S. (2013) Cyanobacterial conversion of carbon dioxide to 2,3-butanediol. *Proc. Natl. Acad. Sci. U.S.A.*, **110** (4), 1249–1254.
112. Xiong, W., Morgan, J.A., Ungerer, J. *et al.* (2015) The plasticity of cyanobacterial metabolism supports direct CO_2 conversion to ethylene. *Nat. Plants*, **1** (5), 15053.
113. Li, H. and Liao, J.C. (2013) Engineering a cyanobacterium as the catalyst for the photosynthetic conversion of CO_2 to 1,2-propanediol. *Microb. Cell Fact.*, **12** (1), 4.
114. Wang, B., Pugh, S., Nielsen, D.R. *et al.* (2013) Engineering cyanobacteria for photosynthetic production of 3-hydroxybutyrate directly from CO_2. *Metab. Eng.*, **16**, 68–77.
115. Lan, E.I., Chuang, D.S., Shen, C.R. *et al.* (2015) Metabolic engineering of cyanobacteria for photosynthetic 3-hydroxypropionic acid production from CO_2 using *Synechococcus elongatus* PCC 7942. *Metab. Eng.*, **31**, 163–170.
116. Wang, Y., Sun, T., Gao, X. *et al.* (2015) Biosynthesis of platform chemical 3-hydroxypropionic acid (3-HP) directly from CO2 in cyanobacterium Synechocystis sp. PCC 6803. *Metab. Eng.*
117. Liu, X., Sheng, J., and Iii, R.C. (2011) Fatty acid production in genetically modified cyanobacteria. *Proc. Natl. Acad. Sci. U.S.A.*, **108** (17), 6899–6904.
118. Gao, Z., Zhao, H., Li, Z. *et al.* (2012) Photosynthetic production of ethanol from carbon dioxide in genetically engineered cyanobacteria. *Energy Environ. Sci.*, **5** (12), 9857–9865.
119. Varman, A.M., Yu, Y., You, L., and Tang, Y.J. (2013) Photoautotrophic production of D-lactic acid in an engineered cyanobacterium. *Microb. Cell Fact.*, **12**, 117.
120. Niederholtmeyer, H., Wolfstädter, B.T., Savage, D.F. *et al.* (2010) Engineering cyanobacteria to synthesize and export hydrophilic products. *Appl. Environ. Microbiol.*, **76** (11), 3462–3466.
121. Lan, E.I. and Liao, J.C. (2012) ATP drives direct photosynthetic production of 1-butanol in cyanobacteria. *Proc. Natl. Acad. Sci. U.S.A.*, **109** (16), 6018–6023.
122. Shen, C.R., Lan, E.I., Dekishima, Y. *et al.* (2011) Driving forces enable high-titer anaerobic 1-butanol synthesis in *Escherichia coli*. *Appl. Environ. Microbiol.*, **77** (9), 2905–2915.
123. Lan, E.I. and Liao, J.C. (2011) Metabolic engineering of cyanobacteria for 1-butanol production from carbon dioxide. *Metab. Eng.*, **13** (4), 353–363.
124. Tokuyama, K., Ohno, S., Yoshikawa, K. *et al.* (2014) Increased 3-hydroxypropionic acid production from glycerol, by modification of central metabolism in *Escherichia coli*. *Microb. Cell Fact.*, **13**, 64.
125. Chu, H.S., Kim, Y.S., Lee, C.M. *et al.* (2015) Metabolic engineering of 3-hydroxypropionic acid biosynthesis in

Escherichia coli. Biotechnol. Bioeng., **112** (2), 356–364.

126. Begemann, M.B., Zess, E.K., Walters, E.M. *et al.* (2013) An organic acid based counter selection system for cyanobacteria. *PLoS One*, **8** (10), e76594.
127. Atsumi, S., Hanai, T., and Liao, J.C. (2008) Non-fermentative pathways for synthesis of branched-chain higher alcohols as biofuels. *Nature*, **451** (7174), 86–89.
128. Li, X., Shen, C.R., and Liao, J.C. (2014) Isobutanol production as an alternative metabolic sink to rescue the growth deficiency of the glycogen mutant of *Synechococcus elongatus* PCC 7942. *Photosynth. Res.*, **120** (3), 301–310.
129. Zhang, S. and Bryant, D.A. (2011) The tricarboxylic acid cycle in cyanobacteria. *Science*, **334** (6062), 1551–1553.
130. Tamagnini, P., Axelsson, R., Lindberg, P. *et al.* (2002) Hydrogenases and hydrogen metabolism of cyanobacteria. *Microbiol. Mol. Biol. Rev.*, **66** (1), 1–20.
131. Dutta, D., De, D., Chaudhuri, S., and Bhattacharya, S.K. (2005) Hydrogen production by Cyanobacteria. *Microb. Cell Fact.*, **4**, 36.
132. Lindblad, P. (1999) Cyanobacterial H2 metabolism: knowledge and potential/strategies for a photobiotechnological production of H2. *Biotechnol. Appl.*, **16** (3), 141–144.
133. Ducat, D.C., Sachdeva, G., and Silver, P.A. (2011) Rewiring hydrogenase-dependent redox circuits in cyanobacteria. *Proc. Natl. Acad. Sci. U.S.A.*, **108** (10), 3941–3946.
134. McNeely, K., Xu, Y., Bennette, N. *et al.* (2010) Redirecting reductant flux into hydrogen production via metabolic engineering of fermentative carbon metabolism in a cyanobacterium. *Appl. Environ. Microbiol.*, **76** (15), 5032–5038.
135. Stripp, S.T., Goldet, G., Brandmayr, C. *et al.* (2009) How oxygen attacks [FeFe] hydrogenases from photosynthetic organisms. *Proc. Natl. Acad. Sci. U.S.A.*, **106** (41), 17331–17336.
136. Ducat, D.C., Avelar-Rivas, J.A., Way, J.C., and Silver, P.A. (2012) Rerouting carbon flux to enhance photosynthetic productivity. *Appl. Environ. Microbiol.*, **78** (8), 2660–2668.

16
Host Organisms: Algae

Elizabeth A. Specht, Prema S. Karunanithi, Javier A. Gimpel, William S. Ansari, and Stephen P. Mayfield

16.1
Introduction to Algae as an Industrial Organism

16.1.1
Algal Diversity

The term *algae* designates a wide range of photosynthetic organisms found throughout all aquatic environments, both freshwater and marine, as well as wet terrestrial environments like soil and symbiotically within lichens. They range from large macroalgae – such as seaweed and kelp – to both eukaryotic and prokaryotic microalgae. Red algae, brown algae, diatoms, cyanobacteria (blue-green algae), and green algae are all classified in distinct phyla with sometimes hundreds of genera within them [1]. More than half a million species have been discovered, but very few have been cultured in the laboratory and even fewer have been genetically characterized. This vast, untapped reserve of biological diversity has spawned much interest in natural products discovery, as algae produce a large number of unique molecules as secondary metabolites [2]. Beyond metabolic diversity, algae's vast range of ecological niches provides potential for identifying tolerance and resistance traits that are crucial for large-scale growth and robustness.

Some species of algae are already grown at industrial scale, mostly for harvesting food additives like agar and carrageenan or for direct consumption for their metabolites like omega-3 fatty acids and antioxidant carotenoids. In some cultures, algae have long played a notable role in cuisine, mostly the seaweed macroalgae. Recently, significant investments have been made in exploring the potential of microalgae for biofuels production, and several commercial and academic pilot facilities exist for developing the growth, harvesting, and lipid extraction technologies that will be required for industrial scale production.

Industrial Biotechnology: Microorganisms, First Edition. Edited by Christoph Wittmann and James C. Liao.

16.1.2 Scalability and Growth Rate

One of the most advantageous traits of algae is their potential for growth at agricultural scale, but with much higher biomass accumulation per acre than that of terrestrial plants and crops. On the macroscopic scale, kelp is well known for its extraordinary growth rate – up to half a meter a day. Microalgae also exhibit short doubling times, and cultures of eukaryotic green algae – the species most explored for large-scale biofuel production – can be grown to saturated density and harvested in as little as a week. Furthermore, algae are not subject to growing seasons to the same extent as most terrestrial plants, making year-round growth possible. Most commercial pilot facilities are situated on nonarable land, often in desert areas, where they do not compete with food crops for land or resources.

16.1.3 Genetic Malleability

Although relatively few of the hundreds of thousands of algal species have been genetically transformed, there has been rapid progress in the past decade of research. Some algae, mostly in the cyanobacterial groups, are naturally competent for transformation [3]. Extensive research has been done with these species – notably in the *Synechococcus*, *Synechocystis*, and *Anabaena* genera – to identify neutral sites and develop transformation vectors with appropriate gene regulatory elements [4–6]. Section 16.2 focuses on the genetic engineering of both the chloroplast and nuclear genomes of green algae, particularly the model species *Chlamydomonas reinhardtii*. This is an active area of research, and new genetic engineering tools – including reporter genes, gene targeting methods, synthetic promoters and regulatory elements, and high-expression transgene cassettes – are being published each year. Recently, there has been a strong push for tools with broad host species compatibility, such that the advances pioneered in the model organisms can be applied to strains more suitable for industrial-scale production.

16.2 Algal Genetic Engineering

At present, the species of green algae most commonly and easily transformed is *C. reinhardtii*. This model organism has been studied for decades, and thus the largest set of genetic tools exists for it. Our detailed discussion of algal genetic engineering focuses primarily on this organism, but recently many of these methods have been adapted to other species. Current progress toward transforming other species is covered in Section 16.2.3. Green algae contain three genomes – nuclear, chloroplast, and mitochondrial – all of which have been transformed; however, almost all genome engineering has been targeted to the nuclear (Section 16.2.1) and the chloroplast (Section 16.2.2) genomes.

16.2.1 Nuclear Genetic Engineering

The complete nuclear genome of *C. reinhardtii* was made available in 2007 [7]. While transformation of this genome is facile, there are several considerations that impact transgene expression from the nuclear genome; so it has not been as well exploited for transgenics as the chloroplast genome. Codon usage plays an important role in transgene expression, as the nuclear genome is highly guanine–cytocine (GC) rich (64% GC) [8]. Codon optimization to reflect codon bias in the nuclear genome can substantially increase transgene expression [9, 10], and several additional advances have been made to increase expression (see Section 16.2.1.3). A distinct advantage of nuclear expression is the potential to add posttranslational modifications, including glycosylation, which is not possible with chloroplast expression. Other potential benefits include targeting to cellular components and easier protein purification through secretion in media from cell-wall-deficient strains [11].

16.2.1.1 Nuclear Transformation Methods

Nuclear transformation in *C. reinhardtii* was first demonstrated in the early 1980s using the yeast *ARG4* gene to complement an *arg7* mutation [12]. Methods of nuclear transformation include particle bombardment or biolistic transformation [13, 14], agitation with glass beads in cell-wall-deficient strains [15], electroporation [16], agitation with silicon-carbide whiskers [17], and biologically induced gene transfer with *Agrobacterium tumefaciens* [18, 19]. *C. reinhardtii* is generally not susceptible to viral infection, so gene transfer using viral agents has not been possible. Electroporation is commonly used for nuclear transformation due to its high transformation efficiency, up to over 10 000 transformants per microgram of DNA [16]. Recently, droplet electroporation on microfluidic chips has been used to improve transformation efficiency by three orders of magnitude for wall-less mutants and over two orders of magnitude for wild-type cells compared to bulk phase electroporation [20]. Translocation by cell-penetrating peptides (CPPs), in particular, the artificially synthesized pVEC (peptide vascular endothelial cadherin), is another potential method for successful delivery of macromolecules like DNA into cell-wall-intact strains of *C. reinhardtii* [21].

16.2.1.2 Regulation of Nuclear Transgene Expression

Integration of foreign DNA into the nuclear genome primarily occurs by nonhomologous end joining [22, 23] because the rate of homologous recombination is very low under normal growth conditions [24–26]. Packaging of transgenes in distinct chromosomal domains [27] and chromatin compaction through histone modifications such as methylation on nucleosomes [28] are thought to contribute to epigenetic silencing, although the full range of silencing mechanisms is still under investigation. For nonsilenced genes, nuclear expression is regulated predominantly at the transcriptional level, necessitating well-characterized robust promoters [27].

Nuclear promoters used to express transgenes include *β_2TUB* [29], *NIT1* [30], *PSAD* (subunit of photosystem I) [31], *LHCB1* (light harvesting complex of photosystem II) [32], and *LIP* (light-inducible protein gene) from *Dunaliella* sp. [33]. The *HSP70A* (heat shock protein 70A) promoter acts as an expression enhancer [34], and *HSP70A-RBCS2* (RuBisCO small subunit) promoter fusions are the most effective to date [35–39]. Recently, *NOS* (nopaline synthase from *Agrobacterium*) and *CaMV35S* (cauliflower mosaic virus 35S) have been used to drive transgene expression, although transformation efficiency is much lower than that of the *HSP70A-RBCS2* chimeric promoter [40].

16.2.1.3
Tools for Increased Nuclear Gene Expression

Several strategies have been pioneered for evading gene silencing to achieve higher nuclear transgene expression, with varied success. Endogenous heat shock factor 1 (*HSF1*) increases expression of transgenes by counteracting epigenetic silencing through scaffold formation and recruitment of RNA Pol II to *HSP70A* promoters [39]. Insertion of one or more copies of the intron sequences from the *RBCS2* gene [41, 42] and use of linearized tandem vectors for coexpression [38] also significantly increase transgene expression. Recently, the viral 2A self-cleaving peptide from the foot-and-mouth disease virus has been used to drive high expression through transcriptional fusion to the selectable marker while creating distinct polypeptides. This strategy is particularly effective with the zeocin resistance gene *ble* because the gene product binds the antibiotic stoichiometrically instead of enzymatically inactivating it, necessitating high expression of the resistance gene (and, thus, of the gene of interest) to survive on zeocin [43]. Transcriptional terminators have also been shown to influence transgene expression. The *psaD* terminator paired with its respective *psaD* promoter increased accumulation of the recombinant protein butyrylcholinesterase to 0.4% [44].

16.2.1.4
Selection and Screening Methods

The first selection markers were genes used to rescue auxotrophic mutants. *ARG7* for arginine biosynthesis [13] and *NIT1* for nitrate reductase [45] are well-known examples. Antibiotic-resistance-conferring genes were developed later, and include genes *sh-ble* for zeocin and phleomycin [46], *CRY1* for emetine [47], *aadA* for spectinomycin [23], *aphVIII* for kanamycin, paromomycin, and neomycin [48], and *aph7* for hygromycin [49]. Most transformation vectors include a selectable marker, but cotransformations can also be done using a selectable marker plasmid and a separate gene of interest plasmid; typically, about a quarter of the selected transformants receive both plasmids.

A handful of nuclear reporters are available for a variety of chemical or fluorescent readouts. The endogenous arylsulfatase gene ARS2, which is tightly regulated and normally only expressed in sulfur starvation, cleaves a chromogenic compound for easy detection [29] and has been used to monitor inducible promoter activity [50]. Luciferase-encoding genes from *Renilla* [32] and *Gaussia* [51] have both been used, but *Gaussia* luciferase may be more sensitive in *Chlamydomonas* [52]. The *RSP3* gene encoding a radial spoke protein required for flagellar motility was used as a reporter in mutant strains lacking endogenous *RSP3* that are paralyzed under aerobic conditions [53]. A suite of codon-optimized fluorescent protein genes have been characterized in *C. reinhardtii* as well, including several green fluorescent protein (GFP) variants, mCherry, tdTomato, and mCerulean [54]. Many common reporters are listed in Table 16.1.

16.2.1.5
Gene Targeting and Genome Editing

Reverse genetics is critical for determining gene function by knocking out endogenous genes and observing phenotypic results. Since homologous recombination is a rare event [26], targeted knockout is not yet possible; so knockouts are obtained through insertional mutagenesis [56]. However, genes can be knocked down using RNA interference (RNAi) due the presence of the required machinery, including Dicer and Argonaute proteins, in algae [57]. The two types of RNA involved in RNAi-mediated gene silencing are small interfering RNAs (siRNAs) and microRNAs (miRNAs). siRNAs are processed from double-stranded RNA (dsRNA) created through sense/antisense pairing or hairpin formation by inverted repeats [58] and have been used to knock down gene expression [59]. Artificial microRNAs (amiRNAs) were created to mimic endogenous single-stranded transcripts of naturally occurring miRNAs [60], promoting gene knockdown [61].

Table 16.1 Reporter genes for nuclear expression.

Reporter	Gene	Source	References	Notes
Arylsulfatase	*ARS2*	*Chlamydomonas reinhardtii*	Davies *et al.* [29] and de Hostos *et al.* [55]	Excreted into media and cleaves chromogenic substrates
Green fluorescent protein	*cGFP*	*Aequorea victoria*	Fuhrmann *et al.* [9]	GFP codon optimized for *C. reinhardtii*
Gaussia luciferase	*G-Luc*	*Gaussia princeps*	Shao and Bock [51]	Codon-optimized luciferase
mCherry; tdTomato	*mCherry; tdTomato*	*Discosoma striata*	Rasala *et al.* [54]	RFP variants optimized for *C. reinhardtii*
Venus; mCerulean	*Venus; mCerulean*	*Aequorea victoria*	Rasala *et al.* [54]	GFP variants optimized for *C. reinhardtii*

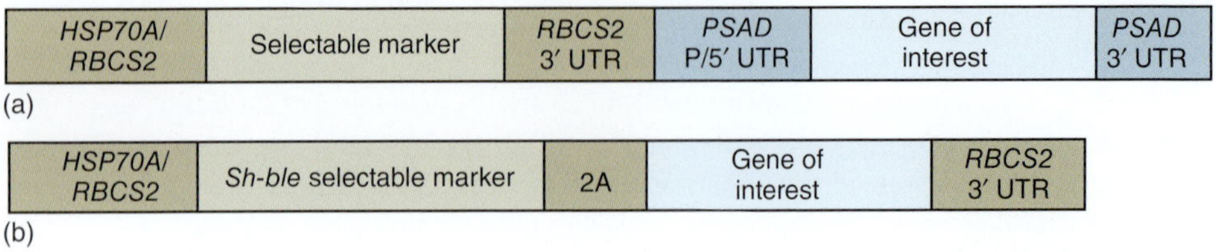

Figure 16.1 Schematic of two typical expression vectors for *Chlamydomonas* nuclear transformation. (a) A standard nuclear expression vector, including a selectable marker cassette and a gene of interest cassette, each with their own promoters, 5′ UTRs and 3′ UTRs. (b) Vector in which the gene of interest has been transcriptionally fused to zeocin resistance by the 2A self-cleaving peptide to drive high gene of interest expression. UTR, untranslated region.

Genome editing is also not routinely possible in the nuclear genome. Recently, zinc finger nucleases have been employed in *C. reinhardtii* [62], but the process for constructing them for each target site requires significant effort and is costly. Other strategies such as transcription-activator-like (TAL) effector nucleases and clustered regularly interspaced short palindromic repeats (CRISPR)/CRISPR-associated (Cas) genome editing have not yet been demonstrated in algae, although multiple groups are working on developing these technologies for algal systems (Figure 16.1).

16.2.2
Chloroplast Genetic Engineering

Due to the endosymbiotic evolutionary history of the chloroplast, it has its own complete genetic system, including a circular chromosome, and transcription and translation machinery including all tRNAs and prokaryotic-like ribosomes. Over time, many genes that originated in the chloroplast genome have moved into the nuclear genome, but their gene products are still targeted for chloroplast localization; there is substantial cross-talk between the nuclear and plastid genomes. In contrast to most terrestrial plants that contain many chloroplasts per cell, *C. reinhardtii* contains a single large chloroplast that occupies over 40% of cell volume [8]. The chloroplast genome is a circular 203 kb chromosome encoding 99 genes [63]. Greater than 20% of the genome consists of repetitive DNA and different classes of short dispersed regions between the coding regions containing genes for photosynthetic proteins and transcriptional–translational machinery [63]. Unlike the GC-rich nuclear genome, the chloroplast genome is only about 35% GC [63]. The chloroplast contains about 80 copies of its genome, and stable transformation requires that all copies contain the transgenic construct or else reversion may be observed [8]; this stable state is referred to as *homoplasmy.*

16.2.2.1
Chloroplast Transformation Methods

Particle bombardment remains the most effective method for transforming the chloroplast of *C. reinhardtii,* and was first demonstrated in the 1980s [64].

DNA-coated gold or tungsten particles are bombarded at high velocity onto a plate with a lawn of algae. The particles penetrate both the cell wall and the chloroplast membranes, where the foreign DNA can access the chloroplast genome [64]. The process is performed by a biolistic gene gun or in a vacuum chamber. Transformation by agitation with glass beads in cell-wall-deficient strains has also been used, albeit with lower transformation efficiency compared to microparticle bombardment [65]. Integration of transgenes occurs readily through homologous recombination, requiring flanking regions of endogenous sequence homologous to target insertion site [64]. The minimal flanking sequence around the recombinant gene should be about 1 kb to ensure successful integration [66]. Since homologous recombination occurs in the chloroplast, knock down of endogenous genes is possible in addition to targeted gene insertion.

16.2.2.2 Gene Regulation in the Chloroplast

Unlike the nuclear genome, in the chloroplast most genes are predominantly regulated at the posttranscriptional level [67]. The 5′ and 3′ untranslated regions (UTRs) largely influence translation rate [68] and mRNA stability [69], respectively. Gene expression in the chloroplast is regulated by posttranscriptional processes influencing RNA stability, RNA processing, *cis*- and *trans*-splicing, and translation [70].

Thus far, most studies have been devoted to identifying and utilizing endogenous chloroplast promoters and UTRs for heterologous gene expression [68]. The most effective promoter and UTR for heterologous gene expression is from the *psbA* gene (encoding the D1 protein of Photosystem II (PSII)), but transgene product accumulation is only significant in *psbA*-deficient strains because of auto-attenuation of translation in the presence of an intact photosystem. However, photosynthesis can be restored by reintroducing the *psbA* coding region under the control of different regulatory elements, allowing transgene expression to be maintained [71]. Other promoters and UTRs from *atpA*, *rbcL*, and *psbD* have been used to drive heterologous gene expression but with varied success [72]. However, because of high recombination rates within the chloroplast, there is some risk of instability from introducing transgenes surrounded by endogenous regulatory sequences. Because of this and the auto-attenuation concerns, novel synthetic UTRs have been derived using discrete elements extracted from high-expressing members of a mutant UTR library [73].

16.2.2.3 Chloroplast Selection and Reporters

Chloroplast transformants can be conveniently selected by restoring photosynthesis in photosystem knockout strains; and *atpB* [64], *rbcL* [74], and *psbH* [75] have all been used as selectable markers in this manner. Antibiotic resistance is also commonly used for selection and can be conferred through heterologous gene expression or induced point mutations in endogenous genes. The *aphA6* gene (from *Acinetobacter baumannii*) for kanamycin resistance [75] and *aadA*

Table 16.2 Reporter genes for chloroplast expression.

Reporter	Gene	Source	References
Renilla luciferase	*R-Luc*	*Renilla reniformis*	Minko *et al.* [80]
GFP	*CrGFP*	*Aequorea victoria*	Franklin *et al.* [81]
Bacterial luciferase	*luxCt*	*Vibrio harveyi*	Mayfield and Schultz [82]
β-Glucuronidase (GUS)	*uidA*	*Escherichia coli*	Klein *et al.* [83]

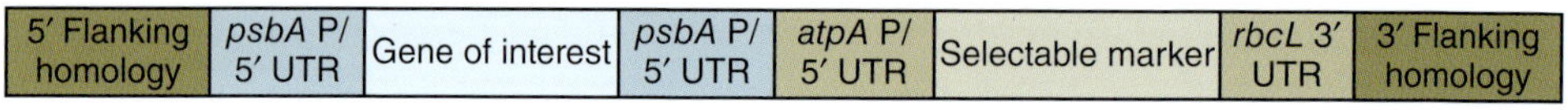

Figure 16.2 Schematic of a typical *Chlamydomonas* chloroplast transformation vector. The gene of interest and the selectable marker are expressed in two separate cassettes. Two sites are commonly targeted: the 3HB site is a neutral site in the chloroplast genome that tolerates transgene insertion, and the *psbA* site is replaced by a transgene for high expression in *psbA* knockouts. The flanking homology regions contain sequence homologous to the chloroplast genome at the intended site of integration. UTR, untranslated region.

(from *Escherichia coli*) for spectinomycin and streptomycin resistance [76] have been used in multiple algal species. Point mutations in the 16S and 23S ribosomal RNAs confer resistance to streptomycin and erythromycin, respectively [77], and mutations in the endogenous *psbA* gene can confer resistance to herbicides like atrazine [78, 79]. Chloroplast reporters, like those used in the nucleus, include fluorescent proteins and luciferase variants (see Table 16.2). However, the same gene cannot be used in both compartments due to the stark difference in codon bias (Figure 16.2).

16.2.3 Other Transformable Species

While *C. reinhardtii* has been the most popular species for pioneering genetic engineering work, many species – encompassing a broad swath of the diversity among microalgae – have, in fact, been successfully transformed. Most of these species have only been transformed in the nuclear genome, as penetrating the chloroplast often proves more difficult and requires accurate plastid genome sequence to obtain flanking regions for integration by homologous recombination.

16.2.3.1 Parameters for Transformation Optimization

Electroporation is generally the first technique tried in most algae for nuclear transformation, while chloroplast transformation often requires a more aggressive approach like biolistic particle transformation. With both of these methods, many parameters can be optimized to increase transformation efficiencies for each species. With electroporation, the voltage, resistance, and current can all be specified, as well as the buffers in which the cells are transformed and subsequently recovered. With particle bombardment, the particle size and material are variable, as well as factors affecting the velocity of the particles such as the vacuum strength, gas pressure, and even the distance between the biolistic gun and the agar plate.

16.2.3.2 Species Successfully Transformed

For some species, notably blue-green algae (cyanobacteria), DNA is taken up directly from the media or through bacterial conjugation, as they are naturally competent for transformation. Even diatoms – which are generally surrounded by a tough silica coat – have been successfully penetrated and transformed after some troubleshooting. At least six species of diatoms have been transformed – all in the nuclear genome [84] – including *Phaeodactylum tricornutum* [85] and *Thalassiosira weissflogii* [86]. Several species of *Chlorella* have been transformed as well, most notably *Chlorella vulgaris* [87]. Green algae in the *Dunaliella* genus, which have significant commercial potential and have been transformed in both the nuclear and chloroplast genomes, include *Dunaliella salina* [88, 89] and *Dunaliella tertiolecta* [90]. Two species of dinoflagellates have been transformed using silicon carbide whiskers [91], as well as two species of red algae, including chloroplast transformation in *Porphyridium* spp. [92, 93]. These species represent widely diverse phyla and genera, and this broad range of transformability is a testament to the potential for applying the genetic engineering tools described in Sections 16.2.1 and 16.2.2 to species beyond *C. reinhardtii.*

16.3 Therapeutic and Nutraceutical Applications

Algae are already harvested for natural products, including astaxanthin from *Haematococcus* for supplements and fish feed, and whole-cell *Chlorella* and *Spirulina* to be blended into specialty drinks and supplements. It is also likely that one of the first commercial applications of engineered algae will be for the therapeutic or nutraceutical industries. These products have much higher value per unit mass than biofuels or bulk animal feed, so their production is economically viable even with relatively low expression levels of recombinant proteins. It is hoped that investment in these specialty products will spur further advancements in algal genetic engineering and in the downstream engineering needed to harvest and process at scale. These advancements can, in turn, eventually be applied to larger systems such as algal biofuel production, thereby essentially subsidizing the sizable development costs of such a platform.

16.3.1 Protein Therapeutics

Protein therapeutics comprise the fastest-growing sector of the booming pharmaceuticals industry [94], and current production platforms have their limitations. For example, mammalian cell culture is expensive and susceptible to contamination; prokaryotic systems cannot properly fold many complex human proteins; and yeast often hyperglycosylate proteins, making them unsuitable for many therapeutic applications. In the past two decades, interest in terrestrial crop plant medicinal production ("molecular pharming") has waxed and waned. Hostility toward this approach usually derives from concerns of gene flow or gene escape into other crops, leading to contamination of the food supply.

Besides the lack of pollen and the ability to grow in enclosed systems – alleviating gene flow concerns – algae have several other advantages over terrestrial plant-derived vaccines. They have a much more rapid strain generation time, allowing transformation and verification of a transgenic strain in a matter of weeks instead of months for crop plants. They also accumulate biomass more quickly per unit area and can be harvested continually instead of seasonally. Finally, growth conditions can be better controlled, leading to less fluctuation in transgenic protein product content than is observed in crop plates from soil and climate variations.

Several human genes have been transformed into algal species, predominantly but not exclusively into *C. reinhardtii.*

16.3.1.1 Human Antibodies and Antibody–Drug Conjugates

Human antibodies are large, complex proteins that require proper folding and disulfide bond formation and precise incorporation of multiple chains, presenting difficulty for many production platforms. However, *C. reinhardtii* chloroplasts have been engineered to produce human antibodies or parts thereof since 2003, when a monoclonal antibody was produced as a single chain instead of a separate heavy and light chain. The antibody folded properly within the chloroplast and recognized its target antigen, herpes simplex virus [95]. A few years later, the first antibody composed of separate heavy and light chains was produced inside algal chloroplasts; the chains assembled into full-length, functional antibodies *in vivo* [96].

Antibody–drug conjugates are a class of molecules in which a humanized or recombinant human antibody is attached to a drug molecule, which is often a small-molecule toxin. Several of these drugs have been approved as cancer treatments, wherein the antibody specifically recognizes cancer cells and is endocytosed, releasing the toxin and triggering cell death. Recently, algal chloroplasts have been shown to be a suitable platform for producing a novel class of antibody–drug conjugates in which the toxin is a protein that is genetically linked to the antibody, not a chemically conjugated small molecule. These toxins include exotoxin A [97] and gelonin [98], both of which inhibit eukaryotic ribosomes, so they cannot be produced within the normal eukaryotic platforms that are required for proper folding of the antibody portion of the molecule. The chloroplast, in contrast, is not susceptible to these toxins because of the prokaryotic nature of its ribosomes, so it is a unique host in that it has the complexity to make antibodies and yet the simplicity of a bacterial platform to evade the effects of the toxin domains. These molecules have indeed been made in *C. reinhardtii* and tested in mice, exhibiting reduced tumor size and improved mouse survival over controls [97, 98].

16.3.1.2
Other Chloroplast-Produced Protein Therapeutics

Several human therapeutic proteins in addition to antibodies have been produced in algae. A human metallothionein protein – one of a class of proteins with metal-binding capacity and an ability to capture oxidative radicals – was introduced into the *C. reinhardtii* chloroplast and appeared to confer increased resistance to UV stress, although only temporarily [99]. Vascular endothelial growth factor (VEGF) stimulates angiogenesis and its upregulation is observed in many cancers, but it also plays a role in wound healing and is decreased in emphysema patients. VEGF was produced in *C. reinhardtii* at 2% total soluble protein, and was shown to be biologically active in a receptor-binding assay [100]. In the same study, high-mobility group protein B1 (HMGB1) – which is also involved in wound healing – was accumulated at similar levels and was able to recruit fibroblasts, demonstrating correct biological activity. Interestingly, erythropoietin was not well expressed from the *C. reinhardtii* chloroplast although it has been expressed from the nuclear genome of another green alga (see Section 16.3.1.3), and insulin – the first approved recombinant protein drug – was also difficult to detect [100], indicating that there is still room for improvement in factors affecting recombinant protein expression in algae.

16.3.1.3
Nuclear-Expressed Protein Therapeutics

Glycosylation can impact protein folding and stability, and is often essential for protein function or recognition. Proteins synthesized in the chloroplast are not glycosylated, necessitating nuclear expression for all therapeutics that require glycosylation, which constitute about 40% of all currently approved protein therapeutics [94]. Although nuclear transgene products do not accumulate to very high levels at present (see Sections 16.2.1.2 and 16.2.1.3), there are some examples of success. The hormone erythropoietin, which regulates blood cell differentiation and is used in anemia treatment, was expressed from the *C. reinhardtii* nuclear genome and secreted into the culture media, accumulating to 100 $\mu g\,l^{-1}$ [42]. In another species of green algae, *Chlorella ellipsoidea*, an antimicrobial peptide was expressed from the nuclear genome and exhibited antimicrobial activity against both gram-negative and gram-positive bacteria as well as a fungus [101, 102].

16.3.2 Nutraceuticals and Nutritional Supplements

The term *nutraceutical* – a portmanteau of "nutrition" and "pharmaceutical" – can refer to any supplement, health product, or formulated food intended to provide prevention against chronic disease or promote general well-being. In this sense, they are often considered similar to supplements. Both of these categories have current and potential future utility for industrial-scale algal production of wild-type and/or engineered strains.

16.3.2.1 Omega-3 Fatty Acids

One of the most popular supplements is derived from fish oil or krill oil, delivering a high concentration of omega-3 fatty acids – in particular, eicosapentaenoic acid (EPA) and docosahexaenoic acid (DHA) – which are essential fatty acids and have notable importance for brain and cardiovascular health. Many algae, at the bottom of the food chain for these fish and krill, produce EPA [103] and DHA [104] themselves and are currently being grown for the purpose of supplementation [105]. However, there has been significant interest in engineering other algal species – ones that are more amenable to large-scale growth and lipid extraction – to produce these valuable fatty acids [106]. Often, only one or a few additional lipid biosynthetic enzymes need to be introduced [107], so this endeavor is also a useful exercise in metabolic engineering for aspirations of modifying more complex pathways (see biofuel lipid engineering in Section 16.4.1).

16.3.2.2 Source of Organic Selenium

Selenium, once thought to be toxic, is now known to be required in trace amounts by humans. Selenium in the form of selenocysteine is incorporated into at least 25 human proteins, several of which have only recently been discovered [108], and selenium must be consumed through the diet for normal function of these proteins. Organic selenium is less toxic and more readily absorbed than inorganic selenium, so the optimal source is to consume organisms that produce selenoproteins themselves. It was long thought that plants – unlike bacteria, archaea, and animals – did not contain any selenoproteins [108]. However, they have now been definitively identified in *C. reinhardtii* by mass spectrometry, confirming that the opal codon UGA can indeed code for selenocysteine in some algae [109]. Attempts to increase organic selenium concentrations in *C. reinhardtii* by introducing human selenoprotein transgenes have shown promising, but not conclusive, initial results [110].

16.3.2.3 Carotenoids from Wild-Type and Engineered Algae

D. salina and *Haematococcus pluvialis* are the main sources for commercial production of nonsynthetic beta-carotene [111] and astaxanthin, respectively. Other microalgae also show great potential for producing other relevant carotenoids such as lutein. These carotenoids are mainly used as animal feed additives and in supplements for human consumption [112, 113] and could become valuable coproducts in the biofuel industry. Natural beta-carotene can reach a price between \$300 and 3000 kg^{-1} [114, 115], while astaxanthin can be sold for \$2500 – 10 000 kg^{-1} [114, 116].

Despite their high value, there are few reports on metabolic engineering for improved carotenoid production in algae. Several attempts have been made to engineer *C. reinhardtii* to produce astaxanthin by introducing carotenoid biosynthesis genes [117 – 119], but these have been met with limited success. *H. pluvialis* has been stably transformed with an herbicide-resistant version of its phytoene desaturase gene (*pds*), the second step of carotenoid synthesis, for enhanced astaxanthin accumulation. This approach led to 26% higher astaxanthin accumulation 48 h after induction with high-light when compared to the wild-type strain [120].

More recently, the phytoene synthase gene (*psy*), the enzyme for the committed step in carotenoid synthesis, has been transformed into *C. reinhardtii*, resulting in increased carotenoid accumulation. Transgenic strains overexpressing *psy* from *D. salina* and *Chlorella zoofingensis* accumulated 2.6- and 2.2-fold more lutein than the controls, respectively [121, 122]. RNAi technology has also been used to modify the carotenoid profile of *C. reinhardtii* and *D. salina* by silencing their *pds* genes. Reductions of up to 93% and 72% of *pds* mRNA were achieved in *Chlamydomonas* and *Dunaliella*, respectively. However, the carotenoid content of *C. reinhardtii* did not change significantly and the effect on *D. salina* carotenoid content was not measured [123, 124].

16.3.3 Recombinant Vaccines

Significant headway has been made in the past decade on expressing pathogen subunits in microalgae for low-cost (and perhaps eventually orally deliverable) vaccines. Algae exhibit several advantages over other expression platforms for recombinant vaccine production. Unlike mammalian cells, algae are not susceptible to viral infection that could be passed to a patient; viral filtering is one of the most costly steps of vaccine purification from animal cell platforms. In fact, the species *C. reinhardtii* in which most vaccine antigen expression has been performed to date does not have a single identified viral pathogen – making it quite unique as a host.

16.3.3.1 Vaccine Antigens Produced in Algae

The first recombinant vaccine antigens were produced in algae in 2003: a foot-and-mouth disease viral protein in the *C. reinhardtii* chloroplast [125] and the hepatitis B surface antigen in *D. salina* [126]. The hepatitis B surface protein, the antigen in the first-ever Food and Drug Administration (FDA)-approved recombinant vaccine, has also been produced via nuclear transformation in *C. reinhardtii*

[127] and in *P. tricornutum* [128]. With these exceptions, almost all other vaccine antigens have been produced in *C. reinhardtii* chloroplasts, including classical swine fever virus protein E2 [129], human glutamic acid decarboxylase 65 implicated in diabetes prevention [130], white spot syndrome virus VP28 protein [131], *Staphylococcus aureus* D2 protein [132], and a mutant version of the human papillomavirus E7 protein [133].

In addition, *C. reinhardtii* has garnered much attention for its unique suitability for producing malarial proteins in the chloroplast. Many of the surface proteins of malaria's causative pathogen, *Plasmodium*, are heavily disulfide bonded and also lack glycosylation [134]. Disulfide bond formation is facilitated within the chloroplast compartment, and chloroplast-synthesized proteins lack posttranslational modifications, making algal chloroplasts the only platform yet demonstrated to produce several properly folded native *Plasmodium* proteins [135]. The genes encoding *MSP1* from two different *Plasmodium* species and the protein *AMA1* from *Plasmodium berghei* were expressed in the *C. reinhardtii* nuclear genome and targeted to chloroplast starch granules via fusion to the major granule-associated protein [136]. The *Plasmodium falciparum* surface proteins *Pfs*25, *Pfs*28, and *Pfs*48/45 – which are present in the sexual reproductive stages of the species most lethal to humans and must be accessible for reproduction and maturation within the mosquito gut – have also been expressed in the chloroplast [135, 137, 138]. These antigens are suitable for a transmission-blocking vaccine because antibodies raised against them provide minimal protection to the individual, but they block transmission at the mosquito's next feeding.

16.3.3.2
Animal Efficacy Studies

While most of these studies provided immunological data demonstrating proper folding or biological activity *in vitro*, at least six of them have been tested in animals. The classical swine fever virus vaccine was purified and injected subcutaneously, where it triggered a systemic immune response [129]. An *S. aureus* subunit vaccine was able to protect 80% of mice from a lethal challenge with the pathogen, even when administered orally instead of parenterally (see Section 16.3.3.3) [132]. The malarial proteins *MSP1* and *AMA1* were able to reduce parasite load and prolong survival in mice exposed to the species of *Plasmodium* that infects rodents, *P. berghei* [136]. The first study of algal-produced *Pfs*25 found that the sera of mice immunized with the purified protein was able to drastically reduce the number of oocysts in mosquito midguts when fed in a blood meal, indicating that the sexual reproduction was halted by the presence of the antibodies raised against the algal-produced antigen [135]. Mice also produced systemic immunoglobulin G (IgG) antibodies to these antigens when injected with purified protein, and mucosal antibodies in the gut when fed whole transgenic algae [137]. Recently, both purified protein and crude algal extracts from *C. reinhardtii* expressing a human papillomavirus antigen were able to confer 70% survival after tumor challenge in mice [133]. All of these are promising preliminary results, and add authenticity to the *in vitro* data, indicating that algal-produced antigens exhibit superb biological activity.

16.3.3.3
Recombinant Oral Vaccines

Recombinant oral vaccines present a variety of advantages over injectable vaccines. Oral immunization can induce mucosal immunity, which provides protection at the site of entry for almost all pathogens. With particular significance in areas with limited financial resources, medical personnel, and infrastructure, oral vaccines can be much cheaper because the antigens may not require extensive purification, and they do not require skilled medical staff for safe injection. For some diseases, such as malaria, the reduced costs of production, processing, shipping, and administration using a freeze-dried transgenic algal oral vaccine are nearly a necessity to become a viable option for the at-risk population for the disease. Two studies tested the stability of antigens within lyophilized algae from 6 to 20 months at room temperature or higher and suggest that this may indeed be possible [132, 137].

Oral immunization with the classical swine fever vaccine produced no response [129], but oral administration of the starch-granule-bound malarial antigens protected mice nearly as well from *P. berghei* challenge as parenteral administration [136]. In order to increase both uptake by the intestinal epithelial cells and immunogenicity, antigens can be fused to subunits from enteric pathogens that allow binding to and uptake by cells in the gut. One such subunit is the beta subunit of cholera toxin, CtxB, which binds to the GM1 ganglioside on intestinal cells. The *S. aureus* fusion to CtxB was fed to mice in the form of lyophilized whole algal cells; this oral route of administration elicited both a mucosal response – measured by IgA titers in fecal samples – and a systemic IgG response to both the *S. aureus* and the CtxB domains [132].

However, the same route of administration with a malarial antigen fused to the CtxB subunit was unable to elicit systemic immunity to the malarial antigen, although systemic CtxB immunity was triggered. Mucosal response was observed for both antigen subunits [137]. The design of this chimeric molecule differed from the *S. aureus* construct in the presence of a furin protease cleavage site between the two domains. It is possible that after internalization and cleavage, the two halves experienced different intracellular fates, leading to different antigen presentation. It may also be that the malarial antigen is inherently less immunogenic than that from *S. aureus*, in which case stronger adjuvants may be needed.

16.4
Bioenergy Applications

Much of the recent interest in algae as a host organism for industrial-scale production revolves around its potential for biofuel production. Ancient algae are the source of the biomass that converted into today's petroleum, so many algae naturally produce the types of hydrocarbons that are suitable for fungible, or "drop-in," fuel alternatives. Algae are considered a promising alternative to other plant-based biofuels for several reasons. Unlike alternatives like corn ethanol or sugarcane ethanol, algal biofuel production will not compete with the food supply, alleviating food-versus-fuel pricing and availability concerns. Algal biofuel crops also will not compete with food crops for land because algae can be grown in outdoor ponds on nonarable land. Finally, the high biomass accumulation per acre compared to other potential fuel crops means that far less land is needed to produce a given volume of fuel.

16.4.1 Altering Lipid Metabolism

Algal lipid metabolism is believed to be largely similar to that of higher plants in that the same enzymes and subcellular compartments are used in the generation of the primary fatty acid types. The primary fatty acids palmitic, stearic, and stearidonic acid are formed in the plastid by the action of the fatty acid synthase (FAS) enzyme and stearoyl acyl-carrier protein (ACP) desaturase, and then transported to the endoplasmic reticulum for further modification and incorporation into triacylglycerol (TAG) molecules [139]. The primary fatty acids of algae are photosynthetic and cellular membrane lipids, but there are also a number of other products such as wax esters, sterols, prenyls, and other hydrocarbons [140]. The traditional means of obtaining high levels of storage lipids or polysaccharides has been through the induction of stress response via changes in nutrients or growth conditions [141, 142]. Metabolic engineering has more recently been used with some success to induce accumulation of both of these compounds.

16.4.1.1 Changes in Nutrient Availability

The most effective and well-known nutrient modification for altering lipid metabolism and storage is the limitation of nitrogen. When nitrogen is limited in the media, algae are unable to continue protein synthesis and therefore growth and replication slow and eventually halt. Photosynthesis will continue, but the algae respond by dramatically increasing their lipid content using the energy provided by photosynthesis or a carbon source such as sugars or acetate [143–145]. Silicon, a nutrient vital to the growth of diatoms, also mimics the effects of nitrogen starvation when it becomes limited [146], as does phosphorus [147]. Iron has also been shown to affect lipid levels, but inversely: excess iron in the media positively influences both biomass and lipid content [145, 148–150]. Carbon dioxide supplied in high levels also supports more robust cell growth and the accumulation of lipids, as carbon dioxide is required for photosynthesis and also for the formation of acetyl-CoA, a substrate of fatty acid biosynthesis [151].

16.4.1.2
Changes in Temperature

Lipids are known to play a role in regulating the fluidity of membranes in response to temperature, with more unsaturated lipids being required to maintain fluidity at low temperatures [152]. Increased polyunsaturated fatty acid levels have been associated with growth in low temperatures for algae [153]. It appears that these modifications are particularly important in the photosynthetic membranes of algae as an adaptation to the cold; and if more saturated fatty acids are desirable, then the algae must be shifted to higher growing temperatures [154].

16.4.1.3
Changes in Light Intensity

Growth of cultures in various light conditions can also affect the proportion of total lipids to membrane lipids, although these trends are more difficult to interpret due to large variations between species. Some studies suggest that low-light levels tend to favor the production of photosynthetic membrane lipids, while high-light levels favor more production of nonpolar storage lipids such as TAG [155, 156]. On the other hand, high-light levels can reduce some polyunsaturated fatty acid levels, but high-light irradiance can favor the production of polyunsaturated fats in the polar membrane lipids, possibly due to a role in the mitigation of free oxygen radicals from photosynthesis under high-light conditions [157, 158].

16.4.1.4
Genetic Manipulation of Lipid Content

Since the majority of lipid regulatory proteins appear to be nuclear-encoded, a primary limitation of successful genetic manipulation of lipid metabolism involves obtaining high-level recombinant protein expression from the nucleus (see Sections 16.2.1.2 and 16.2.1.3). Early attempts to increase lipid metabolism aimed at increasing the levels of substrates available for lipid production. Algae, like higher plants, utilize malonyl-CoA to synthesize lipids, and malonyl-CoA is used to add to lipid chains two carbons at a time. The first attempts to increase lipid accumulation were therefore focused on increasing malonyl-CoA levels by expressing the gene encoding acetyl-CoA carboxylase (ACCase) [159]. Increasing ACCase levels by two- to threefold in the diatom *Cyclotella cryptica* did not result in any increase in lipid levels, indicating that the rate-limiting step of lipid production in algae is not the production of lipid biosynthesis substrates. The lessons learned from early attempts like these seemed to point to regulatory proteins as being the primary bottlenecks in the production of lipids from algae.

Among the simplest modifications capable of producing increased lipid accumulation was a mutation of the sta6 gene that eliminated accumulation of this starch biosynthesis enzyme in *C. reinhardtii*. This mutant exhibited much higher levels of lipid production under nitrogen starvation, presumably due to the inability of this mutant to store energy as starch [160]. One of the most impressive genetic manipulations applied to microalgae for the purpose of biofuel production involves the expression of two heterologous thioesterases in *P. tricornutum*, which resulted in the production of increased proportions of shorter chain fatty acids [161]. Producing shorter chain fatty acids is a crucial advance in the production of biodiesel with excellent cold flow properties. Relatively few of the many lipid biosynthesis metabolic modifications pioneered in higher plants have been transferred to algae so far, but a dramatic expansion in the field of algal metabolic engineering is anticipated in the coming years due to improved genetic tool availability.

16.4.2 Increasing Photosynthetic Efficiency

Photosynthesis typically only captures the energy from about 1% of incident light, leaving ample room for improvement of biomass yields, culture density, and growth rates. Several strategies for increasing photosynthetic efficiency have been employed in algae, as strains with modified photosynthetic apparatuses are quick to make and can be rapidly screened and characterized, at least compared to terrestrial plants.

16.4.2.1 Improving Light Utilization

Microalgae have evolved large light-harvesting complexes (LHCs) for maximizing light absorption in low-light environments. In laboratory culture where light is saturating, excess energy is dissipated through heat and fluorescence quenching in the LHCs. Energy that cannot be dissipated results in direct photodamage and the production of reactive oxygen species, leading to photoinhibition. The large size of the LHCs also limits light penetration into the culture, reducing the achievable cell density [162].

To overcome this, a single RNAi construct has been introduced to silence all 20 LHC protein isoforms of *C. reinhardtii*. These cells have lower mRNA and protein accumulation for all LHC genes, and 68% less chlorophyll than the parental strain, resulting in 290% higher light transmittance in the culture. Furthermore, they present less fluorescence quenching, which leads to an increase in photosynthetic quantum yield. Under high-light conditions, transformed cells were less susceptible to photoinhibition and grew at a faster rate; however, they did not reach a higher cell density [163], which is considered a highly desirable trait for large-scale algae production. Similar results were achieved by downregulating LHC expression using a translational repressor of the LHC protein family, NAB1 [164, 165]. A constitutive version of NAB1 was overexpressed in *C. reinhardtii*, producing a similar but less striking phenotype than the RNAi strain: chlorophyll was only reduced by 20% [166]. Nonetheless, overexpressing a single repressor is notably easier than silencing 20 LHC isoforms.

16.4.2.2 Improving Photosynthetic Efficiency

Photoinhibition caused by excess sunlight is regarded as a major limiting factor for primary productivity in shallow production ponds under high irradiances [167].

One of the main targets of photodamage is PSII, the multiprotein complex that performs the light-driven oxidation of water. Degradation of the D1 subunit of PSII is significantly increased when light is in excess [168]. However, it is possible to select for mutant versions of the algal D1 protein that can evolve up to approximately 4.5 more oxygen *in vivo* under high-light conditions (50% midday sunlight) compared to the control [169]. This was achieved by transforming error-prone polymerase chain reaction (PCR)-amplified D1 coding sequences followed by selection under ionizing radiation. Unfortunately, the mutant strains perform slightly worse under laboratory light conditions (10% midday sunlight) [169], suggesting that this strategy may not translate to increased biomass yield for commercial algae production.

The feasibility of using heterologous genes as interchangeable parts (biobricks) for synthetic biology in *C. reinhardtii* chloroplasts has also been tested using the gene coding for the D1 protein (*psbA*) [170]. This work showed that introducing fully heterologous genes is not the best approach for engineering the chloroplast, but that heterologous coding sequences driven by endogenous regulatory regions are a viable method for modifying the algal photosynthetic machinery [170]. Furthermore, two isoforms of the D1 proteins from *Synechoccocus* sp. PCC 7942 (cyanobacteria) expressed in *C. reinhardtii* reconstituted the low-light and high-light phenotypes associated with each D1 isoform. Interestingly, *C. reinhardtii* expressing the cyanobacterial low-light isoform yielded 11% more dry weight biomass than the strains expressing the high-light isoform or the endogenous D1 protein, which is a highly desirable trait for biofuel production [171].

16.4.2.3
Expanding the Photosynthetically Active Spectrum

Microalgae can only utilize the light spectra between 400 and 700 nm for photosynthesis, also known as *photosynthetically active radiation* (PAR). Expanding this spectrum could enhance primary productivity without additional inputs. Some newly discovered cyanobacterial species can synthesize two novel types of light-harvesting pigments, chlorophyll-d and chlorophyll-f, which can absorb light frequencies of up to 750 nm. It has been proposed that inclusion of these pigments by metabolic engineering and synthetic biology into microalgae and plants would increase the amount of photon flux available for photosynthesis by 19% [172], but this has not yet been demonstrated experimentally.

16.4.3
Modifying Carbon Assimilation

In algae grown phototrophically, all new biomass – including lipids – derives from the fixation of CO_2 into ribulose-1,5-biphosphate (RuBP) to form 3-phosphoglycerate, catalyzed by the enzyme ribulose-1,5-biphosphate carboxylase/oxygenase (RuBisCO). Several other enzymes are also required to regenerate RuBP in the Calvin cycle. Considerable amounts of ATP and NADPH are required for this process, and are supplied by the light-driven activity of photosystems I and II [173]. Several studies have shown that the activity of RuBisCO is the major bottleneck for carbon flux through the Calvin cycle when CO_2 is not enriched in the media and under high-light or high-temperature conditions and all of these conditions are present in industrial-scale ponds in desert areas [173, 174].

16.4.3.1
Engineering RuBisCO

RuBisCO is the first enzyme in the CO_2 fixation pathway and is regarded as a "slow" catalyst; large amounts of RuBisCO are required for a sustainable carboxylation rate, and it has an affinity for oxygen, which is used in a counterproductive competing reaction [175]. *C. reinhardtii* is an ideal host for engineering RuBisCO, since there are RuBisCO-deficient strains that can complete their life cycle heterotrophically [175]. The small and large sub-units of *C. reinhardtii* RuBisCO (*rbcS* and *rbcL* genes) were mutated to resemble spinach RuBisCO, and then transformed into an *rbcS-rbcL*-deficient algae strain, resulting in a holoenzyme that resembled its plant counterpart. Compared to the *C. reinhardtii* RuBisCO, the mutant version had a 17% increase of the *in vitro* CO_2/O_2 specificity factor (Ω) and a 53% decrease of the V_{max} of carboxylation (Vc) [176]. Similarly, the *rbcS* genes of *Arabidopsis* and sunflower were introduced into *C. reinhardtii* while preserving the endogenous large subunit gene (*rbcL*). The Ω was improved by up to 11% while maintaining the V_{max}; however, the cells displayed lower growth rates and lacked pyrenoids, presumably due to mistargeting of RuBisCO [177].

C. reinhardtii rbcL has also been subjected to PCR-based gene shuffling, with oligonucleotides representing the natural diversity of this gene. Three rounds of gene shuffling and three rounds of strain selection resulted in a RuBisCO with up to 20% and 56% increase in Ω and Vc, respectively. Some of the enriched mutations were then incorporated into the tobacco *rbcL* and resulted in 14% Ω and 15% Vc increases [178]. Another interesting strategy is to tune RuBisCO abundance according to the environmental culture conditions to optimize carbon, energy, and nitrogen utilization. *C. reinhardtii* strains with varying levels of RuBisCO have been engineered by manipulating the *rbcL* mRNA maturation factor MRL1. RuBisCO could be reduced to 15% of wild-type levels while maintaining phototrophic growth [179]. An inducible promoter for MRL1 could potentially be used to tune RuBisCO accumulation according to culture conditions like light intensity or CO_2 concentration.

16.5
Other Industrial Applications

Algae are being explored for many additional industrial-scale applications beyond therapeutics and biofuels. Some of these involve alternative uses for algal biomass left over after lipid extraction, while others require genetically engineered strains.

16.5.1 Animal Feedstocks

As much as two-thirds of the grain produced in the United States is used to feed livestock. To reduce costs and increase sustainability of livestock production, there is recent interest in using fast-growing, high-protein algal species as animal feed. Some species are already used extensively as aquaculture feed [180], and algal facilities for biofuels or other industrial uses will soon produce a large excess of residual biomass. Some of this biomass can be added back into the media for subsequent cultures to recycle the nutrients, but it is also a rich source of protein that is valuable as a feed additive for both livestock and farmed fish.

16.5.1.1 Algal Protein Extracts as Animal Feedstocks

After biofuel-relevant lipids have been extracted from algal biomass, the resulting protein-rich algal paste is a valuable coproduct for animal feeds. It has been suggested that this coproduct could subsidize the current cost of algal-based biofuels [181]. Nutritional analysis indicates that algal protein provides a balanced source of amino acids, suitable for many animals from mammals to fish [182]. However, this coproduct makes some chemical extraction methods for lipids unfeasible, as they may leave residual harmful solvents in the algal paste. Mechanical extraction, which is a lower yielding method of extracting fuel lipids, could be used instead, but a cost–benefit analysis will need to be incorporated into these considerations.

16.5.1.2 Engineered Algae to Enhance Immune Function

Recently, algae have been genetically engineered to express proteins from mammalian colostrum, the mother's first milk, which contains many proteins that stimulate immune function and protection in the offspring [71]. In large-scale animal rearing, offspring are usually weaned from the mother's milk or given reconstituted formula at a young age, leaving their immune systems inadequately developed. As a result, low-dose antibiotic administration has become standard practice in commercial farming operations, accounting for the vast majority of all antibiotics sold. Unsurprisingly, widespread antibiotic resistance has emerged and is often directly traceable to livestock farms [183–185].

As an alternative approach, critical colostrum proteins can be introduced into feed or formula using engineered algae, perhaps reducing the need for low-dose antibiotic administration. The colostrum protein mammary-associated serum amyloid, M-SAA, has been expressed from the *C. reinhardtii* chloroplast genome and accumulates to nearly 10% of total soluble protein [71]. This protein stimulates mucin production in the gut of offspring, protecting them from enteric disease. Bioactive peptides derived from chimeric milk proteins have also recently been produced in *C. reinhardtii* chloroplasts [186]. The ability to introduce multiple transgenes into algae allows multiple critical colostrum proteins to be introduced using a single algal strain, and this work is already under way (M. Tran, personal communication).

16.5.2 Industrial Enzymes

Large quantities of relatively pure enzymes are required for many industrial processes, from food processing to paper milling to biofuels. With the recent push toward green manufacturing, many processes that have previously utilized energy-intensive steps like heat and pressure exposure, or steps with harsh chemicals like strong acids or bases, are turning to biological alternatives involve enzymes found in nature. Because of their potential for inexpensive, large-scale growth, algae are being investigated as host organisms for producing these types of enzymes.

16.5.2.1 Enzymes for Processing Cellulosic Biofuels

While algae have attracted attention from biofuels developers and researchers for their ability to accumulate high lipid levels, there may also be a role for them in downstream processing of other biofuel feedstocks. Cellulosic biofuels use the large quantities of carbohydrate polymers in terrestrial plant tissue (namely, cellulose, but also other components of the associated matrix) as a source of sugars for fermentation into ethanol. Some cellulosic crops are grown for the explicit purpose of serving as biofuel feedstocks, such as high-yielding grasses like switchgrass, but another source is residual plant matter left over after harvesting food crops.

For any of these feedstocks, a mix of enzymes must be added to effectively break down the cellulose, hemicellulose, and lignin into their constituent fermentable sugars because no single enzyme can completely degrade these polymers [187]. While some work has focused on engineering cellulosic crops to contain more easily degraded polymers [188], this strategy is limited because of the essential structural role they serve within plants; therefore, even engineered feedstocks require additional degradation enzymes. These enzymes include cellobiohydrolase, endoglucanase, beta-glucosidase, and xylanase. Xylanase – which is currently used in other industries like paper milling – has been produced in multiple algal species, including *D. salina* [90] and *C. reinhardtii* [43], and can even be expressed from the nuclear genome and targeted for secretion into the culture media, further reducing production costs of these enzymes.

16.5.2.2 Enzymes for Food Processing

Xylanase is also used in animal feeds, along with several other enzymes to enhance digestibility and maximize nutritional and caloric yield, including mannanase, alpha-galactosidase, and phytase. Phytase releases absorbable inorganic phosphate from phytic acid, one of the main storage forms of phosphorus in grains. As phytic acid, it is not bioavailable to nonruminant livestock. This not only makes it difficult for the animals to obtain sufficient phosphorus from a grain-based diet but it also contributes to environmental pollution because the phosphorus is excreted as waste, and fuels toxic microbial blooms downstream in the runoff. In fact, an "environmentally friendly" engineered pig that produces phytase in its saliva was created and tested [189], but funding was withdrawn before the regulatory approval process was completed.

Alternatively, these enzymes can be added to animal feed by mixing with edible transgenic algae. Alpha-galactosidase from two genetic sources, as well as two phytase enzymes, have been expressed and shown to be bioactive in both *D. salina* and *C. reinhardtii* [90, 190]. In fact, chickens fed algae-producing phytase have been shown to have lower phytic acid content in their manure [190]. Mannanase was also produced in these algal species but did not exhibit bioactivity, which may be due to the lack of glycosylation because it was produced within the chloroplast [90].

16.6 Industrial-Scale Algal Production

Growing algae at industrial scale requires addressing a host of concerns that are not encountered at the bench. The media inputs must be minimized to remain cost-effective; evaporative water losses must be taken into account; energy-efficient harvesting methods become critical; and crop protection strategies to evade grazers and pathogens must be developed. Laboratory-scale growth ranges from less than a milliliter, for cultures grown in a 96-well-plate high-throughput format, to around 20 l. In enclosed environments, algae can be grown in 100 l bags or equivalent contained systems. Pilot-scale ponds range from hundreds to tens of thousands of liters, providing some information about concerns that appear at scale, especially with outdoor growth. But industrial production will likely occur in ponds harboring hundreds of thousands to millions of liters, with some applications – such as high-value therapeutics or vaccines – suitable for thousand-liter fermenters. Christenson and Sims [191] have compiled a list of all commercial facilities to date, including the location, production type (pond, raceway, tubular photobioreactor, etc.), and scale.

16.6.1 Enclosed Photobioreactors and Fermenters

At present, much of the preliminary culturing, harvesting, and analysis of algae is being done in enclosed bioreactors. This is due in part to tight regulations governing outdoor cultivation of genetically engineered strains, but this contained growth also exhibits distinct advantages, particularly with regard to parameter control and optimizing yields.

16.6.1.1
Optimizing Photobioreactor Design

Enclosed systems alleviate the challenges associated with pathogens and predators; allow for tighter controls on parameters such as media component concentrations, light intensity, and temperature; and are ideal for transgenic strains that may pose an environmental threat if they escape into the environment. However, they are much more expensive to build and maintain than open ponds, so these systems are more suitable for high-value products like therapeutics or nutritional supplements.

As mentioned in Section 16.4.2, photosynthetic organisms typically have a relatively narrow range of light intensity in which they perform best: too much light causes photoinhibition, and too little limits growth. This makes the design of photobioreactors especially critical. The most common design is tubular, although flat panels, manifolds, and serpentine designs have also been tested [192, 193]. Besides expense, there are several drawbacks of enclosed bioreactors including biofouling or growth on the walls limiting light penetrance, and buildup of dissolved oxygen, both of which negatively impact productivity [193].

16.6.1.2
Modifying Trophic Conversion Pathways for Heterotrophic Growth

Growing microalgae under heterotrophic or mixotrophic conditions (in the presence of an organic carbon source) has several advantages. Fermentation systems have already been widely studied and successfully applied in industry, and the culture conditions are highly controlled and reproducible. Moreover, heterotrophic cultures of microalgae achieve higher cell densities, thus resulting in lower harvesting costs [192, 194]. For example, heterotrophic cultivation of *Chlorella protothecoides* for biofuel production resulted in fourfold higher lipid accumulation than autotrophic cultures [195].

Some algal species are naturally heterotrophic, meaning that they can survive without light if provided an appropriate carbon source, so these are candidates for growth in existing industrial fermentation facilities. Indeed, many of the high-expressing transgenic *C. reinhardtii* strains are necessarily nonphotosynthetic (see Section 16.2.2.2). However, many microalgae species are strict autotrophs – a problem that could be circumvented by genetic engineering. *Volvox carteri, C. reinhardtii*, and the diatoms *P. tricornutum* and *Cylindrotheca fusiformis* have been transformed with a hexose transporter (*HUP1*), resulting in strains that can survive on glucose in the dark [196–199]. Despite glucose incorporation, the extent of conversion to full heterotrophy is variable between these four algae. Interestingly, transformed *C. reinhardtii* cells produce 150% more hydrogen than the parental strain upon induction in the presence of glucose [198]. Trophic conversion is a good proof of concept for microalgae metabolic engineering, but despite the advantages of heterotrophic culture, it might not to be optimal for large-scale algae bioproduct production. Additional costs of adding a carbon source and the requirement for enclosed bioreactors given the higher risk of contamination are major drawbacks compared to phototrophic systems.

16.6.2 Open Pond Growth

Realistically, open ponds will be necessary for growing algae at the cost and scale necessary to compete with cheap commodities like fuel and animal feed. However, they come with a profound set of unique challenges – both biological and abiotic. Fortunately, progress is being made on these fronts due to the availability of functional commercial and academic pilot-scale open pond facilities.

16.6.2.1 Media and Water Optimization

The ideal large-scale culture media contains minimal quantities of the required salts and trace elements, and no carbon source – which reduces cost as well as contamination risk. Used media after harvest is often recycled, but may eventually reduce productivity and invite contamination [200]. One cost-cutting and environmentally beneficial proposal is to use diluted wastewater as a feedstock, since it contains high levels of nitrogen and phosphorus [201, 202]. This would serve the dual purpose of supporting algal growth and removing these excess nutrients from the wastewater, which otherwise serves as fertilizer for unwanted algal growth in the open ocean, ultimately resulting in oxygen-depleted dead zones. Water consumption is a significant concern due to evaporative loss from pond surfaces, which can also affect the nutrient concentrations and pH, as can dilution of the culture from rainfall.

Another concern in large-volume open ponds is achieving adequate mixing. Mixing serves two purposes: it increases gas exchange – allowing atmospheric CO_2 to enter the media and oxygen to escape – and it keeps algae in circulation from the bottom of the pond to the surface, modulating light exposure to prevent photoinhibition while maintaining productivity. In smaller ponds, lifters blowing forced air can be used to mix the culture; however, this is energy intensive, so paddlewheels are typically used in large raceway ponds [203].

16.6.2.2 Efficient Harvesting Methods

On a small scale, centrifugation is typically used to separate algal biomass from the culture media. However, due to the huge volume of water relative to algal biomass (typically, only a few grams of dried biomass per liter of culture), large-scale harvesting is a difficult, expensive, and energy-intensive process, and centrifugation is not a viable option. One strategy is to use a filamentous algal species; indeed, this attribute has been used in large-scale *Spirulina* growth for years [204]. Flocculants can also be used to induce aggregation of individual cells. Chemical flocculants – which utilize ions, metal precipitation, or charged polysaccharide or polyacrylamide polymers – are effective but expensive and can be difficult to remove from the algal biomass [205].

Bioflocculants are also being explored, including the introduction of bacteria [206] and filamentous fungi [207, 208] to induce algal aggregation. Some algal species, including the green algae *D. salina* [209] and *C. vulgaris* [210] and the diatom *P. tricornutum* [211], autoflocculate in response to high pH and/or depletion of CO_2, which may serve as a useful feature for self-harvesting batch cultures that naturally aggregate upon reaching saturation.

16.7 Conclusions and Potential of Algal Platforms

Significant progress is being made each year in the areas of algal genetic engineering and tool development, allowing for rapid advancement in achieving the complex engineering of production strains with multiple desirable traits. Bioprospecting for novel species in environments of interest for commercial growth – coupled with the power of inexpensive full-genome sequencing and assembly – also holds promise for discovering new strains and traits that are adaptable to meet the needs of industrial-scale growth, efficient harvesting, and resistance to biological and abiotic stresses. It is likely that algal platforms will not be utilized for a single product, but that a single strain and a single harvest will contribute to multiple processing streams, akin to large-scale agriculture and animal husbandry. Algae exhibit several unique advantages over currently available industrial hosts, making them valuable additions to the commercial organism repertoire as fast-growing, inexpensive, highly tractable bioproduction platforms.

References

1. Norton, T.A., Melkonian, M., and Andersen, R.A. (1996) Algal biodiversity. *Phycologia*, **35** (4), 308–326.
2. Metting, F. Jr., (1996) Biodiversity and application of microalgae. *J. Ind. Microbiol.*, **17** (5-6), 477–489.
3. Barten, R. and Lill, H. (1995) DNA, uptake in the naturally competent cyanobacterium, *Synechocystis* sp. PCC 6803. *FEMS Microbiol. Lett.*, **129** (1), 83–87.
4. Thiel, T. and Poo, H. (1989) Transformation of a filamentous cyanobacterium by electroporation. *J. Bacteriol.*, **171** (10), 5743–5746.
5. Grigorieva, G. and Shestakov, S. (1982) Transformation in the cyanobacterium Synechocystis sp. 6803. *FEMS Microbiol. Lett.*, **13** (4), 367–370.
6. Xu, Y., Alvey, R.M., Byrne, P.O., Graham, J.E., Shen, G., and Bryant, D.A. (2011) *Photosynthesis Research Protocols*, Springer, pp. 273–293.
7. Merchant, S.S., Prochnik, S.E., Vallon, O., Harris, E.H., Karpowicz, S.J., Witman, G.B., Terry, A., Salamov, A., Fritz-Laylin, L.K., Marechal-Drouard, L. *et al.* (2007) The *Chlamydomonas* genome reveals the evolution of key animal and plant functions. *Science*, **318** (5848), 245–251.
8. Franklin, S.E. and Mayfield, S.P. (2004) Prospects for molecular farming in the green alga *Chlamydomonas reinhardtii*. *Curr. Opin. Plant Biol.*, **7** (2), 159–165.
9. Fuhrmann, M., Oertel, W., and Hegemann, P. (1999) A synthetic gene coding for the green fluorescent protein (GFP) is a versatile reporter in *Chlamydomonas reinhardtii*. *Plant J.*, **19** (3), 353–361.
10. Neupert, J., Karcher, D., and Bock, R. (2009) Generation of *Chlamydomonas* strains that efficiently express nuclear transgenes. *Plant J.*, **57** (6), 1140–1150.

11. Lauersen, K.J., Berger, H., Mussgnug, J.H., and Kruse, O. (2013) Efficient recombinant protein production and secretion from nuclear transgenes in *Chlamydomonas reinhardtii*. *J. Biotechnol.*, **167** (2), 101–110.
12. Rochaix, J.D. and Vandillewijn, J. (1982) Transformation of the green-alga *Chlamydomonas-reinhardii* with yeast DNA. *Nature*, **296** (5852), 70–72.
13. Debuchy, R., Purton, S., and Rochaix, J.D. (1989) The argininosuccinate lyase gene of *Chlamydomonas reinhardtii*: an important tool for nuclear transformation and for correlating the genetic and molecular maps of the ARG7 locus. *EMBO J.*, **8** (10), 2803–2809.
14. Mayfield, S.P. and Kindle, K.L. (1990) Stable nuclear transformation of *Chlamydomonas reinhardtii* by using a *C. reinhardtii* gene as the selectable marker. *Proc. Natl. Acad. Sci. U.S.A.*, **87** (6), 2087–2091.
15. Kindle, K.L. (1990) High-frequency nuclear transformation of *Chlamydomonas reinhardtii*. *Proc. Natl. Acad. Sci. U.S.A.*, **87** (3), 1228–1232.
16. Shimogawara, K., Fujiwara, S., Grossman, A., and Usuda, H. (1998) High-efficiency transformation of *Chlamydomonas reinhardtii* by electroporation. *Genetics*, **148** (4), 1821–1828.
17. Dunahay, T.G. (1993) Transformation of *Chlamydomonas reinhardtii* with silicon-carbide whiskers. *Biotechniques*, **15** (3), 452–460.
18. Kumar, S.V., Misquitta, R.W., Reddy, V.S., Rao, B.J., and Rajam, M.V. (2004) Genetic transformation of the green alga – *Chlamydomonas reinhardtii* by *Agrobacterium tumefaciens*. *Plant Sci.*, **166** (3), 731–738.
19. Pratheesh, P.T., Vineetha, M., and Kurup, G.M. (2013) An efficient protocol for the Agrobacterium-mediated genetic transformation of microalga *Chlamydomonas reinhardtii*. *Mol. Biotechnol.*, **56**, 507–515.
20. Qu, B., Eu, Y.J., Jeong, W.J., and Kim, D.P. (2012) Droplet electroporation in microfluidics for efficient cell transformation with or without cell wall removal. *Lab Chip*, **12** (21), 4483–4488.
21. Suresh, A. and Kim, Y.C. (2013) Translocation of cell penetrating peptides on *Chlamydomonas reinhardtii*. *Biotechnol. Bioeng.*, **110** (10), 2795–2801.
22. Tam, L.W. and Lefebvre, P.A. (1993) Cloning of flagellar genes in *Chlamydomonas reinhardtii* by DNA insertional mutagenesis. *Genetics*, **135** (2), 375–384.
23. Cerutti, H., Johnson, A.M., Gillham, N.W., and Boynton, J.E. (1997) A eubacterial gene conferring spectinomycin resistance on *Chlamydomonas reinhardtii*: integration into the nuclear genome and gene expression. *Genetics*, **145** (1), 97–110.
24. Gumpel, N.J., Rochaix, J.D., and Purton, S. (1994) Studies on homologous recombination in the green alga *Chlamydomonas reinhardtii*. *Curr. Genet.*, **26** (5-6), 438–442.
25. Nelson, J.A.E. and Lefebvre, P.A. (1995) Targeted disruption of the Nit8 gene in *Chlamydomonas reinhardtii*. *Mol. Cell. Biol.*, **15** (10), 5762–5769.
26. Zorin, B., Hegemann, P., and Sizova, I. (2005) Nuclear-gene targeting by using single-stranded DNA avoids illegitimate DNA integration in *Chlamydomonas reinhardtii*. *Eukaryotic Cell*, **4** (7), 1264–1272.
27. Cerutti, H., Johnson, A.M., Gillham, N.W., and Boynton, J.E. (1997) Epigenetic silencing of a foreign gene in nuclear transformants of Chlamydomonas. *Plant Cell*, **9** (6), 925–945.
28. Casas-Mollano, J.A., Jeong, B.R., Xu, J., Moriyama, H., and Cerutti, H. (2008) The MUT9p kinase phosphorylates histone H3 threonine 3 and is necessary for heritable epigenetic silencing in Chlamydomonas. *Proc. Natl. Acad. Sci. U.S.A*, **105** (17), 6486–6491.
29. Davies, J.P., Weeks, D.P., and Grossman, A.R. (1992) Expression of the arylsulfatase gene from the beta 2-tubulin promoter in *Chlamydomonas reinhardtii*. *Nucleic Acids Res.*, **20** (12), 2959–2965.
30. Ohresser, M., Matagne, R.F., and Loppes, R. (1997) Expression of the arylsulphatase reporter gene under

the control of the nit1 promoter in *Chlamydomonas reinhardtii. Curr. Genet.*, **31** (3), 264–271.
31. Fischer, N. and Rochaix, J.D. (2001) The flanking regions of PsaD drive efficient gene expression in the nucleus of the green alga *Chlamydomonas reinhardtii. Mol. Genet. Genomics*, **265** (5), 888–894.
32. Fuhrmann, M., Hausherr, A., Ferbitz, L., Schodl, T., Heitzer, M., and Hegemann, P. (2004) Monitoring dynamic expression of nuclear genes in *Chlamydomonas reinhardtii* by using a synthetic luciferase reporter gene. *Plant Mol. Biol.*, **55** (6), 869–881.
33. Park, S., Lee, Y., Lee, J.H., and Jin, E. (2013) Expression of the high light-inducible *Dunaliella* LIP promoter in *Chlamydomonas reinhardtii. Planta*, **238**, 1147–1156.
34. Schroda, M., Blocker, D., and Beck, C.F. (2000) The HSP70A promoter as a tool for the improved expression of transgenes in Chlamydomonas. *Plant J.*, **21** (2), 121–131.
35. Lodha, M., Schulz-Raffelt, M., and Schroda, M. (2008) A new assay for promoter analysis in *Chlamydomonas* reveals roles for heat shock elements and the TATA box in *HSP70A* promoter-mediated activation of transgene expression. *Eukaryotic Cell*, **7** (1), 172–176.
36. Wu, J.X., Hu, Z.L., Wang, C.G., Li, S.F., and Lei, A.P. (2008) Efficient expression of green fluorescent protein (GFP) mediated by a chimeric promoter in *Chlamydomonas reinhardtii. Chin. J. Oceanol. Limnol.*, **26** (3), 242–247.
37. Qiu, Z.Y., Wang, L.H., and Zhou, Q. (2013) Effects of bisphenol A on growth, photosynthesis and chlorophyll fluorescence in above-ground organs of soybean seedlings. *Chemosphere*, **90** (3), 1274–1280.
38. Heitzer, M. and Zschoernig, B. (2007) Construction of modular tandem expression vectors for the green alga *Chlamydomonas reinhardtii* using the Cre/lox-system. *Biotechniques*, **43** (3), 324, 326, 328 passim.
39. Strenkert, D., Schmollinger, S., and Schroda, M. (2013) Heat shock factor 1 counteracts epigenetic silencing of nuclear transgenes in *Chlamydomonas reinhardtii. Nucleic Acids Res.*, **41** (10), 5273–5289.
40. Diaz-Santos, E., de la Vega, M., Vila, M., Vigara, J., and Leon, R. (2013) Efficiency of different heterologous promoters in the unicellular microalga *Chlamydomonas reinhardtii. Biotechnol. Progr.*, **29** (2), 319–328.
41. Lumbreras, V.S.D. and Purton, S. (1998) Efficient foreign gene expression in *Chlamydomonas reinhardtii* mediated by an endogenous intron. *Plant J.*, **14**, 441–447.
42. Eichler-Stahlberg, A., Weisheit, W., Ruecker, O., and Heitzer, M. (2009) Strategies to facilitate transgene expression in *Chlamydomonas reinhardtii. Planta*, **229** (4), 873–883.
43. Rasala, B.A., Lee, P.A., Shen, Z.X., Briggs, S.P., Mendez, M., and Mayfield, S.P. (2012) Robust expression and secretion of Xylanase1 in *Chlamydomonas reinhardtii* by fusion to a selection gene and processing with the FMDV 2A peptide. *PLoS One*, **7** (8), e43349.
44. Kumar, A., Falcao, V.R., and Sayre, R.T. (2013) Evaluating nuclear transgene expression systems in *Chlamydomonas reinhardtii. Algal Res.*, **2** (4), 321–332.
45. Kindle, K.L., Schnell, R.A., Fernandez, E., and Lefebvre, P.A. (1989) Stable nuclear transformation of Chlamydomonas using the Chlamydomonas gene for nitrate reductase. *J. Cell Biol.*, **109** (6, Pt. 1), 2589–2601.
46. Stevens, D.R., Rochaix, J.D., and Purton, S. (1996) The bacterial phleomycin resistance gene *ble* as a dominant selectable marker in *Chlamydomonas. Mol. Gen. Genet.*, **251** (1), 23–30.
47. Nelson, J.A.E., Savereide, P.B., and Lefebvre, P.A. (1994) The Cry1 gene in *Chlamydomonas reinhardtii* – Structure and use as a dominant selectable marker for nuclear transformation. *Mol. Cell. Biol.*, **14** (6), 4011–4019.
48. Sizova, I., Fuhrmann, M., and Hegemann, P. (2001) A *Streptomyces rimosus* aphVIII gene coding for a new type phosphotransferase provides stable

antibiotic resistance to *Chlamydomonas reinhardtii*. *Gene*, **277** (1-2), 221–229.
49. Berthold, P., Schmitt, R., and Mages, W. (2002) An engineered Streptomyces hygroscopicus aph 7″ gene mediates dominant resistance against hygromycin B in *Chlamydomonas reinhardtii*. *Protist*, **153** (4), 401–412.
50. Quinn, J.M., Kropat, J., and Merchant, S. (2003) Copper response element and Crr1-dependent Ni^{2+}-responsive promoter for induced, reversible gene expression in *Chlamydomonas reinhardtii*. *Eukaryotic Cell*, **2** (5), 995–1002.
51. Shao, N. and Bock, R. (2008) A codon-optimized luciferase from *Gaussia princeps* facilitates the in vivo monitoring of gene expression in the model alga *Chlamydomonas reinhardtii*. *Curr. Genet.*, **53** (6), 381–388.
52. Ruecker, O., Zillner, K., Groebner-Ferreira, R., and Heitzer, M. (2008) *Gaussia-luciferase* as a sensitive reporter gene for monitoring promoter activity in the nucleus of the green alga *Chlamydomonas reinhardtii*. *Mol. Genet. Genomics*, **280** (2), 153–162.
53. Sun, X., Perera, S., Haas, N., Lefebvre, P.A., and Silflow, C.D. (2013) Using an RSP3 reporter gene system to investigate molecular regulation of hydrogenase expression in *Chlamydomonas reinhardtii*. *Algal Res.*, **2** (4), 341–351.
54. Rasala, B.A., Barrera, D.J., Ng, J., Plucinak, T.M., Rosenberg, J.N., Weeks, D.P., Oyler, G.A., Peterson, T.C., Haerizadeh, F., and Mayfield, S.P. (2013) Expanding the spectral palette of fluorescent proteins for the green microalga *Chlamydomonas reinhardtii*. *Plant J.*, **74** (4), 545–556.
55. de Hostos, E.L., Togasaki, R.K., and Grossman, A. (1988) Purification and biosynthesis of a derepressible periplasmic arylsulfatase from *Chlamydomonas reinhardtii*. *J. Cell Biol.*, **106** (1), 29–37.
56. Gonzalez-Ballester, D., Pootakham, W., Mus, F., Yang, W., Catalanotti, C., Magneschi, L., de Montaigu, A., Higuera, J.J., Prior, M., Galvan, A. *et al.* (2011) Reverse genetics in *Chlamydomonas*: a platform for isolating insertional mutants. *Plant Methods*, 7, 24.
57. Casas-Mollano, J.A., Rohr, J., Kim, E.J., Balassa, E., van Dijk, K., and Cerutti, H. (2008) Diversification of the core RNA interference machinery in *Chlamydomonas reinhardtii* and the role of DCL1 in transposon silencing. *Genetics*, **179** (1), 69–81.
58. Schroda, M. (2006) RNA silencing in *Chlamydomonas*: mechanisms and tools. *Curr. Genet.*, **49** (2), 69–84.
59. Fuhrmann, M., Stahlberg, A., Govorunova, E., Rank, S., and Hegemann, P. (2001) The abundant retinal protein of the *Chlamydomonas* eye is not the photoreceptor for phototaxis and photophobic responses. *J. Cell Sci.*, **114** (21), 3857–3863.
60. Cerutti, H., Ma, X., Msanne, J., and Repas, T. (2011) RNA-mediated silencing in Algae: biological roles and tools for analysis of gene function. *Eukaryotic Cell*, **10** (9), 1164–1172.
61. Zhao, T., Wang, W., Bai, X., and Qi, Y. (2008) Gene silencing by artificial microRNAs in *Chlamydomonas*. *Plant J.*, **58**, 157–164.
62. Sizova, I., Greiner, A., and Hegemann, P. (2013) Zinc finger nucleases generate DNA double strand brakes and modification in *Chlamydomonas reinhardtii*. *FEBS J.*, **280**, 60.
63. Maul, J.E., Lilly, J.W., Cui, L.Y., dePamphilis, C.W., Miller, W., Harris, E.H., and Stern, D.B. (2002) The *Chlamydomonas reinhardtii* plastid chromosome: Islands of genes in a sea of repeats. *Plant Cell*, **14** (11), 2659–2679.
64. Boynton, J.E., Gillham, N.W., Harris, E.H., Hosler, J.P., Johnson, A.M., Jones, A.R., Randolph-Anderson, B.L., Robertson, D., Klein, T.M., Shark, K.B. *et al.* (1988) Chloroplast transformation in *Chlamydomonas* with high velocity microprojectiles. *Science*, **240** (4858), 1534–1538.
65. Kindle, K.L., Richards, K.L., and Stern, D.B. (1991) Engineering the chloroplast genome – Techniques and capabilities for chloroplast transformation in

Chlamydomonas-reinhardtii. *Proc. Natl. Acad. Sci. U.S.A.*, **88** (5), 1721–1725.
66. Dauvillee, D., Hilbig, L., Preiss, S., and Johanningmeier, U. (2004) Minimal extent of sequence homology required for homologous recombination at the psbA locus in *Chlamydomonas reinhardtii* chloroplasts using PCR-generated DNA fragments. *Photosynth. Res.*, **79** (2), 219–224.
67. Zerges, W. (2000) Translation in chloroplasts. *Biochimie*, **82** (6-7), 583–601.
68. Marin-Navarro, J., Manuell, A.L., Wu, J., and Mayfield, S.P. (2007) Chloroplast translation regulation. *Photosynth. Res.*, **94** (2-3), 359–374.
69. Herrin, D.L. and Nickelsen, J. (2004) Chloroplast RNA processing and stability. *Photosynth. Res.*, **82** (3), 301–314.
70. Rochaix, J.D. (1996) Post-transcriptional regulation of chloroplast gene expression in *Chlamydomonas reinhardtii*. *Plant Mol. Biol.*, **32** (1-2), 327–341.
71. Manuell, A.L., Beligni, M.V., Elder, J.H., Siefker, D.T., Tran, M., Weber, A., McDonald, T.L., and Mayfield, S.P. (2007) Robust expression of a bioactive mammalian protein in *Chlamydomonas* chloroplast. *Plant Biotechnol. J.*, **5** (3), 402–412.
72. Fletcher, S.P., Muto, M., and Mayfield, S.P. (2007) *Transgenic Microalgae as Green Cell Factories*, vol. **616**, Springer, pp. 90–98.
73. Specht, E.A. and Mayfield, S.P. (2013) Synthetic oligonucleotide libraries reveal novel regulatory elements in *Chlamydomonas* chloroplast mRNAs. *ACS Synth. Biol.*, **2** (1), 34–46.
74. Chen, H.C. and Melis, A. (2013) Marker-free genetic engineering of the chloroplast in the green microalga *Chlamydomonas reinhardtii*. *Plant Biotechnol. J.*, **11** (7), 818–828.
75. Bateman, J.M. and Purton, S. (2000) Tools for chloroplast transformation in *Chlamydomonas*: expression vectors and a new dominant selectable marker. *Mol. Gen. Genet.*, **263** (3), 404–410.
76. Goldschmidt-Clermont, M. (1991) Transgenic expression of aminoglycoside adenine transferase in the chloroplast: a selectable marker of site-directed transformation of chlamydomonas. *Nucleic Acids Res.*, **19** (15), 4083–4089.
77. Harris, E.H., Burkhart, B.D., Gillham, N.W., and Boynton, J.E. (1989) Antibiotic resistance mutations in the chloroplast 16S and 23S rRNA genes of *Chlamydomonas reinhardtii*: correlation of genetic and physical maps of the chloroplast genome. *Genetics*, **123** (2), 281–292.
78. Przibilla, E., Heiss, S., Johanningmeier, U., and Trebst, A. (1991) Site-specific mutagenesis of the D1 subunit of photosystem II in wild-type *Chlamydomonas*. *Plant Cell*, **3** (2), 169–174.
79. Erickson, J.M., Rahire, M., Bennoun, P., Delepelaire, P., Diner, B., and Rochaix, J.D. (1984) Herbicide resistance in *Chlamydomonas reinhardtii* results from a mutation in the chloroplast gene for the 32-kilodalton protein of photosystem II. *Proc. Natl. Acad. Sci. U.S.A*, **81** (12), 3617–3621.
80. Minko, I., Holloway, S.P., Nikaido, S., Carter, M., Odom, O.W., Johnson, C.H., and Herrin, D.L. (1999) Renilla luciferase as a vital reporter for chloroplast gene expression in Chlamydomonas. *Mol. Gen. Genet.*, **262** (3), 421–425.
81. Franklin, S., Ngo, B., Efuet, E., and Mayfield, S.P. (2002) Development of a GFP reporter gene for *Chlamydomonas reinhardtii* chloroplast. *Plant J.*, **30** (6), 733–744.
82. Mayfield, S.P. and Schultz, J. (2004) Development of a luciferase reporter gene, luxCt, for *Chlamydomonas reinhardtii* chloroplast. *Plant J.*, **37** (3), 449–458.
83. Klein, U., De Camp, J.D., and Bogorad, L. (1992) Two types of chloroplast gene promoters in *Chlamydomonas reinhardtii*. *Proc. Natl. Acad. Sci. U.S.A*, **89** (8), 3453–3457.
84. Gong, Y., Hu, H., Gao, Y., Xu, X., and Gao, H. (2011) Microalgae as platforms for production of recombinant proteins and valuable compounds: progress and prospects. *J. Ind. Microbiol. Biotechnol.*, **38** (12), 1879–1890.
85. Apt, K.E., Kroth-Pancic, P.G., and Grossman, A.R. (1996) Stable nuclear

transformation of the diatom *Phaeodactylum tricornutum*. *Mol. Gen. Genet.*, **252** (5), 572–579.

86. Falciatore, A., Casotti, R., Leblanc, C., Abrescia, C., and Bowler, C. (1999) Transformation of nonselectable reporter genes in marine diatoms. *Mar. Biotechnol.*, **1** (3), 239–251.
87. Chow, K.-C. and Tung, W. (1999) Electrotransformation of *Chlorella vulgaris*. *Plant Cell Rep.*, **18** (9), 778–780.
88. Feng S, Xue L, Liu H, Lu P. Improvement of efficiency of genetic transformation for *Dunaliella salina* by glass beads method. *Mol. Biol. Rep.* 2009;**36**(6):1433–1439.
89. Tan, C., Qin, S., Zhang, Q., Jiang, P., and Zhao, F. (2005) Establishment of a micro-particle bombardment transformation system for *Dunaliella salina*. *J. Microbiol.*, **43** (4), 361–365.
90. Georgianna, D.R., Hannon, M.J., Marcuschi, M., Wu, S.Q., Botsch, K., Lewis, A.J., Hyun, J., Mendez, M., and Mayfield, S.P. (2013) Production of recombinant enzymes in the marine alga *Dunaliella tertiolecta*. *Algal Res.-Biomass Biofuels Bioprod.*, **2** (1), 2–9.
91. Te, M.R. and Miller, D.J. (1998) Genetic transformation of dinoflagellates (*Amphidinium* and *Symbiodinium*): expression of GUS in microalgae using heterologous promoter constructs. *Plant J.*, **13** (3), 427–435.
92. Lapidot, M., Raveh, D., Sivan, A., Arad, S.M., and Shapira, M. (2002) Stable chloroplast transformation of the unicellular red alga Porphyridium species. *Plant Physiol.*, **129** (1), 7–12.
93. Minoda, A., Sakagami, R., Yagisawa, F., Kuroiwa, T., and Tanaka, K. (2004) Improvement of culture conditions and evidence for nuclear transformation by homologous recombination in a red alga, *Cyanidioschyzon merolae* 10D. *Plant Cell Physiol.*, **45** (6), 667–671.
94. Walsh, G. (2010) Biopharmaceutical benchmarks 2010. *Nat. Biotechnol.*, **28** (9), 917–924.
95. Mayfield, S.P., Franklin, S.E., and Lerner, R.A. (2003) Expression and assembly of a fully active antibody in algae. *Proc. Natl. Acad. Sci. U.S.A.*, **100** (2), 438–442.
96. Tran, M., Zhou, B., Pettersson, P.L., Gonzalez, M.J., and Mayfield, S.P. (2009) Synthesis and assembly of a full-length human monoclonal antibody in algal chloroplasts. *Biotechnol. Bioeng.*, **104** (4), 663–673.
97. Tran, M., Van, C., Barrera, D.J., Pettersson, P.L., Peinado, C.D., Bui, J., and Mayfield, S.P. (2013) Production of unique immunotoxin cancer therapeutics in algal chloroplasts. *Proc. Natl. Acad. Sci. U.S.A.*, **110** (1), E15–E22.
98. Tran, M., Henry, R.E., Siefker, D., Van, C., Newkirk, G., Kim, J., Bui, J., and Mayfield, S.P. (2013) Production of anti-cancer immunotoxins in algae: ribosome inactivating proteins as fusion partners. *Biotechnol. Bioeng.*, **110** (11), 2826–2835.
99. Zhang, Y.K., Shen, G.F., and Ru, B.G. (2006) Survival of human metallothionein-2 transplastomic *Chlamydomonas reinhardtii* to ultraviolet B exposure. *Acta Biochim. Biophys. Sin.*, **38** (3), 187–193.
100. Rasala, B.A., Muto, M., Lee, P.A., Jager, M., Cardoso, R.M., Behnke, C.A., Kirk, P., Hokanson, C.A., Crea, R., Mendez, M. *et al.* (2010) Production of therapeutic proteins in algae, analysis of expression of seven human proteins in the chloroplast of *Chlamydomonas reinhardtii*. *Plant Biotechnol. J.*, **8** (6), 719–733.
101. Bai, L.L., Yin, W.B., Chen, Y.H., Niu, L.L., Sun, Y.R., Zhao, S.M., Yang, F.Q., Wang, R.R., Wu, Q., Zhang, X.Q. *et al.* (2013) A new strategy to produce a defensin: stable production of mutated NP-1 in nitrate reductase-deficient *Chlorella ellipsoidea*. *PLoS One*, **8** (1), e54966.
102. Chen, Y., Wang, Y., Sun, Y., Zhang, L., and Li, W. (2001) Highly efficient expression of rabbit neutrophil peptide-1 gene in *Chlorella ellipsoidea* cells. *Curr. Genet.*, **39** (5-6), 365–370.
103. Wen, Z.Y. and Chen, F. (2003) Heterotrophic production of eicosapentaenoic acid by microalgae. *Biotechnol. Adv.*, **21** (4), 273–294.

104. Kuratko, C.N. and Salem, N. (2013) Docosahexaenoic acid from algal oil. *Eur. J. Lipid Sci. Technol.*, **115**, 965–976.
105. Winwood, R., Jacobsen, C., Skall Nielsen, N., Frisenfeldt Horn, A., and Moltke, S. (2013) *Food Enrichment with Omega-3 Fatty Acids*, pp. 389–404.
106. Hamilton, M.L., Haslam, R.P., Napier, J.A., and Sayanova, O. (2013) Metabolic engineering of *Phaeodactylum tricornutum* for the enhanced accumulation of omega-3 long chain polyunsaturated fatty acids. *Metab. Eng.*, **22C**, 3–9.
107. Vaezi, R., Napier, J.A., and Sayanova, O. (2013) Identification and functional characterization of genes encoding omega-3 polyunsaturated Fatty Acid biosynthetic activities from unicellular microalgae. *Mar. Drugs*, **11** (12), 5116–5129.
108. Gromer, S., Eubel, J.K., Lee, B.L., and Jacob, J. (2005) Human selenoproteins at a glance. *Cell. Mol. Life Sci.*, **62** (21), 2414–2437.
109. Fu, L.H., Wang, X.F., Eyal, Y., She, Y.M., Donald, L.J., Standing, K.G., and Ben-Hayyim, G. (2002) A selenoprotein in the plant kingdom. Mass spectrometry confirms that an opal codon (UGA) encodes selenocysteine in *Chlamydomonas reinhardtii* gluththione peroxidase. *J. Biol. Chem.*, **277** (29), 25983–25991.
110. Hou, Q., Qiu, S., Liu, Q., Tian, J., Hu, Z., and Ni, J. (2013) Selenoprotein-transgenic *Chlamydomonas reinhardtii*. *Nutrients*, **5** (3), 624–636.
111. Raja, R., Hemaiswarya, S., and Rengasamy, R. (2007) Exploitation of *Dunaliella* for *β*-carotene production. *Appl. Microbiol. Biotechnol.*, **74** (3), 517–523.
112. Guedes, A.C., Amaro, H.M., and Malcata, F.X. (2011) Microalgae as sources of carotenoids. *Mar. Drugs*, **9** (4), 625–644.
113. Dufosse, L., Galaup, P., Yaron, A., Arad, S.M., Blanc, P., Chidambara Murthy, K.N., and Ravishankar, G.A. (2005) Microorganisms and microalgae as sources of pigments for food use: a scientific oddity or an industrial reality? *Trends Food Sci. Technol.*, **16** (9), 389–406.
114. Rosenberg, J.N., Oyler, G.A., Wilkinson, L., and Betenbaugh, M.J. (2008) A green light for engineered algae: redirecting metabolism to fuel a biotechnology revolution. *Curr. Opin. Biotechnol.*, **19** (5), 430–436.
115. Borowitzka, M.A. (2013) High-value products from microalgae—their development and commercialisation. *J. Appl. Phycol.*, 1–14.
116. Del Campo, J.A., García-González, M., and Guerrero, M.G. (2007) Outdoor cultivation of microalgae for carotenoid production: current state and perspectives. *Appl. Microbiol. Biotechnol.*, **74** (6), 1163–1174.
117. Fukusaki, E., Nishikawa, T., Kato, K., Shinmyo, A., Hemmi, H., Nishino, T., and Kobayashi, A. (2003) Introduction of the archaebacterial geranylgeranyl pyrophosphate synthase gene into *Chlamydomonas reinhardtii* chloroplast. *J. Biosci. Bioeng.*, **95** (3), 283–287.
118. León, R., Couso, I., and Fernández, E. (2007) Metabolic engineering of ketocarotenoids biosynthesis in the unicelullar microalga *Chlamydomonas reinhardtii*. *J. Biotechnol.*, **130** (2), 143.
119. Wong, K.H. (2006) Transgenic *chlamydomonas reinhardtii* as an experimental system to study the regulation of carotenoid biosynthesis in green microalgae. The University of Hong Kong.
120. Steinbrenner, J. and Sandmann, G. (2006) Transformation of the green alga *Haematococcus pluvialis* with a phytoene desaturase for accelerated astaxanthin biosynthesis. *Appl. Environ. Microbiol.*, **72** (12), 7477.
121. Cordero, B.F., Couso, I., Leon, R., Rodriguez, H., and Vargas, M.A. (2011) Enhancement of carotenoids biosynthesis in *Chlamydomonas reinhardtii* by nuclear transformation using a phytoene synthase gene isolated from Chlorella zofingiensis. *Appl. Microbiol. Biotechnol.*, **91** (2), 341–351.
122. Couso, I., Vila, M., Rodriguez, H., Vargas, M.A., and Leon, R. (2011) Overexpression of an exogenous

phytoene synthase gene in the unicellular alga *Chlamydomonas reinhardtii* leads to an increase in the content of carotenoids. *Biotechnol. Progr.*, **27** (1), 54–60.
123. Sun, G., Zhang, X., Sui, Z., and Mao, Y. (2008) Inhibition of pds gene expression via the RNA interference approach in *Dunaliella salina* (Chlorophyta). *Mar. Biotechnol.*, **10** (3), 219–226.
124. Vila, M., Couso, I., and Leon, R. (2008) Carotenoid content in mutants of the chlorophyte *Chlamydomonas reinhardtii* with low expression levels of phytoene desaturase. *Process Biochem.*, **43** (10), 1147–1152.
125. Sun, M., Qian, K.X., Su, N., Chang, H.Y., Liu, J.X., and Chen, G.F. (2003) Foot-and-mouth disease virus VP1 protein fused with cholera toxin B subunit expressed in *Chlamydomonas reinhardtii* chloroplast. *Biotechnol. Lett*, **25** (13), 1087–1092.
126. Geng, D.G., Wang, Y.Q., Wang, P., Li, W.B., and Sun, Y.R. (2003) Stable expression of hepatitis B surface antigen gene in *Dunaliella salina* (Chlorophyta). *J. Appl. Phycol.*, **15** (6), 451–456.
127. Soria-Guerra, R.E., Ramírez-Alonso, J.I., Ibáñez-Salazar, A., Govea-Alonso, D.O., Paz-Maldonado, L.M.T., Bañuelos-Hernández, B., Korban, S.S., and Rosales-Mendoza, S. (2013) Expression of an HBcAg-based antigen carrying angiotensin II in *Chlamydomonas reinhardtii* as a candidate hypertension vaccine. *Plant Cell Tissue Organ Cult.*, **116**, 133–139.
128. Hempel, F., Lau, J., Klingl, A., and Maier, U.G. (2011) Algae as protein factories: expression of a human antibody and the respective antigen in the diatom *Phaeodactylum tricornutum*. *PLoS One*, **6** (12), e28424.
129. He, D.M., Qian, K.X., Shen, G.F., Zhang, Z.F., Li, Y.N., Su, Z.L., and Shao, H.B. (2007) Recombination and expression of classical swine fever virus (CSFV) structural protein E2 gene in *Chlamydomonas reinhardtii* chroloplasts. *Colloids Surf., B*, **55** (1), 26–30.
130. Wang, X., Brandsma, M., Tremblay, R., Maxwell, D., Jevnikar, A.M., Huner, N., and Ma, S. (2008) A novel expression platform for the production of diabetes-associated autoantigen human glutamic acid decarboxylase (hGAD65). *BMC Biotech.*, **8**, 87.
131. Surzycki, R., Greenham, K., Kitayama, K., Dibal, F., Wagner, R., Rochaix, J.D., Ajam, T., and Surzycki, S. (2009) Factors effecting expression of vaccines in microalgae. *Biologicals*, **37** (3), 133–138.
132. Dreesen, I.A.J., Charpin-El Hamri, G., and Fussenegger, M. (2010) Heat-stable oral alga-based vaccine protects mice from *Staphylococcus aureus* infection. *J. Biotechnol.*, **145** (3), 273–280.
133. Demurtas, O.C., Massa, S., Ferrante, P., Venuti, A., Franconi, R., and Giuliano, G. (2013) A Chlamydomonas-derived human papillomavirus 16 E7 vaccine induces specific tumor protection. *PLoS One*, **8** (4), e61473.
134. Gowda, D.C. and Davidson, E.A. (1999) Protein glycosylation in the malaria parasite. *Parasitol. Today*, **15** (4), 147–152.
135. Gregory, J.A., Li, F., Tomosada, L.M., Cox, C.J., Topol, A.B., Vinetz, J.M., and Mayfield, S. (2012) Algae-produced Pfs25 elicits antibodies that inhibit malaria transmission. *PLoS One*, **7** (5), e37179.
136. Dauvillee, D., Delhaye, S., Gruyer, S., Slomianny, C., Moretz, S.E., d'Hulst, C., Long, C.A., Ball, S.G., and Tomavo, S. (2010) Engineering the chloroplast targeted malarial vaccine antigens in Chlamydomonas starch granules. *PLoS One*, **5** (12), e15424.
137. Gregory, J.A., Topol, A.B., Doerner, D.Z., and Mayfield, S. (2013) Alga-produced cholera toxin-Pfs25 fusion proteins as oral vaccines. *Appl. Environ. Microbiol.*, **79** (13), 3917–3925.
138. Jones, C.S., Luong, T., Hannon, M., Tran, M., Gregory, J.A., Shen, Z., Briggs, S.P., and Mayfield, S.P. (2013) Heterologous expression of the C-terminal antigenic domain of the malaria vaccine candidate Pfs48/45 in

the green algae *Chlamydomonas reinhardtii*. *Appl. Microbiol. Biotechnol.*, **97** (5), 1987–1995.
139. Khozin-Goldberg, I. and Cohen, Z. (2011) Unraveling algal lipid metabolism: recent advances in gene identification. *Biochimie*, **93** (1), 91–100.
140. Guschina, I.A. and Harwood, J.L. (2006) Lipids and lipid metabolism in eukaryotic algae. *Prog. Lipid Res.*, **45** (2), 160–186.
141. Roessler, P.G. (1990) Environmental control of glycerolipid metabolism in microalgae: commercial implications and future research directions. *J. Phycol.*, **26** (3), 393–399.
142. Yu, W.L., Ansari, W., Schoepp, N.G., Hannon, M.J., Mayfield, S.P., and Burkart, M.D. (2011) Modifications of the metabolic pathways of lipid and triacylglycerol production in microalgae. *Microb. Cell Fact.*, **10**, 91.
143. Li, Y.Q., Horsman, M., Wang, B., Wu, N., and Lan, C.Q. (2008) Effects of nitrogen sources on cell growth and lipid accumulation of green alga *Neochloris oleoabundans*. *Appl. Microbiol. Biotechnol.*, **81** (4), 629–636.
144. Dean, A.P., Sigee, D.C., Estrada, B., and Pittman, J.K. (2010) Using FTIR spectroscopy for rapid determination of lipid accumulation in response to nitrogen limitation in freshwater microalgae. *Bioresour. Technol.*, **101** (12), 4499–4507.
145. Chen, M., Tang, H., Ma, H., Holland, T.C., Ng, K.Y., and Salley, S.O. (2011) Effect of nutrients on growth and lipid accumulation in the green algae *Dunaliella tertiolecta*. *Bioresour. Technol.*, **102** (2), 1649–1655.
146. Roessler, P.G. (1988) Effects of silicon deficiency on lipid-composition and metabolism in the diatom cyclotella-cryptica. *J. Phycol.*, **24** (3), 394–400.
147. Li, X., Hu, H.Y., Gan, K., and Sun, Y.X. (2010) Effects of different nitrogen and phosphorus concentrations on the growth, nutrient uptake, and lipid accumulation of a freshwater microalga *Scenedesmus* sp. *Bioresour. Technol.*, **101** (14), 5494–5500.
148. Liu, Z.Y., Wang, G.C., and Zhou, B.C. (2008) Effect of iron on growth and lipid accumulation in *Chlorella vulgaris*. *Bioresour. Technol.*, **99** (11), 4717–4722.
149. Sriharan, S., Bagga, D., and Nawaz, M. (1991) The effects of nutrients and temperature on biomass, growth, lipid production, and *fatty acid* composition of *Cyclotella cryptica* Reimann, Lewin, and Guillard. *Appl. Biochem. Biotechnol.*, **28–29**, 317–326.
150. Yeesang, C. and Cheirsilp, B. (2011) Effect of nitrogen, salt, and iron content in the growth medium and light intensity on lipid production by microalgae isolated from freshwater sources in Thailand. *Bioresour. Technol.*, **102** (3), 3034–3040.
151. Yoo, C., Jun, S.Y., Lee, J.Y., Ahn, C.Y., and Oh, H.M. (2010) Selection of microalgae for lipid production under high levels carbon dioxide. *Bioresour. Technol.*, **101**, S71–S74.
152. Vigh, L. and Joo, F. (1983) Modulation of membrane fluidity by catalytic-hydrogenation affects the chilling susceptibility of the blue-green-alga, anacystis-nidulans. *FEBS Lett.*, **162** (2), 423–427.
153. Sato, N. and Murata, N. (1980) Temperature shift-induced responses in lipids in the blue-green alga, *Anabaena variabilis* the central role of diacylmonogalactosylglycerol in thermo-adaptation. *Biochim. Biophys. Acta*, **619** (2), 353–366.
154. Lynch, D.V. and Thompson, G.A.J. (1982) Low temperature-induced alterations in the chloroplast and microsomal membranes of *Dunaliella salina*. *Plant Physiol.*, **69** (6), 1369–1375.
155. Khotimchenko, S.V. and Yakovleva, I.M. (2005) Lipid composition of the red alga *Tichocarpus crinitus* exposed to different levels of photon irradiance. *Phytochemistry*, **66** (1), 73–79.
156. Sukenik, A., Carmeli, Y., and Berner, T. (1989) Regulation of fatty acid composition by irradiance level in the Eustigmatophyte *Nannochloropsis* sp. *J. Phycol.*, **25** (4), 686–692.
157. Orcutt, D.M. and Patterson, G.W. (1974) Effect of light intensity upon lipid composition of *Nitzschia*

closterium (*Cylindrotheca fusiformis*). *Lipids*, **9** (12), 1000–1003.

158. Constant A.G. and Bloch, K. (1967) Effect of light intensity on lipid composition of *Eugena Gracilis*. *J. Biol. Chem.*, **242** (15), 3538–3542.
159. Dunahay, T.G., Jarvis, E.E., Dais, S.S., and Roessler, P.G. (1996) Manipulation of microalgal lipid production using genetic engineering. *Appl. Biochem. Biotechnol.*, **57-58**, 223–231.
160. Wang, Z.T., Ullrich, N., Joo, S., Waffenschmidt, S., and Goodenough, U. (2009) Algal lipid bodies: stress induction, purification, and biochemical characterization in wild-type and starchless *Chlamydomonas reinhardtii*. *Eukaryotic Cell*, **8** (12), 1856–1868.
161. Radakovits, R., Eduafo, P.M., and Posewitz, M.C. (2011) Genetic engineering of fatty acid chain length in *Phaeodactylum tricornutum*. *Metab. Eng.*, **13** (1), 89–95.
162. Ort, D.R. and Melis, A. (2011) Optimizing antenna size to maximize photosynthetic efficiency. *Plant Physiol.*, **155** (1), 79–85.
163. Mussgnug, J.H., Thomas-Hall, S., Rupprecht, J., Foo, A., Klassen, V., McDowall, A., Schenk, P.M., Kruse, O., and Hankamer, B. (2007) Engineering photosynthetic light capture: impacts on improved solar energy to biomass conversion. *Plant Biotechnol. J.*, **5** (6), 802–814.
164. Mussgnug, J.H., Wobbe, L., Elles, I., Claus, C., Hamilton, M., Fink, A., Kahmann, U., Kapazoglou, A., Mullineaux, C.W., Hippler, M. *et al.* (2005) NAB1 is an RNA binding protein involved in the light-regulated differential expression of the light-harvesting antenna of *Chlamydomonas reinhardtii*. *Plant Cell*, **17** (12), 3409–3421.
165. Wobbe, L., Blifernez, O., Schwarz, C., Mussgnug, J.H., Nickelsen, J., and Kruse, O. (2009) Cysteine modification of a specific repressor protein controls the translational status of nucleus-encoded LHCII mRNAs in Chlamydomonas. *Proc. Natl. Acad. Sci. U.S.A.*, **106** (32), 13290–13295.
166. Beckmann, J., Lehr, F., Finazzi, G., Hankamer, B., Posten, C., Wobbe, L., and Kruse, O. (2009) Improvement of light to biomass conversion by de-regulation of light-harvesting protein translation in *Chlamydomonas reinhardtii*. *J. Biotechnol.*, **142** (1), 70–77.
167. Ritchie, R. and Larkum, A. (2012) Modelling photosynthesis in shallow algal production ponds. *Photosynthetica*, **50** (4), 481–500.
168. Keren, N. and Krieger-Liszkay, A. (2011) Photoinhibition: molecular mechanisms and physiological significance. *Physiol. Plant.*, **142** (1), 1–5.
169. Rea, G., Lambreva, M., Polticelli, F., Bertalan, I., Antonacci, A., Pastorelli, S., Damasso, M., Johanningmeier, U., and Giardi, M.T. (2011) Directed evolution and in silico analysis of reaction centre proteins reveal molecular signatures of photosynthesis adaptation to radiation pressure. *PLoS One*, **6** (1), e16216.
170. Gimpel, J.A. and Mayfield, S.P. (2013) Analysis of heterologous regulatory and coding regions in algal chloroplasts. *Appl. Microbiol. Biotechnol.*, **97** (10), 4499–4510.
171. Vinyard, D.J., Gimpel, J., Ananyev, G.M., Cornejo, M.A., Golden, S.S., Mayfield, S.P., and Dismukes, G.C. (2013) Natural variants of photosystem II subunit D1 tune photochemical fitness to solar intensity. *J. Biol. Chem.*, **288** (8), 5451–5462.
172. Chen, M. and Blankenship, R.E. (2011) Expanding the solar spectrum used by photosynthesis. *Trends Plant Sci.*, **16** (8), 427–431.
173. Raines, C.A. (2011) Increasing photosynthetic carbon assimilation in C-3 plants to improve crop yield: current and future strategies. *Plant Physiol.*, **155** (1), 36–42.
174. Ducat, D.C. and Silver, P.A. (2012) Improving carbon fixation pathways. *Curr. Opin. Chem. Biol.*, **16** (3-4), 337–344.
175. Whitney, S.M., Houtz, R.L., and Alonso, H. (2011) Advancing our understanding and capacity to engineer

nature's CO_2-sequestering enzyme, Rubisco. *Plant Physiol.*, **155** (1), 27–35.
176. Spreitzer, R.J., Peddi, S.R., and Satagopan, S. (2005) Phylogenetic engineering at an interface between large and small subunits imparts land-plant kinetic properties to algal Rubisco. *Proc. Natl. Acad. Sci. U.S.A.*, **102** (47), 17225–17230.
177. Genkov, T., Meyer, M., Griffiths, H., and Spreitzer, R.J. (2010) Functional hybrid Rubisco enzymes with plant small subunits and algal large subunits engineered rbcS cDNA for expression in Chlamydomonas. *J. Biol. Chem.*, **285** (26), 19833–19841.
178. Zhu, G.H., Kurek, I., and Liu, L. (2010) *Chloroplast: Basics and Applications*, vol. **31**, Springer, pp. 307–322.
179. Johnson, X. (2011) Manipulating RuBisCO accumulation in the green alga, *Chlamydomonas reinhardtii. Plant Mol. Biol.*, **76** (3-5), 397–405.
180. Benemann, J.R. (1992) Microalgae aquaculture feeds. *J. Appl. Phycol.*, **4** (3), 233–245.
181. Milledge, J.J. (2011) Commercial application of microalgae other than as biofuels: a brief review. *Rev. Environ. Sci. Biotechnol.*, **10** (1), 31–41.
182. Becker, E.W. (2007) Micro-algae as a source of protein. *Biotechnol. Adv.*, **25** (2), 207–210.
183. Mathew, A.G., Cissell, R., and Liamthong, S. (2007) Antibiotic resistance in bacteria associated with food animals: a United States perspective of livestock production. *Foodborne Pathog. Dis.*, **4** (2), 115–133.
184. Peak, N., Knapp, C.W., Yang, R.K., Hanfelt, M.M., Smith, M.S., Aga, D.S., and Graham, D.W. (2007) Abundance of six tetracycline resistance genes in wastewater lagoons at cattle feedlots with different antibiotic use strategies. *Environ. Microbiol.*, **9** (1), 143–151.
185. Alexander, T.W., Yanke, J.L., Reuter, T., Topp, E., Read, R.R., Selinger, B.L., and McAllister, T.A. (2011) Longitudinal characterization of antimicrobial resistance genes in feces shed from cattle fed different subtherapeutic antibiotics. *BMC Microbiol.*, **11** (1), 19.
186. Campos-Quevedo, N., Rosales-Mendoza, S., Paz-Maldonado, L.M.T., Martínez-Salgado, L., Guevara-Arauza, J.C., and Soria-Guerra, R.E. (2013) Production of milk-derived bioactive peptides as precursor chimeric proteins in chloroplasts of *Chlamydomonas reinhardtii. Plant Cell Tissue Organ Cult.*, 1–9.
187. Brown, T.R. and Brown, R.C. (2013) A review of cellulosic biofuel commercial-scale projects in the United States. *Biofuels, Bioprod. Biorefin.*, **7** (3), 235–245.
188. Abramson, M., Shoseyov, O., and Shani, Z. (2010) Plant cell wall reconstruction toward improved lignocellulosic production and processability. *Plant Sci.*, **178** (2), 61–72.
189. Forsberg, C., Phillips, J., Golovan, S., Fan, M., Meidinger, R., Ajakaiye, A., Hilborn, D., and Hacker, R. (2003) The Enviropig physiology, performance, and contribution to nutrient management advances in a regulated environment: the leading edge of change in the pork industry. *J. Anim. Sci.*, **81** (14, Suppl. 2), E68–E77.
190. Yoon, S.M., Kim, S.Y., Li, K.F., Yoon, B.H., Choe, S., and Kuo, M.M. (2011) Transgenic microalgae expressing *Escherichia coli* AppA phytase as feed additive to reduce phytate excretion in the manure of young broiler chicks. *Appl. Microbiol. Biotechnol.*, **91** (3), 553–563.
191. Christenson, L. and Sims, R. (2011) Production and harvesting of microalgae for wastewater treatment, biofuels, and bioproducts. *Biotechnol. Adv.*, **29** (6), 686–702.
192. Chen, C.Y., Yeh, K.L., Aisyah, R., Lee, D.J., and Chang, J.S. (2011) Cultivation, photobioreactor design and harvesting of microalgae for biodiesel production: a critical review. *Bioresour. Technol.*, **102** (1), 71–81.
193. Mata, T.M., Martins, A.A., and Caetano, N.S. (2010) Microalgae for biodiesel production and other applications: a review. *Renewable Sustainable Energy Rev.*, **14** (1), 217–232.

194. Chen, G.Q. and Chen, F. (2006) Growing phototrophic cells without light. *Biotechnol. Lett*, **28** (9), 607.
195. Miao, X. and Wu, Q. (2006) Biodiesel production from heterotrophic microalgal oil. *Bioresour. Technol.*, **97** (6), 841–846.
196. Hallmann, A. and Sumper, M. (1996) The Chlorella hexose/H+ symporter is a useful selectable marker and biochemical reagent when expressed in Volvox. *Proc. Natl. Acad. Sci. U.S.A.*, **93** (2), 669–673.
197. Zaslavskaia, L.A., Lippmeier, J.C., Shih, C., Ehrhardt, D., Grossman, A.R., and Apt, K.E. (2001) Trophic conversion of an obligate photoautotrophic organism through metabolic engineering. *Science*, **292** (5524), 2073–2075.
198. Doebbe, A., Rupprecht, J., Beckmann, J., Mussgnug, J.H., Hallmann, A., Hankamer, B., and Kruse, O. (2007) Functional integration of the HUP1 hexose symporter gene into the genome of *C. reinhardtii*: impacts on biological H(2) production. *J. Biotechnol.*, **131** (1), 27–33.
199. Fischer, H., Robl, I., Sumper, M., and Kröger, N. (1999) Targeting and covalent modification of cell wall and membrane proteins heterologously expressed in the diatom *Cylindrotheca fusiformis* (Bacillariophyceae). *J. Phycol.*, **35** (1), 113–120.
200. Moheimani, N.R. and Borowitzka, M.A. (2007) Limits to productivity of the alga *Pleurochrysis carterae* (Haptophyta) grown in outdoor raceway ponds. *Biotechnol. Bioeng.*, **96** (1), 27–36.
201. Park, J.B. and Craggs, R.J. (2011) Algal production in wastewater treatment high rate algal ponds for potential biofuel use. *Water Sci. Technol.*, **63** (10), 2403–2410.
202. Park, J.B., Craggs, R.J., and Shilton, A.N. (2011) Wastewater treatment high rate algal ponds for biofuel production. *Bioresour. Technol.*, **102** (1), 35–42.
203. Borowitzka, M. and Moheimani, N. (2013) in *Algae for Biofuels and Energy*, Developments in Applied Phycology, vol. **5** (eds M.A. Borowitzka and N.R. Moheimani), Springer, Netherlands, pp. 133–152.
204. Vonshak, A., Boussiba, S., Abeliovich, A., and Richmond, A. (1983) Production of Spirulina biomass: maintenance of monoalgal culture outdoors. *Biotechnol. Bioeng.*, **25** (2), 341–349.
205. Vandamme, D., Foubert, I., and Muylaert, K. (2013) Flocculation as a low-cost method for harvesting microalgae for bulk biomass production. *Trends Biotechnol.*, **31** (4), 233–239.
206. Powell, R.J. and Hill, R.T. (2013) Rapid aggregation of biofuel-producing algae by the bacterium Bacillus sp. strain RP1137. *Appl. Environ. Microbiol.*, **79** (19), 6093–6101.
207. Zhang, J. and Hu, B. (2012) A novel method to harvest microalgae via co-culture of filamentous fungi to form cell pellets. *Bioresour. Technol.*, **114**, 529–535.
208. Zhou, W., Min, M., Hu, B., Ma, X., Liu, Y., Wang, Q., Shi, J., Chen, P., and Ruan, R. (2013) Filamentous fungi assisted bio-flocculation: a novel alternative technique for harvesting heterotrophic and autotrophic microalgal cells. *Sep. Purif. Technol.*, **107**, 158–165.
209. Besson, A. and Guiraud, P. (2013) High-pH-induced flocculation–flotation of the hypersaline microalga *Dunaliella salina*. *Bioresour. Technol.*, **147**, 464–470.
210. Vandamme, D., Foubert, I., Fraeye, I., Meesschaert, B., and Muylaert, K. (2012) Flocculation of *Chlorella vulgaris* induced by high pH: role of magnesium and calcium and practical implications. *Bioresour. Technol.*, **105**, 114–119.
211. Spilling, K., SeppSepp, I., Fraeye, I., Meesschaert, B., and Muylaert, K. (2011) Flocculation of *Chlorella vulgaris* induced by high pH: rolulation. *J. Appl. Phycol.*, **23** (6), 959–966.

Part VI
Eukaryotic Workhorses: Complex Cells Enable Complex Products

Industrial Biotechnology: Microorganisms, First Edition. Edited by Christoph Wittmann and James C. Liao.

17
Host Organisms: Mammalian Cells

Jennifer Pfizenmaier and Ralf Takors

17.1
Introduction

Mammalian cells play an important role in the production of biopharmaceuticals, that is, they are of key interest for producing medical drugs based on biotechnology. Biopharmaceutical market segments can be divided into nine classes [1]: monoclonal antibodies (mABs), hormones, growth factors, fusion proteins, cytokines, blood factors, therapeutic enzymes, recombinant vaccines, and anticoagulants. With respect to sales, the four top ranking candidates, namely mABs, hormones, growth factors, and fusion proteins count for more than 75% of total market shares, with mABs showing the strongest market dynamics, for instance, representing the biggest fragment (~38%) of U.S. sales (total: US$ 53.8 billion) in 2011. In 2007, worldwide biologics sales already achieved US$ 94 billion. This number significantly increased during the past years, with currently approved 200 biopharmaceuticals for U.S. and European markets [2], reported US$ 120 billion global sales in 2012 and expected U.S. sales of US$ 150 billion in 2015 [3].

These biologics are typically produced by Escherichia *coli*, yeasts, or mammalian cells. While *E. coli* and Saccharomyces *cerevisiae* strains play important roles in the formation of hormones such as insulin or even growth factors (e.g., to treat growth deficiency in children), the production of mABs and fusion proteins is dominated by mammalians. Protein therapeutics require for the application of biologically active forms. This includes the achievement of correct glycosylation patterns, which is the result of the addition of carbohydrate moieties to amino acid residues after correct folding of the complex protein structure. The prokaryote *E. coli* lacks the capability to synthesize correct glycosylated functions. Furthermore, activities for humanizing yeast glycosylation patterns are still ongoing. Consequently, rodent- and human-derived cell lines are favored for the production of recombinant therapeutic proteins [4].

A variety of different cell lines is used today. One group consists of tissue derivatives, originally not showing amplified protein secretion [5]. Examples are Chinese hamster ovary (CHO) [6] or kidney cells like baby hamster kidney (BHK)

Industrial Biotechnology: Microorganisms, First Edition. Edited by Christoph Wittmann and James C. Liao.

or human embryonic kidney (HEK 293) [7]. Besides, other cell lines that suggested native advantages for protein formation and secretion are applied. Among those is the immortal HeLa cell line, which was isolated from the cervical cancer patient "Helen Lane" (correct name: Henrietta Lacks) in 1951. Other examples are HepG2 (a human liver carcinoma cell line) or murine myeloma cells NS0 (typically originating from the plasmacytoma tumor MOPC21) [8]. Recently, engineered human cells such as the neuronal cell line AGE1.HN® [9] or PER.C6® (derived from human embryonic retinal cells) were considered for production as well.

In 2007, nearly 70% of recombinant protein therapeutics was produced in CHO cells [4]. This outstanding role of CHO is expected to persist because of the impressive productivity increases that were achieved during the past years in biopharmaceutical production. While approximately 50 mg product per liter were typically reached by 2×10^6 cells ml^{-1} in batch processes in 1986 [10], current mAB titers of fed-batch processes are typically in the range of 2–5 g l^{-1} after 12 days of cultivation and can be further increased to 5–8 g l^{-1} by thorough systems analysis [11]. Aggarwal [1] and Walsh [2] outlined that mABs are still ranked as number one biologics products on sale – and they are likely to continue. Among the 58 new biopharmaceuticals that gained approval between 2006 and 2010, were 13 mABs [2]. Notably, about half of the newly approved drugs are classified as biosimilars, which outlines the basic need to improve existing cells and processes to withstand the increasing competition in the developing markets.

For the sake of brevity and considering given scientific and economic developments, this contribution focuses on CHO cells as one of the major players in mammalian-cell-based biopharmaceutical production. Unless otherwise stated, these cells are discussed and – where necessary – compared to alternative mammalian production systems.

17.2
Basics of Cellular Structure and Metabolism

One of the characteristic features of mammalian cells in contrast to bacteria is their compartmented structure. Due to the existence of various cell organelles, for example, nucleus, mitochondrion, endoplasmic reticulum (ER), and Golgi apparatus, mammalian cells are divided into compartments with specific functions. Nevertheless, several transport mechanisms form connections between the different compartments and show that local processes are part of a global cellular network. This chapter focuses on the basics of cellular structure and metabolism in the context of the global cellular network.

17.2.1
Cellular Structure

In comparison to bacteria, CHO cells are growing slower and are significantly larger in size. When CHO cells are cultivated in suspension, they show growth rates between 0.025 and 0.045 h^{-1} and cellular diameters of 11 – 16 μm, which are 10-fold higher than those of bacteria. Therefore, the size of bacteria lies even more in the range of mammalian cell organelles, which are further characterized in the following abstracts.

The genetic information of mammalian cells is almost exclusively located in the nucleus (Figure 17.1). This cell organelle is bounded by a double membrane which contains nuclear pore complexes as the connection to the cytoplasm and it is studded with ribosomes [12]. Replication and transcription of DNA are cellular processes that take place in the nucleus. DNA replication occurs during the mammalian cell cycle, which can be divided into different phases. The gap between mitosis and the initiation of DNA replication is termed G_1*-phase,* while the gap between DNA synthesis (S-phase) and mitosis (M-phase) is called G_2*-phase* [13]. In order to pass the different phases of the cell cycle, mammalian cells require cyclin-dependent kinases (Cdks). These Cdks (Cdk1, Cdk2, Cdk4, and Cdk6) are serine/threonine protein kinases which are only catalytically active if regulatory molecules termed *cyclins* are bound. Cyclins received their name as they are synthesized and degraded during each cell cycle [14].

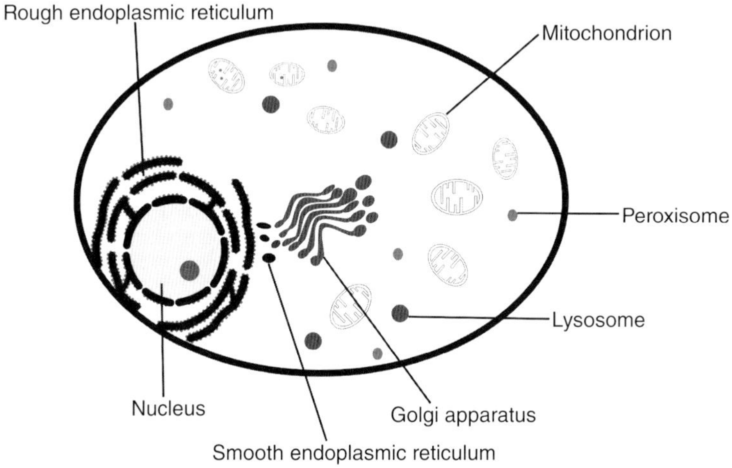

Figure 17.1 Scheme of the compartmentalization of mammalian cells. The nucleus, host of genetic information, is the cell organelle where DNA replication and transcription of DNA occur. A small genome is as well found in mitochondria where important metabolic pathways, for example, TCA cycle, oxidative phosphorylation, and β-oxidation, take place. A variety of acid hydrolases enables the degradation of all kinds of macromolecules in the cell's lysosomes. The degradation of hydrogen peroxide and β-oxidation of fatty acids occurs in the peroxisomes. The rough endoplasmic reticulum (studded with ribosomes) represents the site for protein synthesis, processing, and translocation, while the smooth endoplasmic reticulum is often involved in assembly of vesicles for transport to the Golgi apparatus. Glycosyltransferases, kinases, and proteases modify proteins or lipids on their way through the Golgi apparatus until they are finally delivered to their cellular destinations.

In contrast to bacteria, transcribed units of mammalian cells are significantly longer than the final mRNA, and therefore a specific processing of the transcribed RNA is required before it is transported into the cytoplasm for translation. A methylated guanylate residue called *cap* is added to the 5′ end, while a poly-adenylate segment is added to the 3′ end of the primary transcripts [15]. The eukaryotic primary RNA transcripts consist of coding (exons) and noncoding segments (introns). During a process called *splicing,* introns are removed from the primary RNA transcript in order to receive the final mRNA [15]. Nuclear pore complexes transport mRNAs into the cytoplasm, where translation is located.

Small molecular complexes named ribosomes catalyze and regulate this process of protein synthesis. Mammalian ribosomes consist of ribosomal RNAs (rRNAs) and proteins and can be divided into a large (60S) and a small (40S) subunit. Together these two subunits form an 80S complex, which is responsible for the decoding of mRNAs and the formation of peptide bonds to generate polypeptides. As rRNAs catalyze, this process of peptide bond formation ribosomes are defined as ribozymes [16]. Besides the cytoplasm, ribosomes are also found in the matrix of mitochondria.

The mitochondrion (Figure 17.1) is the only cell organelle beside the nucleus that contains genomic information. Proteins of the complexes of the electron transport chain, tRNAs and rRNAs are encoded by the mitochondrial DNA [17]. A smooth outer mitochondrial membrane and an extensively folded inner membrane give the organelle its characteristic structure with intermembrane space and mitochondrial matrix [18]. While voltage-dependent anion channels (porins) in the outer mitochondrial membrane enable small ions and metabolites to enter the intermembrane space, the inner mitochondrial membrane forms a permeability barrier [18, 19]. As a result, more than 30 carriers in the inner mitochondrial membrane are required to mediate metabolite transport between the intermembrane space and the mitochondrial matrix [18]. The large number of carriers shows as well the importance of intracellular exchange of metabolites between cytoplasm and mitochondria. The tricarboxylic acid cycle (TCA), oxidative phosphorylation, β-oxidation of fatty acids, and parts of amino acid metabolism are mitochondrial pathways which, on the one hand, require the allocation of substrates and, on the other hand, again provide substrates for metabolic pathways that are localized in the cytoplasm. But while mitochondria are the main site for energy production in the cell, they initiate apoptosis as well in response to a variety of death stimuli, for example, activation of oncogenes, DNA damage, or ultraviolet radiation [18, 20].

The ER is a dynamic membrane structure that is constantly rearranged along the cytoskeleton and performs many different functions [21]. Based on differences in morphology, two types of ER have been described. The rough endoplasmic reticulum (RER) is studded with ribosomes and forms a sheetlike cellular structure (Figure 17.1), while the smooth endoplasmic reticulum (SER) shows no connecting ribosomes and has a more tubular structure [22]. The RER represents the site for protein synthesis, processing, and translocation and therefore is found in all different cell types, while the abundance of the SER varies depending on the cell type [21]. The ER domain representing the connecting part between ER and Golgi apparatus is the only type of SER that is found in all cells. In this transitional ER, protein-containing vesicles are assembled in order to transport these proteins to the Golgi apparatus [21]. In general, the ER interacts with almost all cellular compartments, for example, mitochondria, lysosomes, and peroxisomes. Typical processes that occur in the ER are the modification and folding of proteins that are part of the secretory pathway as well as lipid synthesis [22]. N-linked glycosylation and disulfide bond formation are examples of modifications in the ER, influencing protein folding and vice versa [23]. A quality control mechanism in the ER validates the folding of the proteins before they are packaged into vesicles to be transported to the Golgi apparatus [24].

The exchange of cargo coming from the ER and the Golgi apparatus occurs at the *cis*-Golgi network. The Golgi apparatus (Figure 17.1) is built up by stacks of about four to six flattened cisternae and is functionally divided into *cis*-, medial-, and *trans*-Golgi networks. While the *cis*-Golgi network can be defined as the entering point for Golgi substrates, the *trans*-Golgi network serves as the sorting module to send cargo to various intra- or extracellular destinations, for example, lysosomes or plasma membrane, via specific carriers [25]. Due to the characteristic organization of the Golgi apparatus, proteins or lipids pass the *cis*-, medial-, and *trans*-Golgi networks in order to be modified by glycosyltransferases, kinases, and proteases. These enzymes are located in different cisternae [25]. As the Golgi apparatus does not have a quality control mechanism like the ER, this might benefit a correct and efficient posttranslational modification.

For the intracellular digestion of macromolecules, eukaryotic cells possess organelles called *lysosomes* (Figure 17.1). These cell organelles are built up by a membrane that enables the formation of an acidic compartment [26]. Hydrogen ions (H^+) are actively pumped into the lysosomes via V-class H^+ ATPases [27], while accompanying chloride ions (Cl^-) enter the lysosomes via Cl^- channel proteins [28]. The presence of H^+ and Cl^- leads to the acidic pH value inside the lysosomes and provides the optimal conditions for acid hydrolases like phosphatases, nucleases, glycosidases, proteases, peptidases, and lipases [26]. Due to these various types of acid hydrolases, lysosomes are capable of coping with all kinds of macromolecules [26]. The degradation of proteins has been studied in comparison to the growth rate for antibody-producing myeloma cells. From 224 quantified proteins, 90% showed a prolonged half-life compared to the doubling time of myeloma cells. The turnover rates for 15% of the proteins were in the range of less than 10% of specific growth rate. Based on this experimental data, protein dynamics were not significantly affected by the turnover rate as they are markedly decelerated with regard to the growth rate [29].

Peroxisomes are cell organelles with a diameter of 0.1 – 1.0 μm and are bounded by one triple-layered membrane, which is nevertheless thinner than most of the other cellular membranes, for example, the membrane of lysosomes [30, 31]. Originally, only the enzymes oxidase and catalase were known to exist in the matrix of peroxisomes and therefore it was suggested that peroxisomes are organelles of hydrogen peroxide production and degradation [30]. With the identification of various other enzymes like dehydrogenases or acyl transferases, it became apparent that peroxisomal functions encompassed far more than formation and degradation of hydrogen peroxide [30]. In addition, the composition of the matrix enzymes was shown to vary depending on the cell type and environment [31]. Nevertheless, β-oxidation of fatty acids and the degradation of hydrogen peroxide by catalase are highly conserved peroxisomal functions in mammalian cells [31]. As mentioned previously, β-oxidation occurs in mitochondria as well and therefore it is one example of analogous pathways in different cellular compartments [32]. Chain shortening of saturated and unsaturated fatty acids occurs both in peroxisomes and mitochondria but substrates like, for example, prostaglandins, long-chain dicarboxylic, and very long chain monocarboxylic fatty acids are almost exclusively β-oxidized in peroxisomes. So although mitochondria and peroxisomes contain partially similar enzymes for β-oxidation, they have a different range of substrates and possess some distinct physiological functions [32].

17.2.2 Metabolism

Although mammalian cells and bacteria significantly differ in their cellular structure, they share a very similar central metabolism. The metabolic pathway glycolysis (Embden–Meyerhof–Parnas pathway), pentose-phosphate pathway, TCA cycle, oxidative phosphorylation, catabolism of fats, and fermentative and anaplerotic pathways termed *fueling reactions* provide precursor metabolites that are required for the synthesis of the so-called building blocks, for example, amino acids, nucleotides, or fatty acids. Beside cofactors, the cell uses 12 precursor metabolites (glucose-6-P, fructose-6-P, glyceraldehyde-3-P, 3-P-glycerate, phosphoenolpyruvate, pyruvate, ribose-5-P, erythrose-4-P, acetyl-CoA, 2-oxoglutarate, succinyl-CoA, and oxaloacetate) to produce a variety of building blocks. Finally, these building blocks serve as substrates for the synthesis of macromolecules like DNA, RNA, proteins, or lipids [33]. The compartmentalization of mammalian cells is not only a structural characteristic but also results in a separation of metabolic pathways. The TCA cycle, for example, is located in mitochondria. Due to the close connections between different metabolic pathways, transport mechanisms are required to transport substrates of the TCA cycle into mitochondria and to take synthesized precursor metabolites to cellular locations. The citrate/malate antiport and pyruvate/H^+-symport are only two examples of a variety of transport mechanisms between cytosol and mitochondria [18].

The metabolites glutamine and glucose are the main substrates of mammalian cell culture. Nevertheless, even if they are not limited in oxygen, proliferating mammalian cells do not completely oxidize their main substrates to CO_2 but produce the fermentative product lactate. This observation, termed *Warburg effect,* is hypothesized to be a result of the needs of proliferating cells for precursor metabolites instead of maximum ATP formation via complete oxidation of glucose and glutamine to CO_2 [34]. Although glutamine is not one of the essential amino acids of mammals, it is required in the medium of cell cultures [5]. Mammalian cells possess the ability to catabolize glutamine across different pathways. During this process, termed glutaminolysis, glutamine is completely oxidized to CO_2, incompletely oxidized to aspartate or alanine, or it is catabolized to lactate [35]. Each of these catabolic pathways leads to a different number of synthesized ATP molecules per mole of glutamine. According to an approximation by Glacken [35], 24–27 ATP molecules would be synthesized via complete oxidation of glutamine, which would provide about three- to fourfold more ATP molecules than the two other catabolic pathways. The complete oxidation of substrates is connected to the TCA cycle and electron transport chain. The formation of ATP per reduced oxygen atom by the electron transport chain is characterized by the P/O ratio. The theoretical optimum for the P/O ratio of mammalian cells was determined to be 3. Considering the energy dependence of ATP transport into the cytosol, the P/O ratio of 3 is lowered to 2.5 [36, 37]. Another important terminology regarding cellular metabolism is the "adenylate energy charge," which was originally defined as $([ATP] + 0.5 [ADP]) \times ([ATP] + [ADP] + [AMP])^{-1}$ by Atkinson and Walton [38]. Significant variations in adenylate energy charge during a cell's lifespan are unusual as cells stabilize the energy charge at a high level (>0.9) as long as possible [39]. Different nucleoside triphosphates are utilized depending on metabolic pathway. Cytidine triphosphate (CTP) is used in lipid synthesis, guanosine-5'-triphosphate (GTP) in protein synthesis, and uridine-5'-triphosphate (UTP) in polysaccharide synthesis [39]. Although originally defined for adenine nucleotides, the energy charge can as well be utilized for other nucleotides considering the fact that relative concentrations of mono-, di-, and trinucleotides are maintained [39].

17.3 The Genome of CHO Cells

CHO cells were isolated in 1957 [6] and have been used as production hosts since the approval of a recombinant cell line producing human tissue plasminogen activator in 1987 [4]. About 20 years later, the complete nucleotide sequence of mitochondrial DNA (mtDNA) was published [40]. In accordance with other mammalian cells, this approximately 16.3 kb circular hamster mtDNA encodes 22 tRNAs, 2 rRNAs, and 13 proteins which are involved in the electron transport chain [40]. The first publicly available sequence of the nuclear CHO genome was a 2.45 Gb draft genomic sequence for the cell line CHO-K1 [41]. *De novo* gene prediction programs and homology-based methods led to the prediction of 24 383 genes, while common databases and bioinformatic tools (KEGG, Swiss-Prot, TrEMBL, GO, and InterPro) were used for functional annotation of finally 83% of the predicted genes [41]. In 2013, two additional draft genomic sequences were published. Lewis *et al.* [42] sequenced the genome of a female Chinese hamster and assembled a 2.4 Gb draft genomic sequence comprising 24 044 genes. The genomes of six additional CHO cell lines were sequenced and compared to that of the female Chinese hamster. In the end, hamster genes missing in the CHO cell lines were as well as more that 3.7 million single-nucleotide polymorphisms (SNPs), 551 240 indels, and 7063 copy number variations were identified [42]. Brinkrolf *et al.* [43], on the other hand, performed flow cytometric cell sorting in order to isolate individual chromosomes of the Chinese hamster strain 17A/GY, followed by next-generation sequencing and assembly of draft genomic sequences for separated Chinese hamster chromosomes. The total genome size accounted for 2.33 Gb [43].

Beside the assembly of draft genomic sequences and the annotation of protein encoding genes, sequencing of different CHO cell line transcriptomes initiated the identification of CHO-specific miRNAs. This type of small noncoding RNA molecules has become more and more important since the discovery of lin-4 in 1993, which is essential for correct larval development of *Caenorhabditis elegans* [44, 45]. The small noncoding RNA molecule let-7 was the second miRNA that was identified in *C. elegans* [46] and was shown to be highly conserved across species like vertebrate, ascidian, hemichordate, mollusc, annelid, and arthropod [47]. At the moment, the database miRBase (release 20.0) comprises 9076 precursor and 11 717 mature sequences from mammalian cells, including 307 CHO specific mature miRNAs [48, 49]. miRNAs are defined as noncoding small RNA molecules with a length of approximately 22 nucleotides regulating gene expression by degradation of mRNA or inhibition of translation [50]. Biogenesis of miRNAs starts with transcription of pri-miRNAs by RNA polymerase II/III. In the nucleus, the microprocessor complex Drosha/DSCR8 shortens and processes pri-miRNAs into hairpin-loop structures of approximately 70 nt by the use of its RNA-binding and RNAse abilities [50]. The nuclear export of this pre-miRNAs into the cytoplasm is mediated by Exportin-5 [50]. In the cytoplasm, the loop structures of pre-miRNAs are removed by the Dicer/TRBP complex in order to form approximately 22 nt RNA duplexes [50]. Thereafter, the mature miRNA strand becomes part of the RNA-induced silencing complex (RISC), which leads to degradation or inhibition of translation of target mRNAs [50]. The importance of miRNAs for the regulation of gene expression is enhanced by the fact that one miRNA is able to bind to numerous target mRNAs and that miRNAs are involved in almost all crucial cellular processes [50].

17.4 Molecular Biology Tools

The production of therapeutic proteins in mammalian cells requires the expression of product-specific genes in the host cell line. The dihydrofolate reductase (DHFR) and the glutamine synthetase (GS) expression vector systems are most frequently used in CHO production cell lines [51]. The host cells are usually deficient in endogenous DHFR or GS in order to facilitate the identification of vector-containing clones. As glutamine is an essential amino acid in cell culture, clones which contain the GS vector are able to synthesize glutamine from glutamate and can therefore grow in glutamine-free medium [51]. The enzyme DHFR catalyzes the reduction of folate to dihydrofolate and thereafter to tetrahydrofolate, which is an important one-carbon donor for the synthesis of amino acids and nucleotides. The synthesis of glycine or thymidine, for example, requires the one-carbon donor tetrahydrofolate. Therefore, cells require the DHFR gene from the expression system in order to survive in media free from glycine, hypoxanthine, and thymidine [52]. Depending on the expression system, the genes of interest are linked to a selection marker (e.g., DHFR or GS). Both the gene of interest and selection marker are usually under the control of individual promoters. A weak promoter (e.g., SV40) is often used to control the expression of DHFR or GS in order to reduce promoter interference with the promoter of the genes of interest. Instead, strong viral promoters (e.g., from human cytomegalovirus) are chosen for the genes of interest [51]. Expression vector constructs are introduced into host cell lines via transfection. Lipofection, electroporation, and nucleofection are often-used transfection methods in this context [53, 54]. After transfection, low concentrations of specific enzyme inhibitors (e.g., methotrexate (MTX) for DHFR or methionine sulfoximine (MSX) for GS) are added to the cultivation medium so that only cells which contain DNA of the expression vector system survive [4]. Thereafter, the received transfectants are cultivated in the presence of high levels of enzyme inhibitor (e.g., MTX) in order to increase the selection pressure. During this phase, only cells that are able to synthesize a high amount of enzyme (e.g., DHFR) survive. The overproduction of the specific enzyme is often linked to gene amplification so that this most likely results in an increased expression of the gene of interest [4]. During the past years several approaches for site-specific gene integration systems were published and the specific integration at transcription-active sites (hot spots) was discussed in order to simplify the screening process for good producers. Kito *et al.*, for example, used the Cre/loxP recombination system and GFP as a reporter protein to identify high-producing cell lines [55], while Wiberg *et al.* mediated site-specific recombination via FRT/Flp-In recombinase system, thereby reducing genomic position effects [56]. Homologous recombination can be facilitated by the activation of cellular repair mechanisms due to the introduction of site-specific DNA double-strand breaks (DSBs) by custom-designed zinc finger nucleases (ZFNs). In order to cope with this genomic damage, the cell either ligates the resulting ends of the strands (with mutagenic or nonmutagenic result) or applies homologous recombination to repair the damage. Further ZFNs can be utilized to integrate extrachromosomal segments at specific sites via the occurring homologous recombination [57]. The knockout of specific genes in CHO cells has as well been mediated by transfection of the cells with target-specific ZFNs [58, 59].

17.5 Kinetics of Growth and Product Formation

Mammalian cells used for the production of therapeutic proteins descend from rodent (mouse or hamster) or human tissues. Dependent on their origin, they are supposed to be either "natively" specialized for protein secretion (such as myeloma cells like NS0) or show rather low native protein secretion activities like ovary (CHO) and kidney (BHK or HEK 293) cells [5]. Because of this pedigree, it may not surprise that the first recombinant protein blockbuster (annual sales >$1 billion) was produced by adherent cells, namely, erythropoietin (Epogen; Amgen, Thousand Oaks, CA, USA) in adherent CHO, cultivated in roller bottles. Today, suspension cultures are often preferred because production capacities can be scaled up to large volumes such as 20 000 l. In the group of suspended cells, CHO and mouse-myeloma-derived NS0 [60] are the key players in addition to BHK, HEK-293, and PER-C6 production hosts. It is noteworthy that the originally adherent cells need to be adapted to single-cell suspension growth prior to their application as recombinant protein producers. While this process was tedious and difficult in the early days of cell culture technology (in the 1980s), commercial media compositions ease transition today [10].

Presumably because of their origin, cell-specific, recombinant protein formation in mammals occurs partially growth-coupled with a significant, even dominating fraction of growth-decoupled protein production [61]. To capitalize this property, temperature shift-down approaches (37 to 28–31 °C) aim at arresting cells in the G1 phase, thus exploiting their increased metabolic activity for improved recombinant protein production [62]. Wurm [10] outlined that suspension cultures typically achieved approximately 10 pg cell^{-1} day^{-1} in 1986 and were expected to perform approximately 90 pg cell^{-1} day^{-1} in 2004. Recently, 30–35 pg cell^{-1} day^{-1} were published for CHO producers using chemically defined media [63], 55 pg cell^{-1} day^{-1} for a CHO antibody high producer [64], and even greater than 100 pg cell^{-1} day^{-1} for a CHO derivative with overexpressed taurine transporter [65]. The disclosures give rise to the assumption that best performance data of industrial systems are even higher. Related product formation models consider the growth-decoupled nature of therapeutic protein formation by assuming constant, cell-specific product formation rates that may consider inhibiting impacts of lactate or ammonia [66–68]. Typically, these models focus on the metabolic availability of internal substrates like amino acids excluding putative regulations at the transcriptional or posttranslational level or along the protein secretion machinery.

To achieve highest productivities in production scale, suspension processes are performed either in fed-batch or in cell-retention mode, the latter usually termed perfusion cultures. Figure 17.2 represents an exemplary, simplified scheme of some substrate, product, and by-product kinetics observed in suspended fed-batch cultures. Roughly, the process can be divided into the parts (I) and (II) describing cell growth on glucose and glutamine (sometimes called glucoglutaminolysis, [69], phase I) and glucose-based production (phase II). Characteristic of the phase (I) is the accumulation of lactate, a well-known

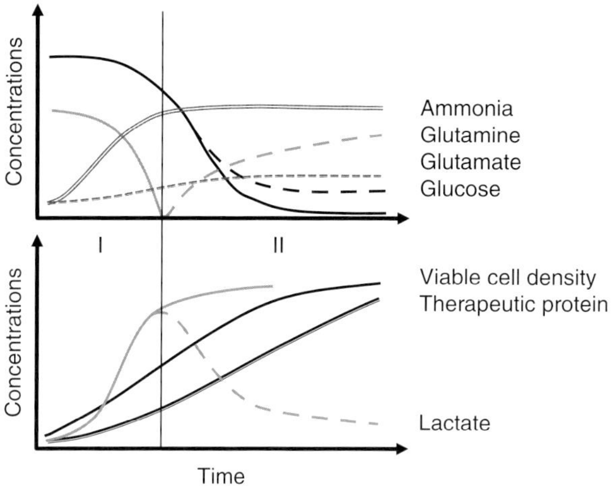

Figure 17.2 Exemplified time series of some substrates, products, and by-products of a CHO fed-batch culture feeding glucose (and glutamine) and producing the therapeutic protein. For the sake of brevity, dynamics of uptake and release of other amino acids were excluded. Roughly two phases can be identified (I and II) that distinguish by lactate, glutamine, and ammonia kinetics. Dotted lines outline putative variations of the said components. It is assumed that a CHO DHFR production system is used, which explains the slight accumulation of glutamate. If CHO producers possess the GS genotype, glutamine would be synthesized internally via glutamine synthetase using glutamate as external substrate. HPLC, high-performance liquid chromatography; HPLC-MS/MS, high-performance liquid chromatography-tandem mass spectrometry; GC-MS, gas chromatography-mass spectrometry; NMR, nuclear magnetic resonance.

phenomenon for actively proliferating cells [70, 71]. Lactate formation basically mirrors the excess of nicotineamide adenine dinucleotide (NADH) due to high glucose uptake rates, relatively low pyruvate drain into mitochondrial TCA (citrate cycle), and high lactate dehydrogenase (ldhA) activity in the cytoplasm. If NADH and/or glucose supply becomes limiting, lactate formation can stop or the acid can even be consumed again, partially substituting glucose uptake [72]. The latter is of importance, as significant growth reductions (25–50%) were observed in the presence of lactate (50–60 mM) for hybridoma or CHO cultures [73, 74]. However, lactate consumption (after previous formation) coincided with reduced cell-specific antibody formation in CHO cultures [75].

Ammonia (NH_4^+) often accumulates in phase (I) together with the consumption of glutamine. Mammalian cells are able to produce glutamine via GS, an ATP-dependent enzyme with relatively low, native activity [5]. To prevent glutamine shortage, the di-amino acid is fed to the cultures. However, because equilibrated feeding compared to cellular needs is hard to install, the surplus of glutamine is converted to glutamate, which serves as an intracellular amino donor finally releasing NH_3. The inhibitory impact of ammonia is known to be more sensitive for cell growth than lactate [76]. Growth inhibition occurs at moderate single-digit millimolar levels and may even affect cell-specific productivity [77]. To prevent the accumulation of nonwanted by-products, several feeding strategies partially replacing the classical substrates have already been developed [78–80].

Mammalian growth usually shows respiratory quotients close to 1, that is, carbon dioxide formation (CER) equals oxygen uptake rate (OUR). With respect to the relatively poor mass transfer conditions installed in cell culture reactors (compared to microbial conditions), dissolved CO_2 (dCO_2) levels can accumulate during fed-batch processes. While optimum dCO_2 concentrations appear to be cell-specific, growth inhibiting partial pressures of approximately 100 mbar had been identified for, for example, hybridoma cells [81].

17.6 Intracellular Metabolome Analysis

Within the concert of omics-technologies, metabolomics surely play an important role in mirroring the key elements of metabolic control and for giving insight into the "downstream" consequences of "upstream" regulations occurring, for example, at the transcriptional or posttranslational level. Therefrom, the necessity to focus system studies on the metabolome was outlined early, introducing metabolic fingerprinting or profiling approaches [82, 83] together with the analysis of metabolite dynamics [84].

Early sample protocols for metabolome analysis concentrated on yeasts [85], outlining the need to wash cells appropriately after filtration to eliminate the nonwanted carryover of media components. de Koning and van Dam [86] further developed yeast-based sampling protocols by introducing cell quenching for rapid metabolism inactivation. Sample suspension was sprayed into a precooled methanol solution – an approach that turned out to find numerous followers applying the same protocol for microbial and mammalian cells. The attractiveness of cold-methanol quenching (typically: 60% v/v methanol in water at −40 °C) is obviously given by its easy applicability to stop intracellular metabolism quickly [87, 88] – a prerequisite for analyzing metabolism dynamics uncoupled from nonwanted sampling-imposed impacts. While the quick inactivation of metabolism is certainly of outmost importance for highly active microbial metabolism showing glycolytic pool turnover times in second and subsecond scales, pool turnover times are lower in mammalian cells, which can be estimated with respect to intracellular pool sizes and fluxes published by [89–91]. Consequently, this phenomenon somewhat alleviates sample processing compared to the strict limitation in time when analyzing microbes.

Beside studies of Wittmann *et al.* [92], Wellerdiek *et al.* [93] and Villas-Boas *et al.* [94] revealed that cold methanol quenching causes nonwanted metabolite leakage in bacteria such as *E. coli* or *Corynebacterium glutamicum* and (less pronounced) in yeasts. Leaking metabolites were also detected for CHO cells applying cold methanol quenching [95]. As a result of their comparative study, the authors outlined that quenching with 5 °C NaCl (0.9% w/v) should be preferred because metabolite leakage was prevented and ATP conversion obviously halted.

On the other hand, Sellick *et al.* [96–98] studied the applicability of cold-methanol quenching for GS-CHO cells, finally concluding that CHO quenching in −40 °C cold 60% methanol supplemented with 0.85% (w/v) ammonium bicarbonate (pH 7.4) resulted in representative metabolite patterns.

An alternative to cell quenching was published by Wiendahl *et al.* [99]. The authors presented a micro heat exchanger, quickly sampling the cell suspension via a cooling device (at −4 °C) into a vacuum chamber. Furthermore, the fast filtration technique was proposed by Volmer *et al.* [100]. Provided that cells were washed with an iso-osmotic NaCl solution at 0 °C, CHO DP12 cells showed physiological adenylate energy charge levels. Recently, Wurm and Zeng [101] presented a mechanical disruption technique using micro nozzles. Computational fluid dynamic (cfd) simulations anticipated that mammalian cell disruption should quickly occur as a first step of automated intracellular metabolome analysis.

Compared to suspended CHO, metabolome analysis of adherent cells appears to be easier as cell/medium separation can be quickly performed by medium suction off the six-well plates used for cultivating the same [102, 103]. Next, washing with iso-osmotic salt solution is typically performed. Then, cells are prepared for *in situ* extraction, simply adding the appropriate extractant to the plates (Figure 17.3).

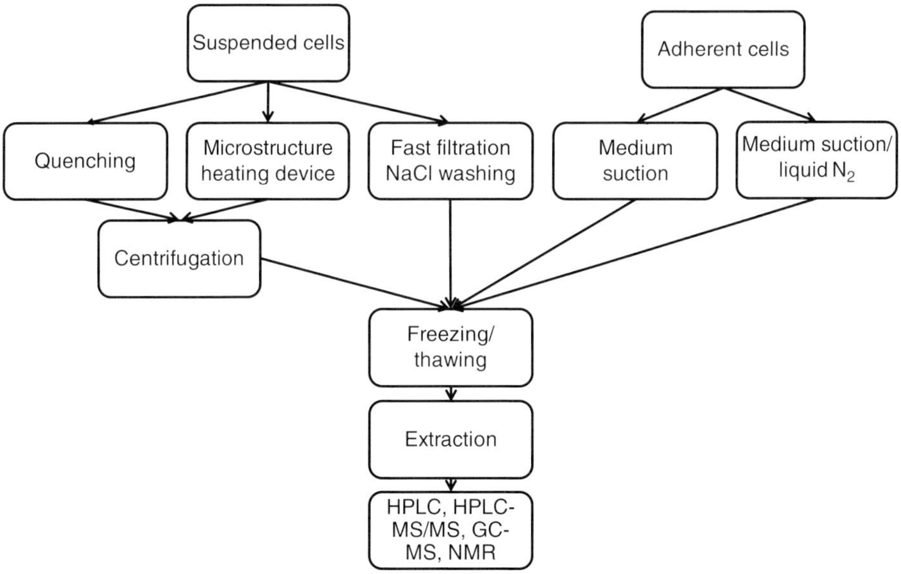

Figure 17.3 Overview of sample processing protocols to access intracellular metabolome patterns. Metabolism of suspended cells is rapidly halted by cooling using either quenching fluids or micro heat exchanger. Alternatively, fast filtration that prevents additional centrifugation for cell/liquid separation may be used. The growth medium of adherent cells (grown, e.g., in six-well plates) can be easily sucked (also in combination with liquid nitrogen addition for metabolism stop). Typically, cells are frozen prior to extraction which is followed by metabolite analytics.

To extract intracellular metabolites, numerous methods have been evaluated in the past for microbial and mammalian applications. Common extractants are perchloric acid, formic acid, potassium hydroxide, several detergents, or organic solvents such as methanol, ethanol, acetonitrile, or chloroform. A comprehensive study was performed by Maharjan and Ferenci [104] for gram-negative *E. coli.* Testing acid or alkaline conditions, membrane permeabilization by freezing in methanol, high-temperature extraction with ethanol or methanol, or chloroform-methanol treatment, the authors outlined the sensitivity of the extraction pattern based on the applied method. While cold methanol extraction was favored, no phosphorylated glycolytic compounds or nucleoside phosphates (like ATP, ADP, or AMP) could be detected. A similar study was performed by Villas-Boas *et al.* [105] focusing on yeast. A basic problem is the instability of reduced redox equivalents such as NADH and NADPH under acid conditions and the instability of their oxidized counterparts NAD^+ and $NADP^+$ at alkaline pH. Hence, measuring valid intracellular levels for each species still remains a challenge. Ritter *et al.* [103] tested several approaches for their applicability using mammalian cells (here: Madin–Darby canine kidney, MDCK). They concluded that methanol/chloroform or methanol boiling should be favored. Focusing on the adherent β-cell line INS-1, Lorenz *et al.* [106] supported this finding by outlining the quality of a 9 : 1 methanol:chloroform extractant mixture that allowed metabolic fingerprinting via HPLC-MS. It is of note that they quenched cellular metabolism by adding liquid nitrogen directly to the culture dishes. On the other hand, Danielsson *et al.* [107] developed a sample processing protocol for adherent clonal rat insulinoma cells (INS-1832/13), finally concluding that extraction with 82% (v/v) methanol is applicable for metabolome analysis.

All extraction methods have in common that intracellular metabolite patterns are measured by subsequent analysis typically applying HPLC, HPLC-MS/MS, or GC-MS protocols described elsewhere. Furthermore, it may be interesting to emphasize that organelle-specific metabolome approaches are still rare, yet one may say even not existing so far. One exception is the fluorescence resonance energy transfer (FRET) approach published by Imamura *et al.* [108]. The authors managed to visualize intracellular ATP levels in mitochondria, indicating surprisingly low levels compared to the cytoplasm. In general, it will be attractive to see how image-based techniques will develop for metabolome analysis in mammalian cells [109].

Presumably because of the sophisticated technologies needed to achieve valid data of intracellular metabolic patterns, metabolic fingerprinting and profiling approaches are mainly applied for extracellular studies so far, that is, for the analysis of the medium composition as a function of cultivation time. Schaub *et al.* [11] demonstrated that these data are a highly valuable part of multivariate data analysis for successful strain and process optimization. Metabolic profiling may also play an important role in elucidating basic apoptosis mechanisms, as shown by Chong *et al.* [110] who succeeded in identifying metabolic apoptosis inducers (such as oxidized glutathione, AMP, and GMP). Recently, Vernardis *et al.* [111]

showed that metabolic shifts of BHK cells cultivated in perfusion reactors result in similar (mostly intracellular and extracellular) metabolite patterns independent of the cultivation scale. Sellick *et al.* [98] successfully applied intra- and extracellular metabolic profiling for optimizing nutrient feeding during antibody production with CHO cells.

Other tools for detailed intracellular studies are metabolic flux analysis (MFA) or flux balance analysis (FBA). MFA is applicable for determined or overdetermined systems, that is, when the degree of freedom, given as the number of fluxes to be calculated minus the number of independent balance equations, is ≤ 0. Usually, MFA requires the measurement of the majority of amino acid exchange rates plus glucose uptake, biomass, product, and CO_2 formation to determine small- to mid-size stoichiometric models of central metabolism [112]. More detailed models can be identified if ^{13}C labeling information, originating from the catabolism of labeled substrates (such as glucose), is tracked in metabolites or in the cell mass for flux estimations [113]. If measured data are not sufficient to determine models, mathematical linear programming approaches that search for optimum solutions in a parameter space considering given, for example, growth maximization constraint can be applied. The latter is usually applied as FBA. There are numerous, successful applications of MFA and FBA for mammalian cell cultures. Early studies concentrated on hybridoma cultures [114, 115]. Applications range from typical multiple-liter laboratory analysis to mammalian cell screening in 24-well plates with simplified metabolic networks [37]. Recently, excellent reviews were given by Quek *et al.* [69], Niklas *et al.* [116], Boghigian *et al.* [117], Wahrheit *et al.* [118], Ahn and Antoniewicz [119], and Niklas and Heinzle [120]. Because they already cover a large field of MFA applications, only some few items are highlighted in the following.

One particularity of MFA in mammalian cells is its compartmentalization, making it necessary to consider exchange fluxes via shuttle systems for instance between mitochondria and cytoplasm. Obviously, the consideration of an additional compartment increases the model complexity, which may result in an underdetermined system. To enable identifiability of the model, one may simplify the same by ignoring cellular compartments, locating all reactions in a "pseudo-prokaryotic" manner. However, what is the impact on metabolic flux patterns?

Own works [121] studied interleukin 8 producing CHO DP12 cells (biological triplicates) and estimated intracellular flux distributions during growth on glucose and glutamine analyzing amino acids, substrates, biomass, and product formation. The noncompartmented model consisted of 73 transformers (reaction rates and transport step), balancing 53 metabolites. The compartmented approach was built on 95 transformers and 75 metabolites. While most of the simulated fluxes were similar in both approaches, a fraction of approximately 15% of totally available ATP was missing in the compartmented model compared to the noncompartmented approach. This was due to the assumed activity of the citrate-pyruvate shuttle that imports pyruvate (via pyruvate/H^+ symporter) into mitochondria by exporting citrate (via citrate/malate antiporter). In cytoplasm, citrate is further

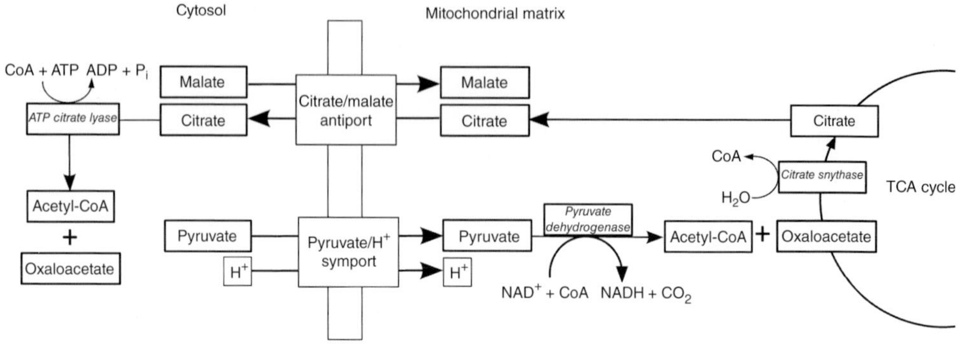

Figure 17.4 Schematic view of citrate availability in mitochondria and in the cytosol. Citrate, which is produced in the TCA cycle inside mitochondria without ATP consumption, is exported into the cytosol using the citrate/malate antiport shuttle. Next, citrate is further catabolized via ATP-dependent citrate lyase in the cytosol to produce acetyl-CoA and oxaloacetate. Hence, cellular compartmentalization finally yields at net ATP need for citrate production in the cytosol, while simplifying models, only concentrating on TCA, ignore this existing ATP demand.

catabolized via ATP-citrate-lyase that demands for ATP (Figure 17.4). In mitochondria, citrate is metabolized in the citric acid cycle without ATP consumption. Hence, a simplified, "prokaryote-like" approach ignores the additional ATP need, finally overestimating ATP availability.

Another hot topic for MFA analysis is the monitoring of changing metabolic flux patterns when mammalian metabolism shifts from lactate production to lactate consumption. CHO perfusion culture studies revealed that during "glucoglutaminolysis," glucose uptake outcompeted glutamine consumption by a factor of 5. Forty-one percentage of glucose was channeled through the pentose-phosphate pathway and the pyruvate drain to lactate formation or to TCA cycle was almost equilibrated [90]. Obviously, this scenario changes significantly when cells are cultivated in the late-nongrowth phase (day 17 of a CHO-fed-batch culture) [122]. The majority of glucose was directed through the pentose-phosphate pathway, generating NADPH, and most of the pyruvate from glycolysis entered the TCA cycle coinciding with no or low-lactate formation. This picture was further fine-tuned by MFA studies of Dean and Reddy [123]. Investigating mAB-producing CHOs, they found that glutamine contributes significantly to anaplerosis and to lactate formation during "glucoglutaminolysis" and that cells shift to glucose-based TCA fueling with decreasing growth rates. Considering that high lactate concentrations may inhibit phosphofructokinase and may force its reconsumption, Mulukutla *et al.* [124] formulated a mechanistic model for mouse myeloma cells (NS0). Following the rationale of the approach, the basic tendencies of the lactate production/consumption shift can be well explained, further outlining the key role of AKT1 and P53 signaling pathway as mediators in the regulatory scheme.

17.7
Proteome and Gene Expression Analysis

Reviewing proteomic and genomic studies on mammalian cells, Gupta and Lee [125] stated that these tools were rarely used, despite their potential for unraveling basic regulatory mechanisms in the cells. One obvious reason is that the DNA sequence of CHO-K1 was published not earlier than 2011 [41]. Another reason is the high degree of complexity that hampered the analysis of the 2.45 Gb genome (coding for ~24,300 genes) more than related studies on, for example, prokaryotes like *E. coli* W3110 with 4.65 Mb coding for approximately 4,300 genes. However, it was already obvious that techniques typically used for the analysis of microbes should also be adaptable to mammalian cells [126].

Indeed, Pascoe *et al.* [75] succeeded in applying proteome analysis for comparing the CHO cultures that consumed lactate (after previous production) with those that did not metabolize it. Among those proteins that significantly alter their expression, the central metabolism enzymes enolase, pyruvate kinase, and glyceraldehyde-3-phosphate dehydrogenase were identified. Besides, the authors argued that process conditions may have a significant impact on the observed differences in the proteome patterns.

The analysis of expression profiles of Schaub *et al.* [127] outlined that transcript studies are well applicable to optimize, for example, media composition for CHO cell lines. The study focused on fed-batch cultures of high and low producers, applying different media. By analyzing the transcript time series, differences between the cell lines were observed and hints for optimizing the synthetic medium could be derived. Moreover, using proprietary CHO microarrays, the transcript courses of an industrial Fc-fusion protein CHO producer were monitored under hyperosmotic stress conditions imposed in batch and fed-batch cultivation [128]. The goal of the study was to identify sensitive gene expression patterns that may be exploited for further increasing cell-specific productivity. It is noteworthy that the latter is typically enhanced under hyperosmotic conditions as a consequence of slowed down or even stopped cell proliferation. The comparison of the resulting gene expression patterns showed a multitude of changing gene transcripts, which had only a small intersection with preliminary studies using hybridoma cells. Therefore, cellular response is obviously very complex and somewhat unique for the cell lines.

This rather general statement is also supported by the multi-omics studies of Dietmair *et al.* [129]. Studying "high" and "low" HEK293 producers via transcriptome, metabolome, and fluxome (FBA) analysis, the high producers revealed (surprisingly) downregulated gene expression profiles together with reduced metabolic activity. Obviously, producer cells managed to synthesize the therapeutic protein more efficiently than the low producers.

Besides, regulatory networking in mammalian cells is even more complex because miRNAs obviously play important, but so far not unraveled, roles. Clarke *et al.* [130] outlined that mRNA processing and protein synthesis are correlated with growth in CHO cells. In addition, miRNAs are key elements of posttranscriptional regulation. Fifty-one miRNAs were found to be associated with increasing growth. Distinct functions, however, are not known for most of them. Therefore, it will certainly be a matter of future studies to elucidate the multifunctional relationship between miRNA and gene expression.

17.8
Improving Cellular Performance by Genetic and Metabolic Engineering

Research activities for improving the performance of production hosts mainly focus on two criteria: (i) the increase of cell-specific productivity and (ii) the optimization of product quality. While item (i) aims at keeping high levels of cell viability, also optimizing protein folding, assembly, and secretion, item (ii) addresses glycoengineering which – in terms of antibody formation – is meant for further improving antibody-dependent cellular cytotoxicity (ADCC) as a property of antibody potency. It is noteworthy that related research is rarely termed *metabolic engineering* (with the exception of the production of heparan sulfate, a precursor of drug heparin in engineered CHO, [131]) but cell or genetic engineering instead. Nevertheless, taking into account one of the key definitions of Bailey [132] defining metabolic engineering as "… the directed improvement of cellular properties through the modification of specific biochemical reactions or introduction of new ones using recombinant DNA technology …," this term may be applied more frequently for the approaches used.

Establishing constant, high levels of cell viability during production processes is of crucial importance. It usually addresses the prevention of apoptosis (the programmed cell death) and/or autophagy (the formation of autophagosomes for replacing organelles). Current studies on autophagy concentrate on external stimuli and internal sequences [3, 133]. Recently, Jardon *et al.* [134] managed to inhibit autophagy by addition of 3-methyl adenine, increasing the tissue-plasminogen activator (t-PA) yield 2.8-fold. On the other hand, for prevention of apoptosis, some tools have already been developed in order to maintain high cell viability. Presumably, the most successful approach considered the overexpression of anti-apoptotic B-cell lymphoma (Bcl) family proteins [135, 136]. Besides, downregulating activities of key members of the pro-apoptotic caspase family were reported [137].

According to the common understanding, mAB production in mammalian cells is not limited by the mRNA levels [51], but by mRNA translation, protein formation, folding, assembling, and/or secretion. Dinnis and James [138] pointed out that significant improvements of the protein production machinery (downstream of transcription) rather demand for a Systems Engineering effort than for alleviating individual steps. A successful example is given by Tigges and Fussenegger [139] who succeeded in expressing the human transcription factor Xbp1 (X-box-binding-protein 1) in CHO-K1 cell lines, finally achieving increased productivities by a multitarget approach. Further findings illustrating the regulatory complexity inside the protein formation machinery were given by

O'Callaghan *et al.* [140]. Their analysis of seven CHO producer cell lines based on data-driven, in-depth protein formation models revealed that sufficient light chain (and also heavy chain) mRNA was produced but downstream protein formation was hampered for individual reasons in the cells.

Barron *et al.* [141] were among the first who addressed miRNAs for evaluating their potential to amplify protein formation. Because temperature shift-down is a well-known approach for controlling cell proliferation (coinciding with increased productivity), the authors monitored related miRNA levels. Of the four miRNAs that were significantly downregulated, miR-7 was selected for dosage increase via transfection into CHO. As a result, cell proliferation was surprisingly blocked and increased productivity observed. In addition, other miRNAs were found to play an important role in protein formation. Comprehensive miRNA screening studies of Strotbek *et al.* [142] identified two human-derived candidates, namely, miR-557 and miR-1287, which increased cell viability and productivity for IgG-producing CHO. Recently, Druz *et al.* [143] investigated the impact of silencing mmu-miR-466h-5p in CHO. The pro-apoptotic role of this miRNA had been identified before. Therefore, reducing the intracellular level of mmu-miR-466h-5p should result in a delayed apoptosis start. Indeed, anti-miR-466h CHO showed the expected phenotype of extended viability together with a delayed activation of caspase-3/7, an important player of the apoptosis cascade.

Hence, the interplay of miRNA levels with protein formation surely represents a highly attractive research field for improving cellular performance, at the same time asking for basic research to unravel underlying regulatory mechanisms.

As outlined earlier, by-products such as lactate can inhibit cell-specific productivity as well. Consequently, multiple efforts were made to reduce lactate production either by increasing the drain from the precursor pyruvate [144, 145] or by reducing ldh activity [146, 147].

Besides, Tabuchi and Sugiyama [65] recently achieved maximum cell-specific mAB productivities greater than 100 pg $cell^{-1}$ day^{-1} with a DHFR-deficient CHO DUK-XB11 cell line overexpressing the taurine transporter TAUT and the alanine aminotransferase 1 ALT1 for amplifying the internal pyruvate and glutamate supply.

Glycosylation is one of the most important posttranslational modifications of therapeutic proteins as it affects biological activity, *in vivo* stability, and immunogenicity. In the field of glycoengineering, glycosylation is modified in order to improve therapeutic proteins. The introduction of new N-linked glycosylation consensus sequences in the proteins Mpl ligand, leptin, and recombinant human erythropoietin (rHuEPO) resulted in hyperglycosylated proteins which showed increased *in vivo* activity and prolonged half-life. The effect of darbepoetin alfa (DA), a variant of rHuEPO which contained two additional sialic acid containing N-linked carbohydrates, was validated in clinical tests. [148]. A huge increase in half-life was as well shown for hyperglycosylated interferon-α variants [149]. Defucosylation of the Fc regional oligosaccharides of antibodies improved ADCC, which is important for the clinical efficacy in anticancer therapy [150]. Therefore, an enhancement of ADCC supports the development of new therapeutic antibodies required for anticancer therapy.

17.9 Outlook

With the publication of the first CHO genome sequence (CHO-K1) [41], surely a new era of systems-based analysis of mammalians has started. Before that, proprietary microarrays were predominately used, inherently limiting the cost-intensive applications to industrial research. Now, making use of next-generation sequencing for DNA analysis, the access to sequence-based studies, including transcript and miRNA approaches, is much more simplified giving rise to the assumption that Systems Biology studies will intensify in the future [126, 151]. It is likely that this systems view will unravel so far uncovered regulatory mechanisms, finally also enabling Metabolic Engineering approaches for cell improvements.

Another impulse can be expected from Synthetic Biology. Already, artificial regulatory circuits and switches are successfully implemented in mammalian cells [152–155]. It is likely that some of the approaches will find their way into production systems to enhance the formation of therapeutic proteins.

References

1. Aggarwal, S.R. (2012) What's fueling the biotech engine – 2011 to 2012. *Nat. Biotechnol.*, **30**, 1191–1197.
2. Walsh, G. (2010) Biopharmaceutical benchmarks 2010. *Nat. Biotechnol.*, **28**, 917–924.
3. Butler, M. and Meneses-Acosta, A. (2012) Recent advances in technology supporting biopharmaceutical production from mammalian cells. *Appl. Microbiol. Biotechnol.*, **96**, 885–894.
4. Jayapal, K.R., Wlaschin, K.F., Hu, W.S., and Yap, M.G.S. (2007) Recombinant protein therapeutics from CHO cells – 20 years and counting. *Chem. Eng. Prog.*, **103**, 40–47.
5. Seth, G., Hossler, P., Yee, J.C., and Hu, W.S. (2006) Engineering cells for cell culture bioprocessing – physiological fundamentals. *Adv. Biochem. Eng. Biotechnol.*, **101**, 119–164.
6. Puck, T.T., Cieciura, S.J., and Robinson, A. (1958) Genetics of somatic mammalian cells. III. Long-term cultivation of euploid cells from human and animal subjects. *J. Exp. Med.*, **108**, 945–956.
7. Graham, F.L., Smiley, J., Russell, W.C., and Nairn, R. (1977) Characteristics of a human cell line transformed by DNA from human adenovirus type 5. *J. Gen. Virol.*, **36**, 59–74.
8. Shulman, M., Wilde, C.D., and Kohler, G. (1978) A better cell line for making hybridomas secreting specific antibodies. *Nature*, **276**, 269–270.
9. Niklas, J., Schrader, E., Sandig, V., Noll, T. *et al.* (2011) Quantitative characterization of metabolism and metabolic shifts during growth of the new human cell line AGE1.HN using time resolved metabolic flux analysis. *Bioprocess. Biosyst. Eng.*, **34**, 533–545.
10. Wurm, F.M. (2004) Production of recombinant protein therapeutics in cultivated mammalian cells. *Nat. Biotechnol.*, **22**, 1393–1398.
11. Schaub, J., Clemens, C., Kaufmann, H., and Schulz, T.W. (2012) Advancing biopharmaceutical process development

by system-level data analysis and integration of omics data. *Adv. Biochem. Eng. Biotechnol.*, **127**, 133–163.
12. Berezney, R. (1991) The nuclear matrix: a heuristic model for investigating genomic organization and function in the cell nucleus. *J. Cell. Biochem.*, **47**, 109–123.
13. Nasmyth, K. (1996) Viewpoint: putting the cell cycle in order. *Science*, **274**, 1643–1645.
14. Malumbres, M. and Barbacid, M. (2005) Mammalian cyclin-dependent kinases. *Trends Biochem. Sci*, **30**, 630–641.
15. Darnell, J.E. Jr. (1982) Variety in the level of gene control in eukaryotic cells. *Nature*, **297**, 365–371.
16. Doudna, J.A. and Rath, V.L. (2002) Structure and function of the eukaryotic ribosome: the next frontier. *Cell*, **109**, 153–156.
17. Scheffler, I.E. (2001) A century of mitochondrial research: achievements and perspectives. *Mitochondrion*, **1**, 3–31.
18. Passarella, S., Atlante, A., Valenti, D., and de Bari, L. (2003) The role of mitochondrial transport in energy metabolism. *Mitochondrion*, **2**, 319–343.
19. Mannella, C.A. (1992) The 'ins' and 'outs' of mitochondrial membrane channels. *Trends Biochem. Sci*, **17**, 315–320.
20. Shi, Y. (2001) A structural view of mitochondria-mediated apoptosis. *Nat. Struct. Biol.*, **8**, 394–401.
21. Voeltz, G.K., Rolls, M.M., and Rapoport, T.A. (2002) Structural organization of the endoplasmic reticulum. *EMBO Rep.*, **3**, 944–950.
22. Chen, S., Novick, P., and Ferro-Novick, S. (2013) ER structure and function. *Curr. Opin. Cell Biol.*, **25**, 428–433.
23. Braakman, I. and Bulleid, N.J. (2011) Protein folding and modification in the mammalian endoplasmic reticulum. *Annu. Rev. Biochem.*, **80**, 71–99.
24. Niwa, M. and Zhong, W. (2012) Membrane compartments and organelles: new tricks by old dogs and new kids on the block. *Curr. Opin. Cell Biol.*, **24**, 449–452.
25. Nakamura, N., Wei, J.H., and Seemann, J. (2012) Modular organization of the mammalian Golgi apparatus. *Curr. Opin. Cell Biol.*, **24**, 467–474.
26. Bainton, D.F. (1981) The discovery of lysosomes. *J. Cell Biol.*, **91**, 66s–76s.
27. Winchester, B.G. (2001) Lysosomal membrane proteins. *Eur. J. Paediatr. Neurol.*, **5** (Suppl. A), 11–19.
28. Stevens, T.H. and Forgac, M. (1997) Structure, function and regulation of the vacuolar (H+)-ATPase. *Annu. Rev. Cell Dev. Biol.*, **13**, 779–808.
29. Yee, J.C., Jacob, N.M., Jayapal, K.P., Kok, Y.J. *et al.* (2010) Global assessment of protein turnover in recombinant antibody producing myeloma cells. *J. Biotechnol.*, **148**, 182–193.
30. Masters, C. and Crane, D. (1996) Recent developments in peroxisome biology. *Endeavour*, **20**, 68–73.
31. Smith, J.J. and Aitchison, J.D. (2013) Peroxisomes take shape. *Nat. Rev. Mol. Cell Biol.*, **14**, 803–817.
32. Poirier, Y., Antonenkov, V.D., Glumoff, T., and Hiltunen, J.K. (2006) Peroxisomal beta-oxidation – a metabolic pathway with multiple functions. *Biochim. Biophys. Acta*, **1763**, 1413–1426.
33. Nielsen, J. and Jewett, M.C. (2008) Impact of systems biology on metabolic engineering of *Saccharomyces cerevisiae. FEMS Yeast Res.*, **8**, 122–131.
34. Vander Heiden, M.G., Cantley, L.C., and Thompson, C.B. (2009) Understanding the Warburg effect: the metabolic requirements of cell proliferation. *Science*, **324**, 1029–1033.
35. Glacken, M.W. (1988) Catabolic control of mammalian-cell culture. *Bio/Technology*, **6**, 1041–1050.
36. Hinkle, P.C. (2005) P/O ratios of mitochondrial oxidative phosphorylation. *Biochim. Biophys. Acta*, **1706**, 1–11.
37. Balcarcel, R.R. and Clark, L.M. (2003) Metabolic screening of mammalian cell cultures using well-plates. *Biotechnol. Progr.*, **19**, 98–108.
38. Atkinson, D.E. and Walton, G.M. (1967) Adenosine triphosphate conservation in metabolic regulation. Rat liver citrate cleavage enzyme. *J. Biol. Chem.*, **242**, 3239–3241.

39. Ataullakhanov, F.I. and Vitvitsky, V.M. (2002) What determines the intracellular ATP concentration. *Biosci. Rep.*, **22**, 501–511.
40. Partridge, M.A., Davidson, M.M., and Hei, T.K. (2007) The complete nucleotide sequence of Chinese hamster (*Cricetulus griseus*) mitochondrial DNA. *DNA Seq.*, **18**, 341–346.
41. Xu, X., Nagarajan, H., Lewis, N.E., Pan, S. *et al.* (2011) The genomic sequence of the Chinese hamster ovary (CHO)-K1 cell line. *Nat. Biotechnol.*, **29**, 735–741.
42. Lewis, N.E., Liu, X., Li, Y., Nagarajan, H. *et al.* (2013) Genomic landscapes of Chinese hamster ovary cell lines as revealed by the Cricetulus griseus draft genome. *Nat. Biotechnol.*, **31**, 759–765.
43. Brinkrolf, K., Rupp, O., Laux, H., Kollin, F. *et al.* (2013) Chinese hamster genome sequenced from sorted chromosomes. *Nat. Biotechnol.*, **31**, 694–695.
44. Lee, R.C., Feinbaum, R.L., and Ambros, V. (1993) The *C. elegans* heterochronic gene lin-4 encodes small RNAs with antisense complementarity to lin-14. *Cell*, **75**, 843–854.
45. Wightman, B., Ha, I., and Ruvkun, G. (1993) Posttranscriptional regulation of the heterochronic gene lin-14 by lin-4 mediates temporal pattern formation in *C. elegans*. *Cell*, **75**, 855–862.
46. Reinhart, B.J., Slack, F.J., Basson, M., Pasquinelli, A.E. *et al.* (2000) The 21-nucleotide let-7 RNA regulates developmental timing in *Caenorhabditis elegans*. *Nature*, **403**, 901–906.
47. Pasquinelli, A.E., Reinhart, B.J., Slack, F., Martindale, M.Q. *et al.* (2000) Conservation of the sequence and temporal expression of let-7 heterochronic regulatory RNA. *Nature*, **408**, 86–89.
48. Kozomara, A. and Griffiths-Jones, S. (2011) miRBase: integrating microRNA annotation and deep-sequencing data. *Nucleic Acids Res.*, **39**, D152–D157.
49. Hackl, M., Jakobi, T., Blom, J., Doppmeier, D. *et al.* (2011) Next-generation sequencing of the Chinese hamster ovary microRNA transcriptome: Identification, annotation and profiling of microRNAs as targets for cellular engineering. *J. Biotechnol.*, **153**, 62–75.
50. Barron, N., Sanchez, N., Kelly, P., and Clynes, M. (2011) MicroRNAs: tiny targets for engineering CHO cell phenotypes? *Biotechnol. Lett.*, **33**, 11–21.
51. Birch, J.R. and Racher, A.J. (2006) Antibody production. *Adv. Drug Delivery Rev.*, **58**, 671–685.
52. Hamlin, J.L. and Ma, C. (1990) The mammalian dihydrofolate reductase locus. *Biochim. Biophys. Acta*, **1087**, 107–125.
53. Li, F., Vijayasankaran, N., Shen, A.Y., Kiss, R. *et al.* (2010) Cell culture processes for monoclonal antibody production. *mAbs*, **2**, 466–479.
54. Lattenmayer, C., Loeschel, M., Schriebl, K., Steinfellner, W. *et al.* (2007) Protein-free transfection of CHO host cells with an IgG-fusion protein: selection and characterization of stable high producers and comparison to conventionally transfected clones. *Biotechnol. Bioeng.*, **96**, 1118–1126.
55. Kito, M., Itami, S., Fukano, Y., Yamana, K. *et al.* (2002) Construction of engineered CHO strains for high-level production of recombinant proteins. *Appl. Microbiol. Biotechnol.*, **60**, 442–448.
56. Wiberg, F.C., Rasmussen, S.K., Frandsen, T.P., Rasmussen, L.K. *et al.* (2006) Production of target-specific recombinant human polyclonal antibodies in mammalian cells. *Biotechnol. Bioeng.*, **94**, 396–405.
57. Durai, S., Mani, M., Kandavelou, K., Wu, J. *et al.* (2005) Zinc finger nucleases: custom-designed molecular scissors for genome engineering of plant and mammalian cells. *Nucleic Acids Res.*, **33**, 5978–5990.
58. Cost, G.J., Freyvert, Y., Vafiadis, A., Santiago, Y. *et al.* (2010) BAK and BAX deletion using zinc-finger nucleases yields apoptosis-resistant CHO cells. *Biotechnol. Bioeng.*, **105**, 330–340.
59. Liu, P.Q., Chan, E.M., Cost, G.J., Zhang, L. *et al.* (2010) Generation of a triple-gene knockout mammalian cell line using engineered zinc-finger nucleases. *Biotechnol. Bioeng.*, **106**, 97–105.

60. Barnes, L.M., Bentley, C.M., and Dickson, A.J. (2000) Advances in animal cell recombinant protein production: GS-NS0 expression system. *Cytotechnology*, **32**, 109–123.
61. Altamirano, C., Cairo, J.J., and Godia, F. (2001) Decoupling cell growth and product formation in Chinese hamster ovary cells through metabolic control. *Biotechnol. Bioeng.*, **76**, 351–360.
62. Kumar, N., Gammell, P., and Clynes, M. (2007) Proliferation control strategies to improve productivity and survival during CHO based production culture: a summary of recent methods employed and the effects of proliferation control in product secreting CHO cell lines. *Cytotechnology*, **53**, 33–46.
63. Huang, Y.M., Hu, W., Rustandi, E., Chang, K. *et al.* (2010) Maximizing productivity of CHO cell-based fed-batch culture using chemically defined media conditions and typical manufacturing equipment. *Biotechnol. Progr.*, **26**, 1400–1410.
64. Templeton, N., Dean, J., Reddy, P., and Young, J.D. (2013) Peak antibody production is associated with increased oxidative metabolism in an industrially relevant fed-batch CHO cell culture. *Biotechnol. Bioeng.*, **110**, 2013–2024.
65. Tabuchi, H. and Sugiyama, T. (2013) Cooverexpression of alanine aminotransferase 1 in Chinese hamster ovary cells overexpressing taurine transporter further stimulates metabolism and enhances product yield. *Biotechnol. Bioeng.*, **110**, 2208–2215.
66. Dorka, P., Fischer, C., Budman, H., and Scharer, J. (2009) Metabolic flux-based modeling of mAb production during batch and fed-batch operations. *Bioprocess. Biosyst. Eng.*, **32**, 183–196.
67. Xing, Z., Bishop, N., Leister, K., and Li, Z.J. (2010) Modeling kinetics of a large-scale fed-batch CHO cell culture by Markov chain Monte Carlo method. *Biotechnol. Progr.*, **26**, 208–219.
68. Nolan, R.P. and Lee, K. (2011) Dynamic model of CHO cell metabolism. *Metab. Eng.*, **13**, 108–124.
69. Quek, L.E., Dietmair, S., Kromer, J.O., and Nielsen, L.K. (2010) Metabolic flux analysis in mammalian cell culture. *Metab. Eng.*, **12**, 161–171.
70. Tsao, Y.S., Cardoso, A.G., Condon, R.G., Voloch, M. *et al.* (2005) Monitoring Chinese hamster ovary cell culture by the analysis of glucose and lactate metabolism. *J. Biotechnol.*, **118**, 316–327.
71. Mulukutla, B.C., Khan, S., Lange, A., and Hu, W.S. (2010) Glucose metabolism in mammalian cell culture: new insights for tweaking vintage pathways. *Trends Biotechnol.*, **28**, 476–484.
72. Li, J., Wong, C.L., Vijayasankaran, N., Hudson, T. *et al.* (2012) Feeding lactate for CHO cell culture processes: impact on culture metabolism and performance. *Biotechnol. Bioeng.*, **109**, 1173–1186.
73. Ozturk, S.S., Riley, M.R., and Palsson, B.O. (1992) Effects of ammonia and lactate on hybridoma growth, metabolism, and antibody production. *Biotechnol. Bioeng.*, **39**, 418–431.
74. Lao, M.S. and Toth, D. (1997) Effects of ammonium and lactate on growth and metabolism of a recombinant Chinese hamster ovary cell culture. *Biotechnol. Progr.*, **13**, 688–691.
75. Pascoe, D.E., Arnott, D., Papoutsakis, E.T., Miller, W.M. *et al.* (2007) Proteome analysis of antibody-producing CHO cell lines with different metabolic profiles. *Biotechnol. Bioeng.*, **98**, 391–410.
76. Schlaeger, E.J. and Schumpp, B. (1989) in *Advances in Animal Cell Biology and Technology for Bioprocesses* (eds R.E. Spier, J.B. Griffiths, J. Stephenne, and P.J. Crooy), Butterworths, Sevenoaks, pp. 386–396.
77. Hansen, H.A. and Emborg, C. (1994) Influence of ammonium on growth, metabolism, and productivity of a continuous suspension Chinese hamster ovary cell culture. *Biotechnol. Progr.*, **10**, 121–124.
78. Altamirano, C., Paredes, C., Illanes, A., Cairo, J.J. *et al.* (2004) Strategies for fed-batch cultivation of t-PA producing CHO cells: substitution of glucose and glutamine and rational design of

culture medium. *J. Biotechnol.*, **110**, 171–179.

79. Genzel, Y., Ritter, J.B., Konig, S., Alt, R. *et al.* (2005) Substitution of glutamine by pyruvate to reduce ammonia formation and growth inhibition of mammalian cells. *Biotechnol. Progr.*, **21**, 58–69.
80. Altamirano, C., Illanes, A., Becerra, S., Cairo, J.J. *et al.* (2006) Considerations on the lactate consumption by CHO cells in the presence of galactose. *J. Biotechnol.*, **125**, 547–556.
81. deZengotita, V.M., Schmelzer, A.E., and Miller, W.M. (2002) Characterization of hybridoma cell responses to elevated pCO(2) and osmolality: intracellular pH, cell size, apoptosis, and metabolism. *Biotechnol. Bioeng.*, **77**, 369–380.
82. Fiehn, O. (2001) Combining genomics, metabolome analysis, and biochemical modelling to understand metabolic networks. *Comp. Funct. Genomics*, **2**, 155–168.
83. Fiehn, O. (2002) Metabolomics--the link between genotypes and phenotypes. *Plant Mol. Biol.*, **48**, 155–171.
84. Buchholz, A., Hurlebaus, J., Wandrey, C., and Takors, R. (2002) Metabolomics: quantification of intracellular metabolite dynamics. *Biomol. Eng*, **19**, 5–15.
85. Saez, M.J. and Lagunas, R. (1976) Determination of intermediary metabolites in yeast. Critical examination of the effect of sampling conditions and recommendations for obtaining true levels. *Mol. Cell. Biochem.*, **13**, 73–78.
86. de Koning, W. and van Dam, K. (1992) A method for the determination of changes of glycolytic metabolites in yeast on a subsecond time scale using extraction at neutral pH. *Anal. Biochem.*, **204**, 118–123.
87. Theobald, U., Mailinger, W., Reuss, M., and Rizzi, M. (1993) In vivo analysis of glucose-induced fast changes in yeast adenine nucleotide pool applying a rapid sampling technique. *Anal. Biochem.*, **214**, 31–37.
88. Theobald, U., Mailinger, W., Baltes, M., Rizzi, M. *et al.* (1997) In vivo analysis of metabolic dynamics in *Saccharomyces cerevisiae*: I. Experimental observations. *Biotechnol. Bioeng.*, **55**, 305–316.
89. Maier, K., Hofmann, U., Reuss, M., and Mauch, K. (2008) Identification of metabolic fluxes in hepatic cells from transient ^{13}C-labeling experiments: part II. Flux estimation. *Biotechnol. Bioeng.*, **100**, 355–370.
90. Goudar, C., Biener, R., Boisart, C., Heidemann, R. *et al.* (2010) Metabolic flux analysis of CHO cells in perfusion culture by metabolite balancing and 2D [^{13}C, ^{1}H] COSY NMR spectroscopy. *Metab. Eng.*, **12**, 138–149.
91. Lu, S.B., Sun, X.M., and Zhang, Y.X. (2005) Insight into metabolism of CHO cells at low glucose concentration on the basis of the determination of intracellular metabolites. *Process Biochem.*, **40**, 1917–1921.
92. Wittmann, C., Kromer, J.O., Kiefer, P., Binz, T. *et al.* (2004) Impact of the cold shock phenomenon on quantification of intracellular metabolites in bacteria. *Anal. Biochem.*, **327**, 135–139.
93. Wellerdiek, M., Winterhoff, D., Reule, W., Brandner, J. *et al.* (2009) Metabolic quenching of *Corynebacterium glutamicum*: efficiency of methods and impact of cold shock. *Bioprocess Biosyst. Eng.*, **32**, 581–592.
94. Villas-Boas, S.G., Moxley, J.F., Akesson, M., Stephanopoulos, G. *et al.* (2005) High-throughput metabolic state analysis: the missing link in integrated functional genomics of yeasts. *Biochem. J*, **388**, 669–677.
95. Dietmair, S., Timmins, N.E., Gray, P.P., Nielsen, L.K. *et al.* (2010) Towards quantitative metabolomics of mammalian cells: development of a metabolite extraction protocol. *Anal. Biochem.*, **404**, 155–164.
96. Sellick, C.A., Hansen, R., Maqsood, A.R., Dunn, W.B. *et al.* (2009) Effective quenching processes for physiologically valid metabolite profiling of suspension cultured Mammalian cells. *Anal. Chem.*, **81**, 174–183.
97. Sellick, C.A., Hansen, R., Stephens, G.M., Goodacre, R. *et al.* (2011)

Metabolite extraction from suspension-cultured mammalian cells for global metabolite profiling. *Nat. Protoc.*, **6**, 1241–1249.

98. Sellick, C.A., Croxford, A.S., Maqsood, A.R., Stephens, G. *et al.* (2011) Metabolite profiling of recombinant CHO cells: designing tailored feeding regimes that enhance recombinant antibody production. *Biotechnol. Bioeng.*, **108**, 3025–3031.
99. Wiendahl, C., Brandner, J.J., Kuppers, C., Luo, B. *et al.* (2007) A microstructure heat exchanger for quenching the metabolism of mammalian cells. *Chem. Eng. Technol.*, **30**, 322–328.
100. Volmer, M., Northoff, S., Scholz, S., Thute, T. *et al.* (2011) Fast filtration for metabolome sampling of suspended animal cells. *Biotechnol. Lett.*, **33**, 495–502.
101. Wurm, M. and Zeng, A.P. (2012) Mechanical disruption of mammalian cells in a microfluidic system and its numerical analysis based on computational fluid dynamics. *Lab Chip*, **12**, 1071–1077.
102. Hofmann, U., Maier, K., Niebel, A., Vacun, G. *et al.* (2008) Identification of metabolic fluxes in hepatic cells from transient ^{13}C-labeling experiments: part I. Experimental observations. *Biotechnol. Bioeng.*, **100**, 344–354.
103. Ritter, J.B., Genzel, Y., and Reichl, U. (2008) Simultaneous extraction of several metabolites of energy metabolism and related substances in mammalian cells: optimization using experimental design. *Anal. Biochem.*, **373**, 349–369.
104. Maharjan, R.P. and Ferenci, T. (2003) Global metabolite analysis: the influence of extraction methodology on metabolome profiles of *Escherichia coli*. *Anal. Biochem.*, **313**, 145–154.
105. Villas-Boas, S.G., Hojer-Pedersen, J., Akesson, M., Smedsgaard, J. *et al.* (2005) Global metabolite analysis of yeast: evaluation of sample preparation methods. *Yeast*, **22**, 1155–1169.
106. Lorenz, M.A., Burant, C.F., and Kennedy, R.T. (2011) Reducing time and increasing sensitivity in sample preparation for adherent mammalian cell metabolomics. *Anal. Chem.*, **83**, 3406–3414.
107. Danielsson, A.P., Moritz, T., Mulder, H., and Spegel, P. (2010) Development and optimization of a metabolomic method for analysis of adherent cell cultures. *Anal. Biochem.*, **404**, 30–39.
108. Imamura, H., Nhat, K.P., Togawa, H., Saito, K. *et al.* (2009) Visualization of ATP levels inside single living cells with fluorescence resonance energy transfer-based genetically encoded indicators. *Proc. Natl. Acad. Sci. U.S.A.*, **106**, 15651–15656.
109. Okumoto, S. (2010) Imaging approach for monitoring cellular metabolites and ions using genetically encoded biosensors. *Curr. Opin. Biotechnol.*, **21**, 45–54.
110. Chong, W.P., Yusufi, F.N., Lee, D.Y., Reddy, S.G. *et al.* (2011) Metabolomics-based identification of apoptosis-inducing metabolites in recombinant fed-batch CHO culture media. *J. Biotechnol.*, **151**, 218–224.
111. Vernardis, S.I., Goudar, C.T., and Klapa, M.I. (2013) Metabolic profiling reveals that time related physiological changes in mammalian cell perfusion cultures are bioreactor scale independent. *Metab. Eng.*, **19**, 1–9.
112. Zamorano, F., Wouwer, A.V., and Bastin, G. (2010) A detailed metabolic flux analysis of an underdetermined network of CHO cells. *J. Biotechnol.*, **150**, 497–508.
113. Metallo, C.M., Walther, J.L., and Stephanopoulos, G. (2009) Evaluation of 13C isotopic tracers for metabolic flux analysis in mammalian cells. *J. Biotechnol.*, **144**, 167–174.
114. Zupke, C. and Stephanopoulos, G. (1995) Intracellular flux analysis in hybridomas using mass balances and in-vitro C-13 nmr. *Biotechnol. Bioeng.*, **45**, 292–303.
115. Bonarius, H.P., Ozemre, A., Timmerarends, B., Skrabal, P. *et al.* (2001) Metabolic-flux analysis of continuously cultured hybridoma cells using (13)CO(2) mass spectrometry in combination with (13)C-lactate nuclear magnetic resonance spectroscopy

and metabolite balancing. *Biotechnol. Bioeng.*, **74**, 528–538.
116. Niklas, J., Schneider, K., and Heinzle, E. (2010) Metabolic flux analysis in eukaryotes. *Curr. Opin. Biotechnol.*, **21**, 63–69.
117. Boghigian, B.A., Seth, G., Kiss, R., and Pfeifer, B.A. (2010) Metabolic flux analysis and pharmaceutical production. *Metab. Eng.*, **12**, 81–95.
118. Wahrheit, J., Nicolae, A., and Heinzle, E. (2011) Eukaryotic metabolism: measuring compartment fluxes. *Biotechnol. J.*, **6**, 1071–1085.
119. Ahn, W.S. and Antoniewicz, M.R. (2012) Towards dynamic metabolic flux analysis in CHO cell cultures. *Biotechnol. J.*, **7**, 61–74.
120. Niklas, J. and Heinzle, E. (2012) Metabolic flux analysis in systems biology of Mammalian cells. *Adv. Biochem. Eng. Biotechnol.*, **127**, 109–132.
121. Meinhold, H. (2012) Metabolic flux analysis in mammalian cells: impact of compartments on intracellular flux patterns in CHO DP12 cells. Master thesis. Institute of Biochemical Engineering, University of Stuttgart, Germany.
122. Sengupta, N., Rose, S.T., and Morgan, J.A. (2011) Metabolic flux analysis of CHO cell metabolism in the late non-growth phase. *Biotechnol. Bioeng.*, **108**, 82–92.
123. Dean, J. and Reddy, P. (2013) Metabolic analysis of antibody producing CHO cells in fed-batch production. *Biotechnol. Bioeng.*, **110**, 1735–1747.
124. Mulukutla, B.C., Gramer, M., and Hu, W.S. (2012) On metabolic shift to lactate consumption in fed-batch culture of mammalian cells. *Metab. Eng.*, **14**, 138–149.
125. Gupta, P. and Lee, K.H. (2007) Genomics and proteomics in process development: opportunities and challenges. *Trends Biotechnol.*, **25**, 324–330.
126. Griffin, T.J., Seth, G., Xie, H., Bandhakavi, S. *et al.* (2007) Advancing mammalian cell culture engineering using genome-scale technologies. *Trends Biotechnol.*, **25**, 401–408.
127. Schaub, J., Clemens, C., Schorn, P., Hildebrandt, T. *et al.* (2010) CHO gene expression profiling in biopharmaceutical process analysis and design. *Biotechnol. Bioeng.*, **105**, 431–438.
128. Shen, D., Kiehl, T.R., Khattak, S.F., Li, Z.J. *et al.* (2010) Transcriptomic responses to sodium chloride-induced osmotic stress: a study of industrial fed-batch CHO cell cultures. *Biotechnol. Progr.*, **26**, 1104–1115.
129. Dietmair, S., Hodson, M.P., Quek, L.E., Timmins, N.E. *et al.* (2012) A multi-omics analysis of recombinant protein production in Hek293 cells. *PLoS One*, **7**, e43394.
130. Clarke, C., Henry, M., Doolan, P., Kelly, S. *et al.* (2012) Integrated miRNA, mRNA and protein expression analysis reveals the role of post-transcriptional regulation in controlling CHO cell growth rate. *BMC Genomics*, **13**, 656.
131. Baik, J.Y., Gasimli, L., Yang, B., Datta, P. *et al.* (2012) Metabolic engineering of Chinese hamster ovary cells: towards a bioengineered heparin. *Metab. Eng.*, **14**, 81–90.
132. Bailey, J.E. (1991) Toward a science of metabolic engineering. *Science*, **252**, 1668–1675.
133. Gump, J.M. and Thorburn, A. (2011) Autophagy and apoptosis: what is the connection? *Trends Cell Biol.*, **21**, 387–392.
134. Jardon, M.A., Sattha, B., Braasch, K., Leung, A.O. *et al.* (2012) Inhibition of glutamine-dependent autophagy increases t-PA production in CHO cell fed-batch processes. *Biotechnol. Bioeng.*, **109**, 1228–1238.
135. Figueroa, B., Sauerwald, T.M., Oyler, G.A., Hardwick, J.M. *et al.* (2003) A comparison of the properties of a Bcl-x(L) variant to the wild-type anti-apoptosis inhibitor in mammalian cell cultures. *Metab. Eng.*, **5**, 230–245.
136. Carlage, T., Kshirsagar, R., Zang, L., Janakiraman, V. *et al.* (2012) Analysis of dynamic changes in the proteome of a Bcl-XL overexpressing Chinese hamster ovary cell culture during exponential and stationary phases. *Biotechnol. Progr.*, **28**, 814–823.

137. Sung, Y.H., Lee, J.S., Park, S.H., Koo, J. *et al.* (2007) Influence of co-down-regulation of caspase-3 and caspase-7 by siRNAs on sodium butyrate-induced apoptotic cell death of Chinese hamster ovary cells producing thrombopoietin. *Metab. Eng.*, **9**, 452–464.
138. Dinnis, D.M. and James, D.C. (2005) Engineering mammalian cell factories for improved recombinant monoclonal antibody production: lessons from nature? *Biotechnol. Bioeng.*, **91**, 180–189.
139. Tigges, M. and Fussenegger, M. (2006) Xbp1-based engineering of secretory capacity enhances the productivity of Chinese hamster ovary cells. *Metab. Eng.*, **8**, 264–272.
140. O'Callaghan, P.M., McLeod, J., Pybus, L.P., Lovelady, C.S. *et al.* (2010) Cell line-specific control of recombinant monoclonal antibody production by CHO cells. *Biotechnol. Bioeng.*, **106**, 938–951.
141. Barron, N., Kumar, N., Sanchez, N., Doolan, P. *et al.* (2011) Engineering CHO cell growth and recombinant protein productivity by overexpression of miR-7. *J. Biotechnol.*, **151**, 204–211.
142. Strotbek, M., Florin, L., Koenitzer, J., Tolstrup, A. *et al.* (2013) Stable microRNA expression enhances therapeutic antibody productivity of Chinese hamster ovary cells. *Metab. Eng.*, **20C**, 157–166.
143. Druz, A., Son, Y.J., Betenbaugh, M., and Shiloach, J. (2013) Stable inhibition of mmu-miR-466h-5p improves apoptosis resistance and protein production in CHO cells. *Metab. Eng.*, **16**, 87–94.
144. Irani, N., Wirth, M., van Den Heuvel, J., and Wagner, R. (1999) Improvement of the primary metabolism of cell cultures by introducing a new cytoplasmic pyruvate carboxylase reaction. *Biotechnol. Bioeng.*, **66**, 238–246.
145. Kim, S.H. and Lee, G.M. (2007) Functional expression of human pyruvate carboxylase for reduced lactic acid formation of Chinese hamster ovary cells (DG44). *Appl. Microbiol. Biotechnol.*, **76**, 659–665.
146. Jeon, M.K., Yu, D.Y., and Lee, G.M. (2011) Combinatorial engineering of ldh-a and bcl-2 for reducing lactate production and improving cell growth in dihydrofolate reductase-deficient Chinese hamster ovary cells. *Appl. Microbiol. Biotechnol.*, **92**, 779–790.
147. Zhou, M., Crawford, Y., Ng, D., Tung, J. *et al.* (2011) Decreasing lactate level and increasing antibody production in Chinese Hamster Ovary cells (CHO) by reducing the expression of lactate dehydrogenase and pyruvate dehydrogenase kinases. *J. Biotechnol.*, **153**, 27–34.
148. Elliott, S., Lorenzini, T., Asher, S., Aoki, K. *et al.* (2003) Enhancement of therapeutic protein in vivo activities through glycoengineering. *Nat. Biotechnol.*, **21**, 414–421.
149. Ceaglio, N., Etcheverrigaray, M., Kratje, R., and Oggero, M. (2008) Novel long-lasting interferon alpha derivatives designed by glycoengineering. *Biochimie*, **90**, 437–449.
150. Natsume, A., Niwa, R., and Satoh, M. (2009) Improving effector functions of antibodies for cancer treatment: enhancing ADCC and CDC. *Drug Des. Dev. Ther.*, **3**, 7–16.
151. Datta, P., Linhardt, R.J., and Sharfstein, S.T. (2013) An 'omics approach towards CHO cell engineering. *Biotechnol. Bioeng.*, **110**, 1255–1271.
152. Bacchus, W., Lang, M., El-Baba, M.D., Weber, W. *et al.* (2012) Synthetic two-way communication between mammalian cells. *Nat. Biotechnol.*, **30**, 991–996.
153. Auslander, S. and Fussenegger, M. (2013) From gene switches to mammalian designer cells: present and future prospects. *Trends Biotechnol.*, **31**, 155–168.
154. Muller, K., Engesser, R., Metzger, S., Schulz, S. *et al.* (2013) A red/far-red light-responsive bi-stable toggle switch to control gene expression in mammalian cells. *Nucleic Acids Res.*, **41**, e77.
155. Karlsson, M., Rebmann, B., Lienemann, P.S., Sprossmann, N. *et al.* (2013) Pharmacologically controlled protein switch for ON-OFF regulation of growth factor activity. *Sci. Rep.*, **3**, 2716.

18
Industrial Microorganisms: *Saccharomyces cerevisiae* and other Yeasts

Diethard Mattanovich, Brigitte Gasser, Michael Egermeier, Hans Marx, and Michael Sauer

Yeast was the first microbial host used by mankind for biotransformation of raw materials, and it marked the early developments of industrial biotechnology. Initially, *Saccharomyces cerevisiae* and closely related species were used because of their high fermentative capacity and based on the vast experience from alcoholic beverage production. While a high fermentation rate is favorable for the production of bioethanol and other primary metabolites, it implicates disadvantages for growth-coupled production. Consequently, a number of other yeasts have been developed for the production of biofuels, biochemicals, lipids, or recombinant proteins.

Modern cell factory engineering allows for the design of new metabolic pathways, expansion of the product portfolio, utilization of new substrates, and engineering of complex cellular features for high productivity and robustness (Figure 18.1). This chapter focuses on recent developments in the application of yeasts in biotechnology. For a more comprehensive treatise of the topic, the reader is referred to recent reviews as those cited here.

18.1
Industrial Application of Yeasts

Yeast has been used by mankind since several thousand years, exploiting its high fermentative capacity to convert sugar into ethanol and CO_2 very efficiently. This so-called Crabtree positive phenotype enabled the production of alcoholic beverages and dough leavening for baking bread. We know today that *S. cerevisiae* (baker's yeast, the yeast *par excellence*) has an outstanding ability to ferment at extremely high rates under aerobic and anaerobic conditions [1]. *S. cerevisiae* technology marks the earliest technologies both for fermentative and respiratory bioprocesses.

Modern fermentation technology emerged from beer and wine production, and it is based on batchwise processing of highly concentrated media where most of the sugar taken up by the cells is efficiently channeled into the metabolic

Industrial Biotechnology: Microorganisms, First Edition. Edited by Christoph Wittmann and James C. Liao.

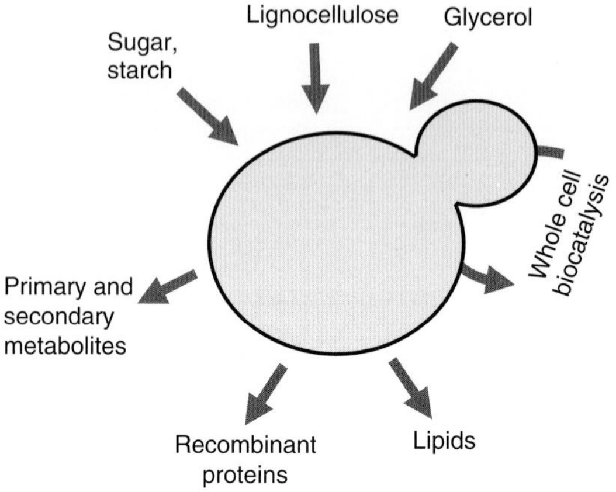

Figure 18.1 Substrates and products of yeast bioprocesses.

by-product ethanol rather than into biomass growth. The exceptionally high fermentation rate of *S. cerevisiae* explains its extraordinary suitability for metabolite production. Today, bioethanol produced with yeast is the largest fermentation product worldwide [2], and yeast technology for production of other metabolites is developed by metabolic engineering [3], as outlined in Section 18.2.

Biomass production, on the other hand, relies on efficient substrate utilization into growth, thus avoiding the accumulation of by-products. Also, yeast technology pioneered process development, to avoid ethanol formation as a major hurdle for efficient production of baker's yeast. As we know today, the key metabolic switch is the flux distribution at the pyruvate branch point, leading either to ethanol or to the tricarboxylic acid (TCA) cycle, providing energy for growth [4]. The problem was solved by process control, limiting the substrate availability by feeding sugar at low rates – the "Zulaufverfahren" developed in 1933 [5], today known as fed-batch technology.

The peculiar metabolic features of *S. cerevisiae* explain why this yeast is especially suitable for overproduction of metabolites, while other yeasts with predominantly respiratory phenotype have advantages for the production of biomass-related products such as recombinant proteins. Natural limitation of sugar uptake limits overflow metabolism in these yeasts, channeling substrate into growth-related products. Table 18.1 provides an overview of yeasts used in biotechnology.

While *S. cerevisiae* was the first yeast used for recombinant protein production, today mainly Crabtree negative yeasts such as *Pichia pastoris*, *Hansenula polymorpha*, or *Yarrowia lipolytica* are employed, not least as they feature a higher secretory efficiency than *S. cerevisiae* (see Section 18.3, and the chapter dedicated to *P. pastoris*).

S. cerevisiae and other yeasts are valuable platform hosts for biorefinery applications where sugar, starch, or lignocellulose are converted to biofuels or value-added chemicals [15]. Beside bioethanol, biologically produced lipids represent

Table 18.1 Yeasts used in biotechnology.

Species	Main applications	Comments	References
Saccharomyces cerevisiae	Bioethanol, polymer precursors, recombinant proteins	Exceptionally high fermentation rate, limitations in protein secretion	[6–8]
Pichia pastoris (*Komagataella* sp.)	Recombinant proteins	High production of recombinant proteins	[9]
Hansenula polymorpha (*Ogatea parapolymorpha*)	Recombinant proteins, xylose utilization	Thermotolerant	[9]
Yarrowia lipolytica	Lipids, citric acid, recombinant proteins	Salt tolerant, utilization of crude glycerol	[6, 9, 10]
Pichia stipitis (*Scheffersomyces stipitis*)	Xylose fermentation	Efficient xylose utilization	[6]
Candida flareri (*Candida famata*)	Riboflavin production	Salt tolerant, issues with genetic stability	[11]
Kluyveromyces marxianus	Biofuels, enzymes, high temperature xylose fermentation	Thermotolerant, xylose utilization	[12, 13]
Kluyveromyces lactis	Enzymes, organic acids	Well characterized, established in food industry, potential model organism for human genetics	[9, 14]

a potential future energy source with high perspectives. Oleaginous yeasts such as *Y. lipolytica* are explored as hosts for production of such lipid-based fuels (Section 18.4). Lignocellulose, by far the most abundant carbohydrate-based biomass feedstock, is built of hexoses and pentoses, with glucose and xylose as the most abundant sugar monomers. For efficient utilization, hosts are required to efficiently metabolize all of these sugars. Section 18.5 summarizes the metabolic engineering approaches enabling *S. cerevisiae* to utilize xylose and arabinose, and the alternative approach using natural pentose utilizing yeasts as hosts for lignocellulose-based bioprocesses.

18.2 Baker's Yeast as Versatile Host for Metabolic Engineering

As outlined above, *S. cerevisiae* is one of the most employed microbial cell factories. Due to its high fermentative capacity and ease of cultivation and genetic manipulation, it developed into one of the most employed engineered cell factories. Among the very successful engineering strategies are the provision of baker's yeast with the ability to use xylose for ethanol production [16, 17] and the production of the antimalarial drug precursor artemisinin [18]. This exemplifies beautifully some basic concepts of industrial microbiology. On the one hand, a mass product with low value but extremely high volume is produced; on the other hand a pharmaceutical product with high value and comparably low volume is produced. On the one hand, the substrate range of the organism is engineered; on the other hand, a plant-derived metabolic pathway is recombinantly introduced into yeast. This shows the great flexibility of this host and the technology.

Butanol is an attractive alternative energy fuel owing to several advantages over ethanol. While *Clostridia* are natural producers of this compound, considerable effort has been put into also developing *S. cerevisiae* to produce butanol and related substances. One reason is again the great success of the established ethanol process; another reason is the high solvent tolerance of this species.

First attempts for the production of 1-butanol by *S. cerevisiae* relied on the introduction of enzymes from native producers to convert acetoacetyl-CoA into 1-butanol. Acetoacetyl-CoA is a natural metabolite of yeast, derived from acetyl-CoA, which itself is a central metabolite that serves as a key precursor in the biosynthesis of various biomolecules [8]. Another approach makes use of the endogenous Ehrlich pathway for isobutanol synthesis. Significant improvements in butanol production were finally reached, eliminating competitive pathways, and particularly taking care to maintain cofactor balance in producing strains [8].

Acetyl-CoA has already been mentioned as a key precursor for many industrially relevant compounds. For example, it is a direct precursor for the mevalonate pathway to obtain isoprenoids. It is also a key precursor for malonyl-CoA, yielding the production of fatty acids (biodiesel) and polyketides [19]. The challenge of engineering the acetyl-CoA availability in yeast lies in its compartmentalization. While acetyl-CoA is readily available in the mitochondrium, the cytosolic pool is low. The cytosolic pool of acetyl-CoA is fed from acetate, which is activated by a bond to coenzyme A at the expense of 1 ATP. It becomes therefore obvious that any metabolic pathway using cytosolic acetyl-CoA aiming at mass production is energetically detrimental and inefficient – if not recombinantly redesigned [20].

Other constraints are important for more complex products, for which mass is not central, but value. For example, central nervous system stimulants are a new class of substances addressed by production with engineered baker's yeast. Expressing the biosynthetic pathways for the opioids thebaine and hydrocodone, and parts of the morphine pathway in yeast, a first step is taken for easy production of opiates [21, 22]. This opens the possibility for the development of new painkillers with less addictive potential. However, it clearly is a new technology that could be abused with many negative consequences – so some contemplation about how to control these developments seems advisable [23]. Not only are narcotics in the center of interest but stimulants such as caffeine and theobromine have also recently been produced with genetically engineered *S. cerevisiae* strains [24].

Synthetic biology aids metabolic engineering approaches and lifts the technology to a new level [25]. New biological parts generate new possibilities and many of them are tested in baker's yeast as common host. For example, cell-surface display of functional proteins opens an entirely new universe to engineer efficient cell factories. Tanaka and Kondo report about the combination of metabolic engineering of target pathways within cells with the efficient display of enzymes to utilize cellulose on the cell membrane outside of the cells to construct a yeast strain which is able to produce ethanol directly from cellulose [26]. One promising approach is the expression of bacteria-derived minicellulosomes on the surface of yeast. Interestingly, the most successful approaches are based on synthetic yeast consortia using various minicellulosome-displaying yeast strains – the optimal composition of enzymatic activities is obtained by changing the combination ratio of the yeast populations [26, 27].

S. cerevisiae laboratory strains have been selected for unicellular growth and multiple genetic markers, mainly auxotrophies, to enable easy selection. It has become evident recently that auxotrophic markers impair specific growth rate, chronological lifespan [28], and tolerance to high ethanol concentrations [29]. Therefore, it appears advisable to use wild-type strains for industrial production rather than (auxotrophic) laboratory strains. For many other yeast species, this is less of a concern as most of their laboratory strains are nearly wild type.

18.3 Protein Production in Yeasts

Apart from metabolites, yeasts are also commonly used for the production of recombinant proteins. Indeed, the first recombinant product ever produced in a eukaryote was human interferon in *S. cerevisiae* in 1981 [30], which was soon followed by hepatitis B surface antigen [31] and human insulin [32]. Since then, many more recombinant proteins produced with yeasts reached the market, the share being approximately 20% of biopharmaceuticals approved in the United States and the European Union [33, 34]. These products mainly include hormones, growth factors, vaccines, and peptides. So far, no yeast-derived antibody or scaffold protein has reached the market, but there are several in the development and clinical pipelines. Especially small binding moieties such as nanobodies and single-chain Fv fragments can be produced in high levels in yeast expression systems. Yeasts are also attractive platforms for antibody surface libraries, enabling high diversity through mating of individual light chain and heavy chain libraries. Industrial enzymes are another example of proteins, where large amounts are needed at low costs. The need for enzymes becomes more and more driven by the production of biofuels lately, but comprises all classes of food and feed enzymes, processing enzymes, enzymes for various industrial applications, as well as pharmaceutically active enzymes [35–38]. Yeasts seem especially suitable for the production of enzymes derived from fungal, plant, and other eukaryotic origin; however, there is no information available on their market share in this sector.

The use of yeasts as expression hosts for protein production and their benefits and pitfalls have been discussed in reviews [39–42]. Along with *S. cerevisiae*, alternative yeasts were developed as protein production hosts, many of them showing higher secretory capacity. These include the methylotrophic species

Candida boidinii, *H. polymorpha* (syn. *Ogataea parapolymorpha*), and *P. pastoris* (syn. *Komagataella* sp.), the lactose utilizing yeast *Kluyveromyces lactis*, *Arxula adenivorans*, *Y. lipolytica*, and *Schizosaccharomyces pombe* [14, 33, 39, 43–49]. Comprehensive overviews about different yeast expression systems including available promoters, vectors, secretion signals, and auxotrophic mutants is given in [40, 41]. Ideally, one would test production of a certain protein in several expression platforms; however, broad host range elements for yeasts are scarce.

While many proteins can be produced at high levels with these expression systems, especially the production of complex, often multimeric human proteins turned out to be a challenge, as their overproduction often leads to misfolded products and triggers stress in the host cells, which consequently results in low product yields. In particular, bottlenecks during oxidative protein folding and passage through the secretion pathway were observed for some recombinant secretory proteins. Strategies to overcome these limitations in order to improve secretory production in yeasts are summarized in [50–52].

The application of yeasts and filamentous fungi for the production of recombinant antibodies and fragments thereof are summarized [53, 54]. As highlighted there, glycoengineering of these microbial hosts to humanize N-glycosylation of antibodies (and other biopharmaceutical proteins) was a tremendous advance toward microbial production of mAbs and other proteins intended for human therapy [55–57]. Glycoengineering was initially started in *S. cerevisiae*, but was pioneered in *P. pastoris*, finally allowing production of human-like sialylated *N*-glycans [58]. Engineering of O-glycosylation was mainly featured in *Ogatea minuta* [59] and later on also in *P. pastoris* [60]. An alternative strategy for preventing yeast-type high-mannose-containing *N*-glycans in *S. cerevisiae* was recently reported [61].

18.4
Lipid Production in Yeasts

Yeasts can convert sugars and other renewable substrates to lipids, which can be further processed to chemicals of higher value or via transesterification to liquid transportation fuels that are currently produced from fossil resources or plant oils [62].

Oleaginous yeasts are able to accumulate lipids up to 80% of their cellular dry weight and comprise genetically very diverse species of both ascomycete and basidiomycete yeasts including the genera *Yarrowia*, *Candida*, *Rhodotorula*, *Rhodosporidium*, *Cryptococcus*, *Trichosporon*, and *Lipomyces* [63]. Many of these species are genetically poorly characterized, which impairs the design and implementation of cell engineering strategies. Lipid accumulation begins in the presence of high concentrations of various carbon sources, if another element is limited in the media. For example, nitrogen limitation inhibits the TCA cycle at the stage of isocitrate. As a consequence, citrate accumulates and is

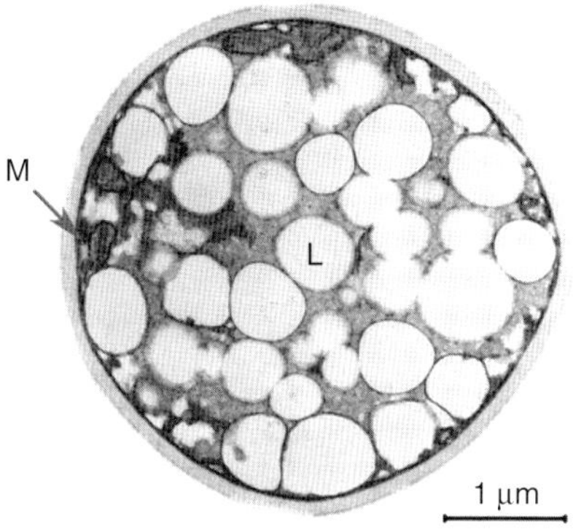

Figure 18.2 *Candida curvata* grown with limiting nitrogen. Total lipid content approximately 40%. M, mitochondrion and L, lipid droplets. (From [64] with permission, ©2007 Elsevier B.V.)

cleaved by ATP-citrate lyase (ACL), yielding oxaloacetate and acetyl-CoA – the precursor molecule for fatty acid synthesis [63]. Through the Kennedy pathway the excess acetyl-CoA is directed into the generation of neutral lipids, which are stored in the form of triacylglycerol (TAGs) as so-called lipid bodies (or lipid droplets – Figure 18.2) [64, 65]. ACL is missing in non-oleaginous yeasts like *S. cerevisiae*, explaining why these yeasts do not accumulate large excess of lipids. In *Y. lipolytica*, native ACL activity is obviously rate limiting, as recombinant expression of mouse-ACL resulted in threefold increased lipid accumulation [66].

The economic feasibility of a lipid production process is determined by the cost of the raw materials as well as the fermentation processes itself. Glucose is the most commonly employed carbon source, but substrates like xylose, lactose, mannose, glycerol, mannitol, L-arabinose, ethanol, as well as wastewaters derived from animal fat treatment and olive oil mills are considered for single-cell oil production in yeast (reviewed in [67]).

The general ability to accumulate lipids varies between yeast species and is dependent on the substrate, nutrient limitation, and the environmental conditions, such as pH and temperature [62]. While *Lipomyces starkeyi, Rhodosporidium toruloides, and Trichosporon cutaneum* grown on glucose reach total lipid contents between 40% and 65% [68–70], *Y. lipolytica* accumulates only up to 20% of lipids [71]. On the other hand, lipid content of *Y. lipolytica* cultivated on glycerol increases to 40% [72]. For further readings, we recommend the reviews by Papanikolaou and Aggelis [73, 74].

Y. lipolytica serves as a model organism for the fundamental understanding and optimization of lipogenesis [63, 75, 76]. Genetic engineering approaches, like the elimination of the β-oxidation pathway, led to an increase in lipid production in *Y. lipolytica.* By metabolic engineering of the lipid synthesis pathway, Tai and Stephanopoulos [77] achieved a total lipid content of 62% in *Y. lipolytica* grown on glucose.

Genetic manipulation of oleaginous yeast will not only further improve the yields and feasibility of single-cell oil production processes but also the generation of novel lipids with high value. Often-named examples are fatty acids similar to natural products like cocoa butter or fish oil, which are becoming increasingly expensive due to high demands in food and pharmaceutical industries [67, 76, 78, 79].

18.5 Pentose-Utilizing Yeasts

Lignocellulosic biomass, such as wheat straw, corn stover, sugarcane bagasse, or wood from spruce, poplar, or beech, is a source of fermentable sugars not directly competing with food or feed production. The interest in pentose-utilizing yeasts has arisen from the development of second-generation biorefineries using not only the cellulose but also the hemicellulose fraction for production of biofuels and biochemicals. For applications in biorefineries, not only the pentose utilization capacity of the strain but also the tolerance toward inhibitors (acetate, hydroxymethylfurfural, 2-furfural, formate) derived from the lignocellulose pretreatment and hydrolysis is of major importance [80]. The most advanced development in this field is the fermentative production of ethanol for fuel purposes, but the developed principles and techniques can give directions for future production strains, products, and processes [16].

18.5.1 Engineered *S. cerevisiae*

Pentose sugars like xylose and arabinose, derived from lignocellulose, are not naturally utilized by the most prominent ethanol producer *S. cerevisiae.* To channel xylose into the metabolic network of *S. cerevisiae,* there are two possible ways. A bacterial pathway uses xylose isomerase converting xylose into xylulose, which is further converted by xylulose kinase into xylulose-5-phosphate, thereby entering the nonoxidative branch of the pentose phosphate cycle. The fungal pathway uses a xylose reductase that is NADPH dependent to convert xylose into xylitol, which is further converted to xylulose by a NAD^+-dependent xylitol dehydrogenase. An imbalance of the cofactors can thus lead to the accumulation of xylitol primarily under anaerobic conditions, as preferred for ethanol production. Channeling arabinose into the pentose phosphate cycle requires also either a bacterial pathway (consisting of an isomerase, a kinase, and an epimerase) or a fungal pathway which also uses an imbalanced set of cofactors leading to the accumulation of arabitol. However, not only is the conversion of the pentose sugars a challenge but also the uptake of the sugars itself. Native transporters of *S. cerevisiae* favor hexoses over pentoses, thereby hindering the simultaneous consumption and fermentation of a mixture of sugars. This problem has been successfully tackled by evolutionary engineering approaches [81]. Metabolic engineering efforts to circumvent the issues of cofactor imbalance and uptake have been reviewed recently in [82, 83]. Furthermore, as already mentioned, the era of synthetic biology offers new possibilities to broaden the substrate and product spectrum of recombinant *S. cerevisiae.* Latest developments in the field have been concisely reviewed by

Tsai *et al.* [84]. Ongoing development in the industrialization of xylose-fermenting yeast has been reviewed in [16].

18.5.2 Yeasts that Naturally Utilize Xylose and Other Pentoses

Native pentose-utilizing yeasts are an upcoming alternative to *S. cerevisiae* in industrial applications. These species possess the ability to grow on pentose sugars. Under fermentative conditions, they accumulate other by-products like xylitol and arabitol, besides ethanol. Furthermore, if the substrate is a mix of hexose and pentose sugars, the same diauxic behavior as described for *S. cerevisiae* is observed. To overcome these obstacles, Harner *et al.* [85] have described genetic improvement strategies for *Pachysolen tannophilus*, *Schefferomyces shehatae*, *Schefferomyces stipitis*, and *Spathaspora passalidarium*. However, some of the pentose-utilizing yeasts offer additional native positive traits for industrial application in a biorefinery. *Kluyveromyces marxianus* and *O. parapolymorpha* (*H. polymorpha*) are of special interest due to their temperature tolerance (up to 50 °C), making them ideal candidates for a simultaneous saccharification and fermentation (SSF) process. *Sugiyamaella (Candida) lignohabitans* shows a remarkable tolerance to a broad pH range [86] and *Zygoascus hellenicus* LK-5G is acid and salt tolerant [87].

18.6 Conclusions

Yeasts are a phylogenetically diverse group of eukaryotic single-cell organisms. As can be seen from the examples above, there is no such thing as the "one" yeast that is best suited for all applications. Nevertheless, *S. cerevisiae* is still the main workhorse, due to its high fermentative capacity and its long tradition in the biotechnology and food industry, which resulted in the availability of many tools for genetic and metabolic engineering. Huge efforts have been made to establish platform strains, for example, strains accumulating precursor metabolites such as acetyl-coA, which are then readily available for production of a wide variety of derived products. Many engineering or screening efforts have been dedicated to the utilization of alternative carbon sources, among them pentoses or complex substrates such as lignocellulose. For other types of products, like lipids or biodiesel, oleaginous yeasts provide excellent production systems, while for recombinant proteins methylotrophic yeasts seem to be particularly attractive. It becomes obvious that a first step into a bright future of yeast biotechnology is a proper appreciation of the biodiversity of yeasts with their varied physiologic and metabolic properties. In a second step, the development of dedicated synthetic biology tools, allowing for metabolic and cell engineering will help optimize this promising class of host organisms for industrial needs. Without doubt, yeasts will be successful hosts in the biotech industry also in future.

Conflict of Interest

The authors declare not to have any conflict of interest.

References

1. Piskur, J., Rozpedowska, E., Polakova, S., Merico, A. *et al.* (2006) How did *Saccharomyces* evolve to become a good brewer? *Trends Genet.*, **22**, 183–186.
2. Mussatto, S.I., Giuliano, D., Guimarães, P.M.R., Silva, J.P.A. *et al.* (2010) Technological trends, global market, and challenges of bio-ethanol production. *Biotechnol. Adv.*, **28**, 817–830.
3. Borodina, I. and Nielsen, J. (2014) Advances in metabolic engineering of yeast *Saccharomyces cerevisiae* for production of chemicals. *Biotechnol. J.*, **9**, 609–620.
4. Frick, O. and Wittmann, C. (2005) Characterization of the metabolic shift between oxidative and fermentative growth in *Saccharomyces cerevisiae* by comparative ^{13}C flux analysis. *Microb. Cell Fact.*, **4**, 30.
5. Intl. Yeast Co. Ltd. (1933) Verfahren zur Herstellung von Hefe nach dem Zulaufverfahren. German Patent DE583760 C.
6. Steensels, J., Snoek, T., Meersman, E., Picca Nicolino, M. *et al.* (2014) Improving industrial yeast strains: exploiting natural and artificial diversity. *FEMS Microbiol. Rev.*, **38**, 947–995.
7. Krivoruchko, A. and Nielsen, J. (2015) Production of natural products through metabolic engineering of *Saccharomyces cerevisiae. Curr. Opin. Biotechnol.*, **35**, 7–15.
8. Kuroda, K. and Ueda, M. (2016) Cellular and molecular engineering of yeast *Saccharomyces cerevisiae* for advanced biobutanol production. *FEMS Microbiol. Lett.*, **363**, fnv247.
9. Wagner, J.M. and Alper, H.S. (2016) Synthetic biology and molecular genetics in non-conventional yeasts: current tools and future advances. *Fungal Genet. Biol.*, **89**, 126–136.
10. Barth, G. (2013) *Yarrowia Lipolytica*, Springer, Berlin, Heidelberg.
11. Dmytruk, K.V. and Sibirny, A.A. (2012) Candida famata (Candida flareri). *Yeast*, **29**, 453–458.
12. Fonseca, G.G., Heinzle, E., Wittmann, C., and Gombert, A.K. (2008) The yeast *Kluyveromyces marxianus* and its biotechnological potential. *Appl. Microbiol. Biotechnol.*, **79**, 339–354.
13. Morrissey, J.P., Etschmann, M.M.W., Schrader, J., and de Billerbeck, G.M. (2015) Cell factory applications of the yeast *Kluyveromyces marxianus* for the biotechnological production of natural flavour and fragrance molecules. *Yeast*, **32**, 3–16.
14. Rodicio, R., Rosaura, R., and Heinisch, J.J. (2013) Yeast on the milky way: genetics, physiology and biotechnology of *Kluyveromyces lactis*. *Yeast*, **30**, 165–177.
15. Buschke, N., Schäfer, R., Becker, J., and Wittmann, C. (2013) Metabolic engineering of industrial platform microorganisms for biorefinery applications – Optimization of substrate spectrum and process robustness by rational and evolutive strategies. *Bioresour. Technol.*, **135**, 544–554.
16. Sànchez Nogué, V. and Karhumaa, K. (2015) Xylose fermentation as a challenge for commercialization of lignocellulosic fuels and chemicals. *Biotechnol. Lett.*, **37**, 761–772.
17. Kim, S.R., Park, Y.-C., Jin, Y.-S., and Seo, J.-H. (2013) Strain engineering of *Saccharomyces cerevisiae* for enhanced xylose metabolism. *Biotechnol. Adv.*, **31**, 851–861.
18. Paddon, C.J. and Keasling, J.D. (2014) Semi-synthetic artemisinin: a model for the use of synthetic biology in pharmaceutical development. *Nat. Rev. Microbiol.*, **12**, 355–367.
19. Krivoruchko, A., Zhang, Y., Siewers, V., Chen, Y. *et al.* (2015) Microbial

acetyl-CoA metabolism and metabolic engineering. *Metab. Eng.*, **28**, 28–42.

20. Kozak, B.U., van Rossum, H.M., Luttik, M.A.H., Akeroyd, M. *et al.* (2014) Engineering acetyl coenzyme A supply: functional expression of a bacterial pyruvate dehydrogenase complex in the cytosol of *Saccharomyces cerevisiae*. *MBio*, **5**, e01696–14.
21. Galanie, S., Thodey, K., Trenchard, I.J., Filsinger Interrante, M. *et al.* (2015) Complete biosynthesis of opioids in yeast. *Science*, **349**, 1095–1100.
22. Ehrenberg, R. (2015) Engineered yeast paves way for home-brew heroin. *Nature*, **521**, 267–268.
23. Oye, K.A., Lawson, J.C.H., and Bubela, T. (2015) Drugs: regulate "home-brew" opiates. *Nature*, **521**, 281–283.
24. Jin, L., Bhuiya Wadud, M., Li, M., Liu, X. *et al.* (2014) Metabolic engineering of *Saccharomyces cerevisiae* for caffeine and theobromine production. *PLoS One*, **9**, e105368.
25. Marx, H., Pflügl, S., Mattanovich, D., and Sauer, M. (2016) in *Synthetic Biology* (eds A. Glieder, C. Kubicek, D. Mattanovich, B. Wiltschi, and M. Sauer), Springer, pp. 255–280.
26. Tanaka, T. and Kondo, A. (2015) Cell-surface display of enzymes by the yeast *Saccharomyces cerevisiae* for synthetic biology. *FEMS Yeast Res.*, **15**, 1–9.
27. Fan, L.-H., Zhang, Z.-J., Yu, X.-Y., Xue, Y.-X. *et al.* (2012) Self-surface assembly of cellulosomes with two miniscaffoldins on *Saccharomyces cerevisiae* for cellulosic ethanol production. *Proc. Natl. Acad. Sci. U.S.A.*, **109**, 13260–13265.
28. Mülleder, M., Capuano, F., Pir, P., Christen, S., Sauer, U. *et al.* (2012) A prototrophic deletion mutant collection for yeast metabolomics and systems biology. *Nat. Biotechnol.*, **30**, 1176–1178.
29. Swinnen, S., Goovaerts, A., Schaerlaekens, K., Dumortier, F., Verdyck, P. *et al.* (2015) Auxotrophic mutations reduce tolerance of *Saccharomyces cerevisiae* to very high levels of ethanol stress. *Eukaryotic Cell*, **14**, 884–897.
30. Hitzeman, R.A., Hagie, F.E., Levine, H.L., Goeddel, D.V. *et al.* (1981) Expression of a human gene for interferon in yeast. *Nature*, **293**, 717–722.
31. Valenzuela, P., Medina, A., Rutter, W.J., Ammerer, G. *et al.* (1982) Synthesis and assembly of hepatitis B virus surface antigen particles in yeast. *Nature*, **298**, 347–350.
32. Stepién, P.P., Brousseau, R., Wu, R., Narang, S. *et al.* (1983) Synthesis of a human insulin gene. VI. Expression of the synthetic proinsulin gene in yeast. *Gene*, **24**, 289–297.
33. Takegawa, K., Tohda, H., Sasaki, M., Idiris, A. *et al.* (2009) Production of heterologous proteins using the fission-yeast (*Schizosaccharomyces pombe*) expression system. *Biotechnol. Appl. Biochem.*, **53**, 227–235.
34. Ferrer-Miralles, N., Domingo-Espín, J., Corchero, J.L., Vázquez, E. *et al.* (2009) Microbial factories for recombinant pharmaceuticals. *Microb. Cell Fact.*, **8**, 17.
35. Borrelli, G.M. and Trono, D. (2015) Recombinant lipases and phospholipases and their use as biocatalysts for industrial applications. *Int. J. Mol. Sci.*, **16**, 20774–20840.
36. Spohner, S.C., Müller, H., Quitmann, H., and Czermak, P. (2015) Expression of enzymes for the usage in food and feed industry with *Pichia pastoris*. *J. Biotechnol.*, **202**, 118–134.
37. Ilmén, M., den Haan, R., Brevnova, E., McBride, J. *et al.* (2011) High level secretion of cellobiohydrolases by *Saccharomyces cerevisiae*. *Biotechnol. Biofuels*, **4**, 30.
38. Kirk, O., Borchert, T.V., and Fuglsang, C.C. (2002) Industrial enzyme applications. *Curr. Opin. Biotechnol.*, **13**, 345–351.
39. Meehl, M.A. and Stadheim, T.A. (2014) Biopharmaceutical discovery and production in yeast. *Curr. Opin. Biotechnol.*, **30**, 120–127.
40. Celik, E. and Calık, P. (2012) Production of recombinant proteins by yeast cells. *Biotechnol. Adv.*, **30**, 1108–1118.
41. Mattanovich, D., Branduardi, P., Dato, L., Gasser, B. *et al.* (2012) Recombinant protein production in yeasts. *Methods Mol. Biol.*, **824**, 329–358.

42. Corchero, J.L., Gasser, B., Resina, D., Smith, W. *et al.* (2013) Unconventional microbial systems for the cost-efficient production of high-quality protein therapeutics. *Biotechnol. Adv.*, **31**, 140–153.
43. van Ooyen, A.J.J., Dekker, P., Huang, M., Olsthoorn, M.M.A. *et al.* (2006) Heterologous protein production in the yeast *Kluyveromyces lactis*. *FEMS Yeast Res.*, **6**, 381–392.
44. Böer, E., Piontek, M., and Kunze, G. (2009) Xplor® 2—an optimized transformation/expression system for recombinant protein production in the yeast *Arxula adeninivorans*. *Appl. Microbiol. Biotechnol.*, **84**, 583–594.
45. Gellissen, G., Kunze, G., Gaillardin, C., Cregg, J. *et al.* (2005) New yeast expression platforms based on methylotrophic *Hansenula polymorpha* and *Pichia pastoris* and on dimorphic Arxula adeninivorans and *Yarrowia lipolytica* – A comparison. *FEMS Yeast Res.*, **5**, 1079–1096.
46. Liu, H.-H., Ji, X.-J., and Huang, H. (2015) Biotechnological applications of *Yarrowia lipolytica*: past, present and future. *Biotechnol. Adv.*, **33**, 1522–1546.
47. Madzak, C., Gaillardin, C., and Beckerich, J.-M. (2004) Heterologous protein expression and secretion in the non-conventional yeast *Yarrowia lipolytica*: a review. *J. Biotechnol.*, **109**, 63–81.
48. Ahmad, M., Hirz, M., Pichler, H., and Schwab, H. (2014) Protein expression in *Pichia pastoris*: recent achievements and perspectives for heterologous protein production. *Appl. Microbiol. Biotechnol.*, **98**, 5301–5317.
49. Gasser, B., Prielhofer, R., Marx, H., Maurer, M. *et al.* (2013) *Pichia pastoris*: protein production host and model organism for biomedical research. *Future Microbiol.*, **8**, 191–208.
50. Idiris, A., Tohda, H., Kumagai, H., and Takegawa, K. (2010) Engineering of protein secretion in yeast: strategies and impact on protein production. *Appl. Microbiol. Biotechnol.*, **86**, 403–417.
51. Hou, J., Tyo, K.E.J., Liu, Z., Petranovic, D. *et al.* (2012) Metabolic engineering of recombinant protein secretion by *Saccharomyces cerevisiae*. *FEMS Yeast Res.*, **12**, 491–510.
52. Delic, M., Göngrich, R., Mattanovich, D., and Gasser, B. (2014) Engineering of protein folding and secretion-strategies to overcome bottlenecks for efficient production of recombinant proteins. *Antioxid. Redox Signal.*, **21**, 414–437.
53. Gasser, B. and Mattanovich, D. (2007) Antibody production with yeasts and filamentous fungi: on the road to large scale? *Biotechnol. Lett.*, **29**, 201–212.
54. Spadiut, O., Capone, S., Krainer, F., Glieder, A. *et al.* (2014) Microbials for the production of monoclonal antibodies and antibody fragments. *Trends Biotechnol.*, **32**, 54–60.
55. De Pourcq, K., De Schutter, K., and Callewaert, N. (2010) Engineering of glycosylation in yeast and other fungi: current state and perspectives. *Appl. Microbiol. Biotechnol.*, **87**, 1617–1631.
56. Hamilton, S.R. and Gerngross, T.U. (2007) Glycosylation engineering in yeast: the advent of fully humanized yeast. *Curr. Opin. Biotechnol.*, **18**, 387–392.
57. Laukens, B., De Visscher, C., and Callewaert, N. (2015) Engineering yeast for producing human glycoproteins: where are we now? *Future Microbiol.*, **10**, 21–34.
58. Hamilton, S.R., Davidson, R.C., Sethuraman, N., Nett, J.H. *et al.* (2006) Humanization of yeast to produce complex terminally sialylated glycoproteins. *Science*, **313**, 1441–1443.
59. Kuroda, K., Kobayashi, K., Kitagawa, Y., Nakagawa, T. *et al.* (2007) Efficient antibody production upon suppression of *O*-mannosylation in the yeast *Ogataea minuta*. *Appl. Environ. Microbiol.*, **74**, 446–453.
60. Hamilton, S.R., Cook, W.J., Gomathinayagam, S., Burnina, I. *et al.* (2013) Production of sialylated *O*-linked glycans in *Pichia pastoris*. *Glycobiology*, **23**, 1192–1203.
61. Parsaie Nasab, F., Aebi, M., Gesche, B., and Frey, A.D. (2013) A combined system for engineering glycosylation

efficiency and glycan structure in *Saccharomyces cerevisiae. Appl. Environ. Microbiol.*, **79**, 997–1007.
62. Lennen, R.M. and Pfleger, B.F. (2013) Microbial production of fatty acid-derived fuels and chemicals. *Curr. Opin. Biotechnol.*, **24**, 1044–1053.
63. Beopoulos, A., Cescut, J., Haddouche, R., Uribelarrea, J.-L. *et al.* (2009) *Yarrowia lipolytica* as a model for bio-oil production. *Prog. Lipid Res.*, **48**, 375–387.
64. Czabany, T., Athenstaedt, K., and Daum, G. (2007) Synthesis, storage and degradation of neutral lipids in yeast. *Biochim. Biophys. Acta*, **1771**, 299–309.
65. Zweytick, D., Athenstaedt, K., and Daum, G. (2000) Intracellular lipid particles of eukaryotic cells. *Biochim. Biophys. Acta, Rev. Biomembr.*, **1469**, 101–120.
66. Zhang, H., Zhang, L., Chen, H., Chen, Y.Q. *et al.* (2014) Enhanced lipid accumulation in the yeast *Yarrowia lipolytica* by over-expression of ATP:citrate lyase from *Mus musculus. J. Biotechnol.*, **192** (Pt. A), 78–84.
67. Ageitos, J.M., Vallejo, J.A., Veiga-Crespo, P., and Villa, T.G. (2011) Oily yeasts as oleaginous cell factories. *Appl. Microbiol. Biotechnol.*, **90**, 1219–1227.
68. Calvey, C.H., Su, Y.-K., Willis, L.B., McGee, M. *et al.* (2016) Nitrogen limitation, oxygen limitation, and lipid accumulation in *Lipomyces starkeyi. Bioresour. Technol.*, **200**, 780–788.
69. Hu, C., Wu, S., Wang, Q., Jin, G. *et al.* (2011) Simultaneous utilization of glucose and xylose for lipid production by *Trichosporon cutaneum. Biotechnol. Biofuels*, **4**, 25.
70. Hu, C., Zhao, X., Zhao, J., Wu, S. *et al.* (2009) Effects of biomass hydrolysis by-products on oleaginous yeast *Rhodosporidium toruloides. Bioresour. Technol.*, **100**, 4843–4847.
71. Lazar, Z., Dulermo, T., Neuvéglise, C., Crutz-Le Coq, A.-M. *et al.* (2014) Hexokinase – A limiting factor in lipid production from fructose in *Yarrowia lipolytica. Metab. Eng.*, **26**, 89–99.
72. Papanikolaou, S. and Aggelis, G. (2002) Lipid production by *Yarrowia lipolytica* growing on industrial glycerol in a single-stage continuous culture. *Bioresour. Technol.*, **82**, 43–49.
73. Papanikolaou, S. and Aggelis, G. (2011) Lipids of oleaginous yeasts. Part I: biochemistry of single cell oil production. *Eur. J. Lipid Sci. Technol.*, **113**, 1031–1051.
74. Papanikolaou, S. and Aggelis, G. (2011) Lipids of oleaginous yeasts. Part II: technology and potential applications. *Eur. J. Lipid Sci. Technol.*, **113**, 1052–1073.
75. Beopoulos, A., Mrozova, Z., Thevenieau, F., Le Dall, M.-T. *et al.* (2008) Control of lipid accumulation in the yeast *Yarrowia lipolytica. Appl. Environ. Microbiol.*, **74**, 7779–7789.
76. Ledesma-Amaro, R. and Nicaud, J.-M. (2016) *Yarrowia lipolytica* as a biotechnological chassis to produce usual and unusual fatty acids. *Prog. Lipid Res.*, **61**, 40–50.
77. Tai, M. and Stephanopoulos, G. (2013) Engineering the push and pull of lipid biosynthesis in oleaginous yeast *Yarrowia lipolytica* for biofuel production. *Metab. Eng.*, **15**, 1–9.
78. Béligon, V., Christophe, G., Fontanille, P., and Larroche, C. (2016) Microbial lipids as potential source to food supplements. *Curr. Opin. Food Sci.*, **7**, 35–42.
79. Xue, Z., Sharpe, P.L., Hong, S.-P., Yadav, N.S. *et al.* (2013) Production of omega-3 eicosapentaenoic acid by metabolic engineering of *Yarrowia lipolytica. Nat. Biotechnol.*, **31**, 734–740.
80. Demeke, M.M., Dietz, H., Li, Y., Foulquié-Moreno, M.R. *et al.* (2013) Development of a D-xylose fermenting and inhibitor tolerant industrial *Saccharomyces cerevisiae* strain with high performance in lignocellulose hydrolysates using metabolic and evolutionary engineering. *Biotechnol. Biofuels*, **6**, 89.
81. Wisselink, H.W., Toirkens, M.J., Wu, Q., Pronk, J.T. *et al.* (2009) Novel evolutionary engineering approach for accelerated utilization of glucose, xylose, and arabinose mixtures by engineered *Saccharomyces cerevisiae* strains. *Appl. Environ. Microbiol.*, **75**, 907–914.
82. Zhang, G.-C., Liu, J.-J., Kong, I.I., Kwak, S. *et al.* (2015) Combining C6 and C5 sugar metabolism for enhancing microbial bioconversion. *Curr. Opin. Chem. Biol.*, **29**, 49–57.

83. Kricka, W., Fitzpatrick, J., and Bond, U. (2014) Metabolic engineering of yeasts by heterologous enzyme production for degradation of cellulose and hemicellulose from biomass: a perspective. *Front. Microbiol.*, **5**, 174.

84. Tsai, C.-S., Kwak, S., Turner, T.L., and Jin, Y.-S. (2015) Yeast synthetic biology toolbox and applications for biofuel production. *FEMS Yeast Res.*, **15**, 1–15.

85. Harner, N.K., Wen, X., Bajwa, P.K., Austin, G.D. *et al.* (2015) Genetic improvement of native xylose-fermenting yeasts for ethanol production. *J. Ind. Microbiol. Biotechnol.*, **42**, 1–20.

86. Bellasio, M., Mattanovich, D., Sauer, M., and Marx, H. (2015) Organic acids from lignocellulose: *Candida lignohabitans* as a new microbial cell factory. *J. Ind. Microbiol. Biotechnol.*, **42**, 681–691.

87. Kodama, S., Nakanishi, H., Thalagala, T.A.T.P., Isono, N. *et al.* (2013) A wild and tolerant yeast suitable for ethanol fermentation from lignocellulose. *J. Biosci. Bioeng.*, **115**, 557–561.

19
Industrial Microorganisms: *Pichia pastoris*

Diethard Mattanovich, Michael Sauer, and Brigitte Gasser

Pichia pastoris was developed as a production system for recombinant proteins in the 1980s. Its ability to grow on methanol, as well as on glucose and other carbon sources, has major implications for their biotechnological applications. Methanol utilization has led to the evolution of strong, tightly regulated promoters, which are used for the expression of heterologous genes. Limited glucose uptake and efficient respiration allow for high biomass yield, and thus high yield of biomass-related products such as recombinant proteins. The production of primary metabolites, on the other hand, is hampered by the strict control of glucose uptake and the efficient respiratory metabolism which is optimized for biomass production. Thus, the physiology of *P. pastoris*, which evolved in its natural habitat, dictates the efficiency of its major applications in industrial biotechnology. Most efforts in metabolic and cell engineering were dedicated to boost the main application, which is the production of recombinant (secretory) proteins. In recent years, *P. pastoris* was also applied to produce chemicals, with most success so far in the manufacture of complex secondary metabolites.

19.1
Physiology and Genetics of *Pichia pastoris*

19.1.1
Methylotrophic Yeasts

Methylotrophic microorganisms can utilize reduced single-carbon molecules such as methanol as their sole carbon and energy source. The use of methanol and other one-carbon molecules as sole carbon source is a metabolic challenge for living organisms, as all C–C bonds need to be closed enzymatically by the cellular metabolism. Besides some gram-negative proteobacteria and gram-positive bacteria [1], a class of yeasts [2] has developed methylotrophy, facing two challenges: oxidation of methanol for energy production and to provide a reactive molecule for C–C bond formation, and formation of these C–C bonds. For both yeast and bacterial pathways, this reactive molecule is formaldehyde [3], a toxic

Industrial Biotechnology: Microorganisms, First Edition. Edited by Christoph Wittmann and James C. Liao.

intermediate which is then either dissimilated to CO_2 or assimilated to biomass constituents.

In nature, methanol is produced either by oxidation of methane (e.g., by methano-oxidizing bacteria in the rhizosphere of plants) or by the decomposition of pectin and lignin (which contain high amounts of methyl ester and methoxyl groups) during growth or decay of plants [4, 5].

Methylotrophic yeasts do not belong to a distinct phylogenetic clade, but are spread in several genera. The main described habitat of methylotrophic yeasts is tree sap. Most natural isolates were found in tree exudates or decaying wood [6, 7]. No information could be found on the chemical composition of these tree saps. However, it can be assumed that wound exudates are rich in carbohydrates and thus a good substrate for proliferation of yeasts. In contrast, xylem sap is poor in carbohydrates, containing mainly inorganic ions. It may be speculated that methanol (from pectin and lignin degradation) diffuses into these saps in certain growth phases of the plant, and represents a (probably fluctuating) carbon source for methylotrophic microorganisms. This would explain the different growth phenotypes of methylotrophic yeasts, which are prepared to live on very low as well as high concentrations of sugars, and on methanol.

Biotechnological interest in methylotrophic yeasts was aroused in the early 1970s when methanol was cheaply available and regarded as nearly unlimited on the basis of fossil resources, leading to the development of processes for production of single-cell protein from methanol (after oxidation of methane). The oil crisis in the mid-1970s marked the end of single-cell protein projects; however, the genetic characterization of methylotrophic yeasts led to their "secondary" use as protein production platforms.

19.1.2
Taxonomy and Natural Isolates

P. pastoris was first described by Guilliermond as *Zygosaccharomyces pastori* [8], and named *P. pastoris* by Phaff [9]. The first isolate worldwide was from the exudate of a chestnut tree in France (strain CBS704, NRRL Y-1603, [8]). For long, only few isolates existed, of which the strain NRRL Y-11430 (CBS7435), isolated from black oak tree in California, was developed as a platform for single-cell proteins by Philips Petroleum. This strain then served as the basis for the development of the protein production system [10].

Based on ribosomal gene sequence data, *P. pastoris* was reclassified in 1995 into a new phylogenetically distinct genus, *Komagataella* [11], which was later split into several species based on 26S rRNA sequencing data [12]. Among the methylotrophic yeasts, *Komagataella* is only distantly related and it is even assumed that *Komagataella* evolved from a different ancestor than other methylotrophic yeasts [13]. Currently, six members of the genus *Komagataella* are described (*K. pastoris, K. phaffii, K. pseudopastoris, K. populi, K. ulmi,* and *K. kurtzmanii*) [14], whereof both *K. pastoris* and *K. phaffii* are among the established *Pichia* protein production platforms [15]. Based on the rules of biological taxonomy, the former type strain of the species *P. pastoris* CBS704 (NRRL Y-1603) remains the type strain of the genus *Komagataella* and the species *K. pastoris*. Consequently, the American isolates mentioned above are classified into the species *K. phaffii*, as, according to Kurtzman [12] they are different enough in their 26S rRNA sequence to justify this distinction. Also, their chromosomal structure differs to some extent, as outlined in Section 19.1.3. It should be noted, however, that the members of the genus *Komagataella* are phenotypically very similar and cannot easily be distinguished from one another by fermentation and assimilation tests routinely employed in yeast taxonomy, so that classification is mainly based on sequence comparisons of a limited number of genes. To avoid confusion and to include all strains employed in biotechnology, we use the established name *P. pastoris* here as a synonym for all *Komagataella* species.

19.1.3
Genomics

So far, genome sequences of *P. pastoris* strains have been published by De Schutter *et al.* [16], Mattanovich *et al.* [17], and Küberl *et al.* [18]. Genome sizes of the three different sequenced strains are very similar (9.5 Mbp), as well as the number of genes (~5300). Also, the coding sequences are very similar. In average, they are up to 91.5% identical at the amino acid sequence level. Noncoding sequences are more divergent, although promoters and terminators function in all strains tested so far. Sequencing confirmed the presence of four chromosomes, as identified by pulsed field gel electrophoresis [19] and a chromosome rearrangement between *K. pastoris* and *K. phaffii*. Genome sequences of different strains, up to date annotation, and many features are publicly available at www.pichiagenome.org [20], hosted by the Austrian Centre of Industrial Biotechnology and the University of Natural Resources and Life Sciences Vienna.

19.1.4
Physiology

The natural habitat of *P. pastoris* is rather poor in free monomeric carbon sources like sugars or sugar alcohols. Consequently, its metabolism is adapted to this scarce environment, different from yeasts like *Saccharomyces cerevisiae* which thrive in environments of high sugar concentrations. This becomes evident by the presence of high-affinity transporters for glucose (Gth1 (PAS_chr1-3_0011) being the major high-affinity glucose transporter [21]). Thus, even low concentrations of these carbon sources can be utilized with reasonable efficiency at low external concentrations. Due to the presence of four putative H^+/glycerol symporters enabling constitutive salt-independent active glycerol transport, *P. pastoris* is able to grow on glycerol as efficiently as on glucose. The maximum specific glucose uptake rate is $0.35\,g\,g^{-1}\,h^{-1}$ at $\mu_{max} = 0.19\,h^{-1}$ on glucose-containing mineral media, and the specific glycerol uptake rate is $0.37\,g\,g^{-1}\,h^{-1}$ on mineral media [17]. Compared to these data, maximum specific methanol uptake rates are markedly lower. There is a great variability of data in literature (e.g., [22] and references therein). Based on reported μ_{max} between 0.05 and $0.07\,h^{-1}$ [23, 24], a maximum specific methanol uptake rate of $0.15-0.20\,g\,g^{-1}\,h^{-1}$ is estimated. It should be noted, however, that methanol is taken up into cells by

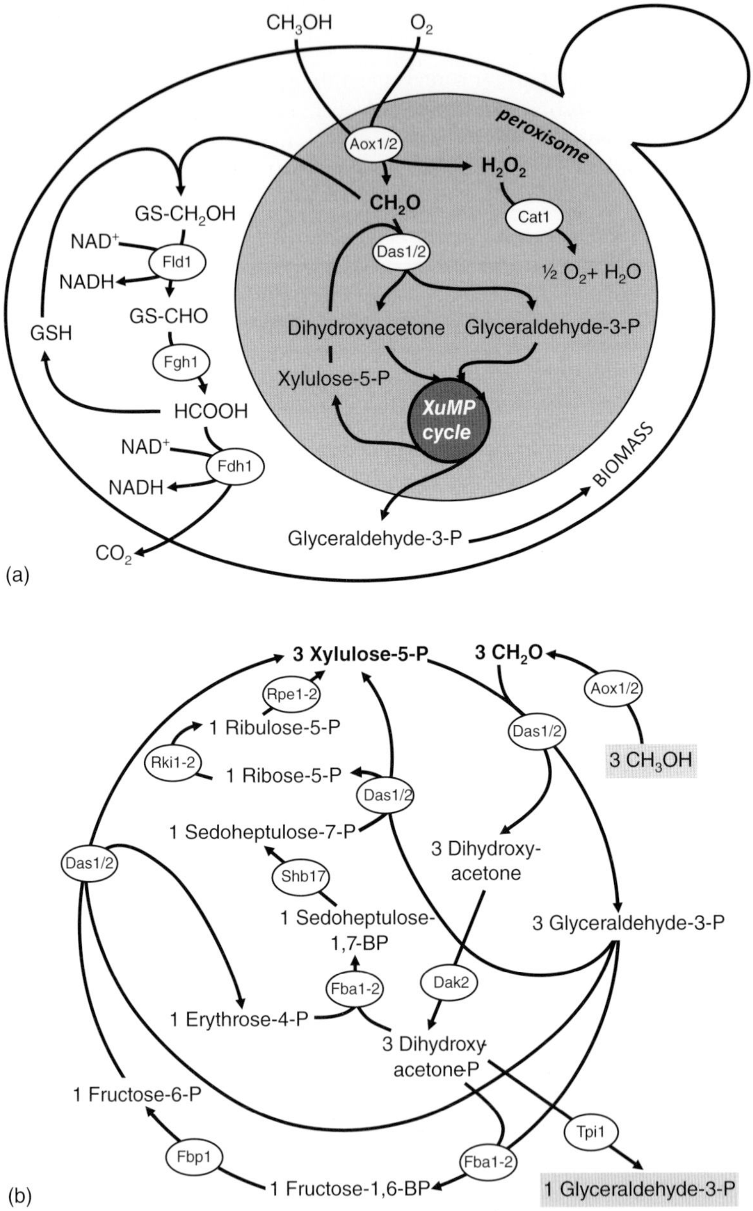
CH3OH
O2
peroxisome
Aox1/2
H2O2
CH2O
Cat1
GS-CH2OH
NAD+
Fld1
NADH
Das1/2
½ O2+ H2O
GS-CHO
GSH
Dihydroxyacetone
Glyceraldehyde-3-P
Fgh1
Xylulose-5-P
XuMP cycle
HCOOH
NAD+
Fdh1
NADH
BIOMASS
Glyceraldehyde-3-P
CO2
3 Xylulose-5-P
3 CH2O
Rpe1-2
Aox1/2
1 Ribulose-5-P
Das1/2
Rki1-2
1 Ribose-5-P
3 CH3OH
Das1/2
1 Sedoheptulose-7-P
3 Dihydroxy-acetone
Das1/2
Shb17
1 Sedoheptulose-1,7-BP
3 Glyceraldehyde-3-P
Fba1-2
Dak2
1 Erythrose-4-P
3 Dihydroxy-acetone-P
1 Fructose-6-P
Tpi1
Fbp1
Fba1-2
1 Fructose-1,6-BP
1 Glyceraldehyde-3-P

(a)

(b)

diffusion without a specific transporter, so that uptake kinetics is controlled by concentration gradients and by the following intracellular utilization (as reported for other methylotrophic yeasts [25]) rather than by kinetics of a transporter protein. Adapted to its natural environment, *P. pastoris* is also able to assimilate the methyl-pentose rhamnose (a constituent of pectin chains) presumably using the alternative fungal pathway of L-rhamnose metabolism [26, 27], while xylose utilization has only been described for *K. pseudopastoris*.

The absence of low-affinity, high-velocity glucose transporters leads to a rather low and stringently regulated maximum specific growth rate. This explains one marked difference of this typical Crabtree-negative yeast to *S. cerevisiae*: limited substrate uptake leads to only low overflow metabolism and thus few by-products; in other words, *P. pastoris* does not follow the "make–accumulate–consume" concept of Crabtree-positive yeasts [28]. Besides ethanol, which is produced at much lower levels than by *S. cerevisiae*, the other main metabolic by-products are arabitol [29], acetate, acetaldehyde, and pyruvate [30]. This metabolic feature has significant advantages for the production of biomass-related products, such as typically recombinant proteins, as even simple batch processes lead to high biomass accumulation, which makes both small-scale screening and bioreactor cultivation easier than, for example, for *S. cerevisiae*. On the contrary, the overflow metabolism of *S. cerevisiae* may constitute a major advantage for the fast production of metabolites, leading to high specific and volumetric productivities. In this respect, the advantage of *P. pastoris* of a low-overflow metabolism may turn into a disadvantage for the production of (at least primary) metabolites.

19.2 Methylotrophic Metabolism

The metabolization of methanol in yeasts begins with its oxidation to formaldehyde, catalyzed by an alcohol oxidase (AOX) that uses molecular oxygen as electron acceptor. From there, metabolism branches either to dissimilation via formate to CO_2 with the production of two NADH as reduction equivalents, or to assimilation in the xylulose 5-phosphate cycle [31]. Oxidation and assimilation of methanol take place in specialized organelles, the peroxisomes, while dissimilation is a cytosolic process. Figure 19.1 illustrates all these steps

Figure 19.1 Metabolic pathways of methanol utilization in *P. pastoris*. (a) Overview of assimilation and dissimilation pathways. (b) Xylulose monophosphate (XuMP) cycle for methanol assimilation. Metabolites entering and leaving the peroxisomes are marked in gray boxes. GSH: reduced glutathione; GS-CH_2OH: *S*-(hydroxymethyl)glutathione; GS-CHO: *S*-formylglutathione; Cat1: catalase; Fld1: formaldehyde dehydrogenase; Fgh1: *S*-formylglutathione hydrolase; Fdh1: formate dehydrogenase; Aox1/2: alcohol oxidase 1 and 2; Das1/2: dihydroxyacetone synthase 1 and 2; Dak2: dihydroxyacetone kinase; Tpi1: triosephosphate isomerase; Fba1-2: fructose 1,6-bisphosphate aldolase; Fbp1: fructose 1,6-bisphosphatase; Shb17: sedoheptulose 1,7-bisphosphatase; Rki1-2: ribose 5-phosphate ketol-isomerase; and Rpe1-2: D-ribulose 5-phosphate 3-epimerase.

and their intracellular localization. Beside formaldehyde, hydrogen peroxide is a second toxic product of the AOX reaction, which is detoxified by a highly expressed peroxisomal catalase. Compartmentalization of these reactions into peroxisomes is crucial not only to protect the cytosol from reactive oxygen species but also because peroxisomal decomposition of H_2O_2 is preventing otherwise high-energy-consuming cellular scavenging reactions [32].

19.2.1
Dissimilation of Methanol to Generate Energy

For energy production, formaldehyde is initially trapped by the abundant metabolite glutathione to form nontoxic *S*-(hydroxymethyl)glutathione, which is successively oxidized to CO_2 by NAD^+-dependent formaldehyde dehydrogenase (FLD, [33]), *S*-formylglutathione hydrolase (FGH), and NAD^+-dependent formate dehydrogenase (FDH, [34]). Condensation of formaldehyde to glutathione is assumed to occur spontaneously, as no homolog of the glutathione-dependent formaldehyde-activating enzyme (Gfa) present in some methylotrophic bacteria [35] was found in the genome of *P. pastoris* and related methylotrophic yeasts. Currently, there is no information whether *P. pastoris* also uses alternative means of formaldehyde oxidation via methyl formate that were observed in other methylotrophic yeasts such as *Candida boidinii* [36] and *Pichia methanolica* [37]. NAD^+ accepts the electrons which are further transferred to mitochondria via the malate–aspartate shuttle [38, 39], where they are used for ATP production in the respiratory chain.

19.2.2
Assimilation of Methanol to Form Biomass

For assimilation, formaldehyde is transferred to xylulose 5-phosphate by the specialized transketolase dihydroxyacetone synthase (DAS), yielding glyceraldehyde 3-phosphate (GAP) and dihydroxyacetone (DHA). DHA is then phosphorylated by dihydroxyacetone kinase DAK to form dihydroxyacetone phosphate (DHAP). Xylulose 5-phosphate is regenerated by the so-called xylulose monophosphate cycle. While these reactions have been believed for many years to be carried out through the pentose phosphate pathway (PPP), Russmayer *et al.* [38] elucidated that the xylulose monophosphate cycle of methanol assimilation utilizes a duplicated methanol inducible enzyme set (Fba1-2, Tal1-2, Rki1-2, Rpe1-2) and is entirely localized to peroxisomes. Furthermore, Russmayer *et al.* [38] showed that this pathway includes sedoheptulose 1,7-bisphosphate, following a reaction with high similarity to the Calvin cycle for ribulose 5-phosphate regeneration for CO_2 assimilation in plants. Similar to plant chloroplasts, all enzymes for the xylulose monophosphate cycle are localized in organelles (the peroxisomes), thus concentrating enzymes and intermediate metabolites in subcellular compartments. In total, 3 mol of methanol (C1) and ATP are needed to form 1 mol of GAP. Depending on the energy status of the cell and the growth rate, the split ratio between the assimilatory and the dissimilatory pathway is variable [40, 41].

19.2.3
Implications for Production Processes

Methanol assimilation is surprisingly efficient, given the fact that all C–C bonds of the entire biomass grown on methanol are formed *de novo*. The first step, methanol oxidation, however, is a thermodynamically inefficient reaction. The reaction energy is not recovered as the electrons are transferred to molecular oxygen instead of redox equivalents for further metabolic use. Consequently, the reaction energy dissipates as metabolic heat. The thermodynamic inefficiency of methanol oxidation has several disadvantages for large-scale cultivation: heat dissipation leads to a strong increase of cooling demand during cultivation, which causes higher process costs compared to standard cultivation of yeasts on glucose. Direct transfer of electrons to oxygen increases the oxygen demand, reaching the limits of aeration in stirred tank bioreactors. As the reaction energy of methanol oxidation is not used for cellular metabolism, the biomass yield on methanol is much lower, about one half than would be theoretically possible.

These disadvantages can be overcome by employing a co-feeding strategy of, for example, glycerol or sorbitol and methanol at substrate-limiting conditions [42]. Metabolic flux analyses revealed that in co-feeds methanol serves mainly for energy production, while glucose or glycerol is the main carbon source. The advantages of co-feeding strategies are a higher biomass yield, lower oxygen demand, and lower metabolic heat; however, they come at the disadvantage of a more complex process regime.

AOX is encoded by two genes (*AOX1* and *AOX2*) in *P. pastoris*, which share 93% and 97% identity on the nucleotide and amino acid sequence level, but have different expression strengths [43]. *AOX1* knockout strains still grow on methanol but with a lower μ_{max}. This phenotype is called mut^{s} (for methanol utilization slow), and leads to a limitation of methanol and oxygen consumption and heat dissipation, often applied in co-feeding strategies as outlined above.

Another, process-based strategy to limit methanol utilization is the limitation of oxygen supply to the bioreactor. As AOX requires methanol and oxygen as substrates, both are rate limiting for the methanol oxidation reaction. Thus, methanol uptake can be controlled by a limitation of oxygen supply. Controlling biomass growth by oxygen limitation, while keeping methanol concentration high, was shown to increase titer and quality of recombinant protein [22, 44].

19.3
Application for the Production of Recombinant Proteins

19.3.1
Promoters

The starting point for the development of *P. pastoris* as a protein production system was the discovery that methylotrophic yeasts accumulate large quantities of AOX when growing on methanol, but essentially none when growing on glucose or glycerol. The need for high amounts of AOX is due to its low efficiency.

After cloning of the *AOX1* promoter it was proven that it is highly regulated and leads to strong transcription on methanol [45]. Consequently, the first *P. pastoris* expression system was based on the *AOX1* promoter. In the following, other methanol-controlled promoters were cloned and characterized as well, such as the *FLD1* promoter [46]. Although the *DAS* promoter was also characterized initially [45], it was not used. High P_{AOX1}-based expression levels have been reported both intracellularly (e.g., *Hevea brasiliensis* hydroxynitrile lyase accumulated to 22% of cell dry weight [47]; tetanus toxin fragment C accumulated up to 27% of total cell protein [48]) and secreted to the culture supernatant. Using the initial expression system, secreted product levels of, for example, >1.5 $g\,l^{-1}$ of the potent thrombin inhibitor hirudin [49], >1.7 $g\,l^{-1}$ tick anticoagulant peptide (TAP) [50], >3.5 $g\,l^{-1}$ of different single-chain Fv antibody fragments [22, 51] or up to 14.8 $g\,l^{-1}$ gelatins [52] were obtained in fed-batch cultivations. Secretion levels could be further enhanced by engineering of the protein folding and secretion machinery (see Section 19.3.6) or by the use of advanced expression systems (see Sections 19.3.1–19.3.4). Despite early successes, some technological limitations of the methanol-based production system have delayed its wide spread in industrial production. Methanol metabolism requires significantly more oxygen than carbohydrate substrates, and releases high metabolic heat, both of which reach the technical limitation of large-scale stirred tank fermenters. In addition, the large amounts of methanol require special safety precautions to prevent the accumulation of flammable or explosive methanol/air mixtures [53]. Therefore, strong constitutive promoters were cloned as well, and the GAP dehydrogenase promoter was characterized as a highly suitable alternative for protein production in *P. pastoris* [54]. Table 19.1 summarizes promoters that were established in the first two decades of *P. pastoris* platform development.

Table 19.1 First generation of *P. pastoris* promoters used for recombinant protein production.

Name	Relative efficiency	Inducing compound	References
P_{AOX1} Alcohol oxidase	High	Methanol	[45, 55]
P_{GAP} Glyceraldehyde 3-phosphate dehydrogenase	High	Constitutive	[54]
P_{FLD1} Formaldehyde dehydrogenase	High	Methanol or methylamine	[46]
P_{PEX8} Peroxisomal matrix protein	Moderate	Methanol or oleate	[56]
P_{YPT1} GTPase involved in secretion	Low	Constitutive	[57]

Recent systems biology studies of the methanol lifestyle of *P. pastoris* have revealed which genes are strongly upregulated by methanol on the transcriptional and translational level [26, 38]. Gasser *et al.* [58] summarized the 45 most suitable strong methanol-regulated promoters. The top five are derived from the genes encoding early steps of methanol utilization (*AOX1* and *AOX2*, *DAS1* and *DAS2*, and *FDH1*). While being strongly repressed in the presence of glucose or glycerol surplus, these genes are derepressed under glucose-limiting conditions so that even without methanol distinct expression levels are obtained [26, 59, 60]. If this is undesired, promoters of genes further downstream in the methanol metabolism should be used (*FLD1, FBA1-2, SHB17, FGH1, DAK2*). Vogl and coworkers [61] evaluated expression levels and regulation of methanol-inducible promoters for the co-expression of several genes of a biosynthetic pathway.

As noted in Section 19.3.1, the use of methanol in large-scale fermentation has some severe restrictions. Recent alternative developments are based on systems biology identification of promoters and on the development of mutated and synthetic promoters. Stadlmayr *et al.* [62] employed transcriptomic data to identify genes with very high expression levels during growth on glucose, glycerol, and methanol and tested their promoters for heterologous protein production. Among the 15 selected promoters, 5 were of strength similar to that of P_{GAP} and considered to be constitutive. In accordance with its function, the *THI11* promoter was shown to be regulated by thiamine availability, leading to an induction strategy where the batch culture is started with thiamine which is consumed to a level allowing strong gene expression in the fed-batch phase. In a dedicated study to find promoters for a most simple induction strategy, Prielhofer *et al.* [21] compared transcriptomes of glycerol batch with glucose-limited cultures and identified the promoter of the high-affinity glucose transporter gene (*GTH1*) P_{G1} as a very efficient, tightly regulated promoter that allowed methanol-free induction. Expression levels of several heterologous genes were at least twofold higher compared those of P_{GAP}. Hartner *et al.* [63] developed a mutant library of the *AOX1* promoter by deletion and/or multiplication of putative transcription factor binding sites, resulting in a collection of promoters ranging from 6% to 160% of the native *AOX1* promoter strength, and including deregulated variants that are independent of methanol induction. Qin *et al.* [64] prepared a GAP promoter variant library by random mutagenesis. Synthetic core promoters were developed for *P. pastoris* based on natural yeast core promoters, with the main advantage of a shorter length compared to native promoters [65]. Table 19.2 lists some of the most important new promoters for heterologous gene expression in *P. pastoris*.

19.3.2
Protein Secretion

As outlined in Section 19.3.1, recombinant protein is either produced in the cytosol or secreted to the culture supernatant when a suitable secretion signal is fused to the 5′ end of the gene of interest. The most widely used secretion signal is the mating factor α (MFα) signal of *S. cerevisiae*, which was shown to

Table 19.2 New generation of *P. pastoris* promoters.

Name	Relative strength	Induction strategy	References
P_{THI11}	High	Depletion of thiamine	[62]
P_{GTH} (P_{G1}) High affinity glucose transporter	High	Slow glucose feed	[21]
P_{AOX1} library Variants of alcohol oxidase promoter	Low to high	Methanol addition or constitutive	[63]
P_{GAP} library Variants of GAP dehydrogenase promoter	Low to high	Constitutive	[64]
Synthetic	Low to high	Constitutive	[65]

be more effective than the native *P. pastoris* leader sequence of acid phosphatase [66]. The complex proteolytic processing of the MFα leader may lead to incomplete removal of N-terminal Glu–Ala dipeptides, which can be overcome by eliminating these from the leader sequence, thus fusing the Lys–Arg recognition sequence of the Kex2 protease directly to the N-terminus of the gene of interest [67, 68]. In the majority of cases, secretory production of recombinant proteins is preferred as it leads in most cases to medium to high concentrations of a quite pure product in the culture supernatant. In this respect, *P. pastoris* was shown to secrete only low levels of endogenous proteins (e.g., 20–40 proteins), in particular when cultivated under suitable bioreactor conditions [17, 69, 70]. As long as cells are viable and intact, usually also the proteolytic activity in the supernatant is low (often even below the detection limit, e.g., [71]), correlating nicely with the low number of proteases found in culture supernatants in the secretome studies.

19.3.3
Vectors and Selection Markers

In general, yeast vectors are either self-replicating in the yeast cell, or they are constructed so that the recombinant DNA is integrated into the host genome. Self-replicating plasmid vectors must contain sequences of the host (or a related organism) which direct DNA replication and are derived from centromeres (CENs) or other autonomously replicating sequences (ARSs). Interestingly, the centromeric regions of *P. pastoris* have not yet been identified. *P. pastoris* ARS-based vectors have been developed [10, 72]; however, they have low replicative stability as all yeast ARS vectors. Without selection pressure, the replicative rate of plasmid loss ranges from 50% to 70% plasmid-free cells after 10 generations.

Genetic stability can be achieved by integration of the recombinant DNA fragments into the genome of *P. pastoris*, mostly by homologous recombination. Therefore, they must be flanked on one or both sides by a homologous sequence of the host, indicating the desired integration position. The minimum required length of this integration sequence is host dependent, for *P. pastoris* in a range of 150 bp. When the vector contains only one flanking homolog, which is cut by a restriction enzyme prior to transformation, mainly integration by gene addition is achieved. In contrast, two flanking regions with homologies to the host genome lead to gene replacement at the integration site. The most frequently used integration sites are the *AOX1* promoter, the *AOX1* terminator, or the *GAP* promoter site. Integrative vectors can be rather easily assembled as desired, so that the required elements, promoter and transcription terminator, integration site, and selection marker, can be combined as required [73].

Over the past 30 years, a wide range of selection markers have become available for *P. pastoris*. Recessive markers, which require a mutation in the host genome that is complemented by the marker, mostly derive from amino acid or nucleotide synthesis pathways, as traditionally applied for other yeasts or bacteria as well (Table 19.3). The disadvantage of these auxotrophic markers is the possibility

Table 19.3 Recessive selection markers used for *P. pastoris*.

Biosynthetic marker	Gene	Useful property	References
HIS4	Histinidol dehydrogenase		[10]
ARG4	PR-amidoimidazole-succinocarboxamide synthase		[74]
ADE1	Argininosuccinate lyase	Pink color of *ade1* strains	[75, 76]
URA3	Orotidine 5′-phosphate decarboxylase	5-FOA selection for Ura^- strains	[75]
URA5	Orotate phosphoribosyl transferase	5-FOA selection for Ura^- strains	[77]
ADE2	Phosphoribosylamino-imidazole carboxylase	Pink color of *ade2* strains	[76]
HIS1, HIS2, HIS5, HIS6	Histidine biosynthesis genes		[78]
ARG1, ARG2, ARG3	Arginine biosynthesis genes		[78]
MET2	Homoserine *O*-transacetylase	Sc*MET2* complements *Pp met2*	[79]
FLD1	Formaldehyde dehydrogenase	Selection of high gene copy numbers on formaldehyde containing media	[80]

Table 19.4 Dominant selection markers used for *P. pastoris*.

Resistance	Gene	Selective drug	References
Kan^R	Neomycin phosphotransferase	G418/geneticin	[82]
Zeo^R	Sh*Ble* gene	Zeocin (*E. coli* and yeast)	[83]
Bsd^R	Blasticidin S deaminase	Blasticidin (*E. coli* and yeast)	www.invitrogen.com
HygMX	Hygromycin phosphotransferase	Hygromycin (*E. coli* and yeast)	[84]
KanMX	Neomycin phosphotransferase	Kanamycin (*E. coli*) G418 (yeast)	[85]
NatMX	*sat1* – nourseothricin acetyltransferase	Nourseothricin (*E. coli* and yeast)	[86]
ARS	*S. cerevisiae* ARR3	Arsenic	[86]

of cross-feeding between marker positive cells and marker negative ones which may take up the missing nutrient when it is released by the positive cells. This is mainly critical in high-cell-density liquid cultures. To overcome this, metabolic auxotrophies concerning the utilization of a nutrient rather than the synthesis of a metabolite are currently being developed, such as the deficiency to utilize glycerol by knocking out *GUT1*-encoding glycerol kinase [81]. Similarly, *P. pastoris FLD1* may be applied as a marker in a respective knockout strain, which allows for selection of positive transformants with high copy numbers on the otherwise toxic substance formaldehyde [80]. Dominant markers have the advantage of being applicable to any strain without prior mutation, as the selection principle is expressed once the marker gene is present. Mostly these are antibiotic resistance markers, but resistance genes against other toxic substances may also be applied (Table 19.4).

19.3.4 Integration of Multiple Gene Copies

Gene expression levels correlate with gene copy numbers. To an upper limit, this correlation also holds for the production levels of the encoded proteins. This correlation was shown to range high up to gene copy numbers of 30 for proteins that are localized to the cytosol [48, 87, 88], while for several secreted proteins a so-called plateauing effect was observed. This means that there is a beneficial effect of gene copy number on product levels observed up to a certain limit, where no more increase in productivity can be gained by increasing the gene copy number further. This upper limit is distinct for each individual protein and expression system used (see review by Aw and Polizzi [89]), for example, being 11–12 copies for insulin precursor [90, 91], about 6–7 copies for human serum albumin [88] or *Citrobacter amalonaticus* phytase [92], but being only 2 copies for human trypsinogen or *Rhizopus oryzae* lipase [93, 94]. Secretory proteins have to pass several potentially rate-limiting steps before being released to the culture supernatant (see Section 19.3.6). Therefore, it appears plausible that at higher gene copy numbers other processes than transcription become saturated and define an upper level of secretion capacity for each individual protein. Indeed, intracellular product accumulation or induction of the unfolded protein response (UPR) pathway, which becomes activated upon accumulation of noncorrectly folded secretory proteins, was frequently observed in multicopy strains [91, 93, 95, 96].

The first strategy to derive clones with multiple gene copies integrated into the genome was based on constructing vectors containing concatemer repeats of the gene of interest [48]. Therefore, the gene expression cassette was cloned in tandem several times in the expression vector. This strategy bears several disadvantages. Concatemer cloning is laborious and time consuming, and leads only to a limited number of repeats, around four to six. While it was expected that this strategy enabled the construction of expression strains with defined gene copy number, it was found that due to homologous recombination during genome integration copies may be lost and a range of different copy numbers is attained [93].

With the availability of antibiotic selection markers, it became possible to select for higher copy numbers of the vector by applying high concentrations of the antibiotic. This strategy is especially suitable for markers with dose-dependent effect, such as G418, Zeocin, or hygromycin. With increasing antibiotic concentration, the number of clones decreases and the chance of high copy clones increases. Alternatively, auxotrophic selection markers with truncated weak promoter regions [76] or the *FLD* marker [80] can be used to select for multicopy transformants. However, this is always a random process and requires a significant screening effort to identify high-producing clones.

The third strategy to raise gene copy numbers is based on integration of the expression cassette into a gene locus which appears naturally in multiple copies. Yeast genomes contain multiple copies of the ribosomal gene cluster containing genes encoding for 25S, 5.8S, and 18S ribosomal RNA, divided by a nontranscribed intergenic spacer (NTS). In *P. pastoris*, loci of these rDNA clusters can be found on each of the four chromosomes [16], in total having approximately 20–30 copies [18, 88]. It was shown that high copy number integration in rDNA of *S. cerevisiae* depends on post-transformational amplification based on selection with the inefficient *LEU2*d gene [97]. This genomic plasticity was utilized to expand gene copy numbers after transformation by growing clones under high selection pressure with other markers. Thus, not only can an initial high copy integration event be selected for (as in strategy 2) but good producer clones can also be further enhanced in gene copy number by post-transformational copy number increase [88, 98–101]. Post-transformational vector amplification has also been described after genome integration into the single-copy AOX-TT locus and successive selection on increasing concentrations of Zeocin [102].

19.3.5 Metabolic Engineering to Enhance Productivity of Recombinant Proteins

Metabolic engineering has proven its power for enabling and enhancing the production of cellular metabolites [103]. However, it has been applied less frequently to improve the production of complex multimeric molecules such as proteins. Metabolic limitations of protein production have been discussed in the past (for *P. pastoris*, see [104]). Common patterns of metabolic analyses of producing strains, compared to wild-type strains, are a reduction in biomass yield and an increase in TCA cycle flux [30, 59, 105, 106], which has been attributed to an increased energy demand for protein synthesis and folding in the recombinant strains. Similarly, a recombinant protein producing strain grown on methanol directed the flux from formaldehyde more toward dissimilation and energy production than to assimilation [41]. Heyland *et al.* [104] also discussed the potential extra demand on NADH for amino acid synthesis as a pulling force of metabolism toward the TCA cycle.

In 2010, the first metabolic model of *P. pastoris* was published [107] and applied to predict gene knockout and overexpression mutations that may have a beneficial effect on recombinant protein production [30]. Over 50% of these single-gene mutations increased production of a human enzyme between 20% and 50%, partly pointing toward NADPH production as the target. The oxidizing branch of the PPP was among the most successful overexpression targets, with glucose 6-phosphate dehydrogenase (*ZWF1*) and most importantly, 6-phosphogluconate lactonase (*SOL3*) as successful targets. After single overexpression, combined enhancement of the first four PPP enzymes was also evaluated. Double overexpression of *ZWF1* and *SOL3* gave the best results, with a two- to threefold increase in the yield of the recombinant protein [108]. The positive impact on recombinant protein production correlated with the increase in glucose flux entering the PPP and thus points to an increased demand for NADPH regeneration in overproducing strains. NADPH provides reduction equivalents for the synthesis of several amino acids, which may explain a higher demand, especially for proteins containing such amino acids at higher frequencies than the *P. pastoris* total proteome. Another engineered pathway that was successfully predicted to enhance protein production was the deletion of fermentation by knocking out pyruvate decarboxylase (*PDC1*) or alcohol dehydrogenase (*ADH2*) [30]. This points to a demand for higher energy efficiency by utilizing more of the carbon source in the TCA cycle. As outlined in Section 19.1.4, the lower fermentative capacity of *P. pastoris* is a principle advantage for recombinant protein production compared to *S. cerevisiae* and can be further enhanced by metabolic engineering.

19.3.6
Engineering Protein Folding and Secretion Pathways to Enhance Productivity

P. pastoris is most favorably employed to produce secreted recombinant proteins. The major advantage is that the product is derived in a quite pure form after cell removal, so that recovery and purification are much easier than from cell lysates. Secondly, proteins that are naturally secreted in their native host often require the passage through the secretory pathway of the recombinant host for correct folding and processing. Most human proteins of potential therapeutic value are secretory proteins (antibodies, hormones, cytokines, enzymes), as well as many technical enzymes (e.g., cellulases, proteases, lipases, etc.), so that the secretory route is of high practical value.

In yeasts, secretory proteins are translocated to the endoplasmic reticulum (ER), where they are folded with the support of chaperones, then travel on to the Golgi system for processing, and further on to the exterior via secretory vesicles. Misfolded fractions of recombinant protein are recognized by quality control systems and may be degraded either by ER-associated degradation (ERAD) or in the vacuole. Glycosylation is initiated in the ER and further processed in the Golgi. The entire secretory pathway of eight yeast species has been reviewed and characterized on a genomic level recently [109].

The most commonly used secretion signal is the MFα pre-pro leader derived from *S. cerevisiae*. After cleaving off the pre-part in the ER, the pro-sequence becomes N-glycosylated and is required for further transport of the protein to the Golgi, where it is cleaved off by endopeptidase Kex2. Despite some drawbacks such as the sometimes observed incorrect processing of the pro-sequences from the protein (leaving unwanted amino acids at the N-terminus of the product), this is still the most efficient leader which outperforms many others (reviewed by Gasser *et al.* [110]). Solutions to overcome the processing problems are overexpression of the processing enzyme Kex2 [111, 112], the use of engineered variants of the MFα leader [113, 114], or the use of alternative leader sequences. While some products can be produced using their native leader sequences (see review by Gasser *et al.* [110]), these mostly failed for other types of products. Thus, *P. pastoris*–derived leader sequences were investigated, and lists of predicted *P. pastoris* leaders were provided by De Schutter *et al.* [16] and Massahi and Calik [115]. Based on these predictions or on actual secretome studies, some novel secretion leaders were introduced in the past years [116–118].

The ER is endowed with chaperones, protein disulfide isomerases and glycosylation enzymes, as well as their accessory proteins. Disulfide bond formation has frequently been observed to constitute a major bottleneck, and the overexpression of *PDI1* alone or in combination with its oxidase *ERO1* was shown to enhance productivity of several proteins in *P. pastoris* [95, 119–123]. Also, the reinforcement of other ER-related processes was assessed with variable success [124, 125]. In particular, overproduction of Kar2 yielded unpredictable results, which can be explained by its cellular function as a signal molecule of the UPR (for details, see Puxbaum *et al.* [126]). Targeted induction of the UPR by overexpression of its transcriptional activator Hac1 [127] poses another possibility of enhancing the folding and secretion capacity, showing beneficial effects on productivity of several recombinant secretory proteins and membrane proteins [92, 96, 121, 127–129]. Misfolded proteins are recognized by the ER quality control machinery, which targets them for retrotranslocation to the cytosol for subsequent degradation by the proteasome. While it has been shown that ERAD is induced by some recombinant proteins [130–132], and that chemical inhibition of proteasome activity increases the yield of recombinant antibody Fab fragments [130], there is no genetic evidence so far that disruption of ERAD function is a general means of enhancing recombinant protein secretion levels. After passing cellular quality control, the protein needs to reach its final destination traveling through the vesicular pathway. Enhancing the targeting components of the secretory machinery (through overexpression of *SEC4* or *SSO2* – both encoding proteins involved in the fusion of secretory vesicles to the plasma membrane) was shown to be beneficial for the secretion of Fab and glucoamylase, respectively [133, 134]. To enhance the release of the product to the cell exterior, cells engineered for a loosened cell wall by knocking out the main cross-linking enzyme Gas1 [85] or other cell-wall-modifying enzymes [135] were proved to be beneficial. Unfortunately, even correctly folded proteins can be subjected to degradation through missorting from the Golgi to the vacuole or by encountering proteases located late in the vesicular pathway (yapsins). To prevent vacuolar degradation, mutants devoid of the main vacuolar proteases encoded by *PEP4* and *PRB1* are in use for many years [136]. Also, strains disrupted for other cellular proteases (Kex1, Sub2, Yps1) have proved to be beneficial to prevent degradation or contamination of some recombinant products [137–141]. Recently, we generated strains deleted for components of the vacuolar protein sorting machinery, an approach previously shown to be beneficial for several other yeasts (reviewed by Idiris *et al.* [142]), which yielded enhanced secretion levels of recombinant antibody fragments and enzymes (Lucas Marsalek *et al.*, in preparation). Impairment of the vacuolar sorting receptor Vps10 was another strategy to prevent missorting of recombinant proteins [113].

As can be seen, most of these cell engineering approaches have targeted reactions closely related to protein folding and secretion based on educated knowledge. Two studies employed genome-scale screening approaches, using either a cDNA overexpression library or a random knockout strategy to select for cells with higher secretory productivity [73, 135]. In both cases, improvement of production of secretory model proteins could be achieved successfully. Interestingly, while several systems biology studies have been performed to characterize the physiology and cellular reactions of *P. pastoris* during production of recombinant proteins [41, 59, 96, 106, 128, 132, 143–146], so far there are only three reports of employing these systems-level data to engineer cells aiming at enhanced productivity [134, 144, 147].

19.3.7 Protein Glycosylation and Glycoengineering

Typical *N*-glycans of *P. pastoris* consist of two *N*-acetyl glucosamine (GlcNAc) and 9–16 mannose residues [148], and are much shorter than those reported for *S. cerevisiae* [149]. The conversion of these glycans to human-like *N*-glycans was achieved by metabolic engineering of *P. pastoris* [148, 150]. Thereby, a yeast-specific mannosyltransferase was deleted and the respective glycosyltransferases for addition of GlcNAc, galactose, and sialic acid, and where needed the pathways for their synthesis were added to attain fully human-like, sialylated *N*-glycans.

P. pastoris *O*-glycans consist of short chains of 1–4 mannose residues on a serine or threonine. Typically, only a small fraction of the potential O-glycosylation sites are actually occupied [151]. The same authors also demonstrated further reduction of site occupancy by partial deletion of protein mannosyltransferase (PMT) genes. Deletion of specific mannosyltransferases and overexpression of an α-1,2 mannosidase led to single mannose *O*-glycans which could be further engineered to contain GlcNAc, galactose, and sialic acid, thus resembling human-like *O*-glycans [152].

It was demonstrated that glycoengineered proteins can be produced in the pilot scale [153]; however, they have not reached the biopharmaceutical market to date. New strategies and perspectives of *P. pastoris* glycoengineering have been reviewed recently [126].

19.3.8 Recombinant Proteins on the Market

Since the first product, the recombinant kallikrein inhibitor ecallantide (KALBITOR®, Dyax, USA), was approved by the FDA in 2009, several more followed, mainly being peptides and small proteins. Antibodies and fragments thereof are currently in clinical trials awaiting approval [154]. Several more biopharmaceutical products such as insulin, interferon-alpha 2b, and hepatitis B vaccine are on the market in India and other Asian countries [155]. Apart from biopharmaceuticals, a range of industrial enzymes, among them phytase, nitrate reductase, or lipase, are commercially available. Direct information on the production process is mostly not available, but it can be concluded from journal and patent literature that mostly the *AOX1* promoter is used in these cases, and other expression systems like the GAP promoter are applied as well. In total, approximately 70 products are on the market, and a list can be found at http://www.pichia.com/science-center/commercialized-products/ [156].

19.4 Application of *P. pastoris* for Metabolite Production

Initially, the biotechnological applications of *P. pastoris* have focused on biomass and recombinant protein production, having the lead position in the latter field today. More recently the production of metabolites in *P. pastoris* has gained significant interest. The first example was *S*-adenosyl-L-methionine (SAM) [157], which was accomplished by overexpression of *S. cerevisiae* SAM synthase to enhance the native pathway in *P. pastoris*. Deletion of cystathionine-beta-synthase further enhanced the SAM titers.

Production of the carotenoid lycopene was accomplished by overexpression of three genes encoding carotenogenic enzymes from *Erwinia* sp. to divert the ergosterol synthesis pathway from farnesyl pyrophosphate (FPP) toward lycopene [158]. By addition of the lycopene-beta-cyclase gene from *Ficus carica,* beta-carotene [159] and astaxanthin [160] production could also be accomplished. The sterol pathway with FPP as a key intermediate is regarded as a hub for the synthesis of many different terpenoids, or more generally isoprenoids. As recent examples, the sesquiterpenoid (+)-nootkatone, an aroma compound of citrus fruits, was produced in *P. pastoris* [161] as well as dammarenediol II, the precursor of ginsengosides, the active compounds in Ginseng [162].

Fatty acid synthesis has been engineered as well. Two pathways to very long chain polyunsaturated fatty acids were realized in *P. pastoris* to demonstrate their feasibility for future reengineering in oilseed crops [163]. Fatty acids and their esters are also interesting potential biofuels. Fatty acid esters with branched chain alcohols are potential low-viscosity biodiesels, and were successfully synthesized in *Escherichia coli* and *P. pastoris* by metabolic engineering [164].

Engineering of the synthesis pathway to vitamins has been demonstrated in *P. pastoris* as well. Like many other yeasts, *P. pastoris* lacks the pathway to biotin, which was introduced by overexpressing four genes from brewing yeast to develop a biotin prototrophic strain [165]. Riboflavin synthesis is naturally upregulated in methanol grown cells, as AOX requires FAD as a cofactor. Further constitutive upregulation of *RIB1*, the first step of riboflavin synthesis, leads to accumulation of free riboflavin, which is additionally enhanced by co-overexpression of all six pathway genes by exchanging their promoters *in situ* with the GAP promoter [166].

As an example of a polymeric high-molecular-weight product hyaluronic acid production was achieved by overexpression of hyaluronan synthase 2 (*xhasA2*) and UDP-glucose dehydrogenase (*xhasB*) from *Xenopus laevis* [167], whereby gram per liter amounts of high-molecular-weight (2.5 MDa) hyaluronic acid were achieved.

It becomes obvious that the published cases of metabolic engineering in *P. pastoris* target more complex molecules rather than simple fermentation by-products. As an exception, production of xylitol via arabitol was achieved by overexpression of the DalD gene from *Klebsiella pneumoniae* encoding D-xylulose-forming D-arabitol dehydrogenase and the xylitol dehydrogenase (XDH) gene from *Gluconobacter oxydans* [168]. Yield and productivity, however, were 10-fold below minimum levels of economical production.

Typical for a Crabtree-negative yeast, glucose uptake is rate limited by a limited set of glucose transporters with high affinity and rather low maximum transport rates [17, 21], contrary to *S. cerevisiae* which expresses numerous low-affinity, high-velocity transporter genes. This indicates that *S. cerevisiae* is more suitable to produce primary, mainly fermentative metabolites, while *P. pastoris* may be an interesting alternative for products of more complex pathways. Recently, overexpression of xylose isomerase and xylulokinase followed by evolutionary engineering led to an expansion of the substrate spectrum of *P. pastoris*, also enabling xylose utilization [169]. In addition, glycerol as alternative substrate is more efficiently utilized than by *S. cerevisiae*, due to the expression of several glycerol transporter genes.

19.5 Conclusion

P. pastoris has reached maturity for production of biopharmaceuticals and technical enzymes. Current literature shows that it is the most frequently used yeast platform for these purposes. Fundamental knowledge of physiology, genomics and cell biology is necessary to understand limitations and benefits, and to take full advantage of the capability of the production system. As an example, central carbon metabolism has major implications for the production process. Strict control of substrate uptake and efficient respiratory metabolism enable efficient utilization of carbon sources for biomass formation and protein production, which can be further enhanced by directing metabolic fluxes to required pathways like those replenishing reduction equivalents and energy. While this efficiency is beneficial for protein production, it hampers the fast uptake and conversion of substrates to primary metabolites as required for bulk chemical production. Detailed understanding of pathways, their fluxes, and regulation can, however, enable the future development of growth-decoupled production also in *P. pastoris*, thus potentially expanding its future palette of applications in industrial biotechnology.

References

1. Anthony, C. (1986) Bacterial oxidation of methane and methanol. *Adv. Microb. Physiol.*, **27**, 113–210.
2. Wegner, G.H. and Harder, W. (1987) Methylotrophic yeasts – 1986. *Antonie Van Leeuwenhoek*, **53**, 29–36.
3. Yurimoto, H., Kato, N., and Sakai, Y. (2005) Assimilation, dissimilation, and detoxification of formaldehyde, a central metabolic intermediate of methylotrophic metabolism. *Chem. Rec.*, **5**, 367–375.
4. Macdonald, R. and Fall, R. (1993) Detection of substantial emissions of methanol from plants to the atmosphere. *Atmos. Environ.*, **27**, 1709–1713.
5. Nakagawa, T., Miyaji, T., Yurimoto, H., Sakai, Y., Kato, N. *et al.* (2000) A methylotrophic pathway participates in pectin utilization by *Candida boidinii*. *Appl. Environ. Microbiol.*, **66**, 4253–4257.
6. Peter, G., Tornai-Lehoczki, J., and Dlauchy, D. (2009) *Ogataea populialbae* sp. nov., a yeast species from white poplar. *FEMS Yeast Res.*, **9**, 936–941.

7. Phaff, H.J. and Knapp, E.P. (1956) The taxonomy of yeasts found in exudates of certain trees and other natural breeding sites of some species of *Drosophila. Antonie Van Leeuwenhoek*, **22**, 117–130.
8. Guilliermond, A. (1920) *Zygosaccharomyces pastori*, nouvelle espèce de levures copulation hétérogamique. *Bull. Soc. Mycol. Fr.*, **36**, 203–211.
9. Phaff, H., Miller, M., and Shifrine, M. (1956) The taxonomy of yeasts isolated from *Drosophila* in the Yosemite region of California. *Antonie Van Leeuwenhoek*, **22**, 145–161.
10. Cregg, J.M., Barringer, K.J., Hessler, A.Y., and Madden, K.R. (1985) *Pichia pastoris* as a host system for transformations. *Mol. Cell. Biol.*, **5**, 3376–3385.
11. Yamada, Y., Matsuda, M., Maeda, K., and Mikata, K. (1995) The phylogenetic relationships of methanol-assimilating yeasts based on the partial sequences of 18S and 26S ribosomal RNAs: the proposal of *Komagataella* gen. nov. (Saccharomycetaceae). *Biosci. Biotechnol., Biochem.*, **59**, 439–444.
12. Kurtzman, C. (2005) Description of *Komagataella phaffii* sp. nov. and the transfer of *Pichia pseudopastoris* to the methylotrophic yeast genus *Komagataella. Int. J. Syst. Evol. Microbiol.*, **55**, 973–976.
13. Ito, T., Fujimura, S., Uchino, M., Tanaka, N., Matsufuji, Y. *et al.* (2007) Distribution, diversity and regulation of alcohol oxidase isozymes, and phylogenetic relationships of methylotrophic yeasts. *Yeast*, **24**, 523–532.
14. Naumov, G.I., Naumova, E.S., Tyurin, O.V., and Kozlov, D.G. (2013) *Komagataella kurtzmanii* sp. nov., a new sibling species of *Komagataella* (*Pichia*) *pastoris* based on multigene sequence analysis. *Antonie Van Leeuwenhoek*, **104**, 339–347.
15. Kurtzman, C.P. (2009) Biotechnological strains of *Komagataella* (*Pichia*) *pastoris* are *Komagataella phaffii* as determined from multigene sequence analysis. *J. Ind. Microbiol. Biotechnol.*, **36**, 1435–1438.
16. De Schutter, K., Lin, Y.C., Tiels, P., Van Hecke, A., Glinka, S. *et al.* (2009) Genome sequence of the recombinant protein production host *Pichia pastoris. Nat. Biotechnol.*, **27**, 561–566.
17. Mattanovich, D., Graf, A., Stadlmann, J., Dragosits, M., Redl, A. *et al.* (2009) Genome, secretome and glucose transport highlight unique features of the protein production host *Pichia pastoris. Microb. Cell Fact.*, **8**, 29.
18. Kuberl, A., Schneider, J., Thallinger, G.G., Anderl, I., Wibberg, D. *et al.* (2011) High-quality genome sequence of *Pichia pastoris* CBS7435. *J. Biotechnol.*, **154**, 312–320.
19. Ohi, H., Okazaki, N., Uno, S., Miura, M., and Hiramatsu, R. (1998) Chromosomal DNA patterns and gene stability of *Pichia pastoris. Yeast*, **14**, 895–903.
20. *Pichia pastoris* Genome Browser www.pichiagenome.org (accessed 4 March 2016).
21. Prielhofer, R., Maurer, M., Klein, J., Wenger, J., Kiziak, C. *et al.* (2013) Induction without methanol: novel regulated promoters enable high-level expression in *Pichia pastoris. Microb. Cell Fact.*, **12**, 5.
22. Khatri, N.K. and Hoffmann, F. (2006) Impact of methanol concentration on secreted protein production in oxygen-limited cultures of recombinant *Pichia pastoris. Biotechnol. Bioeng.*, **93**, 871–879.
23. Pla, I.A., Damasceno, L.M., Vannelli, T., Ritter, G., Batt, C.A. *et al.* (2006) Evaluation of Mut^+ and Mut^S *Pichia pastoris* phenotypes for high level extracellular scFv expression under feedback control of the methanol concentration. *Biotechnol. Progr.*, **22**, 881–888.
24. Jungo, C., Marison, I., and von Stockar, U. (2007) Mixed feeds of glycerol and methanol can improve the performance of *Pichia pastoris* cultures: a quantitative study based on concentration gradients in transient continuous cultures. *J. Biotechnol.*, **128**, 824–837.
25. Brinkmann, U., Mueller, R.H., and Babel, W. (1990) The growth rate-limiting reaction in methanol-assimilating yeasts. *FEMS Microbiol. Rev.*, **7**, 261–265.
26. Prielhofer, R., Cartwright, S.P., Graf, A.B., Valli, M., Bill, R.M. *et al.* (2015)

Pichia pastoris regulates its gene-specific response to different carbon sources at the transcriptional, rather than the translational, level. *BMC Genomics*, **16**, 167.

27. Watanabe, S., Saimura, M., and Makino, K. (2008) Eukaryotic and bacterial gene clusters related to an alternative pathway of nonphosphorylated L-rhamnose metabolism. *J. Biol. Chem.*, **283**, 20372–20382.
28. Hagman, A., Sall, T., Compagno, C., and Piskur, J. (2013) Yeast "make-accumulate-consume" life strategy evolved as a multi-step process that predates the whole genome duplication. *PLoS One*, **8**, e68734.
29. Carnicer, M., Baumann, K., Töplitz, I., Sánchez-Ferrando, F., Mattanovich, D. *et al.* (2009) Macromolecular and elemental composition analysis and extracellular metabolite balances of *Pichia pastoris* growing at different oxygen levels. *Microb. Cell Fact.*, **8**, 65.
30. Nocon, J., Steiger, M.G., Pfeffer, M., Sohn, S.B., Kim, T.Y. *et al.* (2014) Model based engineering of *Pichia pastoris* central metabolism enhances recombinant protein production. *Metab. Eng.*, **24**, 129–138.
31. Yurimoto, H., Oku, M., and Sakai, Y. (2011) Yeast methylotrophy: metabolism, gene regulation and peroxisome homeostasis. *Int. J. Microbiol.*, **2011**, 101298.
32. van der Klei, I.J., Yurimoto, H., Sakai, Y., and Veenhuis, M. (2006) The significance of peroxisomes in methanol metabolism in methylotrophic yeast. *Biochim. Biophys. Acta*, **1763**, 1453–1462.
33. Allais, J., Louktibi, A., and Baratti, J. (1983) Oxidation of methanol by the yeast, *Pichia pastoris*, purification and properties of the formaldehyde dehydrogenase. *Agric. Biol. Chem.*, **47**, 1509–1516.
34. Allais, J., Louktibi, A., and Baratti, J. (1983) Oxidation of methanol by the yeast *Pichia pastoris*. Purification and properties of the formate dehydrogenase. *Agric. Biol. Chem.*, **47**, 2547–2554.
35. Goenrich, M., Bartoschek, S., Hagemeier, C.H., Griesinger, C., and Vorholt, J.A. (2002) A glutathione-dependent formaldehyde-activating enzyme (Gfa) from *Paracoccus denitrificans* detected and purified via two-dimensional proton exchange NMR spectroscopy. *J. Biol. Chem.*, **277**, 3069–3072.
36. Yurimoto, H., Lee, B., Yasuda, F., Sakai, Y., and Kato, N. (2004) Alcohol dehydrogenases that catalyse methyl formate synthesis participate in formaldehyde detoxification in the methylotrophic yeast *Candida boidinii*. *Yeast*, **21**, 341–350.
37. Sakai, Y., Murdanoto, A.P., Sembiring, L., Tani, Y., and Kato, N. (1995) A novel formaldehyde oxidation pathway in methylotrophic yeasts: methylformate as a possible intermediate. *FEMS Microbiol. Lett.*, **127**, 229–234.
38. Russmayer, H., Buchetics, M., Gruber, C., Valli, M., Grillitsch, K. *et al.* (2015) Systems-level organization of yeast methylotrophic lifestyle. *BMC Biol.*, **13**, 80.
39. Cavero, S., Vozza, A., del Arco, A., Palmieri, L., Villa, A. *et al.* (2003) Identification and metabolic role of the mitochondrial aspartate-glutamate transporter in *Saccharomyces cerevisiae*. *Mol. Microbiol.*, **50**, 1257–1269.
40. Solà, A., Jouhten, P., Maaheimo, H., Sánchez-Ferrando, F., Szyperski, T. *et al.* (2007) Metabolic flux profiling of *Pichia pastoris* grown on glycerol/methanol mixtures in chemostat cultures at low and high dilution rates. *Microbiology*, **153**, 281–290.
41. Jorda, J., Jouhten, P., Camara, E., Maaheimo, H., Albiol, J. *et al.* (2012) Metabolic flux profiling of recombinant protein secreting *Pichia pastoris* growing on glucose:methanol mixtures. *Microb. Cell Fact.*, **11**, 57.
42. Ramon, R., Ferrer, P., and Valero, F. (2007) Sorbitol co-feeding reduces metabolic burden caused by the overexpression of a *Rhizopus oryzae* lipase in *Pichia pastoris*. *J. Biotechnol.*, **130**, 39–46.
43. Cregg, J.M., Madden, K.R., Barringer, K.J., Thill, G.P., and Stillman, C.A.

(1989) Functional characterization of the two alcohol oxidase genes from the yeast *Pichia pastoris*. *Mol. Cell. Biol.*, **9**, 1316–1323.

44. Charoenrat, T., Ketudat-Cairns, M., Stendahl-Andersen, H., Jahic, M., and Enfors, S.O. (2005) Oxygen-limited fed-batch process: an alternative control for *Pichia pastoris* recombinant protein processes. *Bioprocess. Biosyst. Eng.*, **27**, 399–406.
45. Tschopp, J.F., Brust, P.F., Cregg, J.M., Stillman, C.A., and Gingeras, T.R. (1987) Expression of the lacZ gene from two methanol-regulated promoters in *Pichia pastoris. Nucleic Acids Res.*, **15**, 3859–3876.
46. Shen, S., Sulter, G., Jeffries, T.W., and Cregg, J.M. (1998) A strong nitrogen source-regulated promoter for controlled expression of foreign genes in the yeast *Pichia pastoris*. *Gene*, **216**, 93–102.
47. Hasslacher, M., Schall, M., Hayn, M., Bona, R., Rumbold, K. *et al.* (1997) High-level intracellular expression of hydroxynitrile lyase from the tropical rubber tree *Hevea brasiliensis* in microbial hosts. *Protein Expression Purif.*, **11**, 61–71.
48. Clare, J.J., Rayment, F.B., Ballantine, S.P., Sreekrishna, K., and Romanos, M.A. (1991) High-level expression of tetanus toxin fragment C in *Pichia pastoris* strains containing multiple tandem integrations of the gene. *Biotechnology (N.Y.)*, **9**, 455–460.
49. Rosenfeld, S.A., Nadeau, D., Tirado, J., Hollis, G.F., Knabb, R.M. *et al.* (1996) Production and purification of recombinant hirudin expressed in the methylotrophic yeast *Pichia pastoris*. *Protein Expression Purif.*, **8**, 476–482.
50. Laroche, Y., Storme, V., De Meutter, J., Messens, J., and Lauwereys, M. (1994) High-level secretion and very efficient isotopic labeling of tick anticoagulant peptide (TAP) expressed in the methylotrophic yeast, *Pichia pastoris*. *Biotechnology (N.Y.)*, **12**, 1119–1124.
51. Damasceno, L.M., Pla, I., Chang, H.J., Cohen, L., Ritter, G. *et al.* (2004) An optimized fermentation process for high-level production of a single-chain Fv antibody fragment in *Pichia pastoris*. *Protein Expression Purif.*, **37**, 18–26.
52. Werten, M., van den Bosch, T., Wind, R., Mooibroek, H., and de Wolf, F. (1999) High-yield secretion of recombinant gelatins by *Pichia pastoris*. *Yeast*, **15**, 1087–1096.
53. Mattanovich, D., Jungo, C., Wenger, J., Dabros, M., and Maurer, M. (2014) Yeast suspension culture, in *Industrial Scale Suspension Culture of Living Cells* (eds H. Meyer and D.R. Schmidhalter), Wiley-Blackwell.
54. Waterham, H.R., Digan, M.E., Koutz, P.J., Lair, S.V., and Cregg, J.M. (1997) Isolation of the *Pichia pastoris* glyceraldehyde-3-phosphate dehydrogenase gene and regulation and use of its promoter. *Gene*, **186**, 37–44.
55. Ellis, S.B., Brust, P.F., Koutz, P.J., Waters, A.F., Harpold, M.M. *et al.* (1985) Isolation of alcohol oxidase and two other methanol regulatable genes from the yeast *Pichia pastoris*. *Mol. Cell. Biol.*, **5**, 1111–1121.
56. Liu, H., Tan, X., Russell, K.A., Veenhuis, M., and Cregg, J.M. (1995) PER3, a gene required for peroxisome biogenesis in *Pichia pastoris*, encodes a peroxisomal membrane protein involved in protein import. *J. Biol. Chem.*, **270**, 10940–10951.
57. Sears, I.B., O'Connor, J., Rossanese, O.W., and Glick, B.S. (1998) A versatile set of vectors for constitutive and regulated gene expression in *Pichia pastoris*. *Yeast*, **14**, 783–790.
58. Gasser, B., Steiger, M.G., and Mattanovich, D. (2015) Methanol regulated yeast promoters: production vehicles and toolbox for synthetic biology. *Microb. Cell Fact.*, **14**, 196.
59. Dragosits, M., Stadlmann, J., Albiol, J., Baumann, K., Maurer, M. *et al.* (2009) The effect of temperature on the proteome of recombinant *Pichia pastoris*. *J. Proteome Res.*, **8**, 1380–1392.
60. Rebnegger, C., Graf, A.B., Valli, M., Steiger, M.G., Gasser, B. *et al.* (2014) In *Pichia pastoris*, growth rate regulates protein synthesis and secretion, mating and stress response. *Biotechnol. J.*, **9**, 511–525.

61. Vogl, T., Sturmberger, L., Kickenweiz, T., Wasmayer, R., Schmid, C. *et al.* (2015) A toolbox of diverse promoters related to methanol utilization: functionally verified parts for heterologous pathway expression in *Pichia pastoris*. *ACS Synth. Biol.*, **5**, 172–186.
62. Stadlmayr, G., Mecklenbrauker, A., Rothmuller, M., Maurer, M., Sauer, M. *et al.* (2010) Identification and characterisation of novel *Pichia pastoris* promoters for heterologous protein production. *J. Biotechnol.*, **150**, 519–529.
63. Hartner, F.S., Ruth, C., Langenegger, D., Johnson, S.N., Hyka, P. *et al.* (2008) Promoter library designed for fine-tuned gene expression in *Pichia pastoris*. *Nucleic Acids Res.*, **36**, e76.
64. Qin, X., Qian, J., Yao, G., Zhuang, Y., Zhang, S. *et al.* (2011) GAP promoter library for fine-tuning of gene expression in *Pichia pastoris*. *Appl. Environ. Microbiol.*, 77, 3600–3608.
65. Vogl, T., Ruth, C., Pitzer, J., Kickenweiz, T., and Glieder, A. (2014) Synthetic core promoters for *Pichia pastoris*. *ACS Synth. Biol.*, **3**, 188–191.
66. Payne, W.E., Gannon, P.M., and Kaiser, C.A. (1995) An inducible acid phosphatase from the yeast *Pichia pastoris*: characterization of the gene and its product. *Gene*, **163**, 19–26.
67. Cabral, K.M., Almeida, M.S., Valente, A.P., Almeida, F.C., and Kurtenbach, E. (2003) Production of the active antifungal *Pisum sativum* defensin 1 (Psd1) in *Pichia pastoris*: overcoming the inefficiency of the STE13 protease. *Protein Expression Purif.*, **31**, 115–122.
68. Schaefer, J.V. and Pluckthun, A. (2012) Engineering aggregation resistance in IgG by two independent mechanisms: lessons from comparison of *Pichia pastoris* and mammalian cell expression. *J. Mol. Biol.*, **417**, 309–335.
69. Huang, C.J., Damasceno, L.M., Anderson, K.A., Zhang, S., Old, L.J. *et al.* (2011) A proteomic analysis of the *Pichia pastoris* secretome in methanol-induced cultures. *Appl. Microbiol. Biotechnol.*, **90**, 235–247.
70. Hohlweg, J., Gruber, C., Altmann, F., Valli, M., Mattanovich, D. *et al.* (2015) Secretome diversity of *Pichia pastoris* in the presence of different carbon sources. *Yeast*, **32**, S153–S154.
71. Resina, D., Cos, O., Ferrer, P., and Valero, F. (2005) Developing high cell density fed-batch cultivation strategies for heterologous protein production in *Pichia pastoris* using the nitrogen source-regulated FLD1 promoter. *Biotechnol. Bioeng.*, **91**, 760–767.
72. Liachko, I. and Dunham, M.J. (2014) An autonomously replicating sequence for use in a wide range of budding yeasts. *FEMS Yeast Res.*, **14**, 364–367.
73. Stadlmayr, G., Benakovitsch, K., Gasser, B., Mattanovich, D., and Sauer, M. (2010) Genome-scale analysis of library sorting (GALibSo): Isolation of secretion enhancing factors for recombinant protein production in *Pichia pastoris*. *Biotechnol. Bioeng.*, **105**, 543–555.
74. Cregg, J.M. and Madden, K.R. (1989) Use of site-specific recombination to regenerate selectable markers. *Mol. Gen. Genet.*, **219**, 320–323.
75. Lin Cereghino, G.P., Lin Cereghino, J., Sunga, A.J., Johnson, M.A., Lim, M. *et al.* (2001) New selectable marker/auxotrophic host strain combinations for molecular genetic manipulation of *Pichia pastoris*. *Gene*, **263**, 159–169.
76. Du, M., Battles, M.B., and Nett, J.H. (2012) A color-based stable multi-copy integrant selection system for *Pichia pastoris* using the attenuated ADE1 and ADE2 genes as auxotrophic markers. *Bioeng. Bugs*, **3**, 32–37.
77. Nett, J.H. and Gerngross, T.U. (2003) Cloning and disruption of the PpURA5 gene and construction of a set of integration vectors for the stable genetic modification of *Pichia pastoris*. *Yeast*, **20**, 1279–1290.
78. Nett, J.H., Hodel, N., Rausch, S., and Wildt, S. (2005) Cloning and disruption of the *Pichia pastoris ARG1, ARG2, ARG3, HIS1, HIS2, HIS5, HIS6* genes and their use as auxotrophic markers. *Yeast*, **22**, 295–304.
79. Thor, D., Xiong, S., Orazem, C.C., Kwan, A.C., Cregg, J.M. *et al.* (2005) Cloning and characterization of the *Pichia pastoris MET2* gene as a

selectable marker. *FEMS Yeast Res.*, **5**, 935–942.

80. Sunga, A.J. and Cregg, J.M. (2004) The *Pichia pastoris* formaldehyde dehydrogenase gene (*FLD1*) as a marker for selection of multicopy expression strains of *P. pastoris*. *Gene*, **330**, 39–47.
81. Näätsaari, L., Mistlberger, B., Ruth, C., Hajek, T., Hartner, F.S. *et al.* (2012) Deletion of the *Pichia pastoris* KU70 homologue facilitates platform strain generation for gene expression and synthetic biology. *PLoS One*, **7**, e39720.
82. Scorer, C.A., Clare, J.J., McCombie, W.R., Romanos, M.A., and Sreekrishna, K. (1994) Rapid selection using G418 of high copy number transformants of *Pichia pastoris* for high-level foreign gene expression. *Biotechnology (N.Y.)*, **12**, 181–184.
83. Higgins, D.R., Busser, K., Comiskey, J., Whittier, P.S., Purcell, T.J. *et al.* (1998) Small vectors for expression based on dominant drug resistance with direct multicopy selection. *Methods Mol. Biol.*, **103**, 41–53.
84. Terentiev, Y., Pico, A.H., Boer, E., Wartmann, T., Klabunde, J. *et al.* (2004) A wide-range integrative yeast expression vector system based on *Arxula adeninivorans*-derived elements. *J. Ind. Microbiol. Biotechnol.*, **31**, 223–228.
85. Marx, H., Sauer, M., Resina, D., Vai, M., Porro, D. *et al.* (2006) Cloning, disruption and protein secretory phenotype of the *GAS1* homologue of *Pichia pastoris*. *FEMS Microbiol. Lett.*, **264**, 40–47.
86. Chen, M.T., Lin, S., Shandil, I., Andrews, D., Stadheim, T.A. *et al.* (2012) Generation of diploid *Pichia pastoris* strains by mating and their application for recombinant protein production. *Microb. Cell Fact.*, **11**, 91.
87. Vassileva, A., Chugh, D., Swaminathan, S., and Khanna, N. (2001) Expression of hepatitis B surface antigen in the methylotrophic yeast *Pichia pastoris* using the GAP promoter. *J. Biotechnol.*, **88**, 21–35.
88. Marx, H., Mecklenbräuker, A., Gasser, B., Sauer, M., and Mattanovich, D. (2009) Directed gene copy number amplification in *Pichia pastoris* by vector integration into the ribosomal DNA locus. *FEMS Yeast Res.*, **9**, 1260–1270.
89. Aw, R. and Polizzi, K.M. (2013) Can too many copies spoil the broth? *Microb. Cell Fact.*, **12**, 128.
90. Mansur, M., Cabello, C., Hernandez, L., Pais, J., Varas, L. *et al.* (2005) Multiple gene copy number enhances insulin precursor secretion in the yeast *Pichia pastoris*. *Biotechnol. Lett.*, **27**, 339–345.
91. Zhu, T., Guo, M., Tang, Z., Zhang, M., Zhuang, Y. *et al.* (2009) Efficient generation of multi-copy strains for optimizing secretory expression of porcine insulin precursor in yeast *Pichia pastoris*. *J. Appl. Microbiol.*, **107**, 954–963.
92. Li, C., Lin, Y., Zheng, X., Pang, N., Liao, X. *et al.* (2015) Combined strategies for improving expression of *Citrobacter amalonaticus* phytase in *Pichia pastoris*. *BMC Biotechnol.*, **15**, 88.
93. Hohenblum, H., Gasser, B., Maurer, M., Borth, N., and Mattanovich, D. (2004) Effects of gene dosage, promoters, and substrates on unfolded protein stress of recombinant *Pichia pastoris*. *Biotechnol. Bioeng.*, **85**, 367–375.
94. Camara, E., Albiol, J., and Ferrer, P. (2015) Droplet digital PCR-aided screening and characterization of *Pichia pastoris* multiple gene copy strains. *Biotechnol. Bioeng.*, **113**, 1542–1551
95. Inan, M., Aryasomayajula, D., Sinha, J., and Meagher, M. (2006) Enhancement of protein secretion in *Pichia pastoris* by overexpression of protein disulfide isomerase. *Biotechnol. Bioeng.*, **93**, 771–778.
96. Lin, X.Q., Liang, S.L., Han, S.Y., Zheng, S.P., Ye, Y.R. *et al.* (2013) Quantitative iTRAQ LC-MS/MS proteomics reveals the cellular response to heterologous protein overexpression and the regulation of HAC1 in *Pichia pastoris*. *J. Proteomics*, **91**, 58–72.
97. Lopes, T.S., Hakkaart, G.J., Koerts, B.L., Raue, H.A., and Planta, R.J. (1991) Mechanism of high-copy-number integration of pMIRY-type vectors into the ribosomal DNA of *Saccharomyces cerevisiae*. *Gene*, **105**, 83–90.

98. Lopes, T.S., Klootwijk, J., Veenstra, A.E., van der Aar, P.C., van Heerikhuizen, H. *et al.* (1989) High-copy-number integration into the ribosomal DNA of *Saccharomyces cerevisiae*: a new vector for high-level expression. *Gene*, **79**, 199–206.
99. Le Dall, M.T., Nicaud, J.M., and Gaillardin, C. (1994) Multiple-copy integration in the yeast *Yarrowia lipolytica*. *Curr. Genet.*, **26**, 38–44.
100. Klabunde, J., Diesel, A., Waschk, D., Gellissen, G., Hollenberg, C.P. *et al.* (2002) Single-step co-integration of multiple expressible heterologous genes into the ribosomal DNA of the methylotrophic yeast *Hansenula polymorpha*. *Appl. Microbiol. Biotechnol.*, **58**, 797–805.
101. Rossolini, G.M., Riccio, M.L., Gallo, E., and Galeotti, C.L. (1992) *Kluyveromyces lactis* rDNA as a target for multiple integration by homologous recombination. *Gene*, **119**, 75–81.
102. Sunga, A.J., Tolstorukov, I., and Cregg, J.M. (2008) Posttransformational vector amplification in the yeast *Pichia pastoris*. *FEMS Yeast Res.*, **8**, 870–876.
103. Jullesson, D., David, F., Pfleger, B., and Nielsen, J. (2015) Impact of synthetic biology and metabolic engineering on industrial production of fine chemicals. *Biotechnol. Adv.*, **33**, 1395–1402.
104. Heyland, J., Fu, J., Blank, L.M., and Schmid, A. (2011) Carbon metabolism limits recombinant protein production in *Pichia pastoris*. *Biotechnol. Bioeng.*, **108**, 1942–1953.
105. Heyland, J., Fu, J., Blank, L.M., and Schmid, A. (2010) Quantitative physiology of *Pichia pastoris* during glucose-limited high-cell density fed-batch cultivation for recombinant protein production. *Biotechnol. Bioeng.*, **107**, 357–368.
106. Jorda, J., Rojas, H.C., Carnicer, M., Wahl, A., Ferrer, P. *et al.* (2014) Quantitative metabolomics and instationary 13C-metabolic flux analysis reveals impact of recombinant protein production on trehalose and energy metabolism in *Pichia pastoris*. *Metabolites*, **4**, 281–299.
107. Sohn, S.B., Graf, A.B., Kim, T.Y., Gasser, B., Maurer, M. *et al.* (2010) Genome-scale metabolic model of methylotrophic yeast *Pichia pastoris* and its use for in silico analysis of heterologous protein production. *Biotechnol. J.*, **5**, 705–715.
108. Nocon, J., Steiger, M., Mairinger, T., Hohlweg, J., Russmayer, H. *et al.* (2016) Increasing pentose phosphate pathway flux enhances recombinant protein production in *Pichia pastoris*. *Appl. Microbiol. Biotechnol.*, **100**, 5955–5963.
109. Delic, M., Valli, M., Graf, A.B., Pfeffer, M., Mattanovich, D. *et al.* (2013) The secretory pathway: exploring yeast diversity. *FEMS Microbiol. Rev.*, **37**, 872–914.
110. Gasser, B., Prielhofer, R., Marx, H., Maurer, M., Nocon, J. *et al.* (2013) *Pichia pastoris*: protein production host and model organism for biomedical research. *Future Microbiol.*, **8**, 191–208.
111. Yang, S., Kuang, Y., Li, H., Liu, Y., Hui, X. *et al.* (2013) Enhanced production of recombinant secretory proteins in *Pichia pastoris* by optimizing Kex2 P1′ site. *PLoS One*, **8**, e75347.
112. Sreenivas, S., Krishnaiah, S.M., Govindappa, N., Basavaraju, Y., Kanojia, K. *et al.* (2015) Enhancement in production of recombinant two-chain Insulin Glargine by over-expression of Kex2 protease in *Pichia pastoris*. *Appl. Microbiol. Biotechnol.*, **99**, 327–336.
113. Fitzgerald, I. and Glick, B.S. (2014) Secretion of a foreign protein from budding yeasts is enhanced by cotranslational translocation and by suppression of vacuolar targeting. *Microb. Cell Fact.*, **13**, 125.
114. Lin-Cereghino, G.P., Stark, C.M., Kim, D., Chang, J., Shaheen, N. *et al.* (2013) The effect of alpha-mating factor secretion signal mutations on recombinant protein expression in *Pichia pastoris*. *Gene*, **519**, 311–317.
115. Massahi, A. and Calik, P. (2015) In-silico determination of *Pichia pastoris* signal peptides for extracellular recombinant protein production. *J. Theor. Biol.*, **364**, 179–188.
116. Liang, S., Li, C., Ye, Y., and Lin, Y. (2013) Endogenous signal peptides

efficiently mediate the secretion of recombinant proteins in *Pichia pastoris*. *Biotechnol. Lett.*, **35**, 97–105.
117. Heiss, S., Puxbaum, V., Gruber, C., Altmann, F., Mattanovich, D. *et al.* (2015) Multistep processing of the secretion leader of the extracellular protein Epx1 in *Pichia pastoris* and implications on protein localization. *Microbiology*, **161**, 1356–1368.
118. Govindappa, N., Hanumanthappa, M., Venkatarangaiah, K., Periyasamy, S., Sreenivas, S. *et al.* (2014) A new signal sequence for recombinant protein secretion in *Pichia pastoris*. *J. Microbiol. Biotechnol.*, **24**, 337–345.
119. Sha, C., Yu, X.W., Zhang, M., and Xu, Y. (2013) Efficient secretion of lipase r27RCL in *Pichia pastoris* by enhancing the disulfide bond formation pathway in the endoplasmic reticulum. *J. Ind. Microbiol. Biotechnol.*, **40**, 1241–1249.
120. Delic, M., Rebnegger, C., Wanka, F., Puxbaum, V., Haberhauer-Troyer, C. *et al.* (2012) Oxidative protein folding and unfolded protein response elicit differing redox regulation in endoplasmic reticulum and cytosol of yeast. *Free Radical Biol. Med.*, **52**, 2000–2012.
121. Gasser, B., Maurer, M., Gach, J., Kunert, R., and Mattanovich, D. (2006) Engineering of *Pichia pastoris* for improved production of antibody fragments. *Biotechnol. Bioeng.*, **94**, 353–361.
122. Shen, Q., Wu, M., Wang, H.B., Naranmandura, H., and Chen, S.Q. (2012) The effect of gene copy number and co-expression of chaperone on production of albumin fusion proteins in *Pichia pastoris*. *Appl. Microbiol. Biotechnol.*, **96**, 763–772.
123. Li, Z., Moy, A., Gomez, S.R., Franz, A.H., Lin-Cereghino, J. *et al.* (2010) An improved method for enhanced production and biological activity of human secretory leukocyte protease inhibitor (SLPI) in *Pichia pastoris*. *Biochem. Biophys. Res. Commun.*, **402**, 519–524.
124. Zhang, W., Zhao, H.L., Xue, C., Xiong, X.H., Yao, X.Q. *et al.* (2006) Enhanced secretion of heterologous proteins in *Pichia pastoris* following overexpression of *Saccharomyces cerevisiae* chaperone proteins. *Biotechnol. Progr.*, **22**, 1090–1095.
125. Samuel, P., Prasanna Vadhana, A.K., Kamatchi, R., Antony, A., and Meenakshisundaram, S. (2013) Effect of molecular chaperones on the expression of *Candida antarctica* lipase B in *Pichia pastoris*. *Microbiol. Res.*, **168**, 615–620.
126. Puxbaum, V., Mattanovich, D., and Gasser, B. (2015) Quo vadis? The challenges of recombinant protein folding and secretion in *Pichia pastoris*. *Appl. Microbiol. Biotechnol.*, **99**, 2925–2938.
127. Guerfal, M., Ryckaert, S., Jacobs, P.P., Ameloot, P., Van Craenenbroeck, K. *et al.* (2010) The *HAC1* gene from *Pichia pastoris*: characterization and effect of its overexpression on the production of secreted, surface displayed and membrane proteins. *Microb. Cell Fact.*, **9**, 49.
128. Vogl, T., Thallinger, G.G., Zellnig, G., Drew, D., Cregg, J.M. *et al.* (2014) Towards improved membrane protein production in *Pichia pastoris*: general and specific transcriptional response to membrane protein overexpression. *New Biotechnol.*, **31**, 538–552.
129. Resina, D., Maurer, M., Cos, O., Arnau, C., Carnicer, M. *et al.* (2009) Engineering of bottlenecks in *Rhizopus oryzae* lipase production in *Pichia pastoris* using the nitrogen source-regulated *FLD1* promoter. *New Biotechnol.*, **25**, 396–403.
130. Pfeffer, M., Maurer, M., Stadlmann, J., Grass, J., Delic, M. *et al.* (2012) Intracellular interactome of secreted antibody Fab fragment in *Pichia pastoris* reveals its routes of secretion and degradation. *Appl. Microbiol. Biotechnol.*, **93**, 2503–2512.
131. Whyteside, G., Alcocer, M.J., Kumita, J.R., Dobson, C.M., Lazarou, M. *et al.* (2011) Native-state stability determines the extent of degradation relative to secretion of protein variants from *Pichia pastoris*. *PLoS One*, **6**, e22692.
132. Vanz, A.L., Lunsdorf, H., Adnan, A., Nimtz, M., Gurramkonda, C. *et al.* (2012) Physiological response of *Pichia pastoris* GS115 to methanol-induced

high level production of the Hepatitis B surface antigen: catabolic adaptation, stress responses, and autophagic processes. *Microb. Cell Fact.*, **11**, 103.
133. Liu, S.H., Chou, W.I., Lin, S.C., Sheu, C.C., and Chang, M.D. (2005) Molecular genetic manipulation of *Pichia pastoris SEC4* governs cell growth and glucoamylase secretion. *Biochem. Biophys. Res. Commun.*, **336**, 1172–1180.
134. Gasser, B., Sauer, M., Maurer, M., Stadlmayr, G., and Mattanovich, D. (2007) Transcriptomics-based identification of novel factors enhancing heterologous protein secretion in yeasts. *Appl. Environ. Microbiol.*, **73**, 6499–6507.
135. Larsen, S., Weaver, J., de Sa Campos, K., Bulahan, R., Nguyen, J. *et al.* (2013) Mutant strains of *Pichia pastoris* with enhanced secretion of recombinant proteins. *Biotechnol. Lett.*, **35**, 1925–1935.
136. Gleeson, M.A., White, C.E., Meininger, D.P., and Komives, E.A. (1998) Generation of protease-deficient strains and their use in heterologous protein expression. *Methods Mol. Biol.*, **103**, 81–94.
137. Wu, M., Shen, Q., Yang, Y., Zhang, S., Qu, W. *et al.* (2013) Disruption of *YPS1* and *PEP4* genes reduces proteolytic degradation of secreted HSA/PTH in *Pichia pastoris* GS115. *J. Ind. Microbiol. Biotechnol.*, **40**, 589–599.
138. Silva, C.I., Teles, H., Moers, A.P., Eggink, G., de Wolf, F.A. *et al.* (2011) Secreted production of collagen-inspired gel-forming polymers with high thermal stability in *Pichia pastoris. Biotechnol. Bioeng.*, **108**, 2517–2525.
139. Boehm, T., Pirie-Shepherd, S., Trinh, L.B., Shiloach, J., and Folkman, J. (1999) Disruption of the *KEX1* gene in *Pichia pastoris* allows expression of full-length murine and human endostatin. *Yeast*, **15**, 563–572.
140. Ni, Z., Zhou, X., Sun, X., Wang, Y., and Zhang, Y. (2008) Decrease of hirudin degradation by deleting the *KEX1* gene in recombinant *Pichia pastoris. Yeast*, **25**, 1–8.
141. Salamin, K., Sriranganadane, D., Lechenne, B., Jousson, O., and Monod, M. (2010) Secretion of an endogenous subtilisin by *Pichia pastoris* strains GS115 and KM71. *Appl. Environ. Microbiol.*, **76**, 4269–4276.
142. Idiris, A., Tohda, H., Kumagai, H., and Takegawa, K. (2010) Engineering of protein secretion in yeast: strategies and impact on protein production. *Appl. Microbiol. Biotechnol.*, **86**, 403–417.
143. Baumann, K., Carnicer, M., Dragosits, M., Graf, A.B., Stadlmann, J. *et al.* (2010) A multi-level study of recombinant *Pichia pastoris* in different oxygen conditions. *BMC Syst. Biol.*, **4**, 141.
144. Huangfu, J., Qi, F., Liu, H., Zou, H., Ahmed, M.S. *et al.* (2015) Novel helper factors influencing recombinant protein production in *Pichia pastoris* based on proteomic analysis under simulated microgravity. *Appl. Microbiol. Biotechnol.*, **99**, 653–665.
145. Vanz, A.L., Nimtz, M., and Rinas, U. (2014) Decrease of UPR- and ERAD-related proteins in *Pichia pastoris* during methanol-induced secretory insulin precursor production in controlled fed-batch cultures. *Microb. Cell Fact.*, **13**, 23.
146. Hesketh, A.R., Castrillo, J.I., Sawyer, T., Archer, D.B., and Oliver, S.G. (2013) Investigating the physiological response of *Pichia* (*Komagataella*) *pastoris* GS115 to the heterologous expression of misfolded proteins using chemostat cultures. *Appl. Microbiol. Biotechnol.*, **97**, 9747–9762.
147. Baumann, K., Adelantado, N., Lang, C., Mattanovich, D., and Ferrer, P. (2011) Protein trafficking, ergosterol biosynthesis and membrane physics impact recombinant protein secretion in *Pichia pastoris. Microb. Cell Fact.*, **10**, 93.
148. Hamilton, S.R., Bobrowicz, P., Bobrowicz, B., Davidson, R.C., Li, H. *et al.* (2003) Production of complex human glycoproteins in yeast. *Science*, **301**, 1244–1246.
149. Dean, N. (1999) Asparagine-linked glycosylation in the yeast Golgi. *Biochim. Biophys. Acta*, **1426**, 309–322.
150. Hamilton, S. and Gerngross, T. (2007) Glycosylation engineering in yeast: the

advent of fully humanized yeast. *Curr. Opin. Biotechnol.*, **18**, 387–392.

151. Nett, J.H., Cook, W.J., Chen, M.T., Davidson, R.C., Bobrowicz, P. *et al.* (2013) Characterization of the *Pichia pastoris* protein-*O*-mannosyltransferase gene family. *PLoS One*, **8**, e68325.
152. Hamilton, S.R., Cook, W.J., Gomathinayagam, S., Burnina, I., Bukowski, J. *et al.* (2013) Production of sialylated *O*-linked glycans in *Pichia pastoris*. *Glycobiology*, **23**, 1192–1203.
153. Ye, J., Ly, J., Watts, K., Hsu, A., Walker, A. *et al.* (2011) Optimization of a glycoengineered *Pichia pastoris* cultivation process for commercial antibody production. *Biotechnol. Progr.*, **27**, 1744–1750.
154. Weinacker, D., Rabert, C., Zepeda, A.B., Figueroa, C.A., Pessoa, A. *et al.* (2013) Applications of recombinant *Pichia pastoris* in the healthcare industry. *Braz. J. Microbiol.*, **44**, 1043–1048.
155. Shekhar, C. (2008) Pichia power: India's biotech industry puts unconventional yeast to work. *Chem. Biol.*, **15**, 201–202.
156. Pichia http://www.pichia.com/science‐center/commercialized‐products/ (accessed 4 March 2016).
157. He, J., Deng, J., Zheng, Y., and Gu, J. (2006) A synergistic effect on the production of S-adenosyl-L-methionine in *Pichia pastoris* by knocking in of S-adenosyl-L-methionine synthase and knocking out of cystathionine-beta synthase. *J. Biotechnol.*, **126**, 519–527.
158. Bhataya, A., Schmidt-Dannert, C., and Lee, P.C. (2009) Metabolic engineering of *Pichia pastoris* X-33 for lycopene production. *Process Biochem.*, **44**, 1095–1102.
159. Araya-Garay, J.M., Feijoo-Siota, L., Rosa-dos-Santos, F., Veiga-Crespo, P., and Villa, T.G. (2012) Construction of new *Pichia pastoris* X-33 strains for production of lycopene and beta-carotene. *Appl. Microbiol. Biotechnol.*, **93**, 2483–2492.
160. Araya-Garay, J.M., Ageitos, J.M., Vallejo, J.A., Veiga-Crespo, P., Sanchez-Perez, A. *et al.* (2012) Construction of a novel *Pichia pastoris* strain for production of xanthophylls. *AMB Express*, **2**, 24.
161. Wriessnegger, T., Augustin, P., Engleder, M., Leitner, E., Muller, M. *et al.* (2014) Production of the sesquiterpenoid (+)-nootkatone by metabolic engineering of *Pichia pastoris*. *Metab. Eng.*, **24**, 18–29.
162. Liu, X.B., Liu, M., Tao, X.Y., Zhang, Z.X., Wang, F.Q. *et al.* (2015) Metabolic engineering of *Pichia pastoris* for the production of dammarenediol-II. *J. Biotechnol.*, **216**, 47–55.
163. Kim, S.H., Roh, K.H., Kim, K.S., Kim, H.U., Lee, K.R. *et al.* (2014) Coexpression of multiple genes reconstitutes two pathways of very long-chain polyunsaturated fatty acid biosynthesis in *Pichia pastoris*. *Biotechnol. Lett.*, **36**, 1843–1851.
164. Tao, H., Guo, D., Zhang, Y., Deng, Z., and Liu, T. (2015) Metabolic engineering of microbes for branched-chain biodiesel production with low-temperature property. *Biotechnol. Biofuels*, **8**, 92.
165. Gasser, B., Dragosits, M., and Mattanovich, D. (2010) Engineering of biotin-prototrophy in *Pichia pastoris* for robust production processes. *Metab. Eng.*, **12**, 573–580.
166. Marx, H., Mattanovich, D., and Sauer, M. (2008) Overexpression of the riboflavin biosynthetic pathway in *Pichia pastoris*. *Microb. Cell Fact.*, **7**, 23.
167. Jeong, E., Shim, W.Y., and Kim, J.H. (2014) Metabolic engineering of *Pichia pastoris* for production of hyaluronic acid with high molecular weight. *J. Biotechnol.*, **185**, 28–36.
168. Cheng, H., Lv, J., Wang, H., Wang, B., Li, Z. *et al.* (2014) Genetically engineered *Pichia pastoris* yeast for conversion of glucose to xylitol by a single-fermentation process. *Appl. Microbiol. Biotechnol.*, **98**, 3539–3552.
169. Li, P., Sun, H., Chen, Z., Li, Y., and Zhu, T. (2015) Construction of efficient xylose utilizing *Pichia pastoris* for industrial enzyme production. *Microb. Cell Fact.*, **14**, 22.

Index

Industrial Biotechnology: Microorganisms, First Edition. Edited by Christoph Wittmann and James C. Liao.

d

e

n